The Eponym Dictionary of Birds

THE EPONYM DICTIONARY OF BIRDS

Bo Beolens, Michael Watkins
and Michael Grayson

B L O O M S B U R Y

LONDON · NEW DELHI · NEW YORK · SYDNEY

Dedication
For Suki Bryson (née Crombet-Beolens)
17.03.73–17.03.2013

First published in 2014 by Bloomsbury Publishing Plc
50 Bedford Square, London, WC1B 3DP

www.bloomsbury.com

ISBN (print) 978-1-4729-0573-4
ISBN (epub) 978-1-4729-0574-1

Bloomsbury Publishing, London, New Delhi, New York and Sydney
Bloomsbury is a trade mark of Bloomsbury Publishing Plc

A CIP catalogue record for this book is available from the British Library

This book is produced using paper that is made from wood grown in managed sustainable forests. It is natural, renewable and recyclable. The logging and manufacturing processes conform to the environmental regulation of the country of origin.

Commissioning editor: Nigel Redman
Designed by Mark Heslington Ltd, Scarborough, North Yorkshire.

Printed and bound in Great Britain by CPI Group (UK) Ltd, Croydon

10 9 8 7 6 5 4 3 2 1

Contents

Sources and Acknowledgements

This book is partly based on *Whose Bird?* and to some extent we were able to draw on our own previous research for this new title. In addition, we have also worked on eponym dictionaries for other faunal groups, and some of the individuals listed in these are common to this dictionary. The major change between *Whose Bird?* and this book is the expansion to cover eponymous scientific names rather than just vernacular names. This involved re-visiting many sources and often approaching the research from a different angle. The new challenge was to identify the individuals whose names are honoured in the binomial of a species for which the vernacular name is definitely not an eponym, or the trinomial of an obscure subspecies.

This was best achieved by reading the etymology (if any) in the original description. A certain number of these are freely available on the internet but the majority are not, and therefore much of our research was conducted in libraries. The three main libraries to which we had access were the Natural History Museum libraries in South Kensington and at Tring, and the library of the Zoological Society of London. Here we could read many of the journals and books that we had identified as the original sources of the type descriptions for the taxa concerned. Obituaries published in journals were also very useful as a source of biographical detail. When any source material could not be found there we were able to get help and advice from friends and contacts in many countries, and here we acknowledge them all.

The first group to thank are the organisations and individuals who helped us with *Whose Bird?*, some of whom have helped even more with this book. The second group is those who are actually mentioned in this book and who provided details of their careers and accomplishments for us to incorporate; indeed, many of them more or less wrote their own entries to save us work. Finally, there are a number of people, some of whom come into one or other of the first two categories, who gave us exceptional help, especially concerning third parties:

In Australia, Bob Forsyth; in Belgium, Robert Dowsett; in Estonia, Dr Jevgeni Shergalin; in Germany, Einhard Bezzel, David Conlin and Prof. Dr Jochen Martens; in India, Nirmala Barure (Reddy), Librarian of the Bombay Natural History Society; in the Netherlands, Jan B. Kaiser; in Peru, Manuel A. Plenge; in South Africa, Rick Nuttall, and in Venezuela, John P. Phelps and Margarita Martínez, Curator of the Colección Ornitológica Phelps, who was particularly kind in looking up journals that we could not access.

In the United Kingdom, Dr Anthony Cheke, Charles Gallimore, Michael Irwin, James Jobling, Guy M. Kirwan, A. M. Macfarlane, Amberley M. Moore, Miss O. C. Ward, and Malcolm Ellis, who unfortunately died just before this book went to the publishers.

Introduction

In the beginning

This book grew, in part, out of our experience in writing *The Eponym Dictionary of Mammals* and *The Eponym Dictionary of Reptiles* (both published by Johns Hopkins University Press), and *The Eponym Dictionary of Amphibians* (published by Pelagic Publishing), but primarily because Bo Beolens and Mike Watkins wrote *Whose Bird? – Men and Women Commemorated in the Common Names of Birds* (published by Christopher Helm in 2003). However, this book is profoundly not a second edition of *Whose Bird?* That book was both light-hearted and limited in its scope (it dealt only with common names of full species).

This is a greatly expanded work taking in not only vernacular eponyms but also eponyms that appear in the scientific names of birds, and it covers all genera, species and subspecies that are eponymous. We have not adopted any one taxonomic authority but try, so far as is possible, to include all bird eponyms ever coined, including mistakes and those not currently in use. We include all extant and historically extinct birds and there are even a few fictitious names. We have generally excluded birds that are only found in fossil form. Our 'yardstick' has been that if a researcher or other interested party comes across a bird name that is an eponym (or seeming one) then we should give an explanation if we can.

To give an idea of how much the scope of the work has increased, there are more than 4,100 entries in the book (including some names that appear to be eponyms but are not) covering just over 10,000 genera, species and subspecies. The original work, *Whose Bird?* covered only 1,124 persons and 2,246 taxa.

Who is it for?

Birders often come across bird names that include a person (such names are properly called 'eponyms'), and their curiosity will be aroused just as ours has been. You will know of some of these birds in your home area. You are certain to come across others on a foreign trip or whilst reading the books we birders seem to accumulate in such great numbers. You may even hear of a familiar bird given an unfamiliar person's name and want to know why. If so, then this book is for you. We hope that it will help satisfy your curiosity and answer your questions.

We hope that ornithologists will find this a handy guide, especially to sort out birds named after people with identical surnames or when only forenames are used. It is customary for descriptions of new species or subspecies to give the origin or meaning of the new names coined, but some etymologies give no clear derivation, and some descriptions include no etymology. Our research has filled some of these gaps.

How to use this book

We use a number of abbreviations to describe taxonomic status, etc.

Alt	=	Alternative common name
b.	=	born
d.	=	died
DNF	=	Dates not found
fl.	=	flourished (i.e. active in his/her profession)
JS	=	Junior synonym of
NCR	=	Not currently recognised
NRM	=	Now regarded as monotypic
NPRB	=	Name preoccupied; replaced by
q.v.	=	*quod vide* (refers to an entry in this book of that person)
SII	=	Sometimes included in
sp.	=	species
ssp.	=	subspecies
Syn	=	Synonymous with
NUI	=	Now usually included in

In the text the following abbreviations are widely used:

AMNH	=	American Museum of Natural History, New York
AOU	=	American Ornithologists' Union
ANWC	=	Australian National Wildlife Collection
BOC	=	British Ornithologists' Club
BOU	=	British Ornithologists' Union
BMNH	=	British Museum (Natural History), now the Natural History Museum
BTO	=	British Trust for Ornithology
CSIRO	=	Commonwealth Scientific and Industrial Research Organisation
MNHN	=	Muséum National d'Histoire Naturelle, Paris
RAOU	=	Royal Australasian Ornithologists' Union
RSPB	=	The Royal Society for the Protection of Birds

USNM = Smithsonian Institution (National Museum of Natural History)
UN = United Nations

The book is arranged alphabetically by names of people. Generally, the easiest way to find your bird is to look it up under the name of the person that is apparently embedded in the bird's name – vernacular, scientific or both. We say 'apparently', as things are not always as simple as they seem: in some bird names an apostrophe implying ownership is a transcription error; others are toponyms – named after places not people. Readers should also beware of spelling errors and variations when names are transcribed from another alphabet – eponymous birds' names are sometimes spelt in several different ways, particularly when the original is in a completely different script such as the cyrillic alphabet or Chinese characters. We have included these variations where we have come across them and where we think they may cause confusion, but complete comprehensiveness in respect of such errors and variations is beyond the scope of this book.

Each entry follows a standard format. First, we give the name of the person honoured. Next, there follows a list of 'their' eponymous birds arranged in up to three sections – genera, species and subspecies (within each section birds are listed chronologically by date of description; birds described in the same year are in alphabetical order of scientific names). In every case we give in sequence: English (vernacular) name, scientific name, describers' name(s) and the original description date. Alternative names follow in parentheses preceded by the abbreviation Alt. for common names and Syn. for scientific names, and followed, if relevant, by a coding to denote taxonomic status (e.g. NCR or NRM – see above for abbreviations). Finally, there is a brief biography of that individual.

Do not expect the length of an entry to reflect the fame or importance of the person honoured; often the reverse is the case. As a general rule very famous public figures (Queen Victoria, for example) are only given a brief biography, as authoritative biographies are easily accessible. Very well-known naturalists are also usually given shorter entries where information about them is readily available and our treatment is invariably from an ornithological perspective, emphasising the reasons for the eponyms (when known). We often give educational achievement, career highlights and their most relevant publications, and if they are honoured in other taxa. We also like to include less well-known or unusual facts such as criminality or how they died if it was in bizarre circumstances.

Although we try to use current scientific names, these are not always as universal as one might suppose. In addition to some birds having been reclassified since they were first discovered and described, there are other cases where various authorities do not agree on which name to use. This often applies to generic names, where there may be disagreement on a bird's taxonomic affinities. Current work on DNA sequencing may help standardise this for future taxonomists.

Where there is dispute or uncertainty as to whether a bird has full species status we sometimes add in brackets the name of the species that some authorities regard the bird as a subspecies of. For example, Black-headed Wagtail is rendered *Motacilla (flava) feldegg* because some authorities treat this as a full species – *Motacilla feldegg*, whereas others treat it as a subspecies of Yellow Wagtail *Motacilla flava feldegg*.

To assist you in your search, we have cross-referenced entries by highlighting (in bold) the names of those describers who have their own entries in the book. If a describer's name appears more than once in any entry it is only highlighted on the first occasion. Some birds are named in different ways after the same person and we have also tried to marry these up using cross-references. So, for example, a species named after Queen Victoria might be called Queen Victoria's Bird or Victoria's Bird, or Queen's Bird or even Empress's Bird. Not surprisingly, this is most often the case where aristocratic titles are concerned. For example, the Earl of Derby, whose family name was Stanley, had birds named after him in at least three different ways. If a name comprises more than one part, such as La Farge or del Hoyo, we treat it alphabetically as if it is all one word (in these cases under 'L' or 'D'), but if the eponym uses only the latter part of the name (e.g. *somereni* for Van Someren) then the main entry is given under the initial letter of the first part of name (in this case 'V'), but the name is also listed under 'S' and cross-referenced to 'V'.

Describers and namers

New species are first brought to the notice of the general scientific community in a formal, published description of a type specimen, essentially a dead example of the species, which will eventually be lodged in a scientific collection (although currently DNA and feathers may be acceptable without a full specimen as 'new' species may have remained unknown because of their rarity, and taking a pair may significantly deplete the gene pool). The person who describes the species will generally coin its scientific name, usually in Latin but sometimes in Latinised ancient Greek. Sometimes the 'new' bird is later reclassified, resulting in a change to the scientific name. This frequently applies to generic names (the first part of a scientific name), but specific names (the second part of a scientific name), once proposed, usually cannot be amended or replaced (except to make a gender agreement with the generic name) – there are precise and complicated rules governing name changes. Conventionally, a change of genus from the one used in the original description is indicated by putting parentheses around the describer's name. For example, the Grey Heron was named *Ardea cinerea* by Linnaeus in 1758 and, since that name remains recognised to this day, the bird is officially named *Ardea cinerea* Linnaeus, 1758. Linnaeus also described the Great Bittern as *Ardea stellaris* in the same year, but in this case the species has since been assigned to a different genus, so we now 'officially' record the bird as *Botaurus stellaris* (Linnaeus, 1758), the brackets indicating that the generic name was not the describer's original choice. Since taxonomy seems to change with the speed of light, we decided that we would automatically be wrong in some instances between proof-reading and publication, so we have not put the name of the original describer in parentheses in any entry; hence the normal convention regarding such brackets does not apply here.

There are no agreed conventions for English names and indeed the choice of vernacular names is often controversial. Very often the person who coined the scientific name will also have given it a vernacular name. However, vernacular names are often added afterwards, sometimes much later and frequently by people other than the describer. Indeed, describers are not always the people who coin the name of a bird in its scientific name either. For example, the discoverer or collector may suggest a name to the person writing the formal description. A label might be attached to a specimen by the collector, or even an unrelated third party such as a trading company, which the describer wrongly thinks was intended by the collector to be its name. Sometimes a species may be mentioned in the literature well before it is described formally. An example would be in a memoir of an exploration, or the captain's log of a voyage. If that first reference has a name attached to it the later describer usually honours that suggestion, although this is not a rule. In this book, therefore, when we refer to a bird having been 'named' by someone, we mean that that person proposed the English or scientific name in question. We refer to someone as a 'describer' or 'author' when they were responsible for the formal description.

Birds named after people's titles

A number of birds are named in such a way that it would be very hard to find the relevant biography without a note against the name. An example of this is Rajah's Scops Owl *Otus brookii*. After the entry you will see '(See Brooke)'. This directs you to the entry for Brooke, as this bird is named after Sir Charles Brooke, second Rajah of Sarawak.

Birds named after more than one person

Throughout the text you may come across several different names for the same taxon. In some cases these names are honorifics: for example, Dixon's Rock Ptarmigan is the same species as Reinhardt's Rock Ptarmigan. This peculiarity has sometimes come about through simple mistakes or misunderstandings, such as believing juveniles or females to be a different species from the adult male. In other cases, the same bird was found at about the same time in two different places and only later has it emerged that the same bird has been named twice. A good example is Cabanis's Tiger Heron, which Heine named *Tigrisoma cabanisi* (1859). It later emerged that Swainson had already named this bird

Tigrisoma mexicanum (1834) – the current scientific name. Some of these duplications persist even today, with the same bird being called something different in different places or by different people. Vernacular names vary enormously, with many alternative names favoured in different parts of the English-speaking world.

Why are birds given the names they are?

Unfortunately not all authors explain their choice of name. They may make it perfectly clear who they are honouring, or leave lots of clues, but may well not bother to say *why* they chose to honour them. The connection can be between the describer and the person honoured, or between the bird and the person. Sometimes a name is a *quid pro quo*, honouring the person who honoured the author in a previous name. It can simply be a matter of timing, such as honouring a great hero of the day. Sometimes a myth or historical figure might be chosen as it reflects an aspect of the bird such as colouring or vocalisation. It may be just that an author decides to use a sequence of names from the same source – perhaps characters from Homer; or it may be just a question of fashion – once someone starts using the names of Greek gods, other authors follow suit. This may not be obvious, especially across time or cultures. A lot of research time has been spent trying to establish such cases, not always successfully. Indeed, as set out below, this might even be deliberate obfuscation. Moreover, what is obvious to the describer at the time may well be lost to future generations.

Unidentified persons

Sadly we have not been able to identify everyone whose name appears in that of a bird. There are just 17 names on our list which are not traceable. Some Victorians had the habit of naming birds after distant female relatives or mistresses. This is difficult to track down if the etymology does not explain the choice; and if, for example, the author's mother, wife and sister all share the same forename, the bird could be named after any one or all three! In some cases we know the author has named the bird after a woman, but he has deliberately (gallantly?) withheld her full name; for example, Harry Church Oberholser did not identify the Alma he had in mind for Alma's Thrush *Catharus ustulatus*. Unhappily some ornithologists, such as Mathews, rarely explained in their etymologies just who they were honouring in a name. Such individuals clearly had no regard for those of us doing this kind of research! The 17 missing names are listed below. We would be delighted if anyone can solve them.

> Alfred's Oropendola *Psarocolius angustifrons alfredi* (see under Alfred)
> Alice's Emerald *Chlorostilbon alice* (see under Alice)
> Alma's Thrush *Catharus ustulatus almae* (see under Alma)
> Naked-faced Spiderhunter *Arachnothera clarae* (see under Clara)
> Cuban Solitaire *Myadestes elisabeth* (see under Elisabeth)
> Bahama Woodstar *Calliphlox evelynae* (see under Evelyn)
> Owl genus *Gisella* (see under Gisella)

Flowerpecker sp. *Prionochilus johannae* (see under Johanna)

Black-tailed Trainbearer ssp. *Lesbia victoriae juliae* (see under Julia)

Mountain Scops Owl ssp. *Otus spilocephalus luciae* (see under Lucy)

Western Jackdaw ssp. *Coloeus monedula sophiae* (see under Sophia/Sophie)

Green-breasted Mountaingem *Lampornis sybillae* (see under Sybil)

Vassori's Tanager *Tangara vassorii* (see under Vassori)

Common House Martin ssp. *Hirundo urbica vogti* (see under Vogt)

Tawny Owl ssp. *Strix aluco willkonskii* (see under Willkonski)

Pin-striped Tit-babbler ssp. *Macronus gularis woodi* (see under Wood)

Golden Whistler ssp. *Pachycephala pectoralis youngi* (see under Young)

Gender allocation and deliberate mis-spellings in scientific names

In some cases we know that a bird is named after a man, even though its scientific name is in feminine form – an example is the recently-described *samveasnae*, named after Sam Veasna. This only seems to occur when a name ends in the letter 'a'. Singular Latin nouns ending in 'a' are regarded as feminine – for example, 'mensa' means 'table', and the possessive/genitive case is 'mensae', not 'mensai'. But the masculine ending remains more widespread. For instance, two birds described in the 1920s named after Cervera (a wren and a rail) both have the binomial *cerverai*. There are also cases where female eponyms are wrongly 'latinised' as masculine.

Some 'Latin' rendering of names may confuse because of slight alphabet differences. For example, any name beginning 'Mc' may, in a binomial, be rendered 'mac'. It is wise, therefore, to search for both spellings. Confusion may also arise as other alphabets (such as cyrillic) have fewer or larger numbers of letters, some of which are often interchangeable when written in the modern English version of the Roman alphabet. Examples are V and W, and J and Y, and letter combinations such as Cz and Ts for Czar and Tsar.

By convention, diacritical marks, such as accents in French and the tilda used in Spanish and Portuguese, have to be ignored in scientific names and the phonetic sense of them expressed in other ways. The Scandinavian letters å, ä, ö and ø are normally expressed as aa, ae, oe and oe and the German ö and ü as oe and ue. In the English name, either spelling is acceptable. We have tried to use correct accents in people's names or their book titles etc.

Weighing the evidence

Ultimately, our decisions on what to include in the book depended upon the weight of available evidence. Wherever there is any doubt, we have made this clear. In some cases we have had to reject a possible attribution when the evidence is just too slim. Nevertheless, it is possible that some (hopefully very few) names in the book have been wrongly attributed, and the authors take full responsibility for all errors and omissions. We welcome any corrections and new information so that future editions may be updated. Please send any comments to the authors, care of the publishers.

The problem of John Gould and Richard Bowdler Sharpe

We came across a real problem in deciding whether some of the more puzzling birds on the list were purely imaginary inventions or genuine mistakes. For example, we know of several hummingbirds and waders that were later shown to be hybrids, not new species. Gould published a considerable corpus of plates and, after his death, Sharpe put out many more that Gould had not published, including some that Gould had not even completed. An unfinished Gould engraving may have appeared to Sharpe to have been complete and therefore a new species. We suspect that poorly prepared or incomplete skins could sometimes have deceived one or both of these gentlemen; modern scientific techniques now allow much more precise comparisons.

A good example of this problem was revealed during our efforts to identify 'Conrad' of Conrad's Inca. Gould calls the bird *Bourcieria conradii*. Bourcier described a bird (1847) that he named *Trochilus conradii*. He bought this bird from Parzudaki, who collected it near Caracas. The bird was processed by Leadbeater (the taxidermist) and was part of the Loddiges Collection of hummingbirds and allies. Gould named another bird, Parzudaki's Starfrontlet, after Parzudaki. Unfortunately Parzudaki's first name was not Conrad, and Gould seems to have stuck with the binomial (*conradii*) given to it by Bourcier, not noting, and possibly not even knowing, the identity of Conrad.

Alas, it is very hard to discover how much Sharpe altered Gould's nomenclature when he took over Gould's work after the latter's death. So any conjecture that Sharpe might have been tempted to embellish such a lucrative series of prints would be pure speculation. We know that modern examination has established that a few of Gould's specimens were unique hybrids and not new species as he believed. In a similar vein, not every species named by Lesson existed. He is known to have had a vivid imagination, and is even rumoured to have put bits of skins together to create 'new' birds. There was a period when such fantasies were openly produced to entertain observers, and times when charlatans produced these three-dimensional lies to make money from collectors. Indeed, so widespread was it at one time that a stuffed Duck-billed Platypus was assumed to be a deliberate mix of bird and mammal skins and dismissed as a hoax!

As the great majority of Gould's and Sharpe's taxonomic judgements were perfectly valid and accurate, we have not eliminated anything we considered (however remotely) to be dubious in case readers encounter these names in reproductions of the plates or find references to them elsewhere.

The Dictionary

A

Aagaard

Buffy Fish Owl ssp. *Ketupa ketupu aagaardi* **Neumann**, 1935

Carl Johan Aagaard (1882–1950) was a Danish naturalist who lived in Siam (Thailand) (1910–1932). He wrote *The Common Birds of Bangkok* (1930).

Abadie

Crested Tit ssp. *Lophophanes cristatus abadiei* **Jouard**, 1929
Chestnut-headed Tesia ssp. *Tesia castaneocoronata abadiei* **Delacour** & **Jabouille**, 1930
[Syn. *Oligura castaneocoronata abadiei*]

René Marquis d'Abadie (1895–1971) was a French ornithologist and zoologist. He was in French Indochina (1929). His collection is housed at the Natural History Museum of Nantes.

Abbot

Abbot's Tanager *Thraupis abbas* **Deppe**, 1830
[Alt. Yellow-winged Tanager]

No explanation is given by the author but we believe this may refer to Abbot Lawrence (1792–1855), a merchant, manufacturer, diplomat, statesman and philanthropist who may have had some links with the Deppe brothers, one of whom, Wilhelm, first described the tanager. Abbot Lawrence supported the natural sciences in general and the work of Louis Agassiz in particular. He could easily have met or known of Ferdinand Deppe (1794–1861), Wilhelm Deppe's younger brother, who collected for the Berlin Museum in Mexico (1824–1827) with William Bullock and Count von Sack (who trained Deppe to prepare skins). Ferdinand Deppe returned to Mexico (1828–1829) with a botanist friend and collected botanical and zoological specimens as well as native artifacts. He collected a little in California and also in Hawaii (1830). He was also an artist and gardener (he has a rose named after him), and is honoured in several plant names. The majority of his bird specimens were studied and catalogued by Lichtenstein, although very poorly and with little acknowledgement of Deppe. Most of Deppe's collections were only studied much later. Wilhelm Deppe was the accountant of the Zoological Department at the Berlin Museum; in his description of the tanager (1830) he did no more than copy Lichtenstein's notes.

Abbott, J. R.

Abbott's Babbler *Malacocincla abbotti* **Blyth**, 1845

Lieutenant-Colonel J. R. Abbott (1811–1888) was Principal Assistant Commissioner at Nimarr on the island of Ramree in the Arakan province of Burma (Myanmar) (1837–1845). As a captain stationed on Ramree he discovered the babbler.

Abbott, W. L.

Vanga genus *Abbottornis* **Richmond**, 1897 NCR
[Now in *Leptopterus*]

Abbott's Booby *Papasula abbotti* **Ridgway**, 1893
Abbott's Sunbird *Cinnyris abbotti* Ridgway, 1894
[Syn. *Cinnyris sovimanga abbotti*]
Barbet sp. *Melanobucco abbotti* Richmond, 1897 NCR
[JS *Lybius leucocephalus albicauda*]
Abbott's Starling *Cinnyricinclus femoralis* Richmond, 1897
Simeulue Serpent Eagle *Spilornis abbotti* Richmond, 1903
Abbott's Cuckooshrike *Coracina abbotti* **Riley**, 1918
[Alt. Pygmy Cuckooshrike, Celebes Mountain Greybird]

Malagasy Sacred Ibis ssp. *Threskiornis bernieri abbotti* Ridgway, 1893
Abbott's White-throated Rail *Dryolimnas cuvieri abbotti* Ridgway, 1894 EXTINCT
Bluethroat ssp. *Luscinia svecica abbotti* Richmond, 1896
Abbott's Blue-rumped Parrot *Psittinus cyanurus abbotti* Richmond, 1902
Black-naped Monarch ssp. *Hypothymis azurea abbotti* Richmond, 1902
Hooded Pitta ssp. *Pitta sordida abbotti* Richmond, 1902
Abbott's Lesser Sulphur-crested Cockatoo *Cacatua sulphurea abbotti* **Oberholser**, 1917
Northern Potoo ssp. *Nyctibius jamaicensis abbotti* Richmond, 1917
Nicobar Megapode ssp. *Megapodius nicobariensis abbotti* Oberholser, 1919
Abbott's Andaman Parakeet *Psittacula alexandri abbotti* Oberholser, 1919
[Alt. Red-breasted Parakeet ssp.]
Antillean Piculet ssp. *Nesoctites micromegas abbotti* **Wetmore**, 1928

William Louis Abbott (1860–1936) was a naturalist and collector. Initially qualifying as a physician at the University of Pennsylvania and working as a surgeon at Guy's Hospital in London, he decided not to pursue medicine but used his private wealth to engage in scientific exploration. As a student (1880) he had collected in Iowa and Dakota and in Cuba and San Domingo (1883), in the company of Joseph Krider, son of the taxidermist John Krider. He went to East Africa (1887), spending two years there. He studied the wildlife of the Indo-Malayan region (1891), using his Singapore-based ship *Terrapin*, and made large collections of mammals from South-East Asia for the USNM in Washington D.C. He switched to Thailand (1897) and spent 10 years exploring and collecting in and around the China Sea. He provided much of the Kenya material in the USNM and was the author of *Ethnological Collections in the United States National Museum from Kilima-Njaro, East Africa* (1890/91). He returned to Haiti and San Domingo (1917), exploring the interior and discovering more new birds. He retired to Maryland but continued the study of birds all his life. Amongst other taxa two mammals and two reptiles are named after him.

Abdim

Stork genus *Abdimia* **Bonaparte**, 1855 NCR
[Now in *Ciconia*]

Abdim's Stork *Ciconia abdimii* **Lichtenstein**, 1823
[Alt. White-bellied Stork]

Bey El-Arnaut Abdim (1780–1827) was a Turkish governor of Dongola in Sudan (1821–1827). He was of great assistance to Rüppell (q.v.) on his North African expedition.

Abdulali

Nicobar Scops Owl *Otus alius* Rasmussen, 1998

Red-vented Bulbul ssp. *Pycnonotus cafer humayuni*
    Deignan, 1951
Besra ssp. *Accipiter virgatus abdulalii* **Mees**, 1981

Shri Humayun Abdulali (1914–2001) was an Indian ornithologist, a cousin of Salim Ali (q.v.). He became Honorary Secretary of the Bombay Natural History Society (1950) and, whilst there, catalogued their collection of bird skins. He was a prolific contributor to their journal, writing c.300 papers. He led two expeditions to the Andaman and Nicobar Islands (1964 and 1966). An amphibian is named after him. Rasmussen explained: '… *alius*, which is Latin for "other" (this being another scops owl from the Nicobar Islands), encapsulates the family name of Mr Humayun Abdulali, who first collected this species, and contributed a great deal to Indian ornithology …'

Abe

Willow Tit ssp. *Poecile montanus abei* Mishima, 1961 NCR
[NUI *Poecile montanus restrictus*]

Very possibly named after Yoshio Abe (1883–1945), who was Professor of Zoology at Karahuto Normal University. He was the first Japanese scientist to study and publish on kinorhynchs (microscopic marine invertebrates) (1930), and one, *Dracoderes abei*, was named after him as late as 1990. A mammal and an amphibian are named after him.

Abeillé

Hummingbird genus *Abeillia* **Bonaparte**, 1850

Abeillé's Grosbeak *Hesperiphona abeillei* **Lesson**, 1839
[Alt. Hooded Grosbeak]
Abeillé's Hummingbird *Abeillia abeillei* Lesson & **De Lattre**,
    1839
[Alt. Emerald-chinned Hummingbird]
Abeillé's Oriole *Icterus abeillei* Lesson, 1839
[Alt. Black-backed Oriole; Syn. *Icterus galbula abeillei*]
Brown Tanager *Orchesticus abeillei* Lesson, 1839
Black-capped Sparrow *Arremon abeillei* Lesson, 1844

Doctor Abeillé and his wife Félice (q.v.) Olymp were French naturalists and collectors who lived in Bordeaux. There is an illustration by Gould of a hummingbird he called Abeillé's Flutterer *Myiabeillia typica*, which appears nowhere else but is the same bird as Abeillé's Hummingbird. Lesson specifically stated that the Brown Tanager is named after the Doctor.

Abert

Abert's Towhee *Melozone aberti* **Baird**, 1852
[Syn. *Pipilo aberti*]

James William Abert (1820–1897) was a Major in the US Army and an ornithologist. He graduated from Princeton (1838) and then entered the US Military Academy. He transferred to the Corps of Topographical Engineers (1843). He was with Frémont's third expedition (1845), whose assignment was 'to make reconnaissance southward and eastward along the Canadian River through the country of Kiowa and Comanche'. Frémont, however, chose to take his main party on to California and gave command of the Canadian River mission to Abert. The expedition followed the headwaters of the Canadian through the breaks in eastern New Mexico and into the Texas Panhandle. In his report Abert described in detail the geology, flora and fauna of the Canadian Valley and mapped the area. He accompanied General Kearny's Army of the West to New Mexico (1846) while continuing his studies of natural science. Afterwards he visited each of the Rio Grande pueblos, then went to Washington to submit his report to Congress. In the Civil War he served in the Shenandoah Valley (1861–1862). He was promoted to Major (1863) and assigned to the US Army Corps of Engineers. Soon afterwards he was severely injured in a fall from his horse, which led to his resignation from the army (1864). He taught English literature at the University of Missouri (1877–1879). Despite their value, his western frontier journals lay almost forgotten in government files until 1941, when Carroll first published the 1845 report in the Panhandle-Plains Historical Review. Keleher published Abert's New Mexico report (1962). John Galvin, a Californian historian, edited special publications of the journals under the title: *Through the Country of the Comanche Indians in 1845* (1967 and 1970). They featured illustrations of Abert's watercolours, many of which were obtained from his descendants. Abert also collected birds for Baird, including the towhee. A mammal is named after him.

Abingdon

Golden-tailed Woodpecker *Campethera abingoni* **A. Smith**, 1836

Montague Bertie, 5th Earl of Abingdon (1784–1854). A law graduate from Oxford University, he held the office of High Steward of Abingdon and Lord-Lieutenant of Berkshire. Smith appears to have missed the *d* from Abingdon's name when spelling the binomial.

Abravaya

Rufous-breasted Hermit ssp. *Glaucis hirsutus abrawayae* **Ruschi**, 1973

Paul Abravaya (b.1945) is an American zoologist. He was in Brazil (1969–1974) working as a field zoologist for Augusto Ruschi. His job was to collect small mammals and assist in fieldwork involving plants and animals. Back in the USA he earned a Masters degree in Biology and published several research papers on small mammals. Later he owned and ran a business called 'Tropical Ecotours' whose principal goal was to educate tourists on the ecology of Costa Rica, Panama and Brazil. He retired (2008) and now spends his time in the field of nature photography. A genus of mammals is named after him. Ruschi used a *w* instead of a *v*; the trinomial should really be *abravayae*.

Aceval

Swallow-tailed Cotinga *Psaliurus acevalianus* **Bertoni**, 1901 NCR
[JS *Phibalura flavirostris*]

Emilio Aceval Merín (1853–1931) was a Paraguayan statesman and president of his country (1898–1902).

Acis

Common Paradise Kingfisher ssp. *Tanysiptera galatea acis* **Wallace**, 1863

Acis, in Greek mythology, was a shepherd whom the nymph Galatea loved. Polyphemus killed him out of jealousy, but Galatea turned his blood into the Acis River.

Acteon

Grey-headed Kingfisher ssp. *Halcyon leucocephala acteon* **Lesson**, 1830

Actaeon, in Greek mythology, was a hunter who made the mistake of encountering the goddess Diana when she was nude. She turned him into a stag and his own dogs tore him to pieces.

Ada

Pompadour Green Pigeon ssp. *Treron pompadora ada* **Meise**, 1930 NCR; NRM

Miss Ada Geertsema (DNF) was the sister of Lieutentant-Colonel Coen Geertsema, a military aide to Prince Bernhard of the Netherlands.

Adalbert

Adalbert's Eagle *Aquila adalberti* **Brehm**, 1861
[Alt. Spanish Imperial Eagle]

Adalbert Wilhelm Georg Ludwig, Prince of Bavaria (1828–1875), son of Ludwig I, King of Bavaria, and husband to Infanta Amalia Philippina of Spain. He appears to have led a singularly untroubled and unadventurous life.

Adametz

Chubb's Cisticola ssp. *Cisticola chubbi adametzi* **Reichenow**, 1910
White-throated Greenbul ssp. *Phyllastrephus albigularis adametzi* Reichenow, 1916 NCR
[JS *Phyllastrephus poensis*]

Major Karl Moritz Ernst Gustav Wilhelm von Adametz (b.1877) was a German army officer (1896–1919) in German colonial Cameroon (1906–1916). He crossed into Spanish Guinea (1916) and was interned (WW1).

Adams, A. L.

Adams's Snowfinch *Montifringilla adamsi* Adams, 1859
[Alt. Tibetan Snowfinch]

Sand Lark ssp. *Calandrella raytal adamsi* **Hume**, 1871

Andrew Leith Adams (1826–1882) was a physician, naturalist and geologist. He was an army surgeon in India (c.1848) with the 22nd Foot (Cheshire Regiment). He wrote an influential report on the cholera epidemic in Malta, where the regiment was stationed (c.1865). He became an academic after retiring from the army, being variously Professor of Zoology at the Royal College of Science in Ireland and Professor of Natural History at Trinity College, Dublin. He wrote *Wanderings of a Naturalist in India, the Western Himalaya and Cashmere* (1867) and *Notes of a Naturalist in the Nile Valley and Malta* (1871). Adams wrote the snowfinch description (1859), accidentally naming it after himself by making reference in print to a previously unpublished description by Frederick Moore of the East India Company's London Museum.

Adams, C. F.

Oriental Magpie Robin ssp. *Copsychus saularis adamsi* **Elliot**, 1890

Charles Francis Adams (1857–1893) was an American taxidermist and collector. He graduated from the University of Illinois (1883). He travelled widely, collecting in Florida, Borneo (1887), Galapagos Islands and New Zealand, where he worked as a taxidermist for the Auckland Museum (1884–1887).

Adams, E.

White-billed Diver *Gavia adamsii* **G. R. Gray**, 1859
[Alt. Yellow-billed Loon]

Edward Adams (1824–1856) was an Arctic explorer (1849–1856), as surgeon and naturalist aboard HMS *Enterprise*. Part of this trip was to search the Bering Sea looking for the missing Franklin expedition. *Enterprise* became locked in by ice in Lancaster Sound and failed to find any sign of Franklin. Adams sailed as ship's surgeon aboard HMS *Hecla* (1856) for the West Africa station. His health was not good after his experiences in the Arctic and he succumbed to typhus and was buried in Sierra Leone.

Adams, P. F.

Sacred Kingfisher ssp. *Todiramphus sanctus adamsi* **Mathews**, 1916 NCR
[NUI *Todiramphus sanctus vagans*]

The original description has no etymology so we can only speculate on Adams's identity. There are a number of possibilities, but the one we favour is Philip Francis Adams (1828–1901), a surveyor, viticulturist and astronomer who was on Lord Howe Island (where the holotype was collected) to observe the transit of Venus (1882).

Adamson

Stout-billed Cuckooshrike ssp. *Coracina caeruleogrisea adamsoni* **Mayr** & **Rand**, 1936

Charles Thomas Johnston Adamson (1901–1978) emigrated from England to Australia (1923). He worked as a sheep shearer and in the sugarcane fields until moving to Papua (1926) to prospect for gold. He was on the Archbold Papua Expedition (1933–1934) and was co-leader of a second expedition (1936). He was a patrol and police officer (1935–1939), and served in the Royal Australian Navy (WW2) in the North Atlantic, the Indian Ocean and finally off Papua. He owned and ran a plantation in Papua (1945–1964), retiring to live in Cooktown, Queensland (1964–1978). When his health failed, he shot himself.

Adanson

African Blue Quail *Coturnix adansonii* **E. Verreaux** & **J. Verreaux**, 1851

Michel Adanson (1727–1806) was a French botanist. He was in West Africa (Senegal) as a bookkeeper for Compagnie des Indes (1748–1754). He collected specimens of all kinds, and after returning to France (1754) wrote two books: *Histoire Naturelle du Sénégal* (1757) and *Familles Naturelles des Plantes* (1763). He lost his position and his income as a result of the French Revolution (1789) and was supported by his servants. He eventually died in penury, leaving a last wish that a wreath representing the 58 plant families that he had named be placed on his grave. Many plant and animal taxa are named after him including generic name of the baobab tree, *Adansonia*.

Adda

Booted Racket-tail ssp. *Ocreatus underwoodii addae* **Bourcier**, 1846

Mrs Adda Wilson (DNF) was the wife of William Savory Wilson (q.v.), who was a cloth merchant and financier. They lived in Paris (1840s) and returned to Philadelphia (1853). He was a brother of Wilson, E. (q.v.) and Wilson, T. B. (q.v.).

Adela

Adela's Hillstar *Oreotrochilus adelae* **Gould**
[Alt. Wedge-tailed Hillstar *Oreotrochilus adela* **d'Orbigny** & **Lafresnaye**, 1838]

Only in Gould do we find this common name. It is so similar to *Oreotrochilus adela* (Wedge-tailed Hillstar) that Gould probably just made a mistake in the spelling, and so others assumed that *adelae* referred to a woman.

Adelaide

Adelaide Rosella *Platycercus adelaidae* **Gould**, 1841 NCR

This bird (often regarded as a hybrid between the Crimson and Yellow Rosellas, and indeed these two birds are commonly considered one species, *Platycercus elegans*) was not named directly after a person, but after the city of Adelaide, founded 1837, which, in its turn was named after Queen Adelaide (Adelheid von Sachsen-Meiningen), a German princess (1792–1849) who married (1818) William Henry of Clarence (1764–1837), later the British King William IV (1830–1837).

Adelaide (Swift)

Adelaide's Warbler *Dendroica adelaidae* **Baird**, 1865

Adelaide Swift (DNF) was the daughter of Robert Swift, an American financier who was patron of a number of collecting trips and obtained the first specimen of the warbler. He collected for the USNM in Puerto Rico, the type locality, during a visit there (1865) with George Latimer.

Adelbert

Adelbert's Bowerbird *Sericulus bakeri* **Chapin**, 1929
[Alt. Fire-maned Bowerbird]

This bowerbird is confined to the Adelbert Mountains in New Guinea, and doubtless derives its vernacular name from the area rather than directly after a person. We believe that the mountains were named after Adelbert von Chamisso (1781–1838), who is also known as Louis Charles Adelaide de Chamisso. Chamisso was the botanist aboard the *Rurik*, captained by Otto von Kotzebue, which explored the South Seas (1816–1823). The crew also conducted the first-ever hydrographical, botanical, and ethnological studies on the nearby Marshall Islands. He kept a diary, *Reise um die Welt mit der Romanzoffischen Entdeckungs-Expedition* (1836) ('Voyage Around the World with the Romanzov Discovery Expedition'), which became a classic of its kind. He was a German poet, playwright, linguist and naturalist, born in France at the Château de Boncourt. He served as a page at the court of William II of Prussia and, after army service and travels, became keeper of the Royal Botanical Gardens. He edited the *Musenalmanach* (1804–1806). His sentimental poetic cycle *Frauenliebe und Leben* (1830) was set to music by Schumann. *Peter Schlemihls wundersame Geschichte* (1814), his tale of a man who sold his shadow to the devil, has become legendary.

Adelbert, M.

Buff-throated Sunbird *Chalcomitra adelberti* **Gervais**, 1834

Vice-Admiral Marie-Charles Adelbert le Barbier de Tinan (1803–1876) was a French explorer and shell collector. He was a career sailor who attained very high office, culminating in being part of the ruling council of Algeria.

Adélie

Adélie Penguin *Pygoscelis adeliae* **Hombron & Jacquinot**, 1841

Adélie Dumont d'Urville (1798–1842) was the wife of Admiral Jules-Sebastien-César Dumont d'Urville, the French explorer who first found the penguin. He also named the Adélie Coast of Antarctica after her and other places in and around Antarctica, where he explored in the *Astrolabe* (1820s). New Zealand's 'Noises' islands are named following his remark about their shape. '*Voilà,*' exclaimed d'Urville when he first saw the clumpy little group of islands. '*C'est noisettes*', which is French for lamb chops!

Adendorff

Adendorff's Clapper Lark *Mirafra apiata adendorffi* **JA Roberts**, 1919 *NCR*
[NUI *Mirafra apiata apiata*]

We know that Adendorff was a South African collector who worked with the describer, but Roberts gave no etymology in his description. We think the person in question may have been Johannes Hendrik Georg Adendorff (1876–1918).

Admiral

Rock-loving Cisticola ssp. *Cisticola emini admiralis* **Bates**, 1930
[Alt. Lazy Cisticola; Syn. *Cisticola aberrans admiralis*]

(See **Lynes**)

Adolf Friedrich

Bustard sp. *Otis adolfi-friederici* **Neumann**, 1907 NCR
[Alt. Kori Bustard; JS *Ardeotis kori struthiunculus*]
Apalis sp. *Apalis adolphi-friederici* **Reichenow**, 1908 NCR
[Alt. Mountain Masked Apalis; JS *Apalis personata*]
Barbet sp. *Gymnobucco adolfi-friederici* Reichenow, 1908 NCR
[Alt. Sladen's Barbet; JS *Gymnobucco sladeni*]
Francolin sp. *Francolinus adolfi-friederici* Reichenow, 1908 NCR
[Alt. Red-winged Francolin; JS *Scleroptila levaillantii kikuyuensis*]
Puffback sp. *Dryoscopus adolfi-friederici* Reichenow, 1908 NCR
[Alt. Pink-footed Puffback; JS *Dryoscopus angolensis nandensis*]
Sunbird sp. *Nectarinia adolfi-friederici* Reichenow, 1908 NCR
[Alt. Red-chested Sunbird; JS *Cinnyris erythrocercus*]

Forest Wood-hoopoe ssp. *Phoeniculus castaneiceps adolfi-friederici* Reichenow, 1908 NCR
[JS *Phoeniculus castaneiceps brunneiceps*]

Fiery-breasted Bush-shrike ssp. *Malaconotus cruentus adolfi-friederici* Reichenow, 1908 NCR; NRM
Meyer's Parrot ssp. *Poicephalus meyeri adolfi-friederici* **Grote**, 1926 NCR
[JS *Poicephalus meyeri meyeri*]

Adolf Friedrich Albrecht Heinrich, Duke of Mecklenburg-Schwerin (1873–1969) was an explorer and colonial politician in Africa, and (1949–1952) first President of German Olympic Committee. He conducted scientific research on the African Rift Valley and crossed Africa from East to West (1907–1908). He led another expedition to Lake Chad and upper reaches of the Congo River and Nile in Sudan (1910–1911) He was the last colonial Governor of Togo (1912–1914). After WW1 he became Vice President, German Colonial Society for South West Africa. He wrote *Von Kongo zum Niger und Nil* (1912). A mammal, an amphibian and two reptiles are named after him.

Adolph

Adolph's Hermit *Phaethornis adolphi* **Gould**, 1857
[Alt. Boucard's Hermit, Syn. *Phaethornis striigularis adolphi*]

(See Adolphe **Boucard**)

Adolphine

Adolphina's Myzomela *Myzomela adolphinae* **Salvadori**, 1876
[Alt. Mountain Myzomela]

Adolphine Susanna Wilhelmina Bruijn *née* Adolfina van Rennesse van Duivenbode (1844–1919) was the wife of Anton August Bruijn (q.v.).

Aëdon

Thick-billed Warbler *Iduna aedon* **Pallas**, 1776
[Syn. *Acrocephalus aedon*]
Northern House Wren *Troglodytes aedon* **Vieillot**, 1809

In Greek mythology Aëdon was the wife of Zethus, king of Thebes. She killed her son, Itylus, and was then changed into a nightingale (or a goldfinch, in some versions).

Aegolius

Owl genus *Aegolius* **Kaup**, 1829

A mythical Cretan, one of several thieves who entered the sacred cave of bees in Crete in order to steal honey. There they perceived the cradle of the infant god Zeus, which was not meant to be seen by mortals. As no-one was allowed to be killed on that sacred spot, Aegolius and his fellow thieves were instead metamorphosed into birds.

Agassiz, A. E.

Darwin's Nothura ssp. *Nothura darwinii agassizii* **Bangs**, 1910
Tuamotu Reed Warbler ssp. *Acrocephalus atyphus agassizi* **Wetmore**, 1919 NCR
[NUI *Acrocephalus atyphus atyphus*]

Alexander Emanuel Agassiz (1835–1910) was born in Switzerland but emigrated to the US with his eminent palaeontologist father, Louis Agassiz (1849), and made a fortune out of copper mining. He graduated from Harvard (1855) and took a second degree (BSc.) (1857) after studying engineering and chemistry. He joined the US Coast Survey (1859) as an assistant, becoming a specialist marine ichthyologist. He worked (1860–1866) as an assistant at the Museum of Natural History that his father had founded at Harvard. He had become involved as an investor in a copper mining venture in Michigan and (1866) became the treasurer of the enterprise. After a struggle, he made the company prosperous, merged and acquired other companies and expanded the conglomerate of which he became president (1871–1910). He returned to Harvard (1870s) to pursue his interests in natural history, giving $500,000 for the Museum of Comparative Zoology there and being its Curator (1874–1885). He visited Peru and Chile (1875) to look at the copper mines and to survey Lake Titicaca. He helped in the examination and classification of the specimens collected by Wyville Thomson on the *Challenger* Expedition and took part in three dredging expeditions (1877–1880) on the Coast Survey's vessel *Blake*. He published much on marine zoology in bulletins and two books: *Seaside Studies in Natural History*, co-written with his stepmother Elizabeth Cary Agassiz (1865) and *Marine Animals of Massachusetts Bay* (1871). He died at sea aboard *Adriatic*.

Aglae

Blue-winged Minla ssp. *Minla cyanouroptera aglae* **Deignan**, 1942

Dr Stella Maria Aglae Deignan *née* Leche (1901–1993) was an American anthropologist and the wife of Herbert Girton Deignan (q.v.).

Aglaia

Tanager genus *Aglaia* **Lesson**, 1838 NCR
[Now in *Tangara*]

Rose-throated Becard *Pachyramphus aglaiae* **Lafresnaye**, 1839

Aglaia was one of the three Graces [daughters of Zeus & Eurynome, Aglaia was Splendour, Euphrosyne was Mirth, and Thalia was Good Cheer] in Greek mythology. Jobling (1991) and Wynne (1969) believe they are named after Aglaé Brelay (fl.1839), wife of French ornithologist and collector Charles Brelay.

Agnete

La Palma Blackbird *Turdus merula agnetae* H. Volsøe, 1949 NCR
[JS *Turdus merula cabrerae*]

Mrs Agnete Volsøe *née* Nielsen (b.1902) was a librarian and wife of Dr Svend Helge Volsøe (1908–1968), a Danish ornithologist who classified many of the subspecies of Canary Island birds.

Aguirre

Thekla Lark ssp. *Galerida theklae aguirrei* **Cabrera**, 1922 NCR
[JS *Galerida theklae erlangeri*]

Manuel Aguirre de Cárcer y de Tejada (1882–1969) was a Spanish diplomat and colonial administrator.

Aharoni

Aharoni's Eagle Owl *Bubo bubo interpositus* **Rothschild** & **Hartert**, 1910
Lesser Short-toed Lark ssp. *Calandrella rufescens aharonii* Hartert, 1910
Shore Lark ssp. *Eremophila alpestris aharonii* **Neumann**, 1934

Professor Dr J. Israel ben A. Aharoni (1882–1946) was born in Lithuania, educated in Prague, and became a zoologist in Israel. He was inducted into the Turkish Army (WW1) and became its official zoologist. He was sent to Damascus (1915) to establish a zoological museum there. Returning to Palestine he became a government zoologist under the British Mandate. When the Hebrew University was founded he became Curator of the Zoological Museum and lectured there until his death. During his numerous research expeditions throughout the Middle East, he assembled an animal collection that is still preserved in a church museum in the Old City of Jerusalem. He is most famous for finding the Golden Hamster *Mesocricetus auratus* (1939) near Mount Aleppo in Syria – a mother with a litter of 10, which he kept as pets and their progeny became the parents of *all* pet hamsters around the world. He wrote an autobiography, *Memories of a Hebrew Zoologist*.

Ahasver

House Sparrow ssp. *Passer domesticus ahasver* **Kleinschmidt**, 1904 NCR
[JS *Passer domesticus tingitanus*]

Ahasver is a traditional name for the Wandering Jew, a figure from medieval Christian folklore.

Aidem

Honeycreeper genus *Aidemedia* James & **Olson**, 1991 EXTINCT

Mrs Joan Aidem was a resident of Molokai and pioneer collector of fossil birds in the Hawaiian Islands (1971). The authors state that: 'The unusual terminal orthography results from our inability to resist creating a palindrome.'

Aigner

Red-wattled Lapwing ssp. *Vanellus indicus aigneri* **Laubmann**, 1913

Aigner (fl.1913) was a German taxidermist at the Munich Museum.

Aiken

Dark-eyed Junco ssp. *Junco hyemalis aikeni* **Ridgway**, 1873
Aiken's Screech Owl *Megascops kennicottii aikeni* **Brewster**, 1891
[Alt. Western Screech Owl ssp.]

Charles Edward Howard Aiken (1850–1936) was born in Vermont. He went to his father's ranch in Turkey Creek,

Colorado Springs (1871), and there he began to study and collect birds, describing their habits and nests. His work was edited by Thomas Brewer (q.v.) (1872) and published in the *Proceedings of the Boston Society of Natural History*. Aiken opened a taxidermy shop (1874). Much of his time was spent in pursuit of birds. He was described as a man of keen hearing and sight with a sharp eye for plumage. He identified a smaller, darker-marked race of the screech owl, hence the eponym. Aiken also acquired the holotype of the junco subspecies. He was honoured not so much for his contribution to ornithology but because he paid for the skins. Colorado College purchased Aiken's collection of 4,700 specimens (1907). Much had already been sent to the USNM. Other specimens were turned over to the college as he mounted them. Aiken joined the American Ornithologists' Union (1898) and was made an Honorary Life Associate (1926).

Ainley

Ainley's Storm Petrel *Oceanodroma* (*leucorhoa*) *cheimomnestes* Ainley, 1980

Dr David G. Ainley is an American marine biologist, ornithologist and climatologist. He graduated from Dickinson College (1968) and undertook his PhD at Johns Hopkins University (1971). He was Program Director for Marine Studies at Point Reyes Bird Observatory (1977–1995) and since then has been Senior Ecological Associate with a private company. He wrote *The Adélie Penguin: Bellwether of Climate Change* (2001) and co-wrote *Storm-petrels of the Eastern Pacific Ocean* (2011).

Ajax

Painted Quail-thrush *Cinclosoma ajax* **Temminck**, 1835

Ajax is a figure from Greek mythology. He was also known as Telamonian Ajax or Ajax the Great; a hero and legendary king of Salamis who plays an important role in Homer's *Iliad*. A mammal is named after him.

Akeley

Brown-chested Alethe ssp. *Pseudalethe poliocephala akeleyae* **Dearborn**, 1909
Desert Lark ssp. *Ammomanes deserti akeleyi* **D. G. Elliot**, 1897
Olive Ibis ssp. *Bostrychia olivacea akeleyorum* **Chapman**, 1912

Carl Ethan Akeley (1864–1926) was an American taxidermist, explorer and artist (the lark is named after him) and his first wife was Delia Julia Akeley *née* Denning (1875–1970), after whom the alethe is named. The ibis's trinomial, meaning 'of the Akeleys', honours them both. They divorced and he re-married (1924).

Alan

Japanese White-eye ssp. *Zosterops japonicus alani* **Hartert**, 1905

(See **Owston**)

Akyildiz

Graceful Prinia ssp. *Prinia gracilis akyildizi* Watson, 1961

Zubeyir Akyildiz (DNF) of the Turkish Forest Department, Ankara, shared in the describer's explorations of southern Turkey.

Albert (Geoffroy Saint-Hilaire)

Woodpecker sp. *Venilia albertuli* **Bonaparte**, 1850 NCR
[Alt. Blood-coloured Woodpecker; JS *Veniliornis sanguineus*]

Albert Geoffroy Saint-Hilaire (1835–1919) was a French zoologist. He became joint Director of the Jardin d'Acclimatation, Paris (1859). As Albert was only 15 when Bonaparte described the bird and the suffix 'uli' means little, we think he may have intended the name as some kind of tribute to Albert's father, Isidore Geoffroy St Hilaire.

Albert, Prince

Albert's Lyrebird *Menura alberti* **Bonaparte**, 1850
[Alt. Prince Albert's Lyrebird]
Prince Albert's Curassow *Crax alberti* **Fraser**, 1852
[Alt. Blue-billed Curassow]

Prince Albert's Riflebird *Ptiloris magnificus alberti* **D. G. Elliot**, 1871
[Alt. Magnificent Riflebird ssp.]

Prince Albert, the Prince Consort (1819–1861), was the husband of Queen Victoria. He was a keen innovator, seeker of knowledge and interested in all of the sciences. So although the birds were named in his honour, this may not have been purely an acknowledgement of his social standing.

Albert, King

Yellow-crested Helmet-shrike *Prionops alberti* **Schouteden**, 1933

King Albert I (1875–1934) was King of the Belgians (1909–1934). Schouteden was a Belgian zoologist who studied the fauna of the Congo, so he honoured his king.

Albert (Meek)

White-eye sp. *Zosterops alberti* **Rothschild** & **Hartert**, 1908 NCR
[JS *Zosterops ugiensis*]

Collared Kingfisher ssp. *Todiramphus chloris alberti* Rothschild & Hartert, 1905
Asian Koel ssp. *Eudynamys orientalis alberti* Rothschild & Hartert, 1907
Lewin's Rail ssp. *Lewinia pectoralis alberti* Rothschild & Hartert, 1907
Striated Thornbill ssp. *Acanthiza lineata alberti* **Mathews**, 1920
Northern Fantail ssp. *Rhipidura rufiventris albertorum* Hartert, 1924 NCR
[JS *Rhipidura rufiventris setosa*]

(See **Meek**) The plural trinomial *albertorum* used for the fantail honours Albert F. Eichhorn (q.v.) as well as Albert Meek.

Albertina (Marschall)

Bay-headed Tanager ssp. *Tangara gyrola albertinae*
Pelzeln, 1877

Countess Albertina Marschall (DNF) was a daughter of Count August F. de Marschall.

Albertina (Schlegel)

Albertina's Myna *Streptocitta albertinae* **Schlegel**, 1866
[Alt. Bare-eyed Myna, Schlegel's Myna]

Albertina Catharina Petronella Schlegel *née* Pfeiffer (b.1829) was the second wife of German ornithologist Hermann Schlegel (q.v.). She was Lidth de Jeude's (q.v.) step-daughter and Schlegel knew her well before they married (1869) as he dedicated the myna to her as being 'one of the aimiable daughters of our dead friend, Professor van Lidth de Jeude.'

Albertis

Black-billed Sicklebill *Drepanornis albertisi* **P. L. Sclater**, 1873
Papuan Mountain Pigeon *Gymnophaps albertisii* **Salvadori**, 1874
Crowned Pigeon sp. *Goura albertisii* Salvadori, 1876 NCR
[Alt. Southern Crowned Pigeon; JS *Goura scheepmakeri*]
Mountain Owlet-nightjar *Aegotheles albertisi* P. L. Sclater, 1874

(See **D'Albertis**)

Albin

Great Curassow *Crax albini* **Lesson**, 1831 NCR
[JS *Crax rubra*]

Eleazar Albin (c.1690–c.1759) was a British painter, naturalist and author. He was born Eleazar Weiss in Germany but moved to England (1707) and anglicised his name by translating it. He wrote the 3-volume *A Natural History of Birds* (1731–1738).

Alcasid

Brown Tit Babbler ssp. *Macronus striaticeps alcasidi*
Du Pont & **Rabor**, 1973

Dr Godofredo L. Alcasid (DNF) was Director of the Philippine National Museum. He co-wrote 'A new race of the naked-faced spider-hunter (*Arachnothera clarae*) from Luzon' (1968).

Alcide

Hummingbird genus *Alcidius* **Boucard**, 1895 NCR
[Now in *Oreotrochilus*]

(See **D'Orbigny**)

Alcippe

Babbler (Pellorneidae) genus *Alcippe* Blyth, 1844

Alcippe was the daughter of Ares, the god of war in Greek mythology. Alcippe was also the name of one of the seven daughters of Alcyoneus, who were transformed into kingfishers.

Alcock

Alcock's Snake Eagle *Spilornis cheela rutherfordi* **Swinhoe**, 1870
[Alt. Crested Serpent Eagle]

Sir Rutherford Alcock (1809–1897) spent most of his career in Asia. He qualified in medicine, then served as a surgeon in the marine brigade in the Carlist Wars (1836). He became Deputy Inspector-General of Hospitals, (1836–1837). He was appointed Consul at Fuchow in China (1844), and was promoted to the Consulate at Shanghai (1845–1846). When Japan opened up to Westerners again (1858) Alcock was appointed to be Consul-General in Japan and he became Minister Plenipotentiary (1859). There was still much resentment in Japan against foreigners and Alcock's native interpreter was murdered at the gates of the Legation (1860). Then Ronin (unemployed members of the old Samurai class) attacked the Legation (1861), which he and his staff repelled. Soon afterwards he returned to England on leave and published his experiences as *Capital of the Tycoon* (1863). He returned to Japan (1863) and gathered English, American, Dutch and French warships to attack and destroy shore batteries as punishment for the previous attacks on Western ships. He was transferred back to China (1865) and was the British Minister in Peking (Beijing) until his retirement (1871). It was Alcock who persuaded the Japanese Court to adopt driving on the left like the British.

Aldrich

Black-capped Chickadee ssp. *Poecile atricapillus aldrichi*
Braund & **McCullagh**, 1940 NCR
[JS *Poecile atricapillus atricapillus*]

Dr John Warren Aldrich (1906–1995) was an American ornithologist and ecologist in the US Fish & Wildlife Service. His bachelor's degree was awarded by Brown University (1928), and his master's (1933) and doctorate (1937) by Western Reserve University, Cleveland.

Aldrovandi

Oriental Hobby *Falco aldrovandii* **Temminck**, 1822 NCR
[JS *Falco severus*]

Dr Ulisse Aldrovandi (1522–1605) was an Italian naturalist. He studied law and medicine, graduating as a physician at Bologna (1553). He was accused of heresy before the Inquisition (1549) but was able to clear himself. He was appointed Professor of Philosophy and Lecturer on Botany at the University of Bologna (1551), becoming Professor of Natural History (1561). He became the first Director of Bologna's botanical garden (1568). He was instrumental in the founding of Bologna's public museum. He willed his huge collection of natural history specimens to the Senate of Bologna, but these were gradually distributed among a variety of institutions. A reptile is named after him.

Aldunate

> Grey-hooded Sierra Finch *Chlorospiza aldunatei* **C. Gay**, 1847 NCR
> [JS *Phrygilus gayi*]

General José Santiago Aldunate Toro (1796–1864) was a Chilean army officer and politician. He joined the army led by Bernardo O'Higgins (1810) and proved to be an outstanding soldier in the wars of liberation against Spain (1810–1824). He was a minister in President Bulnes's government (1842) and a senator for Valparaiso (1842–1852), being also Mayor of Valparaiso (1845) and Director of the Military Academy (1847).

Alecto

> Shining Flycatcher *Myiagra alecto* **Temminck**, 1827

> Palm Cockatoo ssp. *Probosciger aterrimus alecto* Temminck, 1835 NCR
> [JS *Probosciger aterrimus goliath*]

Alecto was a character from Greek mythology, one of the three Furies. According to Hesiod, she was the daughter of Gaea fertilised by the blood spilled from Uranus when Cronus castrated him. Four mammals are named after her.

Alexander, A. H.

> Hummingbird sp. *Aphantochroa alexandri* **Boucard**, 1891 NCR
> [JS *Heliodoxa xanthogonys*]

A. H. Alexander (fl.1878) was an American taxidermist in West Hoboken who styled himself 'Professor'. He published what we assume is a trade catalogue, *Alexander's Taxidermist Depot, where Everything a Taxidermist Needs can be Obtained* (1877). His shop was burgled (1878) by a rival Canadian taxidermist who was careless enough to drop a letter there with his name on it!

Alexander, A. M.

> Alexander's Ptarmigan *Lagopus lagopus alexandrae* **Grinnell**, 1909
> [Alt. Willow Ptarmigan ssp.]
> White-breasted Nuthatch ssp. *Sitta carolinensis alexandrae* Grinnell, 1926

Miss Annie Montague Alexander (1867–1950) founded the Museum of Zoology at California University. She studied both nursing and music but had to abandon both due to eyestrain and headaches. She first became interested in palaeontological work while attending John C. Merriam's lectures at the University of California (1900). Thereafter she was closely associated with the Department and Museum of Paleontology, as well as the Museum of Vertebrate Zoology. She offered to finance Merriam's fossil-hunting expeditions provided she could participate, and she continued to give financial support as well as participating in many field expeditions including Alaska (1908). Despite financing the expeditions and taking part in the digs she did most of the cooking too! She continued taking part right up until she died of a stroke (1950). In an account of the 1905 expedition she wrote: 'We worked hard up to the last. My dear friend Miss Wemple stood by me through thick and thin. Together we sat in the dust and sun, marking and wrapping bones. No sooner were these loaded in the wagon for Davison to haul to Mill City than new piles took their places. Night after night we stood before a hot fire to stir rice, or beans, or corn, or soup, contriving the best dinners we could out of our dwindling supply of provisions. We sometimes wondered if the men thought the fire wood dropped out of the sky or whether a fairy godmother brought it to our door, for they never asked any questions ...'

Alexander, B. F.

> Alexander's Swift *Apus alexandri* **Hartert**, 1901
> [Alt. Cape Verde Swift]
> Alexander's Akalat *Sheppardia poensis* Alexander, 1903
> Barbet sp. *Tricholaema alexandri* **Shelley**, 1903 NCR
> [JS *Tricholaema frontata*]
> Alexander's Kestrel *Falco alexandri* **Bourne**, 1955
> [Syn. *Falco tinnunculus alexandri*]

> Pale-fronted Negrofinch ssp. *Nigrita luteifrons alexanderi* **Ogilvie-Grant**, 1907
> Cut-throat Finch ssp. *Amadina fasciata alexanderi* **Neumann**, 1908
> Lesser Honeyguide ssp. *Indicator minor alexanderi* **C. H. B. Grant**, 1915
> Yellow-bellied Eremomela ssp. *Eremomela icteropygialis alexanderi* **W. L. Sclater** & **Mackworth-Praed**, 1918
> Baglafecht Weaver ssp. *Ploceus baglafecht alexandri* W. L. Sclater, 1925 NCR
> [JS *Ploceus baglafecht eremobius*]

Captain Boyd Francis Alexander (1873–1910) was an ornithologist and traveller in Africa. He was educated at Radley College (1887–1891) and joined the army (1893). He made a comprehensive study of nesting habits and migration of the birds at Rye, Sussex (1896). He led a scientific expedition to the Cape Verde Islands (1897), where he made an extensive collection of native bird species and (1898) went on his first African journey to the Zambezi and Kafuk rivers. He was appointed to the Gold Coast constabulary (1900) and took part in the relief of Kumasi. He explored Lake Chad (1904–1905) with his younger brother Captain Claud Alexander, who died (1904) of fever (Nigeria). He then made a geographical survey of West Africa (1905–1906). He spent some time on the island of Fernando Po (Bioko), and many of the birds, which he described, have *poensis* in their binomial, referring to that island. He was a Royal Geographical Society medallist (1908). He continued his African explorations (1908–1910) until local people in the Dar Tama area of Tchad (Chad) killed him. French soldiers recovered his body and he was buried in Maifoni, Nigeria, alongside his brother. He wrote *From the Niger to the Nile* (1907). His collections are still at Tring, BMNH. Two mammals are named after him. (See also **Boyd**, plus **Alexander, B. F.**, **Francis (Alexander)** and **Herbert (Alexander)**)

Alexander, H. G.

> Aberrant Bush Warbler ssp. *Cettia flavolivacea alexanderi* **Ripley**, 1951

Horace Gundry Alexander (1889–1989) was a British ornithologist whose brother was Wilfred Backhouse Alexander (see below). He taught at Woodbrooke College, Birmingham (1919–1944). He spent much time in India and, as he was a Quaker, his views made it easier for him to become a close personal friend of Mahatma Gandhi. He enrolled the young Indira Gandhi in the Delhi Birdwatching Society (1950).

Alexander, W. B.

> Grey-headed Albatross ssp. *Thalassarche chrysostoma alexanderi* **Mathews**, 1916 NCR; NRM
> Maned Duck ssp. *Chenonetta jubata alexanderi* Mathews, 1916 NCR; NRM
> Dove Prion ssp. *Pachyptila desolata alexanderi* Mathews & **Iredale**, 1921 NCR; NRM

Wilfred Backhouse Alexander (1885–1965) was an English zoologist. He was educated at the University of Cambridge and became Assistant Superintendent of the Cambridge Museum of Zoology (1910), also acting as an Assistant Demonstrator in Zoology and Comparative Anatomy. He was an Assistant Naturalist to the Board of Agriculture and Fisheries (1911). He then left for Australia, where he was Assistant at the Western Australian Museum (1912–1915). He accompanied Professor W. J. Dakin on the first Percy Sladen Trust Expedition to the Abrolhos Islands (1913). During the British Association for the Advancement of Science meeting (1914) he travelled throughout Australia on the presidential train. He was Keeper of Biology at the Western Australian Museum (1915–1920) and was seconded to the Council for Scientific and Industrial Research (CSIR) as science abstractor (1916–1919). He was biologist at the Commonwealth Prickly Pear Board in Brisbane (1920–1924) and officer-in-charge (1924–1925). He edited the journal *Emu* (1924–1925). He then worked at the AMNH (1926), writing his best known work *Birds of the Ocean* (1928). This work is generally recognised to be the first ornithological field guide. Alexander was superintendent of the Tees Estuary survey (1929–1930). He became Director of the Oxford Bird Census (later becoming the Edward Grey Institute of Field Ornithology) (1930–1945). He then became their Librarian (1945–1955). The Alexander Library was named after him (1947). He was elected as a Corresponding Fellow of the American Ornithologists' Union (1921) and Fellow of the RAOU (1939). He was awarded the Tucker Medal of the BTO (1955) and the Union Medal of the British Ornithologists' Union (1959).

Alexander (Milligan)

> Yellow-rumped Thornbill ssp. *Acanthiza chrysorrhoa alexanderi* **Mathews**, 1921 NCR
> [JS *Acanthiza chrysorrhoa pallida*]

Alexander William Milligan (1858–1921) was an Australian ornithologist and collector. He moved to Western Australia from Victoria (1897) where he worked as an accountant for the Department of Lands and Surveys, and latterly as a clerk in a firm of lawyers in Perth. He was Honorary Consulting Ornithologist to the Western Australian Museum (1901).

Alexander the Great

> Moustached Parakeet *Psittacula alexandri* **Linnaeus**, 1758
> [Alt. Red-breasted Parakeet]
> Alexandrine Parakeet *Psittacula eupatria* Linnaeus, 1766

Alexander the Great (356–323 BC) was King of Macedonia, and is much too famous to need an extensive write-up here. He is sometimes described as being the founder of aviculture. He stopped the killing of peacocks for food and had them shipped home to enjoy their beauty. He is also credited with the discovery of the Ring-necked Parakeet.

Alexander (Wetmore)

> White-breasted Wood Wren ssp. *Henicorhina leucosticta alexandri* **A. R. Phillips**, 1986

(See **Wetmore**)

Alexandra

> Golden-headed Cisticola ssp. *Cisticola exilis alexandrae* **Mathews**, 1912
> Red-capped Robin ssp. *Petroica goodenovii alexandrae* Mathews, 1912 NCR; NRM
> Zebra Finch ssp. *Taeniopygia guttata alexandrae* Mathews, 1912 NCR
> [JS *Taeniopygia guttata castanotis*]
> Australian Barn Owl ssp. *Tyto delicatula alexandrae* Mathews, 1912 NCR
> [JS *Tyto delicatula delicatula*]

These birds are named after a place – Alexandra, Northern Territory, Australia.

Alexandra (Princess)

> Alexandra's Parrot *Polytelis alexandrae* **Gould**, 1863
> [Alt. Princess Parrot, Princess of Wales Parakeet]

Alexandra (1844–1925), Princess of Wales, later Queen Alexandra, was a member of the British royal family. The parrot, a rare and elusive inhabitant of the desert regions of Australia, was named by John Gould to celebrate her marriage to Edward VII. Considered by many to be one of the most exquisitely coloured and well proportioned of all birds, the Princess Parrot was described by Baldwin Spencer as the most fitting of the Australian birds to bear the name of this illustrious lady.

Alexandre

> Alexandre's Hummingbird *Archilochus alexandri* **Boucier** & **Mulsant**, 1846
> [Alt. Black-chinned Hummingbird]

Dr M. Alexandre (DNF) discovered the species in the Sierra Madre of Mexico and sent the type specimen to Mexico City.

He was a physician who also collected zoological specimens and sent them back to France.

Alexandrov

Northern House Martin ssp. *Delichon urbicum alexandrovi* **Zarudny**, 1916 NCR
[JS *Delichon urbicum meridionale*]
Common Chaffinch ssp. *Fringilla coelebs alexandrovi* Zarudny, 1916
Bearded Reedling ssp. *Panurus biarmicus alexandrovi* Zarudny & **Bilkevitch**, 1916 NCR
[JS *Panurus biarmicus russicus*]
Black Redstart ssp. *Phoenicurus ochruros alexandrovi* Zarudny, 1918 NCR
[JS *Phoenicurus ochruros phoenicuroides*]

Vasily Georgiyevich Alexandrov (1887–1964) was a Russian botanist who was Director of the Botanical Gardens, Tiflis (Tbilisi, Georgia). He was a Professor (1920) at Tiflis and at Tomsk (1927–1929).

Alexina

Broad-tailed Warbler ssp. *Schoenicola brevirostris alexinae* **Heuglin**, 1863
[Alt. Fan-tailed Grassbird]

Mademoiselle Alexina Tinné (1839–1869) was Dutch but a naturalised British subject. She visited Norway, Italy, Constantinople (Istanbul) and Palestine when she was in her early twenties before settling in Cairo. She explored the White Nile (1861–1864); Heuglin (q.v.) was a member of her expedition. She started to explore the Sahara Desert (1869) in company with three other Europeans and a number of Touareg attendants, who shot and killed all four Europeans.

Alfaro

Alfaro's Hummingbird *Saucerottia alfaroana* **Underwood**, 1896 EXTINCT
[Alt. Miravalles Hummingbird; Syn. *Amazilia alfaroana*]

Red-crowned Ant Tanager ssp. *Habia rubica alfaroana* **Ridgway**, 1905

Don Anastasio Alfaro (1865–1951) was an archaeologist, geologist, ethnologist, zoologist and famous Costa Rican writer. From a young age he collected birds, insects, minerals and plants. He took his first degree at the University of Santo Tomás (1883). He urged the president to create a National Museum (1885) and then he dedicated much of his life to it, becoming Director shortly after it was established (1887). He spent his life teaching and exploring as well as continuing to collect, discovering a number of new taxa that carry his name. He wrote a number of books, including one on Costa Rican mammals, and also wrote poetry. He was much admired throughout Europe and the Americas and corresponded with all the leading naturalists of his day. Among other taxa, three mammals and an amphibian are named after him.

Alfred

Alfred's Oropendola *Psarocolius angustifrons alfredi* **Des Murs**, 1856

Unidentified child, assumed to be a male relative of Des Murs, who wrote the scientific description, or perhaps of the Comte de Castlenau who collected the bird. The original description has the words 'pour encourager à la Science un enfant qui nous est cher.'

Alfred Edmund

Merlin sp. *Falco alfrededmundi* **Kleinschmidt**, 1917 NCR
[JS *Falco columbarius*]

(See **Brehm**)

Alfred (Everett)

Flores Scops Owl *Otus alfredi* **Hartert**, 1897

Wallacean Cuckooshrike ssp. *Coracina personata alfrediana* Hartert, 1898
[Alt. Timor Cuckooshrike]

(See **Everett**)

Alfred (Newton)

Bamboo Warbler *Bradypterus alfredi* **Hartlaub**, 1890
[Alt. Newton's Scrub Warbler]

(See **Newton, A.**)

Alfred (Sharpe)

Swift sp. *Cypselus alfredi* **G. E. Shelley**, 1900 NCR
[Alt. Mottled Swift; JS *Tachymarptis aequatorialis*]

(See **Sharpe, A.**)

Alfred (Vincent)

Olive Sunbird ssp. *Cyanomitra olivacea alfredi* Vincent, 1934

Alfred William Vincent (1904–2000), like his brother Col. Jack Vincent (q.v.), was an ornithologist. He wrote 'On the breeding habits of some African birds' in *Ibis* (1947).

Algonda

Comoros Fody *Foudia eminentissima algondae* **Schlegel**, 1867

Algonda Schlegel (fl.1866) was the describer's sister.

Alice

Alice's Emerald *Chlorostilbon alice* **Bourcier** & **Mulsant**, 1848
[Alt. Green-tailed Emerald]

Unknown etymology. It is odd that the scientific name is just 'Alice', and is not in any way latinised.

Alice (Kennicott)

Grey-cheeked Thrush *Turdus aliciae* **Baird**, 1858 NCR
[JS *Catharus minimus*]

Alice Kennicott was living in Illinois (1858). She was a younger sister of Robert Kennicott (q.v.), who was a naturalist who worked for Baird. She was a great shot and collected many specimens for her brother. She married a Chicago physician, Dr Francis Reilly.

Alice (Robinson)

Purple-backed Sunbeam *Aglaeactis aliciae* **Salvin**, 1896

Copper-rumped Hummingbird ssp. *Saucerottia tobaci aliciae* **Richmond**, 1895
[Syn. *Amazilia tobaci aliciae*]

Anita Alice Mathilde Robinson *née* Phinney (b.1860) was the wife of Colonel Wirt Robinson (q.v.) (1864–1929), US explorer and collector.

Alinder

Golden-winged Sunbird ssp. *Drepanorhynchus reichenowi alinderi* **Laubmann**, 1928 NCR
[JS *Drepanorhynchus reichenowi reichenowi*]

Dr Sven Alinder (1900–1928) was a Swedish entomologist and collector in Kenya for the Bavarian State Zoological Society, Munich. Piteå University awarded his doctorate. He wrote *In the Land of the Nando: Travels and Adventures of a Modern Zoologist in Wild Africa*.

Aline (Bourcier)

Emerald-bellied Puffleg *Eriocnemis alinae* **Bourcier**, 1842

Aline Bourcier (DNF) was the wife of the French naturalist Jules Bourcier (q.v.), one-time Consul General to Ecuador.

Aline (Jackson)

Blue-headed Sunbird *Cyanomitra alinae* **F. Jackson**, 1904

Lady Aline Louise Jackson *née* Cooper (d.1966) was the wife of Sir Frederick Jackson, ornithologist and one-time Governor of Uganda.

Alister

King Parrot genus *Alisterus* **Mathews**, 1911
Grassfinch genus *Alisteranus* Mathews, 1912 NCR
[Now in *Poephila*]
Whistler genus *Alisterornis* Mathews, 1912 NCR
[Now in *Pachycephala*]

Nullarbor Quail-thrush *Cinclosoma alisteri* Mathews, 1910

Grey Fantail ssp. *Rhipidura albiscapa alisteri* Mathews, 1911
Azure Kingfisher ssp. *Ceyx azureus alisteri* Mathews, 1912 NCR
[JS *Ceyx azureus ruficollaris*]
Flock Bronzewing ssp. *Phaps histrionica alisteri* Mathews, 1912 NCR; NRM
Tawny Grassbird ssp. *Megalurus timoriensis alisteri* Mathews, 1912

White-quilled Rock Pigeon ssp. *Petrophassa albipennis alisteri* Mathews, 1912 NCR
[JS *Petrophassa albipennis albipennis*]
Australian Little Bittern ssp. *Ixobrychus minutus alisteri* Mathews, 1913 NCR
[JS *Ixobrychus dubius*]

Alister William Mathews (1907–1985) was the son of the describer (q.v.). He became a printer and a teacher of languages.

Alix

Recurve-billed Bushbird *Clytoctantes alixii* **D. G. Elliot**, 1870

Dr Édouard Alix (1823–1893) was Professor of Zoology at the Catholic University, Paris.

Allard

Metaltail sp. *Ornismya allardi* **Bourcier**, 1839 NCR
[Alt. Tyrian Metaltail; JS *Metallura tyrianthina*]

Jean-Baptiste d'Allard (1769–1848) flourished under the first French Empire and constructed (1810) the building in Montbrison that is now a museum named after him. He spent his life collecting 'curiosities' and acquired the collection of Baron Feutrier in which the original specimen of the hummingbird was found. He had no descendant and left the museum and its contents to his fellow townsfolk.

Allen, A. A.

Moustached Antpitta *Grallaria alleni* **Chapman**, 1912

Chestnut-capped Brush Finch ssp. *Arremon brunneinucha alleni* **Parkes**, 1954

Dr Arthur Augustus Allen (1885–1964) was a US ornithologist who became Professor of Ornithology at Cornell University. He undertook fieldwork in Colombia and Panama and collected the antpitta holotype (1911). He wrote *The Golden Plover and Other Birds* (1939). A populariser of bird study, the St Louis Audubon Society reported (1940) 'Dr A. A. Allen drew an audience of 800 for his "intimate glimpses" into the lives of birds'. (See also **Arthur Allen**)

Allen, A. R.

Bustard sp. *Eupodotis alleni* **Meyer de Schauensee**, 1930 NCR
[Alt. Rüppell's Korhaan; JS *Eupodotis rueppelii*]

Alfred Reginald 'Reggie' Allen Jr (1905–1988) was the son of a neuro-surgeon whose first love was music and who was a founding member of the Savoy Opera Company. Reggie was on the (1930) Meyer de Schauensee expedition to South Africa. When he returned to the US he worked in Philadelphia and New York as an advertising copywriter and became general manager of the Philadelphia Orchestra (1935). He worked for Universal Pictures in Hollywood (1939–1941). He served as a Lieutenant-Commander in Air Intelligence (WW2). After the war he returned to Hollywood to work for the J. Arthur Rank organisation until he was appointed (1950s) to work as Assistant General Manager for the

Metropolitan Opera and Director of Operations, the Lincoln Center, New York. He retired (1969) to become Curator of the Gilbert & Sullivan Collection of the Pierpont Morgan Library.

Allen, C. A.

Allen's Hummingbird *Selasphorus sasin* **Lesson**, 1829

Charles Andrew Allen (1841–1930) was an American collector and taxidermist. After serving in the Union Army during the Civil War he returned home and worked at various jobs, including taxidermy. He once worked on fishing boats out of Newfoundland to learn more about Atlantic seabirds. He was an excellent collector, who became well versed in the habits of the birds and other animals. The heavily wooded hills are still just across the road from Allen's old homestead in California. He was also the inventor of the *Allen Hummer*, a birdcall device, and was an authority on bird lore but had little opportunity for education. Many years later Allen's account of the eponymous hummingbird appeared in Bent's *Life Histories of North American Birds*. William Brewster (q.v.), C. Hart Merriam (q.v.) and Major Charles Bendire (q.v.) were among the many eastern-based scientists who bought bird and mammal skins from this Californian collector. Allen obtained the hummingbird (1877) for Henshaw, who named it after him in appreciation. However, the bird had been discovered c.50 years earlier by Lesson whose name took precedence. Nevertheless, the common name persisted.

Allen, G. M.

Golden Palm Weaver ssp. *Ploceus bojeri alleni* **Mearns**, 1911 NCR; NRM

Small Buttonquail ssp. *Turnix sylvaticus alleni* Mearns, 1911 NCR

[JS *Turnix sylvaticus lepurana*]

Black-winged Oriole ssp. *Oriolus nigripennis alleni* **Amadon**, 1953 NCR; NRM

Dr Glover Morrill Allen (1879–1942) was a collector, curator, editor, librarian, mammalogist, ornithologist, scientist, taxonomist, teacher and writer. He was Librarian at the Boston Society of Natural History (1901–1927). He was employed to oversee the mammal collection at the Museum of Comparative Zoology, Harvard (1907), having been awarded his PhD (1904). He was Curator of Mammals (1925–1938) and then Professor of Zoology (1938–1942). He was keen on all vertebrates, particularly birds (editing *Auk* 1939–1942) and mammals (President of the American Society of Mammalogists 1927–1929). He made many collecting trips (1903–1931), variously to Africa, including the Harvard African Expedition to Liberia (1926), Australia, the Bahamas, Brazil, Labrador and the West Indies. He wrote a great many scientific papers and articles and a number of books. Early works include *The Birds of Massachussetts* (1901), where he notes taking a specimen of Passenger Pigeon *Ectopistes migratorius* (1904), maybe the last recorded in the wild.

Allen, J. A.

Scaly-breasted Thrasher genus *Allenia* **Cory**, 1891

White-eyed Towhee ssp. *Pipilo erythrophthalmus alleni* **Coues**, 1871

Allen's Ptarmigan *Lagopus lagopus alleni* **Stejneger**, 1884 [Alt. Willow Ptarmigan ssp.]

Red-shouldered Hawk ssp. *Buteo lineatus alleni* **Ridgway**, 1885

Golden-tailed Sapphire ssp. *Chrysuronia oenone alleni* **D. G. Elliot**, 1888

Hooded Siskin ssp. *Carduelis magellanica alleni* Ridgway, 1899

Golden-olive Woodpecker ssp. *Piculus rubiginosus alleni* **Bangs**, 1902

[Syn. *Colaptes rubiginosus alleni*]

Bananaquit ssp. *Coereba flaveola alleni* **Lowe**, 1912

Joel Asaph Allen (1838–1921) was a US ornithologist who studied under Agassiz and accompanied him to Brazil (1865). He took a number of field trips in North America and (1873) became chief of an expedition sent out by the Northern Pacific Railroad. He was an Assistant in Ornithology at the Museum of Comparative Zoology, Cambridge, Massachusetts (1870). He became Curator of the Department of Mammals and Birds, AMNH (1885–1921). He wrote many scientific papers and edited *Auk* and the *Bulletin of the Nuttall Ornithological Club*. He also wrote a number of monographs including one with Elliott Coues (q.v.). He organised the AOU with Coues and Brewster (q.v.) and became its first President, and was also a founding member of the National Audubon Society. He is remembered in the names of ten mammals and a reptile.

Allen, W.

Allen's Gallinule *Porphyrio alleni* **Thomson**, 1842

Rear-Admiral William Allen (1793–1864) was an English naval officer who was involved in fighting the African slave trade. He led three expeditions to Africa (two 1832, one 1841), all to the Niger. Allen collected the gallinule type specimen at Idda, near the River Niger. Two mammals are named after him.

Allenby

Kittlitz's Plover ssp. *Charadrius pecuarius allenbyi* **Nicoll**, 1921 NCR; NRM

Field Marshall Sir Edmund Henry Hynman, Viscount Allenby of Megiddo (1861–1936), was a career soldier. He served in Africa before and during the Boer War. He commanded the Cavalry Division, the Cavalry Corps, V Corps, and the Third Army on the Western Front. He was Commander in Chief, Egyptian Expeditionary Force (June 1917), most famous for capturing Jerusalem from Turkish occupation (December 1917). He is also known to be the only Christian General to have succeeded in capturing both Jerusalem and the strategic fortress of Acre – a feat beyond the Crusaders such as Richard Coeur-de-Lion and St. Louis. Allenby was extremely interested in bird migration. Among his staff in Palestine and Egypt toward the end of WW1 was Richard Meinertzhagen (q.v.). The two men are known to have been on very good terms and remained friends, even after Allenby sacked Meinertzhagen. He has one mammal named after him.

Alma

Alma's Thrush *Catharus ustulatus almae* **Oberholser**, 1898 NCR
[Alt. Swainson's Thrush; JS *Catharus ustulatus ustulatus*]

Harry Church Oberholser (q.v.) did not identify the 'Alma' he chose to honour in describing this subspecies, and we are not aware that her identity has ever been made known.

Alma (Jønsson)

Seram Masked Owl *Tyto almae* Jønsson *et al.* 2013

Alma Jønsson is the daughter of Knud Andreas Jønsson, the senior describer of the owl.

Almasy

Eurasian Skylark ssp. *Alauda arvensis almasyi* Keve, 1943 NCR
[JS *Alauda arvensis dulcivox*]

Dr György Ede Almásy Graf von Zsadány und Törökszentmiklós (1867–1933) was a Hungarian zoologist, ethnologist and explorer in Turkistan and western China (1900–1906).

Aloysius

Black-and-white Casqued Hornbill ssp. *Bycanistes subcylindricus aloysii* **Salvadori**, 1906 NCR
[JS *Bycanistes subcylindricus subquadratus*]
Speckled Tinkerbird ssp. *Pogoniulus scolopaceus aloysii* Salvadori, 1906 NCR
[JS *Pogoniulus scolopaceus flavisquamatus*]

Luigi Amedeo Giuseppe Maria Ferdinando Francesco di Savoia, Duke of the Abruzzi (1873–1933) was the son of the King of Spain, a member of the royal House of Savoy and a cousin of the King of Italy; 'Aloysius' is a latinisation of 'Luigi'. He was a mountaineer and explorer who made a number of important first ascents in places as far apart as Alaska, the Rwenzori Mountains in Uganda, and K2 in the Karakoram Range. He was Commander-in-Chief, Royal Italian Navy (WW1). He died in Abruzzi Village, Italian Somaliland (c. 90 km north of Mogadishu, in present-day Somalia). A mammal and a reptile are named after him.

Alphéraky

Common Pheasant ssp. *Phasianus colchicus alpherakyi* **Buturlin**, 1904 NCR
[JS *Phasianus colchicus pallasi*]

Sergei Nikolaevich Alphéraky (1850–1918) was a Russian entomologist, lepidopterist, ornithologist and explorer in Central Asia. He studied at Moscow University (1867–1869) and Dresden (1871–1873).

Alphonse

Alphonse's Crow-tit *Paradoxornis alphonsianus* **J. Verreaux**, 1870
[Alt. Ashy-throated Parrotbill]

(See **Milne-Edwards**)

Alström

Alström's Warbler *Seicercus soror* Alström & Olsson, 1999
[Alt. Plain-tailed Warbler]

Dr Per Johan Alström (b.1961), the senior describer, is a Swedish ornithologist and research scientist in the Department of Systematic Zoology at Uppsala University. He is co-author of *Field Guide to the Rare Birds of Britain and Europe* (1991) and the family monograph *Pipits and Wagtails* (2002), as well as being author of a number of articles and a recorder of bird song.

Altenstein

King Eider *Platypus altensteinii* C. L. Brehm, 1824 NCR
[JS *Somateria spectabilis*]

Karl Freiherr vom Stein zum Altenstein (1770–1840) was a statesman at the court of King Friedrich Wilhelm III of Prussia. He was also a historian interested in natural history. He was perhaps responsible for sending the botanist Friedrich Sellow (q.v.) to Brazil.

Althaea

Hume's Lesser Whitethroat *Sylvia althaea* **Hume**, 1878

Althaea was the wife of King Oeneus of Calydon in Greek mythology. *Althaea* is also a genus of 6–12 species of perennial herb.

Altum

Hildebrandt's Francolin ssp. *Pternistis hildebrandti altumi* **G. A. Fischer** & **Reichenow**, 1884
[Syn. *Francolinus hildebrandti altumi*]

Father Dr Johann Bernard Theodor Altum (1824–1900) was a German catholic priest, zoologist and ornithologist. The University of Berlin awarded his doctorate in natural sciences (1855). He became Professor of Natural Sciences at the Eberswalde Forestry School (1869). He wrote *Der Vogel und sein Leben* (1868) wherein he took issue with A. E. Brehm (q.v.) over ornithology. He was the first ornithologist to theorise regarding the relationship between a bird's song and its territory.

Alvarez, J.

Slaty-legged Crake ssp. *Rallina eurizonoides alvarezi* Kennedy & Ross, 1987

Jesus B. Alvarez Jr (DNF) was an untiring campaigner to conserve Philippine fauna and flora. He was Officer-in-Charge and Director (1972) Parks and Wildlife Office as well as Assistant Director of Forestry for Parks. He wrote: *Our National Parks: Treasure Houses of Wildlife and Beauty* (1973).

Alvarez, M.

Flame-coloured Tanager ssp. *Piranga bidentata alvarezi*
A. R. Phillips, 1966

Miguel Alvarez Del Toro (1917–1996) was a life-long naturalist working mostly in Chiapas State, Mexico, having always collected insects and small vertebrates. He collected birds around Mexico City, in Morelos, and Colonia Sarabia, Oaxaca, in the rainforests, for the Academy of Natural Sciences, Philadelphia (1938–1939). He became a taxidermist at the Museum of Natural History in Tuxtla Gutierrez, Chiapas (1942), and collected vertebrates near the city and surveyed more remote areas. He had no formal education yet was a great teacher, scientist and conservationist. He taught at the Chiapas College of Arts and Sciences and at The National Autonomous University of Mexico (which awarded him an honorary doctorate). He was the principal force behind the creation of six protected areas in Chiapas. He published c.40 papers on birds including: *Las Aves de Chiapas* (1971). Four reptiles and an amphibian are named after him.

Amadon

Hawk genus *Amadonastur* Amaral *et al.* 2009

Long-tailed Hawk *Accipiter amadoni* **Wolters**, 1978 NCR
[JS *Urotriorchis macrourus*]

Orange-headed Thrush ssp. *Zoothera citrina amadoni* **Biswas**, 1951
Greater Blue-eared Glossy Starling ssp. *Lamprotornis chalybaeus amadoni* Wolters, 1952
Long-billed Wren-babbler ssp. *Rimator malacoptilus amadoni* **Koelz**, 1954
North Melanesian Cuckooshrike ssp. *Coracina welchmani amadonis* Cain & Galbraith, 1955
Ethiopian Swallow ssp. *Hirundo aethiopica amadoni* **C. M. N. White**, 1956
Thick-billed Siskin ssp. *Carduelis crassirostris amadoni* George, 1964
Philippine Green Pigeon ssp. *Treron axillaris amadoni* **Parkes**, 1965
Tiny Sunbird ssp. *Cinnyris minullus amadoni* **Eisentraut**, 1965
Green-throated Mountaingem ssp. *Lampornis viridipallens amadoni* Rowley, 1968
Little Greenbul ssp. *Andropadus virens amadoni* **Dickerman**, 1997

Dr Dean Amadon (1912–2003) was an American ornithologist and an authority on birds of prey. Hobart College awarded his bachelor's degree (1934) and Cornell his doctorate (1947). He worked at the AMNH, New York (1937–1973) and was Chairman of the Ornithology Department (1957). He was President of the American Ornithogists' Union (1964–1966) and wrote, with Leslie Brown, *Eagles, Hawks and Falcons of the World* (1968) and, with Jean Delacour (q.v.), *Curassows and Related Birds* (1973).

Amalia (Buturlin)

European Stonechat ssp. *Saxicola torquata amaliae* **Buturlin**, 1929 NCR
[JS *Saxicola rubicola*]

Mrs Amalia J. Buturlin *née* Zarin (fl.1929) was the wife of the describer.

Amalia (Dietrich)

Brown Gerygone ssp. *Gerygone mouki amalia* **Meise**, 1931

Frau Amalia Concordia Dietrich (1822–1891) *née* Nelle was German-born but naturalised Australian (c.1863–1871). She was an agent for the Godeffroy Museum in Hamburg (1872–1879).

Amaryllis

Black-tailed Trainbearer *Lesbia amaryllis* **Bourcier**, 1848 NCR
[JS *Lesbia victoriae*]

In Greek mythology Amaryllis was a shepherdess. The name means 'sparkling'.

Amathusia

Hummingbird genus *Amathusia* **Mulsant** & **E Verreaux** & **J Verreaux** 1865 NCR
[Now in *Doricha*]

Blue-cheeked Rosella *Platycercus amathusia* **Bonaparte**, 1850 NCR
[Alt. Pale-headed Rosella; JS *Platycercus adscitus palliceps*]

In Greek mythology Amathusia was an epithet of the goddess of love, Aphrodite.

Amazili

Amazilia Hummingbird *Amazilia amazilia* **Lesson**, 1827

Amazili is an Inca who is the heroine of Jean Marmontel's novel *Les Incas, ou la Destruction de l'Empire du Pérou* (1777).

Ambrosetti

Buffy-fronted Seedeater *Coccothraustes ambrosettianus* **Bertoni**, 1901 NCR
[JS *Sporophila frontalis*]

Dr Juan Bautista Ambrosetti (1865–1917) was an Argentinian archaeologist, anthropologist and naturalist. His first expedition was to Chaco province (1885). He became Director of Zoology at the Entre Rios Province Museum, Parana, and (1903) Professor of Archaeology at the University of Buenos Aires where he established the Museum of Ethnography (1904). He discovered (1908) the previously lost ruins of the Omaguaca civilisation (10th century AD) in Jujuy province.

Amelie

Rosy-throated Longclaw *Macronyx ameliae* de Tarragon, 1845

Louise-Amélie de Turenne (DNF) was the wife of French explorer Marquis Leonce de Tarragon (1813–1897). The Marquis's mother was Amélie Louise Virginie Goislard de Villebresme (1788–1865). The longclaw may be named after either of these women. The original description is silent on the matter, perhaps deliberately so de Tarragon could avoid upsetting either lady.

Amherst, Lady

Lady Amherst's Pheasant *Chrysolophus amherstiae* **Leadbeater**, 1829

Sarah Countess Amherst (1762–1838) was married to William Pitt Amherst, who was the Governor General of Bengal (1822–1828). Lord Amherst was responsible for sending the first specimen of the pheasant to London (1828). It did not survive the journey but, nevertheless, the ornithologist Leadbeater (q.v.) used the specimen for his official description. The first live specimens successfully reached London in July 1869. An orchid is also named after her.

Amphitrite

Storm-petrel sp. *Fregetta amphitrite* **Jardin**, 1858 NCR
[Alt. Polynesian Storm-petrel; JS *Nesofregetta fuliginosa*]

In Greek mythology Amphitrite was a sea-goddess, the daughter of Oceanus and Tethys, and wife of Poseidon.

Amphitryon

White-throated Dipper ssp. *Cinclus cinclus amphitryon* **Neumann** & Paludan, 1937 NCR
[JS *Cinclus cinclus caucasicus*]

In Greek mythology Amphitryon was a Theban general who accidentally killed Electryon, King of Mycenae, his brother-in-law.

Amy

Clamorous Reed Warbler ssp. *Acrocephalus stentoreus amyae* **E. C. S. Baker**, 1922
Eyebrowed Wren-babbler ssp. *Napothera epilepidota amyae* Kinnear, 1925

Mrs Amy Stevens *née* Ellis (1881–1956) was the wife of British explorer Herbert Stevens.

Amytis

Grasswren genus *Amytis* **Lesson**, 1831 NCR
[Now *Amytornis* **Stejneger**, 1885]

Amytis (fl.450 BC) was a Persian princess, described by the Greek historian Dinon as the most beautiful woman in Asia.

Anabel

Anabel's Bluebird *Sialia mexicana anabelae* **A. W. Anthony**, 1889 NCR
[Alt. Western Bluebird ssp.; JS *Sialia mexicana occidentalis*]

Anabel Anthony *née* Klink (1867–1949) was the wife of Alfred Webster Anthony (q. v.), the describer.

Anais

Golden Myna *Mino anais* **Lesson**, 1839

Anais Lesson (1827–1838) was the daughter of René Lesson, the French naturalist, who named the bird after his late child.

Anchieta

Anchieta's Barbet *Stactolaema anchietae* **Bocage**, 1869
Anchieta's Sunbird *Anthreptes anchietae* Bocage, 1878
[Alt. Red-and-blue Sunbird]
Bat-like Spinetail *Chaetura anchietae* **Souza**, 1887 NCR
[JS *Neafrapus boehmi*]

Anchieta's Tchagra ssp. *Bocagia minuta anchietae* Bocage, 1869
[Alt. Marsh Tchagra, Blackcap Bush-shrike]
Black Scimitarbill ssp. *Rhinopomastus aterrimus anchietae* Bocage, 1892
Rufous-naped Lark ssp. *Mirafra africana anchietae* **da Rosa Pinto**, 1967 NCR
[JS *Mirafra africana occidentalis*]

José Alberto de Oliveira Anchieta (1832–1897) was an independent Portuguese naturalist and collector in Africa, particularly Angola and Mozambique. The Portuguese government hired him as a naturalist (1867), but probably also covertly as a secret agent and informer. He collected many types of mammals and reptiles. A mammal, seven reptiles and two amphibians are named after him.

Andarya

Andarya's Bushshrike *Malaconotus andaryae* **F. Jackson**, 1919 NCR
[Probably a hybrid: *Chlorophoneus bocagei* x *C. sulfureopectus*]

Andarya (fl.1919) was an Ugandan collector for Sir Frederick Jackson.

Andersen, Th.

Regal Sunbird ssp. *Cinnyris regius anderseni* **J. G. Williams**, 1950
Usumbara Weaver ssp. *Ploceus nicolli anderseni* Franzmann, 1983

Thorkild Andersen (b.1912) was a Danish sisal planter and collector in Tanganyika (Tanzania) (1947–1967).

Andersen, To.

Pied Bushchat ssp. *Saxicola caprata anderseni* **Salomonsen**, 1953

Torben Andersen (fl.1950) was a Danish businessman and naturalist in the Philippines.

Anderson, J.

Anderson's Bamboo Partridge *Bambusicola fytchii* Anderson, 1871
[Alt. Mountain Bamboo Partridge]
Anderson's Yellow-backed Sunbird *Aethopyga andersoni* **Oates**, 1890 NCR
[Alt. Crimson Sunbird; JS *Aethopyga siparaja*]

Anderson's Bulbul *Pycnonotus xanthorrhous andersoni* **Swinhoe**, 1870
[Alt. Brown-breasted Bulbul ssp.]
Anderson's Silver Pheasant *Lophura nycthemera andersoni* **J. A. Elliott**, 1871 NCR

John Anderson (1833–1900) was a qualified physician who became Professor of Comparative Anatomy at the Medical School in Calcutta and Director of the Indian Museum there (1865). He joined an expedition to Burma and Yunnan in south-west China as naturalist (1868). A second expedition (1875) only collected in Burma. He wrote a zoological account of the two expeditions; the section on birds covers 233 species. R. Bowdler Sharpe (q.v.) gave assistance in the report's preparation by verifying the identifications. Anderson also wrote monographs of two whale genera. He was elected a Fellow of the Royal Society (1879). Three amphibians, three mammals and eight reptiles are also named after him.

Anderson, M. P.

Pale-vented Pigeon ssp. *Patagioenas cayennensis andersoni* **Cory**, 1915

Malcolm Playfair Anderson (1879–1919) was an American zoologist educated at secondary level in Germany, returning to the US to study zoology and graduate at Stanford University (1904). From age 15 he took part in collecting expeditions to Arizona, Alaska and California. He joined the Cooper Ornithological Club (1901) and wrote a number of articles on ornithology, yet did not confine himself to that subject. He was chosen to conduct the Duke of Bedford's Exploration of Eastern Asia for the Zoological Society of London (1904); he took photographs and extensive notes on the collections and wrote several short stories about the people with whom he lived and worked in the Orient. He was again in western China (1909 and 1910). He was in Peru with Osgood (q.v.) (1912). He died (1919) after falling from scaffolding at the shipyards in Oakland, California. Four mammals are named after him.

Anderson, W.

White-browed Shortwing ssp. *Brachypteryx montana andersoni* **Rand** & **Rabor**, 1967

William Anderson (DNF) was a former Comptroller of Silliman University, Negros, Philippines, and aided others in their fieldwork. He owned an aeroplane and twice flew Rabor in order that he could do aerial suveys of prospective collecting regions (1956).

Anderson, Mount

Grey-fronted Honeyeater ssp. *Meliphaga plumula andersoni* **Mathews**, 1912 NCR (originally described as *Ptilotis chrysotis andersoni*)
[JS *Lichenostomus plumulus planasi*]

This bird is named after Mount Anderson in northern Western Australia, which in turn was named after Charles Anderson (1876–1944), a mineralogist, palaeontologist and museum director.

Andersson, C. J.

Bat Hawk ssp. *Macheiramphus alcinus anderssoni* **Gurney**, 1866
Red-capped Lark ssp. *Calandrella cinerea anderssoni* **Tristram**, 1869

Rock Martin ssp. *Ptyonoprogne fuligula anderssoni* **Sharpe** & **Wyatt**, 1887
Yellow White-eye ssp. *Zosterops senegalensis anderssoni* **G. E. Shelley**, 1892
Golden-tailed Woodpecker ssp. *Campethera abingoni anderssoni* **J. A. Roberts**, 1936

Carl Johan Andersson (1827–1867) was a Swedish explorer and collector in Namibia (1851–1867). He wrote: *Lake Ngami or Explorations and Discoveries during Four Years Wandering in the Wilds of South Western Africa* (1856) and *The Okavango River: a Narrative of Travel, Exploration and Adventure* (1861).

Andersson, J. G.

Oriental Magpie ssp. *Pica pica anderssoni* **Lönnberg**, 1923

Johan Gunnar Andersson (1874–1960) was a Swedish archeologist, geologist and paleontologist who was Director of Sweden's National Geological Survey. He was on the Swedish Antarctic Expedition (1901–1903). He made a number of important archaeological and paleontological finds in China, including the first remains of Peking Man.

André

André's Swift *Chaetura andrei* **Berlepsch** & **Hartert**, 1902
[Alt. Ashy-tailed Swift]
Black-chested Tyrant *Taeniotriccus andrei* Berlepsch & Hartert, 1902

Plain Antvireo ssp. *Dysithamnus mentalis andrei* **Hellmayr**, 1906
Little Tinamou ssp. *Crypturellus soui andrei* Brabourne & **Chubb**, 1914

Eugène André (1861–1922) was a French writer and naturalist. He collected in Venezuela (1897–1900). He may have been in Trinidad (c.1915). His wrote: *A Naturalist in the Guianas* (1904).

Andrew

Andrew's Swallow *Hirundo andrewi* **J. G. Williams**, 1966 NCR
[now considered to be aberrant *Pseudhirundo griseopyga*]

Andrew E. Williams is a botanist who co-authored *Field Guide to Orchids of North America* (1983). He also wrote, together with his father, J. G. Williams (q.v.), *A Field Guide to the Orchids of Britain and Europe, with North Africa and Middle East* (1978).

Andrews, C. W.

Andrews's Frigatebird *Fregata andrewsi* **Mathews**, 1914
[Alt. Christmas Frigatebird]

Dr Charles William Andrews (1866–1924) was a British geologist. He was interested in dinosaurs and (1895–1922) published many articles on dinosaur fossils in the *Geological Magazine* and other publications. The fossilised remains of an ancestral cormorant from the Upper Cretaceous of Transylvania has been named after him. He collected on

Christmas Island (1897–1908) when employed by BMNH and whilst commissioned by the Christmas Island Phosphate Company to survey the natural history of the islands. He was the first person to collect Abbott's Booby *Papasula abbotti* there. He wrote *A Monograph of Christmas Island (Indian Ocean)* (1900).

Andrews, R. C.

> Striated Marsh Warbler ssp. *Megalurus palustris andrewsi* **Bangs**, 1921
> Greater Flameback ssp. *Chrysocolaptes lucidus andrewsi* **Amadon**, 1943

Roy Chapman Andrews (1884–1960) was a larger-than-life American who became an explorer, collector and curator whom many believe to have been the real-life *Indiana Jones*. He always maintained that from his earliest childhood he had a desire for travel and adventure. 'I was born to be an explorer,' he wrote (1935) in *The Business of Exploring*. 'There was never any decision to make. I couldn't do anything else and be happy.' He said, too, that his only ambition was to work at the AMNH. He first worked as a taxidermist, and after graduating (1906) went to New York City and applied for a job at the museum. The Director told him there were no jobs but Andrews persisted saying, 'You have to have somebody to scrub floors, don't you?' The director took him on and from this humble beginning he went on to become the museum's most famous explorer. Initially a taxidermist, he developed an interest in whales and travelled to Alaska, Japan, Korea and China to collect various marine mammals. He was naturalist on the USS Albatross voyage to the Dutch East Indies and Borneo (1909–1910). He led an expedition to China and Outer Mongolia (1921–1923) where he collected both live specimens and fossils, including the first eggs to be positively identified as those of a dinosaur. He continued to make further expeditions over a number of years until returning to the USA (1930). He later became Director of AMNH (1934). On retirement (1942) he moved to California and spent the rest of his life writing about his exploits, including his autobiography *Under a Lucky Star*. Three mammals and an amphibian are named after him.

Andromeda

> Andromeda Thrush *Zoothera andromedae* **Temminck**, 1826
> [Alt. Sunda Thrush]

Andromeda, in Greek mythology, was a princess of Ethiopia. As divine punishment for her mother's boasting, she was chained to a rock as a sacrifice to a sea-monster. The hero Perseus rescued and married her.

Angel(a)

> Angela Starthroat *Ornismya angelae* **Lesson**, 1833
> [Alt. Blue-tufted Starthroat; JS *Heliomaster furcifer*]

This bird appears as the Angela Starthroat in Gould illustrations. Lesson originally described the hummingbird as 'L'Angèle (femelle), *Ornismya Angelae*' and we assume he had no specific angel in mind.

Angela (Diaz)

> Angela's Blue Jay *Cyanolyca angelae* **Salvadori** & **Festa**, 1899 NCR
> [Alt. Black-collared Jay; JS *Cyanolyca armillata quindiuna*]

Angela Díaz-Miranda *née* Savignone (b.1860) was the Italian wife of the Spanish Minister to Ecuador, Antonio Manuel Díaz-Miranda y Arango.

Angela (Kepler)

> Elfin-woods Warbler *Dendroica angelae* C. B. Kepler & **Parkes**, 1972
> Hawaiian Crake sp. *Porzana keplerorum* **Olson** & H. F. James, 1991 EXTINCT

Angela Kay Kepler (b.1943) is an Australian-born New Zealander. She is married to the American biologist Cameron B. Kepler (q.v.). The two often write together, both being specialists on Hawaii and its birds and flora. She wrote a *Comparative Study of Todies (Todidae), with Emphasis on the Puerto Rican Tody, Todus mexicanus* (1977) and has also written guidebooks about Hawaii. The crake is named after her and her husband.

Angelina

> Angelina's Scops Owl *Otus angelinae* **Finsch**, 1912
> [Alt. Javan Scops Owl]

Angeline Henriette Caroline Bartels *née* Maurenbrecher (1877–1920) was the wife of the author and zoologist Max Bartels (q.v.). She made a number of fine watercolours of birds that he had in his collection. Both his collection and her watercolours were donated to the Leiden Museum.

Anna (Branicki)

> Booted Racket-tail ssp. *Ocreatus underwoodii annae* **Berlepsch** & **Stolzmann**, 1894
> Fawn-breasted Brilliant ssp. *Phaiolaima rubinoides annae* Stolzmann, 1926

(See Anna **Branicki**)

Anna (d'Essling)

> Anna's Hummingbird *Calypte anna* **Lesson**, 1829

Anna Masséna, Princess d'Essling, Duchess of Rivoli, *née* Debelle (1802–1887), was the wife of Prince Victor Masséna, the son of one of Napoleon's marshals. French naturalist René Primevere Lesson named the bird for her after having discovered the first specimen among several birds collected for Prince Victor's private collection.

Anna (Dole)

> Ula-ai-hawane *Ciridops anna* **Dole**, 1879 EXTINCT

Anna Prentice Dole *née* Cate (1842–1918) was the wife of the US ornithologist Sanford Dole (q.v.), who is more famed for being a judge who was president of the Hawaiian Republic (1893–1898).

Anna (Normann)

White-headed Vanga ssp. *Artamella viridis annae*
Stejneger, 1879

Anna Normann (1852–1914), a schoolteacher from Bergen, Norway, married Leonhard Stejneger (q.v.) (1876). However, the marriage was not a success and the couple proved wholly incompatible. Anna had no desire to go to America so they decided to separate, later securing a divorce (granted 1892).

Anna (Weber)

Anna's Flowerpecker *Dicaeum annae* **Büttikofer**, 1894
[Alt. Golden-rumped Flowerpecker]

Anna Antoinette Weber van Bosse (1852–1942) was a Dutch botanist. With her husband M. C. W. Weber (d.1937), she collected in the East Indies (1888–1890 and 1899–1900). When Anna died their estate at Eerbeek was bequeathed to the Gelders Landschap foundation. Their house is now an adult education centre. Their library and scientific correspondence were bequeathed to the Artis Library, University of Amsterdam, and to the Zoological Museum of the University of Amsterdam (Institute of Taxonomic Zoology).

Anna Marula

Nimba Flycatcher *Melaenornis annamarulae* **Forbes-Watson**, 1970

Anna Marula Forbes-Watson *née* Kofsky (1941–2006) was the wife of Alexander David Forbes-Watson (q.v.).

Annalisa

Snowy-browed Flycatcher ssp. *Ficedula hyperythra annalisa*
Stresemann, 1931

Mrs Anneliese Heinrich *née* Machatchek (DNF) was the first wife of the German entomologist, collector and explorer Gerd Herrmann Heinrich (q.v.).

Anne (Elliot)

Tawny-capped Euphonia *Euphonia anneae* **Cassin**, 1865

Mrs Anne Eliza Elliot *née* Henderson (DNF) was the wife of Daniel Giraud Elliot (q.v.).

Anne (Meinertzhagen)

African Pipit ssp. *Anthus cinnamomeus annae*
Meinertzhagen, 1921
Desert Lark ssp. *Ammomanes deserti annae*
Meinertzhagen, 1923

Mrs Anne Constance Meinertzhagen *née* Jackson (1888–1928) was the second wife of Richard Meinertzhagen (q.v.). She was well known as an ornithologist in her own right and travelled to Denmark (1921), Egypt and Palestine (1923), Madeira (1925) and India (1925–1926), the latter with her husband, collecting and studying the avifauna. She died from a revolver bullet through the head whilst in the company of her husband just three months after the birth of their child.

This was described as an accident although there was no inquest, and it has been claimed both that she committed suicide suffering post-natal depression after the birth of her third child, and that her husband shot her, particularly as the path of the bullet makes suicide very unlikely.

Ansell

Gabon Coucal *Centropus anselli* **Sharpe**, 1874

Cloud Cisticola ssp. *Cisticola textrix anselli* **C. M. N. White** 1960

Henry F. Ansell (d.1875) was a merchant in Gabon who traded in natural history specimens collected there.

Ansorge

White-collared Oliveback *Nesocharis ansorgei* **Hartert**, 1899
Ansorge's Greenbul *Andropadus ansorgei* Hartert, 1907
Ansorge's Robin Chat *Xenocopsychus ansorgei* Hartert, 1907
[Alt. Angola Cave Chat]

Hairy-breasted Barbet ssp. *Tricholaema hirsuta ansorgii*
G. E. Shelley, 1895
Black Bishop ssp. *Euplectes gierowii ansorgei* Hartert, 1899
Woodhouse's Antpecker ssp. *Parmoptila woodhousei ansorgei* Hartert, 1904
Grey Penduline-tit ssp. *Anthoscopus caroli ansorgei* Hartert, 1905
Yellow-bellied Wattle-eye ssp. *Platysteira concreta ansorgei* Hartert, 1905
Greater Swamp Warbler ssp. *Acrocephalus rufescens ansorgei* Hartert, 1906
Ansorge's Crombec *Sylvietta rufescens ansorgei* Hartert, 1907
[Alt. Long-billed Crombec]
Bearded Bulbul ssp. *Criniger barbatus ansorgeanus* Hartert, 1907
White-browed Sparrow Weaver ssp. *Plocepasser mahali ansorgei* Hartert, 1907
Jameson's Firefinch ssp. *Lagonosticta rhodopareia ansorgei* **Neumann**, 1908
Willcocks's Honeyguide ssp. *Indicator willcocksi ansorgei* **C. H. B. Grant**, 1915
African Green Pigeon ssp. *Treron calvus ansorgei* Hartert & **Goodson**, 1918
Rufous-vented Warbler ssp. *Sylvia subcaerulea ansorgei* **Zedlitz**, 1921
Sabota Lark ssp. *Calendulauda sabota ansorgei* **W. L. Sclater**, 1926
Black-chested Prinia ssp. *Prinia flavicans ansorgei* W. L. Sclater, 1927
Double-banded Sandgrouse ssp. *Pterocles bicinctus ansorgei* **C. W. Benson**, 1947
Plain-backed Pipit ssp. *Anthus leucophrys ansorgei* **C. M. N. White**, 1948

Dr William John Ansorge (1850–1913) was an English explorer and collector who was active in Africa in the second half of the 19th century. He wrote *Under the African Sun*

(1899). He also collected a number of new species of fish from the Niger delta. Four mammals, one amphibian and three reptiles are named after him.

Antaios

St Helena Hoopoe *Upupa antaios* **Olson**, 1975 EXTINCT

Olson's etymology says it all: 'In Greek mythology, Antaios (Latin, Anteus) was a giant wrestler, son of Gaea, whose strength was maintained as long as he was in contact with the earth and who was finally vanquished by Hercules. The St Helena hoopoe was likewise a giant of its kind and as necessarily committed to the earth'. That is to say the bird was probably near-flightless.

Anthony, A. W.

Anthony's Flowerpecker *Dicaeum anthonyi* **McGregor**, 1914
[Alt. Flame-crowned Flowerpecker]

Anthony's Vireo *Vireo huttoni obscurus* Anthony, 1890
[Alt. Hutton's Vireo ssp.]
Anthony's Brown Towhee *Melozone crissalis senicula* Anthony, 1895
[Alt. California Towhee ssp.]
Anthony's Green Heron *Butorides virescens anthonyi* **Mearns**, 1895
[Alt. Green Heron ssp.]
Loggerhead Shrike ssp. *Lanius ludovicianus anthonyi* Mearns, 1898
Cactus Wren ssp. *Campylorhynchus brunneicapillus anthonyi* Mearns, 1902
American Dipper ssp. *Cinclus mexicanus anthonyi* **Griscom**, 1930

Alfred Webster Anthony (1865–1939) was an American collector and ornithologist. He was President of the Audubon Society in Portland, Oregon (1904). He collected birds for years in the Tualatin Valley, his specimens now being in the Carnegie Museum in Pittsburgh, Pennsylvania. Anthony was the first to publish a list of birds of Portland and vicinity. Florence Merriam Bailey (q.v.) used Anthony's list in her *Handbook to the Birds of the Western United States* (1902). He wrote 'Field notes on the birds of Washington County, Oregon' in *Auk* (1886). Two mammals and two reptiles are named after him.

Anthony, H. E.

Anthony's Nightjar *Caprimulgus anthonyi* **Chapman**, 1923
[Alt. Scrub Nightjar]

Dr Harold E. Anthony (1890–1970), son of A. W. Anthony (q.v.), was Curator of Mammals at the AMNH. He was a noted collector of animals, especially in the Neotropics, but his prime interest was in mammals. He was President of the American Society of Mammalogists (1935–1937). He took part in a number of collecting expeditions to, for example, Burma and Ecuador. Chapman's dedication of the nightjar's name is interesting: '… the specimen on which this distinct species is based was shot at night by Mr. Harold E. Anthony, Associate Curator of Mammals in the American Museum, on a trail running through open, grassy, arid country near Portovelo.

Mr. Anthony's capture of the type makes it doubly fitting that this new bird should receive his name in recognition of the contributions he is making to Ecuadorian zoogeography through an intensive study of the mammalia of that country'. Three mammals and an amphibian are also named after him.

Antigone

Sarus Crane *Grus antigone* **Linnaeus**, 1758

Two women in Greek mythology share the name. The goddess Hera transformed Antigone into a stork. A less likely candidate is the Antigone who was the daughter of Oedipus by his mother Jocasta.

Antinori

Whistling Cisticola ssp. *Cisticola lateralis antinorii* **Heuglin**, 1867
Black Saw-wing ssp. *Psalidoprocne pristoptera antinorii* **Salvadori**, 1884

Marchese Orazio Antinori (1811–1882) was an Italian zoologist. He travelled, collecting in Ethiopia (1876–1882). He was head of a scientific station in Shoa (1876–1881), which was set up by the Royal Geographic Society of Italy of which he was a founding member. He was a college 'drop-out': he studied the classics but left (1828) without getting a diploma. His family was noble but not well off, so he needed to earn a living. He spent the next ten years pursuing ornithology and taxidermy, moving to Rome (1837) to work for Bonaparte mounting skins, etc. He undertook various writing and curating tasks until, in a time of political turmoil (1848), he was shot in the right arm which nearly had to be amputated. He then taught himself to write and draw with his left hand. He continued to make a living as a taxidermist, travelling to Greece and Turkey (1850s) and eventually to Syria and Egypt. He explored in Sudan until 'the continuous rains, the fevers, the dysentery and lack of drinking water threatened to bury us all'. He continued to travel, notably to Sardinia with Salvadori (1863) and to Tunisia (1866). He travelled with Beccari to the south of Egypt (1870). When he arrived in Ethiopia (1876) he wrote to his friend Doria telling him it was the most wonderful place he had seen and did not leave with the rest of the expedition, preferring to continue his work there for the rest of his life. A mammal is named after him.

Anton

Anthony's Flufftail *Sarothrura affinis antonii* **Madarász** & **Neumann**, 1911
[Alt. Striped Flufftail ssp.]
Bismarck Fantail ssp. *Rhipidura dahli antonii* **Hartert**, 1926

(See **Reichenow**)

Antonia (Perroud)

Sooty Barbthroat *Threnetes antoniae* **Bourcier** & **Mulsant**, 1846 NCR
[JS *Threnetes niger*]

Madame Antonia Perroud (DNF) was the wife of Benoit Philibert Perroud (1796–1878), who was President of the Linnaean Society at Lyon.

Antonia (Ridgway)

Antonia's Cotinga *Carpodectes antoniae* **Ridgway**, 1884
[Alt. Yellow-billed Cotinga]

Antonia Ridgway (DNF) was the sister of Robert Ridgway (q.v.). He refers to her in his original description of the cotinga as 'a dear sister whose death I mourn'.

Antony

Angola Lark ssp. *Mirafra angolensis antonii* **B. P. Hall**, 1958

Refers to Anthony L Archer (see **Archer, A. L.**).

Apetz

Lesser Short-toed Lark ssp. *Calandrella rufescens apetzii* **Brehm**, 1857

Professor Dr Johann Heinrich Gottfried von Apetz (1794–1857) was a German entomologist, although his time at Jena University was spent studying theology and, latterly, languages. He amassed a large collection of beetles from East Asia. He wrote a number of papers including *Beiträge zur Fauna des Osterlandes* (1840). A number of insects are named after him.

Aphrodite

Great Tit ssp. *Parus major aphrodite* **Madarász**, 1901

Aphrodite was the ancient Greek goddess of love and beauty. A reptile is also named after her.

Apolinar

Apolinar's Wren *Cistothorus apolinari* **Chapman**, 1914

Brother Apolinar María (1877–1949) was a missionary Colombian monk and ornithologist. He was Director of the Institute La Salle in Bogota (1914).

Appert

Appert's Greenbul *Xanthomixis apperti* **Colston**, 1972
[Alt. Appert's Tetraka; Syn. *Phyllastrephus apperti*]

Reverend Dr Otto Appert (1930–2012) was a Swiss missionary and amateur naturalist in Madagascar (1959–1966 and 1973–1990). He wrote several books and articles, such as 'Beobachtungen an *Monias benschi* in Südwest-Madagaskar' and 'La répartition geographique des vangides dans la région du Mangoky et la question de leur présence aux différentes époques de l'année' (1968). A reptile is named after him.

Apsley

Bar-breasted Honeyeater ssp. *Ramsayornis fasciatus apsleyi* **Mathews**, 1912
Bar-shouldered Dove ssp. *Geopelia humeralis apsleyi* Mathews, 1912
Little Corella ssp. *Cacatua sanguinea apsleyi* Mathews, 1912
White-bellied Cuckooshrike ssp. *Coracina papuensis apsleyi* Mathews, 1912

As Mathews named several birds from Melville Island, Northern Territory, and as the Apsley Strait separates Melville Island from Bathurst Island, we do not believe that these taxa are named directly after a person. Apsley Strait was named (1818) after the Secretary of State for War and the Colonies and Lord President of the Council, Henry, Baron Apsley, Lord Bathurst. The first inland town in Australia is called Bathhurst after him too.

Arcé

Arce's Tanager *Bangsia arcaei* **P. L. Sclater** & **Salvin**, 1869
[Alt. Blue-and-gold Tanager; Syn. *Buthraupis arcaei*]

Enrique Arcé (DNF) was a Guatemalan collector who was trained by Salvin (q.v.). He moved to Panama (late 1860s) and collected in Costa Rica and Panama. He obtained the tanager holotype at Cordillera del Chucu, Veraguas. He used a number of his relatives as collectors and combined their work into his returns. Among them was a brother, David, and a son, Enriquito (alive in 1937). Arcé also appears in the names of a number of Central American moths and butterflies.

Archbold

Bowerbird genus *Archboldia* **Rand**, 1940

Archbold's Newtonia *Newtonia archboldi* **Delacour** & **Berlioz**, 1931
Dwarf Sparrowhawk *Accipiter archboldi* **Stresemann**, 1932 NCR
[JS *Accipiter nanus*]
Archbold's Nightjar *Eurostopodus archboldi* **Mayr** & Rand, 1937
[Alt. Mountain Eared Nightjar, Cloud Forest Nightjar]
Archbold's Bowerbird *Archboldia papuensis* Rand, 1940
Snow Mountains Robin *Petroica archboldi* Rand, 1940
Archbold's Owlet-nightjar *Aegotheles archboldi* Rand, 1941
[Alt. Eastern Mountain Owlet-nightjar]

Spangled Kookaburra ssp. *Dacelo tyro archboldi* Rand, 1938

Richard A. Archbold (1907–1976) was an American philanthropist who became a zoologist at the AMNH. He was a member of the Madagascar Expedition (1929–1931) funded by his father and went on to finance and lead expeditions, particularly to Australasia, at times accompanied by, among others, Rand (q.v.) and G. H. H. Tate (q.v.). He also set up a permanent research station at Lake Placid in Florida. As well as birds, 26 insects, three spiders, a fish, an amphibian and a mammal are named after him.

Archer, A. L.

Red-billed Oxpecker ssp. *Buphagus erythrorhynchus archeri* Cunningham van Someren, 1984 NCR; NRM

Anthony 'Tony' L. Archer was a professional big-game hunter in Kenya (1956) and was working as a wildlife guide (2000). He was the author of *A Survey of Hunting Techniques and the Results thereof on Two Species of Duiker and the Suni Antelopes in Zanzibar* (1994). His son, Nigel, has followed in his footsteps. (See also **Antony**)

Archer, G. F.

Archer's Robin Chat *Cossypha archeri* **Sharpe**, 1902
[Alt. Archer's Ground Robin; Syn. *Dryocichloides archeri*]
Archer's Buzzard *Buteo (augur) archeri* **W. L. Sclater**, 1918
Archer's Lark *Heteromirafra archeri* **R. S. Clarke**, 1920

Archer's Greywing Francolin *Scleroptila levalliantoides
    archeri* W. L. Sclater, 1927
[Alt. Orange River Francolin; Syn. *Francolinus levalliantoides
    archeri*]
Alpine Swift ssp. *Tachymarptis melba archeri* **Hartert**, 1928
Eurasian Spoonbill ssp. *Platalea leucorodia archeri*
    Neumann, 1928
Common Kestrel ssp. *Falco tinnunculus archeri* Hartert &
    Neumann, 1932

Sir Geoffrey Francis Archer (1882–1964) was an explorer and administrator who was posted to British Somaliland in 1913, serving as Deputy Commissioner, HM Commissioner and finally as Governor (1919–1922). He carried out extensive field observations and collected 3,000 skins and 1,000 clutches of eggs. His observations of Archer's Lark are particularly important as the species was not subsequently seen for many decades until its rediscovery in Ethiopia in the 1970s. Another of his discoveries, Somali Pigeon *Columba oliviae*, was named after his wife Olive (q.v.). Later he became Governor-General of the Sudan (1924–1926). He co-authored, with Eva M. Godman, the 4-volume *Birds of British Somaliland and the Gulf of Aden: their Life Histories, Breeding Habits and Eggs* (1937–1961). F. J. Jackson (q.v.), who was Archer's uncle, collected the robin chat in Uganda.

Archibald, G. W.

Common Crane ssp. *Grus grus archibaldi* Ilyashenko, 2008

Dr George William Archibald (b.1946) is a Canadian ornithologist. Dalhousie University awarded his bachelor's degree (1968) and Cornell University his doctorate (1975). He is co-founder of the International Crane Foundation (1973) of which he was Director (1973–2000). He was the first winner of the Indianapolis Prize (2006). Archibald pioneered several techniques to rear cranes in human care, including having human handlers wear crane costumes to avoid imprinting. Archibald spent three years with a highly endangered Whooping Crane *Grus americana* named Tex, dressed and acting as a male crane – walking, calling, dancing – to shift her into reproductive condition. Through his dedication and the use of artificial insemination, Tex eventually laid a fertile egg.

Archibald (G. Campbell)

Tasmanian Scrubwren ssp. *Tasmanornis humilis archibaldi*
    Mathews, 1922 NCR
[JS *Sericornis humilis humilis*]

Archibald George Campbell (1880–1954) was President of the RAOU (1934–1935). His parents were Archibald James Campbell (q.v.) and Mrs Elizabeth Campbell.

Archibald (J. Campbell)

Brown Thornbill ssp. *Acanthiza pusilla archibaldi* **Mathews**,
    1910

(See **Campbell, A. J.**)

Archilochus

Hummingbird genus *Archilochus* **Reichenbach**, 1854

Archilochus (c.680–c.645 BC) was a poet from the island of Paros in the Archaic period in Greece, celebrated for his versatile and innovative use of poetic meters and as the earliest known Greek author to compose almost entirely on the theme of his own emotions and experiences.

Archimedes

European Stonechat ssp. *Saxicola torquatus archimedes*
    Clancey, 1949
[Frequently included in *Saxicola (torquatus) rubicola*]

Archimedes (c.287–212 BC) was a Greek mathematician, inventor and scientist who was born in Syracuse, Sicily. He is famous for such devices as the Archimedes screw and for running naked through the streets, shouting '*eureka*' after discovering Archimedes's Principle! A Roman soldier killed him when Syracuse was captured.

Arcos

Glowing Puffleg ssp. *Eriocnemis vestita arcosi*
    Schuchmann, Weller & Heynen, 2001
[Trinomial sometimes amended to (feminine) *arcosae*.

Laura Arcos Terán (b.1938) was three times Dean (1988–2002) of the Department of Biology, Catholic University of Ecuador. She graduated (1962) Universidad Complutense de Madrid, Spain, and was awarded her PhD (1971) by the Max Planck Institute in Tübingen, Germany, where she continued to research (1971–1973). She then became a lecturer back at the Department of Biology, Catholic University of Ecuador (1973) for the rest of her career. She has published widely including *El Parque Yasuní Revela sus Secretos* (2003).

Arechavaleta

Antbird sp. *Formicivora arechavaletae* **Bertoni**, 1901 NCR
[Alt. Dusky-tailed Antbird; JS *Drymophila malura*]
Elaenia sp. *Elainea arechavaletae* Bertoni, 1901 NCR
[Alt. Small-billed Elaenia; JS *Elaenia parvirostris*]

Professor José Arechavaleta y Balpardo (1838–1912) was born in Spain but emigrated to Uruguay (1855) where he qualified as a pharmacist (1862). He was Professor of Zoology, Botany and Natural History at the Faculty of Medicine and Director of both Montevideo's Municipal Laboratory (1874–1905) and Museo Nacional (1892–1912).

Argus

Great Argus *Argusianus argus* **Linnaeus**, 1766
Spotted Nightjar *Eurostopodus argus* **Hartert**, 1892

Argus (or Argos) was a 100-eyed watchman in Greek mythology. The God Mercury killed him, after which the goddess Hera placed Argus's eyes into the peacock's train. An amphibian and four reptiles are also named after Argus.

Ariadne

Hummingbird genus *Ariadne* A. Newton, 1867 NCR
[Now in *Amazilia*]

Ariadne in Greek mythology was daughter of King Minos of Crete and his queen, Pasiphaë, daughter of Helios, the Sun-titan. She aided Theseus in overcoming the Minotaur (actually her half-brother) and was the bride of the god Dionysus. An amphibian is named after her.

Ariel

Fairy Martin *Petrochelidon ariel* **Gould**, 1843
Lesser Frigatebird *Fregata ariel* **Gray**, 1845

Ariel Toucan *Ramphastos vitellinus ariel* **Vigors**, 1826
[Alt. Channel-billed Toucan ssp.]

Ariel, a sprite in folklore, is best known as the spirit in Shakespeare's *The Tempest* but also the name of one of the rebel angels in Milton's *Paradise Lost*.

Aristotle

Aristotle's Cormorant *Phalacrocorax aristotelis* **Linnaeus**, 1761
[Alt. European Shag]

Aristotle (384–322 BC) was one of the greatest ancient Greek philosophers, naturalists and historians. His work *Enquiry concerning Animals*, of which 10 books survive, is considered to be the start of descriptive zoology.

Armand

Yellow-streaked Warbler *Phylloscopus armandii* **Milne-Edwards**, 1865

Jean Pierre Armand David. (See **David, Père**)

Arment

Arment's Cowbird *Molothrus armenti* **Cabanis**, 1851
[Alt. Bronzed Cowbird; Syn. *Molothrus aeneus armenti*]

This is probably not an eponym at all. Gustav Haeberlin collected the original specimens of the cowbird (1826). Cabanis described the species much later, basing his choice of scientific name on an unpublished name of Lichtenstein's – 'I(cterus) armenti'. Later authorities assumed the bird had been named after a person called Arment. However, it is more likely that Lichtenstein derived *armenti* from the Latin word *armentum* (a herd of cattle) – a suitable appellation for a *cow*bird.

Arminjon

Trindade Petrel *Pterodroma arminjoniana* **Giglioli** & **Salvadori**, 1869

Vice-Admiral Vittorio Arminjon (1830–1897) was the first Italian to circumnavigate the globe, which he did (1865) as captain of the warship *Magenta*, which, incidentally, is also commemorated in a bird – the Magenta Petrel *Pterodroma magentae*. Enrico Giglioli was in charge of the scientific aspects of the voyage.

Armit

Ashy Robin ssp. *Heteromyias albispecularis armiti* **De Vis**, 1894

Captain William Edington de Margrat Armit (1848–1901) was a policeman, explorer and amateur naturalist. He was born in Belgium but moved to Australia (c.1870). He served as a policeman in Queensland, and (1883) took part in an expedition to New Guinea sponsored by the Melbourne *Argus*. Armit returned to New Guinea (1893) as private secretary to the administrator, Dr William MacGregor (q.v.).

Armour

Yellow Warbler ssp. *Dendroica petechia armouri* **Greenway**, 1933

Allison Vincent Armour (1863–1941) was a meatpacking millionaire from Chicago. He was a generous sponsor of natural history expeditions, archaeological digs, and agricultural research. He regularly cruised the Caribbean on the *Utowana*, a super-yacht of the era. Two reptiles are named after him.

Armstrong, F.

Armstrong's Sandpiper *Tringa guttifer* **Nordmann**, 1835
[Alt. Nordmann's/Spotted Greenshank]

Frank Bradley Armstrong (1863–1915) was a Canadian ornithologist and taxidermist. He was the son of an amateur naturalist who collected in Massachusetts, Mexico and Texas (1890). He wrote essays on ornithology, mammalogy and oology. He sent thousands of specimens to museums in Europe and the US, including the Field Museum in Chicago and the USNM. He also kept a collection of c.800 birds at his home in Brownsville.

Armstrong, J.

Collared Kingfisher ssp. *Todiramphus chloris armstrongi* **Sharpe**, 1892

Lieutenant-Colonel James Armstrong (1846–1923) was a British army officer in India (1874–1899).

Arnaud

Grey-headed Social Weaver *Pseudonigrita arnaudi* **Bonaparte**, 1850

(See **d'Arnaud**)

Arnault

Arnault's Tanager *Tangara arnaulti* **Berlioz**, 1927
[Presumed hybrid *Tangara cayana* x *T. preciosa*]

Dr Charles Arnault (DNF) was a French aviculturist who lived in southern Algeria (1930–1933). The tanager is only known from one aviary specimen and most authorities consider it to be a hybrid.

Arnold

Bar-throated Apalis ssp. *Apalis thoracica arnoldi*
J. A. Roberts, 1936

Dr George Arnold (1881–1962) was an entomologist who was educated in France and Germany. After qualifying as a doctor of science he joined the Department of Cytology and Cancer Research in Liverpool and, as a hobby, worked on Hymenoptera. He became Curator (1911) and later Director of the National Museum in Bulawayo, Southern Rhodesia (Zimbabwe). A reptile is named after him.

Arnot

Arnot's Chat *Myrmecocichla arnotti* **Tristram**, 1869
[Alt. White-headed Black Chat; Syn. *Pentholaea arnotti*]

David Arnot Jr (1822–1894) was a renowned, unscrupulous South African attorney. He lived at Colesberg, and contributed fossil reptiles, mammals, birds and insects to the South African Museum (1858–1868). Although Tristram used a double 't' in the scientific name, Arnot spelt his surname with a single 't'.

Arolas

Handsome Sunbird ssp. *Aethopyga bella arolasi* **Bourns &
Worcester**, 1894

General Juan Arolas (fl.1883–1898) was a Spanish army officer who was in the Sulu Islands, Philippines (1886–1893), being Governor (1891–1893). He was later in Cuba (1896) and became Military Governor of Havana (1898).

Arrhenius

Red-tailed Ant Thrush ssp. *Neocossyphus rufus arrhenii*
Lönnberg, 1917 NCR
[JS *Neocossyphus rufus gabunensis*]

Captain Karl Johan Ludvig Elias Arrhenius (1883–1923) was a Swedish explorer, collector and mercenary soldier who served in the Belgian colonial army in Rwanda and the Congo (1907–1923). He returned to Sweden once, briefly, for a lecture tour (1916).

Arrigoni

Corsican Goshawk *Accipiter gentilis arrigonii* **Kleinschmidt**,
1903
[Alt. Northern Goshawk ssp.]
Sardinian Buzzard *Buteo buteo arigonii* Picchi, 1903 NCR
[Alt. Common Buzzard ssp.; JS *Buteo buteo pojana*]
Spanish Sparrow ssp. *Passer hispaniolensis arrigonii*
Tschusi, 1903
Willow Tit ssp. *Parus atricapillus arrigonii* **von Burg**, 1925

Ettore Conte Arrigoni degli Oddi (1867–1942) was an Italian ornithologist. He wrote *Ornitologia Italiana* (1929).

Arses

Monarch flycatcher genus *Arses* **Lesson**, 1831

Arses was King of the Persians between 338 BC and 336 BC, the youngest son of Artaxerxes III. His unexpected accession came as a result of the murder of his father and most of his family.

Arsinoë

Hummingbird sp. *Ornismya arsinoe* **Lesson**, 1830 NCR
[Alt. Berylline Hummingbird; JS *Amazilia beryllina*]

Common Bulbul ssp. *Pycnonotus barbatus arsinoe*
Lichtenstein, 1823

There were several women – both real and mythological – named Arsinoë. A mythological nymph (Arsinoë was one of the Nysiads), or perhaps the heartless Arsinoë whom Aphrodite turned to stone because of her coldness towards her suitor Arceophon, may have inspired Lesson. Or he may have been thinking of Arsinoë I, a queen of Egypt and first wife of Ptolemy II, who presumably inspired Lichtenstein's trinomial for the bulbul.

Artemisia

Spotted Laughingthrush ssp. *Garrulax ocellatus artemisiae*
David, 1871

David's original description contains no etymology. Artemisia II of Caria and Halicarnassos, who built the Mausoleum – one of the seven wonders of the ancient world – as a monument to her brother Mausolus, may have inspired him.

Arthur (Buxton)

Greater Kestrel ssp. *Falco rupicoloides arthuri* **Gurney**, 1884

Reverend Arthur Fowell Buxton (1850–1881) was an English clergyman who took his BA degree (1876) from Trinity College, Cambridge. He received the holotype from Mombasa and presented it to the Norwich Museum.

Arthur (Goodson)

Yellow-throated Whistler ssp. *Pachycephala macrorhyncha
arthuri* **Hartert**, 1906 NCR
[NUI *P. macrorhyncha calliope*]
Crested Pitohui ssp. *Ornorectes cristatus arthuri* Hartert,
1930
[Syn. *Pitohui cristatus arthuri*]

(See **Goodson**)

Arthur (McConnell)

Black-billed Thrush ssp. *Turdus ignobilis arthuri* **Chubb**,
1914

Arthur Frederick Vavasour McConnell (1903–1961) was a son of Frederick Vavasour McConnell (q.v.) and Helen Mackenzie McConnell.

Arthur (Penard)

Ground Dove sp. *Chaemepelia arthuri* **Bangs** & **T. E. Penard**, 1918
[Alt. Ruddy Ground Dove; JS *Columbina talpacoti talpacoti*]

Slate-coloured Seedeater ssp. *Sporophila schistacea arthuri*
T. E. Penard, 1923 NCR
[JS *Sporophila schistacea longipennis*]

Arthur P. Penard (1880–1932) co-wrote *De Vogels van Guyana* (1908–1910) with his brother Frederick P. Penard (q.v.). Thomas E. Penard, another brother, was the describer.

Arthur Allen

Red-winged Blackbird ssp. *Agelaius phoeniceus arthuralleni* **Dickerman**, 1974

(See **Allen, A. A.**)

Arthus

Golden Tanager *Tangara arthus* **Lesson**, 1832

Arthus Bertrand (DNF) was the French publisher of Lesson's *Illustrations de Zoologie* (1832), which contains the description of the tanager, presumably named by Lesson in gratitude.

Artobolevsky

Great Spotted Woodpecker ssp. *Dendrocopos major artobolevskii* Charlemagne, 1934 NCR
[JS *Dendrocopos major candidus*]

Dr Vladimir Mikhaylovich Artobolevsky (fl.1934–1958) was a Russian zoologist, who was a Professor at Kiev University and at the Zoological Museum of the Ukrainian Academy of Sciences.

Artur

Bronze Sunbird ssp. *Nectarinia kilimensis arturi* **P. L. Sclater**, 1906

Arthur Lutley Sclater (1873–1922) was the son of Sclater (q.v.), who described the sunbird. He was an English tea planter in Ceylon, and one of the first settlers in Southern Rhodesia (Zimbabwe). He served in the 9th Battalion, Imperial Yeomanry, in the Boer War (1899–1902).

Ascalaphus

Owl genera *Ascalaphia/Ascalaphus* **Geoffroy**, 1837 NCR
[Now *Ascalaphia = Bubo / Ascalaphus = Asio*]

Savigny's Eagle Owl *Bubo ascalaphus* **Savigny**, 1809
[Alt. Pharaoh Eagle Owl]

Ascalaphus is a character from Greek mythology. When Persephone was in the underworld, and Pluto gave her permission to return to the upper world provided she had not eaten anything, Ascalaphus declared that she had eaten part of a pomegranate. Demeter punished him by changing him into an owl. According to Ovid, Persephone herself changed him into an owl by sprinkling him with water from the river Phlegethon.

Aschan

Grey-backed Camaroptera ssp. *Camaroptera brevicaudata aschani* **Granvik**, 1934
Mosque Swallow ssp. *Hirundo senegalensis aschani* Granvik, 1934 NCR
[JS *Cecropis senegalensis saturatior*]

Nils Krister 'Kris' Aschan (1904–1984) was a Swedish cavalry officer who became a professional hunter and settled in Kenya. He was there (1926) when the authorities advertised goods for sale by auction that he had not cleared through customs. His clients included the Roosevelt family.

Ash

Ash's Lark *Mirafra ashi* **Colston**, 1982

Dr John Sidney Ash (1925–2014) was an English ornithologist, a leading expert on African birds, with a special interest in the Horn of Africa. He was co-author of *Birds of Somalia* (1998) with John E. Miskell. Ash and Miskell lived in Somalia and its environs over many years and travelled to its remotest areas. They added over 50 first-time records and found one new species (above) and four new subspecies of birds. Ash was the Research Director of the Game Research Association and later Head of the Medical Ecology Division of the US Medical Research Unit in Ethiopia (1969–1977). Thereafter he worked as an ornithologist for the Food and Agriculture Organization and other UN agencies in Somalia, Uganda, Yemen and the Maldives. He was awarded the Bernard Tucker Medal of the BTO (1967) and the Union Medal of the BOU (1997). For 23 years he was Honorary Research Associate of the USNM. He wrote over 350 articles and reports.

Asha

Indian Reef Heron *Egretta asha* **Sykes**, 1832 NCR
[Alt. Western Reef Heron; JS *Egretta gularis schistacea*]

Asha is a Zoroastran divinity, the personification of truth.

Ashby

Gibberbird genus *Ashbyia* **North**, 1911

White-browed Babbler ssp. *Pomatostomus superciliosus ashbyi* **Mathews**, 1911
Black Honeyeater ssp. *Myzomela nigra ashbyi* Mathews, 1912
Little Corella ssp. *Cacatua sanguinea ashbyi* Mathews, 1912 NCR
[JS *Cacatua sanguinea gymnopis*]
Superb Fairy-wren ssp. *Malurus cyaneus ashbyi* Mathews, 1912
Golden Whistler ssp. *Pachycephala pectoralis ashbyi* Mathews, 1912 NCR
[JS *Pachycephala pectoralis pectoralis*]
White-browed Scrubwren ssp. *Sericornis frontalis ashbyi* Mathews, 1912
Southern Figbird ssp. *Sphecotheres vieilloti ashbyi* Mathews, 1912

Pied Currawong ssp. *Strepera graculina ashbyi* Mathews, 1913

Grey Falcon ssp. *Falco hypoleucos ashbyi* Mathews, 1913 NCR; NRM

Rock Calamanthus ssp. *Calamanthus montanellus ashbyi* Mathews, 1922

Red-browed Finch ssp. *Aegintha temporalis ashbyi* Mathews, 1923 NCR

[JS *Neochmia temporalis temporalis*]

Edwin Ashby (1861–1941) was an English naturalist, conchologist and ornithologist. He visited Australia because of poor health (1880s) and migrated to Adelaide (1888) where he joined his cousin's firm of land agents (1890). When he retired (1914) he continued to run a business from home. He was a collector of natural history specimens (eventually donating his collection to the South Australian Museum) and became a world authority on chitons (a type of mollusc), discovering 20 new species. He also wrote extensively on ornithology and Australian plant cultivation – he created a shrub nursery at his home and turned his land into a bird sanctuary. He was a founder member of the South Australian Ornithological Association (1899) and among the first members of the RAOU (1901), becoming President (1926). He was also a member of many overseas organisations. (See also **Edwin**)

Aspasia

Black Sunbird *Leptocoma aspasia* **Lesson & Garnot**, 1828 NCR

[JS *Leptocoma sericea*]

Aspasia (fl.450 BC) was a beautiful and intelligent woman, born in Miletus (Turkey), but spent much of her life in Athens. Here she became the lover of the orator and statesman Pericles. According to Plutarch, her house became an intellectual centre in Athens, attracting many prominent writers and philosophers.

Astley

Astley's Leiothrix *Leiothrix astleyi* **Delacour**, 1921

[Alt. Doubtful Leiothrix; Syn. *Leiothrix lutea astleyi*]

Reverend Hubert Delaval Astley (1860–1925) was President of the Avicultural Society. He had a most suitable address in Herefordshire: 'The Aviary'. He wrote *My Birds in Freedom and Captivity* (1900). The 'species' bearing his name seems to have been based on aviary mutations of the Red-billed Leiothrix.

Atala

Atala's Emerald *Chlorostilbon atala* **Gould**, c.1861 NCR

[JS *Chlorostilbon mellisugus*]

Probably not an eponym; i.e. it should be the Atala Emerald rather than Atala's Emerald, and may be derived from the Greek *atalos*, meaning delicate. Possibly, as a follow-on to Lesson's naming of *Ornismya atala* (1832), which could be after the heroine of a novel by Chateaubriand *Atala ou Les Amours de Deux Savages dans le Desert* (1801) or even after his daughter: see **Cecilia (Gautrau)**.

Athene

Owl genus *Athene* F. Boie, 1822

Athena was the goddess of wisdom in Greek mythology. Her favourite bird was the owl and it is still depicted on the badge of the city of Athens.

Atherton

Blue-bearded Bee-eater *Nyctyornis athertoni* **Jardine & Selby**, 1830

Lieutenant John Atherton (1797–1827) was part of the British Army in India. He served with the 13th Light Dragoons, fought with them at Waterloo and was stationed in Bangalore (1815). He collected the holotype and was the junior author's nephew.

Atkins

White-breasted Nuthatch ssp. *Sitta carolinensis atkinsi* **W. E. D. Scott**, 1890 NCR

[JS *Sitta carolinensis carolinensis*]

John W. Atkins (fl.1900) was honoured by Scott for his '… careful work done … on the birds of that portion of Florida'. He was manager in Key West for the Telegraph and Cable Company and made the first telephone call from the US to Havana (1900). Western Union owned an underwater telephone cable repair vessel named after him.

Atlapetes

Brush Finch genus *Atlapetes* **Wagler**, 1831

Here Atlas (q.v.) is combined with the Greek word for bird, *petes*.

Atlas

Shore Lark ssp. *Eremophila alpestris atlas* **Whitaker**, 1898

Coal Tit ssp. *Periparus ater atlas* **Meade-Waldo**, 1901

Atlas was a Titan in Greek mythology, condemned to stand at the western edge of the world and hold up the sky.

Atlay

Silver Pheasant ssp. *Lophura nycthemera atlayi* **Oates**, 1910 NCR

[JS *Lophura nycthemera rufipes*]

H. Frank Atlay (1863–1923) was manager of ruby and sapphire mines in Burma (Myanmar) (1887–1923).

Atmore

Mountain Chat ssp. *Oenanthe monticola atmorii* **Tristam**, 1869

Cape White-eye ssp. *Zosterops capensis atmorii* **Sharpe**, 1877

W. Atmore (fl.1875) was a collector in South Africa.

Attenborough

Inambari Gnatcatcher *Polioptila attenboroughi* Whittaker et al., 2013

Sir David Frederick Attenborough (b.1926) is famous as a maker of wildlife television programmes. He studied natural sciences at Cambridge and joined a firm of publishers (1950), where he did not stay long before joining the BBC in the early days of its post-war television service. He has been associated with the BBC, first as an employee and later as a freelance journalist, virtually ever since. He rose high in the organisation's ranks, becoming controller of BBC2 and responsible for introducing colour television to Britain, yet his first love was not administration but photojournalism. He has made some of the most stunning series of nature programmes and produced excellent books to accompany them, such as *The Life of Birds* (1998). A mammal and a reptile are also named after him.

Attila

Tyrant Flycatcher genus *Attila* Lesson, 1830

Attila (c.406–453) the Hun, also known as the 'Scourge of God', having become king or general of his people (433), came out of the plains of Central Asia to conquer half the known world. A mammal is named after him.

Atthis

Common Kingfisher *Alcedo atthis* Linnaeus, 1758

Atthis was a young woman from Lesbos and a favourite of the poetess Sappho, who addressed some of her poems to Atthis.

Attwater

Attwater's Prairie Chicken *Tympanuchus cupido attwateri* Bendire, 1893

Henry Philemon Attwater (1854–1931) was a naturalist and conservationist. He was born in Brighton, England, but emigrated to Ontario, Canada (1873), where he farmed and kept bees. He became interested in natural history and, together with John A. Morden, prepared and exhibited natural history specimens (1883). The two men collected specimens in Bexar County, Texas (1884). Attwater was employed to prepare and exhibit natural history specimens in the Texas pavilion at the New Orleans World's Fair (1884–1885). His major contributions were in the areas of ornithology and conservation. His three ornithological papers deal with the nesting habits of 50 species of birds in Bexar County, the occurrence of 242 species of birds near San Antonio, and the deaths of thousands of warblers (1892). He also contributed specimens to the USNM, collected birds for George B. Sennett and provided notes for W. W. Cooke's *Bird Migration in the Mississippi Valley* (1888). He was elected a Director of the National Audubon Society twice (c.1900 and 1905). Through his influence with farmers, Texas Audubon Society had gained affiliation (1910) with the Texas Farmers' Congress, the Texas Cotton Growers' Association and the Texas Corn Growers' Association. He was also active in the promotion of legislation to protect the Mourning Dove *Zenaida macroura*, which was rapidly declining during the early 1900s. His most important conservation works include *Boll Weevils and Birds* (1903), *Use and Value of Wild Birds to Texas Farmers and Stockmen and Fruit and Truck Growers* (1914) and *The Disappearance of Wild Life* (1917). Two mammals are named after him.

Atwood, J.

California Gnatcatcher ssp. *Polioptila californica atwoodi* Mellnick & Rea, 1994

Jonathan L. Atwood describes himself as a teacher, ornithologist and conservation biologist. His first degree was from University of California at Santa Barbara (1974), his MA from California State University, Long Beach (1978), and his PhD at University of California at Los Angeles (1986). He was an independent biological consultant (1977–1986). He taught undergraduate courses in ornithology, vertebrate zoology, general zoology and introductory biology at Department of Biology, University of California, Los Angeles (1980–1986). He was Director of the Avian Science Division of Manomet Centre for Conservation Sciences (1986–1998). He taught at Antioch University (1998–2011) and was appointed (2011) as Science Director, Biodiversity Research Institute, Gorham, Maine. His research interests focus on integrating behavioural studies of rare and endangered bird species with habitat conservation planning. He was honoured because he '… resolved the relationships between the gnatcatcher species *Polioptila melanura* and *P. californica* and has contributed so much to the conservation of California Gnatcatchers in the U.S.'

Atwood, T.

Dominican Macaw *Ara atwoodi* A. H. Clark, 1908 EXTINCT

Thomas Atwood (d.1793) was Chief Judge of Dominica and the Bahamas. He wrote *The History of the Island of Dominica* (1791). He fell on hard times as he died in the King's Bench prison.

Aubry, C. E.

Crow Honeyeater *Gymnomyza aubryana* J. Verreaux & Des Murs, 1860

Charles Eugène Aubry-Lecomte (1821–1879) was a French civil servant – an administrator in New Caledonia. He was also an amateur naturalist who collected wherever he was posted, for example making a collection of the fishes of Gabon (1850s). He was also amongst the first to describe the iboga root. Other taxa including three amphibians and two reptiles are named after him.

Aubry, O.

Aubry's Parrot *Poicephalus gulielmi aubryanus* Souance, 1856 NCR
[Alt. Red-fronted Parrot ssp.; JS *P. gulielmi gulielmi*]

Abbé Octave Aubry was a French parish priest (1759–1785). According to Levaillant (1799), as quoted by Stresemann

(1952), Aubry, who had one of the most numerous bird collections of his time, was vicar of Saint-Louis en l'Isle.

Aucher

Aucher's Grey Shrike *Lanius meridionalis aucheri*
    Bonaparte, 1853
[Alt. Southerm Grey Shrike ssp.]

Pierre Martin Rémi Aucher-Éloy (1792–1838) was a French pharmacist, botanist and collector in Asia Minor (Turkey), the Middle East, Egypt and Persia (Iran) (1830–1838). He lived in Istanbul (1830) and died in Isfahan (Iran). He wrote *Relations de Voyages en Orient de 1830 à 1838* (1843).

Audebert, J. B.

Audebert's Hummingbird *Chlorestes notata* Reich, 1793
[Alt. Blue-chinned Sapphire]

Jean Baptiste Audebert (1759–1800) was a noted miniaturist and nature artist. He produced *Oiseaux Dorés, ou à Reflets Métalliques* with Francois Vieillot (1800), issued in 32 parts over 26 months. He invented his own colour printing process for the 190 engraved plates, some of which are heightened in gold. He also issued a book (1797) on monkeys drawn from life, *Histoire Naturelle des Singes*. When Audebert died in Paris, he left complete material for a work on hummingbirds and other taxa entitled *Histoire des Colibris, Oiseaux-mouches, Jacamars et Promerops* (1802).

Audebert, J. P.

Thick-billed Cuckoo *Pachycoccyx audeberti* **H. Schlegel**, 1879

Josef-Peter Audebert (1848–1933) was a German naturalist who collected in Madagascar (1876–1882). Hermann Schlegel, when in charge of the museum at Leiden, employed him as a collector (1878), so his specimens were all sent to Holland. A mammal is named after him.

Audenet

Coquette sp. *Ornismya audenetii* **Lesson**, 1832 NCR
[JS *Lophornis chalybeus*]

Adolphe Jean Audenet (1800–1872) was a French banker, bibliophile and collector.

Audouin

Audouin's Gull *Larus audouinii* Payraudeau, 1826
[Syn. *Ichthyaetus audouinii*]

Jean Victoire Audouin (1797–1841) was a French naturalist, born in Paris, where he studied medicine, pharmacy and natural history. He was appointed assistant at the Musée National d'Histoire Naturelle in Paris (1825), later becoming Professor of Entomology there (1833). He also wrote a work on the natural history of French coastal waters; *Récherches pour Servir à l'Histoire Naturelle du Littoral de la France*.

Audrey

Rosy Finch ssp. *Leucosticte brandti audreyana* **Stresemann**, 1939

Miss Audrey Harris (b.1907) was in Sikkim (1936) and wrote *Eastern Vistas* (1939).

Audubon

Heron genus *Audubonia* **Bonaparte**, 1855 NCR
[Now in *Ardea*]

Audubon's Warbler *Dendroica auduboni* **Townsend**, 1837
[Syn. *Dendroica coronata auduboni*]
Audubon's Shearwater *Puffinus lherminieri* **Lesson**, 1839

Audubon's Woodpecker *Picoides villosus audubonii*
    Swainson, 1832
[Alt. Hairy Woodpecker ssp.]
Audubon's Oriole *Icterus graduacauda audubonii* **Giraud**, 1841
Northern Fulmar ssp. *Fulmarus glacialis auduboni*
    Bonaparte, 1857
Audubon's Hermit Thrush *Catharus guttatus auduboni*
    Baird, 1864
Audubon's Caracara ssp. *Caracara cheriway audubonii*
    Cassin, 1865
[Alt. Northern Caracara ssp.; (*C. cheriway* often regarded
    as monotypic)]

John James Laforest Audubon (1785–1851) is remembered as the father of US ornithology. He gave several different accounts of his birth, but was the son of a French Naval Captain and a French girl who worked at his sugar plantation in San Domingo (Haiti). Audubon's real mother died within a short time of his birth, so his natural father took him back to France where he was adopted by Captain Audubon and his legal wife. Captain Audubon sent him (1803) to manage his plantation near Philadelphia to avoid conscription into Napoleon's army, where he became an citizen (1812). In Philadelphia, Audubon met and married Lucy, whose support was critical in achieving his success. He succeeded only because he went to England (1826) where his work was appreciated and subscribers made possible the long publication of his 435 prints (1826–1838). Audubon also wrote (1830s) his *Ornithological Biography*, which describes the habits of the birds he drew. Audubon made a trip to North America (1843), his last great adventure prior to his death. He spent weeks in the woods studying birds and mammals; and his spectacular drawings, which were criticised as over-imaginative by some, were scenes he actually witnessed. There are many extensive works about Audubon from which those interested can get a fuller picture of the great man, but a lesser known fact about Audubon was his predilection for eating many of the birds he shot for their skins. For example, he described a Hermit Thrush as 'very fat and delicate eating' and that 'twenty six starlings made a good and delicate supper' whereas grebes were 'extremely fishy, rancid and fat'. Two mammals are also named after him.

August

Sooty-capped Hermit *Phaethornis augusti* **Bourcier**, 1847
[Alt. Sallé's Hermit]

(See **Sallé**)

Augusta

Emerald Dove ssp. *Chalcophaps indica augusta* **Bonaparte**, 1855

Augusta Amélie Maximilienne Jacqueline Principesa Gabrielli (1836–1900) was the daughter of the describer, French ornithologist Prince Charles Bonaparte.

Augusta Victoria

Empress of Germany's Bird-of-Paradise *Paradisaea raggiana augustaevictoriae* **Cabanis**, 1888

Auguste Viktoria of Schleswig-Holstein (1858–1921) was the first wife of Kaiser Wilhelm II (1869–1941), Emperor of Germany and King of Prussia, who abdicated (1918) and went into exile in the Netherlands for the rest of his life. They married (1881) in an eight-hour ceremony that required everyone to remain standing!

Auguste

Réunion Rail *Dryolimnas augusti* Mourer-Chauviré *et al.*, 1999 EXTINCT

Auguste de Villèle (1858–1943) was a French poet and local politician whose interest in the history of Réunion, where he was born, and hospitality made it possible for numerous naturalists to discover and explore the caves of Réunion.

Aurelia

Aurelia's Puffleg *Haplophaedia aureliae* **Bourcier** & **Mulsant**, 1846
[Alt. Greenish Puffleg]

Aurélie Henon *née* Favre (1814–1889) was the wife of the French agronomist Jacques Louis Henon. She painted watercolours to illustrate his botanical work. Gould illustrated the bird and so may have obtained a specimen from Bourcier.

Aurinia

Hummingbird genus *Aurinia* **Mulsant & E. Verreaux**, 1875 NCR
[Now in *Lophornis*]

Aurinia, a priestess held in veneration by the early Germanic tribes, was mentioned by the Roman writer Tacitus.

Aurivillius

Black Cuckoo ssp. *Cuculus clamosus aurivillii* **Sjöstedt**, 1892 NCR
[JS *Cuculus clamosus gabonensis*]

Dr Per Olof Christopher Aurivillius (1853–1928) was a Swedish entomologist. His doctorate was awarded by Uppsala (1880).

He joined the staff of Naturhistoriska Museet, Stockholm (1881), becoming Professor and then Director (1893).

Austen

Austen's Laughingthrush *Trochalopteron austeni* Austen, 1870
[Alt. Brown-capped Laughingthrush; Syn. *Garrulax austeni*]
Austen's Crow-tit *Paradoxornis guttaticollis* **David**, 1871
[Alt. Spot-breasted Parrotbill; (*P. austeni* is a junior synonym)]
Austen's Brown Hornbill *Anorrhinus austeni* **Jerdon**, 1872
Austen's Barwing *Actinodura waldeni* Austen, 1874
[Alt. Walden's/Streak-throated Barwing]
Austen's Spotted Babbler *Stachyris oglei* Austen, 1877
[Alt. Ogle's Spotted/Snowy-throated Babbler]

Red-billed Scimitar Babbler ssp. *Pomatorhinus ochraceiceps austeni* **Hume**, 1881

Lieutenant-Colonel Henry Homersham Godwin-Austen (1834–1923) was a British army topographer, geologist and surveyor. As an officer (1851–1877) he was assigned to several government surveys in northern India, especially in the Himalayas. He explored and surveyed the region of the Karakorum around K2. This Himalayan peak was named Mount Godwin-Austen in his honour. He was also an ornithologist and described several birds himself, among which was the laughingthrush *Garrulax austeni*. He had given the type specimen to Jerdon (q.v.) to describe and, when Austen wrote a formal description he believed that Jerdon had already gone into print using '*austeni*' as the species name. He was mistaken and, as he was the first to use this name in print himself it took precedence and makes it seem that he named if after himself which is, of course, very bad form. He wrote *Birds of Assam* (1870–1878). (See **Godwin**-Austen)

Austin, O. L.

Spectacled Owl ssp. *Pulsatrix perspicillata austini* **Kelso**, 1938 NCR
[JS *Pulsatrix perspicillata saturata*]

Dr Oliver Luther Austin Jr (1903–1988) was an American ornithologist and collector who was Curator of Birds, Florida State Museum (1957–1984), then Curator Emeritus. Harvard awarded his doctorate (1931). He served in the US Navy (WW2), edited *The Auk* (1968–1977) and became Professor of Zoology at Air University, Maxwell Air Force Base. He wrote *Birds of the World* (1961).

Austin, T. P.

Yellow Robin ssp. *Eopsaltria australis austina* **Mathews**, 1914
Barred Cuckooshrike ssp. *Coracina lineata austini* Mathews, 1916 NCR
[JS *Coracina lineata lineata*]

Thomas Phillips Austin (1874–1937) was an Australian grazier, field ornithologist and oologist. He wrote 'The birds of the Cobbora district' in the *Australian Zoologist* (1918).

Austin Roberts

Fawn-coloured Lark ssp. *Calendulauda africanoides austinrobertsi* **C. M. N. White**, 1947

(See **Roberts, J. A.**)

Austin Smith

Lanceolated Monklet ssp. *Micromonacha lanceolata austinsmithi* **Dwight** & **Griscom**, 1924 NCR; NRM
Tolmie Warbler ssp. *Oporornis tolmiei austinsmithi* **A. R. Phillips**, 1947

Austin Paul Smith (1881–1956) was an American ornithologist who published many articles in *Condor* and *Auk*. He was born in Ohio and moved to Brownsville in Texas (early 1900s). He was in Mexico and California (1908) and in Costa Rica (1920 to mid-1940s).

Aveledo

Stripe-breasted Spinetail ssp. *Synallaxis cinnamomea aveledoi* **W. H. Phelps** & **W. H. Phelps Jr**, 1946

Prof. Ramón Aveledo Hostos (1921–2002) was a Venezuelan ornithologist, Curator of the Museo Ornitológico Phelps, Caracas.

Ayesha

Double-spurred Francolin ssp. *Pternistis bicalcaratus ayesha* **Hartert**, 1917

Ayesha (also transcribed as A'ishah, 'A'isha, Aishat, etc.) bint Abu Bakr (613–678) was the Prophet Mohammed's favourite wife, and is a frequent name in Islamic countries.

Aylmer

Scaly Chatterer *Turdoides aylmeri* **Shelley**, 1885

Gerald Percy Vivian Aylmer (1856–1936) was a British explorer in Somaliland. Some of his travels were in the company of E. Lort Phillips (q.v.). He authored *A Recent Journey in Northern Somaliland* (1898).

Aylwin

Red-necked Francolin ssp. *Pternistis afer aylwinae* **C. M. N. White**, 1947 NCR
[JS *Pternistis afer melanogaster*]

Mrs Aylwin Button (DNF) was the wife of Captain Earl Button (q.v.).

Ayres

Bulbul genus *Ayresillas* **J. A. Roberts**, 1922 NCR
[Now in *Phyllastrephus*]

Ayres's Hawk Eagle *Hieraaetus ayresii* **Gurney**, 1862
Ayres's Cisticola *Cisticola ayresii* **Hartlaub**, 1863
[Alt. Wing-snapping Cisticola]
White-winged Flufftail *Sarothrura ayresi* Gurney, 1877

Thomas Ayres (1828–1913) was a British-born collector and naturalist. He went to Pinetown in Natal, South Africa (1850),

and collected birds, which he sent to Gurney in Norwich, England. Gurney published a series of 11 papers in *Ibis* (1859–1873) describing the species that Ayres collected. Ayres visited Australia and tried his luck in the goldfields there (1852), but returned to South Africa to settle in Potchefstroom as a hunter and trader. He was obviously eager to make his fortune out of gold, as (1870s) he was prospecting on the Lydenburg goldfields. He collected birds, beetles, butterflies and moths. His house was named the 'Ark' as it was 'long, low and stuffed with animals and birds'. He was a mentor to the young Roberts' boys (see **Roberts, J. A.**) and accompanied many legendary hunters on expeditions, for example to Mashonaland. Shelley (q.v.) documented the collection of birds made on such expeditions (*Ibis* 1882). After he returned to South Africa, Ayres even operated a brewery for a couple of years, making 'Ayres XX Pale Ale'. Many people spoke highly of this beer, including Captain William Cloudsley Lucas of the Bengal Yeomanry Cavalry, which was stationed at Rustenberg. Lucas wrote to Ayres saying that the beer had cured him of '… nightly sweatings, terrible affections in the lumbar regions, and a chronic costiveness that had lasted eighteen years.' Unfortunately, Ayres had to close down the brewery when the government changed the law and it became illegal for private persons to brew beer on a commercial basis.

Azara

Azara's Sand Plover *Charadrius collaris* **Vieillot**, 1818
[Alt. Collared Plover]
Ivory-billed Aracari *Pteroglossus azara* Vieillot, 1819
Azara's Bittern *Ixobrychus involucris* Vieillot, 1823
[Alt. Stripe-backed Bittern]
Azara's Spinetail *Synallaxis azarae* **d'Orbigny**, 1835
Black-and-white Monjita *Fluvicola azarae* **Gould**, 1839 NCR
[JS *Heteroxolmis dominicana*]
Tanager sp. *Piranga azarae* d'Orbigny, 1839 NCR
[JS *Piranga flava*]
Teal sp. *Anas azarae* **Merrem**, 1841 NCR
[JS *Anas flavirostris*]
Rush Tyrant sp. *Cyanotis azarae* **P. L. Sclater**, 1866 NCR
[JS *Tachuris rubrigastra*]
Macaw sp. *Ara azarae* **Reichenow**, 1881 NCR
[*Ara glaucogularis* (**Dabbene**, 1921) now used as the valid name]

Azara's Conure *Pyrrhura frontalis chiripepe* Vieillot, 1818
[Alt. Maroon-bellied Parakeet ssp.]
Giant Cowbird ssp. *Cassicus japus azarae* Merrem, 1826 NCR
[JS *Molothrus oryzivorus*]
Greyish Saltator ssp. *Saltator coerulescens azarae* d'Orbigny, 1839
Chimango Caracara ssp. *Milvago chimango azarae* **Brodkorb**, 1939 NCR
[JS *Milvago chimango chimango*]
Golden-crowned Warbler ssp. *Basileuterus culicivorus azarae* **Zimmer**, 1949

Félix Manuel de Azara (1746–1811) was born at Barbunales, Aragon, Spain. He was a military officer but also a naturalist and engineer, who distinguished himself in various expeditions. He was appointed a member of the Spanish Commission and sent to South America (1781–1801) to settle the borders between the Portuguese and Spanish colonies. He rose to Brigadier General when in command of the Paraguayan frontier (1781–1801). While there he turned his attention to the study of mammals, as an observer of the life and habits of quadrupeds in general. His observations, to which he added a lot of hearsay, were not always favourably received, but today his perspicacity as a student of the life of South American mammals is generally acknowledged. He also extended this to birds. Before leaving South America, he sent many notes and observations to his brother, who was then Spanish Ambassador in Paris. Moreau de Saint-Méry published them under the title of *Essai sur l'Histoire Naturelle des Quadrupèdes du Paraguay* (1801). He also wrote *Apuntamientos para la historia natural de los páxaros del Paraguay y río de la Plata* (1802–1805) and *Voyage dans l'Amérique Méridionale depuis 1781 jusqu'en 1801* (1809). Five mammals, an amphibian and a reptile are named after him.

Aziz

> Desert Lark ssp. *Ammomanes deserti azizi* **Ticehurst & Cheesman**, 1924

Abdul Aziz bin Abd ur-Rahman al Su'ud (1876–1953) was King of Saudi Arabia (1932–1953).

Aztec

> Song Sparrow ssp. *Melospiza melodia azteca* **Dickerman**, 1963 NCR
> NUI *Melospiza melodia mexicana*]

Dickerman named this subspecies after the Aztec Indians.

B

Babault

Brown-capped Babbler ssp. *Pellorneum fuscocapillus babaulti* **T. Wells**, 1919
Yellow-streaked Greenbul ssp. *Phyllastrephus flavostriatus babaulti* **Berlioz**, 1936 NCR
[JS *Phyllastrephus flavostriatus graueri*]

Guy Babault (1888–1963) was a French traveller, naturalist, conservationist and collector. He collected in British East Africa and in India and Ceylon (Sri Lanka). He wrote about his extensive collecting missions, among others, *Chasses et Recherches Zoologiques en Afrique Orientale Anglaise* (1917) and *Recherches Zoologiques dans les Provinces Centrales de l'Inde et dans les Régions Occidentales de l'Himalaya* (1922). Many of the animal specimens he collected can be seen in the Bourges Museum, gifted on his return from another trip to East Africa (1927). At least one book was written about his journeys: *Voyage de M. Guy Babault dans l'Afrique Orientale Anglaise 1912–1913.* A mammal is named after him.

Bachman

Bachman's Sparrow *Peucaea aestivalis* **Lichtenstein**, 1823
[Syn. *Aimophila aestivalis*]
Bachman's Warbler *Vermivora bachmanii* Audubon, 1833
EXTINCT
Bachman's Oystercatcher *Haematopus bachmani* Audubon, 1838
[Alt. American Black Oystercatcher]

Dr John Bachman (1790–1874) was a Lutheran minister from Charleston, South Carolina, and a close friend of Audubon (q.v.), whom he greatly aided. Bachman had 8 children by his first wife, Harriet, and we think none by his second wife, Maria. One of his daughters, also called Maria, married (1837) John Woodhouse Audubon (1812–1862), the younger son of John James Audubon. Though a slaveholder himself, Bachman wrote *The Unity of the Human Race* (1850) in which he argued that both master and slave were the same species. Audubon wrote 'My friend Bachman has the merit of having discovered this pretty little warbler …' Two mammals are named after him.

Bäckström

Grey-flanked Cinclodes ssp. *Cinclodes oustaleti baeckstroemii* **Lönnberg**, 1921

Dr Kare Bäckström (fl.1916) was a Swedish zoologist and entomologist with the Swedish Pacific Expedition (1916–1917).

Bacmeister

Lesser Spotted Woodpecker ssp. *Dendrocopos minor bacmeisteri* **Kleinschmidt**, 1916 NCR
[JS *Dendrocopos minor hortorum*]

Walther Kauzmann Bacmeister (1873–1966) was a German lawyer and ornithologist. He fought on the eastern and western fronts (WW1). He was a senior prosecutor in Stuttgart (1928–1938) and occasionally (WW2) was called out of retirement to act as a judge. He specialised in studying the spread of the native birds of south-west Germany even during active service. He wrote *Goethes Beziehungen zur Ornithologie* ('Goethe's Relations with Ornithology') (1918).

Baddeley

Spike-heeled Lark ssp. *Chersomanes albofasciata baddeleyi* **Clancey**, 1959 NCR
[NUI *C. albofasciata albofasciata*]
Peters's Twinspot ssp. *Hypargos niveoguttatus baddeleyi* **Wolters**, 1972 NCR
[JS *Hypargos niveoguttatus macrospilotus*]

M. O. E. Baddeley (b.1934) was a South African collector and taxidermist at the Durban Museum (1952). He was in Mozambique (1960).

Baedeker

Boreal Owl *Nyctale baedeckeri* C. L. Brehm, 1855 NCR
[JS *Aegolius funereus*]

Friedrich Wilhelm Justus Baedeker (1788–1865) was a German pharmacist in Witten, Westphalia, where he owned the town's first pharmacy (1811–1850). He was also an ornithologist and oologist and illustrated plates in a number of Brehm's works. He was part of the Baedeker publishing family, famous for its guidebooks.

Baer, G. A.

Tumbes Hummingbird *Leucippus baeri* Simon, 1901
Tucuman Mountain Finch *Compsospiza baeri* Oustalet, 1904
Short-billed Canastero *Asthenes baeri* Berlepsch, 1906
Crimson-fronted Cardinal *Paroaria baeri* Hellmayr, 1907

Baer's Woodnymph *Thalurania furcata baeri* Hellmayr, 1907
[Alt. Fork-tailed Woodnymph ssp.]
Plumbeous Pigeon ssp. *Patagioenas plumbea baeri* Hellmayr, 1908
Baer's Foliage-gleaner *Syndactyla dimidiata baeri* Hellmayr, 1911
[Alt. Planalto Foliage-gleaner ssp.]

Gustave-Adolphe Baer (1839–1918) was a French naturalist who collected in Brazil and Peru. Hellmayr (q.v.) wrote 'An account of the birds collected by Mons. G. A. Baer in the state of Goyaz, Brazil' in *Novitates Zoologicae* (1908).

Baer, K. E.

Baer's Pochard *Aythya baeri* **Radde**, 1863

Karl Ernst von Baer (Karl Maksimovich) (1792–1876) was a versatile and well-travelled Estonian of German extraction, a naturalist and explorer of Siberia, Novaya Zemlya and the Caspian Sea region. He graduated (1814) as a physician and later took further training in anatomy, eventually joining the staff of Königsberg University (1817) and becoming Professor of Zoology (1821). He was the Director of the Zoological Museum of Königsberg, which he himself had established, then Professor of Anatomy (1826) where he began his embryo research. He is known as the father of Estonian science and world embryology, not only discovering the egg cell but also that embryos have similar developmental stages in virtually all animals – these are known as the Baer Laws. He was also one of the co-founders of the Russian Geographical Society and edited a number of publications on geography. He worked in Austria and Germany before settling in Russia as Head of the Anthropological Museum of the St Petersburg Academy of Sciences and the Director of the Department of Foreign Literature of the Library there. He was also active in St Petersburg in research on geography, ichthyology, ethnography, anthropology and craniology. He worked briefly in the Ministry of Public Education (1862–1867). He is further remembered in 'Baer's Rule', which is about how riverbanks are symmetrical because of the rotation of the earth. He was a contemporary of Darwin (q.v.), with whom he corresponded, and spent the last years of his life writing critiques of Darwin's theories on evolution. Seven different geographical objects on different continents bear the name of Baer, and there is a street named after him in Tartu and his portrait graces the Estonian 2-Kroon banknote.

Bafirawar

Bafirawari Flycatcher *Bradornis pallidus bafirawari* **Bannerman**, 1924
[Alt. Pale Flycatcher ssp.; Syn. *Melaenornis pallidus bafirawari*]

Bafirawar (fl.1924) was a Kenyan collector for Sir Frederick Jackson (q.v.).

Bährmann

Common Nightingale ssp. *Luscinia megarhynchos baehrmanni* **Eck**, 1975 NCR
[NUI *Luscinia megarhynchos megarhynchos*]

Dr Rudolf Bährmann (b.1932) is a German entomologist and ecologist who was Professor at the Institute of Ecology, University of Jena, Germany.

Bailey, A. M.

Bailey's Sparrow *Xenospiza baileyi* **Bangs**, 1931
[Alt. Sierra Madre Sparrow]

Red-billed Chough ssp. *Pyrrhocorax pyrrhocorax baileyi* **Rand** & **Vaurie**, 1955

Dr Alfred Marshall Bailey (1894–1978) was a member of the expedition to Abyssinia (Ethiopia) (1926–1927) organised by the Field Museum of Natural History, Chicago, in conjunction with the Chicago Daily News. He worked at the Louisiana State Museum (1916–1919), the Field Museum (1926–1927), and the Denver Museum of Natural History (1936–1969). He was part of the survey of Alaska undertaken by the US Fish & Wildlife Service (1919–1921). He was the Director of the Chicago Academy of Science (1927–1936). Bailey was a leading light in the American Ornithologists' Union and a notable early photographer and cinematographer. He wrote many articles such as 'Birds of Arctic Alaska' (1948) and books such as his major (co-written) work *The Birds of Colorado* (1965). A mammal is named after him.

Bailey, F. A.

Bailey's Chickadee *Poecile gambeli baileyae* **Grinnell**, 1908
[Alt. Mountain Chickadee ssp.]

Mrs Florence Augusta Bailey *née* Merriam (1863–1948) was an early proponent of the use of binoculars rather than the shotgun in ornithology, evidenced by her first book *Birds Through an Opera Glass* (1889). She was also vehemently opposed to the fashion of decorating women's hats with feathers or even whole birds. She was a younger sister of Clinton Hart Merriam (q.v.), first chief of the United States Biological Survey. She married Vernon Bailey (1899), a pioneering naturalist in his own right, who worked for her brother with whom she was living at the time. She also wrote *A-Birding on a Bronco* and then *Birds of Village and Field* (1898). Her *Handbook of Birds of the Western United States* (1902) was described in *Condor* as 'the most complete textbook of regional ornithology which has ever been published'. Bailey was the first woman to become a fellow of the AOU (1929) and to receive its Brewster Medal (1931), for her comprehensive *Birds of New Mexico* (1928). Her last book was *Among the Birds in the Grand Canyon National Park* (1939).

Bailey, F. M.

Whiskered Yuhina ssp. *Yuhina flavicollis baileyi* **E. C. S. Baker**, 1914 NCR
[JS *Yuhina flavicollis rouxi*]
Rufous Sibia ssp. *Heterophasia capistrata bayleyi* **Kinnear**, 1939

Lieutenant-Colonel Frederick Marshman Bailey (1882–1967) was a soldier and a naturalist. He travelled in Tibet and parts of Central Asia (1903–1909). His experiences in Asia and outstanding military record – he fought at Gallipoli and in France (WW1) – made him the prime candidate for a secret mission into Central Asia (1918–1921) to discover what was happening in Bolshevik Russia. He wrote *Mission to Tashkent* (1946), his recollections of the experience, which were were little short of incredible, as he became a member of the Bolshevik secret police and was given the task of hunting down a British officer called Bailey who was thought to be hiding in the region! He later became a political officer in Sikkim (1921–1928). He wrote his memoirs, *No Passport to Tibet* (1957). Four reptiles and a mammal are named after him. Kinnear gave no etymology, but he did try to change the subspecies name to *baileyi* (1944) – but convention means that the original misspelling has to stand.

Bailey, W. T.

Papuan Frogmouth ssp. *Podargus papuensis baileyi* **Mathews**, 1912
Spangled Drongo ssp. *Dicrurus bracteatus baileyi* Mathews, 1912

W. T. Bailey (fl.1887) was an Australian collector in Queensland.

Baillieu

Palila *Loxioides bailleui* **Oustalet**, 1877

Pierre Etienne Theodore Baillieu (1829–1900) was French consul in Hawaii and a collector of natural history specimens.

Baillon

Baillon's Crake *Porzana pusilla* **Pallas**, 1776
[*Porzana bailloni* (**Vieillot** 1819) is a junior synonym]
Baillon's Aracari *Pteroglossus bailloni* Vieillot, 1819
[Alt. Saffron Toucanet; Syn. *Baillonius bailloni*]
Baillon's Shearwater *Puffinus bailloni* **Bonaparte**, 1857
[Alt. Tropical Shearwater]

Louis Antoine François Baillon (1778–1851) was a French naturalist and collector. He worked as an assistant at the MNHN, Paris (1798–1799). He wrote *Catalogue des Mammifères, Oiseaux, Reptiles, Poissons et Mollusques Testacés Marins Observés dans l'Arrondissement d'Abbeville* (1833). His father, Jean François Emmanuel Baillon (1742–1802), was also an amateur naturalist.

Baily

Grey Peacock Pheasant ssp. *Polyplectron bicalcaratum bailyi* **P. R. Lowe**, 1925 NCR
[NUI *Polyplectron bicalcaratum bicalcaratum*]

William Shore-Baily (d.1932) was a British aviculturist and ornithologist. Aged 23 he went to California, where he enjoyed both birdwatching and duck shooting. He also visited Brazil, Norway, Spain and other parts of Europe. He began keeping birds in captivity (1910), and building a collection of c.1,000 aviary specimens. The Avicultural Society awarded him 36 medals for his achievements in the captive husbandry and breeding of birds.

Baird

Baird's Sparrow *Ammodramus bairdii* **Audubon**, 1844
Baird's Flycatcher *Myiodynastes bairdii* **Gambel**, 1847
Banded Prinia *Prinis bairdi* **Cassin**, 1855
Baird's Sandpiper *Calidris bairdii* **Coues**, 1861
Baird's Wren *Troglodytes pacificus* Baird, 1864
[Alt. Pacific Wren; Syn. *Nannus pacificus*]
Baird's Trogon *Trogon bairdii* **Lawrence**, 1868

Peg-billed Finch *Acanthidops bairdi* **Ridgway**, 1882
Cozumel Vireo *Vireo bairdi* Ridgway, 1885
Baird's Creeper *Oreomystis bairdi* **Stejneger**, 1887
[Alt. Akikiki, Kauai Creeper]

Baird's Cormorant *Phalacrocorax pelagicus resplendens* Audubon, 1838
[Alt. Pelagic Cormorant ssp.]
Cuban Ivory-billed Woodpecker *Campephilus principalis bairdii* Cassin, 1864 EXTINCT
Acorn Woodpecker ssp. *Melanerpes formicivorus bairdi* Ridgway, 1881
Baird's Junco *Junco phaeonotus bairdi* Ridgway, 1883
[Alt. Yellow-eyed Junco ssp.]

Spencer Fullerton Baird (1823–1887) was an American zoologist and giant of American ornithology. He organised expeditions with the steamer *Albatross*. Baird was Assistant Secretary and then Secretary (1878) of the Smithsonian. He wrote *Catalogue of North American Birds* (1858). The young Baird became a friend of John James Audubon (q.v.) and sent him specimens. He is commemorated in the names of five mammals, two reptiles and an amphibian.

Baker, E. C. S.

Rufous-headed Parrotbill *Psittiparus bakeri* **Hartert**, 1900
[Syn. *Paradoxornis bakeri*]
Baker's Yuhina *Yuhina bakeri* **Rothschild**, 1926
[Alt. White-naped Yuhina, Chestnut-headed Yuhina]

Red-faced Liocichla ssp. *Liocichla phoenicea bakeri* Hartert, 1908
Great Crested Tern ssp. *Sterna bergii bakeri* **Mathews**, 1912 NCR
[JS *Thalasseus bergii velox*]
Baker's Cuckoo *Cuculus canorus bakeri* Hartert, 1912
[Alt. Common Cuckoo ssp.]
Streak-breasted Scimitar Babbler ssp. *Pomatorhinus ruficollis bakeri* **Harington**, 1914
Ruddy-breasted Crake ssp. *Porzana fusca bakeri* Hartert, 1917 NCR
[NUI *Porzana fusca fusca*]
Grey Peacock Pheasant ssp. *Polyplectron bicalcaratum bakeri* **P. R. Lowe**, 1925
Alpine Swift ssp. *Tachymarptis melba bakeri* Hartert, 1928

Edward Charles Stuart Baker (1864–1944) was a policeman in colonial India (1883–1912). He was also an amateur ornithologist, oologist and collector. Baker was a productive author, writing *The Indian Ducks and their Allies* (1908), *Game Birds of India, Burmah and Ceylon* (1921), *Fauna of British India: Birds* (1922), *Mishni the Man-eater* (1928), *The Nidification of the Birds of the Indian Empire* (1932) and *Cuckoo Problems* (1942).

Baker, G.

Baker's Bowerbird *Sericulus bakeri* **Chapin**, 1929
[Alt. Fire-maned/Adelbert's/Beck's Bowerbird]

George Fisher Baker Jr (1878–1937) was a US banker whose father had been co-founder of The First National Bank of New York. He became a Trustee of the AMNH.

Baker, J. R.

Baker's Pigeon *Ducula bakeri* **Kinnear**, 1928
[Alt. Vanuatu Imperial Pigeon]

John Randal Baker (1900–1984) was a British zoologist who collected in the New Hebrides (Vanuatu) (1922–1924). He was a member of the Zoology Department of Oxford University and was in Sydney (1933) in connection with the Oxford University Exploration Club expedition to Espiritu Santo, the largest of the islands in the New Hebrides. He met Alan John 'Jock' Marshall (q.v.) and recruited him to join the expedition. Years later when Marshall went to Oxford, he lodged in Baker's house. His works include a series of papers under the title 'The seasons in a tropical rain-forest (New Hebrides)'.

Baldwin

Northern House Wren ssp. *Troglodytes domesticus baldwini* **Oberholser**, 1934 NCR
[JS *Troglodytes aedon aedon*]

Samuel Prentiss Baldwin (1868–1938) was a lawyer and an amateur ornithologist who wrote a number of articles (1920s–1930s) including *The Marriage Relations of the House Wren* (1921) and *Measurements of Birds* (1931) for the Cleveland Museum of Natural History.

Balfour

Socotra Sunbird *Chalcomitra balfouri* **P. L. Sclater** & **Hartlaub**, 1881

Sir Isaac Bayley Balfour (1853–1922) was a Scottish botanist who became Regius Keeper of the Royal Botanic Garden Edinburgh (1890–1922). He explored a number of islands collecting specimens as he went, notably Rodrigues (1874), and made the first botanical study of Socotra (1879–1880) where the expedition collected the sunbird. Socotra contains one of the richest and best-preserved dry tropical floras in the world. Many plants and a reptile are named after him.

Ball

Andaman Scops Owl *Otus balli* **Hume**, 1873

Valentine Ball (1843–1895) was an Irish naturalist who travelled and collected in India. He was naturalist on an expedition there (1876) and sent his specimens to the Natural History Department, National Museum of Ireland. There is a memoir by him appended to the report on the *Scientific Results of the Second Yarkand Mission* (1886). He also translated Tavernier's journals of his travels to India a century before.

Ballion

Himalayan Rubythroat ssp. *Luscinia pectoralis ballioni* **Severtsov**, 1873

Ernst von Ballion (1816–1901) was a Russian entomologist, specialist in *Coleoptera*, and collector.

Ballivian

Stripe-faced Wood Quail *Odontophorus balliviani* **Gould**, 1846

José Ballivián y Segurola (1805–1852) was the 11th President of Bolivia (1841–1847). He was a military commander in the war with Spain, and became president after the breakdown of the Peru-Bolivia confederation. He defended Bolivia against Peruvian incursion and was known as a reformer, but was forced from office (1847).

Ballmann

Ballmann's Malimbe *Malimbus ballmanni* **Wolters**, 1974
[Alt. Gola Malimbe]

Dr Peter Ballmann (b.1941) is a German geoscientist who studies fossil birds and avian osteology. He studied geology in Germany and at Leiden, writing his dissertation on fossil birds. He collected a few birds, some of which he could not identify, whilst he was working (1971) as a field pedologist in the Ivory Coast. These he took to the König Museum in Bonn. Among the collection was the eponymous malimbe. Ballmann presented the specimen to the König Museum. He later worked in Costa Rica (1973–1974), Saudi Arabia (1981–1984) and Swaziland (1985). He regards himself as an amateur, as he wrote to us 'I have met real ornithologists and bird-watchers and it impressed me most that the good ones could identify the birds just by sight and did not have to shoot them first.' Ballman published extensively on palaeo-ornithology (1960s–1970s). His works include 'Die Vögel aus der altburdigalen Spaltenfüllung von Wintershof (West) bei Eichstätt in Bayern' (1969) and 'A new species of fossil barbet (Aves: Piciformes) from the Middle Miocene of the Noerdlinger Ries (southern Germany)' (1983).

Balsac

Crested Lark ssp. *Galerida cristata balsaci* **Dekeyser** & **Villiers**, 1950
Eurasian Spoonbill ssp. *Platalea leucorodia balsaci* **Naurois** & **Roux**, 1974
Three-banded Courser ssp. *Rhinoptilus cinctus balsaci* **Erard**, Hemery & Pasquet, 1993

Professor Henri Heim de Balsac (1899–1973) was a French zoologist. Primarily a mammologist, he worked on biological methods for pollution control. Most of his career was spent in France but he also managed a project at the Moroccan Institute of Scientific Research. He co-wrote *Les Oiseaux du Nord-Ouest de l'Afrique* (1962) with Mayaud (q.v.). (See also **Heim**)

Balston, R. J. & W. E.

Madagascar Swift *Apus balstoni* **E. Bartlett**, 1880

White-browed Scrubwren ssp. *Sericornis frontalis balstoni* **Ogilvie-Grant**, 1909
Australian Yellow White-eye ssp. *Zosterops luteus balstoni* **Ogilvie-Grant**, 1909

Richard James Balston (1839–1916) was a wealthy English businessman (papermaker) who became a patron of the sciences. The Madagascar Swift is named after him and the other two birds named after his brother William Edward Balston (1848–1918), who was also a successful businessman and interested in ornithology. Two mammals are named after William.

Baltimore

Baltimore Oriole *Icterus galbula* **Linnaeus**, 1758

George Calvert, 1st Baron Baltimore (1579–1632), was an English statesman and landowner who obtained a grant for Maryland, north of the Potomac (1632). The city of Baltimore in Maryland is named after him. It is said that the bird received its name because the colours of the male resembled those on his coat-of-arms.

Balzan

Fork-tailed Woodnymph ssp. *Thalurania furcata balzani*
E. L. Simon, 1896

Dr Luigi Balzan (1865–1893) was an Italian naturalist, entomologist and explorer. Aged twenty he went to South America, becoming Professor of Natural Sciences in Asunción, Paraguay. A reptile and an amphibian are named after him.

Bambara

Bambara Cliff Chat *Thamnolaea cinnamomeiventris bambarae* **Bates**, 1928
[Alt. Mocking Cliff Chat ssp.; Syn. *Myrmecocichla c. bambarae*]

The Bambara are a people of the French Sudan (Mali); their language is also known as Bambara.

Bamberg

Eurasian Jay ssp. *Garrulus glandarius bambergi* **Lönnberg**, 1909 NCR
[JS *Garrulus glandarius brandtii*]

Otto Bamberg (1871–1942) was a German oologist and collector in the Caucasus (1895) and northern Mongolia (1908).

Bancroft

Bancroft's Screech Owl *Megascops kennicottii cardonensis* Huey, 1926
[Alt. Western Screech Owl ssp.]
Bancroft's Night Heron *Nyctanassa violacea bancrofti* Huey, 1927
[Alt. Yellow-crowned Night Heron ssp.]

Griffing Bancroft Jr (1907–1998) was an ornithologist and journalist who wrote *The Flight of the Least Petrel* (1932). He was also an egg collector who donated 30,000 birds' eggs to the AMNH (1941). He was instrumental in setting up the Sanival reserve in Florida and is also noteworthy as one of the first people to be awarded the Medal of Freedom, the highest honour for a civilian in the USA, for his work in the field of psychological warfare. His third wife, Jane Eads Bancroft, said of him in an interview (1998): 'He's a very

interesting person. He's really an ornithologist and he's written several books about birds. When he was very young, he went on all these birding expeditions with his father, Griffing Bancroft Sr, in California, and they went around in Baja California and around various mountainous regions. They collected birds' eggs – in those days, it was all right to do it – for the San Diego Museum. At one point, before WW2, they had one of the largest private collections of bird eggs in the country. They have two birds named after them. That was his main interest at first'. Bancroft's grandfather (Hubert Howe Bancroft 1832–1918) is renowned for his 65-volume history of the US, which he sold (1905) to the University of California for $150,000.

Bangs

Tanager genus *Bangsia* **T. E. Penard**, 1919

Santa Marta Antpitta *Grallaria bangsi* **J. A. Allen**, 1900

Grey-breasted Wood Wren ssp. *Henicorhina leucophrys bangsi* **Ridgway**, 1903
Bay-headed Tanager ssp. *Tangara gyrola bangsi* **Hellmayr**, 1911
White-bearded Manakin ssp. *Manacus manacus bangsi* **Chapman**, 1914
Ruddy Kingfisher ssp. *Halcyon coromanda bangsi* **Oberholser**, 1915
European Cuckoo ssp. *Cuculus canorus bangsi* Oberholser, 1919
Olive-backed Woodcreeper ssp. *Xiphorhynchus triangularis bangsi* Chapman, 1919
White-eyed Foliage-gleaner ssp. *Automolus leucophthalmus bangsi* **Cory**, 1919 NCR
[JS *Automolus leucophthalmus leucophthalmus*]
Greater Antillean Grackle ssp. *Quiscalus niger bangsi* **J. L. Peters**, 1921
Brown Prinia ssp. *Prinia polychroa bangsi* **La Touche**, 1922
Mangrove Black Hawk ssp. *Buteogallus subtilis bangsi* Swann, 1922
[Syn. *B. anthracinus bangsi*, if *subtilis* not recognised as a species]
Orange-billed Nightingale Thrush ssp. *Catharus aurantiirostris bangsi* **Dickey** & **Van Rossem**, 1925
White-tipped Dove ssp. *Leptotila verreauxi bangsi* Dickey & Van Rossem, 1926
Bangs's Black Parrot *Coracopsis nigra libs* Bangs, 1927
[Alt. Lesser Vasa Parrot ssp.]
Bangs's Sparrow *Amphispiza bilineata bangsi* **Grinnell**, 1927
[Alt. Black-throated Sparrow ssp.]
Chinese Nuthatch ssp. *Sitta villosa bangsi* **Stresemann**, 1929
Yellow-breasted Crake ssp. *Porzana flaviventer bangsi* Darlington, 1931
Least Grebe ssp. *Tachybaptus dominicus bangsi* **Van Rossem** & Hachisuka, 1937

Outram Bangs (1862–1932) was an American zoologist, born in Watertown, Massachusetts. The family spent a year in England (1873). Bangs attended Harvard (1880–1884). He began (1890) a systematic study of the mammals of eastern North America and wrote over 70 books and articles, the majority on mammals. His collection of over 10,000 mammalian skins and skulls, including over 100 type specimens, was presented to Harvard College (1899). Bangs was appointed Assistant in Mammalogy at Harvard and became Curator of Mammals at the Harvard Museum of Comparative Zoology (1900). He also collected bird specimens. He visited Jamaica (1906) and collected c.100 birds there but his trip was cut short by dengue fever. His collection of over 24,000 bird skins was presented to the Museum of Comparative Zoology (1908) and he went on to increase it. He went to Europe (1925), visiting museums and ornithologists and arranging scientific exchanges. He was a member of the American Ornithologists' Union and wrote articles for its journal, *Auk*. A mammal in named after him.

Banks, E. H.

Sunda Bush Warbler ssp. *Cettia vulcania banksi* **Chasen**, 1935

Edward H. 'Bill' Banks (1903–1988) was, as a British colonial administrator, District Officer in Sarawak, a zoologist, naturalist and Curator of Sarawak Museum, Kuching (1925–1945). He was interned at Batu Lintang during the Japanese occupation (1942–1945) (WW2). He wrote *A Naturalist in Sarawak* (1949) and retired the following year.

Banks, J.

Banks's Black Cockatoo *Calyptorhynchus banksii* **Latham**, 1790
[Alt. Banksian Cockatoo, Red-tailed Black Cockatoo]

Banks's Dove Prion *Pachyptila desolata banksi* **A. Smith**, 1840 NCR; NRM

Sir Joseph Banks (1743–1820) was a highly influential English botanist and explorer. He was born in London and studied at Oxford but left without graduating (1765) and started regularly visiting the reading room of the British Museum, which is where and when he probably first met Daniel Solander (q.v.), a pupil of Linnaeus (q.v. under Linné). Banks made three voyages of note, on HMS *Niger* to Newfoundland and Labrador (1776), during which he collected plants. Banks, accompanied by Solander, was the chief naturalist on James Cook's (q.v.) expedition round the world in HMS *Endeavour* (1768–1771), while on HMS *Sir Lawrence* he led his own expedition to the Hebrides, Iceland and the Orkney Islands (1772). During his last voyage he explored the island of Staffa, writing the first description of Fingal's Cave. He is perhaps best known for founding and stocking Kew Gardens, the foremost botanical collection and research institution in the world. He was an important patron of science and (1778) and became President of the Royal Society for 41 years. It was Banks who had the idea of transporting breadfruit from Tahiti to Jamaica and arranged for Captain James Bligh commanding HMS *Bounty* to achieve this – his failure is well known! Banks founded the African Association, and the Australian colony of New South Wales owes its origin mainly to him.

Banks, R. C.

Yellow Warbler ssp. *Dendroica aestiva banksi* **Browning**, 1994

Dr Richard Charles Banks (b.1931) studied wildlife conservation and zoology at universities in Ohio and California, being awarded his doctorate by Berkeley (1961). He became Curator of the San Diego Natural History Museum (1962), and worked for the Patuxent Wildlife Research Center (PWRC) at the USNM with responsibility for research on systematics, nomenclature and distribution of birds, primarily North American, until retirement (2002). He is Curator Emeritus with the PWRC and a Research Associate at USNM. He was President of the Wilson Ornithological Society (1991–1993) and of the AOU (1994–1996), and is an Honorary Member of the Cooper Ornithological Society. He was (2010) Chairman of the Check-list Committee of the AOU and Editor of *Proceedings of the Biological Society of Washington*. He named a bird after his father Clinton Banks (see **Clinton**).

Bannerman

Storm-petrel genus *Bannermania* **Mathews** & **Iredale**, 1915 NCR
[Now in *Oceanodroma*]

Bannerman's Shearwater *Puffinus bannermani* Mathews & Iredale, 1915
Bannerman's Turaco *Tauraco bannermani* **G. L. Bates**, 1923
Bannerman's Weaver *Ploceus bannermani* **Chapin**, 1932
Bannerman's Sunbird *Cyanomitra bannermani* **C. H. B. Grant & Mackworth-Praed**, 1943

Cape Verde Buzzard *Buteo buteo bannermani* Swann, 1919
[Alt. Common Buzzard ssp.]
Cream-coloured Courser ssp. *Cursorius cursor bannermani* **Rothschild**, 1923 NCR
[NUI *Cursorius cursor cursor*]
Little Swift ssp. *Apus affinis bannermani* **Hartert**, 1928
Bannerman's Pipit *Anthus similis bannermani* G. L. Bates, 1930
[Alt. Long-billed Pipit ssp.]
Bannerman's Paradise-flycatcher *Terpsiphone batesi bannermani* Chapin, 1948
[Alt. Bates's Paradise-flycatcher ssp.]

Dr David Armitage Bannerman (1886–1979) was a British ornithologist on the staff of the BMNH. He was Chairman of the British Ornithologists' Club (1932–1935) and became Honorary President of the Scottish Ornithologists' Club (1959). He was an early leader in the conservation movement. His numerous publications include the standard multi-volume works *The Birds of Tropical West Africa* and *The Birds of the Atlantic Islands*. (Bannerman met a young District Commissioner in West Africa who was a keen birdwatcher, who told him that he took the former work with him on Safari, to which news Bannerman exclaimed 'Not all 8 volumes?' The young man replied 'It only means another porter'). Malcolm Ogilvie, a leading British ornithologist, told us: 'I met him a number of times before he died. He retired here [to Scotland] after a long and very productive life, producing books on West Africa, the Canaries, the Azores, the British Isles, Cyprus, etc., etc., all lavishly produced, with specially commissioned paintings by George Lodge and David Reid-Henry paid for by Bannerman himself who had a private income.'

Baptista

Sapphire-vented Puffleg ssp. *Eriocnemis luciani baptistae* Schuchmann, Weller & Heynen, 2001

Dr Luis Felipe Baptista (1941–2000) was Chair and Curator of Ornithology and Mammalogy at the California Academy of Sciences, San Francisco, for nearly 20 years. His prime area of interest was avian vocalisations, particularly 'dialects', which he studied in the field from Alaska to Costa Rica. He was a great bird mimic too. He died of a heart attck while feeding birds in Golden Gate Park – where he could identify 150 species on calls alone. A symposium on bird song was held in his memory. He wrote solely or jointly 127 publications.

Baraka

Sunbird sp. *Nectarinia barakae* **Sharpe**, 1902 NCR
[Alt. Purple-breasted Sunbird; JS *Nectarinia purpureiventris*]

Green Combec ssp. *Sylvietta virens baraka* Sharpe, 1897
Somali Tit ssp. *Parus thruppi barakae* **F. Jackson**, 1899
[Alt. Acacia Tit, Northern Grey Tit]
Northern Olive Thrush ssp. *Turdus abyssinicus baraka* Sharpe, 1903
Evergreen-forest Warbler ssp. *Bradypterus lopezi barakae* Sharpe, 1906
Scaly-breasted Illadopsis ssp. *Illadopsis albipectus barakae* F. Jackson, 1906

Baraka (d.1911) was an African hunter who collected specimens for Sir Frederick Jackson in Uganda and Kenya. He was skilled at skinning. Baraka also means 'blessed' in several languages and it is difficult to know whether this particular African employee is commemorated in all of the above birds, particularly when, from the Sufi, it is sometimes translated as 'the breath, or essence of life from which the evolutionary process unfolds'. However, all the birds were collected within a few years of each other in the appropriate areas and described by just two men, so it seems likely that they are all eponymous.

Barau

Barau's Petrel *Pterodroma baraui* **Jouanin**, 1964

Armand Barau (1921–1989) was co-author with Nicolas Barré and Christian H. Jouanin of *Oiseaux de la Réunion* (1982), the first serious study of the birds there. He was an agronomist, landowner and amateur ornithologist in Réunion, and (1962–1989) was President of the Centre d'Essai de Recherche et de Formation (CERF), an organisation concerned with sugar production. Jouanin encouraged him to look for the Mascarene Black Petrel that had not been collected since the 19th century (it was not rediscovered until 1970).

Barbara

Lesser Treeswift ssp. *Hemiprocne comata barbarae* **J. L. Peters**, 1939 NCR
[JS *Hemiprocne comata major*]

Barbara Lawrence (1909–1997) worked at the Museum of Comparative Zoology, Harvard University, as Museum Assistant to the director Dr Thomas Barbour (q.v.) and the Curator of Mammals Dr Glover Allen (q.v.). She collected birds in the Philippines.

Barbarita

Orange-billed Nightingale Thrush ssp. *Catharus aurantiirostris barbaritoi* **Aveledo** & **Ginés**, 1952

Dr Xaviero Barbarita (fl.1952) was a Venezuelan zoologist, explorer and collector.

Barberena

Grey-hooded Flycatcher *Hemitriccus barberenae* **Bertoni**, 1901 NCR
[Indeterminate: JS of *Mionectes rufiventris*]

Dr Santiago Ignacio Barberena (1851–1916) was a Salvadorian lawyer, teacher, engineer, historian and essayist. He held doctorates in jurisprudence (University of El Salvador, 1876) and Engineering (University of San Carlos, Guatemala, 1877). He worked as a teacher (1871–1916) including at the University of El Salvador, as Professor of Physics and Topography. He worked as a surveyor in establishing the boundaries between Mexico and Guatemala (1878–1881) and between El Salvador and Honduras (1886). As an archaeologist he undertook the excavation of Mayan sites in Honduras as well as El Salvador. He wrote *History of El Salvador* (1917).

Barbero

Stygian Owl ssp. *Asio stygius barberoi* **Bertoni**, 1930

Dr Andrés José Camilo Barbero Crosa (1877–1951) was a Paraguayan scientist, physician, pharmacologist and philanthropist.

Barbour

Eastern Violet-backed Sunbird ssp. *Anthreptes orientalis barbouri* **Friedmann**, 1931 NCR; NRM
White-crowned Shama ssp. *Copsychus stricklandii barbouri* **Bangs** & **J. L. Peters**, 1927
Grey Jay ssp. *Perisoreus canadensis barbouri* W. S. Brooks, 1920 NCR
[JS *Perisoreus canadensis nigricapillus*]

Dr Thomas Barbour (1884–1946) was an American zoologist. He graduated from Harvard (1906) and obtained his PhD there (1910). He became an Associate Curator of Reptiles and Amphibians at the Harvard Museum of Comparative Zoology, and was its Director (1927–1946). He became Custodian of the Harvard Biological Station and Botanical Garden, Soledad, Cuba (1927). He was Executive Officer in charge of Barro Colorado Island Laboratory, Panama (1923–1945). During his time at the museum he explored in the East Indies, the West Indies, India, Burma, China, Japan, and South and Central America. He was famously jovial good company and would invite all and sundry to eat and converse next door to his office in the 'Eateria' in which his secretary, Helen Robinson, prepared the food for his many guests. Something of an all-rounder, he wrote many articles and books, including *The Birds of Cuba* (1923) and *Naturalist at Large* (1943). He also co-wrote *Checklist of North American Amphibians and Reptiles*. His special area of interest was the herpetology of Central America and 24 reptiles are named after him, as are two mammals and four amphibians.

Barbosa

> Yellow-bellied Hyliota ssp. *Hyliota flavigaster barbozae*
> **Hartlaub**, 1883

(See **Bocage**)

Barej

> Baillon's Crake ssp. *Porzana pusilla bareji* Dunajewski, 1937
> NCR
> [JS *Porzana pusilla pusilla*]

Tomasz Barej (1860–1918), also known as Thomas Barey ,was a Polish forester and collector for the Branicki Museum, Warsaw, in Transcaucasia (1887), Turkmenistan and Iran (1889–1890), Uzbekistan, Kyrgyzstan and Kazakhstan (1893–1894). He was exiled to Siberia (1895) and worked at the museum in Kherson in southern Ukraine (1898). He collected the crake holotype (1894).

Barge

> Curaçao Barn Owl *Tyto bargei* Hartert, 1892
> [Syn. *Tyto alba bargei*]

Dr Charles Augustinius Henri 'Harry' Barge (1844–1919) was Governor (1890–1901) of the Dutch West Indies, of which Curaçao formed part.

Barkly

> Seychelles Black Parrot *Coracopsis nigra barklyi* **E. Newton**, 1867
> [Alt. Lesser Vasa Parrot ssp.]

Sir Henry Barkly (1815–1898) was a British politician and colonial governor. He was a Member of Parliament (1845–1848) and Governor of British Guiana (Guyana) (1848–1853), Jamaica (1853–1856), Victoria, Australia (1856–1863), Mauritius (1863–1870), and Cape Colony, South Africa (1870–1877). Two South African towns, Barkly East and West, are named after him.

Barlow, C.

> Barlow's Chickadee *Poecile rufescens barlowi* Grinnell, 1900
> [Alt. Chestnut-backed Chickadee ssp.]

Chester Barlow (1874–1902) was the Assistant Cashier at the Santa Clara Valley Bank and an amateur ornithologist and oologist. He was first Secretary (for a decade) of the Cooper Ornithological Society, despite his youth. He was the author of many articles (1892–1902) in *The Naturalist, The Oologist* and especially in *The Nidiologist* and *Condor*. These included a paper on 'The Pileolated and Yellow Warblers' (1893) and 'Some additions to Van Denburgh's list of land birds of Santa Clara Co., California' (1900).

Barlow, C. S.

> Barlow's Lark *Calendulauda barlowi* **J. A. Roberts**, 1937

> Tractrac Chat ssp. *Cercomela tractrac barlowi*
> J. A. Roberts, 1937
> Pink-billed Lark ssp. *Spizocorys conirostris barlowi*
> J. A. Roberts, 1942
> Spike-heeled Lark ssp. *Chersomanes albofasciata barlowi*
> **C. M. N. White**, 1961

Charles Sydney 'Punch' Barlow (1905–1979) was a South African businessman. The fifth edition of Roberts' *Birds of South Africa* carries this dedication to Barlow: 'C. S. "Punch" Barlow, best known as one of South Africa's foremost businessmen, was an ardent conservationist and one of the founder members of the John Voelcker Bird Book Fund that initially raised the money for the publication of *The Birds of Southern Africa*. He was a friend of the original author, the late Dr Austin Roberts, with whom he went on bird discovery expeditions ... Among his major interests were his collection of ornithological books, concern for the expansion of knowledge and conservation of habitats. He is remembered as a great entrepreneur, philanthropist, sportsman and lover of nature. Barlow was also Chairman of the Barlow Rand Group (now defunct), which was one of South Africa's largest Mining Houses.'

Barnard, E.

> Parrot genus *Barnardius* **Bonaparte**, 1854

> Barnard's Parakeet *Barnardius barnardi* **Vigors** & **Horsfield**, 1827
> [Alt. Mallee Ringneck; Syn. *Barnardius zonarius barnardi*]

Edward Barnard FRS, FLS, FRHS (1786–1861) was a zoologist, botanist and horticulturalist. He was an official of the British Colonial Office (1804), being appointed British Agent to the Colony of New South Wales (1822). He became Agent-General for the Crown Colonies (1833) with residence in London. He was a Colonisation Commissioner (1835) and headed the Commissioners Finance Committee, which secured the financial loans for South Australia from the British Treasury. He gave the type specimen of the parakeet to the Linnean Society of London, of which he was a fellow.

Barnard, H. G.

> Barnard's Wagtail *Motacilla barnardi* **North**, 1906 NCR
> [Alt. Eastern Yellow Wagtail; JS *Motacilla tschutschensis simillima*]

> Barnard's Brush-turkey *Alectura lathami purpureicollis*
> **Le Souef**, 1898
> [Alt. Australian Brush-turkey ssp.]
> Forest Kingfisher ssp. *Todiramphus macleayii barnardi*
> **A. J. Campbell**, 1911 NCR
> [JS *Todiramphus macleayii incinctus*]

Little Bronze Cuckoo ssp. *Chrysococcyx minutillus barnardi* **Mathews**, 1912

Henry 'Harry' Greensill Barnard (1869–1966) was an Australian zoologist, naturalist and grazier from Queensland. His father was an avid collector and his egg collection, now part of the BMNH collection at Tring, was regarded as the best in the southern hemisphere. He also amassed a fine insect collection, especially moths, butterflies and beetles. His mother was a talented artist who painted many of his specimens. His father trained his sons as collectors including getting aboriginal Australians to tutor them in local lore. Harry was, aged 19, allowed to accompany the government expedition to a mountain range – the expedition leader named a geographical feature Barnard's Spur in Harry's honour. He collected in New Guinea (1894) and Cape York (1896). A reptile and a mammal are named after him.

Barnes, C. S.

Spinetail genus *Barnesia* **Bertoni**, 1901 NCR
[Now in *Synallaxis*]

Charles Stanley Barnes (fl.1900) was an Englishman in Paraguay, where he explored and mapped with Bertoni (1893).

Barnes, H. E.

Barnes's Wheatear *Oenanthe finschii barnesi* **Oates**, 1890
[Alt. Finsch's/White-backed Wheatear ssp.]

Henry Edwin Barnes (1845–1895) was a professional British soldier. He was originally apprenticed to his father, a cabinet-maker, but ran away to join the army. He rose through the ranks and was eventually commissioned. He was posted to Aden (1866). He wrote on Indian and Arabian birds, including such articles as 'On the birds of Aden' and 'List of birds noted at Aden and its vicinity' (1893) and the *Handbook to the Birds of the Bombay Presidency* (1885).

Barnes, P.

Layard's Warbler ssp. *Sylvia layardi barnesi* **Vincent**, 1948
[Syn. *Parisoma layardi barnesi*]

Philip de Villiers Barnes (1883–1951) was Ranger at Giant's Castle Game Reserve, Natal (1911). He was appointed as Assistant Conservator (1913), finally retiring as Conservator (1947). His son, Bill, became Ranger in his turn (1956). Bill was the person who confirmed that Bearded Vultures *Gypaetus barbatus* drop bones from a great height in order to crack them open for their marrow.

Barnés, V.

Yellow-shouldered Blackbird *Agelaius barnesi* 1945 NCR
[Error for *Agelaius xanthomus monensis* Barnés, 1945]
Crested Bobwhite ssp. *Colinus cristatus barnesi* **Gilliard**, 1940

Dr Ventura Barnés Jr (b.1910) was a Puerto Rican agronomist. He was the first Curator of the Phelps Ornithological Collection, Caracas, Venezuela. He travelled extensively in Venezuela and other parts of northern South America. The natural science museum in Bayamón, San Juan, is named after him.

Barnes, W.

Barnes's Astrapia *Astrarchia barnesi* **Iredale**, 1948
[Alt. Barnes's Long-tail; hybrid = *Astrapia stephaniae* x *A. mayeri*]

Wilfred Barnes (fl.1930–1970) was an Australian naturalist working at the Australian Museum who collected on several expeditions.

Barolo

Barolo Shearwater *Puffinus baroli* **Bonaparte**, 1857
[Alt. Macaronesian Shearwater]

Marchese Carlo Tancredi Falletti di Barolo (1782–1838) was an Italian philanthropist who was Mayor of Turin. He promoted children's welfare, training and education. He and his French wife, Marquise Giulia Vittorina Colbert di Maulévrier, were joint founders (1834) of a religious teaching order, 'The Congregation of the Sisters of Saint Anne'. They were the last of their line, but his name continues in the wine Barolo, which he had originally produced in the early 1800s, and his estate is still the location of the vineyard and winery.

Baron

Violet-throated Metaltail *Metallura baroni* **Salvin**, 1893
Baron's Spinetail *Cranioleuca baroni* Salvin, 1895

White-tipped Sicklebill ssp. *Eutoxeres aquila baroni*
    E. Hartert & **C. Hartert**, 1894 NCR
[JS *Eutoxeres aquila heterurus*]
Yellow-breasted Brush Finch ssp. *Atlapetes latinuchus baroni* Salvin, 1895
Baron's Hermit *Phaethornis longirostris baroni* Hartert, 1897
[Alt. Long-billed Hermit ssp.]

Oscar Theodor Baron (1847–1926) was a German engineer in Peru and Ecuador, and an amateur ornithologist and collector. He was interested in wildlife, particularly insects, from the age of 13 while a student at the Gymnasium in Neustadt. He became a sailor, visiting India, China, Australasia and Indonesia. He was shipwrecked on Java, where he contracted scurvy. He shipped to the west coast of America and became a surveyor. He was then (1876) employed loading schooners. He travelled in California collecting insects and was then employed as a location engineer on the railroads there and in Mexico. He visited Ecuador and Peru (1890s) and began collecting hummingbirds for his own cabinet. He moved back to Germany (1893) and kept bees. He wrote a paper entitled 'Notes on the localities visited by O. T. Baron in Northern Peru and on the Trochilidae found there' (1897).

Barraband

Australian parrot genus *Barrabandius* **Bonaparte**, 1850 NCR
[Now in *Polytelis*]

Barraband's Parrot *Pyrilia barrabandi* **Kuhl**, 1820
[Alt. Orange-cheeked Parrot; Syn. *Pionopsitta barrabandi*]
Barraband's Parakeet *Polytelis swainsonii* **Desmarest**, 1826
[Alt. Superb Parrot, Scarlet-breasted Parakeet]

Jacques Barraband (1767–1809) has been called the 'Audubon of France'. His watercolours of flowers and birds, engraved by Langlois, resulted from (1801–1804) a direct commission from Napoleon Bonaparte. François Levaillant, an adventurer and natural historian, who was one of the first to record sightings of Australian parrots and wrote *Histoire Naturelle des Perroquets*, employed Barraband whose skill was greatly esteemed and his name lent prestige to Levaillant's sumptuous work. Barraband's brilliantly coloured engravings of exotic birds coupled with Levaillant's precise descriptions represent the height of realistic bird art of the time. He also illustrated Levaillant's books on toucans and bee-eaters, and on *Birds of Paradise* (1806). He became Professor at the École des Beaux Arts in Lyon (1807). William Swainson wrote of him that he was 'the first artist who ventured to represent the varied attitudes of birds'.

Barral

Sunangel sp. *Heliangelus barrali* **Mulsant** & **J. Verreaux**, 1872 NCR
[Status uncertain: hybrid of *Heliangelus amethysticollis* x *Eriocnemis cupreoventris*?]

Luísa Margarida Borges de Barros Portugal, Comtesse de Barral (1816–1891), was the wife of Eugène, Comte de Barral, and governess to the princesses of Brazil, the daughters of Emperor Dom Pedro II, whose mistress she may have been. The binomial, although in the masculine form, is intended to honour her.

Barratt

Barratt's Warbler *Bradypterus barratti* **Sharpe**, 1876
[Alt. African Scrub Warbler]

Fred A. Barratt (c.1847–1875) was a collector in the Transvaal. Sharpe described the warbler from a specimen he collected there and sent to BMNH (1874).

Barringer

Argus Bare-eye *Phlegopsis barringeri* **Meyer de Schauensee**, 1951 NCR
[Probably a hybrid: *Phlegopsis nigromaculata* x *P. erythroptera*]

Brandon Barringer (1899–1991) was a US banker and a trustee of The Academy of Natural Sciences, Philadelphia.

Barros

Yellow-bridled Finch ssp. *Melanodera xanthogramma barrosi* **Chapman**, 1923

Rafael Barros Valenzuela (b.1890) of Rio Blanco, Chile, discovered the finch. His researches added much to our knowledge of Chilean bird life. Among his many publications were 'Algunas observaciones sobre nidificación y postura de aves' (1925) and *Los Loros Chilenos en la Obra de Molina* (1935).

Barrot

Barrot's Fairy *Heliothryx barroti* **Bourcier**, 1843
[Alt. Purple-crowned Fairy]

Theodore Adolphe Barrot (1801–1870) was a French diplomat who served in Colombia (1831–1835) and the Philippines (1835–1838). He became a senator (1864) under Napoleon III in the Second Empire. He wrote *Unless Haste is Made: A French Sceptic's Account of the Sandwich Islands* (1836). He had the distinction that, although not of noble blood, he became a Knight of the Royal Illustrious Order of St Januarius, which was founded in Naples in the 16th century by the Habsburg King of the Kingdom of the Two Sicilies.

Barrow

Barrow's Goldeneye *Bucephala islandica* **Gmelin**, 1789

Barrow's Korhaan/Bustard *Eupodotis senegalensis barrowii* **J. E. Gray**, 1829
[Alt. White-bellied Bustard ssp.]

Sir John Barrow (1764–1848) was Secretary to the Admiralty and a founder of the Royal Geographical Society. He was a great advocate of Arctic exploration. Several geographical features in the North West Passage to the Pacific Ocean bear his name, such as Barrow Point in Alaska and the Barrow Strait of northern Canada. Although Barrow visited Greenland as a boy, he never actually visited the places named after him. He visited China (1792) as part of the embassy led by Lord McCartney, and wrote an account of the journey, *Travels in China* (1804). McCartney later sent him to South Africa to mediate between the Boers and the Kaffirs (as the South African Dutch settlers and the local black population were referred to 200 years ago). He laid the foundation stone for the monument erected in his honour, a tower 100-feet high at The Hoad at Ulverston, in Cumbria (completed 1850). He was created a baronet.

Barrowclough

Coraya Wren ssp. *Pheugopedius coraya barrowcloughianus* **Aveledo** & Perez, 1994

Dr George F. Barrowclough (b.1948) is Associate Curator, AMNH and Adjunct Professor at City University of New York. The University of Minnesota awarded his PhD (1980). His research concentrates on systematics, population genetics, and conservation of birds. He has published widely. He has travelled to Venezuela a number of times. (See also **George Barrowclough**)

Bartels

Javan Hawk Eagle *Nisaetus bartelsi* **Stresemann**, 1924

Bartels's Nightjar *Caprimulgus pulchellus bartelsi* **Finsch**, 1902
[Alt. Salvadori's Nightjar ssp.]
Bartels's Wood Owl *Strix* (*leptogrammica*) *bartelsi* Finsch, 1906
Edible-nest Swiftlet ssp. *Aerodramus fuciphagus bartelsi* Stresemann, 1927 NCR
[JS *Aerodramus fuciphagus fuciphagus*]
Chestnut-bellied Partridge ssp. *Arborophila javanica bartelsi* **Siebers**, 1929 NCR
[NUI *Arborophila javanica javanica*]

Small Buttonquail ssp. *Turnix sylvaticus bartelsorum*
Neumann, 1929
Asian Palm Swift ssp. *Cypsiurus balasiensis bartelsorum*
R. K. Brooke, 1972

Max Eduard Gottlieb Bartels (1871–1936) was a Dutch plantation owner and naturalist who lived in Java (1896–1936). The
buttonquail and palm swift are named after Bartels and his
three sons, Dr Max Bartels Jr (1902–1943), Dr Ernst Bartels
(b. 1904), and Hans Bartels (1906–1997). Three mammals are
named after him.

Bartlett, A. D.

Cuckoo sp. *Cuculus bartletti* **Layard**, 1854 NCR
[Alt. Lesser Cuckoo; JS *Cuculus poliocephalus*]

Bartlett's Bleeding-heart *Gallicolumba crinigera bartletti*
P. L. Sclater, 1863
[Alt. Mindanao Bleeding-heart ssp.]

Abraham Dee Bartlett (1812–1897) was a taxidermist and
zoologist who was the Superintendent of the Gardens of the
Zoological Society of London (London Zoo) (1859–1897).
Charles Darwin (q.v.) mentioned Bartlett in *The Origin of
Species*, and again in *The Descent of Man and Selection in
Relation to Sex* (1896) where he wrote 'I asked Mr. Bartlett, of
the Zoological Gardens, who has had very large experience
with birds, whether the male tragopan … was polygamous,
and I was struck by his answering, I do not know, but should
think so from his splendid colours.' Bartlett published many
articles on his work and experiences at the London Zoo,
including, for example, 'Notes on the breeding of several
species of birds in the Society's Gardens during the year
1867' (1868). His son was an ornithologist (see below).

Bartlett, E.

Bartlett's Tinamou *Crypturellus bartletti* **P. L. Sclater** &
Salvin, 1873

Bartlett's Emerald *Amazilia lactea bartletti* **Gould**, 1866
[Alt. Sapphire-spangled Emerald ssp.]
Red-headed Malimbe ssp. *Malimbus rubricollis bartletti*
Sharpe, 1890
Blue-diademed Motmot ssp. *Momotus momota bartletti*
Sharpe, 1892 NCR
[JS *Momotus momota ignobilis*]
White-chinned Woodcreeper ssp. *Dendrocincla merula
bartletti* **Chubb**, 1919

Edward Bartlett (1836–1908) was a pioneering ornithologist
and herpetologist in Borneo. He was the son of Abraham Dee
Barlett (q.v.). He was with H. B. Tristram (q.v.) in Palestine
(1863–1864) and also collected in Amazonian Peru (1865–
1869). He served as Curator at Maidstone Museum
(1875–1890). He left for Borneo the following year and
collected there, becoming Curator of Sarawak Museum
(1893–1897). He wrote a paper in the *Journal of the Asiatic
Society* entitled: 'The crocodiles and lizards of Borneo in the
Sarawak Museum, with descriptions of supposed new
species, and the variation of colours in the several species
during life' (1895). At the time of his death he had not quite
finished his most notable work; *Monograph of the Weaver
Birds (Ploceidae) and Arboreal and Terrestrial Finches*, five
parts of which were published (1888–1889). A reptile is
named after him.

Bartlett, H. H.

Nepal House Martin ssp. *Delichon nipalense bartletti* **Koelz**,
1952 NCR
[NUI *Delichon nipalense nipalense*]

Dr Harley Harris Bartlett (1886–1960) was an American botanist, ethnographer and collector. He graduated from Harvard
(1908) and joined the US Plant Industry Bureau in Washington
(1909). He became Assistant Professor of Botany, University
of Michigan (1915), Professor (1921), Head of the Botany
Department (1922–1947) and Director of the Botanical
Garden (1919–1955). He collected in Sumatra (1918 and 1926)
and the Philippines (1935 and 1940) plus collecting expeditions to Tibet, Malaysia, Formosa (Taiwan), Guatemala,
British Honduras (Belize), Panama, Haiti, Argentina, Uruguay
and Chile.

Bartlett, R. A.

Black-capped Chickadee ssp. *Poecile atricapillus bartletti*
Aldrich & Nutt, 1939

Captain Robert 'Bob' Abram Bartlett (1875–1946) was a
Newfoundlander (Canadian) Arctic explorer, sealer, collector
and naturalist. He gained his master's ticket (1904) in Halifax.
He commanded the *Roosevelt* on Robert E. Peary's North
Pole Expedition (1909) and the *Karluk* in Vilhjamar Stefans-
son's Canadian Arctic Expedition (1913). He bought the
schooner *Ethie M. Morrissey* (1925) and (1926–1946) made
about twenty voyages to the Arctic collecting specimens,
helping in archaeological surveys, correcting geographical
charts, and capturing animals for zoos. He also fished and
hunted commercially. He died in New York from pneumonia.

Barton

Sclater's Whistler ssp. *Pachycephala soror bartoni*
Ogilvie-Grant, 1915
Yellow-legged Flyrobin ssp. *Microeca griseoceps bartoni*
Ogilvie-Grant, 1915 NCR
[JS *Microeca griseoceps griseoceps*]

Captain Francis Rickman Barton (1865–1947) was a colonial
administrator in New Guinea whose main interests were in
botany, anthropology and early pioneering photography. He
went to Papua (1899) as Commandant of the Armed Native
Constabulary. He was Resident Magistrate of the Central
Division (1902–1904) and was Acting Administrator for the
colony (1904–1907). He went on leave (1907) and resigned
from his post (1908), becoming Vizier or First Minister in
Zanzibar (1908–1913). He accompanied Captain Charles
Monckton on some of his expeditions. He is noted for taking a
remarkable series of photographs of the native tribes (1897–
1907) and may have shot some early cinema film (1904). He
corresponded widely with anthropologists and photographic
enthusiasts. He became President of the Royal Geographical
Society (1920). A mammal is named after him.

Bartram

Sandpiper genus *Bartramia* **Lesson**, 1831

Bartram's Sandpiper *Bartramia longicauda* **Bechstein**, 1812
[Alt. Upland Sandpiper, Bartramian Tattler]
Bartram's Vireo *Vireo bartramii* **Swainson**, 1832 NCR
[Alt. Red-eyed Vireo; JS *Vireo olivaceus*]

William Bartram (1739–1823) was an American naturalist, explorer and botanist, often called the 'Grandfather of American Ornithology', perhaps because he was the protector of Alexander Wilson (q.v.), who was known as the 'Father of American Ornithology'. He accompanied his father John Bartram (America's 'first' botanist) on a number of expeditions, to the Catskill Mountains (1753–1754), New York and Connecticut (1755) and Florida (1765). William remained in the south drawing flora, collecting botanical specimens, becoming an accomplished ornithologist and befriending both colonial planters and Native American tribes. During the American War of Independence he joined the Georgia Militia (1776). He returned to his home in Pennsylvania (1777) and wrote about his travels. He was offered (1782) the Chair of Botany at the University of Pennsylvania, but never lectured there. As well as befriending Alexander Wilson he took his nephew, Thomas Say (q.v.), under his wing. Bartram assembled a *Catalogue of Birds of North America*. An amphibian is named after him.

Bartsch

Marianas Swiftlet *Aerodramus bartschi* **Mearns**, 1909

Paul Bartsch (1871–1960) was born in Poland but educated in the USA. He joined the Smithsonian staff (1896) and stayed until retiring (1942). He was part of an expedition to the Philippines (1907–1910) aboard the ship *Albatross* among many other expeditions. He taught at a number of universities and was a world authority on molluscs. He also organised Washington's first Boy Scout group, and was a keen member of the DC branch of the National Audubon Society. His account of the Philippines expedition was published in *Copeia* (1941). Two reptiles are named after him.

Barttelot

Blue-shouldered Robin Chat ssp. *Cossypha cyanocampter barttelotti* **G. E. Shelley**, 1890

Major Edmund Musgrave Barttelot (1859–1888) of the Royal Fusiliers joined the army (1879) and served in India. He commanded the rear column of Stanley's expedition for the relief of Emin Pasha (q.v.). Reports varied, and some circulated which claimed he had gone mad. He was killed, possibly by an African who objected to Barttelot interfering in a native festival. The family was dogged by bad luck: Barttelot's elder brother was killed in action in the Boer War (1900), his nephews were both killed (WW1) (1915 and 1918), and his great-nephew killed in action too (WW2) (1944).

Baruffi

Oriole sp. *Oriolus baruffi* **Bonaparte**, 1850 NCR
[Alt. Western Oriole; JS *Oriolus brachyrynchus*]

Abbé Giuseppe Baruffi (1809–1875) was a priest, traveller and populariser of scientific studies.

Bassett

White-eye sp. *Zosterops bassetti* **Sharpe**, 1894 NCR
[Alt. Ashy-bellied White-eye; JS *Zosterops citrinella albiventris*]

Surgeon Rear-Admiral Dr Sir Percy William Bassett-Smith (1861–1927) was a naturalist and naval surgeon. He qualified as a physician (1882) and served in the Naval Medical Service (1883–1921). He was on board HMS *Rambler* (1884–1885) and on HMS *Penguin* (1891–1893) and collected zoological specimens during voyages in the Dutch East Indies (Indonesia). He was appointed lecturer on tropical medicine at the Royal Navy Hospital, Haslar (1900), and was Professor of Clinical Pathology at the Royal Naval Hospital, Greenwich (1912–1921). After he retiring he practised as a consultant in London.

Basulto

Timberline Wren ssp. *Thryorchilus browni basultoi* **Ridgway**, 1908
[*T. browni* often regarded as monotypic]

Francisco Basulto (fl.1908) was a Cuban who was in Costa Rica and helped Ridgway (q.v.) during his collecting expedition. Basulto collected the holotype.

Batchelder

Batchelder's Woodpecker *Picoides pubescens leucurus* **Hartlaub**, 1852
[Alt. Downy Woodpecker ssp.; *P. pubescens oreoecus* Batchelder, 1889 is a junior synonym]

Charles Foster Batchelder (1856–1954) was an American artist and amateur ornithologist, and a leading light in the Nuttall Ornithological Club. Among other articles he wrote an account of the club and information on the nesting habits of a number of birds.

Batcheldor/Batchelor

Brown Skua ssp. *Catharacta antarctica batchelori* **Mathews**, 1929 NCR

N. Batcheldor (fl.1866) was a collector for the Melbourne Museum. Mathews made a mistake in the spelling his name for the trinomial.

Bateman

Kalij Pheasant sp. *Gennaeus batemani* **Oates**, 1906 NCR
[JS *Lophura leucomelanos lathami*]

Hill Partridge ssp. *Arborophila torqueola batemani* **Ogilvie-Grant**, 1906

A. C. Bateman (1870–1925) was a British boundary commissioner and officer in the Burma Police. He collected in the Chin Hills (1905).

Bates, G. L.

Bee-eater sp. *Merops batesiana* **Sharpe**, 1900 NCR
[Alt. Blue-headed Bee-eater; JS *Merops muelleri*]
Bates's Swift *Apus batesi* Sharpe, 1904
Bates's Nightjar *Caprimulgus batesi* Sharpe, 1906
Bates's Sunbird *Cinnyris batesi* **Ogilvie-Grant**, 1908
Bates's Weaver *Ploceus batesi* Sharpe, 1908
Crombec sp. *Sylvietta batesi* Sharpe, 1908 NCR
[Alt. Lemon-bellied Crombec; JS *Sylvietta denti*]
Bates's Paradise-flycatcher *Terpsiphone batesi* **Chapin**, 1921

Grey-throated Rail ssp. *Canirallus oculeus batesi* Sharpe, 1900 NCR; NRM
Blackcap Illadopsis ssp. *Illadopsis cleaveri batesi* Sharpe, 1901
Congo Serpent Eagle ssp. *Dryotriorchis spectabilis batesi* Sharpe, 1904
African Yellow Warbler ssp. *Iduna natalensis batesi* Sharpe, 1905
Grey Ground Thrush ssp. *Zoothera princei batesi* Sharpe, 1905
Many-coloured Bush-shrike ssp. *Chlorophoneus multicolor batesi* Sharpe, 1908
African Cuckoo Hawk ssp. *Aviceda cuculoides batesi* Swann, 1920
Brown-backed Woodpecker ssp. *Dendropicos obsoletus batesi* **W. L. Sclater**, 1921 NCR
[JS *Dendropicos obsoletus obsoletus*]
Bates's White-spotted Flufftail *Sarothrura pulchra batesi* **Bannerman**, 1922
Mountain Robin Chat ssp. *Cossypha isabellae batesi* Bannerman, 1922
White-bellied Kingfisher ssp. *Corythornis leucogaster batesi* Chapin, 1922 NCR
[NUI *Corythornis leucogaster leucogaster*]
Fine-spotted Woodpecker ssp. *Campethera punctuligera batesi* Bannerman, 1923 NCR
[NUI *Campethera punctuligera punctuligera*]
Rufous-naped Lark ssp. *Mirafra africana batesi* Bannerman, 1923

George Latimer Bates (1863–1940) was born in Illinois, and travelled in West Africa (1895–1931). He wrote a *Handbook of the Birds of West Africa* (1930) and a number of articles, notably *Birds of the Southern Sahara and Adjoining Countries* (1933). He also left an unpublished manuscript of *Birds of Arabia*, subsequently utilised by Meinertzhagen (q.v.) for his 1950s work on the subject. Several plants, four mammals, three amphibians and a reptile are named after him.

Bates, H. W.

Bates's Honeyeater *Melidectes ochromelas batesi* **Sharpe**, 1886

Henry Walter Bates (1825–1892) was an English explorer and naturalist. He was the first to describe animal mimicry for science. Forbes (q.v.), who collected the honeyeater and sent it to Sharpe (q.v.), requested it be named after Bates. Bates would have sympathised with the explorer's privations as he had explored the Amazon with Wallace (q.v.), where

they collected 14,000 specimens – 8,000 of which were new to science (1848–1859). He was largely self-taught, having left school at 12 and been apprenticed to a hosier, yet 10 years later published his first scientific paper. His most famous publication was *The Naturalist on the River Amazons*, subtitled *A Record of the Adventures, Habits of Animals, Sketches of Brazilian and Indian Life, and Aspects of Nature under the Equator, during Eleven Years of Travel* (1863). After returning from the expedition he worked as Assistant Secretary of the Royal Geographical Society (1864).

Bathilda

Starfinch genus *Bathilda* **Reichenbach**, 1863 NCR
[Now in *Neochmia*]

Bathilda Aloise Leonie, Comtesse de Cambaceres (1840–1861), was one of the daughters of C. L. Bonaparte (q.v.).

Bathoen

Spike-heeled Lark ssp. *Chersomanes albofasciata bathoeni* **Paterson**, 1958

Kgosi Bathoen II Seepapitso Gaseitsewe (1908–1990) was Paramount Chief and King of the Bangwaketse, Botswana. He became king (1916) when his father was assassinated, assumed power after a period of regency (1928), eventually abdicating (1969).

Batty

Brown-backed Dove *Leptotila battyi* **Rothschild**, 1901

Joseph H. Batty (1846–1906) was a collector in Latin America (1896–1906). In the *Bulletin of the American Museum of Natural History* (1903) there was an article by J. A. Allen (q.v.) entitled 'List of mammals collected by Mr. J. H. Batty in New Mexico and Durango, with descriptions of new species and subspecies'. He wrote another article a few years later on Batty's specimens from other parts of Mexico. In the 1950s there was a paper in *Auk* on the birds he collected in Panama. Recently, it has been suggested by Storrs Olson (q.v.) in *American Museum Novitates* (2008) that Batty sometimes falsified data on his labels.

Baucis

Hummingbird genus *Baucis* **Reichenbach**, 1854 NCR
[Now in *Abeillia*]

In Greek mythology, Baucis and her husband Philemon were an elderly, impoverished couple in Phrygia who used their best efforts to entertain the gods Zeus and Hermes, who were travelling incognito. As a reward, the gods transformed their humble dwelling into a temple for which they were made the priest and priestess. When they died, they were metamorphosed into an intertwined pair of trees.

Baud

Blue-headed Pitta *Hydrornis baudii* **S. Müller** & **Schlegel**, 1839
[Syn. *Pitta baudii*]

Jean Chrétien Baron Baud (1789–1859) started his working life as an official of the Dutch East India Company. He was Governor-General of the Dutch East Indies (1833–1836) and became the Dutch Minister for the Colonies (1842).

Baudin

Baudin's Black Cockatoo *Calyptorhynchus baudinii* **Lear**, 1832
[Alt. Long-billed Black Cockatoo]
Kangaroo Island Emu *Dromaius baudinianus* S. A. Parker, 1984 EXTINCT

Nicolas Thomas Baudin (1754–1803) was a French seafarer and explorer. He set out (1800) from Le Havre with two corvettes, the *Géographe* and the *Naturaliste*, carrying five zoologists, 17 other scientists, three artists and two astronomers among others. His task was to map the coast of Australia. The *Géographe* crawled back to Le Havre (1804) with a living cargo of 72 birds and other animals but no captain, Baudin having died of tuberculosis in Mauritius. However, he did leave an artistic legacy: 46 watercolours and drawings amassed on the expedition to Australia. After the expedition's official artists decamped en route, it was his decision to replace them with assistant gunners Petit (q.v.) and Lesueur (q.v.). 'It will be seen from the work of these two young men whether my choice was good or bad,' he wrote in his log. The results were stunning. A reptile and an amphibian are named after him.

Bauer

Bauer's Parakeet *Barnardius zonarius* Shaw, 1805
[Alt. Port Lincoln Parrot, Australian Ringneck]

Bar-tailed Godwit ssp. *Limosa lapponica baueri* **Naumann**, 1836

Ferdinand Lucas Bauer (1760–1826) was an Austrian illustrator who sailed with Flinders (q.v.) on the *Investigator* to Australia (1801). He had previously worked as a botanical illustrator at the University of Vienna. He spent 18 months on a botanical trip to Greece, Italy, Crete and Cyprus (1784). He then moved to England where he converted his c.1,500 sketches into paintings, although his work of 966 watercolours was only published posthumously (1840). Fortunately, Joseph Banks recommended him for the *Investigator* voyage. He produced the first-ever picture of a live (captured) koala. He returned to Austria (1814) and worked until his death as an illustrator for authors of botanical works in London and Vienna. Tragically many of his personal field drawings and plant collections were lost (WW2).

Baumann

Baumann's Olive Greenbul *Phyllastrephus baumanni* **Reichenow**, 1895

Ernst Baumann (1863–1895) was a German ornithologist who started out training for a commercial career, but then switched to natural history and studied privately at the Zoological Museum, Berlin. He was sent to the German research station at Miss-Höhe, Togo (1892) and explored extensively there (1892–1895). He was on leave in Germany when he died from recurrent malaria. An amphibian is named after him.

Baumgart

Burnt-necked Eremomela ssp. *Eremomela usticollis baumgarti* **Reichenow**, 1905 NCR
[JS *Eremomela usticollis usticollis*]

Dr Baumgart (fl.1904) was a German veterinary surgeon and collector in South West Africa (Namibia) attached to the German army during the Herero war. He sent the holotype (1904) to Reichenow (q.v.).

Baur

Galapagos Mockingbird ssp. *Mimus parvulus bauri* **Ridgway**, 1894

George Herman Carl Ludwig Baur (1859–1898) was an osteologist and testudinologist. He studied in Munich (1882–1884), left for America, and was at Yale (1884–1890). He travelled to the Galapagos Islands (1891) where he made an extensive ornithological collection. He was Assistant Professor of Paleontology, University of Chicago (1893–1898). He died after having been committed to a lunatic asylum. Three reptiles are named after him.

Bayer

White-eye sp. *Zosterops bayeri* **Lönnberg**, 1917 NCR
[Alt. African Yellow White-eye; JS *Zosterops senegalensis jacksoni*]

Dr Leon Bayer (fl.1914) was a collector and big game hunter in East Africa.

Bayley, E. C.

Andaman Treepie *Dendrocitta bayleii* **Tytler**, 1863
[Binomial often given as *bayleyi*]

Sir Edward Clive Bayley (1821–1894) was an archaeologist, lawyer and statesman who was an admistrator in India (1842–1851) before returning to England. He was called to the Bar (1857) and went back to India, becoming a judge there (1859). He acted as temporary Foreign Secretary (1861) and Home Secretary (1862–1872). He was a member of the Supreme Council (1873–1878) before retiring to England. He published a number of works on Indian history and antiquities.

Beal

Beal's Storm-petrel *Oceanodroma leucorhoa beali* **Emerson**, 1906 NCR
[Alt. Leach's Storm-petrel ssp.; JS *Oceanodroma leucorhoa leucorhoa*]

Professor Foster Ellenborough Lascelles Beal (1840–1916) worked for the Biological Survey of the US Department of Agriculture. He worked on birds common to agricultural and horticultural areas and was described as an 'economic ornithologist' by the journal *Condor*. For example, he investigated the diet of the Cedar Waxwing *Bombycilla cedrorum* and found that the species had very little impact on horticulture, no doubt saving it from farmers' wrath. He was a very hard worker and was described as 'having kept his eyes to the microscope more hours per day, and more days, than anyone', hence his study of the stomach contents of 37,000 birds! His favourite saying was 'ignorance is better than error'. He wrote many detailed articles on bird diets.

Bean

Cozumel Wren *Troglodytes aedon beani* **Ridgway**, 1885

Dr Tarleton Hoffman Bean (1846–1916) was an American ichthyologist who worked with the US Fish Commission (1874) after being a high school teacher. He studied medicine at George Washington University (1874–1876). He joined the US National Museum (1877) becoming Curator of Fishes (1879). He discovered many new fish species and became the first Superintendent of the New York Aquarium. He co-wrote many papers and treatises, the most significant being the classic *Oceanic Ichthyology* (1896). A genus and at least 10 species of fish also bear his name.

Beatrice

Grasshopper Sparrow ssp. *Ammodramus savannarum beatriceae* **Olson**, 1980

Mrs Annie Beatrice Wetmore, *née* van der Biest Thielen (1910–1997), was the wife of Alexander Wetmore, who collected the holotype.

Beatty

Bridled Quail Dove ssp. *Geotrygon mystacea beattyi* Danforth, 1938 NCR; NRM
Lesser Masked Weaver ssp. *Ploceus intermedius beattyi* **Traylor**, 1959

Harry Andrew Beatty (1902–1989) was a biologist who was born in St Croix, US Virgin Islands, and brought up on his father's sugarcane plantation. He was educated in Massachusetts and after qualifying he worked on the staff of the Health Department in St Thomas, Virgin Islands, on epidemiology (1933–1940), and later on a project to try and eliminate cattle fever ticks from the deer population (1940–1945). He was based in Liberia (1945) in charge of a malaria control project. He collected in Puerto Rico (1944) for the Philadelphia Academy of Natural Sciences, in various West Africa countries (1948–1952) for the Peabody Museum at Yale, and in Suriname for the Field Museum, Chicago (1960–1962). A reptile is also named after him.

Beaudouin

Beaudouin's Snake Eagle *Circaetus beaudouini* **J. Verreaux & Des Murs**, 1862
[Alt. Beaudouin's Harrier Eagle]

M. Beaudouin (fl.1860) was a professional collector employed by Verreaux (q.v.). He collected in Portuguese Guinea (Guinea-Bissau).

Beaufort

Beaufort's Black-capped Lory *Lorius lory viridicrissalis* De Beaufort, 1909
Large Wren-babbler ssp. *Napothera macrodactyla beauforti* **Voous**, 1950

Professor Lieven Ferdinand de Beaufort (1879–1968) was a Dutch zoologist whose main interests were first fish and then birds. As a student he participated (1902–1903) in the first scientific expedition to New Guinea, headed by the geographer Professor A Wichman. He undertook a second voyage to the Dutch East Indies (1909–1910). He regularly published (1904–1921) about birds collected in New Guinea by him, his friend (later Professor) Cosquino de Bussy, and others. In his early years he was an assistant of Max Weber (q.v.), whom he succeeded as Director of the Zoological Museum of the University of Amsterdam (1922–1949), and was Extraordinary Professor in Zoogeography (1929–1949). He was one of the founders (1901) of the *Nederlandsche Ornithologische Vereeniging* (the Dutch Ornithological Society), becoming Secretary (1911–1924) and Chairman (1924–1956). He was also a member of the editorial board of the society's journal *Ardea* (1924–1956). He edited the 11-volume *Fishes of the Indo-Australian Archipelago* (1911–1962). A mammal and a reptile are named after him.

Beauharnais

Curl-crested Aracari *Pteroglossus beauharnaesii* **Wagler**, 1832

Either after Eugene Beauharnais (1781–1824), Viceroy of Italy and step-son of Napoleon I, or Auguste Beauharnais (1810–1835), who was his son. The elder was the brother-in-law of Charles Lucien Bonaparte.

Beaulieu

Australasian Lark ssp. *Mirafra javanica beaulieui* **Delacour**, 1932
Streak-breasted Scimitar Babbler ssp. *Pomatorhinus ruficollis beaulieui* Delacour & **J. Greenway**, 1940
Black-throated Parrotbill ssp. *Suthora nipalensis beaulieui* **Ripley**, 1953

Louis Henri André David-Beaulieu (1896–1969) was French colonial administrator in Indo-China and a naturalist. He wrote *Les Oiseaux du Tranninh* (1944).

Beauperthuy

Ocellated Woodcreeper ssp. *Xiphorhynchus ocellatus beauperthuysii* **Pucheran & Lafresnaye**, 1850

Dr Louis Daniel Beauperthuy Desbonnes (1807–1871) was a French microbiologist and pioneer in tropical medicine who discovered the cause of yellow fever and the cure for leprosy. After he qualified as a physician in Paris (1837), MNHN, Paris, sent him to open a medical centre in Venezuela (1838), where he settled permanently (1840).

Beavan

Beavan's Bullfinch *Pyrrhula erythaca* **Blyth**, 1862
[Alt. Grey-headed Bullfinch]

Beavan's Rufous-vented Tit *Periparus rubidiventris beavani* **Jerdon**, 1863
Beavan's Wren-warbler *Prinia rufescens beavani* **Walden**, 1866
[Alt. Rufescent Prinia ssp.]
Ashy Drongo ssp. *Dicrurus leucophaeus beavani* Walden, 1867 NCR
[JS *Dicrurus leucophaeus longicaudatus*]

Captain Robert Cecil Beavan (1841–1870) wrote *The Avifauna of the Andaman Islands* (1867) and a series of papers in *Ibis* mostly as 'notes on various Indian birds' (1865–1868). He served with Tytler (q.v.) in the Andaman Islands. His health was poor and he was twice invalided home to England, on the second occasion dying at sea.

Beccari

Blue Flycatcher sp. *Cyornis beccariana* **Salvadori**, 1868 NCR
[Alt. Mangrove Blue Flycatcher; JS *Cyornis rufigastra*]
Beccari's Scrubwren *Sericornis beccarii* Salvadori, 1874
[Alt. Tropical Scrubwren, Little Sericornis]
Beccari's Ground Dove *Gallicolumba beccarii* Salvadori, 1876
[Alt. Bronze Ground Dove; Syn. *Alopecoenas beccarii*]
Beccari's Scops Owl *Otus beccarii* Salvadori, 1876
[Alt. Biak Scops Owl]
Sumatran Cochoa *Cochoa beccarii* Salvadori, 1879

Red-backed Buttonquail ssp. *Turnix maculosus beccarii* Salvadori, 1875
Beccari's Pygmy Parrot *Micropsitta pusio beccarii* Salvadori, 1876
[Alt. Buff-faced Pygmy Parrot ssp.]
Victoria Crowned Pigeon ssp. *Goura victoria beccarii* Salvadori, 1876
Fire-breasted Flowerpecker ssp. *Dicaeum ignipectus beccarii* **H. C. Robinson** & **Kloss**, 1916

Dr Odoardo Beccari (1843–1920) was an Italian botanist. He explored the Arfak Mountains during extensive zoological exploration with D'Albertis (1872–1873), recorded in *Wanderings in the Great Forests of Borneo*. He also explored and collected in the Celebes (Sulawesi) and Sumatra, where he found the Titan Arum or Corpse Flower *Amorphophallus titanum*, the world's largest flower. Seeds of it were sent to the Royal Botanic Gardens at Kew and were successfully grown, flowering for the first time in cultivation (1889). He also collected in Ethiopia. Four mammals, six reptiles and an amphibian are named after him.

Bechstein

Bechstein's Violet-necked Lory *Eos squamata riciniata* Bechstein, 1811
Marsh Sandpiper ssp. *Tringa stagnatilis bechsteini* **Zarudnyi** & Smirnov, 1923 NCR; NRM

Johann Mathaeus Bechstein (1757–1822) was a German scientist. He was Director of the Ducal Academy of Forestry (1810–1818). He adopted Ludwig Bechstein (1810) who became famous for writing fairy tales. He himself wrote *Gemeinnutzige Naturgeschichte Deutschlands* (1789) and (as published posthumously in English) *The Natural History of Cage Birds: Their Management, Habits, Food, Diseases, Treatment, Breeding, and the Methods of Catching Them* (1837). A mammal is named after him.

Beck

Beck's Petrel *Pseudobulweria becki* **R. C. Murphy**, 1928
Beck's Bowerbird *Sericulus bakeri* **Chapin**, 1929
[Alt. Adelbert's/Baker's/Fire-maned Bowerbird]

Nelson's Sparrow ssp. *Ammodramus nelsoni becki* **Ridgway**, 1891 NCR
[NUI *Ammodramus nelsoni nelsoni*]
Warbler Finch ssp. *Certhidea olivacea becki* **Rothschild**, 1898
Burrowing Owl ssp. *Athene cunicularia becki* Rothschild & **Hartert**, 1902 NCR
[JS *Athene cunicularia hypugaea*]
Island Leaf Warbler ssp. *Phylloscopus poliocephalus becki* Hartert, 1929
Midget Flowerpecker ssp. *Dicaeum aeneum becki* Hartert, 1929
Glossy Swiftlet ssp. *Collocalia esculenta becki* **Mayr**, 1931
Pacific Robin ssp. *Petroica multicolor becki* Mayr, 1934
Hill-forest Honeyeater ssp. *Meliphaga orientalis becki* **Rand**, 1936
Island Thrush ssp. *Turdus poliocephalus becki* Mayr, 1941

Rollo Howard Beck (1870–1950) was an American collector, particularly of seabirds. He became interested in natural history as a schoolboy, but left aged 14, working in orchards while continuing his interest encouraged by professional ornithologists, and joined the AOU (1894). He collected for the Museum of the California Academy of Sciences and then for the AMNH, and was well known to Murphy (q.v.) who 'held him in high regard' and named the petrel after him. He spent time in the Galápagos (1897–1898), Alaska (1911) and New Guinea (1928). His wife Ida usually accompanied him. One trip lasted five years, while they explored the South American coast, and another in the South Seas lasted nearly ten, when they were part of the Sanford-Whitney Expedition. Three reptiles are named after him.

Becker, J.

Bahia Tyrannulet *Phylloscartes beckeri* Gonzaga & Pacheco, 1995

Professor Johann Becker (1932–2004) was a Brazilian zoologist, working first as a researcher then Curator at the National Museum of Rio de Janeiro. He gave valuable taxonomic advice to the describers of the tyrannulet. As a student, (1951) he was admitted as apprentice in the Division of Insects at the National Museum. He graduated with an initial degree of BSc. (1954) and became zoologist researcher and later a professor of that museum, being a specialist in genetics, evolution and invertebrate palaeontology. He had collected insects, especially beetles, ever since he was a student and, after his death, his entomological collection c.14,000 specimens was presented to the Zoology Department of Universidade Estadual de Feira de Santana. An insect and an amphibian are named after him.

Becker, R. H.

Becker's Burrowing Owl *Athene cunicularia beckeri* **Cory**, 1915 NCR
[JS *Athene cunicularia grallaria*]
Southern House Wren ssp. *Troglodytes musculus beckeri* Cory, 1916 NCR
[JS *Troglodytes musculus musculus*]

Robert H. Becker (fl.1920) was a naturalist and archaeologist who joined the Field Museum, Chicago (1912), collecing for them in Brazil and Bolivia (1913–1920), including the two eponymous taxa.

Bedford

Bedford's Paradise-flycatcher *Terpsiphone bedfordi* **Ogilvie-Grant**, 1907
Little Rush Warbler *Bradypterus bedfordi* Ogilvie-Grant, 1912 NCR
[JS *Bradypterus baboecala msiri*]

Tit-hylia ssp. *Pholidornis rushiae bedfordi* Ogilvie-Grant, 1904
Eurasian Nuthatch ssp. *Sitta europaea bedfordi* Ogilvie-Grant, 1909

Herbrand Arthur Russell, 11th Duke of Bedford (1858–1940), was President of the Zoological Society of London (1899–1936). He was also a trustee of BMNH. Two mammals are named after him.

Bedout

Masked Booby ssp. *Sula dactylatra bedouti* **Mathews**, 1913

Not named directly after a person, but after Bedout Island (eastern Indian Ocean, off north-west Australia). However, the island is, in turn, named after Rear-Admiral Jacques Bedout (1751–1818) of the French Royal Navy.

Beebe

Rosy Thrush Tanager ssp. *Rhodinocichla rosea beebei* **W. H. Phelps** & **W. H. Phelps Jr**, 1949

Dr Charles William Beebe (1877–1962) was a zoologist, marine biologist, conservationist, explorer and writer. He began his working life looking after the birds at the Bronx Zoo (New York) but became Curator of Ornithology, New York Zoological Society (1899–1952), and Director, Department of Tropical Research (1919). He was greatly interested in deep-sea exploration and made a number of descents in the Bathysphere including (1934) a record one of 923 metres (3,028 feet) off Nonsuch Island, Bermuda. He set up a camp (1942) at Caripito in Venezuela for jungle studies and (1950) bought 92 hectares (228 acres) of land in Trinidad and

Tobago, which became New York Zoological Society's Tropical Research Station (Asa Wright Nature Centre). He married Helen Elswyth Thane Ricker (1900–1981), who wrote romantic novels (pen name Elswyth Thane). Many of his writings were popular books on his expeditions and he made enough money from them to finance his later expeditions. His book *The Bird, Its Form and Function* (1906) presented technical information about bird biology and evolution in a way that was accessible to the general public and made enough money to finance his later expeditions. He made various collecting trips to bring live birds back to the zoo. Perhaps his most outstanding ornithological work is the 4-volume *A Monograph of the Pheasants* (1918–1922). He retired to Trinidad. A number of other taxa including several fish and two amphibians are named after him.

Beechey

Beechey's Jay *Cyanocorax beecheii* **Vigors**, 1829
[Alt. Purplish-backed Jay]

Captain Frederick William Beechey (1796–1856) was a noted geographer and mapmaker. He led an expedition to the Pacific and the Bering Strait in HMS *Blossom* (1825–1828), which made significant discoveries in the Arctic, California and in the Pacific islands. During that voyage Beechey took a formal pardon to Adams, the last survivor of the Mutiny on the *Bounty*, on Pitcairn Island. In *The Zoology of Captain Beechey's Voyage* (1839) the section on ornithology was by Vigors with coloured plates by George B. Sowerby. There is a further book, *The Botany of Captain Beechey's Voyage* (1841). Very many specimens were collected so it is unsurprising that one would be named for the captain. During the voyage (1825) he named Point Barrow after Sir John Barrow (q.v.) of the British Admiralty. He made further voyages such as that on HMS *Sulphur* (1836). A mammal is named after him.

Beehler

Island Thrush ssp. *Turdus poliocephalus beehleri* **Ripley**, 1977

Dr Bruce Beehler (b.1951) is an ornithologist and vice-president of Conservation International's Melanesian Centre for Biodiversity Conservation. Princeton awarded his master's degree and doctorate. He is an expert on New Guinea birds and led the expedition (2005) to the Foja Mountains, Papua, which made many new discoveries including the Wattled Smoky Honeyeater *Melipotes carolae*, named after his wife. He wrote *The Birds of New Guinea* (1985).

Beesley

Beesley's Lark *Chersomanes (albofasciata) beesleyi* **C. W. Benson**, 1966
[Alt. Pygmy Spike-heeled Lark; Syn. *Chersomanes albofasciata beesleyi*]

John S. S. Beesley (b. 1925) is an ornithologist and a former bird pest research officer who worked in Tanzania (1956–1971) and was the first Conservator of Arusha National Park. He moved to Botswana (1972) and co-wrote 'The status of the birds of Gaborone and its surroundings' (1976). He later moved back to Tanzania (until 1984), and retired to Tunbridge Wells, Kent. He discovered the lark near Mt Meru, Tanzania (1965).

Behn

Behn's Thrush *Turdus subalaris* **Seebohm**, 1877
[Alt. Eastern Slaty Thrush; Syn. *Turdus nigriceps subalaris*]
Plain-winged Antwren *Myrmotherula behni* **Berlepsch** & Leverkühn, 1890

Blue-crowned Trogon ssp. *Trogon curucui behni* **Gould**, 1875
Behn's Parakeet *Brotogeris chiriri behni* **Neumann**, 1931
[Alt. Yellow-chevroned/Yellow-winged Parakeet]
Rufous Treepie ssp. *Dendrocitta vagabunda behni* Steinheimer, 2009

William Friedrich Georg Behn (1808–1878) was a German explorer who is famed for his crossing of South America (1847). He was the Director of the Zoological Museum of the Christian Albrechts University of Kiel (1836–1868). A mammal is named after him.

Behnke

Blue Rock Thrush ssp. *Monticola solitarius behnkei* **Niethammer**, 1943 NCR
[JS *Monticola solitarius solitarius*]

H. Behnke (fl.1942) was a German collector in Crete. Judging by the location and date we assume he was either a member of, or connected with, the German armed forces.

Beick

Great Spotted Woodpecker ssp. *Dendrocopus major beicki* **Stresemann**, 1927
Robin Accentor ssp. *Prunella rubeculoides beicki* **Mayr**, 1927 NCR; NRM
White-throated Redstart ssp. *Phoenicurus schisticeps beicki* Stresemann, 1927 NCR; NRM
Beick's Dipper *Cinclus cinclus beicki* **Meise**, 1928 NCR
[Alt. White-throated Dipper ssp.; JS *Cinclus cinclus przewalskii*]
Tengmalm's Owl ssp. *Aegolius funereus beickianus* Stresemann, 1928
Sinai Rosefinch ssp. *Carpodacus synoicus beicki* Stresemann, 1930
Asian Short-toed Lark ssp. *Calandrella cheleensis beicki* Meise, 1933
Beick's Blood Pheasant *Ithaginis cruentus beicki* **Mayr** & Birckhead, 1937
Chinese Hill Warbler ssp. *Rhopophilus pekinensis beicki* Meise, 1937 NCR
[JS *Rhopophilus pekinensis albosuperciliaris*]
Siberian Rubythroat ssp. *Luscinia calliope beicki* Meise, 1937

Walter Beick (1883–1933) was a Russian of Baltic-German origin educated at St Petersburg, Berlin and Munich and wounded (WW1) while serving in the Russian Imperial Army. Fearful of Bolshevik persecution he fled (1920) to China, where he collected and studied natural history in Kansu (Gansu) province. He joined (1927) the Swiss explorer Wilhelm Filchner (1877–1957) on his expedition to Central Asia. (Many of the taxa bearing his name were collected on that expedition and described in an article in *Journal für Ornithologie* by Stresemann (q.v.), 'Aves Beickianae – Beiträge zur Ornithologie von Nordwest-Kansu nach den Forschungen von Walter Beick (†) in den Jahren 1926–1933'). Several attempts were made on Beick's life whilst in China, where the authorities suspected him of being a spy for Turkestan. He eventually sent all his work to the Zoological Museum in Berlin and, severely depressed and convinced that enemies surrounded him, shot himself. He is commemorated in many other taxa including several insects.

Bejarano

Black-throated Thistletail ssp. *Asthenes harterti bejaranoi* **Remsen**, 1981 [Syn. *Schizoeaca harterti bejaranoi*]

Dr Gastón Bejarano (DNF) was a zoologist and the Director of Forestry and National Parks, Ministry of Agriculture, La Paz, Bolivia. The author wrote 'Prof. Bejarano, often at tremendous personal sacrifice, has furthered the study of Bolivian fauna and flora through his encouragement and aid to visiting scientists.' An amphibian is named after him.

Bel

Bel's Silver Pheasant *Lophura nycthemera beli* **Oustalet**, 1898

Jean Marc Bel (1855–1930) was a French explorer, mining engineer and collector in Annam (Vietnam) (1898). He brought the first specimen of the taxon named for him to Europe. He was in the French Congo (1906–1907). He wrote *Mission au Laos et en Annam* (1898).

Bélanger

White-crested Laughingthrush ssp. *Garrulax leucolophus belangeri* **Lesson**, 1831

Charles Paulus Bélanger (1805–1881) was a French traveller. His voyage is commemorated in his published journal, *Voyage aux Indes-Orientales, par le Nord de l'Europe, les Provinces du Caucase, la Géorgie, l'Arménie et la Perse, suivi de Détails Topographiques, Statistiques et Autres sur le Pégou, les Îles de Java, de Maurice et de Bourbon, sur le Cap-de-Bonne-Espérance et Sainte Hélène, pendant les Années 1825, 1826, 1827, 1828 et 1829*, which has illustrations by Geoffroy Saint-Hilaire (q.v.) of the mammals he collected. His trip was part of the French expedition across Europe to India undertaken in order to make a botanical garden at Pondicherry. The expedition collected vast numbers of specimens of dried and living plants and seeds as well as fish, birds, crustaceans, molluscs and a few mammals. Bélanger became the Director of the Botanical Gardens in Martinique (1853). A mammal is named after him.

Belcher, C.

Australasian Robin genus *Belchera* **Mathews**, 1912 NCR
[Now in *Petroica*]

Slender-billed Prion *Pachyptila belcheri* Mathews, 1912

Green Barbet ssp. *Stactolaema olivacea belcheri*
W. L. Sclater, 1927
Spotted Ground Thrush ssp. *Zoothera guttata belcheri*
C. W. Benson, 1950

Sir Charles Frederic Belcher (1876–1970) was an Australian jurist, ornithologist and founder member of RAOU, who spent most of his life in the service of the British Empire. He was educated at the University of Melbourne and Trinity College, Cambridge, and practised law in Victoria (1902–1907). He was called to the Bar in London (1910) and had his own practice in Geelong (1910–1914). He joined the British Colonial Service in Uganda (1914), was an Assistant Judge, Zanzibar (1920); Attorney-General, Nyasaland (1920–1923); Judge, High Court of Nyasaland (1924–1927); Chief Justice, Supreme Court of Cyprus (1927–30); Chief Justice of Trinidad and Tobago; and President, West India Appeal Court (1930–37). He was Chief Legal Adviser, Civil Affairs Branch, East Africa Command (1942–45). In retirement he settled in South Africa. He wrote *The Birds of the District of Geelong, Australia* (1914).

Belcher, E.

Belcher's Gull *Larus belcheri* **Vigors**, 1829
[Alt. Band-tailed Gull]

Admiral Sir Edward Belcher (1799–1877) was a British explorer. After exploring the Pacific coast of America (1825–1828) he was in command of the *Samarang* and surveyed the coast of Borneo, the Philippine Islands and Formosa (Taiwan) (1843–1846). He also explored the Arctic (1852–1854) searching for Franklin. He was court-martialled (1854) for abandoning three ships during this search, but was acquitted. He was the author of *The Last of the Arctic Voyages; Being a Narrative of the Expedition in HMS Assistance, under the Command of ... in Search of Sir John Franklin, During the Years 1852–53–54 with Notes on the Natural History by Sir John Richardson* (1855). He was promoted admiral (1872). A reptile is named after him.

Belding

Belding's Yellowthroat *Geothlypis beldingi* **Ridgway**, 1882
[Alt. Peninsular Yellowthroat]

Belding's Rail *Rallus longirostris beldingi* Ridgway, 1882
[Alt. Clapper Rail ssp.; Syn. *Rallus obsoletus beldingi*]
Belding's Sparrow *Passerculus sandwichensis beldingi*
Ridgway, 1885
[Alt. Savannah Sparrow ssp.]
Belding's Scrub Jay *Aphelocoma californica obscura*
A. W. Anthony, 1889
Leach's Storm-petrel ssp. *Oceanodroma leucorhoa beldingi*
Emerson, 1906 NCR
[JS *Oceanodroma leucorhoa leucorhoa*]
Belding's Plover *Charadrius wilsonia beldingi* Ridgway, 1919
[Alt. Wilson's Plover ssp.]

Lyman Belding (1829–1917) was a professional bird collector who wrote a series of articles about his trips such as *Collecting in the Cape Region of Lower California, West* (1877) and *A Part of my Experience in Collecting* (1900). A mammal and a reptile are named after him.

Belford

Belford's Honeyeater *Melidectes belfordi* **De Vis**, 1890
[Alt. Belford's Melidectes]

George Belford (d.1906), the son of a Samoan chief, collected in New Guinea (1883) for Sir William McGregor (q.v.), who was Administrator and Lieutenant Governor of British New Guinea (1888–1898). He accompanied many expeditions in New Guinea over two decades including McGregor's own (1889–1898). Mount Belford in New Guinea is named after him.

Belisarius

Tawny Eagle ssp. *Aquila rapax belisarius* **Levaillant**, 1850

Flavius Belisarius (c.500–565) was a celebrated Byzantine general in the reign of Justinian, leading campaigns against the Vandals and the Ostrogoths.

Bell, H. H.

Hummingbird sp. *Thalurania belli* **Verrill**, 1905 NCR
[Alt. Blue-headed Hummingbird; JS *Cyanophaia bicolor*]

Sir Henry Hesketh Joudou Bell (1864–1952) was the Administrator of Dominica and an author. His career in colonial administration began in the Bahamas, but within a few years he was promoted to Dominica. Later he was Commissioner, then first Governor of the Uganda Protectorate (1905–1909), where he was very successful in eradicating sleeping sickness. Port Bell in Kampala is named after him. He was then Governor of the Northern Nigeria Protectorate (1909–1912), Governor of the Leeward Islands (1912–1916), and finally Governor of Mauritius (1916–1924). His written works included memoirs, colonial history and fiction.

Bell, J. G.

Bell's Warbler *Basileuterus belli* **Giraud**, 1841
[Alt. Golden-browed Warbler]
Bell's Vireo *Vireo bellii* **Audubon** 1844
Bell's Sparrow *Artemisiospiza belli* **Cassin**, 1850
[Syn. *Amphispiza belli*]

John Graham Bell (1812–1899) was an American taxidermist who accompanied Audubon (q.v.) on his Missouri River Trip (1843). He taught taxidermy to the young Theodore Roosevelt (q.v.). Bell collected in California (1849–1850) and discovered four new species.

Bell, R. M.

Yellow-bellied Eremomela ssp. *Eremomela griseoflava belli*
C. H. B. Grant & Mackworth-Praed, 1941 NCR
[JS *Eremomela icteropygialis polioxantha*]

R. M. Bell (b.1908) was a British colonial administrator in Tanganyika (Tanzania) (1931–1954). He collected birds in the Liwale District and wrote *The Maji-Maji Rebellion in the Liwale District* (1950).

Bell, T.

Oriental Cuckoo ssp. *Cuculus optatus belli* **Mathews**, 1916
NCR; NRM
Long-tailed Cuckoo ssp. *Urodynamis taitensis belli*
Mathews, 1918 NCR; NRM

Thomas Bell (1839–1929) was an English draper with a shop in Samoa and was an early settler on Kermadec Island (1878), finally leaving for New Zealand (1914).

Bell, W. A.

Bell's Pytilia *Pytilia melba belli* **Ogilvie-Grant**, 1907
[Alt. Green-winged Pytilia ssp.]
Bell's Grass Warbler *Cisticola cantans belli* Ogilvie-Grant, 1908
[Alt. Singing Cisticola ssp.]

W. A. Bell (DNF) was a British subscriber to the Ruwenzori Expedition (1905–1906).

Bellona

Hummingbird genus *Bellona* **Mulsant** & **J. Verreaux**, 1866
NCR
[Now in *Orthorhyncus*]

In Roman mythology Bellona was the goddess of war and sister to Mars. An amphibian is named after her.

Belton

Tropeiro Seedeater *Sporophila beltoni* Repenning & Fontana, 2013

Dr William 'Bill' Henry Belton (1914–2009) was an American diplomat and self-taught ornithologist. Stanford University awarded his bachelor's degree in political science (1935) and he joined the US Foreign Service (1938). During this service (1938–1970) his postings included Cuba, Dominican Republic, South America, Australia and Canada where a neighbour introduced him to birding. He ended his career as Deputy Chief of Mission in Rio de Janeiro. He was particularly interested in bird song and made over one thousand recordings, mainly in Latin America and in Rio Grande do Sul (Brazil) in particular - all after retiring, starting in 1971 and continuing until he was 80. Among his written work was the two-volume *Birds of Rio Grande do Sul, Brazil* (1984 & 1985). The original etymology reads: "We name this seedeater in honor of Dr. William "Bill" Belton, an American diplomat who brilliantly revealed with extreme skill and scientific rigor the fantastic world of the birds of Rio Grande do Sul. He also participated significantly in the establishment of 'conservation units' that today remain as a legacy for conservation of birds and their associated environments".

Bemmelen

Swamphen sp. *Porphyrio bemmeleni* **Büttikofer**, 1889 NCR
[Alt. Purple Swamphen; JS *Porphyrio (porphyrio) indicus*]

Adriaan Anthoni van Bemmelen (1830–1897) was a Dutch naturalist. He became the Director of the Zoological Gardens at Rotterdam (1866). A mammal is named after him.

Bendire

Bendire's Sparrow *Peucaea carpalis* **Coues**, 1873
[Alt. Rufous-winged Sparrow]
Bendire's Thrasher *Toxostoma bendirei* Coues, 1873

Western Screech Owl ssp. *Megascops kennicottii bendirei*
Brewster, 1882
Bendire's Crossbill *Loxia curvirostra bendirei* **Ridgway**, 1884
[Alt. Common/Red Crossbill ssp.]
Merlin ssp. *Falco columbarius bendirei* Swann, 1922 NCR
[JS *Falco columbarius columbarius*]

Major Charles Emil Bendire (1836–1897), born Karl Emil Bender in Germany, was an American oologist, zoologist and army surgeon. He collected birds' eggs (1860s and 1870s) while stationed at frontier posts throughout the Department of Columbia, and was famous for the copious notes he made on everything he observed. Fellow officers sent Bendire feathers and eggs from other posts in the West. He became Honorary Curator of Oology at the Smithsonian (1883) and compiled the two-volume *Life Histories of North American Birds* (1892). He personally oversaw the watercolour illustrations to ensure accuracy. He died of Bright's disease. His remarkable collection of 8,000 eggs is housed at the USNM in Washington. Fans of Westerns might like to know that he also once argued Chief Cochise into a truce. A lake, a mountain in Oregon and a mammal are named after him.

Benedetti

Ochre-breasted Brush Finch ssp. *Atlapetes semirufus benedettii* **W. H. Phelps** & **Gilliard**, 1941

Fulvio L. Benedetti (b.1918) was a collector for Phelps in Venezuela. He collected the eponymous holotype (1940) there. He co-wrote *Las Aves de Margarita con Anotaciones Systematicas* (1940).

Benedict

Cozumel Spindalis *Spindalis zena benedicti* **Ridgway**, 1885

Dr James Everard Benedict (1854–1940) was an American zoologist and collector, whose master's degree and doctorate were awarded by Union College, Schenectady. He was resident naturalist aboard the USS *Albatross* (1882–1886) and was Assistant Curator, Division of Marine Invertebrates, USNM.

Benicke

Long-tailed Skua *Lestris benickii* **C. L. Brehm**, 1924 NCR
[JS *Stercorarius longicaudus*]

Benicke (DNF) was the Town Clerk, probably of Schleswig. He was known to Brehm, who mentioned him in volume 3 of his *Beiträge zur Vögelkunde: in Vollständigen Beschreibungen Mehrer* (1822), as travelling from Schleswig to Rügen to observe the birds.

Benjamin

Purple-bibbed Whitetip *Urosticte benjamini* **Bourcier**, 1851

(See **Leadbeater, J. B.**)

Bennett, E.

Bennett's Woodpecker *Campethera bennetti* **A. Smith**, 1836

Edward Turner Bennett (1797–1836) was a British naturalist. He promoted the setting up of a London entomological club (1822), which developed in association with the Linnean Society into a zoological club and eventually the Zoological Society of London (1826), of which he became first Vice Secretary and then Secretary (1831–1836). Five mammals are named after him.

Bennett, G.

Bennett's Cassowary *Casuarius bennetti* **Gould**, 1858
[Alt. Dwarf Cassowary]
Barred Owlet-nightjar *Aegotheles bennettii* **Salvadori** & **d'Albertis**, 1875

Bennett's Bird-of-Paradise *Drepanornis albertisi cervinicauda* **P. L. Sclater**, 1883
[Alt. Black-billed Sicklebill ssp.]

Dr George Bennett (1804–1893) was a British surgeon, botanist and zoologist. He took passage to the South Seas and Australia as surgeon-naturalist on board the *Sophia*, settling permanently there (1836). He wrote such as: *Wanderings in New South Wales, Batavia, Pedir Coast, Singapore, and China : being the journal of a naturalist in those countries during 1832, 1833, and 1834* (1834) and *Gatherings of a Naturalist in Australasia* (1860). He became the first Curator and Secretary of the Australian Museum (1835) and was an early conservationist. He wrote (1860): 'Many of the Australian quadrupeds and birds are not only peculiar to that country, but are, even there, of comparatively rare occurrence: and ... they are in a fair way of becoming extinct. Even in our own time, several have been exterminated; and unless the hand of man be stayed from their destruction, the *Ornithorhynchus* and the *Echidna*, the *Emeu* and the *Megapodius*, like the Dodo, Moa and *Notornis*, will shortly exist only in the pages of the naturalist. The Author hopes that what he has been induced to say with reference to this important subject will not be without weight to every thoughtful colonist.' The fact that all four survive today may be due to his timely campaign. He spent 50 years unsuccessfully trying to fully understand monotreme and marsupial biology. He was a frequent correspondent of both Gould (q.v.) and Gilbert (q.v.) and offered both advice and support in their various Australian sojourns. A mammal, an amphibian and three reptiles are named after him.

Bennett, K. H.

Bennett's Crow *Corvus bennetti* **North**, 1901
[Alt. Little Crow]

Kenric Harold Bennett (1835–1891) was an Australian collector. His interests were birds, plants and Aboriginal artefacts. His writings include: 'Note on the mode of nidification of a species of *Pachycephala*, supposed to be *P. Gilbertii*, from the interior of N. S. Wales' (1887).

Bennett, L. J.

Collared Kingfisher ssp. *Todiramphus chloris bennetti* Ripley, 1947

Dr Logan Johnson Bennett (1907–1957), who collected the holotype, studied biology for his BSc at Central College, Fayette (1930), and at Iowa State College for his MS in zoology (1932) and for his PhD there (1937). He was a Junior Refuge Manager in Wisconsin (1935) and leader of the Cooperative Wildlife Research Unit at Iowa State College, then head of the Cooperative Unit at Pennsylvania State College (1938–1943). He served in the Navy (1943–1945) as Lieutenant, in charge of a malaria control unit. He became Chief of the Fish and Wildlife Service's Branch of Wildlife Research (1948–1953). He was then Executive Director of the Pennsylvania Game Commission (1953). Ripley (q.v.) writes that Bennett, during his war service in the US Naval Reserve (1944), was stationed on Nissan Island near New Ireland where he made a collection of birds. He wrote numerous scientific papers and several books including *The Blue-winged Teal* (1938). He died of a heart attack when about to return from a convention.

Bensbach

Bensbach's Bird-of-Paradise *Janthothoreax bensbachii* **Sharpe**, 1896 NCR
[Alt. Bensbach's Riflebird; believed to be of hybrid origin]

Jacob Bensbach (DNF) was a Dutch colonial civil servant. He was Assistant Resident to the Governor of Celebes (1885–1888) and Deputy Governor of Ternate in the Moluccas (1888–1893). He made various trips to New Guinea (1889–1893), during which he collected natural history specimens. He presented the only known specimen of this bird-of-paradise to the Leiden Museum (1894). It has been suggested that it represents a hybrid between the Magnificent Riflebird *Ptiloris magnificus* and Lesser Bird-of-Paradise *Paradisaea minor*.

Bensch

Bensch's Monias *Monias benschi* **Oustalet** & **Grandidier**, 1903
[Alt. Subdesert Mesite, Bensch's Mesite]

Jean Henri Émile Bensch (1868–1944) was an administrator in the French colonial regime in Madagascar. He was Administrator of the islands of Saint Pierre and Miquelon (1921–1923) and then Governor (1923–1928). He sent bird specimens to Paris (1899) and there is a later reference (1912) to his collection of insects being examined by a fellow enthusiast.

Benson C. E. & G. H.

Desert Lark ssp. *Ammomanes deserti bensoni* **Meinhertzhagen**, 1933
[Has been regarded as a JS of either *A. deserti geyri* or *A. deserti whitakeri*]
Twite ssp. *Carduelis flavirostris bensonorum* Meinertzhagen, 1934

Constantine Evelyn Benson (1895–1960) was an army Captain (WW1) and an Air Commodore in the Royal Auxiliary Air Force (WW2) and, in between, a Director of Lloyds Bank. His elder brother, Guy Holford Benson (1888–1975) was Director of London Assurance. The twite honours them both.

Benson, C. W.

Weaver genus *Bensonhyphantes* **J. A. Roberts**, 1947 NCR
[Now in *Ploceus*]

Benson's Rock Thrush *Monticola bensoni* **Farkas**, 1971
[NUI *Monticola sharpei;* formerly *Pseudocossyphus bensoni*]

Chapin's Apalis ssp. *Apalis chapini bensoni* **Vincent**, 1935
   NCR
[JS *Apalis chapini strausae*]
Bar-throated Apalis ssp. *Apalis thoracica bensoni*
   C. H. B. Grant & **Mackworth-Praed**, 1937 NCR
[Use of name *bensoni* pre-occupied by Vincent's 1935
   usage: replacement name = *Apalis thoracica whitei*]
East Coast Akalat ssp. *Sheppardia gunningi bensoni*
   Kinnear, 1938
Forest Double-collared Sunbird ssp. *Cinnyris fuelleborni
   bensoni* **J. G. Williams**, 1953
Papyrus Yellow Warbler ssp. *Chloropeta gracilirostris
   bensoni* **Amadon**, 1954
Rattling Cisticola ssp. *Cisticola chiniana bensoni* **Traylor**,
   1964
Blue Vanga ssp. *Cyanolanius madagascarinus bensoni*
   Louette & Herremans, 1982

Constantine 'Con' Walter Benson (1909–1982) was born and educated in England. He studied at Cambridge, achieving an 'athletics blue' before he embarked on a career in the colonial service. He spent 20 years in Nyasaland (Malawi) and was awarded the OBE on his retirement (1965). He was a major figure in Central African ornithology and an ultra-meticulous systematist. He collected in Africa for over 30 years, aided throughout by Jali Makawa (q.v.). He wrote widely, including several books, notably the co-written *Birds of Zambia* (1971) and *Birds of Malawi* (1977) with his wife Mary. He was editor of the *Bulletin of the British Ornithologists' Club* (1969–1974) and catalogued the bird collections in the University Museum, Cambridge. (Also see **Mary (Benson)** and **Mrs Benson**)

Benson, H. C.

Benson's Quail *Callipepla douglasii bensoni* **Ridgway**, 1887
[Alt. Elegant Quail ssp.]

Colonel Harry Coupland Benson (1857–1924) joined the US Army Artillery (1878) and graduated at Westpoint (1882), transferring to the cavalry. He was immediately involved in the campaign to capture Geronimo during which he received international recognition by securing a specimen of an Imperial Woodpecker *Campephilus imperialis*. He became First Lieutenant (1887) and accepted an appointment as a mathematics instructor at West Point. During the Spanish-American war he acted as the Collector of Customs, Cuba, and later in the Philippines becoming a Captain (1897). He named a

number of peaks in Yosemite National Park where he became Acting Superintendent (1905) and collected in the south-west USA (he was stationed at Fort Huachuca in Arizona in the 1880s) and north-west Mexico. After he left his Yosemite post (1908) he moved to Yellowstone National Park, where he became Superintendent. He was assigned as Chief of Staff, Philippines Department (1911), becoming Colonel (1914) before retiring (1915). He collected birds and eggs, sending many to the USNM including the quail holotype to Ridgway. He was especially interested in propagating trout in the lakes and streams of Yosemite National Park and to that end cooperated with the State Fish and Game Commisioners.

Benson, J. P.

Terrestrial Brownbul ssp. *Phyllastrephus terrestris bensoni*
   Van Someren, 1945 NCR
[JS *Phyllastrephus terrestris suahelicus*]
Kenrick's Starling ssp. *Poeoptera kenricki bensoni* Van
   Someren, 1945

J. P. Benson (fl.1961) was an ornithologist who collected (including these holotypes) for many years in Kenya, particularly the Meru-Isiolo area whilst employed by the Department of Agriculture in Kenya (c.1938–1962) when he passed the preliminary oral exam in Swahili.

Benson, R. R.

Black-cheeked Warbler ssp. *Basileuterus melanogenys
   bensoni* Griscom, 1927

Rex R. Benson (fl.1930) was an American collector in Panama (1924–1927) for the describer and for the AMNH, New York. He was Director, the Gromaco Foundation for Natural Science, Costa Rica (1962).

Bent

Bent's Crossbill *Loxia curvirostra benti* **Griscom**, 1937
[Alt. Common/Red Crossbill ssp.]
White-throated Thrush ssp. *Turdus assimilis benti*
   A. R. Phillips, 1991

Arthur Cleveland Bent (1866–1954) was a successful Massachusetts businessman and a renowned amateur ornithologist. He wrote the 26-volume *Life Histories of North American Birds*. Until recently, this massive work comprising over 10,000 pages of information, represented perhaps the single most comprehensive resource on the natural history of North American birds.

Bérard

Bérard's Diving-petrel *Pelecanoides berard* **Gaimard**, 1823
[Alt. Common Diving-petrel; Syn. *Pelecanoides urinatrix
   berard*]

Rear Admiral Auguste Bérard (1796–1852) was a French naval officer and explorer, best known for mapping the coast of Algeria (1837). He also took part in the French circumnavigations (1818–1820, 1822–1825 and 1842–1846) which surveyed in the Pacific and particularly Australasia. A mammal genus is named after him.

Berenguer

Thekla Lark ssp. *Galerida theklae berengueri* **Cabrera**, 1922 NCR
[JS *Galerida theklae erlangeri*]

General Dámaso Berenguer y Fusté, Conde de Xauen (1873–1953), was a Spanish soldier and statesman. He was Minister of War (1918), High Commissioner of Spanish Morocco (1919–1922 and Prime Minister (1930–1931).

Berenice

Hornbill genus *Berenicornis* **Bonaparte**, 1850

Berenice II (c.267–221 BC) was the wife of Ptolemy III of Egypt. According to a myth, she dedicated her hair to Aphrodite in return for her husband's safe return from an expedition. The hair was apparently then stolen from the temple where it had been placed, but it was said to have been carried to the heavens and put among the stars. *Berenicornis* = 'Berenice bird'.

Berezowski

Yellow-breasted Tit ssp. *Cyanistes flavipectus berezowskii* **Pleske**, 1893
[Syn. *Cyanistes cyanus berezowskii*]
Berezowski's Blood Pheasant *Ithaginis cruentus berezowskii* **Bianchi**, 1903

Mikhail Mikhailovitch Berezowski (d.1911) was a Russian explorer, geographer and zoologist, and associate of his more famous contemporary Valentin Bianchi (q.v.). They wrote a paper together, *Aves Expeditionis Potanini per Provinciam Gansu et Confinia 1884–1887* (1905) and discovered and described the Blackthroat *Calliope obscura* (1891). He was in north-west Mongolia (1876–1878), and in Gansu and Szechuan (Sichuan), China (1884–1887) and again (1892–1894). He also led an expedition to Kuçar, Turkestan (1905–1907). He collected many written items as well as natural history specimens and his collection of 1,876 oriental text items are held in the Institute of Oriental Manuscripts in St Petersburg. Many of his ornithological specimens are now at the Zoological Museum of Moscow University. A mammal is named after him.

Berg

Tanager genus *Bergia* **Bertoni**, 1901 NCR
[Now in *Stephanophorus*]

Owl sp. *Nyctale bergiana* Bertoni, 1901 NCR
[Alt. Rusty-barred Owl; JS *Strix hylophila*]

White-eyed Foliage-gleaner ssp. *Automolus leucophthalmus bergianus* Bertoni, 1901 NCR
[JS *Automolus leucophthalmus sulphurascens*]

Dr Frederico Guillermo Carlos (Friedrich Wilhelm Carl) Berg (1843–1902) was a Latvian entomologist and naturalist. After working in commerce, he became a conservator of entomological specimens at the Riga Museum. He joined the National Museum, Buenos Aires (1873) and went on expeditions to Patagonia (1874) and Chile (1879). He worked at the National Museum of Natural History, Montevideo (1890–1892)

and was Director, National Museum, Buenos Aires (1892–1901). A mammal and an amphibian are named after him.

Berger

Boubou sp. *Laniarius bergeri* **Reichenow**, 1911 NCR
[Alt. Slate-coloured Boubou; JS *Laniarius funebris*]

Dr Arthur Berger (1871–1947) was a German physician, hunter, explorer and zoologist. He travelled in the Arctic (1900) and Africa (1908–1909) and visited both India and the USA. A reptile is named after him.

Bergius, C.

Greater Crested Tern *Thalasseus bergii* **Lichtenstein**, 1823
[Alt. Swift Tern]

Carl Heinrich Bergius (1790–1818) was a Prussian pharmacist and botanist who was described by Lichtenstein as a promising young naturalist. He obtained the first specimen of the tern that bears his name near Cape Town where he spent time collecting (1816–1818). He was a cavalryman during the Napoleonic Wars and afterwards studied medicine in Berlin. He got a job as a pharmacist with a company called Pallas and Poleman, Capetown (1815). In his spare time he collected zoological and botanical material for the Berlin Museum. He died of tuberculosis in poverty in Cape Town, deserted by his friends and ignored by his former employers.

Bergius, P.

Common Pheasant ssp. *Phasianus colchicus bergii* **Zarudny**, 1914 NCR
[JS *Phasianus colchicus turcestanicus*]

Dr Peter Jonas Bergius (1730–1790) was a Swedish physician and botanist who studied at Lund University and qualified at Uppsala. He practised in Stockholm, where he became Professor of the Medical College (1761). The botanical gardens at Frescati, just north of Stockholm, are named after him.

Bergman

Eurasian Nuthatch ssp. *Sitta europaea bergmani* **Momiyama**, 1931 NCR
[JS *Sitta europaea clara*]
Black Sunbird ssp. *Leptocoma sericea bergmanii* **Gyldenstolpe**, 1955 NCR
[JS *Leptocoma sericea sericea*]

Dr Sten Bergman (1895–1975) was a Swedish zoologist, biologist, writer, traveller and explorer. He visited Kamchatka (1920–1923), the Kurile Islands (1929–1930) and Korea (1935–1936). He made several visits to New Guinea (1948–1950, 1952–1953 and 1956–1958) and to Japan (1960 and 1962). A mammal is named after him.

Bering

Eider sp. ('Bering Goose') *Anas beringii* **Gmelin**, 1789 NCR
[Alt. King Eider; JS *Somateria spectabilis*]
Bering's Cormorant *Phalacrocorax pelagicus* **Pallas**, 1811
[Alt. Pelagic Cormorant]

Vitus Jonassen Bering (1681–1741) was a great explorer, born in Denmark, who spent his life in the Russian Imperial Service. He enlisted in the Russian navy (1703) and only visited Denmark once again in his life (1715). He was ordered by Tsar Peter the Great to lead an expedition (1725–1730) – the first Kamchatka Expedition, during which (1728) he established that Asia and America are two separate continents. In the second Kamchatka Expedition (1733–1743) he was the first to map the west coast of Alaska (1741). Two ships, *St Peter*, with both Bering and Steller on board, and *St Paul* were used for this expedition. On the way home the *St Peter* ran aground on a desolate island, now called Bering Island, where he and his crew had to spend the winter in crude huts. After nine months a boat was constructed from the wreckage of the *St Peter*, and the survivors arrived in Kamchatka (1742). The graves of Bering and five of his crew were discovered (1991), and after scientific examination of the remains in Moscow they were all returned to and re-buried on Bering Island.

Berla

Berla's Parrotlet *Touit ruficauda* Berla, 1954 NCR
[NUI Golden-tailed Parrotlet; JS *Touit surdus chryseurus*]

Rufous-capped Motmot ssp. *Baryphthengus ruficapillus berlai* **Stager**, 1959 NCR; NRM

Herbert Franzioni Berla (1912–1985) was a Brazilian ornithologist and entomologist who worked for the National Museum in Rio de Janeiro. He made a collecting trip to Pernambuco (1946) where he catalogued c.160 bird species. He made a particular study of mites which live on birds.

Berlandier

Berlandier's Wren *Thryothorus ludovicianus berlandieri* **Baird**, 1858
[Alt. Carolina Wren ssp.]

Jean Louis Berlandier (1805–1851) was a Belgian botanist who went to Mexico (1826) as a collector. The Mexican government employed him (1827–1828) and he stayed on, marrying a local woman, dividing his time between a pharmaceutical business and collecting botanical specimens. The Mexican government employed him again (1834) as an interpreter to General Arista, and in charge of the hospitals at Matamoros during the Mexican War. He was drowned whilst trying to cross the San Fernando River. A reptile and an amphibian are named after him.

Berlepsch

Palmcreeper genus *Berlepschia* **Ridgway**, 1887

Berlepsch's Emerald *Chlorostilbon inexpectatus* Berlepsch, 1879 NCR
[Believed to be based on an aberrant *Chlorostilbon poortmani* specimen]
Esmeraldas Woodstar *Chaetocercus berlepschi* **Simon**, 1889
Brilliant sp. *Heliodoxa berlepschi* **Boucard**, 1892 NCR
[JS *Heliodoxa jacula henryi*]

Hermit sp. *Phaethornis berlepschi* **Hartert**, 1894 NCR
[JS *Phaethornis syrmatophorus*]
Berlepsch's Parotia *Parotia berlepschi* **Kleinschmidt**, 1897
[Alt. Bronze Parotia, Berlepsch's Six-wired Bird-of-Paradise]
Berlepsch's Puffleg *Eriocnemis berlepschi* Hartert, 1897 NCR
[JS *Eriocnemis vestita*]
Berlepsch's Tinamou *Crypturellus berlepschi* **Rothschild**, 1897
Stub-tailed Antbird *Sipia berlepschi* Hartert, 1898
[Syn. *Myrmeciza berlepschi*]
Venezuelan Sylph *Aglaiocercus berlepschi* Hartert, 1898
Scarlet-breasted Dacnis *Dacnis berlepschi* Hartert, 1900
Amazonian Antpitta *Hylopezus berlepschi* **Hellmayr**, 1903
Russet-mantled Softtail *Thripophaga berlepschi* Hellmayr, 1905
Berlepsch's Tody Flycatcher *Todirostrum hypospodium* Berlepsch, 1907 NCR
[Believed to be a synonym of *Poecilotriccus sylvia*]
Harlequin Antbird *Rhegmatorhina berlepschi* **Snethlage**, 1907
Berlepsch's Canastero *Asthenes berlepschi* Hellmayr, 1917

Black-eared Hemispingus ssp. *Hemispingus melanotis berlepschi* **Taczanowski**, 1880
Berlepsch's Spotted Woodcreeper *Xiphorhynchus erythropygius aequatorialis* Berlepsch & Taczanowski, 1884
Beryl-spangled Tanager ssp. *Tangara nigroviridis berlepschi* Taczanowski, 1884
Lined Antshrike ssp. *Thamnophilus tenuepunctatus berlepschi* Taczanowski, 1884
Berlepsch's Conure *Pyrrhura melanura berlepschi* **Salvadori**, 1891
[Alt. Maroon-tailed Parakeet ssp.]
Blossomcrown ssp. *Anthocephala floriceps berlepschi* **Salvin**, 1893
Berlepsch's Ruddy Pigeon *Patagioenas subvinacea berlepschi* Hartert, 1898
Berlepsch's Speckled Mousebird *Colius striatus berlepschi* Hartert, 1899
Blue-grey Tanager ssp. *Thraupis episcopus berlepschi* **Dalmas**, 1900
Berlepsch's Masked Gnatcatcher *Polioptila dumicola berlepschi* Hellmayr, 1901
Tufted Flycatcher ssp. *Mitrephanes phaeocercus berlepschi* Hartert, 1902
Grey Antwren ssp. *Myrmotherula menetriesii berlepschi* Hellmayr, 1903
Golden-headed Manakin ssp. *Dixiphia erythrocephala berlepschi* Ridgway, 1906
Strong-billed Woodcreeper ssp. *Xiphocolaptes promeropirhynchus berlepschi* Snethlage, 1908
Plain-mantled Tit Spinetail ssp. *Leptasthenura aegithaloides berlepschi* Hartert, 1909
Ash-breasted Antbird ssp. *Myrmoborus lugubris berlepschi* Hellmayr, 1910
Bronze-olive Pygmy Tyrant ssp. *Pseudotriccus pelzelni berlepschi* **Nelson**, 1913
Three-striped Flycatcher ssp. *Conopias trivirgatus berlepschi* Snethlage, 1914

Checkered Woodpecker ssp. *Veniliornis mixtus berlepschi*
Hellmayr, 1915
Scaly-breasted Hummingbird ssp. *Phaeochroa cuvierii berlepschi* Hellmayr, 1915
Black-tailed Trainbearer ssp. *Lesbia victoriae berlepschi* Hellmayr, 1915
Short-tailed Antthrush ssp. *Chamaeza campanisona berlepschi* **Stolzmann**, 1926
Glittering-bellied Emerald ssp. *Chlorostilbon lucidus berlepschi* **O. Pinto**, 1938

Hans Hermann Carl Ludwig, Graf von Berlepsch (1850–1915), was a German ornithologist and author. He revised the catalogue of Tschudi's (q.v.) collection in the Natural History Museum at Neuchâtel with Hellmayr (q.v.). He amassed a great collection himself and, after his death, over 50,000 skins, including 300 type specimens, went to the museum at Frankfurt. He did not make the collections himself but employed others, such as Garlepp (q.v.) and Stolzmann (q.v.), to collect for him in Latin America. He named a number of birds after his wife Emma (q.v.).

Berlioz

Berlioz's Black Flycatcher *Melaenornis ardesiacus* Berlioz, 1936
[Alt. Yellow-eyed Black Flycatcher]
Berlioz's Sunbird *Anthreptes pujoli* Berlioz, 1958 NCR
[Alt. Grey-chinned Sunbird; JS *Anthreptes rectirostris*]
Berlioz's Tyrant *Knipolegus subflammulatus* Berlioz, 1959 NCR
[Alt. Plumbeous Tyrant; JS *Knipolegus (signatus) cabanisi*]
Berlioz's Swift *Apus berliozi* **Ripley**, 1965
[Alt. Forbes-Watson's Swift]
Berlioz's Woodnymph *Augasma cyaneoberyllina* Berlioz, 1965 NCR
[Status uncertain, perhaps of hybrid origin]
Berlioz's Xenops *Megaxenops ferrugineus* Berlioz, 1966 NCR
[Alt. Peruvian Recurvebill; JS *Simoxenops ucayalae*]

Berlioz's Silver Pheasant *Lophura nycthemera berliozi* **Delacour** & **Jabouille**, 1928
Madagascan Wood Rail ssp. *Canirallus kioloides berliozi* **Salomonsen**, 1934
Scarlet-browed Tanager ssp. *Heterospingus xanthopygius berliozi* **Wetmore**, 1966

Jacques Berlioz (1891–1975) was a French ornithologist and author who was in charge of the bird department at the Paris Museum (mid-20th century). His publications included 'Étude d'une collection d'oiseaux du Tchad' (1938) and 'Le développement de l'ornithologie et l'industrie plumassière' (1959).

Bernard

Collared Antshrike *Sakesphorus bernardi* **Lesson**, 1844

Captain Bernard (DNF) was a French mariner and collector, but little is known about him.

Berney

Owl genus *Berneyornis* **Mathews**, 1916 NCR
[Now in *Ninox*]

Western Gerygone ssp. *Gerygone culicivora berneyi* Mathews, 1912 NCR
[JS *Gerygone fusca exsul*]
Yellow-throated Miner ssp. *Myzantha flavigula berneyi* Mathews, 1912 NCR
[JS *Manorina flavigula flavigula*]
Red-chested Buttonquail ssp. *Turnix pyrrhothorax berneyi* Mathews, 1916 NCR; NRM

Frederic Lee Berney (1865–1949) was an Australian ornithologist who was President of the RAOU (1934–1935).

Bernier

Malagasy warbler genus *Bernieria* **Pucheran**, 1855

Bernier's Ibis *Threskiornis bernieri* Bonaparte, 1855
[Alt. Madagascar Sacred Ibis, Blue-eyed Ibis]
Bernier's Vanga *Oriolia bernieri* **Geoffroy** Saint-Hilaire, 1838
Bernier's Teal *Anas bernieri* **Hartlaub**, 1860
[Alt. Madagascar Teal]

Chevalier Alphonse Charles Joseph Bernier (1802–1858) was a French naval surgeon. He was also a botanist and collector who spent some time (1831–1834) in Madagascar. He took 198 specimens back to France, where they were catalogued (1835). A reptile is named after him.

Bernstein

Chinese Crested Tern *Thalasseus bernsteini* Schlegel, 1863
[Syn. *Sterna bernsteini*]
Scarlet-breasted Fruit Dove *Ptilinopus bernsteinii* Schlegel, 1863
Bernstein's Coucal *Centropus bernsteini* Schlegel, 1866
[Alt. Lesser Black Coucal, Black-billed Coucal]
Sula Scrubfowl *Megapodius bernsteinii* Schlegel, 1866

Bernstein's Black Lory *Chalcopsitta atra bernsteini* **Rosenberg**, 1861
Bernstein's Red Lory *Eos bornea bernsteini* Rosenberg, 1863 NCR
[NUI *Eos bornea bornea*; Syn. *Eos rubra rubra*]
Red-bellied Pitta ssp. *Erythropitta erythrogaster bernsteini* **Junge**, 1958

Heinrich Agathon Bernstein (1828–1865) was a German, latterly naturalised Dutch, zoologist and collector. Son of a German physician he too studied medicine, but attracted by the lure of the tropics he accepted a post as Director of the Sanatorium of Gadok, Java (1854). Here he became acquainted with the local bird life. He described c.70 bird species, which made him well known in Europe, with the result that (1859) the Dutch Government requested him to undertake an expedition to the Moluccas. He established his base on the island of Ternate (1861–1863), journeying to New Guinea (1864) where he experienced a bloody encounter with native inhabitants (1865), soon after which he contracted yellow fever and died on his ship (April 1865). He bequeathed 2,000 bird skins to the Leiden Museum. He left a diary of his last journey (1864–1865) from Ternate to New Guinea.

Berta

Pitta sp. *Pitta bertae* **Salvadori**, 1868 NCR
[JS *Pitta nympha*?]

Bertha Salvadori (1844–1904) *née* King was the describer's cousin and wife. The holotype of the bird, collected in Sarawak, was lost in the post between Italy and Paris (1873). A coloured picture of the stuffed original is all that remains.

Berthelot

Berthelot's Pipit *Anthus berthelotii* **Bolle**, 1862
[Alt. Canary Island Pipit]

Sabin Berthelot (1794–1880) was the French Consul on Tenerife. He wrote *Histoire Naturelle des Îles Canaries* (1835) and *The Exploration of the Southern Coast of Spain* (1867).

Berthemy

Buffy Laughingthrush *Garrulax berthemyi* **Oustalet**, 1876

Jules François Gustave Berthemy (1826–1902) was a French diplomat. He was in Turkey (1852–1857), being promoted (1855) from Attaché to Second Secretary at the French Embassy in Constantinople (Istanbul). He became Minister and Plenipotentiary in Peking (Beijing) (1863). He started the negotiations (1865) to regularise the position of property bought by French missionaries in China. The matter, known as the 'Berthemy Convention', was finally settled by his successor (1895).

Berthold

Dove genus *Bertholdipelia* **Boetticher**, 1950 NCR
[Now in *Streptopelia*]

Dr Paul Erich Berthold Klatt (1885–1958) was a German zoologist. He graduated at the University of Berlin (1904) and taught at the Berlin Agricultural University (1908–1914). His career was interrupted by military service (WW1), in which he was wounded. He worked at the Institute of Genetics, Berlin (1918–1923) and was Professor at the University of Hamburg (1923–1928), and at the University of Halle (1928–1933), returning to Hamburg (1934–1954).

Bertoni

Bertoni's Antbird *Drymophila rubricollis* Bertoni, 1901

Tody Flycatcher ssp. *Euscarthmus gularis bertonii* **Stolzmann**, 1926 NCR
[JS *Poecilotriccus plumbeiceps*]

Dr Moises Santiago Bertoni (1857–1929) was an Italian Swiss biologist, meteorologist, cartographer, agronomist, ornithologist and naturalist. He studied law but (1875) abandoned it for natural sciences. Shortly after the birth of his son, Arnoldo de Winkelried Bertoni (1878–1973), he left Europe and settled in Misiones, Argentina. He then moved to Paraguay (1886), where he established the Swiss colony 'William Tell' on the Parana River on the border with Argentina. The site is now called Puerto Bertoni. He was the first Director of the College of Agriculture in Asunción. He wrote *A Guide to the Birds of South America* (1879).

Bertram/Bertrand

Bertram's Weaver *Ploceus bertrandi* **G. E. Shelley**, 1893
[Alt. Bertrand's Weaver]

Olive Bush-shrike ssp. *Chlorophoneus olivaceus bertrandi* G. E. Shelley, 1893

Bertram Lutley Sclater (1866–1897) was a captain in the British army, having previously been a lieutenant in the Royal Engineers. He served in Uganda and other parts of Africa and explored and mapped in Nyasaland (Malawi) (1880s–1890s). Shelley was a close associate of Philip Lutley Sclater, Bertram's father. It is unclear why Shelley twice chose the spelling *bertrandi* rather than *bertrami*.

Besserer

Three-striped Warbler ssp. *Basileuterus tristriatus bessereri* **Hellmayr**, 1922

Ludwig Willibald Albrecht Johann Max Karl, Freiherr Besserer von Thalfingen (1857–1948), was a German ornithologist.

Best

Orange-bellied Flowerpecker ssp. *Dicaeum trigonostigma besti* **Steere**, 1890

Named after a son of Judge Best (DNF) of Minneapolis. The only source of this information is a cryptic note on a card in the Richmond Index. The holotype was collected by Steere on Siquijor Island, Philippines (1888).

Bethel

Koklass Pheasant ssp. *Pucrasia macrolopha bethelae* **Fleming**, 1947
[May be a clinal morph of the subspecies *biddulphi*]
Black-faced Laughingthrush ssp. *Trochalopteron affine bethelae* **Rand** & Fleming, 1956

Dr Bethel F. Fleming *née* Harris (b.1901) was a medical missionary in India and wife of Dr Robert Lee Fleming Sr, an ornithologist (co-author of *Birds of Nepal with Reference to Kashmir and Sikkim* 1979) and educational administrator. They were in Nepal initially to watch birds but returned (c.1949–c.1962) and founded a hospital in Kathmandu (1959). The couple's full story can be read in Grace Fletcher's biography, *The Fabulous Flemings of Kathmandu: the Story of Two Doctors in Nepal* (1965).

Bethune

Antarctic Tern ssp. *Sterna vittata bethunei* **W. L. Buller**, 1895
White-fronted Tern ssp. *Sterna striata bethunei* W. L. Buller, 1895 NCR
[Name replaced by *Sterna striata aucklandorna*]

Alexander William Bethune (d.1938) was a New Zealand marine engineer on board (1885) the NZ government vessel *Stella*, which annually provisioned the depots on the subantarctic islands. He was chief engineer on *Tutanekai* (1900) and became Inspector of Machinery for the country and Government Surveyor (1916).

Bettington

Golden Whistler ssp. *Pachycephala pectoralis bettingtoni* **Mathews**, 1920 NCR
[JS *Pachycephala pectoralis fuliginosa*]

Albemarle Brindley Bettington (1877–1943) was an Australian farmer, naturalist and collector at Merriwa, New South Wales.

Beven

Black-bellied Sunbird ssp. *Cinnyris nectarinioides beveni* **Van Someren**, 1930 NCR
[JS *Cinnyris nectarinioides erlangeri*]

Dr John Osmonde Beven (1889–1930) was a physician in Kenya (1921–1927).

Beverley (Irwin)

Collared Sunbird ssp. *Anthreptes collaris beverleyae* **Irwin**, 1961 NCR
[JS *Anthodiaeta collaris zuluensis*]

Mrs Beverley Jeanne Irwin *née* Oosthuizen (1924–1983) was a South African from an Afrikaans family and was the wife of M. P. Stuart Irwin (q.v.).

Beverly (Hilty)

Rio Orinoco Spinetail *Synallaxis beverlyae* Hilty & Ascanio, 2009

Beverly J. Hilty is the wife of the senior author of the description, Steven Hilty, who has written authoritative guides to the birds of Colombia and Venezuela. The co-author of the description, David Ascanio, is a Venezuelan ornithologist and tour leader who discovered the bird.

Bewick

Bewick's Wren *Thryomanes bewickii* **Audubon**, 1827

Bewick's Swan *Cygnus columbianus bewickii* **Yarrell**, 1830
[Alt. Tundra Swan ssp.]
Common Redshank ssp. *Tringa totanus bewickii* Rennie, 1831 NCR
[Invalid, as Rennie's 'type specimen' was apparently a Ruff *Philomachus pugnax*]

Thomas Bewick (1753–1828) was an English ornithologist and engraver, the best-known English illustrator of his generation. His woodcuts of birds, mammals and rural scenes made woodcutting an art form. Audubon met the elderly Bewick on his first trip to England (1827) and honoured him by naming a new wren after him, which was the same size as one of the boxwood blocks Bewick engraved his images on. Audubon wrote: 'A complete Englishman, full of life and energy though now seventy-four, very witty and clever, better acquainted with America than most of his countrymen, and an honor to England.' And '… Thomas Bewick is a son of Nature. Nature alone reared him under her peaceful care, and he in gratitude of heart has copied one department of her works that must stand unrivalled forever'. This latter reference is of course to Bewick's woodcuts of British birds. Shortly before he died, Bewick paid a final visit to Audubon and encountered another visitor, William Swainson (q.v.). It was an informal gathering of the three greatest natural history artists of their age.

Bewsher

Comoro Thrush *Turdus bewsheri* **E. Newton**, 1877

Charles E. Bewsher (fl.1890) was a banker, amateur botanist and malacologist in Mauritius. He visited the island of Anjouan (Comoro Islands) (1876) and made a collection of natural history specimens.

Bianchi

Storm-petrel genus *Bianchoma* **Mathews**, 1943 NCR
[Now in *Oceanodroma*]

Bianchi's Warbler *Seicercus valentini* **Hartert**, 1907

Willow Tit ssp. *Poecile montanus bianchii* **Zarudny** & **Harms**, 1900 NCR
[JS *Poecile montanus borealis*]
Bianchi's Blood Pheasant *Ithaginis cruentus michaelis* Bianchi, 1903
Bianchi's Pheasant *Phasianus colchicus bianchii* Buturlin, 1904
[Alt. Common Pheasant ssp.]
Common Treecreeper ssp. *Certhia familiaris bianchii* Hartert, 1905
Great Grey Shrike ssp. *Lanius excubitor bianchii* Hartert, 1907
Cocoa Thrush ssp. *Turdus fumigatus bianchii* Chrostovski, 1921 NCR
[JS *Turdus fumigatus fumigatus*]

Valentin Lvovich Bianchi (more properly Bianki) (1857–1920) was a Russian zoologist and ornithologist and an associate of Berezowski (q.v.). Together they discovered and described the Blackthroat *Calliope obscura* (1891). He was Curator of the Ornithological Department of the Imperial Academy of Sciences at St Petersburg (1896–1920). He co-wrote *Orthoptera and Pseudoneuroptera of the Russian Empire and Adjacent Countries* (1905). He wrote *Instruction for Collecting Birds, their Eggs and Nests* (1909). This was regarded as an important book in its day, as it showed great attention to method. He was also the author of *Colymbiformes et Procellariiformes*, which is the first volume of *Faune de la Russie et des Pays Limitrophes* (1911 & 1913). His son Vitaly Valentinovich Bianki is a well-known writer on nature and ornithology in present-day Russia.

Bicheno

Finch genus *Bichenoa* **Moulton**, 1923 NCR
[Now in *Taeniopygia*]

Bicheno's Finch *Taeniopygia bichenovii* **Vigors** & **Horsfield**, 1827
[Alt. Double-barred Finch, Owl Finch]

James Ebenezer Bicheno (1785–1851) was the British Colonial Secretary for Van Diemen's Land (Tasmania) (1843–1851). He was a keen amateur botanist and experimented with plants on his farm on the banks of the New Town rivulet. Bicheno was famous for his girth, it being said that 'he could fit three full bags of wheat into his trousers'. He made economic and scientific studies, was called to the Bar (1822), and was secretary of the Linnean Society (1825–1832) as well as being a partner in a Glamorganshire ironworks (1832–1842). He had a number of papers on botany and natural history published in a Linnean Society journal. He also wrote *Observations on the Philosophy of Criminal Jurisprudence with Remarks on Penitentiary Prisons* (1819) and assisted Sir William Jardine in preparing the two volumes of *Illustrations of Ornithology* (1830). Furthermore, he was an early Vice-President of the Royal Society of Tasmania. Bicheno died in Hobart and his substantial library, which he bequeathed to Tasmania, was incorporated into the first Tasmanian Library. A Tasmanian town was named after him the year he died, in recognition of his involvement in its development.

Bicknell

Bicknell's Thrush *Catharus bicknelli* **Ridgway**, 1882

Eugene Pintard Bicknell (1859–1925) was a founder of the AOU. He worked with a firm of bankers, but had a life-long interest in natural history. As a 21-year-old amateur ornithologist he climbed the summit of Slide Mountain in the Catskills, not far from New York City (1881). After he arrived near the top he heard Swainson's Thrushes *Catharus ustulatus* singing and calling. Then he heard an unfamiliar song that was more reminiscent of a Veery *C. fuscescens*. A thrush-sized bird flew across the opening, enabling him a clean shot to collect it. He believed it to be a Grey-cheeked Thrush *C. minimus*, but sent the specimen to Ridgway at the USNM, who classified it as a new subspecies of the Grey-cheeked Thrush, and gave it the common name. It was elevated to full species (1995) based on DNA studies and differences in size, song and zoogeography. He wrote *Review of the Summer Birds of Part of the Catskill Mountains* (1882).

Biddulph

Biddulph's Ground-jay *Podoces biddulphi* **Hume**, 1874
[Alt. Xinjiang Ground-jay]

Koklass Pheasant ssp. *Pucrasia macrolopha biddulphi*
GFL Marshall, 1879
Tawny Owl ssp. *Strix aluco biddulphi* **Scully**, 1881

Sir John Biddulph (1840–1921) was an army colonel and a member of the Yarkand Mission to areas around northern India (1873). He wrote *The Pirates of Malabar* and *An Englishwoman in India Two Hundred Years Ago* (1907).

Biedermann

Eurasian Nuthatch ssp. *Sitta europaea biedermanni*
Reichenow, 1907 NCR
[JS *Sitta europaea asiatica*]
Rock Partridge ssp. *Alectoris graeca biedermanni*
Reichenow, 1911 NCR
[JS *Alectoris graeca saxatilis*]
Grey-headed Woodpecker ssp. *Picus canus biedermanni*
Hesse, 1911 NCR
[JS *Picus canus jessoensis*]

Dr Richard Biedermann-Imhoof (1865–1926) was a Swiss naturalist and collector in the Altai (1908).

Biesenbach

Island Thrush ssp. *Turdus poliocephalus biesenbachi*
Stresemann, 1930 NCR
[NUI *Turdus poliocephalus fumidus*]

L. Biesenbach (fl.1930) was a collector in Java.

Biet

Biet's Laughingthrush *Garrulax bieti* **Oustalet**, 1897
[Alt. White-speckled Laughingthrush]

White-browed Fulvetta ssp. *Fulvetta vinipectus bieti*
Oustalet, 1892

Monsignor Felix Biet (1838–1904) was a French missionary on the Burmese/Chinese border. He was posted to Bhamo (1873) and became Bishop of Diana. Two mammals are named after him.

Bigilale

Streak-headed Munia ssp. *Lonchura tristissima bigilalae*
Restall, 1995
[Possibly a hybrid population]

Ilaiah Bigilale (fl.1995) is a Papuan vertebrate biologist and ornithologist, who is Curator of Birds, National Museum & Art Gallery, Port Moresby, Papua New Guinea.

Bilkevitch

Martin sp. *Clivicola bilkewitschi* **Zarudny**, 1910 NCR
[Alt. Grey-throated Martin; JS *Riparia chinensis*]

Streaked Laughingthrush ssp. *Trochalopteron lineatum*
bilkevitchi Zarudny, 1910
European Greenfinch ssp. *Carduelis chloris bilkevitchi*
Zarudny, 1911
Trumpeter Finch ssp *Bucanetes githagineus bilkewitchi*
Zarudny, 1918 NCR
[JS *Bucanetes githagineus crassirostris*]

S. I. Bilkevitch [sometimes written Bilkevich or Bilkewitch] (fl.1930) was a Russian zoologist who was Director of the Transcaspian Museum (1910–1937).

Billy Payn

Fulvous Babbler ssp. *Turdoides fulva billypayni*
Meinertzhagen, 1939 NCR
[NUI *Turdoides fulva fulva*]

William H. Payn (b.1913) was an ornithologist who wrote extensively about his home county, Suffolk. He co-wrote *Ornamental Waterfowl: a Guide to their Care and Breeding* (1957) and in his own right *The Birds of Suffolk* (1962) and *Oh Happy Countryman: a Suffolk Memoir* (1994). He was in Morocco (1939).

Binford

Scaled Antpitta ssp. *Grallaria guatimalensis binfordi*
Dickerman, 1990

Dr Laurence Charles Binford (1935–2009) was an American ornithologist and collector who was Curator of Birds and Mammals, California Academy of Sciences, San Francisco (1965–1989).

Bingham, C. T.

Bingham's Bulbul *Hypsipetes thompsoni* Bingham, 1900
[Alt. White-headed Bulbul; Syn. *Cerasophila thompsoni*]

Mountain Bulbul ssp. *Ixos mcclellandii binghami* **Hartert**,
1902 NCR
[JS *Ixos mcclellandii tickelli*]
Golden Babbler ssp. *Stachyridopsis chrysaea binghami*
Rippon, 1904

Colonel Charles Thomas Bingham (1848–1908) was an entomologist. He edited the first two volumes of *The Fauna of British India, including Ceylon and Burma* (1897 & 1903).

Bingham, H.

Giant Conebill ssp. *Oreomanes fraseri binghami* **Chapman**,
1919 NCR
[*O. fraseri* NRM, with clinal variation]

Professor Hiram Bingham III (1875–1956) was an American academic, explorer, politician and treasure hunter most famed for bringing the Inca city of Machu Picchu to the attention of the world. He was born in Hawaii, where his father and grandfather were missionaries. He graduated from Yale (1898), took a further degreee at the University of California, Berkeley (1900), and Harvard awarded his PhD (1905). He taught at Yale, Princeton and Harvard (1907). His interest in archaeology took him to Chile (1908) and Peru (1911, 1912 and 1915). He is cited as a possible basis for the character 'Indiana Jones', and his book *Lost City of the Incas* (1948) became a bestseller. He was Lieutenant Governor of Connecticut (1922–1924) and served two terms in the US Senate (1926–1932 and 1932–1933). The etymology says that he was the: '... organizer and leader of the Yale University-National Geographic Society Expedition, to whose vision, energy, and executive ability we owe ... our knowledge of the Inca city of Machu Picchu'. An amphibian is named after him.

Biro

Orange-bellied Fruit Dove ssp. *Ptilinopus iozonus biroi*
Madarász, 1897 NCR
[JS *Ptilinopus iozonus humeralis*]

Dr Lajos Biró (1856–1931), a Hungarian zoologist, ornithologist, entomologist, collector, ethnographer and photographer, was in German New Guinea (1896–1901), Singapore (1898), North Africa (1901–1903), Crete (1906) and Turkey (1925). He worked at the Budapest Natural History Museum (1903–1931). His honorary doctorate was awarded (1926) by University of Sciences, Szeged, Hungary. Two amphibians are named after him.

Birula

Willow Grouse ssp. *Lagopus lagopus birulai* **Serebrovsky**,
1926 NCR
[NUI *Lagopus lagopus koreni*]
Birula's Gull *Larus vegae birulai* **Pleske**, 1928 NCR
[Alt. Vega Gull; NUI *Larus vegae vegae*]

Dr Aleksandr Andreevich Bialynicky-Birula (1864–1938) was Director of the Zoological Museum in Leningrad (St Petersburg). A reptile is named after him.

Bishop, C. R.

Bishop's O'o *Moho bishopi* **Rothschild**, 1893 EXTINCT
[Alt. Molokai O'o]

Charles Reed Bishop (1822–1915) was an American businessman who married Princess Bernice Pauahi of Hawaii. He founded the Bishop Museum on Oahu, the largest museum dedicated to the study of the Pacific, as a tribute to his wife.

Bishop, N. H.

Whistling Warbler *Catharopeza bishopi* Lawrence, 1878

Nathaniel Holmes Bishop (1837–1902) was an American explorer, adventurer and businessman. His first claim to fame was that he hiked across South America at the age of 17! The adventure tale he wrote was published as *The Pampas and Andes: a Thousand Miles' Walk across South America* (1869). He wrote two further books, *Voyage of the Paper Canoe – a Geographical Journey of 2500 miles, from Quebec to the Gulf of Mexico, during the years 1874–1875* (1878) and *Four Months in a Sneak-Box – a Boat Voyage of 2600 Miles down the Ohio and Mississippi Rivers, and along the Gulf of Mexico* (1879). He later developed cranberry plantations in Stafford, New Jersey, becoming so successful that he ended up owning some 60 properties, as well as homes in California and Florida as well as Dover, New Jersey. In his will he funded a public library for the Township of Dover, which was completed some years later (1941). He was clearly interested in ornithology as he wrote, in his last book: '... I peered through the branches of the forest to catch a glimpse of what I had searched for through many hundred miles of wilderness since my boyhood ... the Carolina Parrot.' Lawrence's etymology notes: 'Mr Ober requested that I would bestow the name of our friend Mr Nathaniel H. Bishop on some West India bird of his procuring, if the opportunity offered ... Mr Bishop has had much experience in West Indian exploration, and it was in a great measure due to his influence and representations, that Mr Ober's visit to these islands was determined upon; he also contributed substantial aid, by the donation of instruments, and in other ways.'

Bismarck

Bismarck Hanging Parrot *Loriculus tener* **P. L. Sclater**, 1877
[Alt. Green-fronted Hanging Parrot]

Not named directly after the famous German Chancellor, but after the Bismarck Archipelago off the northeast coast of New Guinea, where the parrot is found. Four mammals are also named after this archipelago.

Biswas

Grey-sided Laughingthrush ssp. *Garrulax caerulatus biswasi* **Koelz**, 1953 NCR
[JS *Garrulax caerulatus livingstoni*]

Dr Biswamoy Biswas (1923–1994) was an Indian ornithologist. Sunderlal Hora, then Director of the Zoological Survey of India (ZSI), awarded him a three-year fellowship (1947) which enabled him to study at the BMNH, the Berlin Zoological Museum under Stresemann, and the AMNH under Ernst Mayr. He was awarded his PhD (1952) at the University of Calcutta. He was part of the famous *Daily Mail* expedition (1954) sent to look for the Yeti around Mount Everest. He was elected Corresponding Fellow of the AOU (1963) and later headed the Bird and Mammal Section of the Zoological Survey of India, as Joint Director until his retirement (1981), remaining Emeritus Scientist (1981–1986). A mammal is named after him.

Blaauw

Hooded Parrot ssp. *Psephotus chrysopterygius blaauwi* **van Oort**, 1910 NCR
[JS *Psephotus dissimilis*]
Olive-backed Oriole ssp. *Oriolus sagittatus blaauwi* **Mathews**, 1912 NCR
[JS *Oriolus sagittatus affinis*]
Partridge Pigeon ssp. *Geophaps smithii blaauwi* Mathews, 1912

Frans Ernst Blaauw (1860–1936) was a Dutch ornithologist and aviculturist who turned his estate into a private zoo. He wrote: *A Monograph of the Cranes* (1897).

Blackburn

Blackburnian Warbler *Dendroica fusca* P. L. S. Müller, 1776

Mrs Anna Blackburn(e) (1726–1793) was a botanist who owned a museum at Fairfield in Lancashire, England. She was a correspondent of Linnaeus and a patron of ornithology in London. Her brother, Ashton Blackburn, collected a specimen of the warbler. Her cousin was Sir Ashton Lever – a collector so enthusiastic that price was no object and he bankrupted himself. Anna never married, but preferred to be called Mrs Blackburn as it gave her more standing and authority.

Blainville

Lowland Peltops *Peltops blainvillii* **Lesson** & **Garnot**, 1827
[Alt. Clicking Shieldbill]

Professor Henri Marie Ducrotay de Blainville (1777–1850) was a French zoologist and anatomist. He was one of Cuvier's (q.v.) bitterest rivals and his successor to the chair of Comparative Anatomy at the MNHN, and in the Collège de France. He applied his principles of anatomy to bird classification and had much influence in establishing skeletal evolution as one determinant of classification. He wrote *Cours de Physiologie Générale et Comparée* (1829). Five mammals and a reptile are named after him.

Blake, E. R.

Blake's Spinetail *Synallaxis courseni* E. R. Blake, 1971
[Alt. Apurímac/Coursen's Spinetail]
Blake's Antpitta *Grallaria blakei* **Graves**, 1987
[Syn. Chestnut Antpitta]

Grey-lined Hawk ssp. *Buteo nitidus blakei* **Hellmayr** & **Conover**, 1949

Emmet Reid Blake (1908–1997) was an American ornithologist who was Emeritus Curator of Birds at the Field Museum, Chicago. He was actively collecting in Mexico (1940s–1950s). He wrote *Birds of Mexico* (1953) and *Manual of Neotropical Birds* (1977). A reptile is named after him.

Blake, H. A.

West Indian Woodpecker ssp. *Melanerpes superciliaris blakei* **Ridgway**, 1886

Sir Henry Arthur Blake (1840–1918) was a British diplomat and colonial administrator, naturalist and botanist. He came from humble beginnings as a draper's assistant at a haberdashery shop, but joined the Irish Constabulary (1859), becoming an inspector and Resident Magistrate of Duff, and was promoted to be Special Resident Magistrate at Tuam (1882). He joined the colonial service and was Governor of the Bahamas (1884–1887), Newfoundland (1887–1889), Jamaica (1889–1898), Hong Kong (1898–1903) and Ceylon (Sri Lanka) (1903–1907). The town of Blaketown in Canada is named after him.

Blakiston

Blakiston's Fish Owl *Bubo blakistoni* **Seebohm**, 1884
[Syn. *Ketupa blakistoni*]
Skylark sp. *Alauda blakistoni* **Stejneger**, 1884 NCR
[Alt. Eurasian Skylark; JS *Alauda arvensis pekinensis*]

Water Pipit ssp. *Anthus spinoletta blakistoni* **Swinhoe**, 1863
Barred Buttonquail ssp. *Turnix suscitator blakistoni* Swinhoe, 1871

Thomas Wright Blakiston (1832–1891) was born in Lymington, Hampshire, England. He was commissioned (1851) and later fought in the Crimean War. He explored western Canada with the Palliser Expedition (1857–1859) and three years later was in China, where he organised an expedition up the Yangtze River, going 900 miles further than any Westerner before him. Most of the next 23 years were spent on business in Japan, but he compiled a catalogue of the birds of northern Japan and is today renowned as one of that country's outstanding naturalists. Blakiston lavished money on his hobby, the study of birds, and he became world-famous in ornithology. He preserved many bird specimens, 1,331 of which are now in the museum attached to the Agricultural Department of Hokkaido University. He moved to the USA (1886) but while visiting San Diego he caught pneumonia and died. He published various essays on birds in Japan and some accounts of his explorations in the Tohoku district, besides the accounts of his expedition along the Yangtze River and his essays on birds in Canada. Blakiston was the first person to establish that animals in Hokkaido, Japan's northern island, have northern Asian affinities and differ in appearance from those in Honshu. As a result of Blakiston's work, the Tsugaru Strait, which divides Hokkaido from Honshu, became known as an important zoogeographical boundary, the 'Blakiston Line'.

Blanc

Green-tailed Sunbird ssp. *Aethopyga nipalensis blanci*
Delacour & Greenway, 1939

François Edmond-Blanc (1908–1996) was a professional hunter and naturalist. He was a leading light in the foundation of the International Professional Hunters' Association (1969). He made a number of zoological collecting expeditions to Indochina (1930s), including one (1936) during which he found the bones and skin of a wild ox called the Kouprey *Bos sauveli*, although he does not normally get any credit for its discovery, nor did he describe it for science. He was employed (1955) on a mission to Cameroon to obtain a specimen of a Pygmy Elephant for the University of Copenhagen. (The existence of Pygmy Elephants is still a debated subject today.) He published a book on hunting, *Histoire Mondiale de la Chasse* (1970).

Blanca

Antioquia Brush Finch *Atlapetes blancae* Donegan, 2007

Blanca Huertas is a Colombian entomologist who is the Curator of Lepidoptera at the BMNH, London (2005). Universidad Pedagógica Nacional, Bogotá, Colombia, awarded her bachelor's degree (2000) and University College London her master's (2007). She is married to the describer, ornithologist Thomas Donegan.

Blancaneaux

Slaty-breasted Tinamou ssp. *Crypturellus boucardi blancaneauxi* Griscom, 1935 NCR
[JS *Crypturellus boucardi boucardi*]

François Blancaneaux (1851–1923) was a French businessman, mahogany contractor, amateur naturalist and collector in British Honduras (Belize) (1878–1923). He conducted a biological survey of the country while looking for a reputed cave-dwelling yeti-like creature.

Blanchard, D.

Rufous-naped Tit ssp. *Periparus rufonuchalis blanchardi* Meinertzhagen, 1938 NCR; NRM

Dean Hobbs Blanchard (1912–1999) came from a wealthy Californian family. He travelled and collected in Ethiopia (1933–1934). The Santa Barbara Museum of Natural History houses his collection. He wrote *Ethiopia: its Culture and its Birds* (1969). He collected the holotype.

Blanchard, P.

Rufous-throated Fulvetta ssp. *Alcippe rufogularis blanchardi* Delacour & Jabouille, 1928 NCR
[JS *Alcippe rufogularis stevensi*]

Paul Marie Alexis Joseph Blanchard de la Brosse (fl.1929) was a French Administrator of the Leased Territory of Kwangchowan, China (1922, 1925–1926) and Governor of Cochinchina (Vietnam) (1926–1929).

Blanchet

Peregrine Falcon ssp. *Falco peregrinus blancheti* Lavauden, 1922 NCR
[JS *Falco peregrinus calidus*]

Alfred Blanchet (1872–1944) was a French lawyer, ornithologist and collector in Tunisia (1914–1917). He wrote (1943) *Les Oiseaux de Tunisie* (1955).

Blanchot

Grey-headed Bush-shrike *Malaconotus blanchoti* J. F. Stephens, 1826

General Émile François Michael Blanchot de Verly (1735–1807) was a soldier who was the French Governor of Senegal (1787–1801 and 1802–1807).

Blanco

Lesser Antillean Pewee ssp. *Contopus latirostris blancoi* Cabanis, 1875

T. Blanco (fl.1874) was an apothecary in San Juan, Puerto Rico.

Blancou

Helmeted Guineafowl ssp. *Numida meleagris blancoui* Grote, 1936 NCR
[JS *Numida meleagris galeatus*]

Lucien Jacques Laurent Blancou (1903–1983) was a French ornithologist who was Inspector of Game, French Equatorial Africa (1949–1953). He is said to have invented (1959) the term 'cryptozoology' when referring to the work of the Belgian-France zoologist Bernard Heuvelmans.

Bland

Black-throated Barbet ssp. *Tricholaema melanocephala blandi* E. L. Phillips, 1897

Ivers Bland (d.1903) was a member of the BOU and was in Somalia (1896–1897) on an expedition with the describer.

Blandin

Bengal Florican ssp. *Houbaropsis bengalensis blandini* Delacour, 1928

J. Blandin (DNF) was the French Resident of Svay Rieng Province, Cambodia, Indochina, where Delacour met him (1928).

Blanding

Blanding's Finch *Pipilo chlorurus* Audubon, 1839
[Alt. Green-tailed Towhee; *Fringilla blandingiana* (Gambel, 1843) is a JS]

Dr William Blanding (1772–1857) was an American physician and chemist who had a collection of birds. He was an active member of the Philadelphia Academy of Natural Sciences and once presented a paper there entitled 'Fecundity of the Bass'. He had many interests, including archaeology and native culture, and it was he who found a site, known as Mulberry, which was once a capital city of Cofitachequi, a chiefdom on the Wateree River in central South Carolina (c.1100–1700). He was also a numismatist who manufactured 'ancient' coins as a hobby. He was a friend and patron of Gambel (q.v.), who named the bird after him. However, Audubon's earlier scientific name has precedence but Blanding is still commemorated in one of the bird's common names. He was a keen amateur herpetologist and two reptiles are named after him.

Blanford

Prinia genus *Blanfordius* **Hume**, 1873 NCR
[Now in *Prinia*]

Blanford's Olive Bulbul *Pycnonotus blanfordi* **Jerdon**, 1862
[Alt. Streak-eared Bulbul]
Blanford's Rosefinch *Carpodacus rubescens* Blanford, 1871
[Alt. Crimson Rosefinch]
Blanford's Bush Warbler *Cettia pallidipes* Blanford, 1872
[Alt. Pale-footed Bush Warbler]
Blanford's Snowfinch *Pyrgilauda blanfordi* Hume, 1876
[Alt. Plain-backed Snowfinch, Blanford's Ground Sparrow]
Blanford's Lark *Calandrella blanfordi* **G. E. Shelley**, 1902
Himalayan Owl *Syrnium blanfordi* **Zarudnyi**, 1911 NCR
[JS *Strix nivicolum nivicolum*]

Yellow-legged Buttonquail ssp. *Turnix tanki blanfordii* **Blyth**, 1863
Plain Prinia ssp. *Prinia inornata blanfordi* **Walden**, 1875
Blanford's Warbler *Sylvia leucomelaena blanfordi* **Seebohm**, 1878
[Alt. Arabian Warbler ssp.]
Great Tit ssp. *Parus major blanfordi* Prazak, 1894
Blanford's Black Saw-wing *Psalidoprocne pristoptera blanfordi* **Blundell** & **Lovat**, 1899

William Thomas Blanford (1832–1905) was an English geologist and zoologist. He studied at the Royal School of Mines (1852–1854) and at the mining school in Freiberg, Saxony, before obtaining a post in the Indian Geological Survey (1854), for which he investigated coalmines at Talchir (1854–1857). He then undertook a geological survey of Burma (1860) and was appointed Deputy Superintendent. He surveyed in Bombay (1862–1866), and later (1871–1872) was appointed to the Persian Boundary Commission. Blanford wrote works on the geology of Abyssinia (Ethiopia) (1870), and India (1879), before settling in London (1881). He edited works for the government on Indian fauna, contributing two volumes on mammals (1888–1891). He was elected to the Royal Society and Royal Geographical Society (1874) and was president of the latter (1888–1890). He wrote: *The Scientific Results of the Second Yarkand Mission: Mammalia* (1879). Seven mammals, three amphibians and 17 reptiles are named after him. His name was often misspelt Blandford.

Blasius

Drab Myzomela *Myzomela blasii* **Salvadori**, 1882
Blasius's Wren-babbler *Ptilocichla mindanensis* Blasius, 1890
[Alt. Striated Wren-babbler, Streaked Ground Babbler]

Scaly-breasted Munia ssp. *Lonchura punctulata blasii* **Stresemann**, 1912

Dr Wilhelm August Heinrich Blasius (1845–1912) was a German ornithologist, as were both his brother Rudolf Blasius (1842–1907) and his father Johann Heinrich Blasius (1809–1870). He was the Director of the Braunschweig Museum (1870), which boasted two specimens of the extinct Great Auk *Alca impennis*. He became Professor of Zoology and Botany in the Ducal Technical University in Braunschweig (1871). He is also famed for having used the *Braunschweigische Anzeigen*, a daily newspaper, to publish descriptions quickly: thus acquiring priority for names which he suggested, over those proposed in the more ponderous and mainly English scientific journals. Apparently, this practice was one motivation for the founding of the BOC and its *Bulletin*.

Blayney

Bush Pipit ssp *Anthus caffer blayneyi* **Someren**, 1919

(See **Percival**)

Blaze

European Goldfinch ssp. *Carduelis carduelis blazei* Ghidini, 1903 NCR
[JS *Carduelis carduelis carduelis*]

Elzéar Jean Louis Joseph Blaze (1786–1848) was a captain in Napoleon's Grande Armée (1807–1814) and published: *Life in Napoleon's Army*, *The Memoirs of Captaine Elzéar Blaze* (1837). He stayed in the French Army until retiring (1830). He summed up his feelings in *La Vie Militaire* with the words 'In the career of glory one gains many things; gout and medals, a pension and rheumatism. … And also frozen feet, an arm or leg the less, a bullet lodged between two bones which the surgeon cannot extract … all of these fatigues experienced in your youth, you pay for when you grow old. Because one has suffered in years gone by, it is necessary to suffer more, which does not seem exactly fair.' He was a great huntsman and wrote a number of books on various aspects of hunting including *Le Chasseur aux Filets ou La Chasse des Dames* (1839), which Ghidini described as 'a captivating volume'. It includes a long, enthusiastic section on Le Chardonneret (Goldfinch).

Bleda

Bristlebill genus *Bleda* **Bonaparte**, 1857

Bleda (c.390–445) was the brother of Attila the Hun.

Blewitt

Blewitt's Owl *Athene blewitti* Hume, 1873
[Alt. Forest Owlet; Syn. *Heteroglaux blewitti*]

Painted Bush Quail ssp. *Perdicula erythrorhyncha blewitti* Hume, 1874

Francis Robert Blewitt (1815–1881) and William Turnbull Blewitt (1816–1889) (and their younger brother Henry) were born in India, where their father, also Francis Robert (1787–1836), was serving (1806) in the 8th Light Dragoons Regiment of British army. Robert (as he was known, presumably to distinguish him from his father) was also a serving officer. William worked as a customs official in the Civil Service in the Punjab, where he died of heart disease. Robert and William were avid amateur ornithologists and oologists and collected specimens. Robert sent both the species above to Hume (q.v.), who named them after his friend whom he described as dedicated and zealous in his ornithological pursuits. Hume clearly honours Robert although several authorities ascribe the owl to William. As William collected and there are cards marked with his initials on specimens, it has been assumed by some that it was he who was commemorated. It is possible that *his* name became associated with the owl in its common name when the brothers' and fellow collectors' names were conflated, particularly as he survived his brother by eight years.

Bley

Bronzewing sp. *Reinwardtoenas bleyi* W. Meyer, 1909
[Alt. New Britain Bronzewing; JS *Henicophaps foersteri*]

Father Bernhard Bley (1879–1962) was a German Jesuit missionary and anthropologist in the Bismarck Archipelago, New Guinea.

Blick

Variable Sunbird ssp. *Cinnyris venustus blicki* **Mearns**, 1915 NCR
[NUI *Cinnyris venustus albiventris*]

Dr John Charles Blick (c.1896–1971) was an American physician. He accompanied his friend Henry Childs Frick (q.v.) on an expedition to Abyssinia (Ethiopia) and Lake Rudolf (1911–1912) that Frick had organised with Mearns (q.v.). The express intension of this expedition was to collect birds and small mammals, and during it Blick collected the type specimen (1912). A mammal is named after him.

Bligh

Bligh's Whistling Thrush *Myophonus blighi* **Holdsworth**, 1872
[Alt. Sri Lanka Whistling Thrush]

Spot-bellied Eagle Owl ssp. *Bubo nipalensis blighi* **Legge**, 1878

Samuel Bligh (DNF) was a British coffee planter in Ceylon (1872–1887) who wrote notes on every bird species there. He discovered the whistling thrush (1868) – the last endemic bird species found on the island until Deepal Warakagoda discovered the Serendib Scops Owl *Otus thilohoffmanni* (2001).

Bliss

Bicoloured Mouse-warbler ssp. *Crateroscelis nigrorufa blissi* **Stresemann** & Paludan, 1934

Robert Woods Bliss (1875–1962) was an American diplomat (1903–1933) and philanthropist, who sponsored the AMNH

New Guinea Expedition. He was one of the founders of the Dumbarton Oaks Research Library and Collection, Washington DC.

Blissett

Blissett's Wattle-eye *Platysteira blissetti* **Sharpe**, 1872
[Alt. Red-cheeked Wattle-eye]

Henry Frederick Blissett (1847–1916) was a British Colonial Administrator in West Africa (1869–1885) and amateur naturalist and collector.

Blood

Blood's Bird-of-Paradise *Paradisaea bloodi* **Iredale**, 1948
[Hybrid of *Paradisaea raggiana* x *P. rudolphi*]

Brown Sicklebill ssp. *Epimachus meyeri bloodi* **Mayr** & **Gilliard**, 1951
Tit Berrypecker ssp. *Oreocharis arfaki bloodi* **Gyldenstolpe**, 1955 NCR; NRM

Major Neptune Newcombe Beresford Lloyd Blood (b.1907) was a police officer in New Guinea, where he spent 40 years and was (1958) the District Commissioner for Rabaul, New Britain. He also had a hand in bringing a rare orchid to notice: a New Guinea variety of orchid was initially described (1932), but it was lost to cultivation when WW2 raged across its habitat. The then Captain Blood, at that time in the Australian Army, took a plant with him whilst escaping from the invading Japanese. After the war was over, he left the police and went to work for Hallstrom (q.v.) at his experimental sheep station in the Papua New Guinea Highlands.

Bloxam

Bloxam's Plantcutter *Phytotoma bloxami* **Jardine** & **Selby**, 1827 NCR
[Alt. Rufous-tailed Plantcutter; JS *Phytotoma rara*]
Tit Tyrant sp. *Sylvia bloxami* **J. E. Gray**, 1831 NCR
[Alt. Tufted Tit Tyrant; JS *Anairetes parulus*]

Greater Patagonian Conure *Cyanoliseus patagonus bloxami* **Olson**, 1995
[Alt. Burrowing Parakeet ssp.]

Reverend Andrew Bloxam (1801–1878) was an English botanist and naturalist on HMS *Blonde* in the Pacific (1824–1826). After returning he was ordained in the Church of England and lived the rest of his life as a parish priest, with a strong interest in mycology and botany. Olson later published his diaries and commentaries (1986–1996).

Blumenbach

Blumenbach's Curassow *Crax blumenbachii* **Spix**, 1825
[Alt. Red-billed Curassow]

Johann Friedrich Blumenbach (1752–1840) was a German anatomist, physician, anthropologist, naturalist, physiologist, historian and bibliographer who many regarded as Linnaeus's (q.v. under Linné) natural successor. He was a Professor of Medicine in the University of Göttingen (1776–1836). He is also considered to be the founder of modern anthropology, and the system for classification of human races he devised (1775) is still of relevance today. He coined the word Caucasian as a description of white people and was the first to use the term 'race' to distinguish between different morphs of humanity. His many publications included *On the Natural Varieties of Mankind* (1775), and *Handbuch der Naturgeschichte* (1779). He also taught both Wied (q.v.) and Humboldt (q.v.). Blumenbach never travelled outside Europe, but others sent him specimens and objects from all over the world.

Blundell

Red-headed Weaver *Anaplectes blundelli* **Ogilvie-Grant**, 1900 NCR
[JS *Anaplectes rubriceps leuconotos*]

Herbert Joseph Weld-Blundell (1852–1935) was in Ethiopia (1898). He was wealthy and lived in a castle in Dorset (1905–1935). He dropped 'Blundell' from his name and became plain Mr Weld. He wrote a number of papers among which are 'A journey through Abyssinia to the Nile' (1900) and 'Exploration in the Abai Basin' (1906). He also had an interest in archaeology and made a journey to Cyrenaica (Libya) (1894) to assess its suitability for excavation.

Bluntschli

Bluntschli's Vanga *Hypositta perdita* D. S. Peters, 1996
[NUI *Oxylabes madagascariensis*]

Hans Bluntschli (1877–1962) was a Swiss anatomist. He studied medicine in Zurich, Munich and Leipzig and was awarded his medical degree by Heidelberg University (1903). After working at the Anatomical Institute in Heidelberg he moved to Zurich to study Old World monkeys (1906). He taught at a number of universities as Professor of Anatomy before being offered the chair of anatomy and embryology in Bern (1933), a post he occupied until retirement (1942). He undertook a number of expeditions including to the Amazon (1912–1913) and to Madagascar (1931–1932); the bird species is known only from two juvenile specimens collected there (1931) by Bluntschli, and now known from genetic analysis to involve two juvenile White-throated Oxylabes *Oxylabes madagascariensis*.

Blyth

Woodpecker genus *Blythipicus* **Bonaparte**, 1854

Blyth's Hornbill *Rhyticeros plicatus* **J. Forster**, 1781
[Alt. Papuan Hornbill; Syn. *Aceros plicatus*]
Blyth's Baza *Aviceda jerdoni* Blyth, 1842
[Alt. Jerdon's Baza, Brown Baza]
Blyth's Jungle Babbler *Trichastoma rostratum* Blyth, 1842
[Alt. White-chested Babbler]
Blyth's Leaf Warbler *Phylloscopus reguloides* Blyth, 1842
[Alt. Blyth's Crowned Leaf Warbler]
Blyth's Cuckoo *Cuculus saturatus* Blyth, 1843
[Alt. Himalayan Cuckoo]
Blyth's Hawk Eagle *Nisaetus alboniger* Blyth, 1845
Blyth's Long-tailed Tit *Aegithalos iouschistos* Blyth, 1845
[Alt. Rufous-fronted Tit]

Blyth's Myna *Sturnia blythii* **Jerdon**, 1845
[Alt. Malabar Starling]
Blyth's Olive Bulbul *Iole virescens* Blyth, 1845
[Alt. Olive Bulbul]
Blyth's Parakeet *Psittacula caniceps* Blyth, 1846
[Alt. Nicobar Parakeet]
Blyth's Frogmouth *Batrachostomus affinis* Blyth, 1847
Blyth's Reed Warbler *Acrocephalus dumetorum* Blyth, 1849
Blyth's Rosefinch *Carpodacus grandis* Blyth, 1849
Blyth's Shrike Babbler *Pteruthius aeralatus* Blyth, 1855
Somali Starling *Onychognathus blythii* **Hartlaub**, 1859
Blyth's Tragopan *Tragopan blythii* Jerdon, 1870
Blyth's Pipit *Anthus godlewskii* **Taczanowski**, 1876
[Alt. Godlewski's Pipit]
Blyth's Kingfisher *Alcedo hercules* **Laubmann**, 1917
Blyth's Yuhina *Yuhina bakeri* **Rothschild**, 1926
[Alt. White-naped/Baker's/Chestnut-headed Yuhina]

Plain Prinia ssp. *Prinia inornata blythi* Bonaparte, 1850
Blyth's Parrotbill *Suthora nipalensis poliotis* Blyth, 1851
[Alt. Black-throated Parrotbill ssp.]
Blyth's Fig Parrot *Psittaculirostris desmarestii blythii* **Wallace**, 1864
[Alt. Large Fig Parrot ssp.]
Black-faced Laughingthrush ssp. *Trochalopteron affine blythii* **J. Verreaux**, 1870
Mangrove Blue Flycatcher ssp. *Cyornis rufigastra blythi* Giebel, 1875
White-browed Rosefinch ssp. *Carpodacus thura blythi* **Biddulph**, 1882

Edward Blyth (1810–1873) was an English zoologist and author. He was Curator of the Museum of the Asiatic Society of Bengal (1842–1864) and wrote its catalogue. He wrote a series of monographs on cranes that were gathered together and published as *The Natural History of Cranes* (1881). Hume said of him: 'Neither neglect nor harshness could drive, nor wealth nor worldly advantages tempt him, from what he deemed the nobler path. [He was] ill paid and subjected … to ceaseless humiliations'. In a similar tribute Arthur Grote wrote: 'Had he been a less imaginative and more practical man, he must have been a prosperous one … All that he knew was at the service of everybody. No one asking him for information asked in vain'. Five mammals, two reptiles and an amphibian are named after him.

Bobrinskoy

Wire-tailed Swallow ssp. *Hirundo smithii bobrinskoii* **Stachanov**, 1930 NCR
[JS *Hirundo smithii filifera*]

Professor Count Nikolay Alekseyevich Bobrinskoy (or Bobrinsky) (1890–1964) was a Russian zoologist. He worked at the Zoological Museum in Moscow (1930) and, despite being an aristocrat, was regarded as one of the most prominent Soviet zoologists. (The present Count, Nicholas, is a geographer and lives in Moscow.) He co-wrote: *Mammals of the USSR* (1944) and *The Animal World and Nature of the USSR* and *Animal Geography* (1951). Two mammals are named after him.

Bocage

Bocage's Akalat *Sheppardia bocagei* **Finsch** & **Hartlaub**, 1870
[Alt. Bocage's Ground Robin]
Bocage's Sunbird *Nectarinia bocagii* **G. E. Shelley**, 1879
Bocage's Weaver *Ploceus temporalis* Bocage, 1880
Bocage's Longbill *Amaurocichla bocagei* **Sharpe**, 1892
[Alt. São Tomé Short-tail]
Bocage's Bush-shrike *Chlorophoneus bocagei* **Reichenow**, 1894
[Alt. Grey-green Bush-shrike]
Bocage's Swee Waxbill *Coccopygia bocagei* G. E. Shelley, 1903
[Alt. Angolan Waxbill; Syn. *Estrilda* (*melanotis*) *bocagei*]
Dwarf Olive Ibis *Bostrychia bocagei* **Chapin**, 1923

Fan-tailed Widowbird ssp. *Euplectes axillaris bocagei* Sharpe, 1871
African Thrush ssp. *Turdus pelios bocagei* **Cabanis**, 1882
African Pipit ssp. *Anthus cinnamomeus bocagii* **Nicholson**, 1884
Black-collared Barbet ssp. *Lybius torquatus bocagei* **Sousa**, 1886
Pale-breasted Illadopsis ssp. *Illadopsis rufipennis bocagei* **Salvadori**, 1903

José Vicente Barbosa du Bocage (1823–1907) was Director of the National Zoological Museum of Lisbon, Portugal, which is now named in his honour. He became known as the father of Angolan ornithology and wrote *Ornithologie d'Angola* (1877–1881). He also collected sponges and other specimens. Six mammals, ten reptiles and four amphibians are named after him. (See also **Barbosa**)

Bock

Dark Hawk Cuckoo *Hierococcyx bocki* **R. G. W. Ramsay**, 1886

Carl Ernst Bock (1809–1874) was a German physician and anatomist. He graduated at the University of Leipzig (1831). During the Polish uprising he served as a physician in both the Polish and Russian armies. He became a private lecturer (1832) and later (1837) was employed in charge of autopsies at Leipzig's hospital. He went on to be (1839) Professor of Pathological Anatomy and then the head of the University's Clinical Department (1850). He wrote a number of books and essays on health issues. Arthur Hay, 9th Marquess of Tweeddale (q.v.), employed him to travel to the Malay Archipelago and collect specimens. Hay described around 40 species that Bock collected. R. G. W. Ramsay (q.v.), the describer, was Hay's nephew and biographer.

Bodaly

Helmeted Guineafowl ssp. *Numida meleagris bodalyae* **Boulton**, 1934 NCR
[JS *Numida meleagris marungensis*]

Mrs Jean Bodaly (fl.1934) and her husband, John, were Canadian Protestant missionaries to Angola. She made collections of birds at Chitau. Boulton's etymology mentions

that she had been '… hostess to many naturalists who have visited Angola.'

Bodin/Bodinus

Bodin's Amazon *Amazona festiva bodini* **Finsch**, 1873
[Alt. Festive Amazon/Parrot ssp.]

Dr Karl-August Heinrich Bodinus (1814–1884) was a German physician and zoologist. He was the Director of the Cologne Zoo, and later (1869–1884) of the Berlin Zoological Garden.

Boehm, E.

Yellow-collared Chlorophonia ssp. *Chlorophonia flavirostris boehmi* Conway, 1962 NCR; NRM

Edward Marshall Boehm (1913–1969) was an American aviculturist and self-taught sculptor famed for his porcelain figures of birds.

Boehm, R.

(See **Böhm, R.**)

Boehme, L. B.

Hawfinch ssp. *Coccothraustes coccothraustes boehmei* **Buturlin**, 1929 NCR
[JS *Coccothraustes coccothraustes nigricans*]
Tawny Pipit ssp. *Anthus campestris boehmii* **Portenko**, 1960
[SII *Anthus campestris campestris*]

Leo Borisovitch Boehme (1895–1954) was keen to be a zoologist and, having graduated in law at Moscow State University, took a job as assistant professor (later professor) in the zoology department of an agricultural college in Vladikavkaz (1924–1938). He was arrested by the KGB and exiled to Kazakhstan, where he became an instructor at the Karaganda experimental agriculture establishment (1938–1946). He returned to Vladikavkaz but was arrested again (1948) and exiled to Novozibkov, where he died. He collected in the Caucasus (1922–1933). Both his son and granddaughter became a professors of zoology.

Boehmer

White-quilled Bustard ssp. *Afrotis afraoides boehmeri* **Hoesch** & **Niethammer**, 1940 NCR
[JS *Afrotis afraoides afraoides*]

Walter Böhmer (DNF) was a German farmer in South West Africa (Namibia).

Boetticher

Bishop genus *Boetticherella* **Wolters**, 1943 NCR
[Now in *Euplectes*]

Eurasian Tree Sparrow ssp. *Passer montanus boetticheri* **Stachanov**, 1933 NCR
[JS *Passer montanus montanus*]
Red-winged Starling ssp. *Lamprotornis morio boetticheri* Wolters, 1952 NCR
[JS *Onychognathus morio rueppellii*]

Hans von Boetticher (1886–1958) was a German ornithologist and entomologist who worked at the natural history museum in Coburg. He wrote books on bird familes, such as *Albatrosse und Andere Sturmvögel* (1955).

Bogdanov

Common Pheasant ssp. *Phasianus colchicus bogdanowi* **Buturlin**, 1904 NCR
[JS *Phasianus colchicus principalis*]
Black Francolin ssp. *Francolinus francolinus bogdanovi* **Zarudny**, 1906

Professor Modest Nikolaevich Bogdanov (1841–1888), a Russian zoologist and ornithologist, was supervisor of the ornithological collection at St Petersburg Museum (1879–1885). He carried out a meticulous inventory, identifying all the species, and produced a monograph, *Shrikes of Russian Fauna and Allies* (1881). He also prepared the first issue of the *Checklist of the Birds of the Russian Empire* (1884). A mammal is named after him.

Bogolubov

Cream-coloured Courser ssp. *Cursorius cursor bogolubovi* **Zarudny**, 1885

Colonel I. A. Bogolyubov (DNF) was a Russian explorer in Transcaspia (1884).

Böhm

Banded Parisoma *Sylvia boehmi* **Reichenow**, 1882
Böhm's Bee-eater *Merops boehmi* Reichenow, 1882
Böhm's Spinetail *Neafrapus boehmi* **Schalow**, 1882
[Alt. Bat-like Spinetail]
Böhm's Flycatcher *Muscicapa boehmi* Reichenow, 1884
Böhm's Flufftail *Sarothrura boehmi* Reichenow, 1900
[Alt. Streaky-breasted Flufftail]

Böhm's Barbet *Trachyphonus darnaudii boehmi* **G. A. Fischer** & Reichenow, 1884
[Alt. d'Arnaud's Barbet ssp.]
Red-necked Spurfowl ssp. *Pternistis afer boehmi* Reichenow, 1885 NCR
[JS *Pternistis afer cranchii*]
White-headed Buffalo Weaver ssp. *Dinemellia dinemelli boehmi* Reichenow, 1885
Grey-backed Fiscal ssp. *Lanius excubitoroides boehmi* Reichenow, 1902

Dr Richard Böhm (1854–1884) was a German traveller and zoologist who worked in Tanzania until his premature death, reportedly murdered by 'the tribes westward of Lake Tanganyika', but actually died of a fever (see also entry for Reichard). His book *Von Sansibar zum Tanganjika* (1888) was published posthumously. A mammal is named after him.

Bohndorff

African Wood Owl ssp. *Strix woodfordii bohndorffi* **Sharpe**, 1884 NCR
[JS *Strix woodfordii nuchalis*]
Capuchin Babbler ssp. *Phyllanthus atripennis bohndorffi* Sharpe, 1884

Spotted-backed Weaver ssp. *Ploceus cucullatus bohndorffi* **Reichenow**, 1887
Green-headed Sunbird ssp. *Cyanomitra verticalis bohndorffi* Reichenow, 1887
Plain-backed Pipit ssp. *Anthus leucophrys bohndorffi* **Neumann**, 1906

Friedrich Bohndorff (b.1848) was a German naturalist, explorer and collector in tropical Africa (1876–1884). He originally trained to be a goldsmith, but went to Cairo (1871) to learn Arabic and then went exploring. He went on a German expedition to Central and East Africa (1880–1882), after which he returned to, and settled in Berlin.

Boie

Boie's Woodpecker *Picus boiei* **Wagler**, 1827 NCR
[JS Cream-backed Woodpecker *Campephilus leucopogon*]
Boie's Honeyeater *Myzomela boiei* **S Müller**, 1843
[Alt. Banda Myzomela]

Heinrich Boie (1794–1827) was a German explorer and zoologist. He studied under Blumenbach (q.v.) at Göttingen and worked as an assistant to Temminck (q.v.) at Leiden. After Kuhl (q.v.) died, Boie replaced him. He went to work in the Dutch East Indies and was at Buitenzorg (Bogor) in Java when he died. He wrote some papers with his brother Friedrich (1798–1870) who was a lawyer, ornithologist, herpetologist and entomologist. One reptile and two amphibians are named after him.

Boinet

Band-bellied Crake ssp. *Porzana paykullii boineti* **Bourret**, 1944 NCR, NRM

Professor Édouard Louis Désiré Boinet (1859–1939) was a French military physician who graduated at Bordeaux (1880). He founded the medical school in Hanoi, Vietnam. He later held chairs at Montpellier and Marseilles.

Boissonneau

Hummingbird genus *Boissonneaua* **Reichenbach**, 1854

Streaked Tuftedcheek *Pseudocolaptes boissonneautii* **Lafresnaye**, 1840
Barn Swallow *Hirundo boissonneauti* **Temminck**, 1840
[JS *Hirundo rustica savignii*]

Auguste Boissonneau (1802–1883) was a French ornithologist and dealer in bird specimens. Among his regular customers was Lafresnaye (q.v.), and Boissonneau named several birds after him. He also corresponded with Temminck (q.v.), who suggested he name a tanager after Rieffer (q.v.). Boissonneau was the first person to call himself an 'oculist' and the first to produce glass eyes for human use.

Bojer

Golden Palm Weaver *Ploceus bojeri* **Cabanis**, 1869

Wenzel Bojer (1800–1856) was a Czech naturalist who collected in tropical Africa. He was Director of the Royal Botanic Gardens at Pamplemousses in Mauritius (1848–1849). He has many plants and a reptile named after him.

Bokermann

Araripe Manakin *Antilophia bokermanni* Coelho & Silva, 1998

Dr Werner Carlos Augusto Bokermann (1929–1995) was a zoologist. He received his doctorate in zoology at the Bioscience Institute, São Paulo University. He became head of the bird section, Fundação Parque Zoológico de São Paulo, and stayed there throughout his working life. Many taxa, including a mammal and thirteen amphibians, are named after him.

Boleslavsky

Red-throated Bee-eater ssp. *Merops bulocki boleslavskii* **Pelzeln**, 1858 NCR
[JS *Merops bulocki bulocki*]

Lt von Boleslavsky (fl.1857) was an Austrian explorer in the Sudan and Somaliland, where he travelled with Heuglin (q.v.).

Bolivar

Stripe-breasted Spinetail ssp. *Synallaxis cinnamomeus bolivari* **Hartert**, 1917
Bananaquit ssp. *Coereba flaveola bolivari* **Zimmer & W. H. Phelps**, 1946
White-crowned Manakin ssp. *Dixiphia pipra bolivari* **Meyer de Schauensee**, 1950

Simón Bolívar (1783–1830) 'The Liberator' was one of the great men of Latin American history and needs no biography here. Many places are named after him, and the bananaquit is named after Ciudad Bolívar, Venezuela (the type locality of the subspecies), whilst the manakin is named after the Bolívar department of Colombia. The spinetail may be directly eponymous though Hartert does not make this explicit.

Bolle

White-headed Wood-hoopoe *Phoeniculus bollei* **Hartlaub**, 1858
Bolle's Pigeon *Columba bollii* **Godman** 1872
[Alt. Bolle's Laurel Pigeon, Dark-tailed Laurel Pigeon]

Carl August Bolle (1821–1909) was a German collector, naturalist and botanist. He wrote *Meiner zweiter Beitrage zur Vogelkunde der Canarischen Inseln* (1857) and was co-author of *Die Wirbeltiere der Provinz Brandenburg* (1886), which includes descriptions of 278 native breeding birds.

Bollons

Antarctic Tern ssp. *Sterna vittata bollonsi* **Mathews & Iredale**, 1913 NCR
[JS *Sterna vittata bethunei*]

Captain John Peter Bollons (1862–1929) was an English born, New Zealand amateur ornithologist, naturalist, ethnographer and mariner who captained government steamers which annually provisioned the subantarctic islands. He went to sea (1876) and first landed in New Zealand as a survivor of the wreck (1881) of the barquentine *England's Glory* near the entrance of Bluff Harbour. After the wreck he joined the *Bluff*

pilot cutter and then transferred to a government ketch, engaged in suppressing seal poaching. He became a master mariner (1892) and was employed (1893) by the Marine Department of the New Zealand Government. He liked to name his children after ships that he had commanded so his son's names included Tutanekai and his daughter's Hinemoa. He died unexpectedly after a hernia operation.

Bolton

Apo Sunbird *Aethopyga boltoni* **Mearns**, 1905

First Lieutenant Edward C. Bolton (d.1906) was a US administrator in the military administration of the Philippines. He enlisted in the US Army (1894) and was Military Governor of Davao District, Mindanao, Philippines (1905). He was assassinated.

Bonaparte

Bulbul genus *Bonapartia* **Buettikofer**, 1896 NCR
[Now in *Pycnonotus*, or *Alcurus* if the latter is separated]

Bonaparte's Flycatching-warbler *Myiodioctes bonapartii* **Audubon** NCR
[Based on a female Canada Warbler *Wilsonia canadensis*]
Bonaparte's Gull *Chroicocephalus philadelphia* **Ord**, 1815
Bonaparte's Sandpiper *Calidris fuscicollis* **Vieillot**, 1819
[Alt. White-rumped Sandpiper]
Bonaparte's Shearwater *Puffinus tenuirostris* **Temminck**, 1835
[Short-tailed Shearwater]
Bonaparte's Euphonia *Euphonia hirundinacea* Bonaparte, 1838
[Alt. Yellow-throated Euphonia]
Bonaparte's Starfrontlet *Coeligena bonapartei* **Boissoneau**, 1840
[Alt. Golden-bellied Starfrontlet]
Bonaparte's Zanger *Basileuterus luteoviridis* Bonaparte, 1845
[Alt. Citrine Warbler]
Bonaparte's Friarbird *Melitograis gilolensis* Bonaparte, 1850
[Alt. White-streaked/Gilolo Friarbird]
Bonaparte's Nightjar *Caprimulgus concretus* Bonaparte, 1850
[Alt. Sunda Nightjar]
Bonaparte's Blackbird *Sturnella superciliaris* Bonaparte, 1851
[Alt. White-browed Blackbird]
Bonaparte's Parakeet *Pyrrhura lucianii* **Deville**, 1851
[Alt. Deville's Conure]
Bonaparte's Barbet *Gymnobucco bonapartei* **Hartlaub**, 1854
[Alt. Grey-throated Barbet]
Bonaparte's Tinamou *Nothocercus bonapartei* **G. R. Gray**, 1867
[Alt. Highland Tinamou]

Bonaparte's Hanging Parrot *Loriculus philippensis bonapartei* **Souancé**, 1856
[Alt. Black-billed Hanging Parrot ssp.]
Red-chested Flufftail ssp. *Sarothrura rufa bonapartii* Bonaparte, 1856

Emperor Fairy-wren ssp. *Malurus cyanocephalus bonapartii* G. R. Gray, 1859

Mistle Thrush ssp. *Turdus viscivorus bonapartei* **Cabanis**, 1860

Prince Charles Lucien Bonaparte (originally Jules Laurent Lucien) (1803–1857) was a nephew of the famous statesman, Emperor Napoleon Bonaparte, and a renowned ornithologist. Bonaparte was much travelled, but spent many years in the United States cataloguing birds, and he has been described as the 'father of systematic ornithology'. Whilst in the US, aged 21, he met Audubon and much admired his work and asked for his observations on the wild turkey, which Audubon gave; in return he put him forward for membership of the Philadelphia Academy of Natural Sciences. He eventually settled in Paris and commenced his *Conspectus Generum Avium*, a catalogue of all bird species. He died before finishing it, but its publication was heralded as a major step forward in accomplishing a complete list of the world's birds. He also wrote *American Ornithology* (1825) and *Iconografia della Fauna Italica – uccelli*, (1832). Swainson described Bonaparte as 'destined by nature to confer unperishable benefits on this noble science'. A mammal is named after him.

Bond

Ashy Drongo ssp. *Dicrurus leucophaeus bondi* **Meyer de Schauensee**, 1937

Asian Fairy-bluebird ssp. *Irena puella bondi* Meyer de Schauensee, 1940 NCR

[JS *Irena puella crinigera*]

Cocoa Thrush ssp. *Turdus fumigatus bondi* **Deignan**, 1951

James Bond (1900–1989) was an American ornithologist educated in England at Harrow and Cambridge, which awarded his bachelor's degree (1922). He worked for a Philadelphia banking firm (1922–1925), resigning to take part in an expedition to the Amazon run by the Philadelphia Academy of Sciences for whom he subsequently worked as an ornithologist and Curator of Birds. His area of expertise was the Caribbean and he wrote *Birds of the West Indies* (1936). Ian Fleming knew of Bond and his work as an ornithologist, and asked him for permission to use his name for his British Secret Service agent hero. Bond agreed and in the first Bond film, *Dr No*, Sean Connery, as 007, can be seen examining a copy of *Birds of the West Indies*. James Bond the ornithologist suffered many years of cancer before eventually succumbing.

Bonelli

Western Bonelli's Warbler *Phylloscopus bonelli* **Vieillot**, 1819

Bonelli's Eagle *Aquila fasciata* Vieillot, 1822

Eastern Bonelli's Warbler *Phylloscopus orientalis* **Brehm**, 1855

Franco Andrea Bonelli (1784–1830) was an ornithologist and collector. He taught at the University of Turin (1809–1811), then became the curator of the museum there and totally re-catalogued the collection according to scientific principles. The collection grew and became one of Europe's greatest assemblages of ornithological specimens during his lifetime. He wrote *Catalogue des Oiseaux du Piémont* (1811).

Bonham

Bonham's Sand Partridge *Ammoperdix bonhami* **G. R. Gray**, 1843 NCR

[Alt: See-see Partridge; JS *Ammoperdix griseogularis*]

Edward Walter Bonham (1809–1886) was educated at Harrow and became a diplomat. He was the British Agent resident at Tabriz in Persia (Iran) (1837). He later became Consul at Calais and transferred to Naples (1859).

Bonhote

Little Crow ssp. *Corvus bennetti bonhoti* **Mathews**, 1912 NCR

[*C. bennetti* NRM]

John Lewis James Bonhote (1875–1922) was a British zoologist, ornithologist and writer. After graduating from Trinity College, Cambridge (1897), he became Private Secretary to the Governor of the Bahamas.where he later collected birds and spiders (1901–1902). He described a number of small mammals (1890s and 1900s) and was involved in a collecting expedition to the Malay States (1901). He wrote *Birds of Britain and Their Eggs* (1907) and a number of reports, scientific papers and articles, such as 'The study of bird migration' in *Strand Magazine* (1912). He worked at the Zoological Gardens at Giza, Egypt (1913–1919), also studying and collecting in the Nile Delta. He played a large part in the re-establishment of its egret colony. He edited the *Journal of the Avicultural Society* and was Secretary, Treasurer and Council Member. He bred captive birds, notably waterfowl. He was one of the secretaries of the Fourth International Ornithological Congress, served in various posts in the BOU, and was a long-time council member of the RSPB. Two mammals are named after him.

Bonvalot

Black-browed Tit *Aegithalos bonvaloti* **Oustalet**, 1892

Laughingthrush sp. *Trochalopteron bonvaloti* Oustalet, 1892 NCR

[Alt. Elliot's Laughingthrush; JS *Trochalopteron elliotii elliotii*]

Chinese Babax ssp. *Babax lanceolatus bonvaloti* Oustalet, 1892

Pierre Gabriel Édouard Bonvalot (1853–1933) was a French legislator, author and explorer in Asia. He was financed by the French government to explore Central Asia (1880) and made several subsequent expeditions, going as far east as China and Vietnam, and championed further French colonisation. He led an expedition to Ethiopia through Sudan (1898), founded and edited a journal, *La France de demain* (1898–1904), served as a Deputy in Parliament (1902–1906) and wrote *Une Lourde Tâche* about colonial government as well as other travelogues.

Booth, B. D. M.

White-quilled Rock Pigeon ssp. *Petrophassa albipennis boothi* **Goodwin**, 1969

Major Brian Derek McDonald Booth (1927–2008) was an English explorer, naturalist and ornithologist and a professional soldier (1945–1963). He led the British Berkou-Ennedi Expedition to the Libyan Desert (1957). In Chad the French

Foreign Legion arrested him for shooting 'the wrong sort of crocodile', which he knew to be rare, that he wanted to present to the BMNH London for its collection. He was fined for not having the right permit. He led the fifth Harold Hall Expedition to Western Australia and Northern Territory, Australia (1968). He later ran the Tryon and Moorland Gallery in London, which specialised in natural history paintings.

Booth, C.

Bee Hummingbird *Orthorhynchus boothi* **Gundlach**, 1856 NCR
[JS *Mellisuga helenae*]

Charles Booth (fl.1856) was a Cuban plantation owner with property in Matanzas province, who befriended Gundlach (1839). Gundlach (q.v.) named the hummingbird for Booth apparently unaware that Lembeye (q.v.) had already done so for Booth's wife Helen (q.v.).

Booth, F.

Sunbird sp. *Chalcomitra boothi* **Reichenow** 1902 NCR
[Alt. Amethyst Sunbird; JS *Chalcomitra amethystina kirkii*]

F. Booth (fl.1902) was a collector in German East Africa (Tanzania).

Borelli

Borelli's Parakeet *Pyrrhura borellii* **Salvadori**, 1894 NCR
[Alt. Maroon-bellied Parakeet; JS *Pyrrhura frontalis chiripepe*]
Becard sp. *Hadrostomus borellianus* **Bertoni**, 1901 NCR
[Alt. Chestnut-crowned Becard; JS *Pachyramphus castaneus*]

Stripe-headed Brush Finch ssp. *Arremon torquatus borellii* Salvadori, 1897
Mottled Owl ssp. *Strix virgata borelliana* Bertoni, 1901

Dr Alfredo Borelli (1857–1943) was an Italian zoologist who worked at the Turin Museum (1900–1913). He spent three years (1893–1896) exploring and collecting in Argentina and Paraguay. An amphibian is named after him.

Borgert

Scarlet-chested Sunbird *Chalcomitra borgerti* **Reichenow & Neumann**, 1905 NCR
[JS *Chalcomitra senegalensis*]

Dr Adolf Hermann Constant Borgert (1868–1954) was a German zoologist, entomologist and botanist who collected in Kenya and Uganda (1904–1905). He was Professor at the University of Freiburg (1919–1928) and Director of Zoology at the University of Bonn (1928–1933). He was interested in plankton and wrote *Die Tripyleen Radiolarien der Plankton-Expedition* (1905).

Boris

Babbler genus *Borisia* **Hachisuka** 1935 NCR
[Now in *Sterrhoptilus*]

Icterine Warbler ssp. *Hippolais icterina borisi* **von Jordans**, 1940 NCR, NRM

Boris III (1894–1943) was Tzar of Bulgaria (1918–1943) and was a keen collector and naturalist.

Borissov

Eurasian Eagle Owl ssp. *Bubo bubo borissowi* **Hesse**, 1915

G. Borissov (fl.1913) was a Russian explorer and collector who was in Sakhalin (1911–1913).

Boros

Desert Lark ssp. *Ammomanes deserti borosi* **Horvath**, 1958 NCR
[NUI *Ammomanes deserti isabellina*]

Dr Istvan Boros (1891–1980) was a Hungarian zoologist and herpetologist. He graduated in geology and natural sciences from Pazmany Peter University, Budapest (1914). The University of Pecs, Hungary, awarded his doctorate (1925). He fought (WW1) in the Austro-Hungarian army, was captured by the Russians (1914), and kept prisoner in camps in Russia (1914–1922). His experiences there, where he was well treated, led him to become a Marxist, and he held a number of positions in the Communist administration in Hungary, culminating in becoming Director of the Hungarian National Museum, Budapest (1949–1960). He was responsible for rebuilding the museum after a large part of the collections were burnt out during the Hungarian uprising (1956).

Borrero

Borrero's Cinnamon Teal *Anas cyanoptera borreroi* Snyder & **Lumsden**, 1951
Straight-billed Woodcreeper ssp. *Dendroplex picus borreroi* **Meyer de Schauensee**, 1959
[SII *Dendroplex picus peruvianus*]

José Ignacio Borrero (1921–2004) was a Colombian ornithologist. He wrote such books as *Avifauna de la Región de Soatá, Departamento de Boyacá, Colombia* (1955), and articles including 'Notes on the breeding behaviour of Reiffer's Hummingbird *Amazilia tzacatl*'.

Boscawen

Arabian Wheatear ssp. *Oenanthe lugentoides boscaweni* **G. L. Bates**, 1937

Lieutenant-Colonel the Honourable Mildmay Thomas 'Tommy' Boscawen (1892–1958) was a British army officer who trekked from India via Tibet to Chinese Turkestan (1920s) and explored in Hadhramaut, Yemen and Socotra (1929–1935). He served in the Rifle Brigade (WW1) and was clearly fearless; Lord Moran in his book *The Anatomy of Courage* (1945) mentions him with these words: 'During an attack on Guillemont in the Battle of the Somme an officer of the Rifle Brigade was crossing the open [ground] under heavy shell fire when he dropped his glove. He walked back two or three yards, picked it up and went on.' He became the largest individual sisal planter in Tanganyika and Kenya. A friend in Yemen gave him an Arabian Oryx *Oryx leucoryx*, which he presented to London Zoo.

Bosch

Banded Pitta sp. *Pitta boschii* **S. Müller** & **Schlegel**, 1845
NCR
[JS *Hydrornis irena irena*]

Lieutenant-General Johannes Graaf van den Bosch (1780–1844) was a Dutch soldier and politician, who was Commissioner-General for the Dutch West Indies (1827–1830) and Governor-General of the Dutch East Indies (Indonesia) (1830–1833). He was a prisoner-of-war in England (1808–1813).

Boschma

Chestnut-breasted Munia ssp. *Lonchura castaneothorax boschmai* **Junge**, 1952

Professor Dr Hilbrand Boschma (1893–1976) was a zoologist, herpetologist and expert on crustaceans. His dissertation was on the neck skeleton of crocodiles, but he turned his attention to invertebrates at National Natural History Museum, Leiden (1922), where he became Director (1933–1958). He and Brongersma, his successor, shared a number of expeditions (1920s and 1930s) to Suriname and the Dutch East Indies (Indonesia). Two reptiles are named after him.

Bostanjoglo

Eurasian Penduline-tit ssp. *Remiz pendulinus bostanjogli* **Zarudny**, 1913 NCR
[Hybrid population of *Remiz pendulinus* and *R. macronyx*]

V. N. Bostanjoglo (fl.1919) was a Russian zoologist and entomologist who explored in Kazakhstan (1907).

Botha

Lark genus *Botha* **G. E. Shelley**, 1902 NCR–
[Now in *Spizocorys*]

Botha's Lark *Spizocorys fringillaris* (**Sundevall**, 1850)

General Louis Botha (1862–1919) was a soldier who became the first Prime Minister of the Union of South Africa (1910–1919), having previously been the Prime Minister of the Transvaal (1907). Roberts proposed the vernacular name when he suggested that the lark belonged in Shelley's new genus *Botha*, and named the bird *Botha difficilis*, but this was never widely accepted. The genus *Botha* was disallowed as it was deemed too similar to *Bothus*, a genus of flounders. However, as is often the case, the common name stuck.

Botta

Botta's Wheatear *Oenanthe bottae* **Bonaparte**, 1854
[Alt. Red-breasted Wheatear]

Paul-Emile Botta (1802–1870) was a traveller, diplomat, archaeologist and physician. He spent a year on board the French ship *Héros* as ship's surgeon and naturalist. The vessel traded on the Californian coast (c.1827) under the command of Captain Auguste Duhaut-Cilly, who recorded expeditions ashore with Botta. He was physician to Mohammed Ali, Pasha of Egypt (1830), and was the French Consul in Alexandria (1833). He wrote *Notes on a Journey in*
Arabia and *Account of a Journey in Yemen* (1841). He was appointed Consular Agent for Iraq (1840) and became French Consul at Mosul (1842). He identified Khorsabad as being the site of biblical Nineveh, the ancient capital of Assyria. Botta excavated there (1843–1846) and discovered a dictionary for Class III cuneiform script dating from the seventh century BC. He wrote about his discoveries in the 5-volume *Monuments de Ninivé, Décoverts et Décrits par Botta, Mésurés et Déssinés per E. Flandin* (1849–1850). Three mammals and a reptile are named after him.

Bottego

Chestnut-naped Francolin ssp. *Francolinus castaneicollis bottegi* **Salvadori**, 1898 NCR
[JS *Pternistis castaneicollis castaneicollis*]

Captain Vittorio Bottego (1860–1897) was an Italian explorer. He was an artilleryman and a skilled horseman who wanted adventure and to become a hero, so he arranged a transfer to Eritrea (1887). He set out on a journey of exploration from Berbera with Captain Matteo Grixoni (1892). They reached the upper flow of the Juba River and penetrated to its source (1893). After parting ways with Grixoni, Bottego reached Daua Parma and discovered the Barattieri waterfalls, finally reaching Brava (September 1893). The expedition lost 35 men en route. Bottego set off again (1895) under the auspices of the Italian Geographical Society, with a contingent of 250 local troops. He later tried crossing Ethiopia and was offered a truce, but turned it down and was killed in fighting. The King of Ethiopia kept his men imprisoned for two years, and only when they were released did word of Bottego's fate reach the Italian colonial regime. A mammal and two reptiles are named after him.

Botteri

Botteri's Sparrow *Peucaea botterii* **P. L. Sclater**, 1858

Matteo Botteri (1808–1877) was an ornithologist, botanist and collector. He was born in Croatia and lived in Mexico, where he founded a museum at Orizaba. He collected the sparrow (1854).

Boucard

Boucard's Hermit *Phaethornis adolphi* **Gould**, 1857
[Alt. Adolph's Hermit]
Boucard's Tinamou *Crypturellus boucardi* **P. L. Sclater**, 1859
[Alt. Slaty-breasted Tinamou]
Boucard's Wren *Campylorhynchus jocosus* P. L. Sclater, 1860
Boucard's Hummingbird *Amazilia boucardi* **Mulsant**, 1877
[Alt. Mangrove Hummingbird]
Boucard's Manakin *Manacus coronatus* Boucard, 1879
[Probably a hybrid: *Manacus manacus* x *Dixiphia erythrocephala*]

Boucard's Antwren *Microrhopias quixensis boucardi* P. L. Sclater, 1858
[Alt. Dot-winged Antwren ssp.]
Boucard's Summer Sparrow *Aimophila ruficeps boucardi* P. L. Sclater, 1867
[Alt. Rufous-crowned Sparrow ssp.]

White-bellied Antbird ssp. *Myrmeciza longipes boucardi* **Berlepsch**, 1888

Adolphe Boucard (1839–1905) was a French naturalist who collected in Mexico and spent c.40 years killing humming-birds for science and the fashion trade. He moved to London (1890) but passed his later years in his villa near Ryde on the Isle of Wight. He was author of *The Hummingbird* (1891). He wrote (1894) that 'Now-a-days the mania of collecting is spread among all classes of society, and that everyone possess, either a gallery of pictures, aquarels, drawings, or a fine library, an album of postage stamps, a collection of embroideries, laces … and such like, a collection of humming-birds should be the one selected by ladies. It is as beautiful and much more varied than a collection of precious stones and costs much less …' A reptile is named after him.

Bouchelle

White-collared Swift ssp. *Streptoprocne zonaris bouchellii* **W. Huber**, 1923

Dr Theodore W. Bouchelle (d.1935) was the surgeon and metallurgist of the Eden Mining Company, Nicaragua. He was a member of the Philadelphia Academy of Sciences, and being resident in Nicaragua (1904) he suggested that the Academy send an expedition, which he would host as well as look after the health of its members (1922).

Bougainville

Guanay Cormorant *Phalacrocorax bougainvillii* **Lesson**, 1837 [Syn. *Leucocarbo bougainvillii*]

Admiral Baron Hyacinthe Yves Philippe Potentien de Bougainville (1781–1846) was a French naval officer in command of the corvette *Espérance*. He took part in a circumnavigation (1824–1826). Two mammals are named after him.

Bouguer

White-tailed Hillstar *Urochroa bougueri* **Bourcier**, 1851

Pierre Bouguer (1698–1758) was a French prodigy in mathe-matics and hydrography, taught by his father who was Royal Professor of Hydrography but who died when Pierre was 15. The prodigy was appointed to his father's post, and won three Grand Prix of the Académie Royale des Sciences (1727–1731), all for shipping navigation aids or improvements. He embarked on an expedition to Peru (1735) organised by the Académie Royale des Sciences, in order to measure the length of a degree of meridian at the equator, with Louis Godin (q.v.) among the scientific staff. He travelled in Peru making measurements (1736–1742). His later work, looking at astronomical light sources, earned him the title of 'father of photometry'. He used candlelight as a comparison and published this method in *Essai d'Optique sur la Gradation de la Lumière* (1729). He also wrote *La Figure de la Terre* (1749) and invented the heliometer (1748). Furthermore, 'Bouguer's law', states 'In a medium of uniform transparency the light remaining in a collimated beam is an exponential function of the length of the path in the medium'.

Boulton

Boulton's Hill Partridge *Arborophila rufipectus* Boulton, 1932 [Alt. Sichuan (Hill) Partridge] Boulton's Puff-back Flycatcher *Batis margaritae* Boulton, 1934 [Alt. Boulton's Batis, Margaret's Batis]

Ruddy Treerunner ssp. *Margarornis rubiginosus boultoni* **Griscom**, 1924 Evergreen Forest Warbler ssp. *Bradypterus lopezi boultoni* **Chapin**, 1948

Wolfrid Rudyerd Boulton (1901–1983) was Curator of Birds at The Field Museum, Chicago. He collected in West Africa, Angola and the Kalahari Desert (1931–1946). Boulton collected the hill partridge holotype in southern Sichuan, China, and named the puff-back flycatcher for his wife Margaret (q.v.). A reptile is named after him.

Bouquet

Bouquet's Parrot *Amazona arausiaca* PLS Muller, 1776 [Alt. Red-necked Amazon/Parrot; *Psittacus bouqueti* (Bechstein, 1811) is a junior synonym]

Lucian Bouquet (DNF) was a French academic, Professor and Director of an academy in Paris (1805).

Bourcier

Hummingbird genus *Bourcieria* **Bonaparte**, 1850 NCR [Now in *Coeligena*]

Bourcier's Hermit *Phaethornis bourcieri* **Lesson**, 1832 [Alt. Straight-billed Hermit] Red-headed Barbet *Eubucco bourcierii* **Lafresnaye**, 1845

Orange-eared Tanager ssp. *Chlorochrysa calliparaea bourcieri* Bonaparte, 1851 Bourcier's Quail Dove *Geotrygon frenata bourcieri* Bonaparte, 1855 [Alt. White-throated Quail Dove ssp.]

Claude Marie Jules Bourcier (1797–1873) was French Consul to Ecuador (1849–1850) and a collector and naturalist who specialised in hummingbirds, writing a great many descrip-tions of new species. He named several *franciae* in their scientific binomials after his daughter Francia (q.v.). A reptile is named after him.

Bourdelle

Ashy Bulbul ssp. *Hemixos flavala bourdellei* **Delacour**, 1926

Professor Edouard Sicaire Bourdelle (1876–1960) was a French veterinary surgeon and zoologist at the MNHN, Paris, where he was Professor of Zoology (1926–1947). He was in charge of the menagerie at Le Jardin des Plantes (1926–1936) and created the Paris zoo at Vincennes (1934).

Bourdillon

Great Eared Nightjar ssp. *Eurostopodus macrotis bourdilloni* **Hume**, 1875 Dark-fronted Babbler ssp. *Rhopocichla atriceps bourdilloni* Hume, 1876

Indian Blackbird ssp. *Turdus simillimus bourdilloni* **Seebohm**, 1881

Thomas Fulton Bourdillon (1849–1930) first arrived in Travancore, India, as a planter (1871) and became Conservator of Forests there (1891–1908). He wrote *The Forest Trees of Travancore* (1908).

Bourke

Bourke's Parrot *Neopsephotus bourkii* **Gould**, 1841

General Sir Richard Bourke KCB (1777–1855) was Lieutenant-Governor of the Eastern District and Acting Governor of the Cape Colony (1825–1828). During that period he issued an important ordinance that improved the legal status of the Hottentot people. He then became Governor of New South Wales (1831–1837). Major Sir Thomas Livingstone Mitchell (q.v.) discovered the parrot (1838) near the mouth of the River Bogan. It may not have been named *directly* after Sir Richard, as it was found on the banks of the river Bogan near Fort Bourke, which had already been named after him; hence it may be named for a place rather than a person. Bourke was a relative of the great Parliamentarian Edmund Burke, with whom he stayed when in London.

Bourne, J. & P.

Striated Fieldwren ssp. *Calamanthus fuliginosus bourneorum* **Schodde** & I. J. Mason, 1999

Jack Bourne and Pat Bourne own and run Bourne's Bird Museum at Bool Lagoon, Naracoorte, South Australia. The Bourne family made a big contribution to South Australian ornithology and collected the holotype of this bird.

Bourne, W. R. P.

Bourne's Heron *Ardea purpurea bournei* Naurois, 1966
[Alt. Purple Heron ssp.]

Dr William (Bill) Richmond Postle Bourne (b.1930) is, he tells us, 'more or less retired' but as an Honorary Research Fellow still works at the Department of Zoology, Aberdeen University. After qualifying as a physician he practised geriatrics and joined the Royal Fleet Auxiliary, serving as a naval surgeon (1983–1991) including during the First Gulf War. He is a prolific author with many publications in scientific journals from 'Notes on autumn migration in the Middle East' (1959) to 'Phalaropes in the Arabian Sea and Gulf of Oman' (1997). His particular areas of interest are birds of the Middle East, birds of islands and seabirds. He collected the heron (1951) during a solitary undergraduate expedition. He skinned it and took it to the UK by which time it stank and was maggoty. The BMNH told him it was just a faded Purple Heron and it was only after Abbé Réné de Naurois collected more examples that it was recognised as being different. It is sometimes referred to as Bourne's Heron but Dr Bourne told us he thinks a better name would be Pale Purple Heron.

Bourns

Flowerpecker genus *Bournsia* **McGregor**, 1927 NCR
[Now in *Dicaeum*]

Bourns's Hanging Parrot *Loriculus philippensis bournsi* McGregor, 1905
[Alt. Philippine Hanging Parrot ssp.]
Variable Dwarf Kingfisher ssp. *Ceyx lepidus bournsii* **Steere**, 1890 NCR
[JS *Ceyx lepidus margarethae*]

Dr Frank Swift Bourns (1866–1935) qualifed as a physician (1890) at the University of Michigan and was an early X-ray specialist. He went to the Philippines with the Steere Expedition (1887–1889 and 1898–1899) and then moved on to Borneo to collect there. He worked for the Health Care Department, Seattle (1907–1910), eventually setting up in private practice. With D. C. Worcester (q.v.) he co-wrote *Contributions to Philippine Ornithology* (1898).

Bourqui

Coqui Francolin ssp. *Francolinus coqui bourquii* Monard, 1934 NCR
[JS *Peliperdix coqui coqui*]

Père Charles Bourqui (fl.1934) was a missionary at Kavangu, Angola (1934).

Bourret

White-throated Needletail ssp. *Hirundapus caudacutus bourreti* David-Beaulieu, 1944 NCR
[JS *Hirundapus caudacutus caudacutus*]

René Leon Bourret (1884–1957) was a French zoologist. He undertook a comprehensive herpetological survey of Vietnam and studied Indochinese fauna (1922–1942). He published *Les Tortues de l'Indochine* (1941), the first detailed monograph to deal with all the chelonians of South-East Asia. Among other taxa, seven reptiles, two amphibians and a mammal are named after him.

Bouvier

Bouvier's Fishing Owl *Scotopelia bouvieri* **Sharpe**, 1875
[Alt. Vermiculated Fishing Owl]
Bouvier's Sunbird *Cinnyris bouvieri* **GE Shelley**, 1877
[Alt. Orange-tufted Sunbird]

Aimé Bouvier (1844–1919) was a French collector and zoologist who was active around the year he became Secretary of the French Zoological Society (1876). He and other committee members were forced to resign (1880) after it was discovered that about 5,000 francs of the society's funds were missing and unaccounted for. He made two trips to Australia to collect birds, the first to Cairns (1884–1885) and the second, with the taxidermist Walter Burton, to north-western Australia (1886). Thereafter he collected on Thursday Island, off Cape York, and at Palmerston in Northern Territory (1886). Many of the skins were presented to the BMNH (1887) and the remainder given to G. M. Mathews (q.v.). Among other taxa a mammal and two reptiles are named after him.

Bowdler

Fernbird genus *Bowdleria* **Rothschild**, 1896

White-bellied Kingfisher ssp. *Corythornis leucogaster bowdleri* **Neumann**, 1908

Pale Flycatcher ssp. *Bradornis pallidus bowdleri* **Collin** &
   Hartert, 1927
Rufous Chatterer ssp. *Turdoides rubiginosa bowdleri*
   Deignan, 1964

(See Richard Bowdler **Sharpe**)

Bowen

Collared Pratincole ssp. *Glareola pratincola boweni*
   Bannerman, 1930 NCR
[NUI *Glareola pratincola fuelleborni*]
Spike-heeled Lark ssp. *Chersomanes albofasciata boweni*
   Meyer de Schauensee, 1931

Professor Wilfrid Wedgwood Bowen (1899–1987) was born in
Barbados and became Professor of Biology at Dartmouth
College, US. He was in the Sudan (1922–1931) and wrote
Catalogue of Sudan Birds (1926).

Bower

Shrike-thrush genus *Bowyeria* **Mathews**, 1912 NCR
[Now in *Colluricincla*]

Bower's Shrike-thrush *Colluricincla boweri* **E. P. Ramsay**,
   1885
[Alt. Stripe-breasted Shrike-thrush]
Boobook Owl sp. *Spiloglaux boweri* Mathews, 1913 NCR
[Alt. Southern Boobook; JS *Ninox boobook lurida*]

Red-backed Fairy-wren ssp. *Malurus cruentatus boweri*
   E. P. Ramsay 1887 NCR
[JS *Malurus melanocephalus cruentatus*]
Double-eyed Fig Parrot ssp. *Opopsitta diophthalma boweri*
   Mathews, 1915 NCR
[JS *Cyclopsitta diophthalma macleayana*]
Fernwren ssp. *Oreoscopus gutturalis boweri* Mathews,
   1916 NCR; NRM
Australasian Figbird ssp. *Sphecotheres vieilloti boweri*
   Mathews, 1916 NCR
[JS *Sphecotheres vieilloti flaviventris*]
Long-toed Stint ssp. *Calidris subminuta boweri* Mathews,
   1916 NCR; NRM
Yellow Chat ssp. *Epthianura crocea boweri* Mathews, 1922
   NCR
[JS *Epthianura crocea crocea*]

Captain Thomas Henry Bowyer-Bower (1862–1886) was the
English-born Curator of Ornithology at the Western
Australian Museum (c.1886). He was a naturalist and
collector, especially of the avifauna of Australia, and sent
many specimens of rare and new forms to the BMNH and the
Zoological Society of London. He collected in Queensland
(1884 and 1885) and went to north-western Australia and
Northern Territory (1886) with the taxidermist Walter Burton,
where he died of typhoid. A total of 192 species from the
(1886) trip were sent to the BMNH. After his death Bower's
mother sent his collection to G. M. Mathews (q.v.), who catalogued it.

Bowie

Silvereye sp. *Zosterops bowiae* Horne, 1907 NCR
[JS *Zosterops lateralis lateralis*]

Helen Bowie (fl.1907) was an Australian ornithologist and
founder member of the RAOU. Her uncle, the aviculturist and
ethnologist Dr George Horne, described the bird. She was an
army nurse in France (WW1) in the unit of which Horne was
the senior officer.

Bowman

Black-spotted Bare-eye ssp. *Phlegopsis nigromaculata*
   bowmani **Ridgway**, 1888

David Bowman Riker (1861–1954) was an American whose
father, Clarence B. Riker (q.v.), is said to have inspired Frank
Michler Chapman (q.v.) with his accounts of collecting birds
in the Amazon Basin. David's brother, Herbert Riker (q.v.
under **Herbert**), also has a bird named after him, as does their
mother Jessie (q.v.). The whole Riker family was based near
Santarem (1884–1887), Amazonas, Brazil.

Bowyer

Shrike-thrush genus *Bowyeria* **Mathews**, 1912 NCR
[Now in *Colluricincla*]

(See under Bowyer-**Bower**)

Boyce

Oriental Stork *Ciconia boyciana* **Swinhoe**, 1873

Robert Henry Boyce (1834–1909) was a British (Carlow,
Republic of Ireland) architect and engineer, at one time Prin-
cipal Surveyor of HM Diplomatic and Consular Buildings in
China. He designed, among other buildings, the British
Embassy in Cairo.

Boyd

Boyd's Shearwater *Puffinus boydi* **Mathews**, 1912
[Alt. Cape Verde Little Shearwater; Syn. *Puffinus baroli*
   boydi]

Pink-footed Puffback ssp. *Dryoscopus angolensis boydi*
   Bannerman, 1938

(See **Alexander**)

Boyer

Boyer's Cuckooshrike *Coracina boyeri* **G. R. Gray**, 1846
[Alt. White-lored Cuckooshrike]

Joseph Emmanuel Prosper Boyer (b.1815) was a French sea
captain who explored in the Pacific with Dumont d'Urville.
When he was a young officer (1840) they visited New Zealand
and Boyer assisted with survey work there and during the
rest of the voyage.

Boyle

Chestnut-bellied Flowerpiercer ssp. *Diglossa gloriosissima*
   boylei **Graves**, 1990

Howarth Stanley Boyle (1894–1951), along with Leo E. Miller
(q.v.), collected specimens of the flowerpiercer (1915). He
was collecting in North America and publishing the results
(1900s). He collected in Central America (1911) and spent two

years (1914–1915) undertaking fieldwork in a number of countries in South America (Colombia, Bolivia and Argentina) as assistant to Miller for the AMNH. He left AMNH (1917) to join the navy at the Naval Base Hospital. He published widely (1915–1956) both alone and with a number of co-authors. A reptile is named after him.

Brabourne

Hummingbird genus *Brabournea* **Chubb**, 1916 NCR
[Now in *Leucippus*]

Brabourne's Emerald *Agyrtrina versicolor brabournii* **Bangs & T. E. Penard**, 1918 NCR
[Alt. Versicoloured Emerald ssp.; JS *Amazilia versicolor versicolor*]

Lt Wyndham Wentworth Knatchbull-Hugessen, 3rd Baron Brabourne of Brabourne (1885–1915), was a British ornithologist and author killed in action with the Grenadier Guards. He co-wrote *The Birds of South America* (1912).

Brace

Brace's Emerald *Chlorostilbon bracei* **Lawrence**, 1877 EXTINCT
[Alt. Brace's Hummingbird]

Lewis Jones Knight Brace (1852–1938) was a botanist and bird collector from the Bahamas. He also sent botanical specimens to the New York Botanical Gardens. Brace's Emerald is known only from a single male specimen taken by Brace on New Providence Island, in the Bahamas group, on 13 July 1877. Formerly viewed as a vagrant specimen (or a subspecies) of the Cuban Emerald *C. ricordii*, fossil remains found on New Providence (1982) indicate that it was a full species which had existed on the island since the Pleistocene.

Brack

Band-tailed Sierra Finch ssp. *Phrygilus alaudinus bracki* **O'Neill & T. A. Parker**, 1997

Dr Antonio (José) Brack Egg (b.1940) is a Peruvian biologist who was the first Minister for the Environment in Peru (2008–2011), during which time he brought another four million hectares under protection. He has written c.30 books about conservation, biodiversity and environmental education including *Ecología del Perú* (2000). An amphibian is named after him.

Bradfield

Bradfield's Swift *Apus bradfieldi* **J. A. Roberts**, 1926
Bradfield's Hornbill *Tockus bradfieldi* J. A. Roberts, 1930

Bradfield's Lark *Calendulauda sabota naevia* **Strickland**, 1853
[Alt. Sabota Lark ssp.]
Hartlaub's Francolin ssp. *Pternistis hartlaubi bradfieldi* J. A. Roberts, 1928
Speckled Pigeon ssp. *Columba guinea bradfieldi* J. A. Roberts, 1931 NCR
[JS *Columba guinea phaeonota*]

Rupert Dudley Bradfield (1882–1949) was a South African farmer, naturalist and collector, who lived in Namibia for most of his life with his wife, Marjorie. Bradfield collected the first specimen of the hornbill near their farm, Quickborn, at the Waterberg, near Okahandja, Namibia. Bradfield named a race of another species of swift after his wife Marjory (q.v.) (1935). A reptile is named after him.

Bradshaw

Sunbird sp. *Cinnyris bradshawi* **Sharpe**, 1898 NCR
[Alt. Amethyst Sunbird; JS *Chalcomitra amethystina*]

Karoo Long-billed Lark *Certhilauda subcoronata bradshawi* Sharpe, 1904

Dr Benjamin Frederick Bradshaw (d.1883) was a British physician for a steamship company, then left (1870) to prospect and practise in the diamond fields of South Africa (1871). He was an amateur naturalist, which led him to take employment with explorer George Copp Westbeech (1872–1877), giving him the opportunity to travel into the interior. He returned to England (1880) before becoming Surgeon, Bechuanaland (Botswana) Border Police. He investigated the connection between the tsetse fly and sleeping sickness, writing a number of scientific papers on the issue (1881). He was also an excellent cook.

Brahma/Brama

Woodpecker genus *Brahmapicus* **Malherbe**, 1849 NCR
[Now in *Dinopium*]

Spotted Owlet *Athene brama* **Temminck**, 1821

Brahma, or Brama, is the supreme spirit in the Hindu religion.

Brandt

Brandt's Cormorant *Phalacrocorax penicillatus* Brandt, 1837
Brandt's Mountain Finch *Leucosticte brandti* **Bonaparte**, 1850
[Alt. Black-headed Mountain Finch]

Brandt's Jay *Garrulus glandarius brandtii* **Eversmann**, 1842
[Alt. Eurasian Jay ssp.]
Marsh Tit ssp. *Poecile palustris brandtii* **Bogdanov**, 1879 NCR
[Type specimen lost; name regarded as indeterminate]
Common Pheasant ssp. *Phasianus colchicus brandti* **Rothschild**, 1901 NCR
[JS *Phasianus colchicus mongolicus*]

Johann Friedrich [Fedor Fedorovich] von Brandt (1802–1879) was a German zoologist, surgeon, pharmacologist and botanist who moved to Russia (1831). He explored Siberia and was founding Director of the Zoological Museum of the Academy of Science, St Petersburg. He described several birds from western North America, including the eponymous cormorant. He produced works on systematics, zoogeography, comparative anatomy and the palaeontology of mammals. He co-wrote the two-volume *Medical Zoology* (1829–1833) and solely wrote *Descriptiones et Icones Animalium Rossicorum Novorum vel minus Rite Cognitorum* (1836) dealing in particular with Russia. Other taxa such as fish, insects, five mammals and a reptile are named after him.

Branicki

Branicki's Hummingbird[2] *Heliodoxa branickii* **Taczanowski**, 1874
[Alt. Rufous-webbed Brilliant]
Grey-mantled Wren[2] *Odontorhynchus branickii* Taczanowski & **Berlepsch**, 1885
Branicki's Conure[4] *Leptositta branickii* **Berlepsch & Stolzmann**, 1894
[Alt. Golden-plumed Parakeet]
Branicki's Ibis[5] *Theristicus branickii* Berlepsch & Stolzmann, 1894
[Alt. Andean Ibis; Syn. *Theristicus melanopis branickii*]
Tanager[3] sp. *Chlorochrysa hedwigae* Berlepsch & Stolzmann, 1901 NCR
[Alt. Orange-eared Tanager; JS *Chlorochrysa calliparaea*]

Branicki's Tinamou[1] *Nothoprocta ornata branickii* Taczanowski, 1875
[Alt. Ornate Tinamou ssp.]
Branicki's Blue-and-black Tanager[2] *Tangara vassorii branickii* Taczanowski, 1882
Booted Racket-tail[6] ssp. *Ocreatus underwoodii annae* Berlepsch & Stolzmann, 1894

The first generation of the family were Alexander, Count Branicki[1] (1821–1877), Hieronim Florian Radziwill Konstanty Count Branicki[2] (1824–1884) and his wife Jadwiga (Hedwiga) Potocka, Countess Branicka[3] (1827–1916) and the youngest brother Count Wladyslaw (Ladislas) Michael Branicki[4] (1848–1914). The next generation is represented by the son of Hieronim and Jadwiga, Xavier, Count Branicki[5] (1864–1926) and Countess Anna Branicka[6] (1876–1953), who was a daughter of Count Wladyslaw (Ladislas) Michael Branicki. The superscript numbers refer to which family member eponymised in which birds. The wealthy aristocratic Polish Branicki family was very interested in hunting and Alexander and Konstanty started hunting abroad (1863), visiting Egypt, Sudan, Palestine and Algeria. Zoologists such as Taczanowski (q.v.) and Waga (q.v.) accompanied them and they were persuaded to take a more scientific approach to their favourite pastime (1867). They employed Konstanty Jelski (q.v.) to collect for them in South America. They also employed Jean Stanislas Stolzmann (q.v.), Mikhail Ivanovich Jankowski (q.v.) and Jan Kalinowski (q.v.) as collectors. This resulted in the creation of the Branicki Museum in Poland. Poland was then part of the Imperial Russian Empire and they feared that the museum in St Petersburg might be covetous and transfer the collections. So Konstanty decided to protect it by establishing a private museum, which was achieved (1887) after his death by his nephew Wladyslaw and his own son Xavier. Following the Treaty of Versailles (1919) Poland became independent and the collections were transferred to the state.

Brasher

Brasher's Warbler *Basileuterus culicivorus brasierii* **Giraud**, 1841
[Alt. Golden-crowned Warbler ssp. Trinomial often amended to *brasherii*]

Philip Marston Brasher (d.1880) of Brooklyn was a Wall Street broker and an amateur ornithologist and skilled taxidermist. His close friend Giraud (q.v.) referred to him as a man of 'lively and companionable manners'. During the first half of the 19th century Brasher, sometimes with Giraud, made a collection of bird specimens from Long Island of which Giraud made great use. Brasher was a member of the New York State Legislature and had real estate dealings in St. Lawrence Co., NY, where the towns of Brasher and Brasher Falls are named after him. By his second wife he had a son, Rex Brasher (1869–1960), who was inspired to out-do Audubon in the painting of birds as Audubon had snubbed his father (1840)! Giraud made a spelling error in his original description, misnaming 'Brasher' as 'Brasier'; the trinomial is often amended to *brasherii*.

Brass

Brass's Friarbird *Philemon brassi* **Rand**, 1940

Dr Leonard John Brass (1900–1971) was an Australian plant collector, trained at the Queensland Herbarium. He took part in several of the Archbold Expeditions to Papua New Guinea. He succeeded Rand (q.v.), who was the first Resident Biologist of the Archbold Biological Station (established 1941). He became a naturalised US citizen (1947), but returned to Australia after retiring. A mammal is named after him.

Braun

Braun's Bush-shrike *Laniarius brauni* **Bannerman**, 1939
[Alt. Orange-breasted Bush-shrike]

Buff-throated Apalis ssp. *Apalis rufogularis brauni* **Stresemann**, 1934

Rudolf H. Braun (b.1908) was a German collector who was active in Angola (1942) and southern Africa. A botanical collection he made is held in Swaziland. A reptile is named after him.

Braune

Bronzed Drongo ssp. *Dicrurus aeneus braunianus* **Swinhoe**, 1863

George C. P. Braune (fl.1860) was an assistant to Robert Swinhoe, British Consul to Formosa (Taiwan).

Brauner

Blue Tit ssp. *Cyanistes caeruleus brauneri* **Moltchanov**, 1916 NCR
[JS *Cyanistes caeruleus satunini*]

Aleksandr Aleksandrovich Brauner (1857–1941) was a Ukrainian zoologist, taxonomist and archaeologist, and Professor of Zoology at Odessa (1915).

Brazza

Brazza's Martin *Phedina brazzae* **Oustalet**, 1886
[Alt. Brazza's Swallow, Congo Martin]

Brazza's Thick-billed Cuckoo *Pachycoccyx audeberti brazzae* Oustalet, 1886

Jacques (or Giacomo) C. Savorgnan de Brazza (1859–1888) was the younger brother of Count Pierre Paul François

Camille Savorgnan de Brazza (1852–1905), the distinguished French explorer. Jacques accompanied his brother exploring the Congo, and while his elder brother became Governor of the French Congo, Jacques was the Director of Le Mission de L'ouest Africain. He amassed a fish collection, mainly from the Ogowe River, that was sent to Paris (c.1885).

Bredo

Ruwenzori Turaco ssp. *Ruwenzorornis johnstoni bredoi* **Verheyen**, 1947

Papyrus Yellow Warbler ssp. *Chloropeta gracilirostris bredoi* **Schouteden**, 1955 NCR

[JS *Chloropeta gracilirostris bensoni*]

Hans J. Bredo (1903–1991) was a Belgian government entomologist who collected in the Congo (1930s). He wrote a number of scientific papers such as 'La lutte biologique et son importance économique au Congo Belge' (1934). His major achievement was a scientific explanation for locusts swarming (i.e. a succession of environmental conditions affecting the behaviour of these usually non-gregarious insects).

Brehm (fund)

Rufous-tailed Hummingbird ssp. *Amazilia tzacatl brehmi* Walter & Schuchmann, 1999

The Brehm Fund for International Bird Conservation (established 1982) is named after its founder Wolf W. Brehm. The fund is administered by the Alexander Koenig Research Institute and Museum of Zoology, Bonn, where the junior author works.

Brehm, A. E. & C. L.

Biak Monarch *Monarcha brehmii* **Schlegel**, 1871
[Syn. *Symposiachrus brehmii*]
Brehm's Tiger Parrot *Psittacella brehmii* Schlegel, 1871
Iberian Chiffchaff *Phylloscopus brehmii* **Homeyer**, 1871 NCR
[*P. ibericus* is now regarded as the correct name for this taxon]

Pallid Swift ssp. *Apus pallidus brehmorum* **Hartert**, 1901
Green Woodpecker ssp. *Picus viridis brehmi* **Kleinschmidt**, 1919 NCR
[JS *Picus viridis viridis*]

Alfred Edmund Brehm (1829–1884) was a German traveller, collector, architect and zoologist. He collected in Egypt and Sudan (1847–1852), Spain (1856), Scandinavia (1860) and Ethiopia (1862), where he and his wife contracted malaria. He wrote *Das Leben der Vögel* (1861). His most famous work was the multi-volume *Brehms Tierleben* (1864–1869). He became the first Director of the Hamburg Zoo (1863–1867), then moving to Berlin where he opened the original Berlin Aquarium but then quarrelled with the supervisory board and resigned (1874). He and Finsch (q.v.) travelled to Siberia (1876). The Emperor of Austria honoured him by granting him a peerage (1878). His stepbrother Oskar joined him in Africa but was drowned (1850) while bathing in the Nile. Alfred died at home in Germany after his malaria recurred during a US lecture tour. He is still remembered in Germany as 'Tiervater

Brehm' (Zoo-father Brehm). The monarch, parrot and chiffchaff are all named after him. Dr Christian Ludwig Brehm (1787–1864) was the father of Alfred and of Thekla Brehm (q.v.) and the woodpecker is named after him. He wrote *Beiträge zur Vogelkunde* (1820–1822). The swift subspecies is named after father and son.

Brehmer

Blue-headed Wood Dove *Turtur brehmeri* **Hartlaub**, 1865

Brehmer (DNF) was a nineteenth-century German trader in Gabon.

Brelay

Flowerpiercer sp. *Uncirostrum brelayi* **Lafresnaye**, 1839 NCR
[Alt. Cinnamon-bellied Flowerpiercer; JS *Diglossa baritula*]

Charles Brelay (1791–1857) was a French wholesale hair merchant in Bordeaux and a naturalist and collector.

Brenchley

Chestnut-bellied Imperial Pigeon *Ducula brenchleyi* **G. R. Gray**, 1870

Grey Fantail ssp. *Rhipidura albiscapa brenchleyi* **Sharpe**, 1879

Julius Lucius Brenchley (1816–1873) was an English traveller. He co-wrote *A Journey to Great-Salt-Lake City* (1861) with Jules Remy, and was sole author of *Jottings during the Cruise of the Curacoa among the South Sea Islands in 1865* (1865). During his voyage to the Solomon Islands (1865) he collected over 1,000 objects, many of which form the Brenchley Collection in the British Museum. He has some gardens named after him in Maidstone, Kent, UK, where he was a museum benefactor. A reptile is named after him.

Brewer, C.

Pearly-vented Tody Tyrant ssp. *Hemitriccus margaritaceiventer breweri* **W. H. Phelps Jr**, 1977

Charles Brewer-Carías (b.1939) is a Venezuelan dentist, who, after 20 years in practice, became an explorer and speleologist. He is well known to the indigenous people of the Venezuelan highlands, as he fixed their teeth while exploring their territory. He was shot by thieves (2003) but, despite some loss of flexibility in his shoulder, still explores. The world's largest quartzite cave is named after him, as are two amphibians.

Brewer, T. M.

Brewer's Blackbird *Euphagus cyanocephalus* **Wagler**, 1829
Brewer's Duck *Anas breweri* **Audubon**, 1838 NCR
[Hybrid of Mallard *A. platyrhynchos* x Gadwall *A. strepera*]
Brewer's Sparrow *Spizella breweri* **Cassin**, 1856
Black-headed Bee-eater *Merops breweri* Cassin, 1859

Dr Thomas Mayo Brewer (1814–1880) was an American naturalist and ornithologist. He was very politically active – a family trait as his father took part in the 'Boston Tea Party'. He had contempt for those studying birds in the field, preferring to study museum collections. He also defended the House Sparrow *Passer domesticus* against moves to eliminate this introduced species from the USA. Audubon wrote 'I have named this Duck after my friend Thomas M. Brewer of Boston, as a mark of the estimation in which I hold him as an accomplished ornithologist'. Brewer's only daughter, Lucy (q.v.), was named after Audubon's wife. In turn George Lawrence, her father's friend, named a hummingbird for Lucy. Brewer co-wrote the *History of North American Birds* and alone wrote *North American Oology* (1857). Two mammals are named after him.

Brewster

Buzzard genus *Brewsteria* **Maynard**, 1896 NCR
[Now in *Buteo*]

Brewster's Warbler *Vermivora leucobronchialis* Brewster, 1874 NCR
[Hybrid of *Vermivora chrysoptera* x *V. cyanoptera*]
Least Poorwill *Siphonorhis brewsteri* **Chapman**, 1917

Brewster's Screech Owl *Megascops kennicottii bendirei* Brewster, 1882
[Alt. Western Screech Owl ssp.]
Brewster's Booby *Sula leucogaster brewsteri* **Goss**, 1888
[Alt. Brown Booby ssp.]
Warbling Vireo ssp. *Vireo gilvus brewsteri* **Ridgway**, 1903
Brewster's (Snowy) Egret *Egretta thula brewsteri* Thayer & **Bangs**, 1909
Brewster's Woodpecker *Melanerpes uropygialis brewsteri* Ridgway, 1911
[Alt. Gila Woodpecker ssp.]
Greyish Saltator ssp. *Saltator coerulescens brewsteri* Bangs & **T. E. Penard**, 1918
Willow Flycatcher ssp. *Empidonax traillii brewsteri* **Oberholser**, 1918
Brewster's Green Conure *Aratinga holochlora brewsteri* **Nelson**, 1928
[Alt. Green Parakeet ssp.]

William Brewster (1851–1919) was an ornithologist, born in South Reading (Wakefield), Massachusetts. After Cambridge High School graduation (1869) an eye problem prevented him entering Harvard. Subsequently, he devoted his attention exclusively to ornithology, becoming Assistant in charge of the collection of birds and mammals in the Boston Society of Natural History (1880) and Curator of Ornithology at the Museum of Comparative Zoology, Cambridge (1885). He wrote *Birds of the Cape Regions of Lower California* (1902) and *Birds of the Cambridge Region of Massachusetts* (1906). The eminent ornithologist Ludlow Griscom (q.v.) wrote, 'Having now spent some thirteen years in studying Brewster's field work and records, it is my humble opinion that he was one of the greatest and most naturally gifted field ornithologists that America has ever produced.'

Brian

Flycatcher genus *Briania* **Chasen** & **Kloss**, 1930 NCR
[Now in *Muscicapella*]

(See Brian Houghton **Hodgson**)

Briceno

Flammulated Treehunter ssp. *Thripadectes flammulatus bricenoi* **Berlepsch**, 1907

Salomón Briceño Gabaldón (1826–1912) was a Venezuelan collector for, among others, Walter Rothschild (q.v.). A mammal and an amphibian are named after him.

Bridges

Bridges's Gull *Larus bridgesii* **Fraser**, 1845 NCR
[Alt. Grey Gull; JS *Leucophaeus modestus*]
Bridges's Woodhewer *Drymornis bridgesii* **Eyton**, 1849
[Alt. Scimitar-billed Woodcreeper]
Bridges's Antshrike *Thamnophilus bridgesi* **P. L. Sclater**, 1856
[Alt. Black-hooded Antshrike]

Bridges' Parrot *Pionus maximiliani siy* **Souancé**, 1856
[Alt. Scaly-headed Parrot ssp.; *Pionus bridgesi* (**Boucard**, 1891) is a junior synonym]
Dusky-legged Guan ssp. *Penelope obscura bridgesi* **G. R. Gray**, 1860

Dr Thomas Charles Bridges (1807–1865) was a botanist, traveller and collector in tropical America (1822–1865). The specimens were sent back to Europe for identification. A mammal is named after him.

Brigida

Brigida's Woodcreeper *Hylexetastes brigidai* **Cardoso da Silva**, **Novaes** & Oren, 1995

Manoel Santa Brigida is a Brazilian ornithologist who collected the holotype. He worked as Senior Taxidermist in the Ornithology Section of the Museu Paraense Emilio Goeldi in the last two decades of the 20th century. According to the *Bulletin* of the BOC 'During this time, he contributed tirelessly to the expansion of Museu Goeldi's bird collections and, consequently, to our knowledge of the systematics and distribution of Amazonian birds'.

Brisson

Asity genus *Brissonia* **Hartlaub**, 1860 NCR
[Now in *Philepitta*]

Ultramarine Grosbeak *Cyanocompsa brissonii* **Lichtenstein**, 1823

Mathurin Jacques Brisson (1723–1806) was a French physician and ornithologist. He was educated in philosophy and theology but abandoned both in favour of natural history, and became a Curator at the French Académie des Sciences. However, he was denied access to their natural history collection after a dispute with Buffon and turned, for the rest of his life, to the study of physics. He wrote *Regne Animal* (1756) and the six-volume *Ornithologie ou Méthode contenant la Division des Oiseaux en Ordres* (1760).

Bristol

Rufous Treepie ssp. *Dendrocitta vagabunda bristoli* **Paynter**, 1961

Dr Melvin Lee Bristol (b.1936) is an American botanist and collector, a Research Fellow at Harvard where he gained his bachelor's degree. He was in India, Nepal and Pakistan (1957–1958). His doctorate was in ethnobotany and he studied primitive agriculture in Colombia (1958–1968), taught at the University of Hawaii, and studied the ethnobotany of medicinal plants in Samoa. He and his wife have (1969) run a horticultural farm (speciality daylilies) in Connecticut.

Britton

African Citril ssp. *Serinus citrinelloides brittoni* Traylor, 1970
[Syn. *Crithagra citrinelloides brittoni*]

Peter Leslie Britton (1943–2009) was born in London and during a long stay at Yala, Kenya, was the first person who recognised that the local population of *Serinus citrinelloides* in western Kenya represented an unnamed subspecies. He wrote a number of articles on East African birds such as: 'The status and breeding behaviour of East African Lari' (1974) and co-wrote longer works such as, with Leslie Brown, *The Breeding Seasons of East African Birds* (1980). He emigrated to Australia and published on Australian species such as for *Sunbird*: 'Winter mixed-species flocks at Charters Towers, north Queensland' (1997). He died in Tasmania in 2009.

Broadbent

Honeyeater genus *Broadbentia* **Mathews**, 1913 NCR
[Now in *Lichenostomus*]

Rufous Bristlebird *Dasyornis broadbenti* McCoy, 1867

Broadbent's Ground Thrush *Zoothera lunulata cuneata*
De Vis, 1889
[Alt. Bassian Thrush/Olive-tailed Thrush ssp.]

Kendall Broadbent (1837–1911) was an Australian taxidermist. Whilst collecting (1880s and 1890s) he kept detailed diaries, now at the Queensland Museum, which employed him as a collector. A mammal is named after him.

Broderip

Oriole genus *Broderipus* **Bonaparte**, 1854 NCR
Oriole genus *Broderipornis* **Mathews**, 1930 NCR
[Both now in *Oriolus*]

Black-naped Oriole ssp. *Oriolus chinensis broderipi*
Bonaparte, 1850

William John Broderip (1789–1859) was a British lawyer (called to the Bar in 1817), conchologist and naturalist. He was the Thames Police Court Magistrate (1822–1846) and held the same position in Westminster (1846–1856) but had to resign as increasing deafness meant he could not hear the evidence! He was a founding fellow of the Zoological Society of London (1826).

Brodie, B. C.

Collared Owlet *Glaucidium brodiei* **E. Burton**, 1836

Sir Benjamin Collins Brodie (1783–1862) was a surgeon as was the author and they were friends. He was President of the Royal College of Surgeons (1844) and was 'Sergeant-surgeon' to Queen Victoria. Burton wrote in his etymology that he dedicated the bird '… in token of high respect and ancient friendship.'

Brodie, N.

Monarch sp. *Monarcha brodiei* **E. P. Ramsay**, 1879 NCR
[Alt. Solomons Monarch; JS *Symposiachrus barbatus*]

Neil Brodie (d.1894) was master and part owner of the schooner *Ariel* that traded between Australia and the Solomon Islands. It was wrecked on Bougainville Island (1880). The crew got ashore, but natives poisoned the mate and a crewman (according to a contemporary newspaper report). A passing vessel rescued Brodie and the rest of the crew. He was lost at sea whilst in command of his schooner *Renard*, which foundered in a tropical storm.

Brodkorb

Brodkorb's Swift *Chaetura nubicola* Brodkorb, 1938 NCR
[Alt. Chestnut-collared Swift; JS *Streptoprocne rutila*]

Common Blackbird ssp. *Turdus merula brodkorbi* **Koelz**, 1939 NCR
[JS *Turdus merula intermedius*]
Buff-breasted Flycatcher ssp. *Empidonax fulvifrons brodkorbi* **A. R. Phillips**, 1966
Rusty Sparrow ssp. *Aimophila rufescens brodkorbi* A. R. Phillips, 1966 NCR
[JS *Aimophila rufescens rufescens*]

Professor William Pierce Brodkorb (1908–1992) was an American zoologist and ornithologist who was Assistant Curator of Birds at the Museum of Zoology in Michigan (1936–1946). Later he became Assistant Professor of Biological Sciences at the University of Florida's Department of Zoology. He amassed an enormous collection, presented to the University of Florida, of 12,500 skeletons of birds from 129 families. He published a *Catalogue of Fossil Birds*. He was a prime mover behind what has become the Society for Avian Paleontology and Evolution (SAPE).

Brook

Brook's (Rainbow) Lorikeet *Trichoglossus haematodus brooki* **Ogilvie-Grant**, 1907
[NUI *Trichoglossus haematodus nigrogularis*]
Andean Guan ssp. *Penelope montagnii brooki* **Chubb**, 1917

Edward Jonas Brook (1865–1924) inherited wealth and devoted his time to fox-hunting and aviculture. He was a very keen parrot breeder, and a member of the Parrot Society wrote of the Black Lory *Chalcopsitta atra*: 'This was the first time the species had been bred in England since E. J. Brook was successful in 1909'. Brook was particularly successful in keeping birds-of-paradise, and at one time owned the world's finest collection of foreign birds. Living examples of the lorikeet were collected by Walter Goodfellow (q.v.) and maintained by Brook at Hoddom Castle, Scotland. Chubb (q.v.) gives no etymology in his description of the guan, but Goodfellow, in Brook's employ at the time (1914–1915), collected the type specimen in Ecuador.

Brooke, C. J.

Brooke's Scops Owl *Otus brookii* **Sharpe**, 1892
[Alt. Rajah Scops Owl]

Sir Charles Johnson Brooke (1829–1917) was born Charles Johnson, taking the name Brooke when he became the second 'White Rajah' of Sarawak (1868–1917) on the death of his uncle, James Brooke, the original 'White Rajah' (see **Rajah** and **Brooke, J.**). Rajah James invited Wallace (q.v.) to collect in Sarawak, but it is not clear if Wallace accepted the invitation although he clearly visited Sarawak as it inspired his 'Sarawak Law'. Sir Charles was a notable eccentric who lost an eye in a riding accident and replaced it with a false one which had originally been destined for a stuffed albatross. A mammal is named after him.

Brooke, C. V.

Banded Broadbill ssp. *Eurylaimus javanicus brookei*
H. C. Robinson & Kloss, 1919

Sir Charles Vyner de Windt Brooke (1874–1963), the great-nephew of James Brooke (q.v.), the first 'White Rajah', was Raja Muda of Sarawak and subsequently the third and final 'White Rajah' of Sarawak (1917–1946). He grew up there but was educated in England, graduating from Cambridge before returning to Sarawak (1911). A boom in oil and rubber gave him the revenue to modernise the local institutions and introduce (1924) a version of British law. He banned missionaries and fostered local traditions. He was in Australia (1942–1945) during the Japanese occupation, but resumed as Rajah for a few months (1946), then ceded Sarawak to Britain as a crown colony and retired to London.

Brooke, J.

Grey-and-buff Woodpecker ssp. *Hemicircus concretus brookeanus* **Salvadori**, 1868 NCR
[JS *Hemicircus concretus sordidus*]

Sir James Brooke (1803–1868), the first 'White Rajah' of Sarawak, is believed to have been the model for the eponymous hero of Joseph Conrad's novel *Lord Jim*. He worked for the Honourable East India Company and was near-fatally wounded in the Anglo-Burmese War (1825). He was inspired by the example of Sir Stamford Raffles and resolved to emulate him. He helped put down a rebellion and, with the blessing of the Sultan of Brunei, became Governor and Rajah of Sarawak (1841). He successfully suppressed the locals' propensity for piracy and headhunting. Three reptiles are named after him.

Brooke, R. K.

Levaillant's Cisticola ssp. *Cisticola tinniens brookei*
Herremans *et al.* 1999

Richard Kendall Brooke (1930–1996) was an Indian-born (to British parents) South African ornithologist. Rhodes University College, Grahamstown, awarded his BA in classical languages (1949) as he had originally intended to join the church. He spent his early career in administration in the Southern Rhodesia civil service (Zimbabwe) (1951–1971). He became a scientific officer at the Durban Natural History Museum (1972–1976), then became a clerk on the railways as he was recovering from Paradichlorobenzene poisoning (an insecticide used in the preservation of museum exhibits). He took a position at the Percy Fitzpatrick Institute of African Ornithology at the University of Cape Town where he remained for the rest of his life (1977–1996). His main interests were in distribution, systematics and behaviour particularly of swifts. He surveyed the birds of Kafue National Park, Zambia (1964), resulting in him being co-author of *Birds of Zambia* (1971). He also wrote 321 scientific papers, 248 popular articles and his own obituary!

Brooks, A. C.

Savannah Sparrow ssp. *Passerculus sandwichensis brooksi*
L. B. Bishop, 1915
Saw-whet Owl ssp. *Aegolius acadicus brooksi*
J. H. Fleming, 1916
Evening Grosbeak ssp. *Hesperiphona vespertina brooksi*
Grinnell, 1917

Allan Cyril Brooks (1869–1946) was born at Etawa, India, where his father, William E. Brooks (q.v.), was a civil engineer. Aged four, Allan was sent home to England and remained there until his father migrated to Canada to farm at Milton, Ontario (1881). Even as a small boy Allan was interested in natural history and was encouraged by his father's ornithologist friends. In England, the taxidermist John Hancock taught him how to blow eggs, collect butterflies and recognise plants. He was taught to prepare skins by Thomas McIlwraith (1885), the veteran birdman of eastern Canada. By age 18 he was a marksman and skinner. Brooks senior moved to British Columbia (1887) and Allan devoted every hour he could spare from farm work to search for and observe wildlife, hunting for and preparing specimens for museums and private collectors. Brooks junior received his first major contract (1906) to illustrate Dawson and Bowles's *Birds of Washington*. He also conserved songbirds near his home by the now-discredited method of shooting predators, including many birds of prey. Grinnell's (q.v.) study of grosbeaks was prompted when Brooks drew his attention to variations in his own collection made in British Columbia.

Brooks, W. E.

Brooks's Nuthatch *Sitta cashmirensis* W. E. Brooks, 1871
[Alt. Kashmir Nuthatch]
Brooks's Leaf Warbler *Phylloscopus subviridis*
W. E. Brooks, 1872
Brooks's Niltava *Cyornis poliogenys* W. E. Brooks, 1879
[Alt. Pale-chinned Flycatcher, Brooks's Blue Flycatcher]

William Edwin Brooks (1829–1899) worked as assistant to his father, W. A. Brooks of Newcastle-on-Tyne, who was the chief engineer on the Tyne docks. He went to India as civil engineer on the East India Railway (1856) and published several papers in Allan Hume's (q.v.) journal *Stray Feathers*.

Subsequently (1881) he moved to Canada where he collected specimens for the British Museum. He was the father of Allan Brooks (see above).

Brooks, W. S.

Sharpbill ssp. *Oxyruncus cristatus brooksi* **Bangs** & **Barbour**, 1922

Winthrop Sprague Brooks (1887–1965) was a collector and zoologist, Custodian of Bird's Eggs and Nests, Harvard Museum of Comparative Zoology (1928–1934). He spent time in Eastern Siberia and Alaska (1913–1914) on the Harvard University polar bear hunting expedition led by John Eliot Thayer. He spent several months collecting birds in the Falkland Islands (1917) and travelled in Australia (1926). Two reptiles are named after him.

Brooksbank

Red-rumped Wheatear ssp. *Oenanthe moesta brooksbanki* **Meinhertzhagen**, 1923

George Brooksbank (1863–1934) was a British schoolmaster who was Assistant Headmaster at Aysgarth School, Yorkshire. Meinertzhagen was a pupil there (1886–1889) and later he and Brooksbank became friends and used to go fishing together (1920s).

Brown, E. J.

Brown's Tern *Sternula antillarum browni* **Mearns**, 1916
[Alt. Least Tern ssp.]

Edward Johnson Brown (c.1872–c.1930) collected widely in the US for the USNM in the last few decades of the 19th and first few decades of the 20th centuries. The etymology reads: 'This form is named in honor of Mr. Edward Johnson Brown, who has contributed so largely to the collection of birds in the United States National Museum, from both the Atlantic and Pacific coasts of North America.' He published at least two articles in *Auk* (1894), e.g. 'The Black Tern at Washington, D. C.'

Brown, G.

Brown's Long-tailed Pigeon *Reinwardtoena browni* **P. L. Sclater**, 1877
[Alt. Pied Cuckoo Dove]
Brown's Monarch *Monarcha browni* **E. P. Ramsay**, 1883
[Alt. Kulambangra Monarch; Syn. *Symposiachrus browni*]

The Reverend George Brown (1835–1917) was a Methodist missionary to Melanesia. He was the first Christian missionary to have landed in New Ireland (1875), to the east of the mainland of Papua New Guinea. He spent much time exploring in the company of Cockerell (q.v.). He wrote an autobiography (1905). In the name *Ceyx lepidus sacerdotis* (New Britain Variable Kingfisher) the subgeneric name, which means priest, is a reference to Reverend George Brown, as this form of kingisher was described by Ramsay from a specimen from Brown's collection (1882). A mammal is named after him.

Brown, H.

Steller's Jay ssp. *Cyanocitta stelleri browni* **A. R. Phillips**, 1950 NCR
[JS *Cyanocitta stelleri macrolopha*]

Herbert Brown (1848–1913) was an amateur naturalist. He moved to Tucson, Arizona (1873), to prospect in the mountains, where he nearly died of thirst and survived a number of narrow escapes from Apaches. He worked as journalist, editor and newspaper proprietor in Tucson. He was President of the Audubon Society of Arizona, Curator of the University of Arizona Mineral Museum (1893–1913) and Clerk to the Superior Court of Pima County. A reptile is named after him.

Brown, R.

Brown's Rosella *Platycercus venustus* **Kuhl**, 1820
[Alt. Northern Rosella, Brown's Parakeet]
Brown's (Fairy) Wren *Malurus brownii* **Vigors** & **Horsfield**, 1827 NCR
[Alt. Red-backed Fairy-wren; JS *Malurus melanocephalus*]

Green Rosella ssp. *Platycercus caledonicus brownii* Kuhl, 1820

Robert Brown (1773–1858) was a botanist, educated at Aberdeen and Edinburgh, who studied the flora of Scotland (1791). He was an army official in Ireland (1795) and naturalist to Captain Flinders's (q.v.) Australasian expedition (1801–1805) on board *Investigator*. He was librarian to the Linnean Society and to Sir Joseph Banks (q.v.) and wrote *Prodromus Florae Novae Hollandiae et insula Van-Diemen* (1810).

Brown, W. W.

Timberline Wren *Thryorchilus browni* **Bangs**, 1902

Mountain Elaenia ssp. *Elaenia frantzii browni* Bangs, 1898
Stripe-tailed Yellow Finch ssp. *Sicalis citrina browni* Bangs, 1898
Slaty Vireo ssp. *Vireo brevipennis browni* A. H. Miller & **M. S. Ray**, 1944
Worthen's Sparrow ssp. *Spizella wortheni browni* Webster & Orr, 1954

Wilmot W. Brown Jr (c.1878–c.1953) was a US naturalist and professional collector in Central America, visiting Colombia for the Bangs Collection (1888–1899), Tobago (1892), Panama (1900–1904), and Mexico on the Thayer Expedition (1906–1909 and 1922). In all he collected for over 60 years. He was a trained biological collector who was aware of taxonomic issues and he prepared museum specimens in the field as well as making observations of the behaviour of the species he collected. Bangs (q.v.) wrote an article: 'Remarks on a collection of birds made by W. W. Brown Jr, at David and Divala, Chiriqui'.

Browning

Common Paradise Kingfisher ssp. *Tanysiptera galatea browningi* **Ripley**, 1983
Marsh Wren ssp. *Cistothorus palustris browningi* **Rea**, 1986
Mangrove Vireo ssp. *Vireo pallens browningi* **A. R. Phillips**, 1991

Dr M. Ralph Browning is an American ornithologist and taxonomist who worked at the USNM's Bird and Mammal Laboratories (1966–1994) through the US Geological Survey and the US Fish and Wildlife Service.

Bruce, H. J.

Bruce's Scops Owl *Otus brucei* **Hume,** 1872
[Alt. Pallid Scops Owl]

Brown-cheeked Fulvetta ssp. *Alcippe poioicephala brucei* Hume, 1870

The Reverend Henry James Bruce (1835–1909) was an American missionary in India (1862–1909). He was born at Hardwick, Massachusetts, and died at Panchgani (India). He published several books in the Marathi language.

Bruce, J.

Bruce's Green Pigeon *Treron waalia* **Meyer,** 1793
African barbet sp. *Pogonias brucii* **Rüppell,** 1837 NCR
[Alt. Black-billed Barbet; JS *Lybius guifsobalito*]

James Bruce (1730–1794) was a British explorer who discovered the source of the Blue Nile (1770). He was, for the time, a huge man of 6'4". He was trained in law in Edinburgh. He travelled in Europe before being appointed as Consul in Algiers (1862). He had many adventures but, after Dr Samuel Johnson and Boswell had taken against him, he was generally not believed in England, although in France he was much admired for his travels and achievements. Only after the death of his second wife (1790) did he start to write up his notes and diaries, writing *Travels to Discover the Source of the Nile* (1790). Later generations found that his 'unbelievable' adventures, which he had been accused of inventing, were in fact largely true. A mammal is named after him.

Bruch

Gull genus *Bruchigavia* **Bonaparte,** 1857 NCR
[Now in *Chroicocephalus*]

Carl Friedrich Bruch (1789–1857) was a German zoologist and ornithologist who proposed a trinomial system for naming races of birds. His name is here combined with the Latin word *gavia*, an unidentified seabird.

Brügel

Asian Barred Owlet ssp. *Glaucidium cuculoides bruegeli* Parrot, 1908

Dr Karl Brügel (fl.1900) was a German collector in Borneo (1908–1909) and Siam (Thailand).

Bruijn

Bruijn's Pygmy Parrot *Micropsitta bruijnii* **Salvadori,** 1875
[Alt. Red-breasted/Mountain Pygmy Parrot]
Torrent-lark *Grallina bruijni* Salvadori, 1876
[Syn. *Pomareopsis bruijni*]
Rajah Lory *Chalcopsittacus bruijnii* Salvadori, 1878 NCR
[Alt. Black Lory; JS *Chalcopsitta atra insignis*]
Bruijn's Bird-of-Paradise *Drepanornis bruijnii* **Oustalet,** 1880
[Alt. Pale-billed Sicklebill]

Bruijn's Brush-turkey *Aepypodius bruijnii* Oustalet, 1880
Bruijn's Riflebird *Craspedophora bruijni* **Büttikofer,** 1894 NCR
[Believed to be a hybrid: *Seleucidis melanoleucus* x *Ptiloris magnificus*]

Anton August (Antonie Augustus) Bruijn (1842–1890) was a Dutch plumassier or feather merchant, who was the son-in-law of Duivenbode (q.v.). As a young man he entered the service of the Royal Dutch Navy, and sailed to the Dutch East Indies. He married (1865) Adolphine van Rennesse van Duivenbode (see under Adolphina) on Ternate, in the Maluku Islands (Moluccas). He became a trader in bird-of-paradise plumes and other curiosities of natural history, and exhibited many specimens during the large international Colonial Trade Exhibition in Amsterdam (1883). He also supplied material to Salvadori (q.v.) as well as helping Beccari (q.v.) and D'Albertis (q.v.) when they were collecting in Arfak, New Guinea. Two mammals and a reptile are named after him.

Brunel

Firefinch ssp. *Lagonosticta rhodopareia bruneli* Erard & Roche, 1977 NCR
[Alt. Chad Firefinch; JS *Lagonosticta umbrinodorsalis*]

J. Brunel (DNF) was a French ornithologist. He spent some time in Vietnam and wrote articles such as 'Les oiseaux de la région du Lang-Bian, massif montagneux de la chaine annamitique' (1978).

Brunhilda

Waxbill genus *Brunhilda* **Reichenbach,** 1862 NCR
[Now in *Estrilda*]

Common Paradise Kingfisher ssp. *Tanysiptera galatea brunhildae* Jany, 1955

Brunhilda (or Brynhildr) was a shieldmaiden and valkyrie in Norse mythology and is a central character in Wagner's *Der Ring des Nibelungen*. She is the Valkyrie of the title of the second opera in the Ring cycle, *Die Walküre*, and appears in the third and fourth operas, *Siegfried* and *Die Götterdämmerung*. An amphibian is named after her.

Brünnich

Brünnich's Guillemot *Uria lomvia* **Linnaeus,** 1764
[Alt. Thick-billed Murre]
Kittiwake sp. *Rissa brunnichii* **J. F. Stephens,** 1826 NCR
[Alt. Black-legged Kittiwake; JS *Rissa tridactyla*]

Morten Thrane Brünnich (1737–1827) was Bird Curator in Copenhagen and worked on guillemots. He is described by the Danish bibliographer Jean Anker as the 'founder of Danish faunistic zoology' and was a highly praised systematist. He wrote: *Ornithologia Borealis* (1764), which was a description of a private collection of birds of Denmark, Iceland and other parts of northern Europe.

Bryan, E. H.

Bryan's Shearwater *Puffinus bryani* Pyle *et al.*, 2011

Hawaii Elepaio ssp. *Chasiempis sandwichensis bryani* H. D. Pratt, 1979

Edwin Horace Bryan Jr (1898–1985) went to Hawaii at his uncle's invitation to attend college (1916) where he studied entomology, later teaching the course there. He went on to be Curator at the Bishop Museum, Honolulu (1919–1968). He wrote a number of books including *Ancient Hawaiian Life* (1938) and *Panala'au Memoirs* (1974). He wrote that W. A. Bryan (below) on meeting him said: 'If we are to masquerade around here under the same surname we might as well get acquainted.' They never found common ancestors but became good friends. The shearwater was collected on Midway Atoll (1963) but lay misidentified in a museum drawer until Pyle of the Institute for Bird Populations noted differences in measurement and it was confirmed a new species through DNA analysis. It was presumed extinct but later rediscovered (2012).

Bryan, W. A.

Yellow Bittern ssp. *Ixobrychus sinensis bryani* Seale, 1901 NCR; NRM

Prof. William Alanson Bryan Jr (1874–1942) was an American zoologist, Curator of Ornithology at the Bishop Museum, Honolulu (1901–1907), and a member of the faculty of the College of Hawaii (1907–1919). He visited Easter Island and Chile (1920). He was Director, Los Angeles County Museum of History, Science, and Art (1921–1940). He wrote *Natural History of Hawaii* (1916).

Bryant, E. B.

Magenta-throated Woodstar *Calliphlox bryantae* **Lawrence**, 1867

Mrs Elizabeth Brimmer Bryant *née* Sohier (1823–1916) was the wife (1848) of US ornithologist Dr Henry Bryant (below).

Bryant, H. P.

Bryant's Savannah Sparrow *Passerculus sandwichensis alaudinus* **Bonaparte**, 1853
Bryant's Golden Warbler *Dendroica petechia bryanti* **Ridgway**, 1873
[Alt. Mangrove Warbler ssp.]
Red-winged Blackbird ssp. *Agelaius phoeniceus bryanti* Ridgway, 1887
Bryant's Grassquit *Tiaris olivaceus bryanti* Ridgway, 1898
[Alt. Yellow-faced Grassquit ssp.]

Dr Henry Perkins Bryant (1820–1867) of Boston, Massachusetts, graduated from Harvard (1840) and was a physician and naturalist who trained in Paris and served a year in the French army in North Africa before returning to Boston. As he was in poor health and did not practise medicine he spent his time bird collecting instead, travelling widely in North America and the Caribbean. During the American Civil War he was the surgeon of the 20th Massachusetts Regiment. Lafresnaye's (q.v.) collection was put up for sale (1866) by his widow and Bryant bought the lot and donated it to the Boston Natural History Society. A mammal is named after him.

Bryant, W. P. E.

Bryant's Cactus Wren *Campylorhynchus brunneicapillus bryanti* **A. W. Anthony**, 1894

Walter Pierce E. Bryant (1861–1905) was an American ornithologist, Bird Curator at the California Academy of Sciences. He was founding President of the California Ornithological Club.

Buchanan, A.

Southern Grosbeak Canary *Serinus buchanani* **Hartert**, 1919
[Syn. *Crithagra buchanani*]

Bush Petronia ssp. *Petronia dentata buchanani* Hartert, 1921 NCR; NRM
[Syn. *Gymnoris dentata*]
Fulvous Chatterer ssp. *Turdoides fulva buchanani* Hartert, 1921
Pale Crag Martin ssp. *Ptyonoprogne obsoleta buchanani* Hartert, 1921
Vieillot's Barbet ssp. *Lybius vieilloti buchanani* Hartert, 1924

Captain Angus Buchanan (1886–1954) was a Scottish explorer in northern Africa. As well as collecting bird and mammals he made film and photographic records of his explorations. Buchanan once reported (1922) that he saw a caravan leave Tabelot, Niger, with 7,000 camels and 1,100 men. The train stretched 6 miles from front to back.

Buchanan, F. H.

Rufous-fronted Prinia *Prinia buchanani* **Blyth**, 1844
Buchanan's Bunting *Emberiza buchanani* Blyth, 1845
[Alt. Grey-necked/Grey-hooded Bunting]

Dr Francis Hamilton Buchanan (1762–1829) was a Scottish physician. He was Assistant Surgeon on board a man-of-war, but had to retire through ill health. His health stayed poor for some years, but (1794) he took a job as Surgeon for the East India Company in Bengal. The voyage to India seemed to restore his health fully and, once there, he started collecting plants in Pegu, Ava and the Andaman Islands. He presented these specimens, with drawings, to Sir Joseph Banks (q.v.). He also studied fish. He worked for the Board of Trade in Calcutta (1798), then (1800–1807) continued to study plants and animals in Mysore until his return to Britain. Buchanan collected in both India and Nepal. He returned to Calcutta to become Superintendent of the botanical gardens there, and later became Surgeon to the Governor-General of India. He returned to Britain (1815) following the death of his elder brother, being heir to the family seat at Leney, Perthshire, following which he changed his name to Hamilton, his mother's maiden name. He spent the rest of his life at Leney, 'improving' the gardens by planting many exotics there. He wrote *Travels in the Mysore* and *A History of Nepal*.

Buchanan, W.

Thornbill sp. *Acanthiza buchanani* **Vigors** & **Horsfield**, 1827 NCR
[Indeterminate: possibly based on a female Australian chat (*Epthianura* sp.)]

Walter Buchanan (1787–1856) was an English businessman, naturalist and collector. He was active as a magistrate in Middlesex and prominent in the management of the county lunatic asylum at Hanwell. He was a Fellow of the Linnean Society (1817–1856).

Buchen, W. & M. L.

Souimanga Sunbird ssp. *Cinnyris sovimanga buchenorum* **J. G. Williams,** 1953

Walther Buchen (1887–1961), was an American advertising executive, collector in East Africa and a Trustee of Chicago Natural History Museum. With his wife Margaret L. Buchen (1890–1971) *née* Head he sponsored and led an expedition to the Upper Nile in Kenya and Uganda (1952).

Buckley, C.

Buckley's Ground Dove *Columbina buckleyi* **P. L. Sclater** & **Salvin,** 1877
[Alt. Ecuadorian Ground Dove]
Buckley's Cotinga *Laniisoma buckleyi* P. L. Sclater & Salvin, 1880
[Alt. Andean Laniisoma; Syn. *Laniisoma elegans buckleyi*]
Buckley's Mountain Hummingbird *Pinarolaema buckleyi* **Gould,** 1880 NCR
[Alt. Buckley's Violet-ear; believed to represent an aberrant *Colibri coruscans*]
Buckley's Forest Falcon *Micrastur buckleyi* Swann, 1919
[Alt. Traylor's Forest Falcon]

Black-crowned Tityra ssp. *Tityra inquisitor buckleyi* Salvin & **Godman,** 1890
Marbled Wood Quail ssp. *Odontophorus gujanensis buckleyi* **Chubb,** 1919

Clarence Buckley (d.1889) was a collector of natural history specimens in Ecuador and Bolivia. He remained in South America until c.1880, but details of his death appear to be unrecorded. Buckley collected for a number of artists and ornithologists, including Gould. The British Museum houses a large fish collection made by him, including several eponymous species. There are also over 80 species of birds collected by him in BMNH. Four amphibians and a reptile are named after him.

Buckley, T. E.

Buckley's Lark *Mirafra rufocinnamomea buckleyi* **G. E. Shelley,** 1873
[Alt. Flappet Lark ssp.]
White-throated Francolin ssp. *Peliperdix albogularis buckleyi* **Ogilvie-Grant,** 1892

Thomas Edward Buckley (1846–1902) was a zoologist and palaeontologist. He made four collecting trips to Africa (1872–1889). He also collected c.350 nineteenth-century books on Central and Southern Africa and bequeathed them to the Cambridge University Library.

Buddingh

Asity genus *Buddinghia* **Schlegel** & **Pollen,** 1867 NCR
[Now in *Philepitta*]

(See **Cornelia**)

Buesing

Blue-capped Tanager ssp. *Thraupis cyanocephala buesingi* **Hellmayr** & **Seilern,** 1913

Adolf Freiherr von Büsing-Orville (1867–1948) was a very wealthy German industrialist.

Buffon

Long-winged Harrier *Circus buffoni* **Gmelin,** 1788
Starling sp. *Oriolus buffonianus* Shaw, 1809 NCR
[Alt. White-shouldered Starling; JS *Sturnia sinensis*]
Buffon's Macaw *Ara ambiguus* **Bechstein,** 1811
[Alt. Great Green Macaw]
Buffon's Skua *Stercorarius longicaudus* **Vieillot,** 1819
[Alt. Long-tailed Skua/Jaeger]
Buffon's Plumeleteer *Chalybura buffonii* **Lesson,** 1832
[Alt. White-vented Plumeleteer, Buffon's Hummingbird]
Cockatoo sp. *Plictolophus buffoni* **Finsch,** 1867 NCR
[Alt. Yellow-crested Cockatoo; JS *Cacatua sulphurea parvula*]

Buffon's Green Turaco *Tauraco persa buffoni* Vieillot, 1819
[Alt. Guinea Turaco ssp.]
Buffon's Piculet *Picumnus exilis buffoni* **Lafresnaye,** 1845
[Alt. Golden-spangled Piculet ssp.]

Count Georges Louis Leclerc de Buffon (1707–1788) is one of the giants of zoology and widely written about, hence the brevity of this entry. He is most famous for having developed the species concept, the basis of all taxonomy. Louis XV appointed him to the Academy of Sciences (1734) and later Director of the Jardin du Roi (1739), which Buffon transformed into the present Jardin des Plantes, Paris. He envisaged that his *Histoire Naturelle Générale et Particulière* (1749–1804) would encompass 50 volumes, but 'only' 36 had been produced by the time of his death, and a further 8 posthumously. Two mammals are named after him.

Bugun

Bugun Liocichla *Liocichla bugunorum* Athreya, 2006

The Bugun tribe live close to the sanctuary on the Indian/Chinese border where Ramana Athreya, an astronomer, first saw the bird (1996). There were initially problems over recognition as no type specimen exists as the describer considering it too rare to justify taking a type, so he captured and photographed one and took a few feathers as a DNA sample, then released it.

Bulger

Grey Fantail ssp. *Rhipidura albiscapa bulgeri* **Layard,** 1877

Lieutenant-Colonel George Ernest Bulger (1830–1881) was a British soldier in the 10th Regiment of Foot (1847–1876). He served in South Africa (1860–1864) and later in India and wrote *Notes of a Tour from Bangalore to Calcutta: thence to Delhi, and, Subsequently, to British Sikkim during the Early Part of 1867* (1869). He was also an amateur ornithologist, writing 'List of birds obtained in Sikkim, eastern Himalayas, between March and July 1867' in *Ibis* (1869).

Bull

Brown-capped Vireo ssp. *Vireo leucophrys bulli*
    J. S. Rowley, 1968

Daniel 'Dan' Bernard Bull, who lived at La Mesa, California, was a great friend of the describer who wrote that Bull '… has devoted a lifetime to the study of ornithology and oology'.

Buller, K. G.

White-breasted Whistler ssp. *Pachycephala lanioides bulleri* **Mayr**, 1954 NCR
[JS *Pachycephala lanioides carnavoni*]

Kenneth Gordon Buller (1915–1995) was an English-born collector and taxidermist who moved to Australia (1926). He worked for the Western Australian Zoological Garden (1939–1941), then as a collector for the Water Supply Department on the Canning Stock Route (1943–1945). He collected for the AMNH (1947–1948), then at Western Australian Museum, Perth, until retirement (1948–1976). He was bitten by a Western Tiger Snake (1951) that was in a sack sent (by rail) to the museum and left in the basement apparently containing bottled lizard specimens.

Buller, W. L.

Parakeet genus *Bulleria* **Iredale** & **Mathews**, 1926 NCR
[Now in *Cyanoramphus*]

Buller's Gull *Chroicocephalus bulleri* **Hutton**, 1871
[Alt. Black-billed Gull; Syn. *Larus bulleri*]
Buller's Shearwater *Puffinus bulleri* **Salvin**, 1888
Buller's Albatross *Thalassarche bulleri* **Rothschild**, 1893

Sir Walter Lawry Buller (1838–1906) was a New Zealand lawyer and ornithologist, author of *A History of the Birds of New Zealand*. He was also a government minister (1896–1899). Although born in New Zealand, he achieved his ambition to go 'home' and so he died in England, a country gentleman. He tried to sell Rothschild a specimen of the now-extinct North Island Laughing Owl *Sceloglaux albifacies rufifacies* for a large sum of money. Rothschild examined the bird and realised that it was partly faked, as its tail came from an owl of a different genus. Rothschild pointed all this out publicly and loudly, and Buller was humiliated. (See **Lawry**)

Bullock, D. S.

Thorn-tailed Rayadito ssp. *Aphrastura spinicauda bullocki*
    Chapman, 1934

Professor Dr Dillman Samuel Bullock (1878–1971) was an agriculturist, naturalist and Methodist missionary. He graduated from Michigan State University (1902) and worked for the American Missionary Society, Bunster Agricultural School in Angol, Chile (1902–1912). He taught back in the US (1912–1921) before returning to South America. For a year he worked as Agricultural Commissioner, Department of Agriculture, at the US embassy, Buenos Aires. The etymology reads: 'With much pleasure I name this form for its collector, Mr. D. S. Bullock, of Angol, Chile, as a slight recognition of the service he has rendered ornithology in Chile'. He wrote a series of papers, mostly on the birds of Chile, (1920s–1940s) including 'North American bird migrants in Chile'.

Bullock, W.

Red-throated Bee-eater *Merops bulocki* **Vieillot**, 1817
[Syn. *Merops bullockii* **A Smith**, 1834]
Bullock's Oriole *Icterus bullockii* **Swainson**, 1827
Magpie-jay sp. *Calocitta bullockii* **Wagler**, 1827 NCR
[Alt. White-throated Magpie-jay; JS *Calocitta formosa*]

William Bullock (1773–1849) was an English goldsmith, traveller and amateur naturalist. He established a travelling museum in London, with 3,000 skins and 32,000 other curiosities that he had collected but later auctioned off (1819). He travelled with his son, William Bullock Jr (who was a mine agent and artist who also collected), and Ferdinand Deppe in the southern US and Mexico where he bought up abandoned silver and gold mines. He collected a number of specimens new to science. His friend Swainson (q.v.), to whom he sold many specimens, named the oriole in his and his son's honour from a bird that Bullock Sr collected when visiting one of his mines near Mexico City.

Bulwer, H.

Bulwer's Pheasant *Lophura bulweri* **Sharpe**, 1874

Sir Henry Ernest Gascoyne Bulwer (1836–1914) was a career diplomat who became Governor of Labuan, Borneo (1871–1875), then Lieutenant-Governor of Natal (1875–1880), at which time his secretary was Henry Rider Haggard, author of *King Solomon's Mines* and *She*. He was High Commissioner in Cyprus (1886–1892). He was related to Sir Henry Bulwer-Lytton, the Victorian novelist.

Bulwer, J.

Petrel genus *Bulweria* **Bonaparte**, 1843
Petrel genus *Pseudobulweria* **Mathews**, 1936

Bulwer's Petrel *Bulweria bulwerii* **Jardine** & **Selby**, 1828
Olson's Petrel *Bulweria bifax* **Olson**, 1975 EXTINCT
[Alt. St Helena Bulwer's Petrel]

The Reverend James Bulwer (1794–1879) was an English collector, naturalist and conchologist. He spent several winters travelling in Spain, Portugal and Madeira, and collected the eponymous petrel in Madeira (1825).

Bunge

Siberian Jay ssp. *Perisoreus infaustus bungei* **Sushkin** &
    B. Stegmann, 1929 NCR
[JS *Perisoreus infaustus yakutensis*]

Dr Alexander Georg von Bunge (1803–1890) was a Baltic German botanist who explored in Siberia (1827) and Mongolia (1831). He became Professor of Botany at the University of Tartu (1842–1844).

Buquet

Buquet's Puffleg *Eriocnemis luciani* **Bourcier**, 1847
[Alt. Sapphire-vented Puffleg]

(See **Lucian (Buquet)**)

Burbidge

Burbidge's Blue-backed Parrot *Tanygnathus sumatranus burbidgii* **Sharpe**, 1879

Reddish Scops Owl ssp. *Otus rufescens burbidgei* **Hachisuka**, 1934 NCR

[Origin of only specimen disputed; may belong in nominate race]

Frederick William Burbidge (1847–1920) was an English explorer, discoverer, botanist, artist and lithographer of plants. He was Gardener at Kew Gardens (1868–1870). He was sent to Borneo by James Veitch & Sons, and took part in a botanical expedition to the Sulu Archipelago. He is well known for his exploration of the Kinabalu region, where he collected many orchids (1877–1879) and a small collection of birds incuding the parrot. He wrote *Gardens of the Sun: A Naturalist's Journal of Borneo and the Sulu Archipelago* (1880). He later became Director of the Botanical Gardens at Trinity College, Dublin. He named a number of plants to honour his wife.

Burchell

Burchell's Gonolek *Laniarius atrococcineus* Burchell, 1822
[Alt. Crimson-breasted Shrike]

Turaco sp. *Corythaix burchelli* **A. Smith**, 1831 NCR
[Alt. Purple-crested Turaco; JS *Tauraco porphyreolophus*]

Burchell's (Glossy) Starling *Lamprotornis australis* A. Smith, 1836

Burchell's Courser *Cursorius rufus* **Gould**, 1837

Burchell's Coucal *Centropus burchellii* **Swainson**, 1838

Burchell's Bustard *Neotis burchellii* **Heuglin**, 1867 NCR
[Unique specimen; race or synonym of *Neotis denhami*]

Burchell's Sandgrouse *Pterocles burchelli* **W. L. Sclater**, 1922

William John Burchell (1781–1863) was an English explorer-naturalist who lived on St Helena (1805–1810) as a merchant and then as the local schoolmaster and official botanist. He went to the Cape of Good Hope (1810), undertaking a major exploration of the South African interior (1811–1815) during which he travelled more than 7,000 km through largely unexplored country, documenting this adventure in the two-volume *Travels in the Interior of Southern Africa* (1822 and 1824). He was renowned as a meticulous collector, botanist and artist. He returned to London (1815) to work on his collections. He spent two months in Lisbon (1825) and then went to Brazil, where he collected extensively before returning to home (1830). He became increasingly reclusive and in his last two years became seriously ill, eventually committing suicide. Among other taxa a reptile and a mammal are named after him.

Bureau

Long-tailed Tit ssp. *Aegithalos caudatus bureaui* **Jouard**, 1929 NCR
[JS *Aegithalos caudatus taiti*]

Short-toed Treecreeper ssp. *Certhia brachydactyla bureaui* Jouard, 1929 NCR
[JS *Certhia brachydactyla megarhyncha*]

Dr Louis Marcellin Bureau (1847–1936) was a French physician (graduated Paris, 1877) and zoologist. He was Director of the Natural History Museum in Nantes (1882–1919). After he retired he visited Tunisia (1922).

Buresch

Crested Tit ssp. *Lophophanes cristatus bureschi* **von Jordans**, 1940

Dr Ivan Yosipov Buresch (1885–1980) was a Bulgarian zoologist, Director of the Royal (later National) Museum of Natural History and Professor at the Institute of Zoology, Sofia (1914–1959). Of Czech descent, he graduated in natural science at Charles University, Prague (1909), and undertook postgraduate study at Munich University. An amphibian is also named after him.

Burg

Coal Tit ssp. *Periparus ater burgi* **Jouard**, 1928 NCR
[JS *Periparus ater ater*]

Gustave von Burg (1871–1927) was a Swiss ornithologist who served as President of the Swiss Ornithological Society (1909–1913).

Bürgers

Bürgers' Hawk *Erythrotriorchis buergersi* **Reichenow**, 1914
[Alt. Chestnut-shouldered Goshawk]

Papuan Needletail ssp. *Mearnsia novaeguineae buergersi* Reichenow, 1917

Bürgers's Ground Parrot *Psittacella brehmii buergersi* Reichenow, 1918 NCR
[Alt. Brehm's Tiger Parrot; JS *Psittacella brehmii pallida*]

Papuan Scrubwren ssp. *Sericornis papuensis buergersi* **Stresemann**, 1921

Black Pitohui ssp. *Pitohui nigrescens buergersi* Stresemann, 1922

Bürgers's Blue-collared Parrot *Geoffroyus simplex buergersi* **Neumann**, 1922

Chestnut-backed Jewel-babbler ssp. *Ptilorrhoa castanonota buergersi* **Mayr**, 1931

Theodore Joseph Bürgers (1881–1954) was a German physician and zoologist who participated in the German Sepik Expedition, New Guinea. He supervised the collecting of 3,100 specimens of 240 species of birds, which were sent to the Humboldt University Museum in Berlin. He became Professor of Hygiene and Bacteriology in Düsseldorf (1923) and later in Göttingen. Other New Guinea fauna, including a moth, two reptiles and a mammal are named after him.

Burhan

Togian Hawk Owl *Ninox burhani* Indrawan & Somadikarta, 2004

Burhan is a local resident of Benteng village where this owl was first observed; the name honours his knowledge of the birdlife of Togian. He is one of several villagers interested in studying and conserving local birds. The owl is endemic to the Togian Islands near Sulawesi where the holotype was collected in a local swamp (2001). The describers said at the time that the unexpected discovery of a new endemic species, on an island group so close to the eastern peninsula of Sulawesi, offers hope that further endemic vertebrates may be found. But they warned: '... the discovery provides additional justification to halt forest clearance, which is increasing in the Togian Islands'.

Burke

Green-crowned Warbler *Seicercus burkii* **E. Burton**, 1836

William Augustus Burke (1769–1837) was a colonial civil servant, Inspector General of Hospitals in Bengal (1825). He made an extensive collection of bird specimens from northern India.

Burleigh

Burleigh's Carolina Wren *Thryothorus ludovicianus burleighi* **Lowery**, 1940
Blue Jay ssp. *Cyanocitta cristata burleighi* G. Bond, 1962 NCR
[JS *Cyanocitta cristata bromia*]
Tufted Flycatcher ssp. *Mitrephanes phaeocercus burleighi* **A. R. Phillips**, 1966
Northern Rough-winged Swallow ssp. *Stelgidopteryx serripennis burleighi* A. R. Phillips, 1986

Thomas Dearborn Burleigh (1895–1973) was an American ornithologist and collector. He served in the US Army in France in WW1. He took a master's degree in forestry at the University of Washington (1920) and worked for the US Fish & Wildlife Service (1930–1961). He wrote *Georgia Birds* (1958).

Burmeister

Rough-legged Tyrannulet *Phyllomyias burmeisteri* **Cabanis & Heine**, 1859
Burmeister's Seriema *Chunga burmeisteri* **Hartlaub**, 1860
[Alt. Black-legged/Lesser Seriema]
Slender-tailed Woodstar *Microstilbon burmeisteri* **P. L. Sclater**, 1887

Karl Hermann Konrad Burmeister (1807–1892) was a German ornithologist, Professor and Director of the Institute of Zoology of Martin Luther University at Halle Wittenberg, Germany (1837–1861). He was in the Prussian civil service but got his release by claiming that a persistent stomach complaint was caused by arsenic emissions in the museum and the drinking water in Halle, which had high natrium sulphate content. He sent many specimens to the zoological collections at the institute. These were largely obtained during his two expeditions to South America, Brazil (1850–1852) and La Plata region of Argentina (1857–1860). Subsequently, he was resident in Argentina (1861–1892). He was founding Director of the Institute at the Museo Nacional in Buenos Aires, remaining in post until retirement (1880). He wrote *Reise nach Brasilien* (1853). A mammal, a reptile and two amphibians are named after him.

Burnes

Prinia genus *Burnesia* **Jerdon**, 1863 NCR
[Now in *Prinia*]

Rufous-vented Prinia *Laticilla burnesii* **Blyth**, 1844
[Syn. *Prinia burnesii*]

Sir Alexander 'Bokhara' Burnes (1804–1841) was a British civil servant in Afghanistan (1839–1841) and deputy envoy to the court of Shah Shujah. He was also known as a great womaniser and lavish entertainer, which was not appreciated by the local people. He died during an uprising there (1841) having ordered his men to fire on the attacking mob.

Burnier

Kilombero Weaver *Ploceus burnieri* N. E. Baker & E. M. Baker, 1990

Eric Burnier is a Swiss physician and amateur naturalist who worked at the Ifakara, Tanzania medical research station. He was the first to draw attention to this weaver. He also discovered the Algerian Nuthatch *Sitta ledanti* but someone else got the honour (see under Ledant).

Burrough

Lesser Yellow-headed Vulture *Cathartes burrovianus* **Cassin**, 1845

Dr Marmaduke Burrough (1767–1844) was an American physician and naturalist, US Consul in Calcutta (1828–1830) and Vera Cruz, Mexico (1840). Like Cassin (q.v.) he was a member of the Academy of Natural Sciences of Philadelphia, and he presented the Academy with a specimen (1844) of the vulture that he collected in Mexico. He collected the second rhinoceros to be brought alive to the US (1830). It was of the Indian species, and was displayed in Philadelphia before being sold.

Burton, C.

Yellow Thornbill ssp. *Acanthiza nana burtoni* **Mathews**, 1920 NCR
[JS *Acanthiza nana nana*]

Charles Burton (fl.1896) was a British naturalist and collector. His father was the taxidermist Henry James Burton (q.v.).

Burton, E.

Spectacled Finch *Callacanthis burtoni* **Gould**, 1838

Major Edward Burton (1790–1867) was an army surgeon. He was stationed at Chatham, England (1829–1837), and wrote *A Catalogue of the Collection of Mammalia and Birds in the Museum at Fort Pitt, Chatham* (1838). He wrote a paper on fishes that Cuvier had described. Two reptiles and a mammal are named after him.

Burton, H. J.

Firecrown sp. *Eustephanus burtoni* **Boucard**, 1891 NCR
[Alt. Green-backed Firecrown; JS *Sephanoides sephaniodes*]

Henry James Burton (DNF) was a British taxidermist and natural history dealer in London (1868–1886).

Burton, P. J. K.

Blue-capped Kingfisher ssp. *Halcyon hombroni burtoni* duPont, 1976 NCR
[JS *Actenoides hombroni hombroni*]

Dr Philip John Kennedy Burton (b.1936) is an ornithologist and wildlife artist. He graduated BSc (Zoology) from University College London (1959) and took a postgraduate certificate in education (1960). He was a founding member of the Society of Wildlife Artists (1964). He obtained his PhD from the University of London (1969). He began work at the BMNH Bird Section (1967) as Senior Scientific Officer, then Principal Scientific Officer (1971) in charge of the anatomical collections until early retirement (1988). He took part in several expeditions to Panama, Guyana, Sarawak and elsewhere. He has illustrated a number of books including *Collins Field Guide to Bird Nests, Eggs and Nestlings* (1998), and *Raptors of the World* (2001).

Burton, R. F.

Burton's Grosbeak *Serinus burtoni* **G. R. Gray**, 1862
[Alt. Thick-billed Seedeater; Syn. *Crithagra burtoni*]

Sir Richard Francis Burton (1821–1890) was a noted British explorer, linguist, author and devotee of erotica. He began his career as an army officer but is best remembered as an explorer who, among much more, searched for the source of the Nile. He was appointed British Consul in Trieste (at that time part of the Austro-Hungarian Empire) (1872) and lived there for the rest of his life. He was fluent in over 20 languages and devoted his time to literature, translating the *Kama Sutra* (1883), *The Arabian Nights* (1885) and *The Perfumed Garden* (1886) into English. Immediately after his death his widow burned all his papers – an action that was condemned as one of the greatest acts of literary vandalism of all time. A mammal and a reptile are named after him.

Bury

Yemen Warbler *Sylvia buryi* **Ogilvie-Grant**, 1913
[Alt. Yemen Parisoma; Syn. *Parisoma buryi*]

Southern Grey Shrike ssp. *Lanius meridionalis buryi* J. R. Lorenz & **Hellmayr**, 1901
Streaked Scrub Warbler ssp. *Scotocerca inquieta buryi* Ogilvie-Grant, 1902
Buff-spotted Flufftail ssp. *Sarothrura elegans buryi* Ogilvie-Grant, 1908 NCR
[NUI *Sarothrura elegans elegans*]

George Wyman Bury (1874–1920) was an English naturalist, explorer, political officer and Arabist. He wrote *The Land of Uz* (1911) and *Arabia Infelix or the Turks in Yamen* (1915). His career came to an early end due to an unjust charge of corruption. He was closely associated with the 'Arab Revolt'. A reptile is named after him.

Bustamente

Plushcap genus *Bustamantia* **Bonaparte**, 1844 NCR
[Now in *Catamblyrhynchus*; the spelling *Bustamentia* can also be found]

General Anastasio Bustamante y Oseguera (1780–1853) was a physician and army officer who was President of Mexico (1830–1832, 1837–1839 and 1839–1841). He first came to power by leading a coup. He was deposed twice and twice exiled to Europe.

Butler, A. L.

Nicobar Sparrowhawk *Accipiter butleri* **Gurney**, 1898
Hoopoe sp. *Upupa butleri* **Madarász**, 1911 NCR
[Alt. Eurasian Hoopoe; JS *Upupa epops*]

Stone Partridge ssp. *Ptilopachus petrosus butleri* **W. L. Sclater** & **Praed**, 1920 NCR
[JS *Ptilopachus petrosus petrosus*]
Butler's Rock Dove *Columba livia butleri* **Meinertzhagen**, 1921
Arabian Bustard ssp. *Ardeotis arabs butleri* **Bannerman**, 1930

Arthur Lennox Butler (1873–1939) was born in Karachi, the son of Edward Arthur Butler (q.v.). He went to Ceylon (Sri Lanka) (1891) to become a tea-planter, but became a professional collector instead. He was Curator of the State Museum at Kuala Lumpur, Malaya (1898), and Superintendent of Game Preservation in the Sudan (1901–1915), then went to England and studied hummingbirds. Four reptiles and an amphibian are named after him.

Butler, E. A.

Hume's Owl *Strix butleri* **Hume**, 1878

Lieutenant Colonel Edward Arthur Butler (1842–1916) was a British ornithologist and collector. He arrived in India aged 24, after three years service in Gibraltar. Hume (q.v.), whom he met quite soon after his arrival, was able to use his influence to get Butler on a cable-laying ship called *Amberwitch*, on a trip from Karachi to the Gulf of Oman and back. After service in South Africa in the first Boer War (1880) his regiment returned to the UK and he retired from the army (1883). After his wife died (1912) his health declined and he became depressed and committed suicide (1916). He wrote *Birds of Sind, Cutch and Neighbourhood* (1879).

Butler, W. H.

Butler's Corella *Cacatua pastinator butleri* J. Ford, 1987
[Alt. Western Corella]

Dr William Henry 'Harry' Butler (b.1930) was born in Perth and trained as a teacher. He began working for corporate and government bodies as an environmental consultant and collector (1963) and undertook a major study of Western Australian animals. He collected more than 2,000 examples of mammals, 14 new to science. A passionate conservationist, he presented the popular ABC television series 'In the Wild' (from 1976). He received the Australian of the Year award (1979) and was awarded an honorary Doctorate of

Science by the Edith Cowan University in Perth (2003). One mammal and six reptiles are named after him.

Büttikofer

Thicketbird/Grassbird genus *Buettikoferia* **Madarász**, 1902 NCR
[NPRB *Buettikoferella*]
Thicketbird/Grassbird genus *Buettikoferella* **Stresemann**, 1928

Büttikofer's Warbler *Buettikoferella bivittata* **Bonaparte**, 1850
[Alt. Buff-banded Thicketbird, Buff-banded Grassbird]
Büttikofer's Babbler *Pellorneum buettikoferi* **Vorderman**, 1892
[Alt. Sumatran Babbler]
Apricot-breasted Sunbird *Cinnyris buettikoferi* **Hartert**, 1896

Water Thick-knee ssp. *Burhinus vermiculatus buettikoferi* **Reichenow**, 1898
Little Spiderhunter ssp. *Arachnothera longirostra buettikoferi* **E. Oort**, 1910
White-bellied Woodpecker ssp. *Dryocopus javensis buettikoferi* **Richmond**, 1912 NCR
[NUI *Dryocopus javensis javensis*]
Buffy Fish Owl ssp. *Ketupa ketupu buettikoferi* **Chasen**, 1935 NCR
[JS *Ketupa ketupu minor*]

Johan Büttikofer (1850–1929) was a Swiss zoologist. He made two collecting trips to Liberia (1879–1882 and 1886–1887), but curtailed the second due to ill health. He did, however, make one more trip there (1888). He accompanied Nieuwenhuis to Borneo (1893–1894) where they explored and collected. He was the Director of the Rotterdam Zoo (1897–1924). He wrote 'Zoological researches in Liberia. A list of birds, collected by the author and Mr F. X. Stampfli during their last sojourn in Liberia'. A number of fish, three mammals and two reptiles are named after him.

Büttner

Guinea Turaco ssp. *Tauraco persa buettneri* **Reichenow**, 1891 NCR
[JS *Tauraco persa persa*]

Dr Oscar Alexander Richard Büttner (1858–1927) was a German collector and botanist in Africa. He collected in the Congo (1884–1886), Togo (1890–1892) and Ghana for the Humboldt University Natural History Museum, Berlin, where he became Professor (1893). He is commemorated in the scientific names of a reptile, a mammal and all sorts of taxa from insects to lichens.

Button

Whyte's Barbet ssp. *Stactolaema whytii buttoni* **C. M. N. White**, 1945
Stierling's Wren Warbler ssp. *Calamonastes stierlingi buttoni* C. M. N. White, 1947 NCR
[NUI *Calamonastes stierlingi stierlingi*]

Captain Earl L. Button (b.1913) was a colonial administrator in Northern Rhodesia (Zambia) (1937) until independence

(1964), when he continued as an employee of the Zambian government.

Buturlin

Common Pheasant ssp. *Phasianus colchicus buturlini* **A. H. Clark**, 1907 NCR
[JS *Phasianus colchicus karpowi*]
Corn Bunting ssp. *Emberiza calandra buturlini* **Johansen**, 1907
Grey Partridge ssp. *Perdix perdix buturlini* **Zarudny** & **Loudon**, 1907 NCR
[JS *Perdix perdix robusta*]
Goldcrest ssp. *Regulus regulus buturlini* Loudon, 1911
Common Treecreeper ssp. *Certhia familiaris buturlini* Banjkowski, 1912 NCR
[JS *Certhia familiaris caucasica*]
Lesser Spotted Woodpecker ssp. *Dendrocopos minor buturlini* **Hartert**, 1912
Eyebrowed Thrush ssp. *Turdus obscurus buturlini* **Domaniewski**, 1918 NCR; NRM
Eurasian Nuthatch ssp. *Sitta europaea buturlini* **Momiyama**, 1931 NCR
[JS *Sitta europaea amurensis*]
Siberian Jay ssp. *Perisoreus infaustus buturlini* **Stachanov**, 1932 NCR
[JS *Perisoreus infaustus sibericus*]
Eurasian Oystercatcher ssp. *Haematopus ostralegus buturlini* **Dementiev**, 1941

Dr Sergei Aleksandrovich Buturlin (1872–1938) was a Russian zoologist, explorer and aristocrat, who started out studying law but gave it up for zoology. He explored and collected in Russia, its offshore islands, Siberia and the extreme east of the country (1890–1925). He joined the Zoological Museum of the University of Moscow (1918). He described over 200 new species or subspecies of birds and wrote extensively, such as *The Birds of the Simbirsk Government* (1906). He donated part of his collection (12,000 bird skins) to the Moscow Zoological Museum.

Buvry

Hawfinch ssp. *Coccothraustes coccothraustes buvryi* **Cabanis**, 1862

Dr Louis Leopold Buvry (b.1822) was a German zoologist and explorer in Algeria. He lived in Braunschweig (1857–1874).

Buxton, E. C.

Oriental White-eye ssp. *Zosterops palpebrosus buxtoni* **Nicholson**, 1879

Edmund Charles Buxton (1813–1878) was a British entomologist, hunter, traveller and collector in Tanganyika (Tanzania) (1871) and Sumatra (1876). He died of fever in West Africa.

Buxton, E. N.

African Babbler sp. *Crateropus buxtoni* **Sharpe**, 1891 NCR
[Alt. Brown Babbler; JS *Turdoides plebejus cinerea*]

Sir Edward North Buxton (1840–1924) was a British hunter, conservationist and Liberal politician. He set up the Society for the Preservation of the Wild Fauna of the Empire (1903), of which Oldfield Thomas of the British Museum became an ardent supporter. An antelope is named after him.

Buxton, N. G.

Eurasian Skylark ssp. *Alauda arvensis buxtoni* **J. A. Allen**, 1905 NCR
[JS *Alauda arvensis pekinensis*]

Norman G. Buxton (DNF) was an American explorer and collector in Alaska (1897–1898) and was on the Jesup North Pacific Expedition to Siberia (1900–1902).

Buys

Bennett's Woodpecker ssp. *Campethera bennettii buysi* **Winterbottom**, 1966 NCR
[JS *Campethera bennettii capricorni*]

Peter J. Buys is a South African taxidermist, herpetologist and ornithologist who was Curator, National Museum of Namibia (1957–1987). He co-wrote *Snakes of Namibia* (1983).

Bynoe

Black-tailed Whistler ssp. *Pachycephala melanura bynoei* **Mathews**, 1918 NCR
[JS *Pachycephala melanura robusta*]

Benjamin Bynoe (1804–1865) was a British naval surgeon. He was appointed as Assistant Surgeon on HMS *Beagle* (1831).

His superior, Robert McCormick, was so annoyed that Darwin, instead of himself, was treated as the ship's naturalist that he resigned from the expedition (1832), returning to England. Consequently Bynoe was promoted to Surgeon and served for the rest of that voyage (1836). He was given the same position on the *Beagle*'s third voyage (1837–1843). He was a great success as a naturalist and collector of both mammals and birds. He wrote the first description of the birth of marsupials. He was greatly used by John Gould in the preparation of his *Birds of Australia*, and it was Bynoe who first collected the finch *Chloebia gouldiae* which Gould named in honour of his wife. Bynoe Harbour in Australia was named after him (1839), as are a mammal and reptile.

Byron

Apapane *Nectarinia byronensis* **Bloxam**, 1826 NCR
[JS *Himatione sanguinea*]

Burrowing Parrot ssp. *Cyanoliseus patagonus byroni* **J. E. Gray**, 1831 NCR
[Name viewed as invalid and replaced by *C. patagonus bloxami*]

Admiral George Anson Byron 7th Baron Byron (1789–1868) inherited his title on the death of his cousin, the poet Lord Byron (1824). He joined the Royal Navy (1800) and became a Post Captain (1814). He was on HMS *Blonde*, accompanying the bodies of the monarchs of Hawaii back to their homeland (1824) – they had died of measles during a state visit.

C

Cabanis

Flycatcher-shrike genus *Cabanisia* **Bonaparte**, 1854 NCR
[Now in *Hemipus*]

Cabanis's Tiger Heron *Tigrisoma mexicanum* **Swainson**, 1834
[Alt. Bare-throated Tiger Heron; *Tigrisoma cabanisi* (**Heine**, 1859) is a JS]
Grey-throated Warbling Finch *Poospiza cabanisi* Bonaparte, 1850
Cabanis's Emerald *Chlorostilbon auratus* Cabanis & **Heine**, 1860 NCR
[Believed to be an aberrant specimen of *Chlorostilbon poortmani*]
Cabanis's Wren *Thryothorus modestus* Cabanis, 1861
[Alt. Modest Wren, Plain Wren]
Cabanis's Thrush *Turdus plebejus* Cabanis, 1861
[Alt. Mountain Thrush]
Cabanis's Tanager *Tangara cabanisi* **P. L. Sclater**, 1868
[Alt. Azure-rumped Tanager]
Cabanis's Bunting *Emberiza cabanisi* **Reichenow**, 1875
[Alt. Cabanis's Yellow Bunting]
Cabanis's Greenbul *Phyllastrephus cabanisi* **Sharpe**, 1882
Plumbeous Tyrant *Knipolegus cabanisi* Schulz, 1882
[Syn. *Knipolegus signatus cabanisi*]
Black-capped Social Weaver *Pseudonigrita cabanisi* **Fischer** & Reichenow, 1884
Cabanis's Spinetail *Synallaxis cabanisi* **Berlepsch** & Leverkuhn, 1890
Long-tailed Fiscal *Lanius cabanisi* **Hartert**, 1906

Green Kingfisher ssp. *Chloroceryle americana cabanisii* **Tschudi**, 1846
Abyssinian Scimitarbill ssp. *Rhinopomastus minor cabanisi* **Filippi**, 1853
Great Spotted Woodpecker ssp. *Dendrocopos major cabanisi* **Malherbe**, 1854
Cabanis's Woodpecker *Picoides villosus hyloscopus* Cabanis & Heine, 1863
[Alt. Hairy Woodpecker ssp.]
Green Violetear ssp. *Colibri thalassinus cabanidis* Heine, 1863
Fuscous Flycatcher ssp. *Cnemotriccus fuscatus cabanisi* **Léotaud**, 1866
Shiny Cowbird ssp. *Molothrus bonariensis cabanisii* **Cassin**, 1866
Cabanis's Ground Sparrow *Melozone biarcuata cabanisi* P. L. Sclater & **Salvin**, 1868
[Alt. Prevost's Ground Sparrow ssp.]

Lesser Masked Weaver ssp. *Ploceus intermedius cabanisii* **W. K. H. Peters**, 1868
Buff-browed Foliage-gleaner ssp. *Syndactyla rufosuperciliata cabanisi* **Taczanowski**, 1875
Cabanis's Warbler *Basileuterus culicivorus cabanisi* Berlepsch, 1879
[Alt. Golden-crowned Warbler ssp.]
Livingstone's Turaco ssp. *Tauraco livingstonii cabanisi* Reichenow, 1883
Scaly-breasted Munia ssp. *Lonchura punctulata cabanisi* Sharpe, 1890
Squirrel Cuckoo ssp. *Piaya cayana cabanisi* **J. A. Allen**, 1893

Jean Louis Cabanis (1816–1906) was the most influential European ornithologist of his day. After he completed his studies at the Humboldt University of Berlin (1835–1839) he travelled (until 1841) to North America, where he built a rich collection. He became an assistant in the Museum of Natural History, Berlin, and followed Martin Lichtenstein (q.v.) as Director (1857). He founded (1853) and edited (1853–1893) the *Journal für Ornithologie*. Although he never visited Africa, collectors sent him skins at the Berlin Zoology Museum where he became Senior Custodian (1850) and Head of the Bird Section (1893). He catalogued Heine's (q.v.) huge collection at Halberstadt. Reichenow (q.v.), who succeeded him at the museum, was his son-in-law (see **Maria (Reichenow)**). References to Cabani, Cabini and Caboni in old vernacular bird names all seem to be misspellings of Cabanis.

Cabot

Cabot's Tern *Thalasseus acuflavidus* Cabot, 1847
[Syn. *Thalasseus sandvicensis acuflavidus*]
Cabot's Wren *Thryothorus albinucha* Cabot, 1847
[Alt. White-browed Wren]
Cabot's Tragopan *Tragopan caboti* **Gould**, 1857
[Alt. Chinese Tragopan]

Cozumel Bananaquit *Coereba flaveola caboti* **Baird**, 1873

Dr Samuel Cabot (1815–1885) was a physician, ornithologist and the Curator of the Department of Ornithology at Boston's Society of Natural History. He made an important expedition to the Yucatán Peninsula (1840s), where he discovered several new species. His collection of 3,000 mounted birds is now at the Museum of Comparative Zoology, Harvard.

Cabra

Grass Owl sp. *Strix cabrae* **Dubois**, 1902 NCR
[Alt. African Grass Owl; JS *Tyto capensis*]

Lieutenant-General Alphonse Cabra (1862–1932) was a Belgian army officer (1878–1924) and surveyor and explorer in the Congo, where he settled the limits of the Portuguese enclave of Cabinda (Angola) (1897–1899), the boundaries with the rest of Angola (1901–1902) and French Congo (1903). He carried out an 'inspection' (1905–1906) from Mombasa to Boma, accompanied by his wife, who became the first European woman to traverse Africa from East to West. He was invalided home to Belgium (1906), serving in various military posts there including commanding the fortress of Namur (WW1), eventually retiring as Commander of the 2nd Army Corps.

Cabrera, Anatael

Common Blackbird ssp. *Turdus merula cabrerae* **Hartert**, 1901

Dr Anatael Cabrera y Diaz (1866–1943) was a Spanish physician, entomologist and naturalist who lived in the Canary Islands. He wrote *Catálogo de las Aves del Archipiélago Canario* (1893).

Cabrera, Angel

Coal Tit ssp. *Periparus ater cabrerae* **Witherby**, 1928 NCR
[JS *Periparus ater vieirae*]

Dr Angel Cabrera Latorre (1879–1960) was one of the foremost Spanish zoologists of his era. He graduated from Madrid University with a doctorate (1900). He worked at Madrid Natural History Museum (1900–1925), becoming Chief Curator of Mammals. He was Director, Department of Paleontology, National Museum of Argentina (1925–1947) His expeditions collected both live and fossil specimens in Patagonia and Catamarca, and around Buenos Aires. He spent the rest of his life in Argentina. He co-wrote *Mammals of South America* (1940). Two mammals are named after him.

Cadwalader

Hooded Tinamou ssp. *Nothocercus nigrocapillus cadwaladeri* **Carriker**, 1933
Great Spotted Woodpecker ssp. *Dendrocopos major cadwaladeri* **Meyer de Schauensee**, 1934 NCR
[JS *Dendrocopos major mandarinus*]
Buckley's Cotinga ssp. *Laniisoma buckleyi cadwaladeri* Carriker, 1936
[Alt. Andean Laniisoma ssp.]
Barred Antshrike ssp. *Thamnophilus doliatus cadwaladeri* Bond & Meyer de Schauensee, 1940

Charles Meigs Biddle Cadwalader (1885–1959) was an American philanthropist who was Director of the Academy of Natural Sciences, Philadelphia (1937–1951). The mineral cadwaladerite is named after him.

Caesar

Swamphen genus *Caesarornis* **Reichenbach**, 1852 NCR
[Now in *Porphyrio*]

Caesar was a title of the Roman emperors and does not refer to any one of them in particular.

Cahn

Ornate Melidectes ssp. *Melidectes torquatus cahni* Mertens, 1923

Paul Cahn (1876–1941) was a German zoologist at the Senckenberg Institute, Frankfurt (1903–1941).

Cahoon

Brown-throated Wren ssp. *Troglodytes brunneicollis cahooni* **Brewster**, 1888

John Cyrus Cahoon (1863–1891) was an American ornithologist, oologist and collector. He died when he fell from a cliff while collecting in Newfoundland (1891).

Cailliaud

Cailliaut's Woodpecker *Campethera cailliautii* **Malherbe**, 1849
[Alt. Green-backed Woodpecker]

Frederic Cailliaud (1787–1869) was a French naturalist, geologist and mineralogist. He collected in Egypt and Sudan (1815–1822) while working for the ruler of Egypt, Mohamed Ali Pasha. He was dispatched to discover Roman emerald mines and was successful, bringing back many gems to the ruler. He also discovered a number of ancient Roman and Egyptian sites. In the desert he carefully mapped each oasis. On his return to France he was appointed Curator of the Natural History Museum at Nantes, to which he had sent many of the items he had collected earlier. So far as we can tell the spelling of the binomial with a 't' rather than a 'd' was an error, as Malherbe himself later referred to the species in his monograph of woodpeckers as 'cailliardii' by which time the misspelling had already been accepted and therefore took preference.

Cairns

Cairns's Warbler *Dendroica caerulescens cairnsi* **Coues**, 1897
[Alt. Black-throated Blue Warbler ssp.]

John Simpson Cairns (1862–1895) was an American ornithologist who conducted field studies in the late 19th century near the Craggy Gardens Recreation Area in North Carolina. His correspondence with William Brewster (q.v.) was published in the *North Carolina Historical Review*. Brewster and Cairns were instrumental in proving the palaeontologist Edward Cope's hypothesis that much of the fauna in the Black Mountains was identical to that occurring 1,600 km further north. To quote a local travel guide: 'Within five years after the Civil War, however, a number of zoologists visited the Appalachians to do survey work. These included ... John Simpson Cairns, a local ornithologist who accidentally shot and killed himself while collecting birds north of Balsam Gap.'

Calaby

Australian Masked Owl ssp. *Tyto novaehollandiae calabyi*
    I. J. Mason, 1983

Dr John Henry Calaby (1922–1998) was an Australian biologist, mammalogist and ornithologist. After graduating he worked in the (WW2) munitions industry, only beginning his zoology career afterwards (1945). He was the first Curator of Mammals for the Australian National Wildlife Collection, CSIRO, Discovery Centre, Canberra (1950–1987). Dr Brian Walker, Chief, CSIRO Wildlife and Ecology, once said, 'John is a great collector of wildlife literature and has compiled a treasure trove of wildlife books and journals. These formed the basis of Wildlife and Ecology's library – one of the leading ecological libraries in the world, and definitely the frontrunner in the southern hemisphere.' He wrote *A Field Guide to Australian Birds* (1969). Many taxa including two mammals are named after him.

Caley

Shrike-thrush genus *Caleya* **Mathews**, 1913 NCR
[Now in *Colluricincla*]

George Caley (1770–1829) was a British botanist who explored in Australia. He was self-taught, becoming a labourer at Kew where Sir Joseph Banks appointed him to be a botanical collector in New South Wales (1798–1810). He went with Grant's expedition to Westernport (1801). He tried to cross the Blue Mountains (1804) and named Mount Banks. He visited Norfolk Island and Hobart (1805). He became the Curator of the Botanic Gardens on St Vincent, West Indies (1816–1822).

Callewaert

Helmeted Guineafowl ssp. *Numida meleagris callewaerti*
    Chapin, 1932 NCR
[JS *Numida meleagris galeatus*]

Monsignor Richard Callewaert (1866–1943) was a Belgian missionary and amateur naturalist who collected many different taxa in the Belgian Congo (DRC). We think we was the Callewaert who wrote the tract 'Rapport sur l'éducation des Mulâtres', which appeared in *Compte-rendu du Congrès International pour l'Étude des Problèmes Résultant du Mélange des Races* (1935). A mammal is named after him.

Calliope

Siberian Rubythroat *Luscinia calliope* **Pallas**, 1776
Calliope Hummingbird *Stellula calliope* **Gould**, 1847

Yellow-throated Whistler ssp. *Pachycephala macrorhyncha calliope* **Bonaparte**, 1850

Calliope was the Muse of Epic Poetry in Greek mythology. Calliope (Calliopeia), the 'fair-voiced' and the eldest Muse, is shown crowned in gold and holding a writing tablet, sometimes a roll of paper or book. Calliope is known for taking a fancy to Achilles and for having taught him how to cheer his friends by singing at banquets. Zeus called on her to mediate between Aphrodite and Persephone in their quarrel over the possession of Adonis. She settled the dispute by giving them equal access, providing Adonis with some badly needed free time to himself. She was the mother of Orpheus.

Calthrop

Layard's Parakeet *Psittacula calthropae* **Blyth**, 1849
[Alt. Emerald-collared Parakeet]
Pipit sp. *Anthus calthropae* **Layard**, 1867 NCR
[Alt. Short-tailed Pipit; JS *Anthus brachyurus*]

Barbara Anne Calthrop (DNF) became the first wife of Edgar Layard (1845). They had one son (1848), who collected with his father. She was the daughter of the Reverend John Calthrop, Vicar and Squire of Gosberton, near Spalding, Lincolnshire. Blyth originally mis-transcribed Layard's notes as '*calthrapae*'. The misspelling '*calthorpae*' is also commonly seen.

Camargo

Sharp-billed Treehunter ssp. *Heliobletus contaminatus camargoi* Cardoso da Silva & Stotz, 1992
Great-billed Hermit ssp. *Phaethornis ochraceiventris camargoi* **Grantsau**, 1988 NCR
[JS *Phaethornis (malaris) margarettae*]

Dr Hélio Ferraz de Almeida Camargo (1922–2006) was a Brazilian ornithologist.

Camburn

Golden Weaver sp. *Hyphantornis camburni* **Sharpe**, 1890 NCR
[Alt. Holub's Golden Weaver; JS *Ploceus xanthops*]
Paradise-flycatcher sp. *Tchitrea camburni* **Neumann**, 1908 NCR
[Alt. Bedford's Paradise-flycatcher; JS *Terpsiphone bedfordi*]

Claude Fowler Camburn (fl.1902) was a taxidermist and skinner for Witherby in Kenya (1901) and the Congo (1906). He was in the party led by Mackinder that collected on and climbed Mount Kenya (1899). He was a partner in a nursery garden in Kent until the partnership was dissolved (1918).

Camerano

Camerano's Shrike Babbler *Pteruthius aeralatus cameranoi* **Salvadori**, 1879
[Alt. Blyth's Shrike Babbler ssp.]

Professor Dr Lorenzo Camerano (1856–1917) was a herpetologist and entomologist. He started out as a painter at Turin Natural History Museum (1873), but became fascinated by zoology, becoming a student, graduating from the University of Turin (1878) and becoming Professor (1880). He spent a short time at the University of Cagliari before returning to Turin as Professor of Comparative Anatomy until 1915. He was elected an Italian senator (1909). He wrote *Monografia degli Ofidi Italiani* (1891). An amphibian is named after him.

Campbell

Campbell Shag *Leucocarbo campbelli* **Filhol**, 1878
Campbell Teal *Anas nesiotis* **J. H. Fleming**, 1935

These birds are named after Campbell Island, New Zealand, not directly after a person. Campbell Island was discovered (January 1810) by Captain Frederick Hasselburg of the sealing brig *Perseverance* who named it after his employers, Robert Campbell & Co. of Sydney.

Campbell, A.

Blue Flycatcher sp. *Nitidula campbelli* **Jerdon** & **Blyth**, 1861 NCR
[Alt. Pygmy Flycatcher; JS *Muscicapella hodgsoni*]

Dr Archibald Campbell (1805–1874) was part of the Bengal Medical Service.

Campbell, A. J.

Woodswallow genus *Campbellornis* **Mathews**, 1912 NCR
[Now in *Artamus*]

Campbell's Robin *Petroica boodang campbelli* **Sharpe**, 1898
[Alt. Scarlet Robin ssp.]
Shining Flycatcher ssp. *Myiagra alecto campbelli* Mathews, 1912 NCR
[JS *Myiagra alecto wardelli*]
Spotless Crake ssp. *Porzana plumbea campbelli* Mathews, 1914 NCR
[JS *Porzana tabuensis tabuensis*]
Noisy Scrub-bird ssp. *Atrichornis clamosus campbelli* Mathews, 1916 NCR; NRM
New Holland Honeyeater ssp. *Phylidonyris novaehollandiae campbelli* Mathews, 1923
Striated Pardalote ssp. *Pardalotus striatus campbelli* Mathews, 1924 NCR
[JS *Pardalotus striatus substriatus*]

Archibald James ('A.J.') Campbell (1853–1929) was a career civil servant (1869–1914), being a 'weigher' in the Department of Trade and Customs (1872). He was an active member of the Field Naturalists' Club of Victoria and a founder member (1901) and later President (1909 and 1928) of the RAOU. He was also a founder member (1905) of the Bird Observers' Club. He wrote widely on Australian birds, including popular newspaper articles, and *Nests and Eggs of Australian Birds* (1900). (See **Archibald (Campbell)**)

Campbell, C. W.

Campbell's (Hill) Partridge *Arborophila campbelli* **H. C. Robinson**, 1904
[Alt. Malaysian Partridge; Syn. *Arborophila orientalis campbelli*]

Charles William Campbell (1861–1927) was a member of the British Consular Service in China. He travelled from China (1889) and spent two months in (North) Korea, publishing a report of this journey (1891). A mammal is named after him.

Campbell, R. W.

Campbell's Fairy-wren *Chenorhamphus campbelli* **Schodde** & Weatherly, 1982
[Syn. *Malurus campbelli, Malurus grayi campbelli*]

Robert Watt Campbell is a Scottish ornithologist. He was a bander for the Australian Bird-Banding Scheme and the fairy-wren was discovered (1980) when he netted in the remote Mount Bosavi area in eastern New Guinea. No specimen was taken at that time, though Campbell took measurements and photographs of five individuals (1980–1981).

Campbell, W. A.

Coqui Francolin ssp. *Francolinus coqui campbelli* **J. A. Roberts**, 1928 NCR
[JS *Peliperdix coqui coqui*]

Colonel William A. Campbell (1880–1962) was a South African soldier, businessman and a big game hunter in Natal, South Africa, where he managed estates (1906–1962). He was a member of the National Parks Board. A reptile is named after him.

Campo

Canyon Towhee ssp. *Melozone fusca campoi* **R. T. Moore**, 1949

Professor Rafael Martin del Campo y Sánchez (1910–1987) was a herpetologist who was Director of Zoological Taxonomy at the Institute of Biology at City University in Mexico City. Four reptiles are named after him.

Canace

Spruce Grouse ssp. *Falcipennis canadensis canace* **Linnaeus**, 1766

In Greek mythology, Canace committed incest with her brother Macaraeus and, after the birth of their child, threw herself on the sword that her father had thoughtfully provided.

Candé

White-whiskered Spinetail *Synallaxis candei* **d'Orbigny** & **Lafresnaye**, 1838
Candé's Manakin *Manacus candei* **Parzudaki**, 1841
[Alt. White-collared Manakin]

Admiral Antoine Marie Ferdinand de Maussion de Candé (1801–1867) was a French explorer of South America. He became Governor of Martinique (1859–1864). Other taxa including some gastropods are named after him.

Candida

White-bellied Emerald *Amazilia candida* **Bourcier** & **Mulsant**, 1846

White Tern ssp. *Gygis alba candida* **Gmelin**, 1789

Candida is the name of a saint first referred to by this name in the sixteenth century (as a translation into Latin of 'St Wite', which is much older). It is also a woman's name in a play by George Bernard Shaw but the origin of the name is the Latin word *candidus* (= shining white). It is unlikely that its usage in scientific names is eponymous.

Canfield

Canfield's Quail *Callipepla californica canfieldae* **Van Rossem**, 1939
[Alt. California Quail ssp.]

May Canfield (1879–1938) lived in California and collected birds. She was the aunt of Lawrence M. Huey, who was the Curator of Birds at the San Diego Natural History Museum (1900s). She donated specimens to the museum.

Canivet

Canivet's Emerald *Chlorostilbon canivetii* **Lesson**, 1832
[Alt. Fork-tailed Emerald]

Emmanuel Canivet de Carentan (DNF) was a French collector and ornithologist. Together with the botanist and geologist Pierre Botard (1789–1859) he wrote (pre-1828) a manual on the preparation of animal, plant and other specimens, published as the *Manual de Naturalista Dissector* (1832), one of the classic works on taxidermy. Canivet also wrote *Catalogue des Oiseaux du Département de la Manche* (1843). Elsewhere, Canivet is described simply as a 'curiosity dealer'.

Canning

Andaman Crake *Rallina canningi* **Blyth**, 1863

Charles John Earl Canning (1812–1862) was Governor General of India (1856–1862).

Cantor

Babbler sp. *Alcippe cantori* F. Moore, 1854 NCR
[Alt. Sooty-capped Babbler; JS *Malacopteron affine*]
Puff-backed Bulbul *Microtarsus cantori* F. Moore, 1854 NCR
[JS *Pycnonotus eutilotus*]
Yellow-bellied Bulbul *Criniger cantori* F. Moore, 1854 NCR
[JS *Alophoixus phaeocephalus*]

Dr Theodore Edward (Theodor Edvard) Cantor (1809–1860) was the Danish Superintendent of the European Asylum at Bhowanipur, Calcutta, part of the Bengal Medical Service, which in turn was part of the Honourable East India Company. Cantor was also an amateur zoologist. He was interested in tropical fish, and (c.1840) the King of Siam (Thailand) gave him some bettas (fighting fish). He published an article about them, and this led to the so-called 'betta fever', a fashion for keeping these fish. He also wrote 'List of Malayan birds collected by Theodore Cantor, Esq. MD, with descriptions of the imperfectly known species' (1854) and two similar works on reptiles and fish. Two mammals and twelve reptiles are named after him.

Canute

Knot *Calidris canutus* **Linnaeus**, 1758
[Alt. Red Knot]

Canute (965–1035) was King of Denmark, England and Norway. Most schoolboys will know the tale of his attempt to command the tide, and his pragmatism in making sure the command was for it to go in the direction it was already heading! He is reputed to have regarded the Knot as well worth eating! It is conjectured that 'knot' may be a corruption of the Danish version of his name, Knut.

Capelle

Large Green Pigeon *Treron capellei* Temminck, 1823

Godert Alexander Gerard Philip, Baron van der Capellen (1778–1848), was a member of the Dutch judiciary and a diplomat. He was Governor-General of the Dutch East Indies (1815–1825) when the species bearing his name was discovered. He was a Privy Councillor and Minister of the Interior (1809–1810), becoming Dutch envoy to Great Britain (1838) and attending Queen Victoria's coronation. He served as Lord Chamberlain to the Dutch King William II.

Capello

Common Fiscal ssp. *Lanius collaris capelli* **Bocage**, 1879

Vice-Admiral Hermenegildo Carlos de Brito Capello (1839–1926) was a Portuguese naval officer. He served at sea and was based in Angola (1860–1863 and 1866–1869). He was leader of a number of expeditions in Angola, including one that traversed Africa to the sea in Mozambique (1878–1886). He wrote *De Angola à Contracosta* (1886).

Cardona

Saffron-breasted Redstart *Myioborus cardonai* **Zimmer** & **W. H. Phelps**, 1945
[Alt. Guaiquinima Whitestart]

Tepui Spinetail ssp. *Cranioleuca demissa cardonai* Phelps & **Dickerman**, 1980

Captain Felix Cardona Puig (1903–1982) was a Spanish botanist, explorer and cartographer, notably of the upper reaches of the Caroni, Paragua, Caura and Ventuari rivers in Venezuela. He sailed the world in a steamship (1922–1925) to obtain his masters certificate, during which time he was told that the rivers of Venezuela were full of gold and diamonds – so when his ship docked in Venezuela (1925) he left it to explore. Going back to Spain he worked for the Ministry of Works there, but returned to Venezuela and mapped there (1930s–1960s). He also collected plants and birds for William Phelps, which Phelps donated to AMNH, New York. The Tepui Spinetail ssp. is named after Felix and also his twin sons (b.1937) Jordi and Heinz, both of whom were engineers and members of the Frontier Commission of Venezuela.

Cardoso

Tapajós Scythebill *Campylorhamphus cardosoi* Portes *et al.*, 2013

Dr José Maria Cardoso da Silva is a biologist and biogeographer, who is an executive vice-president of Conservation International and a professor at two universities, Universidade Federal do Amapá (Tropical Diversity) and Universidade

Federal do Pará (Zoology). The Universidade Federal do Pará awarded his bachelor's degree (1986), the Universidade de Brasília his master's (1989) and the Zoological Museum, University of Copenhagen, his doctorate (1995), after which he did post-doctoral research at Museu Paraense Emílio Goeldi (1996).

Carlo

Common Swift ssp. *Apus apus carlo* **Kollibay**, 1905 NCR
[JS *Apus apus apus*]
Common Kestrel ssp. *Falco tinnunculus carlo* **Hartert &
Neumann**, 1907 NCR
[JS *Falco tinnunculus rufescens*]
Red-billed Firefinch ssp. *Lagonosticta senegala carlo*
Zedlitz, 1910 NCR
[JS *Lagonosticta senegala brunneiceps*]
Graceful Prinia ssp. *Prinia gracilis carlo* Zedlitz, 1911

(See Carlo Freiherr von **Erlanger**)

Carlotta

(See **Charlotte (Bonaparte)**)

Carman

Socorro Towhee *Pipilo carmani* **Lawrence**, 1871 NCR
[JS *Pipilo maculatus socorroensis*]

Dr B. F. Carman (fl.1884) was an American physician who was US Vice-Consul at Mazatlan, Mexico (1864–1884).

Carmela

Zitting Cisticola ssp. *Cisticola juncidis carmelae* **Orlando**,
1937 NCR
[JS *Cisticola juncidis juncidis*]

Madame Carmela Orlando (1902–1958) was the wife of Dr Carlo Orlando (q.v.).

Carmelita

Carmelita's Antpitta *Grallaria guatimalensis carmelitae*
Todd, 1915
[Alt. Scaled Antpitta ssp.]

Myrtle Carmelita Carriker (b.1893) *née* Flye was the wife of Melbourne Armstrong Carriker Jr (q.v.). She was the daughter of an American engineer who became a coffee planter at Santa Marta, Colombia, which is where she met her husband while on a collecting expedition (1911). The herpetologist Alexander Grant Ruthven (1882–1971) visited the family and collected the holotype there when collecting amphibians and reptiles. (The book he wrote about the herpetology of the area has a description of the locality by M. A. Carricker Jr., who also later supplied paratypes of the antpitta to US museums.) She returned to the US (1927) and was later divorced (1941). An amphibian is named after her.

Carmichael

Black-tailed Forest Rail ssp. *Rallina mayri carmichaeli*
Diamond, 1969

Dr Leonard Carmichael (1898–1973) was an American educator and psychologist. Tufts University awarded his bachelor's degree (1921) and Harvard his doctorate (1924). He taught at Princeton (1924–1926), at Brown University (1927–1936), University of Rochester (1937–1938) and was then President of Tufts University (1938–1952). He was Secretary of the Smithsonian (1953–1964) and became Vice-President for Research and Exploration, National Geographic Society. A lunar crater is named after him.

Carmichael Low

Sardinian Warbler ssp. *Sylvia melanocephala
carmichaellowi* **Clancey**, 1947 NCR
[NUI *Sylvia melanocephala melanocephala*]

Dr George Carmichael Low (1872–1952) was a researcher in tropical medicine, joining the newly founded London School of Tropical Medicine (1899). He led an expedition (1901–1902) to the Caribbean, helping with malaria eradication projects. He also led a team that studied sleeping sickness at Lake Victoria (1902), then becoming Superintendent of the Albert Dock Hospital, London. He worked again at the London School of Tropical Medicine, becoming Senior Physician. He was co-founder of The Royal Society of Tropical Medicine and Hygiene (1907) and was President (1929–1933). Throughout his career he was interested in ornithology. He compiled the section on birds in 'List of the Vertebrated Animals Exhibited in the Gardens of the Zoological Society of London, 1828–1927' (1929).

Carmina

White-throated Manakin ssp. *Corapipo gutturalis carminae*
V. Barnés, 1955 NCR; NRM

Carmina Barnés (DNF) was the wife of Dr Ventura Barnés Jr (q.v.).

Carmiol

Plumeleteer sp. *Chalybura carmioli* **Lawrence**, 1865 NCR
[Alt. Bronze-tailed Plumeleteer; JS *Chalybura urochrysia*]
Carmiol's Vireo *Vireo carmioli* **Baird**, 1866
[Alt. Yellow-winged Vireo]
Carmiol's Tanager *Chlorothraupis carmioli* Lawrence, 1868

Julian Carmiol (originally Carnigohl) (1807–1885) and Franz Carmiol were father and son. Julian was a Swiss scientist, collector, restaurateur and merchant who went to Costa Rica with his children shortly after being widowed. He explored there (1863) under the aegis of the USNM, and went on their Alaskan expedition (1863) as well as to London (1867). He spent the rest of his life in Costa Rica on the outskirts of the capital where his orchard and garden was home to many native plant species. Specimens of molluscs and other invertebrates, which he collected independently, were sent to European institutions. He was very interested in botany, particularly orchids, and a flower is named after him. The vireo and plumeleteer are named after him (though Lawrence mistakenly called him Julian Carniol); the tanager is after his son Franz, also a collector.

Carnaby

Carnaby's Black Cockatoo *Calyptorhynchus latirostris*
Carnaby, 1948
[Alt. Short-billed Black Cockatoo]

Ivan Clarence Carnaby (1908–1974) was a farmer from Western Australia and the describer. He was a keen entomologist, botanist and ornithologist and collected, notably with H. Steedman (1937–1939). He wrote a number of papers on birdlife in south-western Australia in the journals *Emu*, *Western Australian Bird Notes* and *Western Australian Naturalist*.

Carnap

Northern Crombec ssp. *Sylvietta brachyura carnapi*
Reichenow, 1900

Lieutenant Ernst K. G. H. L. von Carnap-Querheimb (1863–1945) was a German army officer. He was in German East Africa (Tanzania) (1885), Togoland (Benin) (1896), and German Cameroons (1898) where he collected natural history specimens.

Carnegie

Rufous-bellied Euphonia ssp. *Euphonia rufiventris carnegiei*
Dickerman, 1988

This bird is not named directly after Andrew Carnegie but after the natural history museum that he founded in Pittsburgh.

Carol (Andersson)

Grey Penduline-tit *Anthoscopus caroli* **Sharpe**, 1871

Karl Johan Andersson (1827–1867) was a Swedish explorer, hunter, trader, naturalist and ornithologist in South West Africa (Namibia) (1850–1855 and 1856–1867). He studied at Lunds Universitet (1847–1849). He wrote a great deal about his travels, such as *Lake Ngami, or Explorations and Discoveries in the Wilds of Southern Africa* (1856). The name *caroli* is a derivative of Carolus, the latinised version of Charles/Karl.

Carol (Beehler)

Wattled Smoky Honeyeater *Melipotes carolae* Beehler *et al.*, 2007

Carol Beehler is the wife of one of the describers. The etymology reads 'The specific epithet honors Carol Beehler, wife of the senior author, acknowledging her long and unstinting support and her personal commitment to biodiversity studies in New Guinea.'

Carol (Bonaparte)

Bronze-tailed Comet *Polyonymus caroli* **Bourcier**, 1847
Brown-eared Woodpecker *Campethera caroli* **Malherbe**, 1852

(See Charles **Bonaparte**)

Carol (Norman)

Cuckoo sp. *Coccystes caroli* **G. C. Norman**, 1888 NCR
[Alt. Levaillant's Cuckoo; JS *Clamator levaillantii*]

Charles Loyd Norman (1833–1889) was an English banker and father of ornithologist G. C. Norman, the describer.

Carol (Rothschild)

Crested Lark ssp. *Galerida cristata caroli* **Hartert**, 1904 NCR
[JS *Galerida cristata altirostris*]

Nathaniel Charles Rothschild (1877–1923) was a banker and entomologist. His elder brother was Lord Lionel Walter Rothschild, and his daughter was Miriam Rothschild, the famous parasitologist. He worked with Karl Jordan, and together they discovered the rat flea *Xenopsylla cheopis* which was the main vector of bubonic plague in Egypt. He suffered from encephalitis and committed suicide. A mammal is named after him.

Carola (Countess)

Spotted Imperial Pigeon *Ducula carola* **Bonaparte**, 1854

(See **Charlotte (Bonaparte)**)

Carola (Queen)

Carola's Parotia *Parotia carolae* **A. B. Meyer**, 1894
[Alt. Queen Carola's Parotia, Queen Carola's Six-wired Bird-of-Paradise]

(See under **Queen Carola**)

Caroline

Numfor Paradise Kingfisher *Tanysiptera carolinae*
Schlegel, 1871

The island of Numfor, off the coast of New Guinea, was once considered to be part of the Caroline Islands archipelago which Schlegel may have had in mind when he named this bird. However, he originally capitalised *Carolinae* and in his day this usually indicated an eponym. He made no further indication, so it remains a mystery.

Caroline (Ash)

Scaly Babbler ssp. *Turdoides squamulata carolinae* **Ash**, 1981

Caroline Ash is the daughter of the British ornithologist John Ash (q.v.).

Caroline (Brandt)

Hutton's Vireo ssp. *Vireo huttoni carolinae* H. W. Brandt, 1938

Mrs Carrie M. Brandt (fl.1937) was the wife of the describer.

Caroline (Erlanger)

Thekla Lark ssp. *Galerida theklae carolinae* **Erlanger**, 1897

Caroline von Bernus, Baronin von Erlanger (1843–1918), was the describer's mother.

Caroline (Junge)

MacGregor's Honeyeater ssp. *Macgregoria pulchra carolinae* Junge, 1939
Variable Pitohui ssp. *Pitohui kirhocephalus carolinae* Junge, 1952

Carolina Lydia Junge *née* Slingervoet Ramondt (b.1904) was the wife of the describer.

Caroline (La Touche)

Caroline's Shortwing *Brachypteryx leucophris carolinae* **La Touche**, 1898
[Alt. Lesser Shortwing ssp.]

Mrs Caroline La Touche (DNF) was the wife of John David Digues La Touche (q.v.).

Caroline (Rozendaal)

Tanimbar Bush Warbler *Cettia carolinae* F. G. Rozendaal, 1987

Caroline Rozendaal is the wife of Dutch ornithologist Frank G. Rozendaal (1957–2013). A mammal is also named after her.

Carolyn

Spot-breasted Oriole ssp. *Icterus pectoralis carolynae* **Dickerman**, 1981

Carolyn Lyell Dobson *née* Campbell, formerly Dickerman, is the ex-wife of the describer. The University of Minnesota, where her husband was a member of the faculty, awarded her bachelor's degree (1958).

Carp

Carp's Black Tit *Parus carpi* **Macdonald** & **B. P. Hall**, 1957

Bernhard Carp (1901–1966) was a Dutch-born South African businessman and Cape Town naturalist who sponsored many collecting expeditions, particularly to Namibia, by the Zoological Museum of Amsterdam University. The businessman and hunter bought important mammal collections in South Africa and donated them to the museum. Political considerations made it difficult, but after some time the board of the university decided to accept the very important collection, which included rare lagomorph skins and skulls. Carp wrote an autobiography entitled *Why I Chose Africa*. A reptile is named after him.

Carpenter

Philippine Coucal ssp. *Centropus viridis carpenteri* **Mearns**, 1907
Graceful Prinia ssp. *Prinia gracilis carpenteri* **Meyer de Schauensee** & **Ripley**, 1953

William Dorr Carpenter (1879–1958) was an American naturalist.

Carr

Stripe-breasted Spinetail ssp. *Synallaxis cinnamomea carri* **Chapman**, 1895

Albert B. Carr (fl.1898) had a cocoa plantation in Trinidad and was Chapman's (q.v.) host (1895). He survived a bite by a bushmaster *Lachesis* (1898). He was an active member of the Trinidad Naturalists' Club.

Carriker

Carriker's Quail Dove *Geotrygon carrikeri* **Wetmore**, 1941
[Alt. Tuxtla/Veracruz Quail Dove]
Carriker's Antpitta *Grallaria carrikeri* **Schulenberg** & Williams, 1982
[Alt. Pale-billed Antpitta]

Rufous-breasted Antthrush ssp. *Formicarius rufipectus carrikeri* **Chapman**, 1912
Yellow-tailed Oriole ssp. *Icterus mesomelas carrikeri* **Todd**, 1917
Burrowing Owl ssp *Athene cunicularia carrikeri* **Stone**, 1922
Carriker's Conure *Aratinga wagleri minor* Carriker, 1933
[Alt. Scarlet-fronted Parakeet ssp.]
Bolivian Brush Finch ssp. *Atlapetes rufinucha carrikeri* **Bond** & **Meyer de Schauensee**, 1939
Carriker's Mountain Tanager *Dubusia taeniata carrikeri* Wetmore, 1946
[Alt. Buff-breasted Mountain Tanager ssp.]
White-mantled Barbet ssp. *Capito hypoleucus carrikeri* **Graves**, 1986
Black-headed Nightingale Thrush ssp. *Catharus mexicanus carrikeri* **A. R. Phillips**, 1991

Melbourne Armstrong Carriker Jr (1879–1965) was one of the great early naturalists (primarily an ornithologist and entomologist) of Central and northern South America. His was the first modern systematic publication of the birds of Costa Rica (1910), in which he listed 713 species for the country. He greatly enhanced the bird collections of the Carnegie Museum (Pittsburgh) and the USNM. He is the subject of a biography written by his son Professor Melbourne Romaine Carriker (q.v.). A mammal and an amphibian are named after him.

Carruthers

Babbler sp. *Crateropus carruthersi* **Ogilvie-Grant**, 1907 NCR
[Alt. Arrow-marked Babbler; JS *Turdoides jardineii tanganjicae*]
Sunbird sp. *Anthothreptes carruthersi* Ogilvie-Grant, 1907 NCR
[Alt. Western Violet-backed Sunbird; JS *Anthreptes longuemarei angolensis*]
Carruthers's Cisticola *Cisticola carruthersi* Ogilvie-Grant, 1909

Yellow-breasted Tit ssp. *Cyanistes flavipectus carruthersi* **Hartert**, 1917
[Syn. *Cyanistes cyanus carruthersi*]

Alexander Douglas Mitchell Carruthers (1882–1962) was an explorer and naturalist. He was educated at Haileybury and

Trinity College, Cambridge, then trained in land survey and taxidermy. He went with the British Museum expedition to Ruwenzori (1905–1906) and travelled in Russian Turkestan and the borders of Afghanistan (1907–1908), where he did research on wild sheep. He also explored the deserts of Outer Mongolia and the upper Yenisey River (1910) with John H. Miller and Morgan Philips Price. Carruthers was honorary secretary of the Royal Geographical Society (1916–1921). His publications include *Unknown Mongolia* (1913), *Arabian Adventure* (1935) and *Beyond the Caspian, A Naturalist in Central Asia* (1949). Two mammals are named after him.

Carter

Monarch Flycatcher genus *Carterornis* **Mathews**, 1912
[Sometimes merged with *Monarcha*]

Carter's Desertbird *Eremiornis carteri* **North**, 1900
[Alt. Spinifexbird]
Carter's Albatross *Thalassarche carteri* **Rothschild**, 1903
[Alt. Indian Yellow-nosed Albatross/Mollymawk]

White-plumed Honeyeater ssp. *Lichenostomus penicillatus carteri* **A. J. Campbell**, 1899
Australian Reed Warbler ssp. *Acrocephalus australis carterae* Mathews, 1912
Elegant Parrot ssp. *Neophema elegans carteri* Mathews 1912
Welcome Swallow ssp. *Hirundo neoxena carteri* Mathews, 1912
Shining Bronze Cuckoo ssp. *Chrysococcyx lucidus carteri* Mathews, 1913 NCR
[JS *Chrysococcyx lucidus plagosus*]
Red-capped Parrot ssp. *Purpureicephalus spurius carteri* Mathews, 1915 NCR; NRM
Western Bowerbird ssp. *Chlamydera guttata carteri* Mathews, 1920

Thomas Carter (1863–1931) was an English ornithologist who collected and wrote extensively on birds in Western Australia, but was born and died in Yorkshire, England. He worked in his father's business and made several ornithological trips, including one to Iceland, but mostly to Australia. He went to Western Australia (1887), Jackeroo, Boolathanna Station (1887–1889) and Point Cloates (1889–1902). During these journeys he identified 180 bird species and collected 170. He returned to England (1903) with 500 bird skins, which are now in the AMNH, New York. He settled in Western Australia (1904–1913), making ornithological observations which later appeared in *Emu*. Thereafter, he moved back to England (1914) but made several more visits to Western Australia to look for birds (1916–1928). He was a founding member of the RAOU. The Reed Warbler ssp. is probably named after his wife, Annie Ward (b.1869), who was the love of his life and her rejections of his proposal when she was only 17 led to him leaving for Australia. When he sold his properties and returned to England he again proposed and was accepted and they married that year (1903). They returned together to Australia (1904) but returned to England with their three children (1914) where she and the children remained.

Carteret

Slender-billed Cuckoo Dove ssp. *Macropygia amboinensis carteretia* **Bonaparte**, 1854

Vice-Admiral Philip Carteret (1733–1796) was a British naval officer (1747–1794). He was on HMS *Dolphin* with Captain John Byron for their circumnavigation (1764–1766), and again circumnavigated the globe (1766–1769) in command of HMS *Swallow*. During the latter he discovered Pitcairn Island (1767) and the islands now eponymously named Carteret islands. He wrote *Carteret's Voyage Round the World, 1766–1769* (unpublished until 1965). A reptile is named after him.

Casares

Black-banded Woodcreeper ssp. *Dendrocolaptes picumnus casaresi* Steullet & Deautier, 1950

Dr Jorge Casares (DNF) was an Argentinian ornithologist. He wrote the biography *W. H. Hudson, Argentine Ornithologist* (1930).

Cassin

Flycatcher-thrush genus *Cassinia* **Hartlaub**, 1860 NCR
[Now in *Stizorhina*]
African Hawk Eagle genus *Cassinaetus* **W. L. Sclater**, 1922 NCR
[Now placed in either *Aquila* or *Spizaetus*]

Cassin's Auklet *Ptychoramphus aleuticus* **Pallas**, 1811
Cassin's Kingbird *Tyrannus vociferans* **Swainson**, 1826
Cassin's Tern *Sterna hirundinacea* **Lesson**, 1831
[Alt. South American Tern]
Cassin's Jay *Gymnorhinus cyanocephalus* **Wied-Neuwied**, 1841
[Alt. Pinyon Jay]
Cassin's Sparrow *Aimophila cassinii* **Woodhouse**, 1852
Cassin's Finch *Carpodacus cassinii* **Baird**, 1854
Cassin's Honeyguide *Prodotiscus insignis* Cassin, 1856
[Alt. Western Green-backed Honeyguide]
Hummingbird sp. *Trochilus cassini* **Gould**, 1857 NCR
[Alt. Black-chinned Hummingbird; JS *Archilochus alexandri*]
Cassin's Vireo *Vireo cassinii* **Xanthus**, 1858
Cassin's Malimbe *Malimbus cassini* **D. G. Elliot**, 1859
[Alt. Black-throated Malimbe]
Cassin's Flycatcher *Muscicapa cassini* **Heine**, 1860
[Alt. Cassin's Grey Flycatcher]
Dusky-faced Tanager *Mitrospingus cassinii* **Lawrence**, 1861
Golden-collared Woodpecker *Veniliornis cassini* **Malherbe**, 1862
Cassin's Spinetail *Neafrapus cassini* **P. L. Sclater**, 1863
Cassin's Hawk Eagle *Aquila africana* Cassin, 1865
[Syn. *Spizaetus africanus*]
Cassin's Dove *Leptotila cassinii* Lawrence, 1867
[Alt. Grey-chested Dove]
Baudo Oropendola *Psarocolius cassini* **Richmond**, 1898

Cassin's Hermit *Phaethornis longirostris cassinii* Lawrence, 1866 NCR
[Alt. Long-billed Hermit ssp.; JS *Phaethornis longirostris cephalus*]

Cassin's Bullfinch *Pyrrhula pyrrhula cassinii* Baird, 1869
[Alt. Eurasian Bullfinch ssp.]
Cassin's Screech Owl *Megascops guatemalae cassini*
   Ridgway, 1878
White-crested Hornbill ssp. *Tropicranus albocristatus*
   cassini **Finsch**, 1903
Chestnut-backed Antbird ssp. *Poliocrania exsul cassini*
   Ridgway, 1908
[Syn. *Myrmeciza exsul cassini*]

John Cassin (1813–1869) was a Quaker businessman and (unpaid) Curator of Ornithology at the Academy of Natural Sciences of Philadelphia for over a quarter of a century. He is regarded as one of the giants of American ornithology. He described 193 species of birds, many from his expeditions around the world including one to Japan. Cassin accompanied Admiral Perry on this latter historic voyage, serving as the official ornithologist. Among his written works are *Illustrations of the Birds of California, Texas, Oregon, British and Russian America* (1865) and *Birds of Chile* (1855). He joined the lithographer J. T. Bowen (1835) and was seconded from Bowen's company to work on railway surveys (1850s). Bowen, who died (1856) during the production of the railroad surveys, had made Cassin's drawings into lithographs. Cassin assumed the presidency of Bowen's company, but the story that he married Bowen's widow, Lavinia, appears to be wrong as there is no evidence that he divorced his wife, Hannah (married 1837), who was buried next to him (1888). Cassin suffered for 20 years from slow arsenic poisoning and finally died of it. This was the effect of handling bird skins, impregnated with arsenic as a preservative, in the days before rubber gloves.

Castelnau

White-tufted Sunbeam *Aglaeactis castelnaudii* **Bourcier** &
   Mulsant, 1848
Plain-breasted Piculet *Picumnus castelnau* **Malherbe**, 1862
Castelnau's Antshrike *Thamnophilus cryptoleucus*
   Menegaux & **Hellmayr** 1906

Amazonian Royal Flycatcher ssp. *Onychorhynchus*
   coronatus castelnaui **Deville**, 1849
Wedge-billed Woodcreeper ssp. *Glyphorynchus spirurus*
   castelnaudii **Des Murs**, 1856

Francis Louis Nompar de Caumont, Comte de Laporte de Castelnau (1810–1880), was a career diplomat and naturalist. He was born in London, studied natural science in Paris, and then led a French scientific expedition to study the lakes of Canada, the USA and Mexico (1837–1841). He led an expedition (1843–1847) that was the first to cross South America from Peru to Brazil, following the watershed between the Amazon and the Río de la Plata systems, collecting specimens all the way. Following this he assumed several diplomatic posts including Consul-General in Melbourne (1862) and then as the French Consul there (1864–1877). Soon after his return to France he undertook another long voyage of exploration with two botanists and a taxidermist. He retired (1877) and died in Melbourne (1880). A tree genus, many insects, fishes and a reptile are named after him.

Castillo

Plain Thornbird ssp. *Phacellodomus inornatus castilloi*
   Phelps Jr & **Aveledo**, 1987

Dr Rafael Castillo is an eminent Venezuelan physician, a keen naturalist in general and, according to the etymology, '… a lover of birds in particular'.

Castor

Merganser sp. *Mergus castor* **Linnaeus**, 1766 NCR
[Alt. Goosander/Common Merganser; JS *Mergus*
   merganser]

In Greek mythology Castor and Pollux were the twin sons of Tyndarus, King of Sparta (although Pollux's father was said to be the god Zeus). They were regarded as patrons of sailors and athletes.

Castro

Crested Goshawk ssp. *Accipiter trivirgatus castroi* Manuel
   & **Gilliard**, 1952

Arturo P. Castro (b.1914) was a Filipino collector who was employed at the Philippines National Museum, Manila (1941–1949) and thereafter as a laboratory technician. He was on Hoogstraal's Chicago-Philippine expedition (1946–1947).

Catesby

Catesby's Ground Dove *Columbina passerina* **Linnaeus**,
   1758
[Alt. Common Ground Dove]
Catesby's Towhee-bird [archaic] *Pipilo erythrophthalmus*
   Linnaeus, 1758
[Alt. Eastern Towhee]
Catesby's Cowpen-bird [archaic] *Molothrus bonariensis*
   Gmelin, 1789
[Alt. Shiny Cowbird]
Laughing Gull *Atricilla catesbaei* **Bonaparte**, 1854 NCR
[JS *Leucophaeus atricilla*]

Catesby's Tropicbird *Phaethon lepturus catesbyi* **J. F.**
   Brandt, 1839
[Alt. White-tailed Tropicbird ssp.]

Mark Catesby (1683–1749) was an English naturalist, artist and traveller. He made two journeys to the Americas (1712–1719 and 1722–1726). He refers to the colonies as the Carolinas as they were then known. He pre-dates Linnaeus, Bartram and Audubon, all of whom were influenced by him. Lewis and Clark consulted Catesby's work on their cross-country expedition (1804–1806). This work was *The Natural History of Carolina, Florida and the Bahama Islands: Containing the Figures of Birds, Beasts, Fishes, Serpents, Insects and Plants* (1731–1743) wherein he described the Eastern Towhee, naming it after its call. He wrote 'It is a solitary Bird; and one seldom sees them but in pairs. They breed and abide all the Year in Carolina in the shadiest Woods.' Audubon, in a letter to Bartram, wrote 'Let me know if you have ever seen the nest of Catesby's Cowpen-bird. I have every reason to believe that the bird never builds a nest but, like the cuckoo of Europe, drops its eggs into the nests of other birds.' Bartram, in his book on his travels, remarked 'Catesby's ground doves are also here in abundance: they are remarkably beautiful, about the size of a sparrow, and their soft and plaintive cooing perfectly enchanting.' During his travels, Catesby observed that birds migrate. This discovery was entirely contrary to the then prevailing view that birds hibernated in caves or under ponds in the winter. He published these observations in an essay (1747) 'On the passage of birds'. He also described trafficking in Ivory-billed Woodpecker *Campephilus principalis* bills among Native Americans (1731). He observed the similarity in the features of the Native Americans and peoples of Asiatic origin and was the first person to hypothesise the existence in the distant past of a landbridge between Asia and the Americas. He also mentioned the trends in animal size and species diversity as he progressed farther south. He used to ship his snake specimens back to England in jars of rum, which sailors sometimes drank, ruining his specimens! Many plant and animal taxa, including an amphibian and two reptiles, are named after him.

Catharine (Hellmayr)

Bay-headed Tanager ssp. *Tangara gyrola catharinae* **Hellmayr**, 1911

Mrs Catharine 'Kate' Hellmayr (fl.1944) was the describer's wife.

Catharine (Sallé)

Bee Hummingbird sp. *Ornismya catharinae* **Sallé**, 1849 NCR
[JS *Mellisuga minima*]

Madame Catharine Sallé (DNF) was the mother of the describer.

Catherine

Song Thrush ssp. *Turdus ericetorum catherinae* **Clancey**, 1938 NCR
[JS *Turdus philomelos clarkei*]

Mrs Catherine Clancey (DNF) of Glasgow was the mother of the describer.

Cave

Black Wood-hoopoe sp. *Scoptelus cavei* **Macdonald**, 1946 NCR
[Alt. Black Scimitarbill; JS *Rhinopomastus aterrimus notatus*]

Cinnamon Bracken Warbler ssp. *Bradypterus cinnamomeus cavei* Macdonald, 1939
Clapperton's Francolin ssp. *Pternistis clappertoni cavei* Macdonald, 1940 NCR; NRM
Chestnut-headed Sparrow Lark ssp. *Eremopterix signatus cavei* **C. H. B. Grant** & **Mackworth-Praed**, 1941 NCR
[JS *Eremopterix signatus harrisoni*]
Cave's Lark *Calendulauda barlowi cavei* Macdonald, 1953
[Alt. Barlow's Lark ssp.]

Colonel Francis Oswin Cave (1897–1974) served in the Rifle Brigade, retiring as an Honorary Colonel. He wrote, with James D. Macdonald, the classic *Birds of the Sudan – Their Identification and Distribution* (1955). He wrote other articles including 'Notes on birds from the southern Sudan' (1974). Cave also collected in the Sudan, as evidenced by a number of holotypes in the BMNH. Cave became a Catholic missionary in Sudan (1960s) and abandoned serious ornithology.

Cavendish

Common Waxbill ssp. *Estrilda astrild cavendishi* **Sharpe**, 1900

Henry 'Harry' Sheppard Hart Cavendish, 6th Baron Waterpark (1876–1948), was a British explorer and big-game hunter. He was in Portuguese East Africa (Mozambique) (1898). He fought in both the Boer War (1899–1902) and WW1 (1914–1918). A mammal is named after him.

Cayley

Painted Finch genus *Cayleyna* **Iredale**, 1930 NCR
[Now in *Emblema*]

Tree Martin ssp. *Petrochelidon nigricans cayleyi* **Mathews**, 1913 NCR
[JS *Petrochelidon nigricans nigricans*]

Neville William Cayley (1887–1950) was an Australian ornithologist, artist, and President of the RAOU (1936–1937). He wrote *What Bird is That?* (1931).

Cecil (Kloss)

Black-crested Bulbul ssp. *Pycnonotus flaviventris caecilii* **Deignan**, 1948

(See **Kloss**)

Cecil (Mathews)

Kingfisher genus *Cecilia* **Mathews**, 1918 NCR
[Also *Ceciliella* Strand, 1928 (replacement name for the preoccupied *Cecilia*) – now in *Halcyon*]

Cecilia's Rosella *Platycercus eximius cecilae* Mathews, 1911
[Alt. Eastern Rosella ssp.]
Torresian Crow ssp. *Corvus orru cecilae* Mathews, 1912
Partridge Pigeon ssp. *Geophaps smithii cecilae* Mathews, 1912 NCR
[JS *Geophaps smithii smithii*]

Marion Cecil Mathews *née* White (1865–1938) (formerly Wynne; she was a wealthy widow with two children when Mathews met her) was the wife of Gregory Macalister Mathews (q.v.). They married in 1902.

Cecile

Cecile's Woodpecker *Veniliornis kirkii cecilii* **Malherbe**, 1849
[Alt. Red-rumped Woodpecker ssp.]

Cécile Malherbe (DNF) was the daughter of Alfred Malherbe (q.v.).

Cecil(ia)

Black Crowned Crane ssp. *Balearica pavonina ceciliae* **Mitchell**, 1904

Mary Rothes Margaret Cecil *née* Tyssen-Amherst, 2nd Baroness Amherst of Hackney (1857–1919), was an Egyptologist who wrote *Bird Notes from the Nile* (1904). She was one of the few women of her day to hold a peerage of the United Kingdom in her own right.

Cecilia (Gautrau)

Cecilia's Dove *Metriopelia ceciliae* **Lesson**, 1845
[Alt. Bare-faced Ground Dove]

Cécile-Estelle-Atala Gautrau (1819–1845) was the daughter of the describer, René Primavere Lesson by his first wife, Zöe Massiou, who died in childbirth (1819).

Cecilia (Torres)

Alagoas Tyrannulet *Phylloscartes ceciliae* Teixeira, 1987

Cecilia Torres (1952–1985) was stated in the description to be the wife of the Brazilian ornithologist Dante Luiz Martins Teixeira. She had been a PhD student in Edinburgh.

Celaeno

Crimson-collared Grosbeak *Rhodothraupis celaeno* **Deppe**, 1830

In Greek myth Celaeno was a daughter of Atlas who was turned into a star (one of the Pleiades) on her death. The name means 'the dark one'. A mammal is also named after Celaeno.

Celestino, A.

Small Buttonquail ssp. *Turnix sylvaticus celestinoi* **McGregor**, 1907

Andres Celestino (fl.1903) was a collector for John White-head (q.v.) (1895), assistant to McGregor (q.v.) in the Philippines, and father of Manuel Celestino (q.v.).

Celestino, M.

Amethyst Brown Dove ssp. *Phapitreron amethystinus celestinoi* Manuel, 1936 NCR
[JS *Phapitreron amethystinus amethystinus*]
Coppersmith Barbet ssp. *Megalaima haemacephala celestinoi* **Gilliard**, 1949

Manuel Celestino (fl.1949) was a collector and plant-hunter in the Philippines and son of Andres Celestino (q.v.).

Cerruti

Grey-necked Bunting ssp. *Emberiza buchanani cerrutii* **De Filippi**, 1863

Commendatore Marcello Cerruti (1808–1896) was an Italian diplomat originally on behalf of the Kingdom of Sardinia. He held a number of consular posts before leading the Italian expedition to Persia (Iran) (1862). He went to Persia as Minister Plenipotentiary and Italian Ambassador to the court of the Shah. De Filippi was responsible for the scientific department of the expedition and wrote up the expedition report (1865) in *Note di un Viaggio in Persia*. Cerruti was subsequently Minister Plenipotentiary in Berne (1867), Washington (1868), The Hague (1869) and Madrid (1869). He retired (1870) and became a Senator of the Kingdom of Italy.

Cervera

Cervera's Wren *Ferminia cerverai* **Barbour**, 1926
[Alt. Zapata Wren, Cuban Marsh Wren]
Cervera's Rail *Cyanolimnas cerverai* Barbour & **J. L. Peters**, 1927
[Alt. Zapata Rail]

Fermín Zanón Cervera (1875–1944) was a Spanish soldier, landowner, entomologist and naturalist, who worked for the Cuban Ministry of Agriculture and Agronomy. He fought in the Cuban War of Independence (1898) on the Spanish side, and then returned to Spain in their Civil Guard. He returned to Cuba (1904) and married there. He also visited Mexico. He spent a number of years working for the American naturalist Dr Thomas Barbour (q.v.), who recommended he visit the Zapata Swamp to collect. Cervera discovered the wren (1926) and the rail (1927) in the locality of Santo Tomás. He not only knew all the calls of birds but could imitate them, and he recognised that there were unnamed species in the swamp simply because he heard calls that were new to him. He was also an entomologist and several insects are named after him.

Cetti

Warbler genus *Cettia* **Bonaparte**, 1834

Cetti's Warbler *Cettia cetti* **Temminck**, 1820

Fr Francesco Cetti (1726–1778) was an Italian Jesuit priest, zoologist and mathematician who wrote the *Storia Naturale di Sardegna*. The second volume (1776) deals with birds in Sardinia.

Ceyx

Kingfisher genus *Ceyx* Lacépède, 1799

In Greek mythology Ceyx was the son of the Morning Star but was drowned. He was metamorphosed into a kingfisher, as was his wife Alcyone, who was a daughter of Aeolus, god of the winds.

Chabarov

Oriental Greenfinch ssp. *Carduelis sinica chabarovi* **B. Stegmann**, 1929
[SI *Carduelis sinica ussuriensis*]

(See Khabarov)

Chabert

Chabert Vanga *Leptopterus chabert* **S. Müller**, 1776

Although sometimes called 'Chabert's Vanga', this bird is not named after a person at all, but reflects the local (Malagasy) name for the bird: *tcha-chert-be*.

Chabot

Congo Moorchat ssp. *Myrmecocichla tholloni chaboti*
Menegaux & **Berlioz**, 1923 NCR; NRM

Comte Jacques de Rohan-Chabot (1889–1958) was a French
explorer who led the Rohan-Chabot Expedition to Angola and
Rhodesia (Zimbabwe/Zambia) (1912–1914). A reptile is named
after him.

Challenger

Challenger's Lory *Eos histrio challengeri* **Salvadori**, 1891
[Alt. Red-and-blue Lory ssp.]

This race of lory was named after HMS *Challenger*, the lead
ship in an oceanographic expedition to the Pacific (1860–
1868), which explored the area of Nenusa Island where the
bird was supposedly found (the location has been ques-
tioned). A reptile is also named after this ship.

Chamberlain

Chamberlain's Ptarmigan *Lagopus muta chamberlaini*
A. H. Clark, 1907
[Alt. Rock Ptarmigan ssp.]

Frederick Morton Chamberlain (1867–1921) was naturalist
aboard the US Fisheries steamer *Albatross* (1907). He studied
icthyology in Indiana before joining the US Fish Commission
(1897). He participated in various expeditions to Alaska, the
Bering Sea, Hawaii and the South Pacific. He was appointed
as naturalist (1903), taking charge of all the collecting. After
the last cruise to the Philippines he resigned from the Fish
Commission (1911) to become agent for Alaskan salmon fisheries. He was an innovator, designing and constructing
collecting equipment and using photography in the study of
fisheries.

Chamnong

Lesser Coucal ssp. *Centropus bengalensis chamnongi*
Deignan, 1955 NCR
[NUI *Centropus bengalensis bengalensis*]

Chamnong Thepphahatsadin (fl.1955), the Thai district officer
at Khlong Khlung, gave great assistance to Deignan (q.v.).

Chandamony

Chestnut-headed Partridge ssp. *Arborophila cambodiana
chandamonyi* Eames *et al.* 2002

Meas Chandamony (d. 2000) was a Cambodian conservationist from the Department of Forestry and Wildlife who very
sadly contracted malaria and died during the expedition to
the Cardamom Mountains of Cambodia (2000) undertaken by
Flora & Fauna International in conjunction with the Cambodian government.

Chandler

Buff-banded Rail ssp. *Gallirallus philippensis chandleri*
Mathews, 1911 NCR
[JS *Gallirallus philippensis philippensis*]

Australian Logrunner ssp. *Orthonyx temminckii chandleri*
Mathews, 1912 NCR; NRM
Striated Thornbill ssp. *Acanthiza lineata chandleri*
Mathews, 1912 NCR
[JS *Acanthiza lineata clelandi*]
Tawny-crowned Honeyeater ssp. *Glyciphila melanops
chandleri* Mathews, 1912 NCR
[JS *Glyciphila melanops melanops*]

Leslie Gordon Chandler (1888–1980) was an Australian
jeweller, ornithologist, naturalist and vigneron. He was
gassed in WW1, afterwards becoming (1920) press correspondent for the RAOU of which he was a member (1911). He
was instrumental in the creation of a national park in the
Mallee area of Victoria.

Chantre

African Darter ssp *Anhinga rufa chantrei* **Oustalet**, 1882

Ernest Chantre (1843–1924) was a French geologist, archaeologist and anthropologist in the Middle East (1890–1894). He
became Assistant Director, Museum of Lyon (1871) and was
Professor of Anthropology (1908–1924). He excavated the
first Sumerian cuneiform tablets in Turkey (1892–1893).

Chapin

Francolin genus *Chapinortyx* **J. A. Roberts**, 1928 NCR
[Now in *Pternistis*]

Chapin's Spinetail *Telacanthura melanopygia* Chapin, 1915
[Alt. Black Spinetail]
Chapin's Puff-backed Flycatcher *Batis ituriensis* Chapin,
1921
[Alt. Ituri Batis]
Chapin's Apalis *Apalis chapini* **Friedmann**, 1928
Chapin's Flycatcher *Muscicapa lendu* Chapin, 1932
Chapin's Mountain Babbler *Kupeornis chapini* **Schouteden**,
1949
[Syn. *Kupeornis chapini*]
Chapin's Least Honeyguide *Indicator pumilio* Chapin, 1958
[Alt. Dwarf Honeyguide]

Hairy-breasted Barbet ssp. *Tricholaema hirsuta chapini*
Roberts, 1925 NCR
[NUI *Tricholaema hirsuta ansorgii*]
Chapin's (Long-billed) Pipit *Anthus similis chapini* **Grote**,
1937 NCR
[JS *Anthus similis bannermani*]
Rufous-naped Lark ssp. *Mirafra africana chapini*
C. H. B. Grant & **Mackworth-Praed**, 1939
Chapin's Crombec *Sylvietta leucophrys chapini*
Schouteden, 1947
[Alt. White-browed Crombec ssp.]
Stuhlmann's Double-collared Sunbird ssp. *Cinnyris
stuhlmanni chapini* **Prigogine**, 1952
Black-backed Puffback ssp. *Dryoscopus cubla chapini*
Clancey, 1954
Mountain Wagtail ssp. *Motacilla clara chapini* **Amadon**,
1954
Bocage's Akalat ssp. *Sheppardia bocagei chapini*
C. W. Benson, 1955

Chapin's Swift *Schoutedenapus myoptilus chapini*
Prigogine, 1957
[Alt. Scarce Swift ssp.]

Dr James Paul Chapin (1889–1964) was an American ornithologist. He was joint leader of the Lang-Chapin Expedition, which made the first comprehensive biological survey of the Belgian Congo (1909–1915). He was Ornithology Curator for the AMNH and President of the Explorers' Club (1949–1950). He wrote *Birds of the Belgian Congo* (1932), which largely earned him the award of the Daniel Giraud Elliot Gold Medal that year. He famously discovered the Congo Peafowl *Afropavo congensis* after he kept a puzzling feather from a native headdress and was able to match it, quarter of a century later, with the plumes on two dusty specimens in a museum in Belgium that a curator had labelled as juvenile domestic peacocks. Two mammals, three amphibians and four reptiles are named after him.

Chaplin, F. D. P.

Chaplin's Barbet *Lybius chaplini* **S Clarke**, 1920
[Alt. Zambian Barbet]

Sir Francis Drummond Percy Chaplin (1866–1933) was a civil servant who was Administrator in Rhodesia (now Zimbabwe) and Nyasaland (now Malawi) (1914–1923).

Chaplin, L.

Plain-backed Antpitta ssp. *Grallaria haplonota chaplinae*
Robbins & **Ridgely**, 1986

Louise Davis Chaplin *née* Catherwood (1906–1983) came from an upper class Philadelphia family and was described in a newspaper report (1955) with the headline 'Philadelphia socialite finds giving money away hard work'. She co-founded the Catherwood Foundation, which endowed scholarships and financed biological research at the Philadelphia Academy of Sciences and that perhaps explains the describers' statement 'We take pleasure in naming this subspecies after the late Louise Chaplin Catherwood'.

Chapman

Parrot genus *Chapmania* **Ribiero**, 1920 NCR
Parrot genus *Chapmaniana* **Strand**, 1928 NCR
[Both these genera now in *Pyrilia*]

Chapman's Curassow *Crax chapmani* **Nelson**, 1901 NCR
[JS *Crax rubra*]
Chapman's Swift *Chaetura chapmani* **Hellmayr**, 1907
Chapman's Warbler *Xenoligea montana* Chapman, 1917
[Alt. White-winged Warbler]
Chapman's Antshrike *Thamnophilus zarumae* Chapman, 1921
Chapman's Parakeet *Aratinga alticola* Chapman, 1921
[Alt. Mitred Parakeet ssp; Syn. *Aratinga mitrata alticola*]
Chapman's Tyrannulet *Phylloscartes chapmani* **Gilliard**, 1940
[Alt. Chapman's Bristle Tyrant; Syn. *Pogonotriccus chapmani*]

Common Nighthawk ssp. *Chordeiles minor chapmani*
Coues, 1888

Grassland Yellow Finch ssp. *Sicalis luteola chapmani*
Ridgway, 1899
Chapman's Petrel *Oceanodroma leucorhoa chapmani*
Berlepsch, 1906
[Alt. Leach's Storm Petrel ssp.]
Chapman's Ground Cuckoo *Neomorphus rufipennis
nigrogularis* Chapman, 1914 NCR; NRM
[Alt. Rufous-winged Ground Cuckoo ssp]
Chapman's Trogon *Trogon massena australis* Chapman, 1915
[Alt. Slaty-tailed Trogon ssp.]
Plumbeous Pigeon ssp. *Patagioenas plumbea chapmani*
Ridgway, 1916
Rusty-backed Antwren ssp. *Formicivora rufa chapmani*
Cherrie, 1916
Laughing Falcon ssp. *Herpetotheres cachinnans chapmani*
Bangs & T. E. Penard, 1918
Chapman's Cacique *Amblycercus holosericeus australis*
Chapman, 1919
[Alt. Yellow-billed Cacique ssp.]
Chapman's Grosbeak *Pheucticus aureoventris terminalis*
Chapman, 1919
[Alt. Black-backed Grosbeak ssp.]
Slaty Spinetail ssp. *Synallaxis brachyura chapmani* Bangs
& Penard, 1919 NCR
[JS *Synallaxis brachyura nigrofumosa*]
Chapman's Sapphire-rumped Parrotlet *Touit purpuratus
viridiceps* Chapman, 1929
Ocellated Crake ssp. *Micropygia schomburgkii chapmani*
Naumburg, 1930
Spectacled Owl ssp. *Pulsatrix perspicillata chapmani*
Griscom, 1932
Chestnut-crowned Gnateater ssp. *Conopophaga
castaneiceps chapmani* **Carriker**, 1933
Streaked Flycatcher ssp. *Myiodynastes maculatus
chapmani* **Zimmer**, 1937
Chapman's Conure *Pyrrhura melanura chapmani* **Bond** &
Meyer de Schauensee, 1940
[Alt. Maroon-tailed Parakeet ssp.]
Chapman's Mealy Amazon *Amazona farinosa chapmani*
Traylor, 1948 NCR
[NUI *Amazona farinosa farinosa*]
White-throated Tyrannulet ssp. *Mecocerculus leucophrys
chapmani* **Dickerman**, 1985

Frank Michler Chapman (1864–1945) was Curator of Ornithology for the AMNH, New York (1908–1942). He photographed and collected data on North American birds for c.50 years and did much to popularise birdwatching in the US in the 20th century. He began publishing *Bird Lore* (1899) magazine, which became a unifying national forum for the Audubon movement. He had been an enthusiastic collector but became a leading light in the conservation movement. His interest in protection can be traced to a walk in New York City (1886) where he observed that three-quarters of ladies wearing hats had feathers in them: he managed to identify, among others, American Robin, Scarlet Tanager, Blackburnian Warbler, Cedar Waxwing, Bobolink, Blue Jay, Scissor-tailed Flycatcher, Pine Grosbeak and Red-headed Woodpecker. Chapman sponsored the first national Christmas Bird Count (1900). He wrote *Handbook of Birds of*

Eastern North America (1903), *The Distribution of Bird Life in Colombia* (1917) and *The Distribution of Bird Life in Ecuador* (1926). Two mammals are named after him.

Charles

Charles Mockingbird *Nesomimus trifasciatus* **Gould**, 1837
[Alt. Floreana Mockingbird]
Charles Tree Finch *Camarhynchus pauper* **Ridgway**, 1890
[Alt. Medium Tree Finch, Floreana Tree Finch]

These birds derive their name from Charles Island (Floreana) in the Galápagos. The island was named after King Charles II (1630–1685) of England, Scotland and Ireland.

Charlotte

Charlotte's Bulbul *Iole olivacea charlottae* **Finsch**, 1867
[Alt. Buff-vented Bulbul ssp.; Syn. *Hypsipetes charlottae*]

No etymology was provided by Finsch, but perhaps named after Victoria Elisabeth Augusta Charlotte, Princess of Prussia (1860–1919), eldest daughter of Crown Prince Friedrich of Prussia.

Charlotte (Bonaparte)

Charlotte's Woodpecker *Indopicus carlotta* **Malherbe**, 1854 NCR
[Alt. Crimson-backed Flameback; JS *Chrysocolaptes stricklandi*]

Charlotte Honorine Joséphine Pauline Bonaparte, Comtesse Pietro Primoli de Foglio (1832–1901), was the daughter of the ornithologist Charles Lucien Bonaparte (q.v.). Malherbe wrote: 'Je remercie madame la princesse Charlotte Bonaparte, comtesse Primoli, d'avoir bien voulu me permettre de decorer de son nom ce brillant oiseau de Ceylan ...' (Also see **Carola (Countess)**)

Charlotte (Machatschek)

Sulawesi Swiftlet *Aerodramus sororum* **Stresemann**, 1931
Sulawesi Myzomela *Myzomela chloroptera charlottae* Stresemann, 1932 NCR
[JS *Myzomela chloroptera chloroptera*]

Anneliese Heinrich *née* Machatschek (DNF) was the second wife of Gerd Heinrich (q.v.), and she and her younger sister Liselotte (Lotte) Machatschek were in Celebes (Sulawesi) (1930–1932) with Gerd, who later had an affair with Lotte. The sisters are commemorated collectively in the name of the swiftlet (*sororum* means 'of the sisters' in Latin).

Charlotte (McGregor)

Charlotte's Towhee *Melozone crissalis carolae* **McGregor**, 1899
[Alt. California Towhee ssp.; Syn. *Pipilo crissalis carolae*]

Mrs Charlotte Crittenden McGregor (1841–1893) was the describer's mother.

Charlton

Chestnut-necklaced Hill Partridge *Arborophila charltonii* Eyton, 1845
[Alt. Eyton's Hill Partridge]

Lieutenant-Colonel Andrew Charlton (1803–1888) of the East India Company's 74th Bengal Native Infantry was the discoverer of the tea-plant growing wild in Assam (1834) where he was on secondment with the Assam Light Infantry and where he was appointed to be Magistrate (1836). He collected on the Malay Peninsula, where he found the type specimen of the partridge.

Chasen

Chasen's Frogmouth *Batrachostomus chaseni* **Stresemann**, 1937
[Alt. Palawan Frogmouth]

Red-headed Trogon ssp. *Harpactes erythrocephalus chaseni* **Riley**, 1934
Common Iora ssp. *Aegithina tiphia chaseni* Stresemann, 1938 NCR
[JS *Aegithina tiphia aequanimis*]
Chasen's White-crowned Forktail *Enicurus leschenaulti chaseni* **Meyer de Schauensee**, 1940
Yellow-bellied Prinia ssp. *Prinia flaviventris chaseni* **Deignan**, 1942 NCR
[JS *Prinia flaviventris latrunculus*]
Chasen's (Brown) Wood Owl *Strix leptogrammica chaseni* **Hoogerwerf** & De Boer, 1947

Frederick Nutter Chasen (1896–1942) was an English zoologist. He became Assistant Curator of the Raffles Museum (1921), then Director (1932–1942). He was a well-known authority on Malaysian birds and mammals and published many scientific publications on these topics. He also co-authored *The Birds of the Malay Peninsula* (4 vols, 1927–1939) with Herbert C. Robinson (q.v.). Chasen perished at sea when fleeing Singapore during WW2. A reptile is named after him.

Chaulet

Chaulet's Yellow Cissa *Cissa hypoleuca chauleti* Delacour, 1926
[Alt. Indochinese Green Magpie ssp.]

Mr Chaulet (DNF) was resident in Huê province of what is now Vietnam. He was in overall charge of the forestry in the region. Delacours etymology states: 'Named in honour of M. Chaulet, who procured us the specimen'.

Chauvin

Madagascar Owl *Asio chauvini* Lamberton, 1927 NCR
[JS *Asio madagascariensis*]

Charles Herschell-Chauvin (1875–1945) was a French collector and photographer in Madagascar (1903–1927) and had a business selling wildlife specimens – he sold material to the Archbold Expedition.

Chávez

Bushy-crested Jay ssp. *Cyanocorax melanocyaneus chavezi* **W. de W. Miller** & **Griscom**, 1925

Diocletiano Chávez (1843–1936) was a Nicaraguan zoologist and collector. He founded the natural history collections in the Museo Nacional, Managua.

Cheesman

Desert Lark ssp. *Ammomanes deserti cheesmani*
Meinertzhagen, 1923

Colonel Robert Ernest Cheesman (1878–1962) was a British army officer, field naturalist, explorer and ornithologist. He was Private Secretary to Sir Percy Sykes, High Commissioner in Iraq. Cheesman mapped the Arabian coast from Uqair to the head of the Gulf of Salwa (1921). He travelled (1923–1924) to Hufuf and mapped 240 km of desert, identified the site of ancient Gerra, and corrected maps of the wadi system. He spent several months at al-Ahsa collecting specimens of local plants, insects and other animals, and drawing up maps. He wrote an account of that journey, *In Unknown Arabia* (1926). He went on to serve as British Consul for northwest Ethiopia (1925–1934) and in that period explored and mapped Lake Tana and the Blue Nile, a journey he described in *Lake Tana and the Blue Nile: An Abyssinian Quest* (1936). He said of his explorations of the Blue Nile 'It proved to be a venture not to be lightly undertaken, and I understood why the secrets of the Nile Valley remained so long unrevealed.' He was given the Gill Memorial Award by the Royal Geographical Society. Two mammals are named after him.

Chernel

Woodlark ssp. *Lullula arborea cherneli* Prazak 1895 NCR
[JS *Lullula arborea arborea*]

Stefan Chernel von Chernelháza (1865–1922) joined the Hungarian Ornithological Centre (1893) and became its Director (1916).

Cherrie

Cherrie's Tanager *Ramphocelus costaricensis* Cherrie, 1891
Cherrie's Swift *Cypseloides cherriei* **Ridgway**, 1893
[Alt. Spot-fronted Swift]
Cherrie's Antwren *Myrmotherula cherriei* **Berlepsch** & **Hartert**, 1902
Orinoco Softtail *Thripophaga cherriei* Berlepsch & Hartert, 1902
Chestnut-throated Spinetail *Synallaxis cherriei*
Gyldenstolpe, 1930

Greater Antillean Elaenia ssp. *Elaenia fallax cherriei* **Cory**, 1895
Cherrie's Nighthawk *Chordeiles minor aserriensis* Cherrie, 1896
[Alt. Common Nighthawk ssp.]
Yellow-browed Sparrow ssp. *Ammodramus aurifrons cherriei* **Chapman**, 1914
Yellow-olive Flycatcher ssp. *Tolmomyias sulphurescens cherriei* Hartert & **Goodson**, 1917

George Kruck Cherrie (1865–1948) was a naturalist and ornithologist who accompanied Theodore Roosevelt on a trip in Brazil (1913) to find the source of one of the tributaries of the Amazon. He was Assistant Curator, Department of Ornithology, The Field Museum, Chicago (1890s). He also collected extensively in Costa Rica (1894–1897) and, with his wife Stella M. Cherrie, in Colombia (1898). In *Through the Brazilian Wilderness* (1914) Roosevelt described him as an '… efficient and fearless man; and willy-nilly he had been forced at times to vary his career by taking part in insurrections. Twice he had been behind the bars in consequence, on one occasion spending three months in a prison of a certain South American state, expecting each day to be taken out and shot. In another state he had, as an interlude to his ornithological pursuits, followed the career of a gun-runner, acting as such off and on for two and a half years. The particular revolutionary chief whose fortunes he was following finally came into power, and Cherrie immortalized his name by naming a new species of antthrush after him – a delightful touch, in its practical combination of those not normally kindred pursuits, ornithology and gun-running.' Cherrie wrote 'A contribution to the ornithology of the Orinoco region' (1916), *Dark Trails* (1930) as well as *Adventures of a Naturalist.* A mammal and a reptile are named after him.

Chiaradia

Little Owl *Athene chiaradiae* **Giglioli**, 1900 NCR
[JS *Athene noctua noctua*]

Commendatore Emidio Chiaradia (b.1839), who collected the holotype, was an Italian lawyer, politician and a member of the Italian parliament (1883–1900).

Chico

Chico's Tyrannulet *Zimmerius chicomendesi* Whitney et al., 2013

Chico Mendes (born Francisco Alves Mendes Filho, 1944–1988), who was murdered by a rancher, was a Brazilian rubber tapper, trade union leader, environmentalist and rainforest activist. He was born at a time when schools were prohibited on rubber plantations (for fear that the peasants might learn to read and do arithmetic!) and he only learned to read when he was 18. The Chico Mendes Institute for Conservation of Biodiversity is named in his honour.

Children

Children's Warbler *Sylvia childrenii* **Audubon**, 1831 NCR
[Alt. Yellow Warbler; JS *Dendroica aestiva*]

John George Children (1777–1852) was a British entomologist and scientist who was very interested in electricity and published notes on it (1808–1813). He became a Fellow of the Royal Society, London (1807) and later Secretary. He visited Pennsylvania (1802) and was in Spain and Portugal (1808–1809) and his diary records include details of the Peninsular War and observations of minerals, etc. He worked in the British Museum (1816–1840). Audubon collected and described the bird in his *Ornithological Biography or an Account of the Habits of the Birds of the United States of America* (1831), writing that he named it after Children 'as a tribute of sincere gratitude for the unremitted kindness which he has shewn me'. The mineral childrenite is named after him, as is a reptile.

Chivers

White-quilled Bustard ssp. *Afrotis afraoides chiversi*
   J. A. Roberts, 1933 NCR
[JS *Afrotis afraoides afraoides*]

U. B. Chivers (DNF) collected the holotype (1932).

Cholmley

Sand Partridge ssp. *Ammoperdix heyi cholmleyi* **Ogilvie-Grant**, 1897

Alfred John Cholmley (1845–1932) was a British traveller and collector. He accompanied J. Theodore and Mabel Bent on their journey to the Sudan (1897). He became Deputy Lord Lieutenant of the East Riding of Yorkshire (1900).

Christian, E. J.

Great Crested Grebe ssp. *Podiceps cristatus christiani*
   Mathews, 1911 NCR
[JS *Podiceps cristatus australis*]

E. J. Christian (DNF) was an Australian farmer and naturalist in Victoria. He wrote 'Notes on the Black-tailed Native-hen' (1909).

Christian (Lafresnaye)

Plain-brown Woodcreeper ssp. *Dendrocincla lafresnayei
   christiani* **Bangs** & **T. E. Penard**, 1919 NCR
[JS *Dendrocincla fuliginosa ridgwayi*]

Lieutenant-Colonel Christian André de Lafresnaye (b.1844) was the youngest son of Baron Nöel Fréderic Armand André de Lafresnaye. (See **Lafresnaye**)

Christian Ludovic

Merlin *Falco christiani-ludovici* **Kleinschmidt**, 1917 NCR
[JS *Falco columbarius pallidus*]

Dr Christian Ludwig Brehm (1787–1864) was the father of Alfred Edmund Brehm (q.v.) and of Thekla Brehm (q.v.). He wrote *Beiträge zur Vogelkunde* (1820–1822).

Christiana

Black Sunbird ssp. *Leptocoma sericea christianae* **Tristram**,
   1889

Mrs Christiana Heawood *née* Tristram (d.1950) was the describer's daughter.

Christina

Fork-tailed Sunbird *Aethopyga christinae* **Swinhoe**, 1869

Christina Swinhoe *née* Stronach (d.1914) was the wife of Robert Swinhoe.

Christoph

Hooded Crow ssp. *Corvus cornix christophi* Alpheraky, 1910
   NCR
[JS *Corvus cornix sharpii*]

Hugo Theodor Christoph (1831–1894) was a German-born Russian entomologist, collector and explorer. He was curator of the private entomological collection belonging to Grand Duke Nikolai Mikhailovich Romanoff (1880).

Christy

Blue Swallow sp. *Hirundo christyi* **Sharpe**, 1906 NCR
[JS *Hirundo atrocaerulea*]

Dr Cuthbert Christy (1863–1932) qualified as a physician at Edinburgh. He travelled in the West Indies and South America (1890s), subsequently joining the army as Medical Officer. He was in northern Nigeria (1898–1900) and later in Uganda and the Congo (1902–1903). He served in Africa and Mesopotamia (Iraq) (WW1). After the war he explored in the Sudan, Nyasaland (Malawi) and Tanganyika (Tanzania), and was a member of a League of Nations commission enquiring into slavery and forced labour in Liberia. He was Director of the Congo Museum, Tervuren, Belgium. He was on a zoological expedition to the Congo (1932) when he was gored by a buffalo and killed. A mammal, three reptiles and two amphibians are named after him.

Chrostowski

Giant Antshrike *Batara chrostowskii* **Stolzman**, 1926 NCR
[JS *Batara cinerea*]

Lined Antshrike ssp. *Thamnophilus tenuepunctatus
   chrostowskii* **Domaniewski**, 1925 NCR
[JS *Thamnophilus tenuepunctatus tenuifasciatus*]

Tadeusz Chrostowski (1878–1923) was a Polish ornithologist, explorer and collector. He studied mathematics and physics at Moscow State University but did not graduate, as he was arrested as a revolutionary and exiled to Siberia for three years. After being released he became an officer in the Russian Army and served in Manchuria during the Russo-Japanese War (1904–1905). He returned to Poland (1907) to prepare to leave for Brazil to become an apiarist. He arrived in there (1910) to find that bee-keeping was not a commercial proposition, so went on a collecting trip round the Iguaçu river basin, then went back to Poland (1911). He returned to Brazil (1913), settled in Curitiba and conducted a systematic bird survey. Back in Poland (1915) he re-joined the Russian Army and, for a time, worked at the Zoological Museum, Russian Academy of Sciences, St Petersburg. He returned to Poland (1918) and worked at the National Zoological Museum, Warsaw. He went back to Brazil for a third time (1921) and died of malaria at Pinheirinhos.

Chubb

Snipe genus *Chubbia* **Mathews**, 1913 NCR

Chubb's Cisticola *Cisticola chubbi* **Sharpe**, 1892

Red-capped Crombec ssp. *Sylvietta ruficapilla chubbi*
   Ogilvie-Grant, 1910
Chubb's Twinspot *Mandingoa nitidula chubbi* Ogilvie-Grant,
   1912
[Alt. Green-backed Twinspot ssp.]
Rufous-brown Solitaire ssp. *Cichlopsis leucogenys chubbi*
   Chapman, 1924

Ernest Charles Chubb (1884–1972) was an ornithologist who became Curator of the Museum in Durban, South Africa. His father Charles (1851–1924), a fellow ornithologist, was a Curator at the British Museum but was knocked down and killed by a car as he left the premises (1924). Chubb senior wrote *The Birds of British Guiana*, based on the collection of Frederick Vavasour McConnell and, with Lord Brabourne, *The Birds of South America*. Ernest became President of the Southern Africa Association for Advancement of Science (1945). He wrote a paper entitled 'Record of nesting of skimmer at St Lucia' (1943). The crombec and twinspot are named after Ernest, but the solitaire and snipe genus were named after Charles. It seems to us highly probable that the cisticola was named after Charles, as we cannot believe that Sharpe would have named it after a boy who was only 8 years old at the time.

Chun

> Diego Garcia Turtle Dove *Homopelia chuni* **Reichenow**, 1900 NCR
> [Hybrid population of Madagascar Turtle Dove subspecies: *Nesoenas picturata picturata* x *N. p. comorensis*]

Dr Carl Chun (1852–1914) was a marine biologist and Professor of Zoology at the University of Leipzig (1892–1914). He led the German Deep Sea Expedition (1898–1899) to the subantarctic seas. Many marine taxa are named after him.

Cicely

> Common Hawk Cuckoo ssp. *Hierococcyx varius ciceliae* **W. W. A. Phillips**, 1949

Mrs Cicely G. Lushington *née* Kershaw (b.1902) was an ornithologist and author who was born in Ceylon (Sri Lanka) and lived there until moving to Maidstone, England (1949). She wrote a number of books, including *Familiar Birds of Ceylon* (1925) (under her maiden name), and *Bird Life in Ceylon* (1949), which she illustrated too.

Cicero

> Auklet genus *Ciceronia* **Reichenbach**, 1852 NCR
> [Now in *Aethia*]

Marcus Tullius Cicero (106–43BC) was a Roman lawyer, orator, statesman and (63 BC) consul.

Circe

> Circe Hummingbird *Cynanthus latirostris* **Swainson**, 1827
> [Alt. Broad-billed Hummingbird]

> Squirrel Cuckoo ssp. *Piaya cayana circe* **Bonaparte**, 1850

In Greek mythology, Circe was an enchantress and daughter of the sun god Helios. She could turn humans into animals, making Picus a woodpecker.

Clancey

> Corn Bunting ssp. *Emberiza calandra clanceyi* **Meinertzhagen**, 1947
> [SII *Emberiza calandra calandra*]
> African Stonechat ssp. *Saxicola torquatus clanceyi* **M. Latimer**, 1961

> Buffy Pipit ssp. *Anthus vaalensis clanceyi* **Winterbottom**, 1963
> Green-winged Pytilia ssp. *Pytilia melba clanceyi* **Wolters**, 1963 NCR
> [JS *Pytilia melba citerior*]

Dr Phillip Alexander Clancey (1917–2001) was born and educated in Scotland. When young he showed artist ability and great interest in birds. He served in the British Army (WW2), fighting in the Sicilian and Italian campaigns. Immediately after the war, he became Curator of the Natural History Museum in Scarborough, Yorkshire. He accompanied Richard Meinertzhagen (q.v.) on an ornithological expedition through southern Arabia, East and South Africa (1948–1949). He settled in South Africa (1950), becoming the Director of the Natal Museum in Pietermaritzburg, Natal, then took over the Directorship of the Durban Museum and Art Gallery (1950). He wrote widely, including *Birds of Natal and Zululand* (1964) and *The Rare Birds of Southern Africa* (1985).

Clapperton

> Clapperton's Francolin *Pternistis clappertoni* **Children** & **Vigors**, 1826

Hugh Clapperton (1788–1827) was an explorer who was born in Annan, Dumfries and Galloway, Scotland. He served in both the merchant navy and the Royal Navy, travelling to India and Canada. He took part in an expedition that sought the source of the River Niger. Clapperton reached Lake Chad (1823), but on a second expedition to the same area died of fever near Sokoto (Nigeria). His travelling companion Major Denham (q.v.) returned with records of the expedition, which were published as the *Journal of a Second Expedition into the Interior of Africa* (1829).

Clara

> Naked-faced Spiderhunter *Arachnothera clarae* **Blasius**, 1890

No etymology given; perhaps a female relative of Blasius's.

Clara (Lamb)

> Orange-fronted Parakeet ssp. *Aratinga canicularis clarae* **R. T. Moore**, 1937

Mrs Clara Lamb (DNF) was the wife of Chester Converse Lamb (q.v.), a professional collector who worked for Moore (1933–1955) and who collected the holotype.

Clara (Stein)

> Wattled Ploughbill ssp. *Eulacestoma nigropectus clara* **Stresemann** & Paludan, 1934 NCR; NRM
> Snowy-browed Flycatcher ssp. *Ficedula hyperythra clarae* **Mayr**, 1944

Mrs Clara Stein (DNF) was the wife of Georg Hermann Wilhelm Stein (q.v.). She accompanied him on his expedition to the Dutch East Indies (1931–1932) and took part in collecting and study. A mammal is named after her.

Clarisse

Longuemare's Sunangel *Heliangelus clarisse* **Longuemare**,
1841

Clarisse Parzudaki (fl.1840) was the wife of the French
collector Charles Parzudaki (q.v.).

Clark, A. H.

White-backed Woodpecker ssp. *Dendrocopos leucotos
clarki* **Buturlin**, 1908 NCR
[JS *Dendrocopos leucotos leucotos* or *ussuriensis*]
Oriental Greenfinch ssp. *Carduelis sinica clarki* **Kuroda** &
Mori, 1920 NCR
[JS *Carduelis sinica kawarahiba*]

Austin Hobart Clark (1880–1954) was an American zoologist,
biologist, ornithologist and entomologist, who graduated
from Harvard (1903). He worked at the USNM (1908–1950)
and is perhaps best known for his evolutionary theory of
'zoogenesis'. He discussed this theory in his book *The New
Evolution: Zoogenesis* (1930) which argued that the major
types of life on earth evolved separately and independently
from all the others.

Clark, C. C.

Burmese Yuhina ssp. *Yuhina humilis clarki* **Oates**, 1894

C. C. S. Clark (d.1926) was a British civil servant with the India
Public Works Department (1882–1918).

Clark, H. M.

Clark's Screech Owl *Megascops clarkii* **L. Kelso** & E Kelso,
1935
[Alt. Bare-shanked Screech Owl]

Harry M. Clark (DNF) was a US naturalist of Gretna, Kansas,
who was of great help to the senior describer at the beginning of his studies.

Clark, J. H.

Clark's Grebe *Aechmophorus clarkii* **G. N. Lawrence**, 1858

Lieutenant John Henry Clark (1830–1885) was an American
surveyor, naturalist and collector. He was a student of Baird
(q.v.) at Dickinson College (c.1844). He was a zoologist on the
US/Mexican Border Survey (1850–1855), during which period
(1851) he collected the type specimen of a ground snake,
which was one of about 100 new vertebrate species he
collected with Schott (q.v.). Under the auspices of the USNM
he conducted the Texas Boundary Survey (1860). Two reptiles
and an amphibian are named after him.

Clark, M. H.

Marsh Wren ssp. *Cistothorus palustris clarkae* **Unitt**,
Messer & Thery, 1996

Mary Hollis Clark (1921–2010) was an Americn naturalist,
philanthropist and trustee of San Diego Natural History
Museum. The University of Georgia awarded her bachelor's
degree in business administration (1942). She joined the
board of the museum (1964) and became Emeritus Trustee
(1994).

Clark, M. S.

Parrot genus *Clarkona* **Mathews**, 1917 NCR
[Now in *Psephotus*]

Matthew Symonds Clark (1839–1920) was an aviculturist and
ornithologist who emigrated from England to Australia (1850)
and was a founder member of the South Australian Ornithological Association.

Clark, W.

Clark's Nutcracker *Nucifraga columbiana* **A. Wilson**, 1811

Captain William Clark (1770–1838) was a military man and
explorer who later became Governor of Missouri Territory.
The Lewis & Clark Expedition (1804–1806; see under **Lewis**)
crossed the American continent to the Pacific. Clark was
struggling down the canyon of the Salmon River (22 August
1805), testing out the Indians' warning that it could not be
navigated, when he noticed a new bird. Thomas Jefferson,
who initiated the expedition, had the first specimen sent to
Alexander Wilson (q.v.) for description, who later named it
after Clark. Clark is remembered in many other taxa including
a fish.

Clarke, G. V.

Clarke's Weaver *Ploceus golandi* **S. Clarke**, 1913

Brigadier-General Goland Vanholt Clarke (1875–1944) was an
ornithologist and collector in Africa and the Middle East. He
collected the weaver and sent it to his elder brother, Colonel
Stephenson Robert Clarke (1862–1948), who wrote the
description. One of the colonel's great-grandsons provided
us full details about the brothers, and Louis C. G. Clarke (see
Ludovic).

Clarke, S. R.

Zebra Waxbill ssp. *Amandava subflava clarkei* **G. E. Shelley**,
1903
Blood Pheasant ssp. *Ithaginis cruentus clarkei* **Rothschild**,
1920

Colonel Stephenson Robert Clarke (1862–1948) – sometimes
erroneously Robert Stephenson Clarke – was a naturalist,
botanist, great traveller and keen hunter of big game. He
wrote descriptions of a great number of birds. He bought
Borde Hill House, Sussex, England (1893), and created the
famous garden there which remains in the care of his
descendants. The garden is based on plant and seed collections from China, Burma, Tasmania, the Himalayas and the
Andes, brought there by collectors including Francis
Kingdon-Ward (q.v.). A mammal is named after him. (See
Stephenson)

Clarke, W. E.

Song Thrush ssp. *Turdus philomelos clarkei* **Hartert**, 1909
Brown Skua ssp. *Catharacta lonnbergi clarkei* **Mathews**,
1913 NCR
[JS *Stercorarius antarcticus lonnbergi*]

William Eagle Clarke (1853–1938) was a British ornithologist. He was Curator of Leeds Museum (1884–1887) and the Natural History Department, Royal Scottish Museum (1888–1921), becoming Keeper (1909) at the latter.

Claud

Black-shouldered Nightjar *Caprimulgus claudi*
B. Alexander, 1907 NCR
[JS *Caprimulgus nigriscapularis*]

Hill Babbler ssp. *Pseudoalcippe abyssinica claudei*
B. Alexander, 1903
[Trinomial sometimes amended to *claudi*]

Captain Claud Alexander (1878–1904) died during the Niger to Nile Expedition (1904). He was a brother of Captain Boyd Alexander (q.v.), Major Robert Alexander and Herbert Alexander.

Claude

Bar-throated Apalis ssp. *Apalis thoracica claudei*
W. L. Sclater, 1910
Dwarf Cassowary ssp. *Casuarius bennetti claudii* **Ogilvie-Grant**, 1911 NCR
[*C. bennetti* currently regarded as monotypic]
King Bird-of-Paradise ssp. *Cicinnurus regius claudii* Ogilvie-Grant, 1915 NCR
[JS *Cicinnurus regius regius*]
Dusky Babbler ssp. *Turdoides tenebrosa claudei*
Bannerman, 1919 NCR; NRM

(See **Grant, C. H. B.**)

Claudia (Hartert)

Neotropical Palm Swift genus *Claudia* **Hartert**, 1892 NCR
[Now in *Tachornis*]

Claudia's Sunangel *Heliangelus claudia* Hartert, 1898 NCR
[Only known from trade skins; aberrant *Heliangelus clarisse*?]
Claudia's Leaf Warbler *Phylloscopus claudiae* **La Touche**, 1922

Claudia Bernadine Elisabeth Hartert *née* Reinard (1863–1958) was an ornithologist and wife of Ernst Johann Otto Hartert (q.v.); she published several articles jointly with him, some describing new species. She moved from England to Holland (1939). (See also **Reinard**)

Claudia (Hubbard)

Ruddy Kingfisher ssp. *Halcyon coromanda claudiae*
Hubbard & Du Pont, 1974

Claudia Leigh Hubbard (DNF) was the wife of the senior describer, John Patrick Hubbard, and an avid devotee of kingfishers. She was Secretary of the New Mexico Ornithological Society in the 1970s. She and her husband co-wrote *Birds of New Mexico's Park Lands* (1979).

Clay

African Green Pigeon ssp. *Treron calvus clayi* **C. M. N. White**, 1943 NCR
[JS *Treron calvus salvadorii*]
Red-winged Francolin ssp. *Scleroptila levaillantii clayi* C. M. N. White, 1944 NCR
[JS *Scleroptila levaillantii kikuyuensis*]

Gervas Charles Robert Clay (1907–2009) worked for the Provincial Administration in Northern Rhodesia (Zambia) (1930–1964). His first interest was gamebird shooting, but he developed a general interest in birds. He was deaf to bird song, the result of too much quinine taken as a malarial prophylactic, but touring (frequently by bicycle) he discovered much of interest in the Isoka region. He was the first to explore the eastern highlands in Zambia (1942–1943), where he found ten species new to the country's avifauna. He wrote a biography of the Litunga (king) of Barotseland, Lewanika, entitled *Your Friend, Lewanika* (1968).

Cleaver

Blackcap Illadopsis *Illadopsis cleaveri* **G. E. Shelley**, 1874

William Cleaver (fl.1870) was resident in the Gold Coast (Ghana), where he was the agent of Messrs. F. and A. Swanzy of London. He was for a time held captive in the Ashanti War (1873). He was described as being 'a tiny man with a head that instantly attracts attention.' Shelley named the bird after him 'in acknowledgment of his courtesy to me during my recent visit to Cape-Coast Castle.'

Cleaves

Gorgeted Woodstar ssp. *Chaetocercus heliodor cleavesi*
R. T. Moore, 1934

Margaret Forbes Moore *née* Cleaves (DNF) was the second wife (1922) of the describer, Robert Thomas Moore (q.v.). (See Margaret (Moore))

Cleghorn

Prinia sp. *Franklinia cleghorniae* **Blyth**, 1867 NCR
[Alt. Rufous-fronted Prinia; JS *Prinia buchanani*]

Mrs Marjory Isabella Cleghorn *née* Cowan (1838–1887) was the wife of Scottish botanist and Conservator of Forests in India, Hugh Francis Clarke Cleghorn. He was a colleague of Jerdon (q.v.).

Cleland, D. M. & R.

Carola's Parotia ssp. *Parotia carolae clelandiae* **Gilliard**, 1961
[Alt. Queen Carola's Parotia ssp.]

Brigadier Sir Donald Mackinnon Cleland (1901–1975) was the Australian Administrator of Papua New Guinea (1953–1966). He and his wife, Lady Rachel Cleland *née* Evans (1906–2002), are both honoured even although the binomial is in the feminine and implies that only she is intended. He fought in Libya, Greece and Syria (WW2). He stayed on in Port Moresby in retirement, died there, and was accorded a state funeral.

She was active in promoting schools and welfare in PNG, staying on for three years after her husband's death and making a number of visits there from Australia until her death.

Cleland, J. B.

Lewin's Rail ssp. *Lewinia pectoralis clelandi* **Mathews**, 1911

Brown Songlark ssp. *Cincloramphus cruralis clelandi* Mathews, 1912 NCR; NRM

Crested Bellbird ssp. *Oreoica gutturalis clelandi* Mathews, 1912 NCR

[JS *Oreoica gutturalis gutturalis*]

Dusky Miner ssp. *Manorina flavigula clelandi* Mathews, 1912 NCR

[JS *Manorina flavigula obscura*]

Peaceful Dove ssp. *Geopelia placida clelandi* Mathews, 1912

Spotted Bowerbird ssp. *Chlamydera maculata clelandi* Mathews, 1912 NCR; NRM

Striated Thornbill ssp. *Acanthiza lineata clelandi* Mathews, 1912

Tasmanian Boobook ssp. *Ninox novaeseelandiae clelandi* Mathews, 1913 NCR

[JS *Ninox novaeseelandiae leucopsis*]

Painted Firetail ssp. *Emblema pictum clelandi* Mathews, 1914 NCR; NRM

Varied Honeyeater ssp. *Meliphaga versicolor clelandi* Mathews, 1915 NCR

[JS *Lichenostomus versicolor versicolor*]

Yellow Thornbill ssp. *Acanthiza nana clelandi* Mathews, 1920 NCR

[JS *Acanthiza nana nana*]

Red Wattlebird ssp. *Anthochaera carunculata clelandi* Mathews, 1923

Professor Sir John Burton Cleland (1878–1971) was an Australian pathologist, mycologist and ornithologist who qualified as a physician in Sydney (1900). He wrote *The History of Ornithology in South Australia* (1936).

Clemence

Blue-throated Hummingbird *Lampornis clemenciae* **Lesson**, 1829

Olive-backed Sunbird ssp. *Cinnyris jugularis clementiae* Lesson, 1827

Marie Clemence Lesson *née* Dumont de Sainte-Croix (d.1834) was the second wife of French ornithologist René Primavere Lesson. She died of cholera.

Clement

Grey-capped Woodpecker ssp. *Dendrocopos canicapillus clementii* **LaTouche**, 1919 NCR

[JS *Dendrocopos canicapillus scintilliceps*]

Father Columban Clement (fl.1918) was a Belgian Franciscan missionary to China.

Clements

Iquitos Gnatcatcher *Polioptila clementsi* **Whitney** & Alonso, 2005

Dr James (Jim) Franklin Clements (1927–2005) was a naturalist, entrepreneur, adventurer and diplomat. Clements not only has the distinction of having a species named after him but a whole system of classifying the world's bird species. His doctoral thesis was written (1975) at the California Western University when he was in his forties, and became the checklist that has, so far, been published in six editions as *Birds of the World: a Check List* (later *The Clements Checklist of the Birds of the World*). He spent his early years in an orphanage before joining the merchant marine (aged 15) and, when old enough, he joined the navy and served on an aircraft carrier in the Philippine Sea. After military service he took a degree at the University of Minnesota. During the Korean War he again saw military service, this time in the US Air Force. When he left the Air Force he became a partner in a printing company in California. This allowed him sufficient time to take on his doctoral thesis. He retired (1988) and founded Ibis Publishing Company, publishing natural history titles. In retirement he devoted time to raising funds for conservation projects in South America and museum exhibits in the US. The President of Malawi appointed him Honorary Consul for the State of California (1986). He also managed to find time to serve on numerous committees including as President of the San Diego Museum of Natural History and the Explorers Club of San Diego and Los Angeles. He died of myloid leukaemia.

Cleopatra

Little Green Bee-eater ssp. *Merops orientalis cleopatra* **Nicoll**, 1910

Cleopatra (69–30 BC) was Queen of Egypt (52–30 BC). For the full story we recommend Shakespeare's *Antony and Cleopatra*. Suitably, a reptile is named Cleopatra's Asp after her.

Clifford

Silver Pheasant *Gennaeus cliffordi* **Oates**, 1904 NCR

[Hybrid population of *Lophura nycthemera* subspecies]

Lieutenant-Colonel Richard Clifford (b.1877) was in the British Army in India (1898–1922).

Clifton

Blue-winged Kookaburra ssp. *Dacelo leachii cliftoni* **Mathews**, 1912 NCR

[JS *Dacelo leachii occidentalis*]

W. Clifton (fl.1912) was an Australian field naturalist.

Clinton

Northern Cardinal ssp. *Cardinalis cardinalis clintoni* **R. C. Banks**, 1963

Clinton Seeger Banks (1900–1995) was the father of Dr Richard C. Banks (q.v.). Banks named it in appreciation of all the encouragement his father gave him over many years.

Clio

Yellow-throated Whistler ssp. *Pachycephala macrorhyncha clio* **Wallace**, 1863

In Greek mythology Clio was the muse of history.

Cloos

Cape Bunting ssp. *Emberiza capensis cloosi* **Hoesch & Niethammer**, 1940 NCR
[JS *Emberiza capensis bradfieldi*]

Dr Hans Cloos (1885–1951) was a German geologist and explorer (before 1914) in the Dutch East Indies (Indonesia) and South West Africa (Namibia). He was Professor of Geology at the University of Breslau (Wroclaw, Poland) (1919–1926) and Professor of Geology at the University of Bonn (1926–1951).

Clot Bey

Clot Bey's Lark *Ramphocoris clotbey* **Bonaparte**, 1850
[Alt. Thick-billed Lark]

Antoine-Barthélmy Clot (1793–1868) was a French physician who became the head of the Egyptian army medical corps, establishing the first hospital there (1827). He is considered to be the founder of modern medicine in Egypt and has streets named after him in Cairo. He was resident in Egypt and was awarded the honorary title of 'Bey' by Mohammed Ali Pasha (who called him Klute Bey), for his services to Egyptian medicine (1832).

Clotho

Hummingbird genus *Clotho* **Mulsant**, 1876 NCR
[Now in *Eupherusa*]

In Greek mythology Clotho was the youngest of the three Fates. She was responsible for spinning the thread of human life.

Clunie

Long-legged Thicketbird ssp. *Megalurulus rufus cluniei* Kinsky, 1975
[Alt. Long-legged Warbler; Syn. *Ortygocichla rufa cluniei*]

Dr Fergus Clunie was Curator of Rouse Hill House and Farm in New South Wales, Australia. He was Director of the Fiji National Museum, Suva, where he is now a Research Associate, having also been a curator in Northland, New Zealand. He wrote *Birds of the Fiji Bush* (1984).

Coats

Goldcrest ssp. *Regulus regulus coatsi* **Sushkin**,1904

Mr Coats was a collector at Sayan, Siberia (1903). We can find nothing more about him.

Cobb

Cobb's Wren *Troglodytes cobbi* **Chubb**, 1909

Arthur Frederick Cobb (b.1877) was a farmer in the Falkland Islands. He was the manager of Bleaker Island farm when he wrote *Wild Life in the Falklands* (1910), which includes a note on the decline in numbers of the Striated Caracara *Phalcoboenus australis*. He wrote *Birds of the Falkland Islands – A Record of Observation with the Camera* (1933). The wren is a Falkland Islands endemic that Cobb collected. He wrote evocatively, once describing the sound of massed Rockhopper Penguins '… as if thousands of wheelbarrows, all badly in need of greasing, are being pushed at full speed.'

Coburn

Redwing ssp. *Turdus iliacus coburni* **Sharpe**, 1901

Frederick Coburn (1814–1914) was a British taxidermist in Birmingham (1893) who collected in Iceland (1899).

Cochrane

Black Sunbird ssp. *Leptocoma sericea cochrani* **Stresemann** & Paludan, 1932

Commodore Henry L. Cochrane (1871–1949) was an ornithologist and an officer in the Royal Navy who was an early proponent of naval aviation. He was seconded to the Royal Australian Navy and served on the Commonwealth's Navy Board (1917–1918). He wrote *Hooded Dotterels* (1917), and met Stresemann (q.v.) at the 8th International Ornithological Congress in Oxford (1934).

Cockerell

New Britain Friarbird *Philemon cockerelli* **P. L. Sclater**, 1877
Cockerell's Fantail *Rhipidura cockerelli* **E. P. Ramsay**, 1879
[Alt. White-winged Fantail]
Cockerell's Honeyeater *Trichodere cockerelli* **Gould**, 1869
[Alt. White-streaked Honeyeater]

James T. Cockerell (1847–1895) was an Australian collector (1865–1891). He visited both Samoa (1874) and the Bismarck Archipelago (1876) in the company of the Reverend George Brown (q.v.). He collected entomological specimens, particularly stingless bees.

Codrington

Codrington's Indigobird *Vidua codringtoni* **Neave**, 1907
[Alt. Twinspot Indigobird, Green Indigobird]

Robert Edward Codrington (1869–1908) was colonial administrator in Northern Rhodesia (Zambia) (1898–1907). He wrote several articles for the *Geographical Journal*, such as 'A voyage on Lake Tanganyika' (1902).

Coe, M.

Coe's Honeyguide *Melignomon eisentrauti* Louette, 1981
[Alt. Yellow-footed/Eisentraut's Honeyguide]

Dr Malcolm Coe (b. 1930) is a tropical ecologist and naturalist. He worked at the University of Nairobi (1956–1968) and then Department of Zoology, Animal Ecology Research Group, St Peters College, Oxford University, until his retirement (1968–1995). His fieldwork experience (both faunal and floral) includes eastern, central and southern Africa, as well as India. He has also taken part in a number of expeditions; to

Mount Kenya (1957); Liberia (1964 & 1966); the South Turkana Expedition, Kenya (1968–1970); the Royal Society Aldabra Expeditions to the Indian Ocean (1970–1982); the Kora Research Project, Kenya (1982–1985) and the Mkomazi Ecological Research Programme, Tanzania (1990–1997). He was awarded the Royal Geographical Society Busk Medal (1988) as well as the Zoological Society of Southern Africa Gold Medal (1989). Serle collected the first specimen of the honeyguide (1956) but considered it to be an immature Zenker's Honeyguide *Melignomon zenkeri*. Coe saw the bird in the Nimba research compound (1964). He said: 'It flew from the forest edge where it appeared to be searching amongst the leaves and branches of a large climbing *Combretum*. I watched it several times during the day and felt sure it was a honeyguide but one that was quite unknown to me'. Coe wrote this up but did not seek publication. Forbes-Watson collected a series of this honeyguide at Mt Nimba and Colston wrote a formal description, submitting it to the *Bulletin of the British Ornithologists' Club*, when he took responsibility for the Nimba collection at the BMNH. Based partly on a specimen collected by Eisentraut, Louette published the first description of the honyeguide, naming it *Melignomon eisentrauti* (1981) and pre-empting Colston whose own paper was in press and was published soon after. Colston proposed Coe's name for its common name.

Coe, W.

Grey-throated Babbler ssp. *Stachyris nigriceps coei* **Ripley**, 1952
[Sometimes merged with *Stachyris nigriceps nigriceps*]

Dr William Robertson Coe (1869–1960) was an American ornithologist, Curator and Professor of Ornithology (1910–1926) at the Peabody Museum, Yale.

Coffin

Black-throated Bobwhite ssp. *Colinus nigrogularis coffini* **Nelson**, 1932 NCR
[JS *Colinus nigrogularis nigrogularis*]

Howard Earle Coffin (1873–1937) was an American automobile engineer and manufacturer, land developer and aviculturist. He financed a project to collect Ocellated Turkeys *Meleagris ocellata* and other Neotropical gamebirds, and attempt to acclimatise them at his winter home on Sapelo Island, Georgia.

Cognacq

Scaly-breasted Partridge ssp. *Arborophila chloropus cognacqi* **Delacour** & **Jabouille**, 1924
[Alt. Green-legged Hill Partridge ssp.]

Dr Maurice Cognacq (fl.1926) was a creole from the French territory of Guadeloupe who was Governor of Cochinchina, French Indochina (Vietnam) (1921–1926). His administration was regarded as particularly oppressive.

Cohuatl

Hummingbird sp. *Trochilus cohuatl* De La Llave, 1833 NCR
[JS *Calothorax lucifer*]

In Aztec myth Cohuatl is the serpent as a personification of the earth and representing wisdom.

Colclough

Buttonquail genus *Colcloughia* **Mathews**, 1917 NCR
[Now in *Turnix*]

Fork-tailed Swift *Micropus colcloughi* Mathews, 1915 NCR
[JS *Apus pacificus*]

Northern Scrub Robin ssp. *Drymodes superciliaris colcloughi* Mathews, 1914
Collared Kingfisher ssp. *Todiramphus chloris colcloughi* Mathews, 1916

Michael Joseph Colclough (b.1875) was an Australian collector who was taxidermist at the Queensland Museum, Brisbane (1915). He collected the holotype of the scrub robin (1910).

Cole, C. F.

Wattlebird genus *Coleia* **Mathews**, 1912 NCR
Wattlebird genus *Colena* Mathews, 1931 NCR
[Both these now in *Anthochaera*]

Grey Butcherbird ssp. *Cracticus torquatus colei* Mathews, 1912 NCR
[JS *Cracticus torquatus leucopterus*]
Black Currawong ssp. *Strepera fuliginosa colei* Mathews, 1916
Eastern Rosella ssp. *Platycercus eximius colei* Mathews, 1917 NCR
[JS *Platycercus eximius eximius*]

Charles Frederick Cole (1875–1959) was an Australian civil servant, horticulturist and collector.

Cole, G. L.

Bustard sp. *Otis colei* **A. Smith**, 1831 NCR
[Nomen nudum; species now *Neotis ludwigii* Rüppell, 1837]

General Sir Galbraith Lowry Cole (1772–1842) was a distinguished British soldier. He was commissioned (1778) and served in the West Indies, Ireland and Egypt. He was commander of the 4th Division in the Duke of Wellington's army in the Peninsular War and fought in many of the decisive battles of it from Maida (1806) to Toulouse (1814). He was Member of Parliament at Westminster for Fermanagh (1803–1823), and Governor of Mauritius (1823–1828) and of Cape of Good Hope (1828–1833). Although Smith's name *Otis colei* pre-dates Rüppell's *Otis Ludwigii* it is regarded as a *nomen nudum* – i.e. it is not recognised as a valid scientific name because it was not published with an adequate description.

Colenso

Auckland Islands Shag *Leucocarbo colensoi* **W. L. Buller**, 1888
[Syn. *Phalacrocorax colensoi*]

Rev. William Colenso (1811–1899) was a British-born naturalist, ethnologist, philologist and missionary. He was apprenticed to a printer (1826) and took a job in New Zealand

(1834) working for the Church Missionary Society printing copies of Christian texts in the Maori language. Other works followed, including a Maori text of the Treaty of Waitangi (1840). (At the signing of the Treaty, his cautious representations to Lieutenant Governor William Hobson that many Maori were unaware of the meaning of the treaty were brusquely set aside. His observations recorded at the time were published as *The Authentic and Genuine History of the Signing of the Treaty of Waitangi* [1890], the most reliable contemporary European account of the event.) His enthusiasm for natural history was boosted by the brief visit of Charles Darwin (q.v.) on the *Beagle* (1835). The New South Wales government gave him some training (1838) in botany. He collected across North Island until 1852, and became an ordained minister (1844). He was suspended as a deacon (1852) when he fathered a child by a Maori girl, thus becoming a figure of ridicule among the Maori community, whom he had enjoined against sin. He took up politics (1858) and was elected to the General Assembly (1861). He published a large number of scientific papers and was commissioned by the General Assembly to produce a Maori dictionary (1865), but only a section was published.

Coles

Coles's Lorikeet *Trichoglossus colesi* **Le Souëf**, 1910 NCR
[JS *Trichoglossus* (*haematodus*) *moluccanus*]

Clifford Coles (1876–1949) was an Australian natural history dealer, furrier and aviculturist.

Collar

Sangihe Scops Owl *Otus collari* **Lambert** & Rasmussen, 1998

Dr Nigel J. Collar is Leventis Fellow in Conservation Biology at BirdLife International, where he has worked for three decades. The etymology reads: 'this species is named after our friend and colleague Dr Nigel J Collar in recognition of his numerous contributions to the important field of bird conservation. His work has stimulated enormous interest in threatened birds, and has encouraged a conservation ethic and philosophy amongst a generation of amateur and professional ornithologists and birdwatchers.' He is the author of many papers and books including co-writing *Threatened Birds of Africa and related islands* (1985), *Threatened Birds of the Americas* (1992) and *Threatened Birds of Asia* (2001). He also wrote the section on thrushes in the *Handbook of the Birds of the World*.

Collerwart

Yellow-browed Camaroptera ssp. *Camaroptera superciliaris collerwarti* Lletget, 1943 NCR; NRM

Transcription or spelling error. (See **Callewaert**)

Collett

White-cheeked Starling *Poliopsar colletti* **Sharpe**, 1888 NCR
[JS *Spodiopsar cinereaceus*]

Cape Longclaw ssp. *Macronyx capensis colletti* Schou, 1908

Collett's Blue-breasted Quail *Excalfactoria chinensis colletti* **Mathews**, 1912
[Syn. *Coturnix chinensis colletti*]
Rufous Whistler ssp. *Pachycephala rufiventris colletti* Mathews, 1912 NCR
[JS *Pachycephala rufiventris falcata*]
Silver-backed Butcherbird ssp. *Cracticus argenteus colletti* Mathews, 1912

Professor Robert Collett (1842–1913) was a Norwegian zoologist and amateur photographer. He was the Director of the Museum of Natural History at Christiania (1882–1913) and Professor of Zoology (1884–1913). He also described the Hooded Parakeet *Psephotus dissimilis* (1898) based on a collection made by the Norwegian ornithologist Knut Dahl in north and north-west Australia. A mammal, an amphibian and two reptiles are named after him.

Collie

Collie's Magpie-jay *Calocitta colliei* **Vigors**, 1829
[Alt. Black-throated Magpie-jay]

Lieutenant Dr Alexander Collie (1793–1835) was the naval surgeon and naturalist on an expedition (1825–1828) led by Captain Frederick Beechey (q.v.) on HMS *Blossom*, which made some significant ornithological findings during the voyage from Chile to Alaska. Collie collected many specimens that did not survive the return journey to England in good condition, but he made some coloured drawings of birds he thought were new and also took extensive notes. From these, the British ornithologist Nicholas Vigors included a chapter on ornithology in his *Zoology of Captain Beechey's Voyage* (1839), naming several species new to science. Collie also collected many live birds that went on to be exhibited in London Zoo. Collie went to Perth as a colonial administrator where he died before Vigors's work was published. When aboard HMS *Sulphur* he discovered what is now the Collie River in Western Australia. A town in Australia, a mammal and a reptile are also named after him.

Collin

Little Pied Flycatcher ssp. *Ficedula westermanni collini* **Rothschild**, 1925

Albert Collin (1875–1952) was a British ornithologist and shipbroker who lived and worked at Kotka, Finland.

Collingwood

Yellow-breasted Flycatcher ssp. *Tolmomyias flaviventris collingwoodi* **Chubb**, 1920
[SII *Tolmomyias flaviventris aurulentus*]
White-browed Crake ssp. *Poliolimnas cinereus collingwoodi* **Mathews**, 1926 NCR; NRM
[Syn. *Porzana cinerea*]

Captain Collingwood 'Cherry' Ingram (1880–1981) was an ornithologist and botanist who was in Patagonia (1901), Japan (1907) and places as widely spread as Portugal, Corsica and New Zealand. He was a leading member of the Royal Horticultural Society. In addition to much writing on botany he published a number of books on ornithology, including *The Birds of the Riviera* (1926), *In Search of Birds* (1966) and *The Migration of the Swallow* (1970). He is probably best remembered for his garden at Benenden, England, where he had a collection of ornamental Japanese cherry trees: hence his nickname. He started this garden immediately after WW1 in which he fought in the Royal Flying Corps. He was a man of great heart. On his deathbed he asked a friend and fellow enthusiast called Hardy to take seedlings of one of his treasured plants. Hardy did this and reported to Ingram that they had been safely dug up. Ingram replied, 'If I get better, I want them back!'

Collins, A. M.

Longbill sp. *Macrosphenus collinsi* **Riley**, 1924 NCR
[Alt. Yellow Longbill; JS *Macrosphenus flavicans flavicans*]

Yellow-breasted Warbling Antbird ssp. *Hypocnemis subflava collinsi* **Cherrie**, 1916

Major Alfred M. Collins (1876–1951) was an American sportsman, big-game hunter, explorer and sponsor of expeditions, including the Collins-Day South American Expedition (1914–1915), which included Cherrie (q.v.). Collins visited British East Africa (1911–1912) and in north-east Siberia (1918).

Collins, H. B.

Prairie Warbler ssp. *Dendroica discolor collinsi* H. H. Bailey, 1930 NCR
[JS *Dendroica discolor paludicola*]
Chestnut-crowned Warbler ssp. *Seicercus castaniceps collinsi* **Deignan**, 1943

Dr Henry Bascom Collins Jr (1899–1987) was an American ethnologist and archaeologist. He joined the USNM, Division of Ethnology (1924), and became Senior Ethnologist with the US Bureau of Ethnology (1939–1965) and Senior Scientist in the USNM Office of Anthropology (1965), retiring as Archaeologist Emeritus (1967). During his career he carried out a number of important archaelogical excavations, particularly of the Inuit culture in Alaska. Deignan named the warbler 'in recognition of his interest and studies in the fields of Indo-Chinese ethnology and anthropology.'

Colls

Colls's Forest Robin *Erythropygia leucosticta collsi* **B. Alexander**, 1907
[Alt. Forest Scrub Robin ssp.; Syn. *Cercotrichas leucosticta collsi*]

Howard Colls (DNF) seems to have been collecting in West Africa (c.1906–1907). Boyd Alexander (q.v.) gave no reason for naming the scrub robin after Colls.

Colomann

African Penduline-tit *Anthoscopus colomanni* **Madarász**, 1910 NCR
[JS *Anthoscopus (caroli) sylviella*]

(See **Kittenberger** – Colomann is a latinised form of the forename Kálmán; see also **Katona**)

Colston

Colston's Bulbul *Xanthomixis apperti* Colston, 1972
[Alt. Appert's Tetraka; Syn. *Phyllastrephus apperti*]

Forest Scrub Robin ssp. *Erythropygia leucosticta colstoni* Tye, 1991
Dusky Crested Flycatcher ssp. *Elminia nigromitrata colstoni* **Dickerman**, 1994
Tropical Shearwater ssp. *Puffinus bailloni colstoni* Shirihai & Christie, 1996
[Possibly inseparable from *Puffinus bailloni dichrous*]

Peter Robert Colston (b.1935) is a field ornithologist who was Senior Curator at the BMNH where until retirement (1995) he was responsible for the largest scientific collection of bird skins in the world. Shortly after joining the museum in 1961 he was dispatched to Andalucia where he learnt to collect and prepare his first study skins. He made a series of museum expeditions to Australia, sponsored by Major Harold Hall (q.v.), an Australian philanthropist interested in natural history. On the first expedition (1962–1963) he collected extensively in eastern Australia and was responsible for finding and obtaining the first specimens of an undescribed Australo-Papuan babbler in southern Queensland, now known as White-throated Babbler *Pomatostomus halli*. He collected in SWAustralia (1966), then Africa (1970) where he joined the Royal Geographical Society's expedition to northern Turkana, Kenya. Subsequently he became the museum's specialist on African birds, publishing descriptions of four new species. He contributed to *An Atlas of Speciation in African Birds* (1970) and, with K. Curry-Lindahl, co-wrote *The Birds of Mount Nimba, Liberia* (1986). He served as a member of the BOU Records Committee and subsequently co-authored *A Field Guide to the Rare Birds of Britain & Europe*, with Ian Lewington and Per Alström. Always a wader enthusiast, he also published *The Waders of Britain and Europe, North Africa and the Middle East* with Philip Burton. He made private trips to China (1989–1995), together with Per Alström and Urban Olsson, which resulted in the discovery of three previously unknown leaf warblers. Michael Walters of the BMNH (Tring) coined the vernacular name Colston's Bulbul in his *Complete Birds of the World* (1980).

Coltart

Grey-throated Babbler ssp. *Stachyris nigriceps coltarti* **Harington**, 1913
Blue-eared Kingfisher ssp. *Alcedo meninting coltarti* **E. C. S. Baker**, 1919

Dr Henry Neville Coltart (1873–1922) was a British physician and oologist who went to India (1899–1913) as medical officer for a tea-planting company in Assam, and then was in general practice in England (1913–1922).

Comer

Gough Moorhen *Gallinula comeri* **J. A. Allen** 1892

George Comer (1858–1937) was a US seaman and collector who travelled the South Atlantic. He served on whaling ships, achieving the rank of captain, and voyaged to both the Arctic (1875–1919) and Antarctic. When whaling became unprofitable he trapped for furs with the Inuit. He took up photography (1893) and his photographs, together with his observations, form a valuable ethnographic account. He also collected artifacts that he sent to the AMNH, New York. It was written of him: 'he traveled to the uttermost parts of the globe … he was not only interested in capturing whales but also in all the forms of life and mysteries of nature with which he came in contact. A natural student he soon acquired knowledge of a language of the Eskimos, and having overcome the native taboos was able to learn much of Eskimo history and customs'.

Commerson

Commerson's Scops Owl *Otus commersoni* **Oustalet**, 1896
EXTINCT
[Alt. Mauritius Owl; JS *Mascarenotus sauzieri*]

Rufous Hornero ssp. *Furnarius rufus commersoni* **Pelzeln**, 1868

Philibert Commerson (1727–1773) was known as 'doctor, botanist and naturalist of the King'. He accompanied the French explorer Louis Antoine de Bougainville (q.v.) on his round-the-world expedition (1766–1769) on *La Boudeuse* and *L'Etoile*. Commerson was primarily a botanist but he has a wide diversity of animal species named after him, including several fish and two mammals. He also discovered the vine *Bougainvillea* (1760s), naming it after the expedition leader.

Comrie

Curl-crested Manucode *Manucodia comrii* **P. L. Sclater**, 1876

Dr Peter Comrie (1832–1882) was a British naval surgeon who collected the holotype while serving on board HMS *Basilisk* (1871–1875). He collected widely including fauna, flora and anthropological artifacts. He was tried on five counts before a court-martial for insubordination (1875) and, three being proven, forfeited three years seniority and was discharged.

Condamine

Buff-tailed Sicklebill *Eutoxeres condamini* **Bourcier**, 1851

Charles Marie de la Condamine (1701–1774) was a French scientist and traveller. He was originally a soldier. He travelled to North Africa, Palestine, Cyprus and Constantinople (Istanbul) (1730) and started a journey (1735) to Peru. The expedition had two other scientists, with Louis Godin (q.v.) as leader. Its main purpose was to measure a degree of meridian at the equator. There seems to have been considerable friction between them – they travelled by different routes and all eventually met at Quito and then disagreed about the mathematics. They returned to France independently and by different routes – de la Condamine's being a 4-month raft journey down the Amazon. He finally returned to France (1745) after a journey that lasted 10 years.

Condon

White-browed Scrubwren ssp. *Sericornis maculata condoni* **Mathews**, 1942 NCR
[JS *Sericornis frontalis mellori*]

Herbert Thomas Condon (1912–1978) was an Australian ornithologist and nomenclaturist, who joined the South Australian Museum, Adelaide (1929), and was Curator of Birds and Reptiles (1938–1976). He served with the RAAF (WW2). He was President, RAOU (1961–1962). He wrote *Field Guide to the Hawks of Australia* (1949).

Confucius

House Sparrow ssp. *Passer domesticus confucius* **Bonaparte**, 1853 NCR
[JS *Passer domesticus indicus*]

Confucius or Kong Qiu (551–479 BC) was a Chinese philosopher and great teacher.

Congreve

Marsh Tit ssp. *Poecile palustris congrevei* **Kinnear**, 1928 NCR
[JS *Poecile palustris palustris*]

Major William Maitland Congreve (1883–1967) was a British army officer, oologist and collector. He was in Spitsbergen (1923 and 1932) and Transylvania (1928 and 1934). He wrote *An Oologist in Spitsbergen* (1924).

Conigrave

Shrike-thrush genus *Conigravea* **Mathews**, 1913 NCR
[Now in *Colluricincla*]

Fairy Martin ssp. *Petrochelidon ariel conigravi* Mathews, 1912 NCR; NRM
Little Shrike-thrush ssp. *Colluricincla parvula conigravi* Mathews, 1912 NCR
[JS *Colluricincla megarhyncha parvula*]
Papuan Frogmouth ssp. *Podargus papuensis conigravi* Mathews, 1912 NCR
[JS *Podargus papuensis papuensis*]

Charles Price Conigrave (1882–1961) was an Australian zoologist, explorer, photographer and collector. He was a Senior Assistant at the Western Australian Museum (1901) and leader of the Kimberley Exploring Expedition (1911–1912). He wrote *Walk-about* (1938).

Connaught

Duchess of Connaught's Lorikeet *Charmosyna margarethae* **Tristram**, 1879
[Alt. Duchess Lorikeet]

Princess Louise Margaret Alexandra of Prussia, Duchess of Connaught (1860–1917). Tristram writes that he is naming the bird 'in compliment to the bride of H.R.H. the Duke of Connaught.'

Conover

Conover's Dove *Leptotila conoveri* **Bond** & **Meyer de Schauensee**, 1943
[Alt. Tolima Dove]

Ashy-headed Green Pigeon ssp. *Treron phayrei conoveri* Rand & R. Fleming, 1953

[Syn. *Treron pompadora conoveri*]
Singing Quail ssp. *Dactylortyx thoracicus conoveri* Warner & Harrell, 1957

Henry Boardman Conover (1892–1950) was a soldier and amateur ornithologist. He served in the US Army (WW1). All through his life he was a field sports enthusiast. He became interested in scientific ornithology (c.1920), particularly in relation to gamebirds. He was a Trustee of The Field Museum, Chicago, donating his collection of ornithological texts to them. He made significant contributions to Hellmayr's *The Catalogue of Birds of the Americas* wrote a number of articles such as 'A new species of rail from Paraguay' (1934). A mammal is named after him.

Conrad (Festa)

Chiguanco Thrush ssp. *Turdus chiguanco conradi* **Salvadori & Festa**, 1899

Conrado Festa (fl.1899) was an Italian lawyer who was a generous donor of Orthoptera to the Turin Zoological Museum. His son was the junior author, Dr E. Festa.

Conrad (Loddiges)

Conrad's Inca *Coeligena torquata conradii* **Bourcier**, 1847
[Alt. Collared Inca ssp.]

Joachim Conrad Loddiges (1738–1826) was a gardener who established a plant nursery in Hackney, London. His son George Loddiges (q.v.) took over and raised its profile. George was also interested in hummingbirds. His son was Conrad Loddiges II (1821–1865). It is unclear whether Bourcier, who described the hummingbird under the name *Trochilus conradii*, intended to honour George Loddiges's father or son.

Conrad, P.

Streak-eared Bulbul ssp. *Pycnonotus blanfordi conradi* **Finsch**, 1873

Captain Paul Conrad (1836–c.1873) was a German who was in the East Indies (1870–1873). A reptile is also named after him.

Conrads

Yellow-mantled Widowbird ssp. *Euplectes macroura conradsi* **Berger**, 1908
Green-winged Pytilia ssp. *Pytilia melba conradsi* **Grote**, 1922 NCR
[JS *Pytilia melba belli*]

Father H. A. Conrads (1874–1940) was a German Jesuit missionary to Tanganyika (Tanzania). He was resident on Ukerewe Island, Lake Victoria, and made an entomological collection there (1907).

Consobrinorum

Pale-bellied White-eye *Zosterops consobrinorum* **Meyer**, 1904

Consobrinorum means 'of the cousins'. See **Sarasin** for the cousins concerned.

Constant

Constant's Starthroat *Heliomaster constantii* **De Lattre**, 1843
[Alt. Plain-capped Starthroat]

Charles Constant (1820–1905) was a French collector and taxidermist. He is also commemorated in a name of a fossil.

Constantia (Davenport)

Narina Trogon ssp. *Apaloderma narina constantia* **Sharpe & Ussher**, 1872

Mrs Constance Julia Davenport (d.1944) was the daughter of Herbert Taylor Ussher (q.v.).

Constantia (Ripley)

Whiskered Yuhina ssp. *Yuhina flavicollis constantiae* **Ripley**, 1953

Mrs Constance Baillie Ripley *née* Rose (1877–1961) was the mother of S. Dillon Ripley.

Conti

Yellow-bellied Warbler ssp. *Abroscopus superciliaris contii* **de Schauensee** 1946 NCR
[JS *Abroscopus superciliaris superciliaris*]

Niccolò de' Conti (1385–1469) was a Venetian merchant and traveller who visited Syria, Persia (Iran), India, Burma (Myanmar), Vietnam, Malaya, Sumatra, Java, Aden (Yemen), Somalia and Egypt and probably China (1414–1444).

Contino

Glossy-black Thrush ssp. *Turdus serranus continoi* **Fraga & E. Dickinson**, 2008

Francisco Contino (fl.1970) is described by the authors as '… an amateur ornithologist and bird illustrator from Jujuy, Argentina, who wrote several ornithological papers, one in English, and a field guide to the birds of north-west Argentina.'

Contreras, G.

Rufous-browed Peppershrike ssp. *Cyclarhis gujanensis contrerasi* Taczanowski, 1879

Gregorio Contreras (fl.1879), who lived at Cutervo, greatly helped Stolzmann (q.v.), who collected the holotype, during his travels in Peru (1879).

Contreras, J.

Common Miner ssp. *Geositta cunicularia contrerasi* Nores & Yzureta, 1980

Professor Dr Julio Rafael Contreras Roque (b.1933) is an Argentine biologist who was at the Argentine Museum of Natural Sciences, Buenos Aires, and President, Buenos Aires Foundation of Natural History. He was Director of the Biological Station, Isla Victoria, Bariloche (1975). Since 2003 he has lived in Paraguay, where he is member, Scientific Society and Academy of Paraguayan history. He teaches and researches at National University of Pilar where he is director of the Research Institute Bioecología and Subtropical Felix de Azara. A mammal and an amphibian are named after him.

Convers

Convers' Thorntail *Discosura conversii* **Bourcier** & **Mulsant**, 1846
[Alt. Green Thorntail; Syn. *Popelairia conversii*]

M. Convers (DNF) was a French naturalist who collected in Colombia in the middle of the 19th century, but little seems to be known of him.

Cook, J.

Petrel genus *Cookilaria* **Bonaparte**, 1856 NCR
[Now in *Pterodroma* – subgenus]

Cook's Cockatoo *Calyptorhynchus cookii* **Vigors** & **Horsfield**, 1826 NCR
[Alt. Glossy Black Cockatoo; JS *Calyptorhynchus lathami*]
Cook's Petrel *Pterodroma cookii* **G. R. Gray**, 1843
Cook's Parakeet *Cyanoramphus cookii* GR Gray, 1859
[Alt. Norfolk Island Parakeet]

Captain James Cook (1728–1779) was one of the most famed explorers of all time, so there are many biographies to consult. He commanded HMS *Endeavour* on his first expedition, the achievements of which included the discovery of eastern Australia (1770) and much of New Zealand. He commanded HMS *Resolution* on his last two expeditions. He was killed in a skirmish with natives in Hawaii.

Cook, J. P.

Brown Prinia ssp. *Prinia polychroa cooki* **Harington**, 1913
Pacific Swift ssp. *Apus pacificus cooki* Harington, 1914
Red-naped Bush-shrike ssp. *Laniarius ruficeps cooki* **van Someren**, 1919 NCR
[JS *Laniarius ruficeps rufinuchalis*]

John Pemberton Cook (1865–1924) worked for the Bombay Burma Teak Corporation (1883–1913). He left India for Kenya (1913) to run a coffee plantation. A mammal is named after him.

Cook, S. E.

Iberian Magpie *Cyanopica cooki* **Bonaparte**, 1850
[Alt. Azure-winged Magpie; Syn. *Cyanopica cyanus cooki*]

Captain Samuel Edward Cook (1787–1856) was a British geologist, naturalist and collector. After service in the Royal Navy he lived in Spain (1829–1832). He changed his name to Widdrington (1840) and inherited estates through his mother belonging to that family. He was High Sheriff of Northumberland (1854). He wrote *Sketches in Spain* (1834). A tree genus is named after him.

Cooke

White-breasted Nuthatch ssp. *Sitta carolinensis cookei* **Oberholser**, 1917 NCR
[NUI *Sitta carolinensis carolinensis*]

Wells Woodbridge Cooke (1858–1916) was an American ornithologist. He graduated from Ripon College, Wisconsin (1879) with a bachelor's degree and was awarded a master's degree (1882). He taught in various schools founded specifically to educate Native Americans (1879–1885), and at the University of Vermont (1885–1893), being Professor of Agriculture (1886–1893). He also taught at the Agricultural College of Colorado (1893–1900) and the State College of Pennsylvania (1900–1901). He joined the US Biological Survey in Washington DC (1901) and worked there on bird migration and distribution for the rest of his life. He died of pneumonia.

Coomans

Rusty-bellied Fantail ssp. *Rhipidura teysmanni coomansi* **van Marle**, 1940
[SII *Rhipidura teysmanni toradja*]

Dr Louis Coomans de Ruiter (1898–1972) was a Dutch ornithologist, botanist, collector and etymologist who was an administrator in the East Indies (Indonesia) (1921–1936 and 1948–1952).

Coombs

Pale Chanting Goshawk ssp. *Melierax poliopterus coombsi* **J. A. Roberts**, 1931 NCR
[JS *Melierax canorus*]

Mr Cecil Henry Coombs (1887–1953) was a South African businessman.

Cooper, J. G.

Pacific Screech Owl *Megascops cooperi* **Ridgway**, 1878

Cooper's Tanager *Piranga rubra cooperi* Ridgway, 1869
[Alt. Summer Tanager ssp.]

James Graham Cooper (1830–1902), son of W. C. Cooper (q.v.), was a US ornithologist, malacologist and surgeon. Spencer F. Baird (q.v.), Assistant Secretary of the USNM, helped get him work with the Pacific railroad survey parties in the Washington Territory. Along with Baird he wrote a book on the birds of California: *Ornithology, Volume I, Land Birds* (1870). He also wrote *Botanical Report: Explorations and Surveys for a Railroad Route from the Mississippi River to the Pacific Ocean* (1860). A mammal is named after him.

Cooper, W. C.

Cooper's Buzzard *Buteo jamaicensis* **Gmelin**, 1788
[Alt. Red-tailed Hawk]
Cooper's Hawk *Accipiter cooperii* **Bonaparte**, 1828
Cooper's Flycatcher *Contopus cooperi* **Swainson**, 1832
[Alt. Olive-sided Flycatcher]
Cooper's Sandpiper *Tringa cooperii* **Baird**, 1858 NCR
[Presumed hybrid: *Calidris ferruginea* x *C. acuminata*]

William C. Cooper (1798–1864) was a New York conchologist and collector. He studied zoology in Europe (1821–1824), then collected in Nova Scotia, Bahamas and Kentucky. He was one of the founders of the AMNH in New York and the first American member of the Zoological Society of London. He was also the father of Dr James G. Cooper (q.v.).

Coopmans

Coopmans's Tyrannulet *Zimmerius minimus* **Chapman**, 1912
[Syn. *Zimmerius chrysops minimus*]

Paul Coopmans (1962–2007) was a Belgian field ornithologist who graduated at Antwerp and emigrated to Ecuador. He was a leading expert on Ecuadorian birds and popular bird tour leader with an encyclopaedic knowledge of songs and calls. The vernacular name was coined very recently when the bird was split from Golden-faced Tyrannulet, *Z. chrysops*.

Coppinger

Madagascar Turtle Dove ssp. *Nesoenas picturata coppingeri* **Sharpe**, 1884
Common Diving-petrel ssp. *Pelecanoides urinatrix coppingeri* **Mathews**, 1912

Dr Richard William Coppinger (1847–1910) was a naval surgeon and naturalist. Born in Dublin, he studied medicine at Queen's University and entered the medical department of the navy. He served as a naturalist on HMS *Alert* (1878–1882) during a voyage of exploration that took the ship to Patagonia, Polynesia and the Mascarene Islands. He was Inspector-General, Hospitals and Fleets (1901–1904). An amphibian is also named after him.

Coquerel

Mayotte Sunbird *Cinnyris coquerellii* **Hartlaub**, 1860
Coquerel's Coua *Coua coquereli* **Grandidier**, 1867

Dr Charles Coquerel (1822–1867) was a surgeon in the French Imperial Navy and an entomologist who was on collecting expeditions to Madagascar and neighbouring islands. He was the first to identify the screwworm fly (1858) and gave it the name *hominivorax*, which literally means 'man-eater'. Many of the specimens that he collected were only studied after his death. He was the author of many scientific papers, such as 'Orthoptères de Bourbon et de Madagascar' and 'Sur les Monandroptères et Raphiderus'. Two mammals are named after him.

Cora

Cora's Sheartail *Thaumastura cora* **Lesson** & **Gamot**, 1827
[Alt. Peruvian Sheartail]

Cora was a mythological Inca princess of the sun. She appears in a novel, *Les Incas, ou la Destruction de l'Empire du Pérou* (1777), by Jean Francois Marmontel, which is the subject of an opera libretto written by Gudmund Göran Adlerbeth called *Cora och Alonzo*.

Corinna

Black Sunbird ssp. *Leptocoma sericea corinna* **Salvadori**, 1878

Corinna was a celebrated poetess of Tanagra, Greece. Traditionally attributed to the 6th century BC, some scholars argue that she lived much later, around 200 BC.

Cornelia

Cornelia's Eclectus Parrot *Eclectus roratus cornelia* **Bonaparte**, 1850
[Alt. Eclectus Parrot ssp.]

Cornelia Schlegel *née* Buddingh (d.1864) was the first wife (1837) of the German zoologist Hermann Schlegel (1804–1884). Bonaparte wrote: 'I have named this beautiful bird after H. Schlegel's virtuous and talented wife …' – she clearly had an admirer! (See **Buddingh**)

Cornwall

Grey-crowned Babbler ssp. *Pomatostomus temporalis cornwalli* **Mathews**, 1912 NCR
[NUI *Pomatostomus temporalis temporalis*]
Silvereye ssp. *Zosterops lateralis cornwalli* Mathews, 1912
Tawny Frogmouth ssp. *Podargus strigoides cornwalli* Mathews, 1912 NCR
[JS *Podargus strigoides strigoides*]

Edward Mayler Cornwall (1861–1937) was an Australian naturalist, collector and photographer. He was on the first scientific expedition to King Island, Bass Strait (1877) and collected in the far north of Queensland for several years (1880s). He managed a number of businesses in Cairns and Mackay (Queensland). He was the first president of the Field Naturalists Club of Cairns.

Correia

Common Moorhen ssp. *Gallinula chloropus correiana* **R. C. Murphy** & **Chapin**, 1929 NCR
[NUI *Gallinula chloropus chloropus*]
Fan-tailed Gerygone ssp. *Gerygone flavolateralis correiae* **Mayr**, 1931

José G. Correia (1881–1954) was born in Faial in the Azores. He was the cooper aboard the *Daisy*, an old whaler hired by AMNH to undertake a biological survey of South Georgia (1912). The expedition was led by R. C. Murphy and, during the voyage, he taught Correia to collect and prepare bird skins; Correia duly became a taxidermist at New Bedford in Massachusetts. He was sent back to South Georgia (1914) by Murphy, and subsequently went on many AMNH collecting expeditions. He returned to the Azores (1921–1922 and 1928) with his wife, Virginia Correia. They were both members of the Thorne-Correia Expedition to Fernando Po, Principe and Sao Tomé (1928–1929) and visited the Pearl Islands, Panama (1941). The AMNH retain his collected material and his unpublished journals.

Cory

Buzzard genus *Coryornis* **Ridgway**, 1925 NCR
[Now in *Buteo*]

Cory's Shearwater *Calonectris borealis* Cory, 1881
[Syn. *Calonectris diomedea borealis*]
Cory's Least Bittern *Ixobrychus neoxenus* Cory, 1886 NCR
[Variant morph of *Ixobrychus exilis*]
Ochre-browed Thistletail *Schizoeaca coryi* **Berlepsch**, 1888

Clapper Rail ssp. *Rallus longirostris coryi* **Maynard**, 1887

Lesser Antillean Bullfinch ssp. *Loxigilla noctis coryi* Ridgway, 1898

Red-legged Thrush ssp. *Turdus plumbeus coryi* **Sharpe**, 1902

Grey-crowned Palm Tanager ssp. *Phaenicophilus poliocephalus coryi* **Richmond** & **Swales**, 1924

Charles Barney Cory (1857–1921) was an American ornithologist. He donated his collection of 19,000 bird specimens to the Field Museum, Chicago, in exchange for departmental status for ornithology and his appointment as Lifetime Curator without residence obligations. The Department of Ornithology, with Curator Cory and Assistant Curators Cherrie and Dearborn, remained separate for six years. He wrote *The Birds of Haiti and San Domingo* (1885), *The Birds of the West Indies* (1889) and *The Birds of Illinois and Wisconsin* (1909). While he was, undoubtedly, a great ornithologist he was not a conservationist. He was asked to a meeting of the DC Audubon Society; an invitation (1902) which he refused, saying 'I do not protect birds, I kill them!'

Corydon

Broadbill genus *Corydon* Hewitson, 1869

In classical literature, Corydon is a stock name for a shepherd or rustic in pastoral poems and fables, such as those of Virgil and Theocritus.

Cosens

Red-throated Wryneck ssp. *Jynx ruficollis cosensi* **C. H. B. Grant**, 1915 NCR
[NUI *Jynx ruficollis ruficollis*]

Cardinal Woodpecker ssp. *Dendropicos fuscescens cosensi* C. H. B. Grant, 1915 NCR
[JS *Dendropicos fuscescens lafresnayi*]

Lieutenant Colonel Gordon Philip Lewes Cosens (1884–1928) was in Kenya (1912–1913). He transferred to the Egyptian army (1913). A mammal is named after him.

Costa

Costa's Hummingbird *Calypte costae* **Bourcier**, 1839

Louis Marie Panteleon Costa, Marquis de Beau-Regard (1806–1864), was a Sardinian aristocrat and collector and an accomplished amateur ornithologist, archaeologist and historian. He specialised in hummingbirds and had a notable collection, in recognition of which Bourcier named this species.

Costello

Costello's Booby *Papasula abbotti costelloi* Steadman, Schubel & Pahlavan, 1988 EXTINCT

Lou Costello (born Louis Francis Cristillo) (1906–1959) was half of the American comedy duo 'Abbott & Costello', stars of burlesque, radio and film (1935–1956) and TV (1960s–1990s). The authors apparently could not resist the temptation of adding *costelloi* to *abbotti* (although they are aware that *abbotti* was named after W. L. Abbott, q.v., rather than Bud Abbott).

Cottam

Steller's Jay ssp. *Cyanocitta stelleri cottami* **Oberholser**, 1937 NCR
[JS *Cyanocitta stelleri macrolopha*]

Dr Clarence Cottam (1899–1973) was an American naturalist, biologist and conservationist, who was with the US Fish and Wildlife Service (1929–1954), having previously (1927–1929) been an instructor at Brigham Young University where he gained his bachelor's and master's degrees (1926 and 1927). His doctorate was awarded by George Washington University (1936). After retirement from the US Fish and Wildlife Service he was Dean, College of Agriculture, Brigham Young University (1954–1955) and, finally, the first Director of the Welder Wildlife Foundation (1953–1973).

Couch

Couch's Kingbird *Tyrannus couchii* **Baird**, 1858

Couch's Jay *Aphelocoma ultramarina couchii* Baird, 1858
[Alt. Mexican/Grey-breasted Jay ssp.]

General Darius Nash Couch (1822–1897) was a soldier and administrator. He was also an explorer, taking leave of absence to lead a zoological expedition in Mexico. Geisler wrote of the garter snake that Couch collected in 1853, '... named in honor of its indefatigable discoverer, Lt. D. N. Couch, who, at his own risk and cost, undertook a journey into northern Mexico, when the country was swarming with bands of marauders, and made large collections in all branches of zoology ...' He was commissioned into the US Army (1846) and sent to Mexico, where he fought at the Battle of Buena Vista (1847). He bought all of Berlandier's papers, herbarium and collections from his widow (1853). He returned to Washington DC (1854) and in the next year he resigned his army commission and became a merchant and manufacturer in New York and Massachusetts (1855–1861). At the outbreak of the American Civil War he rejoined the army as a colonel then became brigadier-general of volunteers. He offered to resign on the grounds of ill health (1863) but was persuaded to stay by being promoted to major general! He was posted to Pennsylvania and put in charge of all the ceremonies associated with the consecration of the National Cemetery at Gettysburg (1865), the occasion of Abraham Lincoln's Gettysburg Address. After the Civil War he appears to have again resigned from the army and was Collector of the Port of Boston (1866–1867), President of a Virginia mining and manufacturing concern (1867–1877), and an administrator in the state of Connecticut (1877–1884). Two reptiles and an amphibian are named after him.

Coues

Coues's Flycatcher *Contopus pertinax* **Cabanis** & **Heine**, 1859
[Alt. Greater Pewee]

Coues's Arctic Redpoll *Carduelis hornemanni exilipes* Coues, 1862
[Alt. Hoary Redpoll ssp.]

Coues's Gadwall *Anas strepera couesi* Streets, 1876 EXTINCT

Rock Sandpiper ssp. *Calidris ptilocnemis couesi* **Ridgway**, 1880

Cactus Wren ssp. *Campylorhynchus brunneicapillus couesi* **Sharpe**, 1882

Dr Elliott B. Coues (pronounced 'cows'; 1842–1899) was a US Army Surgeon and one of the founders of the American Ornithologists' Union. Coues collected a warbler and asked Baird (q.v.) to name it in honour of his sister, Grace Darlington Coues (q.v.), hence Grace's Warbler. He is also the person who named the sandpiper after Baird. Amongst other works he wrote *Handbook of Field and General Ornithology* (1890), in which he set out in meticulous detail how to 'collect' and preserve birds, *A Checklist of North American Birds* (1873) and the 5-volume *Key to North American Birds* (1872–1903). Three mammals are named after him.

Coulon

Coulon's Macaw *Primolius couloni* **P. L. Sclater**, 1876
[Alt. Blue-headed Macaw; Syn. *Propyrrhura couloni*]

Paul Louis de Coulon (1804–1894) was a Swiss naturalist, the joint founder, with Louis Agassiz, of the Natural History Museum of Neuchâtel, to which he bequeathed money.

Coultas

White-winged Fantail ssp. *Rhipidura cockerelli coultasi* **Mayr**, 1931

Torrent Flyrobin ssp. *Monachella muelleriana coultasi* Mayr, 1934

Uniform Swiftlet ssp. *Aerodramus vanikorensis coultasi* Mayr, 1937

Pacific Baza ssp. *Aviceda subcristata coultasi* Mayr, 1945

Manus Monarch ssp. *Monarcha infelix coultasi* Mayr, 1955
[Syn. *Symposiachrus infelix coultasi*]

Dr William F. (Bill) Coultas (b.1899) collected widely in the Pacific (1929–1943). He led the Whitney South Seas Expedition (1930–1931); his journals chronicling this project are in the AMNH. He also collected in central Africa (1938). He was an early conservationist, commenting (1930) that the flying fox was becoming quite uncommon in Guam because of the proliferation of firearms. As a Lieutenant Commander in the US Naval Intelligence service he commanded an Amphibious Scout School in New Guinea (1943).

Count Raggi

Count Raggi's Bird-of-Paradise *Paradisaea raggiana* **P. L. Sclater**, 1873
[Alt. Raggiana Bird-of-Paradise]

Marchese Francesco Raggi (1807–1887) was an Italian naturalist who explored and collected in Peru in the mid-19th century. Luigi Maria D'Albertis (q.v.), who obtained the type specimen, requested that the new species be named after Raggi, calling him 'a great lover of natural history, and especially of ornithology.'

Coursen

Coursen's Spinetail *Synallaxis courseni* **E. R. Blake**, 1971
[Alt. Apurímac Spinetail]

Charles Blair Coursen (1899–1974) was an American businessman and field ornithologist. He graduated (1922) in business administration and biology from the University of Chicago. He later joined the newly founded General Biological Supply House and became its president (1930–1964) until retirement. He co-wrote *Birds of the Chicago Region* (1934) and wrote *Birds of the Orland Wildlife Refuge* (1947). The General Biological Supply House published *Turtox News* in which he published an article (1962) 'Kitchen botany', aimed at encouraging school children to become interested in vegetables.

Courtois

Courtois's Laughingthrush *Garrulax courtoisi* Menegaux, 1923
[Alt. Blue-crowned Laughingthrush; formerly *Garrulax galbanus courtoisi*]

Orange-headed Thrush ssp. *Zoothera citrina courtoisi* **Hartert**, 1919

The Reverend Fréderic Courtois (1860–1928) was a French missionary to China (1901–1928). He was an amateur naturalist who became Director of the Natural History Museum in Sikawei, near Shanghai (1903). He wrote *Les Oiseaux du Musée de Zi-Ka-Wei* (1912). An amphibian is named after him.

Courtot

Crested Lark ssp. *Galerida cristata courtoti* **Lavauden**, 1926 NCR
[JS *Galerida cristata alexanderi*]

Lieutenant-Colonel Victor-Paul Courtot (fl.1925) was a French army officer and explorer in the Sahara. He led the Mission Tunis-Tchad (1925), which proceeded from the Gulf of Sirte to the Gulf of Benin.

Coutelle

Water Pipit ssp. *Anthus spinoletta coutellii* **Audouin**, 1826

Colonel Jean Marie Joseph Coutelle (1748–1835) was a French scientist, engineer and pioneer balloonist as First Officer of Company of Aeronauts (1794), with orders to develop balloons to help the French Revolutionary Armies. He was part of Napoleon's invasion of Egypt (1798).

Cowan

Plain Martin ssp. *Riparia paludicola cowani* **Sharpe**, 1882

The Reverend William Deans Cowan (1844–1923) was a missionary in Madagascar for ten years (late 19th century). He made a geological expedition in south-central Madagascar and was the author of *The Bara Land: A Description of the Country and People*. A mammal and two amphibians are named after him.

Cowens

Black-streaked Scimitar Babbler ssp. *Pomatorhinus gravivox cowensae* **Deignan**, 1952

Mrs Alice Muriel Sowerby (*née* Cowens) (DNF) was the third wife of Arthur de Carle Sowerby (1885–1954). She met her husband whilst nursing his brother in France after the end of WW2. He asked her to join him in Shanghai, where they married (1946). (His second wife died in 1944, after which he was interned by the Japanese.) They visited England before he retired to Washington DC.

Cox, J. B.

Cox's Sandpiper *Calidris paramelanotos* S. A. Parker, 1982 NCR
[Hybrid: *Calidris melanotos* x *Calidris ferruginea*]

John B. Cox is a British-born Australian ornithologist, resident in Adelaide, who collected two specimens of this bird in South Australia (1973). Shane Parker described them (1982) as a new species, but another birder came forward claiming priority, and there was much argument and debate regarding this at the time. Cox co-wrote *An Annotated Checklist of the Birds of South Australia* (1979).

Cox, P. Z.

Desert Lark ssp. *Ammomanes deserti coxi* **Meinertzhagen**, 1923

Sir Percy Zachariah Cox (1864–1937) was a British diplomat, Agent and Consul at Muscat (1899–1904), Resident and Consul-General in the Persian Gulf (1904–1914), Chief Political Officer in Mesopotamia (1914–1918), Minister at Tehran (1918–1920), and High Commissioner in Iraq (1920–1923).

Coxen

Coxen's Fig Parrot *Cyclopsitta diophthalma coxeni* **Gould**, 1867
[Alt. Double-eyed Fig Parrot ssp.]

Charles Coxen (1809–1879) was born in England, but emigrated to Queensland, Australia (1833) where he collected for the Zoological Gardens, London. He was married to Elizabeth Frances Isaac (1825–1906) and was a brother of Elizabeth Gould (1804–1841), the wife of John Gould. In Australia, Coxen discovered at least 20 species and races of birds new to science. He founded the Queensland Museum (Brisbane) (1862), of which he became Honorary Curator (1871–1876). His wife Elizabeth became the first paid Curator of the Museum, her task being to arrange the shell collection. Apart from his natural history activities he was also a member of the first Queensland Parliament and (1868) became a Crown Lands Commissioner and one of the founders of the Queensland Philosophical Society. Gould described the parrot from a specimen that was sent to him by Waller after the appearance of the original description, which was based on a drawing by Coxen of specimens in his possession. Coxen and his sister Elizabeth travelled with John Gould to Australia. Both Elizabeth and her brother were themselves amateur artists.

Craddock

Black-throated Parrotbill ssp. *Suthora nipalensis craddocki* **Bingham**, 1903

W. H. Craddock (1874–1964) was a British forestry officer in Burma (Myanmar) (1897) and Malaya (Malaysia) (1903).

Cragg

Common Redshank ssp. *Tringa totanus craggi* Hale, 1971

Dr James 'Jim' Birkett Cragg (1910–1996) was a British ecologist, Professor of Zoology, University of Durham (1950–1961) and Director, Merlewood Research Station (1961–1966). He emigrated to Canada, becoming Professor of Biology, University of Calgary (1966–1972), and Killam Professor (1966–1976), retiring as Emeritus Professor of Environmental Science (1976).

Cranbrook (2nd Earl)

Flappet Lark ssp. *Mirafra rufocinnamomea cranbrooki* **B. Alexander**, 1907 NCR
[JS *Mirafra rufocinnamomea tigrina*]

John Stewart Gathorne-Hardy, 2nd Earl of Cranbrook and Lord Medway (1839–1911), was a British politician as a member of the House of Commons (1868–1880 and 1884–1892) Both Cranbrook and Boyd Alexander were officers in the Rifle Brigade, but the author gave no hint as to why he named this subspecies after the Earl

Cranbrook (4th Earl)

Striated Laughingthrush ssp. *Garrulax striatus cranbrooki* **Kinnear**, 1932

John David Gathorne-Hardy, 4th Earl of Cranbrook (1900–1978), was an amateur zoologist who became a Trustee of BMNH, London (1963). He collected in Burma (Myanmar) (1930–1931). Two mammals are named after him.

Cranch

Cranch's Francolin *Pternistis afer cranchii* **Leach**, 1818
[Alt. Red-necked Francolin ssp.]

John Cranch (1758–1816) was a British explorer of tropical Africa and also an accomplished natural historian. He took part in an expedition (1816) led by Captain J. K. Tuckey to discover the source of the Congo River.

Craveri

Craveri's Murrelet *Synthliboramphus craveri* **Salvadori**, 1865

Federico Craveri (1815–1890) was an Italian scientist, chemist and pioneer in the field of meteorology. He was also an explorer, teacher and scholar. He spent 20 years in Mexico (1840–1859), travelling and collecting. He obtained the murrelet, which was named after him and his brother Ettore 20 years later. The 'F. Craveri' Museum of Natural History, Bra, Italy, was founded by Craveri in the brothers' former home and now houses their collections. Strictly speaking, the name should be 'Craveris' Murrelet *Synthliboramphus craveriorum*' because the bird was named for the two brothers.

Crawford

White Tern *Gygis crawfordi* **Nicoll**, 1906 NCR
[JS *Gygis alba*]

Vitelline Warbler ssp. *Dendroica vitellina crawfordi* Nicoll,
1904
Striated Heron ssp. *Butorides striata crawfordi* Nicoll, 1906

James Ludovic Lindsay, 26th Earl of Crawford and 9th Earl of
Balcarres (1847–1913), was an astronomer, collector and
bibliophile. Collections of birds were made during three
cruises of Lord Crawford's yacht *Valhalla*.

Crawfurd, A.

Yellow-bellied Eremomela ssp. *Eremomela icteropygialis
crawfurdi* S. Clarke, 1911 NCR
[NUI *Eremomela icteropygialis griseoflava*]

Stephenson Clarke, who collected (1909) and described the
holotype, gave no etymology or indication as to who Crawfurd was. We think it may be named after A. Crawfurd
(fl.1911), who was a collector in Kenya.

Crawfurd, J.

Woodpecker sp. *Picus crawfurdii* **J. E. Gray**, 1829 NCR
[Alt. White-bellied Woodpecker; JS *Dryocopus javensis*]

Crawfurd's Silver Pheasant *Lophura nycthemera crawfurdii*
J. E. Gray, 1829
[Formerly treated as a race of the Kalij Pheasant,
L. *leucomelanos*]

Dr John Crawfurd (1783–1868) was a British physician,
scholar, administrator and diplomat. He joined the Honourable East India Company and (1810) was with Lord Minto's
expedition to capture Java from the Dutch. Subsequently he
was in Penang and Singapore with Raffles (q.v.). He has been
described as the 'Scotsman who made Singapore British'.
After several years, and a journey back to London, he
returned to India and was employed as a diplomat, heading
missions to Thailand, Burma (Myanmar) and Cochinchina
(Vietnam). He wrote *History of the Indian Archipelago* (1820).
Darwin (q.v.) refers to a Mr Crawfurd who had advised him on
hybrid fowls that he had observed in Indonesia. The same
Crawfurd read a paper at Oxford, 'On the relation of the
domesticated animals to civilisation' (1860).

Crawshay

Red-winged Francolin ssp. *Scleroptila levaillantii crawshayi*
Ogilvie-Grant, 1896

Captain Richard Crawshay (1862–1958) was a hunter and
collector who became an agent for the African Lakes
Company. They sent him to set up a permanent post on the
shores of Lake Mweru (1890). He stayed in Rhodesia (Zambia)
(1890–1891), then abandoned the station and returned to
Blantyre in Nyasaland (Malawi), from where he sent collections of ants to the museums of South Africa. He also
collected butterflies in the Kikuyu Country of British East
Africa (1899–1900). He collected in Tierra del Fuego (1904)
and wrote *The Birds of Tierra del Fuego* (1907). Two mammals
are named after him.

Creagh

Starling sp. *Amydrus creaghi* **Ogilvie-Grant** & **H. O. Forbes**,
1903 NCR
[Alt. Somali Starling; JS *Onychognathus blythii*]

General Garrett O'Moore Creagh (1848–1923) was a British
army officer. He was commissioned into the 95th Regiment of
Foot (1866), was posted to India (1869) and transferred to the
British Indian Army (1870). He fought in the 2nd Anglo-Afghan
War (1879) and was awarded the Victoria Cross for heroism.
He continued to serve in India until sent to China as General
Officer Commanding the Indian contingent during the Boxer
Rebellion (1901). He was Commander-iin-Chief, India
(1909–1914).

Cretzschmar

Cretzschmar's Babbler *Turdoides leucocephala*
Cretzschmar, 1826
[Alt. White-headed Babbler]
Cretzschmar's Bunting *Emberiza caesia* Cretzschmar, 1827

Dr Philipp Jakob Cretzschmar (1786–1845) was a German
physician who taught anatomy at the Senckenberg Institute
in Frankfurt. He was a founder of the Senckenberg Natural
History Museum and the founder and second Director of the
Senckenberg Society for Wildlife Research .

Crishna

Drongo sp. *Edolius crishna* **Gould**, 1836 NCR
[Alt. Hair-crested Drongo; JS *Dicrurus hottentottus*]

Named after the Hindu god Krishna, whose name means
'dark/black'.

Cristina

Cristina's Barbthroat *Threnetes cristinae* Ruschi, 1975 NCR
[Alt. Sooty Barbthroat ssp.; JS *Threnetes niger loehkeni*]

Princess Cristina Maria do Rosário Leopoldina Michaela
Gabriela Rafaela Gonzaga de Orleans-Bragança e Borbon
(b.1950) is a member of the imperial family that ruled Brazil
until 1889. She is very interested in Brazilian forest fauna and
flora and was a member of the expedition during which
Ruschi collected the holotype (1975).

Crockett

Yellow-gaped Honeyeater ssp. *Meliphaga flavirictus
crockettorum* **Mayr** & **de Schauensee**, 1939

Dr Frederick E. Crockett (1907–1979), a physician, and his
wife Charis Crockett *née* Denison (b.1905) were American
explorers. They organised an expedition to the South Seas
and New Guinea (1937–1938). They sailed all the way from
Philadelphia in the 59-foot schooner *Chiva*. Charis wrote *The
House in the Rain Forest* (1942).

Croizat

Blue-winged Minla ssp. *Minla cyanouroptera croizati*
Deignan, 1958 NCR
[NUI *Minla cyanouroptera wingatei*]

African Goshawk ssp. *Accipiter tachiro croizati* Desfayes, 1974

Canyon Wren ssp. *Catherpes mexicanus croizati* **A. R. Phillips**, 1986

White-throated Thrush ssp. *Turdus assimilis croizati* A. R. Phillips, 1991

Leon Camille Marius Croizat (1894–1982) was an Italian botanist, biogeographer and evolutionist. He served in the Italian army (1914–1919), after which he returned to The University of Turin and graduated in law (1920). He hated the Fascists and emigrated to the US (1923), where he took any job going. He painted in watercolours with modest financial success until the Wall Street Crash (1929) killed his market. He tried his luck in Paris, without success, so returned to New York. Merrill (q.v.), then Director of the Arnold Arboretum, Harvard, hired Croizat (1937) as a technical assistant. His drawings were said to be unbelievably accurate. He was sacked for publishing a paper he had written (1946) without his employers' permission. He moved to Caracas, Venezuela, and held a number of academic positions there (1947–1952). He was the botanist on the Franco-Venezuelan expedition to the sources of the Orinoco (1950–1951). He resigned his academic positions (1953) to work full time on biological problems. He and his wife became (1976) the first Directors of Jardin Botanico Xerofito, Coro, which they had founded (1970). It is now named after him, as is a reptile.

Crommelin

Australian Ringneck sp. *Barnardius crommelinae* **Mathews**, 1925 NCR

[JS *Barnardius zonarius*]

Louisa Crommelin Roberta Jowett (*née* Whitwell), Duchess of Bedford (1892–1960), was the wife of Hastings William Sackville Russell, 12th Duke of Bedford (q.v.). Mathews described this form of parakeet from an aviary bird in the collection of the Marquess of Tavistock (courtesy title used by the heir to the Dukedom of Bedford), which was presented to the BMNH upon its death.

Crookshank

Crookshank's White-eye *Zosterops fuscicapilla crookshanki* **Mayr** & **Rand**, 1935

[Alt. Capped White-eye ssp.]

Commander Robert Crookshank (1892–1956) was an officer in the Royal Navy who transferred to the Royal Australian Navy, until retirement (1932). He was in New Guinea (1934) as Captain of the *France*, as part of the Whitney South Seas Expedition.

Crosby

Snowy-breasted Hummingbird ssp. *Saucerottia edward crosbyi* **Griscom**, 1927 NCR

[JS *Amazilia edward margaritarum*]

Captain Maunsell Schieffelin Crosby (1887–1931) was a US naturalist and patron of the sciences. His uncle, Eugene Schieffelin (1827–1906), introduced the Common Starling *Sturnus vulgaris* into the Americas, releasing birds in New

York City (1890–1891). In naming the bird Griscom (q.v.) described Crosby thus: 'friend and choice companion, who assisted in financing my 1927 expedition to eastern Panama, and collected most of the specimens of this new Hummer at my special request'.

Crossin

Crossin's Wren *Hylorchilus navai* Crossin & **Ely**, 1973
[Alt. Nava's Wren]

Chestnut-sided Shrike Vireo ssp. *Vireolanius melitophrys crossini* **A. R. Phillips**, 1991

Richard S. Crossin (1933–2003) was an ornithologist who had a private museum at Tucson, Arizona. He was an active collector in Mexico (1960s–1980s).

Crossley

Madagascar Yellowbrow genus *Crossleyia* **Hartlaub**, 1877

Crossley's Babbler *Mystacornis crossleyi* **Grandidier**, 1870
[Alt. Madagascar Groundhunter, Crossley's Vanga]
Crossley's Ground Thrush *Zoothera crossleyi* **Sharpe**, 1871
Crossley's Ground Roller *Atelornis crossleyi* Sharpe, 1875
[Alt. Rufous-headed Ground Roller]

Alfred Crossley (1829–1888) collected in Madagascar and Cameroon (1870s). Amongst the bird specimens collected by him now in the BMNH one is recorded under E. Crossley. It is not clear if this is a transcription error or whether there were two members of the family collecting. He did not confine his efforts just to collecting birds. A number of butterflies and moths, four reptiles and a mammal are named after him.

Cu

Rusty-capped Fulvetta ssp. *Alcippe dubia cui* Eames, 2002

Dr Nguyen Cu is a Vietnamese ornithologist and collector. He co-wrote *Chim Vietnam* (2000), which is the first illustrated Vietnamese-language guide to his country's avifauna. He works very closely with the describer.

Cuming

Scale-feathered Malkoha *Phaenicophaeus cumingi* **Fraser**, 1839
[Alt. Scale-feathered Cuckoo; Syn. *Lepidogrammus cumingi*]
Cuming's Scrubfowl *Megapodius cumingii* **Dillwyn**, 1853
[Alt. Philippine Scrubfowl/Megapode]

Brown Tit Babbler ssp. *Macronus striaticeps cumingi* **Hachisuka**, 1934 NCR
[NUI *Macronus striaticeps mindanensis*]

Hugh Cuming (1791–1865) was an English naturalist and conchologist, once described as the 'Prince of Collectors'. He collected in the Neotropics (1822–1826 and 1828–1830), Polynesia (1827–1828) and the East Indies (1836–1840). He preceded Darwin in having collected in the Galápagos (1829). His shell collection is housed in the Linnean Library in London. A mammal and seven reptiles are named after him.

Cumming

Kurdish Wheatear *Oenanthe xanthoprymna cummingi* **Whitaker**, 1899 NCR; NRM

W. D. Cumming (d.1925) was a British collector who was a telegraphist at Fao, Iran (1876–1924). He sent several collections of birds to Sharpe and reptiles to Boulenger at the British Museum.

Cunhac

Silver-eared Mesia ssp. *Leiothrix argentauris cunhaci* **H. C. Robinson** & **Kloss**, 1919

Elie J. Cunhac (b.1870) first went to Vietnam (1899) with a surveying expedition. He was French Resident in Dalat, Annam (Vietnam) (1918 to at least 1924), where he suggested the construction of a dam and large lake.

Curzon

Wren-babbler genus *Curzonia* Skinner, 1898 NCR
[Now in *Napothera/Gypsophila*]

George Nathaniel, 1st Marquess Curzon of Kedleston (1859–1925), was a British politician, statesman and Viceroy of India (1899–1905).

Cutting

Nepal House Martin ssp. *Delichon nipalense cuttingi* **Mayr**, 1941

Charles Suydam Cutting (1889–1972) was an American naturalist, ethnologist, collector, patron and Trustee of the AMNH and the Zoological Society in New York. He travelled to Turkestan (1925), Ethiopia (1926–1927) and Burma (Myanmar) and China (1928–1929) with Theodore (q.v.) and Kermit Roosevelt. He was the first westerner to be invited and allowed to visit Lhasa, Tibet (1935). He collected for botanical gardens in the US and England and for museums in the US and India. He visited Lhasa a second time with his wife, Helen McMahon Cutting (1937) and wrote about it in *The Fire Ox and Other Years* (1940). He was also American champion in Real Tennis and took the sport very seriously!

Cuvier, F.

Falcon genus *Cuvieria* **J. A. Roberts**, 1922 NCR
[Now in *Falco*]
African Hobby *Falco cuvierii* **A. Smith**, 1830

Frédéric Cuvier (1773–1838) was a palaeontologist and, like his more famous brother Georges (q.v.), a zoologist. He was the Head Keeper of the menagerie attached to the Museum of Natural History, Paris (1804–1838). He is regarded as the first modern zookeeper in that he believed that animals should receive a good diet and be treated with kindness. The Chair of Comparative Physiology at the museum was especially established for him (1837). He was co-author of the 4-volume *Histoire Naturelles des Mammifères* (1819–1842). He must have felt overshadowed by his older brother, as on his deathbed he asked that his tombstone contain nothing beyond the words 'Brother of Georges Cuvier'. Four mammals are named after him.

Cuvier, G. L.

Becard sp. *Psaris cuvierii* **Swainson**, 1821 NCR
[JS *Pachyramphus viridis*]
Cuvier's Brush-turkey *Talegalla cuvieri* **Lesson**, 1828
[Alt. Red-billed Brush-turkey]
Cuvier's Kinglet *Regulus cuvieri* **Audubon**, 1829 NCR
[Hybrid? *Regulus satrapa* x *R. calendula*]
Cuvier's Rail *Dryolimnas cuvieri* **Pucheran**, 1845
[Alt. White-throated Rail]
Cuvier's Sabrewing *Phaeochroa cuvierii* **de Lattre** & **Bourcier**, 1846
[Alt. Scaly-breasted Hummingbird]

Cuvier's Podargus *Podargus strigoides cuvieri* **Vigors** & **Horsfield**, 1827 NCR
[Alt. Tawny Frogmouth ssp.; JS *Podargus strigoides strigoides*]
Cuvier's Toucan *Ramphastos tucanus cuvieri* **Wagler**, 1827
[Alt. White-throated Toucan ssp.]

Georges Léopold Chrétien Frédéric Dagobert Baron Cuvier (1769–1832), better known by his pen name Georges Cuvier, was a French naturalist; one of the scientific giants of his age. He believed that palaeontological discontinuities were evidence of sudden and widespread catastrophes; that is, that extinctions can happen suddenly. He is also famed for managing to stay in a top government post, as Permanent Secretary of the Academy of Sciences, through three regimes, including Napoleon's. He was born in the Duchy of Württemberg, eastern France, and (1788) took a job as a private tutor to an aristocratic family and moved to Caen, Normandy, staying there until going to Paris (1795) at the invitation of Etienne Geoffroy Saint-Hilaire (q.v.). He taught natural history in a number of establishments before becoming Assistant to the Professor of Animal Anatomy at the MNHN. Among his many achievements, he identified a fossil image from Bavaria as an extinct flying reptile and named it *Pterodactylus*. Audubon (q.v.) said of a kinglet he collected (1812) 'I named this pretty and rare species after Baron Cuvier, not merely by way of acknowledgment for the kind attentions which I received at the hands of that deservedly celebrated naturalist, but as a homage due by every student of nature to one unrivalled in the knowledge of General Zoology'. Cuvier wrote *Tableau Élémentaire de l'Histoire Naturelle des Animaux* (1798), *Mémoires sur les Espèces d'Éléphants Vivants et Fossils* (1800), *Leçons d'Anatomie Comparée* (1801–1805), *Récherches sur les Ossements Fossiles des Quadrupeds* (1812) and *Le Règne Animal Distribué d'Après son Organisation* (1817). An amphibian, three mammals and six reptiles are named after him.

Cyclops

Cyclops Lorikeet *Charmosyna josefinae cyclopum* **Hartert**, 1930
[Alt. Josephine's Lorikeet ssp.]

This race of lorikeet is not named for the Cyclops, the one-eyed giant in Homer's Odyssey, but for a mountain range in New Guinea, after which a mammal and four reptiles are also named.

Czaki

Brown-crested Flycatcher ssp. *Myiarchus tyrannulus czakii*
Stolzmann, 1926 NCR
[JS *Myiarchus swainsoni*]

Dr Joséf Czaki (1857–1946) was a Polish naturalist, collector and physician to the Polish settlement at Colônia Amola Faca, Virmond, Parana, Brazil.

Czarnikow

Capuchin Babbler ssp. *Phyllanthus atripennis czarnikowi*
Ogilvie-Grant, 1907 NCR
[JS *Phyllanthus atripennis bohndorffi*]

Julius Caesar Czarnikow (1838–1909) was a Prussian-born British sugar broker who was one of the subscribers to the Ruwenzori Expedition (1905–1906). He immigrated to England (1854), became a British citizen (1860) and founded (1861) the Czarnikow Group. It is still a large player in the sugar and biofuels markets.

D

Dabbene

Tristan Albatross *Diomedea dabbenena* **Mathews**, 1929
Dabbene's Guan *Penelope dabbenei* **Hellmayr** & **Conover**, 1942
[Alt. Red-faced Guan]

Stripe-capped Sparrow ssp. *Rhynchospiza strigiceps dabbenei* Hellmayr, 1912
Hellmayr's Pipit ssp. *Anthus hellmayri dabbenei* Hellmayr, 1921
Narrow-billed Woodcreeper ssp. *Lepidocolaptes angustirostris dabbenei* **Esteban**, 1948
Buff-fronted Owl ssp. *Aegolius harrisii dabbenei* **Olrog**, 1979

Roberto Dabbene (1864–1938) was an Argentinian ornithologist, the first President of the Asociación Ornitológica del Plata (1916). Among his publications are 'Notas sobre el petrel plateado *Pricoella antarctica*' (1921) and 'Los picaflores de Chile' (1930). A mammal is named after him.

Dabry

Mrs Gould's Sunbird ssp. *Aethopyga gouldiae dabryii* **J. Verreaux**, 1867

Claude-Philibert Dabry de Thiersant (1826–1898) was French consul at Hankow (one of the three cities whose merging formed modern-day Wuhan, China) and an amateur naturalist. He was particularly interested in ichthyology, and at least one fish is named after him. The holotype was collect by a Mr Soubiran for Dabry, who passed it to Verreaux.

D'Achille

Amazonian Parrotlet *Nannopsittaca dachilleae* **O'Neill**, Munn & Franke, 1991

Bárbara D'Achille *née* Bistevins Treinani (1941–1989) was a Latvian-born journalist with Italian nationality who lived in Peru (1961–1989). She wrote on conservation, including the ecological page of the daily paper *El Comercio*. Later she was a consultant for the World Wildlife Fund and US and Canadian development companies. She travelled to remote areas in South America, mainly in Peru. A Peruvian reserve, the *Pampa Galeras – Bárbara D'Achille*, was named in her memory. She wrote *Kuntursuyo – Territory of the Condor : Peru – National Parks and Other Areas of Ecological Conservation* (1996). Terrorists in Huancavelica assassinated her.

Daedalion

Hawk genus *Daedalion* **Savigny**, 1809 NCR
[Now in *Accipiter*]

In Greek mythology Daedalion was a son of Hesperos. He was so grief-stricken at the death of his daughter Chione that he threw himself from the heights of Mount Parnassus. The god Apollo metamorphosed him into a hawk.

Daedalus

Three-striped Warbler ssp. *Basileuterus tristriatus daedalus* **Bangs**, 1908

In Greek mythology Daedalus was an Athenian architect who built the Cretan labyrinth, and fashioned wings from feathers and wax for himself and his son Icarus to escape from Crete. A mammal is also named Daedalus.

Daggett

Red-breasted Sapsucker ssp. *Sphyrapicus ruber daggetti* **Grinnell**, 1901

Frank Slater Daggett (1855–1920) was an American naturalist and ornithologist. Originally a grain merchant in Ohio and Chicago, his main interest was collecting ornithological and entomological specimens. He became Director of the Los Angeles County Museum of History, Science, and Art (1911–1920) and published quite widely, often with Joseph Grinnell (q.v.), such as 'An ornithological visit to Loa, Coronados Islands, Lower California' (1903). He was a long-standing, enthusiastic member of the Cooper Ornithological Club, being President 1901–1903. He is also remembered in the name of the Pleistocene species, Daggett's Eagle *Buteogallus daggetti*.

Dahl

Bismarck Fantail *Rhipidura dahli* **Reichenow**, 1897

Black-tailed Whistler ssp. *Pachycephala melanura dahli* Reichenow, 1897

Professor Dr Karl Friedrich Theodor Dahl (1856–1929) was a German zoologist and arachnologist. He attended the universities of Leipzig, Freiburg and Kiel, achieving the status of 'Privatdozent' (US Associate Professor equivalent). He collected in the Baltic states and the Bismarck Archipelago, New Guinea (1896–1897). He became Curator of Arachnids at the Berlin Natural History Museum (1898) until retirement. He had interests in biogeography and animal behaviour. He wrote 'Das Leben der Vögel auf den Bismarckinseln' (1899). That year he married Maria Grosset (1872–1972), a co-worker at the Zoological Institute of Kiel, who also published several works on spiders and finished editing his book on Melanesia. A reptile is named after him.

D'Albertis

D'Albertis's Bird-of-Paradise *Drepanornis albertisi* **P. L. Sclater**, 1873
[Alt. Black-billed Sicklebill]
D'Albertis's Mountain Pigeon *Gymnophaps albertisii* **Salvadori**, 1874
[Alt. Papuan Mountain Pigeon]
D'Albertis's Grassbird *Megalurus albolimbatus* D'Albertis & Salvadori, 1879
[Alt. Fly River Grassbird]

D'Albertis's Orange-breasted Fig Parrot *Cyclopsitta gulielmitertii suavissima* Salvadori, 1876

Cavaglieri Luigi Maria D'Albertis (1841–1901) was an Italian botanist, ethnologist and zoologist. When in New Guinea (1871–1877) he ventured further than any European, using a steamboat furnished by the New South Wales government, in order to explore and chart the Fly River. He had a very adventurous time, having arrows fired at him by natives. He is reported to have taken a number of human skulls and even a recently severed head of an elderly woman. His behaviour towards the local people probably contributed considerably to their hostility to later European explorers. A mammal and two reptiles are named after him. (See also **Albertis**)

Dalhousie

Long-tailed Broadbill *Psarisomus dalhousiae* **Jameson**, 1835

Christina Ramsay *née* Broun (1786–1839) was the wife of the 9th Earl of Dalhousie (1818–1856). He was one of Wellington's commanders in the Peninsular War and the Battle of Waterloo. He was appointed Lieutenant-Governor of Nova Scotia and was Governor of Canada (1828), then Commander-in-Chief, East Indies (1829–1832).

Dalmas

Dalmas's Barbet *Capito auratus aurantiicinctus* Dalmas, 1900
[Alt. Gilded Barbet ssp.]
Rufous-winged Tanager ssp. *Tangara lavinia dalmasi* **Hellmayr**, 1910

Raymond, Compte de Dalmas (1862–1930), was a wealthy French nobleman, chess master and naturalist. He circumnavigated the world in his yacht *Chazalie*, during which he stayed in Japan (1882–1883) and wrote a book about his experiences there. He was in the Americas (1893–1897), and is recorded collecting at Guira and Yacua on the Venezuelan coast (1895). He also made trips to the Mediterannean, Morocco, Florida and the Caribbean (1899). His initial interest was ornithology, but he developed a passion for invertebrates, becoming an authority on spiders; he made a revision of the Prodidomidae (1918). His other loves were trout fishing and photography. Various marine invertebrates are named after him, a bird named after his wife (see **Emilie, Comptesse de Dalmas**) and several taxa named after the yacht!

Dam

Madagascan warbler genus *Damia* **Schlegel**, 1867 NCR
[Now in *Neomixis*]

(See **Van Dam**)

Dammerman

Sumba Myzomela *Myzomela dammermani* **Siebers**, 1928

Edible-nest Swiftlet ssp. *Aerodramus fuciphagus dammermani* **Rensch**, 1931
Dammerman's Moustached Parakeet *Psittacula alexandri dammermani* **Chasen** & **Kloss**, 1932
[Alt. Red-breasted Parakeet ssp.]
Rufous-sided Honeyeater ssp. *Ptiloprora erythropleura dammermani* **Stresemann** & Paludan, 1934
Sunda Minivet ssp. *Pericrocotus miniatus dammermani* **Neumann**, 1937 NCR; NRM

Dr Karel Willem Dammerman (1888–1951) was a Dutch field zoologist, botanist and collector who worked in the East Indies. He collected beetles from childhood and took his Ph.D (1910), then was appointed Assistant Entomologist of the Botanical Laboratories of the Botanic Gardens, Buitenzorg, Bogor, Java (1910). He became Chief of the Zoological Museum and Laboratory (1919) and ultimately (1932) Director of the Botanical Gardens. He was Chairman of the Society of Nature Protection, Dutch East Indies (1919–1932). On retirement (1939) he returned to Holland and joined the Museum of Natural History, Leiden (1942). He was also Editor of the Botanical Gardens journal *Treubia*. He wrote *Preservation of Wildlife and Nature Reserves in the Netherlands Indies* (1929) and *The Agricultural Zoology of the Malay Archipelago* (1929), as well as an article on the Orang Pendek, a small 'yeti' of Sumatra, still unknown to science. He also wrote (1948) an extensive book on the fauna of Krakatau. He was known for long periods of silence, which caused one of his assistants to name a newly discovered snail *Thiara carolitaciturni*, meaning 'Charles the Silent', a parody on 'William the Silent', a nickname of William I of Orange-Nassau, the leader of the Dutch revolt against the Spanish, assassinated in 1584. An amphibian and a mammal are named after him.

Dammholz

Blackcap ssp. *Sylvia atricapilla dammholzi* **Stresemann**, 1928

Dr Martin Ernst Karl Dammholz (fl.1927) was a German physician, collector and explorer in the Elburz Mountains, Iran (1927). He provided photographs for Gerd Heinrich's (q.v.) *Auf Panthersuche durch Persien* (1933).

Dampier

Variable Goshawk ssp. *Accipiter hiogaster dampieri*
Gurney, 1882
Papuan Hornbill ssp. *Rhyticeros plicatus dampieri* **Mayr**,
1934

Captain William Dampier (1651–1715), English buccaneer, world navigator and explorer, was the first person to circumnavigate the world three times and the first European to sight the Bismarck Archipelago (1699). He joined the Royal Navy (1673) but was discharged almost immediately due to severe illness. He recovered, joined the crew of a buccaneer and for much of his chequered career was a privateer or, from the Spanish perspective, a pirate. Among notable events in his life, he marooned Alexander Selkirk on an island in the Pacific in 1705, only rescuing him four years later, this being the basis of Daniel Defoe's *Robinson Crusoe* (1719).

Danae

Brown-headed Paradise Kingfisher *Tanysiptera danae*
Sharpe, 1880

In Greek mythology Danaë was the daughter of Acrisius, King of Argos. Acrisius was told by an oracle that his daughter's son would cause his death. He thus cast Danaë and her young son Perseus into the sea in a wooden chest, but by the will of the gods the pair survived.

Dane

Streaky-breasted Flufftail ssp. *Sarothrura boehmi danei*
Bannerman, 1920 NCR
[*S. boehmi* usually NRM: the specimen on which *danei* was
based was found at sea off Guinea]

Lieutenant-Commander Arthur Maurice Yate Dane (1885–1953) was a British naval officer who was in West Africa (1920) and later lived in Rhodesia (Zimbabwe). He was captain of HMS *Dwarf*, the ship on which the type specimen of the flufftail was caught.

Danford

Lesser Spotted Woodpecker ssp. *Dendrocopos minor
danfordi* **Hargitt**, 1883

Charles George Danford (1843–1928) was a geologist, palaeontologist, zoologist, artist, traveller and explorer. He was in Asia Minor (Turkey) (1875–1876 and 1879). The Danford Iris was named after his wife, who introduced it to England. A reptile is also named after him.

Daniel

Eurasian Black Vulture ssp. *Aegypius monachus danieli*
Meinertzhagen, 1938 NCR; NRM
[Alt. Cinereous Vulture]

Daniel Meinertzhagen (1875–1898), elder brother of Richard Meinertzhagen, was also an ornithologist.

Danis

Danis's Fulvetta *Fulvetta danisi* **Delacour** & **Greenway**, 1941
[Alt. Indochinese Fulvetta]

Vincent Danis (d.1943) was a French ornithologist who worked at the Natural History Museum in Paris (1938–1940). He wrote 'Etude d'une nouvelle collection d'oiseaux de l'ile Bougainville' (1938) wherein he described new species such as the White-eyed Starling *Aplonis brunneicapillus* collected there (Papua New Guinea). He fought in Norway (1940), joined the Free French forces in London (1940) and was killed in action in Tunisia (1943).

Danjou

Danjou's Babbler *Jabouilleia danjoui* **H. C. Robinson** &
Kloss, 1919
[Alt. Short-tailed Scimitar Babbler]

François André Gustave Abel Danjou (b.1874) was a French diplomat. He was Chancellor in Havana (1901–1905), then Madrid (1905–1906), and held various posts such as Vice-consul in Shanghai, China (1906–1916), and Consul in Singapore (1916–1923). He went on to be French Consul General in Calcutta (1933). While in Singapore he was awarded the Legion d'Honneur.

Dannefaerd

Tomtit ssp. *Petroica macrocephala dannefaerdi* **Rothschild**,
1894

Sigvard Jacob Dannefaerd (1853–1920) was a New Zealand-based Danish collector and photographer who collected in New Zealand and the Chatham Islands (1893–1895) for Rothschild. The Tomtit named for him is from Snares Is.

Daphne

Emerald sp. *Chlorostilbon daphne* **Gould**, 1861 NCR
[Alt. Blue-tailed Emerald; JS *Chlorostilbon mellisugus
phoeopygus*]

In Greek mythology Daphne, when pursued by Apollo, prayed to Gaea (Mother Earth) for aid and was changed into a laurel tree – possibly not the sort of aid she had in mind, although it succeeded in keeping her chaste. A mammal is also named after her.

d'Arnaud

d'Arnaud's Barbet *Trachyphonus darnaudii* **Prevost** & **Des
Murs**, 1847

Joseph Pons d'Arnaud (1811–1884) was a civil engineer, explorer and big-game hunter who was employed by the Egyptian government (1831). His first trip (1840–1841) was an Egyptian expedition looking for the source of the White Nile, which penetrated further than any other previous expedition. He explored in the Sudan (1841–1842) under the leadership of a Turkish officer, Selim Bimbashi, and in the company of Thibaut, a French Muslim who was the French consular agent at Khartoum for c.40 years. He also explored in Ethiopia (1843). He was appointed Lieutenant-Colonel (1856) in an

Egyptian regiment and continued to explore in the Sudan (1860). He wrote *The Aquatic Plants of the Upper Nile* (1860).

Dart

Marsh Tit ssp. *Poecile palustris darti* **Jouard**, 1929 NCR
[JS *Poecile palustris dresseri*]

R. Le Dart (fl.1950) was a French ornithologist and collector. He wrote 'Note sur la capture de la rarissime *Gallinago sabini*' (1952).

Dartmouth

Red-tufted Sunbird ssp. *Nectarinia johnstoni dartmouthi* **Ogilvie-Grant**, 1906

William Heneage Legge (1851–1936), 6th Earl of Dartmouth (1801–1936), sponsored scientific exploration, including an expedition to Mount Ruwenzori. A mammal is named after him.

Darwin

Darwin's Rhea *Rhea pennata* **d'Orbigny**, 1834
[Alt. Lesser Rhea; Syn. *Pterocnemia pennata*]
Darwin's Cactus Ground Finch *Geospiza scandens* **Gould**, 1837
[Alt. Common Cactus Finch]
Darwin's Caracara *Phalcoboenus albogularis* Gould, 1837
[Alt. White-throated Caracara]
Darwin's Large Ground Finch *Geospiza magnirostris* Gould, 1837
Darwin's Large Tree Finch *Camarhynchus psittacula* Gould, 1837
Darwin's Medium Ground Finch *Geospiza fortis* Gould, 1837
Darwin's Small Ground Finch *Geospiza fuliginosa* Gould, 1837
Darwin's Small Tree Finch *Camarhynchus parvulus* Gould, 1837
Darwin's Vegetarian Finch *Platyspiza crassirostris* Gould, 1837
Darwin's Warbler Finch *Certhidea olivacea* Gould, 1837
Darwin's Flycatcher *Pyrocephalus nanus* Gould, 1839
[Syn. *Pyrocephalus rubinus nanus*]
Darwin's Rail *Coturnicops notatus* Gould, 1841
[Alt. Speckled Rail, Speckled Crake]
Darwin's Cocos Island Finch *Pinaroloxias inornata* Gould, 1843
Darwin's Nothura *Nothura darwinii* **Gray**, 1867
[Alt. Darwin's Tinamou]
Darwin's Woodpecker Finch *Camarhynchus pallidus* **P. L. Sclater** & **Salvin**, 1870
Darwin's Sharp-beaked Ground Finch *Geospiza difficilis* **Sharpe**, 1888
Darwin's Large Cactus Finch *Geospiza conirostris* **Ridgway**, 1890
Darwin's Medium Tree Finch *Camarhynchus pauper* Ridgway, 1890
Darwin's Mangrove Finch *Camarhynchus heliobates* **Snodgrass** & **Heller**, 1901

Darwin's Tanager *Thraupis bonariensis darwinii* **Bonaparte**, 1838
[Alt. Blue-and-yellow Tanager ssp.]
Darwin's Koklass Pheasant *Pucrasia macrolopha darwini* **Swinhoe**, 1872
Large Cactus Finch ssp. *Geospiza conirostris darwini* **Rothschild** & **Hartert**, 1899
Scaly-throated Earthcreeper ssp. *Upucerthia dumetaria darwini* **W. E. D. Scott**, 1900 NCR
[JS *Upucerthia dumetaria hypoleuca*]
Green-backed Gerygone ssp. *Gerygone chloronota darwini* **Mathews**, 1912

Charles Robert Darwin (1809–1882) was the prime advocate, together with Wallace (q.v.), of natural selection as the driver of speciation. To quote from his seminal work, *On the Origin of Species by Means of Natural Selection* (1859): 'I have called this principle, by which each slight variation, if useful, is preserved, by the term Natural Selection.' Darwin was naturalist on HMS *Beagle* on her scientific circumnavigation (1831–1836). In South America he found fossils of extinct animals that were similar to extant species. On the Galápagos Islands he noticed many variations among plants and animals of the same general type as those in South America. Darwin collected specimens for further study everywhere he went. On his return to London he conducted thorough research of his notes and specimens. Out of this study grew several related theories: evolution did occur; evolutionary change was gradual taking thousands or even millions of years; the primary mechanism of evolution was 'natural selection'; and the millions of species alive today arose from a single original life form through a branching process called 'specialisation'. However, Darwin held back on publication for many years, not wanting to offend Christians, especially his wife. He is remembered in the names of numerous other taxa including three amphibians, four mammals and nine reptiles.

Daubenton, E.-L.

Daubenton's Parakeet *Psittacula eques* Boddaert, 1783 EXTINCT
[Alt. Réunion Parakeet]

Edmé-Louis Daubenton (1732–1786) was a French naturalist, cousin of Louis Jean-Marie d'Aubenton (q.v.). Buffon (q.v.) engaged him to supervise the colour illustrations of Martinet's engravings for his *Histoire Naturelle* (1749–1789). The name of Daubenton is attached to this taxon because Boddaert's description of *Psittacula eques* was based on one of the plates (collectively known as *Planches Enluminées*) (1783).

Daubenton, L. J.

Daubenton's Curassow *Crax daubentoni* **G. R. Gray**, 1867
[Alt. Yellow-knobbed Curassow]

Dr Louis Jean-Marie d'Aubenton (1716–1800) (more commonly Daubenton) was a French naturalist. His work covered many fields including comparative anatomy, plant physiology, palaeontology, mineralogy and experimental agriculture. He was Professor of Mineralogy at the Jardin des Plantes, Paris, and of Natural History at the School of Medicine. He became first Director of the Museum of Natural History, Paris (1793). His scientific descriptions included 182 species of quadrupeds, for the first section of Georges Buffon's (q.v.) work *Histoire Naturelle Générale et Particulière* (1794–1804). Daubenton was a strange man. Unusually for his day he was vegetarian, once saying: 'It is to be presumed that man, while he lives in a natural state and a graded climate, where the earth spontaneously produces every type of fruit, he feeds himself with these and does not eat animals.' G. R. Gray (q.v.) described the curassow from a specimen bought from Leadbeater (q.v.) some years earlier. It is possibly named after Edmé-Louis Daubenton (q.v.), as the original description is not clear. Three mammals are named after him.

David, A.

David's Hill Partridge *Arborophila davidi* **Delacour**, 1927
[Alt. Orange-necked (Hill) Partridge]

Small Buttonquail ssp. *Turnix sylvaticus davidi* Delacour & **Jabouille**, 1930

André D. David-Beaulieu (b.1896) was a naturalist and civil servant in French Indochina (1940s). He wrote *Les Oiseaux de la Région de Honquan* (1932), *Les Oiseaux du Tranninh* (1944) and *Les Oiseaux de la Province de Savannakhet* (1949).

David, Père

(Père) David's Hill Warbler *Prinia crinigera* **Hodgson**, 1836
[Alt. Striated Prinia, Brown Hill Warbler]
Chinese Beautiful Rosefinch *Carpodacus davidianus* **Milne-Edwards**, 1865
(Père) David's Laughingthrush *Garrulax davidi* **Swinhoe**, 1868
[Alt. Plain Laughingthrush; Syn. *Pterorhinus davidi*]
David's Swan *Cygnus davidi* Swinhoe, 1870 NCR
[indeterminate taxon – see text below]
(Père) David's Snowfinch *Pyrgilauda davidiana* **J. Verreaux**, 1871
[Alt. Small Snowfinch; Syn. *Montifringilla davidiana*]
(Père) David's (Wood) Owl *Strix davidi* **Sharpe**, 1875
[Alt. Sichuan Wood Owl]
(Père) David's Orange-throat *Luscinia pectardens* David, 1877
[Alt. Firethroat]
(Père) David's Tit *Poecile davidi* **Berezowski & Bianchi**, 1891
[Alt. Rusty-breasted Tit]
(Père) David's Fulvetta *Alcippe davidi* **Styan**, 1896
[Alt. Northern Grey-cheeked Fulvetta]
(Père) David's Parrotbill *Neosuthora davidiana* **Slater**, 1897
[Alt. Short-tailed Parrotbill; Syn. *Paradoxornis davidianus*]
(Père) David's Niltava *Niltava davidi* **La Touche**, 1907
[Alt. Fukien/Fujian Niltava]
(Père) David's Bush Warbler *Bradypterus davidi* La Touche, 1923
[Alt. Siberian Bush Warbler]

Brownish-flanked Bush Warbler ssp. *Cettia fortipes davidiana* J. Verreaux, 1870

(Père) David's Blood Pheasant *Ithaginis cruentus sinensis* David, 1873
Rufous-capped Babbler ssp. *Stachyridopsis ruficeps davidi* **Oustalet** 1899

Fr Jean Pierre Armand David (1826–1900) was a French Lazarist priest as well as a fine zoologist. He taught in Savona, and Doria (q.v.) and D'Albertis (q.v.) were among his pupils. He was a missionary to China and the first Westerner to observe many animals, including birds, as well as the Giant Panda *Ailuropoda melanoleuca* and the deer *Elaphurus davidianus* famously named for him. He arrived in China (1862) and started collecting (1863). The French naturalist Alphonse Milne-Edwards (q.v.) classified many of his specimens. He co-wrote *Les Oiseaux de Chine* (1877). Père David collected thousands of specimens and had many plants named after him, as well as an amphibian and a reptile. Swinhoe (q.v.) described the swan from the sole known specimen, since lost. It is variously presumed to be a Snow Goose, a goose/swan hybrid, or a now-extinct species similar to Coscoroba Swan.

David (Harrison)

Eurasian Nuthatch ssp. *Sitta europaea davidi* **J. M. Harrison**, 1955 NCR
[JS *Sitta europaea persica*]

Dr David Lakin Harrison (b.1926) is a physician who joined the Royal Air Force (1953) and served in Iraq (1954), collecting there. After leaving the Air Force he worked extensively in the Middle East and (1970s) was the leader of the Fauna and Flora Survey of Jabal Akhdar (northern Oman). He is Chairman of the Trustees of the Harrison Institute, which is the controlling body of the Harrison Zoological Museum, Sevenoaks, England. He wrote *Footsteps in the Sand* (1959) and co-wrote a revised edition of *The Mammals of Arabia* (1991). He developed an interest in fossil fauna (1985) and in 2010 was researching the Early Eocene in Suffolk, England. Two mammals are named after him. The nuthatch was named after him by his father, James Harrison.

Davidson

Golden-fronted Leafbird ssp. *Chloropsis aurifrons davidsoni* **E. C. S. Baker**, 1920 NCR
[JS *Chloropsis aurifrons frontalis*]

James Davidson (1849–1925) was an ornithologist and British colonial administrator in India (retired 1897). He was a close friend of Hume (q.v.), to whom he loaned his bird collection. He wrote 'Notes on the birds of Khandeish and Nasik' (1895).

David Willard

Sierran Elaenia ssp. *Elaenia pallatangae davidwillardi* **Dickerman & W. H. Phelps Jr**, 1987

(See **Willard**)

Davies

Long-billed Lark sp. *Certhilauda daviesi* **Gunning & J. A. Roberts**, 1911 NCR
[Alt Eastern Long-billed Lark; JS *Certhilauda semitorquata*]

Olive Sunbird ssp. *Cyanomitra olivacea daviesi* **Haagner**, 1907 NCR
[JS *Cyanomitra olivacea olivacea*]
Buffy Pipit ssp. *Anthus vaalensis daviesi* J. A. Roberts, 1914 NCR
[JS *Anthus vaalensis vaalensis*]

Lieutenant Claude Gibney Finch-Davies (1875–1920) was a British soldier, ornithologist and bird painter. Origianlly named 'Davies', he added 'Finch', his wife's maiden name (1916). He joined the Cape Mounted Riflemen (1893) in South Africa and served in the Boer War (1898–1902) and WW1 against German forces in German South West Africa (Namibia). He vandalised a number of books in South African libraries, removing plates of illustrations, and although no charges were made against him his reputation was ruined. He died suddenly of angina. His plates illustrated Horsbrugh's (q.v.) *The Game-Birds and Waterfowl of South Africa* (1912).

Davis

Fairy Tern ssp. *Sternula nereis davisae* **Mathews** & **Iredale**, 1913

Lady Elizabeth Eileen Davis *née* Schischka (fl.1917) was the wife of New Zealand business magnate Sir George Francis Davis.

Davison

Pitta genus *Davisona* **Mathews**, 1934 NCR
[Now in *Hydrornis*]

Davison's Ibis *Pseudibis davisoni* **Hume**, 1875
[Alt. White-shouldered Ibis]
Giant Pitta *Brachyurus davisoni* Hume, 1875 NCR
[JS *Hydrornis caeruleus caeruleus*]
Davison's Leaf Warbler *Phylloscopus davisoni* **Oates**, 1889
[Alt. White-tailed Leaf Warbler]
Silver Pheasant sp. *Gennaeus davisoni* **Ogilvie-Grant**, 1893 NCR
[Hybrid population of *Lophura nycthemera* subspecies]

Crested Serpent Eagle ssp. *Spilornis cheela davisoni* Hume, 1873
Davison's Stripe-throated Bulbul *Pycnonotus finlaysoni davisoni* Hume, 1875
Davison's Sibia *Heterophasia annectans davisoni* Hume, 1877
[Alt. Rufous-backed Sibia ssp.]
Blue-throated Barbet ssp. *Megalaima asiatica davisoni* Hume, 1877
Davison's Brown-eared Bulbul *Hemixus flavala davisoni* Hume, 1877
Siberian Thrush ssp. *Zoothera sibirica davisoni* Hume, 1877
Yellow-breasted Warbler ssp. *Seicercus montis davisoni* **Sharpe**, 1888
Collared Kingfisher ssp. *Todiramphus chloris davisoni* Sharpe, 1892
Grey-throated Babbler ssp. *Stachyris nigriceps davisoni* Sharpe, 1892
Eyebrowed Wren-babbler ssp. *Napothera epilepidota davisoni* Ogilvie-Grant, 1910
Brown-cheeked Fulvetta ssp. *Alcippe poioicephala davisoni* **Harington**, 1915
Pied Triller ssp. *Lalage nigra davisoni* **Kloss**, 1926

William Ruxton Davison (d.1893) was a British ornithologist and Curator of Raffles Museum, in Singapore (1887–1893). He collected for Hume (q.v.) (1870s), during which time (1875) he discovered Gurney's Pitta *Pitta gurneyi*, and later wrote 'A revised list of the birds of Tenasserim' (1878) with Hume. Davison also wrote articles such as 'Notes on the nidification of some Burmese birds'. He wrote descriptions of six birds now in the British Museum, four of which he collected himself in India or the Malay Peninsula. A reptile is named after him.

Dawa

Laced Woodpecker ssp. *Picus vittatus dawae* **Van Tyne** & **Koelz**, 1936 NCR
[JS *Picus xanthopygaeus*]

Noro Surja Dawa (DNF) of Lahul, Himachal Pradesh, India, greatly assisted the describers in their collecting.

Dawson

Rosy Finch ssp. *Leucosticte tephrocotis dawsoni* **Grinnell**, 1913
[Alt. Grey-crowned Rosy Finch ssp.]

William Leon Dawson (1873–1928) was an American Seventh-Day Adventist minister who became a full-time ornithologist and oologist (early 1900s). He founded the Museum of Comparative Oology (1915) (Santa Barbara Museum of Natural History), becoming its first Director (1915–1925). He wrote *Birds of California* (1923).

Day

Black-girdled Barbet *Capito dayi* **Cherrie**, 1916
Day's Elaenia *Elaenia dayi* **Chapman**, 1929
[Alt. Great Elaenia]

Colonel Lee Garnet Day (c.1890–1960) was a New York businessman, banker and philanthropist. He financed expeditions (1915–1920s), including the American Museum expedition to Mount Roraima (1927), which included Chapman (q.v.) and T. D. Carter, who published the ornithological report (1929). The expedition also searched for the missing aviator Paul Redfern, who had been reported as seen living in the jungle. Among other things Day was Chairman of the American Legion's Rehabilitation Committee for New York (1922), advocating better aid for WW1 veterans, and a Trustee of the Field Museum in Chicago. The etymology for the elaenia states that it was named '… after Mr. Lee Garnet Day of New York City … for the support which made our Roraima expedition possible.' A mammal is named after him.

Dearborn

Dearborn's Blue Bunting *Cyanocompsa parellina dearborni* **W. de W. Miller** & **Griscom**, 1931 NCR
[JS *Cyanocompsa parellina parellina*]
Green Shrike Vireo ssp. *Vireolanius pulchellus dearborni* **A. R. Phillips**, 1991

Ned Dearborn (1865–1948) was an American ornithologist. Dartmouth University awarded his bachelor's degree (1891) and New Hampshire State College his master's (1898). He was Assistant Curator of Birds at the Field Museum, Chicago (1901–1909), then worked with the US Biological Survey (1909–1920). He started and ran the Dearborn Fur Farm, Sackets Harbour, New York (1920–1948). He wrote 'Catalogue of a collection of birds from Guatemala' (1907).

Deborah

Rusty-naped Pitta ssp. *Hydrornis oatesi deborah* **B. F. King**, 1978
[Syn. *Pitta oatesi deborah*]

Deborah Bodner is a friend of the describer.

De Carle

Black-streaked Scimitar Babbler ssp. *Pomatorhinus gravivox decarlei* **Deignan**, 1952

(See **Sowerby, A. D. C.**)

Decary

Common Jery ssp. *Neomixis tenella decaryi* **Delacour**, 1931

Raymond Decary (1891–1973) was a colonial administrator in Madagascar (1916–1944). He was a naturalist, botanist, geologist and ethnographer, and was interested in everything to do with Madagascar, contributing over 40,000 specimens of Malagasy flora to the Paris herbarium. He qualified in law (1912) before army service, when he was seriously wounded at the Battle of the Marne (1914) and unable to resume active service. He went to Madagascar (1916) as an officer in the Reserve, thus releasing a fully fit officer for active service. He trained as a colonial administrator (1921), returning to Madagascar (1922). He undertook seven scientific expeditions in the island (1923–1930) and became Director of Scientific Research there (1937). He was again in the French Army in Madagascar (1939–1944), returning to France after the Liberation. Demobilised in 1945, he retired to private life, continuing his research. He wrote *Malagasy Fauna* (1950). Three reptiles and an amphibian are named after him.

Decken

Von der Decken's Hornbill *Tockus deckeni* **Cabanis**, 1868

(See **Von der Decken**)

Decorse

Orange-tufted Sunbird ssp. *Cinnyris osea decorsei* **Oustalet**, 1904

Dr Gaston-Jules Decorse (1873–1907) was an army physician interested in ethnography and linguistics. He travelled in Madagascar, where he collected botanical specimens (1898–1902), and joined a French expedition to Lake Chad (1902–1904). He co-wrote *Rabah et les Arabes du Chari* (1905). Four reptiles are named after him.

Decoux

Red Avadavat ssp. *Amandava amandava decouxi* **Delacour & Jabouille** 1928 NCR
[JS *Amandava amandava punicea*]

Aimé Decoux (d.1961) was an aviculturist near Limoges, France, a friend of the senior author.

De Deken

Black-streaked Scimitar Babbler ssp. *Pomatorhinus gravivox dedekensi* **Oustalet**, 1892

Père Constant De Deken (1852–1896) was a Belgian missionary to China (1870–1880) and on the Upper Mekong River (1892).

Dedem

Streaky-breasted Fantail *Rhipidura dedemi* **Van Oort**, 1911

Grey-headed Woodpecker ssp. *Picus canus dedemi* Van Oort, 1911
[Alt. Sumatran Woodpecker]

Baron Frederik Karel van Dedem (1873–1959) was a Dutch naturalist and collector in Sumatra. The woodpecker may in future be elevated as a full species.

Dedi

Cave Swiftlet ssp. *Collocalia linchi dedii* Somadikarta, 1986

Dedi Ahadiat Somadikarta (1961–1985) was the describer's son.

Defensorum

Gabar Goshawk ssp. *Melierax gabar defensorum* **Meinertzhagen**, 1949 NCR
[JS *Micronisus gabar niger*]

The trinomial *defensorum* means 'of the defenders', a name chosen to honour the Rev. Sir Reginald Stuart Champion (1895–1982), Governor of Aden (1945–1950), and his wife Lady M. Champion.

De Filippi

Lapwing genus *Defilippia* **Salvadori**, 1865 NCR
[Now in *Vanellus*]

Pampas Meadowlark *Sturnella defilippii* **Bonaparte**, 1850
De Filippi's Petrel *Pterodroma defilippiana* **Giglioli &** Salvadori, 1869
[Alt. Defilippe's Petrel, Mas a Tierra Petrel]

Filippo de Filippi (1814–1867) was an Italian doctor, traveller and zoologist. He visited, among other places, Alaska, Mongolia and Turkestan. He succeeded Carlo Gené (q.v.) as Professor of Zoology at the Museum of Natural History in Turin. His efforts to disseminate knowledge of Darwin's evolutionary theory included the presentation of a seminal lecture (1864) 'L'uomo e le scimmie' (Man and the Apes). Filippi was the scientist who accompanied the Duke of Abruzzi's expedition to Alaska, and he also led an expedition (1862) to explore Persia (now Iran). He set out on a government-sponsored scientific voyage aboard *Magenta*, being replaced by Giglioli who published the results. (See also **Philippi**). Two reptiles are named after him.

De Fontaine

Long-tailed Parakeet ssp. *Psittacula longicauda defontainei* **Chasen**, 1934

Percy Mortimer de Fontaine (1875–c.1937) worked c.40 years at the Raffles Museum in Singapore, becoming Chief Taxidermist. He visited the Natuna Islands (1934) and undertook several expeditions in the Malay Archipelago. On a collecting expedition (1914) 100 miles up the Baram River in Borneo, his party heard the war cries of the local people, the Dayaks, and his 100 local helpers melted away. For a week the 17 members of the party barricaded themselves into a small hut, making holes in the planking for their rifles, until the Dayaks moved on. Ironically his father, Captain de Fontaine, was killed (1885) by a similar group of headhunters.

Degen

Flappet Lark sp. *Mirafra degeni* **Ogilvie-Grant**, 1902 NCR
[JS *Mirafra rufocinnamomea*]

J. J. Edward Degen (1852–1922) was born in Basel and died in London. He collected reptiles, mammals and fish in East Africa (c.1895–1905) and was in Ethiopia (1902). After leaving Africa he worked as an articulator/taxidermist at the BMNH, having previously worked for the National Museum in Melbourne (1894). A reptile and an amphibian are named after him.

Degland

Degland's Scoter *Melanitta deglandi* **Bonaparte**, 1850
[Alt. White-winged Scoter]

Dr Côme Damien Degland (1787–1856) was Senior Physician at the Saint Sauveur Hospital in Lille. He was director of their natural history museum where his assembled collection of 1,800 specimens of European birds are still displayed. He wrote *Ornithologie Européenne* (1849).

De Haan

Cream-throated White-eye ssp. *Zosterops atriceps dehaani* Van Bemmel, 1939
Pompadour Green Pigeon ssp. *Treron pompadora dehaani* Van Bemmel & **Voous**, 1951 NCR
[Alt. Grey-cheeked Green Pigeon ssp.; JS *Treron griseicauda wallacei*]

Gunter Adelbert Leonard de Haan (b.1911) was a Javan-born Dutch botanist. He went to the Netherlands to be educated at the Agricultural College, but left for the Dutch East Indies before completing his studies. There he enlisted in the army (1933) but soon bought himself out and joined the Teak Forest Administration (1933–1935). He later became an administrator in the Indonesian Forest Service (1936–1940), which included an exploratory tour in the Moluccas (1937–1938). At the outbreak of war he was suspended, then dismissed, because of his political opinions. After WW2 he was evacuated to Holland but returned (1948) to Halmaheira as a plantation manager, planning to use his spare time to make botanical collections for sale. He made an initial collection of 50 plants, but abandoned the idea as it was poorly received.

He wrote a number of articles including 'Notes on the Invisible Flightless Rail of Halmahera' (1950). During his time at the museum Voous (q.v.) obtained a collection, made by de Haan, of 221 birds from Muna, Buton and the Moluccas.

Deichler

Mistle Thrush ssp. *Turdus viscivorus deichleri* **Erlanger**, 1897
Thekla Lark ssp. *Galerida theklae deichleri* Erlanger, 1899 NCR
[JS *Galerida theklae carolinae*]

Dr Christian Deichler (1876–1954) was a German chemist who was in Tunisia (1896–1899) and the Cameroons.

Deignan

Weaver subgenus *Deignaniplectes* **Wolters**, 1970 NCR
[In *Ploceus*]

Deignan's Babbler *Stachyridopsis rodolphei* Deignan, 1939
[Now often regarded as a synonym of *Stachyridopsis rufifrons rufifrons*]

Hill Blue Flycatcher ssp. *Cyornis banyumas deignani* **de Schauensee**, 1939
Grey-chinned Minivet ssp. *Pericrocotus solaris deignani* **Riley**, 1940
Dollarbird ssp. *Eurystomus orientalis deignani* **Ripley**, 1942 NCR
[Regarded as an intergrade between races *orientalis* and *abundus*]
Asian Barred Owlet ssp. *Glaucidium cuculoides deignani* Ripley, 1948
Puff-throated Babbler ssp. *Pellorneum ruficeps deignani* **Delacour**, 1951
Thick-billed Flowerpecker ssp. *Dicaeum agile deignani* Ripley, 1952 NCR
[JS *Dicaeum agile modestum/pallescens*]
Common Iora ssp. *Aegithina tiphia deignani* **B. P. Hall**, 1957
Chestnut Munia ssp. *Lonchura atricapilla deignani* Parkes, 1958

Herbert 'Bert' Girton Deignan (1906–1968) was a fellow of the John Simon Guggenheim Memorial Foundation (1952) and worked for the USNM (1938–1962). He graduated from Princeton (1928) with a degree in European languages, then (1928–1932) taught English at a school at Chiang Mai, Siam (Thailand). He was an associate of Alexander Wetmore (q.v.), who helped him get a temporary job at the USNM (1933). Deignan then worked at the Library of Congress (1934–1935) before returning to his old job in Thailand (1935–1937) and combining teaching with collecting birds for Wetmore. He returned to the USA and worked at the USNM (1938–1944) then he was in southern Asia as an agent of the Office of Strategic Services (the forerunner to the CIA). He returned to the USNM again (1946) until retiring (1962) to Switzerland. He wrote *The Birds of Northern Thailand* (1945), *Type Specimens of Birds in the United States National Museum* (1961) and *Checklist of the Birds of Thailand* (1963). Deignan described the eponymous babbler along with two other new babblers, in 'Three new birds of the genus *Stachyris*' (1939). He is also remembered in the name of a reptile.

Dejean

Black-bibbed Tit *Parus dejeani* **Oustalet**, 1897 NCR
[JS *Poecile hypermelaenus*]

Père Léonard-Louis Déjean (1846–1906) was a French missionary to China and Tibet (1870–1906). He mainly collected entomological specimens. He died of typhoid.

Delacour

Pheasant genus *Delacourigallus* **Hachisuka** 1941 NCR
[Now in *Lophura*]
Estrildid finch genus *Delacourella* **Wolters**, 1949 NCR
[Now in *Nesocharis*]

Delacour's Little Grebe *Tachybaptus rufolavatus* Delacour, 1932 EXTINCT
[Alt. Alaotra Grebe]
Fulvetta sp. *Alcippe delacouri* Yen, 1935 NCR
[Alt. Yellow-throated Fulvetta; JS *Alcippe cinerea*]

Delacour's Crested Fireback *Lophura ignita macartneyi* **Temminck**, 1813
Delacour's Broadbill *Smithornis capensis delacouri* **Bannerman**, 1923
[Alt. African Broadbill ssp.]
Burmese Shrike ssp. *Lanius collurioides delacouri* **Collin &
Hartert** 1927 NCR
[JS *Lanius collurioides nigricapillus*]
Large-tailed Nightjar ssp. *Caprimulgus macrurus delacouri* Hachisuka 1931 NCR
[JS *Caprimulgus manillensis*]
Hottentot Teal ssp. *Anas punctata delacouri* **Neumann**, 1932 NCR
[JS *Anas hottentota* (monotypic)]
Eyebrowed Wren-babbler ssp. *Napothera epilepidota delacouri* Yen, 1934
Green Jery ssp. *Neomixis viridis delacouri* **Salomonsen**, 1934
White-collared Yuhina ssp. *Yuhina diademata delacouri* Yen, 1935 NCR
[JS *Yuhina diademata ampelina*]
Eurasian Nuthatch ssp. *Sitta europaea delacouri* **Deignan**, 1938 NCR
[JS *Sitta nagaensis*]
Grey-capped Woodpecker ssp. *Dendrocopos canicapillus delacouri* **de Schauensee**, 1938
Black-headed Shrike Babbler ssp. *Pteruthius rufiventer delacouri* **Mayr**, 1941
Yellow-bellied Prinia ssp. *Prinia flaviventris delacouri* Deignan, 1942
Siberian Stonechat ssp. *Saxicola torquata delacouri* David-**Beaulieu** 1944 NCR
[JS *Saxicola stejnegeri*]
Asian Barred Owlet ssp. *Glaucidium cuculoides delacouri* **Ripley**, 1948
Wedge-tailed Green Pigeon ssp. *Treron sphenurus delacouri* **Biswas**, 1950
Long-tailed Widowbird ssp. *Euplectes progne delacouri* Wolters, 1953

Dr Jean Theodore Delacour (1890–1985) was a French-American ornithologist renowned for discovering and rearing some of the rarest birds in the world. He was born in Paris and died in Los Angeles. In France (1919–1920) he created the zoological gardens at Clères and donated them to the French Natural History Museum in Paris (1967). He undertook a number of (particularly pheasant) collecting expeditions to Indochina, particularly Vietnam. He wrote *Birds of Malaysia* (1947) and co-wrote *Birds of the Philippines* (1946) among many other ornithological books, including his memoirs *The Living Air: The Memoirs of an Ornithologist*. A number of taxa are named after him including fish, an amphibian, a reptile and three mammals.

Delafield

Delafield's Ground Warbler *Trichas delafieldii* **Audubon**, 1839 NCR
[Alt. Masked Yellowthroat; JS *Geothlypis aequinoctialis*]

Colonel Joseph Delafield (1790–1875) was an American amateur antiquarian. He had a particular fondness for the preservation of old farmhouses. Audubon says, in *Birds of America*: 'This beautiful little bird I named in honour of Colonel DELAFIELD, President of the Lyceum of Natural History in the city of New York, a gentleman distinguished by his scientific attainments, not less than by those accomplishments and virtues which tend to improve and adorn society.'

Delalande

Delalande's Plovercrest *Stephanoxis lalandi* **Vieillot**, 1818
[Alt. Black-breasted Plovercrest]
Delalande's Coua *Coua delalandei* **Temminck**, 1827 EXTINCT
[Alt. Snail-eating Coua]
Delalande's Antpipit *Corythopis delalandi* **Lesson**, 1830
[Alt. Southern Antpipit]

Delalande's Green Pigeon *Treron calvus delalandii* **Bonaparte**, 1854
[Alt. African Green Pigeon ssp.]
Whiskered Tern ssp. *Chlidonias hybrida delalandii* **Mathews**, 1912

Pierre Antoine Delalande (1787–1823) worked for the MNHN in Paris. He collected with Auguste de Saint-Hilaire (q.v.) in the region around Rio de Janeiro (1816), and with his nephew J. P. Verreaux (q.v.) and Andrew Smith (q.v.) in the African Cape (1818). Later Geoffroy Saint-Hilaire (q.v.) employed him as a taxidermist. An amphibian and four reptiles are named after him.

Delamere

Black-faced Waxbill ssp. *Estrilda erythronotos delamerei* **Sharpe**, 1900
Short-tailed Lark ssp. *Pseudalaemon fremantlii delamerei* Sharpe, 1900
Long-tailed Widowbird ssp. *Euplectes progne delamerei* **G. E. Shelley**, 1903

Hugh Cholmondeley, 3rd Baron Delamere (1870–1931), was a pioneer settler (1903) in East Africa (Kenya). He farmed a huge acreage and his trial and error over many years was the foundation for much of Kenya's later agriculture which learned much from his many failures and eventual success. He was also well known for his antics in personal life as unofficial leader of the Happy Valley Set – white hedonist settlers whose lifestyles apparently degenerated into wife-swapping and drug-taking. Many of his attitudes can be summed up by a much quoted utterance of his that 'I am going to prove to you all that this is a white man's country.' Despite this he admired the Masai.

de Lattre, H.

Hummingbird genus *Delattria* **Bonaparte**, 1850 NCR
[Now in *Lampornis*]

de Lattre's Sabrewing *Campylopterus hemileucurus* **Deppe**, 1830
[Alt. Violet Sabrewing]

Augustine Henri de Lattre (1801–1867), like his younger brother Pierre Adolphe (q.v.), was a French collector, naturalist and artist; he specialised in painting animals, particularly horses. He collected in Mexico with his brother (1838). He painted a portrait of Boston, a famous American thoroughbred (c.1840). (See also **Henry de Lattre**)

de Lattre, P. A.

de Lattre's Coquette *Lophornis delattrei* **Lesson**, 1839
[Alt. Rufous-crested Coquette]
Tawny-crested Tanager *Tachyphonus delatrii* **Lafresnaye**, 1847
Chestnut-capped Warbler *Basileuterus delattrii* **Bonaparte**, 1854
[Alt. Rufous-capped Warbler; Syn. *Basileuterus rufifrons delattrii*]

Thicket Tinamou ssp. *Crypturellus cinnamomeus delattrii* Bonaparte, 1854

Pierre Adolphe de Lattre (1805–1854), like his older brother Henri (q.v.), was a French collector, artist and naturalist. He collected in the Americas (1831–1851), including Mexico with his brother (1838).

de Lautour

de Lautour's Duck *Biziura delautouri* **H. O. Forbes**, 1892
EXTINCT
[Alt. New Zealand Musk Duck]

Dr Harry Archibald de Lautour (1849–1917) was born in West Bengal, India, where his father was a judge. He was educated at Cheltenham College and studied medicine at Kings College, London (1873), before becoming a member of the Royal College of Surgeons (1874). He registered to practise medicine (1875) then joined the Defence Forces of New Zealand, as Surgeon. He moved to Oamaru, New Zealand (1882), and soon rose to Surgeon Major (1888), then Lieutenant Colonel (1895). While there he did microscopic work of scientific value on the diatomaceous deposits of the Oamaru district. He became Principal Medical Officer of the Otago District (1903). Forbes (q.v.), who described the duck, said of de Lautour ... 'This gentleman ... is well known through his papers on the diatomaceous deposits discovered by him in his district ... I have proposed the name of Biziura de

Lautouri, after the gentleman to whom I am indebted for the acquisition of these bones.' The duck is believed to have become extinct around the 16th century.

Delegorgue

Delegorgue's Pigeon *Columba delegorguei* Delegorgue, 1847
[Alt. Eastern Bronze-naped Pigeon]
Harlequin Quail *Coturnix delegorguei* Delegorgue, 1847

Louis-Adulphe Joseph Delegorgue (1814–1850) was a French hunter and naturalist who bequeathed the collections he made in southern Africa (1830–1839) to the BMNH. He described his adventures in a book, *Travels in Southern Africa* (1847). He also collected with Wahlberg (q.v.). Apparently he named both birds after himself, which is, of course, frowned upon.

Delessert

Wynaad Laughingthrush *Garrulax delesserti* **Jerdon**, 1839
Flameback sp. *Indopicus delesserti* **Malherbe**, 1849 NCR
[Alt. Greater Flameback; JS *Chrysocolaptes guttacristatus*]

Adolphe Delessert (1809–1869) was a French naturalist who collected in India (1834–1839), sponsored by his wealthy uncle. He met Jerdon (q.v.) in the Nilgiri hills and gave him a new babbler, which Jerdon described in his honour. He wrote *Souvenirs d'un Voyage dans l'Inde Exécuté de 1834 à 1839* (1843).

del Hoyo

White Hawk ssp. *Leucopternis albicollis delhoyoi* Bahr, 2010

Dr Josep del Hoyo (b.1954) was a Catalan physician (GP) until he became a full-time ornithologist, publisher and conservationist. He founded Lynx Edicions and was senior editor of the 17-volume *Handbook of the Birds of the World* (1992–2013), and served on the Global Council of BirdLife International (2005–2013). Bahr coined the above scientific name as a replacement for *Leucopternis albicollis costaricensis* (Sclater, 1919), reassigning *Leucopternis albicollis* to an enlarged genus *Buteo*, and thus avoiding conflict with *Buteo jamaicensis costaricensis*; but this is not widely accepted.

Dellon

Réunion Fody *Foudia delloni* Cheke & J. P. Hume, 2008
EXTINCT

Charles Gabriel Dellon (1649–1709) was a French physician and traveller who first mentioned this Réunion endemic (1668). It became extinct shortly afterwards. Dellon travelled to Goa (1668) but fell foul of the Inquisition and spent four years in prison and then as a slave in Portuguese galleys. He was released in Lisbon (1677) and returned to France. He wrote of his experiences in *Relation d'un Voyage des Indes Orientales* (1685).

Deluz

White-fringed Antwren ssp. *Formicivora grisea deluzae* **Ménétriés**, 1835

M. Deluz (fl.1834) was a French resident in Rio de Janeiro, Brazil, about whom we can find no more (note the binomial is feminine).

Dement'ev

Eurasian Nightjar ssp. *Caprimulgus europaeus dementievi* **Stegmann**, 1949
Eurasian Sparrowhawk ssp. *Accipiter nisus dementjevi* **Stepanyan**, 1958

Professor Georgii Petrovich Dement'ev (1898–1969) was a Russian zoologist, ornithologist and conservationist. He went to Moscow University (1920) to study under Buturlin (q.v.), with whom he wrote *The Complete Identification Book for the Birds of the USSR* (1928). He became head of the ornithological laboratory of the Zoological Museum, Moscow (1956). After Stalin's death he was one of the first Russian scientists to press for conservation of nature as it was no longer suicidal to not follow the party line. He is perhaps best known for his monumental six-volume work with N. A. Gladkov, *Birds of the Soviet Union* (published in English 1966–1968).

Demery

African Yellow White-eye ssp. *Zosterops senegalensis demeryi* **Büttikofer**, 1890

Archery T. Demery (d.1891) was a West African whose father was Jackson Demery, Büttikofer's collector. Büttikofer sent Archery to Holland to be trained as a natural history collector. He collected with his father in Sierra Leone, Gold Coast (Ghana) and Liberia (1890–1891).

Dendy

Olive-tailed Thrush ssp. *Turdus lunulatus dendyi* **Mathews**, 1912 NCR
[JS *Zoothera lunulata lunulata*]
Tawny Frogmouth ssp. *Podargus strigoides dendyi* Mathews, 1912 NCR
[JS *Podargus strigoides phalaenoides*]
Western Gerygone ssp. *Gerygone culicivora dendyi* Mathews, 1912 NCR
[JS *Gerygone fusca fusca*]

Dr Arthur Dendy (1865–1925) was an English zoologist and a world expert on sponges. His bachelor's degree in zoology (1885), master's (1887) and doctorate (1891) were awarded by Victoria University, Manchester. He worked on the team evaluating the reports of the Challenger Expedition (1885), after which he joined the staff of the British Museum. He was appointed to be Demonstrator and Assistant Lecturer in biology at the University of Melbourne (1888). He was Professor of Biology, Canterbury College, University of New Zealand (1893–1902 and 1903–1905) and held the identical post at University of Cape Town, South Africa. He returned to England (1905) as Professor of Biology, King's College, London. He died following an operation for chronic appendicitis. An amphibian is named after him.

Denham

Denham's Bustard *Neotis denhami* **Children** & **Vigors**, 1826

Lieutenant-Colonel Dixon Denham (1786–1828) was an English soldier who explored extensively across Africa. He wrote about this in *Narrative of Travels and Discoveries in Northern and Central Africa* (1826). His exploits included crossing of the Sahara from Tripoli to Lake Chad with Clapperton (q.v.) and (then as Major Denham) returned with records of the expedition, which were published as the *Journal of a Second Expedition into the Interior of Africa* (1829). He was also a fine illustrator. Denham became Governor-General of Sierra Leone, where he died of fever.

Deninger

Buru Honeyeater *Lichmera deningeri* **Stresemann**, 1912

Chestnut-breasted Malkoha ssp. *Phaenicophaeus curvirostris deningeri* Stresemann, 1912
Island Thrush ssp. *Turdus poliocephalus deningeri* Stresemann, 1912

Professor Dr Karl Deninger (1878–1918) was a German zoologist and explorer who was leader of the Second Freiburg Moluccas Expedition (1911–1912) in which Stresemann, whose first wife was Deninger's sister Elizabeth, participated.

Denise

Ochre-breasted Brush Finch ssp. *Atlapetes semirufus denisei* **Hellmayr**, 1911

Louis Simon Denise (1863–1914) was a French ornithologist, poet, bibliographer and art critic. He was co-founder (1909) of the journal *Revue Française d'Ornithologie*.

Dennis

Pacific Robin ssp. *Petroica multicolor dennisi* Cain & I. C. J. Galbraith, 1955

Geoffrey 'Geoff' F. C. Dennis (1918–1995) was an Australian forestry officer and botanical collector in the Solomon Islands (1946–1995). He co-wrote *Palms of the Solomon Islands* (1985).

Dennistoun

Golden-crowned Babbler *Stachyris dennistouni* **Ogilvie-Grant**, 1895
[Syn. *Sterrhoptilus dennistouni*]

John Dennistoun (DNF) was a member of the firm of Glasgow merchants, J & A Dennistoun. He was based in London and was one of those who provided funds for the holotype collector John Whitehead's expedition to the Philippines.

Dent

Dent's Short-tailed Warbler *Sylvietta denti* **Ogilvie-Grant**, 1906
[Alt. Lemon-bellied Crombec]

Apalis sp. *Apalis denti* Ogilvie-Grant, 1907 NCR
[Alt. Buff-throated Apalis; JS *Apalis rufogularis nigrescens*]

Tit-hylia ssp. *Pholidornis rushiae denti* Ogilvie-Grant, 1907

Captain Richard Edward Dent (b.1882) was a British explorer who collected in tropical Africa (1901 and 1906). He became a game warden in Kenya (c.1928–1931) in charge of fisheries, and was one of the very few people to have seen the Marozi or Spotted Lion *Panthera leo maculatus* (a mysterious big cat which is probably a mutation rather than a valid subspecies). He introduced black bass and tilapia into Lake Naivasha (1928). Four mammals are named after him.

Dente

Roosevelt Stipple-throated Antwren *Epinecrophylla dentei* Whitney *et al.*, 2013

Emilio Dente (DNF) was a highly skilled taxidermist. He was at the Zoology Department, Universidade de São Paulo, Brazil (1947) when the 17-year old Bokermann (q.v.) joined the staff as a supernumerary servant labourer. He took Bokermann, who was indigent, under his wing and fed him, as well as teaching him laboratory techniques. They collected together on a number of expeditions, such as in Pará (1955). Dente was at the Belém Virus Laboratory (1963) and worked at the Adolfo Lutz Institute (1974). An amphibian is named after him.

Deplanche

Deplanche's Lorikeet *Trichoglossus haematodus deplanchii* **J. Verreaux** & **Des Murs**, 1860
[Alt. Coconut/New Caledonia Lorikeet]

Emile Deplanche (1824–1875) was a French naval surgeon. He was appointed (1854) as Principal Surgeon on *Le Rapide* and embarked for Cayenne (French Guiana). Here he survived an outbreak of yellow fever, returning to France to convalesce for six months, and was then posted to Tahiti. From Tahiti he sailed to New Caledonia, where he established his reputation as a botanist and collector. He co-wrote *Essais sur la Nouvelle Calédonie* (1863). A reptile and a tree are named after him.

Deppe

Oahu Ou *Psittirostra psittacea deppei* **Rothschild**, 1905 NCR; NRM EXTINCT
Blue-grey Gnatcatcher ssp. *Polioptila caerulea deppei* **Van Rossem**, 1934

Ferdinand Deppe (1794–1861) was a horticulturist, collector and artist. He arrived in Mexico (1824) with Count von Sack, an irresolute 'expedition leader' who soon returned to Germany while Deppe stayed in Mexico (1827). He made a brief visit home, returning to Mexico with botanist Wilhelm Schiede (until 1836). Many of the specimens went to the Berlin Zoology Museum. Four reptiles and a mammal are named after him.

Déprimoz

Crested Lark ssp. *Galerida cristata deprimozi* **Lavauden** 1924 NCR
[JS *Galerida cristata arenicola*]

Jean Déprimoz (d.1922) was a French taxidermist and collector who in 1914 was in India and Ceylon (Sri Lanka) with Guy Babault (q.v.).

Derby

Hummingbird genus *Derbyomyia* **Bonaparte**, 1854 NCR
[Now in *Eriocnemis*]

Chestnut-tipped Toucanet *Aulacorhynchus derbianus* **Gould**, 1835
Derby's Guan *Oreophasis derbianus* **G. R. Gray**, 1844
[Alt. Horned Guan]
Derbian Screamer *Chauna derbiana* G. R. Gray, 1845 NCR
[JS *Chauna chavaria*]
Derby's Puffleg *Eriocnemis derbyi* **Bourcier** & **de Lattre**, 1846
[Alt. Black-thighed Puffleg]
Derbyan Parakeet *Psittacula derbiana* **L. Fraser**, 1852
[Alt. Lord Derby's Parakeet]
Grey-backed Tailorbird *Orthotomus derbianus* F. Moore, 1855

Pauraque ssp. *Nyctidromus albicollis derbyanus* Gould, 1838
Derby's Flycatcher *Pitangus sulphuratus derbianus* **Kaup**, 1852
[Alt. Great Kiskadee ssp.]

Edward Smith Stanley (1775–1851) was 13th Earl of Derby; two mammals are named after him. (See **Stanley** and **Lord Derby**).

Derby (Town)

Australian Bustard ssp. *Ardeotis australis derbyi* **Mathews**, 1912 NCR; NRM
Australian Darter ssp. *Anhinga novaehollandiae derbyi* Mathews, 1912 NCR; NRM
Galah ssp. *Eolophus roseicapilla derbyana* Mathews, 1912 NCR
[JS *Eolophus roseicapilla kuhli*]
Little Woodswallow ssp. *Artamus minor derbyi* Mathews, 1912
Red-headed Honeyeater ssp. *Myzomela erythrocephala derbyi* Mathews, 1912 NCR
[JS *Myzomela erythrocephala erythrocephala*]
Buff-sided Robin ssp. *Poecilodryas superciliosa derbyii* Mathews, 1913 NCR
[JS (monotypic) *Poecilodryas cerviniventris*]
Western Corella ssp. *Cacatua pastinator derbyi* Mathews, 1916

These birds are all named after the town of Derby, Western Australia.

De Rham

De Rham's Garnet *Lamprolaima rhami* **Lesson**, 1839
[Alt. Garnet-throated Hummingbird]

Henri Casimir de Rham (1785–1873) was a Swiss merchant, diplomat and amateur naturalist who collected in the US. He emigrated from Switzerland to New York (1805), where he started a successful business in the cotton trade, becoming first Swiss Consul there (1832–1846). He founded (1832) the Swiss Benevolent Society of New York. He also had a ship named after him. (See also **Rham**)

Derjugin

Coal Tit ssp. *Periparus ater derjugini* **Zarudny** & **Loudon**, 1903

Professor Dr Konstantin Michailovich Derjugin (1878–1938) was a hydrobiologist at Leningrad State University. A reptile is named after him.

Dernedde

Dernedde's Hummingbird *Uranomitra derneddei* **Simon**, 1911 NCR
[Alt. Violet-crowned Hummingbird; JS *Amazilia violiceps*]

Professor Karl Dernedde (1863–1943) was a German zoologist.

De Roepstorff

Andaman Masked Owl *Tyto deroepstorffi* **Hume**, 1875

Frederick Adolph de Roepstorff (1842–1883) was a member of the Indian Survey and Forests Department of the Indian Civil Service. He was in the Andaman and Nicobar Islands (1875–1883), becoming officer in charge at Nancowry Harbour (1882–1883). He was very interested in anthropology and linguistics and wrote *Vocabulary of the Dialects Spoken in the Nicobar and Andaman Isles* (1875).

De Roo

Blue-headed Sunbird ssp. *Cyanomitra alinae derooi* **Prigogine**, 1975

Antoon Emeric Marcel De Roo (1936–1971) was a Belgian naturalist. He co-wrote 'Contribution à l'étude des ciroptères de la République du Togo' (1969). A mammal and an amphibian are named after him.

De Schauensee

Plain Chachalaca ssp *Ortalis vetula deschauenseei* **Bond**, 1936
De Schauensee's Shrike Babbler *Pteruthius aeralatus schauenseei* **Deignan**, 1946
[Alt. Blyth's Shrike Babbler ssp.]
Blackish Cuckooshrike ssp. *Coracina coerulescens deschauenseei* **J. E. duPont**, 1972

Dr Rodolphe Meyer de Schauensee (1901–1984) was an ornithologist, Curator and Director, Academy of Natural Sciences, Philadelphia. He went on many expeditions, including at least three to Siam (Thailand), where he collected anything and everything. He wrote, among a considerable output, *The Birds of Colombia* (1964), *The Species of Birds of South America with their Distribution* (1966), *A Guide to the Birds of America* (1971) and *The Birds of China* (1984). Two reptiles are named after him. (Also see **Meyer de Schauensee, Rodolphe** and **Schauensee**)

Desfontaines

European Stonechat ssp. *Saxicola torquata desfontainesi* **Blanchet**, 1925 NCR
[JS *Saxicola rubicola*]

René Louiche Desfontaines (1750–1833) was a French botanist and collector in Tunisia and Algeria (1783–1785). He became Professor of Botany at Le Jardin des Plantes, Paris (1786) and later Director, MNHN.

Desgodins

Black-headed Sibia *Heterophasia desgodinsi* **Oustalet**, 1877

Abbé Auguste Desgodins (1826–1913) was a French missionary in Tibet (1855–1870). He created a dictionary, *Dictionnaire Thibétain-Latin-Français* (1899), as well as writing up the mission with his brother Charles H. Desgodins in *La Mission du Thibet de 1855 à 1870* (1872). Other taxa, particularly insects, are named after him.

Desmarest

Brassy-breasted Tanager *Tangara desmaresti* **Vieillot**, 1819
Desmarest's Fig Parrot *Psittaculirostris desmarestii* Desmarest, 1826
[Alt. Large Fig Parrot]

European Shag ssp. *Phalacrocorax aristotelis desmarestii* Payraudeau, 1826

Anselme Gaetan Desmarest (1784–1838) was a French zoologist and palaeontologist. He wrote *Histoire Naturelle des Tangaras, des Manakins et des Todiers* (1805) and *Considérations Générales sur la Classe des Crustacés – et Description des Espèces de ces Animaux, qui Vivent dans la Mer, sur les Côtes, ou dans les Eaux Douces de la France* (1825) and was co-author of the *Dictionnaire des Sciences Naturelles* (insects and crustaceans) (1816–1830). Many other taxa including four mammals are named after him.

Des Murs

Des Murs's Wiretail *Sylviorthorhynchus desmursii* Des Murs, 1847
[Alt. Des Murs's Spinetail]
Des Murs's Antbird *Hylophylax punctulatus* Des Murs, 1856
[Alt. Dot-backed Antbird]

Darjeeling Woodpecker ssp. *Dendrocopos darjellensis desmursi* **J. Verreaux**, 1870
[*D. darjellensis* sometimes regarded as monotypic]

Marc Athanase Parfait Oeillet Des Murs (1804–1894) was a French historian, local politician and amateur ornithologist. He was a magistrate (1830–1838) and became a lawyer (1841) but retired shortly afterwards (1846) to restore Château St Jean, which he had bought (1843) and which he eventually sold (1885) having spent most of his fortune on it. He was Mayor of Nogent-le-Rotrou (1860–1868). He wrote several papers and books including with Jules Verreaux (q.v.) and Florent Prévost (q.v.), his most important ornithological work being *Iconographie Ornithologique* (1849).

Deville

Oropendola sp. *Cassicus devillii* **Bonaparte**, 1850 NCR
[Alt. Olive Oropendola; JS *Psarocolius* (*bifasciatus*) *yuracares*]
Deville's Parakeet *Pyrrhura devillei* Massena & Souancé, 1854
[Alt. Blaze-winged Parakeet]
Trogon sp. *Aganus devillei* **Cabanis** & **Heine**, 1863 NCR
[Alt. Black-throated Trogon; JS *Trogon rufus sulphureus*]
Parakeet sp. *Conurus devillei* G. R. Gray, 1870 NCR
[Invalid name: replaced by *Brotogeris cyanoptera*]
Deville's Hawk Eagle *Spizaetus devillei* **A. J. C. Dubois**, 1874 NCR
[Alt. Black-and-chestnut Eagle; JS *Spizaetus isidori*]
Striated Antbird *Drymophila devillei* Menegaux & **Hellmayr**, 1906

Berylline Hummingbird ssp. *Amazilia beryllina devillei* **Bourcier** & **Mulsant**, 1848
Cinnamon-throated Woodcreeper ssp. *Dendrexetastes rufigula devillei* **Lafresnaye**, 1850
Saffron-billed Sparrow ssp. *Arremon flavirostris devillii* **Des Murs**, 1856

Emile Deville (1824–1853) was a French naturalist who collected in Latin America (1843–1847). By order of King Louis-Philippe, Castelnau (q.v.) accompanied the botanist H. A. Weddell (q.v.), the geologist d'Osery (q.v.), and Deville to Brazil and Peru. An amphibian and other taxa are also named after him.

The suggestions that the birds are named after Charles Joseph Sainte-Claire Deville (1814–1876), a geologist and meteorologist, or his brother Etienne (1818–1881), strike us as tenuous as, although Charles Deville may have been born in St Thomas in the West Indies, where his father was the French Consul, we cannot trace any connection with the South American mainland and he appears to have spent his professional life in France. Etienne was a chemist with no links to natural history.

De Vis

Fairy-wren genus *Devisornis* **Mathews**, 1917 NCR
[Now in *Malurus*]
Flyrobin genus *Devioeca* Mathews, 1925 NCR
[Now in *Microeca*]

De Vis's Thornbill *Acanthiza murina* De Vis, 1897
[Alt. Papuan Thornbill]

De Vis's Bird-of-Paradise *Paradisaea raggiana intermedia* De Vis, 1894
[Alt. Raggiana Bird-of-Paradise ssp.]
Dimorphic Fantail ssp. *Rhipidura brachyrhyncha devisi* **North**, 1898

Charles Walter De Vis (1829–1915) was a British-born cleric who gave up the Church to concentrate on ornithology. He was the first Director of the Queensland Museum (1882–1905), before which he published many popular articles under the pen name 'Thickthorn'. He was a founder member of the Royal Society of Queensland (1884) and its President (1888–1889). He was a founder member and first Vice-president of the Australasian Ornithologists' Union (1901). He described 551 new fossil and living species, including the thornbill. Five reptiles and two mammals are named after him. (See **Vis**)

Devron

Parrot-billed Seedeater ssp. *Sporophila peruviana devronis* **J. Verreaux**, 1852

Monsieur Devron (fl.1850) was a French collector from whom Verreaux (q.v.) acquired the type specimen, but nothing more seems to be known about him.

De Wet

Lark genus *Dewetia* **Buturlin**, 1904 NCR
[Now in *Spizocorys*]

Commandant-General Christiaan Rudolf De Wet (1854–1922) was a General in the Orange Free State Army. He served in both Anglo-Boer Wars (1880–1881 and 1899–1902). He was one of the signatories to the Treaty of Vereeniging (1902) and entered politics after the peace, but was involved in a rebellion (1914), fined and imprisoned.

De Witte

Long-billed Pipit ssp. *Anthus similis dewittei* **Chapin**, 1937
White-throated Francolin ssp. *Peliperdix albogularis dewittei* Chapin, 1937

Dr Gaston-François de Witte (1897–1980) was in the Belgian Congo (Democratic Republic of Congo) (1933–1935 and 1946–1949). He was originally a colonial administrator, but worked as a naturalist, becoming a collector for the Institut des Parcs Nationales Congo-Belge in Tervuren (1938). Two reptiles and five amphibians are named after him.

Dharmakumarsinhji

Small Minivet ssp. *Pericrocotus peregrinus dharmakumari* **Koelz**, 1950 NCR
[JS *Pericrocotus cinnamomeus cinnamomeus*]
Oriental Skylark ssp. *Alauda gulgula dharmakumarsinhjii* **Abdulali**, 1976

Raof Shri Dharmakumarsinhji (1917–1986) was Prince of Bhavnagar State, Gujarat, India, and an ornithologist. He wrote *Birds of Saurashtra* (1957).

Diamond

Diamond's Paradise-crow *Phonygammus keraudrenii diamondi* Cracraft, 1992
[Alt. Trumpet Manucode; Syn. *Manucodia keraudrenii diamondi*]
Mountain Mouse-warbler ssp. *Crateroscelis robusta diamondi* **Beehler** & Prawiradilaga, 2010

Dr Jared Mason Diamond (b.1937) is an American physiologist who is a Professor at the Medical School of University College Los Angeles, California. He has also been a Research Associate in Ornithology and Mammalogy at the Los Angeles County Museum of Natural History (1985). He is interested in nutrition and ornithology. He has published extensively, including detailed studies of the avifauna of Papua New Guinea and the Philippines. His studies on bird diversity in New Guinea's tropical rainforests have made fundamental contributions to our understanding of species's coexistence, altitudinal segregation of montane species, speciation in rainforest environments and bowerbird evolution. He won the Tyler Prize for Environmental Achievement for pioneering work in conservation biology (2001), has a MacArthur Foundation award, and won the Pulitzer Prize (1998) for his book *Guns, Germs and Steel*. A mammal and an amphibian are named after him.

Diana

Sunda Blue Robin *Myiomela diana* **Lesson**, 1831
[Syn. *Cinclidium diana*]

In Roman mythology Diana was the virgin goddess of the chase, associated with wild animals and woodlands. She was also a goddess of the moon. Her name is commemorated in other taxa, including a reptile and two mammals.

Diard

Pheasant genus *Diardigallus* **Bonaparte**, 1856 NCR
[Now in *Lophura*]

Black-bellied Malkoha *Phaenicophaeus diardi* **Lesson**, 1830
Diard's Trogon *Harpactes diardii* **Temminck**, 1832
Diard's Fireback *Lophura diardi* Bonaparte, 1856
[Alt. Siamese Fireback]

Pierre Medard Diard (1795–1863) was a French naturalist and explorer who collected in the East Indies (1827–1848), often in collaboration with Alfred Duvaucel (q.v.). A mammal and a reptile are named after him.

Diaz

Mexican Duck *Anas diazi* **Ridgway**, 1886

Augustin Diaz (1829–1893) was a Mexican explorer, geographer and military engineer. He was Director of the Mexican Geographical and Exploring Commission. A mammal is named after him.

Dickerman

American Dipper ssp. *Cinclus mexicanus dickermani*
    A. R. Phillips, 1966
Lesser Greenlet ssp. *Hylophilus decurtatus dickermani*
    Parkes, 1991
Streak-backed Oriole ssp. *Icterus pustulatus dickermani*
    A. R. Phillips, 1995

Dr Robert William Dickerman (b.1926) is an American physician, ornithologist and collector. He was on the faculty at the University of Minnesota, Museum of Natural History (1963), at the Western Foundation of Vertebrate Zoology (1987) and

at the Museum of Southwestern Biology, University of New Mexico, Albuquerque (1994). He wrote *The Song Sparrows of the Mexican Plateau* (1963).

Dickey, D. R.

Dickey's Jay *Cyanocorax dickeyi* **R. T. Moore**, 1935
[Alt. Tufted Jay]

Dickey's Egret *Egretta rufescens dickeyi* **Van Rossem**, 1926
[Alt. Reddish Egret ssp.]
Common Poorwill ssp. *Phalaenoptilus nuttallii dickeyi*
    Grinnell, 1928
Turquoise-browed Motmot ssp. *Eumomota superciliosa dickeyi* **Griscom**, 1929 NCR
[JS *Eumomota superciliosa apiaster*]
Spot-bellied Bobwhite ssp. *Colinus leucopogon dickeyi*
    Conover, 1932
Highland Guan ssp. *Penelopina nigra dickeyi* Van Rossem, 1934 NCR; NRM

Donald Ryder Dickey (1887–1932) was an American ornithologist, collector and photographer. Born in Iowa he lived most of his life in California. He took 7,000 black-and-white photographs and collected 50,000 specimens of birds and mammals (1908–1923). These form the Donald Ryder Dickey Collection, which his widow presented to University College Los Angeles. His publications were many, the most notable on birds being *The Birds of El Salvador* (1938), co-written with A. J. Van Rossem (q.v.). Two mammals are named after him.

Dickey, F.

Varied Bunting ssp. *Passerina versicolor dickeyae*
    Van Rossem, 1934
Audubon's Oriole ssp. *Icterus graduacauda dickeyae* Van Rossem, 1938

Florence Dickey *née* Murphy (fl.1932) was the widow of Donald Ryder Dickey (see above). She wrote *Familiar Birds of the Pacific Southwest* (1935).

Dickinson, E.

Coral-billed Scimitar Babbler ssp. *Pomatorhinus ferruginosus dickinsoni* Eames, 2002

Edward C. Dickinson (b.1938) is a British businessman (until 1989) and ornithologist, who is now a Research Associate of the National Museum of Natural History, Leiden. He was co-founder of the Trust for Oriental Ornithology (now the Trust for Avian Systematics). He co-wrote *Field Guide to the Birds of Southeast Asia* (1993), *A Guide to the Birds of Philippines* (2000) and *Priority! The Dating of Scientific Names in Ornithology* (2011) as well as being managing editor of the third and fourth editions of the *Howard & Moore Complete Checklist of the Birds of the World* (2003, 2013–14).

Dickinson, J.

Dickinson's Kestrel *Falco dickinsoni* **P. L. Sclater**, 1864

Dr John Dickinson (1832–1863) was an English physician and missionary. He joined David Livingstone but died of blackwater fever in Nyasaland (Malawi). He collected the holotype of the kestrel.

Didi

> Striated Heron ssp. *Butorides striatus didii* **W. W. A. Phillips** & Sims, 1958 NCR
> [JS *Butorides striata albolimbata*]

Al Amir Ibrahim Fa'amuladeri Kilegefa'anu, later known as Ekgamuge Ibrahim Ali Didi (c.1933–1975), was a Maldivian politician and member of the royal family. He was Prime Minister of his country (1953–1957).

Dido

> Southern Marquesan Reed Warbler ssp. *Acrocephalus mendanae dido* **R. C. Murphy** & **Mathrews**, 1928

In Virgil's *Aeneid* Dido was the founder and queen of Carthage. The description gives no reasoning for this name choice.

Dieffenbach

> Dieffenbach's Rail *Gallirallus dieffenbachii* **G. R. Gray**, 1843 EXTINCT

Johann Karl Ernst Dieffenbach (1811–1855) was a German naturalist and physician. He travelled widely in New Zealand, writing *Travels in New Zealand* (1843), which contain descriptions of birds including the rail. He made the first translation into German of Darwin's *The Voyage of the Beagle* (1844). Dieffenbach and James Heberley became the first Europeans to climb Mount Taranaki (1839). He died of typhus.

Diesing

> Plush-crested Jay ssp. *Cyanocorax chrysops diesingii* **Pelzeln**, 1856

Dr Karl Moriz Diesing (1800–1867) was an Austrian zoologist, naturalist and helminthologist. He wrote *Revision der Nematoden* (1861).

Dietrichsen

> Dietrichsen's Lory *Glossopsitta porphyrocephala* Dietrichsen, 1837
> [Alt. Purple-crowned Lorikeet]

Lionel Lorenzo Dietrichsen (1806–1846) was a British ornithologist as well as a 'dealer in patent medicines and perfumery' in London's Oxford Street. He advertised in the *Literary Gazette* (1830) that he had a large quantity of foreign bird skins for sale. He described the lory (1832) as *Psittacus purpureus*, giving rise to the eponym. However, the name was considered invalid having been used earlier by Müller to describe another species, so (1837) Dietrichsen re-described it as *Trichoglossus porphyrocephalus*. He is the same Lionel Dietrichsen reported as having committed suicide by cutting his throat. According to press reports 'He had been engaged in a chancery suit, and had latterly evinced a lowness of spirits and eccentricity of manner resulting, as was supposed from his too close application of study.'

Diggles

> Papuan Pitta ssp. *Erythropitta macklotii digglesi* G. Kreft, 1869
> Diggles's Finch *Poephila cincta atropygialis* Diggles, 1876
> [Alt. Black-throated Finch ssp.]
> Chocolate Diggles's Finch *Poephila cincta nigrotecta* **Hartert**, 1899 NCR
> [NUI *Poephila cincta atropygialis*]

Silvester Diggles (1817–1880) was an Australian piano tuner, artist and teacher of drawing and music, as well as a pioneer entomologist and amateur ornithologist. He wrote *The Ornithology of Australia* (1866–1870), which was illustrated with the help of his niece, Rowena Birkett. There is a published volume of his correspondence with E. P. Ramsay (q.v.).

Dillon

> Dillon's Eagle Owl *Bubo capensis dillonii* **des Murs** & **Prévost**, 1846
> [Alt. Cape Eagle Owl ssp.]

> Speckled Pigeon ssp. *Columba guinea dilloni* **Bonaparte**, 1854 NCR
> [NUI *Columba guinea guinea*]

Dr Léon Richard Quartin-Dillon (d.1840) was a French physician, botanist, hunter and explorer in Ethiopia (1839–1840). He died of an illness and one of his collecting partners, Antoine Petit, was taken and drowned by a crocodile in the Tacazze River (1843). The botanical genus *Quartinia* is named after him.

Dillon Ripley

> Clapper Rail ssp. *Rallus longirostris dillonripleyi* **W. H. Phelps Jr** & **Aveledo**, 1987 NCR
> [NUI *Rallus longirostris phelpsi*]

(See under **Ripley**)

Dillwynn

> Dwarf Kingfisher sp. *Ceyx dillwynni* **Sharpe**, 1868 NCR
> [Alt. Oriental Dwarf Kingfisher; JS *Ceyx erithaca*]

Lewis Llewellyn Dillwyn (1814–1892) was a member of the family that owned the company that manufactured the famous Swansea Pottery, where he started work (1831). He was Mayor of Swansea (1847) and its Member of Parliament (1855–1892). He collected in Borneo (1854) and co-wrote *Contributions to the Natural History of Labuan, and the adjacent coasts of Borneo* (1855).

Dinelli

> Dinelli's Doradito *Pseudocolopteryx dinellianus* **Lillo**, 1905

> Variable Antshrike ssp. *Thamnophilus caerulescens dinellii* **Berlepsch**, 1906

Luis M. Dinelli was a naturalist and collector working in Argentina (1904–1939). He collected over 300 bird skins in Tucumán province (1904) which are now in the National Museum in Buenos Aires. He wrote 'Notas biológicas sobre las aves del noroeste de la República Argentina' (1918).

Dinemelli

Buffalo Weaver genus *Dinemellia* **Reichenbach**, 1863

White-headed Buffalo Weaver *Dinemellia dinemelli*
Rüppell, 1845

Dinemelli (DNF) was a collector in Ethiopia, which is all we know about him.

Diomedes

Albatross genus *Diomedea* **Linnaeus**, 1758

Scopoli's Shearwater *Calonectris diomedea* **Scopoli**, 1769

In Greek mythology Diomedes was a warrior in the Trojan War. On his death, his companions were turned into seabirds.

Dixon, C.

Long-tailed Thrush *Zoothera dixoni* **Seebohm**, 1881

Charles Dixon (1858–1926) was a British oologist, journalist and author. He wrote *Our Rarer Birds: being Studies in Ornithology and Oology* (1888).

Dixon, J. S.

Dixon's Ptarmigan *Lagopus muta dixoni* **Grinnell**, 1909
[Alt. Rock Ptarmigan ssp.]
White-winged Scoter ssp. *Melanitta deglandi dixoni*
W. S. Brooks, 1915 NCR
[JS *Melanitta deglandi deglandi*]

Joseph Scattergood Dixon (1884–1952) was an ornithologist who accompanied Miss Annie Alexander on the expeditions to collect birds in Alaska (1907–1908), which she financed and led. He was employed by the Museum of Vertebrate Zoology at Berkeley as Assistant Curator of Mammals and Economic Ornithology before leaving to work for the National Park Service (1931).

Dobson

Andaman Cuckooshrike *Coracina dobsoni* **Ball**, 1872

George Edward Dobson (1848–1895) was an Irish zoologist. He was an army surgeon (1868–1888) who served in India (1868) and in the Andaman Islands (1872–1876) before returning to England, retiring at the rank of surgeon-major. He was an expert on small mammals, especially Chiroptera and Insectivora. He published several articles and papers, including two on the Andamanese people (1875–1877). He also took their pictures, as he was an early aficionado of photography. He wrote a monograph on Asian bats (1876). He also published a collection of medical hints for travellers. He was appointed as Curator of the Royal Victoria Hospital's museum (1878). Eight mammals and an amphibian are named after him.

Dod

La Selle Thrush ssp. *Turdus swalesi dodae* **Graves** & **Olson**, 1986

Annabelle Jean 'Tudy' Dod *née* Stockton (1913–1997) was an American missionary, ornithologist and educationist in Puerto Rico (1946–1963) and the Dominican Republic (1964–1988). She wrote *Endangered and Endemic Birds of the Dominican Republic* (1992).

Dodge

Bornean Swiftlet *Collocalia dodgei* **Richmond**, 1905
[Syn. *Collocalia linchi dodgei*]

H. D. Dodge (DNF) of Connecticut explored in northern Borneo with George A. Goss (1904). Richmond writes: 'In the small collection of birds presented to the US National Museum by Messrs. Goss and Dodge, and obtained by them during their recent expedition to Mount Kina Balu, north Borneo, is a single example of a small swiftlet that I cannot identify with any described species.'

Dodson, E.

Dodson's Bulbul *Pycnonotus* (*barbatus*) *dodsoni* **Sharpe**, 1895

Southern Grey Shrike ssp. *Lanius meridionalis dodsoni*
Whitaker, 1898 NCR
[JS *Lanius meridionalis algeriensis*]
Familiar Chat ssp. *Cercomela familiaris dodsoni*
J. D. Macdonald, 1953 NCR
[JS *Cercomela familiaris galtoni*]

Edward Dodson (1872–1948), brother of W. Dodson (q.v.), was a taxidermist working at the BMNH. He undertook a number of collecting expeditions in the Middle East, Morocco, Libya and Somalia in the late 19th century. He accompanied Donaldson-Smith (q.v.) on an expedition (1894–1895) to Somaliland (Somalia) during which they succeeded in reaching Lake Rudolph. With the ornithologist Arthur Blayney Percival (q.v.) he took part in a Royal Society expedition to Arabia (1898–1899), in the employ of Joseph Whitaker (q.v.). He also accompanied expeditions to Patagonia (Argentina) (1899) and Tripoli (Libya) (1901–1902). Thereafter his health broke down and he had to abandon further planned trips. He was also a skilled engineer and inventor of the 'Zenith' carburettor and the special laminated wooden propellor for WW1 aircraft. An amphibian is named after him.

Dodson, W.

Spotted Thick-knee ssp. *Burhinus capensis dodsoni*
Ogilvie-Grant, 1899
[Alt. Spotted Dikkop ssp.]

W. Dodson (1873–1948), brother of Edward Dodson (q.v.) was a traveller, collector and taxidermist on the Royal Society expedition to Arabia (1899) with Blayney Percival, as well as in tropical Africa. He died of fever in Aden.

Doering

Andean Tinamou ssp. *Nothoprocta pentlandii doeringi*
Cabanis, 1878

Dr Adolf Döring (1848–1925) was a German zoologist and geologist who lived in Argentina (1872–1925) and became Professor of Zoology, University of Cordoba (1878).

Doerries

Blakiston's Fish Owl ssp. *Bubo blakistoni doerriesi*
   Seebohm, 1895
Eurasian Eagle Owl ssp. *Bubo bubo doerriesi* **Buturlin**, 1910
   NCR
[NPRB *Bubo bubo ussuriensis*]
Grey-capped (Pygmy) Woodpecker ssp. *Dendrocopos
   canicapillus doerriesi* **Hargitt**, 1881
Common Kestrel ssp. *Falco tinnunculus doerriesi* Swann,
   1920 NCR
[JS *Falco tinnunculus tinnunculus*]

Fritz Doerries (1851–1953) was a German lepidopterist,
collector and explorer in Siberia. He normally collected with
his brother.

Doflein

Green Woodpecker ssp. *Picus viridis dofleini* **Stresemann**,
   1919 NCR
[JS *Picus viridis karelini*]

Dr Franz Theodor Doflein (1873–1924) was a German zoologist, ichthyologist and herpetologist. He studied natural
sciences at the universities of Strasbourg and Munich (1893–
1898). He joined the Bavarian State Collection of Zoology,
Munich (1901), and became Professor, Department of
Zoology, University of Freiburg (1912) and at the University of
Breslau (Wroclaw, Poland) (1918). He made a number of
collecting expeditions, to Central America and USA (1898),
China, Japan (where he made a notable collection of marine
specimens in Sagami Bay) and Ceylon (Sri Lanka) (1904–
1905), and Macedonia (1917–1918), when he was attached to
the German Army. He wrote *Lehrbuch der Protozoenkunde*
(1906). An amphibian is named after him.

Doggett, F.

Northern Cassowary ssp. *Casuarius unappendiculatus
   doggetti* **Rothschild**, 1904 NCR; NRM

Frederick Doggett (DNF) in the 1871 census is described as a
naturalist living in Cambridge. He was later described as the
Cambridge animal doctor and taxidermist, and Rothschild
wrote that he 'has charge of the living birds.'

Doggett, W. G.

Amethyst Sunbird ssp. *Chalcomitra amethystina doggetti*
   Sharpe, 1902

Walter Grimwood Doggett (1876–1904) was a collector and
naturalist in East Africa (1899–1903). He was working (1901)
for Sir Harry H. Johnston (q.v.), Special Commissioner for
Uganda, who records being accompanied by Doggett on a
mission to collect the skin of an okapi, and on another occasion the pair of them shooting a 'five-horned giraffe'.
Ogilvie-Grant wrote 'On the birds collected by the late W. G.
Doggett on the Anglo German frontier of Uganda' (1905). He
was drowned in the Kagera River in Uganda. A mammal is
named after him.

Doherty

Chestnut-backed Thrush *Zoothera dohertyi* **Hartert**, 1896
Doherty's Fruit Dove *Ptilinopus dohertyi* **Rothschild**, 1896
[Alt. Red-naped Fruit Dove]
Doherty's Greybird *Coracina dohertyi* Hartert, 1896
[Alt. Pale-shouldered/Sumba Cicadabird]
Doherty's White-eye *Lophozosterops dohertyi* Hartert, 1896
[Alt. Crested White-eye]
Sula Pitta *Erythropitta dohertyi* Rothschild, 1898
[Syn. *Erythropitta erythrogaster dohertyi*]
Doherty's Bush-shrike *Telophorus dohertyi* Rothschild, 1901
Drongo sp. *Dicrurus dohertyi* Hartert, 1902 NCR
[Alt. Hair-crested Drongo; JS *Dicrurus hottentottus
   guillemardi*]

Variable Pitohui ssp. *Pitohui kirhocephalus dohertyi*
   Rothschild & Hartert, 1903
Yellow-bellied Gerygone ssp. *Gerygone chrysogaster
   dohertyi* Rothschild & Hartert, 1903
Doherty's Hanging Parrot *Loriculus philippensis dohertyi*
   Hartert, 1906
[Alt. Philippine Hanging Parrot ssp.]
Rufous-naped Lark ssp. *Mirafra africana dohertyi* Hartert,
   1907 NCR
[JS *Mirafra africana athi*]
Plain Martin ssp. *Riparia paludicola dohertyi* Hartert, 1910
   NCR
[JS *Riparia paludicola ducis*]

William Doherty (1857–1901) was an American collector,
regarded by Rothschild as the best he had ever employed.
Before becoming a collector he travelled (1877) through
Europe, Turkey and into Palestine and Egypt and thence to
Persia (Iran) (1881). He started (1882) collecting entomological specimens seriously, to bolster his finances, and
(1882–1883) he roamed through India, Burma (Myanmar), the
Malay Archipelago and was in the far reaches of Indonesia
(1887). He went to England, to visit Hartert at Tring (1895), and
it was there that he met Lord Rothschild, who recruited him
as a bird collector. An article by B. Verdcourt in *The
Conchologists' Newsletter* (1992) mentions him collecting
molluscs in East Africa, where, in Nairobi, he died of
dysentery. (See also **William Doherty**)

Dohrn, H. W. L.

Dohrn's Flycatcher *Horizorhinus dohrni* **Hartlaub**, 1866
[Alt. Dohrn's Thrush-babbler, Principe Flycatcher-babbler]
Dohrn's Warbler *Acrocephalus brevipennis* Keulemans,
   1866
[Alt. Cape Verde Cane/Swamp Warbler]

Heinrich Wolfgang Ludwig Dohrn (1838–1913) was a German
politician and entomologist. His family came from Pommern,
part of which is now in Poland. He studied entomology at
Stettin (Szczecin) where he graduated (1858). He collected
on the island of Príncipe (1865). His father was Dr Karl
Augustus Dohrn (q.v.) and his younger brother was Felix
Anton Dohrn (1840–1909), the founder of the Stazione
Zoologica di Messina 'Anton Dohrn' and whose godfather
was Felix Mendelssohn, the composer.

Dohrn, K. A.

Dohrn's Hermit *Glaucis dohrnii* **Bourcier** & **Mulsant**, 1852
[Alt. Hook-billed Hermit; Syn. *Ramphodon dohrnii*]

Dr Karl Augustus Dohrn (1806–1892) was a German who inherited a fortune from his grandfather, a sugar, spice and wine entrepreneur in Stettin. He was well able to afford to follow his hobbies, which included entomology. He had at least two sons, Heinrich Wolfgang Ludwig Dohrn (q.v.) and Felix Anton Dohrn.

Dolan

Dolan's Eared Pheasant *Crossoptilon crossoptilon dolani* **de Schauensee**, 1937
[Alt. White Eared Pheasant ssp.]

Brooke Dolan II (1908–1945) was an American naturalist who studied at Princeton and Harvard. His early adult years were spent as an Asian explorer, making several expeditions to Siberia, Sichuan (China), and Tibet. His first Tibetan expedition was with Ernst Schäfer (1931–1932) and he returned to the Tibetan borderlands (1934–1935). The results were published by others as 'Zoological results of the second Dolan expedition to western China and eastern Tibet, 1934–1936' (1938). He joined the Army Air Force after the Japanese bombing of Pearl Harbor. His expedition to Lhasa, Tibet, from India (1942–1943) was as a captain and was funded by the Office of Strategic Services (OSS), since Franklin D. Roosevelt wanted to explore the possibility of moving military supplies to Chiang Kai-shek's Republican Chinese government via Tibet. Lieutenant Colonel Ilya Tolstoy, grandson of the Russian author Leo Tolstoy, accompanied him. Dolan took a signed photograph of Roosevelt as a gift for the young Fourteenth Dalai Lama. After this he joined the United States Military Observer Group in Yunnan. 'He was killed while attempting the rescue of Allied bomber crews downed behind enemy lines in Chongking while on a mission for the OSS' (Hoffman 1983).

Dole

Dole's Flycatcher *Chasiempis dolei* **Stejneger**, 1887 NCR
[Alt. Kauai Elepaio; JS *Chasiempis sclateri*]
Akohekohe *Palmeria dolei* **SB Wilson**, 1891
[Alt. Crested Honeycreeper]

Sanford Ballard Dole (1844–1926) was an American (Hawaiian) politician and judge who was President of the Hawaiian Republic (1893–1898). He and a cousin started the pineapple business in Hawaii, and his family is still well known in the fruit trade.

Dolgushin

Sand Martin ssp. *Riparia riparia dolgushini* **Gavrilov** & Savtchenko, 1991 NCR
[NUI *Riparia riparia riparia*]

Professor Igor Alexandrovich Dolgushin (1908–1966) was a Russian ornithologist and collector in Kazakhstan (1946).

Domaniewski

Puffleg sp. *Vestipedes domaniewskii* **Stolzmann**, 1926 NCR
[Alt. Buff-thighed Puffleg; JS *Haplophaedia assimilis affinis*]

Eurasian Nuthatch ssp. *Sitta europaea domaniewskii* Dunajewski, 1934 NCR
[JS *Sitta europaea europaea*]

Professor Janusz Domaniewski (1891–1954) was a Polish ornithologist, zoogeographer and environmentalist. He was Professor of Zoology at Lublin University and later Director of the National Museum of Zoology in Warsaw.

Dombrain

Australian Masked Owl ssp. *Tyto novaehollandiae dombraini* **Mathews**, 1914 NCR
[JS *Tyto novaehollandiae novaehollandiae*]
Ground Parrot ssp. *Pezoporus wallicus dombraini* Mathews, 1914 NCR
[JS *Pezoporus wallicus wallicus*]
Turquoise Parrot ssp. *Neophema pulchella dombraini* Mathews, 1915 NCR; NRM

Dr Ernest Arthur D'Ombrain (1867–1944) was an Irish/Australian ophthalmologist, ornithologist and collector. He arrived in Australia (1877) and qualified as a physician at the University of Melbourne, then studied in England and Vienna. He practised at Casterton, Victoria (1898–1910), and as an eye specialist in Sydney, NSW (1910–1928). He was a founding member of the RAOU (1901) and its President (1927–1928).

Dombrowski

Dombrowski's Yellow Wagtail *Motacilla flava dombrowskii* **Tschusi**, 1903 NCR
[Inter-racial hybrid of *Motacilla flava* subspecies]

Robert Ritter von Dombrowski (1869–1932) was an Austrian ornithologist at the Natural History Museum, Bucharest (1895–1916).

Domvile

Orange-breasted Green Pigeon ssp. *Treron bicinctus domvilii* **Swinhoe**, 1870

Admiral Sir Compton Edward Domvile (1842–1924) was a British naval officer (1856–1905) who fought against pirates in China (1866–1868). He was Commodore in Jamaica (1882), Director of Naval Ordnance (1891–1894) and Commander-in-Chief, Mediterranean Station (1902–1905).

Don

Scaly Francolin ssp. *Pternistis squamatus doni* **C. W. Benson**, 1939

P. J. Don (b.1895) was a South African laboratory assistant at Witwatersrand (1930). He collected in Nyasaland (Malawi) (1936).

Donaldson Smith

Donaldson Smith's Nightjar *Caprimulgus donaldsoni*
Sharpe, 1895

Donaldson Smith's Sparrow Weaver *Plocepasser
donaldsoni* Sharpe, 1895

Northern Grosbeak Canary *Serinus donaldsoni* Sharpe, 1895
[Syn. *Crithagra donaldsoni*]

Donaldson's Turaco *Tauraco leucotis donaldsoni* Sharpe,
1895
[Alt. White-cheeked Turaco ssp.]

Rüppell's Robin Chat ssp. *Cossypha semirufa donaldsoni*
Sharpe, 1895

Olive Bee-eater ssp. *Merops superciliosus donaldsoni*
Oberholser, 1904 NCR
[JS *Merops superciliosus superciliosus*]

Golden-breasted Starling ssp. *Cosmopsarus regius
donaldsoni* **van Someren**, 1919 NCR
[JS *Lamprotornis regius regius*]

Dr Arthur Donaldson Smith (1864–1939) was a traveller and
big-game hunter of American birth who spent much time in
East Africa. He visited Lake Rudolph (Lake Turkana) (1895 and
1899). He was in Ethiopia (1896) and may have been present
at the Ethiopian victory over the Italians at the Battle of
Adwa. He wrote *Through Unknown African Countries* (1897)
and was later elected a Fellow of the Royal Geographical
Society. A mammal is named after him. Note that his name is
often incorrectly hyphenated in the literature. (See also
Smith, A. D.)

Dora

Arabian Woodpecker *Dendrocopos dorae* **Bates** & **Kinnear**,
1935

Mrs Dora Philby *née* Johnston (1888–1957) was the first wife
of Harry St John Philby (q.v.). Though described by George
Latimer Bates and Norman Boyd Kinnear, it was their occasional co-worker, Philby, who proposed the scientific name
after his wife.

Dorabtata

Alpine Swift ssp. *Tachymarptis melba dorabtatai* **Abdulali**,
1965

Sir Dorabji Tata (1859–1932) was an Indian industrialist and
philanthropist who founded the Tata industrial group.

D'Orbigny

D'Orbigny's Seedsnipe *Thinocorus orbignyianus* **Saint-
Hilaire** & **Lesson**, 1831
[Alt. Grey-breasted Seedsnipe]

D'Orbigny's Chat Tyrant *Ochthoeca oenanthoides* d'Orbigny
& **Lafresnaye**, 1837

Ocellated Piculet *Picumnus dorbignyanus* Lafresnaye, 1845

D'Orbigny's Puffleg *Eriocnemis dorbignyi* **Bourcier** &
Mulsant, 1846 NCR
[JS *Eriocnemis glaucopoides*]

Creamy-breasted Canastero *Asthenes dorbignyi*
Reichenbach, 1853

D'Orbigny's Parakeet *Bolborhynchus orbygnesius* **Souance**,
1856
[Alt. Andean Parakeet]

D'Orbigny's Woodcreeper *Dendrocincla fuliginosa atrirostris*
d'Orbigny & Lafresnaye, 1838
[Alt. Plain-brown Woodcreeper ssp.]

Rusty Flowerpiercer ssp. *Diglossa sittoides dorbignyi*
Boissonneau, 1840

Buff-throated Woodcreeper ssp. *Xiphorhynchus guttatus
dorbignyanus* **Pucheran** & Lafresnaye, 1850

Saffron-billed Sparrow ssp. *Arremon flavirostris dorbignii*
P. L. Sclater, 1856

Alcide Charles Victor Dessalines d'Orbigny (1802–1857) was
a traveller, collector, illustrator and naturalist. He was the
author of *Dictionnaire Universel d'Histoire Naturelle*. His
father Charles-Marie Dessalines d'Orbigny (1770–1856) was
a ship's surgeon. Both he and Alcide studied shells. Alcide
went to the Academy of Science in Paris to pursue his
methodical paintings and classification of natural history
specimens. The MNHN sent him to South America (1826)
where the Spanish briefly imprisoned him, mistaking his
compass and barometer, which had been supplied by
Humboldt (q.v.), for 'instruments of espionage'. After he left
prison, he lived for a year with the Guarani Indians learning
their language. He spent five years in Argentina and then
travelled north along the Chilean and Peruvian coasts, before
moving into Bolivia and returning to France (1834). Once
home he donated thousands of specimens of animals: birds,
fish, reptiles, insects, mammals as well as plants, samples of
rocks, fossils, land surveys, pre-Columbian pottery, etc. to
the MNHN. His fossil collection led him to determine that
there were many geological layers, revealing that they must
have been laid down over millions of years. This was the first
time such an idea was put forward. He is also remembered in
the names of an amphibian, two mammals and five reptiles,
as well as other taxa.

Doreen

New Zealand parrot genus *Doreenia* **Mathews**, 1930 NCR
[Now in *Nestor*]

Doreen Peall (b.1901) was in New South Wales, Australia, in
1934, but we know no more about her.

Doria

Doria's Hawk *Megatriorchis doriae* **Salvadori** & **Albertis**,
1875
[Alt. Doria's Goshawk]

Greater Hoopoe Lark ssp. *Alaemon alaudipes doriae*
Salvadori, 1868

Marchese Giacomo Doria (1840–1913) was an Italian ornithologist who collected in Persia (Iran) with de Filippi (q.v.)
(1862–1863) and in Borneo with Beccari (q.v.) (1865–1866). He
was the first Director of the Natural History Museum in Turin
(1867–1913). Six mammals, three amphibians and eight
reptiles, among other taxa, are named after him.

Doricha

Sheartail genus *Doricha* **Reichenbach**, 1854

Doricha was a Greek courtesan, also known as Rhodopis, who lived in the seventh century BC. According to Herodotus, she was a fellow-slave of the fable-teller Aesop.

Doris

Common Paradise Kingfisher ssp. *Tanysiptera galatea doris*
Wallace, 1862

In Greek mythology, Doris was a sea nymph and mother of the Nereids. Her parents were Oceanus and Tethys.

Dorje

Brown Dipper ssp. *Cinclus pallasii dorjei* **Kinnear**, 1937

Raja Sonam Tobgye Dorje (1896–1953) was the Foreign Minister of the Maharaja of Bhutan.

Dorothy (Bate)

Short-toed Treecreeper ssp. *Certhia brachydactyla dorotheae* **Hartert**, 1904

Dorothy Minola Alice Bate (1879–1951) worked as an archaeozoologist and palaeontologist at the British Museum (1898–1948). She started as a teenager sorting bird skins and ended up in charge of the zoology section at Tring. She is remembered for her work on the interpretation of the Mount Carmel excavation (1930s), and was very interested in the field of climate interpretation. Much of her work was on recently extinct animals, trying to understand why giant or dwarf forms evolved. She wrote many scientific papers, her first being 'A short account of a bone cave in the Carboniferous limestone of the Wye Valley' (1901). She died of a heart attack while suffering from cancer. On her desk at Tring was a list of 'Papers to write'. By the last item in the list she had written 'Swan Song'.

Dorothy (White)

Honeyeater genus *Dorothina* **Mathews**, 1914 NCR
[Now in *Meliphaga*]

Dorothy's Grasswren *Amytornis dorotheae* Mathews, 1914
[Alt. Carpentarian Grasswren, Red-winged Grasswren]

Hooded Parrot ssp. *Psephotellus chrysopterygius dorotheae*
Mathews, 1915 NCR
[JS *Psephotus dissimilis*]

Dorothy E. White (1888–c.1927) was the daughter of the Australian naturalist, surveyor and sheep station owner Henry Luke White (1860–1927) (q.v.). His collection of over 13,000 specimens of birds and their eggs from throughout Australia was donated to the Victoria Museum (1917). It is one of the top three such collections in the world, and the most important egg collection in Australia.

Dorst

Dorst's Cisticola *Cisticola guinea* **Lynes**, 1930
[*Cisticola dorsti* Chappuis & Érard, 1991 is a JS]
Dorst's Tyrannulet *Serpophaga berliozi* Dorst, 1957 NCR
[Considered a synonym of *Myiopagis gaimardii*]

Ecuadorian Ground Dove ssp. *Columbina buckleyi dorsti*
Koepcke, 1962

Professor Dr Jean Dorst (1924–2001) was a French ornithologist and Secretary General of the Charles Darwin Foundation. He studied biology and palaeontology at the Faculty of Sciences of the University of Paris then joined the staff of MNHN (1947), eventually becoming Director (1975) but resigned (1985) in protest of government reforms at the museum. He was administrator of the Oceanographic Institute. He was a member of, and sometimes an office holder in many scientific societies. He wrote *Les Migrations des Oiseaux* (1956), *Les Oiseaux* (1959) and *La Vie des Oiseaux* (1972) among many others. He was also was one of the writers of the documentary Le Peuple Migrateur (Winged Migration) (2001) and the film was dedicated to him.

d'Osery

Casqued Oropendola *Psarocolius oseryi* **Deville**, 1849
d'Osery's Hermit *Phaethornis oseryi* **Bourcier** & **Mulsant**, 1852 NCR
[Alt. White-bearded Hermit; JS *Phaethornis hispidus*]

Eugene Comte d'Osery (1818–1846) was a French traveller and collector. He was killed by Indians while a member of Castelnau's (q.v.) collecting expedition to the source of the Amazon. 'Between 1843 and 1847 Castelnau and his party sailed to Rio de Janeiro on board the French brig *Petit-Thouards*, and began his meteorological, magnetic, botanical, and zoological observations there. The expedition resolved to cross South America from Rio via Minas Gerais, Goiás, and Mato Grosso. They explored the north of Mato Grosso, the Paraguay River as far as Asunción, and from Vila Bela they travelled to Bolivia through Potosí, finally reaching La Paz. From there they journeyed to Lima, where they spent some time, then proceeded to the source of the Amazon from whence they sailed to Pará. Owing to the death of a member of the expedition, d'Osery, who was killed by Indians, a large part of the records were lost ... [but] the minutes were saved, which made the writing of this history of the expedition possible.' D'Osery has plants, fish and other taxa named after him.

Doubleday

Doubleday's Hummingbird *Cyanthus doubledayi* **Bourcier**, 1847
[Formerly *Cynanthus latirostris doubledayi*]

Henry Doubleday (1808–1875), an entomologist and ornithologist, was known as 'the Epping Naturalist'. He was a member of the Entomological Society (1833) and introduced the practice of 'sugaring' to capture moths. He wrote *A Nomenclature of British Birds* (1838) and tried to establish uniformity in entomological nomenclature through his *Synonymic List of British Lepidoptera* (1847–1850). His family also founded the eponymous publishing company. Bourcier entitled his paper describing the bird 'Description de deux nouvelles espèces de Charaxes des Index orientales, de la Collection de M. Henri Doubleday.' And clearly decided to name one after Henry.

Dougall

Dougall's Tern *Sterna dougallii* **Montagu**, 1813
[Alt. Roseate Tern]

Dr Peter McDougall (1777–1814) was a Scottish physician and amateur naturalist from Glasgow. Montagu described the tern from birds collected by McDougall in the Firth of Clyde. When McDougall sent the specimens to Montagu he drew the latter's attention to the difference between these birds and all the other terns that 'swarmed in their company on the same rocky islands'.

Douglas, D.

Douglas's Quail *Callipepla douglasii* **Vigors**, 1829
[Alt. Elegant Quail]

David Douglas (1796–1834) was a British botanist and traveller who collected in North America (1823–1834) and Hawaii (1834). Suffering from bad eyesight he fell into a pit in Hawaii and was gored to death by a feral bull that had been trapped there. The Douglas Fir is named after him, and he introduced the Sitka Spruce and the Lodgepole Pine to the UK. A mammal and a reptile are also named after him.

Douglas, R.

Blue-rumped Pitta ssp. *Hydrornis soror douglasi* **Ogilvie-Grant**, 1910

Robert Douglas (fl.1910) was employed at the International Medical Centre in Shanghai (1910) and collected in southern China.

Dove

Tasmanian Thornbill *Acanthiza dovei* **Mathews**, 1922 NCR
[JS *Acanthiza ewingii*]

Spotted Quail-thrush ssp. *Cinclosoma punctatum dovei* Mathews, 1912

Hamilton Stuart Dove (1864–1941) was an English naturalist and immigrant to Tasmania (1890–1941), where he farmed.

Dow

Dow Tanager *Tangara dowii* **Salvin**, 1863
[Alt. Spangle-cheeked Tanager]

Captain John Melmoth Dow (1827–1892) was an American naturalist and explorer. He collected plants and animals in South and Central America, notably Costa Rica. A Zoological Society of London paper by A Günther (1869) is entitled 'An account of the fishes of the states of Central America, based on collections made by Capt. J. M. Dow, F. Godman, Esq., and O. Salvin, Esq.' Dow was in command of the schooner *Starbuck* (1892–1894). He worked for the American Packet Service and sent plants to Britain. There are a number of plants (particularly orchids) named after him and after his wife or some other female relative.

Dowsett

Brown-headed Apalis ssp. *Apalis alticola dowsetti* **Prigogine**, 1973

Robert Jack Dowsett (b.1942) is an English zoologist and former game warden who has specialised in African ornithology and mammalogy for 50 years. He is author or co-author of several books, including *Checklist of the Birds of the Afrotropical and Malagasy Regions* (1993), *The Birds of Malawi* (2006), *The Birds of Zambia* (2008) and *Priority! The Dating of Scientific Names in Ornithology* (2011), as well as *An Atlas of the Mammals of Malawi* (1988). He is a member of the International Ornithological Congress's Standing Committee on Ornithological Nomenclature. Several species of African butterfly that he discovered have been named after him. This bird name replaced an earlier one whose pre-occupation he had pointed out to the author, but which he himself considers doubtfully recognisable!

Drasche Cuckooshrike sp. *Rectes draschei* **Pelzeln**, 1876 NCR
[JS *Coracina schisticeps*?]

Dr Richard Freiherr von Drasche Wartinberg (1850–1923) was an Austrian geologist, zoologist, explorer and collector. His family were industrialists with a large brick-making business that he headed. He visited Spitsbergen (1873), the Indian Ocean and East Asia, including Réunion, Mauritius and Luzon (Philippines) (1875–1876) and later Japan.

Dresser

Common Eider ssp. *Somateria mollissima dresseri* **Sharpe**, 1871
White-winged Brush Finch ssp. *Atlapetes leucopterus dresseri* **Taczanowski**, 1883
Marsh Tit ssp. *Poecile palustris dresseri* **Stejneger**, 1886
Common Starling ssp. *Sturnus vulgaris dresseri* **Buturlin**, 1904 NCR
[JS *Sturnus vulgaris porphyronotus*]
Eastern Rock Nuthatch ssp. *Sitta tephronota dresseri* **Zarudny** & Buturlin, 1906
Brown Accentor ssp. *Prunella fulvescens dresseri* **Hartert**, 1910

Henry Eeles Dresser (1838–1915) was an English businessman, ornithologist, oologist and traveller. His father was a timber merchant who sent his son to Sweden (1854) to learn Swedish and the timber trade. He worked in this trade until 1863, when he sailed to Texas with a cargo for the Confederate government. He spent 13 months travelling and collecting in southern Texas. He was based at San Antonio (famous for the Alamo), where he shared a house with Heermann (q.v.). He wrote *A History of the Birds of Europe* (1871), *A Monograph of the Meropidae* (1884) and *A Monograph of the Coraciidae* (1893).

Drouhard

Greater Vasa Parrot ssp. *Coracopsis vasa drouhardi* **Lavauden**, 1929

Eugène Jean Drouhard (1874–1945) was an agronomist and botanist who collected in Madagascar and Mozambique (1901–1934), where he was Inspector of Forests (1929). A mammal is also named after him.

Drouyn

White Eared Pheasant ssp. *Crossoptilon crossoptilon drouynii* **J. Verreaux**, 1868

(See **L'Huys**)

Drowne

Drowne's Fantail *Rhipidura drownei* **Mayr**, 1931
[Alt. Brown Fantail]

Dr Frederick Peabody Drowne (1880–1930) was an American physician, traveller and collector. He took his medical degree in Baltimore (1904) and served in the Spanish-American War and WW1. He was active in the Galápagos (1897–1898) and the Solomon Islands (1930). He kept a diary, which was incorporated in 'Review of the ornithology of the Galápagos Islands, with notes on the Webster-Harris Expedition' (1899) by Hartert and Rothschild. He wrote a monograph, *The Reptiles and Batrachians of Rhode Island* (1905). He died of a self-inflicted bullet wound.

Drozdov

Parrotbill genus *Enendrozdovoma* Kashin, 1978 NCR
[Now in *Conostoma*]

Professor Nikolai Nikolaevich Drozdov (b.1937) is a Russian ornithologist, biogeographer, filmmaker, conservationist and broadcaster. He is on the faculty of Moscow State University, which awarded his bachelor's degree (1963) and his master's (1968).

Drummond, J. M.

Swamp Harrier ssp. *Circus approximans drummondi* **Mathews** & **Iredale**, 1913 NCR; NRM

James Mackay Drummond (1869–1940) was a New Zealand journalist and naturalist. He co-wrote *The Animals of New Zealand* (1904).

Drummond, T.

Drummond's Snipe *Scolopax drummondii* **Swainson** & **J. Richardson**, 1832 NCR
[Alt. Wilson's Snipe; JS *Gallinago delicata*]

Thomas Drummond (1793–1835) was a Scottish botanist. He was Assistant Naturalist to Dr John Richardson (q.v.), who was Surgeon Naturalist on the 'Northern Land Expeditions' under Sir John Franklin (q.v.) (1819–1822 and 1825–1827). Richardson wrote *Fauna Boreali-American* (1829) in which he said of Drummond: '... to whose unrivalled skill in collecting, and indefatigable zeal, we are indebted for most of the insects, the greater part of the specimens of plants, and a considerable number of the quadrupeds and birds'. Richardson went on to say how Drummond went hundreds of miles on foot through all sorts of perilous country and weather, such privations as anyone with less zeal would have succumbed to. He made a second trip to America (1830–1833), collecting extensively in the south and west of the current US. His ambition was to undertake a complete botanical survey of Texas but he died in Havana while collecting in Cuba.

Dryas

Ultramarine Lorikeet *Coryphilus dryas* **Gould**, 1842 NCR
[JS *Vini ultramarina*]
Arafura Fantail *Rhipidura dryas* Gould, 1843
Spotted Nightingale Thrush *Catharus dryas* Gould, 1855

Blue-breasted Kingfisher ssp. *Halcyon malimbica dryas* **Hartlaub**, 1854
Ring-necked Dove ssp. *Streptopelia capicola dryas* **Grote**, 1927 NCR
[NUI *Streptopelia capicola tropica*]
Forbes's Forest Rail ssp. *Rallicula forbesi dryas* **Mayr**, 1931

Dryas ('oak') was the name of several minor characters in Greek mythology. A dryad is a mythological tree-nymph.

Drygalski

Eaton's Pintail ssp. *Anas eatoni drygalskii* **Reichenow**, 1904

Erich Dagobert von Drygalski (1865–1949) was a German geophysicist, geographer and polar explorer. He studied mathematics at the universities of Königsberg, Berlin, Bonn and Leipzig (1882–1887), finally being awarded a doctorate. He led two expeditions (1891 and 1893), the second lasting through the winter in western Greenland. He became Associate Professor (1898) and then Professor (1899) at Berlin. He led the German South Polar Expedition (1901–1903) with the *Gauss*. He wrote *Die Deutsche Südpolar-Expedition auf dem Schiff Gauss* (1902) as well as twenty books and two atlases documenting the expedition's findings (1905–1931). He became a professor at Munich (1906) and presided over the Geographic Institute that he founded (1910–1949). He continued to explore, including Spitsbergen (1910), North America and Asia. Other taxa are named after him.

Dubois

Réunion Pink Pigeon *Nesoenas duboisi* **Rothschild**, 1907 EXTINCT
Réunion Night Heron *Nycticorax duboisi* Rothschild, 1907 EXTINCT
Réunion Kestrel *Falco duboisi* Cowles, 1994 EXTINCT

Sieur Dubois and Sieur D. B. are the pseudonymous names of a traveller who reached the islands of Madagascar and Réunion at the time of early colonisation by France. He wrote a book *Les Voyages Faits par le Sieur D.B. aux Isles Dauphine ou Madagascar et Bourbon ou Mascarenne, ès années 1669, 70, 71 et 72* (1674) about his journeys and the wildlife he saw including details of several of Réunion's endemic bird species which, like the three eponymous ones, have since become extinct, such as the Réunion Ibis, Réunion Swamphen, and Réunion Rail.

Dubois, A. J. C.

Dubois's Seedeater *Sporophila ardesiaca* A. Dubois, 1894
Dubois's Leaf Warbler *Phylloscopus cebuensis* A. Dubois, 1900
[Alt. Lemon-throated Warbler]

Dubois's Black-headed Weaver *Ploceus melanocephalus duboisi* **Hartlaub**, 1886

Piping Hornbill ssp. *Bycanistes fistulator duboisi*
W. L. Sclater, 1922

Alphonse Joseph Charles Dubois (1839–1921) was a Belgian naturalist, as was his father Charles Fréderic Dubois (1804–1867). They published many works together, including *Les Oiseaux de l'Europe* (1868–1872), which was completed by Alphonse and published after Charles's death. He became Curator of the Department of Vertebrates at the Royal Museum of Natural History, Brussels (1869).

Du Bus

Mountain Tanager genus *Dubusia* **Bonaparte**, 1850

Grey-fronted Dove ssp. *Leptotila rufaxilla dubusi* Bonaparte, 1855

Bernard-Amé-Léonard du Bus de Gisignies (1808–1874) was a Belgian ornithologist and palaeontologist. He wrote *Esquisses Ornithologiques: Descriptions et Figures d'Oiseaux Nouveaux ou Peu Connus* (1843).

Duchaillu

Duchaillu's Yellow-spotted Tinkerbird *Buccanodon duchaillui* **Cassin**, 1855
[Alt. Yellow-spotted Barbet]

Paul Belloni Duchaillu (1835–1903) was a French-born American explorer and anthropologist. He spent much of his childhood in Gabon, West Africa, where his father was a trader. He learned the local languages and became interested in exploring further inland. He went to the US (1852), becoming a citizen. With the support of the Philadelphia Academy of Natural Sciences he put together an expedition to explore Gabon (1855–1859). He collected widely, the specimens including many live birds and other animals, some of them previously unknown to science. He took back gorillas to the US where they had never been seen. He published an account of the expedition, *Explorations in Equatorial Africa* (1861). He also virtually redrew the geographical map of the region. Duchaillu made a second expedition (1863–1865) during which he visited hitherto unknown areas and verified previous reports of 'pygmy' people. His account of the latter expedition was published as *A Journey to Ashango-Land* (1857). He also wrote *Stories of the Gorilla Country* (1867), *Wild Life under the Equator* (1868), *My Apingi Kingdom* (1870) and *The Country of the Dwarfs* (1871). He also travelled in Scandinavia (1871–1878) and wrote about it in *The Land of the Midnight Sun* (1881) and *The Viking Age* (1889).

Duchassain

Duchassain's Hummingbird *Lepidopyga coeruleogularis*
Gould, 1850
[Alt. Sapphire-throated Hummingbird; *Trochilus duchassaini* (**Bourcier**, 1851) is a JS]

We believe 'Duchassain' to be a transcription error and that the bird was named for Placide Duchassaing de Fonbressin (1819–1873), a zoologist who was active at the right time and in the right place. There are records of specimens of marine corals and sponges collected by him on the Panama coast (c.1850). Duchassaing was born on Guadeloupe, to a French-Creole family of planters. He was sent to school in France and went to university in Paris, achieving a doctorate in medicine, geology and zoology. He returned to Guadeloupe as a physician. He travelled widely in the Caribbean to neighbouring islands: Nevis, St Eustatius, St Martin, St Barthélemy, St Croix, Cuba and also Panama, treating people during cholera outbreaks. He returned to France (1867) where he eventually died. Duchassaing was a natural history collector who sent many specimens to Europe. He wrote *Mémoire sur les Coralliaires des Antilles* (1860) and co-wrote *Spongiaires de la Mer Caraibe* (1864). Gould first described the hummingbird from a bird collected by J. Warszewicz (q.v.) in Chiriquí, Panama.

Ducis

Plain Martin ssp. *Riparia paludicola ducis* **Reichenow**, 1908

Duke (hence *ducis*) Adolf Friedrich von Mecklenburg (1873–1969) was a German explorer and colonial politician in Africa and (1949–1952) the first President of the German Olympic Committee. He conducted scientific research on the African Rift Valley and famously crossed Africa from East to West (1907–1908). He led another expedition to Lake Chad and the upper reaches of the Congo River and Nile (Sudan) (1910–1911), about which *Vom Kongo zum Niger und Nil* was written (1912–1914). After WW1 he became Vice President (his brother Johann Albrecht had been President) of the German Colonial Society for South West Africa. A reptile is named after him.

Ducorps

Ducorps's Corella *Cacatua ducorpsii* **Pucheran**, 1853
[Alt. Ducorps's Cockatoo, Solomons Cockatoo]

Louis-Jaques Ducorps (1811–c.1858) was an explorer for the French navy. He was administrator on board the French vessel *L'Astrolabe* during her circumnavigation (1822–1825) under the command of Dumont d'Urville.

Dufresne

Dufresne's Amazon *Amazona dufresniana* Shaw, 1812
[Alt. Blue-cheeked Amazon]
Dufresne's Waxbill *Coccopygia melanotis* **Temminck**, 1823
[Alt. Swee Waxbill, Yellow-bellied Waxbill; Syn. *Estrilda melanotis*]

Louis Dufresne (1752–1832) was a French ornithologist and taxidermist and one of the naturalists on *L'Astrolabe*'s voyage of discovery (1785–1787). This expedition visited Madeira, Tenerife, Trinidad, the coast of Brazil, Cape Horn, the Sandwich Islands (Hawaii) and then to the coast of north-west America and Alaska. Dufresne became a curator at the MNHN, Paris (1793), and he continued to travel on behalf of the museum. A claim to fame was popularising arsenic in the preparation of skins (1802), which helped build the greatest collection of bird specimens in the world at that time. His private collection held over 1,600 birds, 800 eggs and hundreds of other specimens and fossils that he sold to the University of Edinburgh (1819). Dufresne was awarded the

Legion of Honour (1829). He died of lung disease. A mammal is named after him.

Dugand

Dugand's Antwren *Herpsilochmus dugandi* **de Schauensee**, 1945

Straight-billed Woodcreeper ssp. *Dendroplex picus dugandi* **Wetmore** & **W. H. Phelps**, 1946
Ochre-breasted Tanager ssp. *Chlorothraupis stolzmanni dugandi* de Schauensee, 1948
Collared Puffbird ssp. *Bucco capensis dugandi* **Gilliard**, 1949
Rusty-belted Tacapulo ssp. *Liosceles thoracicus dugandi* de Schauensee, 1950

Armando Dugand (1906–1971) was a Colombian naturalist and Director of the Institute of Natural Sciences of the National University of Colombia (1940–1953). He was one of the founders (1940) of the magazine *Caldasia*. He headed several bird collecting expeditions for the institute into the Colombian interior (1940s). A reptile is named after him.

Dugès

Yellow Warbler ssp. *Dendroica aestiva dugesi* Coale, 1887
Rufous-capped Warbler ssp. *Basileuterus rufifrons dugesi* **Ridgway**, 1892

Professor Alfredo Augusto Delsescautz Dugès (1826–1910) was Professor of Natural History, University of Guanajuato, Mexico. He is regarded as being the father of Mexican herpetology, as he was the first to define Mexican herpetofauna in Linnaean terms. Six reptiles are named after him.

Duivenbode

Sunbird genus *Duyvena* Mathews, 1925 NCR
[Now in *Aethopyga*]

Duyvenbode's Sunbird *Aethopyga duyvenbodei* **Schlegel**, 1871
[Alt. Elegant Sunbird]
Duivenbode's Lory *Chalcopsitta duivenbodei* **A. J. C. Dubois**, 1884
[Alt. Brown Lory]
Duivenbode's Riflebird *Paryphephorus duivenbodei* **A. B. Meyer**, 1890 NCR
[Presumed hybrid: *Ptiloris magnificus* x *Lophorina superba*]
Duivenbode's Six-wired Bird-of-Paradise *Parotia duivenbodei* **Rothschild**, 1900 NCR
[Presumed hybrid: *Parotia sefilata* x *Lophorina superba*]
Duivenbode's Bird-of-Paradise *Paradisaea duivenbodei* Menegaux, 1913 NCR
[Presumed hybrid: *Paradisaea minor* x *Paradisaea guilielmi*]

Renesse van Duivenbode was the family and company name of a father and son, merchants based on the island of Ternate in the Moluccas. Interestingly the name (sometimes spelt 'Duyvenbode') means 'pigeon-post messenger' and was an honorific conferred when the family used carrier pigeons to keep in touch with William of Orange during the siege of Leiden. The Duivenbodes were sometimes called the 'King(s) of Ternate' because of their vast wealth. The family member most interested in natural history appears to have been Constantijn Willem Rudolf van Renesse van Duivenbode (b. 1858). When naming the riflebird, Meyer (q.v.) wrote: 'This new species I dedicate to Mr. C. W. R. van Renesse van Duivenbode, of Ternate, to whom science is already indebted for many interesting additions to the Papuan avifauna.' Alfred Russel Wallace (q.v.) visited Ternate and rented a house from M. D. van Renesse van Duivenbode, saying of him that he was '… of an ancient Dutch family, but who was educated in England, and speaks our language perfectly. He was a very rich man, owned half the town, possessed many ships, and above a hundred slaves. He was, moreover, well educated, and fond of literature and science – a phenomenon in these regions'. Like Bruijn (q.v.) the Duivenbodes were heavily involved in the feather trade. It may be that some of these birds are named after Lodewijk Diederik Hendrik Alexander van Renesse van Duivenbode (c.1832–c.1881) who seems to have accompanied Wallace on at least one expedition.

Dulcie

Grassbird genus *Dulciornis* **Mathews**, 1912 NCR
[Now in *Megalurus*]

White-cheeked Honeyeater ssp. *Meliornis nigra dulciei* Mathews, 1911 NCR
[JS *Phylidonyris niger gouldii*]
Tawny Grassbird ssp. *Megalurus alisteri dulciei* Mathews, 1912 NCR
[JS *Megalurus timoriensis alisteri*]
White-faced Storm-petrel ssp. *Pelagodroma marina dulciae* Mathews, 1912

Dulcie Marian Wynne (b.1894) was the stepdaughter of the describer, G. M. Mathews.

Dumas

Buru Thrush *Zoothera dumasi* **Rothschild**, 1898

Mountain Tailorbird ssp. *Phyllergates cuculatus dumasi* **Hartert**, 1899
[Syn. *Orthotomus cuculatus dumasi*]

Johannes Maximiliaan Dumas (d.1917) was a feather merchant, surveyor and self-taught naturalist who collected in Irian Jaya (1899–1917), partly in the service of the botanist Dr K. Heyne's and partly as an indefatigable and intrepid explorer. He also collected in the Moluccas for Everett (q.v.), Stresemann (q.v.) and others. Most specimens he collected went to the Forest Research Institute. He was part of a scientific expedition of Sentani Lake and surrounding area sponsored by the Treub Company and Royal Dutch Geographical Society (1903). He spent seven years (1907–1915) in the interior as naturalist and surveyor on an extended military mapping and collecting expedition.

Duméril

New Zealand Bellbird ssp. *Anthornis melanura dumerilii* **Lesson** & **Garnot**, 1828 NCR
[JS *Anthornis melanura melanura*]
Duméril's Amazilia *Amazilia amazilia dumerilii* Lesson, 1832
[Alt. Amazilia Hummingbird ssp.]

Dr André Marie Constant Duméril (1774–1860) was a zoologist and herpetologist. He qualified as a physician (1793), and became Professor of Anatomy, MNHN, Paris (1801–1812), changing to Professor of Herpetology and Ichthyology (1813–1857). He built up one of the largest natural history collections of the time. Towards the end of his career his son, Auguste Henri André (1812–1870), also a distinguished zoologist, assisted him and later took over his father's professorship (1857). Twenty-one reptiles and six amphibians are named after one or other of them.

Dumont, C. H. F.

Dumont's Myna *Mino dumontii* **Lesson**, 1827
[Alt. Yellow-faced Myna]

Charles Henri Frédéric Dumont de Sainte Croix (1758–1830) was a French lawyer, amateur ornithologist and the father-in-law of the describer.

DuMont, P. A.

Crested Coua ssp. *Coua cristata dumonti* **Delacour**, 1931

Philip Atkinson DuMont (1903–1996) was an American zoologist. Drake University awarded his bachelor's degree (1926). He collected in Madagascar (1929–1931) with one of the Archbold expeditions, seconded from the AMNH, New York. He worked for the US Fish & Wildlife Service (1935–1972).

Duncan

Seedeater genus *Duncanula* **Chubb**, 1921 NCR
[Now in *Catamenia*]

Speckled Rail ssp. *Coturnicops notatus duncani* Chubb, 1916 NCR; NRM
Paramo Seedeater ssp. *Catamenia homochroa duncani* Chubb, 1921

Duncan Vavasour McConnell (b.1908) was a son of Frederick Vavasour McConnell (q.v.) and Helen Mackenzie McConnell *née* Gibson. Chubb (q.v.) was a friend of the McConnell family.

Dunn

Dunn's Lark *Eremalauda dunni* **G. E. Shelley**, 1904

Colonel Henry Nason Dunn (1864–1952) was a British army surgeon who became a big-game hunter. He left diaries to the National Army Museum relating to his time (1897–1906) in Sudan and Somaliland. Dunn himself collected the lark. He also has a mammal named after him.

Dunne

Dunne's Finch *Nesospiza acunhae dunnei* **Hagen**, 1952
[Alt. Inaccessible Island Finch ssp.]

Jacobus Charles Dunne (fl.1938) was a South African vulcanologist and geologist who was on the Norwegian Scientific Expedition to Tristan da Cunha (1937–1938). He wrote *Volcanology of the Tristan da Cunha Group* (1941).

Duperrey

New Guinea Scrubfowl sp. *Megapodius duperryii* **Lesson** & **Garnot**, 1826 NCR
[Alt. Orange-footed Scrubfowl; JS *Megapodius reinwardt*]

Captain Louis Isidore Duperrey (1786–1865) was a French naval officer (1802). He was second in command and hydrologist on board *L'Uranie* during its circumnavigation (1817–1820), then was appointed (1821) to command *La Coquille* for its circumnavigation (1822–1825). A reptile is named after him.

Dupetit-Thouars

White-capped Fruit Dove *Ptilinopus dupetithouarsii* **Neboux**, 1840

Admiral Abel Aubert Dupetit-Thouars (1793–1864) was a French naval officer who explored in the Pacific (1836–1839). In Tahiti he entered into a confrontation with the English missionary and Consul George Pritchard, finally expelling him and establishing a French protectorate over the territory. Relations between France and Great Britain soured because of the 'Pritchard Affair'. (Also see **Thouars**)

Dupont

Becard sp. *Platyrhynchus duponti* **Vieillot**, 1823 NCR
[Alt. Green-backed Becard; JS *Pachyramphus viridis*]
Dupont's Hummingbird *Tilmatura dupontii* **Lesson**, 1832
[Alt. Sparkling-tailed Woodstar]

Monsieur Dupont (1798–1873) was a French dealer in natural history specimens, but nothing more seems recorded about him.

duPont, J. E.

duPont's Blue-backed Parrot *Tanygnathus sumatranus duponti* **Parkes**, 1971
[Alt. Azure-rumped Parrot ssp.]
Rufous-crowned Sparrow ssp. *Aimophila ruficeps duponti* Hubbard, 1975

John Eleuthere duPont (1938–2010) was an American ornithologist. He wrote *South Pacific Birds* (1976), *Philippine Birds* (1971) and co-wrote *Recent Volutidae of the World* (1970). He also published articles such as 'Notes on Philippine birds' with Rodolphe Meyer de Schauensee (q.v.) (1959). He described over 20 new taxa of birds. duPont founded the Delaware Museum of Natural History (1957) and its collection of molluscs, birds and bird books very much reflects his interests. Having been convicted of the third-degree murder of David Schultz, a wrestler (1996), he was sentenced to 40 years' detention in either a prison or a mental hospital. He died in prison.

Dupont, L. P.

Dupont's Lark *Chersophilus duponti* **Vieillot**, 1820

Léonard Puech Dupont (1795–1828) was a French naturalist and collector. Vieillot named the lark after him when Dupont showed it to him. According to Vieillot's label, Dupont collected it in Provence although this may be an error as Dupont went on the Joseph Ritchie expedition to North Africa, which he left, following a dispute, taking 200 specimens with him back to Paris just a year before the lark was named; so the specimen could actually have been collected in North Africa. The species is now known from Spain but has never otherwise been recorded in France.

Duprez

Barbary Partridge ssp. *Alectoris barbara duprezi* **Lavauden**, 1930 NCR
[JS *Alectoris barbara spatzi*]

Captain Maurice Duprez (1891–1943) was a French explorer and army officer (1912–1943) who was wounded in WW1 (1914) and posted to Tunisia (1915–1920). He was a political officer in Algeria and the Sahara (1920–1933 and 1936–1943), serving in Syria (1934–1936). He was a member of the Tunis to Chad Expedition (1925). He died of tuberculosis.

Durnford

Durnford's Oystercatcher *Haematopus durnfordi* **Sharpe**, 1896 NCR
[NUI American Oystercatcher *Haematopus palliatus palliatus*]

Henry Durnford (1848–1878) was a British explorer and collector in Patagonia (1875–1878). He wrote 'Notes on the birds of Central Patagonia' (1877).

Dusit

Puff-throated Babbler ssp. *Pellorneum ruficeps dusiti* **E. Dickinson** & Chaiyaphun, 1970

Dusit Bhanijbhatana (fl.1970) was Director-General of the Royal Thai Forest Department.

Dussumier

Seychelles Sunbird *Cinnyris dussumieri* **Hartlaub**, 1860

Philippine Turtle Dove ssp. *Streptopelia bitorquata dussumieri* **Temminck**, 1823
Shikra ssp. *Accipiter badius dussumieri* Temminck, 1824
Small Buttonquail ssp. *Turnix sylvaticus dussumier* Temminck, 1828

Jean-Jacques Dussumier (1792–1883) was a French merchant, collector, traveller and ship owner. He was most active (1816–1840) in South-East Asia and around the Indian Ocean. He seems to have collected mainly molluscs and fish, a large number being named after him including a whole genus of herrings called *Dussumieria*. A mammal and five reptiles are also named after him.

Duvaucel

Trogon genus *Duvaucelius* **Bonaparte**, 1854 NCR
[Now in *Harpactes*]

Duvaucel's Trogon *Harpactes duvaucelii* **Temminck**, 1824
[Alt. Scarlet-rumped Trogon]
River Lapwing *Vanellus duvaucelii* **Lesson**, 1826

Duvaucel's Barbet *Megalaima australis duvaucelii* Lesson, 1830
[Alt. Blue-eared Barbet ssp.]

Alfred Duvaucel (1796–1824) was a French naturalist who was Cuvier's (q.v.) stepson. He was sent (1817) by Cuvier with Diard (q.v.) to collect in India for the MNHN, Paris. They established a botanical garden in Chanannagar (1818–1819), and collected in Sumatra under contract to Stamford Raffles (q.v.), but were dismissed when Raffles discovered that they were sending most of what they collected back to Paris rather than to him. Duvaucel died in India. Among other taxa, two reptiles and a mammal are named after him.

Dwight

Dwight's Bush Tanager *Chlorospingus dwighti* Underdown, 1931
[Syn. *Chlorospingus flavopectus dwighti*]

Horned Lark ssp. *Eremophila alpestris dwighti* **Stresemann**, 1922 NCR
[JS *Eremophila alpestris leucansiptila*]
Yellowish Flycatcher ssp. *Empidonax flavescens dwighti* **Van Rossem**, 1928 NCR
[JS *Empidonax flavescens salvini*]
Hermit Thrush ssp. *Hylocichla guttata dwighti* **L. B. Bishop**, 1933 NCR
[JS *Catharus guttatus auduboni*]

Dr Jonathan Dwight (1858–1929) was an amateur American ornithologist who amassed over sixty thousand North American bird skins, which he eventually deposited at the AMNH, New York City. He graduated from Harvard (1880) and received his medical degree at Columbia University (1893). He practised medicine for 15 years then turned to the full-time study of birds. As an undergraduate he began collecting eggs and taking notes about his bird sightings. After his training in rifle shooting in the New York National Guard he took to 'collecting' birds, using his surgical skills to prepare the skins. He published many articles such as 'The sequence of plumages and moults of the passerine birds of New York' (1900) and 'The gulls (Laridae) of the world: their plumages, moults, variations, relationships and distribution' (1925). He was also an early leading light in the American Ornithologist's Union, serving as President, Vice-President and Treasurer. He amassed an extensive ornithological library.

Dybowski, B.

Great Bustard ssp. *Otis tarda dybowskii* **Taczanowski**, 1874
Emerald-bellied Puffleg ssp. *Eriocnemis alinae dybowskii* Taczanowski, 1882
Dybowski's Tree Sparrow *Passer montanus dybowskii* **Domaniewski**, 1915
[Alt. Eurasian Tree Sparrow ssp.]

Benedykt (Benoit) Dybowski (1833–1930) was a Polish biologist who was born in Belarus. He was an ardent proponent of Darwin's theory of evolution. He was appointed Adjunct Professor of Zoology in Warsaw (1862), but after the failure of the January 1863 Uprising (against the Russian Empire) he was banished and spent time as a political exile in Siberia.

Here support from the Zoological Cabinet at Warsaw allowed him to undertake investigations into the natural history of Lake Baikal and other parts of the Soviet Far East. He was pardoned (1877) and went to Kamchatka as a physician (1878). He was appointed (1883–1906) to the Chair of Zoology, University of Lemburg, Poland (Lviv, Ukraine) until his retirement. A mammal and two amphibians are named after him.

Dybowski, J.-T.

African warbler genus *Dybowskia* **Oustalet**, 1892 NCR
[Now in *Heliolais*]

Dybowski's Twinspot *Euschistospiza dybowskii* Oustalet, 1892
[Alt. Dybowski's Dusky Twinspot]

Heuglin's Francolin ssp. *Pternistis icterorhynchus dybowskii* Oustalet, 1892 NCR; NRM

Jan (Jean-Thadée) Dybowski (1856–1928) was a botanist and explorer of (especially equatorial) Africa. He led a Congo expedition (1891) and wrote accounts of his travels, to Chad (1893) (*La Route du Tchad*) and the Congo (1912) (*Le Congo Méconnu*). Dybowski established new gardens and plantations in Tunisia and organised schools of agriculture. Later (c.1908) he became French Inspector-General of colonial agriculture. However, the Dybowski family comprised many scientists, including an outstanding arachnologist, so it is very difficult to track down quite what is named after whom. (See also **Xavier,** who was Jan's brother). Dybowski was largely responsible for the isolation and introduction of the psychotropic drug ibogaïne. Dybowski and Landrin isolated the alkaloid, which they named ibogaïne from the bark of the root (1901), and showed it to have the same psychoactive properties as the root itself. Two mammals are named after him.

Dyleff

Levaillant's Cisticola ssp. *Cisticola tinniens dyleffi* **Prigogine**, 1952

Dr Pierre Dyleff (1888–1978) was a Russian emigré pathologist who lived and collected in the Belgian Congo (Democratic Republic of Congo). He was the medical officer for the Bafwaboli Mining Company at Angamu (1930s–1950s).

Dyott

Wattlebird genus *Dyottornis* **Mathews** 1912 NCR
[Now in *Anthochaera*]

Australasian Gannet ssp. *Morus serrator dyotti* Mathews 1913 NCR; NRM
Paradise Riflebird ssp. *Ptiloris paradiseus dyotti* Mathews 1915 NCR; NRM

Robert A. Dyott (fl.1913) was an Australian ornithologist and collector. He was in Tasmania and Queensland (1913–1915).

Dyson

Dyson's Puffbird *Bucco dysoni* **P. L. Sclater**, 1855 NCR
[Alt. White-necked Puffbird; JS *Notharchus hyperrhynchus*]

David Dyson (1823–1856) was an English naturalist and collector and former weaver in a Salford factory, with a passion for entomology, conchology and ornithology. He went to, and collected in, the eastern USA (1843) returning to England with a collection of over 18,000 specimens. He was commissioned by the British Museum and the Earl of Derby (q.v.) to go to British Honduras (Belize) (1844–1845) and was in Venezuela (1846–1847). He became Curator of the Earl of Derby's collection at Knowsley (1848–1851).

Dzieduszycki

Grey-headed Woodpecker ssp. *Picus canus dzieduszyckii* **Domaniewski**, 1925 NCR
[JS *Picus canus canus*]

Włodzimierz Jan Graf Dzieduszycki (1821–1899) was a Polish zoologist and collector who founded the Dzieduszycki Museum in Lviv, Ukraine.

E

Eades

> White-faced Storm-petrel ssp. *Pelagodroma marina eadesi* **Bourne**, 1953
> [Trinomial may be changed to (plural) *eadesorum*]

Reverend Ernest A. D. Eades (c.1915–at least 1953) was a Nazarene missionary to the Cape Verde Islands. Bourne named the petrel after both Mr and Mrs Eades – hence the use of *eadesorum* would be correct – for their help in allowing Bourne to visit the small island of Cima.

Earl

> Earl's Weka *Ocydromus earli* G. R. Gray, 1862 NCR
> [JS *Gallirallus australis australis*]

Percy Earl (DNF) was a naturalist, conchologist, entomologist and ornithologist. He was in New York (1836) before a trip taking him to New Zealand via Colombia and Cape Town. He collected in New Zealand (1842–1844) during which time Edward Shortland showed him the swamp deposit of moa bones recently discovered at Waikouaiti, and Earl made a large collection of Moa remains (1843). He brought back the Weka holotype from New Zealand (1845). He forwarded many specimens to the British Museum including shipping some home on HMS *Erebus* and *Terror*. Gray merely mentions the name Mr Percy Earl at the end of his description, although elsewhere he refers to Earl's description of the habits of other New Zealand birds.

Earl of Derby

> Earl of Derby's Parakeet *Platycercus stanleyii* **Vigors**, 1830 NCR
> [Alt. Stanley Rosella, Western Rosella; JS *Platycercus icterotis*]

(See **Stanley**)

Earle

> Earle's Babbler *Turdoides earlei* **Blyth**, 1844
> [Alt. Striated Babbler]

Willis Earle (DNF) was an English merchant in India (1844). He *might* have been the same person as the Willis Earle who was a wealthy coal merchant living near Liverpool (1810) and who also seems to have been involved in the slave trade. There is a record of him having donated some bird skins from Mauritius to the museum in Calcutta. Strickland (q.v.) described him as a friend, as did Blyth (q.v.) who collected in the Himalayas and around Calcutta.

Eaton, A. E.

> Eaton's Pintail *Anas eatoni* **Sharpe**, 1875
> [Alt. Kerguelen/Southern Pintail]
>
> ---
>
> Fulmar Prion ssp. *Pachyptila crassirostris eatoni* **Mathews**, 1912

Reverend Alfred Edmund Eaton (1845–1929) was an English explorer, entomologist and naturalist who published many scientific papers (1860s–1920). He collected on Kerguelen Island with the Transit of Venus Expedition (1874–1875), and on Madeira and Tenerife (1902).

Eaton, W. F.

> Mottled Owl ssp. *Ciccaba virgata eatoni* **L. Kelso** & E. H. Kelso, 1936 NCR
> [JS *Strix virgata centralis*]

Warren Francis Eaton (1900–1936) was an American naturalist who graduated from Harvard (1922) and worked in Boston (1922) and New York (1923–1936) for a firm of cotton goods factors. He died from appendicitis.

Echo

> Echo Parakeet *Psittacula echo* **A. Newton** & **E. Newton**, 1876
> [Alt. Mauritius Parakeet; Syn. *Psittacula eques echo*]

In Greek mythology, Echo was a mountain nymph who fell in love with the vain youth Narcissus. Since the latter only loved himself, Echo pined away, crying endlessly until all that was left of her was her voice. In another version, Echo would distract and amuse the goddess Hera with long stories while Hera's husband Zeus took advantage of the time to court the other nymphs. When Hera discovered the trickery she punished Echo by taking away her voice, except in foolish repetition of another person's words.

Eck

> Great Tit ssp. *Parus major ecki* **von Jordans**, 1970
> Sharp-tailed Glossy Starling ssp. *Lamprotornis acuticaudus ecki* **Clancey**, 1980
> Coal Tit ssp. *Periparus ater eckodedicatus* **J. Martens**, Tietze & Sun, 2006

Dr Siegfried Eck (1942–2005) was a German ornithologist who was entirely self-taught, never having attended any university. His doctorate was honorary, awarded by the University of Mainz (2002). He was Curator of the ornithological collection, State Natural History Museum, Dresden (1967–2005) and was a Corresponding Fellow, American Ornithologists' Union (1988–2005). He worked on taxonomy and phylogenetics of Palearctic birds by using proportional differences of contour feathers, on species concepts and on the history of ornithology.

Edgar

Black-naped Oriole ssp. *Oriolus chinensis edgari* **Chasen**, 1939 NCR
[JS *Oriolus chinensis maculatus*]

Alexander 'Sandy' Thomson Edgar (1900–1983) was a Scottish-born New Zealand naturalist, conservationist and rubber planter in Malaya (Malaysia) (1920–1961). He lived in Perak (1920–1940) and served in Malayan Local Forces and the Indian Army (1940–1945), including three years as a Japanese prisoner-of-war in Singapore and on the notorious Siam railway (1942–1945). He worked again in Malaya (1946–1959) as advisor to several companies. He farmed in New Zealand (1960–1981). He wrote *Manual of Rubber Planting (Malaya)* (1938).

Edith (Baker)

Makira/San Cristobal Moorhen genus *Edithornis* **Mayr**, 1933
[Now usually included within *Gallinula*]

Common Cicadabird ssp. *Coracina tenuirostris edithae*
Stresemann, 1932
[Alt. Sulawesi Slender-billed Cicadabird]

Edith Brevoort Baker *née* Kane (1884–at least 1930) was the wife (1911) of George Fisher Baker Jr (1878–1937), an American financier who, at the time of their marriage, was Vice President and Director of the First National Bank (his father was President until 1909). He was also an enthusiastic swimmer and yachtsman.

Edith (Cole)

Cole's Crow *Corvus edithae* **E. L. Phillips**, 1895
[Alt. Somali Crow/Dwarf Raven]

Miss Edith Cole (1859–1940) was a British entomologist and botanist. She was in Somaliland (1895) with Mr and Mrs Ethelbert Lort Phillips. The two women collected butterflies and botanical specimens while the men (see **Gillett** and **Phillips, E. E. L.**) collected larger fauna. She discovered the eponymous cacti genus *Edithcolea* there. She is remembered in the names of several other plants. Phillips heard the crow from his tent and immediately recognised it as a new species.

Edith (McGregor)

Edith's Titmouse *Periparus elegans edithae* **McGregor**, 1907
[Alt. Elegant Tit ssp.]

Edith McGregor (fl.1907) was the wife of US ornithologist Richard Crittenden McGregor.

Edmund

Weaver sp. *Othyphantes edmundi* **Madarász**, 1914 NCR
[Alt. Baglafecht Weaver; JS *Ploceus baglafecht*]

Edmund Kovacs (1866–1919) was a Hungarian zoologist. He was in Abyssinia (1910–1919).

Edouard

White-winged Fairy-wren ssp. *Malurus leucopterus edouardi* **A. J. Campbell**, 1901

(See **Edward VII**)

Edouard (Verreaux)

Crested Guineafowl *Guttera edouardi* **Hartlaub**, 1867
[Syn. *Guttera pucherani edouardi*]

Jean Baptiste Edouard Verreaux (See **Verreaux, J. B.**)

Edquist

Paradise Kingfisher genus *Edquista* **Mathews**, 1918 NCR
[Now in *Tanysiptera*]

Alfred George Edquist (1873–1966) was an Australian educationist and collector. He was Supervisor of Nature Study and Science, Education Department of South Australia. He wrote *Nature Studies in Australasia* (1916).

Eduard

Tylas Vanga *Tylas eduardi* **Hartlaub**, 1862

(See **Newton, E.**)

Edward VII

White-winged Fairy-wren ssp. *Malurus leucopterus edouardi* **A. J. Campbell**, 1901

Edward VII (1841–1910) became King of Great Britain and Ireland on the death of his mother, Queen Victoria (1901). (See **Prince of Wales**)

Edward VIII

Edward Lyrebird *Menura novaehollandiae edwardi* Chisholm, 1921
[Alt. Superb Lyrebird ssp.]

Edward, Prince of Wales (1894–1972) became King of Great Britain as Edward VIII (1936) and abdicated later the same year. He lived the rest of his life as the Duke of Windsor.

Edward (Migdalski)

Spotted Dove ssp. *Streptopelia chinensis edwardi* **Ripley**, 1948 NCR
[JS *Spilopelia chinensis suratensis*]

Edward C. Migdalski (1918–2009) was an American ichthyologist, collector and preparator for the fish collection at Yale. After WW2 he was at Bingham Ocean Laboratory at Yale and led a collecting expedition in India and Nepal (1946–1947). He joined the Peabody Museum at Yale (1958). He wrote quite widely, either as sole or co-author, such as *The Fresh and Saltwater Fishes of the World* (1976) and *The Complete Book of Fish Taxidermy* (1981). He was also extremely interested in sports and games, founding the Yale Fishing Club and the Yale Shooting Club, and was fanatical about college gridiron football.

Edward (Wilson)

Edward's Hummingbird *Amazilia edward* **de Lattre** & **Bourcier**, 1846
[Alt. Snowy-bellied Hummingbird; Syn. *Saucerottia edward*]

(See **Wilson, E.**)

Edwards

Edwards's Lorikeet *Trichoglossus capistratus* **Bechstein**, 1811
[Alt. Marigold Lorikeet; Syn. *Trichoglossus haematodus capistratus*]
Great Indian Bustard *Otis edwardsii* **J. E. Gray**, 1831 NCR
[JS *Ardeotis nigriceps*]
Moss-backed Tanager *Bangsia edwardsi* Elliott, 1865
Edwards's Rosefinch *Carpodacus edwardsii* **J. Verreaux**, 1870
[Alt. Dark-rumped Rosefinch]
Swamphen sp. *Porphyrio edwardsi* Elliott, 1878 NCR
[Alt. Purple Swamphen; JS *Porphyrio porphyrio viridis*]
Cape Verde Shearwater *Calonectris edwardsii* **Oustalet**, 1883
Edwards's Fig Parrot *Psittaculirostris edwardsii* Oustalet, 1885
Edwards's Pheasant *Lophura edwardsi* Oustalet, 1896

Henri (father) and Alphonse (son) Milne-Edwards (see **Milne-Edwards**). The swamphen, and *probably* all the birds named by Oustalet, were named after Alphonse Milne-Edwards (1835–1900) but it is not always clear which Edwards is intended.

Edwards, E. P.

Singing Quail ssp. *Dactylortyx thoracicus edwardsi* **Warner** & **Harrell**, 1957

Dr Ernest Preston 'Buck' Edwards (1919–2011) was an American biologist, ornithologist and collector. The University of Virginia awarded his bachelor's degree (1940) and Cornell his master's (1941) and doctorate (1949). He served in the US Army in both WW2 and the Korean War. He taught at the University of Kentucky and the University of the Pacific and led birding tours to Mexico (1950s–early 1964). He was a member of the faculty and taught at Sweet Briar College, Virginia (1965–1990), retiring as Duberg Professor of Ecology. He wrote *Field Guide to the Birds of Mexico* (1972).

Edwards, G.

Australian Parrot sp. *Psittacus edwardsii* **Bechstein**, 1811 NCR
[Alt. Turquoise Parrot; JS *Neophema pulchella*]
Tropicbird sp. *Phaethon edwardsii* **Brandt**, 1838 NCR
[Alt. White-tailed Tropicbird; JS *Phaethon lepturus*]

Curassow sp. *Crax edwardsii* **Reichenbach**, 1850 NCR
[Alt. Great Curassow; JS *Crax rubra*]
White-bearded Manakin *Manacus edwardsi* **Bonaparte**, 1850 NCR
[JS *Manacus manacus*]

George Edwards (1694–1773) was an illustrator, naturalist, and ornithologist. He was Librarian, Royal College of Physicians, London (1733–1764), and corresponded regularly with Linnaeus. He wrote a 4-volume *A Natural History of Birds* (1743–1751). A reptile and mammal are named after him.

Edwin

Silvereye ssp. *Zosterops australasiae edwini* **Mathews**, 1923 NCR
[JS *Zosterops lateralis chloronotus*]

(See **Ashby**)

Egerton

Rusty-fronted Barwing *Actinodura egertoni* **Gould**, 1836

Sir Philip de Malpas Grey-Egerton (1806–1881) was an English palaeontologist and geologist. He travelled in Switzerland where he was introduced to Professor Louis Agassiz, who inspired him to make a study of fossil fish, and his collection of them is now in the British Museum. He was a Member of Parliament (1830–1881).

Egolia

Hummingbird genus *Egolia* **Mulsant**, **E. Verreaux** & **J. Verreaux**, 1865 NCR
[Now in *Calliphlox*/*Philodice*]

Perhaps derived from the Greek mythological character Aegolius (q.v.).

Egusquiza

Euphonia sp. *Euphonia egusquizae* **Bertoni**, 1901 NCR
[Alt. Green-chinned Euphonia; JS *Euphonia chalybea*]

General Juan Bautista Luis Egusquiza Isasi (1845–1902) was a Paraguayan statesman who was President of Paraguay (1894–1898).

Ehrenberg

Ehrenberg's Redstart *Phoenicurus phoenicurus samamisicus* Hablizi, 1783
[Alt. Common Redstart ssp.]
Spotted Thick-knee ssp. *Burhinus capensis ehrenbergi* **Zedlitz**, 1910 NCR
[JS *Burhinus capensis dodsoni*]

Professor Christian Gottfried Ehrenberg (1795–1876) was a German natural scientist. He started studying theology at Leipzig (1815) but changed direction (1817) and went to Berlin to study medicine. He was working on fungi and was a lecturer at the University of Berlin (1820), where he became a Professor of Medicine (1827). He travelled extensively (1820–1825), mainly in the company of his friend Hemprich (q.v.). They travelled widely in north-east Africa and the Middle

East, from Lebanon to Sinai and from the Nile to Abyssinia (Ethiopia). Furthermore, he travelled with Humboldt (q.v.) to Asia (1829) and visited England to meet Darwin (q.v.) at Oxford (1847). He is regarded as the founder of Micropalaeontology. He bequeathed his collection to the Zoology Museum in Berlin. He published a great many articles and books, especially on fungi and corals. He was the first person to establish that the phosphorescence in the sea is caused by the presence of plankton-like micro-organisms. A mammal and an amphibian are named after him.

Ehrenreich

Blue-grey Tanager ssp. *Thraupis episcopus ehrenreichi* **Reichenow**, 1915

Dr Paul Ehrenreich (1856–1914) was a German physician, anthropologist and explorer in Brazil (1884–1889), India and Eastern Asia (1892–1893), and in North America and Mexico (1898 and 1906).

Eichhorn

Eichhorn's Myzomela *Myzomela eichhorni* **Rothschild** & **Hartert**, 1901
[Alt. Crimson-rumped Myzomela, Yellow-vented Myzomela]
Eichhorn's Friarbird *Philemon eichhorni* Rothschild & Hartert, 1924
[Alt. New Ireland Friarbird]

Bronze Ground Dove ssp. *Gallicolumba beccarii eichhorni* Hartert, 1924
Velvet Flycatcher ssp. *Myiagra hebetior eichhorni* Hartert, 1924
Northern Fantail ssp. *Rhipidura rufiventris albertorum* Hartert, 1924 NCR
[JS *Rhipidura rufiventris setosa*]
Russet-tailed Thrush ssp. *Zoothera heinei eichhorni* Rothschild & Hartert, 1924
White-rumped Swiftlet ssp. *Aerodramus spodiopygius eichhorni* Hartert, 1924
Black Sunbird ssp. *Leptocoma sericea eichhorni* Rothschild & Hartert, 1926
Louisiade White-eye ssp. *Zosterops griseotinctus eichhorni* Hartert, 1926
Pied Goshawk ssp. *Accipiter albogularis eichhorni* Hartert, 1926
Solomons Hawk Owl ssp. *Ninox jacquinoti eichhorni* Hartert, 1929

Albert F. Eichhorn (c.1868–1933) was an Australian farmer. He collected in New Guinea (1900), the Solomon Islands (1903 and 1907) and New Ireland and New Britain (1910 and 1920s). Much of his collecting was for or with Albert Meek (q.v.), and the fantail's trinomial *albertorum* [= of the Alberts] commemorates both.

Eisenhofer

Laced Woodpecker ssp. *Picus vittatus eisenhoferi* **Gyldenstolpe**, 1916
[*P. vittatus* sometimes regarded as monotypic]

Emil Eisenhofer (1879–1962) was a German railway engineer who lived in Thailand (1903–1917 and 1930–1962). He was

interned (1917) as Siam declared war on Germany but returned to Germany in 1920. He worked in Turkey for some years (1920s). He worked (1930s and 1940s) on a number of projects, including building a large brewery in Bangkok.

Eisenmann

Azuero Parakeet *Pyrrhura eisenmanni* Delgado, 1985
[NUI Painted Parakeet; Syn. *Pyrrhura picta eisenmanni*]
Inca Wren *Pheugopedius eisenmanni* **T. A. Parker** & **O'Neill**, 1985
[Syn. *Thryothorus eisenmanni*]

Grassland Yellow Finch ssp. *Sicalis luteola eisenmanni* **Wetmore**, 1953
Olivaceous Piculet ssp. *Picumnus olivaceus eisenmanni* **W. H. Phelps** & **Aveledo**, 1966
Collared Inca ssp. *Coeligena torquata eisenmanni* **Weske**, 1985
Fiery-throated Hummingbird ssp. *Panterpe insignis eisenmanni* **Stiles**, 1985
Least Grebe ssp. *Tachybaptus dominicus eisenmanni* **Storer** & Getty, 1985

Dr Eugene Eisenmann (1906–1981) was an American ornithologist who was born in Panama. He graduated in law at Harvard (1930), then practised in New York, but found the job so irksome he retired early (1957) and became a naturalist. He was a Research Associate at the American Museum of Natural History (1957–1981) and was a member of the International Commission on Zoological Nomenclature.

Eisentraut

Eisentraut's Honeyguide *Melignomon eisentrauti* Louette, 1981
[Alt. Yellow-footed/Coe's Honeyguide]

Black-crowned Waxbill ssp. *Estrilda nonnula eisentrauti* **Wolters**, 1964

Professor Dr Martin Eisentraut (1902–1994) was a German zoologist and collector. He was on the staff of the Berlin Zoological Museum, working on bat migration and the physiology of hibernation. His first overseas trip was to West Africa (1938). He left Berlin to become Curator of Mammals at the Stuttgart Museum (1950–1957), then became Director of the Alexander Koenig Museum in Bonn, where he lived for the rest of his life. He made six trips to Bioko and Cameroon (1954–1973). Some of the material collected on these trips is still being studied. Eisentraut wrote many scientific papers and three books, including *Die Wirbeltierfauna von Fernando Poo und Westkamerun* (1973). He also published a slim volume of poems. Four mammals and a reptile are named after him.

Ekman

Hispaniolan Nightjar *Caprimulgus ekmani* **Lönnberg**, 1929
[Syn. *Antrostomus cubanensis ekmani*]

Dr Eric Leonard Ekman (1883–1931) was a Swedish botanist who lived in the West Indies (1917–1931). His passion for botany started when he was still at school and he went on to graduate from Lund University (1907). That year he was offered free passage to Argentina and he spent three months collecting plants in Misiones. He took a post at the Swedish Museum of Natural History in Stockholm (1908), and while there was able to travel and collect widely in Europe. Lund awarded his doctorate (1914). He hoped to collect in Brazil but was assigned stops in Cuba and Hispaniola. His plan to proceed to Brazil was delayed by WW1, plague in Cuba and political unrest in Haiti. He left Haiti for Cuba (1917) and, apart from another short visit to Haiti (1917), spent ten years there. Further disagreements with, and pressure from, the Swedish Royal Academy resulted in him returning to Haiti (1924–1928) followed by a period of intensive fieldwork in the Dominican Republic (although he found time to climb its mountains to discover which was the highest: Pico Duarte). He finally managed to persuade the Academy (1930) to send him to Brazil, but he died of influenza (weakened by pneumonia, malaria and blackwater fever) before he could set sail. There are streets named after him in Santiago and Santo Domingo, and a department of the Botanical Gardens in Cuba also bears his name.

Elbel

Black-crested Bulbul ssp. *Pycnonotus flaviventris elbeli*
Deignan 1954
Moustached Barbet ssp. *Megalaima incognita elbeli*
Deignan, 1956
Puff-throated Babbler ssp. *Pellorneum ruficeps elbeli*
Deignan, 1956

Dr Robert Edwin Elbel (1925–2005) was an American entomologist and herpetologist. The University of Kansas awarded his bachelor's and master's degrees, and the University of Oklahoma his doctorate. He served in the US Navy in WW2 and then for the US Civil Service (1950–1975) including being in Thailand (1951–1953 and 1961–1963) with the US Public Health Service's campaign to combat malaria. A mammal is named after him.

Elbert

Bronze-tailed Glossy Starling ssp. *Lamprotornis chalcurus elberti* **Neumann**, 1920 NCR
[JS *Lamprotornis chalcurus emini*]
Brown Babbler ssp. *Turdoides plebejus elberti* **Reichenow**, 1921 NCR
[JS *Turdoides plebejus cinerea*]
White-faced Cuckoo Dove ssp. *Turacoena manadensis elberti* **Rensch**, 1926 NCR; NRM

Dr Johannes Eugen Wilhelm Elbert (1878–1915) was a German naturalist. He, his wife and his assistant C. Gründler acquired an important collection (1910) during a geographical-botanical expedition in south-east Sulawesi, which he led on behalf of the Frankfurt Society for Geography and Statistics. The main purpose of the expedition was to explore the geographical relationship between the Asian and Australian areas. He explored in the Nusa Tenggara islands in Indonesia e.g. Bali, Lombok, Sumbawa, Salayer, Tukang Besi, Flores and Wetar; and also on the islands off Sulawesi,

such as Muna and Buton. He was also in the Cameroons (1914). Two amphibians are named after him.

Eleanor (Phillips)

Nightjar sp. *Caprimulgus eleanorae* **J. C. Phillips**, 1913 NCR
[Alt. Freckled Nightjar; JS *Caprimulgus tristigma*]

Mrs Eleanor Hayden Phillips *née* Hyde (1880–1975) was the wife (1908) of the describer Dr John Charles Phillips (q.v.) of Harvard University. He collected the bird in an expedition to the Sudanese province of Sennaar (1912–1913) for the Museum of Comparative Zoology, where he was Associate Curator of Birds. Although he qualified as a physician he never practised, was independently wealthy and philanthropic. He made other expeditions including to the Congo (1924). He wrote c.200 articles and books including the 4-volume *A Natural History of the Ducks* (1922–1926).

Eleanor (Semple Galey)

Brown-capped Vireo ssp. *Vireo leucophrys eleanorae*
G. Sutton & **Burleigh**, 1940

Mrs Eleanor Semple Galey (d.<1940) was the daughter of John Bonner Semple of Cornell University. The vireo was named 'in honor and in memory' of her.

Eleonora (of Arborea)

Eleonora's Falcon *Falco eleonorae* **Gené**, 1839

Eleonora of Arborea (c.1350–1404) was the warrior-princess national heroine of Sardinia who passed enlightened legislation to protect birds of prey (although cynics might say that this was to keep them for the aristocracy alone). She died 'during an epidemic of the pest [plague]'. The falcon was first observed in Sardinia in 1830.

Eleonora (van der Schroeff)

Eleonora's Cockatoo *Cacatua galerita eleonora* **Finsch**, 1863
[Alt. Sulphur-crested Cockatoo ssp.]

Maria Eleonora van der Schroeff (1812–1892) was the wife of Dr Gerardus Frederik Westerman (q.v.), the first Director (1843–1890) of Artis, the Amsterdam Zoo. Otto Finsch (q.v.) discovered this cockatoo at the zoo.

Elfrieda

Sabota Lark ssp. *Calendulauda sabota elfriedae* **Hoesch** & **Niethammer**, 1940 NCR
[NUI *Calendulauda sabota waibeli*]

Mrs Elfrieda Hoesch (DNF) was the wife of Dr Walter Hoesch (q.v.).

Elgas

Elgas's Greater White-fronted Goose *Anser albifrons elgasi*
Delacour & Ripley, 1975

Robert 'Bob' Elgas (1920–2010) was honoured, according to the etymology, because he '... has for some time conducted an active survey of these geese and endeavoured to find their nesting grounds'. He wrote the article 'Breeding populations of Tule White-fronted Geese in northwestern Canada' (1970).

Elgin

Elgin's Serpent Eagle *Spilornis elgini* **Blyth**, 1863
[Alt. Andaman Serpent Eagle]

James Bruce 8th Earl of Elgin (1811–1863) was a diplomat and colonial administrator who was Governor of Jamaica (1842–1847), Governor-General of Canada (1847–1854), High Commissioner to China (1857–1860), overseeing the 2nd Opium War and the destruction of the Old Summer Palace in Peking (Beijing), and Viceroy of India (1861–1863). It was his father, the 7th Earl who bought the famous marbles from the Parthenon. According to a contemporary account by Beavan (q.v.), despite Blyth (q.v.) describing the bird, Tytler (q.v.) found it and suggested it be named after Lord Elgin.

Eli

Pygmy Nuthatch ssp. *Sitta pygmaea elii* **A. R. Phillips**, 1986

(See **Ely**)

Elicia

Elicia's Goldentail *Hylocharis eliciae* **Bourcier** & **Mulsant**, 1846
[Alt. Blue-throated Goldentail]

Madame Elicia Alain – we have been unable to discover anything about her.

Elise

Elisa's Flycatcher *Ficedula elisae* **Weigold**, 1922
[Alt. Green-backed Flycatcher; Syn. *Ficedula narcissina elisae*]

Mrs Elise Johanna Marie Weigold *née* Anders (DNF) was the wife of the describer Dr Hugo Max Weigold (q.v.).

Elisabeth

Cuban Solitaire *Myadestes elisabeth* **Lembeye**, 1850

No etymology was given by the author and we have not been able to identify a strong candidate so the identity of the Elisabeth he had in mind may never be known.

Elisabeth (Heine)

Forest Kingfisher ssp. *Todiramphus macleayii elisabeth* **Heine**, 1883
[Trinomial often given as *elizabeth*, but original spelling was with an *s*]

Elisabeth Heine *née* Rimpau (1844–1932) was the wife of the describer, H. Heine (q.v.).

Elisabeth (Kozlova)

Great Grey Owl ssp. *Strix nebulosa elisabethae* **B. Stegmann**, 1925 NCR
[NUI *Strix nebulosa lapponica*]

Ortolan Bunting ssp. *Emberiza hortulana elisabethae* **H. Johansen**, 1944 NCR; NRM

Dr Elizabeth Vladimirovna Kozlova *née* Pushkariova (1892–1975) was a well-known Russian ornithologist whose husband was General Petr Kuzmich Kozlov. She worked at the Department of Ornithology at the St Petersburg Museum (1932–1975). She was a member of the expedition of the Geographical Society to Mongolia (1923–1926) and wrote *Birds of Southwestern Transbaikalia, Northern Mongolia and Central Gobi* (1930) based on the collection of this expedition. She divided a large part of Central Asia into ornithogeographic districts and reached important conclusions about the historic zoogeography, which she summarised in her papers 'Avifauna of the Tibetan Plateau, its genetic relationships and history' (1952) and 'The birds of zonal steppes and deserts of Central Asia' (1975).

Elisabeth (Stresemann)

Miombo Blue-eared Starling *Lamprotornis elisabeth* **Stresemann**, 1924
[Alt. Southern Blue-eared Glossy Starling]

Mrs Elisabeth Stresemann *née* Deniger (fl.1924) was the first wife of the describer, Professor Erwin Stresemann (q.v.).

Elisabeth (van Dedem)

Wakolo Myzomela ssp. *Myzomela wakoloensis elisabethae* **van Oort**, 1911

Elisabeth, Baroness van Dedem (DNF), was daughter of Baron Coenraad Willem van Dedem, a general in the Dutch army, and sister of Baron Frederik Karel van Dedem (1873–1959). She married (1871) William Edward Hartpole Lecky (1938–1903), an Irish historian.

Eliza (Alexander)

Black-crowned Waxbill ssp. *Estrilda nonnula elizae* **B. Alexander**, 1903

Lady Eliza Alexander *née* Speirs (d.1927). The original description contains no etymology so we cannot be sure that we are right, but at the time the description was made she was the widow of Sir Claud Alexander (1831–1899), who was related to the author.

Eliza (Lefèvre)

Mexican Sheartail *Doricha eliza* **Lesson** & **de Lattre**, 1839

Madame Eliza Lefèvre (DNF) was the wife of Dr Amédée Lefèvre (1798–1869), Professor of Zoology at Rochfort (1839).

Elizabeth (Campbell)

Superb Fairy-wren ssp. *Malurus cyaneus elizabethae* **A. J. Campbell**, 1901

Mrs Elizabeth Melrose Campbell *née* Anderson (d.1915) was the wife (1879) of Archibald James Campbell (q.v.).

Elizabeth (Elliott)

Common Diving-petrel ssp. *Pelecanoides urinatrix elizabethae* H. F. I. Elliott, 1954 NCR
[JS *Pelecanoides urinatrix dacunhae*]

Lady Elizabeth Margaret Elliott *née* Phillipson (1912–2007) was the wife (1939) of Sir Hugh Francis Ivo Elliot 3rd Baronet (1913–1989), the describer and holotype collector. He was Administrator of Tristan da Cunha (1950–1953) and President of the British Ornithologists' Union (1975–1979).

Elizabeth (La Touche)

Silver-breasted Broadbill ssp. *Serilophus lunatus elisabethae* **La Touche**, 1921
[Alt. Gould's/Collared Broadbill ssp.]
Vinous-throated Parrotbill ssp. *Sinosuthora webbiana elisabethae* La Touche, 1922

Elizabeth La Touche (DNF) was a daughter of the describer, J. D. D. La Touche (q.v.).

Elizabeth (Macdonald)

Double-banded Sandgrouse ssp. *Pterocles bicinctus elizabethae* **J. Macdonald**, 1954 NCR
[JS *Pterocles bicinctus bicinctus*]

Dr Elizabeth 'Betty' Macdonald (c.1922–c.2002) was a physician and the wife (1938) of James David Macdonald (q.v.). She went on the Hall Expedition to Australia (1962–1963), doubling as catering supremo and medical officer. They settled in Brisbane after his retirement (1969).

Elizabeth (van Someren)

Red-chested Flufftail ssp. *Sarothrura rufa elizabethae* **van Someren**, 1919

Mrs Elizabeth van Someren *née* Cunningham (DNF) was the wife of Dr Victor Gurner Logan van Someren (q.v.).

Ella

Philippine Fairy-bluebird ssp. *Irena cyanogastra ellae* **Steere**, 1890

Mrs Ella Beal (DNF) was a cousin of the describer.

Ellenbeck

Ellenbeck's Honeyguide *Prodotiscus zambesiae ellenbecki* **Erlanger**, 1901
[Alt. Green-backed/Slender-billed Honeyguide ssp.]
Black-faced Sandgrouse ssp. *Pterocles decoratus ellenbecki* Erlanger, 1905
Moorland Francolin ssp. *Scleroptila psilolaema ellenbecki* Erlanger, 1905 NCR
[NUI *Scleroptila psilolaema psilolaema*]

Dr Hans Ellenbeck (fl.1900) was a German physician and collector who was on Erlanger's (q.v.) expedition to Somalia, Abyssinia (Ethiopia) and Sudan (1900–1901).

Ellinor

Black-headed Apalis ssp. *Apalis melanocephala ellinorae* **van Someren**, 1944 NCR
[JS *Apalis melanocephala nigrodorsalis*]

Mrs Ellinor Catherine van Someren *née* Macdonald (1915–at least 1974) was the wife of the describer, Dr G. R. C. van Someren (b.1913), who worked for the East Africa Research Unit in Kenya. She made an outstanding contribution in the field of medical entomology, having worked in the Medical Research Laboratories, Nairobi (1935–1973). She wrote c.40 scientific papers.

Elliot, D. G.

Elliot's Storm-petrel *Oceanites gracilis* Elliot, 1859
[Alt. White-vented Storm-petrel]
Elliot's Woodpecker *Dendropicos elliotii* **Cassin**, 1863
Elliot's Laughingthrush *Trochalopteron elliotii* **J. Verreaux**, 1870
Elliot's Topaz *Crinis chlorolaemus* Elliot, 1870 NCR
[Hybrid: *Anthracothorax nigricollis* x *Chrysolampis mosquitus*]
Kofiau Paradise Kingfisher *Tanysiptera elliotii* **Sharpe**, 1870
Elliot's Pheasant *Syrmaticus elliotii* **Swinhoe**, 1872
Elliot's Bird-of-Paradise *Epimachus elliotii* **Ward**, 1873 NCR
[Hybrid? *Paradigalla carunculata* x *Epimachus fastosus*]
Elliot's Pitta *Hydrornis elliotii* **Oustalet**, 1874
[Alt. Bar-bellied Pitta; Syn. *Pitta elliotii*]
Elliot's Sapphire *Chlorestes subcaerulea* Elliot, 1874 NCR
[Probably an aberrant *Chlorestes notata*]
Elliot's Hummingbird *Amazilia lucida* Elliot, 1877
[Status uncertain: known from a single specimen]
Wine-throated Hummingbird *Atthis elliotii* **Ridgway**, 1878

Australasian Swamphen ssp. *Porphyrio poliocephalus elliotii* **Salvadori**, 1879 NCR
[Alt. Purple Swamphen ssp.; NUI *Porphyrio melanotus samoensis*]
Violet-crowned Hummingbird ssp. *Amazilia violiceps elliotii* **Berlepsch**, 1889
Thekla Lark ssp. *Galerida theklae elliotii* **Hartert**, 1897

Daniel Giraud Elliot (1835–1915) was Curator of Zoology at The Field Museum in Chicago and one of the founders of the American Ornithologists' Union. He was independently wealthy, enabling him to produce a series of bird books illustrated by magnificent colour plates, including his own excellent work, long after most publishers employed smaller formats and cheaper techniques. Elliot could also afford to commission the best bird artists of the day, including Josef Wolf and Josef Smit, both formerly employed by John Gould (q.v.). The lithograph series include works on pittas, pheasants, hornbills and birds of prey. The Daniel Giraud Elliot Medal is awarded by the US National Academy of Sciences '… for meritorious work in zoology or paleontology published in a three- to five-year period.' Two mammals are named after him.

Elliot, W.

Indian Babbler sp. *Turdoides elliotti* **Jerdon**, 1845 NCR
[Alt. Yellow-billed Babbler; JS *Turdoides affinis*]

Sir Walter Elliot (1803–1887) was a career civil servant in the Indian Civil Service, Honourable East India Company, Madras (1821–1860). He was Commissioner for the Administration of the Northern Circars (1845–1854) and a member of the Council of the Governor of Madras (1854–1860). A distinguished Orientalist, his interests included botany, zoology, Indian languages, numismatics, and archaeology. He was a regular correspondent of Charles Darwin (q.v.). His Indian herbarium was given to the Edinburgh Botanic Garden. He retired to Scotland and, despite blindness, worked on local natural history projects. Two mammals and two reptiles are named after him. Jerdon incorrectly spelt Elliot's name with a double 't'.

Elliott-Smith

Splendid Astrapia ssp. *Astrapia splendidissima elliottsmithi* **Gilliard**, 1961 NCR; NRM

Lieutenant-Colonel Sydney Elliott-Smith (1900–1974) was an Australian army officer who commanded a Papuan infantry battalion in WW2 (1944–1946). He was also a magistrate and Civil Administrator Papua (1942), Royal Papuan Constabulary and District Commissioner. He survived a plane crash in Papua New Guinea (1951). The astrapia was named after 'Mr and Mrs Sydney Elliott-Smith', so perhaps the plural form *elliottsmithorum* should have been used. Her name was Myola Elliott-Smith *née* Harboard (1897–1974) and a lake in PNG is named after her.

Ellis, W.

Brush Warbler genus *Ellisia* **Hartlaub**, 1860 NCR
[Now in *Nesillas*]

Reverend William Ellis (1794–1872) was a British Protestant missionary, botanist and gardener. He was ordained as a priest (1818) and became a missionary in Polynesia (1816–1824). As his wife was seriously ill they returned to England where he became Assistant Foreign Secretary, London Missionary Society (1830). His wife died (1835) and he re-married (1837) and lived in Hertfordshire, working as a Congregationalist pastor at Hoddesdon (1847). He resigned from the London Missionary Society as he became ill himself (1848), but recovered (1853) and was appointed as official representative of the London Missionary Society to Madagascar. He was refused permission to land on three occasions (1852–1856) – resulting from French pressure to exclude English-speaking Protestant missionaries from what they regarded as their exclusive sphere of influence. He stayed in Mauritius and eventually was allowed into Madagascar (1861–1865).

Ellis, W. W.

Greater Akialoa *Hemignathus ellisianus* **G. R. Gray**, 1860 EXTINCT
Ellis's Sandpiper *Prosobonia ellisi* **Sharpe**, 1906 EXTINCT
[Alt. White-winged Sandpiper]

William Webb Ellis (d.1786) was a surgeon's mate, artist and collector on board *Resolution* on Captain Cook's (q.v.) third voyage. Against the orders of the Admiralty, he published a two-volume version of the Cook expedition (1782), which contained engravings made from his own drawings. He also corresponded with Sir Joseph Banks (q.v.). He died as a result of a fall from the main mast of a ship in Ostend. We assume him to be a relative (the grandfather perhaps) of the famous William Webb Ellis, who invented Rugby football (1823) by picking up and running with the ball during a game of soccer; however, we have no direct evidence.

Ellsworth

Gentoo Penguin ssp. *Pygoscelis papua ellsworthi* **R. C. Murphy**, 1947

Lincoln Ellsworth (born William Linn Ellsworth) (1880–1951) was an American engineer from a very wealthy family. He explored in both the Arctic and Antarctic. He failed to reach the North Pole by airship with Amundsen (1925) but they flew over it (1926). He made four Antarctic expeditions (1933–1939) using a converted herring boat as his base (called *Wyatt Earp* after the legendary lawman, whom Ellsworth revered). Lots of Antarctica features are named Ellsworth after him.

Elpenor

Ascension Flightless Crake *Mundia elpenor* Olson, 1973
EXTINCT
[Syn. *Atlantisia elpenor*]

Elpenor was a member of Odysseus's crew in Greek mythology. On the island of Aeaea, he fell to his death from the roof of a house. Like Elpenor, the crake was stranded on an island, fell down a hole and perished (probably in the 18th century). Olson wrote in his etymology: 'The Ascension rail was also stranded on an island and upon falling off the lip of the fumarole, descended straight to the bowels of the earth and was not known again until its shades were stirred up by inquiring mortals invading its underworld tomb.'

Elphinstone

Elphinstone's Pigeon *Columba elphinstonii* **Sykes**, 1832
[Alt. Nilgiri Wood Pigeon, Spotted Wood Pigeon]

The Hon. Mountstuart Elphinstone (1779–1859) went to India (1795) as an employee of the East India Company, but became a diplomat in Poona and other Indian 'residencies', and was British Governor of Bombay (1819–1827). Although a diplomat, he gained a reputation for courage in various military actions. He returned to England (1827) and was twice offered the post of Governor-General of India, but he declined for reasons of poor health. In retirement he wrote a history of India.

Elsa

Chestnut-capped Brush Finch ssp. *Arremon brunneinucha elsae* **Parkes**, 1954

Dr Elsa Guerdrum Allen (1888–1969) of Cornell University was an ornithologist and the wife of Arthur A. Allen. She published a number of books and papers including 'John Abbot: pioneer naturalist of Georgia' (1957) and 'A history of American ornithology before Audubon' (1979).

Elsey

Black-fronted Dotterel genus *Elseyornis* **Mathews**, 1914

Pale-headed Rosella ssp. *Platycercus adscitus elseyi*
Mathews, 1912
[Hybrid *P. adscitus adscitus* x *P. adscitus palliceps*]

Dr Joseph Ravenscroft Elsey (1834–1858) was a surgeon, explorer and physician who qualified in London (1855). He was Assistant Naturalist on the North Australian exploring expedition led by Augustus Gregory (1855–1856). He was then on St Kitts, West Indies (1857–1858). His pastoral property, Elsey Station, was the setting for Mrs Aeneas Gunn's book on tropical outback life in Australia, *We of the Never-Never*. A genus of Australian turtles is named after him.

Elwes

Elwes's Crake *Amaurornis bicolor* **Walden**, 1872
[Alt. Black-tailed Crake; *Porzana elwesi* **Hume**, 1875 is a JS]
Elwes's Eared Pheasant *Crossoptilon harmani* Elwes, 1881
[Alt. Tibetan/Harman's Eared Pheasant]

Horned Lark ssp. *Eremophila alpestris elwesi* **Blanford**, 1872
Oriental White-eye ssp. *Zosterops palpebrosus elwesi*
E. C. S. Baker, 1922 NCR
[JS *Zosterops palpebrosus palpebrosus*]

Henry John Elwes (1846–1922) was a wealthy English collector and illustrator, mainly of plants but also of birds, butterflies and moths. After five years in the Guards he travelled very widely, over much of Europe, Asia Minor, India, Tibet, Mexico, North America, Chile, Russia, Siberia, Formosa, China and Japan. Elwes discovered many plants, some named after him. His home was Colesbourne Park near Cheltenham, UK, where he lived for 47 years (1875–1922). He created a collection of about 140 different varieties of snowdrops, including one named after him. He wrote *Memoirs of Travel, Sport and Natural History* and an article on 'The geographic distribution of Asiatic birds' (1873). Most of his writing was on birds, but he also wrote the 7-volume *Trees of Great Britain and Ireland* (1906–1913). He wrote a number of other monographs, such as one on lilies (1880).

Ely

Pygmy Nuthatch ssp. *Sitta pygmaea elii* **A. R. Phillips**, 1986

Charles Adelbert Ely (b.1933) is a biologist who was at Fort Hays State University, where he was Professor of Zoology. He co-wrote *Birds of Kansas* (2011).

Emerson

Yellow Rail ssp. *Coturnicops noveboracensis emersoni*
H. H. Bailey, 1935 NCR
[JS *Coturnicops noveboracensis noveboracensis*]

William Otto Emerson (1856–1940) was an American ornithologist, artist, collector for the California Academy of Sciences and a founder of the Cooper Ornithological Club.

Emil

Ornate Honeyeater ssp. *Melidectes torquatus emilii*
A. B. Meyer, 1886

Count Emil Turati (1858–1938) was a collector and a banker in Milan, and was a relative (probably son) of Conte Ercole Turati (q.v.).

Emile

Hummingbird genus *Emilia* **Mulsant & E. Verreaux**, 1866
NCR
[Now in *Lepidopyga*]

Ruddy Cuckoo Dove *Macropygia emiliana* Bonaparte, 1854

François-Charles-Émile Parzudaki *né* Fouqueux (1829–1862), who was possibly a relation of Charles Parzudaki (q.v.) was a French traveller, naturalist and collector of bird skins. He certainly used the name Parzudaki before 1858, when he was given permission to adopt the name Fouqueux-Parzudaki. He may have taken over the Parzudaki natural history business in Paris after Parzudaki's death (1847) and wanted to use the same name. He sold a number of specimens to P. L. Sclater (q.v.) that were collected by Emile Parzudaki on an expedition to Brazil and Colombia (1854). He collected with Sallé (q.v.) in Mexico (c.1855–1856) and they wrote *Catalogue des Oiseaux de Méxique* (1862). He was described by Bonaparte (q.v.) as being a 'jeune naturaliste voyageur' and remarked that he had had the cuckoo dove brought to his attention by Emile.

Emilia

Plain Antvireo ssp. *Dysithamnus mentalis emiliae* **Hellmayr**, 1912
Emilia's Antwren *Microrhopias quixensis emiliae* **Chapman**, 1921
[Alt. Dot-winged Antwren ssp.]

(See **Snethlage**)

Emilie, Comtesse de Dalmas

Tanager sp. *Calliste emiliae* **Dalmas**, 1900 NCR
[Alt. Rufous-winged Tanager; JS *Tangara lavinia*]

Emilie, Comtesse de Dalmas (DNF) was the wife of Raymond, Compte de Dalmas (1862–1930). (See **Dalmas**)

Emilie (Galichon)

Green Hermit ssp. *Phaethornis guy emiliae* **Bourcier & Mulsant**, 1846

Madame Emilie Galichon (DNF) was presumably a relation of Emile-Louis Galichon (1829–1875), a noted collector, art historian and editor of the *Gazette des Beaux-Arts*.

Emilie (Hose)

Mountain Black-eye *Chlorocharis emiliae* **Sharpe**, 1888

Mrs Emilie Ellen 'Poppy' Hose *née* Ravn (b.1879) was the wife of Dr Charles Hose (q.v.).

Emily (Sharpe)

Western Negrofinch ssp. *Nigrita canicapillus emiliae*
Sharpe, 1869
[Alt. Grey-headed Nigrita ssp.]
Common Paradise Kingfisher ssp. *Tanysiptera galatea
emiliae* Sharpe, 1871

Mrs Emily Eliza Sharpe *née* Burrows (1843–1928) was the
wife (1867) of R. B. Sharpe (q.v.).

Emin

Warbler genus *Eminia* **Hartlaub**, 1881

Chestnut Sparrow *Passer eminibey* Hartlaub, 1880
Emin's Shrike *Lanius gubernator* Hartlaub, 1882
Pratincole sp. *Glareola emini* **G. E. Shelley**, 1888 NCR
[Alt. Rock Pratincole; JS *Glareola nuchalis nuchalis*]
Rock-loving Cisticola *Cisticola emini* **Reichenow**, 1892
[Syn. *Cisticola aberrans emini*]

Emin's Weaver *Ploceus baglafecht emini* Hartlaub, 1882
[Alt. Baglafecht Weaver ssp.]
Spotted Creeper ssp. *Salpornis spilonotus emini* Hartlaub,
1884
d'Arnaud's Barbet ssp. *Trachyphonus darnaudii emini*
Reichenow, 1891
Red-rumped Swallow ssp. *Cecropis daurica emini*
Reichenow, 1892
Black-billed Turaco ssp. *Tauraco schuettii emini*
Reichenow, 1893
Brown-crowned Tchagra ssp. *Tchagra australis emini*
Reichenow, 1893
Red-bellied Paradise-flycatcher ssp. *Terpsiphone rufiventer
emini* Reichenow, 1893
Red-winged Pytilia ssp. *Pytilia phoenicoptera emini* **Hartert**,
1899
Speckle-fronted Weaver ssp. *Sporopipes frontalis emini*
Neumann, 1900
Arrow-marked Babbler ssp. *Turdoides jardineii emini*
Neumann, 1904
Black Scimitarbill ssp. *Rhinopomastus aterrimus emini*
Neumann, 1905
Emin's Black-collared Lovebird *Agapornis swindernianus
emini* Neumann, 1908
Chestnut-bellied Sandgrouse ssp. *Pterocles exustus emini*
Reichenow, 1919 NCR
[JS *Pterocles exustus olivascens*]
Bronze-tailed Starling ssp. *Lamprotornis chalcurus emini*
Neumann, 1920
Stone Partridge ssp. *Ptilopachus petrosus emini* Neumann,
1920 NCR
[NUI *Ptilopachus petrosus petrosus*]
Red-tailed Greenbul ssp. *Criniger calurus emini* **Chapin**,
1948

Emin Pasha (1840–1892), whose real name was Isaak Eduard
Schnitzer (q.v.), was a German explorer and administrator in
Africa. He made important contributions to the geographical
knowledge of the Sudan and central Africa. He became a
physician in Albania, which was then a part of the Ottoman
Empire. The people there called him *Emin*, meaning 'faithful

one'. He was appointed as Medical Officer (1876) on the staff
of General Charles G. Gordon, British Governor-General and
Administrator of the Sudan. Gordon, who became world
famous by being killed by the Mahdi at Khartoum (1885),
appointed Emin, with the title of Bey, to be the *Pasha*
(governor) of the southern Sudanese province of Equatoria
(1878). Emin then began his explorations, as well as being
active as a naturalist and collector. As a ruler, Emin's claim to
fame was that he abolished slavery in the territories which
he commanded. A Sudanese uprising (1885) forced him to
retreat into what is now Uganda. A search party, led by Henry
Morton Stanley, on what was to be his last African expedition, reached Emin (1888). Emin joined the German East
Africa Company (1890), which controlled what is now
Tanzania. He led an expedition to the upper Congo River
region but was beheaded by slave traders in the region of
Lake Tanganyika. Two mammals and two reptiles are named
after him.

Emin Bey

Emin Bey's Sparrow *Passer eminibey* **Hartlaub**, 1880
[Alt. Chestnut Sparrow]

(See **Emin**)

Emlen

'Emlen's Mockingbird' *Mimus polyglottos* **Linnaeus**, 1758
[Northern Mockingbird – see text]
Pewee sp. *Myiochanes emleni* **W. Stone**, 1931 NCR
[Alt. Greater Pewee; JS *Contopus pertinax minor*]

John Thompson Emlen (1908–1997) was an American zoologist and ornithologist who was Professor of Zoology,
Department of the University of Wisconsin (1946–1974). He
apparently studied one particular individual Northern Mockingbird *Mimus polyglottos* which came to be known as
'Emlen's Mockingbird' (not therefore the name of a new
species or subspecies). An amphibian is named after him.

Emma (Berlepsch)

Emma's Parrotlet *Touit emmae* **Berlepsch**, 1889 NCR
[Alt. Spot-winged Parrotlet; JS *Touit stictopterus*]

Long-tailed Sylph ssp. *Aglaiocercus kingii emmae*
Berlepsch, 1892

Emma Karoline Wilhelmine Gräfin von Berlepsch *née* von
Bülow (b.1855) was the wife (1881) of the German ornithologist Hans Graf von Berlepsch (q.v.).

Emma (Jabouille)

Purple-throated Sunbird ssp. *Leptocoma sperata emmae*
Delacour & **Jabouille**, 1928

Madame Emma Jabouille (NDF) was the wife of the French
ornithologist and civil servant Pierre Jabouille (q.v.).

Emma (Salvadori)

Emma's Conure *Pyrrhura emma* **Salvadori**, 1891
[Alt. Venezuelan Parakeet; Syn. *Pyrrhura leucotis emma*]

Mrs Emma Salvadori *née* Jourdain (1846–1928) was the wife of Alfredo Salvadori and the sister-in-law of Conte Tommaso Salvadori (q.v.), who named the bird after her.

Emmott

Splendid Fairy-wren ssp. *Malurus splendens emmottorum* **Schodde** & I. J. Mason, 1999

Angus James Emmott (b.1962) and his wife Karen (*emmottorum* = of the Emmotts) are Australian farmers, naturalists and photographers; he wrote *Snakes of Western Queensland: a Field Guide* (2009). Two reptiles are named after them.

Emperor

Emperor Penguin *Aptenodytes forsteri* **G. R. Gray**, 1844

Although there are occasional references to 'Emperor's Penguin', these are all mistakes for Emperor Penguin, which is named for its size and importance as being the largest penguin, and not after any specific emperor. Interestingly two other penguins which might have a possessive attached (King Penguin *Aptenodytes patagonicus* and Victoria Penguin *Eudyptes pachyrhynchus*) never seem to be referred to except by their correct name.

Emperor of Germany

Emperor of Germany's Bird-of-Paradise *Paradisaea guilielmi* **Cabanis**, 1888
[Alt. Emperor Bird-of-Paradise]

Kaiser Wilhelm II (1859–1941) was the last German Emperor and King of Prussia (reigned 1888–1918). He was an energetic promoter of the arts, sciences, public education and social welfare. However, as war leader he was considered ineffective and lost the support of the army, leading to his abdication (November 1918) and exile in the Netherlands. A mammal is named after him.

Empress

Empress Hummingbird *Heliodoxa imperatrix* **Gould**, 1856
[Alt. Empress Brilliant]

Gould wrote: 'I am desirous of dedicating this new Humming Bird to the Empress of the French'. (See **Eugenie**).

Empress of Germany

Empress of Germany's Bird-of-Paradise *Paradisaea raggiana augustaevictoriae* **Cabanis**, 1888
[Alt. Raggiana Bird-of-Paradise ssp.]

Auguste Viktoria of Schleswig-Holstein (1859–1921) was the first wife of Kaiser Wilhelm II (see **Emperor of Germany**).

Enendrozdov

Parrotbill genus *Enendrozdovoma* Kashin, 1978 NCR
[Now in *Conostoma*]

(See **Drozdov**)

Enewton

Mascarene White-eye sp. *Zosterops enewtoni* **Hartlaub**, 1877 NCR
[Alt. Reunion Grey White-eye; JS *Zosterops borbonicus*]

(See **Newton, E.**)

Engelbach

Little Cuckoo Dove ssp. *Macropygia ruficeps engelbachi* **Delacour**, 1928
Black-headed Sibia ssp. *Heterophasia desgodinsi engelbachi* Delacour, 1930
Engelbach's Silver Pheasant *Lophura nycthemera engelbachi* Delacour, 1948

Dr Pierre Engelbach (1890–1961) was an amateur French ornithologist, who was a resident of Kampot, Cambodia; he was the medical officer of the provinces of Pakse and Saravane (1924–1946) and accompanied Delacour on his sixth expedition to Indochina (Laos) (1931–1932). He visited Bokor, an area close to his home in Kampot (1935 and 1936). He carried out a limited ornithological survey of the Cardamom Mountains in Cambodia (1944), which he wrote up as 'Notes de voyage dans les monts des Cardamomes (Cambodge)' (1952). Earlier he wrote a 'Note sur quelques oiseaux du Cambodge' (1938), apparently unpublished.

Érard

Erard's Lark *Heteromirafra sidamoensis* Érard, 1975 NCR
[Alt. Sidamo/Archer's Lark; JS *Heteromirafra archeri*]
Erard's Lark, *Mirafra degodiensis* Érard, 1975 NCR
[JS Gillett's Lark *Mirafra gilletti*]

Christian Érard is a Professor and Assistant Director at the Laboratoire de Zoologie (Mammifères & Oiseaux) of the MNHN, Paris. He is a specialist in birds of tropical wet forests. He spent a total of 53 months in the field (1968–1987), making five visits to French Guiana and 17 trips to various African countries, which included over two years in Gabon. He has published many scientific papers, several books and more recently an online paper (2002). His most notable publications included a collaborative work *Oiseaux de Guyane* (1992). He is also a president of the Société d'Etudes Ornithologiques de France (SEOF) and editor of *Revue d'Ecologie: La Terre et la Vie*.

Erckel

Erckel's Francolin *Pternistis erckelii* **Rüppell**, 1835
[Syn. *Francolinus erckelii*]

Theodor Erckel (1811–1897) was a German taxidermist who became a servant and helper to Rüppell (q.v.) (1825) and accompanied him on his second journey to Ethiopia (1830). They returned to Europe (1834) and Erckel worked as Rüppell's assistant at the institute at Frankfurt University. The institute's Director, Cretzschmar (q.v.), and Rüppell quarrelled to such an extent that Rüppell withdrew (1844) and Erckel then took over as curator of the collections (1844–1880). He was elected a life member of the Senckenberg Society for

Wildlife Research (1891), which Cretzschmar founded. Rüppell named the francolin in recognition of Erckel's loyal service.

Erebus

Glossy Black Cockatoo ssp. *Calyptorhynchus lathami erebus* **Schodde** & I. J. Mason, 1993

In Greek mythology, Erebus was the son of Chaos and the personification of darkness and the gate to Hades – but this cockatoo is named after Mount Erebus in central Queensland, which in turn is named after the primordial deity.

Eriksson

Spike-heeled Lark ssp. *Chersomanes albofasciata erikssoni* **Hartert**, 1907

Axel Wilhelm Eriksson (1846–1901) was a Swedish pioneer settler, trader, ornithologist, collector and naturalist in South West Africa (Namibia) (1866–1901).

Erlanger

Apalis sp. *Apalis erlangeri* **Reichenow**, 1905 NCR
[Alt. Red-fronted Warbler; JS *Urorhipis rufifrons smithi*]
Crombec sp. *Sylvietta erlangeri* Reichenow, 1905 NCR
[Alt. Somali Crombec; JS *Sylvietta isabellina*]
Eremomela sp. *Eremomela erlangeri* Reichenow, 1905 NCR
[Alt. Yellow-vented Eremomela; JS *Eremomela flavicrissalis*]
Seedeater sp. *Poliospiza erlangeri* Reichenow, 1905 NCR
[Alt. Stripe-breasted Seedeater; JS *Crithagra reichardi striatipecta*]
Erlanger's Boubou *Laniarius erlangeri* Reichenow, 1905 NCR
[Alt Black/Somali Boubou; JS *Laniarius nigerrimus*]
Erlanger's Lark *Calandrella erlangeri* **Neumann**, 1906
Southern White-faced Owl *Ptilopsis erlangeri* **Ogilvie-Grant**, 1906
Western Black-headed Batis *Batis erlangeri* Neumann, 1907

Lanner Falcon ssp. *Falco biarmicus erlangeri* **Kleinschmidt**, 1901
Emerald-spotted Wood Dove ssp. *Turtur chalcospilos erlangeri* Reichenow, 1902 NCR; NRM
Eurasian Scops Owl ssp. *Otus scops erlangeri* **Tschusi**, 1904 NCR
[JS *Otus scops scops*]
Thekla Lark ssp. *Galerida theklae erlangeri* **Hartert**, 1904
Banded Martin ssp. *Riparia cincta erlangeri* Reichenow, 1905
Bearded Scrub Robin ssp. *Erythropygia quadrivirgata erlangeri* Reichenow, 1905 NCR
[JS *Erythropygia quadrivirgata quadrivirgata*]
Black-bellied Sunbird ssp. *Cinnyris nectarinioides erlangeri* Reichenow, 1905
Broad-ringed White-eye ssp. *Zosterops poliogastrus erlangeri* Reichenow, 1905 NCR
[Alt. Montane White-eye ssp,; JS *Zosterops poliogastrus poliogastrus*]
Grey-backed Camaroptera ssp. *Camaroptera brevicaudata erlangeri* Reichenow, 1905

Moorland Chat ssp. *Pinarochroa sordida erlangeri* Reichenow, 1905 NCR
[JS *Pinarochroa sordida sordida*]
White-bellied Bustard ssp. *Eupodotis senegalensis erlangeri* Reichenow, 1905
Pale Flycatcher ssp. *Bradornis pallidus erlangeri* Reichenow, 1905
Pale Prinia ssp. *Prinia somalica erlangeri* Reichenow, 1905
Common Waxbill ssp. *Estrilda astrild erlangeri* Reichenow, 1907 NCR
[JS *Estrilda astrild peasei*]
Spotted Creeper ssp. *Salpornis spilonotus erlangeri* Neumann, 1907
Black-crowned Tchagra ssp. *Tchagra senegalus erlangeri* Neumann, 1907 NCR
[JS *Tchagra senegalus habessinicus*]
White-browed Sparrow Weaver ssp. *Plocepasser mahali erlangeri* Reichenow, 1907 NCR
[JS *Plocepasser mahali propinquatus*]
Brubru ssp. *Nilaus afer erlangeri* **Hilgert**, 1907 NCR
[JS *Nilaus afer minor*]
Brown-hooded Kingfisher ssp. *Halcyon albiventris erlangeri* Neumann, 1908 NCR
[NUI *Halcyon albiventris orientalis*]
Rüppell's Griffon (Vulture) ssp. *Gyps rueppellii erlangeri* **Salvadori**, 1908
Chestnut-bellied Sandgrouse ssp. *Pterocles exustus erlangeri* Neumann, 1909
Hadada Ibis ssp. *Bostrychia hagedash erlangeri* Neumann, 1909 NCR
[NUI either *B. hagedash nilotica* or *B. h. brevirostris*]
Speckled Mousebird ssp. *Colius striatus erlangeri* **Zedlitz**, 1910
[SII *Colius striatus leucotis*]
Common Quail ssp. *Coturnix coturnix erlangeri* Zedlitz, 1912
Blackstart ssp. *Cercomela melanura erlangeri* Neumann & Zedlitz, 1913 NCR
[NPRB *Cercomela melanura neumanni*]
Lesser Honeyguide ssp. *Indicator minor erlangeri* Zedlitz 1913 NCR
[JS *Indicator minor diadematus*]
White-rumped Shrike ssp. *Eurocephalus ruppelli erlangeri* Zedlitz, 1913 NRM
[Alt. Northern White-crowned Shrike]
Collared Pratincole ssp. *Glareola pratincola erlangeri* Neumann, 1920
Barn Owl ssp. *Tyto alba erlangeri* **W. L. Sclater**, 1921

Baron Carlo von Erlanger (1872–1904) was a German collector from Ingelheim in the Rhineland. He travelled in Tunisia (1893 and 1897) and wrote two trip reports. He visited Abyssinia (Ethiopia) and Somaliland (1900–1901), accompanied for part of the time by Oskar Neumann (q.v.). He named c.40 new avian taxa as well as having almost as many named after him. He died in a car accident. Two mammals, a reptile and an amphibian are named after him.

Ermak

Common Crossbill ssp. *Loxia curvirostra ermaki* Kozlova, 1930 NCR
[JS *Loxia curvirostra curvirostra*]

Ermak (or Yermak) (d.1585) was a Cossack leader and Russian folk hero. He conquered the khanate of Sibir (the root of the name Siberia) (1582). He appears to have drowned, weighed down by his armour, whilst swimming across the Wagay River. His life and death have been the subjects of numerous Russian books, songs, and paintings.

Ermann

Laughing Dove ssp. *Spilopelia senegalensis ermanni* **Bonaparte**, 1856
[Syn. *Streptopelia senegalensis ermanni*]

Professor Georg Adolf Erman(n) (1806–1877) was a German physicist, explorer and plant-collector in Central Asia and Siberia. He was appointed Professor of Physics at Berlin in 1839.

Ernest (Hartert)

Barn Owl ssp. *Tyto alba ernesti* **Kleinschmidt**, 1901
Ernest's Grand Munia *Lonchura grandis ernesti* **Stresemann**, 1921
[Alt. Great-billed Manakin ssp.]
White-crowned Wheatear ssp. *Oenanthe leucopyga ernesti* **Meinertzhagen**, 1930
Black Monarch ssp. *Monarcha axillaris ernesti* **Rothschild**, 1931 NCR
[JS *Monarcha* (or *Symposiachrus*) *axillaris axillaris*]

(See **Ernst (Hartert)** and **Hartert, E.**)

Ernest (Hose)

Peregrine Falcon ssp. *Falco peregrinus ernesti* **Sharpe**, 1894

Ernest Shaw Hose (1873–1968) was in the Malayan Civil Service resident in Sarawak, Borneo (1892–1919), and as a planter, claiming to have been the first person to grow rubber there (1902). Like his cousin, Dr Charles Hose, he was a naturalist. He collected the type specimen of the Peregrine subspecies on Mount Dulit, Sarawak. He retired to Norfolk.

Ernest (Mayr)

Green-backed Robin ssp. *Pachycephalopsis hattamensis ernesti* **Hartert**, 1930

(See **Mayr, E.**)

Ernest (Saunders)

Moorland Chat ssp. *Pinarochroa sordida ernesti* **Sharpe**, 1900

Ernest H. Saunders (DNF) was working for the Australian Museum in Sydney, collecting on Lord Howe Island (1887). He was a collector and taxidermist on the Mackinder (q.v.) expedition to Mount Kenya (1899).

Ernst (Bartels)

Javan Needletail sp. *Hirundapus ernsti* M. Bartels Jr, 1931 NCR
[Alt. Silver-backed Needletail. Now regarded as a synonym of *Hirundapus cochinchinensis*]

Dr Ernst Bartels (1904–1976) was a Dutch naturalist. (See **Bartels**)

Ernst (Hartert)

Eurasian Treecreeper ssp. *Certhia familiaris ernsti* **Kuroda**, 1924 NCR
[JS *Certhia familiaris daurica*]
Lesser Spotted Woodpecker ssp. *Dendrocopos minor ernsti* **Domaniewski**, 1933 NCR
[JS *Dendrocopos minor colchicus*]
Marsh Tit ssp. *Poecile palustris ernsti* **Yamashina**, 1933

(See **Ernest (Hartert)** and **Hartert, E.**)

Ernst Mayr

Ebony Myzomela ssp. *Myzomela pammelaena ernstmayri* **Meise**, 1929
[Alt. Bismarck Black Myzomela ssp.]
Tawny-breasted Parrotfinch ssp. *Erythrura hyperythra ernstmayri* **Stresemann**, 1938 NCR
[JS *Erythrura hyperythra microrhyncha*]
Nepal House Martin ssp. *Hirundo cuttingi ernstmayri* **Wolters**, 1952 NCR
[Unnecessary replacement name for *Delichon nipalense nipalense*]
Arabian Golden-winged Grosbeak ssp. *Rhynchostruthus socotranus ernstmayri* Wolters, 1949 NCR
[JS *Rhynchostruthus percivali*]

(See **Mayr, E.**)

Erwin

Slaty-backed Flycatcher *Muscicapa erwini* **Wolters**, 1950 NCR
[Unnecessary replacement name for *Ficedula hodgsonii*]

Northern Puffback ssp. *Dryoscopus gambensis erwini* **Sassi**, 1923
[Alt. Ghazal Puffback, when merged with *D. gambensis malzacii*]
Glossy Swiftlet ssp. *Collocalia esculenta erwini* Collin & **Hartert**, 1927
Oriental White-eye ssp. *Zosterops palpebrosus erwini* **Chasen**, 1934 NCR
[JS *Zosterops palpebrosus auriventer*]
Lemon-bellied White-eye ssp. *Zosterops intermedia erwini* **Meise**, 1941 NCR
[NPRB *Zosterops chloris mentoris*]

(See **Stresemann**)

Esacus

Thick-knee genus *Esacus* **Lesson**, 1831

In classical mythology Esacus (or Aisakos) was a son of Priam, King of Troy. He was metamorphosed into a seabird or shorebird. The Greek word *aisakos* is the name of a shorebird of some kind, but its true identity is uncertain.

Escherich

Common Bulbul ssp. *Pycnonotus barbatus escherichi*
  Grote, 1922 NCR
[JS *Pycnonotus* (*barbatus*) *tricolor*]

Dr Karl Escherich (1871–1951) was a German physician, ento-mologist and collector who graduated in medicine (1893) and took a doctorate in zoology (1896). He became Professor of Forest Zoology at Tharandt Forest School (1907). He was Professor of Applied Zoology at Ludwig-Maximilians Univer-sity, Munich (1914) and Rector of the University (1933–1936). He visited the USA (1911) and was in the Cameroons (1912). He was an early supporter of Adolph Hitler and took part in the abortive putsch (1923), but after that appears to have tried to distance himself although still a member of the Nazi party.

Espinach

Spot-breasted Oriole ssp. *Icterus pectoralis espinachi*
  Ridgway, 1882

Ramón Espinach (fl.1880) was a Costa Rican landowner and rancher at La Palma de Nicoya. He apparently owned the oriole as a pet – it was reputed to have a wonderful voice, running through the scales perfectly!

Esteban

Chaco Earthcreeper ssp. *Tarphonomus certhioides estebani*
  Wetmore & J. L. Peters, 1949
[Syn. *Upucerthia certhioides estebani*]
Great Rufous Woodcreeper ssp. *Xiphocolaptes major*
  estebani Cardoso da Silva, **Novaes** & Oren, 1991

Dr Juan G. Esteban (fl.1940) was an Argentinean zoologist at the Instituto Miguel Lillo, Tucumán.

Estelle

Estella's Hillstar *Oreotrochilus estella* **d'Orbigny** &
  Lafresnaye, 1838
[Alt. Andean Hillstar]

Estelle-Marie d'Orbigny (1801–1893) was the elder sister of the senior describer, Alcide Dessalines d'Orbigny (q.v.). However, as the scientific name is not in the genitive, the word *estella* may be meant to convey the quality of the bird and that it actually means something like 'Starlike Hillstar'. It is only in Gould (q.v.) that we find 'Estella's'; everywhere else it is just 'Estella', and Gould may have been mistaken in thinking that it was eponymous – though it seems coinci-dental that a relative of the describer should bear a name so similar to the binomial.

Esther

Mountain Serin *Serinus estherae* **Finsch**, 1902

Esther Finsch (fl.1924) was the daughter of the describer, Otto Finsch (q.v.).

Estudillo

Estudillo's Curassow *Crax estudilloi* G. A. Allen, Jr *et al.*,
  1977 NCR
[Hybrid: *Crax* sp. x *Crax fasciolata*?]

Dr Jesus Estudillo Lopez (1933–2010) was a Mexican avicul-turist, zoologist and veterinary surgeon. He was internationally recognised for his skill in breeding threatened bird species in captivity, including the Resplendent Quetzal *Pharomacrus mocinno*, at his facility, Vida Silvestre, in Ixta-paluca, México. He wrote 'Notes on rare cracids in the wild and in captivity' (1986).

Etchécopar

African Barred Owlet ssp. *Glaucidium capense etchecopari*
  Erard & F. Roux, 1983
[Sometimes treated as a race of Chestnut Owlet,
  G. castaneum]

Robert Daniel Etchécopar (1905–1990) was a French ornithol-ogist who co-wrote, with F. Hüe, *Les Oiseaux du Nord de l'Afrique* (1964), *Les Oiseaux du Proche et du Moyen Orient* (1970) and *Les Oiseaux de Chine, de Mongolie et de Corée* (1976). He originally studied law and intended to follow his father into the family business as a notary. Instead he went to Paris where Jacques Berlioz (q.v.) at the MNHN welcomed him. He joined the staff of the Société Ornithologique de France (1935), devoting 30 years of his life to it, becoming secretary-general (1944). He became Director of the Ringing (banding) Centre of MNHN until retirement (1955–1976).

Ethel

Gerygone/Peep-warbler genus *Ethelornis* **Mathews**, 1912
  NCR
[Now in *Gerygone*]

Grey Butcherbird ssp. *Cracticus torquatus ethelae*
  Mathews, 1912 NCR
[JS *Cracticus torquatus leucopterus*]
Grey-fronted Honeyeater ssp. *Lichenostomus plumulus*
  ethelae Mathews, 1912 NCR
[Intergrade between *L. plumulus plumulus* and *L. plumulus*
  graingeri]
Rufous Fieldwren ssp. *Calamanthus campestris ethelae*
  Mathews, 1912 NCR
[NUI *Calamanthus campestris campestris*]
Silver Gull ssp. *Larus novaehollandiae ethelae* Mathews
  1912 NCR
[JS *Chroicocephalus novaehollandiae novaehollandiae*]
Painted Firetail ssp. *Emblema pictum ethelae* Mathews,
  1914 NCR; NRM
Mulga Parrot ssp. *Psephotus varius ethelae* Mathews 1917
  NCR; NRM

(See **White, S. A. & E. R.** and **Rosina**)

Eugen

Yellow-whiskered Greenbul ssp. *Andropadus latirostris*
  eugenius **Reichenow**, 1892 NCR
[NUI *Andropadus latirostris latirostris*; Syn. *Eurillas*
  latirostris latirostris]

(See **Kretschmer**)

Eugene

Blue Whistling Thrush ssp. *Myophonus caeruleus eugenei* **Hume**, 1873

(See **Oates**)

Eugenie

Hummingbird genus *Eugenia* **Gould**, 1856 NCR
[Now in *Heliodoxa*]

Eugenie's Fruit Dove *Ptilinopus eugeniae* Gould, 1856
[Alt. White-headed Fruit Dove]

The Empress Eugénie de Montijo (1826–1920) was a Spanish countess who married Napoleon III and reigned in France (1853–1870). She was the first person to have an asteroid named after her. She also has a very large diamond appropriately named in her honour, since she was renowned for her lavish and stylish lifestyle. (See also **Empress**).

Euler

Euler's Flycatcher *Lathrotriccus euleri* **Cabanis**, 1868
Pearly-breasted Cuckoo *Coccyzus euleri* Cabanis, 1873

Carl Hieronymus Euler (1834–1901) was a farmer, amateur ornithologist and Swiss consul in Rio de Janeiro, Brazil (1897–1901). Ornithological knowledge of the region of the north Paraíba valley and its c.360 species is largely due to survey work undertaken by him (1867–1900) and through his studies on the breeding behaviour of Brazilian birds. He deposited his collections in the Berlin Museum. Together with his comments on nests and eggs, they largely derive from the 'Boa Fazenda Vale Cantagalo', which was his property. Unfortunately, about 40% of the avifauna described by Carl Euler can no longer be found there.

Eurynome

Scale-throated Hermit *Phaethornis eurynome* **Lesson**, 1832

In Greek mythology Eurynome was the mother of the three Graces.

Eustace

Laura's Woodland Warbler ssp. *Phylloscopus laurae eustacei* **C. W. Benson**, 1954

Major William Eustace Poles (1902–1990) served in Burma (1943–1945). After WW2 he worked for the Game & Tsetse Department of Northern Rhodesia (Zambia). The Scottish ecologist Frank Fraser Darling accompanied Poles on a trek through Northern Rhodesia. As they rested on the top of Muchinga escarpment, Poles set out to 'create a view', which they could look at before they moved on. Fifty Africans were set to work felling trees, and 'after about two hours our felling revealed a very fine view indeed'.

Eva

Rainbow Starfrontlet ssp. *Coeligena iris eva* **Salvin**, 1897

Salvin provided no etymology, and the identity of 'Eva' is unknown.

Eva (Meise)

Sulawesi Myzomela ssp. *Myzomela chloroptera eva* **Meise**, 1929

Mrs Eva Meise (b.1907) was the wife of the describer Wilhelm Meise (q.v.).

Evelina

Evelina's Puffleg *Eriocnemis evelinae* **E. Hartert** & C Hartert, 1894 NCR
[Alt. Glowing Puffleg; JS *Eriocnemis vestita smaragdinipectus*]

The Hon. Charlotte Louise Adela Evelina Rothschild (1873–1947) was a sister of Walter Rothschild (q.v.).

Evelyn

Bahama Woodstar *Calliphlox evelynae* Bourcier, 1847

Bourcier (q.v.) called this hummingbird the 'Troch[ilus] d'Evelyn', but gave no etymology to say who he was referring to.

Everett

Everett's Bulbul *Hypsipetes everetti* **Tweeddale**, 1877
[Alt. Yellowish Bulbul; Syn. *Ixos everetti*]
Everett's Flowerpecker *Dicaeum everetti* **Sharpe**, 1877
[Alt. Brown-backed Flowerpecker]
Spot-throated Flameback *Dinopium everetti* Tweeddale, 1878
Everett's White-eye *Zosterops everetti* Tweeddale, 1878
Chestnut-crested Yuhina *Yuhina everetti* Sharpe, 1887
Bornean Piculet sp. *Sasia everetti* **Hargitt**, 1890 NCR
[Alt. Rufous Piculet; JS *Sasia abnormis*]
Everett's Thrush *Zoothera everetti* Sharpe, 1892
Everett's Spiderhunter *Arachnothera everetti* Sharpe, 1893
[Alt. Bornean Spiderhunter; Syn. *Arachnothera affinis everetti*]
Bush Warbler sp. *Androphilus everetti* **Hartert**, 1896 NCR
[Alt. Chestnut-backed Bush Warbler; JS *Bradypterus castaneus castaneus*]
Everett's Monarch *Monarcha everetti* Hartert, 1896
[Alt. White-tipped Monarch; Syn. *Symposiachrus everetti*]
Everett's Hornbill *Rhyticeros everetti* **Rothschild**, 1897
[Alt. Sumba Hornbill; Syn. *Aceros everetti*]
Everett's (Scops) Owl *Otus alfredi* Hartert, 1897
[Alt. Flores Scops Owl]
Gerygone sp. *Gerygone everetti* Hartert, 1897 NCR
[Alt. Plain Gerygone; JS *Gerygone inornata*]
Russet-capped Tesia *Tesia everetti* Hartert, 1897
Everett's Buttonquail *Turnix everetti* Hartert, 1898
[Alt. Sumba Buttonquail]

Everett's Blue-backed Parrot *Tanygnathus sumatranus everetti* Tweeddale, 1877
White-bellied Munia ssp. *Lonchura leucogastra everetti* Tweeddale, 1877
Philippine Scops Owl ssp. *Otus megalotis everetti* Tweeddale, 1879
Black-bibbed Cicadabird ssp. *Coracina mindanensis everetti* Sharpe, 1893

Philippine Green Pigeon ssp. *Treron axillaris everetti*
Rothschild, 1894

Rusty-breasted Whistler ssp. *Pachycephala fulvotincta everetti* Hartert, 1896

Mountain Tailorbird ssp. *Phyllergates cuculatus everetti* Hartert, 1897

Philippine Hawk Owl ssp. *Ninox spilocephala everetti* Sharpe, 1897 NCR
[JS *Ninox philippensis reyi*]

Black-backed Fruit Dove ssp. *Ptilinopus cinctus everetti* Rothschild, 1898

Elegant Pitta ssp. *Pitta elegans everetti* Hartert, 1898 NCR
[NUI *Pitta elegans concinna*]

Island Leaf Warbler ssp. *Phylloscopus poliocephalus everetti* Hartert, 1899

Eastern Koel ssp. *Eudynamys orientalis everetti* Hartert, 1900

Brush Cuckoo ssp. *Cacomantis variolosus everetti* Hartert, 1925
[Syn. *Cacomantis sepulcralis everetti*; or included in *Cacomantis variolosus sepulcralis*]

Australasian Barn Owl ssp. *Tyto alba everetti* Hartert, 1929 NCR
[JS *Tyto delicatula delicatula*]

Alfred Hart Everett (1848–1898) was a British civil servant who worked as an administrator in North Borneo. He was also a naturalist who collected for wealthy patrons such as the Marquess of Tweeddale and Walter Rothschild, who in turn named various species after Everett. He was engaged by the Royal Society (1878–1879) to explore 'the Caves of Borneo' in search of remains of ancient humans. It is believed that a jawbone from an orang-utan, which he found in a cave, may have been used in the 'Piltdown Man' hoax. Everett was interested in all aspects of natural history and anthropology. His death made the front page of the *Sarawak Gazette*. Five mammals, four amphibians and three reptiles are named after him.

Evermann

Evermann's Rock Ptarmigan *Lagopus muta evermanni*
D. Elliot, 1896

Dr Barton Warren Evermann (1853–1932) was a school-teacher (1876–1886) and a student at Indiana University, where he was awarded his bachelor's degree (1886), master's (1888) and doctorate (1891). He worked for the Bureau of Fishes in Washington (1891–1914) in various roles that he combined with lecturing on zoology at Cornell (1900–1903), Yale (1903–1906), and, later, Stanford, after he became Director of the Museum, California Academy of Sciences (1914). A reptile is also named after him.

Eversmann

Eversmann's Booted Warbler *Iduna caligata* **Lichtenstein**, 1823
[Alt. Booted Warbler; Syn. *Hippolais caligata*]

Eversmann's Redstart *Phoenicurus erythronotus* Eversmann, 1841
[Alt. Rufous-backed Redstart]

Eversmann's Pigeon *Columba eversmanni* **Bonaparte**, 1856
[Alt. Pale-backed Pigeon, Yellow-eyed Pigeon]

Eversmann's Warbler *Phylloscopus borealis* **Blasius**, 1858
[Alt. Arctic Warbler]

Eurasian Eagle Owl ssp. *Bubo bubo eversmanni* Dementiev, 1931 NCR
[JS *Bubo bubo turcomanus*]

Alexander Eduard Friedrich Eversmann (Eduard Aleksandrovich Eversmann) (1794–1860) was a pioneer Russian, of German extraction, physician and entomologist and Professor of Zoology at Kazan in Russia. After being educated in Germany he worked for two years as a physician (1818–1820) at an arms factory in Zlatoust. He became disenchanted with medicine and followed his fascination for zoology, eventually becoming Professor of Zoology and Botany at the University in Kazan. He travelled in, and wrote about remote areas of the Russian Empire in Asia – visiting Bukhara (1820), the Kirghiz steppes (1825), the Urals (1827), Orenburg and Astrakhan provinces (1829), and finally through the Caucasus (1830). He concentrated on Lepidoptera and during his travels collected widely, giving detailed scientific descriptions of many mammals, birds and insects, and becoming recognised as the greatest expert on the fauna of south-east Russia. Two mammals and a reptile are named after him.

Ewing

Ewing's Thornbill *Acanthiza ewingii* **Gould**, 1844
[Alt. Tasmanian Thornbill]

Ewing's Fruit Dove *Ptilinopus regina ewingii* Gould, 1842
[Alt. Rose-crowned Fruit Dove ssp.]

Reverend Thomas James Ewing (c.1813–1882) was born in England, moved to Tasmania (1833) and was admitted to holy orders (1838). He became headmaster of the Queen's Orphan Schools, as well as a keen amateur naturalist and collector. When John Gould (q.v.) stayed in Tasmania, Reverend Ewing helped him with his researches into the local avifauna. In recognition, Gould named the dove and thornbill after him. He was accused of serious misconduct with one of the older girls in his charge (1841), and although eventually exonerated of criminal behaviour, he was considered to have been imprudent and was deprived of his headmastership (1844), but was retained as Chaplain (1844–1863). He was in England in 1846–1847 and returned there for good in 1863, filling a number of ecclesiastical posts up to the time of his death. Ewing wrote 'List of birds of Tasmania'. An amphibian is named after him.

Exton

Yellow-fronted Tinkerbird ssp. *Pogoniulus chrysoconus extoni* **Layard**, 1871

Dr Hugh Exton (1833–1903) was an English physician, anthropologist, geologist and collector, who emigrated to South Africa (1861). He was the first President of the South African Geological Society.

Eyerdam

White-bellied Cuckooshrike ssp. *Coracina papuensis
eyerdami* **Mayr**, 1931

Walter Jakob Eyerdam (1892–1974) was an American cooper,
gold prospector, conchologist, naturalist and collector in
Alaska, Siberia (Kamchatka and Lake Baikal), Haiti (1927), the
Solomon Islands (1929–1930) with the Whitney South Pacific
Expedition of the AMNH, New York and Chile (1939, 1957 and
1959). The Field Museum, Chicago, houses his personal
collection of c.58,000 shells.

Eyles

Eyles's Harrier *Circus eylesi* **Scarlett**, 1953 EXTINCT

James 'Jim' Roy Eyles (1926–2004) was an amateur archaeologist and the major excavator of the Wairau Bar site in New
Zealand. When he was still a schoolboy (1939), he discovered
a moa-hunter's burial site near Blenheim, which included
human bones, tools, necklaces and a moa's egg from which
the contents had been extracted. Ronald Scarlett (q.v.),
assisted by Eyles, excavated the site (1942). Articles
mentioning Eyles are scattered in records of the Canterbury
Museum and the Royal Society of New Zealand library. He
was inaugural Director of the Nelson Provincial Museum,
though said not to be academically inclined.

Eyton

Eyton's Whistling Duck *Dendrocygna eytoni* Eyton, 1838
[Alt. Plumed Whistling Duck]
Eyton's Hill Partridge *Arborophila charltonii* Eyton, 1845
[Alt. Chestnut-necklaced (Hill) Partridge]
Spectacled Spiderhunter *Arachnothera eytoni* **Salvadori**,
1874 NCR

[Unnecessary replacement name for *Arachnothera
flavigaster*]

Buff-throated Woodcreeper ssp. *Xiphorhynchus guttatus
eytoni* **P. L. Sclater**, 1854

Thomas Campbell Eyton (1809–1880) was an English naturalist. He matriculated from St John's College, Cambridge
(1828) and corresponded with Agassiz and Darwin (q.v.),
although he opposed Darwinism. He wrote *History of the
Rarer British Birds* (1836), *A History of the Oyster and the
Oyster Fisheries* (1858) and *Osteologia Avium* (1871–1878).
Though it may seem that Eyton named the whistling duck
after himself, the name *eytoni* was coined by Gould. However,
it was not 'officially' published until Eyton produced *A Monograph on the Anatidae, or Duck Tribe* (1838).

Ezra

Ezra's Sunbird *Aethopyga nipalensis ezrai* **Delacour**, 1926
[Alt. Green-tailed Sunbird ssp.]

Alfred Aaron D. Ezra (1872–1955) was an aviculturist who
achieved the first recorded captive breeding in Britain of
many species, including the Rothschild's Myna *Leucopsar
rothschildi* (1931). He was born in India and privately
educated in Calcutta, the son of an immensely wealthy
merchant. The family made its money from exporting cotton
and, it is rumoured, opium. He came to Europe (1912), travelling overland through the Pamirs and Turkestan. He settled in
England and lived at Foxwarren Park in Surrey where his
extensive collection of birds included Pink-headed Ducks
Rhodonessa caryophyllacea (now extinct). On his death his
great friend Jean Delacour (q.v.) wrote that a 'happy, prosperous and delightful era in aviculture had come to an end,
the likes of which would probably not be seen again'.

F

Faber, D. A.

Hair-crested Drongo ssp. *Dicrurus hottentottus faberi*
Hoogerwerf, 1962

Dirk A. Faber (d.1960) was a Dutch soil scientist. He was Assistant Soil Surveyor at the Institute of Soil Research, part of the Agricultural Research Station at Bogor, Indonesia. The etymology states: "The new race is named to the honor of the late Mr. D. A. Faber, who accompanied the Prinsen Island expedition as a very ardent student in soil-conservation and who was a good comrade". He went on a number of expeditions, including the one to Prinsen Island (near Java) (1951), with Hoogerwerf, who describes in his book *Udjung Kulon, the Land of the Last Javan Rhinoceros*, how the two of them explored and searched together for the Javan Rhinoceros. Faber made a list of the plants and took soil and rock samples. He was taken seriously ill in 1956 and returned to the Netherlands, but never fully recovered.

Faber, F.

Long-tailed Duck *Clangula faberi* **C. L. Brehm**, 1824 NCR
[JS *Clangula hyemalis*]

Frederik Faber (1795–1828) was a Danish soldier, ornithologist and explorer. He collected in Iceland (1819–1821) and was Regimental Quartermaster, Schleswig Cavalry, at Horsens (1821–1828).

Fabricius

Great Black-backed Gull *Larus fabricii* **C. L. Brehm**, 1830
NCR
[JS *Larus marinus*]

Bishop Otto Fabricius (1744–1822) was a Danish missionary, explorer and naturalist in Greenland (1768–1773). He wrote *Fauna Groenlandica* (1780).

Faenorum

Desert Lark ssp. *Ammomanes deserti faenorum*
Meinertzhagen, 1951 NCR
[JS *Ammomanes deserti insularis*]

Lieutenant-Colonel Sir William Rupert Hay (1893–1962) and Lady Sybil Edith Hay (b.1894) *née* Abram (married 1925) as '*faenorum*' means 'of the hays'. Hay was a soldier who became an administrator and political officer in British Protectorates in Arabia, before and after serving in the Indian subcontinent. He was in charge of the largely Kurdish

area of Northern Iraq (1918–1920) and produced a book on his experiences: *Two Years in Kurdistan*. His job was to establish and maintain British rule in the area after WW1 and break-up of the Ottoman Empire. He was Chief Commissioner of Baluchistan (Pakistan) (1943–1946), then Political Officer at Bahrain (1946–1953) and responsible for what was called the Trucial States (UAE, Oman and Qatar). Twelve days before Meinertzhagen's (q.v.) description of this race of lark, Ripley (q.v.) described the same form as *Ammomanes deserti insularis* ['*insularis*' being Latin for 'of the island', which in this case is Bahrain] so Ripley's description has priority.

Fagan

Yemen Accentor *Prunella fagani* **Ogilvie-Grant**, 1913
[Alt. Arabian Accentor]

Red-bellied Malimbe ssp. *Malimbus erythrogaster fagani*
Ogilvie-Grant, 1907 NCR; NRM
Sickle-winged Guan ssp. *Chamaepetes goudotii fagani*
Chubb, 1917
Red-bellied Paradise-flycatcher ssp. *Terpsiphone rufiventer fagani* **Bannerman**, 1921

Charles Edward Fagan (1855–1921) was a British conservationist who was Assistant Secretary (1889–1919), then first Secretary, of the Natural History Departments of the British Museum (1919–1921). He had joined the museum aged 18 (1873) as a 'Second-class Assistant'. He took great interest in the Bird Room and enabled its expansion and development. His chief interest was in the preservation of native fauna at home and in the colonies. He was the British representative on the International Committee for the Protection of Nature (1913). He collected in New Guinea (1910–1912) on the BOU expedition under Ogilvie-Grant (q.v.). A mammal is named after him.

Fahrettin

Snow Finch ssp. *Montifringilla nivalis fahrettini* Watson, 1961
[Alt. White-winged Snowfinch ssp.]

Fahrettin Ozgecil (DNF) of the Turkish Forest Department, Antalya, accompanied the describer in western Turkey.

Fairbank

Kerala Laughingthrush *Trochalopteron fairbanki* **Blanford**,
1869
[Syn. *Strophocincla fairbanki, Garrulax fairbanki*]

Rev. Samuel Bacon Fairbank (1822–1898) was an American missionary in India (1846–1852). Illinois College awarded his BA (1842) and MA (1845), after which he attended Andover Theological Seminary (1845) where he was ordained DD (Doctor of Divinity) and where he became convinced his calling was as a missionary. He married Abbie Allen (1845) and the couple travelled to India (1846). After her death (1852) he went back to the USA, married Mary Ballentine (1856) and returned to India (1857), where he remained for the rest of his life. He lived in the Marathi district and spent much time and effort on education and improving the subsistence agriculture of the area. His main natural history interests were malacology and ornithology; he was a regular correspondent of Hume (q.v.) and a contributor to *Stray Feathers*. He died during a train journey in South India, possibly due to heatstroke.

Fairchild

Antipodes Island Parakeet *Pezoporus fairchildii* **J. Hector**, 1895 NCR
[JS *Cyanoramphus unicolor*]

Captain John Fairchild (1835–1898) was a New Zealander who was master on government steamers, and supply vessels to subantarctic islands. He rediscovered the parakeet (1886), which had originally been described without a type locality, by Lear (1831), on Antipodes Island. Fairchild died whilst superintending the loading of some iron rails when he was struck on the head by a chain. He is commemorated in other taxa such as the New Zealand Electric Ray.

Fairfax

Golden Bowerbird ssp. *Prionodura newtoniana fairfaxi* **Mathews**, 1915 NCR; NRM
Spotted Catbird ssp. *Ailuroedus melanotis fairfaxi* Mathews, 1915 NCR
[JS *Ailuroedus melanotis maculosus*]

Sir James Oswald Fairfax (1863–1928) was an Australian lawyer and newspaper proprietor. He died whilst playing a round of golf in Sydney.

Falck

Chukar Partridge ssp. *Alectoris chukar falki* **Hartert**, 1917

Dr Johan Peter Falck (1732–1774) was a Swedish physician and botanist who studied under Linnaeus (q.v. under Linné) at Uppsala and then went to Russia (1763), where he became Professor of Medicine and Botany at the medical college in St Petersburg (1765). He was a member of Pallas's (q.v.) expedition (1768–1774). He committed suicide.

Falkenstein

Falkenstein's Greenbul *Chlorocichla falkensteini* **Reichenow**, 1874
[Alt. Yellow-necked Greenbul]

Familiar Chat ssp. *Cercomela familiaris falkensteini* **Cabanis**, 1875

Variable Sunbird ssp. *Cinnyris venustus falkensteini* **G. A. Fischer** & Reichenow, 1884

Julius Falkenstein (1842–1917) was a German physician who acted as photographer for the German Africa Society's Loango Expedition to the west coast of Africa (1873–1876). The expedition brought back one of the first live gorillas to appear in Europe, which was exhibited at the Berlin Aquarium. He published, inter alia, *Afrikanisches Album*, which included 72 original photographs of the Loango coast.

Falla

Falla's Petrel *Pterodroma occulta* Imber & Tennyson, 2001
[Alt. Vanuatu Petrel]

Fairy Prion ssp. *Pachyptila turtur fallai* **Oliver**, 1930 NCR
[JS *Pachyptila turtur turtur*]

Sir Robert Alexander Falla (1901–1979) was an ornithologist and museum administrator in New Zealand. He was assistant zoologist with the British, Australian and New Zealand Antarctic Research Expedition led by Sir Douglas Mawson (1929–1931). He was Director of the Dominion Museum in Wellington (1947–1966). A reptile is named after him. (See also **Lalfa**)

Fannin

Fannin's Heron *Ardea herodias fannini* **Chapman**, 1901
[Alt. Great Blue Heron ssp.]

John [Jack] Fannin (1837–1904) was a Canadian naturalist and taxidermist. He was the first Curator (1886–1904) of the Provincial Museum of Natural History and Anthropology (Royal British Columbia Museum). He started the museum's collection of birds, other animals, plants and historical artefacts from British Columbia. He collected and preserved specimens from 1870 onwards, and published the first checklist of birds for British Columbia (1891), covering 307 species. He updated it as *A Preliminary Catalogue of the Collections of Natural History and Ethnology in the Provincial Museum, Victoria B.C* (1898). He noted (1894) 'Mr. Charles deB. Green, who spends a good deal of his spare time in making collections for the Museum, writes me from Kettle River, Okanogan district, British Columbia, to the effect that while climbing to an osprey's nest he was surprised to find his actions resented by not only the ospreys but also by a pair of Canada geese (*Branta canadensis*), the latter birds making quite a fuss all the time Mr. Green was in the tree.'

Fanny

Fanny's Woodstar *Myrtis fanny* **Lesson**, 1838
[Alt. Purple-collared Woodstar]
Green-crowned Woodnymph *Thalurania fannyi* **de Lattre** & **Bourcier**, 1846

Mrs Frances 'Fanny' Wilson *née* Stokes (DNF) was the wife of the British collector, Edward Wilson (q.v.). Her brother-in-law was Dr T. B. Wilson (q.v.). (See also **Frances (Wilson)**)

Fargo

Rufous-fronted Thornbird ssp. *Phacellodomus rufifrons fargoi* **Brodkorb**, 1935 NCR
[NUI *Phacellodomus rufifrons sincipitalis*]
Red-crested Finch ssp. *Coryphospingus cucullatus fargoi* Brodkorb, 1938

William Gilbert Fargo (1867–1957) was an American civil engineer, surveyor, palaeozoologist, malacologist and ornithologist who was Honorary Curator of Birds at the Museum of Zoology, University of Michigan.

Farkas

Farkas's Rock Thrush *Monticola sharpei bensoni* Farkas, 1971
[Alt. Benson's Rock Thrush; formerly *Pseudocossyphus bensoni*]

Dr Tibor B. Farkas (1921–1996) was a Hungarian-born South African ornithologist. Budapest University, where he taught (1952–1957), awarded his MSc in Zoology (1952), his teaching degree (1953) and PhD (1956). After research (1957–1958) at the University of Vienna he emigrated to South Africa and was an ornithologist for what is now Nature Conservation in Pretoria (1958–1966). After research in Germany (1967–1968) he worked at the Percy Fitzpatrick Institute (1969–1972). He led the Ornithology Department of the National Museum of Bloemfontein until retirement (1973–1986). He published c.60 papers, such as: 'The birds of Korannaberg, eastern Orange Free State, South Africa' (1988).

Farley

Boreal Chickadee ssp. *Poecile hudsonicus farleyi* **W. E. Godfrey**, 1951

Frank La Grange Farley (1870–1949) was a Canadian ornithologist and naturalist. He worked in an Ontario bank (1890–1892), then went to Red Deer, North-West Territories, Canada, as a homesteader (1892–1907). He moved to Camrose, Alberta (1907), and joined a real estate brokerage business. He married twice – to sisters! He wrote *Birds of the Battle River Region* (1932).

Farquhar

Chestnut-bellied Kingfisher *Todiramphus farquhari* **Sharpe**, 1899

Pacific Imperial Pigeon ssp. *Ducula pacifica farquhari* Sharpe, 1900 NCR
[NUI *Ducula pacifica pacifica*]

Admiral Sir Arthur Murray Farquhar (1855–1937) served in the Royal Navy. During the Franco-Prussian War (1870) he commanded HMS *Zealous* in the Pacific, and spent a lot of time in San Francisco, keeping in touch with London by telegraph in case of new orders. He was in the New Hebrides (Vanuatu) (1894).

Fatima Lima

Inambari Woodcreeper *Lepidocolaptes fatimalimae* Rodrigues *et al.*, 2013

Dr Maria de Fátima Cunha Lima is a Brazilian ornithologist at Universidade Federal de Pernambuco, Brazil. She was an assistant of the late Fernando da Costa Novaes (q.v.) at the Museu Paraense Emílio Goeldi in the 1980s.

Fatio

Eurasian Treecreeper ssp. *Certhia familiaris fatioi* **Jouard**, 1929 NCR
[JS *Certhia familiaris macrodactyla*]

Dr Victor Fatio de Beaumont (1838–1906) was a Swiss zoologist and physiologist. The University of Leipzig awarded his doctorate. He contracted typhoid (1861), causing him to forget nearly all his physiological knowledge. He went to Paris and studied zoology under Henri Milne-Edwards (q.v.) at MNHN (1862) and became an expert on phylloxera, a disastrous infection of grapevines, writing *État de la question phylloxérique en Europe en 1877* (1878). His major work was the 6-volume *Faune des Vertébrés de la Suisse* (1869–1904).

Faustino

Cream-bellied Fruit Dove ssp. *Ptilinopus merrilli faustinoi* Manuel, 1936

Dr Leopoldo Alcarez Faustino (1892–1935) was a Filipino zoologist. He attended the University of the Philippines but left without graduating (1912). He went to Columbus, Ohio, and worked as a messenger boy for Western Union, saving enough money to enroll in Ohio State University, where he graduated in engineering (1917). He returned to Manila (1918) and was employed as an assayer by the Bureau of Science and became lecturer in metallurgy, University of the Philippines (1920). Sponsored by the government he returned to Ohio State University (1921–1922) for degrees in mining engineering, moving to Stanford University for a master's degree (1922) and a doctorate (1924). He underwent field training with the US Geological Survey (1921–1924) and returned to the Philippines (1924). He became a lecturer in geology at the University of the Philippines (1926–1930). He was Acting Director of the National Museum (1930), then Director (1933) and chief of the division in mineral resources, Department of Agriculture and Commerce (1934). He wrote a number of short works such as *Coral reefs of the Philippine Islands* (1931). He died of cancer.

Faxon

Hermit Thrush ssp. *Catharus guttatus faxoni* **Bangs** & **T. E. Penard**, 1921

Dr Walter Faxon (1848–1920) was an American ornithologist and carcinologist, who worked at Harvard's Museum of Comparative Zoology. He proved that Brewster's Warbler is actually a hybrid. A peony was named after him by his botanist brother, Charles.

Fea

Fea's Thrush *Turdus feae* **Salvadori**, 1887
[Alt. Grey-sided Thrush]
Fea's Petrel *Pterodroma feae* Salvadori, 1899
[Alt. Cape Verde Petrel]
Honeyguide sp. *Indicator feae* Salvadori, 1901 NCR
[Alt. Spotted Honeyguide; JS *Indicator maculatus maculatus*]
São Tomé White-eye *Zosterops feae* Salvadori, 1901
[SII Principe White-eye *Zosterops ficedulinus*]

Black-throated Parrotbill ssp. *Suthora nipalensis feae* Salvadori, 1889
African Scops Owl ssp. *Otus senegalensis feae* Salvadori, 1903

Leonardo Fea (1852–1903) was an Italian explorer, painter and naturalist. He was Assistant at the Natural History Museum in Genoa and liked exploring far-off, little-known countries. He visited Burma (Myanmar), islands in the Gulf of Guinea, and the Cape Verde Islands. Salvadori (q.v.), who described all the birds, was a friend. Two mammals, four amphibians and five reptiles are named after him.

Featherston

Corncrake *Rallus featherstonii* **W. L. Buller**, 1865 NCR
[JS *Crex crex*]
Featherston's Shag *Phalacrocorax featherstoni* W. L. Buller, 1873
[Alt. Pitt Island Shag]

Dr Isaac Earl Featherston (1813–1876) was an English physician, newspaperman and politician. He qualified in medicine at the University of Edinburgh (1836), then became Surgeon Superintendent on the New Zealand Company ship *Olympus* (1841). Arriving in Wellington, he was shocked by the conditions there. He felt that the New Zealand Company deliberately misled would-be migrants and later fought a duel over the issue with Colonel William Wakefield, the company's principal agent. He practised medicine in Wellington for a number of years, during which time he became Editor of the *Wellington Independent* (1845). He took up politics to encourage devolution and was elected unopposed as the first Superintendent of Wellington Province (1853–1858), and later re-elected (1861–1870). The people he helped govern knew him as 'The Little Doctor'. He also took part in national politics and was a member of the House of Representatives (1853–1870). He was something of an arbitrator between Maori and settlers. Buller worked for him (1864) as his assistant, so his bird names honoured his erstwhile boss. Featherston became New Zealand's first agent general in London (1871) and died at his home in Sussex.

Fedden

White-bellied Woodpecker ssp. *Dryocopus javensis feddeni* **Blyth**, 1863

Francis Fedden (1839–1887) was a British surveyor, geologist, conchologist and collector. He graduated from the Royal School of Mines, London. Then the Geological Survey of India employed him (1860–1887), first in Burma (1869–1864),

then central India, where he eventually became Superintendent. He wrote: 'The geology of the Káthiáwár Peninsula in Guzerat' (1884). He died suddenly of heart disease. His widow gave his shell collection to the National Museum of Wales (1894).

Fedjuschin

Black Grouse ssp. *Lyrurus tetrix fedjuschini* Charlemagne, 1934 NCR
[JS *Lyrurus tetrix tetrix*]

Professor A. V. Fedjuschin (b.1892) was a Russian zoologist who was at the University of Minsk (1926–1957). He co-wrote *The Birds of White Russia* (1967).

Feilden

Feilden's Falconet *Polihierax insignis* **Walden**, 1871
[Alt. White-rumped Falcon]

Colonel Henry Wemyss Feilden (1838–1921) was a noted British natural historian and explorer who fought in the American Civil War for the Confederate side. He was a member of the Zoological Society of London. He visited the Faeroe Islands (1872) to look for a reported Great Auk *Pinguinus impennis*, despite it having been presumed extinct (1844). He was a contributor of various notes and regional lists, which appeared in early issues of *Ibis*, such as: 'Ornithological notes from Natal' by Majors E. A. Butler, H. W. Feilden and Captain S. G. Reid, in 1882, who 'found [them]selves condemned to a life of comparative idleness for months at Newcastle, owing to the unexpectedly peaceful results of the operations against the Boers ...' He also published a number of articles in the *Zoologist*, such as: 'The nest of the alligator' (1870).

Feldegg

Black-headed Wagtail *Motacilla* (*flava*) *feldegg* **Michahelles**, 1830

Feldegg's Falcon *Falco biarmicus feldeggii* **Schlegel**, 1843
[Alt. Lanner Falcon ssp.]

Colonel Baron Christoph Fellner von Feldegg (1780–1845) was an Austrian army officer, naturalist and ornithologist who collected in the Balkans. He fought in the Napoleonic wars and was made a baron (1817) because of his gallantry. He served in Dalmatia, becoming Colonel and CO of the 6th Battalion of Chasseurs. He served for a while with Michahelles (q.v.). His collection numbered 4,549 birds and 3,000 molluscs.

Félice (Abeillé)

Go-away-bird sp. *Chizaeris feliciae* **Lesson**, 1839 NCR
[Alt. Grey Go-away-bird; JS *Corythaixoides concolor*]

Copper-rumped Hummingbird ssp. *Saucerottia tobaci feliciae* Lesson, 1840
[Syn. *Amazilia tobaci feliciae*]
Violet-bellied Hummingbird ssp. *Damophila julie feliciana* Lesson, 1844

Félice Olymp (DNF) was the wife of Doctor Abeillé. (See **Abeillé**)

Félice (Hombron)

Golden Fruit Dove *Columba felicia* **Hombron** & **Jacquinot**, 1841 NCR
[JS *Ptilinopus luteovirens*]

Madame Félice Hombron (DNF) was the wife of Dr Jacques Hombron. (See **Hombron**)

Fenichel

Rufous-collared Monarch ssp. *Arses insularis fenicheli* **Mathews**, 1930 NCR; NRM

Sámuel Fenichel (1868–1893) was a Hungarian archaeologist, entomologist and ethnologist at the Hungarian National Museum (1888). He collected in New Guinea (1891–1893), where he died of gall bladder disease.

Fennell

Alpine Accentor ssp. *Prunella collaris fennelli* **Deignan**, 1964

Chester 'Chet' M. Fennell (1914–1972) was an American physician, medical researcher and collector. He was in the US army in Korea (1947–1948) and was in Taiwan (1959). He wrote 'Some observations on the birds of southern Korea' (1952).

Fenwick

Fenwick's Antpitta *Grallaria fenwickorum* Barrera & Bartels, 2010
[Alt. Antioquia Antpitta, Urrao Antpitta; Syn. *Grallaria urraoensis*]

Named after Dr George Fenwick, his wife Rita and their children Cyrus, Sarah and Rachel. George Fenwick is the President of the American Bird Conservancy. He received a PhD in Pathobiology from Johns Hopkins University. He created the American Bird Conservancy (1994) and directed the creation of the Latin American Bird Reserve Network. Rita was American Bird Conservancy (ABC)'s Vice-President for Development. George said of the antpitta's naming: 'I am deeply honored by this naming. I know it reflects in equal parts on the contributions of both my family and the ABC organization, both of which have sought to further bird conservation efforts in Colombia'. The evaluation was remarkable, as it was one of the first times that a new species had been described from a captured individual, ringed, measured, photographed, sampled for DNA, and then released. Shortly after the description using the scientific name *Grallaria fenwickorum* was published, a second description using the name *Grallaria urraoensis* appeared. Some question if the first description is valid.

Ferdinand, Glen

Yellow-rumped Thornbill ssp. *Acanthiza chrysorrhoa ferdinandi* **Mathews**, 1916 NCR
[Intergrade population of *Acanthiza chrysorrhoa* subspecies]

Named after the type locality, Glen Ferdinand in the Musgrave Ranges of central Australia.

Ferdinand, Tsar

Bananal Antbird *Cercomacra ferdinandi* **Snethlage**, 1928

Tsar Ferdinand of Bulgaria (1861–1948) was an ornithologist, botanist, author and collector of German origin (the same family as Prince Albert of Saxe-Coburg-Gotha, Queen Victoria's Prince Consort). He became regent of Bulgaria (1887) and later reigned as monarch (1908–1918). He abdicated after Bulgaria's defeat in WW1 and returned to Coburg, where he lived out his life.

Fermín

Zapata Wren genus *Ferminia* **Barbour**, 1926

(See **Cervera**)

Fernandez

Juan Fernández Firecrown *Sephanoides fernandensis* **P. P. King**, 1831
Juan Fernández Tit Tyrant *Anairetes fernandezianus* **Philippi**, 1857
Juan Fernández Petrel *Pterodroma externa* **Salvin**, 1875

These birds are not named directly after a person but after the three Juan Fernández Islands off the coast of Chile, to which they are all endemic. The islands were named after Juan Fernández de Quiros, a Portuguese navigator in the employ of the Spanish crown, who discovered them (1570s). The major island's claim to fame is that it was where Alexander Selkirk was marooned (1704) and became the real-life 'Robinson Crusoe' (see **Dampier**). It is now more often known as Isla Robinson Crusoe. Fernández also established a Christian settlement at Espiritu Santo, Vanuatu.

Fernandina

Fernandina's Flicker *Colaptes fernandinae* **Vigors**, 1827
[Alt. Cuban Flicker]
Yellow-headed Warbler *Teretistris fernandinae* **Lembeye**, 1850

José María de Jesús Damiano de Herrera y Herrera, Conde de la Fernandina (1788–1864), was a Spanish nobleman and head of a notable Cuban landowning family, which settled there mid-18th century. He succeeded to the title on his father's death (1819) and was loaded with honours and responsibilities by King Fernando VII, who also created him permanent royal councillor in Havana. Vigors says in his etymology 'I have named this species in honour of the Conde de Fernandina, at the express desire of Mr MacLeay, who had received various marks of attention and assistance in his Zoological pursuits from that nobleman, a resident and extensive proprietor in the Island of Cuba'. (As Fernandina was once another name for Cuba, as Vigors himself also points out, we first believed that the name was a toponym rather than an eponym.)

Fernando

Common Tailorbird ssp. *Orthotomus sutorius fernandonis*
  Whistler, 1939

E. C. Fernando (c.1900–1966) was a hunter and collector for the Colombo Museum, Ceylon (Sri Lanka) (1931). He worked as taxidermist and collector for the Field Museum, Chicago. He tanned the pelt of the largest leopard ever shot in Sri Lanka – 2.7 m long. A mammal is named after him. One of his 10 children, Dr Henry Fernando (1928–2008), was an entomologist with the Department of Agriculture in Colombo.

Ferrari-Pérez

Black-chested Sparrow sp. *Amphispiza ferrariperezi*
  Ridgway, 1886 NCR
  [JS *Peucaea humeralis*]

Professor Dr Fernando Ferrari-Pérez (1857–1933) was a Mexican zoologist and chemist who was naturalist for the Geographical and Exploring Commission, Republic of Mexico, which was established (1877) to catalogue the natural history and resources of Mexico. He wrote 'Catalogue of animals collected by the Geographical and Exploring Commission of the Republic of Mexico' (1886). Ferrari-Pérez is also credited with introducing cinema into Mexico by purchasing the rights for exploitation of the Frères Lumière patent for the country. A reptile was named after him.

Ferret

African Paradise-flycatcher ssp. *Terpsiphone viridis ferreti*
  Guérin-Méneville, 1843

Colonel Pierre Victor Adolphe Ferret (1814–1882) was a French army officer and explorer. He was in Abyssinia (1839–1843) and was Chief of Staff, Oran Province, Algeria (1866). He co-wrote: *Voyage en Abyssinie 1839–1843* (1847).

Ferrier

Rufous Scrub-bird ssp. *Atrichornis rufescens ferrieri*
  Schodde & Mason, 1999

Dr Simon Ferrier is an Australian entomologist and environmentalist. The University of Queensland awarded his bachelor's degree (1978) and the University of England NSW his doctorate (1984). He worked the Department of Environment and Climate Change, New South Wales (1985–2008) and then joined CSIRO, Canberra.

Ferry

Bananaquit ssp. *Coereba flaveola ferryi* **Cory**, 1909

John Farwell Ferry (1877–1910) was an American field naturalist, ornithologist and collector who graduated from Yale (1901) and worked for the US Biological Survey. He collected in Costa Rica, Panama, Venezuela and the Netherlands West Indies for the Field Museum, Chicago (1908–1910). He died of acute pneumonia.

Festa

Crested Lark ssp. *Galerida cristata festae* **Hartert**, 1922
Willow Tit ssp. *Parus atricapillus festae* **von Burg**, 1925 NCR
[JS *Poecile montanus montanus*]

Dr Enrico Festa (1868–1938) worked for the Zoological Museum of the University of Turin. He travelled and collected in Panama and Ecuador (1895–1898). He visited Palestine (1905) and the Lebanon and (early 1920s) took an expedition to Cyrenaica in Libya. Six reptiles and three amphibians are named after him.

Festetich

Fig Parrot sp. *Cyclopsittacus festetichi* Madarász, 1902 NCR
  [Alt. Double-eyed Fig Parrot; JS *Cyclopsitta diophthalma*]

Rudolf Count Festetics von Tolna (1865–1943) was a Hungarian traveller in the Pacific (1893–1900). He and his wife cruised in their own yacht *Tolna* and had a very adventurous time. The Count published an account of their voyage *Chez les Cannibales: Huit Ans de Croisière dans l'Océan Pacifique à Bord du Yacht 'Le Tolna'* (1903).

Fiedler

Social Flycatcher ssp. *Myiozetetes similis fiedleri*
  Dunajewski 1939 NCR
  [JS *Myiozetetes similis similis*]

Arkady Adam Fiedler (1894–1985) was a Polish writer, journalist and adventurer. He travelled in Mexico, Indochina, Brazil, Madagascar, West Africa, Peru (1934), USA and Canada (1965). He was an officer in the Polish army and took part in the Greater Poland Uprising (1918) and in WW2 was with the UK-based Polish armed forces.

Field

Greater Kestrel ssp. *Falco rupicoloides fieldi* **D. G. Elliot**, 1897

Marshall Field (1835–1906) was an American business magnate, founder of Marshall Field & Co., philanthropist and a major patron of the Columbian Museum of Chicago, which changed its name to the Field Museum in his honour (1905) and to better reflect its focus on natural sciences.

Figgins

Long-tailed Ground Dove ssp. *Uropelia campestris figginsi*
  Oberholser, 1931

Jesse Dade Figgins (1867–1944) was an American ornithologist and archaeologist, who was Director, Colorado Museum of Natural History (1910–1935), having previously been employed both by the USNM and the AMNH. After leaving Colorado he became affiliated to the University of Kentucky. He wrote *Birds of Kentucky* (1945).

Filchner

Daurian Redstart ssp. *Phoenicurus auroreus filchneri*
  Parrot, 1907 NCR
  [JS *Phoenicurus auroreus auroreus*]

Dr Wilhelm Filchner (1877–1957) was a German scientist. He explored in Russia (1898), Tibet and China (1903–1905) and Spitsbergen (1910), and was the leader of the German South Polar Expedition (1911–1912). He travelled and explored in

Nepal and Tibet (1939) and was in India when WW2 started. He was interned (1940–1946) and later stayed on, living in Poona (Pune). He wrote his autobiography *Ein Forscherleben* (1950). The Filchner Ice Shelf in Antarctica is named after him.

Filewood

> Crimson Rosella ssp. *Platycercus elegans filewoodi*
> **McAllan** & M. D. Bruce, 1989 NCR
> [NUI *Platycercus elegans elegans*]

Lionel Winston Charles Filewood (b.1936) is an Australian biologist who worked (1970s and 1980s) for the Department of Agriculture, Stock and Fisheries, Papua New Guinea. He co-wrote *Scientific Names used in Birds of New Guinea and Tropical Australia* (1978). An Australian court sentenced him to 9 months imprisonment for child pornography offences (2010).

Filhol

> Southern Rockhopper Penguin ssp. *Eudyptes chrysocome filholi* **Hutton**, 1879

Professor Dr Antoine Pierre Henri Filhol (1843–1902) was a French physician, malacologist, palaeontologist, speleologist and zoologist. He qualified as a physician (1873) and was doctor and naturalist on the French Expedition to Campbell Island, New Zealand (1874) to observe the Transit of Venus. He took part in a number of maritime expeditions on board *Travailleur* (1880–1882) and *Talisman* (1883). He was Professor of Comparative Animal Anatomy at the MNHN in Paris (1894–1902) and became a member of the Academy of Science (1897) and Chairman of the Zoological Society of France (1898). Among his writings is *Études sur les mammifères fossiles de Sansan* (1891). Filhol Peak on Campbell Island is named after him.

Finlayson

> Finlayson's Bulbul *Pycnonotus finlaysoni* **Strickland**, 1844
> [Alt. Stripe-throated Bulbul]

George Finlayson (1790–1823) was a Scottish surgeon and naturalist. He accompanied the East India Company mission to Siam and Cochinchina (1821–1822) as surgeon and naturalist, returning with it to Calcutta (1823). Finlayson's health had been broken by the trip and he died soon after returning to Britain. He wrote an account of the trip as *The Mission to Siam and Hue 1821–1822* (1826) with a memoir of the author by Raffles (q.v.). A mammal is named after him.

Finn

> Finn's Weaver *Ploceus megarhynchus* **Hume**, 1869
> [Alt. Finn's Baya Weaver, (Indian) Yellow Weaver]

Frank Finn (1868–1932) was a British ornithologist educated at Maidstone Grammar School and Brasenose College, Oxford. He collected in East Africa (1892) and became Assistant Superintendent of the Indian Museum, Calcutta (1894), then Deputy Superintendent (1895–1903). He was the author of many books on the birds of India, including *How to Know*

the *Indian Ducks* (1901), *The Birds of Calcutta* (1901), *How to Know The Indian Waders* (1906), *The Water Fowl of India and Asia* (1909), and *Game Birds of India and Asia* (1911) among others. Hume (q.v.) first described the weaver from a specimen in his own collection made at Kaladhungi (1869). Finn later rediscovered it near Calcutta, and Oates (q.v.) gave it its vernacular name (c.1890). Frank Finn used over 100 of the illustrations from Hume's *The Game Birds of India* (1880) in his *Indian Sporting Birds* (1915).

Finot

> White-throated Robin *Irania finoti* **Filippi**, 1863 NCR
> [JS *Irania gutturalis*]

Auguste Baron Finot (fl.1877) was French Consul to Persia (1863).

Finsch

> Pipipi genus *Finschia* **Hutton**, 1903

> Finsch's Amazon *Amazona finschi* **P. L. Sclater**, 1864
> [Alt. Lilac-crowned Amazon]
> Finsch's White-eye *Zosterops finschii* **Hartlaub**, 1868
> [Alt. Dusky White-eye]
> Finsch's Wheatear *Oenanthe finschii* **Heuglin**, 1869
> Finsch's Flycatcher-thrush *Stizorhina finschi* **Sharpe**, 1870
> [Alt. Finsch's Rufous Thrush, Finsch's Rusty Flycatcher]
> Finsch's Bearded Bulbul *Alophoixus finschii* **Salvadori**, 1871
> Finsch's Conure *Aratinga finschi* **Salvin**, 1871
> [Alt. Crimson-fronted Parakeet]
> Finsch's Parakeet *Psittacula finschii* **Hume**, 1874
> [Alt. Grey-headed Parakeet, Eastern Slaty-headed Parakeet]
> Finsch's Duck *Chenonetta finschi* Van Beneden, 1875
> EXTINCT
> Finsch's Euphonia *Euphonia finschi* Sclater & Salvin, 1877
> Finsch's Francolin *Scleroptila finschi* **Bocage**, 1881
> Finsch's Pygmy Parrot *Micropsitta finschii* **E. P. Ramsay**, 1881
> [Alt. Green Pygmy Parrot]
> Finsch's Imperial Pigeon *Ducula finschii* Ramsay, 1882
> Finsch's Reed Warbler *Acrocephalus rehsei* Finsch, 1883
> [Alt. Nauru Reed Warbler]
> South Island Oystercatcher *Haematopus finschi* **Martens**, 1897
> Finsch's Honeyeater *Lichmera notabilis* Finsch, 1898
> [Alt. Black-chested Honeyeater, Black-necklaced Honeyeater]

> Finsch's Bulbul *Iole olivacea charlottae* Finsch, 1867
> [Alt. Buff-vented Bulbul ssp.]
> Sulawesi Babbler ssp. *Trichastoma celebense finschi* **Walden**, 1876
> Northern Fantail ssp. *Rhipidura rufiventris finschii* Salvadori, 1882
> Red-bellied Pitta ssp. *Erythropitta erythrogaster finschii* Ramsay, 1884
> Lesser Bird-of-Paradise ssp. *Paradisaea minor finschi* **Meyer**, 1885
> Black-billed Turaco ssp. *Tauraco schuettii finschi* **Reichenow**, 1899 NCR
> [JS *Tauraco schuettii schuettii*]

Olive-brown Oriole ssp. *Oriolus melanotis finschi* **Hartert**, 1904

Thick-billed Flowerpecker ssp. *Dicaeum agile finschi* **Bartels**, 1914

Orange-bellied Fruit Dove ssp. *Ptilinopus iozonus finschi* **Mayr**, 1931

Friedrich Hermann Otto Finsch (1839–1917) was a German ethnographer, naturalist and traveller. He visited the Balkans, North America, Lapland, Turkestan and north-west China with Alfred Brehm (q.v.) and also the South Seas, spending nearly a year (1879–1880) on the Marshall Islands. Bismarck appointed him Imperial Commissioner for the German Colony of 'Kaiser-Wilhelm-Land' (1884) in New Guinea. He founded the town of Finschhafen there (1885), which remained the seat of German administration (1885–1918). He was the director of a number of museums at various times, including Bremen where he succeeded Hartlaub (q.v.) as Curator (1884), and Braunschweig. Among other works he co-wrote (with Hartlaub) *Die Vögel Ost-Afrikas*. A mammal and a reptile are named after him.

Fischer, Georg

Fischer's Fruit Dove *Ptilinopus fischeri* Brüggemann, 1876
[Alt. Red-eared Fruit Dove]

Georg Fischer (DNF) was an army surgeon who collected in the East Indies (1873–1896).

Fischer, Gustav

Fischer's Turaco *Tauraco fischeri* **Reichenow**, 1878

Fischer's Greenbul *Phyllastrephus fischeri* Reichenow, 1879

Fischer's Whydah *Vidua fischeri* Reichenow, 1882
[Alt. Straw-tailed Whydah]

Fischer's Sparrow Lark *Eremopterix leucopareia* Fischer & Reichenow, 1884
[Alt. Fischer's Finch Lark]

Fischer's Starling *Spreo fischeri* Reichenow, 1884
[Syn. *Lamprotornis fischeri*]

White-eyed Slaty Flycatcher *Dioptrornis fischeri* Reichenow, 1884

Fischer's Lovebird *Agapornis fischeri* Reichenow, 1887

Crombec sp. *Sylvietta fischeri* Reichenow, 1900 NCR
[Alt. Red-faced Crombec; JS *Sylvietta whytii minima*]

Flappet Lark ssp. *Mirafra rufocinnamomea fischeri* Reichenow, 1878

Yellow-rumped Tinkerbird ssp. *Pogoniulus bilineatus fischeri* Reichenow, 1880

Black-headed Weaver ssp. *Ploceus melanocephalus fischeri* Reichenow, 1887
[SII *Ploceus melanocephalus dimidiatus*]

Blue-headed Coucal ssp. *Centropus monachus fischeri* Reichenow, 1887

Hildebrandt's Francolin ssp. *Pternistis hildebrandti fischeri* Reichenow, 1887 NCR; NRM

Rattling Cisticola ssp. *Cisticola chiniana fischeri* Reichenow, 1891

Fischer's Ground Thrush *Zoothera guttata fischeri* **Hellmayr**, 1901
[Alt. Spotted Ground Thrush ssp. Syn. *Geokichla guttata*]

Pel's Fishing Owl ssp. *Scotopelia peli fischeri* **Zedlitz**, 1908 NCR; NRM

Dr Gustav Adolf Fischer (1848–1886) was a German explorer of East and Central Africa. He was an army physician who (1876) joined the expedition of the Denhardt brothers (Clemens and Gustav) to Wituland (coastal northern Kenya) and explored there again the following year. He explored the Tana River (1878), again together with the Denhardt brothers who were very influential in the region, and stayed on until (1890) the sultanate of Wituland came under British influence, after which the brothers lost all their property without compensation. Fischer stayed in Zanzibar (1878–1882) as a physician. He made (1882) an expedition into the interior of East Africa (Masai lands), sponsored by the Hamburger Geographischen Gesellschaft, then returned to Germany (1883). He undertook his last mission (1885) to meet Emin Pasha (q.v.), who was living in southern Sudan but who fled to Uganda because of a local uprising. Fischer reached Nyanza on Lake Victoria's east coast, but was unable to go further, so returned to Zanzibar. He died in Berlin the year after of a tropical fever. Fischer and Reichenow (q.v.) were very close friends, the latter describing many of the eponymous birds. A mammal is named after him.

Fischer, J. G.

Fischer's Eider *Somateria fischeri* **J. F. Brandt**, 1847
[Alt. Spectacled Eider]

Johann [Grigoriy] Gotthelf Fischer von Waldheim (1771–1853) was a German anatomist, entomologist and palaeontologist who settled in Russia (1804), becoming a professor at Moscow University where he founded (1805) the Moscow Imperial Society of Naturalists. He remained there for the rest of his life. Fischer's major contributions in virtually every field of natural history made him one of the intellectual giants of his age, playing a major role in the development of the natural sciences in Russia. His other areas of interest included the study of molluscs and other marine invertebrates. He was the author of numerous works on palaeontology, geology and entomology, including his 5-volume monograph *Ethnomography of Russia* (1820–1851). He was one of the initiators and founder members of the Moscow Naturalists' Society.

Fisher, A. K.

Mountain Pygmy Owl sp. *Glaucidium fisheri* **Nelson** & **T. S. Palmer**, 1894 NCR
[JS *Glaucidium gnoma gnoma*]

Seaside Sparrow ssp. *Ammodramus maritimus fisheri* **Chapman**, 1899

Song Sparrow ssp. *Melospiza melodia fisherella* **Oberholser**, 1911 NCR
[NUI *Melospiza melodia montana*]

Dr Albert Kenrick Fisher (1856–1948) was an American physician, ornithologist, collector and founder member of the American Ornithologists' Union. He helped found the Branch of Economic Ornithology in the U.S. Department of Agriculture, Division of Entomology (1885), and was involved in the creation of the Bureau of Biological Survey, part of the Fish and Wildlife Service. He was on a number of major expeditions – Death Valley (1891), Harriman Alaska (1899) and Pinchot South Seas Expedition (1929) before retiring (1931). He wrote *The Hawks and Owls of the United States in Their Relation to Agriculture* (1893). An amphibian is named after him.

Fisher, T. H.

Romblon Hawk Owl ssp. *Ninox spilonotus fisheri*
Rasmussen *et al.* 2012

Timothy H. Fisher (1947–2010) was a British ornithologist and pioneer birdwatcher in the Philippines, where he lived (1974–2010). He had a bachelor's degree in biology and was a qualified chartered accountant. He co-wrote *A Photographic Guide to Birds of the Philippines* (2001).

Fisher, W. J.

Fisher's Petrel *Aestrelata fisheri* **Ridgway**, 1883 NCR
[JS Mottled Petrel *Pterodroma inexpectata*]

William James Fisher (1830–1903) was a US Tidal Observer and trader at Kodiak Island, Alaska (1879–1899). He took the type specimen of the petrel (1882), sending it to the USNM, where his collection of artifacts and natural history specimens is known as the Fisher Collection.

Fitzgerald

Rufous-bellied Seedsnipe ssp. *Attagis gayi fitzgeraldi*
Chubb, 1918 NCR
[JS *Attagis gayi gayi*]

Major Edward Arthur Fitzgerald (1871–1930) was an officer in the British army and a mountaineer, traveller and explorer. He led a number of expeditions to New Zealand and South America. He was the first to ascend Aconcagua (1897), the highest mountain in South America. A reptile is named after him.

Fitzpatrick

Sira Barbet *Capito fitzpatricki* Seeholzer *et al.*, 2012

Fulvous Wren ssp. *Cinnycerthia fulva fitzpatricki* **Remsen** & Brumfield, 1998

John Weaver Fitzpatrick (b.1951) is an American ornithologist. Harvard awarded his BA in Biology (1974) and Princeton his PhD (1978). He was then Curator of Birds at the Field Museum, Chicago (1978–1989), during which he led a number of expeditions to Peru, discovering and describing seven new species. He was Executive Director of the Archbold Biological Station in Florida (1988–1995) then became Director of Cornell Laboratory of Ornithology and Professor in ecology and evolutionary biology (1995). He has written a number of articles and papers, as well as contributions to larger works and several books including co-writing *Neotropical Birds: Ecology and Conservation* (1996).

Fitzroy (River)

Fitzroy Cockatoo *Cacatua galerita fitzroyi* **Mathews**, 1912
[Alt. Sulphur-crested Cockatoo ssp.]

Named after the type locality, the Fitzroy River in the north of Western Australia, which was named after Captain Robert FitzRoy, commander HMS *Beagle* on Darwin's voyage and founder of the Meteorological Office, London.

Fitzsimons

Black-throated Canary ssp. *Serinus angolensis fitzsimonsi*
J. A. Roberts, 1932 NCR
[JS *Crithagra atrogularis semideserti*]

Frederick William FitzSimons (1871–1951) was Director, Port Elizabeth Museum and Snake Park (1906–1936). He was a dynamic personality, appointed to run a 'sleepy' museum, and quickly energised it and the local inhabitants. He wrote *Snakes* (1932). A reptile is named after him.

Fjeldså

Brown-backed Antwren *Epinecrophylla fjeldsaai* Krabbe *et al.*, 1999
[Syn. *Myrmotherula fjeldsaai*]

Professor Jon Fjeldså (b.1942) is Professor of Biodiversity at the Zoological Museum, University of Copenhagen, Denmark. Bergen University, Norway awarded his first degree and University of Copenhagen his doctorate (1975); he has taught at the latter ever since, as well as being curator (1971). His main interests are the systematics and biogeography of Andean and African birds. He has built up an avian tissue collection, which (1987) was the second largest in the world, in order to study avian genetics. He has held a number of high offices, being on the Council of BirdLife International and the Danish Representative on the International Ornithological Congress. He has written or co-authored over 100 scientific papers and several books including *Birds of the High Andes* (1990) and *The Grebes: Podicipedidae (Bird Families of the World)* (2004). The etymology for the antwren reads: 'We take the pleasure of naming this species in honor of Prof. Jon Fjeldså of the Zoological Museum, University of Copenhagen. Through his countless publications, most based on results obtained during field trips to the most hostile of environments, he has inspired a large number of biologists to leave their desks and get into the field'.

Fleay

Wedge-tailed Eagle ssp. *Aquila audax fleayi* **Condon** & Amadon, 1954

David Howells Fleay (1907–1993) was an Australian naturalist who pioneered the captive breeding of endangered species, being the first to breed captive platypus. He worked in his father's chemist shop, then as a teacher, but went on to study for a degree in botany, zoology and education at Melbourne University (1927–1931). He became passionately involved in conservation. He was the last person to photograph a thylacine (Tasmanian 'tiger', in captivity in Hobart Zoo); it bit him on the buttock and he proudly carried the scar for life. Asked

(1934) to design the 'native animals' section for Melbourne Zoo, where he worked four years developing captive breeding programmes. He was involved in captive breeding for several decades (1940s and 1950s). He co-founded the Wildlife Preservation Society of Queensland (1962) and created a conservation park, now named after him, that was sold to the Queensland government, and worked there for the rest of his life. An amphibian is named after him.

Fleck

Fleck's Coucal *Centropus senegalensis flecki* **Reichenow**, 1893
[Alt. Senegal Coucal ssp.]
Long-billed Crombec ssp. *Sylvietta rufescens flecki* Reichenow, 1900

Dr Eduard Fleck (1843–at least 1911) was a German ornithologist who explored and collected in German South West Africa (Namibia) (1889–1892). He sent his specimens to Reichenow (q.v.) in Berlin. He was elected a member of the Deutsche Ornithologische Gesellschaft (1894) and published extensively. However, after a few years the flow of publications abruptly ceased when (c.1895) he moved to the village of Azuga in the Transylvanian Alps (then part of the Austro-Hungarian Empire, now in Romania). There, with an old schoolfriend, he set up a factory to make Portland cement and (1911) worked as director and continued private study.

Flegel

West African Seedeater *Poliospiza flegeli* **Hartert**, 1886 NCR
[JS *Crithagra canicapilla*]

Eduard Robert Flegel (1855–1886) was a Baltic-German (born Lithuania) explorer. He was originally a merchant and worked in Lagos (1875–1879), where he joined an expedition to explore the River Niger. He then explored in West Africa (1880–1882) and led the Niger Expedition (1885–1886) for the German African Society. Some of his writings were collected and published posthumously as *Vom Niger-Benüe; Briefe aus Afrika* (1890).

Fleming, C.

Fulmar Prion ssp. *Pachyptila crassirostris flemingi* Tennyson & Bartle, 2005

Sir Charles Alexander Fleming (1916–1987) was a New Zealand ornithologist, palaeontologist and environmentalist. He was President, the Orniithological Society of New Zealand (1948–1949). He was a mentor and inspiration to the describers and an expert on the natural history of the Auckland Islands.

Fleming, J. H.

Fleming's Grouse *Dendragapus obscurus flemingi* **Taverner**, 1914 NCR
[Alt. Dusky Grouse; JS *Dendragapus obscurus richardsonii*]

James Henry Fleming (1872–1940) was a Canadian ornithologist who developed an early interest in birds and by 1888 was a member of the Royal Canadian Institute. He became an Associate Member of the AOU (1893) and was later its President (1932–1935). The National Museum of Canada made him Honorary Curator of Ornithology (1913), and many other honours followed worldwide. He was Honorary Curator of the Royal Ontario Museum of Zoology (1927). He collected more than 32,000 specimens and one of the largest personal ornithological libraries of his time, both of which he bequeathed to the Royal Ontario Museum.

Fleming, R. L.

Black-faced Laughingthrush ssp *Garrulax affinis flemingi* **Rand**, 1953 NCR
[JS *Trochalopteron affine affine*]
Rock Bunting ssp. *Emberiza cia flemingorum* **J. Martens**, 1972

Dr Robert Leland Fleming Sr (1905–1987) was at the Field Museum, Chicago. He was in India (1928) and spent many years resident in Nepal. He became Superintendent of Public Schools in India (1949–1953). His son, Dr Robert 'Bob' Leland Fleming Jr, graduated with a bachelor's degree from Albion College (1959) and gained a doctorate at Michigan State University (1967) and is a Professor at Future Generations Graduate School. The Rock Bunting is named after both of them and they co-wrote *Birds of Nepal* (1976).

Flemming

Spot-crowned Antvireo ssp. *Dysithamnus puncticeps flemmingi* **Hartert**, 1900

G. Flemming (DNF) was a collector in Ecuador (1899–1900).

Fletcher

Australian Swamphen ssp. *Porphyrio melanotus fletcherae* **Mathews**, 1911 NCR
[JS *Porphyrio melanotus melanotus*]
Superb Fairywren ssp. *Malurus cyaneus fletcherae* Mathews, 1912 NCR
[JS *Malurus cyaneus cyaneus*]

Jane Ada Fletcher (1870–1956) was an Australian schoolteacher in Tasmania, an ornithologist, founder member (1901) of the RAOU and a broadcaster.

Flinders

Flinders's Cockatoo *Calyptorhynchus banksii* **Latham**, 1790
[Alt. Red-tailed Black Cockatoo]
Flinders's Cuckoo *Eudynamys flindersii* **Vigors** & **Horsfield**, 1826 NCR
[Alt. Pacific Koel; JS *Eudynamys orientalis cyanocephalus*]

Matthew Flinders (1774–1814) was an English explorer and navigator who joined the British navy and trained as a navigator, having wanted to be a sailor and explorer ever since reading *Robinson Crusoe*. He sailed to Australia on HMS *Reliance* (1795) as a midshipman, and with George Bass, the ship's physician, explored south of Sydney in a tiny boat called *Tom Thumb*. As a Lieutenant (1798), Flinders was given command of the *Norfolk* and discovered the passage between the Australian mainland and Tasmania named Bass

Strait after his friend, while its largest island would later be named Flinders Island. He returned to England (1800) and married but was sent exploring again (1802). He circumnavigated the mainland of Australia aboard the *Investigator* (1802–1803). The French later captured him in Mauritius, treating him as a spy and holding him captive (1803–1810). He then returned to England a broken man in ill health, but still managed to write the story of his circumnavigation under the title *A Voyage to Terra Australis* published the day he died.

Flint

> Tanager genus *Flintthraupis* Kashin, 1978
> [Now in *Neothraupis*]

Professor Vladimir Evgenevich Flint (1924–2004) was a Russian zoologist, ornithologist and conservationist. He graduated from Moscow State University (1943) and studied the role of small mammals in epidemics for his PhD. He began writing on birds and herpetology, including: *Guide to the Birds of the USSR* (1968) and many of papers on waders. He began work at the Zoological Museum of Moscow University (1969), devoting himself to ornithology. He took part in a number of expeditions to the remote areas of the USSR, such as to the Indigirka River Delta in Siberia (1971).

Florence (Delamere)

> Stone Partridge ssp. *Ptilopachus petrosus florentiae* **Ogilvie-Grant**, 1900 NCR
> [NUI *Ptilopachus petrosus petrosus*]

Lady Florence Delamere (1878–1914) was the wife of Hugh Cholmondeley (see **Delamere**) and the daughter of the 4th Earl of Enniskillin.

Florence (Ingram)

> Black-faced Woodswallow *Artamus florenciae* **Ingram**, 1906 NCR
> [JS *Artamus cinereus melanops*]

Mrs Florence Maude Ingram *née* Laing (DNF) was the wife of the describer Captain Collingwood Ingram (q.v.).

Florence (van Rossem)

> Florence's Hummingbird *Saucerottia florenceae* **Van Rossem** & **Hachisuka**, 1938 NCR
> [Status uncertain. Hybrid involving *Amazilia beryllina*?]

Mrs Florence van Rossem *née* Stevenson (d.1944) was the second wife of the describer Adriaan Joseph van Rossem (q.v.).

Florentino

> West Indian Woodpecker ssp. *Melanerpes superciliaris florentinoi* Garrido, 1966 NCR
> [JS *Melanerpes superciliaris superciliaris*]
> Cuban Pewee ssp. *Contopus caribaeus florentinoi* Regalado Ruiz, 1977 NCR
> [JS *Contopus caribaeus caribaeus*]

Florentino García Montaña (fl.1955–c.1988) was an amateur Cuban naturalist. He co-wrote –*Las Aves de Cuba*–:

–*Especies Endémicas*– (1980) and *Las Aves de Cuba: Subespecies Endémicas* (1987).

Floresi

> Jamaican Hummingbird sp. *Trochilus floresii* **Bourcier** & **Mulsant**, 1846 NCR
> [Alt. Jamaican Mango; JS *Anthracothorax mango*]
> Floresi's Flamebearer *Selasphorus floresii* **Gould**, 1861 NCR
> [Hybrid: Allen's Hummingbird *Calypte anna* x Anna's Hummingbird *Selasphorus sasin*]

Damiano Floresi d'Areais (DNF) was an Italian engineer and a commissioner of a number of mines in Mexico who sent specimens to Gould. He was elected a Fellow of the Geological Society of London, (1833). He concentrated on botanical collecting and a number of Central and South American plants, such as the orchid *Lepanthes floresii*, are named after him. Gould wrote that Floresi died of fever in 'the pestilential region of Panama'.

Flower

> Asian Pied Starling ssp. *Gracupica contra floweri* **Sharpe**, 1897
> Crested Serpent Eagle ssp. *Spilornis cheela floweri* Swann, 1920 NCR
> [JS *Spilornis cheela burmanicus*]
> Chestnut-bellied Sandgrouse ssp. *Pterocles exustus floweri* **Nicoll**, 1921

Captain Stanley Smyth Flower (1871–1946) was Director, Cairo Zoological Gardens, Giza, Egypt (1898–1924). He had previously spent two years as Scientific Adviser to the Siamese government. Flower visited the zoo at Madras (Chennai) as an adviser (1913) and wrote *Zoological Gardens of the World* (1908–1914). An amphibian, three mammals and three reptiles are named after him. The sandgrouse subspecies, from the Nile Valley of Egypt, is now believed to be extinct.

Flückiger

> Woodchat Shrike ssp. *Lanius senator flueckigeri* **Kleinschmidt**, 1907 NCR
> [JS *Lanius senator rutilans*]
> Blue Tit ssp. *Parus caeruleus flueckigeri* **J. M. Harrison**, 1945 NCR
> [JS *Cyanistes caeruleus caeruleus*]

Ernst Flückiger (1875–1957) was a Swiss naturalist. He collected in Algeria (1903 and 1929) and the Balkans (1935). He has at least one plant named after him.

Flynn/Stone

> Blue-rumped Pitta ssp. *Hydrornis soror flynnstonei* Rozendaal, 1993

The eponym in this case is a rare composite of two people, Sean Leslie Flynn (1941–1971) and Dana Stone (1939–1971). Flynn was an actor (following his famous father Errol Flynn into the profession) (1955–1967), then became a freelance photographer (1966) (having tried being a big-game hunter and safari guide in Africa), often working for *Time* magazine. Stone was a CBS cameraman (taught by Flynn), and both were best known for their photojournalism from the Vietnam war. They went missing (1970) after they had left Phnom Penh heading south-east along the Route Coloniale, being taken prisoner by the Viet Cong and later handed over to the Khmer Rouge. Their graves were discovered in 1991.

Fonseca

> Pink-legged Graveteiro *Acrobatornis fonsecai* **Pacheco,**
> **Whitney** & Gonzaga, 1996

Paulo Sergío Moreira da Fonseca is a Brazilian ornithologist who is known to his friends as 'PS'. The original description reads that it is dedicated to: 'our multi-talented friend of many years, not only because he was the first to gasp in wonder at the living bird, but also in recognition of his unending encouragement and deep generosity. PS has contributed much valuable data to our continuing studies of the Brazilian avifauna.'

Fontanier

> Tiny Hawk ssp. *Accipiter superciliosus fontanieri*
> **Bonaparte**, 1853

Henri Victor Fontanier (1830–1870) was in Colombia (c.1850) before becoming the French Consul at Tientsin in China (1870), combining that job with collecting for the Paris Museum. On 21 June 1870 a crowd of locally prominent representatives at Tianjin (Tientsin) marched (1870) on a Roman Catholic orphanage run by French and Belgian nuns (the French Sisters of Charity), accusing them of kidnapping children and taking their eyes to make medicine. The crowd demanded to search for the truth. Fontanier lost his temper and fired into the crowd, killing the District Magistrate's servant. The xenophobic Chinese mob attacked, killing 24 foreigners, including Fontanier and the nuns, mutilating their bodies. Two mammals are named after him. We believe he might have been the son of another French naturalist, Victor Fontanier (1796–1857), a pharmacist who was sent out by the French government as an envoy to the Persian Gulf (1834) and wrote *Voyage in the Indian Archipelago* (1852).

Fooks

> Scaly Laughingthrush ssp. *Trochalopteron subunicolor*
> *fooksi* **Delacour** & **Jabouille**, 1930

Francis 'Frank' Edward Fooks (1892–1967) was a British aviculturist who worked in the livestock department of London department store, Derry & Toms. He was seriously wounded in action (WW1). He became Director of Clères Zoological Park in France (1920–1967), which was virtually emptied of its collections and despoiled during WW2. He returned there (1945) and set about restoring it (1945–1950).

Forbes, H. O.

> Forbes's Mannikin *Lonchura forbesi* **P. L. Sclater**, 1879
> [Alt. New Ireland Mannikin/Munia]
> Forbes's Hawk Owl *Ninox forbesi* P. L. Sclater, 1883
> [Alt. Tanimbar Boobook; Syn. *Ninox squamipila forbesi*]

> Forbes's Forest Rail *Rallina forbesi* **Sharpe**, 1887
> [Alt. Red-backed Forest Rail]
> Forbes's Parakeet *Cyanoramphus forbesi* **Rothschild**, 1893
> [Alt. Chatham Parakeet]
> Forbes's Snipe *Coenocorypha chathamica* Forbes, 1893
> EXTINCT
> Lemon Dove *Haplopelia forbesi* **Salvadori**, 1904 NCR
> [JS *Columba larvata inornata*]

> Black Myzomela ssp. *Myzomela nigrita forbesi* **Ramsay,**
> 1880

Henry Ogg Forbes (1851–1932) was a Scottish explorer and natural history collector who retraced Wallace's (q.v.) footsteps in the Moluccas on an expedition. He wrote *A Naturalist's Wanderings in the Eastern Archipelago* (1885). After a number of ill-fated expeditions in New Guinea, Forbes was appointed Meteorological Observer in Port Moresby. He was later Director of the Canterbury Museum, New Zealand (1890–1893), and Consulting Director on Museums in Liverpool (1911–1932). Three reptiles and two mammals are named after him.

Forbes, V. C. W.

> Apurimac Brush Finch *Atlapetes forbesi* Morrison, 1947

Sir Victor Courtenay Walter Forbes (1889–1958) was a British career diplomat who served in Mexico, Spain and Peru.

Forbes, W. A.

> Forbes's Plover *Charadrius forbesi* **G. E. Shelley**, 1883
> Forbes's Blackbird *Anumara forbesi* **P. L. Sclater**, 1886
> [Syn. *Curaeus forbesi*]
> Forbes's Kite *Leptodon forbesi* Swann, 1922
> [Alt. White-collared Kite]

> Blue-breasted Kingfisher ssp. *Halcyon malimbica forbesi*
> **Sharpe**, 1892
> Black-bellied Firefinch ssp. *Lagonosticta rara forbesi*
> **Neumann**, 1908

William Alexander Forbes (1855–1883) was a British anatomist, collector and zoologist. He became a member of the Zoological Society of London. He collected in the Americas, particularly Brazil, where he conducted a survey of the birds of Pernambuco (1881). He also collected in West Africa (1882), where he died on the Upper Niger River. He wrote valuable papers on bird anatomy: 'Contributions to the anatomy of passerine birds: on the structure of the stomach in certain genera of tanagers' (1880) and 'Report on the anatomy of the petrels (Tubinares), collected during the voyage of HMS Challenger, Zoology of the Challenger Expedition' (1882). Forbes collected the kite (1882) but it apparently lay in the British Museum for many years before Swann recognised it as a new species, naming it after the collector.

Forbes, W. C.

> White-eye sp. *Zosterops forbesi* **Bangs**, 1922 NCR
> [Alt. Everett's White-eye; JS *Zosterops everetti basilanicus*]

> Striated Grassbird ssp. *Megalurus palustris forbesi* Bangs,
> 1919

William Cameron Forbes (1870–1959) was an American banker and diplomat. He was in the Philippines as Commissioner of Commerce and Police (1904–1908), and Governor-General (1908–1912). He was sent (1921) as joint head of the Wood-Forbes Commission to investigate conditions in the Philippines, which concluded that Filipinos were not yet ready for independence. He was a keen naturalist and obtained the holotype of the white-eye there. He was American ambassador to Japan (1930–1932).

Forbes-Watson

Forbes-Watson's Swift *Apus berliozi* **Ripley**, 1965

Black-crowned Sparrow Lark ssp. *Eremopterix nigriceps forbeswatsoni* Ripley & **Bond**, 1966

Alexander 'Alec' David Forbes-Watson (1935–2013) was an ornithologist with great experience of African birds. He was closely involved in Kenya conservation, being a Game Warden there before becoming Curator of Ornithology at Kenya's National Museum (1960). He wrote or co-wrote c.20 articles such as 'Observations at a nest of the Cuckoo-Roller *Leptosomus discolor*' (1967) and 'Notes on birds observed in the Comoros on behalf of the Smithsonian Institution' (1969) as well as being co-author of the *Checklist of Birds of the Afrotropical and Malagasy Regions* (1993). He also discovered and described the Nimba Flycatcher, which he collected in Liberia and named after his former wife Anna Marula (q.v.).

Forbush

Forbush's Sparrow *Melospiza lincolnii striata* **Brewster**, 1889 NCR
[Alt. Lincoln's Sparrow; JS *Melospiza lincolnii gracilis*]

Edward Howe Forbush (1858–1929) was the American ornithologist after whom the Forbush Bird Club was named. He had very little formal education but began learning about ornithology and taxidermy as a child. He was appointed ornithologist to the Massachusetts Board of Agriculture (1903), then State Ornithologist (1908) and became the first Director of the Massachusetts Department of Agriculture's Division of Ornithology (1920). Among his published works are *Useful Birds and their Protection* (1905) and *Birds of Massachusetts and other New England States* (1925). He took part in collecting expeditions to Florida, the Pacific Northwest, Western Canada and Alaska. He was a co-founder of the Naturalists' Exchange, now the Naturalists Club at the Museum of Science in Springfield, Massachusetts. Brewster (q.v.) and Forbush must have known each other well – Forbush succeeded Brewster as President of the Massachusetts Audubon Society in 1914.

Ford, H. A.

Sarus Crane ssp. *Grus antigone fordi* M. D. Bruce & McAllan, 1989 NCR
[JS *Grus antigone gillae*]

Dr Hugh Alistair Ford (b.1946) is a British ornithologist and ecologist who emigrated to Adelaide, Australia (1973) and moved (1977) to the University of New England. He was editor of *Emu* (1981–1985) and wrote *Ecology of Birds: an Australian Perspective* (1989).

Ford, J. R.

Kimberley Honeyeater *Meliphaga fordiana* **Schodde**, 1989

Channel-billed Cuckoo ssp. *Scythrops novaehollandiae fordi* Mason, 1996
Chestnut Quail-thrush ssp. *Cinclosoma castanotum fordianum* Schodde & Mason, 1999

Dr Julian Ralph Ford (1932–1987), an Honorary Associate, Western Australian Museum, was an ornithologist and herpetologist. He received degrees in chemistry and zoology (1953) and a doctorate (1984) from the University of Western Australia. He worked for Shell as an industrial chemist (1954–1960), then was a Senior Lecturer in Chemistry, Curtin University, Western Australia (1987). He co-wrote 'Northern extension of the known range of the Brush Bronzewing' (1959). A reptile is named after him.

Forns

Oriente Warbler *Teretistris fornsi* Gundlach, 1858

Dr Ramon M. Forns (DNF) was Principal of the 'Santa Teresa' School and is known to have made collections of Cuban birds (1850s–1870s), sending many to the USNM.

Forrer

Forrer's Hummingbird *Amazilia forreri* **Boucard**, 1893 NCR
[Alt. Amazilia Hummingbird; JS *Amazilia amazilia*]

Yellow-green Vireo ssp. *Vireo flavoviridis forreri* **Madarász**, 1885
Black-headed Siskin ssp. *Carduelis notata forreri* **Salvin** & **Godman**, 1886

Alphonse Forrer (1836–1899) was born in England but moved to the USA, where he joined the Union army at the outbreak of the American Civil War (1861). He was employed by the British Museum (1865) to collect zoological specimens in western Mexico and the US, continuing to collect in Mexico (1880s). He also supplied specimens to other museums and made four trips to Europe. A reptile and an amphibian are named after him.

Forrest, G.

Laughingthrush sp. *Ianthocincla forresti* **Rothschild**, 1921 NCR
[Alt. Assam Laughingthrush; JS *Trochalopteron chrysopterum woodi*]
Sichuan Leaf Warbler *Phylloscopus forresti* Rothschild, 1921

Black-naped Monarch ssp. *Hypothymis azurea forrestia* **Oberholser**, 1911
White-bellied Woodpecker ssp. *Dryocopus javensis forresti* Rothschild, 1922
Spotted Dove ssp. *Streptopelia chinensis forresti* Rothschild, 1925 NCR
[NUI *Spilopelia chinensis tigrina*]

Golden-breasted Fulvetta ssp. *Lioparus chrysotis forresti*
Rothschild, 1926
Slender-billed Scimitar Babbler ssp. *Pomatorhinus superciliaris forresti* Rothschild, 1926

George Forrest (1873–1932) was a British collector, primarily of botanical specimens, but he would collect anything from butterflies to birds. He started work in a chemist's shop, where he learned about the medicinal qualities of plants and how to dry and preserve them. He went to Australia at the height of the (1891) gold rush and spent 10 years panning for gold before returning to Britain (1902). He was employed as a clerk at the Royal Botanic Gardens in Edinburgh (1903). He went to Talifu, China (1904), then (1905) set out for north-west Yunnan, close to the border with Tibet. At Tzekou his party was based at a French mission, and after a collecting foray, Forrest and his team of 17 collectors returned to the mission to find that the locals had turned on the foreigners. Of his group, only Forrest escaped with his life. Continuing to collect despite this tragedy, and despite being struck down by malaria, he and his new team of collectors amassed a very considerable weight of specimens that he took back to Britain (1906). During his life he made six further expeditions to Yunnan, Sichuan, Tibet, and Upper Burma. He discovered more than 1,200 plant species new to science as well as many birds and mammals. He died from a heart attack in China while still collecting. Three mammals are named after him.

Forrest

Australasian Lark ssp. *Mirafra javanica forresti* **Mayr** & McEvey, 1960

Named after the type locality, the Forrest River in northern Western Australia. The river is named after John Forrest (q.v.).

Forrest, J.

Forrest's Honeyeater *Lichenostomus virescens forresti*
Ingram, 1906
[Alt. Singing Honeyeater ssp.]

The Rt. Hon. Sir John Forrest (1847–1918) was an Australian explorer and statesman who was first Prime Minister of Western Australia (1890–1901). He led an expedition in search of Ludwig Leichhardt (1869) and another (1870) from Perth (Western Australia) to Adelaide (South Australia) around the Great Australian Bight. He crossed the Great Victoria Desert (1874) from Champion Bay, Western Australia, to the Overland Telegraph Line, Northern Territory, during which time he collected botanical specimens. He became Deputy Surveyor-General for Western Australia before becoming Secretary to the Treasury and Prime Minister, thereafter (1901) holding other ministerial posts. He died at sea. Two towns, Forrestdale and Glen Forrest, are named after him, as well as a mammal.

Forskål

African Grey Hornbill ssp. *Tockus nasutus forskalii*
Ehrenberg, 1833 NCR
[NUI *Tockus nasutus nasutus*]

Peter Forsskål (1736–1763) was a Swedish traveller and naturalist who was at Uppsala University (1751) and studied at Georg-August University in Göttingen (1756). He was Professor of Natural History, Copenhagen University (1760). He died of the plague in Yemen while participating in the Danish expedition to Arabia (1761–1763). His journals and notes were published posthumously (1775). A reptile is named after him.

Forsten

Forsten's Scrubfowl *Megapodius forsteni* **G. R. Gray**, 1847
[Alt. Forsten's Megapode]
Forsten's Lorikeet *Trichoglossus forsteni* **Bonaparte**, 1850
[Alt. Sunset Lorikeet; Syn. *Trichoglossus haematodus forsteni*]
Forsten's Oriole *Oriolus forsteni* Bonaparte, 1850
[Alt. Grey-collared Oriole]
Purple-bearded Bee-eater *Meropogon forsteni* Bonaparte, 1850
White-bellied Imperial Pigeon *Ducula forsteni* Bonaparte, 1854

Forsten's Pitta *Pitta sordida forsteni* Bonaparte, 1850
[Alt. Hooded Pitta ssp.]

Eltio Alegondas Forsten (1811–1843) collected in the East Indies (1838–1843). He was primarily a botanist and was interested in the pharmaceutical properties of plants, on which he wrote at least one scientific paper, 'Dissertatio botanico-pharmaceutico-medica inauguralis de cedrela febrifuga' (1836) published at Leiden. We assume that he was Dutch. Three reptiles are named after him.

Forster, J. G. A.

Blue Duck *Malacorhynchus forsterorum* **Wagler**, 1832 NCR
[JS *Hymenolaimus malacorhynchos*]
South Island Tomtit *Miro forsterorum* **G. R. Gray**, 1843 NCR
[JS *Petroica macrocephala*]
Forster's Dove *Gallicolumba ferruginea* J. G. A. Forster, 1844
EXTINCT
[Alt. Tanna Ground Dove]

Johann Georg Adam Forster (1754–1794), known as George, sailed on several expeditions with his father Johann Reinhold Forster (q.v.) aboard the *Resolution*, notably on James Cook's second voyage to the Pacific. He was elected a Fellow of the Royal Society (1775) for his share in the description of the flora of the South Seas. Later in his life, he became Professor of Natural History at Wilna and Librarian at Mainz. The dove was only known from two specimens, which are both now lost; the only remaining records are paintings (one sketched by Forster). Tanna is a Vanuatu island. A mammal is named after him. The two birds with the binomial *forsterorum* are named after both father and son.

Forster, J. R.

Forster's Caracara *Phalcoboenus australis* **Gmelin**, 1788
[Alt. Striated Caracara]
Broad-billed Prion *Pachyptila forsteri* **Latham**, 1790 NCR
[JS *Pachyptila vittata*]

Blue Duck *Malacorhynchus forsterorum* **Wagler**, 1832 NCR
[JS *Hymenolaimus malacorhynchos*]
Forster's Tern *Sterna forsteri* **Nuttall**, 1834
South Island Tomtit *Miro forsterorum* **G. R. Gray**, 1843 NCR
[JS *Petroica macrocephala*]
Emperor Penguin *Aptenodytes forsteri* G. R. Gray, 1844
Forster's Petrel/Shearwater *Puffinus gavia* J. R. Forster,
1844
[Alt. Fluttering Shearwater]
Fan-tailed Gerygone *Petroica forsteri* G. R. Gray, 1860 NCR
[JS *Gerygone flavolateralis*]

Buff-banded Rail ssp. *Gallirallus philippensis forsteri*
Hartlaub, 1852 NCR
[JS *Gallirallus philippensis ecaudatus*]
Silver Gull ssp. *Chroicocephalus novaehollandiae forsteri*
Mathews, 1912
Sacred Kingfisher ssp. *Todiramphus sanctus forsteri*
Mathews & **Iredale**, 1913 NCR
[JS *Todiramphus sanctus vagans*]

Johann Reinhold Forster (1729–1798) was a clergyman in Danzig (now Gdansk, Poland). He became a naturalist and accompanied James Cook (q.v.) on his second voyage around the world (1772–1773), which extended further into Antarctic waters than anyone previously. He discovered five new species of penguin. However, he gained a reputation as a constant complainer and troublemaker. His complaints about Cook continued after his return and became public, destroying Forster's career in England. He went to Germany and became a Professor of History and Mineralogy. Unpleasant and troublesome to the end, Forster refused to relinquish his notes of the voyage. They were not found and published until c.50 years after his death (which explains Forster's posthumous naming of the shearwater). His son Johann George Adam (q.v.) was also on Cook's voyage as an artist. A mammal is named after him. Two taxa called *forsterorum* are named after both father and son.

Förster

New Britain Bronzewing *Henicophaps foersteri* **Rothschild**
& **Hartert**, 1906
Förster's Honeyeater *Melidectes foersteri* Rothschild &
Hartert, 1911
[Alt. Huon Melidectes/Honeyeater]

Friedrich Förster (1865–1918) was a German botanist and collector. He described two new birds-of-paradise with Rothschild (1906).

Forsyth

Alpine Chough ssp. *Pyrrhocorax graculus forsythi*
Stoliczka, 1874

[Alt. Yellow-billed Chough ssp.] Sir Thomas Douglas Forsyth (1827–1886) originally went to India with the Honourable East India Company. He was Commissioner of the Punjab (1862–1872) and was sent to Yarkand (1870) to visit Yakub Beg, then ruler of independent Chinese Turkestan. The mission failed, so he made another expedition, the famous Second Yarkand Expedition (1873), in which Dr. Ferdinand Stoliczka (q.v.)

played an important part. Forsyth was instrumental in preventing a war between British India and Burma (Myanmar) (1875). He wrote *Report of a Mission to Yarkund in 1873* (1875). A reptile is also named after him.

Fortich

Striated Wren-babbler ssp. *Ptilocichla mindanensis fortichi*
Rand & Rabor, 1957

Ismael Fortich (fl.1950) was a Filipino landowner at Sierra Bullones, Bohol. He donated the building materials for a new school (1953).

Forwood

White Wagtail ssp. *Motacilla alba forwoodi* **Ogilvie-Grant** &
H. O. Forbes, 1899 NCR
[NUI *Motacilla alba alba*]

Sir William Bower Forwood (1840–1928) was an English shipping magnate, yachtsman, politician and philanthropist. He joined his father's cotton business (1859) and ran it (1862) with his brother, Arthur. They made a fortune from speculation, exploiting the futures markets and running the Union blockade of the Confederacy. He was a Director of the Cunard Line (1888–1923) and of the Bank of Liverpool (1887–1928).

Fosse

Square-tailed Nightjar *Caprimulgus fossii* **Hartlaub**, 1857
[Alt. Gabon/Mozambique Nightjar]

W. Fosse (DNF) collected in the area of Gabon. Nothing seems to be recorded about him.

Foster

Guira Tanager ssp. *Hemithraupis guira fosteri* **Sharpe**, 1905

William T. Foster (1867–1915) was a field naturalist and collector. He was a seaman who became a shopkeeper at Sapucay, Paraguay (1894–1915) and regularly supplied bird and mammal specimens to the British Museum.

Foster Smith

Whistling Heron ssp. *Syrigma sibilatrix fostersmithi*
Friedmann, 1949

Foster D. Smith Jr (fl.1930–c.2006) was in Venezuela (1947–1948), where he collected the holotype of the heron. He co-wrote papers with Herbert Friedmann, including: 'A contribution to the ornithology of northeastern Venezuela' (1950) and 'A further contribution to the ornithology of northeastern Venezuela' (1955).

Fowler

Ochraceous Bulbul ssp. *Alophoixus ochraceus fowleri*
Amadon & Harrisson, 1957

James Alexander Fowler III (d.2003) was an American ornithologist and herpetologist at the AMNH, New York (1952). Columbia awarded his PhD (1961), following which he joined the faculty of the State University of New York until retirement (1985). He was honoured because, when he was a graduate student at Columbia University, he had spent much time arranging all the bird holotypes in taxonomic order, with file cards for each one, so greatly facilitating future research. He was still studying and teaching when he suffered a fatal cerebral haemorrhage.

Fox, T. V.

Fox's Swamp Warbler *Calamornis foxi* **W. L. Sclater**, 1927 NCR
[Alt. Greater Swamp Warbler; JS *Acrocephalus rufescens ansorgei*]
Fox's Weaver *Ploceus spekeoides* **C. H. B. Grant** & **Mackworth-Praed**, 1947

T. V. Fox (1879–1918) collected birds, including Fox's Weaver, for Stephenson Robert Clarke (q.v.) in Uganda (1910–1913).

Fox, W. J.

Rusty-backed Spinetail ssp. *Cranioleuca vulpina foxi* **Bond** & **de Schauensee**, 1940

William J. Fox (fl.1940) was librarian at the Academy of Natural Sciences, Philadelphia.

Frade

Sabota Lark ssp. *Calendulauda sabota fradei* da Rosa Pinto, 1963 NCR
[JS *Calendulauda sabota suffusca*]
Príncipe Seedeater ssp. *Crithagra rufobrunnea fradei* Naurois, 1975

Dr Fernando Frade Viegas da Costa (1898–1983) was a Portuguese zoologist, ornithologist, naturalist and explorer who worked in Portugal's overseas territories: Timor, Mozambique, Angola, Portuguese Guinea, São Tomé and Príncipe, Cape Verde and Goa. He became Professor of Zoology and Director of the Lisbon Zoo. He wrote *Protecção à Fauna em Mozambique* (1953).

Fraga

Rufous-banded Miner ssp. *Geositta rufipennis fragai* Nores & **Yzurieta**, 1986

Dr Rosendo M. Fraga is an Argentinean ornithologist. The University of California awarded his doctorate. He worked at the USNM's Tropical Research Institute, Panama (1989). He wrote 'Phylogeny and behavioural evolution in the family Icteridae' (2008).

Fraith

Laysan Apapane *Himatione fraithii* **Rothschild**, 1892 EXTINCT
[Scientific name sometimes corrected to *Himatione freethii*]

(See **Freeth**)

Frances (Cole)

Frances's Sparrowhawk *Accipiter francesiae* **A. Smith**, 1834
[formerly *Accipiter francesii*]

Lady Henrietta Frances Cole *née* Harris (c.1785–1848) was wife to Sir Galbraith Lowry Cole, Governor of the Cape Colony, and a patroness of science. Sir Andrew Smith named the sparrowhawk after her and wrote: 'To Lady Frances Cole I am indebted for the only specimen I possess of this apparently undescribed species – and the name it bears in the South African Museum is an indication of the high respect entertained for Her Ladyship as a well known and zealous Patroness of Science.' She apparently had the holotype collected in Madagascar.

Frances (Grayson)

Red-breasted Chat ssp. *Granatellus venustus francescae* **Baird**, 1865

Frances Jane Grayson *née* Timmons (d.1908) was the wife of American ornithologist and collector Andrew Jackson Grayson (q.v.). After Grayson's death she married Dr G. B. Crane of St Helena, California.

Frances (Wilson)

Mrs. Wilson's Tanager *Tangara larvata franciscae* **P. L. Sclater**, 1856
[Alt. Golden-hooded Tanager ssp.]

(see **Fanny**)

Francia

Francia's Azure-crown *Amazilia franciae* **Bourcier** & **Mulsant**, 1846
[Alt. Andean Emerald; Syn. *Chrysuronia franciae*]

Francia Bourcier (DNF) was the daughter of Jules Bourcier (q.v.), who was an amateur naturalist and the French Consul to Ecuador (1849–1850).

Francis (Alexander)

Bioko Trogon sp. *Heterotrogon francisci* **B. Alexander**, 1903 NCR
[Alt. Bar-tailed Trogon; JS *Apaloderma vittatum*]

Lieutenant Colonel Boyd Francis Alexander (1834–1917) was the father of Captain Boyd Alexander (q.v.).

Francis, H. F. & W. F.

Livingstone's Flycatcher ssp. *Erythrocercus livingstonei francisi* **W. L. Sclater**, 1898

Capt. H. F. Francis (d.1901) and his brother W. F. Francis (d.1900) were British explorers and collectors in Mozambique (1898); both served and were killed in the Boer War. Sclater's brief original description does not specify which brother is being honoured in the bird's name.

Franck

Pied Bushchat ssp. *Saxicola caprata francki* **Rensch**, 1931

Paul Friedrich Franck (fl.1940) was a German big-game hunter and taxidermist at Zoological Museum Buitenzorg

(Bogor), Java (1922–1940). He died (c.1943) in a WW2 prisoner-of-war camp in India.

Frank

Rufous-tailed Bush-hen ssp. *Amaurornis olivacea frankii* **Schlegel**, 1879 NCR
[Alt. Pale-vented Bush-hen; JS *Amaurornis moluccana moluccana*]

Gustav Adolph Frank (1809–1880) was a German collector and animal dealer in Amsterdam with worldwide trade connections.

Franklin, James

Prinia genus *Franklinia* **Jerdon**, 1863
[Now in *Prinia* – may be recognised as a subgenus]

Franklin's Nightjar *Caprimulgus monticolus* J. Franklin, 1831
[Syn. *Caprimulgus affinis monticolus*]
Golden-throated Barbet *Megalaima franklinii* **Blyth**, 1842
[Syn. *Psilopogon franklinii*]
Franklin's Prinia *Prinia hodgsonii* Blyth, 1844
[Alt. Grey-breasted Prinia, Franklin's Wren Warbler]

Plain Prinia ssp. *Prinia inornata franklinii* Blyth, 1844

Major James Franklin (1783–1834) served in the British Army in India, where he made the first military survey. He was also an ornithologist, geologist and author. His younger brother was Sir John Franklin (q.v.). Alhough most authorities list Jerdon (q.v.) as responsible for *Franklinia*, he himself credits Blyth (q.v.) with coining the name.

Franklin, John

Franklin's Gull *Larus pipixcan* **Wagler**, 1831
[Syn. *Leucophaeus pipixcan*]

Franklin's Grouse *Falcipennis canadensis franklinii* **Douglas**, 1829
[Alt. Spruce Grouse ssp.]

Sir John Franklin (1786–1847) was an officer in the Royal Navy and well-known explorer for the North-West Passage. The youngest of 12 boys, Franklin joined the navy in his youth and spent the rest of his life in its service. His first Arctic voyage (1818) was commanding a vessel trying to reach the North Pole. He attempted to find the North-West Passage, the sea route across the Arctic to the Pacific Ocean (1819). He returned empty-handed after two years amid rumours of starvation, murder and cannibalism. He was appointed as Governor of Van Diemen's Land (Tasmania) (1837) and the following year played host to John Gould (q.v.) for three months, and to Eliza Gould even longer as she was pregnant. Lady Franklin accompanied Gould on one of his collecting trips. Franklin was involved in several more voyages to the north, before disappearing (1845) in another attempt to cross the Arctic by sea. A search was undertaken under the command of Sir Clements Robert Markham. Log books were later found confirming that he had died along with most of his officers and crew – in his case probably from lead poisoning, ironically leaching from the solder used to seal canned provisions that were then considered the height of technology. A mammal is named after him.

Franks

Franks's Guillemot *Uria francsii* **Sabine**, 1817
[Alt. Brünnich's Guillemot/Thick-billed Murre; JS *Uria lomvia*]

A man named Frederick Franks (DNF) sent a specimen of the guillemot to William Leach (q.v.), who exhibited it at the Linnean Society, London, unaware that it had been described by Linnaeus as *Alca lomvia*.

Frantzius

Fiery-billed Aracari *Pteroglossus frantzii* **Cabanis**, 1861
Frantzius's Nightingale Thrush *Catharus frantzii* Cabanis, 1861
[Alt. Ruddy-capped Nightingale Thrush]
Prong-billed Barbet *Semnornis frantzii* **P. L. Sclater**, 1864
Mountain Elaenia *Elaenia frantzii* **G. N. Lawrence**, 1865

Silver-throated Tanager ssp. *Tangara icterocephala frantzii* Cabanis, 1861
Highland Tinamou ssp. *Nothocercus bonapartei frantzii* G. N. Lawrence, 1868

Alexander von Frantzius (1821–1877) was a German physician. He and fellow doctor Carl Hoffmann (q.v.) went to Costa Rica (1853) to explore and collect mainly botanical specimens. Frantzius became a successful businessman and the eventual owner of a drugstore known popularly as 'Botica Francesa', which is currently one of the largest private organisations in Costa Rica. Advertisements in the press of the time said that the drugstore was 'managed by naturalists Frantzius and Zeledón' (q.v.). An extinct volcano in Costa Rica is also named after him.

Fraser, L.

Flycatcher genus *Fraseria* **Bonaparte**, 1854

Fraser's Sunbird *Deleornis fraseri* **Jardine** & **Selby**, 1843
[Alt. Scarlet-tufted Sunbird; Syn. *Anthreptes fraseri*]
Fraser's Rufous Thrush *Stizorhina fraseri* **Strickland**, 1844
[Alt. Rufous Flycatcher-thrush; Syn. *Neocossyphus fraseri*]
Fraser's Forest Flycatcher *Fraseria ocreata* Strickland, 1844
Fraser's Eagle Owl *Bubo poensis* L Fraser, 1854
Fraser's Spinetail *Cranioleuca antisiensis* **P. L. Sclater**, 1859
[Alt. Line-cheeked Spinetail]
Giant Conebill *Oreomanes fraseri* P. L. Sclater, 1860
Fraser's Warbler *Basileuterus fraseri* P. L. Sclater, 1884
[Alt. Grey-and-gold Warbler]

Black-crowned Tityra ssp. *Tityra inquisitor fraserii* **Kaup**, 1852
Tambourine Dove ssp. *Turtur tympanistria fraseri* Bonaparte, 1855 NCR; NRM
Band-tailed Barbthroat ssp. *Threnetes ruckeri fraseri* **Gould**, 1861 NCR
[JS *Threnetes ruckeri ruckeri*]

Louis Fraser (1819–?1884) was a British zoologist and collector. He was also variously a curator, explorer, zookeeper, consul, author, dealer and taxidermist. He was employed first as office boy, then Clerk, Assistant Curator and finally Curator to the Museum of the Zoological Society of London (1832–1841 and 1842–1846). In between

(1841–1842) he collected in West Africa as the official naturalist on the Niger expedition. He collected in North Africa (1847) (self-financed), in the Bights of Benin (Nigeria) (1851–1853) and Ecuador and Guatemala for P. L. Sclater (q.v.) (1857–1860). He wrote the 14-part *Zoologica Typica* (1845–1849) and *Catalogue of the Knowsley Collection* (1850) as well as around 40 papers in the *Proceedings of the Zoological Society of London* (1839–1866). He took charge of Lord Derby's zoological collections at Knowsley (1848–1850) before becoming Vice-consul at Whydah [Ouida] to the Kingdom of Dahomey (1850–1853). He also tried to establish himself as a natural history dealer. He was last heard of collecting in Florida (1883); we can find no record of his death. A mammal is named after him. (See also **Fraser, T.**)

Fraser, M.

Inaccessible Island Finch ssp. *Nesospiza acunhae fraseri* Ryan, 2008

Mike Fraser is a writer and naturalist who works for the RSPB as Conservation Officer, Lothians and Borders. He was formerly warden on the Isle of May. In 1982 he was invited at a couple of days' notice to join the Denstone Expedition to Inaccessible Island. He worked in Tristan da Cunha and Gough Island under the auspices of the Percy FitzPatrick Institute of African Ornithology, Cape Town University. He has written three books including, with his wife Liz, a botanical and wildlife illustrator, *The Smallest Kingdom*, about plant collecting in the Cape.

Fraser, T.

Fraser's Hermit *Glaucis fraseri* **Gould**, 1861 NCR
[Alt. Band-tailed Barbthroat; JS *Threnetes ruckeri*]

Gould states that a T. Fraser had collected the hummingbird in Ecuador. Since Louis Fraser (q.v.) was collecting in Ecuador for P. L. Sclater at the time, we conjecture that Gould made a mistake in the transcription and it is named after Louis Fraser.

Fraser, W. T.

Stork-billed Kingfisher *Pelargopsis fraseri* **Sharpe**, 1870 NCR
[JS *Pelargopsis capensis capensis*]

William T. Fraser (d.1880) grew up in Surabaya, Java, where his father was a merchant. His interest in natural history was evidenced by his correspondence with the Zoological Society of London, e.g. (1869) on the disputed existence of rhinoceros on Borneo (since confirmed). He became the British Consul in Batavia (Jakarta), Java.

Fratrisregis

Spot-winged Grosbeak ssp. *Mycerobas melanozanthos fratrisregis* **Deignan**, 1943 NCR; NRM

'*Fratris regis*' is Latin for 'King's brother's'. Phrabat Somdet Phra Pinklao Chaoyuhua (1808–1866) was (1851) Vice, or second, king of Siam (Thailand), ruling at the same time as his brother King Rama IV. He was very interersted in western culture and spoke English. He had a scientific interest in the natural history of his country.

Frauenfeld

Imperial pigeon sp. *Carpophaga frauenfeldi* **Pelzeln**, 1865 NCR
[Alt. Pacific Imperial Pigeon; JS *Ducula pacifica*]

Georg Ritter von Frauenfeld (1807–1873) was an Austrian zoologist, entomologist, malacologist and explorer. He was on the Novara Expedition (1857–1859), which circumnavigated the globe. He was a curator of invertebrates at the Vienna Natural History Museum (1852).

Frazar

Hispaniolan Flycatcher sp. *Contopus frazari* **Cory**, 1883 NCR
[Alt. Hispaniolan Pewee; JS *Contopus hispaniolensis*]

Frazar's Green Heron *Butorides virescens frazari* **Brewster**, 1888
Frazar's Oystercatcher *Haematopus palliatus frazari* Brewster, 1888
[Alt. American Oystercatcher ssp. possibly a hybrid population of *H. palliatus* x *H. bachmani*]
Greater Yellowlegs ssp. *Tringa melanoleuca frazari* Brewster, 1902 NCR
[*T. melanoleuca* NRM]
Frazar's Vireo *Vireo huttoni cognatus* **Ridgway**, 1903
[Alt. Hutton's Vireo ssp.]

Marston Abbott Frazar (1860–1925) was a taxidermist and collector for Sennett (q.v.) in the 1880s. He was interested in natural history from the age of 16 and was elected a member of the Nuttall Ornithological Club. He made his first collecting trip (1881) to San Domingo for Cory (q.v.) and his second (1886–1887) to Lower California for Brewster (q.v.). He wrote *An Ornithologist's Summer in Labrador* (1887). He set up as a taxidermist (1889), buying an existing business in Boston, and later developed a fur business too. He published some papers (mostly with the Nuttall Club) often as Abbott M. Frazar (1876–1881).

Frederick (McConnell)

Ant-shrike genus *Frederickena* **Chubb**, 1918

(see **McConnell**)

Frederick (Penard)

Greyish Mourner ssp. *Rhytipterna simplex frederici* **Bangs & T. E. Penard**, 1918

Frederick Paul Penard (1876–1909) was born in Suriname to Jewish parents of French extraction who had made a fortune trading timber; they moved to the USA (1891). He was co-author with his brother Arthur Philip Penard (1880–1932) of the 2-volume *De Vogels van Guyana* (1908 and 1910); both were interested in birds from an early age and accompanied their father on trips into the rainforest. They left school very young (9 and 11 respectively) because 'the first symptoms of a terrible disease were manifested' and they were to spend the 'rest of their lives in seclusion'. They suffered from leprosy and both were sadly blinded by the disease, but still persevered until they completed their book and continued to study the local indigenous people. Another brother, Thomas Edward Penard (1878–1936) (q.v.), was the junior describer.

Freer

Freer's Blue-backed Parrot *Tanygnathus sumatranus freeri* **McGregor**, 1910

Paul Caspar Freer (1862–1912) was Professor of Chemistry at the University of Michigan and later Director of the Government Scientific Laboratories, Manila, and Dean of the Philippine Islands Medical School. He edited the *Philippine Journal of Science*, wherein McGregor published his description.

Freeth

Laysan Apapane *Himatione freethii* **Rothschild**, 1892 EXTINCT

[Original spelling of scientific name = *Himatione fraithii*]

Captain George Douglas Freeth (DNF) was the American Governor of Laysan (1892) in the Hawaiian Islands. Earlier (1890) he had been granted the right by the Kingdom of Hawaii to mine guano on Laysan in return for a royalty, but this badly affected the island's ecosystem. Rothschild originally used the incorrect spelling *fraithii* as the binomial, and many authorities believe this spelling must take priority as the valid scientific name.

Freire

Pale-tailed Barbthroat ssp. *Threnetes loehkeni freirei* **Ruschi**, 1976 NCR

[NUI *Threnetes niger loehkeni*]

Paulo Reglus Neves Freire (1921–1997) was a Brazilian social reformer, humanist and educationalist. He came from an impoverished background, but rose to qualify as a lawyer and to have an enormous influence on educational thinking and practice. After the military coup in Brazil (1964) he was arrested and imprisoned as a traitor, but then exiled to Bolivia from whence he moved to Chile (1964–1969). His most famous book is *Pedagogy of the Oppressed* (1968). He was a visiting professor at Harvard (1969–1970) and then moved to Switzerland until he was allowed to return to Brazil (1980) and became Secretary of Education in São Paulo state (1988).

Fremantle

Short-tailed Lark *Pseudalaemon fremantlii* **E. L. Phillips**, 1897

Major Guy Fremantle (b.1867) was in the British Army in Somaliland, and collected there (1896). A man of the same rank and name, whom we believe to be him, was also reported as in the Coldstream Guards and sailing from East London to London (1902) after the end of the Boer War, in which he served.

Fresnay

White-flanked Antwren ssp. *Myrmotherula axillaris fresnayana* **D'Orbigny**, 1835

(See **Lafresnaye**)

Freycinet

Dusky Megapode *Megapodius freycinet* **Gaimard**, 1823
Guam Flycatcher *Myiagra freycineti* **Oustalet**, 1881 EXTINCT

Captain Louis Claude Desaules de Freycinet (1779–1842) was a French navigator who was involved, with Baudin (q.v.), on board *Casuarina* and *Le Geographe*, in mapping the Western Australian coast north of Perth (1803). He explored in the Pacific (1817–1820) on *L'Uranie* and, after she was wrecked, he bought *La Physicienne*. The flycatcher was collected on Freycinet's expedition to Guam. Oustalet named a race of Rufous Fantail collected at the same time *Rhipidura rufifrons uraniae*, after the ship. An amphibian is also named after Freycinet.

Frick

Frick's Weaver *Othyphantes fricki* **Mearns**, 1913 NCR
[Alt. Baglafecht Weaver; JS *Ploceus baglafecht reichenowi*]

Jameson's Firefinch ssp. *Lagonosticta rhodopareia fricki* Mearns, 1913 NCR
[JS *Lagonosticta rhodopareia rhodopareia*]
Yellow White-eye ssp. *Zosterops senegalensis fricki* Mearns, 1913 NCR
[JS *Zosterops abyssinicus flavilateralis*]
Rattling Cisticola ssp. *Cisticola chiniana fricki* Mearns, 1913
Frick's Sombre Greenbul *Andropadus importunus fricki* Mearns, 1914 NCR
[JS *Andropadus importunus insularis*]
Northern Grey Tit *Parus thruppi fricki* Mearns, 1914 NCR
[Alt. Acacia Tit; JS *Parus thruppi thruppi*]
Northern Brownbul ssp. *Phyllastrephus strepitans fricki* Mearns, 1914 NCR; NRM
Moorland Francolin ssp. *Francolinus africanus fricki* **Friedmann**, 1928 NCR
[JS *Scleroptila psilolaemus*]
Orange-breasted Bush-shrike ssp. *Chlorophoneus sulfureopectus fricki* Friedmann, 1930 NCR
[JS *Chlorophoneus sulfureopectus similis*]

Childs Frick (1883–1965) was the son of Henry Clay Frick, co-founder of the United States Steel Corporation and whose Manhattan mansion overlooking Central Park now houses the world-renowned Frick Collection of paintings, sculpture and decorative art. Childs did not follow his father into industry, but became a brilliant scientist in his own right. He was Honorary Curator of the Department of Vertebrate Palaeontology at the Princeton Museum of Natural History, which he often supported with his own funds. He excavated for fossils in North America and undertook an expedition to Ethiopia and Kenya, cataloguing the birds recorded there (1937). Among other taxa an extinct sabre-toothed 'cat' and a fossil gastropod are named after him. On his death he bequeathed the world's largest collection of mammalian fossils, over 250,000 specimens, to the AMNH in New York

City. It was his efforts, more than anyone else's, which were responsible for our knowledge of prehistoric North American camels, rhinoceroses, antelope-like animals, deer and carnivores.

Friederichsen

Black Bishop ssp. *Euplectes gierowii friederichseni*
G. A. Fischer & **Reichenow**, 1884

The original description has no etymology and no clue in the rest of the text. We think Fischer and Reichenow may have had Ludwig Friederichsen (1841–1915) in mind. He was a German cartographer, ethnologist and collector who was in Tanganyika (Tanzania) in 1884.

Friedmann

Friedmann's Lark *Mirafra pulpa* Friedmann, 1930
[Alt. Friedmann's Bush Lark]

Violet-backed Starling ssp. *Cinnyricinclus leucogaster friedmanni* **Bowen**, 1930 NCR
[NUI *Cinnyricinclus leucogaster leucogaster*]
Double-toothed Barbet ssp. *Lybius bidentatus friedmanni* **Bannerman**, 1933 NCR
[NUI *Lybius bidentatus aequatorialis*]
Speckled Tinkerbird ssp. *Pogoniulus scolopaceus friedmanni* Bannerman, 1933 NCR
[JS *Pogoniulus scolopaceus flavisquamatus*]
Osprey ssp. *Pandion haliaetus friedmanni* Wolfe, 1946 NCR
[JS *Pandion haliaetus haliaetus*]
Banded Broadbill ssp. *Eurylaimus javanicus friedmanni* **Deignan**, 1947 NCR
[JS *Eurylaimus javanicus pallidus*]
Virginia Rail ssp. *Rallus limicola friedmanni* **Dickerman**, 1966

Dr Herbert Friedmann (1900–1987) was an American ornithologist and collector. Cornell University awarded his PhD (1923). He spent the next two years on a postdoctoral fellowship, devoted to fieldwork in Argentina (1923–1924) and Africa (1924–1925). Among other works he wrote a classic monograph, *The Cowbirds: A Study in the Biology of Social Parasitism* (1929) and co-wrote sections of the 11-volume *The Birds of Middle and North America* (1901–1950). He contributed to the ornithological collection at Cornell University and became Curator of Birds at the Department of Zoology, USNM (1929–1958). He was later the Director of the Los Angeles County Museum (1961–1970). He died of cancer.

Fries

Large Scimitar Babbler ssp. *Pomatorhinus hypoleucos friesi* **Delacour**, 1927 NCR
[JS *Pomatorhinus hypoleucos brevirostris*]

Jules Fries (fl.1928) was French Resident-Superior in Annam (Vietnam) (1927–1928).

Frobenius, J.

South American Gull sp. *Larus frobenii* **Philippi** & Landbeck, 1861 NCR
[Alt. Belcher's Gull; JS *Larus belcheri*]

Common Miner ssp. *Geositta cunicularia frobeni* Philippi & Landbeck, 1864

J. Frobenius (fl.1860) was a collector in Chile (1853).

Frobenius, L.

Spotted-backed Weaver *Ploceus cucullatus frobenii* **Reichenow**, 1923
[Alt. Village Weaver ssp.]

Dr Leo Viktor Frobenius (1873–1938) was a German ethnologist, archaeologist and philosopher. He first travelled in the Congo (1904) and (until 1918) concentrated on western and central Sudan. He became Honorary Professor at the University of Frankfurt (1932) and Director of the Municipal Ethnographic Museum (1935).

Fromm

Black-bellied Seedcracker ssp. *Pyrenestes ostrinus frommi* Kothe, 1911 NCR
[*P. ostrinus* NRM]
Brimstone Canary ssp. *Serinus sulphuratus frommi* Kothe, 1911 NCR
[JS *Crithagra sulphurata sharpii*]
Crimson-rumped Waxbill ssp. *Estrilda rhodopyga frommi* Kothe, 1911 NCR
[JS *Estrilda rhodopyga centralis*]
Helmeted Guineafowl ssp. *Numida meleagris frommi* Kothe, 1911 NCR
[JS *Numida meleagris marungensis*]

Captain Paul Fromm (1864–1940) was a German army officer in German East Africa (Tanzania) (1888–1889 and 1909–1911) and German South West Africa (Namibia) (1904–1905).

Fruiti

Peregrine Falcon ssp. *Falco peregrinus fruitii* **Momiyama**, 1927
[Trinomial sometimes corrected to *furuitii*]

I. Furuiti (DNF) was a collector in the Volcano Islands, Japan. Momiyama (q.v.) unfortunately missed a letter *u* out of the binomial, and this original misspelling has priority.

Frutuoso

São Miguel Scops Owl *Otus frutuosoi* Rando *et al.*, 2013 EXTINCT

Gaspar Frutuoso (c.1522–c.1591) was a 16th-century Portuguese priest, humanist and Azorean historian from the Azores island of São Miguel. The University of Salamanca awarded an arts certificate (1548) and he was later ordained (1554). He became a parish priest in the Azores (1558–1560), then returned to Salamanca where he obtained his doctorate (1563). He was vicar in Ribeira Grande (1565–1591). He wrote detailed histories and geographies of the Azores, the Canary Islands and to a lesser extent the Cape Verde Islands. This was published in six volumes as *Saudades da Terra*.

Fry

Spotted Owlet ssp. *Athene brama fryi* **E. C. S. Baker**, 1919
NCR
[JS *Athene brama brama*]

Colonel Arthur Brownfield Fry (1873–1954) was a British army surgeon in India (1899–1930). He saw action in Waziristan (1900–1902), took part in the Tibet Mission (1903–1904) and in WW1 served in France and Mesopotamia (Iraq). He was Professor of Hygiene at the Calcutta School of Tropical Medicine (1922–1925).

Fuertes

Fuertes's Oriole *Icterus fuertesi* **Chapman**, 1911
[Alt. Ochre Oriole]
Fuertes's Parrot *Hapalopsittaca fuertesi* Chapman, 1912
[Alt. Indigo-winged Parrot]

Fuertes's Red-tailed Hawk *Buteo jamaicensis fuertesi*
Sutton & **Van Tyne**, 1935

Louis Agassiz Fuertes (1874–1927) was an ornithologist and a painter of birds. He showed great interest and skill from a very early age. He met Elliott Coues (q.v.) (1894) and showed him some of his paintings. He was greatly encouraged by Coues's comments on his work and this may have finally decided him to concentrate on painting birds as a career. He went on collecting trips to Florida (1898), Alaska (1899) and Texas and New Mexico (1901). In addition to his travels within the USA, he also visited the Bahamas, the Canadian prairies and Rockies, Mexico and Jamaica, all before the outbreak of WW1. He also travelled to Abyssinia (1926) with W. H. Osgood (q.v.). He lectured on ornithology at Cornell University (1923–1927) until his death in a road traffic accident.

Fuggles-Couchman

Green-backed Camaroptera ssp. *Camaroptera brachyura fugglescouchmani* **Moreau**, 1939
Brown Woodland Warbler ssp. *Phylloscopus umbrovirens fugglescouchmani* Moreau, 1941

Norman Robin Fuggles-Couchman (1907–1994) was an ornithologist and collector who was British Commissioner for Agriculture in Tanganyika (Tanzania) (1930–1962). He was a member of the British Ornithologists' Union for over 60 years. He wrote *Agricultural Change in Tanganyika, 1945–1960* (1964). At the thanksgiving service following his death, a swift entered the church and flew about for five minutes before exiting.

Fullagar

Masked Booby ssp. *Sula dactylatra fullagari* R. O'Brien & Davies, 1991 NCR
[JS *Sula dactylatra tasmani*]

Dr Peter J. Fullagar (b.1938) is an English-born Australian ornithologist who worked for CSIRO (1964–1992). The University of London awarded both his bachelor's degree (1960) and his doctorate (1964). Since retiring (1992) as Senior Research Scientist he has been an Honorary Research Fellow working as Honorary Curator, Wildlife Sound Library and Bioacoustics Laboratory, Australian National Wildlife Collection. He is an expert on the avifauna of Lord Howe Island.

Fülleborn

Forest Double-collared Sunbird *Cinnyris fuelleborni*
Reichenow, 1899
Black Tit sp. *Parus fuelleborni* Reichenow, 1900 NCR
[Alt. White-winged Black Tit; JS *Parus leucomelas insignis*]
Fülleborn's Alethe *Pseudalethe fuelleborni* Reichenow, 1900
[Alt. White-chested Alethe]
Fülleborn's Boubou *Laniarius fuelleborni* Reichenow, 1900
Fülleborn's Longclaw *Macronyx fuelleborni* Reichenow, 1900
Pennant-winged Nightjar *Caprimulgus fuelleborni* Reichenow, 1900 NCR
[JS *Macrodipteryx vexillarius*]
Weaver sp. *Ploceus fuelleborni* Reichenow, 1900 NCR
[Alt. Bertram's Weaver; JS *Ploceus bertrandi*]

African Dusky Flycatcher ssp. *Muscicapa adusta fuelleborni* Reichenow, 1900
Green-backed Woodpecker ssp. *Campethera cailliautii fuelleborni* **Neumann**, 1900 NCR
[JS *Campethera cailliautii nyansae*]
Collared Praticole ssp. *Glareola pratincola fuelleborni* Neumann, 1910

Dr Friedrich Fülleborn (1866–1933) was a German physician. He served as a military doctor with the German Army in East Africa (1896–1910). He was an expert on tropical diseases and parasitology. He became Director of the Hamburg Institute for Marine and Tropical Diseases (1930). An amphibian and a reptile are named after him.

Fuller

Sparrow Weaver genus *Fullerellus* **Oberholser**, 1945 NCR
[Now in *Plocepasser*]

Arthur Bennett Fuller (1893–1976) was an American zoologist who collected in Kenya (1930) and was Curator, Cleveland Museum of Natural History (1931–1945).

Fytche

Mountain Bamboo Partridge *Bambusicola fytchii* **J Anderson**, 1871

Lieutenant-General Albert Fytche (1820–1892) was Chief Commissioner in Burma (1867–1871). At that time Burma (Myanmar) was coming under the influence of the Viceroy in Delhi. He wrote *Burma Past and Present, with Personal Reminiscences of the Country* (1878).

G

Gabb

Hispaniolan Kingbird *Tyrannus gabbii* **G. N. Lawrence**, 1876
[Syn. *Tyrannus caudifasciatus gabbii*]

Professor William More Gabb (1839–1878) was an American invertebrate palaeontologist. Philadelphia Central High School awarded his first degree (1857) and later his MA. He became James Hall's student and assistant in order to acquire knowledge of geology and became an active member of the Philadelphia Academy of Natural Sciences (1860). He worked on the geological survey of California, under Professor Whitney (1861), during which time he classified all the Cretaceous and Tertiary fossils (1862–1865). He wrote the relevant chapter in *Geological Survey of California* (1864) as well as the entire second volume. He surveyed Santo Domingo for the Santo Domingo Land and Mining Company (1868–1872). He went to Costa Rica on government contract and undertook a topographical and geological survey of that country assisted by Zeledón (q.v.) and wrote *On the Topography and Geology of Santo Domingo* (1873). He contracted malaria and returned to Philadelphia (1876), but tried to return to work a mining claim in Costa Rica until ill health forced him to return home, where he prepared all his notes for publication shortly before dying. A mammal is named after him.

Gabriel

Fairy sp. *Ornismya gabriel* **de Lattre**, 1843 NCR
[Alt. Purple-crowned Fairy; JS *Heliothryx barroti*]

(See **Prêtre**)

Gabriele

Olivaceous Warbler ssp. *Acrocephalus dumetorum gabrielae* **Neumann**, 1934 NCR
[JS *Iduna pallida elaeica*]
European Stonechat ssp. *Saxicola torquata gabrielae* Neumann & Paludan, 1937 NCR
[JS *Saxicola rubicola*]
Striped Honeyeater ssp. *Plectorhyncha lanceolata gabrielae* **Mathews** & Neumann, 1939 NCR; NRM

Dr Gabriele Neuhaüser (b.1911) was a German collector and zoologist. She learnt how to prepare specimens for museums while undertaking her doctoral thesis (1933). She then travelled to Turkey on a collection expedition, consolidating her reputation as a competent museum collector, so she was asked (1936) by the AMNH to collect in Queensland (1937), often alone. She married a man called John Scott; after WW2 she found collecting no longer financially viable and became a librarian in Brisbane.

Gabrielle

Green Magpie sp. *Cissa gabriellae* **Oglivie-Grant**, 1906 NCR
[Alt. Indochinese Green Magpie; JS *Cissa hypoleuca*]

Yellow-spotted Barbet ssp. *Buccanodon duchaillui gabriellae* **Bannerman**, 1924 NCR; NRM

Mrs Gabrielle Maud Vassal (1880–1959), wife of the French physician Dr J. J. Vassal (q.v.), was an English travel writer and amateur naturalist. She followed her husband to his postings in the French Colonies, culminating in the French Congo. She wrote *Three Years in Vietnam (1907–1910)*, also called *On and Off Duty in Annam* (1910), *In and Around Yunnan Fu* (1922) and *Life in French Congo* (1925). She collected during these postings, sending specimens back to the BMNH. A mammal is named after her.

Gabrielson

Gabrielson's Rock Ptarmigan *Lagopus muta gabrielsoni* **Murie**, 1944

Ira Noel Gabrielson (1889–1977) was an American biologist and conservationist. He graduated BA (biology) from Morningside College, Iowa, then became a schoolmaster in Marshalltown (1912–1915). He joined the Bureau of Biological Survey working in the western states on economic ornithology, rodent control and game management (1915–1946). He became Chief of the Bureau (1935) and, when this was combined with the Bureau of Fisheries, forming the United States Fish and Wildlife Service (1940), Gabrielson became Director. He retired (1946) but became President of the Wildlife Management Institute (1946–1970). He helped found the International Union for Conservation of Nature and Natural Resources (1948) and organise the World Wildlife Fund (1961). He received many honours including the Audubon Conservation Award. He wrote four books including *Wildlife Management* (1951) and co-authored another six, including *The Birds of Alaska* (1959). Popularly known as 'Mr Conservation', his collection of 9,000 bird skins in now at the USNM.

Gadd

White-winged Snowfinch ssp. *Montifringilla nivalis gaddi*
Zarudny & **Loudon**, 1904
Rock Pigeon ssp. *Columba livia gaddi* Zarudny & Loudon,
1906
Coal Tit ssp. *Periparus ater gaddi* Zarudny, 1911

G. G. Gadd (fl.1902) was a Finnish explorer who was in Persia (Iran) with Zarudny (q.v.) (1903–1904). He wrote a number of papers (1902–1911) including one on cicadas (1909) published by the St Petersburg Nature Society.

Gadow

Gadow's Sunbird *Nectarinia kilimensis gadowi* **Bocage**,
1892
[Alt. Bronze Sunbird ssp.]

Dr Hans Friedrich Gadow (1855–1928) was a German zoologist who spent most of his adult life in England having been invited, after graduating from Heidelberg, by Albert Günther (q.v.) to work on the BMNH's *Catalogue of Birds*, particularly volumes VIII and IX. His main contribution to ornithology was a method of taxonomy based on comparisons of 40 characteristics, which he set out in a paper to the Zoological Society of London, 'On the classification of birds' (1892). He became Curator (1884) of the Strickland Collection at Cambridge University as well as lecturing on the morphology of vertebrates. He settled in England, marrying an English woman who accompanied him on two journeys in Spain, which resulted in his *In Northern Spain* (1897). Gadow contributed to Alfred Newton's (q.v.) *A Dictionary of Birds* (1893–1896). He also contributed the article 'Bird' to the *Encyclopaedia Britannica*. He wrote a number of scientific descriptions. An amphibian and three reptiles are named after him.

Gaetke

Bluethroat ssp. *Luscinia svecica gaetkei* **Kleinschmidt**,
1904 NCR
[JS *Luscinia svecica svecica*]

Heinrich Gätke (1814–1897) was a German ornithologist and artist who lived on Heligoland (1841–1897) and pioneered migration studies there.

Gaffney

Buff-rumped Warbler ssp. *Phaeothlypis fulvicauda gaffneyi*
Griscom, 1927 NCR
[JS *Phaeothlypis fulvicauda leucopygia*]

Brigadier-General Dale Vincent Gaffney (1894–1950) was an officer in the US Army Air Force (1917–1950). He served in a number of postings, including Panama (1927), Bermuda and Alaska, where he was commanding officer, Alaska wing (1948–1950). He assisted the collector Rex Benson (q.v.) in the latter's explorations of western Panama.

Gaige

Spotted Towhee ssp. *Pipilo maculatus gaigei* **Van Tyne** &
G. Sutton, 1937

Professor Frederick McMahon Gaige (1890–1976) was an American entomologist, herpetologist and botanist who was Director of the Zoological Museum, University of Michigan. His wife, Helen Beulah Thompson Gaige, was a well-known herpetologist – and in their honour the American Society of Ichthyologists and Herpetologists makes an annual award to a graduate student of herpetology.

Gaikwar

Crombec sp. *Sylvietta gaikwari* **Sharpe**, 1901 NCR
[Alt. Somali Crombec; JS *Sylvietta isabellina*]

Sir Sayaji Rao, Maharajah Gaekwar of Baroda (Gujarat) (1863–1939), was an Indian princely ruler. His rule (1875–1939) raised Baroda from chaos to a model state. He was also in Somaliland (Somalia) (1900).

Gaimard

Gaimard's Cormorant *Phalacrocorax gaimardi* **Lesson** &
Garnot, 1828
[Alt. Red-legged Cormorant]
Forest Elaenia *Myiopagis gaimardii* **D'Orbigny**, 1840

Joseph (possibly Jean) Paul Gaimard (1793–1858) was a French naval surgeon, explorer and naturalist. He voyaged to Australia and the Pacific (1817–1819) aboard the *Uranie* keeping a journal, *Journal du Voyage de Circumnavigation, tenu par Mr Gaimard, Chirurgien à Bord de la Corvette L'Uranie*. Unfortunately, further entries were lost when the ship was wrecked off the Falklands. Gaimard continued on board *Physicienne*, the ship that rescued the expedition and was been purchased as a replacement. He was on the *Astrolabe*, under the command of Dumont d'Urville, when it visited New Zealand (1826). He led an expedition on the *Récherche* to northern Europe (1838–1840), visiting Iceland, the Faeroes, northern Norway, Archangel and Spitsbergen. His contemporary, zoologist Henrik Krøyer, who went with Gaimard to Spitsbergen (1838), described him thus; 'he was of medium build, with curly black hair and a rather unattractive face, but with a charming and agreeable manner'. He was something of a dandy and, in Iceland, gave out many sketches of himself! A number of fish species and a mammal and a reptile are named after him.

Gain

Antarctic Tern ssp. *Sterna vittata gaini* **R. C. Murphy**, 1938

Dr Louis Gain (fl.1910) was a French zoologist and botanist working at MNHN, Paris. He was on the Second French Antarctic Expedition (1908–1910).

Gairdner

Gairdner's Woodpecker *Picoides pubescens gairdnerii*
Audubon, 1839
[Alt. Downy Woodpecker ssp.]

Dr Meredith Gairdner (d.1837) was a naturalist from Edinburgh, Scotland. He left for Canada (1832) employed by the Hudson's Bay Company, and was stationed on the Columbia River (c.2 years). There he studied natural history, collecting materials to prepare a monograph of the vast riparian area. He also witnessed an eruption of Mount St Helens. Unfortunately he contracted tuberculosis when he was sent to treat an outbreak. The disease left him weak and he left for Hawaii, hoping the climate would help, but it proved fatal. He left bequests to help poor Hawaiian children. He wrote *Observations During a Voyage from England to Fort Vancouver, on the North-west Coast of America* (1834). Audubon (q.v.) wrote '… a specimen of a Woodpecker sent from the Columbia river by Dr Meredith Gairdner to Professor Jameson of Edinburgh, who kindly lent it to me for the purpose of being described … I hope to be able to give a figure of this species at the end of the present work.' A fish was named after him.

Gajdusek

Gajdusek's Hooded Mannikin *Lonchura spectabilis gajduseki* **Diamond**, 1967
[SI *Lonchura spectabilis wahgiensis*]

Dr D. Carleton Gajdusek (1923–2008) was an American physician who was co-recipient of the Nobel Prize for Medicine (1976). The University of Rochester, New York, awarded his first degree (1943) and he went on to achieve a medical degree at Harvard (1946). After postgraduate work at Columbia, Caltech and Harvard he was drafted serving as a research virologist at the Walter Reed Medical Graduate School. He worked at the Pasteur Institute in Tehran (1952–1953) on epidemiological problems in isolated populations. He was Visiting Investigator at Walter & Eliza Hall Institute of Medical Research in Melbourne (1954), where he began the work that culminated in the Nobel Prize (on kuru, the first human prion disease demonstrated to be infectious, which Gajdusek associated with the practice of funerary cannibalism). He was also the first person to recognise the potential zoogeographical significance of the Karimui Basin as a tropical enclave within the Highlands of New Guinea. Gajdusek became head of laboratories for virological and neurological research at the National Institutes of Health (1958), and was inducted to the National Academy of Sciences (1974) in the discipline of microbial biology. He took 56 children to the USA ostensibly to further their education, but was convicted (1997) of child molestation. After serving his sentence he spent the rest of his life in Europe, openly admitting molesting boys and his approval of incest in the TV documentary *The Genius and the Boys* (2009).

Galatea

Common Paradise Kingfisher *Tanysiptera galatea* **G. R. Gray**, 1859

Galatea was a sea-nymph in Greek mythology.

Galbraith

Buff-breasted Wren ssp. *Thryothorus leucotis galbraithii* **J. L. M. Lawrence**, 1861

John R. Galbraith (fl.1865) was a taxidermist in New York (1865) who worked with James McLeannan (q.v.) in the 1850s and collected in Panama (1860–1861).

Gale

Australian Masked Owl ssp. *Tyto novaehollandiae galei* **Mathews**, 1914

Captain Gale (fl.1914) may have been the master of the *James Nicol Fleming*, a clipper that traded between England and New Zealand (1875).

Galen

Hummingbird genus *Galenia* **Mulsant, E. Verreaux & J. Verreaux**, 1866 NCR
[Now in *Boissonneaua*]

Galen (AD 129–217), also known as Aelius Galenus or Claudius Galenus, was a celebrated physician and author of Greek origin who practised in Rome. He was court physician to a number of the emperors including Marcus Aurelius, Commodus and Septimus Severus.

Galindo

Greenish Puffleg ssp. *Haplophaedia aureliae galindoi* **Wetmore**, 1967 NCR
[JS *Haplophaedia aureliae caucensis*]

Dr Pedro Galindo (DNF), a Panamanian entomologist, collected with and for Alexander Wetmore (q.v.) in Panama (1962–1965). He became Director of the Gorgas Memorial Laboratory (1973).

Galinier

Abyssinian Catbird *Parophasma galinieri* **Guérin-Méneville**, 1843
[Alt. Juniper Babbler]

Captain Joseph Germain Galinier (1814–1888) was a French army officer. He helped map and survey the Tigre and Simen areas of Abyssinia (1840–1842) and co-wrote the 3-volume *Voyage en Abyssinie dans les Provinces du Tigré, du Samen et de l'Amhara* (1847–1848).

Gallardo

Hooded Grebe *Podiceps gallardoi* Rumboll, 1974

Angel Gallardo (1867–1934) and his grandson José María Gallardo (1925–1994) are both honoured. Angel was Director of the Argentine Museum of Natural Sciences, one of the great Argentine natural historians, and a passionate advocate of education. He later went into politics and was fervently anti-communist and pro-Mussolini, having met the man. He wrote *Memories for my Children and my Grandsons*. José Maria also became director of the museum and wrote *Anfibios y Reptiles* (1994) and had two reptiles named after him.

Galton

Familiar Chat ssp. *Cercomela familiaris galtoni* **Strickland**, 1853

Sir Francis Galton (1822–1911) was an English anthropologist, traveller and explorer. He was in Austria and the Balkans

(1840), in Egypt, Sudan and Syria (1845), and Damaraland (Namibia) (1851). He was greatly influenced by his cousin, Charles Darwin (q.v.). *The Origin of Species* (1859) led him to study eugenics, and he is regarded as its founder.

Gambel

Gambel's Quail *Callipepla gambelii* Gambel, 1843
Gambel's Chickadee *Poecile gambeli* **Ridgway** 1886
[Alt. Mountain Chickadee]

Gambel's Sparrow *Zonotrichia leucophrys gambelii* **Nuttall**, 1840
[Alt. White-crowned Sparrow ssp.]
White-fronted Goose ssp. *Anser albifrons gambelli* **Hartlaub**, 1852
Gambel's (Loggerhead) Shrike *Lanius ludovicianus gambeli* Ridgway 1887 NCR
[JS *Lanius ludovicianus excubitorides*]

Dr William Gambel (1821–1849) was an American physician, naturalist and collector, notably travelling with Thomas Nuttall (q.v.) on a collecting trip to North Carolina (1838). He was back in Philadelphia briefly (1839) before going to collect minerals in New Jersey and then accompanied Nuttall to Cambridge, Massachussets, where Nuttall was lecturing on botany at Harvard and Gambel acted as his apprentice. Gambel set out on his own (1840) to go to California (1841–1842), He became short of money and so joined the US Navy as a clerk on USS *Cyane*. He served on a number of vessels (1842–1845) during which time he collected on the Pacific coasts of both North and South America, visited the Sandwich Islands (Hawaii) and returned to Philadelphia via Cape Horn (1845). He studied medicine at the University of Pennsylvania (1845–1848) and served as Recording Secretary at the Philadelphia Academy of Natural Sciences (1848–1849). He decided to set up in medical practice in California and left by the overland route. He joined a party of the famous 49-ers on their way to the Californian Goldrush. Most of the party died trying to cross the Sierra Nevada Mountains in midwinter. He made it across into California but caught typhoid and died. The controversy over Gambel's Quail is not resolved. Gambel collected a specimen (1842), labelling it 'Gambel's Quail' and sent it home, possibly to his friend Nuttall, who, he believed mistakenly, had given it a formal description. Despite the mistake the name became official. To add to the general sense of confusion, John Cassin (q.v.) also claimed that he named it in Gambel's honour. A reptile and an oak are also named after him.

Gamble

Rufous-naped Whistler ssp. *Aleadryas rufinucha gamblei* **Rothschild**, 1897

Robert Gamble (DNF) was a collector in New Guinea.

Ganesh

Square-tailed Bulbul *Hypsipetes ganeesa* **Sykes**, 1832

Ganesh is a Hindu god revered as the Remover of Obstacles.

Ganier

Cliff Swallow ssp. *Petrochelidon pyrrhonota ganieri* **A. R. Phillips**, 1986

Albert Franklin Ganier Sr (1883–1973) was an American civil engineer who graduated from Purdue University (1908) and worked for the Nashville, Chattanooga and St Louis Railway (1908–1948). He had a life-long interest in ornithology and was a co-founder of the Tennessee Ornithological Society (1915). He left an important photographic archive.

Garbe

Garbe's Antwren *Myrmotherula longipennis garbei* H. von **Ihering**, 1905
[Alt. Long-winged Antwren ssp.]

P. Ernest William Garbe (1853–1925) was a naturalist collecting in Brazil at the end of the 19th century. He became zoologist for the Zoological Museum of the University of São Paulo (MZUSP). He also collected the first specimen of the rare bat species *Lichonycteris obscura*. Two dragonfiles are named after him

Garden

Night Heron sp. *Nycticorax gardeni* **Gmelin**, 1789 NCR
[Alt. Black-crowned Night Heron; JS *Nycticorax nycticorax hoactli*]

Dr Alexander Garden (1730–1791) was a Scottish physician and botanist who lived and practised in the Carolinas (1752–1785). He had supported George III during the American War of Independence and returned to Britain after it. The gardenia is named after him.

Gardener

Gardener's Bowerbird *Amblyornis macgregoriae* **De Vis**, 1890
[Alt. MacGregor's Bowerbird]

This name error is sometimes encountered; it is not 'Gardener's Bowerbird' but a 'Gardener Bowerbird'. All members of the genus *Amblyornis* are gardener bowerbirds because the males' bowers are surrounded by 'gardens' that are decorated with objects that the male collects.

Gardner

Blood Pheasant *Phasianus gardneri* **Hardwicke**, 1827 NCR
[JS *Ithaginis cruentus cruentus*]

The Honourable Edward Gardner (1784–1861) was a botanist and a British diplomat. He was Commissioner of Kumaon (1814) and first Resident in Nepal (1820–1829) after the Anglo-Nepalese War (1814–1816) (since which Gurkhas have served in the British Army).

Garlepp

Cochabamba Mountain Finch *Compsospiza garleppi* **Berlepsch**, 1893
Hermit sp. *Phaethornis garleppi* **Boucard**, 1893 NCR
[Alt. Planalto Hermit; JS *Phaethornis pretrei*]

Garlepp's Tinamou *Crypturellus atrocapillus garleppi*
Berlepsch, 1892
[Alt. Black-capped Tinamou ssp.]
Garlepp's Torrent Duck *Merganetta armata garleppi*
Berlepsch, 1894
Puna Rhea *Rhea pennata garleppi* **Chubb**, 1913
[Alt. Lesser Rhea ssp.]

Gustav Garlepp (1862–1907) was a German who collected in Latin America (1883–1897), sometimes with his brother Otto Garlepp (1864–1959). He was a bank employee who first went to South America to collect insects for the Natural History Museum in Dresden, but also collected birds for Berlepsch, including 40 new species, and eggs for Maximilian Küschel. He landed in Pará, Brazil, and crossed the continent to Peru, where he stayed for four years. After a short break in Germany he went back, this time to Bolivia. When he returned to Germany (1892) he had amassed 1,530 bird specimens. He returned to Bolivia (1893) together with his wife and his brother. He visited Germany for the last time (1900), presenting 3,000 specimens of 600 species of birds at the Annual Meeting of the German Ornithologists' Society, Leipzig. He was murdered during a collecting expedition in Paraguay. His collection is now in the Senckenberg Museum, Frankfurt. A mammal is named after him.

Garman

Common Gallinule ssp. *Gallinula galeata garmani*
J. A. Allen, 1876

Dr Samuel Walton Garman (1843–1927) was an American naturalist who took part in an expedition to the American West (1868). He graduated in Illinois (1870), became a school-teacher, and was Professor of Natural Science at a seminary in Illinois (1871–1872). He became Louis Agassiz's special student (1872) and (1873) was Assistant Director, Herpetology and Ichthyology Section, Museum of Comparative Zoology, Harvard. He hunted fossils in Wyoming with Cope (q.v.) (1872) and was in South America with Alexander Agassiz (q.v.) and surveyed Lake Titicaca (1874). He wrote *Reptiles of Easter Island* (1908). Four reptiles and an amphibian are named after him.

Garnot

Peruvian Diving-petrel *Pelecanoides garnotii* **Lesson**, 1828

Prosper Garnot (1794–1838) was a French naval surgeon, naturalist and collector who worked closely with Lesson (q.v.). They were both on board *La Coquille* during its circumnavigation of the world (1822–1825). Lesson and Garnot co-authored the zoological section of the voyage's report, *Voyage Autour du Monde Exécuté par Ordre du Roi sur la Corvette La Coquille Pendant les Années 1822–1825*, which was published in Paris (1828–1832). A grass genus is named after him.

Garrett

Society Islands Reed Warbler ssp. *Acrocephalus caffer garretti* Holyoak & Thibault, 1978 EXTINCT
[Alt. Garrett's Reed Warbler; Leeward Is. Warbler *A. musae garretti*]

Andrew Garrett (1823–1887) was an American naturalist, ichthyologist and malacologist who collected in Polynesia for the Godeffroy Museum, Hamburg (1861–1885).

Garrod

Grey-backed Storm-petrel genus *Garrodia* **W. A. Forbes**, 1881

Alfred Henry Garrod (1846–1879) was an English anatomist and zoologist. He contributed to descriptions of specimens obtained by the *Challenger* expedition (1872–1876). He died of tuberculosis.

Garza

Mexican Chickadee ssp. *Poecile sclateri garzai*
A. R. Phillips, 1986

Aldegundo Garza de León (b.1939) is a Mexican businessman and ornithologist who founded the Museum of Birds of Mexico (1993). The library at the museum is named after Allan R. Phillips, who described this subspecies and mentioned Garza in *The Known Birds of North and Middle America*.

Gasquet

Marshbird sp. *Xanthornus gasquet* **Quoy** & **Gaimard**, 1824 NCR
[Alt. Yellow-rumped Marshbird; JS *Pseudoleistes guirahuro*]

General Baron Jean Gasquet (1764–1819) was a French army officer (1781–1811). His nephew was the surgeon, naturalist and explorer Joseph Gaimard (q.v.).

Gaudichaud

Gaudichaud's Kingfisher *Dacelo gaudichaud* **Quoy** & **Gaimard**, 1824
[Alt. Rufous-bellied Kookaburra]

Charles Gaudichaud-Beaupré (1789–1854) was born in Angoulème, France. He studied pharmacy and became a dispenser in the French Navy (1810). He took part in several major expeditions as a naturalist; aboard the *Uranie* and *Physicienne* (1817–1820) circumnavigation, to eastern and western South America with the *Herminie* (1831–1833), and circumnavigating again on *Bonite* (1836–1837). After the *Bonite* expedition, Gaudichaud was appointed professor in pharmacy, in Paris, was attached to the MNHN and worked on the botanical collections from his expeditions. Two reptiles and an amphibian are named after him.

Gaughran

Red-headed Parrotfinch ssp. *Erythrura cyaneovirens gaughrani* **duPont**, 1972

Dr. James 'Jim' Alan Gaughran (b.1932) is an American waterpolo player (1956 Olympic Games) of Stanford University and was head coach of Stanford's waterpolo team (1969–1973). He was with duPont on the (1970) expedition to Western Samoa during which the parrotfinch holotype was collected.

Gaumer

Yucatan Swift *Chaetura vauxi gaumeri* **G. N. Lawrence**, 1882
[Alt. Vaux's Swift ssp.]
Caribbean Dove ssp. *Leptotila jamaicensis gaumeri*
Lawrence, 1885

Bright-rumped Attila ssp. *Attila spadiceus gaumeri* **Salvin & Godman**, 1891

George Franklin Gaumer (1850–1929) collected in the Americas. Kansas University awarded his first degree (1876) and masters (1893). He collected (mostly birds) in Cuba (1878), Yucatan, Mexico (1878–1881) and south-west US (1881–1884). He was professor of natural sciences at the University of North Mexico, after which he moved permanently to Yucatan, practising medicine. His later collections, with his two sons, were mostly botanical. He wrote *Monografía de los Mamíferos de Yucatán* (1917). Many plants and a mammal are named after him.

Gavrilov

Pale Sand Martin ssp. *Riparia diluta gavrilovi* Loskot, 2001

Professor Eduard I. Gavrilov is a Russian ornithologist at the Institute of Zoology, Almaty, Kazakhstan. He wrote *Fauna and Distribution of the Birds of Kazakhstan* (1999).

Gay, C.

Gay's Seedsnipe *Attagis gayi* **Geoffroy Saint-Hilaire & Lesson**, 1831
[Alt. Rufous-bellied Seedsnipe]
Grey-hooded Sierra Finch *Phrygilus gayi* **Gervais**, 1834
Velvetbreast sp. *Lafresnaya gayi* **Bonaparte**, 1850 NCR
[No type locality known; Alt. Mountain Velvetbreast; Syn *Lafresnaya lafresnayi saul*]

Claude Gay (1800–1873) was a French botanist. He went to Chile (1828–1932), then returned to France. He returned to Chile (1834–1840) and was later in the USA (1859–1860) studying American mining techniques. He wrote many works including his 24-volume magnum opus *Historia Física y Política de Chile* (1843–1851). An amphibian is named after him.

Gay, F.

Oahu Elepaio *Chasiempis gayi* **S. B. Wilson**, 1891 NCR
[JS *Chasiempis ibidis*]

Francis Gay (DNF) was an American explorer and collector in the Hawaiian Islands with the Rothschild Expedition. His family initiated sugarcane cultivation, and introduced the mangosteen tree in Hawaii.

Gayet

Sultan Tit ssp. *Melanochlora sultanea gayeti* **Delacour & Jabouille**, 1925

Victor Gayet-Laroche (DNF) was Director of the railway system in Tonkin, French Indochina (Vietnam). He collected the holotype (1924) and gave it to Delacour (q.v.).

Gazzola

Cuckooshrike genus *Gazzola* **Bonaparte**, 1850 NCR
[Now in *Coracina*]

Giovanni Battista Conte Gazzola (1757–1834) was an Italian scientist, inventor, palaeobiologist and collector. He made a special study of fossil fish in the Verona museum. His own collection was split up and some of it is in Paris after he 'donated' specimens to Napoleon Bonaparte.

Gedge

Clapperton's Francolin ssp. *Pternistis clappertoni gedgii* **Ogilvie-Grant**, 1891

Ernest Gedge (1862–1935) went to Assam, India, to work on a tea estate (1879). He joined the Imperial British East Africa Company (1888) and was in the party Sir Frederick Jackson (q.v.) led to Uganda (1889). He was a special correspondent for *The Times* (1892–1893) and was in England (1894–1896). He prospected for gold and other minerals in many countries – in the Yukon, Canada (1898–1899), Southern Rhodesia (Zimbabwe) (1900–1902), and Kenya (1902–1905). He visited Russia (1907) and returned to East Africa (1909). He was in Malaya, Borneo and Java (1912–1913) and Mexico (1914). He was regarded as too old to fight in WW1 so he served as a policeman instead. After WW1 he made a trip to Australia (1933) but otherwise stayed in Europe and the Mediterranean.

Geisler

Brown-capped Jewel-babbler *Ptilorrhoa geislerorum* **A. B. Meyer**, 1892
[Alt. Dimorphic Jewel-babbler]

Geislers' Catbird *Ailuroedus buccoides geislerorum* A. B. Meyer, 1891
[Alt. White-eared Catbird ssp.]
Black-billed Sicklebill ssp. *Drepanornis albertisi geisleri* A. B. Meyer, 1893

Bruno Geisler (1857–1945) and his brother Herbert G. Geisler were German taxidermists. They collected together, mainly for the museum in Dresden, but also for the dealer Wilhelm Schlüter (q.v.), in Ceylon (Sri Lanka) and Java (1887–1989), then in New Guinea (1890–1892). Bruno later became a curator at Dresden museum where A. B. Meyer (q.v.) was Professor. Bruno also illustrated works by Meyer and others. *Drepanornis albertisi geisleri* is named after Bruno and the other two birds after both brothers. Two amphibians are named after Bruno.

Gené

Slender-billed Gull *Chroicocephalus genei* Breme, 1839
Rufous-tailed Antbird *Drymophila genei* **De Filippi**, 1847

Carlo Giuseppe Gené (1800–1847) was an Italian author and naturalist. He studied at the University of Pavia and later (1828) became Assistant Lecturer in Natural History there. He collected in Hungary (1829) and made four trips to Sardinia (1833–1838), mostly collecting insects. On Bonelli's (q.v.) death (1830) Gené succeeded him as Professor of Zoology and Director of the Royal Zoological Museum in Turin, and De Filippi (q.v.), in turn, succeeded him. An amphibian is named after him.

Genestier

Rusty-capped Fulvetta ssp. *Alcippe dubia genestieri*
Oustalet, 1897
[Syn. *Schoeniparus dubius genestieri*]

Père Annet Genestier (1856–1937) was French priest,
ordained in 1885, and a botanist. He was a missionary to
Qiunatong, China, on the Tibetan border (1885–1937). Francis
Kingdon-Ward visited him (1913) and was thankful for a good
meal!

Gengler

Common Chaffinch ssp. *Fringilla coelebs gengleri*
Kleinschmidt, 1909

Dr Josef Gengler (1863–1931) was a German physician and
ornithologist.

Genin

White-bellied Emerald ssp. *Amazilia candida genini* **Meise**,
1938

Auguste Genin (1862–1931) was a Franco-Mexican industrialist, entrepreneur, archaeologist, ethnographer and
collector.

Gentry

Ancient Antwren *Herpsilochmus gentryi* **B. Whitney** &
J. Alvarez-Alonso, 1998

Alwyn H. Gentry (1945–1993) was a US botanist. He wrote
extensively (1970s–1990s) on biodiversity in Neotropical rainforests. For some years he was Senior Curator at the
Missouri Botanical Gardens. Over a 25-year research career
he published c.200 scientific papers and collected c.80,000
plant specimens. His major work was *Field Guide to the Families and Genera of Woody Plants of Northwest South
America*. He died in the plane crash that also killed Ted
Parker (q.v.). An amphibian is named after him.

Geoffroy

Parrot genus *Geoffroyus* **Bonaparte**, 1850

Geoffroy's Dove *Claravis godefrida* **Temminck**, 1811
[Alt. Purple-winged Ground Dove]
Geoffroy's Parrot *Geoffroyus geoffroyi* **Bechstein**, 1811
[Alt. Red-cheeked Parrot]
Rufous-vented Ground Cuckoo *Neomorphus geoffroyi*
Temminck, 1820
Geoffroy's Dotterel/Plover *Charadrius leschenaultii* **Lesson**,
1826
[Alt. Greater Sand Plover]
Geoffroy's Wedgebill *Augastes geoffroyi* **Bourcier**, 1843
[Alt. Wedge-billed Hummingbird]

Geoffroy's Blood Pheasant *Ithaginis cruentus geoffroyi*
J. Verreaux, 1867

Étienne Geoffroy Saint-Hilaire (1772–1844) was a French
naturalist. He trained for the Church but abandoned it to
become Professor of Zoology (aged 21!), when the Jardin du
Roi was renamed Le Musée National d'Histoire Naturelle. In
his *Philosophie Anatomique* (1818–1822) and other works, he
expounded the theory that all animals conform to a single
plan of structure. This was strongly opposed by Cuvier (q.v.),
who had been his friend, and (1830) a widely publicised
debate between the two took place. Despite their differences, the two men did not become enemies; they respected
each other's research, and Geoffroy gave one of the orations
at Cuvier's funeral (1832). Modern developmental biologists
have confirmed some of Geoffroy Saint-Hilaire's ideas. His
son, Isidore Geoffroy Saint-Hilaire (q.v. under **Isidor**), was
also a zoologist, and was an authority on deviation from
normal structure. He succeeded to his father's professor-
ships. Sixteen mammals and a reptile are named after
Étienne and two mammals after Isidore. *Godefrida* is Latin for
Geoffrey.

George Barrowclough

Scaled Flowerpiercer ssp. *Diglossa duidae
georgebarrowcloughi* **Dickerman**, 1987

(See **Barrowclough**)

Gerasimov

Rock Ptarmigan ssp. *Lagopus muta gerasimovi* Redkin, 2005

Nikolai Nikolaevich Gerasimov is a Russian ornithologist and
conservationist working in the Russian Far East since 1992
with his wife Alevtina Ivanova, re-establishing the Aleutian
Canada Goose *Branta hutchinsii leucopareia*. He and Yuri N.
Gerasimov wrote a number of papers together during the
1990s and 2000s such as: 'Observations of the spring migration of divers and seaducks along the western coast of
Kamchatka, Russia' (1997).

Gerbe

Eurasian Treecreeper ssp. *Certhia familiaris gerbei* **Jouard**,
1930 NCR
[JS *Certhia familiaris macrodactyla*]

Dr Jean-Joseph Zéphirin Gerbe (1810–1890) was a French
naturalist and doctor of science. He completed *Ornithologie
Européenne* (1867), the work that Côme Damien Degland
(q.v.) was revising at the time of his death. He also published
a number of articles mostly on ornithological subjects and a
French translation of Alfred Edmund Brehm's (q.v.) works
under the title *La Vie des Animaux Illustrée*. A mammal is
named after him.

Gerchner

Redwing ssp. *Turdus iliacus gerchneri* **Zarudny**, 1918 NCR
[JS *Turdus iliacus iliacus*]
Eurasian Treecreeper ssp. *Certhia familiaris gerchneri*
Charlemagne, 1928 NCR
[JS *Certhia familiaris familiaris*]

W. J. Gerchner (DNF) was a German ornithologist in the
Ukraine.

Gerd

Lesser Spotted Woodpecker ssp. *Dendrocopos minor gerdi*
E. Dickinson, Frahnert & Roselaar, 2009 NCR
[JS *Dendrocopos minor buturlini*]

(See **Heinrich**)

Gerhard

African Yellow White-eye ssp. *Zosterops senegalensis
gerhardi* Elzen & König, 1983

Gerhard Nikolaus is a German ornithologist who collected the holotype and lived for an extended period in Sudan, gaining an extensive knowledge of the avifauna of that country. He wrote *Distribution Atlas of Sudan's Birds with Notes on Habitat and Status* (1987).

Germain/German

Germain's Peacock Pheasant *Polyplectron germaini* **D. G. Elliot**, 1866
Germain's Swiftlet *Aerodramus germani* **Oustalet**, 1876
[Alt. Oustalet's Swiftlet, German's Swiftlet]
Sooty-headed Bulbul ssp. *Pycnonotus aurigaster germani*
Oustalet, 1878

Louis Rodolphe Germain (1827–1917) was a veterinary surgeon in the French colonial army, serving in Indochina (Vietnam) (1862–1867), and went to New Caledonia (1975–1878). He made zoological collections (including the first specimen of the pheasant) in his spare time, donating them to the MNHN. A mammal is named after him. The use of *germani* by Oustalet (q.v.) appears to be a spelling error.

Gertrude (Legendre)

Silver-eared Mesia ssp. *Leiothrix argentauris gertrudis*
Ripley, 1948 NCR
[JS *Leiothrix argentauris vernayi* – itself SI *Leiothrix
argentauris argentauris*]

Mrs Gertrude Legendre *née* Sanford (1902–2000) was an American socialite (daughter of a congressman and rug magnate), explorer, big-game hunter, environmentalist, plantation owner and spy! After her education she spent six years (1923–1929) travelling the world and hunting big game in Africa, and North America. Shortly after exploring in Abyssinia for the AMNH (1929) she married her co-leader Sydney J. Legendre (1903–1948) and accompanied him on his expeditions to Indochina (1931–1932) and Assam (1946). During WW2 she worked for the OSS and was the first American woman captured on the Western Front, being held prisoner for six months before escaping to Switzerland. She wrote two autobiographies (1948 and 1987).

Gertrude (Whitney)

Niau Kingfisher *Todiramphus gertrudae* **R. C. Murphy**, 1924
[Syn. *Todiramphus gambieri gertrudae*]

Gertrude Vanderbilt Whitney (1875–1942) was an American philanthropist and well known both as a socialite and as a sculptress (working for a time in Paris with Rodin). She founded the Whitney Museum. (See **Whitney, G.**)

Gervais

Magpie robin genus *Gervaisia* **Bonaparte**, 1854 NCR
[Now in *Copsychus*]

François Louis Paul Gervais (1816–1879) was a zoologist, palaeontologist, and anatomist. He was Blainville's (q.v.) student and succeeded him as Professor of Comparative Anatomy, MNHN, Paris (1868), where he had been Assistant (1835–1845). He was Professor of Zoology and Comparative Anatomy in the Science Faculty at Montpellier (1845–1868), and Head of the Faculty (1856). Four mammals, a reptile, and a fish genus are named after him.

Gestro

Gestro's Fruit Dove *Ptilinopus ornatus gestroi* **Salvadori** &
Albertis, 1875
[Alt. Ornate Fruit Dove ssp.]

Raffaello Gestro (1845–1936) was an Italian entomologist, one of the most important Italian coleopterists. He was Deputy Director, and then Director, of the Genoa Natural History Museum, and President of the Entomological Society of Italy. He published 147 papers on Coleoptera (beetle) taxonomy and described 936 new insect species, many based on specimens from other Italian collectors, including Fea (q.v.), D'Albertis (q.v.) and Beccari (q.v.). Many insects are named after him.

Getty

Junin Tapaculo *Scytalopus gettyae* Hosner *et al.*, 2013

Caroline Marie Getty (b.1957) is a member of the American dynasty which is famous for its oil wealth and family squabbles that end in litigation. She is a granddaughter of Jean Paul Getty, who founded Getty Oil. She is a conservationist who has served on many boards including the National Fish and Wildlife Foundation and is a director of the Worldwide Fund for Nature.

Geyr

Desert Lark ssp. *Ammomanes deserti geyri* **Hartert**, 1924

Professor Dr Hans Freiherr Geyr von Schweppenburg (1884–1963) was an ornithologist and explorer who worked in the Arctic (1907–1908), Sudan (1913) and Algerian Sahara (1913–1914). A reptile is named after him.

Ghiesbreght

White Hawk ssp. *Leucopternis albicollis ghiesbreghti* **Du
Bus**, 1845

Auguste Boniface Ghiesbreght (1810–1893) was a Belgian botanist and zoologist who collected in tropical America. He started his career by accompanying his lifelong friend J. J. Linden (q.v.) on his two expeditions to Latin America (1835–1837 and 1837–1841). He later returned to Brazil. Several plants, particularly orchids, are named after him.

Ghigi

Ghigi's Grey Peacock Pheasant *Polyplectron bicalcaratum ghigii* **Delacour** & **Jabouille**, 1924

Professor Alessandro Ghigi (1875–1970) was Chancellor of Bologna University (1931–1943). His major interest was breeding hybrids, so is best described as a geneticist. He was a friend of Delacour (q.v.), who gave him pheasant specimens he had collected on his various expeditions to Vietnam. Ghigi developed pheasant hybrids and mutations in captivity, including one known as Ghigi's Golden Pheasant (*Chrysolophus pictus* mut. *luteus*). He was one of the founders (1911) of the *Rivista Italiana di Ornitologia*.

Gibbs

Vanikoro White-eye *Zosterops gibbsi* Dutson, 2008

David Gibbs is a British ecologist, naturalist and ornithologist. He discovered the white-eye in 1994, and Dutson collected the first specimens three years later. Gibbs co-wrote *Pigeons and Doves: a Guide to the Pigeons and Doves of the World* (2001).

Gibson

Red-billed Emerald *Chlorostilbon gibsoni* **L. Fraser**, 1840

Gibson (DNF) was employed by Loddiges (q.v.) as a collector in Colombia.

Gibson, J. D.

Gibson's Albatross *Diomedea gibsoni* Robertson & Warham, 1992
[Alt. Antipodean Albatross; Syn. *Diomedea exulans gibsoni*]

John Doug Gibson (c.1925–1984) was an Australian amateur ornithologist deeply interested in seabirds, their ecology and distribution, and a pioneer of albatross studies. He was a founder member of the Illawarra Bird Observers Club, publisher of his *The Birds of the County of Camden (Including the Illawarra Region)*. He wrote numerous articles (1950s–1970s), published in several Australasian journals. Together with members of the New South Wales Albatross Study Group, Gibson suggested, in an article in *Notornis* (1967), a unified system for describing the plumage of the great albatrosses. Known as the 'Gibson Plumage Index' (GPI) or 'Gibson Code', the system gives numerical values to particular degrees of coloration on the back, head, inner wing and tail. Gibson's Albatross has been elevated from a race of the Wandering Albatross *D. exulans*, although species status remains controversial.

Gibson-Hill

Orange-headed Thrush ssp. *Zoothera citrina gibsonhilli* **Deignan**, 1950

Dr Carl Alexander Gibson-Hill (1911–1963) was a British physician, naturalist and ornithologist. He qualified as a physician in London (1938) before becoming Resident Medical Officer on Christmas Island and the Cocos Islands (1938–1941). He transferred to Singapore as a health officer and as Assistant Curator, Raffles Museum, four days before Singapore fell to the Japanese. He was interned as a prisoner-of-war in Changi Prison (1942–1945). After a stint in the Falkland Islands (1945–1946) he returned to Singapore as Curator of Zoology, Raffles Museum (1947), and became the museum's last expatriate director (1956–1963), dying shortly before he was due to retire.

Gierow

Gierow's Bishop *Euplectes gierowii* **Cabanis**, 1880
[Alt. Black Bishop]

Paul Gierow (DNF) was a German naturalist who collected in Angola (late 1870s).

Giffard

White-crowned Robin Chat ssp. *Cossypha albicapillus giffardi* **Hartert**, 1899
Sun Lark ssp. *Galerida modesta giffardi* Hartert, 1899 NCR
[NUI *Galerida modesta modesta*]

Colonel William Carter Giffard (1859–1921) was in the Gold Coast (Ghana) (1897–1898). He collected mostly bird specimens, which he sent to Walter Rothschild's (q.v.) museum at Tring. A mammal is named after him.

Gifford

Gifford's Ground Finch *Camarhynchus giffordi* **Swarth**, 1929 NCR
[Aberrant *Camarhynchus pallidus* or hybrid x *Certhidea olivacea*]

Professor Edward Winslow Gifford (1887–1959) was an American herpetologist, archaeologist, ethnographer and anthropologist. After graduating from high school (1904) he was appointed as an assistant at the California Academy of Sciences, at first in conchology but later in ornithology. He was on the Academy's expedition to the Galapagos Islands (whence came the only known specimen) (1905–1906). He observed and later described the use of a thorn or twig by the Pallid Tree (or Woodpecker) Finch *Camarhynchus pallidus* to pry out insects from the bark of trees. The scepticism that greeted this was finally overcome 35 years later when filmed. He joined the University of California at Berkeley as Assistant Curator (1912) and was promoted to Associate Curator (1915), then Lecturer (1920). He undertook field studies in the Galapagos again (1913–1919) and then to Tonga (1920–1921). He developed the collection of the Lowie Museum of Anthropology (now Phoebe A. Hearst Museum of Anthropology) in Berkeley into a major one, becoming full Professor in 1945. He sailed to the Fiji Islands (1947) on board *Thor I* and led two further expeditions to New Caledonia and Yap in Micronesia (1947–1956).

Gigas

Giant Coua *Coua gigas* Boddaert, 1783
Laughing Kookaburra *Dacelo gigas* Boddaert, 1783 NCR
[JS *Dacelo novaeguineae*]
Pitta sp. *Pitta gigas* Temminck, 1823 NCR
[Alt. Giant Pitta; JS *Hydrornis caeruleus caeruleus*]

Giant Hummingbird *Patagona gigas* Vieillot, 1824
Mottle-backed Elaenia *Elaenia gigas* P. L. Sclater, 1871
Waterfall Swift *Hydrochous gigas* **Hartert & A. L. Butler,** 1901
[Alt. Giant Swiftlet; Syn. *Collocalia gigas*]
Atitlan Grebe *Podilymbus gigas* **Griscom,** 1929 EXTINCT
[Alt. Giant Pied-billed Grebe]

Great Thrush ssp. *Turdus fuscater gigas* L. Fraser, 1841
Hooded Mountain Tanager ssp. *Buthraupis montana gigas* Bonaparte, 1851
Red-legged Honeycreeper ssp. *Cyanerpes cyaneus gigas* Thayer & Bangs, 1905
Dollarbird ssp. *Eurystomus orientalis gigas* Stresemann, 1913
Altamira Oriole ssp. *Icterus gularis gigas* Griscom, 1930 NCR
[NUI *Icterus gularis gularis*]
Wallace's Owlet-nightjar ssp. *Aegotheles wallacii gigas* **Rothschild,** 1931

Gigas was a giant in Greek mythology, the child of Uranos and Gaea. The name is applied to taxa that are 'giants' of their kind including six reptiles, four mammals and two amphibians.

'Gigi' (G. G.)

Cape Penduline-tit ssp. *Anthoscopus minutus gigi* **Winterbottom,** 1959

Gerald Graham ('G. G.') Smith (1892–1976) was a botanist and naturalist who joined the Board of the East London Museum (1940), becoming Chairman (1942) before retiring (1973). He co-wrote *The Wild Flowers of the Tsitsikama National Park* (1967).

Giglioli

Bunting sp. *Emberiza gigliolii* **Swinhoe,** 1867 NCR
[Alt. Meadow Bunting; JS *Emberiza cioides castaneiceps*]

Asian Rosy Finch ssp. *Leucosticte arctoa gigliolii* **Salvadori,** 1869

Prof. Dr Enrico Hillyer Giglioli (1845–1909) was an Italian zoologist, anthropologist, photographer and ornithologist who graduated from the University of Pisa (1864). He started teaching zoology at the University of Florence (1869) and was Director, Royal Zoological Museum, Florence. He succeeded de Filippi (q.v.) on the *Magenta* expedition after de Filippi's death from cholera in Hong Kong (1867), and wrote up the expedition's results and report. An amphibian and a mammal are named after him.

Gil

Iberian Azure-winged Magpie ssp. *Cyanopica cooki gili* **Witherby,** 1923 NCR; NRM
[Syn. *Cyanopica cyanus cooki*]

Dr Gil Augusto Lletget (1889–1946) was a Spanish ornithologist.

Gilbert (Nkwocha)

Gilbert's Mountain Babbler *Kupeornis gilberti* **Serle,** 1949
[Alt. White-throated Mountain Babbler]

Gilbert Nkwocha (DNF) was an African collector who worked as a bird-skinner for William Serle in West Africa (1944–1955).

Gilbert, J.

Whistler genus *Gilbertornis* **Mathews,** 1912 NCR
[Now in *Pachycephala*]

Gilbert's Whistler *Pachycephala inornata* **Gould,** 1841

Philippine Megapode ssp. *Megapodius cumingii gilbertii* **G. R. Gray,** 1862
Brown Noddy ssp. *Anous stolidus gilberti* Mathews, 1912 NCR
[JS *Anous stolidus pileatus*]

John Gilbert (c.1812–1845) was an English naturalist and explorer in south-western Western Australia. He was the principal, and assiduous, collector of birds for Gould (q.v.) (1840–1842) who, despite high expectations, poorly served Gilbert, often leaving him with insufficient funds and equipment and barely acknowledging his huge contribution of specimens, descriptions and detailed observations. Gilbert's specimens were sent back to Europe on board *Beagle* and *Napoleon*. Gilbert once stayed with a settler, Mrs Brockman, who wrote: 'He used to go out after breakfast. ... .. and we seldom saw him until late afternoon, when he would come in with several birds and set busily to work to skin and fill them out before dark. ... .. He was an enthusiast at his business, never spared himself, and often came in quite tired out from a long day's tramp after some particular bird, but as pleased as a child if he succeeded in shooting it.' His employer the Zoological Society of London trained him as a taxidermist. He was Curator of the Shropshire and North Wales Natural History Society in Shrewsbury until 1837. He left for Australia (1838) becoming naturalist on Ludwig Leichhardt's expedition to Port Essington (1844–1845). At the Gulf of Carpentaria Aborigines speared him to death (1845). Two of the expedition, aboriginals themselves, probably caused the attack, having treated the people they met very badly including raping a local aboriginal woman. Leichhardt's account says Gilbert heard the noise of the attack in the night and rushed from his tent with his gun only to receive a spear in the chest. The only words he spoke were: 'Charlie, take my gun, they have killed me' before dropping lifeless to the ground. A reptile and two mammals are named after him.

Giles

Giles's Pale-bellied Tapaculo *Scytalopus griseicollis gilesi* Donegan & Avendaño-C., 2008
Giles's Antpitta *Grallaria milleri gilesi* Salaman, Donegan & Prys-Jones, 2009
[Alt. Brown-banded Antpitta ssp.]

O. A. Robert Giles is a British sponsor of research and conservation with strong links to the NGO ProAves Colombia. The antpitta lay unregarded in the British Museum (Natural

History) (1878–2001). It is thought that the bird's habitat has already been destroyed and that it may have become extinct before being recognised.

Gill, E. L.

Swainson's Francolin ssp. *Pternistis swainsonii gilli* **J. A. Roberts**, 1932 NCR
[NUI *Pternistis swainsonii swainsonii*]
Karoo Long-billed Lark ssp. *Certhilauda subcoronata gilli* J. A. Roberts, 1936

Edwin Leonard Gill (1877–1956) worked in various museums in the United Kingdom before being appointed Director, South African Museum Cape Town (1925). He implemented and carried through a programme of modernisation and expansion. The museum suffered financially (WW2) and Gill decided to ameliorate the situation by resigning (1942) so that his salary would not be a burden. He was a skilled hobby taxidermist at home, where his sister helped him. An amphibian is named after him.

Gill, H. B.

Sarus Crane ssp. *Grus antigone gillae* **Schodde**, Blackman & Haffenden, 1989

Mrs H. B. Gill (fl.1989) is an Australian field ornithologist. She discovered the species in Australia (1969) and prepared the holotype.

Gillett

Gillett's Lark *Mirafra gilletti* **Sharpe**, 1895

Major Frederick Alfred Gillett (1872–1944) was in Somaliland (Somalia) (1895) on an expedition, which included not only his sister, Miss Gillett, but Mr and Mrs Ethelbert Lort Phillips – a grand Victorian outing! We surmise that Gillett was related to Phillips (q.v.) and was his heir, as he changed his name to Frederick Alfred Lort Phillips (1926). He settled in the UK and wrote his memoirs: *The Wander Years: Hunting and Travel in Four Continents* (1931).

Gilliard

Gilliard's Honeyeater *Melidectes whitemanensis* Gilliard, 1960
[Alt. Bismarck Honeyeater, Bismarck Melidectes]

Greater Flowerpiercer ssp. *Diglossa major gilliardi* **Chapman**, 1939
Elegant Tit ssp. *Periparus elegans gilliardi* **Parkes**, 1958
Chestnut-backed Jewel-babbler ssp. *Ptilorrhoa castanonota gilliardi* **Greenway**, 1966
Marquesan Swiftlet ssp. *Aerodramus ocistus gilliardi* **Somadikarta**, 1994

Ernest Thomas Gilliard (1912–1965) was an American ornithologist closely associated with the AMNH in New York. He took part in the New Britain Expedition and an exploration of the Whiteman Mountains (1958–1959), hence the scientific name of the honeyeater. He wrote chiefly on birds of New Guinea, including *Living Birds of the World* (1959) and the posthumously published *Birds of Paradise and Bower Birds*

(1969) and the co-written *Handbook of New Guinea Birds* (1968). See also **Thomas (Gilliard)**. An amphibian and a mammal are named after him.

Gilman

Gilman's Screech Owl *Megascops kennicottii gilmani* **Dickey**, 1916 NCR
[Alt. Western Screech Owl ssp.; JS *Megascops kennicottii aikeni*]

Marshall French Gilman (1871–1944) was a well-respected, all-round Californian naturalist who studied the wildlife of a number of sites, including Death Valley, where he was caretaker (1930s). A number of plants are named after him including at least one genus. He was a member of the Cooper Ornithological Society (1901–1944).

Gindi

Buff-crested Bustard *Lophotis gindiana* **Oustalet**, 1881

Abdou Gindi (DNF) was an Egyptian explorer and collector in Zanzibar, Abyssinia (Ethiopia) and Somaliland (Somalia).

Ginés

Bluish Flowerpiercer ssp. *Diglossa caerulescens ginesi* **W. H. Phelps** & **W. H. Phelps Jr**, 1952

Hermano Ginés or Brother Ginés (1912–2011) was born Pablo Mandazen Soto, in Navarra, Spain. He was a notable Venezuelan naturalist who founded Sociedad de Ciencias Naturales La Salle. He studied theology in Barcelona and Belgium, and natural sciences for his bachelor's degree in Colombia (1936). He then started to teach natural sciences (1939) in Caracas, Venezuela. He led 28 scientific expeditions and wrote c.40 scientific papers. Two amphibians are named after him, as is the research ship *Hermano Ginés*.

Giraud

Giraud's Oriole *Icterus chrysater giraudii* **Cassin**, 1848
[Alt. Yellow-backed Oriole ssp.]
Horned Lark ssp. *Eremophila alpestris giraudi* **Henshaw**, 1884
Ladder-backed Woodpecker ssp. *Picoides scalaris giraudi* **Stone**, 1920 NCR
[NUI *Picoides scalaris cactophilus*]
Olive Warbler ssp. *Peucedramus taeniatus giraudi* **Zimmer**, 1948

Jacob Post Giraud Jr (1811–1870) was an American naturalist and treasurer of the New York Lyceum of Natural History. He wrote *The Birds of Long Island* (1844).

Gironière

Black-chinned Fruit Dove ssp. *Ptilinopus leclancheri gironieri* **J. Verreaux** & **Des Murs**, 1862

Dr Paul Proust de la Gironière (1797–1862) was a French physician resident in the Philippines (1820–1861), where he was a pioneering agriculturist. He founded the town of Jalajala. He wrote *Aventures d'un Gentilhomme Breton aux Îles Philippines* (1855).

Gisella

Owl genus *Gisella* **Bonaparte**, 1854 NCR
[Now in *Aegolius*]

There is no etymology given in either 1854 publication announcing this genus – we have found no obvious candidate.

Giulianetti

Island Leaf Warbler ssp. *Phylloscopus poliocephalus giulianettii* **Salvadori**, 1896
Tawny-breasted Honeyeater ssp. *Xanthotis flaviventer giulianettii* **Mayr**, 1931 NCR
[Viewed as an intergrade population between races *saturatior* and *visi*]

Amadeo Giulianetti (1869–1901) was an Italian naturalist, ethnologist, explorer and collector who went to New Guinea as assistant to Loria (q.v.) (1889), then returned to Europe (1895). He was in Port Moresby again (1896) with William MacGregor (q.v.) and was appointed Travelling Government Agent for the collecting of natural history specimens, and also appointed (1897) Government Agent for the Interior. He became administrator of Mekeo (1898) and Delena district (1898–1901).

Gladkov

Warbler genus *Gladkovia* Kashin, 1977 NCR
[Now in *Cettia*]

Eurasian Eagle Owl ssp. *Bubo bubo gladkovi* Zaletaev, 1962 NCR
[JS *Bubo bubo turcomanus*]

Professor Nikolay Alekseyevich Gladkov (1905–1975) was a Russian ornithologist. He joined the Natural Science Department of Moscow University (1926) and was concurrently made chief of the Biological Station in Staro-Pershino. He became a researcher at the Zoological Museum of Moscow Lomonosov University (1934), eventually becoming Director (1964–1969). During his Red Army service (WW2) he was captured by the Germans and held in a camp in France until liberated by American forces (1944). He returned to the Zoological Museum (1947) and became Chief of the Ornithological Division, leaving (1954) for the Zoogeography Department of Moscow University. He remained as Professor of Zoogeography there, combining this with being Director, Zoological Museum of Moscow Lomonosov University (1964–1969). He co-wrote, with Dementi'ev (q.v.), the monumental *Birds of the Soviet Union* (English translation 1966–1968).

Glanville

African Black Swift ssp. *Apus barbatus glanvillei* **C. W. Benson**, 1967

R. R. Glanville (b.1901) was a British senior agricultural officer in Sierra Leone (1924–1944) and principal agricultural officer in Nigeria (1944–1950). He wrote *Agricultural Survey of the Existing and Potential Rice Lands in the Swamp Areas of the Little Scarcies, Great Scarcies, Port Loko and Rokel Rivers* (1931).

Glaszner

Eurasian Jay ssp. *Garrulus glandarius glaszneri* **Madarász**, 1902

Károly Glaszner (DNF) was a Hungarian ornithologist who collected in Cyprus (1901–1904).

Glauert

Sooty Tern ssp. *Onychoprion fuscatus glauerti* **Mathews**, 1922 NCR
[JS *Onychoprion fuscatus serratus*]
Little Shearwater ssp. *Puffinus assimilis glauerti* Mathews, 1936 NCR
[JS *Puffinus assimilis tunneyi*]
Singing Honeyeater ssp. *Lichenostomus virescens glauerti* Mathews, 1942 NCR
[JS *Lichenostomus virescens virescens*]

Ludwig Glauert (1879–1963) was born in England and trained as a geologist. He moved to Perth, Western Australia (1908), and joined the geological survey as a palaeontologist. He volunteered at the Western Australian Museum (1908–1910), joining the permanent staff (1910) as Scientific Assistant, then Keeper of Geology and Ethnology (1914). He worked on the Margaret River caves (1909–1915), studying remains from the Pleistocene. He served in the Australian Army (1917–1919) and then studied Australian material in the British Museum before returning to Perth (1920) as keeper, Western Australian Museum's biological collections, becoming Curator (1927) and Director (1954). His interests were legion – he was the leading authority on Western Australian reptiles, used his own money to buy books for the museum, and helped with the taxidermy. He retired (1956) but went on working on reptiles and scorpions. A reptile and two amphibians are named after him.

Gmelin

Hummingbird genus *Gmelinius* **Boucard**, 1894 NCR
[Now in *Cyanophaia*]

Common Pheasant ssp. *Phasianus colchicus gmelini* **Buturlin**, 1904 NCR
[JS *Phasianus colchicus torquatus*]
Swift Parrot ssp. *Lathamus discolor gmelini* **Mathews**, 1923 NCR; NRM

Johann Friedrich Gmelin (1748–1804) belonged to a well-known family of German naturalists. He was Professor of Medicine, Georg-August University in Göttingen. Other members of his family were with Pallas (q.v.) on one of his expeditions. He published the 13th edition of Linnaeus's *Systema Naturae* (1788–1796). Three reptiles and a mammal are named after him.

Godeffroy

Yap Monarch *Monarcha godeffroyi* **Hartlaub**, 1868
Marquesan Kingfisher *Todiramphus godeffroyi* **Finsch**, 1877

Johann Caesar Godeffroy (1813–1885) was a member of a German trading house importing copra from the Pacific and was interested in ornithology. His family was French but moved to Germany to avoid religious persecution. His fleet of ships traded largely in the Pacific, where he used them as floating collection bases, with paid collectors on board to search for zoological specimens as well as trading commercially. He established c.45 trading posts and bought land and property, laying the foundations of German colonial power. He used his natural history collection to found a museum in Hamburg (1860), naming it after himself. He sent collectors to the Pacific and employed well-known naturalists in Hamburg, such as Otto Finsch (q.v.) and Gustav Hartlaub (q.v.). Eventually he neglected commerce and the company went bankrupt (1879). A reptile is named after him.

Godefrida

Purple-winged Ground Dove *Claravis godefrida* **Temminck**, 1811

(See **Geoffroy**)

Godfrey, R.

Barratt's Warbler ssp. *Bradypterus barratti godfreyi* **J. A. Roberts**, 1922

Dr R. Godfrey (1872–1948) was a Scottish physician and priest who was a missionary to South Africa (1907–1946).

Godfrey, W. E.

Varied Thrush ssp. *Ixoreus naevius godfreii* **A. R. Phillips**, 1991

W. Earl Godfrey (1910–2002) was a Canadian ornithologist, Curator of Ornithology at the National Museum of Canada (1947–1976). The American Birding Association gave him the Ludlow Griscom Award (2000). He wrote *The Birds of Canada* (1966).

Godin, I.

Ground Dove sp. *Chamaepelia godinae* **Bonaparte**, 1855
NCR
[Alt. Ruddy Ground Dove; JS *Columbina talpacoti*]

Isabel Godin des Odonais (1728–1792) was a well-educated woman from a wealthy family who was the wife of Jean Godin des Odonais (q.v.) – she was 14 when they married. She was a remarkable woman: setting off to find her husband, whom she had not seen for 18 years (because of colonial politics her husband could not cross from French Guiana into Ecuador to re-join her), she attempted, with other members of her familly, to row 450 miles up the Amazon and cross the Andes. Her servants abandoned them, and the rest of her party perished, and she went on alone wearing her brother's boots. She was helped by some indigenous people and emerged alone after many weeks the lone survivor of a party of 42 people! She was eventually reunited with her husband after twenty years apart. They spent the rest of their lives together in a small French town. Prince Charles Bonaparte said he had '... given Madame Godin's name to a remarkable species of South American bird, the *Chamaepelia godinae*, consecrated,' he said, 'to the memory, which can never be too much honoured, of Isabel Godin des

Odonais, who, alone and abandoned, travelled across the American continent in its greatest width, sustained by her greatness of soul and her martyrdom to duty.'

Godin, J.

Godin's Puffleg *Eriocnemis godini* **Bourcier**, 1851
[Alt. Turquoise-throated Puffleg]

Jean Louis Godin (1704–1760) was a French astronomer who worked in South America (1735). He was present in a minor role as a member of the French government-sponsored expedition to tropical America (1735–1738). After being with the expedition in Peru and Ecuador he was offered a post at Lima, Peru, as Professor of Mathematics (1738–1751). He lost most of his money speculating and returned to Europe where he took the post of president of Cadiz midshipman's college. He was ennobled by the Spanish king for his work pertaining to the earthquake in Lisbon. Bourcier named two other hummingbirds after members of that expedition in the same paper that describes the puffleg. He was cousin to Jean Godin des Odonais (1712–1792), who was a French naturalist and professor in Quito, Ecuador, and married an heiress (1743) (q.v. **Godin, I.** above), resigned his chair and gave his time to natural science and Indian languages. Godin des Odonais explored in Ecuador and northern Peru, and collected a herbarium containing over 4,000 plant species. He also made drawings of c.800 species of animals. Later he spent 15 years exploring Cayenne and Brazilian Guiana, north of the Amazon. He donated his botanical collections to the MNHN, where they are still preserved. He published widely on both the plants and animals of northern South America, and also wrote dictionaries of South American languages. So it is possible that the bird is named after Godin des Odonais, but Jean Louis Godin is more likely. Only six specimens of this hummingbird are known, and it has not been recorded since the 19th century and a search (1980) for it was unsuccessful; it may be extinct.

Godlewski

Godlewski's Bunting *Emberiza godlewskii* **Taczanowski**, 1874
Godlewski's Pipit *Anthus godlewskii* Taczanowski, 1876
[Alt. Blyth's Pipit]

Wiktor Ignacy Aleksandrovich Godlewski (1831–1900) was a Polish collector. He was exiled in Siberia after the failed Polish uprising (1863) in Ukraine. Along with other Polish naturalists who were also exiles – including Benedykt Dybowski (q.v.), who trained him as a collector, and Michal Jankowski (q.v.)—he settled in a village by Lake Baikal (1867). He studied the local fauna (1867–1877) thanks to the Zoological Cabinet in Warsaw, as the exiles were refused help by the Eastern-Siberian Department of Imperial Russian Geographical Society. They sent their material to the Cabinet, and Taczanowski (q.v.) negotiated the sale of part of it to the museums of Western Europe. The funds obtained, together with donations from the Branicki (q.v.) brothers, enabled the exiles' survival and continued study. Most of their research on the lake took place in winter when they studied ice that Godlewski himself chiselled out, thanks to his technical abilities and great physical strength. They also explored the region (1876) around the Angara River source and Kosogol Lake (Khubsugul), Mongolia – leading to the realisation that the Lake Baikal fauna was unique. They were allowed to return to Poland in 1876. Several taxa by Lake Baikal were named after Godlewski.

Godman, A.

Lord Howe Pigeon *Columba vitiensis godmanae* **Mathews**, 1915 EXTINCT
[Alt. Metallic Pigeon ssp.]

Dame Alice Mary Godman (1869–1944) was Deputy President of the British Red Cross Society and the second wife of Frederick DuCane Godman (q.v.). No skins or specimens of the Lord Howe Pigeon were ever obtained for scientific study, and it is only known from paintings and written accounts.

Godman, F. DuC.

Godman's Euphonia *Euphonia godmani* **Brewster**, 1889
[Alt. Scrub Euphonia; Syn. *Euphonia affinis godmani*]

Golden-crowned Warbler ssp. *Basileuterus culicivorus godmani* **Berlepsch**, 1888
Northern Bobwhite ssp. *Colinus virginianus godmani* **Nelson**, 1897
Godman's Fig Parrot *Psittaculirostris desmarestii godmani* **Ogilvie-Grant**, 1911
[Alt. Large Fig Parrot ssp.]

Dr Frederick DuCane Godman (1834–1919) was a British naturalist (particularly entomologist and ornithologist) who, with his friend Osbert Salvin (q.v.) (and 18 others), co-founded the BOU and compiled the massive *Biologia Centrali Americana* (1888–1904). They presented their joint collection to the BMNH periodically (1885–1870). Godman qualified as a lawyer, but was wealthy and did not need to work, so devoted his life to ornithology. He visited Norway, Russia, the Azores, Madeira, the Canary Islands, India, Egypt, South Africa, Guatemala, British Honduras (Belize) and Jamaica, sometimes with Salvin. The trip to India (1886) was with H. J. Elwes (q.v.) to whose sister Edith he had been married to until she died (1875). He re-married (1891) Sir Francis Chaplin's (q.v.) sister Alice and she accompanied him on his African and West Indies travels. His daughter Eva Mary Godman co-wrote the 4-volume *The Birds of British Somaliland and the Gulf of Aden: their Life Histories, Breeding Habits and Eggs* (1937–1961) with Sir Geoffrey F. Archer (q.v.)—she was killed (1965) by a vehicle when she crossed a street to post a letter. The Godman-Salvin Medal, a prestigious award of the BOU, is named after them. Five reptiles, three mammals and an amphibian are named after him.

Godwin-Austen

Godwin-Austen's Laughingthrush *Trochalopteron austeni* Godwin-Austen, 1870
[Alt. Brown-capped Laughingthrush]
Godwin-Austen's Wren-babbler *Spelaeornis chocolatinus* Godwin-Austen & **Walden**, 1875
[Alt. Long-tailed Wren-babbler, Naga Wren-babbler]

Assam Laughingthrush ssp. *Trochalopteron chrysopterum godwini* **Harington**, 1914
Streak-breasted Scimitar Babbler ssp. *Pomatorhinus ruficollis godwini* **Kinnear**, 1944

(See **Austen**)

Goeldi

Goeldi's Antbird *Akletos goeldii* **Snethlage**, 1908
[Syn. *Myrmeciza goeldii*]

Orange-fronted Yellow Finch ssp. *Sicalis columbiana goeldii* **Berlepsch**, 1906
Amazonian Trogon ssp. *Trogon ramonianus goeldii* **Ridgway**, 1911 NCR
[JS *Trogon ramonianus crissalis*]

Emil August Goeldi (1859–1917) was a Swiss zoologist. He went to Brazil (1880), first working at the Museu Nacional. Later he reorganised the newly founded (1866) Pará Museum of Natural History and Ethnography now bearing his name: Museu Paraense Emílio Goeldi. He became well known for his studies of Brazilian birds and mammals. Goeldi returned to Switzerland (1907) to teach biology and physical geography at the University of Bern (1907–1917). He wrote *Aves do Brasil* (1894) and *Die Vogelwelt des Amazonensstromes* (1901). He was also a self-confessed racist who did not like Brazilians and worked in a Swiss enclave! Two mammals and two amphibians are named after him.

Goering

Pale-headed Jacamar *Brachygalba goeringi* **P. L. Sclater** & **Salvin**, 1869
Slaty-backed Tanager *Hemispingus goeringi* P. L. Sclater & Salvin, 1871
[Alt. Slaty-backed Hemispingus]

Christian Anton Goering (1836–1905) was a German artist and naturalist who studied under C. L. Brehm (q.v.) and Burmeister (q.v.). He became Curator at the Zoological Museum of the University of Halle (1854) and accompanied Burmeister on his expedition to Brazil, Uruguay and Argentina (1856–1859). He worked in Venezuela (1866–1872) collecting for the British Museum.

Goethals

Goethals's Hummingbird *Goethalsia bella* **Nelson**, 1912
[Alt. Pirre Hummingbird, Rufous-cheeked Hummingbird]

George Washington Goethals (1858–1928) was a US Army Colonel. He was the Chief Engineer of the Panama Canal (1907–1914) and personally directed its construction. He completed the project six months ahead of schedule and $23 million below budget. A crane operator on the canal said of him 'Few men could have stood the amount of work he put on himself. Men broke down; men went crazy; men took to drink. The colonel kept as keen as a brier'. Goethals became Governor of the Panama Canal Zone (1914–1916).

Goffin

Goffin's Cockatoo *Cacatua goffiniana* **Roselaar & Michels**, 2004
[Alt. Tanimbar Corella; Syn. *Cacatua goffini*]

Goffin's Barbet *Trachylaemus purpuratus goffinii* Goffin, 1863
[Alt. Yellow-billed Barbet ssp.; Syn. *Trachyphonus purpuratus goffinii*]

Andreas Leopold Goffin (1837–1863) was a Belgian ornithologist at Rijksmuseum voor Natuurlijke Historie, Leiden. Otto Finsch's original etymology names it after Herr A. Goffin 'out of friendship'. However, it was later (2000) shown that the description of this species was based on two specimens that actually belonged to a different cockatoo species, *Cacatua ducorpsii*. Therefore, *Cacatua goffini* became a synonym for *Cacatua ducorpsii* and left the Tanimbar species without a proper scientific name. Roselaar & Michels provided the name *Cacatua goffiniana* 'to respect the original intent of Otto Finsch to dedicate a species of cockatoo to him [Goffin]'. Other references say the name honoured Finsch's 'girlfriend' but we cannot tell if this was some sort of veiled reference to a 'special' male friend.

Goldie

Goldie's Lorikeet *Psitteuteles goldiei* **Sharpe**, 1882
Goldie's Bird-of-Paradise *Paradisaea decora* **Salvin & Godman**, 1883

Painted Quail-thrush ssp. *Cinclosoma ajax goldiei* **E. Ramsay**, 1879
Goldie's Jungle Hawk Owl *Ninox theomacha goldii* **Gurney**, 1883
Brown Cuckoo Dove ssp. *Macropygia amboinensis goldie* **Salvadori**, 1893 NCR
[JS *Macropygia amboinensis cinereiceps*]

Andrew Goldie (1840–1891) was born in Scotland and died in Port Moresby, Papua New Guinea. He moved to New Zealand (1862), working there as a nurseryman (c.1873). He then travelled to Britain (1874) and Melbourne (1875) before beginning his explorations in New Guinea (1876). Whilst collecting there (1877) he discovered and named the Goldie River and discovered traces of gold. He then explored the south coast of New Guinea (1878), naming the Blunden River, Milport Harbour and Glasgow Harbour. Goldie returned to Sydney (1878) to have gold samples assayed. He bought land near Hanubada, New Guinea, and established a trading store (1878). He named the Redlick group of islands and discovered Teste Island in Freshwater Bay (1879). He was given £400 as compensation by the government (1886) when they decided to remove European settlement from the Hanubada area, and he used it to purchase 50 suburban acres and three town allotments on which he built Port Moresby's first store (1897). It has been suggested that many of the scientific discoveries claimed by Goldie were actually made by his associate, Carl von Hunstein (q.v.).

Goldman

Goldman's Hummingbird *Goldmania violiceps* **Nelson**, 1911
[Alt. Violet-capped Hummingbird]
Goldman's Quail Dove *Geotrygon goldmani* Nelson, 1912
[Alt. Russet-crowned Quail Dove]

Goldman's Warbler *Dendroica coronata goldmani* Nelson, 1897

[Alt. Yellow-rumped Warbler ssp.]
Elegant Trogon ssp. *Trogon elegans goldmani* Nelson, 1898
Buff-collared Nightjar ssp. *Caprimulgus ridgwayi goldmani* Nelson, 1899 NCR
[JS *Caprimulgus ridgwayi ridgwayi*]
Song Sparrow ssp. *Melospiza melodia goldmani* Nelson, 1899
Blue-diademed Motmot ssp. *Momotus lessonii goldmani* Nelson, 1900
Thicket Tinamou ssp. *Crypturellus cinnamomeus goldmani* Nelson, 1901
Yellow Rail ssp. *Coturnicops noveboracensis goldmani* Nelson, 1904
Goldman's Yellowthroat *Geothlypis beldingi goldmani* **Oberholser**, 1917
[Alt. Belding's Yellowthroat ssp.]
Botteri's Sparrow ssp. *Peucaea botterii goldmani* **A. R. Phillips**, 1943

Major Edward Alphonso Goldman (1873–1946) was a field naturalist and mammalogist who was born in Illinois. Nelson (q.v.) hired him (1892) to assist his biological investigations of California and Mexico, and then as Field Naturalist and eventually Senior Biologist with the United States Bureau of Biological Survey. He spent c.14 years collecting all over Mexico. Their biological exploration and collecting expeditions (1892–1906) are said to have been: 'among the most important ever achieved by two workers for any single country'. They investigated in every Mexican state, collecting 17,400 mammals and 12,400 birds, and amassing an enormous fund of information of the country's natural history. The best account of their work is Goldman's *Biological Investigations in Mexico* (1951). Goldman also had an honorary position with the USNM, as an Associate in Zoology (1928–1946). He was part of the Biological Survey of Panama (1911–1912) during the canal's construction. His results were published in *The Mammals of Panama* (1920). He was President of the Biological Society of Washington (1927–1929) and of the American Society of Mammalogists (1946). He assisted the United States government in negotiating with Mexico to protect migratory birds (1936). Goldman's bibliography includes over 200 titles. He named over 300 forms of mammal, most of them subspecies. Eight mammals, an amphibian and a reptile as well as Goldman Peak in Baja California bear his name.

Goliath

Goliath Heron *Ardea goliath* **Cretzschmar**, 1829
Goliath Coucal *Centropus goliath* **Bonaparte**, 1850
Goliath Imperial Pigeon *Ducula goliath* **G. R. Gray**, 1859

Palm Cockatoo ssp. *Probosciger aterrimus goliath* **Kuhl**, 1820

Goliath of Gath (about 1,030 BC) was a Philistine warrior of giant size who was killed with a slingshot by David, later King of the Jews (see 1 Samuel 17.4, Bible). The name is applied to species that are giants of their kind, including three mammals, an amphibian and a reptile.

Golz

Common Nightingale ssp. *Luscinia megarhynchos golzii*
  Cabanis, 1873
[Formerly *Luscinia megarhynchos hafizi*]
Yellow-breasted Apalis ssp. *Apalis flavida golzi*
  G. A. Fischer & **Reichenow**, 1884

Dr W. H. T. Golz (1825–1898) was a German ornithologist.

Gomes

Rufous-naped Lark ssp. *Mirafra africana gomesi*
  C. M. N. White, 1944

António de Figueiredo Gomes e Sousa (1896–1973) was a Brazilian botanist and collector who worked for White (q.v.) in Angola (1943).

Gonzales

Scarlet Minivet ssp. *Pericrocotus speciosus gonzalesi*
  Ripley & **Rabor**, 1961

Rodolfo B. Gonzales (DNF) was a Filipino ornithologist and collector, who was Rabor's assistant at Silliman University, Negros, Philippines and became Profesor and Chairman of the Biology Department for 35 years. He wrote 'A study of the breeding biology and ecology of the Monkey-eating Eagle' (1968). The university's natural history museum is now named after him.

Good

Common Bulbul ssp. *Pycnonotus barbatus goodi* **Rand**, 1955
  NCR
[NUI *Pycnonotus barbatus inornatus*]

Dr Albert Irwin Good (1884–1975) was an American priest, naturalist and Presbyterian missionary to the Cameroons (1909–1949). He collected entomological and ornithological specimens for the Carnegie, Cleveland and Stanford University museums.

Goodall

Dusky-tailed Canastero ssp. *Asthenes humicola goodalli*
  Marin, Kiff & Peña, 1989

William Jack Davies Goodall (1892–1980) was a British businessman resident in Chile. He originally studied naval architecture but emigrated to Chile (1912) to work for a nitrate company. Here he worked with Alfred Johnson, who became his lifelong friend and partner in various businesses in Chile, and co-author of *Las Aves de Chile* (1946).

Goodenough

Red-capped Robin *Petroica goodenovii* **Vigors** & **Horsfield**, 1827

Samuel Goodenough (1743–1827) was a naturalist and botanist as well as a clergyman who became Bishop of Carlisle (1808–1827). He was the first Treasurer and later Vice-President of the Linnaean Society of London. His main interests were seaweeds and sedges. The genus of Australian plants, *Goodenia*, is named after him.

Goodfellow

Myna genus *Goodfellowia* **Hartert**, 1903 NCR
[Now in *Basilornis*]

Goodfellow's White-eye *Apoia goodfellowi* Hartert, 1903
[Alt. Black-masked White-eye, Mindanao White-eye; Syn. *Lophozosterops goodfellowi*]
Goodfellow's Jungle Flycatcher *Rhinomyias goodfellowi*
  Ogilvie-Grant, 1905
[Alt. Slaty-backed Jungle Flycatcher]
Goodfellow's Kingfisher *Ceyx goodfellowi* Ogilvie-Grant, 1905 NCR
[Alt. Variable Dwarf Kingfisher; JS *Ceyx lepidus margarethae*]
Taiwan Firecrest *Regulus goodfellowi* Ogilvie-Grant, 1906
[Alt. Flamecrest]
Goodfellow's Lory *Eos goodfellowi* Ogilvie-Grant, 1907 NCR
[Believed to be based on immature Red Lory, *Eos bornea*]

Black-billed Thrush ssp. *Turdus ignobilis goodfellowi*
  Hartert & **Hellmayr**, 1901
White-browed Shortwing ssp. *Brachypteryx montana goodfellowi* Ogilvie-Grant, 1912
Dwarf Cassowary ssp. *Casuarius bennetti goodfellowi*
  Rothschild, 1914 NCR; NRM
Rufous-banded Owl ssp. *Ciccaba albitarsis goodfellowi*
  Chubb, 1916 NCR
[JS *Strix albitarsis albitarsis*]
Hooded Pitta ssp. *Pitta sordida goodfellowi* **C. M. N. White**, 1937

Walter Goodfellow (1866–1953) was a British ornithologist and explorer. He discovered the Mikado Pheasant *Syrmaticus mikado* in Formosa and led the British Ornithologists' Union's Expedition (1909–1911) to New Guinea. He collected live birds for aviculturists such as Mrs Johnstone (q.v.). His first expedition was to Colombia and Ecuador (1898–1899) during which he and Claude Hamilton (q.v.) collected c.4,000 skins covering 550 species. His last expedition (1936) was to Melville Island. He was awarded the British Ornithologists' Union Medal (1912). Goodfellow's collection was later acquired by the British Museum. Two mammals are named after him.

Goodson

Goodson's Pigeon *Patagioenas goodsoni* **Hartert**, 1902
[Alt. Dusky Pigeon]
Hartert's Leaf Warbler *Phylloscopus goodsoni* Hartert, 1910

Rufous-capped Babbler ssp. *Stachyridopsis ruficeps goodsoni* **Rothschild**, 1903
Buff-banded Rail ssp. *Gallirallus philippensis goodsoni*
  Mathews, 1911
Golden Tanager ssp. *Tangara arthus goodsoni* Hartert, 1913
Goodson's Pipit *Anthus leucophrys goodsoni*
  Meinertzhagen, 1920
[Alt. Plain-backed Pipit ssp.]
Bismarck Whistler ssp. *Pachycephala citreogaster goodsoni*
  Rothschild & Hartert, 1924
Eclectus Parrot ssp. *Eclectus roratus goodsoni* Hartert, 1924 NCR
[JS *Eclectus roratus solomonensis*]

Mackinlay's Cuckoo Dove ssp. *Macropygia mackinlayi goodsoni* Hartert, 1924

Grey-cheeked Green Pigeon ssp. *Treron pompadora goodsoni* Hartert, 1927 NCR

[JS *Treron griseicauda wallacei*]

Black Cicadabird ssp. *Coracina melas goodsoni* Mathews, 1928

Little Shrike-thrush ssp. *Colluricincla megarhyncha goodsoni* Hartert, 1930

Arthur Thomas Goodson (1873–1931) was a native of Tring, Hertfordshire, UK. He was thus ideally placed to take employment at the Rothschild Museum (1893), which is now incorporated in the Natural History Museum, Tring. He became Assistant Ornithologist to Hartert (q.v.) at the museum. (See **Arthur (Goodson)**)

Goodwin

Cook Islands Fruit Dove ssp. *Ptilinopus rarotongensis goodwini* Holyoak, 1974

Richard Patrick 'Derek' Goodwin (1920–2008) was a British ornithologist, conservationist, collector and aviculturist. After WW2 service in the Royal Artillery, he joined the Natural History Museum, London (1945). He was a member of the third Harold Hall Australian Expedition (1965). He wrote *Pigeons and Doves of the World* (1977), *Estrildid Finches of the World* (1984) and *Crows of the World* (1986).

Goossens

African Munia sp. *Pseudospermestes goossensi* **A. Dubois**, 1905 NCR

[Alt. Black-and-white Mannikin; JS *Lonchura bicolor poensis*]

E. Goossens (1866–1906) was a Belgian missionary to the Congo (1903–1906).

Gordius

Common Pheasant ssp. *Phasianus colchicus gordius* Alphéraky & **Bianchi**, 1907 NCR

[JS *Phasianus colchicus zarudnyi*]

In Greek mythology Gordius was the King of Phrygia (Turkey) who devised a most complicated knot. In 333 BC, while wintering in Phrygia, Alexander the Great attempted to untie the knot. When he could not find the end of the knot to unbind it, he sliced it in half with a stroke of his sword, producing the required result (the so-called 'Alexandrian solution').

Gordon, C.

Red-breasted Swallow ssp. *Cecropis semirufa gordoni* **Jardine**, 1852

Surgeon-General Sir Charles Alexander Gordon (1821–1899) was a British army surgeon in India at the battle of Maharajport (1843) and during and after the Indian Mutiny (1857–1859). He was medical officer in charge of an expeditionary force in West Africa (1848) and China (1860–1861). He was in Paris as the War Department's Medical Commissioner to the French army (1870–1871) and was at a number of

battles in the Franco-Prussian War and the siege of Paris. He was in India again (1870s) and visited Burma (Myanmar) as recorded in his book *Our Trip to Burmah, with Notes on that Country, in December 1874* (1875).

Gordon, J.

Tristan Thrush ssp. *Nesocichla eremita gordoni* Stenhouse, 1924

John Gordon McHaffie-Gordon (1875–1938) was a Scottish lepidopterist, ornithologist, conchologist and naturalist. He served with the Black Watch (WW1). He was very interested in the island of Tristan da Cunha and co-wrote an article on Tristan's birds and their eggs.

Gordon

Helmeted Friarbird ssp. *Philemon buceroides gordoni* **Mathews**, 1912

The original description has no etymology. We do not know who Gordon was. Two possible contenders for the honour are John Grant Gordon (1858–1951), who was an Australian businessman, and Leslie Gordon Chandler (1888–1980), an Australian businessman and ornithologist.

Gore

Woodpecker sp. *Gecinus gorii* **Hargitt**, 1887 NCR

[Alt. Scaly-bellied Woodpecker; JS *Picus squamatus flavirostris*]

Colonel St George Corbet Gore (1849–1913) of the Royal Engineers was Surveyor-General of India (1899–1904). While still a Lieutenant he was deputed (1878) by Colonel Everest (after whom Mount Everest is named) to explore and map the Pishin valley in the North-West Province of India (Baluchistan in Pakistan). As a Captain he shot a specimen of the woodpecker on the banks of the Helmand River (1884), Afghanistan. A reptile is named after him.

Gorgo

Sylph sp. *Cynanthus gorgo* **Mulsant, E. Verreaux** & **J. Verreaux**, 1866 NCR

[Alt. Long-tailed Sylph; JS *Aglaiocercus kingi*]

Gorgo, Queen of Sparta, was the wife of Leonidas, King of Sparta, who, with his bodyguard of 300, died at the Battle of Thermopylae (BC 480). According to Plutarch, when asked by a woman from Attica, 'Why are you Spartan women the only ones who can rule men?' Queen Gorgo replied: 'Because we are also the only ones who give birth to men.'

Gosling

Gosling's Bunting *Emberiza goslingi* **B. Alexander**, 1906

[Syn. *Emberiza tahapisi goslingi*]

Gosling's Apalis *Apalis goslingi* **B. Alexander**, 1908

Freckled Nightjar ssp. *Caprimulgus tristigma goslingi* B. Alexander, 1907 NCR

[JS *Caprimulgus tristigma sharpei*]

Captain George Bennet Gosling (1872–1906) was an explorer and zoologist. He followed the Uele River, exploring and collecting from the Niger to the Nile (1905–1906) on Boyd Alexander's (q.v.) expedition, during which he died of black-water fever at Niangara on the Uele River. A mammal is named after him.

Gosse, P. H.

Gosse's Macaw *Ara gossei* **Rothschild**, 1905 EXTINCT
[Alt. Jamaican Red Macaw]
Rufous-tailed Flycatcher *Tyrannula gossii* **Bonaparte**, 1851 NCR
[JS *Myiarchus validus*]

Yellow-breasted Crake ssp. *Porzana flaviventer gossii* Bonaparte, 1856

Philip Henry Gosse (1810–1888) was an English naturalist, a Plymouth Brethren missionary and a science populariser. He is best known for his attempt to reconcile biblical literalism with uniformitarianism, trying to reconcile biblical ideas on creation with the growing evidence of geological eras of millions of years duration, in his work *Omphalos: An Attempt to Untie the Geological Knot*. This advanced the theory that fossils had never lived, but God inserted them into rocks during the Creation as a way of making the world appear older than it was. Gosse's only child famously wrote *Father and Son* – his birth was noted by Gosse in his diary as 'Received green swallow from Jamaica. E delivered of a son'. He was clearly a man engrossed in his work! Gosse made a career of writing textbooks on a wide range of subjects, from Jamaican natural history to marine biology. Various taxa are named after him including a reptile.

Gosse, P. H. G.

Pampa Finch *Embernagra gossei* **Chubb**, 1918 NCR
[JS *Embernagra platensis olivascens*]

Dr Philip Henry George Gosse (1879–1959) was the grandson of his namesake above, an English physician (1907) in general practice (retired 1930), author and, at the age of only 16, explorer in the Andes with the Fitzgerald expedition (1896–1897). He served in France and India (WW1) with the Royal Army Medical Corps. He was married thrice and divorced twice. He was fascinated by pirates and wrote a number of books on their history as well as 'Notes on the natural history of the Aconagua Valley' (1899) and *The Birds of the Balearic Islands* (1926).

Götz

Common Kingfisher ssp. *Alcedo atthis goetzii* **Laubmann**, 1923 NCR
[JS *Alcedo atthis bengalensis*]

Dr Wilhem J. H. Götz (1902–1979) of Munich was a German zoologist and ornithologist who was a friend of Laubmann (q.v.). He joined the staff of the State Natural History Museum, Stuttgart (1925) eventually becoming Director.

Goudot

Goudot's Guan *Chamaepetes goudotii* **Lesson**, 1828
[Alt. Sickle-winged Guan]
Shining-green Hummingbird *Lepidopyga goudoti* **Bourcier**, 1843

Justin-Marie Goudot (d.1845) was a French zoologist who worked in Colombia (1822–1843), described by some as a 'furious collector of birds'. Goudot began collecting bird skins for the Paris museum (1828) and also collected mammals, reptiles, insects and other invertebrates, and botanical material. He was the first Head of the Zoology Department of the Colombian National Museum, having been asked by Cuvier (q.v.) and Humboldt (q.v.), on behalf of the new Vice-President, Francisco Antonio Zea, to create the department (1822). He wrote a paper on the Mountain Tapir 'Nouvelles observations sur le Tapir Pinchaque' (1843). A reptile is named after him.

Gould, E.

Finch genus *Gouldaeornis* **Mathews**, 1923 NCR
[Now in *Erythrura/Chloebia*]

Mrs Gould's Sunbird *Aethopyga gouldiae* **Vigors**, 1831
[Alt. Blue-throated Sunbird]
Gouldian Finch *Erythrura gouldiae* **J Gould**, 1844
[Syn. *Chloebia gouldiae*]

Elizabeth Gould (1804–1841) was the wife of John Gould (q.v.). She was taught lithography by Edward Lear (q.v.), and went on to use this skill to create illustrations from John Gould's drawings. While on an Australian expedition (1841) English ornithologist John Gould came across what many believe to be the most beautiful finch in the world, and he named it 'The Lady Gouldian' to honour his artist wife although she bore no actual aristocratic title.

Gould, J.

Thorntail genus *Gouldia* **Bonaparte**, 1850 NCR
[NPRB *Popelairia* **Reichenbach**, 1854]

Gould's Blue-wren *Malurus cyaneus* **Ellis**, 1782
[Alt. Superb Fairy-wren]
Gould's Albatross *Thalassarche chrysostoma* **J. R. Forster**, 1785
[Alt. Grey-headed Albatross]
Gould's Coquette *Lophornis gouldii* **Lesson**, 1832
[Alt. Dot-eared Coquette]
Gould's Broadbill *Serilophus lunatus* Gould, 1834
[Alt. Silver-breasted Broadbill]
Gould's Parrotbill *Paradoxornis flavirostris* Gould, 1836
[Alt. Black-breasted Parrotbill]
Gould's Frogmouth *Batrachostomus stellatus* Gould, 1837
Gould's Toucanet *Selenidera gouldii* **Natterer**, 1837
Gould's Shearwater *Puffinus assimilis* Gould, 1838
[Syn. Little Shearwater]
Orange-bellied Parrot *Nanodes gouldii* **Ewing**, 1842 NCR
[JS *Neophema chrysogaster*]
Gould's Petrel *Pterodroma leucoptera* Gould, 1844
[Alt. White-winged/Sooty-capped Petrel]
Gould's Storm-petrel *Fregetta tropica* Gould, 1844
[Alt. Black-bellied Storm-petrel]
Gould's Zebra Dove *Geopelia placida* Gould, 1844
[Alt. Peaceful Dove]
Tern sp. *Sterna gouldi* Reichenbach, 1845 NCR
[Alt. Sooty Tern; JS *Onychoprion fuscatus serratus*]
Gould's Jewelfront *Heliodoxa aurescens* Gould, 1846

Gould's Violetear *Colibri coruscans* Gould, 1846
[Alt. Sparkling Violetear]
White-eye sp. *Zosterops gouldi* Bonaparte, 1851 NCR
[Alt. Silvereye; JS *Zosterops lateralis chloronotus*]
Gould's Inca *Coeligena inca* Gould, 1852
Silver Gull *Gelastes gouldi* Bonaparte, 1854 NCR
[NPRB *Chroicocephalus novaehollandiae forsteri*]
Black Bittern *Ardetta gouldi* Bonaparte, 1855 NCR
[JS *Dupetor flavicollis australis*]
Gould's Euphonia *Euphonia gouldi* **P. L. Sclater**, 1857
[Alt. Olive-backed Euphonia]
Gould's Emerald *Chlorostilbon elegans* Gould, 1860
EXTINCT?
Gould's Woodnymph *Augasma smaragdinea* Gould, 1860
[Status uncertain: possibly a hybrid, but parentage
unknown]
Gould's Woodstar *Chaetocercus decorata* Gould, 1860 NCR
[Hybrid: *Chaetocercus mulsant* x *Chaetocercus heliodor*]
Gould's Plumeleteer *Chalybura urochrysia* Gould, 1861
[Alt. Bronze-tailed Plumeleteer]
Gould's Fulvetta *Schoeniparus brunneus* Gould, 1863
[Alt. Dusky Fulvetta, Brown-capped Fulvetta; Syn. *Alcippe
brunnea*]
Gould's Bronze Cuckoo *Chrysococcyx russatus* Gould, 1868
[Syn. *Chrysococcyx minutillus russatus*]
Gould's Shortwing *Brachypteryx stellata* Gould, 1868
[Syn. *Heteroxenicus stellatus*]
Gould's Sunangel *Heliangelus micraster* Gould, 1872
[Alt. Little Sunangel, Flame-throated Sunangel]
Gould's Tanager *Tangara gouldi* Sclater, 1886 NCR
[Hybrid: *Tangara desmaresti* x *Tangara cyanoventris*]
Scrubtit *Acanthornis gouldi* **Mathews**, 1916 NCR
[JS *Acanthornis magna magna*]

Green-tailed Trainbearer ssp. *Lesbia nuna gouldii* **Loddiges**,
1832
Gould's Red-tailed Black Cockatoo *Calyptorhynchus banksii
macrorhynchus* Gould, 1842
Plain-backed Pipit ssp. *Anthus leucophrys gouldii* **L. Fraser**,
1843
Gould's Harrier *Circus approximans gouldi* Bonaparte, 1850
[Alt. Swamp Harrier ssp.; NRM]
Gould's (Wild) Turkey *Meleagris gallopavo mexicana* Gould,
1856
Song Sparrow ssp. *Melospiza melodia gouldii* **Baird**, 1858
Gould's Manucode *Phonygammus keraudrenii gouldii*
G. R. Gray, 1859
[Alt. Trumpet Manucode ssp.]
Gould's Hermit *Phaethornis griseogularis zonura* Gould,
1860
[Alt. Grey-chinned Hermit ssp.]
Spectacled Monarch ssp. *Monarcha trivirgatus gouldii*
Gray, 1861
[Syn. *Symposiachrus trivirgatus gouldii*]
Chestnut Thrush ssp. *Turdus rubrocanus gouldi* **J. Verreaux**,
1870
Stork-billed Kingfisher ssp. *Pelargopsis capensis gouldi*
Sharpe, 1870
White-cheeked Honeyeater ssp. *Phylidonyris niger gouldii*
Schlegel, 1872
Australian Reed Warbler ssp. *Acrocephalus australis gouldi*
AJC Dubois, 1901

Varied Triller ssp. *Lalage leucomela gouldi* Mathews, 1912
NCR
[JS *Lalage leucomela rufiventris*]
Brown Gerygone ssp. *Wilsonavis richmondi gouldiana*
Mathews, 1920 NCR
[JS *Gerygone mouki richmondi*]

John Gould (1804–1881) was the son of a gardener at Windsor Castle who became an illustrious British ornithologist, artist and taxidermist. Gould was born in Dorset, England, and became acknowledged around the world as 'The Bird Man'. He was employed as a taxidermist by the newly formed Zoological Society of London and travelled widely in Europe, Asia and Australia. He was arguably the greatest and certainly the most prolific publisher and original author of ornithological works in the world. In excess of 46 volumes of reference work were produced by him in colour (1830–1881). He published 41 works on birds, with 2,999 remarkably accurate illustrations by a team of artists, including his wife. His first book, on Himalayan birds, was based on skins shipped to London, but later he travelled to see birds in their natural habitats. Gould and his wife, Elizabeth (q.v. under **Gould, E.**), arrived on board *Parsee* in Australia (1838) to spend 19 months studying and recording the natural history of the continent. By the time they left Gould had not only recorded most of Australia's known birds, and collected information on nearly 200 new species, but he had also gathered data for a major contribution to the study of Australian mammals. His best known works include: *The Birds of Europe*, *The Birds of Great Britain*, *The Birds of New Guinea* and *The Birds of Asia*. He also wrote monographs on the Odontophorinae, Trochilidae and Pittidae. Gould was commercially minded and pandered to Victorian England's fascination with the exotic, particularly hummingbirds, with which he is particularly associated. His superb paintings, and prints of these and other birds were greatly sought after, so much so that he had trouble keeping up with the demand. The large corpus of unpublished and unfinished work, which he left at his death, supports this. Five mammals, two reptiles and an amphibian are named after him.

Gounelle

Broad-tipped Hermit *Anopetia gounellei* **Boucard**, 1891
[Syn. *Phaethornis gounellei*]

Edmond Gounelle (1850–1914) was a French entomologist and naturalist who collected in Brazil (1887 and 1892–1893).

Gourdin

Yellow-vented Bulbul ssp. *Pycnonotus goiavier yourdini*
G. R. Gray, 1847
[Trinomial sometimes amended as *gourdini*]

Ensign Jean-Marie Gourdin (b.1814) was a French naval officer. He was on board the *Astrolabe* in the Pacific and Antarctic (1837–1839). Gray (q.v.) originally used the spelling *Yourdini*, presumably in error. Although many later authors corrected this to *gourdini*, the original spelling has priority.

Govinda

Black Kite ssp. *Milvus migrans govinda* **Sykes**, 1832

In Hindu mythology Govinda, the cow-finder, is another name for Krishna.

Goyder

Goyder Grasswren *Amytornis goyderi* **Gould**, 1875
[Alt. Eyrean Grasswren]

George Woodroffe Goyder (1826–1898) was an explorer and Surveyor-General of South Australia, who emigrated from England (1848). He researched rainfall and devised 'Goyder's Line of Rainfall', which is a line drawn on the map of South Australia that proved to be remarkably reliable. South of this line rainfall could be deemed as reliable enough for most agricultural activities; north of it only grazing was recommended.

Graba

Atlantic Puffin ssp. *Fratercula arctica grabae* **C. L. Brehm**, 1831

Carl Julian von Graba (1799–1874) was a German-speaking Danish lawyer and traveller who visited the Faeroes (1828). He wrote *Tagebuch, Geführt auf einer Reise nach Färö in Jahre 1828* (1830). In his day Schleswig and Holstein were regarded as duchies controlled by Denmark and were ceded to Germany by Denmark after the second Schleswig War (1864).

Graber

Sedge Wren ssp. *Cistothorus platensis graberi* **Dickerman**, 1975

Richard Rex Graber (1924–1997) was an American ornithologist and ecologist. He wrote extensively on Illinois's birds.

Grace

Grace's Warbler *Dendroica graciae* **Baird**, 1865

Grace Darling Page, *née* Coues (1847–1925) was the sister of Dr Elliot Coues (q.v.). She married Charles Albert Page (1868), the United States Ambassador to Switzerland. After his death she married (1884) Dana Estes, a publisher. The warbler was collected by Coues, who asked Baird (q.v.) to name it after his sister, who was 18 at the time.

Grace Anna

White-edged Oriole *Icterus graceannae* **Cassin**, 1867

Grace Anna Lewis (1821–1912) was an American teacher and botanist from a Quaker family in Pennsylvania. She came under the influence of a fellow Quaker, John Cassin (q.v.), who introduced her to the Academy of Natural Sciences in Philadelphia and also introduced her to Baird (q.v.) at the USNM.

Graeffe

Fiji Whistler *Pachycephala graeffii* **Hartlaub**, 1866

Purple-capped Fruit Dove ssp. *Ptilinopus porphyraceus graeffei* **Neumann**, 1922 NCR
[Believed to be a hybrid population between *Ptilinopus porphyraceus porphyraceus* and *P. porphyraceus fasciatus*]

Dr Eduard Graeffe (1833–1916) was born in Switzerland, died in part of the Austro-Hungarian Empire (Slovenia) but regarded himself as an Austrian. He was a physician, author, zoologist and ornithologist. Godeffroy (q.v.) employed him (1860) to organise his personal collection and to start his personal museum. He travelled to the Pacific (1861) as a paid collector for Godeffroy and (1862–1870) visited nearly all the Islands between longitudes 170° and 180°E. He returned to Hamburg (1871) with a new collection to add to Godeffroy's museum, and published *Travels in the Interior of Viti Levu*. Godeffroy's business was in a bad state (1874) and he could no longer afford to pay Graeffe, who took the position of an inspector at the Austrian Zoological Station in Trieste.

Graells

Lesser Black-backed Gull ssp. *Larus fuscus graellsii* **A. E. Brehm**, 1857

Prof. Dr Mariano de la Paz Graells y de la Aguera (1809–1898) of Madrid was a botanist, entomologist and malacologist. He qualified in medicine and natural sciences at the University of Barcelona, where he was first an Associate and later full Professor of Physics and Chemistry. There was an epidemic in Barcelona (1835) which he was prominent in combating. He moved to Madrid (1837) as Professor of Zoology at the Museum of Natural Sciences and Director of the Botanical Gardens. He joined a scientific expedition to the Pacific (1845), establishing the Spanish Natural History Society facilities to allow for the acclimatisation of tropical plants. The phylloxera outbreak virtually wiped out European vineyards in the latter part of the 19th century, and Graells, as a senior man in the Council of Agriculture, had to deal with its effect in Spain. He was a founding member of the Spanish Academy of Exact Sciences and was honoured in Spain and by several foreign governments. Wallace (q.v.) mentions him in connection with plants collected in Spain. He wrote 'Subfamilia felina fauna mastodologica' (1897). A mammal is named after him.

Graf

Grafs Finch *Pooecetes gramineus* **Gmelin**, 1789
[Alt. Vesper Sparrow, Bay-winged Bunting]

The 'Grafs Finch' appears in Audubon's monumental *Ornithological Biography* as a picture. Graf is a mistake as Audubon clearly states that he called the bird 'The Grass Finch or Bay-winged Bunting'.

Graf, H.

White-collared Starling genus *Grafisia* G. L. Bates, 1926

Dr H. Graf (fl.1909) was a German herpetologist who collected in the Cameroons.

Graham

Grey-necked Wood Rail ssp. *Aramides cajanea grahami*
Chubb, 1919 NCR
[JS *Aramides cajanea cajanea*]

Sir Ronald William Graham (1870–1949) was a British diplomat. He was Assistant Under-Secretary, Foreign Office (1916–1919), Minister to the Netherlands (1919–1921) and Ambassador to Italy (1921–1933).

Grandidier

Madagascar Spinetail *Zoonavena grandidieri* **J. Verreaux**, 1867

Alfred Grandidier (1836–1921) was a French explorer, geographer and ornithologist who collected in Madagascar (1865) and who wrote *Histoire Naturelle des Oiseaux de Madagascar* (1876) as part of his 40-volume *L'Histoire Physique, Naturelle et Politique de Madagascar* in collaboration with others such as Alphonse Milne-Edwards (q.v.) and Leon Vaillant (q.v.). He recovered bones that turned out to be *Aepyornis maximus* – the extinct Elephant Bird (1866). The mineral grandidierite, which is found in Madagascar, is named after him, as is Mont Alfred Grandidier in the French Antarctic, plus eight reptiles, three mammals and an amphibian.

Grant, C. H. B.

Seedcracker sp. *Pyrenestes granti* **Sharpe**, 1908 NCR
[Alt. Lesser Seedcracker; JS *Pyrenestes minor*]

Grant's Wood-hoopoe *Phoeniculus damarensis granti*
Neumann, 1903
[Alt. Violet Wood-hoopoe ssp.]
African Green Pigeon ssp. *Treron calvus granti*
Van Someren, 1919
Double-banded Courser ssp. *Rhinoptilus africanus granti*
W. L. Sclater, 1921
Olive Sunbird ssp. *Cyanomitra olivacea granti* **Vincent**, 1934
Evergreen Forest Warbler ssp. *Bradypterus lopezi granti*
C. W. Benson, 1939
Crested Malimbe ssp. *Malimbus malimbicus granti*
Bannerman, 1943 NCR
[JS *Malimbus malimbicus malimbicus*]
African Paradise-flycatcher ssp. *Terpsiphone viridis granti*
J. A. Roberts, 1948
Bocage's Akalat ssp. *Sheppardia bocagei granti* **Serle**, 1949
Yellow-fronted Canary ssp. *Serinus mozambicus granti*
Clancey, 1957
[Syn. *Crithagra mozambica granti*]

Captain Claude Henry Baxter Grant (1878–1958) was a British ornithologist and collector. He collected in southern Africa (1903–1907) and in South America (1908–1910). He served with the Rifle Brigade and the East African Expeditionary Force (WW1). He was Editor of the *Bulletin of the British Ornithologists' Club* (1935–1940) and was co-author of *African Handbook of Birds* (1952) with C W Mackworth-Praed (q.v.). Three mammals are named after him.

Grant, J. A.

Crested Francolin ssp. *Dendroperdix sephaena grantii*
Hartlaub, 1866

Colonel James Augustus Grant (1827–1892) was a Scottish naturalist and explorer. After completing his education at Marischal College, Aberdeen, he joined the British army and served in India during the Sikh Wars (1849) and the Indian Mutiny (1857–1858), during which he was wounded. He spent considerable time (1860–1863) in Africa with John Hanning Speke (q.v.), searching for the source of the Nile. He never saw the source, as he was unable to walk for six months because of debilitating leg ulcers. He kept a record of the journey and published it as *A Walk across Africa* (1864), in which he described 'the ordinary life and pursuits, the habits and feelings of the natives'. He served as an intelligence officer in the Abyssinian campaign (1868), retiring with the rank of Lieutenant Colonel. Two mammals are named after him.

Grant, R.

Raggiana Bird-of-Paradise ssp. *Paradisaea raggiana granti*
North, 1906

Robert Grant (fl.1906) was assistant taxidermist of the Australian Museum, Sydney.

Grant-Mackie

Grant-Mackie's Wren *Pachyplichas jagmi* Millener, 1988
EXTINCT

Dr Jack A. Grant-Mackie (b.1932) is a palaeontologist. He taught at the University of Auckland, New Zealand, in the Department of Geology where the Triassic-Jurassic marine faunas of the south-west Pacific, especially molluscs, are his principal research interest. Since retiring (1998) he is still an honorary research associate there, although he hopes to ease himself out over the next 12 months (2013). His secondary interest is New Zealand fossil birds, especially of the Quaternary. He co-wrote 'The Jurassic palaeobiogeography of Australasia' in *Memoirs of the Australasian Association of Palaeontologists* (2000). The scientific honorific is a 'joke' based on Grant-Mackie's initials, perpetrated by Millener, one of his PhD students, who wrote the description – but it is nonetheless valid.

Grant (Ogilvie-)

Painted Honeyeater genus *Grantiella* **Mathews**, 1911

Bronze Ground Dove *Phlegaenas granti* **Salvadori**, 1893
NCR
[JS *Gallicolumba beccarii solomonensis*]
Nazca Booby *Sula granti* **Rothschild**, 1902
Grant's Bluebill *Spermophaga poliogenys* Ogilvie-Grant,
1906
[Alt. Grant's Red-headed Bluebill]
Ant Thrush sp. *Neocossyphus granti* **B. Alexander**, 1908
NCR
[Alt. White-tailed Ant Thrush; JS *Neocossyphus poensis praepectoralis*]

Silver Pheasant *Gennaeus granti* **Oates**, 1910 NCR
[JS *Lophura nycthemera rufipes*]
Southern White-faced Owl *Ptilopsis granti* **Kollibay**, 1910
Grant's Starling *Aplonis mystacea* Ogilvie-Grant, 1911
[Alt. Yellow-eyed Starling]
Flycatcher sp. *Bradyornis granti* **Bannerman**, 1911 NCR
[Alt. Pale Flycatcher; JS *Bradornis pallidus parvus*]

Solomon Hawk Owl ssp. *Ninox jacquinoti granti* **Sharpe**,
1888
Eurasian Sparrowhawk ssp. *Accipiter nisus granti* Sharpe,
1890
Black Dwarf Hornbill ssp. *Tockus hartlaubi granti* **Hartert**,
1895
Eyebrowed Wren-babbler ssp. *Napothera epilepidota granti*
Richmond, 1900
Olive-green Camaroptera ssp. *Camaroptera chloronota
granti* Alexander, 1903
Green-backed Guan *Penelope jacquacu granti* **Berlepsch**,
1908
Rifleman ssp. *Acanthisitta chloris granti* **Mathews &
Iredale**, 1913
Rufous Fantail ssp. *Rhipidura rufifrons granti* Hartert, 1918
Pale-billed Scrubwren ssp. *Sericornis spilodera granti*
Hartert, 1930
Leaden Honeyeater ssp. *Ptiloprora plumbea granti* **Mayr**,
1931
Mottle-breasted Honeyeater ssp. *Meliphaga mimikae granti*
Rand, 1936
Uniform Swiftlet ssp. *Aerodramus vanikorensis granti* Mayr,
1937 NCR
[JS *Aerodramus vanikorensis yorki*]

(See **Ogilvie-Grant**)

Grantsau

Sincorá Antwren *Formicivora grantsaui* Gonzaga,
Carvalhães & Buzzetti, 2007

Rolf Grantsau (b.1928) is a German-born ornithologist, naturalist and artist, who moved to Brazil (1962). He was a
member of two Brazilian expeditions to the Antarctic. He
wrote *The Complete Guide to the Identification of the Birds of
Brazil* (2006). He was the first to notice the distinctiveness of
this antwren and collected the first specimen (1965). A
carnivorous plant and a beetle are named after him.

Granvik

African Green Pigeon ssp. *Treron calvus granviki* **Grote**,
1924
[Sometimes treated as a JS of *Treron calvus gibberifrons*]
Grey Apalis ssp. *Apalis cinerea granviki* Grote, 1927 NCR
[JS *Apalis cinerea cinerea*]
Moustached Grass Warbler ssp. *Melocichla mentalis
granviki* **Grant** & Mackworth-Praed, 1941 NCR
[JS *Melocichla mentalis amauroura*]

Dr Sven Hugo Granvik (b.1889) was a Swedish zoologist and
collector in Kenya and Uganda (1925–1927). Lund University
awarded his doctorate (1923).

Grassmann

Hazel Grouse ssp. *Tetrastes bonasia grassmanni* **Zedlitz**,
1920 NCR
[JS *Tetrastes bonasia volgensis*]

Lieutenant Wilhelm Grassmann (d.1918) was in the German
army in Poland.

Grauer

Grauer's Warbler genus *Graueria* **Hartert**, 1908

Grauer's Cuckooshrike *Coracina graueri* **Neumann**, 1908
Grauer's Scrub Warbler *Bradypterus graueri* Neumann,
1908
[Alt. Grauer's Swamp Warbler]
Grauer's Warbler *Graueria vittata* Hartert, 1908
[Alt. Grauer's Forest Warbler]
Grauer's Broadbill *Pseudocalyptomena graueri* **Rothschild**,
1909
[Alt. African Green Broadbill]

Streaky Seedeater ssp. *Serinus striolatus graueri* Hartert,
1907
[Syn. *Crithagra striolata graueri*]
Black-headed Waxbill ssp. *Estrilda atricapilla graueri*
Neumann, 1908 NCR
[JS *Estrilda kandti kandti*]
Grauer's Yellow-bellied Wattle-eye *Dyaphorophyia concreta
graueri* Hartert, 1908
Many-coloured Bush-shrike ssp. *Chlorophoneus multicolor
graueri* Hartert, 1908
Rwenzori Double-collared Sunbird ssp. *Cinnyris stuhlmanni
graueri* Neumann, 1908
Yellow-streaked Greenbul ssp. *Phyllastrephus flavostriatus
graueri* Neumann, 1908
Dusky Twinspot ssp. *Euschistospiza cinereovinacea graueri*
Rothschild, 1909
Village Weaver ssp. *Ploceus cucullatus graueri* Hartert,
1911
[SII *Ploceus cucullatus cucullatus*]
Abyssinian Owl ssp. *Asio abyssinicus graueri* **Sassi**, 1912
Grauer's Ground Thrush *Zoothera camaronensis graueri*
Sassi, 1914
[Alt. Black-eared Ground Thrush ssp.]
Tawny-flanked Prinia ssp. *Prinia subflava graueri* Hartert,
1920
African Scops Owl ssp. *Otus senegalensis graueri* **Chapin**,
1930 NCR
[NUI *Otus senegalensis senegalensis*]

Rudolf Grauer (1870–1927) was an Austrian explorer and
zoologist who collected extensively during an expedition to
the Belgian Congo (1909) and again (1910–1911) on an expedition paid for by the Austrian Imperial Museum. His research
focused on the Albertine Rift. He suffered from actinomy-
cosis contracted in Africa and eventually succumbed to it.
He is also commemorated in the names of other taxa
including two mammals, an amphibian and two reptiles.

Graves

Fulvous Wren ssp. *Cinnycerthia fulva gravesi* **Remsen** &
Brumfield, 1998

Dr Gary R. Graves is an American ornithologist who is Curator
of Birds at the USNM. The University of Arizona, Little Rock,
awarded his bachelor's degree (1976), Louisiana State
University his master's (1980) and Florida State University his
doctorate (1983). He was appointed Honorary Professor at
the Centre for Macroecology, Evolution and Climate, University of Copenhagen, for a 5-year period (2011). His primary
interests centre on the ecology, biogeography, and evolution
of birds and he is one of the co-organisers of the Bird 10K
genome project which seeks to obtain whole genome
sequences of the 10,400+ living species of birds. One of the
most recent of his many papers was the co-authored 'An
update of Wallace's zoogeographic regions of the world'
(2013).

Gray, G. R.

Parrot genus *Graydidascalus* **Bonaparte**, 1854

Gray's Robin *Turdus grayi* Bonaparte, 1838
[Alt. Clay-coloured Robin, Clay-coloured Thrush]
Gray's Tanager *Piranga rubriceps* G. R. Gray, 1844
[Alt. Red-hooded Tanager]
Gray's Shag *Leucocarbo chalconotus* G. R. Gray, 1845
[Alt. Bronze Shag; Syn. *Phalacrocorax chalconotus*]
Gray's Brush-turkey *Macrocephalon maleo* **S. Müller**, 1846
[Alt. Maleo]
Gray's Greybird *Coracina schisticeps* G. R. Gray, 1846
[Alt. Grey-headed Cuckooshrike, Black-tipped Cicadabird]
Gray's Sapphire *Hylocharis grayi* **de Lattre** & **Bourcier**, 1846
[Alt. Blue-headed Sapphire]
Gray's Lark *Ammomanopsis grayi* **Wahlberg**, 1855
Gray's Lory *Trichoglossus coccineifrons* G. R. Gray, 1858
NCR
[Probable hybrid: *Trichoglossus haematodus* x *Chalcopsitta
sintillata*]
Gray's Goshawk *Accipiter henicogrammus* G. R. Gray, 1860
[Alt. Moluccan Goshawk]
Kingfisher sp. *Sauropatis grayi* **Cabanis** & **Heine**, 1860 NCR
[Alt. Collared Kingfisher; JS *Todiramphus chloris sordidus*]
Gray's Oriole *Oriolus phaeochromus* G. R. Gray, 1861
[Alt. Dusky-brown Oriole, Halmahera Oriole]
Gray's Warbler *Locustella fasciolata* G. R. Gray, 1861
[Alt. Gray's Grasshopper Warbler]
Broad-billed Fairy-wren *Chenorhamphus grayi* **Wallace**,
1862
[Syn. *Malurus grayi*]
Gray's Honeyeater *Xanthotis polygrammus* G. R. Gray, 1862
[Alt. Spotted Honeyeater]
Pearl-bellied White-eye *Zosterops grayi* Wallace, 1864

Correndera Pipit ssp. *Anthus correndera grayi* Bonaparte,
1850
Great Argus ssp. *Argusianus argus grayi* **D. G. Elliot**, 1865
Gray's Piping Guan *Pipile cumanensis grayi* **Pelzeln**, 1870
[Alt. Blue-throated Piping Guan ssp.]
Bare-faced Curassow ssp. *Crax fasciolata grayi* **Ogilvie-
Grant**, 1893

George Robert Gray (1808–1872) was an Assistant Keeper
(Ornithology) at the British Museum's Department of Zoology
and the younger brother of John Edward Gray (q.v.). He wrote
Genera of Birds (1844) and *A Fasciculus of the Birds of China*
(1871), which had illustrations by Swainson (q.v.). With his
brother he wrote *Catalogue of the Mammalia and Birds of
New Guinea in the Collection of the British Museum* (1859).
The parrot genus name *Graydidascalus* means 'Gray the
teacher'. A mammal is named after him.

Gray, J. E.

Gray's Malimbe *Malimbus nitens* J. E. Gray, 1831
[Alt. Blue-billed Malimbe]
Gray's Pond Heron *Ardeola grayii* **Sykes**, 1832
[Alt. Indian Pond Heron]

John Edward Gray (1800–1875) was a British ornithologist
and entomologist. He started at the British Museum (1824)
with a temporary appointment at 15 shillings a day, but rose
to Curator of Birds (1840–1874) and then Head of the Department of Zoology. Gray published descriptions of a large
number of animal species, including many Australian reptiles
and mammals, and was the leading authority on many
reptiles. He was also an ardent philatelist and claimed that
he was the world's first stamp collector. He worked at the
museum with his brother George Robert Gray (q.v.), and
together they published a *Catalogue of the Mammalia and
Birds of New Guinea in the Collection of the British Museum*
(1859). He wrote *Gleanings from the Menagerie and Aviary at
Knowsley Hall* (1846–1850), which was illustrated by Lear
(q.v.). Gray suffered a severe stroke (1869), paralysing his
right side, including his writing hand, yet he continued to
publish to the end of his life by dictating to his wife, Maria
Emma, who had always worked with him as an artist and
occasional co-author. Twenty-three reptiles, nine mammals
and an amphibian are named after him. (We have assigned
most of the 'Gray' birds under the entry for G. R. Gray, as he
wrote many more of the descriptions. It is, however, very
difficult to be completely sure, as not all etymologies make it
clear which brother is honoured.)

Graydon

Chestnut-necklaced Hill Partridge ssp. *Arborophila
charltonii graydoni* **Sharpe** & **Chubb**, 1906

P. N. Graydon (DNF) was a British tobacco planter in
Sandakan, North Borneo (1890–1902).

Grayson

Grayson's Dove *Zenaida graysoni* **G. N. Lawrence**, 1871
[Alt. Socorro Dove]
Socorro Mockingbird *Mimus graysoni* Lawrence, 1871
[Syn. *Mimodes graysoni*]

Cinnamon Hummingbird ssp. *Amazilia rutila graysoni*
G. N. Lawrence, 1867
Grayson's Oriole *Icterus pustulatus graysonii* **Cassin**, 1867
[Alt. Tres Marias Oriole, Streak-backed Oriole ssp.]
Grayson's Bobwhite *Colinus virginianus graysoni*
G. N. Lawrence, 1867
[Alt. Northern Bobwhite ssp.]

Ladder-backed Woodpecker ssp. *Picoides scalaris graysoni*
Baird, 1874
Grayson's Thrush *Turdus rufopalliatus graysoni* **Ridgway**,
1882
[Alt. Rufous-backed Thrush ssp.]
Great-tailed Grackle ssp. *Quiscalus mexicanus graysoni*
P. L. Sclater, 1884
Elf Owl ssp. *Micrathene whitneyi graysoni* Ridgway, 1886
Tropical Parula ssp. *Parula pitiayumi graysoni* Ridgway,
1887
Grayson's Parrotlet *Forpus cyanopygius insularis* Ridgway,
1888
[Alt. Tres Marias Parrotlet, Mexican Parrotlet ssp.]

Andrew Jackson Grayson (1819–1869) was an American ornithologist and artist. Appropriately for a birdman, his father's middle name was Wren. He was considered to be the most accomplished bird painter in North America of his time, and was often referred to as 'the Audubon of the West'. When he began painting western American birds (1853) there was no systematic avifaunal record from the Sierra Nevada to the Pacific Ocean. Grayson regarded his *Birds of the Pacific Slope* as a completion of Audubon's *Birds of America*, which did not include all the western species. The USNM recruited Grayson as a field ornithologist and he became one of their principal collectors for California, Mexico and its offshore islands. He discovered many new species, some of which were named after him. He also wrote species accounts, recorded scientific data and published articles on travel and natural history. However, his greatest achievements were his paintings, which are virtually colour-perfect and depict their subjects in natural settings and behaviour. In 16 years he painted over 175 bird portraits, of which 156 survive, preserved in a single collection. Ill-luck often beset Grayson in his latter years. He was shipwrecked; his son was mysteriously murdered; he was bankrupted; and he contracted yellow fever on a field expedition, from which he died. A mammal is named after him.

Grebnitzki

Common Rosefinch ssp. *Carpodacus erythrinus grebnitskii*
Stejneger, 1885

Nikolai A. Grebnitzki (DNF) was a Russian administrator in Siberia and a collector who was in Kamtchatka (1885) and collected molluscs on Bering Island with Stejneger (q.v.).

Green

Scrubtit ssp. *Acanthornis magna greeniana* **Schodde** &
Mason, 1999

Dr. Robert 'Bob' Geoffrey Green (b.1925) was a Tasmanian farmer. He was very interested in ornithology and photography, becoming a professional wildlife photographer (1953) and honorary ornithologist at the Queen Victoria Museum and Art Gallery, Launceston, Tasmania (1959). He sold his farm (1960) and joined the museum staff, becoming Curator (1962). The University of Tasmania awarded him an honorary doctorate (1987). A reptile is named after him.

Greenewalt

Mountain Velvetbreast ssp. *Lafresnaya lafresnayi
greenewalti* **W. H. Phelps** & **W. H. Phelps Jr**, 1961
Brown Violetear ssp. *Colibri delphinae greenwalti* Ruschi,
1962 NCR; NRM

Crawford Hallock Greenwalt (1902–1993) was an American businessman, pioneer high-speed photographer and conservationist. He worked for the DuPont Corporation (1948–1967), rising to President of the Board (1962–1967). He became a trustee of the AMNH, New York (1954). Greenwalt photographed hummingbirds in Brazil and Ecuador and designed equipment to photograph them in flight. He also wrote award-winning books and articles based on his avian researches, including *Hummingbirds* (1960) and *Bird Song: Acoustics and Physiology* (1968).

Greenway, J. C.

Red-winged Laughingthrush ssp. *Trochalopteron formosum
greenwayi* **Delacour** & **Jabouille**, 1930
Greenway's Blood Pheasant *Ithaginis cruentus holoptilus*
Greenway, 1933
[SI *Ithaginis cruentus rocki*]

Dr James Cowan Greenway Jr (1903–1989) was an American collector and ornithologist. He accompanied his close friend Delacour (q.v.) on expeditions to Vietnam (1929–1930 and 1938–1939), a major objective of which was to search for pheasants in general, and Edwards's Pheasant *Lophura edwardsi* in particular. They co-wrote 'VIIe Expedition Ornithologique en Indochine française' (1940). Greenway was an author and editor of *Check-list of Birds of the World* (started 1931) and he co-produced the second edition of the first volume. He worked with Mayr (q.v.) and Griscom (q.v.) and (1959) became Curator of Birds at the Museum of Comparative Zoology, Harvard. He is well known for his *Extinct and Vanishing Birds of the World* (1967). Two reptiles are named after him.

Greenway, P.

Bearded Scrub Robin ssp. *Erythropygia quadrivirgata
greenwayi* **Moreau**, 1938

Dr Percy James Greenway (1897–1980) was a South African botanist and collector at the East African Agricultural Research Station (1927–1950) and East African Herbarium (1950–1958). The University of Witwatersrand awarded his doctorate (1954).

Gregorjew

(See **Grigorjew**)

Greschik

Great Rosefinch ssp. *Carpodacus rubicilla greschiki* Keve,
1943 NCR
[JS *Carpodacus severtzovi*]

Dr Eugen Greschik (b.1887) was a Hungarian zoologist at the National Museum in Budapest. He was the first editor (1929) of the Hungarian quarterly magazine *Kogsag* (White Egret). He wrote 'Ueber den Bau der Milz einiger Vögel' (1916).

Greta

Scrub Honeyeater ssp. *Meliphaga albonotata gretae*
Gyldenstolpe & **Gilliard**, 1955 NCR; NRM
[Taxon originally described as *Meliphaga montana gretae*]

Sofia Margareta Emilie 'Greta', Countess Gyldenstolpe *née* Heijkenskjöld (1894–1991), was in New Guinea (1951) with her husband (1917), Count Nils Gyldenstolpe (1886–1961) (q.v.)

Grey, G.

Grey's Fruit Dove *Ptilinopus greyii* **Bonaparte**, 1857
[Alt. Red-bellied Fruit Dove]
Kakapo *Strigops greyii* **G. R. Gray**, 1862 NCR
[JS *Strigops habroptilus*]

North Island Weka *Gallirallus australis greyi* **W. L. Buller**, 1888

Sir George Grey (1812–1898) was a soldier, explorer, colonial governor, premier and scholar. He explored Western Australia on government-financed expeditions, to Hanover Bay and Shark Bay (1837–1839). On the first expedition he was speared by an Aborigine, whom he afterwards shot; nevertheless, he championed the cause of assimilation, and respect for the aboriginal people was a trait he shared with Gould (q.v.), who was a frequent correspondent. He became Governor of New Zealand (1845), where his greatest success was his management of Maori affairs. He scrupulously observed the terms of the Treaty of Waitangi, assuring Maoris that their land rights were fully recognised. He became Governor of the Cape Colony (1853) and High Commissioner for South Africa. He sought to convert the frontier tribes to Christianity, to 'civilise' them. Grey supported mission schools and built a hospital for African patients. When he returned to New Zealand he was elected to parliament, but he remained a keen naturalist and botanist, and established extensive collections and important libraries at Cape Town and Auckland. He wrote books on Australian aboriginal vocabularies and his Western Australian explorations, and also took a scholarly interest in the Maori language and culture. Two mammals and a reptile are named after him.

Grigorjew

Turtle Dove ssp. *Turtur communis grigorjewi* **Zarudny** & **Loudon**, 1902 NCR
[JS *Streptopelia turtur arenicola*]

Alexander W. Grigorjew (DNF) was Secretary of the Imperial Geographical Society, St Petersburg, Russia. The describers originally used the spelling *gregorjewi* in error.

Grill

Black Coucal *Centropus grillii* **Hartlaub**, 1861
[Alt. Black-chested Coucal]

Johan Wilhelm Grill (1815–1864) was a Swedish zoologist who worked at the Swedish Academy of Veterinary Medicine in Stockholm.

Grimm

Grey Shrike sp. *Lanius grimmi* **Bogdanov**, 1881 NCR
[Alt. Steppe Grey Shrike; JS *Lanius* (*meridionalis*) *pallidirostris*]

Dr Oscar A. Grimm (1845–1921) was a Russian Professor of Zoology, Leningrad University. He was an expert on the fauna of the Caspian Sea. He wrote 'On the scientific exploration of the Caspian Sea' (1876).

Grimwood

Grimwood's Longclaw *Macronyx grimwoodi* **C. W. Benson**, 1955

Swamp Flycatcher ssp. *Muscicapa aquatica grimwoodi* **Chapin**, 1952

Major Ian Robert Grimwood (1912–1990) was Chief Game Warden of Kenya (1959–1964). He had a great part in rescuing some of the last wild Arabian oryx to establish a captive-breeding group and save them from extinction. In Africa he guided Colonel Charles A. Lindbergh, the famous airman who made the first solo transatlantic flight. Grimwood was awarded the World Wildlife Fund gold medal (1972). The citation was '… for his dedication to the conservation of wildlife in Africa, Asia and Latin America; his contribution to the establishment of national parks and game reserves in Kenya, Pakistan and Peru, and to the survival of endangered species such as oryx and vicuña.'

Grinda

Grinda's Bushtit *Psaltriparus minimus grindae* **Ridgway**, 1883
[American Bushtit ssp.]

Francisco C. Grinda (DNF) was a Mexican landowner at La Paz, California. Lyman Belding (q.v.) collected the bird and suggested that Ridgway (q.v.) name it after Grinda, who had been so helpful to him. Ridgway, in his description, quotes Belding's letter to him.

Grinnell, G.

Grinnell's Waterthrush *Parkesia noveboracensis notabilis* **Ridgway**, 1881 NCR; NRM
[Alt. Northern Waterthrush ssp.; Syn. *Seiurus noveboracensis notabilis*.]

George Bird Grinnell (1849–1938) was an American ornithologist, publisher and conservationist. He attended a school run by John James Audubon's (q.v.) widow Lucy. Despite a poor degree (Yale, 1870), he talked his way onto a fossil-collecting expedition (1870) and then served as the naturalist on Custer's expedition to the Black Hills (1874). He returned to Yale and gained a PhD in palaeontology (1880). He was very interested in Native Americans and got on so well with them that many tribes had names for him. The Pawnee called him 'White Wolf' and eventually adopted him into their tribe. The Gros Ventre called him 'Gray Clothes', the Black Feet 'Fisher Hat' and the Cheyenne called him 'Wikis', which means 'bird', observing that he came and went with the seasons. He wrote 26 books including *The Cheyenne Indians – Their History and*

Ways of Life (1923). He was also editor of *Forest and Stream*, the leading natural history magazine in North America. He was the founder of the Audubon Society (1886) and began publication of its *Audubon Magazine* the next year. In only three months more than 38,000 people joined the society. Overwhelmed by the response, Grinnell had to disband the group (1888) (although it was subsequently reconstituted). He was also an early member of the American Ornithologists' Union from shortly after its foundation (1883). He became an advisor to Theodore Roosevelt (q.v.) whom he had convinced of the need for hunting regulation and habitat conservation, and Glacier National Park was established through his efforts. The National Wildlife Confederation Hall of Fame citation reads: 'To accomplish his goal of ensuring effective enforcement of game laws, Grinnell advocated a game warden system to be financed by small fees from all hunters. ..... a revolutionary concept that would become a corner-stone of game management. Realizing that the enforcement of game laws was the solution to only half a problem, Grinnell turned his attention to habitat conservation. In 1882, he began an editorial effort to persuade America to manage timberlands efficiently to yield a sustained "crop". He was also drawn to the plight of Yellowstone National Park, launching a campaign to expose Federal neglect and ensure the park against commercialisation. Grinnell's efforts attracted the admiration and support of Theodore Roosevelt, an avid reader of *Forest and Stream*. Before he ascended to the presidency, Roosevelt launched his career as a conser-vationist by joining Grinnell's battle for Yellowstone. When Roosevelt became president in 1901, the conservation philos-ophy first formulated by George Bird Grinnell became the basis of the American conservation program.'

Grinnell, J.

Ruby-crowned Kinglet ssp. *Regulus calendula grinnelli*
   W. Palmer, 1897
Grinnell's Jay *Cyanocitta stelleri carbonacea* J. Grinnell, 1900
[Alt. Steller's Jay ssp.]
Northern Pygmy Owl ssp. *Glaucidium californicum grinnelli*
   Ridgway, 1914
Grinnell's Screech Owl *Megascops kennicottii quercinus*
   J. Grinnell, 1915 NCR
[Alt. Western Screech Owl ssp.; JS *Megascops kennicottii bendirei*]
Red-winged Blackbird ssp. *Agelaius phoeniceus grinnelli*
   Howell, 1917
Grinnell's Shrike *Lanius ludovicianus grinnelli* **Oberholser**, 1919
[Alt. Loggerhead Shrike ssp.]
Mountain Chickadee ssp. *Poecile gambeli grinnelli*
   Van Rossem, 1928 NCR
[JS *Poecile gambeli baileyae*]
Grinnell's Crossbill *Loxia curvirostra grinnelli* **Griscom**, 1937
[Alt. Common/Red Crossbill ssp.]
House Finch ssp. *Carpodacus mexicanus grinnelli*
   R. T. Moore, 1939 NCR
[JS *Carpodacus mexicanus frontalis*]

Joseph Grinnell (1877–1939) – a distant cousin of George Grinnell (q.v.) – was instrumental in shaping the philosophy of the United States National Park System. He was Director of the Museum of Vertebrate Zoology at Berkeley (1908–1939) and an 'incomparable authority on the birds and mammals of the West Coast'. When he died, A. H. Miller (q.v.), with whom he had often written, took his post. It was Grinnell who coined the expression 'niche' in the context of 'competitive exclusion' and the expression is now used universally for almost any situation where species (including people) have established an exclusive position. He was also editor of *Condor* for over 30 years, and co-author with Miller of *The Distribution of the Birds of California* (1944).

Griscom

Griscom's Antwren *Myrmotherula ignota* Griscom, 1929
[Alt. Moustached Antwren]

Great Curassow ssp. *Crax rubra griscomi* **Nelson**, 1926
Unicoloured Jay ssp. *Aphelocoma unicolor griscomi*
   Van Rossem, 1928
Grey-throated Chat ssp. *Granatellus sallaei griscomi*
   Van Rossem, 1934 NCR
[JS *Granatellus sallaei boucardi*]
Black-headed Siskin ssp. *Carduelis notata griscomi* Van
   Rossem, 1938 NCR
[JS *Carduelis notata forreri*]
House Finch ssp. *Carpodacus mexicanus griscomi*
   R. T. Moore, 1939
Eastern Meadowlark ssp. *Sturnella magna griscomi*
   Van Tyne & Trautman, 1941
Colima Pygmy Owl ssp. *Glaucidium palmarum griscomi*
   Moore, 1947
Western Wood Pewee ssp. *Contopus sordidulus griscomi*
   Webster, 1957
[SII *Contopus sordidulus sordidulus*]

Ludlow Griscom (1890–1959) was an American ornithologist. Columbia University awarded his bachelor's degree (1912), then he studied at Cornell under Arthur Allen (q.v.) and Louis Agassiz Fuertes (q.v.) and was awarded his masters (1915). He taught there and at Virginia University whilst studying for his doctorate, which he did not complete. He joined the staff at AMNH (1916–1927) as Assistant Curator. He took a post at Harvard's Museum of Comparative Zoology (1927) as Research Curator of Zoology becoming Research ornitholo-gist (1948–1955) but took early retirement, being in poor health after a stroke (1950). He kept notes on the birds he observed on field trips throughout the state of Massachu-setts (1930s–1950s) and undertook fieldwork variously in Canada, Belize, Panama, Guatemala and Mexico. He was an early advocate of identifying birds through binoculars by field marks, rather than by shooting them and 'collecting' speci-mens. He was a pioneer of conservation and in many ways born before his time. For example, while he was in Guatemala collecting birds for the AMNH (1930s), he noted that coffee growers left much of the natural forest to shade their plants and 'in such growth, the avifauna was little, if any, different from its original condition'; something that went otherwise unnoticed for decades. The Conventions of the American

Birding Association have bestowed (since 1980) an award named the 'Ludlow Griscom Distinguished Birder Award'. Griscom himself was an inveterate lister.

Griswold

Chestnut-hooded Laughingthrush ssp. *Garrulax treacheri griswoldi* **J. L. Peters**, 1940

John Augustus 'Gus' Griswold Jr (1912–1991) was an aviculturist and ornithologist. He was a field collector for the Harvard University Museum of Comparative Zoology, leading expeditions to Siam (Thailand), Peru and Mexico. He became Curator of Birds, Philadelphia Zoological Gardens (1947). A reptile and an amphibian are named after him.

Griveaud

Comoro Green Pigeon *Treron griveaudi* **C. W. Benson**, 1960

African Palm Swift ssp. *Cypsiurus parvus griveaudi* C. W. Benson, 1960
Frances's Sparrowhawk ssp. *Accipiter francesii griveaudi* C. W. Benson, 1960

Dr Paul Griveaud (1907–1980) was an entomologist who worked for many years at the Entomological Laboratory, ORSTOM, Antananarivo, Madagascar. He co-wrote *La Protection des Richesses Naturelles, Archéologiques et Artistiques à Madagascar* (1968). He has a reptile and a mammal named after him.

Grombchevsky

Himalayan Snowcock ssp. *Tetraogallus himalayensis grombczewskii* **Bianchi**, 1898

General Bronislav Ludvigovich Grombchevsky (1855–1926) was a Polish explorer in the Russian service in Central Asia (1888–1892).

Grönvold

Gull-billed Tern ssp. *Gelochelidon nilotica groenvoldi* **Mathews**, 1912

Henrik Grönvold (1858–1940) was a Danish artist, taxidermist and naturalist. He worked as a draughtsman for the Danish Artillery (1880–1892), then left Denmark for America. He travelled via England and chanced upon employment at the BMNH in London as a preparator of bird skeletons (1892–1940), except when he accompanied Ogilvie-Grant (q.v.) on an expedition to the Savage Islands (1895). He was highly regarded as an artist and did work for Lord Rothschild (q.v.), Captain George Shelley (q.v.), Sir Walter Lawry Buller (q.v.) and the authority on mammals, Michael Rogers Oldfield Thomas. (See **Henrik**)

Gross

Clapper Rail ssp. *Rallus longirostris grossi* **Paynter**, 1950
[Syn. *Rallus crepitans grossi*]

Dr Alfred Otto Gross (1883–1970) was an American ornithologist whose doctorate was awarded (1912) by Harvard. He taught at Bowdoin College as an instructor (1912–1913) and Professor of Biology (1913–1953). He was the ornithologist on Donald MacMillan's Arctic Expeditions (1934 and 1937).

Grosvenor

Bismarck Thicketbird *Megalurulus grosvenori* **Gilliard**, 1960

Dr Gilbert Hovey Grosvenor (1875–1966) is considered to have been the father of photojournalism. Amherst College, Massachusetts, awarded his bachelor's degree (1897). He was President of the National Geographic Society (1920–1954) and the first full-time editor of the *National Geographic Magazine* (1899–1954). Gilliard wrote: 'It gives me great pleasure to name this new thicket warbler in honor of Dr Gilbert Grosvenor as a token of my gratitude to him and to the National Geographic Society for their generous support of my explorations.' He has a sandstone arch in Utah named after him.

Grote

Bishop genus *Groteiplectes* **Wolters**, 1943 NCR
[Now in *Euplectes*]

Black-browed Fulvetta *Alcippe grotei* **Delacour**, 1936

Fischer's Greenbul ssp. *Phyllastrephus fischeri grotei* **Reichenow**, 1910 NCR; NRM
Green-winged Pytilia ssp. *Pytilia melba grotei* Reichenow, 1919
Hildebrandt's Francolin ssp. *Pternistis hildebrandti grotei* Reichenow, 1919 NCR; NRM
African Dusky Flycatcher ssp. *Muscicapa adusta grotei* Reichenow, 1921 NCR
[JS *Muscicapa adusta pumila*]
Vinaceous Dove ssp. *Streptopelia vinacea grotei* Reichenow, 1926 NCR; NRM
Madagascar Bulbul ssp. *Hypsipetes madagascariensis grotei* **Friedmann**, 1929
Yellow-fronted Canary ssp. *Serinus mozambicus grotei* **W. L. Sclater** & **Mackworth-Praed**, 1931
[Syn. *Crithagra mozambica grotei*]
Bluethroat ssp. *Luscinia svecica grotei* Dement'ev 1932 NCR
[JS *Luscinia svecica volgae*]
Olive-flanked Robin Chat ssp. *Cossypha anomala grotei* Reichenow, 1932
African Pipit ssp. *Anthus cinnamomeus grotei* **Niethammer**, 1957

Herman Grote (1882–1951) was a sisal planter in German East Africa (1908–1911). He was working in Russia when WW1 started and was interned at Orenburg as an enemy alien (1915–1918). This enabled him to learn Russian and he later translated many Russian ornithological papers and studies into German. He worked at the Zoological Museum in Berlin (1945–1951).

Gruber

Gruber's Hawk *Onychotes gruberi* **Ridgway**, 1870
[JS Hawaiian Hawk *Buteo solitarius* **Peale**, 1848]

Ferdinand Gruber (1830–1907) was Curator of the Gold Park Museum in San Francisco. Gruber's name became attached to this hawk when Ridgway (q.v.) named a 'new' species that was thought to come from California, but it was actually the Hawaiian Hawk.

Gruchet

Réunion Owl *Mascarenotus grucheti* Mourer-Chauviré *et al.*, 1994 EXTINCT

Harry Gruchet (d.2013) was formerly a curator at Muséum d'Histoire Naturelle de Saint-Denis, Réunion. He was one of those who suggested where to look for fossils.

Grumm-Grzhimaylo

White-winged Snowfinch ssp. *Montifringilla nivalis groumgrzimaili* **Zarudny** & **Loudon**, 1904

Grigory Y. Grumm-Grzhimaylo (1860–1936) was a Russian entomologist who travelled widely in Central Asia (1884–1890). On some of his journeys his younger brother, Lieutenant M. Y. Grumm-Grzhimaylo (b.1862), accompanied him. We cannot be sure which brother was honoured, but the elder is more likely as he appears to have been the scientist. A reptile is named after him.

Grün

Little Owl ssp. *Athene noctua grueni* **von Jordans** & **Steinbacher**, 1941 NCR
[JS *Athene noctua vidalii*]

H. Grün (d.1963) was a German collector in Spain and Portugal (1926–1963).

Grzimek

Grzimek's Barbthroat *Threnetes grzimeki* Ruschi, 1973 NCR
[Alt. Rufous-breasted Hermit; JS *Glaucis hirsutus*]

Professor Dr Bernhard Klemens Maria Grzimek (1909–1987) was a German zoologist, author and veterinary surgeon who travelled widely (1950s) in Tanzania. He was Director of the Frankfurt Zoo (1945–1974). When President of the Frankfurt Zoological Society he raised funds for conservation through his television series in Germany 'A Place for Wild Animals'. He expanded the activities of the society in Tanzania and elsewhere and lobbied government to make conservation a priority. He made the film *Serengeti Shall Not Die* (1958).

Guatimozin

Black Oropendola *Psarocolius guatimozinus* Bonaparte, 1853
[Syn. *Gymnostinops guatimozinus*]

Guatimozin, or Cuauhtémoc (c.1495–1522) was the last Aztec emperor, nephew and son-in-law of Montezuma II. Bonaparte (q.v.) probably used this name to indicate the bird's relationship with the Montezuma Oropendola *Psarocolius montezuma*, which Lesson (q.v.) had named (1830).

Guerin

Guerin's Helmetcrest *Oxypogon guerinii* **Boissonneau**, 1840
[Alt. Bearded Helmetcrest]

Grey-headed Woodpecker ssp. *Picus canus guerini* **Malherbe**, 1849

Félix Edouard Guérin-Méneville (1799–1874) was an entomologist. He is most well known for his illustrated work *Iconographie du Règne Animal de G. Cuvier 1829–44*, a complement to Cuvier's (q.v.) and Latreille's work, which lacked illustrations; and also for introducing silkworm breeding to France. There is a Felix-Edouard Guerin-Méneville Collection of Crustacea at the Academy of Natural Sciences, Philadelphia. He made the first description of Rouget's (q.v.) Rail *Rougetius rougetii*. He was an all-rounder, having written scientific papers on plants, insects and others. A reptile is named after him.

Guglielm

Yellow-browed Bulbul ssp. *Acritillas indica guglielmi* **Ripley**, 1946
[Syn. *Hypsipetes indicus guglielmi*]

Major-General William Joseph Donovan (1883–1959) commanded the American troops in Ceylon (Sri Lanka) (1944) and was a courageous and successful soldier. He was awarded the Congressional Medal of Honor (WW1) where his actions earned him the nickname 'Wild Bill'. During WW2 he became head of the OSS (Office of Strategic Services), the predecessor of the Central Intelligence Agency. He was an Assistant (1945–1946) to Robert Jackson, Chief American Prosecutor at the Nuremberg War Crimes Trials. He became US Ambassador to Thailand (1953). The word *guglielm* is a latinised version of 'William'.

Guichard

Arabian Partridge ssp. *Alectoris melanocephala guichardi* **Meinertzhagen**, 1951
[Species sometimes regarded as monotypic]

Kenneth Mackinnon Guichard (1914–2002) was a British entomologist and art connoisseur. He worked on the Desert Locust Survey, Oman (1949–1950), and in the Sahara (1952). He was on Socotra Island, Yemen, as entomologist attached to the Middle East Command's expedition (1967). Many of his specimens were stored in the Natural History Museum, London, for c.20 years before examination. He made his living through his flair for spotting good paintings and etchings, buying cheap and selling expensive. A reptile is named after him.

Guiers

African Reed Warbler ssp. *Acrocephalus baeticatus guiersi* Colston & Morel, 1984

It sounds like an eponym, but it is not. It is named after a place, Lac de Guiers, Senegal.

Guilding

Guilding's Amazon *Amazona guildingii* **Vigors**, 1837
[Alt. St Vincent Parrot]

The Reverend Lansdown Guilding (1797–1831) was an amateur naturalist from St Vincent, West Indies, who collected there. He was educated at Oxford, England. He left his collection to the British Museum.

Guillarmod

Yellow Canary ssp. *Serinus flaviventris guillarmodi*
J. A. Roberts, 1936
[Syn. *Crithagra flaviventris guillarmodi*]

Charles Frédéric Jacot-Guillarmod (1912–1979) was a South African entomologist and collector. The University of Pretoria awarded his BSc (1934) and MSc (1936). He was employed by the Division of Entomology of the Department of Agriculture and Forestry, where he was involved in locust research for two years, working mainly in Zululand and the Upington area, then several other posts in the department (1936–1958). He was then at the Albany Museum, Grahamstown, where he was the first Curator of the Department of Invertebrate Zoology (1958–1977) apart from two years spent in the US (1962–1964), returning to become Director (1964). He worked a further two years as a Curator after resigning as Regional Director (1977–1979) but died of a heart attack.

Guillelm

Eurasian Skylark ssp. *Alauda arvensis guillelmi* **Witherby**, 1921

(See **Tait, W.**)

Guillemard

Bar-bellied Cuckooshrike ssp. *Coracina striata guillemardi* **Salvadori**, 1886
Hair-crested Drongo ssp. *Dicrurus hottentottus guillemardi* Salvadori, 1889
Guillemard's Crossbill *Loxia curvirostra guillemardi* **Madarász**, 1903
[Alt. Common/Red Crossbill ssp.]

Dr Francis Henry Hill Guillemard (1853–1933) was a British physician (although he never practised), geographer, author and traveller. He visited Lapland (1873) and southern Africa (1877–1878) and took part in the first Boer War (1881). He visited Madeira and the Canary Islands several times. He made an extensive zoological expedition (1882–1884), basing himself on the yacht *Marchesa*, on which he visited Kamchatka, New Guinea and the major islands of the Malay Archipelago. After his return he wrote *The Cruise of the Marchesa to Kamschatka and New Guinea, with notices of Formosa, Liu-Kiu, and various islands of the Malay Archipelago* (1886). He went to Cyprus (1887) and started the Cyprus Exploration Fund. He finally settled in Cambridge and served as the Cambridge University Press's Geographical Editor. He wrote a number of other works including *The Life of Ferdinand Magellan, and the First Circumnavigation of the Globe, 1480–1521* (1890) and *Malaysia and the Pacific Archipelagos* (1894).

Guillot

Carolina Chickadee ssp. *Poecile carolinensis guilloti* **Oberholser**, 1938 NCR
[JS *Poecile carolinensis agilis*]

James P. Guillot (b.1887) was an American zoologist who was trained as a lawyer and was Secretary, Louisiana Department of Conservation.

Guimet

Guimet's Flutterer *Klais guimeti* **Bourcier**, 1843
[Alt. Violet-headed Hummingbird]

Jean Baptiste Guimet (1795–1871) was a French chemist whose fortune was based on the invention of a blue dye (ultramarine), which made him extremely famous in his time. His son travelled the world collecting art, artefacts and natural exhibits for their own museum.

Guise

Guise's Honeyeater *Ptiloprora guisei* **De Vis**, 1894
[Alt. Rufous-backed/Red-backed Honeyeater]

Reginald Edward Guise (1850–1902) was born in England, became an army officer (1869–1873), then resigned his commission and moved to Australia. He arrived in Sydney (June 1874) and days later took ship to Brisbane in order, possibly, to take up a job as manager of a sheep station. There is a mystery about him as he appears to have been in New Guinea (1883) but was asked to leave (1884), returned, and was officially deported (1885) but returned again (1886). He served there as a Colonial Administrator (1891–c.1894). He wrote a number of reports on native languages and customs for the Royal Anthropological Institute. He settled in New Guinea where his grandson, Sir John Guise, became the first Governor-General of the Independent State of Papua New Guinea.

Güldenstädt

Güldenstädt's Redstart *Phoenicurus erythrogastrus* **Güldenstädt**, 1775
[Alt. White-winged Redstart]

Professor Johann Anton Gueldenstaedt (1745–1781) was a Baltic-German, born in Riga in Latvia, then part of the Russian Empire. He was a physician, natural scientist and traveller. He made several expeditions to the Caucasus and trans-Caucasus regions (1768–1773) for the Imperial Academy of Science in St Petersburg, where he was a Professor (1771). He was the author of diaries containing extensive geographical, biological and ethnographical material on the Caucasas and Ukraine, as commissioned by the Empress Catherine II. During this period Gmelin (q.v.) was his assistant. Pallas (q.v.) published his *Reisen durch Russland und im Caucasischen Gebürge* posthumously (1787–1791). Güldenstädt was the first person to describe a number of birds, including the Ferruginous Duck *Aythya nyroca* (1770). A mammal is named after him.

Gulielmi Terti

William's Fig Parrot *Cyclopsitta gulielmitertii* **Schlegel**, 1866
[Alt. Orange-breasted Fig Parrot]

Willem Alexander Paul Frederik Lodewijk (1817–1890) of the Netherlands. This translates from its latinised form as William Third. (See **William, King of Holland**)

Gulliver

Gulliver's White-eye *Zosterops gulliveri* **Castlenau &
E. P. Ramsay**, 1877 NCR
[Alt. Canary White-eye; JS *Zosterops luteus*]

T. A. Gulliver (1847–1931) was an employee of the Postal &
Telegraph Department in Queensland, Australia and a field
natural history collector (1865–1891).

Gundlach

Cuban Vireo *Vireo gundlachii* **Lembeye**, 1850
Cuban Black Hawk *Buteogallus gundlachii* **Cabanis**, 1855
Gundlach's Mockingbird *Mimus gundlachii* Cabanis, 1855
[Alt. Bahama Mockingbird]
Antillean Nighthawk *Chordeiles gundlachii* **G. N. Lawrence**,
1857
Gundlach's Hawk *Accipiter gundlachi* G. N. Lawrence, 1860

Golden Warbler ssp. *Dendroica petechia gundlachi* **Baird**,
1865
Greater Antillean Grackle ssp. *Quiscalus niger gundlachii*
Cassin, 1867
Northern Flicker ssp. *Colaptes auratus gundlachi* **Cory**, 1886

Juan Cristóbal (*né* Johannes Christoff) Gundlach (1810–1896)
was a German-born naturalised Cuban ornithologist and
collector. He learnt the art of dissection and taxidermy by
watching his older brother, who was a zoology student. An
event that nearly cost him his life ironically allowed him to
follow his chosen profession: while hunting he accidentally
discharged a small gun so close to his nose that he lost his
sense of smell. After that he could calmly dissect, macerate
and clean skeletons without difficulty. He was a Curator at
the University of Marburg and later at the Senckenberg
Museum in Frankfurt. He took part in a collecting expedition
to Cuba (1830), and stayed on and collected there and in
Puerto Rico. He wrote the first major work on the island's
birds, *Ornitología Cubana* (1876). He met American explorer
Charles Wright (q.v.) and explored the virgin forest that
became Alejandro de Humboldt National Park. He was
zealous and single-minded and tended to keep what he
collected, and described it for science himself. Two amphibians, two mammals and a reptile are named after him.

Gunn

Silver Gull ssp. *Chroicocephalus novaehollandiae gunni*
Mathews, 1912
Emu ssp. *Dromiceius novaehollandiae gunni* Mathews,
1922 NCR
[JS *Dromaius novaehollandiae diemenensis* – EXTINCT]

Ronald Campbell Gunn (1808–1881) emigrated (1829) from
Scotland to Australia, where he became a superintendent of
prisons and a police magistrate. However, it is as a botanist
and collector that he is remembered, having corresponded
with many of the greats of natural history of his time such as
J. E. Gray (q.v.). He was elected to the House of Assembly
(1885–1860) and subsequently became Deputy Commissioner
for Crown Lands in northern Tasmania. He edited the *Tasmanian Journal of Natural Science* (1842–1849). He was married
twice and had 12 children. When he died he left his herbarium

to the Royal Society of Tasmania, which transferred to the
National Herbarium, Sydney. William Jackson Hooker said of
him, 'Ronald Campbell Gunn ... to whose labours the Tasmanian Flora is so largely indebted, was the friend and
companion of the late Mr. Lawrence, from whom he imbibed
his love of botany. Between 1832 and 1850, Mr Gunn collected
indefatigably over a great portion of Tasmania ... There are
few Tasmanian plants that Mr Gunn has not seen alive, noted
their habits in a living state, and collected large suites of
specimens with singular tact and judgement. These have all
been transmitted to England in perfect preservation, and are
accompanied with notes that display remarkable powers of
observation, and a facility for seizing important characters in
the physiognomy of plants, such as few botanists possess.'
Various plants and a mammal are named after him.

Gunning

Sunbird genus *Gunningia* **J. A. Roberts**, 1922 NCR
[Now in *Anthreptes*]

Gunning's Robin *Sheppardia gunningi* Haagner, 1909
[Alt. East Coast Akalat]

Dr Jan Willem Bowdewyn Gunning (1860–1913) was a Dutch
physician. He went to South Africa (1884) and became
Director at the Transvaal Museum, Pretoria (1896), which
position he held until shortly before his death. He founded the
Pretoria National Zoo, which apparently started in his
garden. Gunning was co-founder of the African Ornithologists' Union. Austin Roberts (q.v.) worked under him. A
mammal is named after him.

Gunnison

Gunnison Sage Grouse *Centrocercus minimus* Young *et al.*,
2000

This bird is named after Gunnison County in Colorado, USA,
and not directly after John Williams Gunnison (1812–1853),
the American military officer and explorer.

Günther, A. C. L. G.

Myzomela sp. *Myzomela guentheri* **Gadow**, 1884 NCR
[Alt. Black-bellied Myzomela; JS *Myzomela erythromelas*]

Shining-blue Kingfisher ssp. *Alcedo quadribrachys
guentheri* **Sharpe**, 1892

Dr Albert Carl Ludwig Gotthilf Günther (1830–1914) was
primarily one of the giants of herpetology. He recognised
(1867) that the tuatara was not a lizard but an entirely separate order of reptiles. He was educated as a physician in
Germany, joining the British Museum (1856) and becoming
Keeper, Zoological Department (1857). He was naturalised
British (1862), changing his middle names to Charles Lewis.
He became President, Biological Section, British Association
for the Advancement of Science (1880) and was President of
the Linnean Society (1881–1901). He wrote *The Reptiles of
British India*. Three mammals, 26 amphibians and 67 reptiles
are named after him.

Günther (Niethammer)

> Common Linnet ssp. *Carduelis cannabina guentheri*
> **Wolters**, 1953

(See **Niethammer**)

Gurnet

> Thick-billed Seedeater ssp. *Serinus burtoni gurneti*
> **Gyldenstolpe**, 1926 NCR
> [JS *Serinus burtoni kilimensis*]

Dr Victor Gurnet Logan van Someren (See **Van Someren**).

Gurney

> Gurney's Eagle *Aquila gurneyi* **G. R. Gray**, 1861
> Gurney's Ground Thrush *Zoothera gurneyi* **Hartlaub**, 1864
> [Alt. Orange Ground Thrush]
> Gurney's Sugarbird *Promerops gurneyi* **J. Verreaux**, 1871
> Gurney's Pitta *Hydrornis gurneyi* **Hume**, 1875
> [Alt. Black-breasted Pitta; Syn. *Pitta gurneyi*]
> Giant Scops Owl *Mimizuku gurneyi* Tweeddale, 1879
> [Alt. Lesser Eagle Owl]
> Gurney's Buzzard *Buteo poecilochrous* Gurney, 1879
> [Alt. Puna Hawk]

> Pacific Baza ssp. *Aviceda subcristata gurneyi* **E. P. Ramsay**,
> 1882
> Black-necked Grebe ssp. *Podiceps nigricollis gurneyi*
> **J. A. Roberts**, 1919
> Crested Honey-buzzard ssp. *Pernis ptilorhynchus gurneyi*
> **Stresemann**, 1940 NCR
> [JS *Pernis ptilorhynchus ruficollis*]

John Henry Gurney (1819–1890) was a banker in Norwich, England, and an amateur ornithologist who worked at the BMNH. Most of his writing was on the birds of his own county, but he also wrote on collections of African birds, as well as editing the works of others, and he had a particular interest in birds of prey. His son (1848–1922), who shared the same name, was also an ornithologist.

Gustav (Garlepp)

> Gustav's Tanager *Malacothraupis gustavi* **Berlepsch**, 1901
> NCR
> [Alt. Slaty Tanager; JS *Creurgops dentatus*]

> Gustav's Parakeet *Brotogeris cyanoptera gustavi*
> Berlepsch, 1889
> [Alt. Cobalt-winged Parakeet ssp.]

(See **Garlepp**)

Gustav (Schlegel)

> Pechora Pipit *Anthus gustavi* **Swinhoe**, 1863

Gustaaf Schlegel (1840–1903) was a Dutch ornithologist and sinologist, and the son of Hermann Schlegel (q.v.). He made his first trip to China (1857) in order to collect bird specimens. He is credited with being the first European to document the Chinese origins of gunpowder. Gustaaf's magnum opus was his 4-volume Dutch-Chinese dictionary (1882–1891).

Gutiérrez

> Philippine Turtle Dove ssp. *Streptopelia dussumieri*
> *gutierrezi* **Hachisuka**, 1930 NCR
> [JS *Streptopelia bitorquata dusumieri*]

E. Gutiérrez (DNF) was a Filipino plant-hunter in the Philippines.

Guy

> Hummingbird genus *Guyornis* **Bonaparte**, 1854 NCR
> [Now in *Phaethornis*]

> Guy's Hermit *Phaethornis guy* **Lesson**, 1833
> [Alt. Green Hermit, White-tailed Hermit]

J. Guy (DNF) was a French naturalist, about whom little is known.

Gwendoline

> Banded Lapwing ssp. *Zonifer tricolor gwendolenae*
> **Mathews**, 1912 NCR
> [JS *Vanellus tricolor*; NRM]
> Crested Tern ssp. *Thalasseus bergii gwendolenae*
> Mathews, 1912
> [SII *Thalasseus bergii cristatus*]
> White-browed Babbler ssp. *Pomatostomus superciliosus*
> *gwendolenae* Mathews, 1912 NCR
> [JS *Pomatostomus superciliosus superciliosus*]

Gwendoline Carter (DNF) was a daughter of Thomas Carter (q.v.), a British naturalist and pastoralist in Australia.

Gyldenstolpe

> Tupana Scythebill *Campylorhamphus gyldenstolpei* Aleixo
> *et al.*, 2013

> Grey-headed Woodpecker ssp. *Picus canus gyldenstolpei*
> **E. C. S. Baker**, 1918 NCR
> [NUI *Picus canus hessei*]

Count Nils Carl Gustaf Fersen Gyldenstolpe (1886–1961) was a Swedish zoologist and ornithologist who was attached to the Riksmuseum in Stockholm (1914–1961). He travelled in Siam (Thailand) (1911), Central Africa (1921) and New Guinea (1951). An amphibian and a reptile are named after him.

H

Haagner

Sunbird genus *Haagneria* **J. A. Roberts**, 1924 NCR
[Now in *Cyanomitra*]

Haagner's Robin Chat *Cossypha haagneri* **Gunning**, 1901
NCR
[Hybrid: *Cossypha dichroa* x *Cossypha natalensis*]
Whydah sp. *Microchera haagneri* Roberts, 1926 NCR
[Hybrid: *Vidua paradisaea* x *Vidua chalybeata*?]

Alwin Karl Haagner (1880–1962) was a South African ornithologist and naturalist. He was at the Transvaal Museum (1906–1911) and thereafter Director of the Pretoria Zoological Gardens. He wrote *Sketches of South African Bird-Life* (1914).

Haast

New Zealand Rockwren *Xenicus haasti* **W. L. Buller**, 1869
NCR
[JS *Xenicus gilviventris*]
Great Spotted Kiwi *Apteryx haastii* **T. H. Potts**, 1872
Haast's Eagle *Harpagornis moorei* Haast, 1872 EXTINCT

Dr Sir Johann Franz Julius von Haast (1822–1887), was a German-born geologist. He became a naturalised New Zealander, where he worked for the Canterbury provincial government. He was instrumental in founding the Canterbury Museum, becoming its first Director. He described the eagle (1872) from bones discovered a few years earlier at the Glenmark Station (see **Moore, G. H.**). He was the German Consul in New Zealand (1880) and the first New Zealander to be awarded the Royal Geographical Society's Gold Medal, for his work on moas. The Haast Pass in the Southern Alps, the Haast River, and the town of Haast, all in New Zealand, are named after him.

Habel

Large Tree Finch ssp. *Camarhynchus psittacula habeli*
P. L. Sclater & **Salvin**, 1870

Dr Simeon Habel (DNF) was an Austrian physician, traveller and naturalist who explored in Central and South America, spending six months (1868) on the Galapagos Islands. He wrote *The Sculptures of Santa Lucia Cosumalwhuapa in Guatemala* (1878). A reptile is named after him.

Habenicht

Habenicht's Pitta *Erythropitta habenichti* **Finsch**, 1912
[Alt. Red-bellied Pitta; Syn. *Erythropitta erythrogaster habenichti*]

Captain P. Habenicht (fl.1911) of the Imperial German Navy was on the survey vessel *Planet* in German New Guinea, the Bismarck Archipelago and Palau (1907–1911). He and his crew were relieved and returned to Germany (1912). He collected the holotype (1911).

Haberer

Grey-streaked Flycatcher ssp. *Muscicapa griseisticta haberi* **Parrot**, 1907 NCR; NRM

Dr Karl Albert Haberer (1864–1941) was a German naturalist, student of politics, and collector in Japan and China. The Boxer Rebellion (1899) forced him to stay in Peking (Beijing) for longer than he planned, but he bought some 'dragon bones' (fossilised bones used in traditional Chinese medicine) in a market. Among them was a human-like molar that eventually led to the discovery 'Peking Man' (*Homo erectus pekinensis*). He wrote *Schädel und Skeletteile aus Peking* (1902).

Hachisuka

Hachisuka's Sunbird *Aethopyga primigenia* Hachisuka, 1941
[Alt. Grey-hooded Sunbird]

Purple Sunbird ssp. *Cinnyris asiaticus hachisukai* **Delacour** & **Jabouille**, 1928 NCR
[JS *Cinnyris asiaticus intermedius*]
Green Kingfisher ssp. *Chloroceryle americana hachisukai* **Laubmann**, 1941
Mountain Shrike ssp. *Lanius validirostris hachisuka* **Ripley**, 1949
Olive-winged Bulbul ssp. *Pycnonotus plumosus hachisukae* **Deignan**, 1952 NCR
[JS *Pycnonotus plumosus insularis*]

Masauji Hachisuka, 18th Marquis Hachisuka (1903–1953), was a Japanese ornithologist who studied biology at Cambridge. He led a 50-man expedition to Mount Apo's summit, Mindanao, Philippines (1929), and established a bird-collecting centre at Galong. He wrote *A Handbook of the Birds of Iceland* (1927) and *Birds of the Philippine Islands* (1931–1935).

Hachlow

Horned Lark ssp. *Eremophila alpestris hachlowi* **Meise**, 1932 NCR

[JS *Eremophila alpestris brandti*]

Professor Vitaly Andreyevich Khokhlov (1890–1983) was a Russian ornithologist, palaeontologist and comparative anatomist who was at Tomsk University (1924–1928). He was marginalised after expressing his belief in the existence of the Yeti.

Hadden

Bougainville Bush Warbler *Cettia haddeni* **LeCroy** & Barker, 2006

[Alt. Odedi]

Don Hadden is a New Zealand retired schoolteacher, ornithologist and wildlife photographer. He lives in an Aboriginal community near Alice Springs, Northern Territory, Australia. He lived in Bougainville, Papua New Guinea, where he taught at Panguna International School (1976–1980 and 1999–2002). He is the author of several books including *Birds and Bird Lore of Bougainville and the North Solomons* (2004) and *Birds of the Outback* (2010). He put in much effort trying to find the warbler after local people reported its existence.

Hadi

Abyssinian Ground Thrush ssp. *Zoothera piaggiae hadii* **J. D. Macdonald**, 1940

Muhammad Abdel Hadi (DNF) was a Sudanese collector and taxidermist at the Sudan Government Museum (1940).

Haeberlin

Emerald sp. *Chlorostilbon haeberlini* **Reichenbach**, 1855 NCR

[Alt. Red-billed Emerald; JS *Chlorostilbon gibsoni chrysogaster*]

Ernst Haeberlin (1819–1895) was a German palaeontologist and collector in Santa Marta, Colombia (c.1854).

Haffer

Antbird genus *Hafferia* **Isler**, Bravo & Brumfield, 2013 NCR

[JS of *Akletos* (Dunajewski, 1948)]

Campina Jay *Cyanocorax hafferi* Cohn-Haft et al., 2013

Dr Jürgen Haffer (1932–2010) was a German ornithologist, geologist and biogeographer. The University of Göttingen awarded his doctorate (1957). He worked as a field geologist for Mobil Oil and lived in countries as diverse as Iran, Egypt, Norway and Colombia. He devised the Amazonian refugia theory to explain the rapid diversification of Neotropical fauna during the Pleistocene. He was a friend of Erwin Stresemann (q.v.) and worked closely with Ernst Mayr (q.v.), of whom he wrote a full-length biography (2007).

Hafiz

Common Nightingale ssp. *Luscinia megarhynchos hafizi* **Severtzov**, 1873 NCR

[JS *Luscinia megarhynchos golzii*]

Muhammad Shams ud-Din Hafiz (1326–1390), also known as Shamsuddin Muhammad Hafiz, etc., was a Persian lyric poet and Sufi philosopher.

Hagen

Little Green Pigeon ssp. *Treron olax hageni* **Parrot**, 1907 NCR; NRM

Velvet-fronted Nuthatch ssp. *Sitta frontalis hageni* Parrot, 1908 NCR

[JS *Sitta frontalis saturatior*]

Dr Bernhard Hagen (1853–1919) was a German physician and amateur naturalist. After studying medicine at Munich University he was employed by a planting company in Sumatra. Here he made several, mainly zoological, collecting expeditions. The Astrolabe Company employed him in New Guinea (1893–1895), then he returned to Germany (1895), but revisited New Guinea (1905) with his wife. He was a section head (1897–1904) at the Senckenberg Museum in Frankfurt, founding their Ethnology Department. He published widely on zoology, geography and ethnography. Two mammals and a reptile are named after him.

Hagenbeck

Common Pheasant ssp. *Phasianus colchicus hagenbecki* **Rothschild**, 1901

Carl Hagenbeck (1844–1913) was German; a Hamburg-based animal dealer, circus proprietor and founder of Tierpark Hagenbeck (1907). Many consider him the father of the modern zoo, because he introduced 'natural-looking' animal enclosures without bars. He was a pioneer in human tableaux, depicting Samis, Samoans, Nubians and Inuits, etc. in their natural environments.

Hagmann

Antshrike sp. *Sakesphorus hagmanni* **Miranda-Ribeiro**, 1927

[Alt. Glossy Antshrike; JS *Sakesphorus luctuosus*]

Dr Gottfried (Godofredo) A. Hagmann (1874–1946) was a Swiss-born Brazillian zoologist who spent many years as Chief Zoologist, then Director, Museu Paraense Emilio Goeldi, Belém, Brazil (1899–1904). He collected the holotype. He wrote 'Die Eier von *Caiman niger*' (1906). A reptile is named after him.

Hahn

Hahn's Macaw *Diopsittaca nobilis* **Linnaeus**, 1758

[Alt. Red-shouldered Macaw; *Psittacara hahni* Souancé, 1856 is a JS]

Dr Carl Wilhelm Hahn (1786–1835) was a German (Bavarian) zoologist, the son of a palace gardener from Hungary. He called himself a 'nature researcher' and he was very proficient at drawing, but did not colour his illustrations. He moved from Fürth to Nürnberg (Nuremberg) (1816) as there were no competent illustrators/colourists to be found in Fürth. Many of his drawings were of insects, arachnids and fish (being primarily an entomologist), but most of his writings were on spiders, including his *Monagraphie der Spinnen* (1820). Hahn published an ornithological atlas (1834) and wrote *Die Vögel aus Asien, Afrika, America und Neuholland* (1819–1836). He once complained to the Bavarian Public Natural History Society that he had spent all his money in the pursuit of science and art. He died in Nürnberg of a pulmonary infection.

Hall, H.

Hall's Babbler *Pomatostomus halli* Cowles, 1964
[Alt. White-throated Babbler]

Major Harold Wesley Hall (1888–1964) was an Australian zoologist, collector, explorer, philanthropist and entrepreneur who promoted 'Hall's fortified wines'. He supported natural history research, having made a fortune through the Mt Morgan goldfields. J. D. Macdonald (q.v.), head of the Bird Room at the British Museum (Natural History), approached him (1961) for sponsorship. Macdonald secured £25,000 over five years to undertake the museum's 'Harold Hall Expeditions'. Macdonald led the first of five such bird-collecting expeditions to Australia (1963) to rebuild the Australian bird collections as Mathews and Rothschild collections had been sold to American institutions. Other Bird Room staff led the other expeditions, including Mrs Pat Hall (q.v.), who subsequently edited a book on the findings of all five expeditions, and Peter Colston (q.v.).

Hall, P.

Hall's Greenbul *Andropadus hallae* **Prigogine**, 1972 NCR
[Alt. Mrs Hall's Greenbul; Syn. *Eurillas virens hallae*]

Piping Cisticola ssp. *Cisticola fulvicapilla hallae*
  C. W. Benson, 1955
Ochraceous Bulbul ssp. *Alophoixus ochraceus hallae*
  Deignan, 1956
Long-billed Pipit ssp. *Anthus similis hallae* **C. M. N. White**,
  1957 NCR
[NUI *Anthus similis dewittei*]
African Reed Warbler ssp. *Acrocephalus baeticatus hallae*
  C. M. N. White, 1960
Brown-chested Alethe ssp. *Pseudalethe poliocephala*
  hallae **Traylor**, 1961

Beryl Patricia ('Pat') Hall (1917–2010) was an ornithologist and a voluntary assistant, later Honorary Associate, of the BMNH Bird Room (1947–1971). She took part in museum expeditions to South West Africa (Namibia) (1949–1950) with J. D. Macdonald (q.v.) and Colonel Cave (q.v.), to Bechuanaland (Botswana) (1950, 1957 and 1962), and to Angola (1957). She also led the third Harold Hall expedition to Australia (1965). She was co-author of *An Atlas of Speciation in African Passerine Birds* (1970) and edited *Birds of the Harold Hall Australian Expeditions 1962–1970* (1974). She wrote two autobiographical books; *What a Way to Win a War* (1978) and *A Hawk from a Handsaw* (1993). She was awarded the Union Medal of the British Ornithologists' Union (1973). (See **Pat** and **Mrs Hall**)

Hall, R

Fairy-wren genus *Hallornis* **Mathews**, 1912 NCR
[Now in *Malurus*]

Hall's Giant Petrel *Macronectes halli* Mathews, 1912
[Alt. Northern Giant Petrel]

Australasian Bushlark ssp. *Mirafra javanica halli* **Bianchi**,
  1907
King Penguin ssp. *Aptenodytes patagonicus halli* Mathews,
  1911
[*A. patagonicus* sometimes regarded as monotypic]
Little Kingfisher ssp. *Ceyx pusillus halli* Mathews, 1912
Scarlet-chested Parrot ssp. *Neophema splendida halli*
  Mathews, 1916 NCR; NRM

Robert Hall (1867–1949) was an Australian naturalist and ornithologist. He was a founder member of the RAOU (President 1912–1913), Curator of the Tasmanian Museum and Botanical Gardens (1908–1912) and thereafter an orchard grower in Tasmania. Hall was a keen collector and parts of his private bird and egg collections are in the Tasmanian Museum and the Australian National Museum, Melbourne. His Siberian collection with Ernie Trebilcock (1903) is held at the BMNH at Tring, England. He wrote a number of papers including 'Notes on a collection of bird skins from the Fitzroy River, northwestern Australia' (1902).

Hallinan

Scale-throated Earthcreeper ssp. *Upucerthia dumetaria*
  hallinani **Chapman**, 1919 NCR
[NUI *Upucerthia dumetaria hypoleuca*]

Thomas Hallinan (1882–1950) was an American engineer, herpetologist, ornithologist and collector in tropical America and Asia. He worked on the construction and mainanance of the Panama Canal and collected there (1915 and 1924) and in Chile (1917). He wrote 'Notes on some Panama Canal Zone birds' (1924).

Halls

Yariguíes Slate-crowned Antpitta *Grallaricula nana hallsi*
  Donegan, 2008

Alan G. Halls (d.2005) was a British birdwatcher whose death occurred around the time that the bird was discovered. He wrote articles on birds for his local newspaper, *Caversham Bridge*. Donegan wrote that Alan Halls had enthusiasm for tutoring younger birders, including himself.

Hallstrom

Petrel genus *Hallstroma* **Mathews**, 1943 NCR
[Now in *Pterodroma*]

Hallstrom's Bird-of-Paradise *Pteridophora alberti hallstromi*
  Rothschild, 1931 NCR; NRM
[Alt. King of Saxony Bird-of-Paradise ssp.]
Hallstrom's Parrot *Psittacella madaraszi hallstromi* **Mayr** &
  Gilliard, 1951
[Alt. Madarász's Tiger Parrot ssp.]

Sir Edward John Lees Hallstrom (1886–1970) was born in Coonamble, New South Wales, Australia. He was a pioneer of refrigeration, a philanthropist and leading aviculturist. He began work in a furniture factory aged 13, but later opened his own factory to make ice-chests and then wooden cabinets for refrigerators. He eventually designed and manufactured the first popular domestic Australian refrigerator. Hallstrom made generous donations to medical research, children's hospitals and the Taronga Zoo in Sydney, becoming an honorary life director there. He commissioned (1940) the artist Cayley (q.v.) to paint all the Australian parrots. Twenty-nine large watercolours were produced and presented to the Royal Zoological Society of New South Wales. He visited New Guinea, home of his eponymous parrot (1950). There is a research collection of 1,600 rare books on Asia and the Pacific at the University of New South Wales Library known as the Hallstrom Pacific Collection, purchased with funds Hallstrom gave to the Commonwealth government (1948) for the purpose of establishing a library of Pacific affairs and colonial administration. Among the rare books in the collection is John Gould's *Birds of New Guinea*. He established (1950s) a centre in the southern highlands of Papua New Guinea, ostensibly to introduce local sheep farming, but really to have a base from which a well-known collector, Fred W. Shaw Mayer (q.v.) could devote all his time to collecting bird-of-paradise specimens for him and the Taronga Zoo. However, there was corruption which involved trafficking in rare and endangered species during his time as director there and he stands accused of giving in to pressure to appoint the officials who carried out this trade. A mammal is also named after him.

Halsey

Halsey's Warbler *Dendroica nigrescens halseii* **Giraud**, 1841
[Alt. Black-throated Grey Warbler ssp.]

Abraham Halsey (1790–1857) was a cashier in a foundry, then a clerk and cashier in a bank in New York. He was elected as a member of the Brooklyn Lyceum of Natural History on publishing a paper on lichen (1818). He was vice-president (1825–1833) at the same time as Giraud (q.v.) was treasurer of the Lyceum. Giraud made a collection of birds in Texas (1838) and published (1842) a book with plates drawn by Halsey (1841), in recognition of which Giraud named the warbler.

Halsuet

Halsuet's Wren-babbler *Spelaeornis troglodytoides halsueti* **David**, 1875
[Alt. Bar-winged Wren-babbler ssp.]

Halsuet (fl.1850) was a French missionary at Tsinling, China.

Hambroek

Mountain Scops Owl ssp. *Otus spilocephalus hambroecki* **Swinhoe**, 1870

Antonius Hambroek (1607–1661) was a Dutch missionary (1648–1661) to Taiwan, where the owl is found. Zheng Chenggong – the Koxinga (literally 'Lord with the Imperial Surname') who took over the island – martyred him. During the wresting of the island from the Dutch East India Company, Hambroek and his family were taken prisoner and sent to the Governor with a demand that he surrender the garrison. Hambroek was told he would die if he came back with a displeasing answer. As the Governor refused to surrender, Hambroek was executed on his return.

Hamerton

Lesser Hoopoe Lark *Alaemon hamertoni* **Witherby**, 1905

Rufous-tailed Scrub Robin ssp. *Erythropygia galactotes hamertoni* **Ogilvie-Grant**, 1906

Colonel Albert Ernest Hamerton (1873–1959) was an English army medical officer who explored in Somaliland (1904–1906). He was in Uganda and Nyasaland (now Malawi) (1908–1913) and kept notes on the incidence of sleeping sickness there.

Hamilton, A.

Southern Cassowary ssp. *Casuarius casuarius hamiltoni* **Mathews**, 1915 NCR
[JS *Casuarius casuarius johnsonii*; species often regarded as monotypic]

Alexander Greenlaw Hamilton (1852–1941) was an Irish-born schoolteacher (1866–1920), botanist and naturalist. He went to Australia (1866) and became a schoolteacher aged 15! He taught in several schools, becoming headmaster at Mount Kembla Public School (1887) and of Willoughby Public School (1905). He was President of the Linnean Society of New South Wales (1916).

Hamilton, C.

Starfrontlet sp. *Helianthea hamiltoni* **Goodfellow**, 1900 NCR
[Alt. Buff-winged Starfrontlet; JS *Coeligena lutetiae*]

Claud Hamilton (DNF) was an explorer who collected in Colombia and Ecuador (1898–1899).

Hamilton, F.

Hamilton's Pheasant *Lophura leucomelanos hamiltonii* **J. E. Gray**, 1829
[Alt. Kalij Pheasant ssp.]

Dr Francis Hamilton *né* Buchanan (1762–1829) joined the Bengal service of the Honourable East India Company as an assistant surgeon (1815), dropping his family name at the same time. He was often employed on survey work on all manner of subjects including fisheries, and today is regarded as an ichthyologist – a number of fish and a reptile are named after him.

Hamilton, J. E.

Tristan Skua *Stercorarious antarcticus hamiltoni* **Hagen**, 1952
[Alt. Brown Skua ssp.]

Dr James Erik Hamilton (1891–1957) was a British scientist who was resident in the Falkland Islands (1919–1957), being Government Naturalist there (1919–1949). He was seconded to the Discovery Expeditions for 14 years. He wrote a number of papers on the southern sealion and the leopard seal, as well as co-authoring 'The birds of the Falkland Islands' in *Ibis* (1961).

Hamilton (Bartlett Mathews)

Inland Thornbill ssp. *Acanthiza albiventris hamiltoni*
  Mathews, 1911 NCR
[JS *Acanthiza apicalis albiventris*]

Hamilton Bartlett Mathews (1873–1959) was an Australian surveyor, son of Robert Hamilton Mathews (q.v.) and the elder brother of the describer, Gregory Macalister Mathews (q.v.).

Hamilton (Robert Mathews)

Magpie Goose ssp. *Anseranas semipalmata hamiltoni*
  Mathews, 1912 NCR; NRM

Middle name of Robert Mathews (see also **Robert (Mathews)**)

Hamlin

Rennell Shrikebill *Clytorhynchus hamlini* **Mayr**, 1931

Grey-throated White-eye ssp. *Zosterops ugiensis hamlini*
  R. C. Murphy, 1929
Island Leaf Warbler ssp. *Phylloscopus poliocephalus hamlini* Mayr & **Rand**, 1935

Dr Hannibal Hamlin (1904–1982) was a collector in the Pacific, leading the Whitney South Seas Expedition (1927–1930). His namesake ancestor was Abraham Lincoln's Vice President. He graduated from Yale (1927) and its medical school (1936). In Mayr's (q.v.) 'Birds collected during the Whitney South Sea Expedition – 14' (1931) he wrote on the geography of Rennell Island and the ecology of its birdlife. During WW2 he was a naval surgeon, having joined the staff of Massachusetts General Hospital (1939) where he returned after the war, rising to Chief of Neurosurgery (1977–1979) before retiring.

Hammond, W. A.

Hammond's Flycatcher *Empidonax hammondii* **Xantus**, 1858

Dr William Alexander Hammond (1828–1900) was an American physician, naturalist and soldier. He started collecting (c.1847) on the Pacific Railroad Survey for the USNM. The University of New York awarded his MD (1848) and he was in private practice for a short while after his internship, then joined the army as Assistant Surgeon (1849). For the next ten years he served at various frontier stations, collecting all the while. He resigned (1859) from the army and became Professor of Anatomy and Physiology at the University of Maryland. During the Civil War (1861) he re-joined as Surgeon-General of the United States Army and starting making changes in the medical department. He clashed with Edward Stanton, Secretary of War, and was court-marshalled (1864) but later exonerated (1878). He was a lecturer (1865–1867) on nervous and mental diseases at the College of Physicians and Surgeons in New York, then became Professor of Nervous and Mental Diseases at the Bellevue Hospital Medical College (1867–1873) and at the university

(1874). He was one of the founders (1882) of the New York Medical School, later returning to private practice (1888). Hammond collected birds for Baird (q.v.), and arranged for Xantus (q.v.) to work with for him, so Xantus named the flycatcher in gratitude. An amphibian and a reptile are named after him.

Hammond, W. O.

Hammond's Petrel *Puffinus mauretanicus* **P. R. Lowe**, 1921
[Alt. Balearic Shearwater]

William Oxenden Hammond (1817–1903) was a taxidermist and naturalist who lived and worked in Canterbury, England, and inherited the senior partnership in a bank from his father (1863). He was educated at Harrow and Balliol College, Oxford, and briefly served in the 17th Lancers and the Rifle Brigade. He travelled widely in Europe and to Canada and Egypt and collected birds eggs. He was a president of the East Kent Natural History and Scientific Society. He collected birds to paint them and is noted for his watercolour portraits of 374 British birds in 17 volumes. He wanted to portray all the birds found in Britain, but sometimes took shooting expeditions in Europe (i.e. Netherlands) to get specimens. They were never exhibited publicly, yet were particularly admired by Edward Wilson (q.v.), the accomplished bird artist, who commented on their 'refinement, truth and exactness'. Being wealthy, he was a hobbyist rather than a professional artist. He bequeathed his stuffed birds to the Canterbury Museum, to which his descendants sold the portraits (almost miniatures).

Hance

Hance's Thrush *Zoothera dauma hancii* **Swinhoe**, 1863
[Alt. Scaly Thrush ssp. SII *Zoothera dauma dauma*]

Dr Henry Fletcher Hance (1827–1886) was a British diplomat in China (1844–1886). He was a collector and wrote on the flora of Hong Kong.

Handley

Escudo Hummingbird *Amazilia tzacatl handleyi* **Wetmore**, 1963

Charles Overton Handley Jr (1924–2000) was primarily a mammalogist. He graduated from Virginia Tech (1944) and after WW2 service in Europe became a collector of birds for the USNM, making four expeditions to the High Arctic, one to Labrador, one to Guatemala, and a seven-month expedition to the Kalahari Desert in southern Africa. After completing his MSc and PhD at the University of Michigan (1955), he became Curator, Department of Vertebrate Zoology, USNM. He made numerous visits to Panama (1957–1967), compiling an inventory of all its mammals. He made a similar inventory in Venezuela and organised a bat-trapping programme to study their movements. He continued with trips to an island off Panama (1993). He worked at the museum for 53 years. He published 188 scientific papers, including the posthumous 'A new species of three-toed sloth (Mammalia: Xenarthra) from Panama, with a review of the genus *Bradypus*' (2001), as well as two books. The museum created The Handley Memorial

Fund in his honour, which supports research in mammalogy and tropical biology. Six mammals are named after him.

Haniel

> Blood-breasted Flowerpecker ssp. *Dicaeum sanguinolentum hanieli* **Hellmayr**, 1912
> Australian Hobby ssp. *Falco longipennis hanieli* Hellmayr, 1914
> Oleaginous Hemispingus ssp. *Hemispingus frontalis hanieli* Hellmayr & **Seilern**, 1914

Dr Curt Berthold Haniel (1886–1951) was a German zoologist and collector in tropical America and the East Indies (Timor), who worked at Zoologische Staatssammlung, Munich. He wrote 'Variationsstudie an timoresischen Amphidromusarten' (1921).

Hannah

> Blue-necked Tanager ssp. *Tangara cyanicollis hannahiae* **Cassin**, 1865

Mrs Hannah Cassin (c.1806–1888) *née* Wright was the wife (1837) of John Cassin, who described the tanager. Her birth year is uncertain – some sources give c.1813 – but she was said to be 82 when she died – her headstone reveals 15th September as the day, but the year is hard to decipher.

Hannum

> Masked Tityra ssp. *Tityra semifasciata hannumi* **Van Rossem & Hachisuka**, 1937

Robert G. Hannum (fl.1959) was an American zoologist who worked at the Western Foundation of Vertebrate Zoology, Los Angeles, and collected the holotype with Van Rossem in Mexico (1937). He was an inspector for the Bureau of Biological Survey (1939).

Hans Günther

> Eurasian Jay ssp. *Garrulus glandarius hansguentheri* Keve, 1967 NCR
> [JS *Garrulus glandarius anatoliae*]

Professor Günther Niethammer (q.v. under **Günther** and **Niethammer**) and Dr Hans Kumerloeve (q.v.).

Harcourt

> Harcourt's Storm-petrel *Oceanodroma castro* Harcourt, 1851
> [Alt. Madeiran Storm-petrel, Band-rumped Storm-petrel]

Edwin Vernon Harcourt (1825–1891) was an English naturalist. He described the petrel in his book *Sketch of Madeira Containing Information for the Traveller or Invalid Visitor.* Darwin (q.v.), in *The Origin of Species*, refers to him as a man who was 'very knowledgeable about Madeira'.

Hardwick[e], C. B.

> Latham's Snipe *Gallinago hardwickii* **J. E. Gray**, 1831

Captain Charles Browne Hardwick[e] (1788–1880) was a British collector who was one of the first to settle in Tasmania

(1816), where he became a farmer. He served (1807–1813) in the Royal Navy, leaving as a lieutenant. He arrived in Australia (1814) as third officer on the convict transport *General Hewitt.* He then took command of the cutter *Elizabeth*, which traded regularly between Sydney and Van Diemen's Land, and was wrecked in a storm (1816) with no loss of life. He sailed around the coast for five months (1823) and reported that it looked very inhospitable. He discovered the snipe in Tasmania (part of its wintering grounds).

Hardwicke, T.

> Hardwicke's Leafbird *Chloropsis hardwickii* **Jardine & Selby**, 1830
> [Alt. Orange-bellied Leafbird]
> Bay-backed Shrike *Lanius hardwickii* **Vigors**, 1831 NCR
> [JS *Lanius vittatus*]
>
> Brown-capped Woodpecker ssp. *Dendrocopos nanus hardwickii* **Jerdon**, 1845
> [Syn. *Dendrocopos moluccensis hardwickii*, or included in *D. m. nanus*]

Major-General Thomas Hardwicke (1756–1835) served with the Bengal Artillery, which in his day was a regiment in the Bengal Army of the Honourable East India Company. He was an amateur naturalist who is credited with being the first to make the red panda widely known, through a paper which he presented to the Linnean Society of London (1821): 'Description of a new genus … from the Himalaya chain of hills between Nepaul [sic] and the Snowy Mountains'. Two mammals and five reptiles are named after him.

Hardy, E. C.

> Hardy's Lemon-bellied Crombec *Sylvietta denti hardyi* **Bannerman**, 1911

Rear-Admiral E. C. Hardy (1866–1934) was a Royal Navy hydrographer in China (1882) and in British Columbia on board HMS *Egeria* (1900–1901). He was promoted to Vice-Admiral on retirement (1926). He invited Willoughby P. Lowe (q.v.) to accompany him as naturalist on a survey of the West African coast. Lowe obtained the type specimen of the crombec in Sierra Leone.

Hardy, J. W.

> Hardy's Pygmy Owl *Glaucidium hardyi* **Vielliard**, 1989
> [Alt. Amazonian Pygmy Owl]
>
> Red-bellied Woodpecker ssp. *Melanerpes carolinus harpaceus* **W. Koelz**, 1954
> White-throated Jay ssp. *Cyanolyca mirabilis hardyi* **A. R. Phillips**, 1966 NCR; NRM

Dr John William (Bill) Hardy (1930–2012) was an American ornithologist and author. After graduating from the Southern Illinois University he completed an MA at Michigan State University and a PhD at the University of Kansas. Thereafter he was Professor of Biology at Occidental College, Los Angeles (1961–1973). He went on to be Curator (and Curator Emeritus) of Ornithology and Bioacoustics at the Florida Museum of Natural History, at the University of Florida

(1973–1990). Hardy was a pioneer of studies of vocal and social behaviour in jays and parrots, but he is best known for his series of more than two dozen 'sound monographs' issued on audio cassette, including *Sounds of Florida Birds* (1998), and *Voices of the New World Owls* (1999). He was the founder of the Florida Museum of Natural History Bioacoustic Laboratory and Archive, which has grown to become the second largest (in number of bird species represented) in the Western Hemisphere. Koelz's name *harpaceus* is a compound of three people's names: J. W. Hardy, William Pielou (q.v.) and George J. Wallace (q.v.).

Hare

Fawn-coloured Lark ssp. *Calendulauda africanoides harei*
J. A. Roberts, 1917
Cardinal Woodpecker ssp. *Dendropicos fuscescens harei*
J. A. Roberts, 1924 NCR
[NUI *Dendropicos fuscescens fuscescens*]
Large-billed Lark ssp. *Galerida magnirostris harei*
J. A. Roberts, 1924
Ferruginous Lark sp. *Pseudammomanes harei* J. A. Roberts, 1937 NCR
[Alt. Red Lark; JS *Calendulauda burra*]

H. Leighton Hare (fl.1932) was a South African ornithologist and collector, who wrote 'The birds of Philipstown, Cape Province, with notes on their habits' (1915).

Hargitt

Woodpecker sp. *Melanerpes hargitti* **A. J. C. Dubois**, 1899 NCR
[Hybrid between morphs of *Melanerpes cruentatus*]

White-bellied Woodpecker ssp. *Dryocopus javensis hargitti*
Sharpe, 1884
Golden-headed Quetzal ssp. *Pharomachrus auriceps hargitti*
Oustalet, 1891
Fine-spotted Woodpecker ssp. *Campethera punctuligera hargitti* Sharpe, 1902 NCR
[JS *Campethera punctuligera balia*]

Edward Hargitt (1835–1895) was a Scottish ornithologist and landscape painter. He became an expert on woodpeckers and wrote the Picidae in the *Catalogue of the Birds in the British Museum* (1890).

Hargrave

American Crow ssp. *Corvus brachyrhynchos hargravei* **A. R. Phillips**, 1942
Scaled Quail ssp. *Callipepla squamata hargravei* Rea, 1973

Dr Lyndon Lane Hargrave (1896–1978) was an American anthropologist, archaeologist and ornithologist with Arizona National Parks Service, and Director of the Museum of Northern Arizona. He was Research Professor of Ethnobiology, Prescott College, Arizona, when an honorary doctorate was awarded to him by Northern Arizona University (1971). He wrote *Mexican Macaws: Comparative Osteology and Survey of Remains from the Southwest* (1970).

Harington, D. C.

Brown-cheeked Fulvetta ssp. *Alcippe poioicephala haringtoniae* **Hartert**, 1909

Mrs Dorothy Caroline Harington *née* Pepys (1879–1969) was the wife of Lieutenant-Colonel H. H. Harington (below).

Harington, H. H.

Bulbul genus *Haringtonia* **Mathews** & **Iredale**, 1917 NCR
[Now in *Hypsipetes*]

Silver Pheasant sp. *Gennæus haringtoni* **Oates**, 1910 NCR
[Hybrid of *Lophura nycthemera* races]
Harington's Babbler *Stachyridopsis ambigua* Harington, 1915
[Alt. Buff-chested Babbler]

Eurasian Jay ssp. *Garrulus glandarius haringtoni* **Rippon**, 1905
Indian Spot-billed Duck ssp. *Anas poecilorhyncha haringtoni* Oates, 1907
Grey Bushchat ssp. *Saxicola ferreus haringtoni* **Hartert**, 1910 NRM
Rusty-cheeked Scimitar Babbler ssp. *Pomatorhinus erythrogenys haringtoni* **E. C. S. Baker**, 1914 NCR
[JS *Pomatorhinus erythrogenys ferrugilatus*]
Tree Pipit ssp. *Anthus trivialis haringtoni* **Witherby**, 1917
Blunt-winged Warbler ssp. *Acrocephalus concinens haringtoni* Witherby, 1920

Lieutenant-Colonel Herbert Hastings Harington (1868–1916) was an English career soldier and amateur naturalist who was killed in action in Mesopotamia. He became interested in birds when posted to Burma (1890s) and wrote *Harington's Birds of Burma* (1909). He discovered several new bird subspecies including the duck and pipit.

Harlan

Harlan's Hawk *Buteo jamaicaensis harlani* **Audubon**, 1830
[Alt. Red-tailed Hawk ssp.]

Dr Richard Harlan (1796–1843) was a physician. Audubon (q.v.) named this subspecies after his friend, saying upon bestowing the honour: 'This I might have done sooner, had not I waited until a species should occur, which in size and importance should bear some proportion to my gratitude toward that learned and accomplished friend.' Harlan, a young Quaker doctor, was a supporter of Audubon in the bitter feud that developed among members of the Philadelphia Academy of Natural Sciences (1823) between followers of the deceased Alexander Wilson (q.v.) – including George Ord (q.v.), who had taken on the task of finishing Wilson's bird volumes – and Audubon's supporters, who favoured his more 'artistic' style of bird painting. Ord's campaign against Audubon succeeded in denying him membership of the academy for many years. Harlan was also an amateur palaeontologist and helped discover the first American plesiosaur. He also published *Fauna Americana* (1825). Three mammals and an amphibian are named after him.

Harman

Harman's Eared Pheasant *Crossoptilon harmani* **Elwes,**
1881
[Alt. Tibetan Eared Pheasant]

Captain Henry John Harman (1850–1883) was a surveyor and
officer in the British Indian army.

Harmand

White-rumped Falcon ssp. *Polihierax insignis harmandi*
Oustalet, 1876
Lesser Yellownape ssp. *Picus chlorolophus harmandi*
Delacour, 1927 NCR
[NPRB *Picus chlorolophus krempfi*. Probably a JS of
P. chlorolophus annamensis]

Dr François-Jules Harmand (1845–1921) was a French Navy
surgeon-naturalist and explorer in Indochina (1873–1877). He
was Civil Commissioner-General in Tonkin (1883), held
consular posts in Thailand (1881), India (1885) and Chile
(1890), and was French Ambassador to Japan (1894–1906).

Härms

Buff-bellied Pipit ssp. *Anthus rubescens haermsi* **Zarudny,**
1908 NCR
[NUI *Anthus rubescens japonicus*]
Common Reed Bunting ssp. *Emberiza schoeniclus harmsi*
Zarudny, 1911 NCR
[JS *Emberiza schoeniclus pyrrhuloides*]
Tawny Owl ssp. *Strix aluco harmsi* Zarudny, 1911
Rock Sparrow ssp. *Petronia petronia harmsi* Keve, 1948
NCR
[JS *Petronia petronia intermedia*]

Michael Härms (1878–1941) was a Baltic-German zoologist
and ornithologist who collected in Persia (Iran) (1900–1901).
He was curator at the Zoological Museum, Dorpat, Livonia
(Tartu, Estonia) (1922–1939).

Harpaceus

Red-bellied Woodpecker ssp. *Melanerpes carolinus*
harpaceus **Koelz,** 1954

According to a letter written by J. W. Hardy (q.v.) to Wetmore
(q,v,), the word *harpaceus* is a compound of pieces of three
people's names: Hardy's own, Wm. Pielou (q.v.) and George
J. Wallace (q.v.).

Harriet

Mrs Gould's Sunbird ssp. *Aethypyga gouldiae harrietae*
Delacour & J. C. Greenway, 1940 NCR
[JS *Aethopyga gouldiae dabryii*]

Mrs Harriet Greenway *née* Lauder (DNF) was the mother of
J. C. Greenway (q.v.).

Harrington

Chestnut-breasted Malkoha ssp. *Phaenicophaeus*
curvirostris harringtoni **Sharpe,** 1877

Professor Mark Walrod Harrington (1848–1926) was an
American botanist, meteorologist and astronomer. The
University of Michigan awarded his bachelor's degree (1868)
and his master's (1871). He became Professor of Astronomy
in Peking (Beijing) and was Director of the Detroit Observatory (1879–1891). He was the first civilian chief of the United
States Weather Bureau (1891–1895) from which he was
removed because of 'concerns over his health'. He retired
(1899) and one evening left home to attend a dinner – and
was not seen again until he was placed in a mental institution
as 'John Doe' (1907), where his wife and son found him
(1908), but he never recovered sufficiently for him to leave
there.

Harris, C. M.

Galapagos Cormorant *Phalacrocorax harrisi* **Rothschild,**
1898
[Alt. Flightless Cormorant; Syn. *Nannopterum harrisi*]

Charles Miller Harris (DNF) was an English taxidermist and a
collector for Lord Rothschild's Museum at Tring. He was in
the Galapagos Islands (1897–1898) and kept a careful diary of
his experiences there.

Harris, E.

Harris's Hawk *Parabuteo unicinctus* Temminck, 1824
Harris's Sparrow *Zonotrichia querula* **Nuttall,** 1840

Harris's Hawk ssp. *Parabuteo unicinctus harrisi* **Audubon,**
1838
Harris's Woodpecker *Picoides villosus harrisi* Audubon,
1838
[Alt. Hairy Woodpecker ssp.]

Edward Harris (1799–1863) was a farmer landowner, breeder
of horses, and amateur ornithologist who lived a life of
leisure. Audubon (q.v.) named the hawk and woodpecker to
honour his friend, who had helped him financially (1823)
during the preparation of his *Birds of America* when he had
been reduced 'to the lowest degree of indigence.' He had just
experienced a humiliating departure from Philadelphia,
denied membership of the Science Academy (see under
Harlan). Edward Harris bought all of Audubon's paintings and
gave him an extra $100 in appreciation of his talent saying:
'men like you ought not to want for money'. Years later
Audubon said of the incident, 'I would have kissed him, but
that is not the custom in this icy city'. Audubon called Harris
'one of the best friends I have in the world'. A mammal is
named after him.

Harris, R. H.

Cockatoo genus *Harrisornis* **Mathews,** 1914 NCR
[Now in *Calyptorhynchus*]

Dr Ronald Hamlyn-Harris (1874–1953) was an English zoologist, entomologist and director of the Queensland Museum,
Brisbane (1910–1917). His doctorate (1902) was awarded by
Eberhard Karl University, Tübingen, Germany. He went to
Australia (1903) and became a schoolmaster at Toowoomba
Grammar School, Queensland. After resigning from the
museum as his health was failing he became a fruit farmer
and later returned to teaching. He became City Entomologist
for Brisbane (1926–1934) with the brief to control mosquitoes.

He ended his career as a lecturer in zoology at the University in Brisbane (1936–1943). One of his hobbies was to play the role of Charles Dickens at meetings of the Brisbane Dickens Fellowship.

Harrison, E. N.

> Spotted Nightingale Thrush ssp. *Catharus dryas harrisoni*
> **A. R. Phillips** & W Rook, 1963
> Rufous-banded Miner ssp. *Geositta rufipennis harrisoni*
> Marin, Kiff & Pena, 1989

Ed N. Harrison (1914–2002) was a naturalist and ornithologist. He was President of the Cooper Ornithological Society. He established the Western Foundation of Vertebrate Zoology (1956). The WFVZ was formed at a time when many natural history museums were unwilling to add eggs to their holdings, so a respected repository was needed. Harrison himself contributed approximately 11,000 egg sets, 2,000 nests, and 1,700 study skins to the museum. There is an Ed N. Harrison Memorial Scholarship for ornithology students established in his memory. A mammal is named after him.

Harrison, J. J.

> Chestnut-headed Sparrow Lark ssp. *Eremopterix signatus harrisoni* **Ogilvie-Grant**, 1900

Lieutenant-Colonel James Jonathan Harrison (1858–1923) was a British hunter and collector in Somalia, Kenya and Abyssinia (1899–1900). He also travelled in Japan, India and the Americas. He was in the Congo (1904–1905) where he persuaded six 'pygmies' to accompany him to Khartoum and thence to England. These six persons toured (1905–1907) to display themselves as curiosities, and when not on tour lived at Harrison's house in Yorkshire, but eventually returned to the Congo (1908). He wrote *Life among the Pygmies of the Ituri Forest, Congo Free State* (1905).

Harrison, J. M.

> European Greenfinch ssp. *Carduelis chloris harrisoni*
> **Clancey**, 1940
> Eurasian Nuthatch ssp. *Sitta europaea harrisoni* **Voous** &
> **van Marle**, 1953 NCR
> [JS *Sitta europaea caesia*]

Dr James Maurice Harrison (1892–1971) was a British ornithologist and prominent member of many ornithological and zoological societies and organisations worldwide. He was a member of the Harrison shipowning family, J. & C. Harrison, known in the shipping business as 'Hungry Harrison'. He started to read medicine before WW1, but joined the Royal Navy (1914). After service in destroyers with the Dover Patrol, he returned to St. Thomas's Hospital and qualified. He rejoined the navy and was sent to the eastern Mediterranean, where his ship was sunk by German battle cruisers. He had to swim for it to the island of Thasos. He was awarded the Distinguished Service Cross for his actions in caring for the wounded. He was a general practitioner in Sevenoaks, England (1920–1970). He made many collecting expeditions, both at home and abroad, from North Africa to Lapland. He left a massive collection of more than 35,000 skins and

specimens, which formed the basis for the Harrison Zoological Museum Trust. He wrote *The Birds of Kent* (1953). He will always be remembered for his defence of George Bristow concerning others' suspicions about the 'Hastings Rarities', a series of very rare birds supposedly collected in Hastings, UK, but later proved to have come from elsewhere. His son David L. Harrison of the Harrison Institute is a noted zoologist. A mammal is named after him.

Harrison, W. F.

> Olivaceous Woodcreeper ssp. *Sittasomus griseicapillus harrisoni* **G. M. Sutton**, 1955

William Frank Harrison (d.1966) wrote under the pen name 'Pancho'. He was a Canadian school teacher who farmed in southern Tamaulipas, gaining title to the land at El Cielo (1935) after he came across it on a hunting trip (1926). He spent the rest of his life hybridising plants and fruit trees, raising cattle in the cloud forest and welcoming scientists and birders until he was murdered (1966) by jealous lowland farmers. After this the area became a field station dedicated solely for the purpose of conservation and preservation. The etymology reads: 'I take pleasure in naming this new form in honor of Mr. Frank Harrison, who resides at the type locality (Rancho del Cielo, Tamaulipas, Mexico) and who has been unfailingly kind to me and to many of my students and friends who have visited him from time to time.'

Harrisson

> Yellow-eared Spiderhunter ssp. *Arachnothera chrysogenys harrissoni* **Deignan**, 1957

Major Thomas Harnett Harrisson (1911–1976) was a British polymath and an anthropologist, ornithologist, ecologist, explorer, journalist, broadcaster, soldier, ethnologist, museum curator, mass observationist, archaeologist, filmmaker, writer and conservationist. He spent much of his life in Sarawak and during WW2, after having been a radio critic (1942–1944), he was parachuted (1945) into Borneo to organise tribesmen against the Japanese. He was Curator, Sarawak Museum (1947–1966), and undertook pioneering excavations at Niah, Sarawak, discovering a 40,000-year-old skull. After leaving Sarawak he lived in the USA, UK and France before being killed in a motor accident in Thailand. He wrote *Savage Civilisation* (1937). An amphibian is named after him.

Harry White

> Lyrebird genus *Harriwhitea* **Mathews**, 1912 NCR
> [Now in *Menura*]

(See **White, H. L.**)

Hart

> Pink-billed Lark ssp. *Spizocorys conirostris harti*
> **C. W. Benson**, 1964
> Grey-backed Sparrow Lark ssp. *Eremopterix verticalis harti*
> C. W. Benson & **Irwin**, 1965

Dr Robert 'Rob' C. Hart (b.1947) is a Northern Rhodesian (Zambian) limnologist, ornithologist, collector and photographer, who is a Professor in the School of Biological and Conservation Sciences, University of KwaZulu-Natal, Pietermaritzburg.

Hartert

Jery genus *Hartertula* **Stresemann**, 1925

Bolivian Earthcreeper *Tarphonomus harterti* **Berlepsch**, 1892

Dulit Frogmouth *Batrachostomus harterti* **Sharpe**, 1892

Hartert's Hermit *Phaethornis mexicanus* Hartert, 1897
[Alt. Mexican Hermit; Syn. *Phaethornis longirostris mexicanus*]

Hartert's Sunangel *Heliangelus dubius* Hartert, 1897 NCR
[Probably a melanistic variation of *Heliangelus amethysticollis*]

Black-throated Thistletail *Schizoeaca harterti* Berlepsch, 1901

Hartert's Woodstar *Acestrura harterti* **Simon**, 1901 NCR
[Believed to be a hybrid involving *Chaetocercus berlepschi*]

Peruvian Piedtail *Phlogophilus harterti* Berlepsch & **Stolzmann**, 1901

Hartert's White-eye *Zosterops luteirostris* Hartert, 1904
[Alt. Splendid White-eye, Gizo White-eye]

Hartert's Leaf Warbler *Phylloscopus goodsoni* Hartert, 1910

Hartert's Camaroptera *Camaroptera harterti* **Zedlitz**, 1911
[Alt. Angola Camaroptera]

Hartert's Flycatcher *Ficedula harterti* **Siebers**, 1928
[Alt. Sumba Flycatcher]

Hartert's Double-eyed Fig Parrot *Cyclopsitta diophthalma virago* Hartert, 1895

Hartert's Orange-fronted Hanging Parrot *Loriculus aurantiifrons meeki* Hartert, 1895

Eastern Bronze-naped Pigeon ssp. *Columba delegorguei harterti* **Neumann**, 1898 NCR
[JS *Columba delegorguei sharpei*]

Great Spotted Woodpecker ssp. *Dendrocopos major harterti* **Arrigoni**, 1902

Eurasian Skylark ssp. *Alauda arvensis harterti* **J. I. S. Whitaker**, 1904

Common Reed Bunting ssp. *Emberiza schoeniclus harterti* **Sushkin**, 1906

Thick-billed Green Pigeon ssp. *Treron curvirostra harterti* **Parrot**, 1907 NCR
[JS *Treron curvirostra curvirostra*]

Red-billed Helmet-shrike ssp. *Prionops caniceps harterti* Neumann, 1908

Rufous-naped Lark ssp. *Mirafra africana harterti* Neumann, 1908

Banded Broadbill ssp. *Eurylaimus javanicus harterti* **van Oort**, 1909

Frilled Monarch ssp. *Arses telescopthalmus harterti* van Oort, 1909

Olivaceous Piculet ssp. *Picumnus olivaceus harterti* **Hellmayr**, 1909

Black Pitohui ssp. *Pitohui nigrescens harterti* **Reichenow**, 1911

Dusky Myzomela ssp. *Myzomela obscura harterti* **Mathews**, 1911

Great Slaty Woodpecker ssp. *Mulleripicus pulverulentus harterti* Hesse, 1911

Turquoise Flycatcher ssp. *Eumyias panayensis harterti* van Oort, 1911

Ashy-bellied White-eye ssp. *Zosterops citrinella harterti* Stresemann, 1912

Asian Koel ssp. *Eudynamys scolopaceus harterti* **Ingram**, 1912

Northern Golden Bulbul ssp. *Thapsinillas longirostris harterti* Stresemann, 1912

Masked Lapwing ssp. *Lobibyx miles harterti* Mathews, 1912 NCR
[JS *Vanellus miles miles*]

Spotted Nightjar ssp. *Eurostopodus argus harterti* Mathews, 1912 NCR; NRM

White-browed Scrubwren ssp. *Sericornis frontalis harterti* Mathews, 1912

Blyth's Leaf Warbler ssp. *Phylloscopus trochiloides harterti* **E. C. S. Baker**, 1913 NCR
[NPRB *Phylloscopus reguloides assamensis*]

Grey-rumped Treeswift ssp. *Hemiprocne longipennis harterti* Stresemann 1913

Tyrian Metaltail ssp. *Metallura tyrianthina harterti* **Schlüter**, 1913 NCR
[JS *Metallura tyrianthina oreopola*]

Little Tinamou ssp. *Crypturellus soui harterti* **Brabourne** & **Chubb**, 1914

Western Marsh Harrier ssp. *Circus aeruginosus harterti* Zedlitz, 1914

Eurasian Wryneck ssp. *Jynx torquilla harterti* Polyakov, 1915 NCR
[JS *Jynx torquilla torquilla*]

Rufous-crowned Antpitta ssp. *Pittasoma rufopileatum harterti* **Chapman**, 1917

Black Drongo ssp. *Dicrurus macrocercus harterti* Baker, 1918

Pygmy Wren-babbler ssp. *Pnoepyga pusilla harterti* **H. C. Robinson** & **Kloss**, 1918

Common Buzzard ssp. *Buteo buteo harterti* **Swann**, 1919

Tawny Owl ssp. *Strix aluco harterti* **La Touche**, 1919 NCR
[JS *Strix (aluco) nivicolum*]

Brown-eared Bulbul ssp. *Microscelis amaurotis harterti* **Kuroda**, 1922

Micronesian Starling ssp. *Aplonis opaca harterti* **Momiyama**, 1922 NCR
[JS *Aplonis opaca guami*]

African Paradise-flycatcher ssp. *Terpsiphone viridis harterti* **Meinertzhagen**, 1923

Japanese Pygmy Woodpecker ssp. *Dendrocopos kizuki harterti* Kuroda, 1923 NCR
[JS *Dendrocopos kizuki matsudairai*]

Hartert's Pygmy Parrot *Micropsitta bruijnii necopinata* Hartert, 1925
[Alt. Red-breasted Pygmy Parrot ssp.]

White-headed Black Chat ssp. *Pentholaea arnotti harterti* Neunzig, 1926

Clamorous Reed Warbler ssp. *Acrocephalus stentoreus harterti* **Salomonsen**, 1928

Pale Blue Flycatcher ssp. *Cyornis unicolor harterti* H. C. Robinson & **Kinnear**, 1928

Cream-browed White-eye ssp. *Lophozosterops superciliaris hartertianus* **Rensch**, 1928

Hartert's Fairy Lorikeet *Charmosyna pulchella rothschildi* Hartert, 1930

Hartert's Ground Parrot *Psittacella brehmii harterti* **Mayr**, 1931

[Alt. Brehm's Tiger Parrot ssp.]

Papuan Grassbird ssp. *Megalurus macrurus harterti* Mayr, 1931

Shining Bronze Cuckoo ssp. *Chrysococcyx lucidus harterti* Mayr, 1932

Blyth's Hornbill ssp. *Rhyticeros plicatus harterti* Mayr 1934

[Alt. Papuan Hornbill ssp.; Syn. *Aceros plicatus harterti*]

Booted Eagle ssp. *Hieraaetus pennatus harterti* Stegmann, 1935 NCR; NRM

[Syn. *Aquila pennata*]

Eurasian Stone-curlew ssp. *Burhinus oedicnemus harterti* **Vaurie**, 1963

Ernst Johann Otto Hartert (1859–1933) was a German ornithologist and oologist. He wrote *Die Vögel der Paläarktischen Fauna* and undertook many collaborations (1890s–1920s). He travelled extensively, often on behalf of his employer Walter (Lord) Rothschild. He was the ornithological curator of Rothschild's private museum at Tring, which later became an annexe to the BMNH, housing all of the bird skins. An amphibian is named after him.

Harting

Double-banded Courser ssp. *Rhinoptilus africanus hartingi* **Sharpe**, 1893

[Syn. *Smutsornis africanus hartingi*]

James Edmund Fotheringham Harting (1841–1928) was an English solicitor, naturalist and falconer. He was a contributor to, and editor of, *The Field* (1869–1920), and editor of *The Zoologist* (1877–1896).

Hartlaub

Madagascar Starling genus *Hartlaubius* **Bonaparte**, 1853

Hartlaub's Sheathbill *Chionis minor* Hartlaub, 1841

[Alt. Black-faced Sheathbill, Lesser Sheathbill]

Hartlaub's Jay *Cyanocorax melanocyaneus* Hartlaub, 1844

[Alt. Bushy-crested Jay]

Hartlaub's Warbler *Oreothlypis superciliosa* Hartlaub, 1844

[Alt. Crescent-chested Warbler; Syn. *Parula superciliosa*]

Red-headed Barbet *Micropogon hartlaubii* **Lafresnaye**, 1845 NCR

[JS *Eubucco bourcierii*]

São Tomé Scops Owl *Otus hartlaubi* Giebel, 1849

Hartlaub's Gull *Chroicocephalus hartlaubii* **Bruch**, 1853

Turquoise Dacnis *Dacnis hartlaubi* **P. L. Sclater**, 1855

Hartlaub's Sunbird *Anabathmis hartlaubii* Hartlaub, 1857

[Alt. Príncipe Sunbird]

Hartlaub's Duck *Pteronetta hartlaubii* **Cassin**, 1860

Black Dwarf Hornbill *Tockus hartlaubi* **Gould**, 1861

Heart-spotted Woodpecker *Hemicercus hartlaubi* Gould, 1861 NCR

[JS *Hemicircus concretus*]

Hartlaub's Bustard *Lissotis hartlaubii* **Heuglin**, 1863

Hartlaub's Babbler *Turdoides hartlaubii* **Bocage**, 1868

[Alt. Angola Babbler, Southern White-rumped Babbler]

Hartlaub's Scrubfowl *Megapodius eremita* Hartlaub, 1868

[Alt. Melanesian Scrubfowl]

Hartlaub's Francolin *Pternistis hartlaubi* Bocage, 1869

[Alt. Hartlaub's Spurfowl; Syn. *Francolinus hartlaubi*]

Hartlaub's Marsh Widowbird *Euplectes hartlaubi* Bocage, 1878

Hartlaub's Turaco *Tauraco hartlaubi* **G. A. Fischer** & **Reichenow**, 1884

Brown-backed Scrub Robin *Erythropygia hartlaubi* Reichenow, 1891

Cardinal Woodpecker ssp. *Dendropicos fuscescens hartlaubii* **Malherbe**, 1849

Chestnut-winged Starling ssp. *Onychognathus fulgidus hartlaubii* Hartlaub, 1858

Greater Blue-eared Starling ssp. *Lamprotornis chalybaeus hartlaubi* **Neumann**, 1908 NCR

[JS *Lamprotornis chalybaeus chalybaeus*]

Karel Johan Gustav Hartlaub (1814–1900) was an academic and explorer. He was Professor of Zoology at Bremen. He trained as a physician but his hobby was 'exotic' ornithology and he published a number of papers on African birds, including checklists, although he never set foot on the continent. He also founded the *Journal für Ornithologie* with Cabanis. Many people sent him specimens, which he eventually gave to the Hamburg Museum. He described c.30 Southern African birds. He wrote *System der Ornithologie Westafrikas* (1857) and *Die Vögel Madagaskars und benachbarten Inselgruppen* (1897), and co-wrote two books with Finsch including *Die Vögel Ostafrikas* (1870).

Hartley

Grey-throated Babbler ssp. *Stachyris nigriceps hartleyi* **Chasen**, 1935

Air Marshall Sir Christopher Harold Hartley (1913–1998) of the Royal Air Force (1939–1970) retired as Controller of Aircraft. He was as a student at Balliol College, Oxford, which awarded him a bachelor's degree in zoology (1937). He took part as a zoologist in Oxford University Expeditions to Sarawak (1932), Spitsbergen (1933) and Greenland (1937). He then became a schoolmaster at Eton (1937–1939) until called up as a member of the Royal Air Force Volunteer Reserve (1939–1945), transferring to a regular commission in peacetime. After retiring from the RAF he became a Director of Westland Aircraft (1971–1983) and Chairman of British Hovercraft Corporation (1974–1978).

Hartweg

Prevost's Ground Sparrow ssp. *Melozone biarcuata hartwegi* **Brodkorb**, 1938

Dr Norman Edouard 'Kibe' Hartweg (1904–1964) was a herpetologist whose specialty was the distribution and taxonomy of turtles. He worked (1927–1964) at the University of Michigan, where he took his doctorate (1934) and was Assistant Curator, Herpetology (1934), then Curator (1947). Two reptiles and two amphibians are named after him.

Harwood

Harwood's Francolin *Pternistis harwoodi* **Blundell** & **Lovat**, 1899

Leonard C. Harwood (fl.1888–at least 1920) was an English naturalist and taxidermist who had premises in Hammersmith, London. He accompanied Lord Delamere (q.v.) to Kenya (1896–1897) and made a 3-month collecting tour to Somaliland (1897) with Richard Hawker (q.v.). A mammal is named after him.

Hasbrouck

Hasbrouck's Screech Owl *Megascops asio hasbroucki* **Ridgway**, 1914
[Alt. Eastern Screech Owl ssp.]
White-browed Piculet ssp. *Sasia ochracea hasbroucki* **Deignan**, 1947 NCR
[JS *Sasia ochracea reichenowi*]

Dr Edwin Marble Hasbrouck (1866–1956) was an ornithologist and physician who was active in Texas (1880s). Georgetown Medical School awarded his bachelors degree (1895), then his MD (1897). He was a physician attached to a survey team in Texas running telegraph and powerlines. He became Captain in the United States army medical corps (WW1) after which he was in the Veterans Administration as an 'examining surgeon'. He wrote a number of papers including 'Summer birds of Eastland County, Texas' (1889) and others on the Carolina Parakeet *Conuropsis carolinensis* and the Ivory-billed Woodpecker *Campephilus principalis*. Hasbrouck's grandson, David Ervin Cummins of Sarasota, Florida, told us 'I remember, in about nineteen thirty-something, my grandfather taking me down to the Natural History Museum (I think) and showing me his work site and thousands and thousands of drawers of bird skins. He was taking each skin, checking for ID, noting condition and tagging all written by hand. He spent many years in his retirement doing this kind of volunteer work'.

Hass

Large-billed Crow ssp. *Corvus macrorhynchos hassi* **Reichenow**, 1907 NCR
[JS *Corvus macrorhynchos colonorum*]

Walter Hass (DNF) was a German forestry officer at Shantung in China.

Hasselt

Little Pied Flycatcher ssp. *Ficedula westermanni hasselti* **Finsch**, 1898

(See **Van Hasselt**)

Hauchecorn

Barn Owl ssp. *Tyto alba hauchecorni* Kleinschmidt, 1940
[Syn. *Tyto furcata hauchecorni*]

Dr Friedrich Hauchecorne (1894–1938) was Director of Halle Zoo and Cologne Zoo (1929–1938). He wrote: *Führer durch den Zoologischen Garten der Stadt Halle (Saale)* (1929).

Haughton

Nordmann's Greenshank *Totanus haughtoni* **J. Armstrong**, 1876 NCR
[Alt. Spotted Greenshank; JS *Tringa guttifer*]

Professor Rev. Samuel Haughton (1821–1897) was an Irish zoologist and ordained a priest (1847) who was Professor of Geology, Trinity College, Dublin (1851–1881). He is best remembered for devising the humane equations (1866) for execution by hanging, thus ensuring that the condemned person dies of a broken neck at the time of the drop and does not slowly strangle to death. Armstrong wrote: 'I propose to describe this presumably new species under the above name, dedicating it to my valued friend, the Rev. Professor Haughton, of Trinity College, Dublin, whose labors have done so much to enlarge the field of Natural History research.'

Hausburg

Fine-banded Woodpecker ssp. *Campethera tullbergi hausburgi* **Sharpe**, 1900

Campbell B. Hausberg (1873–1941) was a photographer and mountaineer with Mackinder (q.v.), who first climbed Mount Kenya (1899). Hausberg Valley on Mt Kenya is named after him. He settled in Kenya (1904) to grow sisal. His name has been spelt Hausberg and Hausburg, but Hausberg seems the commoner.

Hauxwell

Plain-throated Antwren *Isleria hauxwelli* **P. L. Sclater**, 1857
[Syn. *Myrmotherula hauxwelli*]
Black-banded Crake *Laterallus hauxwelli* Sclater & **Salvin**, 1868 NCR
[Invalid replacement name for *Anurolimnas fasciatus*]
Hauxwell's Thrush *Turdus hauxwelli* **G. N. Lawrence**, 1869
Coquette sp. *Lophornis hauxwelli* **Boucard**, 1892 NCR
[Alt. Festive Coquette; JS *Lophornis chalybeus verreauxii*]
Barbthroat sp. *Threnetes hauxwelli* Boucard, 1895 NCR
[Alt. Pale-tailed Barbthroat; JS *Threnetes leucurus cervinicauda*]

Ochre-bellied Flycatcher ssp. *Mionectes oleagineus hauxwelli* **Chubb**, 1919
White-breasted Wood Wren ssp. *Henicorhina leucosticta hauxwelli* Chubb, 1920

John Hauxwell (DNF) was an English bird collector mentioned by Henry Walter Bates (q.v.) in *The Naturalist on the River Amazons* (1864). Bates writes of being shown how to use 'a blow-gun, by Julio, a Juri Indian, then in the employ of Mr. Hauxwell, an English bird-collector'. We can surmise that some of the above birds were first collected by this method, better preserving the skin than shooting. Chubb (1919) refers to 'the late J. Hauxwell'.

Havell

Havell's Tern *Sterna havelli* **Audubon**, 1838 NCR
[Alt. Forster's Tern; JS *Sterna forsteri*]

Robert Havell Jr (1793–1878) was an English engraver who produced many plates for Audubon's *Birds of America*. He moved (1829) from London to New York to work more closely with Audubon (q.v.). Audubon in his *Ornithological Biography* (1831) wrote: 'I have several reasons for naming this Tern after Mr. ROBERT HAVELL, of Oxford Street, London. In the first place I consider him as one of the best ornithological engravers in England. Secondly, I feel greatly indebted to him for the interest which he has always evinced in my publication, which, I dare venture to assert, is the largest work of the kind that has hitherto appeared, and the engraving of which has cost him much trouble and anxiety. Thirdly, I consider myself entitled to express my gratitude in this manner, the individual on whom I confer the honour being more deserving of it than many to whom similar compliments have been paid.' The species had, however, already been named by Nuttall (q.v.) (1834).

Hawker

Red-tailed Wheatear *Saxicola hawkeri* **Oglivie-Grant**, 1908 NCR
[Alt. Kurdish Wheatear; JS *Oenanthe xanthoprymna*]

Purple Grenadier ssp. *Uraeginthus ianthinogaster hawkeri* **E. L. Phillips**, 1898 NCR; NRM
Marico Sunbird ssp. *Cinnyris mariquensis hawkeri* **Neumann**, 1899 NCR
[JS *Cinnyris mariquensis osiris*]

Richard Macdonnell Hawker (1865–1930) was a British collector in Arabia (1897), Somaliland (1897), Sudan (1900) and Somalia (1901).

Hawkins, R. W.

White-bellied Wren ssp. *Uropsila leucogastra hawkinsi* Monroe, 1963 NCR
[JS *Uropsila leucogastra brachyura* (or of *U. leucogastra australis* if latter race is recognised)]

Roland W. Hawkins (DNF), with Arthur C. Twomey (q.v.), collected the holotype in Honduras (1950). He worked in the Ornithology Department, Carnegie Museum of Natural History (1947–1952).

Hawkins, W.

Hawkins's Rail *Diaphorapteryx hawkinsi* H. O. Forbes, 1892 EXTINCT
[Alt. Giant Chatham Islands Rail]

William Hawkins (DNF) found the remains of the rail in the Chatham Islands. It was long thought that the species went extinct prior to European discovery, but recent evidence suggests that the bird may have survived much later, perhaps into the 1800s. Hawkins was also the first person to discover the Chatham Petrel *Pterodroma axillaris* and was Forbes's guide when he visisted the Chatham Islands (1892).

Hay

Brown Barbet ssp. *Caloramphus fuliginosus hayii* **J. E. Gray**, 1831

Captain Hay (DNF). Unfortunately the only indication that Gray records in his etymology is that the holotype was from

'... the collection of Capt. Hay'. We believe this to be Captain (later Major) W. E. Hay who published on the genus *Bombycistoma* (1841) and who was recorded as shooting in the Tibetan border region (c.1859). He sent specimens to P. L. Sclater (q.v.).

Haynaldi

Yellowish Bulbul ssp. *Hypsipetes everetti haynaldi* **Blasius**, 1890
[Syn. *Ixos everetti haynaldi*]

Dr Lajos Haynaldi (1826–1891) was a Hungarian botanist.

Haynes

Capuchin Babbler *Crateropus haynesi* **Sharpe**, 1871 NCR
[JS *Phyllanthus atripennis rubiginosus*]

Lieutenant-Colonel J. W. Haynes (1839–1914) was a British army officer in the Gold Coast (Ghana) (1871–1874).

Heberer

Blue-eared Barbet ssp. *Megalaima australis hebereri* **Rensch**, 1930 NCR
[NUI *Megalaima australis australis*]

Dr Gerhard Heberer (1901–1973) was a German anthropologist, geneticist, ethnologist and zoologist in the East Indies (Indonesia) (1927). The University of Halle awarded his doctorate (1924). He lectured at Eberhard Karls University, Tübingen (1928–1938) and was Professor of Zoology and Comparative Anatomy (1932). He was an early and enthusiastic member of the Nazi party and was in the SS, working in the Race and Settlement Main Office and was a member of Himmler's SS-German Ancestral Heritage Research Foundation. He was Professor of General Biology and Anthropogeny, Friedrich Schiller University, Jena (1938–1945) and lectured on 'Germanization' to the inmates of Buchenwald concentration camp (1944). He was imprisoned and subjected to denazification (1945–1947) and became Director of the Anthropological Research Centre, Georg-August University, Göttingen (1949–1970).

Heck

Red-lored Amazon *Chrysotis hecki* **Salvadori**, 1891 NCR
[JS *Amazona autumnalis lilacina*]
Curassow sp. *Crax hecki* **Reichenow**, 1894 NCR
[Alt. Great Curassow; JS *Crax rubra rubra*]

Dwarf Cassowary ssp. *Casuarius bennetti hecki* **Rothschild**, 1899
[*C. bennetti* sometimes treated as monotypic]
Heck's Finch *Poephila acuticauda hecki* **Heinroth**, 1900
[Alt. Heck's Grassfinch, Long-tailed Finch ssp.]

Dr Ludwig Franz Friedrich Georg Heck (1860–1951) was a German zoologist. He became Director of Cologne Zoo (1886), and later of Berlin Zoo. He was also interested in ethnology and part of the Nazi movement (1930s). Both his sons also became directors of zoos. Two mammals are named after him.

Hecker

Brünnich's Guillemot ssp. *Uria lomvia heckeri* **Portenko**,
1944
[Alt. Thick-billed Murre ssp.]

Dr Roman Fedorovich Hecher (pronounced and often trans-
literated as 'Gekker') (1900–1991) was a Russian
palaeontologist and geologist who led an expedition to
Wrangel Island for the Academy of Sciences of the USSR
(Russia) (1939). Of German origin, he was otherwise known
as Robert Roman Gustav Wilhelm von Hecker. His doctorate
was awarded in biology (1937) and he was appointed
professor (1944). He dealt with finds of mammoths preserved
in the permafrost.

Hector

Weka ssp. *Gallirallus australis hectori* **F. W. Hutton**, 1874

Dr Sir James Hector (1834–1907) was a Scottish-born Canadian geologist who took his medical degree at Edinburgh
and, as both geologist and surgeon, was part of the Palliser
expedition to western North America (1857–1860). He discovered and named many landmarks in the Rocky Mountains,
including Kicking Horse Pass, the route later taken by the
Canadian Pacific Railway. He returned to Scotland via the
Pacific Coast, the California goldfields, and Mexico. He
became the Director of the Geological Survey of New
Zealand (1865) and eventually the Curator of the Colonial
Museum in Wellington (Museum of New Zealand Te Papa).
He wrote *Outlines of New Zealand Geology* (1886). Two
mammals are named after him.

Hedley

Blue-faced Honeyeater ssp. *Entomyzon cyanotis hedleyi*
Mathews, 1912 NCR
[JS *Entomyzon cyanotis griseigularis*]
Peaceful Dove ssp. *Geopelia placida hedleyi* Mathews,
1912 NCR
[JS *Geopelia placida placida*]
Slender-billed Thornbill ssp. *Acanthiza iredalei hedleyi*
Mathews, 1912

Charles Hedley (1862–1926) was an English malacologist and
chronic asthmatic. His health demanded a warmer, drier
climate, so he moved to New Zealand (1881) and thence to
Australia (1882). He was on the staff of the Queensland
Museum, Brisbane (1889–1891), worked at the Australian
Museum, Sydney (1891–1924), and was leader of their expe-
dition to the Torres Strait (1907) collecting biological and
ethnographic material. After resigning from the Australian
Museum, he became Scientific Director of the Great Barrier
Reef Committee.

Hedwig (Branicki)

Metaltail sp. *Metallura hedvigae* **Taczanowski**, 1874 NCR
[Alt. Fiery-throated Metaltail; JS *Metallura eupogon*]
Orange-eared Tanager *Chlorochrysa hedwigae* **Berlepsch
& Stolzmann**, 1901 NCR
[JS *Chlorochrysa calliparaea*]

(See **Branicki**)

Hedwig (Sauer-Guerth)

African Blue Tit ssp. *Cyanistes teneriffae hedwigii* Dietzen
et al. 2007

Hedwig Sauer-Guerth has been a Technical Assistant at the
Institute of Pharmacy and Molecular Biotechnology, University of Heidelberg, since 1984. She contributed greatly to the
laboratory work of the authors' and many other people's
studies.

Heermann

Heermann's Gull *Larus heermanni* **Cassin**, 1852

Song Sparrow ssp. *Melospiza melodia heermanni* **Baird**,
1858

Dr Adolphus Lewis Heermann (1827–1865) was an army
physician and naturalist. He came to the attention of Spencer
Baird (q.v.) at the USNM, and was assigned to a surveying
party for the Pacific Railroad line. Heermann was especially
interested in collecting birds' eggs, and he is credited with
coining the term 'oology' for the practice. He retired from the
army early due to illness and died two years later in a hunting
accident when he stumbled and his rifle discharged and
killed him. He appears to have looked many years older than
he was – the effect, among other things, of syphilis. A
mammal is named after him.

Heilprin

Heilprin's Jay *Cyanocorax heilprini* **Gentry**, 1885
[Alt. Azure-naped Jay]

Professor Angelo Heilprin (1853–1907) was a Hungarian-born
American geographer, scientist and explorer. He was
Professor of Invertebrate Palaeontology at the Academy of
Natural Sciences in Philadelphia. He wrote *A Complete
Pronouncing Gazetteer or Geographical Dictionary of the
World*, with his son Louis, and *The Geographical and Geolog-
ical Distribution of Animals* (1907). In an article (1882) entitled
'On the value of the 'Neoarctic' as one of the primary zoolog-
ical regions', he tried to show that the Nearctic and
Palaearctic should form one region – the Triarctic (Holarctic).
He went with Peary on his Arctic expedition (1891–1892). He
examined (1902) an erupting volcano, Mount Pelee on the
island of Martinique, at close range. He also founded the
American Alpine Club. He died shortly before he was due to
give the presidential address to the Association of American
Geographers. An amphibian is named after him.

Heim

Crested Tit ssp. *Parus cristatus heimi* **Jouard**, 1929 NCR
[JS *Lophophanes cristatus mitratus*]

(See **Balsac**)

Heine

Black-capped Tanager *Tangara heinei* **Cabanis**, 1850
Heine's Ground Thrush *Zoothera heinei* Cabanis, 1850
[Alt. Russet-tailed Thrush]
Heine's Kingbird *Tyrannus apolites* Cabanis & **Heine**, 1859
NCR

[Hybrid: *Empidonomus varius* x *Tyrannus melancholicus*]
Heine's Hermit *Phaethornis apheles* Heine, 1884 NCR
[Alt. Grey-chinned Hermit; JS *Phaethornis griseogularis zonura*]

Lesser Shrikebill ssp. *Clytorhynchus vitiensis heinei* **Finsch & Hartlaub**, 1870
Lesser Short-toed Lark ssp. *Calandrella rufescens heinei* **Homeyer**, 1873

Ferdinand Heine (1809–1894) was a German landowner and collector. His massive collection is housed by the Museum of Natural Science within the Heineanum Halberstadt Museum in Halberstadt, Saxony, and was recognised as the biggest private collection of birds in the mid-19th century. It has 15,000 books and 27,000 specimens of birds. Jean Louis Cabanis wrote about Heine in *Museum Heineanum, Verzeichniss de Ornithologischensammlung des Oberamtmann Ferdinand Heine auf Gut St. Burchard vor Halberstadt.*

Heineken

Blackcap ssp. *Sylvia atricapilla heineken* **Jardine**, 1830

Dr Karl (also known as Carlos) Heineken (d.1830) was a German physician and naturalist on Madeira (1826–1830). He wrote 'Notice of some of the birds of Madeira' (1829).

Heinrich

Great Shortwing genus *Heinrichia* **Stresemann**, 1931

Heinrich's Brush Cuckoo *Cacomantis heinrichi* Stresemann, 1931
[Alt. Moluccan Cuckoo]
Heinrich's Nightjar *Eurostopodus diabolicus* Stresemann, 1931
[Alt. Satanic/Diabolical Nightjar]
Heinrich's Thrush *Zoothera heinrichi* Stresemann, 1931
[Alt. Geomalia, Sulawesi Mountain Thrush; Syn. *Geomalia heinrichi*]
Heinrich's Robin Chat *Cossypha heinrichi* **Rand**, 1955
[Alt. White-headed Robin Chat, Rand's Robin Chat]

Common Starling ssp. *Sturnus vulgaris heinrichi* Stresemann, 1928 NCR
[JS *Sturnus vulgaris caucasicus*]
Scaly-headed White-eye ssp. *Lophozosterops squamiceps heinrichi* Stresemann, 1931
Sulawesi Woodcock ssp. *Scolopax celebensis heinrichi* Stresemann, 1932
Uniform Swiftlet ssp. *Aerodramus vanikorensis heinrichi* Stresemann, 1932
Sulawesi Thrush ssp. *Cataponera turdoides heinrichi* Stresemann, 1938
Yellow-breasted Greenfinch ssp. *Carduelis spinoides heinrichi* Stresemann, 1940
Lesser Spotted Woodpecker ssp. *Dendrocopos minor heinrichi* **von Jordans**, 1940 NCR
[JS *Dendrocopos minor buturlini*]
Banded Prinia ssp. *Prinia bairdii heinrichi* **Meise**, 1958

Gerd Herrmann Heinrich (1896–1984) was a German entomologist, ornithologist, taxonomist, collector and philanderer!

He joined the German army (1914) as a cavalryman and later (WW1) became a pilot in the German Air Force. In WW2 he served in the Luftwaffe (WW2) as a way of avoiding being arrested by the Gestapo. He and his family abandoned their possessions in East Prussia (1945) to avoid the advancing Russian army and fled westwards. Eventually they emigrated to the USA (1951), becoming American citizens. He explored in Persia (Iran), Sulawesi, Burma (Myanmar), North America, Europe and Africa, usually with his wife Hildegard, collecting mammal and bird skins to sell to museums. He wrote *Der Vogel Schnarch – Zwei Jahre Rallenfang un Urwaldforschung in Celebes* ('The snoring bird – two years of rail trapping and jungle exploration in the Celebes') (1932). During his expeditions he also collected other taxa such as amphibians and reptiles, and particularly ichneumon wasps, writing the definitive book *Burmesische Ichneumoninae* over the course of 40 years. He is the subject of a biographical work by his son Bernd Heinrich, *The Snoring Bird: My Family's Journey through a Century of Biology*. He described 1,479 new species, mostly insects, but discovered at least five species and 40 subspecies of birds. Several insects, two mammals and an amphibian are named after him.

Heinroth

Heinroth's Fantail *Rhipidura matthiae* Heinroth, 1902
[Alt. Matthias Fantail]
Heinroth's Shearwater *Puffinus heinrothi* **Reichenow**, 1919

Heinroth's Reed Warbler *Acrocephalus stentoreus celebensis* Heinroth, 1903
[Alt. Clamorous Reed Warbler ssp.]
Glossy Swiftlet ssp. *Collocalia uropygialis heinrothi* **Neumann**, 1919 NCR
[JS *Collocalia esculenta stresemanni*]
Slender-billed Cicadabird ssp. *Coracina tenuirostris heinrothi* **Stresemann**, 1922
Island Thrush ssp. *Turdus poliocephalus heinrothi* **Rothschild & Hartert**, 1924

Oskar August Heinroth (1871–1945) was very interested in natural history even as a child. He qualified as a medical doctor but after just one year in practice turned his attention to full-time zoology. He worked (unpaid) as an assistant at Berlin Zoo, and only began to be paid after his participation in Menke's two-year expedition on board *Eberhard* to New Guinea and the South Seas. He studied the largest aquaria of the time – London, Naples, Amsterdam and Cologne – and built an innovative version as part of the Berlin Zoo. It was destroyed by allied bombing (1945) but has since been restored and modernised. He wrote *The Birds of Central Europe* with his first wife Magdalena Wiebe. His second wife Katharina (1897–1989) remained a director of the zoo after his death (1945–1956).

Helen (Booth)

Helena's Calypte *Mellisuga helenae* **Lembeye**, 1850
[Alt. Bee Hummingbird]

Helen Booth *née* Fernandez (DNF) was the wife of Charles Booth, who owned a sugar plantation in Matanzas Province, Cuba. The Booths befriended Gundlach (q.v.) when he arrived in Cuba (1839). Gundlach found the hummingbird (1844), whose male is the smallest known bird, and later gave all his notes on it to Lembeye (q.v.) to help him with his book *Aves de la Isla de Cuba* (1850) in which the original description appeared. (See **Booth, C.**)

Helen (Cutting)

Red-headed Trogon ssp. *Harpactes erythrocephalus helenae* **Mayr**, 1941

Mrs Helen Cutting *née* McMahon, formerly Brady (1895–1967), was the wife of Charles Suydam Cutting (1889–1972), a New York financier, explorer, big-game hunter and naturalist who was a Trustee of the AMNH, New York. She often travelled with her husband, and on their trip to Tibet (1937) she became the first American woman to reach Lhasa. She was the third wife (1920) and then widow of James Cox Brady – they married in Westminster Abbey! His fortune was worth £35 million.

Helen (Greenway)

Spot-necked Babbler ssp. *Stachyris strialata helenae* **Delacour** & **Greenway**, 1939

Mrs Helen Greenway (DNF) was the wife of J. C. Greenway (q.v.). An amphibian is named after her.

Helen (Kelsall)

Yellow-crowned Gonolek ssp. *Laniarius barbarus helenae* **Kelsall**, 1913

Helen Kelsall (DNF) was the wife of the describer Colonel Harold Joseph Kelsall (q.v.).

Helen (Lavauden)

Crested Lark ssp. *Galerida cristata helenae* **Lavauden**, 1926

Helen Lavauden (DNF) was the daughter of Louis Lavauden (1881–1935).

Helen (McConnell)

Mrs McConnell's Manakin *Neopipo cinnamomea helenae* **McConnell**, 1911
[Alt. Cinnamon Neopipo ssp]

(See **Mackenzie, H.**)

Helen (of Troy)

Short-crested Monarch *Hypothymis helenae* **Steere**, 1890

Helen of Troy was reputed to be the most beautiful woman in the world.

Helena

Helena's Parotia *Parotia helenae* **De Vis**, 1897
[Alt. Eastern Parotia]

Princess Helena Augusta Victoria (1846–1923) married into the House of Augustenborg and was known as Princess Christian of Schleswig-Holstein. She was the third daughter

of Queen Victoria and Prince Albert. While we believe this to be correct, De Vis's original text has no etymology and one authority (Wynne) states that it was named after Helena Forde (1832–1910), who was the daughter of the Australian oologist A. Scott and one of the illustrators of Krefft's *Mammals of Australia* (1871).

Helene

Princess Helene's Coquette *Lophornis helenae* **de Lattre**, 1843
[Alt. Black-crested Coquette]

Princess Hélène d'Orléans (1814–1858) was the wife of the Duc d'Orléans, who was a patron of natural history expeditions.

Helenor

White-tailed Warbler *Poliolais helenorae* **B. Alexander**, 1903 NCR
[JS *Poliolais lopezi*]

Yellow-bellied Eremomela ssp. *Eremomela icteropygialis helenorae* Alexander, 1899

Helenor Alexander (1869–1898) was the elder sister of Boyd Alexander (q.v.), who commemorated her after her death.

Heliodore

Heliodore's Woodstar *Chaetocercus heliodor* **Bourcier**, 1840
[Alt. Gorgeted Woodstar]

Jules Bourcier (q.v.) originally spelt the binomial *heliodor*, but later amended this to *heliodori* (the original spelling takes priority). The mistaken binomial is derived from the Greek words meaning 'gift of the sun', but Bourcier's amendment rather suggests an eponym – and Heliodore Bourcier (1813–1838), a soldier, has been suggested. The original text of 1840 has no etymology.

Heller

Heller's Ground Thrush *Turdus helleri* **Mearns**, 1913
[Alt. Taita Thrush]
Puna Thistletail *Schizoeaca helleri* **Chapman**, 1923

Pacific Wren ssp. *Troglodytes pacificus helleri* **Osgood**, 1901
[Syn. *Nannus pacificus helleri*]
White-starred Robin ssp. *Pogonocichla stellata helleri* Mearns, 1913
Heller's Francolin *Francolinus hildebrandti helleri* Mearns, 1916 NCR; NRM
[Alt. Hildebrandt's Francolin; Syn. *Pternistis hildebrandti*]

Edmund Heller (1875–1939) was an American zoologist and ornithologist. He collected in the Colorado and Mohave deserts (1896–1897) whilst still a student, and interrupted his studies to spend seven months (1899) in the Galapagos Islands. After graduating from Stanford University (1901) he became Western Field Collector for the Field Colombian

Museum and took part in its African expedition (1907). On his return he was appointed Curator of Mammals at the Museum of Vertebrate Zoology of the University of California (1908), and was a member of the Alexander Alaskan expedition. He was in Africa again with the Smithsonian-Roosevelt and the Rainey Expeditions (1909–1912). He was a member of the Lincoln Ellsworth expedition to British Columbia (1914) and later to Alberta. An expedition was sent (1915) to explore the newly discovered lost city of Machu Picchu, Peru, with Heller employed as naturalist overseeing the collecting of 891 mammals, 695 birds, c.200 fishes and several tanks of reptiles and amphibians. He joined (1916) Roy Chapman Andrews (q.v.) on the AMNH Expedition to China. Rainey became official photographer for the Czech army in Siberia, and invited Heller to Russia. They travelled (1918) by rail across Siberia to the Urals and back. Heller was leader (1919) of the USNM Cape-to-Cairo Expedition and then worked for a short time for the Roosevelt Wild Life Experiment Station, studying large game animals in Yellowstone National Park. Later (1919–1925) he became Assistant Curator of Mammals at the Field Museum under Osgood (q.v.) and visited Peru (1922–1923) and Africa (1923–1926). The trip to Africa was his last and (1928–1935) he became Director of the Milwaukee Zoological Garden, until he became Director of the Fleishhacker Zoo in San Francisco (1935–1939). Three mammals and three reptiles are named after him.

Hellmayr

Chough genus *Hellmayria* Poche, 1904 NCR
[Now in *Pyrrhocorax*]
White-browed Spinetail genus *Hellmayrea* **Stolzmann**, 1926

Red-shouldered Spinetail *Gyalophylax hellmayri* **Reiser**, 1905
Buff-banded Tyrannulet *Mecocerculus hellmayri* Berlepsch, 1907
Streak-capped Spinetail *Cranioleuca hellmayri* **Bangs**, 1907
Hellmayr's Pipit *Anthus hellmayri* **Hartert**, 1909
Santa Marta Antbird *Drymophila hellmayri* **W. E. C. Todd**, 1915
[Syn. *Drymophila caudata hellmayri*]
Nunlet sp. *Nonnula hellmayri* Chrostowski, 1921 NCR
[Alt. Rusty-breasted Nunlet; JS *Nonnula rubecula*]

Familiar Chat ssp. *Cercomela familiaris hellmayri* **Reichenow**, 1902
Marsh Tit ssp. *Poecile palustris hellmayri* **Bianchi**, 1902
Slaty-backed Nightingale Thrush ssp. *Catharus fuscater hellmayri* Berlepsch, 1902
Shining Sunbird ssp. *Cinnyris habessinicus hellmayri* **Neumann**, 1904
Hellmayr's Woodcreeper *Xiphorhynchus elegans insignis* Hellmayr, 1905
[Alt. Elegant Woodcreeper ssp.]
Plain-throated Antwren ssp. *Myrmotherula hauxwelli hellmayri* **Snethlage**, 1906
Chestnut Woodpecker ssp. *Celeus elegans hellmayri* Berlepsch, 1908
Tyrannine Woodcreeper ssp. *Dendrocincla tyrannina hellmayri* **Cory**, 1913
Grey Penduline-tit ssp. *Anthoscopus caroli hellmayri* **J. A. Roberts**, 1914

Grey-fronted Dove ssp. *Leptotila rufaxilla hellmayri* **Chapman**, 1915
Hellmayr's Rusty-faced Parrot *Hapalopsittaca amazonina theresae* Hellmayr, 1915
Plain-winged Antshrike ssp. *Thamnophilus schistaceus hellmayri* Cory, 1916 NCR
[NUI *Thamnophilus schistaceus schistaceus*]
Rusty-margined Flycatcher ssp. *Myiozetetes cayanensis hellmayri* Hartert & **A. Goodson**, 1917
White-backed Fire-eye ssp. *Pyriglena leuconota hellmayri* Stolzmann & **Domaniewski**, 1918
Brown Goshawk ssp. *Accipiter fasciatus hellmayri* **Stresemann**, 1922
Green Kingfisher ssp. *Chloroceryle americana hellmayri* **Laubmann**, 1922
Common Miner ssp. *Geositta cunicularia hellmayri* **J. L. Peters**, 1925
Narrow-billed Woodcreeper ssp. *Lepidocolaptes angustirostris hellmayri* Naumberg, 1925
Slender-billed Xenops ssp. *Xenops tenuirostris hellmayri* W. E. C. Todd, 1925
Slaty-capped Flycatcher ssp. *Leptopogon superciliaris hellmayri* **Griscom**, 1929
[SII *Leptopogon superciliaris superciliaris*]
Red-billed Scythebill ssp. *Campylorhamphus trochilirostris hellmayri* Laubmann, 1930
Eye-ringed Flatbill ssp. *Rhynchocyclus brevirostris hellmayri* Griscom, 1932
Cordilleran Flycatcher ssp. *Empidonax occidentalis hellmayri* **Brodkorb**, 1935
Grey-hooded Attila ssp. *Attila rufus hellmayri* **O. M. Pinto**, 1935
Rufous Gnateater ssp. *Conopophaga lineata hellmayri* O. M. Pinto, 1936 NCR
[JS *Conopophaga lineata lineata*]
Barn Owl ssp. *Tyto alba hellmayri* Griscom & **J. C. Greenway**, 1937
[Syn. *Tyto furcata hellmayri*]
Spot-billed Toucanet ssp. *Selenidera maculirostris hellmayri* Griscom & Greenway, 1937 NCR
[JS *Selenidera gouldii*]
Squirrel Cuckoo ssp. *Piaya cayana hellmayri* Pinto, 1938
Double-collared Seedeater ssp. *Sporophila caerulescens hellmayri* **Wolters**, 1939
Golden-billed Saltator ssp. *Saltator aurantiirostris hellmayri* **Bond** & de Schauensee, 1939
Great-billed Parrot ssp. *Tanygnathus megalorynchos hellmayri* **Mayr**, 1944
Purple Honeycreeper ssp. *Cyanerpes caeruleus hellmayri* **Gyldenstolpe**, 1945

Carl Eduard Hellmayr (1878–1944) was an Austrian zoologist who worked for a decade in the USA. One of his earlier written contributions to science was 'Beiträge zur Ornithologie Nieder Österreichs' (1899). He accepted (1903) a post at the Zoological Institute at Munich (Germany) with the task of reorganising the ornithological department of the Bavarian State Museum, a position he held until 1922, apart from three years (1905–1908) when he studied Neotropical birds at the Rothschild Museum, Tring. He accepted a post (1922) as Associate Curator of Birds of the Field Museum, Chicago, to complete the *Catalogue of the Birds of the Americas*, which his predecessor, Cory (q.v.), had started. He returned to Vienna (1931), but after the Nazis invaded Austria (1938) he was arrested and imprisoned without explanation. All his property was confiscated. After four months he got out to Switzerland, where he remained. His publications (1920s–1960s) include *Birds of Chile* and the co-written *Catalogue of Birds of the Americas and the Adjacent Islands*. (See also **Catharine (Hellmayr)**)

Helme

Ruffed Grouse ssp. *Bonasa umbellus helmei* H. H. Bailey, 1941
[SII *Bonasa umbellus umbellus*]

Arthur Hudson Helme (1860–1947) was a veterinarian, taxidermist and ornithologist on Long Island, New York.

Heloise

Heloise's Hummingbird *Atthis heloisa* **Lesson** & **de Lattre**, 1839
[Alt. Bumblebee Hummingbird]

Heloise (1101–1164) was the niece of Canon Fulbert of Notre Dame, famous for her tragic love affair with the philosopher Peter Abelard. Abelard and Heloise were secretly married but tried to maintain the fiction that they were single. They had a son called Astrolabe. It all turned out badly: Abelard became a Dominican monk and Heloise a nun.

Helymus

Hummingbird genus *Helymus* **Mulsant**, 1875 NCR
[Now in *Heliangelus*]

In Greek mythology, Helymus (or Elymus) was the ancestor of the Elymians, natives of Sicily.

Hemprich

Hemprich's Hornbill *Tockus hemprichii* **Ehrenberg**, 1833
Hemprich's Gull *Ichthyaetus hemprichii* **Bruch**, 1853
[Alt. Sooty Gull]

Cardinal Woodpecker ssp. *Dendropicos fuscescens hemprichii* Ehrenberg, 1833
Siberian Stonechat ssp. *Saxicola maurus hemprichii* Ehrenberg, 1833

Wilhelm Friedrich Hemprich (1796–1825) was a physician, traveller and collector. He co-wrote *Natural Historical Journeys in Egypt and Arabia* (1828) with Ehrenberg (q.v.), whom he had met whilst studying medicine in Berlin. They were invited to serve (1820) as naturalists on an expedition to Egypt and they continued to journey and collect in the region, including the Lebanon and the Sinai Peninsula before returning to Egypt and on to Ethiopia. Hemprich died of fever in the Eritrean port of Massawa. A mammal and two reptiles are named after him.

Hemptinne

Common Pheasant ssp. *Phasianus colchicus hemptinnii* **La Touche**, 1919 NCR
[JS *Phasianus colchicus decollatus*]

Father Anselm de Hemptinne (DNF) was a Belgian Franciscan missionary to China.

Hendee

Silver-eared Laughingthrush ssp. *Trochalopteron melanostigma hendeei* **Bangs** & **Van Tyne**, 1930 NCR
[JS *Trochalopteron melanostigma connectans*]
Blue-naped Pitta ssp. *Hydrornis nipalensis hendeei* Bangs & Van Tyne, 1931

Russell W. Hendee (1899–1929) was an American zoologist. He was in Alaska (1921–1922), working for the Colorado Museum of Natural History. He was a member of the Godman-Thomas expedition to Peru (1925) collecting for the British Museum, and was in Indochina with Theodore Roosevelt (q.v.) (1929). Hendee had a tropical fever and went into a delirium, was hospitalised but committed suicide by throwing himself out of a window. Two mammals are named after him.

Henderson, A. A.

African Scops Owl ssp. *Otus senegalensis hendersonii* **Cassin**, 1852 NCR
[NUI *Otus senegalensis senegalensis*]

Dr Andrew A. Henderson (1816–1875) was an American naval surgeon and naturalist. He collected on the Angolan coast whilst attached to the US brig *Perry* (1850) during the Wilkes Expedition. He was surgeon on board USS *Richmond* during the American Civil War, when it passed the Confederate batteries at Vicksburg on the River Mississippi (1862) and off Port Hudson (1863).

Henderson, A. D.

Short-billed Dowitcher ssp. *Limnodromus griseus hendersonii* **Rowan**, 1932

Archibald Douglas Henderson (1878–1963) was a Canadian ornithologist and oologist. He wrote 'Nesting of the American hawk owl' (1919).

Henderson, G.

Henderson's Ground Jay *Podoces hendersoni* **Hume**, 1871
[Alt. Mongolian Ground Jay]

Lieutenant-Colonel Dr George Henderson (1836–1929) was an English traveller and botanist in the Imperial Medical Service (1859–1889). He qualified as a physician at Aberdeen University (1858) and became Professor of Surgery at Lahore Medical College and Superintendent of the Central Jail in Lahore (1868). He was Director of the Royal Botanic Gardens in Calcutta (Kolkata) (1872) and Professor of Botany at Calcutta University. He was the physician and botanist for the first Yarkand Mission to Central Asia (1870), during which he collected the first specimen of the ground jay (1870). He corresponded with Sir Joseph Hooker at Kew and they regularly sent each other plants for their respective gardens. Henderson wrote *Lahore to Yarkand: Incidents of the Route and Natural History of the Countries Traversed by the Expedition of 1870*, with Allan Hume (1873).

Henderson, J. B.

Yellow-breasted Crake ssp. *Porzana flaviventer hendersoni*
Bartsch, 1917

Dr John Brooks Henderson Jr (1870–1923) was a physician, naturalist and amateur malacologist, specialising in West Indian shells; his collection is in the USNM. He graduated from Harvard (1891) and Columbia Law School (George Washington University) (1893), entering government service as a private secretary. He later (1897) travelled to Europe and Turkey as a civilian observer of the armies of the great European powers. He collected in Cuba on the *Tomas Barrera* expedition (1914) and in Haiti (1917). He was a citizen member of the USNM's Board of Regents (1911–1923). He wrote *The Cruise of the Tomas Barrera* (1916). He owned a yacht, *Eolis*, on which he cruised to collect molluscs off the Florida Keys (1910–1913). Paul Bartsch (q.v.), the describer of the crake, was a close associate of Henderson's and was a guest on the cruises. The original description has no etymology but this association is convincing. A reptile is named after him.

Henderson, R.

Henderson's Mango *Anthracothorax prevostii hendersoni*
Cory, 1887
[Alt. Green-breasted Mango ssp.]

Robert Henderson (DNF) collected birds on the Caribbean islands of Old Providence and St Andrews for a winter (1886–1887), but we know nothing more about him.

Henderson

Henderson Fruit Dove *Ptilinopus insularis* **North**, 1908
Henderson Lorikeet *Vini stepheni* North, 1908
Henderson Petrel *Pterodroma atrata* **Mathews**, 1912
Henderson Reed Warbler *Acrocephalus taiti* **Ogilvie-Grant**, 1913

These birds are named after their breeding place, Henderson Island in the Pitcairn group. Henderson Island, sighted (1606) by Pedro Fernandez de Quiros, a Portuguese navigator, was rediscovered by Captain James Henderson of the British East India merchant ship *Hercules* (1819) and is named after the venerable captain.

Henke

Henke's Sunbird *Leptocoma sperata henkei* **A. B. Meyer**,
1884
[Alt. Purple-throated Sunbird ssp.]
Frilled Monarch ssp. *Arses telescopthalmus henkei*
A. B. Meyer, 1886

Karl Gottlieb Henke (1830–1899) was a German ornithologist who was sickly as a child and did not go to school but went birding instead. He later learned to be a shoemaker and was a self-taught taxidermist who came to the attention of Reichenbach (q.v.), who took him on at the Dresden Zoological Museum (1851) and trained him as a preparator. Henke is best remembered for his three expeditions to Russia (1853–1854, 1857–1862 & 1869–1877). Between trips he lived in his home village, Saupsdorf, and created his own museum of natural sciences. He re-joined the Dresden Museum (1881) and became keeper of ornithology there (1889). He closed down his private museum and moved the majority of his collection to Dresden Zoological Museum, where they can be seen to this day. A. B. Meyer (q.v.) wrote Henke's obituary.

Henri

Eurasian Skylark ssp. *Alauda arvensis henrii* A. Vaucher,
1923 NCR
[JS *Alauda arvensis harterti*]

Henri Vaucher (1856–1910) was a Swiss botanist who was in Morocco (1879–1910).

Henri, Prince

Henri's Laughingthrush *Trochalopteron henrici* **Oustalet**
1892
[Alt. Brown-cheeked Laughingthrush]
Tibetan Snowfinch *Montifringilla henrici* Oustalet, 1892

Long-tailed Rosefinch ssp. *Uragus sibiricus henrici*
Oustalet, 1892
Tibetan Snowcock ssp. *Tetraogallus tibetanus henrici*
Oustalet, 1892
Bar-backed (Hill) Partridge ssp. *Arborophila brunneopectus henrici* Oustalet, 1896
Puff-throated Bulbul ssp. *Alophoixus pallidus henrici*
Oustalet, 1896

Henri, Prince d'Orléans (1867–1901) (See **Prince Henri**)

Henric (Heinrich)

Damar Flycatcher *Ficedula henrici* **Hartert**, 1899

(See **Kühn**)

Henrica

Bolivian Spinetail *Cranioleuca henricae* Maijer & **Fjeldså**,
1997
[Alt. Inquisivi Spinetail]

Henrica G. van der Werff is the mother of the senior author, the late Sjoerd Maijer.

Henrici

Yellow-crowned Barbet *Megalaima henricii* **Temminck**,
1831

Major Henri Albert von Henrici (c.1783–1838) was an Austrian army officer who entered the service of the Dutch in the East Indies (Indonesia). He made a collection of vertebrates in Borneo.

Henrietta (Sanford)

Island Leaf Warbler ssp. *Phylloscopus poliocephalus
henrietta* **Stresemann**, 1931

Mrs Henrietta Edwards Sanford *née* Whitney (1876–1963) was the second wife of Leonard Cutler Sanford (1868–1950) (q.v.). Her younger sister, Sarah Tracy Whitney (1877–1901) had been Sanford's first wife.

Henrietta (Tregellas)

Superb Fairywren ssp. *Malurus cyaneus henriettae* **Mathews**, 1912 NCR
[JS *Malurus cyaneus cyanochlamys*]
Green Rosella ssp. *Platycercus caledonicus henriettae* Mathews, 1915 NCR
[JS *Platycercus caledonicus brownii*]

Henrietta Tregellas (fl.1912) was the wife of the Australian ornithologist Thomas H. Tregellas (q.v.).

Henrik

Rufous-naped Lark ssp. *Mirafra africana henrici* **G. L. Bates**, 1930

(See **Grönvold**)

Henry de Lattre

Henry de Lattre's Cazique *Ornismyia henrica* **Lesson** & **de Lattre**, 1839 NCR
[Also known as *Delattria henrica*; Alt. Amethyst-throated Mountaingem; JS *Lampornis amethystinus*]

(See **de Lattre, H.**)

Henry, J.

Green-crowned Brilliant ssp. *Heliodoxa jacula henryi* **G. N. Lawrence**, 1867

Joseph Henry (1797–1878) was the first Secretary of the Smithsonian Institution (1846–1878). He was apprenticed as a watchmaker and silversmith, aged 13, but became a scientist and engineer. He was given a free education at the Albany Academy (1819–1824) and was appointed Professor of Mathematics and Natural Philosophy there (1826). He invented prolifically, including the electromechanical relay (1835).

Henry, T. C.

Common Nighthawk ssp. *Chordeiles minor henryi* **Cassin**, 1855

Lieutenant-Colonel Thomas Charlton Henry (1825–1877) was an American army surgeon and collector. He obtained specimens of the nighthawk at Fort Webster, New Mexico.

Hens

Woodchat Shrike ssp. *Lanius senator hensii* **Clancey**, 1948 NCR
[JS *Lanius senator senator*]

Dr P. A. Hens (1888–1971) was a Dutch zoologist. A memorial fund was set up at Maastricht, Netherlands, to provide financial support for research in his honour. He wrote 'Avifauna of the Dutch province of Limburg, in addition to a comparison with that of adjacent areas' (1926).

Henshaw

Wrentit ssp. *Chamaea fasciata henshawi* **Ridgway**, 1882
Dark-eyed Junco ssp. *Junco hyemalis henshawi* **A. R. Phillips**, 1962 NCR
[Unnecessary 'replacement' name for *Junco hyemalis cismontanus*]

Henry Wetherbee Henshaw (1850–1930) was a naturalist and ethnologist. He was the naturalist on the Wheeler survey of the American West (1872–1879). He worked for the US Bureau of Ethnology (1879–1893) and edited *American Anthropologist* (1888–1893). He visited Hawaii several times (1894–1904). He joined the US Department of Agriculture (1905), working in the Biological Survey, and became (1910) the official in charge. A reptile is named after him.

Henslow

Henslow's Sparrow *Ammodramus henslowii* **Audubon**, 1829

The Reverend John Stevens Henslow (1796–1861) was an English cleric, geologist and Professor of Botany, Cambridge University. He is honoured because he advised Audubon on booksellers in England when publishing *Birds of America*. Henslow was asked (1831) by Captain FitzRoy of the *Beagle* to recommend a young naturalist to join an expedition around the world without pay. Henslow picked a promising young student of his – Charles Darwin (q.v.). It was to Henslow that Darwin shipped all of the specimens he collected throughout his journey, and to whom he first told his theory of evolution. A devout Christian, Henslow always remained a sceptic, but he was pleased by the impact made by his former student.

Henson

Brown-eared Bulbul ssp. *Microscelis amaurotis hensoni* **Stejneger**, 1892 NCR
[NUI *Microscelis amaurotis amaurotis*]
Marsh Tit ssp. *Poecile palustris hensoni* Stejneger, 1892

Harry Vernon Henson (fl.1900) was an American who worked in Japan for the British trading house Jardine Matheson (1885). His collection of Japanese bird specimens was purchased by the USNM.

Henst

Henst's Goshawk *Accipiter henstii* **Schlegel**, 1873

Gideon van der Henst (DNF) was a former student at the Leiden Museum who assisted Douwe Van Dam (q.v.) on his second expedition to Madagascar (1868–1873).

Hepburn, D.

Spurfowl genus *Hepburnia* **Reichenbach**, 1852 NCR
[Now in *Galloperdix*]

Lieutenant-Colonel D. Hepburn (1788–1851) was an officer in the British army in India.

Hepburn, J. E.

Hepburn's Rosy Finch *Leucosticte tephrocotis littoralis* **Baird**, 1869
[Alt. Grey-crowned Rosy Finch]

James Edward Hepburn (1810–1869) qualified as a lawyer in London and moved to California (1852). He travelled very widely, between San Francisco and Vancouver, and once went as far north as Sitka. He did not confine himself to coastal regions, and made a number forays to the Rockies. His skins and specimens went to various places and people; Jardine (q.v.) and Tristram (q.v.) both had some, and others went to Baird (q.v.) at the USNM. His relatives presented what became the 'Hepburn Collection' to the University of Cambridge Zoological Museum (1870). It consists mostly of material from western North America, but also includes specimens from the Atlantic coast of the USA, from Chile and from some Pacific islands.

Heptner

> Short-toed Eagle ssp. *Circaetus gallicus heptneri*
> **Dementiev**, 1932 NCR; NRM

Vladimir Georgievich Heptner (1901–1975) was a Professor of Biology at the Laboratory of Zoological Geography and Taxonomy, the Vertebrates Zoology Department, Moscow State University. He worked on *Mammals in the Soviet Union* with others over a number of years; so far five volumes have been published. He was widely regarded as the leading mammalogist of the Soviet Union of his day. A mammal is named after him.

Herbert, E. G.

> Herbert's Babbler *Stachyris herberti* **E. C. S. Baker**, 1920
> [Alt. Sooty Babbler]

> Plain Prinia ssp. *Prinia inornata herberti* E. C. S. Baker, 1918
> Oriental Skylark ssp. *Alauda gulgula herberti* **Hartert**, 1923

Edward Greville Herbert (1870–1951) was an English collector and naturalist. He wrote an article on 'Nests and eggs of birds in Central Siam' in the *Journal of the Natural History Society of Siam* (1926).

Herbert (Alexander)

> Herbert's Woodland Warbler *Phylloscopus herberti*
> **B. Alexander**, 1903
> [Alt. Black-capped Woodland Warbler]

> Buff-spotted Woodpecker ssp. *Campethera nivosa herberti*
> B. Alexander, 1908

Herbert Alexander (1874–1946) was an artist and author who lived in Kent, England. His brother was Boyd Alexander (q.v.).

Herbert (Deignan)

> Philippine Trogon ssp. *Harpactes ardens herberti* **Parkes**, 1970

(See **Deignan**)

Herbert (Riker)

> Coraya Wren ssp. *Pheugopedius coraya herberti* **Ridgway**, 1888

Herbert Ashley Riker (d.1937) was an American whose father was the amateur naturalist Clarence B. Riker (q.v.). The family was based near Santarém, Brazil.

Hercules

> Blyth's Kingfisher *Alcedo hercules* **Laubmann**, 1917

> Hooded Butcherbird ssp. *Cracticus cassicus hercules*
> **Mayr**, 1940

Hercules, or in Greek Herakles, was the legendary strong man of classical mythology. He was famous for his twelve Labours, taking time out to be a member of the crew of the *Argo* and the hero of Euripides's tragedy *Alcestis*.

Herero

> Herero Chat *Namibornis herero* **de Schauensee**, 1931

This bird is not named after a person but a whole people. Most of the Herero people live in Namibia and the remainder are in Angola and Botswana.

Heriot

> Blue-breasted Flycatcher *Cyornis herioti* **R. G. W. Ramsay**, 1886

Frederick Maitland-Heriot (1852–1925) was a British collector who was in the Philippines (1884). A man of this name whom we suspect is the same person is recorded as being in Buenos Aires (1894–1898), and as a life member of the Argentine St Andrew's Society of the River Plate (for expatriate Scotsmen).

Herman

> Black-sided Robin ssp. *Poecilodryas hypoleuca hermani*
> **Madarász**, 1894

Dr Ottó Herman (1835–1914) was a Hungarian naturalist, entomologist, archaeologist and ornithologist at the Royal Hungarian Bureau for Ornithology. He wrote *The Method for Ornithophaenology Inaugurated by the Hungarian Central Office of Ornithology* (1905).

Herminier

> Guadeloupe Woodpecker *Melanerpes herminieri* **Lesson**, 1830

(See **L'Herminier, F. L.**)

Hermotimus

> Sunbird genus *Hermotimia* **Reichenbach**, 1854 NCR
> [Now in *Leptocoma*]

Hermotimus was a prophet and divine of Phocis and father of Aspasia, concubine of the Persian king Cyrus the Younger (d.401 BC).

Hernandez

> American White Pelican *Onocrotalus hernandezii* **Wagler**, 1832 NCR
> [JS *Pelecanus erythrorhynchos*]

Dr Francisco Hernandez (1514–1587) was a Spanish physician and botanist who trained at Universidad de Salamanca. He became personal physician to King Philip II of Spain, who ordered him to go to Mexico, where he travelled for several years (1570–1577), visiting many famous Aztec sites, particularly the renowned gardens, which the Spaniards had not yet destroyed. He consulted Aztec physicians on the medicinal qualities of plants. After he returned to Spain he handed in a huge written report, which Philip II put in the Royal Library, Monastery of Escorial, and never had published. The monastery and library were destroyed by fire (1671), but some of Hernandez's work survives through copies others had made. Two reptiles are named after him.

Hernsheim

Purple-capped Fruit Dove ssp. *Ptilinopus porphyraceus hernsheimi* **Finsch**, 1880

Franz Hernsheim (1845–1909) was German Consul in the Marshall Islands (1878–1885) and manager of the Jaluit Trading Company, which administered German possessions in Micronesia, before returning to Germany (1886). He wrote *Südsee-Erinnerungen (1875–1880)* to which Finsch wrote the foreword.

Herran

Herran's Thornbill *Chalcostigma herrani* **de Lattre** & **Bourcier**, 1846
[Alt. Rainbow-bearded Thornbill]

Pedro Alcantara Herran (1800–1872) was a Colombian general, politician and President of New Granada (Colombia) (1841–1844).

Herse

Swallow genus *Herse* **Lesson**, 1837 NCR
[Now in *Hirundo*]

In Greek mythology, Herse was the beautiful daughter of Cecrops, King of Athens.

Hershkovitz

Black Tinamou ssp. *Tinamus osgoodi hershkovitzi* **E. R. Blake**, 1953

Philip Hershkovitz (1900–1997) was an American zoologist. He was with the Field Museum of Natural History in Chicago and a specialist on primates, particularly Neotropical primates. He had a reputation for 'being ornery and combative as well as a loner' yet he was 'very responsive to anyone interested in his work'. He wrote *Living New World Monkeys* (1977). Three mammals are named after him.

Hesse

Grey-headed Woodpecker ssp. *Picus canus hessei* **Gyldenstolpe**, 1916

Dr Erich Hesse (1874–1945) was a German zoologist and ornithologist who worked at the zoological museums in Leipzig and Berlin.

Heude

Heude's Parrotbill *Paradoxornis heudei* **David**, 1872
[Alt. Reed Parrotbill]

Pierre Marie Heude (1836–1902) was a French conchologist and Jesuit missionary who collected and described specimens whilst living in China (late 1800s). Whilst his records were preserved and used by others, his collection was thought lost for over 100 years. However, Australian scientists working in Beijing rediscovered it, still with labels in his handwriting. He wrote *Mémoires concernant l'Histoire Naturelle de l'Empire Chinois, par des Pères de la Compagnie de Jésus*, which was published in a series of instalments, bound in different volumes printed in Shanghai. His best-known work is probably *Conchyliologie fluviatile de la province de Nanking (et la Chine Centrale)*. He died at Zi-ka-wei (Xujiahui, suburb of Shanghai) in China. A mammal is named after him.

Heuglin

Turaco genus *Heuglinornis* **Boetticher** 1935 NCR
[Now in *Tauraco*]

Heuglin's Bustard *Neotis heuglinii* **Hartlaub**, 1859
Heuglin's White-eye *Zosterops poliogastrus* Heuglin, 1861
[Alt. Montane White-eye, Broad-ringed White-eye]
Heuglin's Courser *Rhinoptilus cinctus* Heuglin, 1863
[Alt. Three-banded Courser]
Heuglin's Francolin *Pternistis icterorhynchus* Heuglin, 1863
Heuglin's Robin Chat *Cossypha heuglini* Hartlaub, 1866
[Alt. White-browed Robin Chat]
Heuglin's Wheatear *Oenanthe heuglini* **Finsch** & Hartlaub, 1870
Heuglin's Masked Weaver *Ploceus heuglini* **Reichenow**, 1886
Coucal sp. *Centropus heuglini* **Neumann**, 1911 NCR
[Alt. Blue-headed Coucal; JS *Centropus monachus fischeri*]

Heuglin's Gull *Larus fuscus heuglini* Bree, 1876
[Alt. Lesser Black-backed Gull ssp.]
Rusty Babbler ssp. *Turdoides rubiginosa heuglini* **Sharpe**, 1883
Brown-backed Woodpecker ssp. *Dendropicos obsoletus heuglini* Neumann, 1904
Swallow-tailed Bee-eater ssp. *Merops hirundineus heuglini* Neumann, 1906
Half-collared Kingfisher ssp. *Alcedo semitorquata heuglini* **Laubmann**, 1925 NCR; NRM

Martin Theodor von Heuglin (1824–1876) was a German mining engineer and ornithologist. He was the son of a pastor and a vocal opponent of evolutionary theories. He was born in Ditzingen, where the local school is named after him. His first visit to Africa was to Egypt, where he learnt Arabic (1850) and visited the Sinai Peninsula with Alfred Brehm (q.v.). He then got a job with the Austrian consulate in Khartoum, which allowed him to travel extensively in Abyssinia. He journeyed (1857–1858) to East Africa, an expedition financed by Archduke Ferdinand Max of Austria. Not only did he catch malaria on this trip, but he was lucky to survive after he was speared in the neck by an irate local. He explored (1861) in Abyssinia (Ethiopia), publishing an account of these travels in *Reise nach Abessinien, den Gala-ländern, Ostsudan und Chartum in den Jahren (1861–1862)* to which Brehm wrote the foreword. He subsequently visited other parts of eastern Africa and synthesised his knowledge of birds there in

Ornithologie von Nordost-Afrika (1869–1875). He visited Spitsbergen (1870) and the North Polar Sea (1871) before returning finally to Egypt and Abyssinia (1875–1876). He died of pneumonia and is buried in the Prague cemetery in Stuttgart. Two mammals are named after him.

Heurn

Grand Munia ssp. *Lonchura grandis heurni* **Hartert**, 1932
Rusty Pitohui ssp. *Pitohui ferrugineus heurni* Hartert, 1932 NCR
[JS *Pitohui ferrugineus holerythrus*]
White-faced Robin ssp. *Tregellasia leucops heurni* Hartert, 1932

(See **van Heurn**)

Hewitt

Clapper Lark sp. *Megalophonus hewitti* **J. A. Roberts**, 1926 NCR
[Alt. Eastern Clapper Lark; JS *Mirafra fasciolata*]

Rufous-eared Warbler ssp. *Malcorus pectoralis hewitti* J. A. Roberts, 1932 NCR
[JS *Malcorus pectoralis pectoralis*]
White-throated Canary ssp. *Serinus albogularis hewitti* J. A. Roberts, 1937
[Syn. *Crithagra albogularis hewitti*]

Dr John Hewitt (1880–1961) was the British-born Director of the Albany Museum, Grahamstown, South Africa (1910–1958). He had been Curator of the Sarawak Museum (1905–1908) during which time he collected entomological specimens in Borneo. Thereafter he was Assistant, Lower Vertebrates, Transvaal Museum (1909–1910). There he commenced his systematic work on the South African arachnids. He studied vertebrate zoology and archaeology. Among other works he wrote *A Guide to the Vertebrate Fauna of the Eastern Cape Province. Part 1 Mammals and Birds* (1931). A mammal, two reptiles and five amphibians are named after him.

Hey

Hey's Sand Partridge *Ammoperdix heyi* **Temminck**, 1825
[Alt. Sand Partridge]

Michael Hey (1796–1832) was a German surgeon, collector and alcoholic. Rüppell (q.v.) took him on as a collector on his expedition to Sinai and Nubia (1821–1827) and was infuriated by Hey's drunkenness. His behaviour occasionally jeopardised the entire expedition as, when drunk, he became very negligent, as evidenced in a letter from Rüppell to Cretzschmar (q.v.). Hey went back to Germany after the expedition, but returned shortly afterwards to Egypt. He died in Cairo; happily he and Rüppell were reconciled before his death.

Heywood

Heywood's Plover *Charadrius heywoodi* **G. R. Gray**, 1848 NCR
[Alt. White-fronted Plover; JS *Charadrius marginatus*]

Dr Thomas Richard Heywood Thomson (c.1813–1876) was a British naval surgeon who was a member of the Niger Expedition (1841–1842).

Higham

Pheasant Coucal ssp. *Centropus phasianinus highami* **Mathews**, 1922 NCR
[JS *Centropus phasianinus melanurus*]

Walter Ernest Higham (fl.1936) was a British aviculturist and bird photographer.

Hildebrand

Ashy Bulbul ssp. *Hemixos flavala hildebrandi* **Hume**, 1874

Lieutenant-General C. P. Hildebrand (1833–1888) served in the British Army in India (1849–1883) and collected in Burma.

Hildebrandt

Hildebrandt's Francolin *Pternistis hildebrandti* **Cabanis**, 1878
Hildebrandt's Starling *Lamprotornis hildebrandti* Cabanis, 1878

Johann Maria Hildebrandt (1847–1881) was a German botanist and explorer who collected and travelled in Arabia, East Africa, Madagascar and the Comoro Islands (1872–1881). He was also interested in languages and (1876) published *Zeitschrift für Ethiopia*, which deals with the vocabularies of dialects in the Johanna Islands. He died in Madagascar of yellow fever. His father, who was a painter and entomologist, named a beetle after him. Additionally, three reptiles, two mammals and an amphibian are named after him.

Hildegard (Heinrich)

Black-faced Canary ssp. *Serinus capistratus hildegardae* **Rand** & **Traylor**, 1959
[Syn. *Crithagra capistrata hildegardae*]

Mrs Hildegard Heinrich (fl.1954) was the second wife of Gerd Heinrich (q.v.). She accompanied her husband on a number of his expeditions – to the Celebes (Sulawesi) (1930–1932) and to Angola (1954).

Hildegard (Hinde)

Buff-bellied Warbler *Euprinodes hildegardae* **Sharpe**, 1899 NCR
[JS *Phyllolais pulchella*]

Mrs Hildegard Beatrice Hinde *née* Ginsburg (1871–1959) was a linguist who wrote on East African languages and was the wife of Dr Sidney Langford Hinde (q.v.). She and her husband were in the Congo (1891–1894) and Kenya (1895–1915). Three mammals are named after her.

Hileret

Cordilleran Canastero ssp. *Asthenes modesta hilereti* **Oustalet**, 1904

Clodomiro Hileret (1852–1903) was a French sugar-mill owner at Santa Ana in Tucumán, Argentina. He was known as being very helpful to visiting naturalists, letting them study in his plantations, etc.

Hilgert

Ovampo Sparrowhawk *Accipiter hilgerti* **Erlanger**, 1904 NCR
[JS *Accipiter ovampensis*]
Eurasian Jay *Garrulus hilgerti* **Kleinschmidt**, 1940 NCR
[JS *Garrulus glandarius glandarius*]

Rosy-patched Bush-shrike ssp. *Rhodophoneus cruentus hilgerti* **Neumann**, 1903
[Syn. *Telophorus cruentus hilgerti*]
Brubru ssp. *Nilaus afer hilgerti* Neumann, 1907
[SII *Nilaus afer minor*]
Buff-crested Bustard ssp. *Lophotis gindiana hilgerti* Neumann, 1907 NCR; NRM
Speckled Mousebird ssp. *Colius striatus hilgerti* **Zedlitz**, 1910
Grey Wren Warbler ssp. *Calamonastes simplex hilgerti* Zedlitz, 1912 NCR; NRM
Thekla Lark ssp. *Galerida theklae hilgerti* **Rothschild & Hartert** 1912 NCR
[JS *Galerida theklae superflua*]
Reichenow's Seedeater ssp. *Serinus angolensis hilgerti* Zedlitz, 1912 NCR
[JS *Serinus reichenowi*]
Ring-necked Dove ssp. *Streptopelia capicola hilgerti* Zedlitz, 1913 NCR
[JS *Streptopelia capicola somalica*]
Northern Crombec ssp. *Sylvietta brachyura hilgerti* Zedlitz, 1916 NCR
[JS *Sylvietta brachyura leucopsis*]
Eurasian Jackdaw ssp. *Coloeus monedula hilgerti* Kleinschmidt, 1935 NCR
[JS *Coloeus monedula spermologus*]

Carl Hilgert (1866–1940) was a German collector in North Africa (1896–1901 and 1909–1903) and the Near East (1907).

Hill, G. F.

Black-tailed Whistler ssp. *Pachycephala melanura hilli* **A. J. Campbell**, 1910 NCR
[Intergrade between *P. melanura melanura* and *P. melanura robusta*]
Northern Rosella ssp. *Platycercus venustus hilli* **Mathews**, 1910
Masked Finch ssp. *Poephila personata hilli* Mathews, 1923 NCR
[JS *Poephila personata personata*]

Gerald Freer Hill (1880–1954) was a zoologist whose major research work was a taxonomic study of termites. He was naturalist and photographer of the Commonwealth Government Exploration Party, led by Henry Barclay, which explored Central Australia and the Northern Territory (1911). Alfred Ewart, Government Botanist, Victoria, told Hill that he would be paid five shillings for every new species he collected. Hill collected more than 600 specimens. He stayed in the Northern Territory (1911–1917). He worked at the Australian Institute of Tropical Medicine (1919–1922) and then for CSIR (1928–1941). A reptile is named after him.

Hill, R.

Bahama Mockingbird ssp. *Mimus gundlachii hillii* **W. T. March**, 1864

Richard Hill (1795–1872) was a Jamaican politician, naturalist and author, who co-wrote *The Birds of Jamaica* (1847).

Himalia

Hummingbird genus *Himalia* **Mulsant**, 1876 NCR
[Now in *Ocreatus*]

In Greek mythology, Himalia was a nymph seduced by Zeus. She bore three sons to the god.

Hinde

Hinde's Pied Babbler *Turdoides hindei* **Sharpe**, 1900

Pectoral-patch Cisticola ssp. *Cisticola brunnescens hindii* Sharpe, 1896

Dr Sidney Langford Hinde (1863–1930) was Medical Officer of the Interior in British East Africa and a Captain in the Congo Free State Forces, as well as a naturalist and collector. He was also a Provincial Commissioner in Kenya and collected there. He wrote *The Fall of The Congo Arabs* (1897) and co-wrote *The Last of the Masai* (1901). Three mammals and a reptile are named after him.

Hindwood

Eungella Honeyeater *Bolemoreus hindwoodi* Longmore & Boles, 1983
[Syn. *Lichenostomus hindwoodi*]

Keith Alfred Hindwood (1904–1971) was an Australian businessman and ornithologist. He was President of the RAOU (1944–1946).

Hitchcock

Scaled Flowerpiercer ssp. *Diglossa duidae hitchcocki* **W. H. Phelps** & **W. H. Phelps Jr**, 1948

Dr Charles Baker Hitchcock (1906–1969) was an American geographer, cartographer and explorer. Harvard awarded his bachelor's degree in geology (1928); then he immediately went on the AMNH Tyler Duida expedition to southern Venezuela (1928–1929). He joined the American Geographical Society and worked on *Map of Hispanic America* (1930–1947). Columbia University, New York, awarded his masters (1933). He returned many times to Venezuela as a field scientist for the Phelps family. He was Assistant Director of the American Geographical Society (1943–1952), becoming Director (1953–1967).

Hiwa

Saipan Reed Warbler *Acrocephalus hiwae* Yamashina, 1942

Minori Hiwa (DNF) was the assistant of the describer, Yamashina (q.v.).

Hoa

Vietnamese Cutia ssp. *Cutia legalleni hoae* Eames, 2002

Dinh Thi Hoa is a Vietnamese businesswoman and supporter of ornithology, especially in Vietnam. She was educated at the Moscow State Institute of International Relations in the Soviet Union (Russia) and returned to Hanoi to work for the Ministry of Foreign Affairs Press Centre. She was recommended for a World Bank Scholarship at Harvard Business School from where she graduated (1992) and then worked as a brand manager for Proctor & Gamble in Bangkok (1992–1993). She returned to Hanoi (1994) and established Galaxy Company, a consulting firm of which she is managing director.

Hochstetter, A.

Reischek's Parakeet *Cyanoramphus hochstetteri* **Reischek**, 1889
[Alt. Antipodes Red-crowned Parakeet]

Arthur von Hochstetter (DNF) was the son of C. G. F. Hochstetter (below). Reischek wrote that he was naming the parakeet '... after Arthur von Hochstetter, the son of a sincere friend from whom I received many kindnesses, and who has too soon passed away'.

Hochstetter, C. G. F.

South Island Takahe *Porphyrio hochstetteri* **A. B. Meyer**, 1883

Christian Gottlieb Ferdinand Ritter von Hochstetter (1829–1884) was an Austrian geologist. He studied theology at Tübingen University (1847–1851), gaining a PhD (1852). He joined the Austrian Geological Survey (1853) following an invitation from Professor Wilhelm von Haidinger, and was subsequently appointed Chief Geologist for Bohemia and (1856) admitted as a lecturer to the University of Vienna. With Haidinger's support he found favour at the Austrian Court and was offered the position of Official Geologist to the *Novara* expedition, an Austrian Imperial Frigate scientific circumnavigation that took in Australia and New Zealand (1858–1859). Hochstetter carried out extensive investigations of the geology of various Pacific islands and parts of Asia, before heading south. He wrote many scientific papers but also popular works on botany, geology and palaeontology. After returning to Austria (1860) he married an Englishwoman, received a knighthood from the King of Württemberg, and was appointed Professor of Mineralogy and Geology at the Royal and Imperial Polytechnic Institute in Vienna (1860–1874). He continued to teach, work in the field, write prolifically and was President of the Geographical Society of Vienna (1867–1882). Poor health (diabetes and bronchitis) caused him to relinquish this position. Hochstetter was selected (1872) by the Austrian Emperor Franz Joseph as tutor in natural history to Crown Prince Rudolph. He was (1876–1884) first Superintendent of the Imperial Natural History Museum in Vienna. Obituary notices spoke of his genius, skills as an educator, and breadth of geological researches. An amphibian is named after him.

Hocking

Hocking's Parakeet *Aratinga hockingi* Arndt, 2006 NCR
[Taxonomy disputed; may not be separate from *Aratinga mitrata*]

Koepcke's Screech Owl ssp. *Megascops koepckeae hockingi* Fjeldså *et al.* 2012
[Alt. Apurimac Screech Owl ssp.]

Dr Peter 'Pedro' J. Hocking is a Peruvian zoologist, biologist and ornithologist at the Peruvian Natural History Museum in Lima. He was born in Peru to American missionary parents and studied at Bible College in the USA before returning to Peru (1963), renouncing his American citizenship, and taking Peruvian nationality. He is interested in cryptozoology, for example writing 'Large Peruvian mammals unknown to zoology' (1992). An amphibian is named after him.

Hodge

Andaman Woodpecker *Dryocopus hodgei* **Blyth**, 1860

Captain S. Hodge (1792–1876) was the naval officer who commanded the guardship at Port Blair, Andaman Islands (1860).

Hodgen

Hodgen's Rail *Gallinula hodgeni* **Scarlett** 1955 EXTINCT
[Alt. Hodgens's Waterhen; binomial often amended to *hodgenorum*]

Joseph Hodgen (DNF) and his son, Rob, owned the swamp in Pyramid Valley, North Canterbury, New Zealand, where the bones of this rail were first found by Ron Scarlett (q.v.). Originally described as *Rallus hodgeni*, the American ornithologist Storrs Olson (q.v.) transferred it into the genus *Gallinula* and amended the binomial to the plural form *hodgenorum*. The Hodgens also discovered 17 complete skeletons of the Giant Moa *Dinornis maximus* when they were digging a hole in which to bury a dead horse (1936).

Hodgson, A.

Hodgson's Partridge *Perdix hodgsoniae* **Hodgson**, 1857
[Alt. Tibetan Partridge]

Anne Hodgson *née* Scott (d.1868) was the wife of the British naturalist Brian Hodgson (q.v.), who described the bird and named it after her.

Hodgson, B. H.

White-bellied Redstart genus *Hodgsonius* **Bonaparte**, 1850

Hodgson's Rosefinch *Carpodacus erythrinus* **Pallas**, 1770
[Alt. Common Rosefinch]
Hodgson's Pigeon *Columba hodgsonii* **Vigors**, 1832
[Alt. Speckled Wood Pigeon]
Hodgson's Hawk Eagle *Nisaetus nipalensis* Hodgson, 1836
[Alt. Mountain Hawk Eagle]
Hodgson's Mountain Finch *Leucosticte nemoricola* Hodgson, 1836
[Alt. Plain Mountain Finch]

Hodgson's Fulvetta *Fulvetta vinipectus* Hodgson, 1837
[Alt. White-browed Fulvetta]
Hodgson's Grandala *Grandala coelicolor* Hodgson, 1843
Hodgson's Hawk Cuckoo *Hierococcyx nisicolor* **Blyth**, 1843
[Syn. *Hierococcyx fugax nisicolor*]
Hodgson's Prinia *Prinia hodgsonii* Blyth, 1844
[Alt. Grey-breasted Prinia]
Hodgson's Bushchat *Saxicola insignis* **G. R. Gray** &
   J. E. Gray, 1847
[Alt. White-throated Bushchat]
Broad-billed Warbler *Tickellia hodgsoni* F. Moore, 1854
Hodgson's Redstart *Phoenicurus hodgsoni* F. Moore, 1854
Pygmy Blue Flycatcher *Muscicapella hodgsoni* F. Moore,
   1854
Hodgson's Frogmouth *Batrachostomus hodgsoni* G. R. Gray,
   1859
Hodgson's Treecreeper *Certhia hodgsoni* **W. E. Brooks,** 1871
Hodgson's Tree Pipit *Anthus hodgsoni* Richmond, 1907
[Alt. Olive-backed Pipit]

Hodgson's Imperial Pigeon *Ducula badia insignis* Hodgson,
   1836
[Alt. Mountain Imperial Pigeon ssp.]
Hodgson's Munia *Lonchura striata acuticauda* Hodgson,
   1836
[Alt. White-rumped Munia ssp.]
Hodgson's Pied Wagtail *Motacilla alba alboides* Hodgson,
   1836
Red-headed Trogon ssp. *Harpactes erythrocephalus*
   hodgsonii **Gould**, 1838 NCR
[NUI *Harpactes erythrocephalus erythrocephalus*]
Hodgson's Broadbill *Serilophus lunatus rubropygius*
   Hodgson, 1839
[Alt. Silver-breasted Broadbill ssp.]
White-bellied Woodpecker ssp. *Dryocopus javensis*
   hodgsonii **Jerdon**, 1840
Hodgson's Barbet *Megalaima lineata hodgsoni* Bonaparte,
   1850
[Alt. Lineated Barbet ssp.]
Large-tailed Nightjar ssp. *Caprimulgus macrurus hodgsoni*
   E. C. S. Baker, 1930 NCR
[JS *Caprimulgus macrurus albonotatus*]
Golden Eagle ssp. *Aquila chrysaetos hodgsoni* **Ticehurst**,
   1931 NCR
[JS *Aquila chrysaetos daphanea*]

Brian Houghton Hodgson (1800–1894) was an official of the
East India Company and Assistant Resident in Nepal (1825–
1843), then Darjeeling (1845–1859). He was very interested in
Buddhism, and was among those who introduced it to Britain,
and in the languages of Nepal and northern India. He
amassed 9,512 specimens of birds, belonging to 672 species,
of which 124 had never been described previously. He
described 79 of these, but the rest were described by others .
A fiercely patriotic man, he once said that Cuvier (q.v.), who
had stolen a march on Hardwicke (q.v.) by naming the red
panda *Ailurus fulgens* because Hardwicke's return to
England was delayed, would 'prevent England reaping the
zoological harvest of her own domains'. He described no
fewer than 138 different passerines and lodged many specimens with the BMNH. Five mammals and a reptile are named
after him.

Hoedt

Wetar Ground Dove *Gallicolumba hoedtii* **Schlegel**, 1871

Jungle Hawk Owl ssp. *Ninox theomacha hoedtii* Schlegel,
   1871
Northern Fantail ssp. *Rhipidura rufiventris hoedti* **Büttikofer**,
   1892

Dirk Samuel Hoedt (1813–1893) was a Dutch civil servant,
being Secretary for the Moluccas (1854–1855), where he also
collected (1853–1867).

Hoerning

Eurasian Nuthatch ssp. *Sitta europaea hoerningi*
   Kleinschmidt, 1928 NCR
[JS *Sitta europaea caesia*]

R. Hoerning (1858–1940) was a German field ornithologist
who was active in Thuringia (1893–1928).

Hoesch

Mountain Pipit *Anthus hoeschi* **Stresemann**, 1938

Sabota Lark ssp. *Calendulauda sabota hoeschi* Stresemann,
   1939 NCR
[NUI *Calendulauda sabota waibeli*]
Familiar Chat ssp. *Cercomela familiaris hoeschi*
   Niethammer, 1955 NCR
[JS *Cercomela familiaris angolensis* (and invalid)]
Gray's Lark ssp. *Ammomanopsis grayi hoeschi* Niethammer,
   1955
Tractrac Chat ssp. *Cercomela tractrac hoeschi*
   Niethammer, 1955
Chestnut-backed Sparrow Lark ssp. *Eremopterix leucotis*
   hoeschi **C. M. N. White**, 1959

Dr Walter Hoesch (1896–1961) was a German soldier who
was invalided out of the army (WW1), after which he trained
as a lawyer. He moved to South West Africa (Namibia) (1930)
and bought a farm, but was forced to sell up (1932) after a
disastrous drought. He started to collect specimens for
Professor Karl Jordan, an entomologist at Lord Rothchild's
Tring Museum and (1933) accompanied Jordan on his South
West African tour, and developed relationships with other
zoologists including Stresemann (q.v.). He spent the rest of
his life in South West Africa and made a private collection of
small animals. He wrote 'Die Vögelwelt Deutsches-
Südwestafrikas namentlich des Damara- und Namalandes'
(1940) and about 50 other papers on various scientific
subjects. A reptile and an amphibian are named after him.

Hoevell

Hoevell's Flycatcher *Cyornis hoevelli* **A. B. Meyer**, 1903
[Alt. Blue-fronted Blue Flycatcher]

Baron Gerrit Willem Wolter Carel van Hoevell (1848–1920)
was the Dutch colonial Governor of Celebes (Sulawesi)
(1902). He wrote *Ambon en Meer Bepaaldelijk de Oeliasers*
(1875).

Hoeven

Bare-eyed Rail ssp. *Gymnocrex plumbeiventris hoeveni*
H. K. B. Rosenberg, 1866

Dr Jan van der Hoeven (1801–1868) was a Dutch zoologist and physician who took degrees in physics (1822) and medicine (1824) at Leiden University, then practised as a physician (1824–1826). He became Professor of Zoology and Mineralogy at Leiden University (1826–1868) and was a traditionalist, being one of the last professors at Leiden to teach in Latin.

Hoffmann

Hoffmann's Conure *Pyrrhura hoffmanni* **Cabanis**, 1861
[Alt. Sulphur-winged Parakeet]
Hoffmann's Woodpecker *Melanerpes hoffmannii* Cabanis, 1862

Steely-vented Hummingbird ssp. *Saucerottia saucerrotei hoffmanni* Cabanis & **Heine**, 1860
[Syn. *Amazilia saucerrottei hoffmanni*]
Hoffmann's Antthrush *Formicarius analis hoffmanni* Cabanis, 1861
[Black-faced Antthrush ssp.]
Slaty-tailed Trogon ssp. *Trogon massena hoffmanni* Cabanis & Heine, 1863

Carl H. Hoffmann (1823–1859) was a German physician. He went to Costa Rica (1853) with Alexander von Frantzius (q.v.) and began to explore the country and collect mainly botanical specimens. He was later a physician to the Costa Rican army. A mammal and a reptile are named after him.

Hoffmanns

White-breasted Antbird *Rhegmatorhina hoffmannsi*
Hellmayr, 1907
Hoffmanns's Woodcreeper *Dendrocolaptes hoffmannsi*
Hellmayr, 1909

Cinereous Antshrike ssp. *Thamnomanes caesius hoffmannsi*
Hellmayr, 1906
Ornate Antwren ssp. *Epinecrophylla ornata hoffmannsi*
Hellmayr, 1906
Little Tinamou ssp. *Crypturellus soui hoffmannsi* **Brabourne**
& **Chubb**, 1914 NCR
[JS *Crypturellus soui albigularis*]

Wilhelm Hoffmanns (1865–1909) was a German naturalist who collected in Peru and Brazil (1903–1908). A mammal is named after him.

Höfling

Velvety Black Tyrant ssp. *Knipolegus nigerrimus hoflingi*
Lencioni-Neto, 1996

Dr Elizabeth Höfling is a Brazilian zoologist and Professor at the Department of Zoology, University of São Paulo. Universidade Estadual Paulista Júlio de Mesquita Filho awarded her Bachelor's degree (1973) and Universidade de São Paulo her doctorate in zoology (1979). Her post-doctoral work was at MNHN, Paris, of which she became a correspondent (2004).

Hofmann

Hofmann's Sunbird *Cinnyris hofmanni* **Reichenow**, 1915
[Syn. *Cinnyris shelleyi hofmanni*]

Hofmann (DNF) was a collector in Tanganyika (Tanzania). Reichenow (q.v.) gives no further details in his very brief description.

Holboell

Holboell's Redpoll *Carduelis flammea holboellii* **C. L. Brehm**, 1831
[Alt. Common Redpoll ssp.; NUI *Carduelis flammea flammea*]
Holboell's Grebe *Podiceps grisegena holbollii* **Reinhardt**, 1854
[Alt. Red-necked Grebe ssp.]

Carl Peter Holboell (1795–1856) was a captain in the Danish Royal Navy who served in Greenland and became interested in natural history there. At least one Greenland plant is named after him. Holboell was presumed to have died at sea when the ship on which he was sailing to Greenland from Denmark disappeared without trace.

Holderer

Pheasant sp. *Phasianus holdereri* **Schalow**, 1901 NCR
[Alt. Common Pheasant; JS *Phasianus colchicus strauchi*]

Tibetan Lark ssp. *Melanocorypha maxima holdereri*
Reichenow, 1911

Dr Julius Holderer (1866–1950) was a German entomologist and collector in China and Tibet (1897–1899), after which he became a civil servant and district magistrate until retiring (1931).

Holdsworth

Sri Lanka Scimitar Babbler ssp. *Pomatorhinus melanurus holdsworthi* **Whistler**, 1942

Edmund William Hunt Holdsworth (1829–1915) was a British zoologist who had a particular interest in fisheries. He was in Ceylon (Sri Lanka) (1868) and wrote *Catalogue of the birds found in Ceylon : with some remarks on their habits and local distribution, and descriptions of two new species peculiar to the island* (1872).

Holland

Versicolored Emerald ssp. *Amazilia versicolor hollandi*
W E C. **Todd**, 1913

Dr William Jacob Holland (1848–1932) was a Jamaican-born American Presbyterian minister, entomologist and palaeontologist, and Director of the Carnegie Museum (1898–1922).

Holliday

Cape Batis ssp. *Batis capensis hollidayi* **Clancey**, 1952

Clayton S. Holliday (b.1930) was a South African archaeologist and zoologist at the Rhodes-Livingstone Museum (Livingstone Musem), Northern Rhodesia (Zambia) (1957).

Holmberg

Grey-headed Kite *Micraëtus holmbergianus* **Bertoni**, 1901
NCR
[JS *Leptodon cayanensis*]

Barn Owl ssp. *Tyto alba holmbergiana* Bertoni, 1901 NCR
[JS *Tyto alba tuidara/Tyto furcata tuidara*]

Dr Eduardo Ladislao Holmberg (1852–1937) was an Argentinian zoologist who was Director of the Buenos Aires Botanical Gardens. He was also a novelist, the first South American to write science fiction.

Holmes, D. A.

Scaly-breasted Munia ssp. *Lonchura punctulata holmesi*
Restall, 1992
[SII *Lonchura punctulata nisoria*]

Derek Anthony Holmes (1938–2000) was a field-ornithologist and conservationist who lived and collected in Indonesia. He was the first editor of *Kukila*, the Bulletin of the Indonesian Ornithologists' Union (1985). He wrote *Birds of Java and Bali* (1989).

Holmes, S.

Auckland Snipe *Scolopax holmesi* **Peale**, 1848 NCR
[Alt. Subantarctic Snipe; JS *Coenocorypha aucklandica*]

Dr Silas Holmes (1815–1849) graduated in medicine from Harvard (1836). He became Assistant Surgeon (1838) aboard the brig *Porpoise* (joining at Sydney), and later transferred to the sloop *Peacock*, during the US Exploration Expedition to the Pacific Ocean, Antarctica, Australia, the South Sea Islands, and the north-west coast of America (1838–1842). He also collected some natural history material when the *Porpoise* visited the Auckland Islands. He kept a journal, which is now in Yale's library. Holmes Glacier in Porpoise Bay, Antarctica, is named after him. He was promoted to Passed Assistant Surgeon (1843). He was noted for his ascerbic wit. He drowned when a boat capsized at Mobile Bay, Alabama.

Holroyd

Holroyd's Woodpecker *Micropternus brachyurus holroydi*
Swinhoe, 1870
[Alt. Rufous Woodpecker ssp.; Syn. *Celeus brachyurus
holroydi*]

Captain George Sowley Holroyd (1842–1870) was a British army officer in India with the 73rd Regiment of Foot. He met Swinhoe (q.v.) in Hainan and assisted him in collecting specimens. He died at sea when the regiment was en route to Ceylon (Sri Lanka).

Holst

Yellow Tit *Parus holsti* **Seebohm**, 1894

P. Aug. Holst (d.1895) was a Swedish (or possibly Norwegian) naturalist who was in Japan (1889) and Formosa (Taiwan) (1893–1894). During his time in Taiwan he ascended Mount

Yushan and collected many zoological specimens including the tit. He sent specimens to Boulenger (1880s). His date of death is known by a note taken of his long-gone gravestone in Taiwan. Among other taxa a spider from Taiwan is named after him, as is an amphibian.

Holt, E. G.

Saffron Finch ssp. *Sicalis flaveola holti* **W. de W. Miller**,
1925 NCR
[JS *Sicalis flaveola brasiliensis*]
Pale-browed Treehunter ssp. *Cichlocolaptes leucophrus
holti* Pinto, 1941
Red-legged Honeycreeper ssp. *Cyanerpes cyaneus holti*
Parkes, 1977

Ernest Golsan Holt (1889–1983) was an ornithologist, explorer and naturalist. He had a roving commission from the US Biological Survey as there are reports of his activities in Alabama, Alaska, North Dakota and Virginia. He visited South America a number of times (1926–1947). He was a biologist (1930s) with the US Soil Erosion Service. He co-wrote 'Birds of Autauga and Montgomery counties, Alabama' in *Auk* (1914). An amphibian also named after him. (See **Margaret (Holt)** and **Margaret (Boulton)**)

Holt, H. F. W.

Holt's Mountain Bulbul *Ixos mcclellandii holtii* **Swinhoe**,
1861

Henry Frederick William Holt (d.1890) was in the British consular service in China and a colleague of Robert Swinhoe (q.v.) who introduced him to ornithology and lent him his Chinese taxidermist when he started collecting. He was Acting Interpreter, Ningpo Consulate (1864–1876). He was noted as an orientalist and wrote 'Notes on the Chinese game of chess' (1885), was Honorary Secretary of the Royal Asiatic Society, and made a significant collection of Chinese coins.

Holtemüller

Yellow-necked Spurfowl ssp. *Pternistis leucoscepus
holtemulleri* **Erlanger**, 1904 NCR; NRM

H. Holtemüller (DNF) was a collector and explorer in Abyssinia (Ethiopia) (1903).

Holub

Holub's Golden Weaver *Ploceus xanthops* **Hartlaub**, 1862
[Alt. Large Golden Weaver]

Croaking Cisticola ssp. *Cisticola natalensis holubii* **Pelzeln**,
1882

Emil Holub (1847–1902) was a Bohemian (Czech) naturalist who also studied South African fossils. Like his father, he trained as a physician but was always fascinated by wildlife and foreign lands; his compelling ambition was to follow in the footsteps of David Livingstone. He was a physician, zoologist, botanist, hunter, taxidermist, artist and cartographer, an avid collector of specimens and, above all, a keen observer. His first trip to Africa was in 1872 and he practised medicine to pay his way. He travelled extensively in south-central Africa, gathering varied and valuable natural history material, including c.30,000 specimens! On his return from his first trip he wrote *Seven Years in Africa*. He took another trip (1883) to Africa, which ended in disaster after ten weeks when a number of the party died from malaria and all the equipment was lost. When he returned to Europe he fell upon hard times and was forced to sell much of his collection. He, too, eventually died from malaria, contracted on his second trip.

Hombron

Hombron's Kingfisher *Actenoides hombroni* **Bonaparte**, 1850
[Alt. Blue-capped Kingfisher]

Dr Jacques Bernard Hombron (1798–1852) was a French naval surgeon and naturalist. He was in the Pacific aboard the *Astrolabe*, which sailed with *Zélée* (1837–1840). He wrote *Adventures les Plus Curieuses des Voyageurs* (1847). He described a number of taxa with Jacquinot (q.v.).

Homeyer

White-vented Whistler *Pachycephala homeyeri* **Blasius**, 1890

Great Grey Shrike ssp. *Lanius excubitor homeyeri* **Cabanis**, 1873
Golden Eagle ssp. *Aquila chrysaetos homeyeri* **Severtzov**, 1888
Eurasian Nuthatch ssp. *Sitta europaea homeyeri* **Hartert**, 1892 NCR
[Transitional form: *Sitta europaea europaea/ S. europaea caesia*]

Eugen Ferdinand von Homeyer (1809–1899) was a German ornithologist who was opposed to the views of Charles Darwin (q.v.). He wrote *Verzeichniss der Vögel Deutschlands* (1885).

Hoogerwerf

Hoogerwerf's Pheasant *Lophura hoogerwerfi* **Chasen**, 1939
[Alt. Sumatran Pheasant; Syn. *Lophura inornata hoogerwerfi*]

Island Thrush ssp. *Turdus poliocephalus hoogerwerfi* Chasen, 1939 NCR
[JS *Turdus poliocephalus loeseri*]
Pale Cicadabird ssp. *Coracina ceramensis hoogerwerfi* Jany, 1955

Andries Hoogerwerf (1906–1977) wrote much on Indonesian fauna, particularly that of Java (1940s–1970s). His works include *De avifauna van Tjibodas en omgeving (Java)* (1949) and *Udjung Kulon: The Land of the Last Javan Rhinoceros* (1970). He is believed to have been the only person to photograph the now-extinct Javan Tiger (1938). A mammal is named after him.

Hoogstraal

Philippine Fairy-bluebird ssp. *Irena cyanogastra hoogstraali* **Rand**, 1948

Harold 'Harry' Hoogstraal (1917–1986) was an American expert in the field of medical zoology, parasitology, entomology and ecology. He gained two degrees from the University of Illinois (1938 and 1942), then served in the US Army (1942–1946). After the war he took two further degrees at the London School of Hygiene and Tropical Medicine. He made a number of field trips: to Mexico (1940) and to the Solomon Islands, New Guinea, and New Hebrides (1945) while serving at a military medical research establishment nearby. His next trip was to Mindanao and Palawan (1946–1947), then Africa (1948–1949), culminating in Madagascar and Egypt. He then moved to Egypt to organise and become Head of the Department of Medical Zoology, US Naval Medical Research, for the rest of his life. He was renowned for working 18-hour days and had over 500 publications and edited many more. He was a member of 30 professional societies and a volunteer in 20, collecting numerous professional honors. He amassed the biggest collection of ticks outside of the British Museum, donating them to the USNM. He died in Cairo on his 69th birthday from lung cancer. Three mammals and a reptile are named after him.

Hoopes

Eastern Meadowlark ssp. *Sturnella magna hoopesi* **W. Stone**, 1897

Josiah Hoopes (1832–1904) was an American botanist, oologist and collector. He started (1853) a tree nursery, now one of the largest in the USA.

Hoover

Hoover's Warbler *Dendroica coronata hooveri* **R. C. McGregor**, 1899; NRM
[Alt. Yellow-rumped Warbler; Syn. *Setophaga coronata*]

Theodore 'Tad' Jesse Hoover (1871–1955) was an American mining and metallurgical engineer, naturalist and elder brother of Herbert Hoover, President of the USA. Stanford awarded his bachelor's degree (1901), and after working in business he returned there (1919) as Professor of Mining and Metallurgy, eventually retiring as Emeritus Professor. The Theodore J. Hoover Natural Preserve in California is named after him.

Hopke

Hopke's Cotinga *Carpodectes hopkei* **Berlepsch**, 1897
[Alt. Black-tipped Cotinga]

Gustav Hopke (d.c.1902) was a German field collector who specialised in South American mammals. He collected mostly in Paraguay and Chile (c.1897), and in Colombia where the cotinga occurs. Berlepsch refers (1902) to 'the late' Mr Gustav Hopke.

Hopkinson, E.

Ahanta Francolin ssp. *Pternistis ahantensis hopkinsoni* **Bannerman**, 1934

Dr Emilius Hopkinson (1869–1951) graduated from Trinity College, Oxford, and did his medical training at St Thomas's

Hospital, London. He then served as a medical officer in the Imperial Yeomanry, going to South Africa (1900–1901) where he was awarded the DSO. He spent a decade serving as a medical officer in the Gambia (1901–1911), then being Travelling Commissioner (1911–1929) during which he was awarded CMG (1922). While in the Gambia he published two books on its law and the Mendengo language. His interest in birds extended to aviculture and he wrote the booklet *A List of Birds of Gambia* (1919) and the book *Records of Birds bred in Captivity* (1926). Whenever he returned home on leave he brought live birds with him for London Zoo, but on one such trip during WW1 a submarine sank his ship and his collections were lost. By WW2 he had to stay in the UK and no longer study winter migrations in West Africa. He was a very committed member of the BOC. Bannerman (q.v.) wrote in his etymology: '… I name this new race of Francolin after my friend Dr. E. Hopkinson, C.M.G., D.S.O., who during a long residence in Gambia has done so much to further our knowledge of the birds of that colony.'

Hopkinson, H.

Mountain Bamboo Partridge ssp. *Bambusicola fytchii hopkinsoni* **Godwin-Austen**, 1874

General Henry Hopkinson (1820–1899) was a surveyor and the British Commissioner in Assam, India (1861–1874).

Hopson

African Reed Warbler ssp. *Acrocephalus baeticatus hopsoni* Fry, 1974 NCR
[NUI *Acrocephalus baeticatus cinnamomeus*]

Antony John Hopson (b.1932) was in the Federal Fisheries Service, Nigeria, at the research station on Lake Chad. He wrote *Fisheries of Lake Chad* (1967).

Hopwood

Ashy Drongo ssp. *Dicrurus leucophaeus hopwoodi* **E. C. S. Baker**, 1918

Cyril Hopwood (b.1880) was British forestry officer in the Imperial Forest Service, India (1901–1924) and an oologist. He was in the Chin Hills of Burma (Myanmar) at intervals (1913–1915) and co-wrote 'A list of birds from the North Chin Hills' (1917).

Horice

Hazel Grouse ssp. *Tetrastes bonasia horicei* Hachler, 1950 NCR
[JS *Tetrastes bonasia styriacus*]

Dr Alfred Horice (1865–1945) was a Czech physician and amateur ornithologist who was a founding member of the Czechoslovak Society for Ornithology.

Horn

Fairy Tern ssp. *Sternula nereis horni* **Mathews**, 1912 NCR
[JS *Sternula nereis nereis*]

William Austin Horn (1841–1922) was an Australian politician, pastoralist and mining magnate who financed an expedition to central Australia (1894). He wrote *Bush Echoes* (1901).

Hornby

Hornby's Storm-petrel *Oceanodroma hornbyi* **G. R. Gray**, 1854
[Alt. Ringed Storm-petrel]

Admiral Sir Phipps Hornby (1785–1867) was the British commander-in-chief in the Pacific (1847–1850). He collected animal specimens for his brother-in-law, Edward Smith-Stanley, 13th Earl of Derby (q.v.). After his retirement (1853) Hornby continued to receive honours, being promoted full admiral (1858) and becoming a Knight Grand Cross of the Order of the Bath (1861).

Hornemann

Hornemann's Redpoll *Carduelis hornemanni* **Holboell**, 1843
[Alt. Arctic Redpoll, Hoary Redpoll]

Jens Wilken Hornemann (1770–1841) was a Danish botanist, Professor at the University of Copenhagen, who wrote *Flora Danica* (1806). He and Holboell (q.v.) were friends and Hornemann honoured Holboell by naming a plant after him.

Hornschuch

Hornschuch's Velvet Scoter *Melanitta hornschuchii* **C. L. Brehm**, 1824 NCR
[JS *Melanitta fusca*]

Dr Christian Friedrich Hornschuch (1793–1850) was a German bryologist and botanist who was apprenticed to a pharmacist (1808). He moved to Regensburg (1813) as assistant to the botanist, David Heinrich Hoppe, accompanying him on an expedition to the Adriatic (1816). He later worked for Heinrich Christian Funck and finally moved to Greifswald as a botanical demonstrator at the university, being appointed Assistant Professor of Natural History and Botany, a Director of the University of Greifswald's botanical gardens (1820) and Full Professor (1827).

Horsbrugh, B. R.

Red-necked Falcon *Falco horsbrughi* **Gunning** & **J. A. Roberts**, 1911
[Syn. *Falco chicquera horsbrughi*]

Lieutenant-Colonel Boyd Robert Horsbrugh (1871–1916) was an ornithologist and aviculturist who had remarkable success in breeding and rearing birds in South Africa, collaborating closely with his younger brother C. B. Horsbrugh (q.v.). He served in the Royal Warwickshire Regiment (1893–1895) in Ceylon (Sri Lanka) and in the Army Service Corps (1895–1911 and 1914–1916), seeing active service in Sierra Leone (1898–1899) and in South Africa (1899–1902). He was invalided home to England (1902), but returned after the end of the Boer War to various postings in South Africa (1905–1909). He retired (1911) after two years service in Ireland. He returned to the army (1914), served in France (1915), but his health failed and he was invalided home. He underwent a serious surgical operation (1916) but failed to recover. He wrote *The Game-Birds & Water-Fowl of South Africa* (1912).

Horsbrugh, C. B.

Red-backed Buttonquail ssp. *Turnix maculosus horsbrughi*
W. Ingrams, 1909

Charles Bell Horsbrugh (1874–1952), younger brother of B. R. Horsbrugh (q.v.), was a British aviculturist (1905) and naturalist at the Transvaal Museum, Pretoria, and a collector in Africa, North America and New Guinea. He hunted with his younger brother and took trips into the Veld with him to study birds in their natural habitat.

Horsfield

Horsfield's Babbler *Malacocincla sepiaria* Horsfield, 1821
[Alt. Horsfield's Jungle Babbler]
Horsfield's Bronze Cuckoo *Chrysococcyx basalis* Horsfield, 1821
[Alt. Narrow-billed Bronze Cuckoo]
Horsfield's Goshawk/Sparrowhawk *Accipiter soloensis* Horsfield, 1821
[Alt. Chinese Goshawk/Sparrowhawk]
Horsfield's Hill Partridge *Arborophila orientalis* Horsfield, 1821
[Alt. Grey-breasted Hill Partridge, White-faced Partridge]
Horsfield's Nightjar *Caprimulgus macrurus* Horsfield, 1821
[Alt. Large-tailed Nightjar]
Horsfield's Woodcock *Scolopax saturata* Horsfield, 1821
[Alt. Dusky Woodcock, Javan Woodcock]
Malabar Whistling Thrush *Myophonus horsfieldii* **Vigors**, 1831
Horsfield's Snipe *Scolopax horsfieldii* . **E. Gray**, 1831 NCR
[Alt. Pin-tailed Snipe; JS *Gallinago stenura*]
Indian Scimitar Babbler *Pomatorhinus horsfieldii* **Sykes**, 1832
Falconet sp. *Hierax horsfieldi* **Lesson**, 1843 NCR
[Alt. Black-thighed Falconet; JS *Microhierax fringillarius*]
Horsfield's Cuckoo *Cuculus optatus* **Gould**, 1845
[Alt. Oriental Cuckoo; *Cuculus horsfieldi* is JS]
Horsfield's Thrush *Zoothera horsfieldi* **Bonaparte**, 1857
[Alt. Spot-winged Thrush; Syn. *Zoothera dauma horsfieldi*]

Horsfield's Kalij (Pheasant) *Lophura leucomelanos lathami* Gray, 1829
[Alt. Black-breasted Kalij; *Gennaeus horsfieldii* is JS]
Green-tailed Sunbird ssp. *Aethopyga nipalensis horsfieldi*
Blyth, 1843
Horsfield's Lark *Mirafra javanica horsfieldii* Gould, 1847
[Alt. Horsfield's Bush Lark, Australasian Bush Lark ssp.]

Dr Thomas Horsfield (1773–1859) was an American naturalist. He studied medicine at the University of Pennsylvania, but became an explorer and prolific collector of plants and animals. He began his career in Java while it was under Dutch rule, but when Napoleon Bonaparte annexed Holland the British East India Company was able to take control (1811). Horsfield's poor health made him seek other employment (1819) and he was moved by the company to continue his research under their direction in London, as Curator and then Keeper of the India House Museum. Whilst in Java he became a good friend of Sir Thomas Stamford Raffles (q.v.). He wrote *Zoological Researches in Java and the*

Neighbouring Islands (1824). Five mammals and three reptiles are named after him.

Horus

Horus Swift *Apus horus* **Heuglin**, 1869

Horus, son of Osiris, was one of the gods of ancient Egypt, a falcon deity representing the sky.

Horvath

Crested Bobwhite ssp. *Colinus cristatus horvathi* **Madarász**, 1904

Dr Géza Horváth (1847–1937) was an entomologist. He was Director, Zoology Department, Hungarian Natural History Museum, Budapest (1895–1923), and took a similar post at the University of Budapest (1913). He was editor of *General Catalogue of the Hemiptera*. A reptile is named after him.

Hose

Black Oriole *Oriolus hosii* **Sharpe**, 1892
Hose's Broadbill *Calyptomena hosii* Sharpe, 1892
Hose's Partridge *Rhizothera dulitensis* **Ogilvie-Grant**, 1895
Flowerpecker sp. *Dicaeum hosii* Sharpe, 1897 NCR
[Alt. Scarlet-backed Flowerpecker; JS *Dicaeum cruentatum nigrimentum*]

Giant Pitta ssp. *Hydrornis caeruleus hosei* **E. C. S. Baker**, 1918

Dr Charles Hose (1863–1929) was a naturalist who lived in Sarawak and Malaysia (1884–1907). He was 'Resident of Baram' (1891–1903), then transferred to Sibu (1903–1907). Hose successfully investigated the principal cause of the disease beri-beri. He was also a good cartographer who produced the first reliable map of Sarawak. He sent huge collections of zoological, botanical and ethnographic material to many museums and institutions, including the British Museum and at least four British universities, one of which (Cambridge) awarded him an honorary DSc (1900). He was an extremely bulky man, which meant that when he went to visit the local tribes in their long houses they had to reinforce the floors. He was still remembered (1995) for his extreme size! We are indebted to a member of his family for some reminiscences of him. He successfully put a stop to the head-hunting raids among the various villages by the simple expedient of organising a boat-race *à la* the University Boat Race over a similar distance and a similar sinusoidal course (although there were twenty-two dugouts with crews of 70+) (1899). This satisfied their honour, apparently. Another story concerns a journey on the Trans-Siberian Railway when the train stopped near Lake Baikal and he acquired three live Baikal seals which he put in the luggage rack! Not surprisingly they died. As each succumbed he skinned them on the train, to the interest and surprise of his fellow travellers. He returned to Sarawak several times after retirement, very possibly in connection with the development of the oilfields at Miri (they are still producing). He became an expert on the production of acetone (used in the manufacture of cordite) as he ran a factory for it at Kings Lynn (WW1). The raw materials for making acetone were maize and horse chestnuts. He wrote *Fifty Years of Romance and Research* (1927) and *The Field Book of a Jungle Wallah* (1929). Fort Hose in Sarawak, now a museum, was named after him. Seven mammals and three amphibians are named after him.

Hoskins

Hoskins's Pygmy Owl *Glaucidium hoskinsii* **Brewster**, 1888
[Alt. Baja Pygmy Owl; Syn. *Glaucidium gnoma hoskinsii*]

Francis Hoskins (DNF) of El Triumfo, Baja California, assisted Marston Abbott Frazar (q.v.), who was a paid bird collector for Sennett (q.v.). Frazar collected the owl holotype and requested that it be named after Hoskins.

Houghton

Lowland Akalat ssp. *Sheppardia cyornithopsis houghtoni* **Bannerman**, 1931

Colonel George John Houghton (1873–1947) was a British army surgeon in Sierra Leone (1911–1914) and France (WW1).

House

Black Grasswren *Amytornis housei* **A. W. Milligan**, 1902

Frederick Maurice House (1865–1936), an Australian naturalist who farmed sheep in Western Australia, was a member of F. S. Brockman's 1901 expedition to the Kimberley area.

Houy

Standard-winged Nightjar *Caprimulgus houyi* **Neumann**, 1915 NCR
[JS *Macrodipteryx longipennis*]

Blackcap Babbler ssp. *Turdoides reinwardtii houyi* Neumann, 1915 NCR
[JS *Turdoides reinwardtii stictilaema*]

Dr Reinhard Houy (d.1912) collected in Tanganyika (Tanzania) (1911–1912). He wrote *Beiträge zur Kenntnis der Haftscheibe von Echeneis* (1909). An amphibian is named after him.

Howard

Howard's Sooty Grouse *Dendragapus fuliginosus howardi* **Dickey** & **Van Rossem**, 1923 NCR

Ozra William Howard (1877–1928) was a member of the Cooper Ornithological Club in Los Angeles, California. He personalised his stationery with: 'O W Howard, Field Work a Speciality in Ornithology, Oology and Entomology'. He published a number of articles in *Condor* (1899–1906) including 'Nesting of Belding's Sparrow' (1899) and 'The English Sparrow in the Southwest' (1906). Dickey and van Rossem described this race of grouse in recognition of his 'many years of enthusiastic ornithological work in southern California and Arizona', of his local knowledge of the area, and of his assistance when they collected the bird at Mount Pinos.

Howe

Fantail genus *Howeavis* **Mathews**, 1912 NCR
[Now in *Rhipidura*]

Fieldwren sp. *Calamanthus howei* Mathews, 1909 NCR
[Alt. Rufous Fieldwren; JS *Calamanthus campestris*]

Striated Grasswren ssp. *Amytornis striatus howei* Mathews, 1911 NCR
[JS *Amytornis striatus striatus*]
Grey Currawong ssp. *Strepera versicolor howei* Mathews, 1912 NCR
[JS *Strepera versicolor melanoptera*]
Large-billed Scrubwren ssp. *Sericornis magnirostra howei* Mathews, 1912
Purple-gaped Honeyeater ssp. *Lichenostomus cratitius howei* Mathews, 1912 NCR
[JS *Lichenostomus cratitius occidentalis*]
Jacky Winter ssp. *Microeca fascinans howei* Mathews, 1913 NCR
[JS *Microeca fascinans assimilis*]
Galah ssp. *Eolophus roseicapilla howei* Mathews, 1917 NCR
[Intergrade between *E. r. roseicapilla* and *E. r. albiceps*]

Frank Ernest Howe (1878–1955) was an Australian oologist and field ornithologist whose day job was tailoring.

Howell, A. H.

Howell's Nighthawk *Chordeiles minor howelli* **Oberholser**, 1914
[Alt. Common Nighthawk ssp.]
Howell's Seaside Sparrow *Ammodramus maritimus howelli* **Griscom** & Nichols, 1920 NCR
[JS *Ammodramus maritimus fisheri*]

Arthur Holmes Howell (1872–1940) was an American ornithologist. He joined the Bureau of Biological Survey of the United States Department of Agriculture (1895); ten years later Harry C. Oberholser (q.v.) joined him there. His periodical field service took him to many western states but his most notable work was done in the south. Howell wrote the 579-page *Florida Bird Life* (1932).

Howell, K. M.

Green Barbet ssp. *Stactolaema olivacea howelli* Jensen & Stuart, 1982

Kim Monroe Howell (b.1945) is Professor of Zoology and Marine Biology, University of Dar Es Salaam. He co-wrote *A Field Guide to the Reptiles of East Africa* (2006). A mammal, an amphibian and three reptiles are named after him.

Hoy, C. M.

Light-vented Bulbul ssp. *Otocompsa sinensis hoyi* **Riley**, 1923 NCR
[JS *Otocompsa sinensis sinensis*]

Charles McCauley Hoy (1897–1923), whose parents were missionaries in Japan and China, was an American collector for the USNM in China. He went to Australia on the Abbott Expedition (1919). He died as a result of a ruptured appendix.

He was the first westerner to obtain a specimen of the (probably extinct) Baiji, (Yangtze River Dolphin) when he shot one while duck hunting on Dongting Lake (1916).

Hoy, G. A.

Hoy's Screech Owl *Megascops hoyi* König & **Straneck**, 1989
[Alt. Montane-forest Screech Owl, Yungas Screech Owl]

Rufous-banded Miner ssp. *Geositta rufipennis hoyi*
Contreras, 1980
Hooded Siskin ssp. *Carduelis magellanica hoyi* König, 1981

Gunnar Arthur Hoy (1901–1997) was a Norwegian-born ornithologist of German extraction; his father was German and his mother Swedish. As such, he was required to join the German army after its occupation of Norway (WW2). As a result, after the war he was accused of being a collaborator and was imprisoned. He emigrated from Norway to Argentina (1951) with his wife and six-year old son, and (1954) became a Curator at the Natural History Museum of the University of Salta. Hoy kept in touch with many of the world's leading ornithologists and published a number of articles such as 'Un ave nueva para la Argentina – *Contopus virens*' (1981). König wrote Hoy's obituary.

Hoyer, H.

Razorbill *Alca hoieri* Merrem, 1819 NCR
[JS *Alca torda*]

Dr Henrik Hoyer (d.1615) was a Norwegian physician in the Færœ Islands in the early part of the 17th century.

Hoyer, H. F.

Common Whitethroat ssp. *Sylvia communis hoyeri*
Dunajewski, 1938 NCR
[JS *Sylvia communis communis*]

Dr Henryk Ferdynand Hoyer (1864–1947) was a Polish anatomist and zoologist. He studied at Wroclaw, Strasbourg and Berlin and then was an Assistant at the University of Wurzburg. He moved to the Jagiellonian University, Cracow (1894), as Assistant Professor, becoming Full Professor of Comparative Anatomy (1904). The university bestowed honorary doctorates on him twice (1924 and 1934). With 183 of his colleagues at the university he was arrested and sent to Sachsenhausen concentration camp (1939). He survived and ran a pharmacy in Cracow (1941–1944) in the hospital for prisoners-of-war.

Hoyo

(See **del Hoyo**)

Hoyt

Hoyt's Horned Lark *Eremophila alpestris hoyti* **Bishop**, 1896

William 'Will' Henry Hoyt (1855–1929) was an American collector, artist, ornithologist and taxidermist born in Stamford, Connecticut. He moved to Florida and opened a taxidermy shop (1882–1892). The type series of the lark bearing his name was collected during one of his trips to North Dakota (1895). Bishop wrote his obituary.

Hubbard

Coqui Francolin ssp. *Peliperdix coqui hubbardi* **Ogilvie-Grant**, 1895

Reverend Edward Henry Hubbard (1863–1898) was an English missionary to Uganda (1891–1898). He was accidentally shot by a colleague. He translated a number of books of the New Testament into local East African languages.

Huber, J.

Blackish-grey Antshrike ssp. *Thamnophilus nigrocinereus huberi* **Snethlage**, 1907
Burnished-buff Tanager ssp. *Tangara cayana huberi* **Hellmayr**, 1910

Dr Jacques Huber (1867–1914) was a Swiss botanist. He founded (1895) the Amazonicum Herbarium Musei Paraensis (Goeldi Museum Herbarium).

Huber, W.

Cloud-forest Screech Owl *Megascops huberi*
L. & E. H. Kelso, 1936
[Taxonomy debated – name regarded as invalid by some; others argue it has priority over *Megascops petersoni* (Fitzpatrick & O'Neill, 1986)]

Dr Wharton Huber (1877–1942) was an American zoologist at the Academy of Natural Sciences of Philadelphia. He collected in Nicaragua (1922) and the USA and Mexico (1927–1934).

Hudson

Hudson's Black Tyrant *Knipolegus hudsoni* **P. L. Sclater**, 1872
Hudson's Canastero *Asthenes hudsoni* P. L. Sclater, 1874
[Alt. Hudson's Soft-tail]

William Henry Hudson (1841–1922) was born in Buenos Aires to American parents. He spent his childhood on the pampas but a heart condition forced his emigration to England (1870). He is generally regarded as a British author, naturalist and ornithologist. However, he is best known for his exotic romances, especially *Green Mansions* (1904), which is set in a South American jungle. His memoir *Far Away and Long Ago* (1918) lovingly recalls his childhood. Hudson was a sensitive observer of nature, particularly birds. He hated the vogue for 'collecting' birds and once said of John Gould (q.v.) that his obsession with hummingbirds was not the selfless appreciation of a true lover of nature but no more than a magpie-like addiction: 'He regarded natural history principally as a science of dead animals – a necrology'. Whilst he met Gould only once (c.1875), he wrote a satirical pamphlet about him and described him as a 'pretentious and unscientific ornithologist'. His books describe plants and animals in a highly personal manner with great force and beauty. Other works are *The Purple Land* (1885), *Argentine Ornithology* (1888), *The Naturalist in La Plata* (1892), *A Shepherd's Life* (1910) and *A Hind in Richmond Park* (1922). The novelist John Galsworthy said of him: 'Hudson is … the finest living observer, and the greatest living lover of bird and animal life, and of Nature in her moods.'

Hüe

Thekla Lark ssp. *Galerida theklae huei* **Érard** & Naurois, 1973

François Hüe (1905–1972) was a French ornithologist. He co-wrote several books with R. D. Etchecopar (q.v.), including *The Birds of North Africa from the Canary Islands to the Red Sea* (1967). He was killed in a traffic accident.

Huegel, A. A.

Snares Island Snipe *Coenocorypha huegeli* **Tristram**, 1893

Anatole Andreas Aloys Freiherr von Hügel (1854–1928) was an Austrian-born anthropologist, collector and explorer in Australasia (1874–1878) and son of Karl (q.v. below). His family had moved to England (1867) after his father's retirement. He became the first Curator of the University of Cambridge Museum of Archaeology and Anthropology (1883).

Huegel, K. A.

Tailorbird sp. *Orthotomus huegelii* **Pelzeln**, 1857 NCR
[Alt. Common Tailorbird; JS *Orthotomus sutorius*]

Karl Alexander Anselm Freiherr von Hügel (Baron Charles von Hügel) (1795–1870) was an Austrian army officer, diplomat, botanist and explorer primarily remembered for his travels in northern India (1830s). After studying law he became an officer (1813) in the Austrian Hussars and fought various campaigns against Napoleon. He travelled in Scandinavia, Russia and both France and Italy where he was stationed. He took up residence in Hietzing, Vienna (1824), where he established a botanical garden through which he introduced plants from Australia into European gardens. He was engaged to a Hungarian countess, but when she broke it off to marry the Austrian Chancellor, Charles began a grand tour in Asia and Australia (1831–1836). On his return he wrote the four-volume *Kaschmir und das Reich der Siek*. He also founded the Austrian Imperial Horticultural Society. He married a Scottish woman and retired to England (1867) with his family, including his son Anatole (q.v. above).

Huet, J.

Huet's Fulvetta *Alcippe hueti* **David**, 1874
[Alt. Grey-cheeked Fulvetta; Syn. *Alcippe morrisonia hueti*]

Joseph Huet (1827–1903) was a French zoologist at MNHN, Paris. He was primarily a mammalogist and wrote 'Note sur les carnassiers du genre *Bassaricyon*' (1883). Two mammals are named after him.

Huet, N.

Huet's Parrotlet *Touit huetii* **Temminck**, 1830
[Alt. Scarlet-shouldered Parrotlet]

Nicolas Huet (1770–1830) was a French natural history artist. He was attached to the MNHN in Paris and collaborated closely there with Prêtre (q.v.).

Huey, L. M.

Common Poorwill ssp. *Phalaenoptilus nuttallii hueyi* **Dickey**, 1928
Summer Tanager ssp. *Piranga rubra hueyi* **Van Rossem**, 1938 NCR
[JS *Piranga rubra cooperi*]
Mangrove Warbler ssp. *Dendroica petechia hueyi* Van Rossem, 1947 NCR
[JS *Dendroica petechia castaneiceps*]

Laurence Markham Huey (1892–1963) was an American field collector employed by the San Diego Natural History Museum, where there is a collection named after him. He took a trip to the Coronado Islands (1913) and met Donald R. Dickey (q.v.), with whom he worked for the next decade. Huey took full-time employment at the San Diego Natural History Museum as Curator for the Department of Birds and Mammals (1923–1961). He published a number of papers and articles, among which are 'Field notes: trip to the Coronado Islands, June 1926' and 'Range extension of Pocket Gophers along a new road in the arid southwest' (1941). A mammal is named after him.

Huey, W. S.

Sharp-tailed Grouse ssp. *Tympanuchus phasianellus hueyi* **Dickerman** & Hubbard, 1994 EXTINCT

William 'Bill' Samuel Huey (1925–2010) was Director of the New Mexico Department of Game and Fish. After high school he enlisted (1943) and was a turret gunner (1945) in the US Air Force. After WW2 he earned a BSc in Agriculture at New Mexico A&M (1947). He married that year after an engagement of just a few hours, and became a game warden. He became Chief of Public Affairs, Assistant Director, Director and finally Secretary of the Department. The describers of the grouse wrote: 'We name this extinct subspecies in honor of William S. Huey. ... .. in recognition of his significant role in conserving wildlife and its habitats in New Mexico, North America, and around the world.' Among his written works was *New Mexico Beaver Management* (1956).

Hugh Land

Velasquez's Woodpecker ssp. *Melanerpes santacruzi hughlandi* **Dickerman**, 1987

Dr Hugh Coleman Land (1929–1968) was an American ornithologist and collector in Guatemala. Marshall University, West Virginia, awarded his bachelor's degree (1950), Ohio State University his master's and the University of Oklahoma his doctorate (1960). He was at the Biology Department, Concord College, Athens, West Virginia (1958). He wrote *Birds of Guatemala* (1970), published posthumously. At the time of his death from Hodgkin's disease, he was a full Professor at Northwestern State College.

Hugonis

Rufous-faced Warbler ssp. *Abroscopus albogularis hugonis* **Deignan**, 1938

Dr Hugh McCormick Smith (1865–1941) was a physician and ichthyologist. He joined the US Fish Commission (1886) and spent over four decades in US Government service, becoming the Commissioner of the United States Bureau of Fisheries. He visited 22 countries to study their resources and methods, during which he made extensive collections; the expedition to the Philippines on board USS *Albatross* was particularly productive. He published very widely including *The Fresh Water Fishes of Siam, or Thailand* (1945) – a country where he lived (1923–1925) and collected the warbler (1931). His contribution to natural history was large and 25 species of fish, birds, amphibians, reptiles, invertebrates and plants are named after him. Deignan chose to 'latinise' the name Hugh.

Hull, A. F. B.

Lord Howe Starling *Aplonis fusca hulliana* **Mathews**, 1912 EXTINCT
[Alt. Tasman Starling ssp.]
Flesh-footed Shearwater ssp. *Puffinus carneipes hullianus* Mathews, 1912; NRM

Arthur Francis Basset Hull (1862–1945) was an Australian barrister, ornithologist and philatelist. He was lamed by polio-myelitis (aged 15) and had to wear a surgical boot and use a walking stick thereafter. He practised at the Tasmanian Bar (1891–1892), then moved to New South Wales and worked at the General Post Office in Sydney (1892–1900). He visited Europe and on his return (1903) worked at The Department of Mines until his retirement (1921). He became the honorary ornithologist of the Australian Museum, Sydney.

Hull, G. D.

Galapagos Mockingbird ssp. *Mimus parvulus hulli* **Rothschild**, 1898

Galen Dorens Hull (1865–1909) took a bachelor's degree at Dartmouth College, New Hamshire. He was a collector on the Webster-Harris expedition to the Galápagos Islands (1897–1898), then was involved in mining in North Carolina (1899). He returned to the Galapagos for Rothschild (1907).

Humayun

Red-vented Bulbul ssp. *Pycnonotus cafer humayuni* **Deignan**, 1951

(see **Abdulali**)

Humbert

Humbert's Cardinal *Paroaria humberti* Angelini, 1901 NCR
[Believed to be a melanistic specimen of *Paroaria dominicana*]

Umberto I or Humbert I (1844–1900) was King of Italy (1878–1900). His full name was Umberto Ranieri Carlo Emanuele Giovanni Maria Ferdinando Eugenio of Savoy. He was hated by the political left, especially the anarchists, as he was a hard-line conservative. The Italian-American anarchist, Gaetano Basso, assassinated him.

Humblot

Flycatcher genus *Humblotia* **Milne-Edwards** & **Oustalet**, 1885

Humblot's Flycatcher *Humblotia flavirostris* Milne-Edwards & Oustalet, 1885
[Alt. Grand Comoro Flycatcher]
Humblot's Heron *Ardea humbloti* **Grandidier** & Milne-Edwards, 1885
[Alt. Madagascar Heron]
Humblot's Sunbird *Cinnyris humbloti* Milne-Edwards & Oustalet, 1885

Henri Joseph Léon Humblot (1852–1914) was apparently sent to the Comoro Islands from Madagascar (1884) to ensure that France was recognised as their 'protector' nation. This being achieved, he became 'Resident' (1886). Sometime later his formal power was removed, but he remained the biggest landholder, using slaves to farm his land. He must have been an amateur naturalist as he is credited with the discovery (1885) of the orchid *Angraecum leonis* as described by Reichenbach, and sent various specimens to the orchid dealers Sander & Co.

Humboldt

Humboldt Penguin *Spheniscus humboldti* **Meyen**, 1834
Humboldt's Sapphire *Hylocharis humboldtii* **Bourcier** & **Mulsant**, 1852

Humboldt's Letter-billed Araçari *Pteroglossus inscriptus humboldti* **Wagler**, 1827
Red-necked Spurfowl ssp. *Pternistis afer humboldtii* **W. K. H. Peters**, 1854
Band-tailed Sierra-finch ssp. *Phrygilus alaudinus humboldti* **Koepcke**, 1963

Baron Friedrich Wilhelm Heinrich Alexander von Humboldt (1769–1859) was a Prussian naturalist, explorer and politician. After attending universities at Frankfurt an der Oder and Göttingen (1791) he enrolled at the Freiberg Mining Academy to learn natural history and earth sciences to help him with his intended future travels. To complete his experience he then worked as an Inspector of Mines in Prussia for five years. After two years of disappointments and delays, he explored (1799–1804) in South America, collecting thousands of specimens, mapping and studying natural phenomena. The trip took in parts of Venezuela, Peru, Ecuador, Colombia and Mexico. He returned via the USA, where Thomas Jefferson entertained him at Monticello. He made a journey (1829) of similarly epic proportions, ranging from the Urals east to Siberia. He was also a patron of Louis Agassiz. Humboldt's *Personal Narrative* was inspirational to later travellers in the tropics, notably Darwin (q.v.) and Wallace (q.v.). His most famous writing was the 5-volume work *The Cosmos* (1845–1862). Humboldt did research in many other fields, including astronomy, forestry and mineralogy. The Humboldt Current that runs south-to-north just off the Pacific coast of South America was named after him, as are five mammals and two amphibians.

Hume, A. O.

Hume's Ground Tit *Pseudopodoces humilis* Hume, 1871
[Alt. Hume's Groundpecker, Tibetan Ground Jay]
Hume's Wheatear *Oenanthe albonigra* Hume, 1872
Hume's Babbler *Stachyridopsis rufifrons* Hume, 1873
[Alt. Rufous-fronted Babbler]
Hume's Hawk Owl *Ninox obscura* Hume, 1873
[Syn. *Ninox scutulata obscura*]
Hume's Lark *Calandrella acutirostris* Hume, 1873
[Alt. Hume's Short-toed Lark]
Hume's Wren-babbler *Sphenocichla humei* **Mandelli**, 1873
[Alt. Blackish-breasted Babbler, Sikkim Wedge-billed
    Babbler]
Hume's Blue-throated Barbet *Megalaima incognita* Hume,
    1874
[Alt. Moustached Barbet]
Hume's Parakeet *Psittacula finschii* Hume, 1874
[Alt. Grey-headed Parakeet, Finsch's Parakeet]
Hume's Leaf Warbler *Phylloscopus humei* **W. E. Brooks**,
    1878
Hume's Owl *Strix butleri* Hume, 1878
[Alt. Hume's Tawny Owl]
Hume's Swiftlet *Aerodramus maximus* Hume, 1878
[Alt. Black-nest Swiftlet]
Hume's Whitethroat *Sylvia althaea* Hume, 1878
Hume's Treecreeper *Certhia manipurensis* Hume, 1881
[Alt. Manipur Treecreeper]
Hume's Reed Warbler *Acrocephalus orinus* **Oberholser**,
    1905
[Alt. Large-billed Reed Warbler; Replaces Hume's earlier
    name *Acrocephalus macrorhynchus* (1871), which
    proved to be invalid]

Red-throated Barbet ssp. *Megalaima mystacophanos humii*
    G. F. L. Marshall & **C. H. T. Marshall**, 1870 NCR
[NUI *Megalaima mystacophanos mystacophanos*]
Hume's Nightjar *Caprimulgus europaeus unwini* Hume, 1871
[Alt. Eurasian Nightjar ssp.]
Common Starling ssp. *Sturnus vulgaris humii* W. E. Brooks,
    1876
Hume's White-eye *Zosterops palpebrosus auriventer* Hume,
    1878
[Alt. Oriental White-eye ssp.]
Black-throated Parrotbill ssp. *Suthora nipalensis humii*
    Sharpe, 1883
Hawfinch ssp. *Coccothraustes coccothraustes humii*
    Sharpe, 1886
Red-fronted Rosefinch ssp. *Carpodacus puniceus humii*
    Sharpe, 1888
Hume's White-cheeked Bulbul *Pycnonotus leucogenys
    humii* **Oates**, 1889 NCR
[Hybrid population: *Pycnonotus leucogenys* x *Pycnonotus
    leucotis*]
Checker-throated Woodpecker ssp. *Picus mentalis humii*
    Hargitt, 1889
Collared Kingfisher ssp. *Todiramphus chloris humii* Sharpe,
    1892
White-breasted Woodswallow ssp. *Artamus leucorynchus
    humei* **Stresemann**, 1913
Rufous Woodpecker ssp. *Micropternus brachyurus humei*
    Kloss, 1918

Common Iora ssp. *Aegithina tiphia humei* **E. C. S. Baker**,
    1922
Square-tailed Bulbul ssp. *Hypsipetes ganeesa humii*
    Whistler & **Kinnear**, 1932

Allan Octavian Hume CB (1829–1912) was a famous Theo-
sophist, poet, botanist and writer on Indian birds as well as a
passionate advocate of Indian self-rule. He was born in
London to a radical Member of Parliament, Joseph Hume. He
was sent to sea as a junior midshipman (aged 13) and served
in the Mediterranean aboard HMS *Vanguard*. After education at the East India Company's own school, Haileybury, and
further training in medicine and surgery at University College
Hospital, London, he joined the Bengal Civil Service (1849)
and was appointed to the district of Etawah, Uttar Pradesh.
Here he introduced free primary education and founded a
vernacular newspaper, *Lokmitra* (People's Friend). When the
Indian Mutiny broke out (1857), Hume was heavily involved in
fighting. He caught cholera, recovered, returned to Etawah
with 50 men and occupied the town. He was again involved in
fighting, leading charges against the rebels and, on one
occasion, capturing their artillery. He was made a
Commander of the Order of the Bath (1860) as a reward.
When the Mutiny was over, Hume was accused of being too
lenient with the mutineers – only seven men were hanged in
his administrative area on his judgements. He started *Stray
Feathers*, a quarterly ornithological journal (1872) and wrote
most of it. He travelled very widely throughout India, accumulating an enormous collection of skins. He wrote
Agricultural Reform of India and the three-volume classic
The Game Birds of India Burma and Ceylon. His other ornithological works included *Indian Oology and Ornithology*
(1869), and the 10-volume *The Nests and Eggs of Indian Birds*
(1873–1883). After his retirement and having dabbled briefly
with Theosophy (late 1870s to early 1880s), he was a
co-founder of the Indian National Congress (1885) and
became its General Secretary (1885–1906). He lived in Simla
and suffered the loss of 25 years work when, during the early
spring of 1883, while he and all Europeans were in the plains,
a servant sold all his papers and correspondence (weighing
well over 100 kilos!) as waste paper. This episode killed his
interest in ornithology and he gave his collection to the
BMNH in London. R. B. Sharpe (q.v.) was despatched to
Simla to pack up and escort the collection to England. There
were 47 crates, each weighing c.500 kilograms, which had to
be moved from Hume's museum (which was at about 7,800
feet above sea level in the mountains) to the nearest railway
station, to a port and to London. As a widower back in Britain
from 1894 Hume took up botany and founded the South
London Botanical Institute. Two mammals are named after
him.

Hume, M. A.

Mrs Hume's Pheasant *Syrmaticus humiae* **A. O. Hume**, 1881
[Alt. Bar-tailed Pheasant, Hume's Pheasant]

Mary Ann Hume *née* Grindall (1824–1890) was the wife of
Allan Octavian Hume (q.v.).

Humphreys

Dark-necked Tailorbird ssp. *Orthotomus atrogularis humphreysi* **Chasen** & **Kloss**, 1929

John Lisseter Humphreys (1881–1929) joined the Straits Settlements (Malysia) Civil Service (1905), was British agent at Terengganu (1916–1919), Adviser at Terengganu (1919–1925) and at Kedah (1925–1926), and British Governor of North Borneo (Sabah, Malaysia) (1926–1929). He died whilst on leave in China.

Hunstein

Hunstein's Mannikin *Lonchura hunsteini* **Finsch**, 1886
[Alt. Mottled Munia, White-cowled Munia]

Hunstein's Manucode *Phonygammus keraudrenii hunsteini* **Sharpe**, 1882
[Alt. Trumpet Manucode ssp.]
Hunstein's Magnificent Bird-of-Paradise *Diphyllodes magnificus hunsteini* **A. B. Meyer**, 1885
White-eyed Robin ssp. *Pachycephalopsis poliosoma hunsteini* **Neumann**, 1922

Carl Hunstein (1843–1888) was a German trader in German New Guinea (1885–1888). He was swept out to sea by a tidal wave and drowned when a 20 m high tsunami, caused by the volcanic eruption that completely destroyed the Ritter Islands, reached New Guinea. The Hunstein Mountains and the Hunstein Forest in New Guinea, and various plants bear his name.

Hunt

Blue Bird-of-Paradise ssp. *Paradisaea rudolphi hunti* **Le Soeuf**, 1907 NCR
[JS *Paradisaea rudolphi rudolphi*]

Atlee Arthur Hunt (1864–1935) was an Australian lawyer, public servant and politician who was called to the New South Wales Bar (1892). He was Secretary and permanent head of the Department of External Affairs (1901–1917). After Australia took over responsibility for British New Guinea (Papua New Guinea), he visited the territory and reported (1905) in terms that had an immediate effect on the efficiency of the administration. He visited New Guinea a second time (1919) to inspect the recently mandated German territories. He retired from public service (1930).

Hunter

Hunter's Cisticola *Cisticola hunteri* **G. E. Shelley**, 1889
Hunter's Sunbird *Chalcomitra hunteri* G. E. Shelley, 1889

Henry Charles Vicars Hunter (1861–1934) was both a big-game hunter and naturalist. A mammal is named after him.

Hurley

Yakima Wren *Thryomanes bewickii hurleyi* **S. G. Jewett**, 1944 NCR
[Alt. Bewick's Wren ssp.; JS *Thryomanes bewickii calophonus*]

John B. Hurley (1896–1974) was an American ornithologist and oologist. He wrote 'Annotated list of Yakima County birds' (1921).

Hüsker

Slender-billed Cuckoo Dove ssp. *Macropygia amboinensis hueskeri* **Neumann**, 1922

Dr Carl Hüsker (1849–1928) was a German hydrographer and explorer in the Bismarck Archipelago (1875).

Hutchins

Hutchins's Goose *Branta hutchinsii* **J. Richardson**, 1832
[Alt. Cackling Goose; Syn. *Branta canadensis hutchinsii*]

Dr Thomas Hutchins (c.1742–1790) was an English naturalist and surgeon who worked for the Hudson's Bay Company as Chief Factor in Fort Albany, Ontario (1774–1782). His studies of local fauna and flora included finding edible plants useful for the prevention of scurvy. At the request of the Royal Society he made observations on magnetic declination, the angle between magnetic north and true north.

Hutchinson

Black-and-cinnamon Fantail ssp. *Rhipidura nigrocinnamomea hutchinsoni* **Mearns**, 1907

Wallace Irving Hutchinson (b.1881) worked for the US Forestry Service. He was in the employ of the Philippine Government on Mindanao (1906–1908), where he collected with Mearns (q.v.) and with whom he made the first ascent of Mount Malindang. He wrote 'A Philippine substitute for *Lignum vitae*' (1908).

Hutson

Willcocks's Honeyguide ssp. *Indicator willcocksi hutsoni* **Bannerman**, 1928

Major-General Henry Porter Wolseley Hutson (1893–1991) was a British army officer, and amateur ornithologist, in Nigeria (1926–1928) and India (1943–1945), retiring in 1947. He wrote *The Ornithologists' Guide; Especially for Overseas* (1956).

Hutton, F.

Rail genus *Huttonena* **Mathews**, 1929 NCR
[Now in *Gallirallus* or *Cabalus*]

Hutton's Rail *Gallirallus modestus* Hutton, 1872 EXTINCT
[Alt. Chatham Islands Rail; Syn. *Cabalus modestus*]
Hutton's Fruit Dove *Ptilinopus huttoni* **Finsch**, 1874
[Alt. Rapa Fruit Dove, Long-billed Fruit Dove]
Hutton's Shearwater *Puffinus huttoni* Mathews, 1912

Frederick Wollaston Hutton (1836–1905) was an English geologist and zoologist who settled in New Zealand. He served in the Indian Mercantile Marine, and then in the army (1855–1865). He saw service in the Crimean War and the Indian Mutiny. He wrote *Catalogue of the Birds of New Zealand* (1871). The Royal Society of New Zealand established (1909) the Hutton Memorial Fund in his memory. It awards the Hutton Medal and provides grants for the encouragement of research into the zoology, botany and geology of New Zealand.

Hutton, T.

Afghan Babbler *Turdoides huttoni* **Blyth**, 1847
[Syn. *Turdoides caudata huttoni*]

Grey-necked Bunting ssp. *Emberiza buchanani huttoni* Blyth, 1849 NCR
[NUI *Emberiza buchanani buchanani*]
Black-throated Accentor ssp. *Prunella atrogularis huttoni* Moore, 1854
Mountain Scops Owl ssp. *Otus spilocephalus huttoni* **Hume**, 1870

Captain Thomas Hutton (1806–1875) was an officer in the 37th Bengal Native Infantry who served in the Afghan War (1839–1840). He retired (c.1850) and bought land near Mussoorie, Uttar Pradesh, where he built a 'Manor House'. He sold this (1853) but stayed on in the area. He was a regular correspondent of Allan Octavian Hume (q.v.), who quoted him frequently in his writings. He also started experimental sericulture in Mussoorie and (1872) wrote an appendix to a book on sericulture, giving the names of all the species of Indian silkworms known to him. He was an all-round naturalist and was one of those who introduced the giant African snail to India, although the specimens that he released died from cold in the winter. A mammal is named after him.

Hutton, W.

Hutton's Vireo *Vireo huttoni* **Cassin**, 1851

William Rich Hutton (1826–1901) was an artist, surveyor and civil engineer. He worked for a year in California (1847) as a clerk to his uncle, who was paymaster for the volunteers who formed part of the 1st New York Regiment. He was chosen to assist Lieutenant Ord in the Pueblo Survey (1849). He spent that summer helping with the survey, and sketched and drew scenes in the area in his spare time. Although he was on the move all the time, he supplied the USNM with a number of specimens, including the vireo, which he took near Monterey, California. He returned to Washington DC (1853) and married one of Francis Clopper's daughters and inherited the Clopper estate and woodlands, which now form part of Seneca Creek State Park. Among his engineering projects were the old Cabin John Bridge and the Washington Aqueduct, the Washington Bridge over the Harlem River and the Hudson River Tunnel in New York. He was embarrassed by the honour of having a bird named after him, and said in a letter to a relative '... it goes against my principles to name after individuals unless for important scientific service.' He is the subject of at least two biographies.

Hutz

Olive-winged Bulbul ssp. *Pycnonotus plumosus hutzi* **Stresemann**, 1938
Elegant Pitta ssp. *Pitta elegans hutzi* **Meise**, 1941
[SII *Pitta elegans concinna*]

Baronin Marie-Izabel von Plessen *née* von Jenisch (1906–1971) was the wife (1934) of Baron Viktor von Plessen (1900–1980). The name *hutzi* appears to be from Hütz, which may have been her nickname. One reference calls her 'The courageous and enthusiastic zoologist who accompanied her husband on his last expedition.' This was to the Lesser Sunda Islands, Indonesia (1937–1938). (See **Plessen** & **Marie (von Plessen)**)

Hyder

White-eyed Buzzard *Astur hyder* **Sykes**, 1832 NCR
[JS *Butastur teesa*]

Hyder Ali Khan (1722?–1782) was a Carnatic soldier who ruled Mysore, southern India (1761–1782). He was one of the first commanders to use rockets as weapons of war against European troops.

I

Ibarra

Stripe-headed Sparrow ssp. *Peucaea ruficauda ibarrorum* **Dickerman**, 1987

Professor Jorge Alfonso Ibarra (1921–2000), who was Director of Museo Nacional de Historia Natural, Guatemala, and Vincente Ibarra and his family, advised and helped Dickerman in the field. 'Named in honor of Jorge Ibarra for his efforts in the conservation of Guatemala wildlife, and for "Don" Vicente Ibarra and his family at La Avellana who were my valued mentors and indispensable aides in the field.' A reptile and an amphibian are also named after the professor.

Icelanders

Rock Ptarmigan ssp *Lagopus muta islandorum* **F. Faber**, 1822

This bird is named after the people of Iceland in general.

Ida (Beck)

Northern Marquesan Reed Warbler ssp. *Acrocephalus percernis idae* **R. C. Murphy** & **Mathews**, 1928

Mrs Ida May Beck (1883–1970) *née* Menzies was the wife (1907) of Rollo Beck (q.v.).

Ida (Pfeiffer)

Madagascar Pond Heron *Ardeola idae* **Hartlaub**, 1860

Ida Laura Pfeiffer (1797–1858) was an Austrian traveller and a very remarkable woman. She wrote *Reiser einer Wienerin in das Heilige Land* (1846) about her journey to Egypt and Palestine (1842). She visited and wrote about a great number of countries and (1846–1848) went round the world, publishing an account of it, *Eine Frau fährt um die Welt* (1850). She started another journey (1851) lasting three years, taking in England, South Africa, parts of North and South America which she had missed on her first circumnavigation, Australia and the Malay Archipelago, spending 18 months in the Sunda Islands and the Moluccas. She set off again (1856) from Vienna to explore Madagascar, where she became embroiled in a plot to overthrow the government, leading to her expulsion. Her experiences were recounted in *Reise nach Madagaskar* (1861), together with a biography of her by her son. An amphibian is named after her.

Idalia

Minute Hermit *Phaethornis idaliae* **Bourcier** & **Mulsant**, 1856

In Greek mythology Idalia is an epithet used in relation to the goddess Aphrodite.

Idas

Hummingbird genus *Idas* **Mulsant**, 1876 NCR
[Now in *Lophornis*]

The name Idas appears more than once in Greek mythology. The most famous character bearing the name was one of the Argonauts. However, another Idas was one of the companions of Diomedes, who were transformed into birds.

Ignesti

Indigobird sp. *Hypochera ignestii* **Moltoni**, 1925 NCR
[Alt. Village Indigobird; JS *Vidua chalybeata ultramarina*]

Ugo Ignesti (fl.1930) was an Italian zoologist, philologist and collector in Abyssinia and British Guiana (Guyana) (1931). He wrote: *La Lingua degli Amharic, Trascritta in Caratteri Latini: Grammatica, Esercizi e Vocabolario* (1937).

Ihering

Ihering's Piculet *Picumnus jheringi* **Berlepsch**, 1884 NCR
[Alt. Mottled Piculet; JS *Picumnus nebulosus*]
Ihering's Woodcreeper *Xiphorhynchus juruanus* H. von Ihering, 1905
[Alt. Elegant Woodcreeper; Syn. *Xiphorhynchus elegans juruanus*]
Ihering's Tyrannulet *Phylloscartes difficilis* H. von Ihering & R. von Ihering, 1907
[Alt. Serra do Mar Tyrannulet]
Narrow-billed Antwren *Formicivora iheringi* **Hellmayr**, 1909
Ihering's Antwren *Myrmotherula iheringi* **Snethlage**, 1914

Buff-fronted Owl ssp. *Aegolius harrisii iheringi* **Sharpe**, 1899

Hermann von Ihering (1850–1930) was a German-Brazilian zoologist, malacologist and geologist. He trained as a physician and served in the German army. He settled in Rio Grande do Sul, Brazil (1880). He founded the São Paulo Museum (1894) and spent 22 years as its first Director. He wrote *Catálogos da Fauna Brasileira* (1907) and *As Aves do Brazil* with his son Rudolpho Teodoro Gaspar Wilhelm von Ihering (1883–1939). He returned to Germany (1924) where he died. Three mammals, three reptiles and an amphibian are named after him.

Ijima

Ijima's Leaf Warbler *Phylloscopus ijimae* **Stejneger**, 1892

Ijima's Copper Pheasant *Syrmaticus soemmerringii ijimae* **Dresser**, 1902

Common Sand Martin ssp. *Riparia riparia ijimae* **Lönnberg**, 1908

Japanese Bush Warbler ssp. *Cettia cantans ijimae* **Kuroda**, 1922 NCR
[JS *Cettia diphone cantans*]

Japanese Pygmy Woodpecker ssp. *Dendrocopos kizuki ijimae* **Taka-Tsukasa**, 1922

Varied Tit ssp. *Poecile varius ijimae* Kuroda, 1922 NCR
[JS *Poecile varius varius*]

Dr Isao Ijima (1861–1921) was Professor of Zoology at Tokyo University and first President of the Ornithological Society of Japan. The University of Leipzig awarded his doctorate (1924). A reptile is named after him.

Illiger

Illiger's Macaw *Primolius maracana* **Vieillot**, 1816
[Alt. Blue-winged Macaw]

Yellow-browed Tody Flycatcher ssp. *Todirostrum chrysocrotaphum illigeri* **Cabanis & Heine**, 1859

'Carl' Johann Karl Wilhelm Illiger (1775–1813) was a German zoologist, the first Director of the Zoological Museum of the University of Berlin (1811–1813). Illiger is credited with inventing the word 'Proboscidea' for the order of mammals containing the elephants. He was the author of *Prodromus Systematis Mammalium et Avium* (1811), an overhaul of the Linnean system of classification. A mammal is named after him.

Ilse (Grantsau)

Hyacinth Visorbearer ssp. *Augastes scutatus ilseae* **Grantsau**, 1967

Ilse Grantsau is the describer's wife (1957) and with him co-wrote a book on the medicinal qualities of aloe.

Ilse (Rensch)

Bare-throated Whistler ssp. *Pachycephala nudigula ilsa* **Rensch**, 1928

Mrs Ilse Rensch *née* Maier (1902–1992) was a botanist who was in the Lesser Sunda Islands (1927) with her husband, Bernhard Rensch (q.v.).

Ilya

Bocage's Akalat ssp. *Sheppardia bocagei ilyai* **Prigogine**, 1987

Ilya Romanovich Prigogine (1917–2003) was a Russian/Belgian physical chemist who won the 1977 Nobel Prize in Chemistry. His brother Alexandre Prigogine (q.v.) described the bird.

Imam

Crested Lark ssp. *Galerida cristata imami* **Meinertzhagen**, 1923 NCR
[JS *Galerida cristata tardinata*]

Strictly speaking this species is not named after a person but after a person's title. Yahya Mohammed Hamid ed-Din (1864–1948) became Imam of the Zaydis, i.e. he was effectively ruler of the mountainous areas of North Yemen (1904). The Turks (Yemen was part of the Ottoman Empire) did not recognise his rule. War resulted (1911), the Turks recognised him as ruler, and he became a loyal subject of the Ottoman Empire. At the end of WW1, the Ottoman Empire collapsed and Yahya became ruler, later King (1926), of the newly independent state of North Yemen. There was another war (1934) in which Yemen was comprehensively defeated by Saudi Arabia. There was internal opposition to his rule and an unsuccessful coup d'état (1946), during which Yayha was shot and died; his son succeeded him.

Imelda

Amethyst Brown Dove ssp. *Phapitreron amethystinus imeldae* de la Paz, 1976

Imelda Remedios Visitacion Romualdez Marcos (b.1929) is a Filipina politician who was the wife of former President Ferdinand Marcos of the Philippines. After she and her husband were forced to go into exile (1986), the full extent of her collection of footwear was revealed – around 2,700 pairs. Subsequently she returned to the Philippines and was elected to the Philippine Parliament (1995 & 2010).

Imperator

Variegated Antpitta ssp. *Grallaria varia imperator* **Lafresnaye**, 1842

Dom Pedro II (1825–1891) was the last Emperor of Brazil. He succeeded to the throne (1830) aged 5. He assumed legal powers (1840–1889) but later abdicated, the royal family going into voluntary exile, and Brazil became a republic.

Imperialis

Imperial Pheasant *Lophura imperialis* **Delacour & Jabouille**, 1924 NCR
[Hybrid: *Lophura edwardsi* x *Lophura nycthemera*]

Khai Dinh (1885–1925) was Emperor of Annam (1916–1925). He visited France (1922) and during his reign Western influence on Vietnamese culture began to be noticeable. His tomb in Hue is a mixture of Oriental and European architecture and, as the travel guides would say, 'Well worth a visit'.

Impey

Himalayan Monal *Lophophorus impejanus* **Latham**, 1790
[Alt. Impeyan Pheasant]

Lady Mary Impey (1749–1818) was the wife of Sir Elijah Impey, the first Governor of the State of Bengal. Lady Impey was the first person to keep this pheasant species in captivity at her private zoo in Calcutta. She was unsuccessful in providing live birds to Europe, this only being achieved by Lord Harding (late 1850s) when they became the prized possession of Prince Albert (q.v.). Lady Impey commissioned many paintings to record and identify species that were then unknown in India.

Im Thurn

Im Thurn's Blackbird *Macroagelaius imthurni* **P. L. Sclater**, 1881

[Alt. Golden-tufted Grackle, Tepui Mountain Grackle]

Sir Everard Ferdinand im Thurn (1852–1932) was a British diplomat, colonial administrator and explorer as well as a botanist and anthropologist. He was in British Guiana (Guyana) (1877–1899), was Lieutenant-Governor of Ceylon (Sri Lanka) (1901–1904), and Governor of Fiji (1904–1910). He wrote *Among the Indians of Guiana* (1883) and numerous notes and minutes regarding the Fiji Islands and their history.

Ince

Ince's Paradise-flycatcher *Terpsiphone paradisi incei* **Gould**, 1852

Commander John M. R. Ince (fl.1828–1850) was a British naval officer who served on HMS *Challenger* (1828–1830). He was promoted from Mate to Lieutenant (1841) while serving aboard HMS *Implacable* and in that rank served on board HMS *Fly* in Australian and New Guinea waters (1842–1846). He was later on board HMS *Pilot* (1850). He kept a series of logs for most of his career. Whilst he was on HMS Fly, John MacGillivray (q.v.) was on board as the botanist and natural history collector.

Indira

Bar-winged Wren-babbler ssp. *Spelaeornis troglodytoides indiraji* **Ripley**, Saha & Beehler, 1991

Indira Priyadarshini Gandhi (1917–1984) was Prime Minister of India and a keen member of the Delhi Birdwatching Society. Her own guards assassinated her.

Indranee

Brown Wood Owl ssp. *Strix leptogrammica indranee* **Sykes**, 1832

In Hindu mythology, Indranee (or Indrani) is the wife of Indra, god of war and tempests. She was said to have a thousand eyes.

Inez

Tyrannulet genus *Inezia* **Cherrie**, 1909

Enriqueta Iñez Cherrie (b.1898) was the daughter of American ornithologist George Cherrie (q.v.), who described this genus. A mammal is named after her.

Inglis, C. M.

Manipur Bush Quail ssp. *Perdicula manipurensis inglisi* **Ogilvie-Grant**, 1909

Ashy Prinia ssp. *Prinia socialis inglisi* **Whistler** & **Kinnear**, 1933

Charles McFarlane Inglis (1870–1954) was a Scottish naturalist and planter, who went to India (1888). He was Curator, Darjeeling Museum (1926–1948).

Inglis, J.

Scimitar Babbler sp. *Pomatorhinus inglisi* **Hume**, 1877 NCR
[Alt. Large Scimitar Babbler; JS *Pomatorhinus hypoleucos hypoleucos*]

James Inglis (1845–1908) was a Scottish author, tea and indigo planter, merchant and politician in India. He went to New Zealand aged 19 and joined the gold rush, then went to India (1866) at the instigation of his brother, to work as a tea merchant before becoming an indigo planter. He loved tiger hunting and pig sticking and wrote sporting verses. He wrote, among other works: *Sport and Work on the Nepaul Frontier: or, Twelve Years Sporting Reminiscences of an Indigo Planter* (1878). He moved to Australia (1877) where he was a Commissioner for the Indian government. Hume (q.v.) said of him 'Mr James Inglis has, for some years past, most kindly collected birds for me in the north-eastern corner of the Cachar District … Altogether Mr. Inglis has presented our museum with specimens of 157 species.'

Ingouf

Patagonian Tinamou *Tinamotis ingoufi* **Oustalet**, 1890

Read-Admiral Jules Alexandre Ingouf (1846–1901) was an officer in the French navy (1861–1901) who explored and collected in Argentina. He commanded the *Volage*, which surveyed the Straits of Magellan (1882). He became commandant of the French coastguard service (1899).

Ingram

Black-chinned Honeyeater ssp. *Melithreptus gularis ingrami* **Mathews**, 1912 NCR
[Intergrade between *Melithreptus gularis gularis* and *M. gularis laetior*]

Common Coot ssp. *Fulica australis ingrami* Mathews, 1912 NCR
[JS *Fulica atra australis*]

Crested Finchbill ssp. *Spizixos canifrons ingrami* **Bangs** & **J. C. Phillips**, 1914

White-browed Crake ssp. *Porzana cinerea ingrami* Brasil, 1917 NCR; NRM

Captain Collingwood Ingram (1880–1981) was an ornithologist and botanist whose father, Sir William Ingram (1847–1924), owned the *Illustrated London News*. Collingwood Ingram travelled the world collecting specimens, including visits to Japan and Australia (1907). He was an expert on cherry trees and found a species growing in England that had become extinct in its native Japan, but was successfully reintroduced a few years later. In WW1 he was a fighter pilot in the Royal Flying Corps. He wrote *Isles of the Seven Seas* (1937). A reptile is also named after him.

Ingrams

Lichtenstein's Sandgrouse ssp. *Pterocles lichtensteinii ingramsi* **G. L. Bates** & **Kinnear**, 1937

William Harold Ingrams (1897–1973) was wounded in action in WW1. He became a British colonial administrator in

Zanzibar (1919–1927 and 1932–1933), Mauritius (1927–1933) and Aden Protectorate area (1934–1944). He was a member of the British Control Commission in Germany (1945–1947) and served in the Gold Coast (Ghana) (1947–1948), Gibraltar (1949), Hong Kong (1950) and Uganda (1956) before retiring (1968). He and his wife were the first Europeans to explore parts of the Hadhramaut (1930s) and were jointly awarded the Founder's Medal of the Royal Geographical Society (1940) for their explorations.

Inouye

Inouye's Three-toed Woodpecker *Picoides tridactylus inouyei* **Yamashina**, 1943

Dr Motonori Inouye (1901–1990) was a Japanese forest entomologist. He specialised in the taxonomy and control of conifer-feeding aphids.

Iohannis

Johannes's Tody Tyrant *Hemitriccus iohannis* **E. Snethlage**, 1907

João Baptista de Sá (DNF) was a collector on the Museu Goeldi expedition to the Rio Purús region of Brazil (1904). Snethlage (q.v.) presumably latinised João into *iohannis* – hence 'Johannes' in the bird's common name. There is some speculation that he may have been an intimate friend of Ms Snethlage. Although we can find no confirmation of this, our source wrote 'Snethlage was no nun, nor saint for that matter, and very advanced for her times. She seems to have had a number of paramours during her stint there'. There has also been speculation that the name referred to Schönmann, the other collector on the trip, who prepared specimens, but this seems less likely since his forename was Joseph.

Iphigenia

Eurasian Jay ssp. *Garrulus glandarius iphigenia* **Sushkin** & Ptuschenko, 1914

In Greek mythology, Iphigenia was a daughter of Agamemnon and Clytemnestra. She was to be sacrificed to appease the goddess Artemis and allow the Greek fleet to set sail for Troy. However, Artemis saved the girl and left a deer in her place. Two plays by Euripides (5th century BC) survive and tell her story.

Iradi

Antillean Palm Swift ssp. *Tachornis phoenicobia iradii* **Lembeye**, 1850

Francisco Solá y Iradi (fl.1850) was a Cuban landowner.

Irby

Irby's Long-tailed Tit *Aegithalos caudatus irbii* **Sharpe** & **Dresser**, 1871

Lt.-Col. Leonard Howard Loyd Irby (1836–1905) was a career soldier and amateur ornithologist. Gazetted to the 90th Light Infantry (1854), he served in Crimea including the Siege of Sevastopol (1854–1855). His regiment was posted to China (1857) but the troopship ran aground and was lost off Sumatra

(1857). Rescued 10 days later, greeted with news of the Indian Mutiny, the regiment was ordered to Calcutta. Irby served in India (1857–1860) and was in action at both Cawnpore and Lucknow. He transferred to the 74th Highlanders (1864) and was part of the Gibraltar garrison (1868–1871) before retiring (1874). He wrote *The Ornithology of the Straits of Gibraltar* (1875).

Iredale

Jacana genus *Irediparra* **Mathews**, 1911
Australian Robin genus *Iredaleornis* Mathews, 1912 NCR
[Now in *Heteromyias*]

Slender-billed Thornbill *Acanthiza iredalei* Mathews, 1911
Kingfisher sp. *Alcedo iredalei* **E. C. S. Baker**, 1921 NCR
[Alt. Blyth's Kingfisher; JS *Alcedo hercules*]
South Island Snipe *Coenocorypha iredalei* **Rothschild**, 1921 EXTINCT

Little Penguin ssp. *Eudyptula minor iredalei* Mathews 1911
Crimson Finch ssp. *Neochmia phaeton iredalei* Mathews, 1912 NCR
[JS *Neochmia phaeton phaeton*]
Crested Shrike-tit ssp. *Falcunculus frontatus iredalei* Mathews, 1912 NCR
[JS *Falcunculus frontatus frontatus*]
Lesser Frigatebird ssp. *Fregata ariel iredalei* Mathews, 1914
Black-throated Tit ssp. *Aegithalos concinnus iredalei* E. C. S. Baker, 1920

Tom Iredale (1880–1972) was an English artist and naturalist who spent much of his life in Australia. He started work apprenticed to a pharmacist, became a clerk, and later Secretary to Gregory MacAlister Mathews (q.v.), the Australian ornithologist, after working with him for a number of years at the British Museum. While Mathews is credited as the author of *Birds of Australia*, Iredale is said to have written much of the text. He collected widely in the Kermadec Islands and Queensland. He published extensively on birds, ecology and shells, and was the conchologist at the Australian Museum in Sydney (1922–1924). He is remembered in many gastropod and fish scientific names as well as birds. He also wrote 'popular' science articles for newspapers under various pseudonyms, including *Garrio*, which is Latin for 'I chatter'. The jacana genus *Irediparra* combines Iredale's name with *Parra*, a name used by Linnaeus for the jacanas. (See also **Tomirdus**)

Irena

Fairy-bluebird genus *Irena* **Horsfield**, 1821

Malayan Banded Pitta *Hydrornis irena* **Temminck**, 1836

In Greek mythology Irene (or Eirene) was the goddess of peace.

Irena (Domaniewska)

Piculet sp. *Picumnus irenae* **Domaniewski**, 1925 NCR
[Alt. Ocellated Piculet; JS *Picumnus dorbignyanus jelskii*]

Irena Domaniewska (1892–1984) was a teacher and the wife of Professor Janusz Domaniewski, who described the piculet.

Irene

Irene's Scops Owl *Otus ireneae* **Ripley**, 1966
[Alt. Sokoke Scops Owl, Morden's Scops Owl]

(see **Morden**)

Iris

Iris Lorikeet *Psitteuteles iris* **Temminck**, 1835
Rainbow Pitta *Pitta iris* **Gould**, 1842
Opal-crowned Manakin *Lepidothrix iris* **Schinz**, 1851
Rainbow Starfrontlet *Coeligena iris* Gould, 1854
Iris Glossy Starling *Lamprotornis iris* **Oustalet**, 1879
[Alt. Emerald Starling; Syn. *Coccycolius iris*]

Brown-throated Sunbird ssp. *Anthreptes malacensis iris*
Parkes, 1971

In Greek mythology Iris, an Oceanid, was the messenger of Hera and goddess of the rainbow. The name is usually applied to 'rainbow-coloured' species and includes both a mammal and an amphibian.

Iris (Darnton)

Dollarbird ssp. *Eurystomus orientalis irisi* Deraniyagala, 1951

Mrs Iris Sheila Darnton *née* Moreton (1900–at least 1975) was an intrepid traveller and collector who was a regular visitor to Sri Lanka. She also visited Trinidad (1957) resulting in an article in *Ibis* on the display of manakins. She was a Fellow of the Royal Geographical Society and a talented artist and good photographer. Apparently the naming came about because she observed some unfamiliar birds in Sri Lanka and asked an apparent authority what they were. He assured her they were common, but unsatisfied she informed the Director of the National Museum (Paul Deraniyagala) who 'collected' the birds. He identified them as a new race, which he named after her. She wrote *Jungle Journeys in Ceylon* (1975).

Iris (Priest)

Locust Finch *Paludipasser irisae* **Roberts**, 1932 NCR
[JS *Paludipasser locustella locustella*]

Mrs Iris Priest (DNF) was the wife of Captain Cecil D. Priest (q.v.).

Irving

Grey-crowned Rosy Finch ssp. *Leucosticte tephrocotis irvingi* Feinstein, 1958
[SII *Leucosticte tephrocotis tephrocotis*]

Lieutenant-Colonel Dr Laurence Irving (1895–1979) was an American zoologist, physiologist and Arctic researcher. He served in the US Army in WW1 (1917–1919). Bowdoin College awarded his bachelor's degree, Harvard his master's, and Stanford his doctorate. He taught at Stanford (1925–1931), was Professor of Experimental Biology at University of Toronto (1931–1937), and was Professor of Biology, Swarthmore College (1937–1949) broken by service (WW2) in the US Army Air Corps (1943–1946). He was chief of the physiology section, Arctic Health Research Center, Anchorage, Alaska (1949–1962), and was the founding Director of the Institute of Arctic Biology, University of Alaska Fairbanks (1962–1966).

Irwin

Stierling's Wren Warbler ssp. *Calamonastes stierlingi irwini*
Smithers & **Paterson**, 1956
Sowerby's Barbet ssp. *Buccanodon whytii irwini* **Benson**, 1956 NCR
[Syn. *Stactolaema sowerbyi sowerbyi*]
Rufous-naped Lark ssp. *Mirafra africana irwini* **A.A. Pinto**, 1968 NCR
[NUI *Mirafra africana gomesi*]
Yellow-bellied Waxbill ssp. *Coccopygia quartinia stuartirwini*
Clancey, 1969

Michael Patrick Stuart Irwin (b.1925) is a British ornithologist and natural historian from Ulster who spent 63 years in Africa and now lives in Norfolk. He was associated with Natural History Museum in Bulawayo, Southern Rhodesia (Zimbabwe), for over 40 years, but only joining it officially in 1959. He became Regional Director in Bulawayo (1975). He published extensively, often co-authored articles, e.g. with Stuart Keith and C. W. Benson 'The genus *Sarothrura* (Aves, Rallidae)' (1970), but also longer works alone such as *The Birds of Zimbabwe* (1981), and for many years he was the editor of *Honeyguide*. In a letter to us he said 'And while the Rufous-naped Lark from Angola is now regarded as a synonym, I personally feel it is quite distinct!' (See also **Stuart Irwin**)

Isaacson

Isaacson's Puffleg *Eriocnemis isaacsonii* **Parzudaki**, 1845
NCR
[Hybrid? = *Eriocnemis* sp. x *Heliangelus* sp.]

Joseph P. Isaacson (DNF) was an English naturalist who was Curator of the Liverpool Zoological Gardens (c.1838). The hummingbird that bears his name is known from three old 'Bogotá' specimens. It is generally believed to be of hybrid origin, but could be a valid form awaiting rediscovery.

Isabella

Isabella's Oriole *Oriolus isabellae* **Ogilvie-Grant**, 1894
[Alt. Isabela Oriole, Green-lored Oriole]

This rare oriole was described by Ogilvie-Grant, whose mother's name was Isabela, so it could have been named for her. However, the bird is an inhabitant of Isabela Province, Luzon, which was named (1856) after Queen Isabella of Spain. As it is most often referred to as the Isabela Oriole we believe it was named for the province.

Isabella (Burton)

Mountain Robin Chat *Cossypha isabellae* **G. R. Gray**, 1862

Lady Isabel Burton *née* Arundell (1831–1896) was the wife of Sir Richard Burton (q.v.), the famous English traveller and author. She wrote *The Romance of Isabel, Lady Burton* (1897), an unfinished autobiography, published posthumously.

Isabella (Cortes)

Gorgeted Puffleg *Eriocnemis isabellae* Cortes-Diago *et al.*, 2007

The etymology for this hummingbird, in *Ornitologia Neotropical,* reads: 'We take pleasure in naming this species for Isabella Cortes, daughter of Alexander Cortés-Diago'. Alexander Cortés-Diago and Luis Alfonso Ortega discovered the bird (2005) in Colombia, having three sightings of it during surveys of montane cloud forest in the Serrania del Pinche.

Isabella (Thirion)

White-browed Purpletuft *Iodopleura isabellae* **Parzudaki**, 1847

Isabelle Thirion (fl.1846) was the wife of French naturalist, trader and explorer Eugène Thirion (1813–1879), who collected hummingbirds in South America (1846) and explored the Orinoco and Rio Negro.

Isaure

Bronze-tailed Plumeleteer ssp. *Chalybura urochrysia isaurae* **Gould**, 1861

Isaure, Baroness de Lafresnaye *née* Guéneau de Montbeillard (1802–1893), was the wife of Baron Nöel Fréderic Armand André de Lafresnaye (q.v.).

Isidor(e)

Isidore's Rufous Babbler *Pomatostomus isidorei* **Lesson**, 1827
[Alt. New Guinea Babbler, Papuan Babbler]
Isidor's Eagle *Spizaetus isidori* **Des Murs**, 1845
[Alt. Black-and-chestnut Eagle; Syn. *Oroaetus isidori*]
Blue-rumped Manakin *Lepidothrix isidorei* **P. L. Sclater**, 1852

Great Jacamar ssp. *Jacamerops aureus isidori* **Deville**, 1849

Isidore Geoffroy Saint-Hilaire (1805–1861) was the son of Étienne Geoffroy Saint-Hilaire (q.v.) and continued his father's professorial work. He was an expert on teratology (deviations from normal structure in organisms) and was author of *Histoire Générale et Particulière des Anomalies de l'Organisation Chez l'Homme et les Animaux* (1832–1837).

Isidro

Orange-bellied Flowerpecker ssp. *Dicaeum trigonostigma isidroi* **Rand** & **Rabor**, 1969

Dr Antonio Isidro y Santos (1901–at least 1969) was a Filipino biologist and educator. The University of Chicago awarded his doctorate. He was the founder and first President of Mindanao State University (1961), having been Vice President for Academic Affairs in the University of the Philippines. He wrote many papers and longer works on education.

Isis

Paradise Kingfisher sp. *Tanysiptera isis* **G. R. Gray**, 1860 NCR
[Alt. Common Paradise Kingfisher; JS *Tanysiptera galatea margarethae*]

In Egyptian mythology Isis was the goddess of children and worshipped as the ideal mother and wife. The annual flooding of the Nile River was said to be the result of Isis's tears as she wept for her brother-husband Osiris.

Isleib

Spruce Grouse ssp. *Falcipennis canadensis isleibi* **Dickerman** & Gustafson, 1996

Malcolm E. 'Pete' Isleib (1938–1993) of Cordova, Alaska, was an American ornithologist, commercial fisherman and conservationist. He co-wrote *Birds of the North Gulf Coast-Prince William Sound region, Alaska* (1973). He died in an accident while re-fitting his fishing vessel.

Isler

Antwren genus *Isleria* Bravo, Chesser & Brumfield, 2012

Chapada Flycatcher *Suiriri islerorum* Zimmer, Whittaker & Oren, 2001

Morton L. Isler (b.1929) and Phyllis R. Isler (b.1931) are ornithologists living in Virginia, USA, most well known for their book *The Tanagers* (1987). Both are Research Associates at the USNM. The original etymology for the flycatcher reads: 'We take great pleasure in naming this species after our good friends and colleagues, Morton and Phyllis Isler, in recognition of their numerous contributions to Neotropical ornithology …' Both retired from other professions (1981) to devote themselves full time to ornithology. They continue (2010) to examine species limits in antbirds with an emphasis on the use of vocalisations as taxonomic characters, typically publishing one or two papers a year.

Issel

Fawn-coloured Lark ssp. *Calendulauda africanoides isseli* **Hoesch** & **Niethammer**, 1940 NCR
[JS *Calendulauda africanoides harei*]

Professor Arturo Issel (1842–1922) was an Italian geologist, malacologist and palaeontologist. He was appointed Professor of Geology at the University of Genoa (1866). He conducted marine research along the Eritrean coast (1870s) and wrote *Viaggio nel Mar Rosse e tra I Bogos* (1876). A mammal and a reptile are named after him.

Itys

Seedsnipe genus *Itys* **Wagler**, 1830 NCR
[Now in *Thinocorus*]

Itys is a minor figure in Greek mythology. He was killed by his own mother, Procne, and then fed to his father, the tyrant Tereus of Thrace. This was in revenge for Tereus's rape of Procne's sister, Philomela.

Ivanov

Crested Lark ssp. *Galerida cristata iwanowi* **Zarudny** & Loudon, 1903

General Nikolai Alexandrovich Ivanov (1851–1918) was the Russian Governor-General of Turkistan (1901–1904).

Iwasaki

Japanese Sparrowhawk ssp. *Accipiter gularis iwasakii* Mishima, 1962

Tajuki Iwasaki (DNF) was a Japanese collector.

Izzard

Grey Crested Tit ssp. *Lophophanes dichrous izzardi* **Biswas**, 1955 NCR
[NUI *Lophophanes dichrous dichrous*]

Ralph William Burdick Izzard (1910–1992) was a British journalist who covered the Everest Expedition (1953). He was in Naval Intelligence in WW2; his commanding officer was Ian Fleming, who is reputed to have used, as a model for James Bond's Baccarat game in *Casino Royale*, a game of poker that Izzard played against covert German intelligence agents in a casino in Pernambuco, Brazil. Izzard wrote four books chronicling his experiences in India, Nepal and the Middle East.

J

Jaba

Common Pheasant ssp. *Phasanius colchicus jabae*
Zarudny, 1909 NCR
[JS *Phasanius colchicus bianchii*]

Rittmeister P. M. Schaba (DNF) was a Russian Army officer who was in the Amur region (1910) – *jabae* being a latinised form of his surname.

Jabouille

Scimitar Babbler genus *Jabouilleia* **Delacour**, 1927

Red Junglefowl ssp. *Gallus gallus jabouillei* Delacour &
Kinnear, 1928
Indochinese Cuckooshrike ssp. *Coracina polioptera
jabouillei* Delacour, 1951

Pierre Charles Edmond Jabouille (1875–1947) was a French colonial administrator in Indochina (Resident Superior of Annam) (1905–1933) and also an ornithologist and editor of *L'Oiseau*. He accompanied his friend Delacour (q.v.) on all his expeditions there (1923–1939), and with him wrote the 4-volume *Les Oiseaux de l'Indochine Française* (1931).

Jackson, F. J.

Jackson's Golden-backed Weaver *Ploceus jacksoni*
G. E. Shelley, 1888
[Alt. Golden-backed Weaver]
Black-throated Apalis *Apalis jacksoni* **Sharpe**, 1891
Jackson's Francolin *Pternistis jacksoni* **Ogilvie-Grant**, 1891
Jackson's Hornbill *Tockus jacksoni* Ogilvie-Grant, 1891
[Syn. *Tockus deckeni jacksoni*]
Jackson's Widowbird *Euplectes jacksoni* Sharpe, 1891
Jackson's Pipit *Anthus latistriatus* Jackson, 1899
[Syn. *Anthus cinnamomeus latistriatus*]
African Babbler sp. *Turdinus jacksoni* Sharpe, 1900 NCR
[Alt. Mountain Illadopsis; JS *Illadopsis pyrrhoptera*]
Jackson's Crimsonwing *Cryptospiza jacksoni* Sharpe, 1902
[Alt. Dusky Crimsonwing]
Jackson's Akalat *Sheppardia aequatorialis* Jackson, 1906
[Alt. Equatorial Akalat]

White-headed Wood-hoopoe ssp. *Phoeniculus bollei
jacksoni* Sharpe, 1890
Black-throated Wattle-eye ssp. *Platysteira peltata jacksoni*
Sharpe, 1891 NCR
[JS *Platysteira peltata mentalis*]
Red-faced Crombec ssp. *Sylvietta whytii jacksoni* Sharpe,
1897

Yellow-rumped Tinkerbird ssp. *Pogoniulus bilineatus
jacksoni* Sharpe, 1897
Brown Parisoma ssp. *Sylvia lugens jacksoni* Sharpe, 1899
[Syn. *Parisoma lugens jacksoni*]
Jackson's Sunbird *Nectarinia tacazze jacksoni* **Neumann**,
1899
[Alt. Tacazze Sunbird ssp.]
Jackson's White-eye *Zosterops senegalensis jacksoni*
Neumann, 1899
[Alt. African Yellow White-eye ssp.]
Bocage's Bush-shrike ssp. *Chlorophoneus bocagei jacksoni*
Sharpe, 1901
Lesser Swamp Warbler ssp. *Acrocephalus gracilirostris
jacksoni* Neumann, 1901
Jackson's Black Cuckoo *Cuculus clamosus jacksoni*
Sharpe, 1902 NCR
[Intergrade population between *Cuculus clamosus
clamosus* and *C. clamosus gabonensis*]
Lemon Dove ssp. *Columba larvata jacksoni* Sharpe, 1904
Pink-breasted Lark ssp. *Calendulauda poecilosterna
jacksoni* Ogilvie-Grant, 1913 NCR; NRM
Red-and-yellow Barbet ssp. *Trachyphonus erythrocephalus
jacksoni* Neumann, 1928 NCR
[Intergrade population between *Trachyphonus
erythrocephalus shelleyi* and *T. e. versicolor*]
Jackson's Bustard *Neotis denhami jacksoni* **Bannerman**,
1930
[Alt. Denham's Bustard ssp.]
Western Green Tinkerbird ssp. *Pogoniulus coryphaea
jacksoni* **Sclater**, 1930 NCR
[Invalid name; replaced by *Pogoniulus coryphaea
hildamariae*]

Sir Frederick John Jackson (1859–1929) was an English administrator and explorer, but also a naturalist and keen ornithologist. He led a British expedition to make contact with Emin Pasha (q.v.) after the latter was isolated by the Mahdi's victory in the Sudan. He led another expedition financed by the British East Africa Company (1889) to explore the new Kenya colony, later becoming its first Governor. He was also Governor of Uganda (1911–1918), describing the country as 'a hidden Eden, a wonderland for birds'. He wrote *The Birds of Kenya Colony and the Uganda Protectorate*, published posthumously (1938). Five mammals and four reptiles are also named after him.

Jackson, S. W.

Yellow Robin sp. *Eopsaltria jacksoni* **Le Souef**, 1909 NCR
[Alt. Eastern Yellow Robin; JS *Eopsaltria australis chrysorrhos*]
Gerygone sp. *Pseudogerygone jacksoni* **A. J. Campbell**, 1912 NCR
[Alt. Western Gerygone; JS *Gerygone fusca exsul*]
Rufous Scrub-bird ssp. *Atrichornis rufescens jacksoni* **H. L. White**, 1920 NCR
[NUI *Atrichornis rufescens rufescens*]

Sydney William Jackson (1873–1946) was an Australian oologist, taxidermist and collector. Henry White (q.v.), the describer of the scrub-bird subspecies, bought his egg collection (1907) and employed him as a collector and as curator for the collection (1907–1927).

Jacobi

African Finfoot *Podica jacobi* **Reichenow**, 1906 NCR
[JS *Podica senegalensis camerunensis*]

Island Monarch ssp. *Monarcha cinerascens jacobii* **Neumann**, 1924 NCR
[NUI *Monarcha cinerascens cinerascens*]

Dr Arnold Friedrich Victor Jacobi (1870–1948) was a German zoologist and ethnographer who was Professor at the Technical University and Director, State Museum for Natural History, Dresden (1906–1937). He was in Tonkin (Vietnam) (c.1900). He wrote *Mimikry und Verwandte Erscheinungen* [*Mimicry and related phenomena*] (1913). A reptile is named after him.

Jacobs

Grey-headed Woodpecker ssp. *Picus canus jacobsii* **La Touche**, 1919 NCR
[JS *Picus canus guerini*]

Father Thadée Jacobs (DNF) was a Belgian Franciscan missionary to Hupeh, China.

Jacobson

Golden-bellied Gerygone ssp. *Gerygone modiglianii jacobsoni* **van Oort**, 1909 NCR
[JS *Gerygone sulphurea sulphurea*]
Great Eared Nightjar ssp. *Eurostopodus macrotis jacobsoni* **Junge**, 1936

Edward Richard Jacobson (1870–1944) was a Dutch businessman and skilled amateur naturalist. He managed a trading company in Java, but also lived for some years in Sumatra. He made extensive collections for Dutch museums, leaving his business (1910) to devote himself to natural history. His main interest was entomology, but he collected specimens of other taxa too. He died in an internment camp during the Japanese occupation. Two reptiles and an amphibian are named after him.

Jacot

Pine Grosbeak ssp. *Pinicola enucleator jacoti* Jenks, 1938 NCR
[JS *Pinicola enucleator montanus*]

Western Bluebird ssp. *Sialia mexicana jacoti* **A. R. Phillips**, 1991

Edouard C. Jacot (fl.1938) was an American field naturalist and collector in Arizona (1920s and 1930s). He and Jenks collected together in Arizona (1936 and 1937). He wrote 'Notes on the spotted and flammulated screech owls in Arizona' (1931).

Jacquet

Common Bush Tanager ssp. *Chlorospingus flavopectus jacqueti* **Hellmayr**, 1921

Dr H. Jacquet (DNF) was in Caracas (1920).

Jacquin

Whistling Duck sp. *Anas jacquini* **Gmelin**, 1788 NCR
[Presumed to be a JS of *Dendrocygna arborea*]
Trinidad Piping Guan *Pipile jacquini* **Reichenbach**, 1862 NCR
[JS *Pipile pipile*]

Joseph Franz Edler von Jacquin (1766–1839) was an Austrian analytical chemist and ornithologist who was Professor of Botany and Chemistry, University of Vienna (1797–1838), having inherited the positions from his father. Mozart, a friend of the family, taught his sister piano.

Jacquinot

Solomon Hawk Owl *Ninox jacquinoti* Bonaparte, 1850
Tongan Whistler *Pachycephala jacquinoti* Bonaparte, 1850

It is unclear whether Bonaparte named these birds after Honoré Jacquinot (1815–1887), who was a French naval surgeon and zoologist, or his older brother Charles Hector Jacquinot (1796–1879), a naval officer; especially as they sailed together on Dumont d'Urville's *Astrolabe* expedition (1837–1840). Charles was commander of the expedition corvette *Zélée* and Honoré was naturalist on the same vessel. The voyage visited the Antarctic and Australasia. While anchored off New Zealand, Honoré was able to describe and illustrate 15 species of local mollusc, plus several species of fish and crustacea, and is certainly commemorated in the names of some such taxa. Charles had been an ensign on the *Coquille* on his first trip to Antarctica (1826–1829), after which he was awarded the Cross of Honour. D'Urville named Mount Jacquinot after his best friend Charles and, after d'Urville's death, Charles supervised the publication of the last expedition's narrative and went on to achieve the rank of Vice Admiral. He was said to have been modest, no doubt a result of his request to be buried without military honours. As Hombron (q.v.), who was Honoré's collecting partner, was also honoured by Bonaparte, it seems likely that the birds were named after Honoré.

Jaczewski

Rufous-capped Antshrike ssp. *Thamnophilus ruficapillus jaczewskii* **Domaniewski**, 1925

Professor Tadeusz Jaczewski (1899–1974) was a Polish entomologist who was Professor of Zoology, Warsaw University (1948–1974). He specialised in the study of Hemiptera.

Jagor

Chestnut Munia ssp. *Lonchura atricapilla jagori*
K. E. Martens, 1866

Professor Dr Fedor Jagor (1817–1900) was a German ethnographer and naturalist who travelled in Asia, particularly the Philippines, in the second half of the 19th century, collecting for the Berlin Museum. He wrote *Reisen in den Philippinen* (1873). He described the country thus: 'Few countries in the world are so little known and so seldom visited as the Philippines, and yet no other land is more pleasant to travel in than this richly endowed island kingdom. Hardly anywhere does the nature lover find a greater fill of boundless treasure.' He also wrote about Indonesia and southern Malaya. Two mammals and two reptiles are named after him.

James, A. C.

Sharp-tailed Grouse ssp. *Tympanuchus phasianellus jamesi*
F. C. Lincoln, 1917
Canyon Towhee ssp. *Melozone fusca jamesi*
C. H. Townsend, 1923

Arthur Curtiss James (1867–1941) was an American patron of the sciences, especially in regard of the AMNH, New York. He was immensely rich, having inherited a fortune from his father, and he increased it greatly through investments in mining and railways.

James, C.

Trumpet Manucode ssp. *Phonygammus keraudrenii jamesii*
Sharpe, 1877

Dr C. James (DNF) was a British physician who collected in New Guinea and the Solomon Islands (1875).

James, F.

James's Tchagra *Tchagra jamesi* **G. E. Shelley**, 1885
[Alt. Three-streaked Tchagra]

Frank Linsly James (1851–1890) was an explorer of the Sudan, Somalia, India and Mexico. He published *Experiences and Adventures during Three Winters Spent in the Sudan* (1883) and *The Unknown Horn of Africa – an Exploration from Berbera to the Leopard River*, which was edited by his widow (1890). A wounded elephant killed him.

James, H. B.

James's Flamingo *Phoenicoparrus jamesi* **P. L. Sclater**, 1886
[Alt. Puna Flamingo]

Henry Berkeley James (1846–1892) was a British businessman who spent nearly 20 years in Chile. He began work as a clerk for a company in Valparaíso and at 25 managed a saltpetre mine near the Peruvian border. Shortly after being appointed he narrowly escaped death when an earthquake and tsunami destroyed his home, which was right next to the beach. He was a keen amateur naturalist, first collecting butterflies and moths after a mule journey to the remote Chanchamayo in central Peru, where he also wanted to see the endemic birds. Following the war between Chile and Peru

and Bolivia over saltpetre mining (1879), he returned to England. After marrying, he returned to Chile (1881) and began collecting birds himself, and buying them from other people. He published a 15-page pamphlet on the birds of Chile (1855), retiring to England soon after. He employed a German collector called Carlos Rhaner, who obtained the type specimen of James's Flamingo in Tarapacá, northern Chile. His *New List of Chilean Birds* was finished and published posthoumously by Sclater. Mrs Berkeley James presented the type specimen to the BMNH as part of the whole of James's collection of 1,382 skins and 678 eggs of Chilean birds. A reptile is named after him.

James, H. W.

Grey-backed Cisticola ssp. *Cisticola subruficapilla jamesi*
Lynes, 1930

Hubert William James (1883–c.1976) was an English ornithologist in South Africa and Rhodesia (Zimbabwe), where he was Keeper of Oology, Queen Victoria Museum, Salisbury (Harare) (1965).

Jameson, J. S.

Jameson's Firefinch *Lagonosticta rhodopareia* **Heuglin**, 1868
Jameson's Antpecker *Parmoptila jamesoni* **G. E. Shelley**, 1890
Jameson's Wattle-eye *Platysteira jamesoni* **Sharpe**, 1890

Jameson's Firefinch ssp. *Lagonosticta rhodopareia jamesoni*
G. E. Shelley, 1882

James Sligo Jameson (1856–1888) was an Irish hunter, explorer and naturalist. He collected in Borneo, South Africa, Spain, Algeria, the Rocky Mountains and, finally, the Belgian Congo, where he died of haemorrhagic fever at Bangala, whilst on an expedition with Stanley to rescue Emin Pasha (q.v.). According to his obituary in *The Times* (8 November 1890), he witnessed a cannibal banquet in the Upper Congo and was accused by Stanley of instigating it. He wrote *Story of the Rear Column of the Emin Pasha Relief Expedition*, which was published posthumously (1890).

Jameson, R.

Jameson's Gull *Larus jamesonii* J. Wilson, 1831 NCR
[Alt. Silver Gull; JS *Chroicocephalus novaehollandiae*]

Professor Robert Jameson (1774–1854) was a Scottish mineralogist, geologist and naturalist who became Regius Professor of Natural History at Edinburgh University (1804–1854). He started out as an apprentice to a surgeon with the aim of becoming a ship's surgeon. He changed his mind about medicine and the sea and decided to concentrate on minerals and the land. He was given the job of looking after Edinburgh University's Natural History Collection. He spent a year at Freiburg studying mining (1800). Among his more celebrated pupils at Edinburgh was Charles Darwin (q.v.), although Darwin, who, only 16 at the time, said he found Jameson's lectures boring. Gould (q.v.) illustrated the gull under the name *Xema jamesonii* in *Birds of Australia* (1848). A reptile is named after him.

Jameson, W.

Jameson's Snipe *Chubbia jamesoni* Jardine & **Bonaparte**, 1855
[Alt. Andean Snipe; Syn. *Gallinago jamesoni*]

Ecuadorian Hillstar ssp. *Oreotrochilus chimborazo jamesonii* **W. Jardine**, 1849
Jameson's Brilliant *Heliodoxa jacula jamersoni* **Bourcier**, 1851
[Alt. Green-crowned Brilliant ssp.; spelling of trinomial is often corrected to *jamesoni*]

Dr William Jameson (1796–1873) was a botanist and a Professor at the University of Quito, Ecuador. He prepared an account of Ecuadorian flora, *Synopsis Plantarum Quitensium* (1864–1865). An all-round naturalist, he collected holotypes of birds as well as many plants. He was among the first to descend into the crater of Chimborazo, an extinct volcano in the Ecuadorian Andes.

Jamrach

Cassowary sp. *Casuarius jamrachi* **Rothschild**, 1904 NCR
[Hybrid? = *Casuarius casuarius* x *C. unappendiculatus*]

William Jamrach (fl.1902) was an animal dealer in London, the son of Charles Jamrach (1815–1891), a leading importer and exporter of wild animals, selling to zoos, circus owners, and wealthy collectors. William sold the mounted holotype of the cassowary to Rothschild (q.v.). It came from an unknown location and its taxonomy is still unclear.

Janet

Desert Lark ssp. *Ammomanes deserti janeti* **Meinertzhagen**, 1933 NCR
[NUI *Ammomanes deserti geyri*]

Mrs Janet Wood (1907–1997) was the sister of Theresa Rachel Clay (q.v.), Meinertzhagen's (q.v.) 'companion'.

Jankowski, M. I.

Jankowski's Bunting *Emberiza jankowskii* **Taczanowski**, 1888
[Alt. Rufous-backed Bunting]

Jankowski's Swan *Cygnus columbianus jankowskyi* Alphéraky, 1904 NCR
[Alt. Tundra Swan; NUI *Cygnus columbianus bewickii*]

Mikhail Ivanovich Jankowski (1842–1912) was a Polish nobleman trained as a collector by Dybowski (q.v.). With Dybowski, Godlewski (q.v.) and others he was exiled in Siberia, following the Polish Uprising (1863). Although he was pardoned (1873) he stayed in Manchuria. He continued to collect birds for the Warsaw Zoological Collection and later for the Branicki Museum. After release exiles were obliged to settle in remote areas. With his fellow exiles he undertook a scientific expedition along the Amur River (1872–1874); here they studied the fauna of Lake Baikal and other regions of eastern Siberia with the help of the Zoological Cabinet back in Poland. He settled in Sidemi Bay (1875), which now bears his name, and there created an experimental model farm. He bred a prize-winning strain of horses and domesticated deer, building the first herd in Russia, and cultivated ginseng. He managed the goldfields on Askold Island (1876) and later a printshop. He was a renowned hunter (called 'four-eyes' by the locals) yet advocated conservation of the Taiga and its fauna. His interests were very wide-ranging including astronomy, archaeology (a local prehistoric culture still bears the name 'Jankowski Culture'), agriculture, geography and all natural history. He discovered three birds as well as over 100 species of butterfly, 17 of which carry his name. His son Y. M. Jankowski (q.v.) also had a bird named for him.

Jankowski, Y. M.

Eurasian Magpie ssp. *Pica pica jankowskii* Stegmann, 1928 NCR
[JS *Pica pica anderssoni*]

Yura (George) Michailowitch Jankowski (sometimes 'Yankovsky') (1879–1956) was a hunter who lived in the region of Vladivostock, and then (1923) North Korea. His father was Mikhail Ivanovich Jankowski (q.v.).

Japp

Eastern Clapper Lark ssp. *Mirafra fasciolata jappi* **Traylor**, 1962

Richard G. Japp (fl.1960–1969) was an amateur zoologist and collector in Barotseland, Northern Rhodesia (Zambia). He was local representative of the Witwaterstrand Native Labour Association at Kalabo, Zambia, recruiting labourers for the South African gold mines. He and his wife Hazel were Traylor's (q.v.) hosts and helped him plan his trip. He also collected in Namibia (1969).

Jardine, Capt. W.

Jardine's Parrot *Poicephalus gulielmi* **Sir W. Jardine**, 1849
[Alt. Red-fronted Parrot]

Captain William Jardine (1834–1869) was the son of Sir William Jardine (q.v.), the British naturalist. He was in the Congo (1849) and brought back a live specimen of the 'Jardine's Parrot', which he called 'Congo Jack'.

Jardine, Sir W.

Jardine's Harrier *Circus assimilis* Jardine & **Selby**, 1828
[Alt. Spotted Harrier]
Jardine's Triller *Coracina tenuirostris* Jardine, 1831
[Alt. Common Cicadabird, Slender-billed Cicadabird]
Jardine's Babbler *Turdoides jardineii* **A. Smith**, 1836
[Alt. Arrow-marked Babbler]
Jardine's Hummingbird *Boissonneaua jardini* **Bourcier**, 1851
[Alt. Velvet-purple Coronet]
Jardine's Pygmy Owl *Glaucidium jardinii* **Bonaparte**, 1855
[Alt. Andean Pygmy Owl]
Sunbird sp. *Nectarinia jardinii* **Hartlaub**, 1857 NCR
[JS Purple-banded Sunbird *Cinnyris bifasciatus*?]
Slaty Finch *Spodiornis jardinii* **P. L. Sclater**, 1866 NCR
[JS *Haplospiza rustica*]

Hairy Woodpecker ssp. *Picoides villosus jardinii* **Malherbe**, 1845

Cocoa Woodcreeper ssp. *Xiphorhynchus susurrans jardinei* **Dalmas**, 1900

Sir William Jardine, Seventh Baronet of Applegirth (1800–1874), was a Scottish ornithologist who created the *Naturalist's Library* (1833–1845). His private museum was among the finest of his day. He wrote *Illustrations of Ornithology* (1826) with Prideaux Selby (q.v.), *The Natural History of Hummingbirds* (1833) and *The Natural History of the Nectariniidae* (1834).

Jardine (River)

Black Butcherbird ssp. *Cracticus quoyi jardini* **Mathews**, 1912

White-streaked Honeyeater ssp. *Trichodere cockerelli jardinei* Mathews, 1917 NCR; NRM

We believe these two birds to be named after the Jardine River, Cape York Peninsula, Queensland, which is the type locality for the honeyeater and the area in which the butcherbird occurs. The lack of any etymology (usual for Mathews's descriptions) means that the butcherbird *could* be named after Sir William Jardine (q.v.).

Jarland

Eurasian Eagle Owl ssp. *Bubo bubo jarlandi* **La Touche**, 1921 NCR

[JS *Bubo bubo kiautschensis*]

Surgeon-Major Jarland (fl.1921) served with the French Colonial Army in Yunnan, China (1917–1921). He gave La Touche a pair of nestlings of this owl.

Jeanne

Blue Robin sp. *Heteroxenicus joannae* **La Touche**, 1922 NCR
[Alt. Blackthroat; JS *Luscinia obscura*]

Oriental White-eye ssp. *Zosterops palpebrosus joannae* La Touche, 1921 NCR
[NUI *Zosterops palpebrosus palpebrosus*]

Jeanne La Touche (DNF) was a daughter of J. D. D. La Touche (q.v.).

Jeffery

Fruithunter *Chlamydochaera jefferyi* **Sharpe**, 1887
[Alt. Black-breasted Fruithunter]
Bornean Green Magpie *Cissa jefferyi* Sharpe, 1888
[Syn. *Cissa thalassina jefferyi*]
Luzon Sunbird *Aethopyga jefferyi* **Ogilvie-Grant**, 1894
[Syn. *Aethopyga pulcherrima jefferyi*]
Philippine Eagle *Pithecophaga jefferyi* Ogilvie-Grant, 1896
[Alt. Monkey-eating Eagle]

Jeffery Whitehead (c.1832–1909) was an English stockbroker who financed his explorer and collector son, John Whitehead (q.v.).

Jelski

Golden-collared Tanager *Iridosornis jelskii* **Cabanis**, 1873
Jelski's Metaltail *Metallura jelskii* Cabanis, 1874 NCR
[JS *Metallura phoebe*]
Plain-breasted Earthcreeper *Upucerthia jelskii* Cabanis, 1874
Jelski's Bush Tyrant *Knipolegus signatus* **Taczanowski**, 1875
[Alt. Andean Tyrant]
Jelski's Chat Tyrant *Ochthoeca jelskii* Taczanowski, 1883
Jelski's Tinamou *Crypturellus rubripes* Taczanowski, 1886 NCR
[Alt. Pale-browed Tinamou; JS *Crypturellus transfasciatus*]
Stolzman's Flycatcher *Empidonomus jelskii* **Stolzman**, 1926 NCR
[Alt. Variegated Flycatcher; JS *Empidonomus varius rufinus*]

Jelski's Ocellated Piculet *Picumnus dorbignyanus jelskii* Taczanowski, 1882
[Alt. Ocellated Piculet ssp.]
Jelski's Woodnymph *Thalurania furcata jelskii* Taczanowski, 1874
[Alt. Fork-tailed Woodnymph ssp.]
Montane Foliage-gleaner ssp. *Anabacerthia striaticollis jelskii* Stolzman, 1926 NCR
[Dismissed as discoloured specimen of *Anabacerthia striaticollis montana*]

Konstanty Jelski (1837–1896) was a Russian ornithologist who discovered about 60 new species of birds. He was born in Minsk and studied medicine in Moscow (1853–1856), then natural history in Kiev (1856–1860). He became Curator in the Zoological Museum of Kiev (1862), but had to flee abroad after the January Revolt (1863), in which he participated. Finally, (1865) he went with the French navy to Guyana to become a collector for the Branickis (q.v.). Five years later he moved to Peru for reasons of health. There he became Curator at the Museum of Lima (1874). He was pardoned by the Russian Government (1878) and became Curator of the Krakow Museum (1878–1896). His specimens are displayed in museums in Lima, Paris, Warsaw and elsewhere. A mammal and an amphibian are named after him.

Jenson

Black-headed Antbird ssp. *Percnostola rufifrons jensoni* Capparella, Rosenberg & Cardiff, 1997

Peter Stonewall Jenson (1936–2010) was an American conservationist. Hamline University awarded his bachelor's degrees in archaeology and geology, and the University of Minnesota a master's degree. He founded (1964) Exploraciones Amazónicas (Explorama Tours), Iquitos, Peru, after completing two years Peace Corps service in Peru. He originally went on a short trip to Iquitos and never really left it.

Jentink

Barred Rail *Hypotaenidia jentinki* **Sharpe**, 1893 NCR
[JS *Gallirallus torquatus sulcirostris*]

Hair-crested Drongo ssp. *Dicrurus hottentottus jentincki* **A. G. Vordeman**, 1893

Dr Fredericus Anna Jentink (1844–1913) was the Director of the Dutch National Museum of Zoology at Leiden. He was one of five zoologists chosen by the Third International Congress of Zoology (Leiden, 1895) to deliberate and form a 'codex' on zoological nomenclature, the precursor of today's process. He described and named many animals, particularly mammals, and wrote numerous papers including: 'On two mammals from the Calamianes Islands' (1895) and 'Mammals collected by the members of the Humboldt Bay and the Merauke River Expeditions: Nova Guinea' (1907). More significant, however, were his *Catalogue Ostéologique des Mammifères* (1887) and *Catalogue Systématique des Mammifères* (1892). Four mammals are named after him. (Vorderman appears to have made the fairly common mistake of inserting a *c* into Jentink's name.)

Jerdon

Tree Warbler genus *Jerdonia* **Hume**, 1870 NCR
[Now in *Hippolais*]

Jerdon's Minivet *Pericrocotus erythropygius* Jerdon, 1840
[Alt. White-bellied Minivet]
Jerdon's Baza *Aviceda jerdoni* **Blyth**, 1842
[Alt. Blyth's Baza]
Jerdon's Leafbird *Chloropsis jerdoni* Blyth, 1844
[Alt. Indian Leafbird]
Jerdon's Bushlark *Mirafra affinis* Blyth, 1845
Jerdon's Nightjar *Caprimulgus atripennis* Jerdon, 1845
Jerdon's Leaf Warbler *Phylloscopus griseolus* Blyth, 1847
[Alt. Sulphur-bellied Warbler]
Prinia sp. *Drymoica jerdoni* Blyth, 1847 NCR
[Alt. Jungle Prinia; JS *Prinia sylvatica sylvatica*]
Jerdon's Courser *Rhinoptilus bitorquatus* Blyth, 1848
Indian Black-lored Tit *Parus jerdoni* Blyth, 1856 NCR
[JS *Parus (xanthogenys) aplonotus*]
Jerdon's Babbler *Chrysomma altirostre* Jerdon, 1862
Jerdon's Starling *Acridotheres burmannicus* Jerdon, 1862
[Alt. Vinous-breasted Starling]
Jerdon's Bushchat *Saxicola jerdoni* Blyth, 1867
Babbler sp. *Timalia jerdoni* **Walden**, 1872 NCR
[Alt. Chestnut-capped Babbler; JS *Timalia pileata bengalensis*]

Jerdon's Imperial Pigeon *Ducula badia cuprea* Jerdon, 1840
[Alt. Mountain Imperial Pigeon ssp.]
Eastern Orphean Warbler ssp. *Sylvia crassirostris jerdoni* Blyth, 1847
Rufous Woodpecker ssp. *Micropternus brachyurus jerdonii* **Malherbe**, 1849
Jerdon's Laughingthrush *Trochalopteron cachinnans jerdoni* Blyth, 1851
[Alt. Black-chinned Laughingthrush ssp.]
Grey-hooded Warbler ssp. *Phylloscopus xanthoschistos jerdoni* **W. E. Brooks**, 1871
Rufous-breasted Accentor ssp. *Prunella strophiata jerdoni* W. E. Brooks, 1872
Tickell's Blue Flycatcher ssp. *Cyornis tickelliae jerdoni* Holdsworth, 1872
Jerdon's Munia *Lonchura kelaarti jerdoni* Hume, 1874
[Alt. Black-throated Munia ssp.]

Little Ringed Plover ssp. *Charadrius dubius jerdoni* **W. V. Legge**, 1880

Thomas Claverhill Jerdon (1811–1872) was a British physician with both zoological and botanical interests. He was born in Durham and educated at the University of Edinburgh. He studied medicine and became an Assistant Surgeon in the East India Company. He published *Birds of India* (1862–1864), which according to Darwin (q.v.) was *the* book on Indian birds. He also wrote *Illustrations of Indian Ornithology* and *The Game Birds and Wildfowl of India*, as well as *Mammals of India* and writings on ants among others. A mammal, eight reptiles and four amphibians are named after him.

Jesse

Green-winged Pytilia ssp. *Pytilia melba jessei* **G. E. Shelley**, 1903

William Jesse Sr (1809–?1906) was a British zoologist. When the original nominee, Captain Bevan, fell ill, Jesse was the zoologist sent on the Abyssinian Expedition (1868) because of his experience collecting in South America. Otto Finsch (q.v.) wrote up the bird collection (1870) 'with notes by the collector William Jesse'.

Jessie (Bruce)

Gough Island Bunting *Nesospiza jessiae* **W. E. Clarke**, 1904 NCR
[JS *Rowettia goughensis*]

Mrs Jessie Bruce *née* Mackenzie (1870–1942) was the wife of the Scottish naturalist, oceanographer and Antarctic explorer, Dr William Speirs Bruce (1867–1921). He led the *Scotia* expedition (1902–1904) which collected the bunting holotype. She had been a nurse at the medical practice of William Bruce's father, in London. She was an alcoholic and violent towards her children, and after her husband's death she emigrated to Australia, leaving her children with guardians. Clarke described the bird, an immature *goughensis* with buffy plumage, as a different species.

Jessie (Riker)

Eared Dove ssp. *Zenaida auriculata jessieae* **Ridgway**, 1888

Mrs Jessie Riker (fl.1887) was in the Amazon region of Brazil (1884–1887) with her husband Clarence B. Riker (q.v.) and sons Bowman Riker (q.v.) and Herbert Riker (q.v. under **Herbert**).

Jesup

Yellow-bellied Chat Tyrant ssp. *Ochthoeca diadema jesupi* **J. A. Allen**, 1900

Morris Ketchum Jesup (1830–1908) was an American banker and philanthropist. He part-funded Peary's Arctic expeditions and funded the ethnographic Jesup North Pacific Expeditions (1897–1902). He was President of the AMNH, New York (1881) and one of the founders of the YMCA.

Jewett

American Goldfinch ssp. *Carduelis tristis jewetti* **Van Rossem**, 1943

Hermit Thrush ssp. *Catharus guttatus jewetti* **A. R. Phillips**, 1962

Dr Stanley Gordon Jewett (1885–1955) was an American ornithologist who collected for the Field Museum, Chicago, in Colombia and Venezuela (1910–1911). He worked in the Pacific Northwest of the USA for the US Fish & Wildlife Service on temporary contracts (1910–1916), then permanently (1916–1949). Oregon State College awarded his honorary doctorate (1953).

Jimenez

Ornate Tinamou ssp. *Nothoprocta ornata jimenezi* J. Cabot, 1997

Marcos Jimenez de la Espada (1831–1898) was a Spanish zoologist, explorer and writer. He studied natural sciences in the Universidad Complutense de Madrid (1850–1855), writing a study of amphibian taxonomy. His first job (1857) was as an Assistant Curator in the Natural History Department of the University at the same time undertaking a similar role at the Madrid Natural History Museum. He visited the Americas as part of the Pacific Scientific Commission (1862–1865) and studied the collections (786 species) made there, publishing articles about them (1866–1874) culminating in *Batrachians: Vertebrates from the Pacific trip* (1875). He presented his doctorial thesis years later (1898) just six months before he died. Four amphibians are named after him.

Jin

Indochinese Green Magpie ssp. *Cissa hypoleuca jini* **Delacour**, 1930

Dr Jin-Kwok-Jung Sun (DNF) was a Chinese scientist. He presented the holotype to the museum at Sun Yat-sen University in Canton (Guangzhou).

Jitkow

Common Starling ssp. *Sturnus vulgaris jitkowi* **Buturlin**, 1904 NCR

[Intergrade between *Sturnus vulgaris vulgaris* and *S. vulgaris poltaratskyi*]

Boris Mikhailovich Zhitkov (1872–1943) was an outstanding Russian zoologist and ornithologist. He was a Professor of Vertebrate Zoology at Moscow State University, the first scientific 'projecter' (manager at the planning stage) of the Astrakhan Nature Reserve, and the founder of a scientific school. He was particularly interested in the biology of the extreme northern regions of Russia and undertook expeditions and research in Novaya Zemlya, and the White Sea, River Volga and central Asia. The All-Russian Scientific Research Institute of Fur Farming and Hunting in Kirov (previously Vyatka) was named after him (1973). He co-wrote *The Ornithology of the Commander Islands* (1915). A mammal is also named after him.

Jitnikow

Great Tit ssp. *Parus major jitnikowi* **Zarudny**, 1910 NCR

[JS *Parus major intermedius*; or *Parus cinereus intermedius* if *P. major* is split into three species]

M. J. Zhitnikov (DNF) was a Russian collector in Persia (Iran) (1907). He wrote the article 'Ornithological observations in Atrek River basin: winter of 1898 and spring of 1899' (1900).

Joanna

(See **Jeanne**)

Johanna

Flowerpecker sp. *Prionochilus johannae* **Sharpe**, 1888 NCR

[Alt. Palawan Flowerpecker; JS *Prionochilus plateni plateni*]

Sharpe gave no etymology in his original description of the identity of Johanna. Mrs Sharpe was called Emily, and none of their 10 daughters had a name resembling Joan/Johanna. Perhaps Johanna was a relative of John Whitehead, the collector of the flowerpecker holotype? Or perhaps the bird is named after the wife of one of Sharpe's professional acquaintances such as Mrs Jane Sclater, the wife of Philip Lutley Sclater (q.v.) (see **Johanna (Sclater)** below).

Johanna (Island)

Madagascar Kingfisher ssp. *Corythornis vintsioides johannae* **Meinertzhagen**, 1924

[Syn. *Alcedo vintsioides johannae*]

Named after Johanna (Anjouan) Island in the Comoros. A reptile is also named after this island.

Johanna (Loddiges)

Blue-fronted Lancebill *Doryfera johannae* **Bourcier**, 1847

Jane Cooke *née* Loddiges (b.1812) was the daughter of British nurseryman and hummingbird specialist George Loddiges (q.v.). Both she and her husband, Edward William Cooke, were notable botanical painters.

Johanna (Sclater)

Bronze Ground Dove ssp. *Gallicolumba beccarii johannae* **P. L. Sclater**, 1877

Jane Anne Eliza Sclater *née* Hunter-Blair (1836–1915) was the wife of the describer, Philip Lutley Sclater; *johannae* is a latinisation of Jane.

Johanna (Verreaux)

Johanna's Sunbird *Cinnyris johannae* **E. & J. Verreaux**, 1851

[Alt. Madame Verreaux's Sunbird]

(See under **Madame Verreaux**)

Johannes

Johannes's Tody Tyrant *Hemitriccus iohannis* **E. Snethlage**, 1907

(See under **Iohannis**)

Johannis (= John C.)

Warsangli Linnet *Carduelis johannis* **S. Clarke**, 1919

John P. Stephenson Clarke (b.1896) was a colonial policeman in British Somaliland (Somalia) (1918–1920). He is one of the Clarke family, four of whom are mentioned in this book.

Johannis (= John M.)

Yellow-bibbed Fruit Dove ssp. *Ptilinopus solomonensis johannis* **P. L. Sclater**, 1877

(See John **Murray**)

Johansen, H.

Johansen's Bean Goose *Anser fabalis johanseni* **Delacour**, 1951

Prof. Hans Johansen (1897–1973) was a Danish/Russian ornithologist and collector, resident and explorer in Siberia (1916–1937), Professor at Tomsk University (1931–1937) and Professor and Head of Bird Ringing Centre, Museum of Zoology, University of Copenhagen (1943–1960). His father was Herman Eduardovich Johansen (below).

Johansen, H. E.

Common Cuckoo ssp. *Cuculus canorus johanseni* **Tschusi**, 1903 NCR
[JS *Cuculus canorus canorus*]
Common Starling ssp. *Sturnus vulgaris johanseni* **Buturlin**, 1904 NCR
[JS *Sturnus vulgaris porphyronotus*]

Prof. Herman Eduardovich Johansen (1866–1930) was a Russian/Danish zoologist and Professor at Tomsk University; his son Hans Johansen (q.v.) succeeded him on his death.

John (Waterstradt)

Cinnamon-breasted Whistler *Pachycephala johni* Hartert, 1903

(See **Waterstradt**)

John (Whitehead)

Greater Racket-tailed Drongo ssp. *Dicrurus paradiseus johni* Hartert, 1902

(See **Whitehead, J.**)

Johns

Black-throated Sunbird ssp. *Aethopyga saturata johnsi* **H. C. Robinson** & **Kloss**, 1919

John Francis Johns (1885–at least 1967), was a Student Interpreter in Siam (Thailand) (1907), Vice-Consul (1915) then Consul (1921) Senggora District, Thailand, Acting British Consul, Saigon (Ho Chi Minh City), Vietnam (1919) and Consul-General and Chargé d'Affairs, Bangkok, (1924–1932), retiring on grounds of ill-health. He gave assistance to Kloss (q.v.) during the latter's visit to Saigon. An amphibian is also named after him.

Johnson, A.

Pacific Tuftedcheek *Pseudocolaptes johnsoni* **Lönnberg** & **Rendahl**, 1922
[Syn. *Pseudocolaptes lawrencii johnsoni*]

Black-crested Bulbul ssp. *Pycnonotus flaviventris johnsoni* **Gyldenstolpe**, 1913

Axel Axelson-Johnson (1876–1958) inherited (1910) the company Axel Johnson AB that his father, Axel Johnson, had founded (1873) and which is still active. He also took over his father's position as Consul-General in Sweden for Siam (Thailand) (1911). He was honoured in bird names because he 'at many opportunities kindly has promoted the interests of this museum' (National Museum of Natural History, Stockholm).

Johnson, D.

Large-tailed Nightjar ssp. *Caprimulgus macrurus johnsoni* **Deignan**, 1955

Dr David H. Johnson (b.1912) was an American zoologist. He was a member of the Biological Society of Washington, which published a number of his papers, including 'The spiny rats of the Riu Kiu Islands' (1946). He worked at the USNM as Associate Curator (1941), becoming Curator (1957). He was in charge of the Mammals Division (1948–1965). An extinct mammal is named after him.

Johnson, G. R.

Southern Cassowary ssp. *Casuarius casuarius johnsonii* **F. Müller**, 1866
[Species often viewed as monotypic; taxonomy uncertain]

George Randall Johnson (1833–1919) was a barrister (1861) who played cricket for Cambridge and MCC (Marylebone Cricket Club). He left England (1865) and arrived in Queensland (1866) and then moved to New Zealand (1867) where he became a member of the Legislative Council (1872–1892). He returned to England (1892) and lived in Exeter. He shot a cassowary at Rockingham Bay and presented it to the Australian Museum. Müller's description, based on Johnson's account, ignored an earlier newspaper account (1854) of an Australian cassowary.

Johnson, H. R.

Blackcap Illadopsis ssp. *Illadopsis cleaveri johnsoni* **Büttikofer**, 1889

Dr Hilary Richard Wright Johnson (1837–1901) was the 11th President of Liberia (1884–1892) and the first Liberian president to be born in Africa.

Johnson, N. K.

Johnson's Tody Tyrant *Poecilotriccus luluae* Johnson & Jones, 2001
[Alt. Lulu's Tody Tyrant, Lulu's Tody Flycatcher]

Royal Sunangel ssp. *Heliangelus regalis johnsoni* **Graves** *et al.*, 2011

Professor Ned Keith Johnson (1932–2003) was Professor of Integrative Biology at the University of California, Berkeley, and Curator of Ornithology at its Museum of Vertebrate Zoology. His life-long love of birds started when, aged seven, he saw a Red-shafted Flicker *Colaptes (auratus) cafer* and could not believe how beautiful it was. In his career he collected and prepared over seven thousand specimens. He published his first paper aged 17 (1949) and went on to publish 125 papers and longer works, including co-authoring a *Checklist of North American Birds*. He was honoured '… *in recognition of his many contributions to avian systematics and biogeography.*' He died of cancer after a 15-year battle with the disease.

Johnston

Red-tufted Sunbird *Nectarinia johnstoni* **G. E. Shelley**, 1885
Lemon Dove *Haplopelia johnstoni* G. E. Shelley, 1893 NCR
[JS *Columba larvata larvata*]
Johnston's Turaco *Ruwenzorornis johnstoni* **Sharpe**, 1901
[Alt. Ruwenzori Turaco]

Johnston's Woodpecker *Dendropicos elliotii johnstoni* G. E. Shelley, 1887
[Alt. Elliot's Woodpecker ssp.]
Hildebrandt's Francolin ssp. *Pternistis hildebrandti johnstoni* G. E. Shelley, 1894 NCR, NRM
Yellow-throated Woodland Warbler ssp. *Phylloscopus ruficapilla johnstoni* **W. L. Sclater**, 1927

Sir Harry Hamilton Johnston (1858–1927) was a formidable English explorer and colonial administrator. He was a larger-than-life character and became known as the 'Tiny Giant', as he was just five feet tall. He was an accomplished painter, photographer, cartographer, linguist, naturalist and writer. He began exploring tropical Africa (1882) and met up with Henry Morton Stanley in the Congo (1883). He also travelled in Cameroon, Nigeria, Liberia, Mozambique, Nyasaland (Malawi), Angola, Southern Rhodesia (Zimbabwe), South Africa, Tanzania, Kenya, Uganda and North Africa. He was in East Africa (1884), then joined the colonial service (1885) taking various posts across Africa, in Cameroon, Nigeria, Liberia, Mozambique, Tunisia, Zanzibar and Uganda. He also established a British Protectorate in Nyasaland. Johnston was Queen Victoria's first Commissioner and Consul-General to British Central Africa. A member of the Royal Academy of Art, his paintings of African wildlife are exceptional. He spoke over 30 African languages, as well as Arabic, Italian, Spanish, French and Portuguese. He was knighted (1896) and after retirement (1904) he continued his pursuit of natural history. He discovered more than 100 new birds, reptiles, mammals and insects, the most notable being the okapi *Okapia johnstoni*. He wrote more than 60 books including *The Story of My Life* (1923), and more than 600 monographs and short articles. He made the very first Edison cylinder recordings in Africa, which preserved his squeaky voice for posterity. In addition to the okapi, five other mammals, three amphibians and two reptiles are named after him.

Johnstone, F.

Johnstone's Grassquit *Tiaris bicolor johnstonei* **P. R. Lowe**, 1906
[Alt. Black-faced Grassquit ssp.]

Sir Frederick John William Johnstone (1841–1913) was an English politician, racehorse owner and close friend of the Prince of Wales (q.v. as **Edward VII**). (See also **Laura (Johnstone)**)

Johnstone, M.

Johnstone's Lorikeet *Trichoglossus johnstoniae* **Hartert**, 1903
[Alt. Mindanao Lorikeet, Mount Apo Lorikeet]
Johnstone's Bush Robin *Tarsiger johnstoniae* **Ogilvie-Grant**, 1906
[Alt. Collared Bush Robin; Syn. *Luscinia johnstoniae*]

Scarlet Minivet ssp. *Pericrocotus speciosus johnstoniae* Ogilvie-Grant, 1905

Mrs Marion A. Johnstone (DNF) was a well-known aviculturist who received an award for being the first to breed the lorikeet, which is named after her. Goodfellow (q.v.) sent a bird to her in England (1903), the first of its kind to be imported into Europe.

Johnstone, R. E.

Rainbow Pitta ssp. *Pitta iris johnstoneiana* **Schodde** & Mason, 1999

Dr Ronald 'Ron' Eric Johnstone (b.1949) is an ornithologist, West Australian Museum, Perth. He collected the type specimen. He wrote *Handbook of Western Australian Birds* with G. M. Storr (1998). A mammal and a reptile are named after him.

Johnstone, R. S.

Euler's Flycatcher ssp. *Empidonax euleri johnstonei* **Barbour**, 1911
[Syn. *Lathrotriccus euleri flaviventris*]

Sir Robert Stewart Johnstone (1855–1936) was a barrister and colonial administrator who became Chief Justice, Grenada and the Grenadines (1909). He studied law at Trinity College Dublin and served as a Major in the Manchester Regiment. Barbour (q.v.) wrote that he '… very kindly aided the members of the Museum's expedition while at Grenada.' He was Commissioner for Lagos Western District (1889–1894) and an administrator in the Bahamas (1902–1914). He donated ethnographic items from Nigeria and the West Indies to the British Museum (1923–1929). An amphibian is also named after him. ([The taxonomy of the Grenada population is confusing: if the genus is *Empidonax*, the name *flaviventris* becomes preoccupied and is replaced by *johnstonei*; but if *Lathrotriccus* is retained, the valid name for this taxon is *Lathrotriccus euleri flaviventris*.)

Joicey

Seram Thrush *Zoothera joiceyi* **Rothschild** & **Hartert**, 1921

Belford's Honeyeater ssp. *Melidectes belfordi joiceyi* Rothschild, 1921

James 'Jimmy' John Joicey (1870–1932) was a British entomologist, lepidopterist and general naturalist, who founded the Hill Museum in Surrey. He travelled widely in Europe, Asia and the Americas. He was a lover of orchids, but bankrupted himself in the early 1900s and had to promise to abandon his orchid-collecting habit. He switched to Lepidoptera and by 1930 had amassed 380,000 specimens, only to go bankrupt again owing the then enormous sum of £300,000. His collection went to the BMNH in London.

Jolanda

Rinjani Scops Owl *Otus jolandae* Sangster *et al.*, 2013

Jolanda is the wife of the senior describer, George Sangster.

Joly

White-collared Jay ssp. *Cyanolyca viridicyanus jolyaea* **Bonaparte**, 1852

Professor W. Joly (1812–1885) was a French collector in Peru (1852). He was a Professor at Montpellier University (1839–1872).

Jones, A. E.

Puff-throated Babbler ssp. *Pellorneum ruficeps jonesi* **E. C. S. Baker**, 1920 NCR
[JS *Pellorneum ruficeps punctatum*]
Griffon Vulture ssp. *Gyps indicus jonesi* **Whistler**, 1927 NCR
[JS *Gyps fulvus fulvescens*]

Alexander Edward Jones (1878–1947), the collector of the holotype, was the owner of a tailoring business and an amateur ornithologist. He was an active member of the Bombay Natural History Society (1910–1947). He published a number of papers in the Society's journal, and was working on a comprehensive paper, 'Birds of Simla and adjacent hills' (1948), only part of which was published before he died, the rest being in note form, some, unfortunately in his private shorthand. He gifted (1944) his collection of 3,000 bird skins, most of which he collected himself, to the Society. He was also keen on butterflies and flowers. He died of heart failure.

Jones, H.

Jones's Silver Pheasant *Lophura nycthemera jonesi* **Oates**, 1903

Major Henry Jones (1838–1921) was an independently wealthy artist who specialised in painting pheasants and similar birds. After retiring from army service in India (1881) he worked in the Bird Room of the British Museum. At the turn of the 19th/20th century he produced around 1,200 watercolours of birds which he left to the Zoological Society of London. Nearly 90 years later, Paul A. Johnsgard used many as illustrations for his books *The Pheasants of the World* and *Bustards, Hemipodes, and Sandgrouse: Birds of Dry Places*. Oates (q.v.) wrote that he named the bird '... after my friend Major Henry Jones, who has greatly assisted me in studying and discriminating the various forms of Silver-Pheasants.'

Jones, M. P.

Nubian Nightjar ssp. *Caprimulgus nubicus jonesi* **Ogilvie-Grant & H. O. Forbes**, 1899 NCR
[JS *Caprimulgus nubicus torridus*]

Morris Paterson Jones (b.1847) was a Liverpool lawyer (1869–1921) and magistrate who was Chairman of the Museums Sub-Committee at Liverpool Museum. He was also a leading light in the Powysland Club, which sought to found a museum in North Wales.

Jordans

Lesser Spotted Woodpecker ssp. *Dendrocopos minor jordansi* **Götz**, 1925 NCR
[JS *Dendrocopos minor hortorum*]
Eurasian Thick-knee ssp. *Burhinus oedicnemus jordansi* **Neumann**, 1932 NCR
[NUI *Burhinus oedicnemus oedicnemus*]
Common Whitethroat ssp. *Sylvia communis jordansi* **Clancey**, 1950 NCR
[JS *Sylvia communis communis*]
Common Raven ssp. *Corvus corax jordansi* **Niethammer**, 1953 NCR
[JS *Corvus corax canariensis*]
Crested Lark ssp. *Galerida cristata jordansi* Niethammer, 1955
Eurasian Jay ssp. *Garrulus glandarius jordansi* Keve, 1966 NCR
[Uli *Garrulus glandarius albipectus*]

(See **von Jordans**)

Joret

Joret's Koklass (Pheasant) *Pucrasia macrolopha joretiana* **Heude**, 1883

Reverend Hippolyte Joret (1842–1901) was a Jesuit priest. He was Heude's fellow missionary in China and procured the pheasant holotype. A race of sika deer *Cervus nippon joretianus*, now considered invalid, was named after him. He wrote *Voyage au Hoei-tcheou fou* (1880).

Josefa

Brown-capped Vireo ssp. *Vireo leucophrys josephae* **P. L. Sclater**, 1859

Señora Josefa Borja y Davilos (DNF) lived in Ecuador where this race of vireo is found. The vireo was collected by Louis Fraser (q.v.) on property belonging to her family and was named after her by Sclater (q.v.) on Fraser's particular request.

Josephine (Finsch)

Josephine's Lorikeet *Charmosyna josefinae* **Finsch**, 1873

Josephine Finsch, *née* Wychodil (DNF) was the second wife of the German ornithologist Otto Finsch (q.v.), whom she married (1873) but later divorced. She accompanied her husband on his first trip to the South Seas (1879–1882).

Josephine (Lacroix & Mulsant)

Golden-tailed Sapphire ssp. *Chrysuronia oenone josephinae* **Bourcier** & **Mulsant**, 1848

Madame Joséphine Lacroix (DNF) had a husband, Julien, who was a deputy from Beaujolais in the French parliament (1848) and had been a benefactor to Bourcier (q.v.) and Joséphine Mulsant, the wife of Fleury Mulsant, counsel to the Court of Appeal in Chambéry.

Josephine (McConnell)

Boat-billed Tody Tyrant *Hemitriccus josephinae* **Chubb**, 1914

Pale-breasted Spinetail ssp. *Synallaxis albescens josephinae* Chubb, 1919

Mrs Josephine Vavasour de Laszlo, née McConnell (b.1906), was a daughter of Frederick Vavasour McConnell (q.v.) and Helen Mackenzie McConnell (q.v. under Helen (McConnell) and Mackenzie, H.). Chubb (q.v.) was a friend of the McConnell family and named these birds after their daughter.

Jouanin

Jouanin's Petrel *Bulweria fallax* Jouanin, 1955

Christian Jouanin (b.1925) is a French ornithologist who began his long association with the Paris museum at aged 15 as a pupil of Jacques Berlioz (q.v.), who was in charge of the bird department there. His interests have been divided between hummingbirds and petrels – an apparently strange association, but one shared with Osbert Salvin (q.v.) who was active in the 19th century. Jouanin pointed out the differences between 'his' petrel and Bonaparte's *Pterodoma aterrima* (now *Pseudobulweria aterrima*), describing it as a new species. This event stimulated his interest in the Indian Ocean and he began to search for the breeding grounds of *P. aterrima*, a species that had not been collected since the preceding century. His quest led to the discovery of another unknown gadfly petrel, which he named for Armand Barau (q.v.). Shortly afterwards he made the distinction between the Réunion and Seychelles populations of Audubon's Shearwater *Puffinus lherminieri*, naming the Seychelles representatives *nicolae* after his wife (1947). He participated with Francis Roux in G. E. Maul's (Director of the Funchal Museum in Madeira) expedition to the Selvagens. This resulted a lasting friendship with the Zinos. Jouanin is co-author with Mougin of the chapter on the '*Procellariiformes*' in Peters' *Check-list of the Birds of the World*, Vol 1, second edition (1978). He was General Secretary of the French Societé Nationale de Protection de la Nature (1966–1981). He was Vice-President of the International Union for Conservation of Nature (IUCN) (1970–1975), and a Permanent Executive Committee member of International Ornithological Committee (1970–1978).

Jouard

Willow Tit ssp. *Poecile atricapillus jouardi* **von Burg**, 1925 NCR
[JS *Poecile montanus montanus*]

Henri Jouard (1896–1938) was a French ornithologist and collector who was co-founder of the journal *Alauda*. He wrote 'Sous-espèces nouvelles de passereaux paléarctiques (Paridae et Carthiidae)' (1929).

Jourdain

Red-backed Shrike ssp. *Lanius collurio jourdaini* Parrot, 1910 NCR
[JS *Lanius collurio collurio*]
Red Phalarope ssp. *Phalaropus fulicarius jourdaini* **Iredale**, 1921 NCR; NRM
[Alt Grey Phalarope ssp.]

Reverend Francis Charles Robert Jourdain (1865–1940) was a British ornithologist, oologist, traveller and collector. He led the first Oxford University Expedition to Spitsbergen. He was President of the Oxford Ornithological Society, and a founder of the British Oological Association (1932–1939), now called the Jourdain Society in his memory.

Jourdan

Jourdan's Woodstar *Chaetocercus jourdanii* **Bourcier**, 1839
[Alt. Rufous-shafted Woodstar]

Claude Jourdan (1803–1873) was a French zoologist, geologist and Director of the Museum of Natural History, Lyons (1832–1869). He collected in Trinidad.

Jouy

Jouy's Wood Pigeon *Columba jouyi* **Stejneger**, 1887 EXTINCT
[Alt. Ryukyu Pigeon, Silver-banded Pigeon]
Thrush sp. *Turdus jouyi* Stejneger, 1887 NCR
[Alt. Brown-headed Thrush; JS *Turdus chrysolaus*]

Slaty-breasted Rail ssp. *Gallirallus striatus jouyi* Stejneger, 1887
Rufous-capped Warbler ssp. *Basileuterus rufifrons jouyi* **Ridgway**, 1892
Lesser Goldfinch ssp. *Carduelis psaltria jouyi* Ridgway, 1898
Grey Heron ssp. *Ardea cinerea jouyi* **A. H. Clark**, 1907

Pierre Louis Jouy (1856–1894) was an American diplomat, amateur naturalist and ethnographer. He wrote a handbook of a *Unique Collection of Ancient and Modern Korean and Chinese Works of Art, Procured in Korea During 1883* with Edward Grey (1888). He was collecting in Japan (1881) and Korea (1885–1889), mostly with Dr F. C. Dale. Several species of fish, which he collected whilst in Hong Kong and Shanghai (1882 and 1883), are named after him. Later (c.1892) he collected in Arizona and New Mexico. There are a number of articles relating to his travels in the USNM annual reports of the 1880s.

Juana Phillips

Spotted Owl ssp. *Strix occidentalis juanaphillipsae* **Dickerman**, 1997

Mrs Juana Farfán Bautista de Phillips (fl.1996) is Mexican and was the wife of the American ornithologist Allan R. Phillips (q.v.).

Judd

Song Sparrow ssp. *Melospiza melodia juddi* **L. B. Bishop**, 1896 NCR
[NUI *Melospiza melodia melodia*]

Elmer T. Judd (1866–1942) was an American ornithologist and field naturalist in North Dakota where he was also a prominent Freemason. He collected the holotype of the sparrow with Bishop (q.v.). He wrote and self-published *List of North Dakota Birds: Found in the Big Coulee, Turtle Mountains and Devils Lake Region* (1917).

Judith

Kelp Gull ssp. *Larus dominicanus judithae* Jiguet, 2002

Judith Jiguet (b.1972) is the wife of the describer Frédéric Jiguet, who is Associate Professor at the MNHN, Paris, and Director of Centre de Recherches sur la Biologie des Populations d'Oiseaux. His etymology states: 'This subspecies was named in honour of Judith who supported my one-year long works on seabirds at Kerguelen Islands.' (See **Mélisande**)

Jugurtha

Eurasian Hobby ssp. *Falco subbuteo jugurtha* **Hartert & Neumann**, 1907 NCR
[JS *Falco subbuteo subbuteo*]

Jugurtha (ca.160–104 BC) was King of Numidia (Algeria). He was defeated by the Romans, captured, and put to death.

Julia

Black-tailed Trainbearer ssp. *Lesbia victoriae juliae* **Hartert**, 1899

Hartert left no etymology and there are too many possible candidates to make even an inspired guess as to the 'Julia' in question. Oscar Theodor Baron collected the hummingbird holotype in Peru.

Julia (Tweeddale)

Whitehead's Spiderhunter *Arachnothera juliae* **Sharpe**, 1887

Purple-throated Sunbird ssp. *Leptocoma sperata juliae* **Tweeddale**, 1877

Julia, Marchioness of Tweeddale (1846–1937), was the wife of Arthur Hay, 9th Marquess of Tweeddale (q.v.).

Juliana

Black-backed Monarch *Monarcha julianae* **Ripley**, 1959
[Syn. *Symposiachrus julianae*]

Queen Juliana of the Netherlands (1909–2004) took the throne (1948) and reigned until abdicating (1980) in favour of her daughter, Queen Beatrice. Ripley wrote: 'This species is named, by gracious permission, in honor of Her Majesty, the Queen of the Netherlands', but did not explain further.

Julie

Julie's Hummingbird *Damophila julie* **Bourcier**, 1842
[Alt. Violet-bellied Hummingbird]

Anne-Julie Roncheval (DNF) became the wife of French naturalist Martial Etienne Mulsant (q.v.). Mulsant worked very closely with Bourcier (q.v.) and they jointly described many birds.

Julien

Pearly-breasted Cuckoo *Coccyzus julieni* **G. N. Lawrence**, 1864 NCR
[An earlier name for *Coccyzus euleri* (Cabanis, 1873), but suppressed; was regarded at one time as a JS of *C. americanus*]

Dr Alexis Anastay Julien (1840–1919) was an American geologist. He went to the guano island of Sombrero as the resident chemist (1860), staying in the Lesser Antilles (1860–1864). He sent his collections of birds and land shells to the USNM. He surveyed the islets around Sankt Bartholomeus (Saint Barthélemy) (1862) and visited Curacao, Bonaire and Aruba (1881–1882). The University of New York awarded his doctorate (1882). A reptile is named after him.

Julius

Tawny-breasted Tinamou *Nothocercus julius* **Bonaparte**, 1854

(See **Verreaux**)

June

Blossom-headed Parakeet ssp. *Psittacula roseata juneae* **Biswas**, 1951

June Davies (DNF) was an acquaintance of the describer and lived at Bovey Tracey in Devon, England. The Indian zoologist Biswamoy Biswas (1923–1994), when describing this subspecies of parakeet, said she was '… so much interested in ornithology'.

Junge

Blyth's Hornbill ssp. *Rhyticeros plicatus jungei* **Mayr**, 1937
[Species sometimes regarded as monotypic]
Little Bronze Cuckoo ssp. *Chrysococcyx minutillus jungei* **Stresemann**, 1938
Sulawesi Nightjar ssp. *Caprimulgus celebensis jungei* **Neumann**, 1939
Oriental Dwarf Kingfisher ssp. *Ceyx erithaca jungei* **Ripley**, 1942
[SII *Ceyx erithaca erithaca*]

Dr George Christofell Alexander Junge (1905–1962) was a Dutch ornithologist. He studied zoology at the University of Amsterdam, which awarded his doctorate (1934). Shortly afterwards he was offered the post of head of ornithology at the Rijksmuseum van Natuurlijke Historie, Leiden (1934–1962); this also included mammals until 1950. He wrote many papers particularly on the birds of Indonesia. He became the recognised expert on the birds of New Guinea despite never visiting the island, although he did spend some time in Java (1948–1949). He also studied the birds of Trinidad and Tobago. He died suddenly of a heart attack shortly after he had attended a meeting of the Netherlands Ornithological Union.

K

Kabali

Rufous-naped Lark ssp. *Mirafra africana kabalii*
C. M. N. White, 1943
White-browed Scrub Robin ssp. *Erythropygia leucophrys
kabalii* White, 1944 NCR
[JS *Erythropygia leucophrys munda*]

Kabali Muzeya (DNF) was an African collector employed by
Charles White. He collected for him around Kajilsha and
Isoka in Zambia (1943–1944) and a number of his skins are in
the collection at Livingstone Museum.

Kaeding

Kaeding's Storm-petrel *Oceanodroma leucorhoa kaedingi*
A. W. Anthony, 1898 NCR
[Alt. Leach's Storm-petrel; NUI *Oceanodroma leucorhoa
socorroensis*]

Henry Barroilhet 'H.B.' Kaeding (1877–1913) was a mineralogist and miner and amateur ornithologist who collected
with his brother. He met Chester Barlow (q.v.) and joined the
Cooper Ornithological Society through him. He visited
Mexico, Korea and Nicaragua for work and on private expeditions. He wrote 'In Mexican waters' (1897) in *The Nidologist*,
which included a first description of the eggs of the
Black-vented Shearwater *Puffinus opisthomelas*. Kaeding
also wrote 'Bird life on the Farallone Islands' (1903) and
'Birds from the west coast of Lower California and adjacent
islands' (1905) in *Condor*.

Kaempfer

Kaempfer's Tody Tyrant *Hemitriccus kaempferi* **Zimmer**,
1953
Kaempfer's Woodpecker *Celeus obrieni* Short, 1973
[Alt. Piaui Woodpecker]

Emil Kaempfer (DNF) was a German naturalist who amassed
a large collection of South American fauna for the AMNH
(1926–1931) while collecting birds in eastern Brazil. A catalogue based on his collection was published (1939) as
'Studies of birds from eastern Brazil and Paraguay'. The tody
tyrant was known from only two specimens until rediscovered (1991). Kaempfer collected the woodpecker type
specimen (1926), but it was not recorded again for eighty
years (2006).

Kaestner

Kaestner's Antpitta *Grallaria kaestneri* **Stiles**, 1992
[Alt. Cundinamarca Antpitta]

Peter Graham Kaestner (b.1953) is an American diplomat and
leading 'lister'. Born in Baltimore, Maryland, he graduated in
biology from Cornell University (1976). He then served as a
Peace Corps volunteer, teaching in a secondary school in
Nyankunde Village in Zaïre (Democratic Republic of Congo).
He entered the Foreign Service (1980), the perfect career for
a birder! He has served in India, Papua New Guinea,
Colombia, Malaysia, Namibia, Guatemala and Brazil. As a
birder, he is ranked in the top world listers (number 8 in 2012),
and is recognised by the Guinness Book of Records as
having been the first person to see a representative of each
of the 159 bird families. He discovered the antpitta (1989) in
the mountains of eastern Colombia.

Kalckreuth

Amethyst Sunbird ssp. *Chalcomitra amethystina kalckreuthi*
Cabanis, 1878
[SII *Chalcomitra amethystine kirkii*]

Lieutenant von Kalckreuth (DNF) was a Prussian explorer
and collector in East Africa. He was in Zanzibar (1864) and
travelled with Hildebrandt (q.v.) (1876–1877).

Kalimaya

Slender-billed Miner ssp. *Geositta tenuirostris kalimayae*
Krabbe, 1992

Kalimaya Krabbe (b.1985) is the daughter of the describer,
Danish ornithologist Niels Kaare Krabbe.

Kalinde

Turner's Eremomela ssp. *Eremomela turneri kalindei*
Prigogine, 1958

Kalinde Musika (DNF) was an African hunter employed by
Alexandre Prigogine (q.v.).

Kalinowski

Kalinowski's Tinamou *Nothoprocta kalinowskii* **Berlepsch** &
Stoltzman, 1901 NCR
[Alt. Ornate Tinamou; SII *Nothoprocta ornata branickii*]

Kalinowski's Scrub Blackbird *Dives warczewiczi kalinowskii*
Berlepsch & Stoltzman, 1892

Swallow-tailed Nightjar ssp. *Uropsalis segmentata kalinowskii* Berlepsch & Stolzmann, 1894
White-tipped Dove ssp. *Leptotila ochroptera kalinowskii* Stolzmann, 1926 NCR
[JS *Leptotila verreauxi decipiens*]
Barred Forest Falcon ssp. *Micrastur ruficollis kalinowskii* Dunajewski, 1938 NCR
[JS *Micrastur ruficollis zonothorax*]

Jan Kalinowski (1860–1942) was a Polish zoologist who led an expedition to Peru (1889–1902). Four mammals are named after him.

Kamla

Eyebrowed Jungle Flycatcher ssp. *Rhinomyias gularis kamlae* Leh, 2005

Kamla Sreedharan (DNF) was the wife of field ornithologist Slim Sreedharan, an associate of the Sarawak Museum, who collected the holotype. Her study of Borneo birds was short-lived, sadly cut short by a terminal illness.

Kamol

Black-throated Parrotbill ssp. *Suthora nipalensis kamoli* Eames, 2002

Kamol Komolphalin is a Thai conservationist and bird artist.

Kandt

Kandt's Waxbill *Estrilda kandti* **Reichenow**, 1902
Tinkerbird sp. *Barbatula kandti* Reichenow, 1903 NCR
[Alt. Yellow-rumped Tinkerbird; JS *Pogoniulus bilineatus jacksoni*]

Dr Richard Kandt (1867–1918) was a physician and the first German 'Resident' in Rwanda (1907–1913). He was sent on an expedition by the Kaiser and Bismarck to find the source of the Nile (1898). He wrote *Caput-Nili, eine Empfindsame Reise zu den Quellen des Nils* (1904). A mammal is named after him.

Kannegieter

Bar-bellied Cuckooshrike ssp. *Coracina striata kannegieteri* **Büttikofer**, 1896

Jan Zacharias Kannegieter (1862–1899) was a Dutch entomologist and collector in Ceylon (Sri Lanka) (1889) and in the East Indies; Sumatra (1889 and 1896), Java (1890–1891) and on Bangka Island (1898).

Kapustin

Willow Ptarmigan ssp. *Lagopus lagopus kapustini* **Serebrovsky**, 1926 NCR
[JS *Lagopus lagopus lagopus*]
Great Tit ssp. *Parus major kapustini* **Portenko**, 1954

W. Kapustin (fl.1926) was a Russian collector in Lapland and the Transbaikal region.

Karelin

European Green Woodpecker ssp. *Picus viridis karelini* **Brandt**, 1841
Great Tit ssp. *Parus major karelini* **Zarudny**, 1910

Grigory Silitsch Karelin (1801–1872) was a collector who explored in Central Asia (1832–1840) involving a number of voyages of discovery on the Caspian Sea and penetrating further into Cossack territory than any Russian had done before, culminating in his 'great expedition' (1840). An amphibian is also named after him.

Karlene

Karlene's Warbler *Geothlypis nelsoni karlenae* **R. T. Moore**, 1946
[Alt. Hooded Yellowthroat ssp.]

Karlene Wilhemina Pim *née* Moore (1915–1968) was the daughter of the describer, Robert Thomas Moore (q.v.), who wrote that she 'accompanied him and helped him on his last expedition in Ecuador.'

Karpov

Common Pheasant ssp. *Phasianus colchicus karpowi* **Buturlin**, 1904
Pine Bunting ssp. *Emberiza leucocephalos karpovi* **Zarudny**, 1913 NCR
[JS *Emberiza leucocephalos leucocephalos*]

Professor Vladimir Pavlovich Karpov (1870–1943) was a Russian biologist, Professor of Histology at Moscow State University, and was in Manchuria (1901).

Kasnakow

Babax genus *Kasnakowia* **Bianchi**, 1904 NCR
[Now in *Babax*]

Aleksandr Nikolaevich Kaznakov (1872–1932) was a Russian naturalist who became Director, Caucasus Museum. He accompanied Kozlov (q.v.) on his expedition to Mongolia (1907–1909). A reptile is also named after him.

Kastschenko

Tawny Pipit ssp. *Anthus campestris kastschenkoi* **Johansen**, 1952

Dr Nikolai Feofanovich Kastschenko (1858–1935) was a physician and founding Professor of Zoology at Tomsk University (1888). He described and named a number of mammals, especially marmots. During his time at Tomsk he undertook and encouraged wildlife research in the area and led expeditions into wider Siberia. His actions and studies led to his university being recognised as the centre for zoology in Siberia. He published a number of papers, including one on the results of his 1898 zoological expedition, and wrote the first manual in Russian on biology for medical students. His collections provided the beginning of the museum at the university (now comprising 120,000 zoological specimens). A mammal is named after him.

Katharine (Bowen)

Black-faced Sandgrouse ssp. *Pterocles decoratus*
   katharinae **Bowen**, 1930 NCR
[JS *Pterocles decoratus loveridgei*]

Mrs Katharine Winthrop Bowen *née* Knaebel (1901–1987) was the wife of the describer, Wilfrid Wedgwood Bowen (q.v.).

Katherine (de Vis)

Mountain Thornbill *Acanthiza katherina* **De Vis**, 1905

Katherine Elizabeth de Vis (formerly Luckle) *née* Coulson (DNF) was the second wife of the describer, English naturalist Charles De Vis (q.v.).

Kathleen (Phelps)

Red-banded Fruiteater ssp. *Pipreola whitelyi kathleenae*
   Zimmer & **W. H. Phelps**, 1944

Mrs Kathleen Phoebe Phelps *née* Deery (1908–2001) was the wife of William H. Phelps Jr (q.v.). She collected the holotype.

Kathleen (White)

Boulton's Batis ssp. *Batis margaritae kathleenae* **C. M. N.**
   White, 1941
Crested Guineafowl ssp. *Guttera edouardi kathleenae*
   C. M. N. White, 1943 NCR
[NUI *Guttera pucherani verreauxi*]

Miss Kathleen A. A. White (DNF) was the sister of the describer, Charles Matthew Newton White (q.v.).

Katona

Short-winged Cisticola ssp. *Cisticola brachypterus katonae*
   Madarász, 1904

This is an alternative Hungarian rendering of Kittenberger (q.v.).

Katsumata

Hainan Peacock Pheasant *Polyplectron katsumatae*
   Rothschild, 1906

Indochinese Green Magpie ssp. *Cissa hypoleuca*
   katsumatae Rothschild, 1903

Zensaku Katsumata (DNF) was a Japanese pharmacist who collected birds on Hainan (1902–1903) for Alan Owston (q.v.), the Yokohama dealer who supplied them to Lord Rothschild (q.v.).

Kaup

Lizard Buzzard genus *Kaupifalco* **Bonaparte**, 1854
Tyrant Flycatcher genus *Kaupornis* Bonaparte, 1854 NCR
[Now in *Myiarchus*]

White-browed Hawk *Leucopternis kaupi* Bonaparte, 1850
NCR

[JS *Leucopternis kuhli*]
Kaup's Flycatcher *Arses kaupi* **Gould**, 1851
[Alt. Pied Monarch]

Johann Jakob von Kaup (1803–1873) was a German zoologist, ornithologist and palaeontologist who was the Director of the Grand Duke's natural history 'cabinet' in Darmstadt. He wrote *Classification der Säugethiere und Vögel* (1844). An amphibian is named after him.

Kawall

Kawall's Amazon *Amazona kawalli* **Grantsau** & **Camargo**,
   1989
[Alt. White-faced Amazon]

Nelson Kawall (1928–2006) was a leading Brazilian aviculturist and exponent of captive breeding programmes. He kept and bred budgerigars as a child and was given his first parrot, a Blue-fronted Amazon *Amazona aestiva*, when he was 12 years old. He established himself as a commercial breeder (1968) whose company AVEX was involved in the import and export of parrots. Ornithologist Rolf Grantsau (q.v.) became a friend. He was the first Brazilian breeder to sex his birds by laparoscopy, then karyotype and the first to captive-breed several species. At one time he had 50 different species of parrots in the facilities he constructed in his yard. His collection included a pair of Spix's Macaws *Cyanopsitta spixii*. He stopped trading (1978) because of changes in Brazilian legislation on commercial exploitation of native fauna, transferring the Brazilian species to his house, where he continued his breeding activities privately. While visiting (1968) a fellow bird lover, José Xavier de Mendonça, in Santarém, Pará, he saw for the first time a parrot that displayed characteristics of the Mealy Amazon *Amazona farinosa* but with very evident differences. Taking the bird to São Paulo he spotted another specimen in the aviary of another breeder, Alcides Vertamatti. Kawall always felt that José Xavier de Mendonça was the true discoverer. Kawall discontinued as a breeder (1997) but was important counsellor for many young ornithologists and breeders. He died in late 2006 of a heart attack whilst recovering from pneumonia.

Keartland

Australian chat genus *Keartlandia* **Mathews**, 1917 NCR
[Now in *Epthianura*]

Keartland's Honeyeater *Ptilotula keartlandi* **North**, 1895
[Alt. Grey-headed Honeyeater; Syn. *Lichenostomus
   keartlandi*]

George Arthur Keartland (1848–1926) was an Australian botanist, ornithologist and field naturalist. He was a member of the Calvert expedition (1897–1897), which set out to fill in blanks in the map of Australia. They were lucky that only two members of the expedition died. They had to abandon their equipment and collections, including 3,000 bird skins. The expedition was re-created (1966) but used Land Rovers instead of camels!

Keast

Grey Fantail ssp. *Rhipidura albiscapa keasti* **Ford**, 1981

Dr James Allen Keast (1922–2009) was an Australian ornithologist, Curator of Birds and Reptiles, Australian Museum, Sydney (1955–1960) and a pioneer television presenter (1958–1960). He served in the Australian forces in WW2 in New Guinea and New Britain (1941–1945). The University of Sydney awarded his bachelor's degree (1950) and his master's (1952), and Harvard his doctorate (1955). He was Professor of Biology, Queens University, Ontario, Canada (1962–1989). He wrote *Bush Birds of Australia* (1973).

Keay

Keay's Bleeding-heart *Gallicolumba keayi* **W. E. Clarke**, 1900
[Alt. Negros Bleeding-heart]

William A. Keay (DNF) was an English sugar planter (about 1880–1900) at San Carlos, Negros (Philippines).

Keays

Pygmy Tyrant sp. *Ochthoeca keaysi* **F. M. Chapman**, 1901
NCR
[Alt. Hazel-fronted Pygmy Tyrant; JS *Pseudotriccus simplex*]

Herbert H. Keays (DNF) was a mining engineer who collected in Peru for the AMNH, including a new species of swift (1899). He was associated with the Inca Mine in Peru and eventually settled in Prescott, Arizona, offering to continue as a collector there for J. A. Allen (q.v.). Two mammals are named after him.

Keith

Rattling Cisticola ssp. *Cisticola chiniana keithi* **Parkes**, 1987

George Stuart Keith (1931–2003) was a British/American ornithologist, pioneer sound-recordist, and founding President of the American Birding Association (1970). He was evacuated to Canada during WW2, returning to England (1945) and eventually obtaining a degree in classics at Oxford, after serving in the British army in the Korean War. He became a research associate in the Ornithology Department, AMNH, New York (1958). At the time of his death he had seen over 6,500 different species of birds. (See also **Stuart Keith**).

Kelaart

Yellow-eared Bulbul genus *Kelaartia* **Jerdon**, 1863
[Genus often merged with *Pycnonotus*]

Black-throated Munia *Lonchura kelaarti* Jerdon, 1863
Legge's Hawk Eagle *Nisaetus kelaarti* **Legge**, 1878

Kelaart's Nightjar *Caprimulgus indicus kelaarti* **Blyth**, 1851
[Alt. Jungle Nightjar ssp.]

Lieutenant-Colonel Dr Edward Frederick Kelaart (1819–1860) was a British physician and zoologist born in Ceylon (Sri Lanka). He qualified as a doctor at Edinburgh and also attended classes in medicine in Paris. While at Edinburgh he made his first contribution as a naturalist, delivering a paper (1839) on 'The timber trees of Ceylon'. He was in the Ceylon medical service and also served in Gibraltar (1843–1845). Whilst stationed there he studied the plants of the Rock of Gibraltar and wrote 'Flora Calpensis – a contribution to the botany and topography of Gibraltar and its neighbourhood'. He returned to Ceylon (1849–1854) and wrote *Prodromus Fauna Zeylanica* (1852), the first work to give scientific classifications to the mammals of Ceylon. Also in these five years he produced a great deal of work including publications on geology, mammals, birds, reptiles and the cultivation of cotton. He was appointed Naturalist to the Ceylon Government, which paid £200 per annum (a lot of money in the middle of the 19th century) plus expenses on top of his army pay! One of his tasks as official naturalist was to investigate why the Ceylon pearl fisheries had not produced any profit. He investigated the life history of the pearl oyster and produced no less than four reports on the subject, which were published, some posthumously (1858–1863). When the Governor of Ceylon fell ill, and went home to England on board the *Nubia* (1860), his health was of such concern that Kelaart, accompanied by his wife and five children, was sent along as the Governor's medical attendant. The Governor died two days before *Nubia* arrived at Southampton, and Kelaart died the next day and was buried in Southampton soon after the ship docked. Three mammals, two reptiles and an amphibian are named after him.

Kellen

Babbling Starling *Neocichla kelleni* **Büttikofer**, 1888 NCR
[JS *Neocichla gutturalis gutturalis*]

Pieter Jacob van der Kellen (DNF) of the Dutch Natural History Museum, Leiden, took part in the Dutch Ethnographic Museum's South West Africa expedition (1884–1885) and stayed on in Angola (1885–1888). The leaders of the expedition were D. D. Veth and L. D. Godeffroy. Van der Kellen's job was to act as hunter and to search out and collect 'interesting' animals. He returned to Africa and is recorded as exploring in the area around Mossamedes, Angola (1896), and he also led an expedition for the Mossamedes Company to the Zambezi (1899–1900). Hermann Baum, a botanist, was a member of this expedition and wrote the trip report (1903), and he and van der Kellen published descriptions of a number of African ants. A mammal is also named after him.

Keller

Island Thrush ssp. *Turdus poliocephalus kelleri* **Mearns**, 1905

Fletcher L. Keller (DNF) was a hemp-planter in Davao, Mindanao, Philippines. He accompanied Mearns (q.v.) on an expedition to Mount Apo. A mammal is named after him.

Kelley

Kelley's Tit Babbler *Macronus kelleyi* **Delacour**, 1932
[Alt. Grey-faced Tit Babbler; Genus name often spelt *Macronous*]

William Vallandigham Kelley (1861–1932) was an American philanthropist and sponsor who was a successful businessman in Chicago and a member of one of Darke County's pioneer families. There were a number of Kelley-Roosevelt Field Museum Expeditions (1920s and 1930s), and there were also Delacour expeditions around this time. It is likely that they knew each other and that Kelley sponsored Delacour's work.

Kellogg, C. L.

Kellogg's Rock Ptarmigan *Lagopus muta kelloggae* **Grinnell**, 1910
[SII *Lagopus muta nelsoni*]

Clara Louise Kellogg (1879–1967) was an intrepid naturalist who explored the American West from Alaska to the tip of Baja California for more than half a century, in conjunction with Annie M. Alexander (q.v.). These two women contributed more than 25,000 fossil, plant and animal specimens to the University of California's Natural History Museum. Kellogg had published the second-ever paper on mammalogy in the USA known to be authored by a woman (1910). The Louise Kellogg research award of the Museum of Vertebrate Zoology at the University of California, Berkeley, is given in her memory.

Kellogg, J. P. & L. J.

Hildebrandt's Starling ssp. *Lamprotornis hildebrandti kelloggorum* **Neumann**, 1944 NCR; NRM

John Payne Kellogg (1897–1980) and Mrs Louise Jewett Kellogg *née* Mitchell (1903–1965) lived at Libertyville, Illinois, and were on a shooting safari in Tanganyika (Tanzania) (1931) when they collected the starling holotype.

Kelsall

Fraser's Forest Flycatcher ssp. *Fraseria ocreata kelsalli* **Bannerman**, 1922
Olive-green Camaroptera ssp. *Camaroptera chloronota kelsalli* **W. L. Sclater**, 1927

Colonel Harold Joseph Kelsall (1867–1950) was a British army officer (Royal Artillery), ornithologist and plant-collector in the Straits Settlements, Malaya (Malaysia) (1892), Sierra Leone (1910–1913) and Colombia (1924). He wrote 'Notes on a collection of birds from Sierra Leone' (1914) and named a subspecies for his wife Helen (q.v.).

Kelso

Tropical Screech Owl ssp. *Megascops choliba kelsoi* **Hekstra**, 1982 NCR
[JS *Megascops choliba crucigerus*]

Dr Leon Hugh Kelso (1907–1982) was an American ornithologist, botanist, collector and publisher of the journal *Biological Leaflets*. He joined the Department of Agriculture, US Biological Survey (1930). He and his wife Estelle H. Kelso wrote 'A key to species of American owls' (1934).

Kemp, A. C.

Western Red-billed Hornbill *Tockus kempi* Treca & **Erard**, 2000

Dr Alan Charles Kemp (b.1944) is a freelance ornithological and natural history consultant and guide. Rhodes University, South Africa, awarded both his BSc in zoology and his doctorate (1973). He was at the Department of Birds, Transvaal Museum, Pretoria (1969–2001) as Head Curator (1974) then Museum Manager (1999–2001). He is an Honorary Research Associate, Percy FitzPatrick Institute of African Ornithology, University of Cape Town. He is a leading expert on hornbills and birds of prey, and has written many scientific papers on avifauna including the monograph *The Hornbills* (1995).

Kemp, R.

Flyrobin genus *Kempia* **Mathews**, 1912 NCR
[Now in *Microeca*]

Kemp's Longbill *Macrosphenus kempi* **Sharpe**, 1905

Kemp's Olive-bellied Sunbird *Cinnyris chloropygius kempi* **Ogilvie-Grant**, 1910
Black-backed Butcherbird ssp. *Cracticus mentalis kempi* Mathews, 1912
Black-naped Tern ssp. *Sterna sumatrana kempi* Mathews, 1912 NCR
[JS *Sterna sumatrana sumatrana*]
Blue-winged Kookaburra ssp. *Dacelo leachii kempi* Mathews, 1912 NCR
[JS *Dacelo leachii leachii*]
Little Shearwater ssp. *Puffinus assimilis kempi* Mathews, 1912 NCR
[JS *Puffinus (assimilis) elegans*]
Red-headed Myzomela ssp. *Myzomela erythrocephala kempi* Mathews, 1912 NCR
[JS *Myzomela erythrocephala erythrocephala*]
Rufous Fantail ssp. *Rhipidura rufifrons kempi* Mathews, 1912 NCR
[JS *Rhipidura rufifrons rufifrons*]
Silver-crowned Friarbird ssp. *Philemon argenticeps kempi* Mathews, 1912
Common Waxbill ssp. *Estrilda astrild kempi* **Bates**, 1930

Robert 'Robin' Kemp (b.1871) was originally an accountant who worked for a company building a railway in Sierra Leone. However, he became a naturalist and professional collector and combined his collecting in Sierra Leone (1902–1904) with his day job. He was in New Zealand (1906–1908) and collected in East Africa (1908–1911), in Australia (1912–1914), and in Argentina (1916–1917). Four mammals are named after him. (See also **Robin (Kemp)**)

Kemp, S. W.

White-throated Fantail ssp. *Rhipidura albicollis kempi* **E. C. S. Baker**, 1913 NCR
[NPRB *Rhipidura albicollis stanleyi*]

Dr Stanley Wells Kemp (1882–1945) was a zoologist and anthropologist. He joined the Fisheries Research Section,

Department of Agriculture, Dublin (1903) as Assistant Naturalist. He joined the Indian Museum, Calcutta (1911) as Superintendent, Zoological Section. There he worked very closely with the Scottish zoologist Dr Nelson Annandale (in 1925 he would write Annandale's obituary in the *Records of the Indian Museum*). He was on the Abor Punitive Expedition (1911–1912), during which government scientists made extensive natural history collections including in the Garo Hills. Kemp later joined the Colonial Office (1924) as Director of Research, Discovery Committee, and led the second Antarctic Discovery Expedition (1924) (relating to Whale Fisheries). He was Director of the Plymouth Marine Laboratory (1936–1945). He lost all his personal possessions, his library, and his unpublished works as the result of a German air raid (1941). Three amphibians are named after him and his wife.

Kempff

Grey-throated Leaftosser ssp. *Sclerurus albigularis kempffi* Kratter, 1997

Professor Noel Kempff Mercado (1924–1986) was a Bolivian biologist, ornithologist and conservationist. The University of Santa Cruz awarded his bachelor's degree (1946). He and some colleagues were murdered by a gang of cocaine producers and smugglers, whom they had taken by surprise in a national park – that park in Bolivia is now named after him, as is an amphibian.

Kempton

Black-hooded Thrush ssp. *Turdus olivater kemptoni* **W. H. Phelps** & **W. H. Phelps Jr**, 1955

Dr James Harry Kempton (1891–1970) was an American scientist at the Department of Agriculture, Caracas, Venezuela (1942–1958). The describers noted his 'good humored companionship' during an expedition they shared with him.

Kemsies

Red-tailed Hawk ssp. *Buteo jamaicensis kemsiesi* **Oberholser**, 1959

Dr Emerson Kemsies (1905–1970) was an American ornithologist. He was Curator of Ornithology, University of Cincinnati (1947). He co-wrote *Birds of Southwestern Ohio* (1953).

Kennedy

Cape Batis ssp. *Batis capensis kennedyi* **Smithers** & **Paterson**, 1956

Major-General Sir John Noble Kennedy (1893–1970) was a British army officer who originally joined the navy, but switched services (1914) and fought in France, Flanders and the Middle East during WW1. In WW2 he held a number of important staff positions in London, including being Assistant Chief of the Imperial General Staff (1943–1945). He was Governor of Southern Rhodesia (Zimbabwe) (1947–1953).

Kennicott

Kennicott's Screech Owl *Megascops kennicottii* **D. G. Elliot**, 1867
[Alt. Western Screech Owl]

Kennicott's Arctic Warbler *Phylloscopus borealis kennicotti* **Baird**, 1869

Robert Kennicott (1835–1866) was an American naturalist who founded the Chicago Academy of Sciences, and who explored the American Northwest (1857–1859). At 17 he was sent to study under Dr Jared Potter Kirtland (q.v.) in Cleveland. Kennicott worked for Baird (q.v.) at the USNM, largely helping to classify animals collected on the western frontier by army personnel involved in railroad surveys. Through Baird he went to Canada and met Hudson's Bay's chief trader, Bernard Ross (q.v.), who became a close friend. One of Baird's biographers described Kennicott thus: 'He became the consummate collector, and when more demanding responsibilities intruded upon his direct involvement in collecting and classifying, he became a collector of collectors. Under his training and guidance virtually all the major natural scientists of the nineteenth century developed their enthusiasms and their professional competence.' After a period as Curator in Chicago he left to explore 'Russian America' and spent the rest of his life in Alaska. Kennicott suffered a second and fatal heart attack near Nulato, Alaska, at the age of just 30 (1866). A town in Alaska now bears his name, and he is also commemorated in the names of several fish.

Kenrick

Kenrick's Starling *Poeoptera kenricki* **G. E. Shelley**, 1894

Major Reginald Watkin Edward-Kenrick (DNF) was a British Army officer in Kenya. The type specimen in the BMNH was first discovered and apparently collected by Kenrick in the Usambara Mountains. Kenrick presented it to the museum (1894). The museum also has a specimen of a young female (collected 1879) and it is not entirely clear which of the two birds is Kenrick's specimen.

Kenyon

Kenyon's Shag *Phalacrocorax kenyoni* Siegel-Causey, 1991 NCR
[NUI *Phalacrocorax pelagicus*]

Karl Walton Kenyon (1918–2007) was a wildlife biologist and marine mammalogist. He studied at Pomona College and Cornell University (1930s) and served as an aviator in the US Navy in WW2. He taught at Mills College for two years before joining the US Fish and Wild Life Service in Seattle. He collected the type specimen of the shag in 1959 at Amchitka, Aleutian Islands. The shag was not described until 1991, but its taxonomic status remains in doubt. Rohwer *et al.* in their 'Critical evaluation of Kenyon's Shag' in *Auk* (2000) virtually dismiss it and conclude that it is within the variation possible for the Pelagic Cormorant *Phalacrocorax pelagicus*. Kenyon also carried out a 4,000-mile aerial search for the Caribbean

Monk Seal *Monachus tropicalis* (1973), concluding that the species is extinct as was feared.

Kepler

Kepler's Crake *Porzana keplerorum* **Olson** & H. F. James, 1991 EXTINCT

Dr Cameron B. Kepler and his wife, Angela (q.v.), are both honoured in the binomial. He gained both his bachelor's and master's degrees at the University of California, Santa Barbara, and Cornell awarded his doctorate while he was working at the USNM. His entire working life has been with the US Fish & Wildlife Service.

Keraudren

Keraudren's Manucode *Phonygammus keraudrenii* **Lesson** & **Garnot**, 1826
[Alt. Trumpet Manucode; Syn. *Manucodia keraudrenii*]

Pierre François Keraudren (1769–1858) was a French military physician. He was trained at the School of Naval Medicine, Brest (1782), and was Ship's Surgeon aboard the *Géographe* (1801) under Nicolas-Thomas Baudin (q.v.) on trips to Mauritius, Australia and elsewhere. The manucode must have been collected in New Guinea during one of these voyages. Baudin named Cape Keraudren in Western Australia, now a nature reserve, after him. He was appointed by Napoleon to be the French Navy's Medical Corps Inspector-General (1813–1845). He wrote a description of the cholera outbreak in India (1831).

Kerdel

Green Violetear ssp. *Colibri thalassinus kerdeli* **Aveledo** & Perez, 1991

Dr Francisco Kerdel-Vegas (b.1928) is a Venezuelan physician and dermatologist who graduated (1951) and took a master's degree at New York University (1954). He became a professor of dermatology at Hospital Vargas de Caracas, Venezuela, before becoming a diplomat who was Ambassador to United Kingdom (1987–1992), and to France (1995–1999).

Kermit

Kermit's Antwren *Myrmotherula kermiti* **Cherrie**, 1916 NCR
[Alt. Sclater's Antwren; JS *Myrmotherula sclateri*]

Kermit Roosevelt (1889–1943) was the son of Theodore Roosevelt (1858–1919), the 26th President of the USA, and a keen amateur ornithologist. He accompanied his father on a number of his expeditions, including that to Brazil (1913). Cherrie was on the same expedition and named the bird in honour of his employer's son. Like his father, Kermit was an explorer, hunter and writer. In WW1 he fought in the British Army, transferring to the American Army after the USA entered the war. He then went into business and founded both the Roosevelt Steamship Company and United States Line. He joined the British Army again at the beginning of WW2 and transferred to American forces after Pearl Harbor. He was by then in very bad physical shape and drank heavily.

He was posted to Alaska (1942) but without any real responsibilities or duties, and became very depressed. He committed suicide by shooting himself. The antwren is now considered to be a variant of Sclater's Antwren. (See also **Roosevelt, K.**).

Kerr, E. L.

Choco Tinamou *Crypturellus kerriae* **Chapman**, 1915

Mrs Elizabeth L. Kerr (DNF) was an American who travelled in Colombia (1905). She collected the type specimen of the tinamou. Chapman wrote that her '… work in the Atrato Valley has added materially to our knowledge of the avifauna of that part of Colombia.'

Kerr, J. G.

Pale-crested Woodpecker ssp. *Celeus lugubris kerri* **Hargitt**, 1891

Sir John Graham Kerr (1869–1957) was a Scottish embryologist, zoologist, explorer and collector in Argentina (1889–1891) and Paraguay (1896). His bachelor's degree (1896) was awarded by Cambridge. He was Professor of Zoology at Glasgow University (1902–1935). He resigned his chair upon his election as a Member of Parliament, holding the Combined Scottish Universities' seat until it was abolished (1950).

Kersten

Dark-backed Weaver ssp. *Ploceus bicolor kersteni* **Hartlaub** & **Finsch**, 1870

Dr Otto Kersten (1839–1900) was a German chemist and traveller. He was with Baron von der Decken in the unsuccessful attempt to climb Mount Kilimanjaro (1862). Kersten published six volumes of memoirs (1869–1879). A reptile is also named after him.

Kersting

Slender-billed Cuckoo Dove ssp. *Macropygia amboinensis kerstingi* **Reichenow**, 1897

Dr Otto Kersting (b.1863) was a German botanist and collector in New Guinea (1896) and Togoland (Togo Republic) (1897–1909).

Kessler

Kessler's Thrush *Turdus kessleri* **Przewalski**, 1876
[Alt. White-backed Thrush]

Little Owl ssp. *Athene noctua kessleri* **Semenov**, 1899 NCR
[JS *Athene noctua indigena*]

Karl Theodorovich Kessler (1815–1881) was a Russian/German zoologist and collector who was one of the founders of the St Petersburg Society of Naturalists (1868) and its President (1868–1879). He took part in the (1874) Fedtschensko Expedition to Turkestan and the Aralo-Caspian Expedition (1877). He wrote about fish and other vertebrates in European Russia in his reports on the two expeditions.

Several freshwater fish, such as Kessler's Loach *Nemacheilus kessleri,* are named after him. He was Dean of St Petersburg University (1880).

Kettlewell

Brown Tit Babbler ssp. *Macronus striaticeps kettlewelli* **Guillemard**, 1885

Charles T. Kettlewell (DNF) was a British yachtsman who was both owner and master of the steam yacht *Marchesa*, which cruised the Pacific (1882–1884). He sold her (1891) and replaced her with a newer vessel, which he also named *Marchesa*. Collections of birds were made during the yacht's travels.

Keysser

Cassowary sp. *Casuarius keysseri* **Rothschild**, 1912 NCR
[Alt. Dwarf Cassowary; JS *Casuarius bennetti*]

Large Scrubwren ssp. *Sericornis arfakiana keysseri* **Stresemann**, 1925 NCR
[JS *Sericornis nouhuysi oorti*]
Island Thrush ssp. *Turdus poliocephalus keysseri* **Mayr**, 1931

Dr Christian Gottlob Keysser (1877–1961) was a German ethnologist, linguist and Lutheran missionary to New Guinea (1899–1920). He taught theology (1922–1939) at the University of Erlangen, which awarded him a doctorate (1929).

Keytel

Broad-billed Prion ssp. *Pachyptila vittata keyteli* **Mathews**, 1912 NCR; NRM

Casper Keytel (DNF) was a South African businessman who visited Tristan da Cunha (1907) and lived there (1908–1909) to trade in sheep, cattle and dried fish, and to try to exploit guano deposits on Inaccessible Island.

Khabarov

Oriental Greenfinch ssp. *Carduelis sinica chabarovi* **B. Stegmann**, 1929
[SII *Carduelis sinica ussuriensis*]

Yerofey Pavlovich Khabarov, Boyar of Ilimsk (b.1603), was a Russian adventurer, merchant and explorer. He was in Siberia (1625–1641) and the Amur River area (1649–1653). He came into conflict with the Chinese and fought a local war against them, but a third of his troops mutinied and, heavily outnumbered despite a reinforcement of Cossack cavalry, his forces withdrew. Khabarov was blamed, arrested and sent to Moscow for trial (1655) but was acquitted. He returned to Siberia and was last recorded being threatened with being placed in irons if he refused to lead another expedition to the Amur (1658). The city of Khabarovsk is named after him.

Khama

Grey-backed Sparrow Lark ssp. *Eremopterix verticalis khama* **Irwin**, 1957

Sir Seretse Khama (1921–1980) was King of the Bamangwato (1924–1956). His early adult life was one of controversy, as he married an Englishwoman, Ruth Williams (1948). In the climate of that time an inter-racial marriage caused enormous furore both in the government of South Africa, which had just introduced Apartheid, and among the tribal elders of the Bamangwato. He and his wife were exiled to England (1951–1956). He renounced the throne (1956), returning to Bechuanaland as a private citizen until taking up active politics and becoming Prime Minister of Bechuanaland (1965) and first President of Botswana (1966–1980). He died of pancreatic cancer.

Kibort

Brown Dipper *Cinclus kiborti* **Madarász**, 1903 NCR
[JS *Cinclus pallasii*]

Eurasian Skylark ssp. *Alauda arvensis kiborti* Zaliesski, 1917

M. Y. Kibort (DNF) was a Polish ornithologist and Conservator of Krasnoyarsk Museum, Siberia (1893–1903), where he was in exile.

Kidder

Kerguelen Petrel *Aphrodroma kidderi* **Coues**, 1875 NCR
[JS *Aphrodroma brevirostris*; Syn. *Pterodroma brevirostris*]

Dr Jerome Henry Kidder (1842–1889) was a surgeon in the US Navy, astronomer, naturalist and explorer. Harvard awarded his bachelor's degree (1862) and he then served with the Union army in the American Civil War. His medical degree was awarded by the University of Maryland (1866). He served in Japan (1867) and was on board USS *Swatara* for the expedition to the Kerguelen Islands to observe the Transit of Venus (1874). He returned to Harvard and took a master's degree (1875). Later he served on the US Fish Commission.

Kiener

Racket-tail sp. *Ornismya kieneri* **Lesson**, 1832 NCR
[Alt. Booted Racket-tail; JS *Ocreatus underwoodii*]
Rufous-bellied Hawk Eagle *Lophotriorchis kienerii* **Geoffroy Saint-Hilaire**, 1835
[Syn. *Hieraaetus kienerii*]
Rusty-crowned Ground Sparrow *Melozone kieneri* **Bonaparte**, 1850
Zimmer's Woodcreeper *Dendroplex kienerii* **Des Murs**, 1855

Louis Charles Kiener (1799–1881) was a French conchologist and zoologist. He was curator of the private collection of the Duke of Rivoli (1834).

Kikuchi

Russet Sparrow ssp. *Passer rutilans kikuchii* **Kuroda**, 1924 NCR
[JS *Passer rutilans rutilans*]

Yonetaro Kikuchi (1869–1921) was a collector for the Taipei Museum in Formosa (Taiwan). He collected examples of the Mikado Pheasant *Syrmaticus mikado* for Alan Owston (q.v.). A reptile and a mammal are also named after him.

Kimbutu

Archer's Robin Chat ssp. *Cossypha archeri kimbutui*
Prigogine, 1955

François Kimbutu (DNF) was an African hunter and skinner in the Congo, working for Alexandre Prigogine.

King, B. F.

Black-headed Sibia ssp. *Heterophasia desgodinsi kingi*
Eames, 2002

Benjamin Frank King (b.1937) is an American ornithologist specialising in the study of Asian birds. He is President of KingBird Tours, which organises bird-watching tours in Asia. He was appointed a Field Associate of the AMNH (1986). He is the senior author of *A Field Guide to the Birds of South-East Asia* (1975).

King, G.

Red-tailed Wheatear ssp. *Oenanthe chrysopygia kingi*
Hume, 1871 NCR; NRM

Sir George King (1840–1909) was a British botanist who was the first Superintendent of the Royal Botanic Gardens, Calcutta (1871–1898), and who initiated the Botanical Survey of India (1890).

King, H. H.

Yellow-breasted Barbet ssp. *Trachyphonus margaritatus kingi* **Bowen**, 1931 NCR
[NUI *Trachyphonus margaritatus margaritatus*]

Harold Henry King (1885–1954) was a British government entomologist in Sudan (1906–1932). He was in Northern Rhodesia (Zambia) (1934) to experiment with poison dust as an antidote to swarms of locusts.

King (Island)

Striated Pardalote ssp. *Pardalotus striatus kingi* **Mathews**, 1912 NCR
[JS *Pardalotus striatus striatus*]
Dusky Robin ssp. *Melanodryas vittata kingi* Mathews, 1914
Strong-billed Honeyeater ssp. *Melithreptus validirostris kingi* Mathews, 1915 NCR; NRM
Grey Shrike-thrush ssp. *Colluricincla harmonica kingi* Mathews, 1923 NCR
[JS *Colluricincla harmonica strigata*]
Yellow Wattlebird ssp. *Anthochaera paradoxa kingi* Mathews, 1925

These birds are all named after King Island, which is in the Bass Strait between Tasmania and the Australian mainland. The island is named after Philip Gidley King (1758–1808), the third Governor of New South Wales (see **King, P. P.**).

King Louis (XVI)

King Louis (XVI)'s Starfrontlet *Coeligena iris* **Gould**, 1854
[Alt. Rainbow Starfrontlet]

Louis XVI (1754–1793) was King of France at the time of the French Revolution, which started with the storming of the Bastille on 14 July 1789 and which led him to the guillotine.

King of Holland

King of Holland's Fig Parrot *Cyclopsitta gulielmitertii* **Schlegel**, 1866
[Alt. Orange-breasted Fig Parrot]

Willem III (1817–1890) was King of the Netherlands (1849–1890). The word *gulielmitertii* is a latinised version of 'William the Third'. (See **William, King of Holland, Gulielmi Terti**)

King of Saxony

King of Saxony Bird-of-Paradise *Pteridophora alberti* **Meyer**, 1894

King Friedrich August Albert Anton Ferdinand Joseph Karl Maria Baptist Nepomuk Wilhelm Xavier Georg Fidelis (1828–1902) was King of Saxony (1873–1902). Whilst he was a Catholic king of a country that had a significant Protestant minority, he seems to have been a popular monarch with all parties. He was also a soldier who fought in the Seven Weeks' War (1866), during which time he was Commander-in-Chief of the Saxon Army. During the Franco-Prussian War (1870) he proved to be a successful commander of the IV Army and was made a Field Marshal of the German Reich (1871).

King, P. P.

Long-tailed Sylph *Aglaiocercus kingii* **Lesson**, 1832
Pacific Gull ssp. *Larus pacificus kingi* **Mathews**, 1916 NCR
[JS *Larus pacificus pacificus*]

Rear-Admiral Philip Parker King (1791–1856) was an Australian-born British marine surveyor. He was also a collector who travelled in South America (1827–1832) as Commander of the British South American Survey. His father was Philip Gidley King, the 3rd Governor of New South Wales (see **King (Island)**). The King family returned to England (1807) and Philip Parker King entered the navy as a young man. He was in command of the cutter *Mermaid* (1818) and made a number of discoveries including the Goulburn Islands, which he named after Major Goulburn who was Colonial Secretary in the New South Wales. He carried out the first survey of the Great Barrier Reef (1819). In command of *Bathurst* (1821), he carried out a second survey of the reef and of the Torres Strait. Six reptiles are named after him.

King, S. W.

Nihoa Millerbird *Acrocephalus familiaris kingi* **Wetmore**, 1924

Lieutenant-Commander Samuel Wilder King (1886–1959) was born in Honolulu and studied at the US Naval Academy in Annapolis. He served in the US Navy from 1910 until 1924. In

1934 he was elected as a delegate to Congress, serving until 1943. When the USA joined in WW2 he resigned to accept a Naval Commission, rising to captain and retiring in 1946. He returned to Hawaii, serving in various public offices and being eventually appointed Governor (1953–1957). He was Captain of the USS *Tanager* in 1923 when it undertook a scientific survey of the north-western Hawaiian Islands. Wetmore (q.v.) named the bird in his honour, because King's 'interest and seamanship' contributed much to the success attained by the expedition.

Kingdon-Ward, F.

Ward's Trogon *Harpactes wardi* **Kinnear** 1927

Hoary-throated Barwing ssp. *Actinodura nipalensis wardi* Kinnear, 1932 NCR
[JS *Actinodura waldeni saturatior*]

Captain Francis Kingdon-Ward (1885–1958) was an English botanist, collector and explorer. He wrote *The Land of the Blue Poppy* (1913), and *Plant Collecting on the Edge of the World* (1930). He was one of those intrepid late Victorians who went everywhere, surviving the most enormous perils – in his case storms, torrents and an earthquake measuring over 9.5 on the Richter scale! He travelled repeatedly to Asia, exploring in Assam, Burma (Myanmar), China, India and Tibet. He served in the Indian Army (WW1) and went on to teach jungle survival techniques to Allied Forces (WW2), after previously avoiding capture by the Japanese and having made his way alone through the Burmese jungle to India. He was awarded the Founders' Gold Medal of the Royal Geographical Society (1930). Ward died of a stroke. Kinnear (q.v.) wrote the description of the trogon from a specimen '… collected in Burma by a native for F. Kingdon-Ward'. Two mammals are named after him.

Kinnear

Pale-throated Wren-babbler *Spelaeornis kinneari* **Delacour & Jabouille**, 1930
Arabian Lark *Eremalauda kinneari* **G. L. Bates**, 1934 NCR
[Alt. Dunn's Lark; JS *Eremalauda dunni eremodites*]

Rufous Treepie ssp. *Dendrocitta vagabunda kinneari* **E. C. S. Baker**, 1922
Pin-striped Tit Babbler ssp. *Macronus gularis kinneari* Delacour & Jabouille, 1924
Spectacled Barwing ssp. *Actinodura ramsayi kinneari* Delacour, 1927 NCR
[JS *Actinodura ramsayi yunnanensis*]
White-browed Piculet ssp. *Sasia ochracea kinneari* **Stresemann**, 1929
Bar-tailed Lark ssp. *Ammomanes cinctura kinneari* Bates, 1935 NCR
[JS *Ammomanes cinctura arenicolor*]
Shining Sunbird ssp. *Cinnyris habessinicus kinneari* Bates, 1935
Belford's Honeyeater ssp. *Melidectes belfordi kinneari* **Mayr**, 1936
Red-backed Buttonquail ssp. *Turnix maculosus kinneari* **Neumann**, 1939
White-breasted Nuthatch ssp. *Sitta carolinensis kinneari* **Van Rossem**, 1939

Eurasian Wren ssp. *Troglodytes troglodytes kinneari* **Biswas**, 1955 NCR
[JS *Troglodytes troglodytes nipalensis*]
Blue-winged Leafbird ssp. *Chloropsis cochinchinensis kinneari* **B. P. Hall** & **Deignan**, 1956

Dr Sir Norman Boyd Kinnear (1882–1957) was a Scottish zoologist who was Curator of the Museum of the Bombay Natural History Society (1907–1919). He worked as a volunteer at the Royal Scottish Museum (1905–1907). He joined the Natural History Museum in London (1920) as an Assistant in the Department of Zoology, becoming Keeper of Zoology (1945) and Director of the Museum until his retirement (1947–1950). He was one of the founders of the Society for the History of Natural History, and also active in both the Zoological Society of London and the National Trust. He wrote *The History of Indian Mammalogy and Ornithology* (1952).

Kinnis

Indian Blackbird ssp. *Turdus simillimus kinnisii* **Kelaart**, 1851

Dr John Kinnis (1795–1852) was a British military surgeon-naturalist in Ceylon (Sri Lanka) (1841–1851).

Kirby

Red-winged Warbler *Heliolais kirbyi* **Haagner**, 1909 NCR
[JS *Heliolais erythropterus rhodopterus*]

Speckled Mousebird ssp. *Colius striatus kirbyi* **Sharpe**, 1907 NCR
[JS *Colius striatus minor*]

Major Frederick Vaughan Kirby (d.1945) was a British big-game hunter and Chief Game Conservator in Zululand (1911–1929). He wrote *In Haunts of Wild Game* (1896). His native name was 'Maqaqamba'.

Kirk, J.

Red-rumped Woodpecker *Veniliornis kirkii* **Malherbe**, 1845

John Kirk (DNF) was a collector on Tobago.

Kirk, Sir J.

Kirk's Francolin *Francolinus kirkii* **Hartlaub**, 1868 NCR
[Alt. Crested Francolin; JS *Dendroperdix sephaena rovuma*]

Kirk's Amethyst Sunbird *Chalcomitra amethystina kirkii* **G. E. Shelley**, 1876
Arrow-marked Babbler ssp. *Turdoides jardineii kirkii* **Sharpe**, 1876
Kirk's White-eye *Zosterops maderaspatanus kirki* G. E. Shelley, 1880
[Alt. Grand Comoro White-eye ssp.]
Green-winged Pytilia ssp. *Pytilia melba kirki* G. E. Shelley, 1903 NCR
[JS *Pytilia melba soudanensis*]
Chestnut-fronted Helmet-shrike ssp. *Prionops scopifrons kirki* **W. L. Sclater**, 1924

Sir John Kirk (1832–1922) was a Scottish administrator in Africa. He took part, as a physician and naturalist, on David Livingstone's second Zambezi expedition (1858–1863). Kirk contributed much to the eradication of the slave trade. He was Vice-Consul in Zanzibar (1866) and then Consul-General (1873). During his service he persuaded the sultan to abolish the slave trade (1873) and also to concede mainland territories to the British East Africa Company (1887). Kirk was a polymath whose interests included botany, geography, history, geology, chemistry and photography, as well as the study of Swahili, Arabic, Spanish, Portuguese and French. He was a Fellow of the Royal Botanical Society and collected and sent specimens to the Royal Botanical Gardens at Kew. He also collected the first fish from Lake Nyasa to reach Western science and observed that they were almost all endemic. One amphibian, one reptile and three mammals are named after him.

Kirtland

Kirtland's Owl *Aegolius acadicus* **Gmelin**, 1788
[Alt. Northern Saw-whet Owl; *Nyctale kirtlandii* (Hoy, 1852) is a JS]
Kirtland's Warbler *Dendroica kirtlandii* **Baird**, 1852

Dr Jared Potter Kirtland (1793–1877) was a naturalist, botanist, doctor, legislator, teacher and writer. He also founded the Cleveland Museum of Natural History and the Cleveland Medical College. He was a contemporary of Agassiz, Audubon (q.v.) and other notables who were documenting and classifying the plants and animals of North America at that time. Kirtland was named after his maternal grandfather, Jared Potter, who was himself a distinguished physician and amateur naturalist. He was an accomplished horticulturist who managed a large plantation of white mulberry trees for the rearing of silkworms. He is also credited with originating 26 varieties of cherries and 6 of pears. Kirtland made one important scientific discovery that brought him national attention: he asserted that the freshwater mussels (*Unionacea*) have distinct sexes, which had been mistakenly classified previously as different species. The Assistant-Secretary of the Smithsonian Institution disputed this at first but, when Agassiz proved Kirtland right, he later made up for his early doubts by giving Kirtland's name to a species of warbler which the naturalist had been the first to collect and classify. Three reptiles are named after him.

Kistchinski

Dunlin ssp. *Calidris alpina kistchinskii* **Tomkovich**, 1986

Alexander A. Kistchinski (1937–1980) was a Russian zoologist and ornithologist. He was at Central Laboratory for Nature Conservation, Ministry of Agriculture of the USSR, Moscow (1969). He wrote *Waterfowl in North-east Asia* (1973).

Kistjakovski

Common Rosefinch ssp. *Carpodacus erythrinus kistjakovskii* Charlemagne, 1933 NCR
[JS *Carpodacus erythrinus erythrinus*]

Professor Aleksander Fyodorovich Kistjakovski (b.1904) was a Russian biologist at Kiev State University (1929–1958).

Kittenberger

African Swift sp. *Apus kittenbergeri* **Madarász**, 1910 NCR
[Variously regarded as a JS of *Apus barbatus* or *Tachymarptis aequatorialis*]

Kálmán Kittenberger (1881–1958) was a Hungarian biologist, naturalist, traveller and collector in East Africa (1902–1914), and later in Uganda and the Congo (1925–1929). He wrote *Big Game Hunting and Collecting in East Africa, 1903–1926* (1929). A Wildlife Park in Hungary is named after him. (See also **Colomann** and **Katona**)

Kittlitz

Crake genus *Kittlitzia* **Hartlaub**, 1892 NCR
[Now in *Porzana*]

Kittlitz's Plover *Charadrius pecuarius* **Temminck**, 1823
[Alt. Kittlitz's Sand Plover]
Kittlitz's Murrelet *Brachyramphus brevirostris* **Vigors**, 1829
Kittlitz's Thrush *Zoothera terrestris* Kittlitz, 1830 EXTINCT
[Alt. Bonin Thrush]
Kittlitz's Wood Pigeon *Columba versicolor* Kittlitz, 1832 EXTINCT
[Alt. Bonin Wood Pigeon, Bonin Black Pigeon]
Kittlitz's Starling *Aplonis corvina* Kittlitz, 1833 EXTINCT
[Alt. Kosrae Starling]
Gull sp. *Chroicocephalus kittlitzii* **Bruch**, 1853 NCR
[Alt. Franklin's Gull; JS *Leucophaeus pipixcan*]
Kittlitz's Crake *Porzana monasa* Kittlitz, 1858 EXTINCT
[Alt. Kosrae Crake]
White-eye sp. *Zosterops kittlitzi* **Finsch**, 1880 NCR
[Alt. Grey-brown White-eye; JS *Zosterops cinereus*]

Oriental Greenfinch ssp. *Carduelis sinica kittlitzi* **Seebohm**, 1890
White Tern ssp. *Gygis alba kittlitzi* **Hartert**, 1891 NCR
[JS *Gygis alba candida*]

Friedrich Heinrich Freiherr von Kittlitz (1799–1874) was a Polish-born German artist, naval officer, explorer and ornithologist. He undertook a round-the-world journey, the *Senjawin* Expedition (1826–1829), during which specimens of the now-extinct bird species listed above were collected. Based on this voyage, Kittlitz wrote *Twenty-Four Views of the Vegetation of the Coasts and Islands of the Pacific* (English edition, 1861). He was a friend of Eduard Rüppell (q.v.) and accompanied him to North Africa (1831), but due to poor health he had to return to Germany.

Klaas

Klaas's Cuckoo *Chrysococcyx klaas* **J. F. Stephens**, 1815

Levaillant (q.v.) named Klaas's Cuckoo after his Khoikhoi (then known as Hottentot) servant and collector, who presumably found the bird (1784). No more is known about him.

Klages

Klages's Antbird *Drymophila klagesi* **Hellmayr** & **Seilern**, 1912
[Syn. *Drymophila caudata klagesi*]

Klages's Antwren *Myrmotherula klagesi* **Todd**, 1927

Klages's Coquette *Lophornis chalybeus klagesi* **Berlepsch & Hartert**, 1902
[Alt. Festive Coquette ssp.]
Grey-crowned Flycatcher ssp. *Tolmomyias poliocephalus klagesi* **Ridgway**, 1906
Streak-capped Treehunter ssp. *Thripadectes virgaticeps klagesi* Hellmayr & Seilern, 1912
Black-chested Tyrant ssp. *Taeniotriccus andrei klagesi* Todd, 1925

Samuel M. Klages (1875–1957) was an American collector. He left the United States for South America (1891) and collected in a range of countries: in Venezuela (1898–1913), Trinidad & Tobago (1912–1913), Venezuela again (1913–1914 and 1919), French Guiana (1917–1918) and in Brazil (1919 and 1920–1926). Most of these trips were undertaken for the Bird Section of the Carnegie Museum of Natural History in Pittsburgh. He continued to make trips to South America until 1932. He then lived for a while with a niece, Alva Held, a high school counsellor, her mother and his sister Mary Amelia, in Knoxville. Apparently he annoyed Alva when his hair and moustache dye stained her bathroom sink (!) and he left to set up home in an apartment. He must have made bad investments as he is listed in the (1943) Pittsburgh city Directory as a writer at a Perrysville Avenue address and in receipt of the state dole. He wrote an unpublished romantic novel and offered a publisher an account of his travels, but his manuscripts were all burnt after he died of heart disease, having already been diagnosed as senile. He is also commemorated in the scientific names of a number of lepidoptera.

Klais

Violet-headed Hummingbird genus *Klais* **Reichenbach**, 1854

Kleis was the daughter of Sappho, the Greek poetess from Lesbos.

Klee

Grey Tinamou ssp. *Tinamus tao kleei* **Tschudi**, 1843

E. Klee (b.c.1814) was a Prussian collector in Peru (1838–1842). He was a seaman who had arrived in Peru on board the *Princesse-Louise* and was Tschudi's (q.v.) traveling companion. In addition to collecting, he acted as general handyman and factotum.

Klein

Chukar Partridge ssp. *Alectoris chukar kleini* **Hartert**, 1925
[SII *Alectoris chukar cypriotes*]

Dr Eduard Klein (1864–1943) was an Austrian ornithologist based in Bulgaria.

Kleinschmidt, J. T.

Pink-billed Parrotfinch *Erythrura kleinschmidti* **Finsch**, 1878

Pacific Robin ssp. *Petroica multicolor kleinschmidti* Finsch, 1876
Silktail ssp. *Lamprolia victoriae klinesmithi* **E. P. Ramsay**, 1876

[Trinomial often amended to *kleinschmidti*, but original spelling used here]

Johann Theodor Kleinschmidt (1834–1881) was a German sailor, merchant and animal collector. He was the uncle of Otto Kleinschmidt (q.v.). He emigrated from Germany to the USA (1858) and worked for 12 years for a trading house in St Louis. He fought in the Civil War, achieving the rank of Major. As was very common with immigrants into the USA in the 19th century, he changed his name and became John Theodore Klinesmith. However, the war forced him to cease trading. He went to Melbourne (1870) but failed to find work, so moved on to Fiji where he settled – but in penury when the cotton trade collapsed and he lost everything. J. C. Godeffroy (q.v.) gave him the position (1875) of animal collector and explorer based at his company's establishment on the island of Mioko (in Papua New Guinea between New Britain and New Ireland). From there he made many trips to various islands in the South-West Pacific. He sent numerous bird specimens to Godeffroy along with drawings of the habitat where he collected. On one of his collecting expeditions on Mioko, aborigines murdered him and his two local helpers.

Kleinschmidt, O.

Kleinschmidt's Falcon *Falco kreyenborgi* Kleinschmidt, 1929 NCR
[Alt. Peregrine Falcon; JS *Falco peregrinus cassini*]

Crested Lark ssp. *Galerida cristata kleinschmidti* **Erlanger**, 1899
Willow Tit ssp. *Poecile montanus kleinschmidti* **Hellmayr**, 1900
Eurasian Jay ssp. *Garrulus glandarius kleinschmidti* **Hartert**, 1903 NCR
[JS *Garrulus glandarius fasciatus*]
Rock Pipit ssp. *Anthus petrosus kleinschmidti* Hartert, 1905
Common Cuckoo ssp. *Cuculus canorus kleinschmidti* **Schiebel** 1910 NCR
[JS *Cuculus canorus canorus*]
Corn Bunting ssp. *Emberiza calandra kleinschmidti* Gornitz, 1921 NCR
[JS *Emberiza calandra calandra*]
Grey-headed Sparrow ssp. *Passer griseus kleinschmidti* **Grote**, 1922 NCR
[JS *Passer griseus griseus*]
Barn Owl ssp. *Tyto alba kleinschmidti* **von Jordans**, 1923 NCR
[NUI *Tyto alba alba*]
Eurasian Nuthatch ssp. *Sitta europaea kleinschmidti* Hartert & **Steinbacher**, 1933 NCR
[JS *Sitta europaea amurensis*]
European Stonechat ssp. *Saxicola torquata kleinschmidti* **Meise**, 1934 NCR
[JS *Saxicola stejnegeri*]
Northern Goshawk ssp. *Accipiter gentilis kleinschmidti* **von Jordans**, 1950 NCR
[JS *Accipiter gentilis gentilis*]

Dr Otto Kleinschmidt (1870–1954) – full name Konrad Ernst Adolf Otto Kleinschmidt – was a German Protestant pastor and an accomplished artist. Throughout his life he was an enthusiastic amateur ornithologist and accumulated over

10,000 bird skins. Before becoming a pastor (1897) he had worked as assistant to Count von Berlepsch (q.v.). He founded (1927) the Kirchliche Forschungsheim, a religious research institute in Wittenberg, Saxony, the aim of which was to promote dialogue between theologists and proponents of the theory of evolution. He wrote the 3-volume *Die Vögel der Heimat – Raubvögel, Eulen, Singvögel* (1951) (The Birds of the Homeland).

Klinikowski

Pearly-eyed Thrasher ssp. *Margarops fuscatus klinikowskii* Garrido & **Remsen**, 1996

Ronald Francis Klinikowski (1939–2008) was an American herpetologist who made a number of collecting trips to the Caribbean, including Cuba (1959), Jamaica (1961) and Saint Lucia (1962 and 1963). An amphibian is also named after him.

Kloss

Great Nicobar Serpent Eagle *Spilornis klossi* **Richmond**, 1902
Kloss's Leaf Warbler *Phylloscopus ogilviegranti* **La Touche**, 1922
Black-crowned Fulvetta *Alcippe klossi* **Delacour** & **Jabouille**, 1931
[Syn. *Pseudominla klossi*]

Olive-backed Sunbird ssp. *Cinnyris jugularis klossi* Richmond, 1902
Kloss's Eagle Owl *Bubo coromandus klossii* **H. C. Robinson**, 1911
[Alt. Dusky Eagle Owl ssp.]
Chestnut Forest Rail ssp. *Rallina rubra klossi* **Ogilvie-Grant**, 1913
Kloss's Red-headed Trogon *Harpactes erythrocephalus klossi* Robinson, 1915
Sclater's Whistler ssp. *Pachycephala soror klossi* Ogilvie-Grant, 1915
White-browed Scimitar Babbler ssp. *Pomatorhinus schisticeps klossi* **E. C. S. Baker**, 1917
Sooty-headed Bulbul ssp. *Pycnonotus aurigaster klossi* **Gyldenstolpe**, 1920
Blue-throated Flycatcher ssp. *Cyornis rubeculoides klossi* Robinson, 1921
Kloss's Leaf Warbler ssp. *Phylloscopus ogilviegranti klossi* Riley, 1922
Common Green Magpie ssp. *Cissa chinensis klossi* Delacour & Jabouille, 1924
Shikra ssp. *Accipiter badius klossi* Swann, 1925 NCR
[JS *Accipiter badius poliopsis*]
Hill Prinia ssp. *Prinia superciliaris klossi* **Hachisuka**, 1926
Yellow-browed Tit ssp. *Sylviparus modestus klossi* Delacour & Jabouille, 1930

Cecil Boden Kloss (1877–1949) was an English ethnologist and zoologist. He was a member of the staff of the museum in Kuala Lumpur (1908), for which he travelled extensively as a collector. He started to work under H. C. Robinson (q.v.), who was Curator of Birds for the Federated States Museums, Malaysia (1908). He was the Director of the Raffles Museum in Singapore (1923–1932) and established its *Bulletin* (1928).

Three mammals and four reptiles are named after him. (Also see **Cecil**)

Klossovski

Common Pheasant ssp. '*Phasianus zerafschanicus* sive *klossovskii*' Tarnovski, 1891
[Now *Phasianus colchicus zerafschanicus*]

V. F. Klossovski (DNF) was a Russian resident in Samarkand (1877–1891). The describer, George Tarnovski, introduced this taxon to science in an odd way, naming it as a new species – *zerafschanicus* 'sive' (= or) *klossovskii*. As he placed *zerafschanicus* first, that name has priority.

Knox

Brown Tinamou ssp. *Crypturellus obsoletus knoxi* **W. H. Phelps** Jr, 1976

Dr William Graham Knox (1918–2009) was a New York physician and surgeon. He graduated from Columbia College (1942) and worked in New York City hospitals (1942–1988), interrupted by service in WW2 in the South Atlantic and Pacific theatres of war with the US Navy Medical Corps. He was also Professor of Clinical Surgery, Columbia College of Physicians and Surgeons. The etymology says he was honoured 'in recognition of his friendship and constant interest in ornithological investigations'.

Knudsen

Hawaiian Stilt *Himantopus knudseni* **Stejneger**, 1887
[Alt. Black-necked Stilt; Syn. *Himantopus mexicanus knudseni*]

Valdemar Emil Knudsen (1819–1898) was a Norwegian naturalist who collected on Hawaii. He was also an explorer, gold prospector, rancher and poet as well as royal agent for King Kamehameha. He prospered in California as a merchant during the Gold Rush, then sailed to Hawaii (late 1840s). He was fluent in Hawaiian and acted as an intermediary between native Hawaiians and immigrants from the American mainland. His native name was 'Kanuka'. He collected for his fellow Norwegian ornithologist Stejneger (q.v.).

Kobayashi, K. (Junior)

Japanese Robin ssp. *Luscinia akahige kobayashii* **Momiyama**, 1940 NCR
[JS *Luscinia akahige tanensis*]

Keisuke Kobayashi Jr (1908–2000) was, like his father (q.v.), a Japanese ornithologist and oologist. He wrote *Birds of Japan in Natural Colours* (1956).

Kobayashi, K. (Senior)

Micronesian Myzomela ssp. *Myzomela rubratra kobayashii* **Momiyama**, 1922
Light-vented Bulbul ssp. *Pycnonotus sinensis kobayashii* **Kuroda**, 1930 NCR
[JS *Pycnonotus* (*Otocompsa*) *sinensis orii*]

Keisuke Kobayashi Sr (1879–1933) was a Japanese ornithologist, oologist and collector.

Kobylin

Red-backed Shrike ssp. *Lanius collurio kobylini* **Buturlin**, 1906

A. M. Kobylin (DNF) was a Russian who resided in Transcaucasia for three years (1903–1906) and collected birds, sending their skins to Buturlin (q.v.).

Koch

Koch's Pitta *Erythropitta kochi* Bruggemann, 1876
[Alt. Whiskered Pitta; Syn. *Pitta kochi*]

Bar-bellied Cuckooshrike ssp. *Coracina striata kochii* **Kutter**, 1882

Gottlieb von Koch (1849–1914) was a German collector and taxidermist who was Professor of Zoology at the technical school in Darmstadt. He wrote *Die Gorgoniden des Golfes von Neapel und der Angrenzenden Meeresabschnitte* (1887).

Koelz

Green-tailed Sunbird ssp. *Aethopyga nipalensis koelzi* **Ripley**, 1948
Chestnut-bellied Nuthatch ssp. *Sitta cinnamoventris koelzi* **Vaurie**, 1950

Dr Walter Norman Koelz (1895–1989) was an American zoologist, botanist, anthropologist and collector whose doctorate was awarded by the University of Michigan (1920). He was on the McMillan Expedition to the American Arctic (1925). He joined the Himalayan Research Institute and was in India (1930–1932 and 1933). He explored (1934–1941) Persia (Iran), Nepal and parts of India, collecting c.30,000 bird specimens for the Zoological Museum, University of Michigan, and c.30,000 plants for the university's herbarium.

Koenig

(See **König**)

Koeniswald

Tawny-browed Owl *Pulsatrix koeniswaldiana* **M. Bertoni** & W. Bertoni, 1901

Lesser Woodcreeper ssp. *Picolaptes fuscus koeniswaldianus* Bertoni, 1914 NCR
[JS *Xiphorhynchus fuscus fuscus*]

Dr Gustavo Koeniswald (or Gustav Koenigswald) (fl.1896–at least 1906) was a German author and ornithologist at the Museu Paulista, São Paulo, in the late nineteenth century. He wrote 'Ornithologia Paulista' in the *Journal fur Ornithologie* (1896).

Koepcke

Koepcke's Cacique *Cacicus koepckeae* **Lowery** & **O'Neill**, 1965
[Alt. Selva Cacique]
Sira Curassow *Pauxi* (*unicornis*) *koepckeae* **Weske** & **Terborgh**, 1971
Koepcke's Hermit *Phaethornis koepckeae* **Weske** & **Terborgh**, 1977

Koepcke's Screech Owl *Megascops koepckeae* Hekstra, 1982
[Alt. Maria Koepcke's Screech Owl]

Oasis Hummingbird ssp. *Rhodopis vesper koepckeae* **Berlioz**, 1974

Maria Koepcke (1924–1971) was born Maria Emilia Ana von Mikulicz-Radecki in Leipzig, Germany. She went to Peru (1950) where she joined her husband Hans, who was an ecologist at the Museum of Natural History in San Marcos, where she too began work (1960) as an ornithologist. She came to be known as 'the Mother of Peruvian Ornithology'. Her study of a coastal biome of Peru was a seminal work. She also wrote 'Corte ecológico transversal en los Andes del Peru Central' (1954), 'Die Vögel des Waldes von Zarate' (1958) and *Las Aves del Departmento de Lima* (1964). She was killed in an air crash in Amazonia on Christmas Eve 1971. She described several birds including the Cactus Canastero *Asthenes cactorum*. A mammal is named after her and both a reptile and an amphibian after her and her husband.

Koester

Saw-wing sp. *Psalidoprocne koesteri* **Neumann**, 1933 NCR
[Alt. Black Saw-wing; JS *Psalidoprocne pristoptera reichenowi*]

Groundscraper Thrush ssp. *Psophocichla litsitsirupa koesteri* Neumann, 1929 NCR
[JS *Psophocichla litsitsirupa stierlingi*]
Montane Nightjar ssp. *Caprimulgus ruwenzorii koesteri* Neumann, 1931

Paul Koester (DNF) was a collector in Angola.

Koike

Koike's Blue-naped Parrot *Tanygnathus lucionensis koikei* **Hachisuka**, 1930 NCR
[JS *Tanygnathus lucionensis salvadorii*]

K. Koike (DNF) was a Japanese resident of Davao in the Philippines. He was President of the local Japanese Association. He acted as manager of Marquess Hachisuka's (q.v.) Mount Apo expedition (1929), and also accompanied Goodfellow (q.v.) on his third expedition there.

Kolb

Kolb's Vulture *Gyps coprotheres* **J. R. Forster**, 1798
[Alt. Cape Griffon, Cape Vulture; *Vultur kolbii* (Daudin, 1800) is a JS]

Peter Kolb (1675–1725) was a German polymath who attended Halle University, where he studied philosophy, theology, mathematics and physics. He obtained his master's degree (1703) with a thesis entitled 'De natura cometarum', in which he explained comets as natural phenomena. In 1705 he was sent to the Cape of Good Hope, tasked with making astronomical and meteorological observations. He also made studies of the South African fauna, flora and local peoples. He wrote *The present state of the Cape of Good-Hope – A Particular Account of the Several Nations of the Hottentots* (English translation, 1731). He is also credited with introducing the musical instrument, the goura, to the outside world.

Kolichis

Sunda Bush Warbler ssp. *Cettia vulcania kolichisi*
R. E. Johnstone & Darnell, 1997

Nicholas Kolichis is Managing Director of NK Contractors Pty Ltd. and an Honorary Associate of Western Australian Museum, Perth, which he sponsors and helps support. He is also an ornithologist and naturalist and has published a number of papers and reports, including 'First description of the nest and eggs of the Black Grasswren *Amytornis housei* (Milligan) with notes on breeding', written jointly with R. E. Johnstone (1999).

Kollar

Hoary-throated Spinetail *Synallaxis kollari* **Pelzeln**, 1856

Vincent Kollar (1797–1860) was a German entomologist and collector. He was born in Kranovitz (since 1946 in Poland and called Krzanowice) and died in Vienna where he was a Curator at the Zoological Museum.

Kollibay

Common Swift ssp. *Apus apus kollibayi* **Tschusi**, 1902 NCR
[JS *Apus apus apus*]

Paul Robert Kollibay (1863–1919) was a German notary, ornithologist and collector. He wrote *Die Vogel Der Preussischen Provinz Schlesien* (1906).

Kollmannsperger

Desert Lark ssp. *Ammomanes deserti kollmannspergeri*
Niethammer, 1955

Dr Franz Kollmannsperger (b.1907) was a German zoologist, ecologist and ornithologist at the University of Saarbrücken. He took part in the international Sahara expedition (1953–1954). He wrote *Von Afrika nach Afrika* (1965). A mammal is named after him.

Komarow

Common Pheasant ssp. *Phasianus colchicus komarowii*
Bogdanov, 1886 NCR
[JS *Phasianus colchicus principalis*]

General Aleksandr Vissarionovich Komarov (1830–1904) served in the Russian army in the Caucasus (1855–1883), and was then military governor of the Transcaspian Region (1883–1889). He defeated the Afghans in battle at the Kushka River (1885). He retired (1890) and devoted his time to archaeology, palaeontology and ethnography.

König, A. F.

Swift sp. *Micropus koenigi* **Reichenow**, 1894 NCR
[Alt. Little Swift; JS *Apus affinis galilejensis*]

Common Chaffinch ssp. *Fringilla coelebs koenigi* **Rothschild** & Hartert, 1893 NCR
[JS *Fringilla coelebs africana*]

Barbary Partridge ssp. *Alectoris barbara koenigi*
Reichenow, 1899
Levaillant's Woodpecker ssp. *Picus vaillantii koenigi*
Erlanger, 1899 NCR; NRM
Southern Grey Shrike ssp. *Lanius meridionalis koenigi*
Hartert, 1901
African Grey Woodpecker ssp. *Dendropicos goertae koenigi*
Neumann, 1903
Eurasian Jay ssp. *Garrulus glandarius koenigi* **Tschusi**, 1904
NCR
[JS *Garrulus glandarius cervicalis*]
Eurasian Wren ssp. *Troglodytes troglodytes koenigi*
Schiebel, 1910
[Syn. *Nannus troglodytes koenigi*]
Blackcap ssp. *Sylvia atricapilla koenigi* **von Jordans**, 1923
NCR
[NUI *Sylvia atricapilla pauluccii*]

Professor Alexander Ferdinand König (1858–1940) was a German naturalist born in St Petersburg. He studied zoology at the Universities of Marburg and Greifswald. He came from a wealthy merchant family and so could afford, in later life, to fund expeditions to the Arctic and Africa. He personally visited Egypt and the Sudan. He founded the Museum König, Bonn (1912), with his collections.

König, A. & H.

White-browed Antbird ssp. *Myrmoborus leucophrys*
koenigorum **O'Neill** & **T. A. Parker**, 1997

Arturo H. König (b.1924), an orthodontist, and Helen König *née* Maunsell (b.1930 in Washington State, USA) of Lima, Peru, used to own a plantation near Tingo Maria, Department of Huánaco, but lost it due to the actions of terrorists. The etymology says they ' ... have been of tremendous help to LSUMZ [Louisiana State University Museum of Zoology] personnel for more than three decades.'

König, C.

Saffron Finch ssp. *Sicalis flaveola koenigi* G. A. Hoy, 1978

Prof. Dr Claus König (b.1931) is an ornithologist, a friend of the describer and ' ... a connoisseur of Argentine birds'. The Goethe University at Frankfurt am Main awarded his doctorate (1959), after which he was head of the Ludwigsburg Ornithological Station (1962–1971). He was Professor of Zoology at Stuttgart University and was at the State Natural History Museum in Stuttgart as head of the ornithology section (1971–1996) and as Director of the Museum (1997–2000). Since his retirement he has continued to work there as a volunteer. He is author or co-author of a number of papers including 'Zur Kenntnis der Kleinsäugetiere von Ibiza (Balearen)' (1958) and 'Owl vocalizations as interspecific differentiation patterns and their taxonomical value as ethological isolating mechanisms between various taxa' (2000). He is also senior author of *Owls of the World* (1999 and 2009).

Königsegg

Scops Owl sp. *Scops koenigseggi* **Madarász**, 1912 NCR
[Alt. African Scops Owl; JS *Otus senegalensis senegalensis*]

Clapperton's Francolin ssp. *Pternistis clappertoni*
koenigseggi Madarász, 1914

Fidelis Graf Königsegg (1879–1941) was a Hungarian collector in the Sudan (1912–1914).

Kopsch

Pygmy-goose sp. *Nettapus kopschii* **Swinhoe**, 1873 NCR
[Alt. Cotton Pygmy-goose; JS *Nettapus coromandelianus*]

Henry Charles Joseph Kopsch (1845–1913) was a British civil servant seconded to the Chinese Customs Service (1862–1900), becoming Deputy Commissioner (1867) and Commissioner (1868). He was Statistical Secretary and then Postal Secretary (1891–1897). He wrote a pamphlet on bi-metallism (1903). He was a friend of Swinhoe (q.v.), who also named a race of Sika Deer *Cervus nippon kopschi* after him.

Korejew

Horned Grebe ssp. *Podiceps auritus korejevi* **Zarudny & Loudon**, 1902 NCR
[Alt. Slavonian Grebe; JS *Podiceps auritus auritus*]
Marsh Tit ssp. *Parus communis korejewi* Zarudny & **Härms**, 1902 NCR
[JS *Poecile palustris palustris*]
Saxaul Sparrow ssp. *Passer ammodendri korejewi* Zarudny & Härms, 1902 NCR
[NUI *Passer ammodendri ammodendri*]
Great Bustard ssp. *Otis tarda korejewi* Zarudny, 1905 NCR
[NUI *Otis tarda tarda*]
Water Rail ssp. *Rallus aquaticus korejewi* Zarudny, 1905
Rock Pigeon ssp. *Columba livia korejewi* Zarudny & Loudon, 1906 NCR
[JS *Columba livia neglecta*]
Common Reed Bunting ssp. *Emberiza schoeniclus korejewi* Zarudny, 1907
Northern Hawk Owl ssp. *Surnia ulula korejewi* Zarudnyi & Loudon, 1907 NCR
[JS *Surnia ulula tianschanica*]
Brown Dipper ssp. *Cinclus tenuirostris korrejewi* Zarudny, 1909 NCR
[JS *Cinclus pallasii tenuirostris*]
Eurasian Three-toed Woodpecker ssp. *Picoides tridactylus korejewi* Zarudny & Loudon, 1914 NCR
[JS *Picoides tridactylus tianschanicus*, which is itself sometimes included in *P. tridactylus alpinus*]
Twite ssp. *Carduelis flavirostris korejevi* Zarudny & Härms, 1914
White-winged Woodpecker ssp *Dendrocopos leucopterus korejevi* Zarudny, 1923 NCR; NRM

B. P. Korejew (DNF) was a Russian ornithologist and collector in Turkestan (1902–1905).

Koren

Willow Ptarmigan ssp. *Lagopus lagopus koreni* **Thayer & Bangs**, 1914

Johan Koren (1879–1919) was a Norwegian mariner, explorer, zoologist and collector. He was a deckhand in the Antarctic on the *La Belgica* expedition (1897–1899) – Roald Amundsen was first mate on board the vessel. He collected in Novaya Zemlya (1902–1903) for the Zoological Museum in Christiania (Oslo). He was shipwrecked on an uninhabited island in the Crozet group, Antarctica, when sealing (1906), and was only rescued after three months (1907). Stranded and destitute in Australia, he worked as a labourer for a year to save money for an expedition to Siberia (1908), but was shipwrecked again (1909) whilst moving camp, again rescued, finishing up in Nome, Alaska (1910). He went to northern Siberia (1911) but was shipwrecked for a third time (1912) on Little Diomede Island, Bering Strait. He walked across the ice from there to Alaska (1913), being the first person in recorded history to do so. He was in eastern Siberia (1914–1916) working for a commercial firm in Vladivostok when the Russian revolution broke out (1917). He tried to organise humanitarian relief but his health failed and he died. He was buried in a churchyard which has since been turned into a public park, so his grave is unknown.

Koroviakov

Chukar Partridge ssp. *Alectoris chukar koroviakovi* **Zarudny**, 1914

B. D. Koroviakov (DNF) was a Russian explorer in Persia (Iran).

Korthals

Korthals's Green Pigeon *Treron sphenurus korthalsi* **Bonaparte**, 1855
[Alt. Wedge-tailed Green Pigeon ssp.]

Pieter Willem Korthals (1807–1892) was a Dutch botanist and traveller. He collected mostly in Sumatra (1830s and 1840s) whilst working for the Dutch East India Service (1831–1836). His name is associated with a number of orchids, but his work on *Nepenthes* (pitcher plants) was seminal – he wrote the first monograph on them (1839), naming nine species. He collected specimens of the Giant Pitcher Plant in the highlands of Borneo.

Koslow

(See under **Kozlov**)

Kosswig

Bearded Reedling ssp. *Panurus biarmicus kosswigi* Kumerloeve, 1958

Dr Curt Kosswig (1903–1982) was a German zoologist and geneticist. He was Professor and Director of the Zoological Institute, Istanbul University (1937–1955). He returned to Germany and worked at the Zoological Institute, University of Hamburg (1955–1969). He wrote *Zoogeography of the Near East* (1955).

Kotzebue

Kotzebue's Kittiwake *Rissa kotzebui* **Bonaparte**, 1856
[Alt. Black-legged Kittiwake; JS *Rissa tridactyla*]

Count Otto Evstafievich von Kotzebue (1788–1846) was a Russian naval officer and Estonian explorer. He was born and died in Tallinn, Estonia – then part of Russia. He went with von Krusenstern on board *Nadezhda* on his circumnavigation (1803–1806), and then commanded two voyages around the world. During the first (1815–1818), when captain of the *Rurik*, he checked the location of, and gathered information on, the islands and Pacific coast of Siberia. He also explored the north-west coast of Alaska (a Russian possession at the time), looking for a North-west Passage, and (1816) he discovered and explored Kotzebue Sound. His exploration here was curtailed when he sustained a chest injury during a storm. On the second journey (1823–1826), captaining the *Predpriyatie*, he discovered c.400 islands in the South Seas around Samoa and the Marshall Islands, etc. His expeditions made reports on ethnography, oceanography and natural history. Among other things he conducted the first research on fossil ice in Alaska. He wrote *A Voyage of Discovery* (1821) and *A New Voyage round the World* (1828). He had to retire due to ill health (1830). Mention of the kittiwake is made in the *Journal of the Arctic Expedition* (1881) by John Muir (*The Cruise of the Corwin*): 'Kotzebue's gull, the kittiwake, about the ship; no seals or walrus …'

Kovacs

Roller sp. *Coracias kovacsi* **Madarász**, 1911 NCR
[Alt. Lilac-breasted Roller; JS *Coracias caudatus lorti*]
Weaver sp. *Othyphantes kovacsi* Madarász, 1914 NCR
[Alt. Baglafecht Weaver; JS *Ploceus baglafecht emini*]

Ödön (Edmund) Kovács (1886–1919) was a Hungarian explorer and collector in Abyssinia (Ethiopia) (1910 and 1914–1919).

Kowald

Blue-capped Ifrita *Ifrita kowaldi* **De Vis**, 1890

Charles Kowald (d.1896) was a government agent at Mekeo in British New Guinea (1889–1896), working as a patrol officer with the then British New Guinea Protectorate (PNG). He was a botanist and lepidopterist. He was accidentally killed when a stick of dynamite, which he was holding, exploded.

Kozlov

Roborovski's Rosefinch genus *Kozlowia* **Bianchi**, 1907

Kozlov's Accentor *Prunella koslowi* **Przevalsky**, 1887
[Alt. Mongolian Accentor]
Kozlov's Bunting *Emberiza koslowi* Bianchi, 1904
[Alt. Tibetan Bunting]
Kozlov's Babax *Babax koslowi* Bianchi, 1905
[Alt. Tibetan Babax]

Himalayan Snowcock ssp. *Tetraogallus himalayensis koslowi* Bianchi, 1898

General Petr Kuzmich Kozlov (1863–1935) was a Russian researcher in Central Asia who was one of Przevalsky's (q.v.) companions on his fourth and last expedition. He was sent to Tibet to improve relations there and led the Mongolo-Tibetan (1899–1901 and 1923–1926) and Mongolo-Sichuan (1907–1909) expeditions. He stopped on the Silk Road, just inside the present-day Chinese border with Mongolia, when he

discovered Khara-Khoto, the 'Black City' (1908), which had been described by Marco Polo. It was 'the city of his dreams' and his excavations uncovered many scrolls which he took back for study, along with geographic and ethnographic materials. He wrote *Mongoliya I Kam*. Four mammals and a reptile are named after him.

Kozlova

Willow Ptarmigan ssp. *Lagopus lagopus kozlowae* **Portenko**, 1931
Black-bellied Sandgrouse ssp. *Pterocles orientalis koslovae* **Meinertzhagen**, 1934 NCR
[JS *Pterocles orientalis arenarius*]
Common Reed Bunting ssp. *Emberiza schoeniclus kozlovae* Portenko, 1960 NCR
[JS *Emberiza schoeniclus parvirostris*]

Dr Elizaveta Vladimirovna Kozlova (1892–1975) was a Russian ornithologist at the Zoological Institute, Leningrad (St Petersburg). She was the wife of General Petr Kuzmich Kozlov (q.v.).

Kramer

Kramer's Parakeet *Psittacula krameri* **Scopoli**, 1769
[Alt. Ring-necked Parakeet, Rose-ringed Parakeet]

Wilhelm Heinrich Kramer (d.1765) – also known by the 'latinised' version of his name, Guilielmi Henrici Kramer – was an Austrian naturalist who was originally trained in medicine, which he practised in Bruck (close to Vienna). He wrote *Elenchus Vegetabilium et Animalium per Austriam Inferiorem Observatorum* (1756) (The plants and animals of Lower Austria), one of the first works to adopt Linnean binomial nomenclature.

Krasheninnikov

Rock Ptarmigan ssp. *Lagopus muta krascheninnikowi* Potapov, 1985
[SII *Lagopus muta pleskei*]

Dr Stepan Petrovich Krasheninnikov (1711–1755) was a Russian naturalist, ichthyologist, geographer, and explorer in Siberia. He was a member of Bering's (q.v.) second Kamchatka Expedition (1731–1742). On his return to St Petersburg he wrote his doctoral thesis on ichthyology. Later he became Professor of Natural History at St Petersburg University. His last expedition was to Ladoga and Novgorod (1752), but he died before publishing his observations. A volcano in Kamchatka is named after him.

Krebs

Red-necked Spurfowl ssp. *Pternistis afer krebsi* **Neumann**, 1920 NCR
[JS *Pternistis afer castaneiventer*]

Georg Ludwig Engelhard Krebs (1792–1844) was a German immigrant to South Africa who became a farmer and a prodigious collector. He was trained as an apothecary and, after working in Hamburg, took a four-year contract (1817) with a firm of apothecaries in Cape Town called Pallas and Poleman. Here he succeeded Carl Heinrich Bergius (q.v.), who had lost his job but collected for Lichtenstein (q.v.). Krebs and Bergius collected together until Bergius's death from tuberculosis (1818). Krebs met other well-known naturalists in the Cape, including Brehm (q.v.), von Chamiso (q.v.), and Delalande (q.v.). He wrote to his brother asking him to ask Lichtenstein to see if he would buy specimens. Lichtenstein agreed, and Krebs sent first consignment (1820), followed by a further three large consignments (1821). After finishing his contract with the apothecaries (1821) he worked as a collector. Lichtenstein obtained offical Prussian government authorisation for him to collect for the Natural History Museum, Berlin – hence the book title (1971) about his life: *Ludwig Krebs – Cape Naturalist to the King of Prussia 1820–8.* Krebs opened his own pharmacy in Grahamstown (1826). He was severely stricken with 'rheumatism' (1828), which curtailed much of his active collecting. His last consignment to the Berlin museum (1829) included a complete Bushman (San tribesman) pickled in brine in a barrel, a rhinoceros, an elephant, and a (now extinct) quagga. Krebs succeeded in gaining permanent residence rights to own property, and bought a farm naming it 'Lichtenstein'. He continued to make expeditions into the unknown parts of southern Africa, sending consignments of specimens to Lichtenstein for disposal, normally by public auction. A mammal is also named after him.

Kreczmer

Garden Warbler ssp. *Sylvia borin kreczmeri* Dunajewski 1938 NCR
[JS *Sylvia borin borin*]

Bogdan Kreczmer (b.1908) was a Polish zoologist and collector in Eastern Europe, obtaining the holotype of the warbler in Lithuania (1936).

Krefft

Long-tailed Myna *Mino kreffti* **P. L. Sclater**, 1869
Pitta sp. *Pitta kreffti* **Salvadori**, 1869 NCR
[Alt. Noisy Pitta; JS *Pitta versicolor simillima*]

Johann Ludwig Gerard Krefft (1830–1881) was a German adventurer, artist and naturalist in Australia. He emigrated from Germany to the USA (1851) and worked as an artist in New York, but sailed for Australia (1852) to join the gold rush. He was a miner (1852–1857), then joined the National Museum in Melbourne as a collector and artist. He seems to have had a temper: he feuded with the Museum Trustees and was dismissed (1874). He refused to accept the dismissal and barricaded himself in his office, only to be carried out of the building, still sitting on his chair, deposited in the street and the door locked behind him. He felt he had been hard done by and so set up a rival 'Office of the Curator of the Australian Museum' and successfully sued the trustees for a substantial sum of money. That was the end of his career and he never worked seriously again, but he wrote natural history articles for the Sydney press. He wrote *The Snakes of Australia* (1869) and *The Mammals of Australia* (1871). Three reptiles, three amphibians and a mammal are named after him.

Krempf

Lesser Yellownape ssp. *Picus chlorolophus krempfi* **Delacour** & **Jabouille**, 1931 NCR
[NUI *Picus chlorolophus annamensis*]

Dr Armand Krempf (b.1879) was a French marine biologist. He was the founding Director of the Oceanographic and Fisheries Service, Nha Trang Institute of Oceanography, Indochina (Vietnam). He wrote 'Carcass on coast of Annam, 1883' (1925) about a large, enigmatic 'armour-plated' creature reportedly washed up on the Vietnamese coast. The identity of this oddity remains a mystery.

Kretschmer

Kretschmer's Longbill *Macrosphenus kretschmeri* **Reichenow** & **Neumann**, 1895

Dr Eugen Franz Kretschmer (1868–1894) was a German collector in Kenya. His youth was spent in Russian-occupied Poland. He studied medicine in Munich, Leipzig and Kiel. He was associated with the museum in Kiel (1891–1893), then joined the Berlin Natural History Museum expedition to British East Africa (1894). As he was such a keen ornithologist the museum, to which his small collection went after his death, had great hopes that he would become a major figure in the field. However, shortly after leaving the station at Marangu, Kretschmer and his companion Dr Lent, along with 13 native Kenyans, were murdered. (See also **Eugen**)

Kreyenborg

Pallid Falcon *Falco kreyenborgi* **Kleinschmidt**, 1939 NCR
[Alt. Kleinschmidt's Falcon, Peregrine Falcon; JS *Falco peregrinus cassini*]

Dr Herman Kreyenborg (1889–1963) was a German philologist whose specialities were Indian and Persian languages. He was the librarian at the University of Münster, and a keen falconer.

Krider

Krider's Hawk *Buteo jamaicensis kriderii* **Hoopes**, 1873
[Alt. Red-tailed Hawk ssp.]

John Krider (1819–c.1896) was a Philadelphia taxidermist, gunsmith and professional collector who took two specimens of 'Krider's Hawk' in Iowa (1872). He and Heermann (q.v.) made a bird collecting expedition to Florida (1848). Unusually, Krider published his field notes in books or articles in *Forest and Stream* magazine. He co-wrote *Krider's Sporting Anecdotes, Illustrative of the Habits of Certain Varieties of American Game* (1853). N. H. Bishop (q.v.) recorded that 'Monday, November 9, was a cold, wet day. Mr Knight and the old, enthusiastic gunsmith-naturalist of the city, Mr John Krider, assisted me to embark in my now decked, provisioned, and loaded canoe' in his *Voyage of a Paper Canoe* (1874).

Krieg

Great Antshrike ssp. *Taraba major kriegi* **Laubmann**, 1930 NCR
[JS *Taraba major major*]

Krieg's Conure *Pyrrhura frontalis kriegi* Laubmann, 1932
NCR
[Alt. Maroon-bellied Parakeet; *kriegi* now regarded as an
intermediate population between *Pyrrhura frontalis
frontalis* and *P. frontalis chiripepe*]

Professor Dr Hans Krieg (1891–1975) was the Director of the
Munich Museum and made a number of expeditions to South
America, bringing back important collections from Argentina,
Bolivia, Brazil and Paraguay. A reptile and an amphibian are
named after him.

Krimhild

Waxbill genus *Krimhilda* **Wolters**, 1943 NCR
[Now in *Estrilda*]

In Germanic legend, Krimhild or Kriemhild was the beautiful
heroine in the Nibelungenlied.

Krishnakumarsinhji

Sand Lark ssp. *Calandrella raytal krishnakumarsinhji* **Vaurie
& Dharmakumarsinhji**, 1954

His Highness Raol Shri Krishnakumarsinhji (1919–1965) was
the 27th Maharajah of Bhavnagar and elder brother of the
junior author.

Krishna Raju

Abbott's Babbler ssp. *Malacocincla abbotti krishnarajui*
Ripley & Beehler, 1985

K. S. R. Krishna Raju (d.2002) was an Indian ornithologist who
collected the holotype. He published more than 30 papers on
ornithology (1968–1991). He joined the Bombay Natural
History Society Bird Migration study station in Point Calimere
in Tamil Nadu as a field biologist. He was Honorary Secretary
of Andhra Pradesh Natural History Society. He promoted the
conservation of the Eastern Ghats and undertook at least one
field trip with Ripley (q.v.).

Krone

Restinga Tyrannulet *Phylloscartes kronei* **Willis** & Oniki,
1992

Sigismund Ernst Richard (Ricardo) Krone (1861–1917) was
born in Dresden, Germany, but left there for Brazil (1884),
settling in Iguape where he opened a pharmacy. He was a
keen cave explorer who discovered 41 new caves and
studied every aspect of the Ribeira Valley. He wrote his great
work on the valley's caves, *As Grutas Calcáreas do Vale do
Rio Ribeira de Iguape* (1906), which also explored the
geology, ecology and palaeontology of the area. The original
etymology reads: 'We name this species after Ricardo Krone
of Iguape, who was the premier zoologist of the first agricultural boom of the Ribeira Valley at the turn of the century …'
[i.e. turn of 19th/20th century]. He is also commemorated in
the names of a number of fish such as the cave-dwelling
catfish *Pimelodella kronei*.

Krug

Puerto Rican Screech Owl *Gymnoglaux krugii* **Gundlach**,
1874 NCR
[JS *Megascops nudipes*]

Carl (Karl) Wilhelm Leopold Krug (1833–1898) was a German
businessman who worked in Puerto Rico (1857–1876), where
he became the German Vice Consul, acquired ownership of
the firm he worked for, and married a wealthy landowner's
daughter. His personal hobbies were zoology and botany.
Krug and Gundlach (q.v.), who was Krug's guest, collected
together. A reptile is also named after him.

Krüper

Krüper's Nuthatch *Sitta krueperi* **Pelzeln**, 1863

Theobald Johannes Krüper (1829–1917) was a German ornithologist, entomologist, collector and natural history trader.
He travelled widely in Europe, visiting Lapland, Iceland and
Gotland (in the Baltic Sea, off the coast of Sweden) (1855–
1857); and the Ionian Islands of Greece, and Western Turkey
(1858–1872). Krüper wrote 'Ornithologische Notizen über
Griechenland' (1862) and 'Beitrag zur Ornithologie Klein-
asiens' (1875). He became Director of the Museum of Athens
University (1872) and continued in that post until his death.

Krynicki

Eurasian Jay ssp. *Garrulus glandarius krynicki*
Kaleniczenko, 1839

Professor Johann Krynicki (1797–1838) was a Polish entomologist who was a Professor at Kharkov University in the
Ukraine, where his collection is housed.

Kubary

White-eye genus *Kubaryum* **Momiyama**, 1922 NCR
[Now in *Zosterops*]

Pohnpei Fantail *Rhipidura kubaryi* **Finsch**, 1876
Carolina (Islands) Ground Dove *Gallicolumba kubaryi*
Finsch, 1880
[Alt. White-fronted Ground Dove]
Kubary's Crow *Corvus kubaryi* **Reichenow**, 1885
[Alt. Mariana Crow, Guam Crow]

Jan Stanislav Kubary (1846–1896) was a Polish medical
student whose father was Hungarian and mother German.
He fled Poland because of his participation in the Polish
Revolt (1863) against the Russian occupying forces. He went
to Germany where he met Johann Godeffroy (q.v.),
who owned a natural history trading company in Hamburg.
Godeffroy employed Kubary as a collector, sending him to the
Pacific (1869), where he collected on Tonga, Samoa, the
Marshall Islands and Pohnpei until Godeffroy's company
went bankrupt (1879). After hard times, lacking a regular
income, Kubary collected specimens for the Berlin Ethno-
graphical Museum (1884–1885) and worked as an interpreter
for the Imperial German Navy. He was employed by the
Neuguinea-Kompanie (1887–1895), managing the station of
Konstantinhafen (Papua New Guinea). He then returned to
Pohnpei in the Caroline Islands, where he settled and

acquired a plantation. When his plantation was attacked and devastated by the Spanish authorities as a punitive action for native unrest he committed suicide. His widow, the daughter of an American farmer on Pohnpei, remained there until her death (1937). He wrote (in Polish and German) the three-volume *Ethnographische Beitrage zur Kenntnis des Karolinen Archipels* (1889), as well as a number of articles for the *Journal des Museum Godeffroy*, mostly on ethnography with a particular interest in religious practices.

Kubitschek

Versicolored Emerald ssp. *Amazilia versicolor kubtchecki* **Ruschi**, 1959

Dr Juscelino Kubitschek de Oliveira (1902–1976) was a medic by profession who became President of Brazil (1956–1961). 'Jusselino Kubtcheck' is a variant spelling of his name (used by Ruschi). Before becoming President, he was Mayor of Belo Horizonte, Minas Gerais province. Kubitschek's presidency was marked by a time of political optimism. He launched the 'Plan of National Development', famous for its motto: 'Fifty years of progress in five'. He raised enormous sums of money for capital projects, including construction of the new capital city of Brasília. After a military takeover (1964), Kubitschek was deprived of his political rights and, temporarily, exiled. He returned to Brazil (1967) where he died in a car crash (1976). A genus of Neotropical rodents, *Juscelinomys*, is also named after him.

Kudashev

Common Linnet ssp. *Carduelis cannabina kudashevi* **Portenko**, 1960 NCR
[JS *Carduelis cannabina bella*]

The author gave no indication of who the bird is named after. Circumstantial evidence leads us to believe it was Prince Alexander Koudashev (DNF), who (with Zarudny, q.v.) described the subspecies *Acanthis cannabina taurica* and *A. cannabina persica* (1916); so it seems probable that Portenko (q.v.) had him in mind when describing another subspecies of Linnet, but we have been unable to find more about him or to confirm our assumption. He wrote 'Preliminary list of birds observed in the district of Sochi, Black Sea region' (1916).

Kuehner

Guerrero Brush Finch *Arremon kuehnerii* Navarro-Sigüenza, García-Hernández & Peterson, 2013

Carl R. Kuehner III (b.1963) is a Director of the National Fish and Wildlife Foundation. He is a businessman, a real estate developer in Connecticut and a philanthropist. He graduated with a business degree (1982) and joined his father's company, taking over when his father retired. He is Chief Executive Officer and President, Building and Land Technology Corporation; he was formerly Chairman of BNC Financial Group and the Bank of New Canaan. Philanthropy is something of a family trait as his mother set up a humanitarian charity following the earthquake in Haiti, delivering $30m in aid; his daughter now runs that organisation. He was honoured in the finch's binomial for '…

his longstanding engagement and dedication to fish and wildlife in the United States.'

Kuhl

Kuhl's Sunbird *Aethopyga eximia* **Horsfield**, 1821
[Alt. White-flanked Sunbird]
Kuhl's Lorikeet *Vini kuhlii* **Vigors**, 1824
[Alt. Rimatara Lorikeet]
Kuhl's Ground Thrush *Zoothera interpres* **Temminck**, 1828
[Alt. Chestnut-capped Thrush]
Kuhl's Shearwater *Procellaria kuhli* Boie, 1835 NCR
[Alt. Cory's Shearwater; JS *Calonectris diomedea*]
White-browed Hawk *Leucopternis kuhli* **Bonaparte**, 1850

Kuhl's Cape Parrot *Poicephalus fuscicollis fuscicollis* Kuhl, 1820
[Alt. Brown-necked Parrot ssp.]
Galah ssp. *Eolophus roseicapilla kuhli* **Mathews**, 1912

Dr Heinrich Kuhl (1797–1821) was a German zoologist who became an assistant to Coenraad Temminck (q.v.) at the Leiden Museum. He wrote *Conspectus Psittacorum* (1820). He travelled to Java (1820) with his friend van Hasselt (q.v.) to study the fauna of the Dutch East Indies. After less than a year in Java, Kuhl died in Buitenzorg (Bogor) of a liver infection brought on by the tropical climate and overexertion. Six mammals, three reptiles and an amphibian are named after him.

Kühn

Damar Flycatcher *Ficedula henrici* **Hartert**, 1899
Kühn's Myzomela *Myzomela kuehni* **Rothschild**, 1903
[Alt. Crimson-hooded Myzomela]
Kühn's White-eye *Zosterops kuehni* Hartert, 1906
[Alt. Ambon White-eye]

Drab Whistler ssp. *Pachycephala griseonota kuehni* Hartert, 1898
Orange-banded Thrush ssp. *Zoothera peronii audacis* Hartert, 1899
Papuan Pitta ssp. *Erythropitta macklotii kuehni* Rothschild, 1899
Rufous-sided Gerygone ssp. *Gerygone dorsalis kuehni* Hartert, 1900
Wallacean Drongo ssp. *Dicrurus densus kuehni* Hartert, 1901
Barred Rail ssp. *Gallirallus torquatus kuehni* Rothschild, 1902
Grey-sided Flowerpecker ssp. *Dicaeum celebicum kuehni* Hartert, 1903
Wetar Blue Flycatcher *Cyornis hyacinthinus kuehni* Hartert, 1904
Snowy-browed Flycatcher ssp. *Ficedula hyperythra audacis* Hartert, 1906
Black-faced Cuckooshrike ssp. *Coracina novaehollandiae kuehni* Hartert, 1916 NCR
[JS *Coracina novaehollandiae melanops*]
Large-tailed Nightjar ssp. *Caprimulgus macrurus kuehni* Rothschild & Hartert 1918 NCR
[NUI *Caprimulgus macrurus schlegelii*]

Barn Owl ssp. *Tyto alba kuehni* Hartert, 1929 NCR
[NUI Eastern Barn Owl *Tyto delicatula delicatula*]
Spotted Honeyeater ssp. *Xanthotis polygrammus kuehni*
  Hartert, 1930

Dr Heinrich Kühn (1860–1906) was a German explorer and naturalist, particularly an entomologist, who collected in Sulawesi (1882–1885), and on Damar Island and the Kai Islands (c.1898). Two birds have *audacis* as a trinomial – a zoological gentle joke as 'kühn' is German for bold and 'audacis' is the genitive of the Latin 'audax', the word for bold.

Kuiper

Barred Buttonquail ssp. *Turnix suscitator kuiperi* **Chasen**, 1937 NCR
[NUI *Turnix suscitator suscitator*]

F. J. Kuiper (DNF) was a Dutch ornithologist who collected, principally for van Marle (q.v.), in the East Indies (Indonesia) (1935–1937).

Kulczynski

Blackish-grey Antshrike ssp. *Thamnophilus nigrocinereus kulczynskii* **Domaniewski** & **Stolzmann**, 1922

Dr Władysław Jan Kulczynski (1854–1919) was a Polish zoologist, arachnologist, mountaineer and conservationist who was Professor of Zoology at the Cracow Museum.

Kumagai

Eurasian Nuthatch ssp. *Sitta europaea kumagaii*
  Momiyama, 1928 NCR
[JS *Sitta europaea hondoensis*]

Saburo Kumagai (1904–1954) was a Japanese engineer, ornithologist and collector. He wrote *Avifauna in Tohoku district* (1929).

Kumerloeve

Horned Lark ssp. *Eremophila alpestris kumerloevei*
  Roselaar, 1995

Dr Hans Kumerloeve (1903–1995) was the Director of the Natural History Museum in Dresden and the first director of the State Scientific Museums in Vienna (1936–1945). He and Niethammer carried out the first ornithological study expeditions of northern and western Turkey (1933). His name at birth was Kummerlöwe, which means 'underdeveloped lion' in German and not a good sounding name for a senior SS officer, which he became. He anglicised the spelling of his name after WW2. He was imprisoned for a number of years. He wrote a book on the mammals of Turkey as well as a number of papers on its birds.

Kumlien

Kumlien's Gull *Larus glaucoides kumlieni* **Brewster**, 1883
[Alt. Iceland Gull ssp.]

Thure Ludwig Theodor Kumlien (1819–1888) was a Swedish naturalist who settled in America. He lived in Wisconsin and collected on his own land. He also acted as a collecting agent for various European museums, including the Natural History Museum in Stockholm. He became Professor of Natural Science at Albion College (1867) and Curator at the Milwaukie Museum in Wisconsin (1883). He co-wrote *History of American Birds*. His son Ludwig (1853–1902) was also an ornithologist who took part in the Howgates Arctic Expedition (1877–1878), worked for the US Government in fisheries investigation (1879–1891) and, with N. Hollister, wrote *The Birds of Wisconsin* (1903).

Kuntz

House Swift ssp. *Apus nipalensis kuntzi* **Deignan**, 1958

Captain Robert E. Kuntz (1916–2002) was an American navy parasitologist and medical entomologist (1943–1964) and collector. The University of Michigan awarded his master's degree in zoology, and his doctorate in parasitology. After retiring from the navy he was Chairman of the Department of Parasitology, Southwest Foundation for Research and Education, until retirement (1981). He spent periods of time in Africa, the Middle East and South-East Asia, and was in Formosa (Taiwan) during naval service (1947–1953).

Künzel

White-collared Swift ssp. *Streptoprocne zonaris kuenzeli*
  Niethammer, 1953

Werner Künzel (DNF) was a German sculptor and artist who lived at Trupana, Bolivia (1952).

Kuper

Rufous Fantail ssp. *Rhipidura rufifrons kuperi* **Mayr**, 1931

Henri Kuper (DNF) was a German who became a copra planter and trader on Santa Ana, Solomon Islands, sometime before 1914. He played host to members of the Whitney South Seas Expedition (1927).

Kuroda

Shama genus *Kurodornis* **Hachisuka**, 1941 NCR
[Now in *Copsychus*]

Kuroda's Shelduck *Tadorna cristata* Kuroda, 1917 EXTINCT?
[Alt. Crested Shelduck]

Micronesian Starling ssp. *Aplonis opaca kurodai*
  Momiyama, 1920
Micronesian Myzomela ssp. *Myzomela rubratra kurodai*
  Momiyama, 1922
Izu Thrush ssp. *Turdus celaenops kurodai* Momiyama, 1923
  NCR; NRM
White-backed Woodpecker ssp. *Dendrocopos leucotos*
  kurodae Götz, 1926 NCR
[JS *Dendrocopos leucotos namiyei*]
Collared Treepie ssp. *Dendrocitta frontalis kurodae*
  Delacour, 1927 NCR; NRM
Grey Bunting ssp. *Emberiza variabilis kurodai* Momiyama,
  1927 NCR; NRM
Eurasian Three-toed Woodpecker ssp. *Picoides tridactylus*
  kurodai Yamashina, 1930
[SII *Picoides tridactylus alpinus*]

Grey-capped Pygmy Woodpecker ssp. *Dendrocopos canicapillus kurodai* **La Touche**, 1931 NCR
[NPRB *Dendrocopos canicapillus nagamichii*]
Pacific Swift ssp. *Apus pacificus kurodae* **Domaniewski**, 1933
[SII *Apus pacificus pacificus*]

Dr Nagamichi Kuroda (1889–1978) was a Japanese ornithologist. He published widely including 'A contribution to the knowledge of the avifauna of the Riu Kiu Islands and the vicinity' (1925), 'On a small collection of birds from the Riu Kiu Islands' (1926), *Birds of the Island of Java* (1933–1936) and *Parrots of the World in Life Colours* (1975). Most of his extensive collection of specimens was destroyed in WW2; what survived is now in the Yamashina Institute for Ornithology at Chiba, Japan (see also **Nagamichi**).

Kuser

Kuser's Blood Pheasant *Ithaginis cruentus kuseri* **Beebe**, 1912

Colonel Anthony Rudolph Kuser (1862–1929) of Trenton, New Jersey, was an American of Swiss descent who was an avid pheasant fancier and wealthy patron of the New York Zoological Park. 'Colonel' was an honorary title given for serving as a member of Leon Abbett's staff (Abbett was Governor of New Jersey). Kuser also served Governors Wertz and Griggs. He was President of the South Jersey Gas and Electric Lighting Company, which he took over in 1909, and also Vice-President of the Public Services Corporation of New Jersey. Together with his four brothers he had interests in brewing and an ice-making company. He offered $60,000 for an expedition to conduct research on pheasants (1909) and to write an updated monograph on the family. His only condition was to have William Beebe, who later wrote the required monograph, as the expedition leader. Kuser declared himself willing to pay for all the best artists to do the colour plates, to meet all printing, binding and marketing costs, and even to let the book be published in the name of the New York Zoological Society. Kuser bequeathed money to the New Jersey Audubon Society.

Kutter

Night Heron sp. *Butio kutteri* **Cabanis**, 1881 NCR
[Alt. Malayan Night Heron; JS *Gorsachius melanolophus*]

Dr Otto Friedrich Kutter (1834–1891) was a German surgeon, oologist and collector. He was in the Philippines in 1878.

L

Labat

Labat's Conure *Aratinga labati* Labat, 1724 EXTINCT
[Alt. Guadeloupe Parakeet]

Jean Baptiste Labat (1663–1738) was a French missionary, although British 17th-century biographers describe him as a pirate because he defended Guadeloupe against the English. He was sent by his order (the Dominicans) to the West Indies (1693), first to Martinique and then to Guadeloupe. He performed valuable botanical research and wrote extensively about his travels. His account of a hunting expedition (1696) described how a party of six men captured over 200 Black-capped Petrels *Pterodroma hasitata*, which they hauled from their burrows in Guadeloupe. Apparently he also decimated the flamingo colony on Aves Island (1705). There are no existing specimens of the conure that bears his name, so its taxonomy may never be fully elucidated and its status as a separate species is hypothetical.

Labrador

Ornate Tinamou ssp. *Nothoprocta ornata labradori* J. Cabot, 1997

José Sánchez Labrador (1717–1798) was a jesuit missionary and naturalist. He was a prolific author of catholic doctrine, anthropology, the Guarani language and agriculture as well as natural history. He entered the 'Society of Jesus' (1732) and was sent to South America two years later (1734) where he studied theology at the University of Córdoba (1734–1739). He continued to study and also began to teach theology in Buenos Aires as well as at several missions and eventually to Indians (1760–1767) before being expelled and taking up residence in Ravenna, Italy, where he wrote of his experiences of the people and natural history of South America. He wrote the four-volume *Natural Paraguay* (1771).

Lady Amherst

Lady Amherst's Pheasant *Chrysolophus amherstiae* **Leadbeater**, 1829

(See **Amherst, Lady**)

Lady MacGregor

Lady MacGregor's Bowerbird *Amblyornis macgregoriae* **De Vis**, 1890

(See **MacGregor, M.**)

Lady Ross

Lady Ross's Turaco *Musophaga rossae* **Gould**, 1852

(See **Ross, E.**)

Laenen

Firecrest ssp. *Regulus ignicapilla laeneni* **van Marle** & **Voous**, 1949 NCR
[NUI *Regulus ignicapilla balearicus*]
Blue-naped Mousebird ssp. *Urocolius macrourus laeneni* **Niethammer**, 1955
Eastern Olivaceous Warbler ssp. *Iduna pallida laeneni* Niethammer, 1955
Grey-headed Sparrow ssp. *Passer griseus laeneni* Niethammer, 1955
Yellow-bellied Eremomela ssp. *Eremomela icteropygialis laeneni* Niethammer, 1955 NCR
[JS *Eremomela icteropygialis alexanderi*]

J. R. Laenen (DNF) was a Belgian collector, zoologist and explorer. He collected the firecrest (named after both him and his wife) while making a zoological survey in Algeria (1949), the sparrow, eromomela and warbler at Lake Chad (1954) and the mousebird in Niger (1955). Laenen and Niethammer (q.v.) wrote the article 'Hivernage au Sahara' (1954) and collected together in the Sudan.

La Farge

Scarlet-naped Myzomela *Myzomela lafargei* **Pucheran**, 1853
[Alt. Red-capped Myzomela]

Ensign Antoine Auguste Thérèse Pavin de la Farge (1812–1839) was on board the exploration vessel *La Zélée* in the Pacific (1837–1839). He died at sea aged just 27.

Lafayette

Lafayette's Junglefowl *Gallus lafayetii* **Lesson**, 1831
[Alt. Sri Lanka Junglefowl]

Marie Joseph Paul Yves Roch Gilbert du Mothier, Marquis de Lafayette (1757–1834), was a French nobleman and cavalry officer. He inherited a large fortune but, coming from a long military line, he studied at the military academy in Versailles and became a cavalry captain aged 16 (1773). He bought a ship (1777) and left for America to fight in the revolution against the British. He was appointed a Major-General and was assigned to the staff of George Washington, achieving several notable victories. After the British surrender he returned home, where he cooperated closely with Ambassadors Benjamin Franklin and then Thomas Jefferson on behalf of American interests. In France he worked toward establishing a constitutional monarchy in the period before the French Revolution (1789). He tried unsuccessfully to help the reluctant King and Queen, was denounced as a traitor, and had to flee the country. He returned (1800) and found his fortune had been confiscated, but was later elected to the Chamber of Deputies (1815). As one of its vice-presidents, he worked for Napoleon's abdication after the Battle of Waterloo. Lafayette became a focus of resistance to the Bourbon kings and (1830) led the Revolution dethroning them. He refused the presidency of the new republic, instead helping make Louis Philippe the constitutional monarch of France, which he regretted by the time of his death, preferring the republican option. La Fayette city in California and Fayetteville in North Carolina are named after him.

Lafresnaye

Hummingbird genus *Lafresnaya* **Bonaparte**, 1850

Glossy Flowerpiercer *Diglossa lafresnayii* **Boissonneau**, 1840
Mountain Velvetbreast *Lafresnaya lafresnayi* Boissonneau, 1840
Lafresnaye's Spinetail *Hellmayrea gularis* Lafresnaye, 1843
[Alt. White-browed Spinetail]
Great Iora *Aegithina lafresnayei* **Hartlaub**, 1844
Lafresnaye's Vanga *Xenopirostris xenopirostris* Lafresnaye, 1850
Lafresnaye's Woodcreeper *Xiphorhynchus guttatoides* Lafresnaye, 1850
Lafresnaye's Rail *Gallirallus lafresnayanus* **J. Verreaux** & **Des Murs**, 1860
[Alt. New Caledonian Rail; Syn. *Tricholimnas lafresnayanus*]
Lafresnaye's Piculet *Picumnus lafresnayi* **Malherbe**, 1862

Blue-breasted Bee-eater ssp. *Merops variegatus lafresnayii* **Guérin-Méneville**, 1843
Red-billed Scythebill ssp. *Campylorhamphus trochilirostris lafresnayanus* **d'Orbigny**, 1846
Lafresnaye's Chestnut-fronted Macaw *Ara severus castaneifrons* Lafresnaye, 1847
Cardinal Woodpecker ssp. *Dendropicos fuscescens lafresnayi* Malherbe, 1849
Montane Woodcreeper ssp. *Lepidocolaptes lacrymiger lafresnayi* **Cabanis** & **Heine**, 1859
Plain-brown Woodcreeper ssp. *Dendrocincla fuliginosa lafresnayei* **Ridgway**, 1888

Baron Nöel Fréderic Armand André de Lafresnaye (1783–1861) was a French ornithologist and collector. He wrote with d'Orbigny (q.v.) and described a number of species including parrots and macaws. After his death his bird collection was purchased by the American collector Henry Bryant (q.v.) and donated to the Boston Natural History Society.

Lagden

Lagden's Bush-shrike *Malaconotus lagdeni* **Sharpe**, 1884

Sir Godfrey Yeatman Lagden (1851–1934) was an English diplomat. With letters of introduction to the High Commissioner, Sir Bartle Frere, he sailed for South Africa (1877). He became Chief Clerk to the State Secretary in Transvaal, and acted as Secretary to the Administrator, Sir Owen Lanyon. Subsequently he became a war correspondent for the *Daily Telegraph* during the Egyptian Campaign (1882–1883). He served as a diplomat in Sierra Leone and the Gold Coast (Ghana), and then in Basutoland (Lesotho) (1884) where he became Resident Commissioner (1890). He was appointed Commissioner in Swaziland (1892), and again in Basutoland (1893–1901). He served during the South African War and was repeatedly mentioned in dispatches. He wrote a standard book, *The Basutos: The Mountaineers and their Country; being a Narrative of Events Relating to the Tribe from its Formation early in the Nineteenth Century to the Present Day* (1909). He was also a distinguished hunter and played first-class cricket for the MCC (Marylebone Cricket Club). His family founded the Yeatman Hospital in Sherborne, Dorset, England, the town where he was born.

Laglaize

Versicolored Emerald ssp. *Amazilia versicolor laglaizei* **E. L. Simon**, 1921 NCR
[JS *Amazilia versicolor millerii*]

Léon François Laglaize (DNF) was a French traveller, naturalist and collector in New Guinea (1880–1897) and Venezuela (1898).

Lagrandière

Lagrandiere's Barbet *Megalaima lagrandieri* **J. Verreaux**, 1868
[Alt. Red-vented Barbet]

Vice-Admiral Pierre Paul Marie de Lagrandière (1807–1876) was the French Governor of Cochinchina (Vietnam) (1863–1868). He is best known for having occupied three Cambodian provinces, on instruction from France.

Lahille

Antshrike sp. *Thamnophilus lahilleanus* **Bertoni**, 1901 NCR
[Alt. Tufted Antshrike; JS *Mackenziaena severa*]

Dr Fernand Lahille (1861–1940) was a French physician, marine biologist, ichthyologist and zoologist who went to Argentina (1893). He worked at the La Plata Museum of Natural History (1893–1899). He was the first head of the Fish and Game Department, Ministry of Agriculture, and was Professor of Zoology, University of Buenos Aires (1910). Lahille Island in Antarctica is named after him.

Laing

Marsh Wren ssp. *Cistothorus palustris laingi* Harper, 1926
Northern Goshawk ssp. *Accipiter gentilis laingi* **P. A. Taverner**, 1940

Hamilton Mack Laing (1883–1982) was a Canadian naturalist, ornithologist, photographer and artist. He served with the Royal Flying Corps in WW1. He took part in a number of expeditions, including the Smithsonian Expedition to Lake

Athabasca (1920). He had a nut farm at Comox, Vancouver Island (1922–1950), where there is a Mack Laing Nature Park on land he bequeathed to the community.

Lairet

Green-bellied Hummingbird ssp. *Amazilia viridigaster laireti* **W. H. Phelps** Jr & Aveledo, 1988
[Syn. *Amazilia cupreicauda laireti* (if *cupreicauda* split from *viridigaster*)]

Dr Andrés Lairet (DNF) is a distinguished Venezuelan physician and a 'greatly appreciated' friend of the authors.

Lais

Wailing Cisticola *Cisticola lais* **Hartlaub** & **Finsch**, 1870

Lais, who lived in the 4th century BC, was a celebrated hetaera (educated courtesan) and the mistress of the great Greek artist Apelles.

Lako

Variegated Tinamou ssp. *Crypturellus variegatus lakoi* **Miranda-Ribeiro**, 1938 NCR
[*C. variegatus* NRM]

Károly Lako (1895–1960) was a Hungarian taxidermist who lived in Brazil (1920–1960).

Laland

Black-breasted Plovercrest *Stephanoxis lalandi* **Vieillot**, 1818

(See **Delalande**)

Lalfa

Slender-billed Prion ssp. *Heteroprion belcheri lalfa* **Mathews**, 1939 NCR; NRM
[Syn. *Pachyptila belcheri*]

(See **Falla** – *lalfa* is an anagram of his name)

Lamb

Oaxaca Screech Owl *Megascops lambi* **R. T. Moore** & **J. T. Marshall**, 1959
[Syn. *Megascops cooperi lambi*]

Rufous-crowned Sparrow ssp. *Aimophila ruficeps lambi* **J. Grinnell**, 1926 NCR
[JS *Aimophila ruficeps canescens*]
Lamb's Stygian Owl *Asio stygius lambi* R. T. Moore, 1937
Northern Potoo ssp. *Nyctibius jamaicensis lambi* Davis, 1959
Ladder-backed Woodpecker ssp. *Picoides scalaris lambi* **A. R. Phillips**, 1966 NCR
[NUI *Picoides scalaris scalaris*]
Russet Nightingale Thrush ssp. *Catharus occidentalis lambi* A. R. Phillips, 1969

Chester Converse Lamb (1883–1965) was a collector in Mexico and western USA. The Museum of Vertebrate

Zoology (MVZ), University of California, Berkeley, employed him as a Field Assistant collecting in Lower California and western USA (1925–1932). He rose to become Assistant Curator of Mammals. He then (1932–1955) collected extensively in Mexico, accumulating 40,000 bird skins and many nests, eggs and mammals, for the Moore Lab of Zoology, Occidental College, Los Angeles. In his citation for the Stygian Owl subspecies, Moore said of him: 'He is more than a collector, he has the indefatigable zest of a real student of bird behaviour, which has won him the friendship of every ornithologist ...'

Lambert

Lambert's Fairy-wren *Malurus lamberti* **Vigors** & **Horsfield**, 1827
[Alt. Variegated Fairy-wren]

Aylmer Bourke Lambert (1761–1842) was a wealthy British botanist who spent so much money on acquiring botanical specimens that, on his death, his family found very little capital left. As a result, his collections were sold at auction shortly after his demise and were split up; mainly between The British Museum in London, the Botanic Garden in Geneva, and a collection in the USA which re-acquired specimens collected during the Lewis (q.v.) and Clark (q.v.) expedition. Lambert was elected to the Royal Society in London (1791) and was Vice-President of the Linnean Society (1796–1842). He wrote *A Description of the Genus Pinus*, detailing all the then-known conifers. *Pinus lambertiana* (Sugar Pine) is named after him.

Lambruschini

Gull genus *Lambruschinia* **Salvadori**, 1864 NCR
[Now in *Chroicocephalus*]

Slender-billed Gull *Xema lambruschinii* **Bonaparte**, 1840 NCR
[JS *Chroicocephalus genei*]

Raffaele Lambruschini (1788–1873) was an Italian parliamentarian (1847–1849), senator (1860–1873), priest, agronomist and teacher. He was arrested for political agitation and exiled to Corsica (1812–1814). He was Professor of Agriculture at the University of Pisa (1841). He founded the journal *La Gioventù* (1862).

Lamm

Pernambuco Foliage-gleaner *Automolus lammi* **Zimmer**, 1947

Donald Wakeham Lamm (1914–1996) was an American ornithologist and diplomat who chose his overseas postings to further his ornithological interests. He served in Japan (1936–1942) and was interned after the attack on Pearl Harbour (1941), but was released and was posted to Recife in Brazil (1943–1947), Australia twice (1947–1949 and 1960–1964), Mozambique (1952–1954), and the Gold Coast (Ghana) (1954–1957). He published widely, often with local ornithologists, such as with A. A. da Rosa Pinto (q.v. under Pinto, A.) 'Contribution to the study of the ornithology of Sul do Save

(Mozambique)', which appeared in four parts (1953 and 1960). After retiring from the US State Department's service he continued to travel and research.

Lamont

Brown Quail ssp. *Synoicus ypsilophorus lamonti* **Mayr &
Gilliard**, 1951

William Lamont (DNF) of Lae, New Guinea, was described in the original etymology as 'a true sportsman'. He was a member of the Gilliard Mount Hagen Expedition (1950).

Lampert

Scarlet-chested Sunbird ssp. *Chalcomitra senegalensis
lamperti* **Reichenow**, 1897

Dr Kurt Lampert (1859–1918) was a German zoologist at the State Natural History Museum, Stuttgart (1884), becoming Professor (1890) and finally Director (1894).

Landbeck

Chilean Pelican *Pelecanus landbecki* **Philippi**, 1909 NCR
[Alt. Peruvian Pelican; JS *Pelecanus thagus*]

Plumbeous Rail ssp. *Pardirallus sanguinolentus landbecki*
Hellmayr, 1932

Christian Ludwig Landbeck (1807–1890) was a German ornithologist and collector who settled in Chile (1852–1890) and was Curator of the Santiago de Chile Museum (1859–1854).

Lang

Yellow-billed Oxpecker ssp. *Buphagus africanus langi*
Chapin, 1921
Orange River Francolin ssp. *Scleroptila levaillantoides langi*
J A. Roberts, 1932 NCR
[JS *Scleroptila levaillantoides levaillantoides*]

Herbert Lang (1879–1957) was born in Germany and trained as a taxidermist working at the Natural History Museum, University of Zurich, and at a commercial establishment in Paris. He moved to the USA (1903) and joined the AMNH as a taxidermist. He led the Congo expedition (1909–1915). On returning to New York he became Assistant in Mammalogy, then Assistant Curator (1919). He returned to Portuguese West Africa (Angola) for the AMNH (1925) with Rudyerd Boulton (q.v.). They covered 4,000 miles and collected 1,200 mammal specimens. He stayed on in Africa, taking a job with the Transvaal Museum, South Africa. He made a number of expeditions, including one for the AMNH into the Kalahari Desert. He took over the management of a hotel in Pretoria (1935). Two reptiles are named after him.

Langhorne

Grey Imperial Pigeon ssp. *Ducula pickeringii langhornei*
Mearns, 1905
[Species sometimes regarded as monotypic]

Colonel George Taylor Langhorne (1867–1962) was an American army officer who graduated from West Point (1889) and

was in the Philippines (1901–1902). He served in the Spanish-American War and commanded the 8th Cavalry (1919) in Texas. His cousin was Nancy Witcher Langhorne, later Viscountess Astor, who was the first woman to sit as a Member of Parliament in the British House of Commons.

Langsdorff

Langsdorff's Thorntail *Discosura langsdorffi* **Temminck,**
1821
[Alt. Black-bellied Thorntail]

Langsdorff's Toucanet *Selenidera reinwardtii langsdorffii*
Wagler, 1827
[Alt. Golden-collared Toucanet ssp.]

Baron Georg Heinrich von Langsdorff (otherwise Grigoriy Ivanovich) (1774–1852) was a German physician, botanist, zoologist, traveller, ethnographer and diplomat. He graduated as a Doctor of Medicine at Göttingen University (1797) and was then sent to Portugal. He was elected as a corresponding member of the Academy of Science at St Petersburg (1803) and joined Krusenstern's round-the-world expedition (1803–1806) on board the *Nadezhda*, representing the Academy. He continued to travel widely, to Japan (1804–1805); north-western America (1805–1806); and Kamchatka, Siberia and European Russia (1806–1808). He became Associate Professor of Botany at the Academy, moving to zoology (1809). He wrote *Remarks and Observations on a Voyage Around the World from 1803–1807* (1812) and was elected as an Extraordinary Member of the Academy before being appointed as Russian Consul General in Brazil (1813–1817). He was also Chargé d'Affaires for Russia to Portugal in Rio, where the Portuguese government was in exile during the Napoleonic Wars. Whilst in Brazil he travelled to Minas Gerais with the French scientist Saint-Hilaire (q.v.). He returned to Russia (1821) to organise an expedition in Brazil, which he led (1822–1828), but he caught a tropical fever that caused a psychological breakdown. He returned to Germany (1830), retiring (1831) and living there until his death. One amphibian and one reptile are named after him.

Lansberge, J. W.

Flores Minivet *Pericrocotus lansbergei* **Büttikofer**, 1886
[Alt. Little Minivet]

Johan Wilhelm van Lansberge (1830–1905) was Governor-General of the Dutch East Indies (Indonesia) (1875–1881). He was born in Bogota, Colombia, and brought up in Venezuela where his father R. F. C. van Lansberge (q.v.) was the Dutch Consul-General.

Lansberge, R. F. C.

Grey-capped Cuckoo *Coccyzus lansbergi* **Bonaparte**, 1850

Reinhart Frans Cornelis van Lansberge (1804–1873) was a Dutch administrator in the West Indies. He was Dutch Consul-General in Venezuela and Governor of Curacao (1855–1859) and Suriname (1859–1867); during his administration slavery was abolished (1863). His son J. W. Lansberge (q.v.) had a similar career and interests. One reptile is named after him.

Lantz

Vanga genus *Lantzia* **Hartlaub**, 1877 NCR
[Now in *Schetba*]

Lantz's Sparrowhawk *Accipiter lantzii* **J. Verreaux**, 1866
NCR
[Alt. Madagascar Sparrowhawk; JS *Accipiter
madagascariensis*]
Lantz's Brush Warbler *Nesillas lantzii* **Grandidier**, 1867
[Alt. Subdesert Brush Warbler]

Auguste Lantz (DNF) was a French naturalist. He was a taxidermist at the MNHN, Paris (1862), and Curator of the Natural
History Museum on Réunion (1863–1893). He explored and
collected in Madagascar with Grandidier (q.v.) (1865–1867).

Lanyon

Antioquia Bristle Tyrant *Pogonotriccus lanyoni* **G. R. Graves**,
1988
[Syn. *Phylloscartes lanyoni*]

Yucatan Flycatcher ssp. *Myiarchus yucatanensis lanyoni*
Parkes & **A. R. Phillips**, 1967
Clay-coloured Thrush ssp. *Turdus grayi lanyoni* **Dickerman**,
1981

Dr Wesley Edwin Lanyon (b.1926) is an American ornitholo-
gist. He was an instructor at Miami University in Ohio and
spent five summers as a naturalist with the National Park
Service, as well as writing many articles (particularly on
Myiarchus flycatchers) and several books including a mono-
graph on meadowlarks. He was President of the American
Ornithologists' Union (1976–1978) and is Curator Emeritus of
the Division of Birds, American Museum of Natural History,
New York, having started there as Assistant Curator (1957).
His most well known publication is *Biology of Birds* (1964).

Laperouse

Laperouse's Megapode *Megapodius laperouse* **Gaimard**,
1823
[Alt. Micronesian Megapode]

Jean-François de Galaup, Comte de Lapérouse (1741–1788?),
was a French naval officer who commanded (1784) the
French expedition to the Pacific. There were two ships,
Boussole and *Astrolabe* (we think this was the first of several
ships of this name used in French voyages of exploration), on
the expedition, which arrived in Botany Bay (26 January
1788) only five days after Governor Philip arrived there with
the first convicts from England. The ships anchored in what is
now called Frenchman Bay. Lapérouse took the opportunity
to send his journals, some charts and letters back to Europe
with a British naval ship, which proved to be invaluable in the
light of what was to come. The French expedition sailed from
Botany Bay, never to be seen again by Europeans. An Irish
captain called Peter Dillon found evidence that Vanikoro in
the Santa Cruz Islands (1827) was the site of the double shipwreck. Whether Lapérouse drowned, died of disease or was
killed by the local inhabitants has never been established.

LaPersonne

Streaked Rosefinch ssp. *Carpodacus rubicilloides
lapersonnei* **Hartert** & **Steinbacher**, 1932 NCR
[JS *Carpodacus rubicilloides lucifer*]

V. S. LaPersonne (fl.1900–1934) was a British Indian ornithologist who also visited Iraq, Iran, Kuwait and New Zealand.
His collecting trips to islands in the Persian Gulf (1914 and
1921–1923) were to investigate migration. He wrote a number
of articles including: 'The Common Central Asian Kingfisher
(*Alcedo atthis pallasii* Reichenb.) in Nepal' (1933), all
published in the *Journal of the Bombay Natural History
Society*; he was Assistant Curator of the Society's museum.

Laptev

Chukar Partridge ssp. *Alectoris chukar laptevi* **Dementiev**,
1945 NCR
[JS *Alectoris chukar koroviakovi*]

Professor Dr M. K. Laptev (DNF) was a Russian zoologist and
collector. He led the first scientific expedition to Turkmenistan (1923). He wrote *Account of the Land Fauna of
Vertebrates by Means of the Method of Itinerary Counting*
(1930). He studied the avifauna of the south-east Caspian
(Turkmenistan) as part of the Kara-Kalpak expedition (1932)
and animal husbandry and land cultivation (1940s).

Larisch

Red-capped Robin Chat ssp. *Cossypha natalensis larischi*
Meise, 1958

N. von Larisch (DNF) was a collector in Angola.

La Sagra

La Sagra's Flycatcher *Myiarchus sagrae* **Gundlach**, 1852

Ramón de la Sagra (1798–1871) was a Spanish economist
and botanist who lived and worked in Cuba, and died in Switzerland. He was Director of the Havana Botanical Gardens
(1822–1834). During this time he established a model farm
and collected widely. Later he returned to Spain and devoted
himself to the study of political economy before joining the
French revolution (1854). He was a deputy in the Cortes
(Spanish parliament) (1854–1856). He wrote *Principios de
Botanica Agricola* (1833) and the 13-volume *Historia Física,
Política y Natural de la Isla de Cuba* (1839–1861). Four reptiles
are named after him.

La Selle

La Selle Thrush *Turdus swalesi* **Wetmore**, 1927

The thrush was named after a geographical location; the
type locality was the 'Massif de la Selle' in Haiti.

Lashmar

Mallee Whipbird ssp. *Psophodes leucogaster lashmari*
Schodde & Mason, 1991

Allen Frederick Cooper Lashmar (1917–1993) wrote (1934–
1937) about the birds of Kangaroo Island, where he lived and
where this race of whipbird occurs, e.g. 'Birds noted in

eastern portion of Kangaroo Island' in *South Australian Ornithologist* (1935). Some people called him he 'Father of ornithology on Kangaroo Island'.

Latham

Swift Parrot genus *Lathamus* **Lesson**, 1830

Latham's Myna *Scissirostrum dubium* Latham, 1802
[Alt. Grosbeak Starling, Scissor-billed Starling]
Latham's Cockatoo *Calyptorhynchus lathami* **Temminck,**
1807
[Alt. Glossy Black Cockatoo]
Crested Bunting *Emberiza lathami* **J. E. Gray,** 1831
[Syn. *Melophus lathami*]
Latham's Brush-turkey *Alectura lathami* J. E. Gray, 1831
[Alt. Australian Brush-turkey]
Latham's Snipe *Gallinago hardwickii* J. E. Gray, 1831
[Alt. Japanese Snipe]
Latham's Francolin *Peliperdix lathami* **Hartlaub,** 1854
[Alt. Forest Francolin]
Yellow-throated Scrubwren *Sericornis lathami* **Mathews,**
1912 NCR
[JS *Sericornis citreogularis*]

Grey-headed Chickadee ssp. *Poecile cinctus lathami*
J. F. Stephens, 1817
Kalij Pheasant ssp. *Lophura leucomelanos lathami*
J. E. Gray, 1829
Red-billed Quelea ssp. *Quelea quelea lathamii* **A. Smith,**
1836
Varied Sittella ssp. *Daphoenositta chrysoptera lathami*
Mathews, 1912 NCR
[NUI *Daphoenositta chrysoptera chrysoptera*]

Dr John Latham (1740–1837) was a British physician, naturalist and author. He played a leading role in the formation of the Linnean Society of London (1788) and was a Fellow of the Royal Society. He was a practising physician in Kent, England, until his retirement (1796). He wrote *General Synopsis of Birds* (1781–1785), contributed the descriptions of birds in *The Voyage of Governor Phillip to Botany Bay* (1789), and wrote the *Index Ornithologicus* (1790) and the *General History of Birds* (1821–1828), starting the latter at 81 years old. He designed, sketched and coloured the illustrations himself. He knew all the important English naturalists and collectors of his day, so was able to examine practically all the specimens and drawings of Australian birds which reached England. In his later books he provided the first published descriptions and scientific names of many iconic Australian birds. He has been called the 'grandfather' of Australian ornithology.

Lathbury

Golden-winged Sunbird ssp. *Drepanorhynchus reichenowi lathburyi* **J. G. Williams,** 1956

General Sir Gerald William Lathbury (1906–1978) was an ornithologist and professional soldier in the British Army (1926–1969). In WW2 he commanded paratroops and fought in North Africa, Sicily and Europe. He was wounded and captured at Arnhem but escaped – having persuaded his captors that he was not a Brigadier-General, just a Lance Corporal so was not closely guarded. After WW2 he became Commander-in-Chief, East Africa (1955–1957), Commander-in-Chief, Eastern Command (1960–1961), Quartermaster-General (1961–1965) and Governor of Gibraltar (1965–1969). Before his last official appointment he was a member of the third Harold Hall (q.v.) Australia expedition.

Latimer, G.

Puerto Rican Vireo *Vireo latimeri* **Baird,** 1866

George Latimer (1873–1874) from Pennsylvania was US Consul in Puerto Rico, first in Mayagüez (1835–1838) and then in San Juan (1846–1853), where he made a collection of stone implements and ancient Amerindian remains now at the USNM.

Latimer, M.

Spike-heeled Lark ssp. *Chersomanes albofasciata latimerae*
Winterbottom, 1958 NCR
[JS *Chersomanes albofasciata macdonaldi*]
Cape Longclaw ssp. *Macronyx capensis latimerae* **Clancey,**
1963 NCR
[NUI *Macronyx capensis colletti*]

Miss Marjorie Courtenay-Latimer (1907–2004) was the first Curator of the Museum at East London, South Africa, which was based on the Latimer family collection and holds the only remaining Dodo egg. She was primarily an ichthyologist best remembered for her rediscovery (1938) of the Coelacanth, a fish that had been believed to be extinct for 70 million years. She was highly regarded in her country, and clay casts of her footprints were placed in Heroes Park alongside those of Nelson Mandela and Walter Sisulu (2003).

La Touche

Bunting genus *Latoucheornis* **Bangs,** 1931
[Now often merged with *Emberiza*]

La Touche's Bunting *Latoucheornis siemsseni* Martens,
1906
[Alt. Slaty Bunting, Chinese Blue Bunting; Syn. *Emberiza siemsseni*]

La Touche's Sunbird *Aethopyga christinae latouchii*
H. H. Slater, 1891
[Alt. Fork-tailed Sunbird ssp.]
Mountain Scops Owl ssp. *Otus spilocephalus latouchi*
Rickett, 1900
Large Woodshrike ssp. *Tephrodornis virgatus latouchei*
Kinnear, 1925
Chinese Babax ssp. *Babax lanceolatus latouchei*
Stresemann, 1929
Bianchi's Warbler ssp. *Seicercus valentini latouchei* Bangs,
1929
Dollarbird ssp. *Eurystomus orientalis latouchei* Allison, 1946
NCR
[JS *Eurystomus orientalis calonyx*]
Sooty-headed Bulbul ssp. *Pycnonotus aurigaster latouchei*
Deignan, 1949

John David Digues La Touche (1861–1935) was a French-born, English-educated, Irish ornithologist. He was Inspector of Customs in China (1882–1921) and wrote *A Handbook of the Birds of Eastern China* (1925–1934). He intended to retire to Ireland, but died at sea near Majorca on his way home from Spain. A mammal he collected is named after him, as is a reptile and an amphibian. (See also **Touchen**)

Latreille

Rufous-bellied Seedsnipe ssp. *Attagis gayi latreillii* **Lesson**, 1831

Professor Pierre André Latreille (1762–1833) was a French entomologist at the MNHN, Paris (1798), and Professor of Zoology (1829–1832). He trained as a Roman Catholic priest before the French Revolution, during which he was imprisoned (1793), narrowly avoiding the guillotine.

Laubmann

Golden-bellied Grosbeak ssp. *Pheucticus chrysogaster laubmanni* **Hellmayr & Seilern**, 1915
Middle Spotted Woodpecker ssp. *Dendrocopos medius laubmanni* **Götz**, 1923 NCR
[NUI *Dendrocopos medius caucasicus*]
Blue-eared Kingfisher ssp. *Alcedo meninting laubmanni* **Mathews**, 1925 NCR
[NUI *Alcedo meninting coltarti*]
Common Magpie ssp. *Pica pica laubmanni* **Stresemann**, 1928 NCR
[JS *Pica pica bactriana*]
Red-legged Partridge ssp. *Alectoris rufa laubmanni* **von Jordans**, 1928 NCR
[JS *Alectoris rufa intercedens*]
Chestnut-eared Bunting ssp. *Emberiza fucata laubmanni* **Stachanow**, 1929 NCR
[JS *Emberiza fucata fucata*]
Brown Tinamou ssp. *Crypturellus obsoletus laubmanni* Neumann, 1933 NCR
[JS *Crypturellus obsoletus obsoletus*]
Collared Kingfisher ssp. *Todiramphus chloris laubmannianus* **Grote**, 1933

Dr Alfred Louis Laubmann (1886–1965) was a German veterinary surgeon and ornithologist, and Professor at the Zoological Museum, Munich. Living in eastern Germany immediately after WW2 he was forced into heavy labour in a factory and not allowed to visit the museum where his own library and books were housed. Among other works he wrote *Kingfishers* (1922) and *Vögel* (1930).

Laura (Boulton)

Laura's Woodland Warbler *Phylloscopus laurae* **Boulton**, 1931
[Alt. Mrs Boulton's Woodland Warbler; Syn. *Seicercus laurae*]

(See **Mrs Boulton**)

Laura (Johnstone)

Bananaquit ssp. *Coereba flaveola laurae* **Lowe**, 1908

Lady Laura Caroline Johnstone *née* Russell, Countess of Wilton (1842–1916) was the wife of Sir Frederick Johnstone

(q.v.). Percy Lowe (q.v.), the describer of the bananaquit race, was aboard the Johnstones' yacht *Zenaida* in the West Indies (1908).

Laura Gray

Amethyst Starling ssp. *Cinnyricinclus leucogaster lauragrayae* **Bowen**, 1930 NCR
[Alt. Violet-backed Starling; JS *Cinnyricinclus leucogaster verreauxi*]

Mrs Laura Caroline Gray *née* Sherman (1884–1968) was the wife of Prentiss Nathaniel Gray (1884–1934), a prominent New York banker and big-game hunter. He wrote *African Game-Lands: A Graphic Itinerary in Kenya and along the Livingstone Trail in Tanganyika, Belgian Congo, and Angola, 1929.*

Laurence

Common Raven ssp. *Corvus corax laurencei* **Hume**, 1873

(See **Lawrence, J. L. M.**)

Laurente

White-browed Shortwing ssp. *Heteroxenicus cruralis laurentei* **La Touche**, 1921 NCR
[JS *Brachypteryx montana cruralis*]
Laurente's Bush Warbler *Cettia pallidipes laurentei* La Touche, 1921
[Alt. Pale-footed Bush Warbler ssp.]
Pale-chinned Blue Flycatcher ssp. *Cyornis poliogenys laurentei* La Touche, 1921
Streak-breasted Scimitar Babbler ssp. *Pomatorhinus ruficollis laurentei* La Touche, 1921
Chestnut-crowned Warbler ssp. *Seicercus castaniceps laurentei* La Touche, 1922
Blue-throated Barbet ssp. *Cyanops davisoni laurentii* **T. Wells**, 1923 NCR
[JS *Megalaima asiatica davisoni*]

E. P. Laurente (DNF) was a French Customs Officer in Mengtsz, Yunnan, China, where he assisted La Touche (q.v.) in collecting bird specimens. What he collected elsewhere (Szemao in south Yunnan) were also sent to La Touche, who sent his notes to Lord Rothschild (q.v.).

Laurina

Marchioness Doria's Mesia *Leiothrix argentauris laurinae* **Salvadori**, 1879
[Alt. Silver-eared Mesia ssp.; Syn. *Mesia argentauris laurinae*]

Laura Marulli, Marchesa di Doria (1834–1882), was the daughter of Gennaro Marulli, Duca di San Cesareo, and wife of Marcantonio, Marchese di Doria and Principe de Angri.

Lauterbach

Lauterbach's Bowerbird *Chlamydera lauterbachi* **Reichenow**, 1897
[Alt. Yellow-breasted Bowerbird]

Frilled Monarch ssp. *Arses telescopthalmus lauterbachi*
Reichenow, 1897

Carl Adolph Georg Lauterbach (1864–1937) was a German botanist who collected in northern New Guinea. He wrote widely on orchids and led a scientific expedition to New Guinea (1890–1891).

Lavauden

European Robin ssp. *Erithacus rubecula lavaudeni*
Bannerman, 1926 NCR
[JS *Erithacus rubecula witherbyi*]

Louis Lavauden (1881–1935) was a French zoologist and ecologist. He was in Tunisia and explored in the Sahara (1927), and was the Chief Forestry Officer in Madagascar (1928–1931).

Lavender

Miss Lavender Pease's Cisticola *Cisticola aridulus lavendulae* **Ogilvie-Grant** & Reid, 1901
[Alt. Desert Cisticola ssp.]

Lavender Mary Medlicott *née* Pease (1889–1989) was the daughter of Sir Alfred Edward Pease, the collector of the holotype.

Lavery

Zitting Cisticola ssp. *Cisticola juncidis laveryi* **Schodde** & Mason, 1979

Dr Hugh John Lavery (b.1935) is an Australian environmentalist, naturalist and ornithologist. He wrote *Wild Ducks and other Waterfowl in Queensland* (1971).

Lavinia

Hummingbird genus *Lavinia* **Mulsant** & E. Verreaux, 1877
NCR
[Now in *Metallura*]

In Roman mythology Lavinia was the last wife of Aeneas, who named the city of Lavinium after her.

Lavinia (Bowen)

Lavinia's Tanager *Tangara lavinia* **Cassin**, 1858
[Alt. Rufous-winged Tanager]

Lavinia Brown Bowen (later Jones) *née* Davis (1805–1888) was an American painter of lithographs, famed for her natural history studies. Many of these were coloured by her, but drawn by Titian R. Peale (q.v.), who collected specimens as a member of the government-sponsored United States Exploring expedition (1838–1842). Her first husband John T. Bowen (d.1856) was the respected lithographer of Philadelphia who printed many of Audubon's (q.v.) plates. The colouring of lithographs normally involved teams of colourists, usually women, each assigned to fill in one colour of the print only. However, it seems that Lavinia was often the sole colourist as she was so expert in that art. When J. T. Bowen died, Lavinia operated the company, which changed its name from J. T. Bowen's to Bowen's & Co. She married a second time (c.1859) to Robert Henry Jones of Manchester, England.

Lavrov

Black-headed Gull ssp. *Chroicocephalus ridibundus lavrovi*
Zarudny, 1912 NCR; NRM

Professor Sergei Dmitreivich Lavrov (fl.1929) was a Russian zoologist at Omsk (1925–1929).

Lawes

Lawes's Parotia *Parotia lawesii* **E. P. Ramsay**, 1885
[Alt. Lawes's Six-wired Bird-of-Paradise]

The Reverend William George Lawes (1839–1907) was a missionary in New Guinea. He first went to work (1860) on Savage Island (Niue). After a visit to England he sailed (1874) to New Guinea and settled in Port Moresby, where his party became the first permanent European residents. He left for England on holiday (1878), having produced the first book in a Papuan language and the first printed piece of paper in Papua New Guinea. He was in great demand as his knowledge of Papua was unrivalled and he was liked and trusted by the local tribes. He returned to Port Moresby (1881) and acted as interpreter for the Protectorate proclamation (1884). He founded a new training college at Vatorata (1892), serving there (1892–1902). He retired to Sydney (1906). He wrote *Grammar & Vocabulary of Language Spoken by Motu Tribe (New Guinea)* (1896), and also translated the Bible into Motu. He is credited with collecting *Dendrobium lawesii*, a New Guinean orchid. A reptile and a mammal are named after him.

Lawrence, G. N.

Vireo genus *Lawrencia* **Ridgway**, 1886 NCR
[Now in *Vireo*]
Hummingbird genus *Lawrencius* **Boucard**, 1895 NCR
[Now in *Elvira*]

Lawrence's Goldfinch *Carduelis lawrencei* **Cassin**, 1850
Bare-legged Owl *Margarobyas lawrencii* **P. L. Sclater** & **Salvin**, 1868
[Alt. Cuban Screech Owl; Syn. *Gymnoglaux lawrencii*]
Lawrence's Quail Dove *Geotrygon lawrencii* Salvin, 1874
[Alt. Purplish-backed Quail Dove]
Lawrence's Tuftedcheek *Pseudocolaptes lawrencii*
Ridgway, 1878
[Alt. Buffy Tuftedcheek]
Lawrence's Thrush *Turdus lawrencii* **Coues**, 1880
Hummingbird sp. *Amazilia lawrencei* **D. G. Elliot**, 1889 NCR
[Alt. Green-bellied Hummingbird; JS *Amazilia viridigaster*]
Lawrence's Starfrontlet *Coeligena lawrencei* Boucard, 1892
NCR
[Probably a hybrid: *Coeligena torquata* x *Lafresnaya lafresnayi*]

Lawrence's Crested Flycatcher *Myiarchus tuberculifer lawrenceii* **Giraud**, 1841
[Alt. Dusky-capped Flycatcher ssp.]
Song Wren ssp. *Cyphorhinus phaeocephalus lawrencii*
Lawrence, 1863

Lawrence's Warbler *Helminthophaga lawrencii* Herrick, 1874 NCR

[Hybrid: *Vermivora chrysoptera* x *V. cyanoptera*]

Happy Wren ssp. *Pheugopedius felix lawrencii* Ridgway, 1878

Stripe-headed Sparrow ssp. *Peucaea ruficauda lawrencii* Salvin & **Godman**, 1886

Lawrence's Hummingbird *Cynanthus latirostris lawrencei* **Berlepsch**, 1887

[Alt. Broad-billed Hummingbird ssp.]

Jamaican Oriole ssp. *Icterus leucopteryx lawrencii* **Cory**, 1887

Lawrence's Flycatcher *Lathrotriccus euleri lawrencei* **J. A. Allen**, 1889

[Alt. Euler's Flycatcher ssp.]

Forest Thrush ssp. *Cichlherminia lherminieri lawrencii* Cory, 1891

[If *Cichlherminia* is merged with *Turdus*, the name of this taxon must be changed to *Turdus lherminieri montserrati*]

Lawrence's Swift *Chaetura cinereiventris lawrencei* Ridgway, 1893

[Alt. Grey-rumped Swift ssp.]

George Newbold Lawrence (1806–1895) was a businessman involved with wholesale pharmaceuticals. He was an amateur ornithologist helping on Pacific surveys for Baird (q.v.) and Cassin (q.v.), reporting on waterbirds. He then worked voluntarily at the USNM, describing the birds collected on those surveys. He also acted as agent for John Gould (q.v.) in America and was a frequent correspondent. He left his collection of 8,000 skins to the AMNH. (Although it may appear that Lawrence named the Song Wren subspecies after himself, which would have been a major taxonomic *faux pas*, Sclater bestowed the name in manuscript, but, because Lawrence first formally published the name *lawrencii* in this context, he is officially credited as the describer.)

Lawrence, J. L. M.

Common Raven ssp. *Corvus corax laurencei* **Hume**, 1873

John Laird Mair Lawrence (1811–1879), first Baron Lawrence, was a colonial administrator in India (1829–1859), serving in many posts. After the first Sikh War, he became Commissioner (1846) of the newly acquired territory and (1849) after the complete annexation of the Punjab he and his brother, Sir Henry Montgomery Lawrence, reorganised its administration. The Punjab remained quiet during the Indian Mutiny (1857–1858) and, according to the inscription on his memorial plaque in Westminster Abbey, it was he who maintained the peace. He returned to England (1859) but was persuaded to come out of retirement to become Viceroy of India (1864–1869).

Lawry

Southern Brown Kiwi ssp. *Apteryx australis lawryi* **Rothschild**, 1893

(See **Buller**)

Lawson, M.

African Black Swift ssp. *Micropus apus lawsonae* **J. Vincent**, 1933 NCR

[JS *Apus barbatus*]

Miss M. Lawson (DNF) was a British ornithologist at the British Museum of Natural History (1927–1954).

Lawson, W. J.

Fiscal Flycatcher ssp. *Sigelus silens lawsoni* **Clancey**, 1966

Lawson's Batis *Batis poensis occulta* Lawson, 1984

[Alt. West African Batis, Bioko Batis ssp.]

Walter James Lawson (b.1937) is an amateur ornithologist who was based for some time at the Durban Natural Science Museum. His description of the new batis, 'The West African mainland forest dwelling population of *Batis*: a new species', appeared in *Bulletin of the British Ornithologists' Club*. He also wrote such articles as 'Speciation in the forest-dwelling populations of the avian genus *Batis*' (1986) and 'Systematics and evolution in the savanna species of the genus *Batis*' (1987).

Layard

Babbler genus *Layardia* **Blyth**, 1855 NCR

[Now in *Turdoides*]

Shining Parrot genus *Layardiella* **Mathews**, 1917 NCR

[Now in *Prosopeia*]

Layard's Chat *Cercomela tractrac* **Wilkes**, 1817

[Alt. Tractrac Chat; *Saxicola layardi* (Sharpe, 1876) is a JS]

Layard's Parakeet *Psittacula calthropae* Blyth, 1849

[Alt. Emerald-collared Parakeet]

Layard's Flycatcher *Muscicapa muttui* Layard, 1854

[Alt. Brown-breasted Flycatcher]

Layard's Warbler *Sylvia layardi* **Hartlaub**, 1862

[Alt. Layard's Tit-babbler; Syn. *Parisoma layardi*]

Layard's Black-headed Weaver *Ploceus nigriceps* Layard, 1867

[Alt. Village Weaver; Syn. *Ploceus cucullatus nigriceps*]

Layard's Seedeater *Serinus leucopterus* **Sharpe**, 1871

[Alt. Protea Canary; Syn. *Crithagra leucoptera*]

Layard's White-eye *Zosterops explorator* Layard, 1875

[Alt. Fiji White-eye]

Whistling Fruit Dove *Ptilinopus layardi* **D. G. Elliot**, 1878

[Alt. Velvet Dove, Yellow-headed Dove]

Vanuatu Scrubfowl *Megapodius layardi* **Tristram**, 1879

Large Cuckooshrike ssp. *Coracina macei layardi* Blyth, 1866

Lineated Woodcreeper ssp. *Lepidocolaptes albolineatus layardi* **P. L. Sclater**, 1873

Streaked Fantail ssp. *Rhipidura verreauxi layardi* **Salvadori**, 1877

Common Bulbul ssp. *Pycnonotus barbatus layardi* **Gurney**, 1879

Red-banded Flowerpecker ssp. *Dicaeum eximium layardorum* Salvadori, 1880

Layard's Thrush *Turdus poliocephalus layardi* **Seebohm**, 1891

[Alt. Island Thrush ssp.]

Shining Bronze Cuckoo ssp. *Chrysococcyx lucidus layardi* Mathews, 1912

Lesser Shrikebill ssp. *Clytorhynchus vitiensis layardi* Mayr, 1933

Crested Goshawk ssp. *Accipiter trivirgatus layardi* **Whistler**, 1936

Edgar Leopold Layard (1824–1900) was born in Florence, Italy. He spent ten years in Ceylon (Sri Lanka), where he studied the local fauna. He went to the Cape Colony, South Africa (1854), as a civil servant on the staff of the Governor, Sir George Grey. He became Curator of the South African Museum (1855) in his spare time. Later he was Honorary British Consul in New Caledonia. From here, he and his son Edgar Leopold Calthrop Layard (known as Leopold Layard) (b.1848) made collecting trips all over the South-West Pacific. The name *layardorum* honours both father and son. Layard wrote *The Birds of South Africa* (1867), which was later updated by Richard Bowdler Sharpe (q.v.). Two mammals and a reptile are named after him.

Laybourne

Rufous-crowned Sparrow ssp. *Aimophila ruficeps laybournae* Hubbard, 1975

Roxie Collie Simpson Laybourne (1921–2003) pioneered forensic ornithology to help protect airplanes from collisions with birds. She worked for more than 50 years at the USNM, where she developed expertise in identifying dead birds from their feathers to learn what types of birds struck planes. The *New York Times* once dubbed her the Miss Marple of Eider-down. A plane from Boston flew into a flock of starlings and crashed, killing 62 of 72 people aboard, the deadliest ever bird-related air crash. Laybourne, on assignment to the Federal Aviation Administration, identified the birds involved from charred fragments clogging the engines. She solved c.1,000 cases annually. According to the *Los Angeles Times*: 'Laybourne's work was not completely limited to flight paths. In 1986, she helped the FBI nab a husband who had murdered his wife in Alaska and dumped her body in the ocean. Although the body was never recovered, the victim's down coat washed ashore. Laybourne was able to match the coat feathers – from a Chinese duck – to those found in the back of the husband's van, contributing to the conviction'.

Lazdin

Rock Bunting ssp. *Emberiza cia lasdini* **Zarudny**, 1917 NCR
[JS *Emberiza cia par*]

W. Y. Lazdin (V. J. Lasdin) (DNF) was a Russian explorer, collector and zoologist. He is also described as one of the outstanding explorers of the Pamir-Alai mountains (1915) where, like Zarudny (q.v.), he collected bird skins. He wrote in regard to Caspian Sea ichthyology 'Der Bau und die Entwick-elung [*sic*] des Schadels von Exocoetus' (1913).

Leach, J. A.

Australian chat genus *Leachena* **Mathews**, 1916 NCR
[Now in *Epthianura*]

Southern Boobook ssp. *Spiloglaux boobook leachi* Mathews, 1913 NCR
[JS *Ninox boobook boobook*]

Dr John Albert Leach (1870–1929) was an Australian teacher and naturalist. Melbourne University, where he studied part-time, awarded his bachelor's degree (1904), master's (1906) and doctorate (1912). He was President, RAOU (1922–1924) and editor of *Emu* (1914–1924).

Leach, R. H.

Eastern Ground Parrot ssp. *Pezoporus wallicus leachi* **Mathews**, 1912

Noisy Miner ssp. *Manorina melanocephala leachi* Mathews, 1912

Spotted Pardalote ssp. *Pardalotus punctatus leachi* Mathews, 1912 NCR
[JS *Pardalotus punctatus punctatus*]

Yellow-rumped Thornbill ssp. *Acanthiza chrysorrhoa leachi* Mathews, 1912

Richard Howell Walker Leach (1841–1914) was an English collector in Tasmania (1863).

Leach, W. E.

Leach's Storm-petrel *Oceanodroma leucorhoa* **Vieillot**, 1818
Large-tailed Antshrike *Mackenziaena leachii* Such, 1825
Leach's Kingfisher *Dacelo leachii* **Vigors** & **Horsfield**, 1827
[Alt. Blue-winged Kookaburra]
Leach's Black Cockatoo *Calyptorhynchus leachii* **Gould**, 1842 NCR
[Alt. Glossy Black Cockatoo; JS *Calyptorhynchus lathami*]

William Elford Leach (1790–1836) was a British zoologist. He studied medicine, but he did not practice, instead working at the British Museum. Here he became a world-renowned expert on crustaceans and molluscs, also working on insects, mammals and birds. He wrote *The Zoological Miscellany* (1814). He suffered a nervous breakdown (1821) due to overwork and resigned from the museum (1822). He went to Europe to convalesce, and travelled through France, Greece and Italy where he died of cholera. A mammal and four reptiles are named after him.

Leadbeater, B.

Southern Ground-hornbill *Bucorvus leadbeateri* **Vigors**, 1825
Leadbeater's Cockatoo *Cacatua leadbeateri* Vigors, 1831
[Alt. Major Mitchell's Cockatoo; Syn. *Lophochroa leadbeateri*]
Leadbeater's Brilliant *Heliodoxa leadbeateri* **Bourcier**, 1843
[Alt. Violet-fronted Brilliant]

Benjamin Leadbeater (1773–1851) was an English taxidermist and dealer in natural objects, as well as an ornithologist. His son John Leadbeater (1800–1856) was his partner in the busi-ness located in central London. Vigors (q.v.) described the hornbill and cockatoo from specimens in Leadbeater's collection. Bourcier refers only to 'M. (= Monsieur) Lead-beater', so could be referring to either Benjamin or John. He wrote that Leadbeater possessed one of the best collections of hummingbird specimens. Leadbeater's grandson John (q.v.) was also a taxidermist.

Leadbeater, J.

Helmeted Honeyeater *Ptilotis leadbeateri* **McCoy**, 1867 NCR
[Alt. Yellow-tufted Honeyeater; JS *Lichenostomus melanops cassidix*]
Fig parrot sp. *Cyclopsitta leadbeateri* McCoy, 1875 NCR
[Alt. Double-eyed Fig Parrot; JS *Cyclopsitta diophthalma macleayana*]

John Leadbeater (1832–1888) followed in the footsteps of his grandfather Benjamin (q.v.) and became chief taxidermist at the National Museum, Melbourne (1858). Gould (q.v.) originally named the honeyeater *Ptilotis cassidix*, but in the same year it was named *Ptilotis leadbeateri* by Frederick McCoy; Gould was first so his name has priority, but John Leadbeater is also remembered in the name of Leadbeater's Possum *Gymnobelideus leadbeateri*, which appears on the Coat of Arms of the State of Victoria.

Leake

Western Fieldwren ssp. *Calamanthus montanellus leakei* **Mathews**, 1922 NCR; NRM

Bruce Wyborn Leake (1880–1962) was an Australian farmer, field naturalist and aviculturist. He wrote *Reminiscences* (1961).

Lear

Lear's Macaw *Anodorhynchus leari* **Bonaparte**, 1856
[Alt. Indigo Macaw]
Solomons Cockatoo *Lophochroa leari* **Finsch**, 1863 NCR
[JS *Cacatua ducorpsii*]

Edward Lear (1812–1888) was a poet, traveller and artist. He is famous for his nonsense verse, limericks, stories and songs, such as 'The Owl and the Pussycat', but was also one of the most talented 19th-century illustrators of birds. As a teenager and a largely self-taught artist (1827) he prepared posters, leaflets and advertisements and produced *Illustrations of the Family of the Psittacidae or Parrots* in 1832, when he was only 20. He worked (1832–1837) for Lord Stanley, the 13th Earl of Derby (q.v.), who was President of the Zoological Society of London, including as a tutor to all the household's children. He worked as illustrator for many bird book publishers, particularly John Gould (q.v.), with whom he contributed to *The Birds of Europe* (1832–1837), the *Monograph of the Ramphistidae, or Family of Toucans* (1834–1835) and finally the *Monograph of the Trogonidae, or Family of Trogons* (1838). He is believed to be the first bird illustrator who preferred to draw from live specimens, rather than using skins. Lear spent the latter part of his life in southern Europe (1837–1888) and died in Italy.

Lebrun

Patagonian Yellow Finch *Sicalis lebruni* **Oustalet**, 1891

Édouard A. Lebrun (1852–1904) was a French taxidermist who collected in Argentina (1882).

Leclancher

Leclancher's Bunting *Passerina leclancherii* **Lafresnaye**, 1840
[Alt. Orange-breasted Bunting]
Leclancher's Dove *Ptilinopus leclancheri* **Bonaparte**, 1855
[Alt. Black-chinned Fruit Dove]

Charles René Augustin Leclancher (1804–1857) was a surgeon aboard the French ship *La Favourite* (1841–1844) in the Bay of Bengal, Arabia and Japan and on board the *Bayonnaise* (1847–1850).

Le Conte, J.

Le Conte's Sparrow *Ammodramus leconteii* **Audubon**, 1844

Dr John Le Conte (1818–1891) was a physician and scientist who operated a gunpowder factory for the Confederacy during the American Civil War. He was Professor of Physics and Chemistry at the University of Georgia, and later became President of the University of California, Berkeley (1875–1881). Audubon named the sparrow after his friend, 'Dr Le Conte', which probably refers to him, but might refer to his first cousin John Lawrence Le Conte (q.v.). A mammal is named after him.

Le Conte, J. L.

Le Conte's Thrasher *Toxostoma lecontei* **G. N. Lawrence**, 1851
African Dwarf Kingfisher *Ispidina lecontei* **Cassin**, 1856
[Syn. *Ceyx lecontei*]

Dr John Lawrence Le Conte (1825–1883) was an American biologist and the most important American entomologist of the 19th century. He undertook a number of collecting expeditions to the western USA, one with Louis Agassiz (1848) to the Rocky Mountains, and to Central America including Honduras (1857) and Panama (1867). He also visited Europe, Algeria and Egypt. He founded the Entomological Society of America and was co-founder of the National Academy of Science. He was also a physician during the American Civil War. His father John Eatton Le Conte (1784–1860) was also a naturalist and US Army engineer, and some of his writings are addressed to his son who may also have contributed some illustrations. He was first cousin of Dr John Le Conte (q.v.). Two reptiles and two mammals are named after him.

LeCroy

Black-bibbed Cuckooshrike ssp. *Coracina mindanensis lecroyae* **Parkes**, 1971
Green-backed Robin ssp. *Pachycephalopsis hattamensis lecroyae* Boles, 1989
Orange-bellied Euphonia ssp. *Euphonia xanthogaster lecroyana* **Aveledo** & **Perez**, 1994 NCR
[NUI *Euphonia xanthogaster badissima*]
MacGregor's Bowerbird ssp. *Amblyornis macgregoriae lecroyae* Frith & Frith, 1997
Goldenface ssp. *Pachycare flavogriseum lecroyae* **Beehler** & Prawiradilaga, 2010

Mary Kathryn LeCroy *née* Strother (b.1935) is an ornithologist. She was Thomas Gilliard's (q.v.) assistant (1959–1965), and Senior Scientific Assistant in the Department of Ornithology at the AMNH, where she is now a Research Associate and is completing the publication of annotated lists of the avian types. In this time she has undertaken field studies in New Guinea on bird-of-paradise behaviour and has studied the taxonomy of New Guinea and South-West Pacific birds. She has also done fieldwork on terns on the Museum's Great Gull Island Field Station and in Venezuela. She specialises in the avifauna of New Guinea and is a member of the Standing Committee on Ornithological Nomenclature of the International Ornithological Committee and a Fellow of the American Ornithologists' Union. She also has a feather mite named after her.

Ledant

Algerian Nuthatch *Sitta ledanti* **Vieillard**, 1976

Jean-Paul Ledant (b.1951) is a Belgian ecologist and ornithologist. He has carried out surveys and ecological audits of birds and other taxa in a number of African countries including Chad, Cameroon and Algeria. He wrote, with J. P. Jacob, *Liste rouge des espèces d'oiseaux menacées en Algérie* (1982).

Ledeboer

Hardhead ssp. *Aythya australis ledeboeri* **Bartels** & **P. F. Franck**, 1938 NCR
[*Aythya australis* NRM]

Adriaan Johan Marie Ledeboer (1877–1945) was a Dutch coffee plantation manager on Java (1907–1942). He was also a 'conservationist', although some of his methods would not be approved today: e.g. he 'conserved' deer partly by shooting their natural predators, including the rare Javan Leopard. Ledeboer was interned by the Japanese (1942) and died in the internment camp.

Ledoux

Coal Tit ssp. *Periparus ater ledouci* **Malherbe**, 1845
Lesser Spotted Woodpecker ssp. *Dendrocopos minor ledouci* Malherbe, 1855

Captain L. F. Ledoux (fl.1842–at least 1866) was an engineer in the French army at Bône, Algeria (1842).

Lee

Velasquez's Woodpecker ssp. *Melanerpes santacruzi leei* **Ridgway**, 1885

Dr Thomas Lee (DNF) was a regular member of the group of naturalists used by the US Fish Commission on the cruises of their research vessel *Albatross*. He was aboard when they called at Cozumel Island off the Yucatan Peninsula (1885), and again in the Bahamas and the West Indies (1886). A reptile is named after him.

Le Fol

Lesser Racket-tailed Drongo ssp. *Dicrurus remifer lefoli* **Delacour** & **Jabouille**, 1928

Aristide Eugène Le Fol (1878–1967) was a French colonial administrator. He was Acting Governor, New Caledonia (1896–1897), Resident Superior, first in Cambodia (1927–1929), then Laos (1932–1933).

Le Gallen

Vietnamese Cutia *Cutia legalleni* **H. C. Robinson** & **Kloss**, 1919

Maurice Joseph Le Gallen (1873–1956) was a French colonial administrator. He was Acting Resident in Cambodia (1914), Resident of Tonkin (1915–1916) and Governor-General of Cochinchina (Vietnam) (1916–1921).

Legendre

Green-backed Tit ssp. *Parus monticolus legendrei* **Delacour**, 1927

Marcel Legendre (fl.1927) was a French ornithologist who was editor of the journal *L'Oiseau*. He wrote *Oiseaux de Cage* (1952).

Legge, G.

Short-tailed Pipit ssp. *Anthus brachyurus leggei* **Ogilvie-Grant**, 1906

Captain the Honourable Gerald Legge (1882–1915) was a British army officer and explorer who was on the Ruwenzori Expedition (1905–1906). He was killed in action in WW1.

Legge, W. V.

Fairy-wren genus *Leggeornis* **Mathews**, 1912 NCR
[Now in *Malurus*]

Legge's Flowerpecker *Dicaeum vincens* **P. L. Sclater**, 1872
[Alt. White-throated Flowerpecker]
Legge's Hawk Eagle *Nisaetus kelaarti* Legge, 1878

Legge's Baza *Aviceda jerdoni ceylonensis* Legge, 1876
[Alt. Jerdon's Baza ssp.]
Scarlet Robin ssp. *Petroica boodang leggii* **Sharpe**, 1879
Orange-breasted Green Pigeon ssp. *Treron bicinctus leggei* **Hartert**, 1910
Superb Fairy-wren ssp. *Malurus cyaneus leggei* Mathews, 1912
Whiskered Tern ssp. *Chlidonias hybrida leggei* Mathews, 1912 NCR
[NUI *Chlidonias hybrida javanicus*]
Barred Buttonquail ssp. *Turnix suscitator leggei* **E. C. S. Baker**, 1920
Oriental Scops Owl ssp. *Otus sunia leggei* **Ticehurst**, 1923
Bar-winged Flycatcher-shrike ssp. *Hemipus picatus leggei* **Whistler**, 1939
White-rumped Shama ssp. *Copsychus malabaricus leggei* Whistler, 1941
Grey-breasted Prinia ssp. *Prinia hodgsonii leggei* **G. Watson**, 1986 NCR
[JS *Prinia hodgsonii pectoralis*]

William Vincent Legge (1841–1918) was a professional soldier and a keen collector of bird skins. He was born in Van Diemen's Land (Tasmania) but educated in England, France and Germany. He was commissioned in the Royal Artillery

(1862) and served in Melbourne (1867–1868). From there he went to Ceylon (Sri Lanka) where, as Secretary of the Royal Asiatic Society, he reorganised the museum at Colombo and made a large collection of birds. He was the military commander in Tasmania (1883–1890 and 1898–1904), retiring with the rank of Lieutenant-Colonel. Legge was also a founder of the RAOU (1901) and was its first President. He wrote *History of the Birds of Ceylon* (1880), and *Systematic List of Tasmanian Birds* (1887). He also assisted in the compilation of the *List of Vernacular Names for Australian Birds*. The bird skins he collected in Ceylon were presented to the Hobart Museum in Tasmania (1902).

Legters

Northern Barred Woodcreeper ssp. *Dendrocolaptes sanctithomae legtersi* **Paynter**, 1954 NCR
[JS *Dendrocolaptes sanctithomae sanctithomae*]

David Brainerd Legters (1908–2002) was a collector and an American missionary to Yucatan, Mexico.

Leguat

Leguat's Giant *Leguatia gigantea* **Schlegel**, 1858 (hypothetical)

Leguat's Gelinote *Aphanapteryx leguati* **Milne-Edwards**, 1873 EXTINCT
[Alt. Rodriques Blue Rail]
Leguat's Owl *Mascarenotus murivorus* Milne-Edwards, 1873 EXTINCT
[Alt. Rodrigues Owl; note that *Bubo leguati* (**Rothschild**, 1907) is a JS]
Leguat's Starling *Necrospar leguati* **H. O. Forbes**, 1898 (misidentification)
[Based on an albinistic specimen of *Cinclocerthia gutturalis*]

François Leguat of Bresse (1637–1735) was a French naturalist and explorer. Leguat sailed from Amsterdam (1690) to help establish a colony of French Protestant refugees on the island of Réunion in the Indian Ocean. Instead of Réunion, he and seven companions were left on the nearby uninhabited island of Rodrigues (1691). After spending a year on the island, the homesick group constructed a wooden boat to escape to Mauritius. Leguat wrote *Voyage et Avantures en Deux Îles Desertes des Indies Orientales* (1708), which covered his travels to and within Rodrigues, Mauritius, Java and the Cape of Good Hope. The book is notable in containing natural history observations on the now-extinct fauna of Rodrigues. It is recorded that he 'noted that vast herds of the (now-extinct) Rodrigues tortoise seemed to include certain individuals with the responsibility of sentinels, but he wryly commented that it wasn't clear what predator they were looking for, or how the alarm would be sounded, or what effective response the herd might take'. 'Leguat's Giant', or 'Le géant', was depicted as a sort of giant rail, but actually seems to have been inspired by historical descriptions of flamingos. There is no hard evidence of its existence, unless the discovery of a single *Ciconia* bone on Réunion indicates that a stork species might be behind the *Leguatia* accounts.

Lehmann, J.

Red-necked Spurfowl ssp. *Pternistis afer lehmanni* **J. A. Roberts**, 1931 NCR
[NUI *Pternistis afer castaneiventer*]

J. Lehmann (DNF) was a collector in South Africa (1930).

Lehmann(-Valencia)

Brown-throated Parakeet ssp. *Aratinga pertinax lehmanni* **Dugand**, 1943
[Syn. *Eupsittula pertinax lehmanni*]
Black-throated Tody Tyrant ssp. *Hemitriccus granadensis lehmanni* **Meyer de Schauensee**, 1945
Giant Antpitta ssp. *Grallaria gigantea lehmanni* **Wetmore**, 1945

Professor Frederico Carlos Lehmann-Valencia (1914–1974) was a biologist and conservationist who founded a natural history museum in Colombia, using as its basis his own vast collection (1936). He also founded Museo de Ciencias Naturales, Santa Teresita de Cali, Colombia (1963). Other taxa, including three amphibians and a reptile, are named after him.

Le Hunte

Superb Bird-of-Paradise ssp. *Lophorina superba lehunti* **Rothschild**, 1932 NCR
[JS *Lophorina superba minor*]

Sir George Ruthven Le Hunte (1852–1925) was a British colonial administrator. He was Commissioner for the Western Pacific (1883–1887), President of Dominica (1887–1894), Lieutenant-Governor of British New Guinea (Papua New Guinea) (1898–1903), Governor of South Australia (1903–1909), and Governor of Trinidad & Tobago (1909–1916). The holotype was collected when Le Hunte was Governor of British New Guinea, and he was responsible for forwarding it to the BMNH.

Leigh

Leigh's Tit Warbler *Acanthiza chrysorrhoa leighi* **Ogilvie-Grant**, 1909
[Alt. Yellow-rumped Thornbill ssp.]

William Henry Leigh, 2nd Baron Leigh of Stoneleigh Abbey (1824–1905), was in New South Wales when his son-in-law, the Earl of Jersey, was Governor of New South Wales (1891–1893). Leigh presented two specimens of the thornbill to the British Museum (1903). He was a member of the Privy Council and Lord Lieutenant of Warwickshire (1856–1905).

Lekhakun

Grey-eyed Bulbul ssp. *Iole propinqua lekhakuni* **Deignan**, 1954
Hill Blue Flycatcher ssp. *Cyornis banyumas lekhakuni* Deignan, 1956

Dr Bunsong Lekhakun (1907–1992) (also known as Boonsong Lekagul) was a Thai physician, ornithologist and conservationist. Chulalongkorn University, Bangkok, awarded his medical degree (1933). He established Thailand's first poly-clinic (1935). At first a hunter, he became a conservationist as his awareness grew of how Thailand's forests were becoming fragmented. He founded the Association for the Conservation of Thailand (1993). Two mammals, a reptile and an amphibian are named after him.

Lembeye

Cuban Gnatcatcher *Polioptila lembeyei* **Gundlach**, 1858

Juan Lembeye (1816–1889) was a Cuban ornithologist who wrote *Aves de la Isla de Cuba* (1850). He was the first to describe the Bee Hummingbird *Calypte helenae*, the world's smallest bird.

Lemos

White-chested Swift *Cypseloides lemosi* **Eisenmann &
V. Lehmann**, 1962

Dr Antonio José Lemos-Gúzman (1900–1967) was a Colombian physician and historian, Dean of Cauca University when its Natural History Museum opened (1939) and Governor of Cauca (1961–1962).

Lemprière

Palawan Blue Flycatcher *Cyornis lemprieri* **Sharpe**, 1884
Palawan Hornbill *Anthracoceros lemprieri* Sharpe, 1885 NCR
[JS *Anthracoceros marchei*]

Crested Jay ssp. *Platylophus galericulatus lemprieri*
F. Nicholson, 1883

Edward Philip Lemprière (1851–1891) was in British North Borneo (1879–1883) during which period he collected and sent specimens to Sharpe (q.v.), who described him as 'his friend'. He was Private Secretary to the Governor of British North Borneo (1879–1881) and Assistant Resident of the East Coast (1881–1883). He was Private Secretary to the Governor of South Australia (1886–1890), resigning on the grounds of ill health and returning to England.

Lendl

Eurasian Jay ssp. *Garrulus glandarius lendli* **Madarász**,
1907 NCR
[JS *Garrulus glandarius krynicki*]

Dr Adolf Lendl (1862–1943) was a Hungarian zoologist, politician and collector. He received his doctorate (1887) and became a professor (1888), joining the Zoology Department, National Museum of Hungary, Budapest (1890). He was elected to the Hungarian parliament (1901). He was in Anatolia, Turkey (1906) and Argentina (1909). He was Director, Budapest Zoo (1911–1919).

Lenz

Fantail sp. *Rhipidura lenzi* **Blasius**, 1883 NCR
[Alt. Northern Fantail; JS *Rhipidura rufiventris cinerea*]
Tody Flycatcher sp. *Poecilotriccus lenzi* **Berlepsch**, 1884
NCR
[Alt. Rufous-crowned Tody Flycatcher; JS *Poecilotriccus ruficeps*]

Professor Dr Heinrich Wilhelm Christian Lenz (1846–1913) was a German zoologist and arachnologist who visited Colombia (1884). He was Director of the Lübeck Museum.

Leonardina

Bagobo Babbler genus *Leonardina* **Mearns**, 1905

(See **Wood, L. Sr**)

Leopold

Metallic Pigeon ssp. *Columba vitiensis leopoldi* **Tristram**,
1879

Edgar Leopold Calthrop Layard (See **Layard**)

Leopold II

Bare-faced Go-away-bird ssp. *Corythaixoides personatus
leopoldi* **G. E. Shelley**, 1881
White-bellied Kingfisher ssp. *Corythornis leucogaster
leopoldi* **A. J. C. Dubois**, 1905

Léopold II (1835–1909) was King of the Belgians (1865–1909) and established and ruled the Congo Free State (1885–1908).

Léotaud

Chestnut Woodpecker ssp. *Celeus elegans leotaudi*
Hellmayr, 1906

Antoine Léotaud (1814–1867) was a Trinidadian/French physician, chemist and ornithologist. He studied in France and qualified as a physician (1826–1839). He wrote *Oiseaux de l' Île de la Trinidad, (Antilles)* (1866).

Lepcha

Green-backed Tit ssp. *Parus monticolus lepcharum*
R. Meinertzhagen & A. Meinertzhagen, 1926 NCR
[JS *Parus monticolus yunnanensis*]

The Lepcha, or Róng, are the aboriginal inhabitants of Sikkim, north-east India.

Lerch

Lerch's Woodnymph *Thalurania lerchi* **Mulsant &
J. Verreaux**, 1871 NCR
[Hybrid: *Thalurania furcata* x *Chrysuronia oenone*. Syn.
Timolia lerchi]

Mulsant and Verreaux remarked that the collector of this hummingbird was Dr Lerch (DNF), a French botanical explorer in New Granada (a republic in north-west South America, centred on present-day Colombia) for 20+ years from 1847. Little seems to have been recorded about him.

Lerdo

Hummingbird sp. *Thaurnatias lerdi* d'Oca, 1875 NCR
[JS Berylline Hummingbird *Amazilia beryllina*?]

Sebastián Lerdo de Tejada y Corral (1825–1889) was a Mexican jurist and statesman who was President of Mexico

(1872–1876). The true identity of this taxon, said to come from Veracruz, Mexico, is uncertain.

Lereboullet

> Jamaican Thrush sp. *Turdus lereboulleti* **Bonaparte**, 1854 NCR
> [JS *Turdus jamaicensis*]

Dr Dominique Auguste Lereboullet (1804–1865) was a French physician and Professor of Comparative Anatomy and Zoology at the University of Strasbourg. He died of an apoplectic fit.

Leschenault

> Chestnut-headed Bee-eater *Merops leschenaulti* **Vieillot**, 1817
> Leschenault's Forktail *Enicurus leschenaulti* Vieillot, 1818
> [Alt. White-crowned Forktail]
> Greater Sand Plover *Charadrius leschenaultii* **Lesson**, 1826
> Sirkeer Malkoha *Taccocua leschenaultii* Lesson, 1830
> [Alt. Sirkeer Cuckoo; Syn. *Phaenicophaeus leschenaultii*]

> Brown Fish Owl ssp. *Ketupa zeylonensis leschenaultii* **Temminck**, 1820

Jean Baptiste Louis Claude Theodore Leschenault de la Tour (1773–1826) was a French botanist and naturalist to two Kings of France, Louis XVIII (1814–1824) and Charles X (1824–1830). He wrote one of the first descriptions of coconuts and their oil extraction. He was botanist on the voyage of the *Casuarina*, *Géographe* and *Naturaliste*, (1801–1803) collecting widely in Australia (1801–1802). He also collected in Java (1803–1806) and India (1816–1822). He sent many of the plants and seeds he discovered in India to the French island of Réunion for cultivation, including varieties of cotton and sugarcane. Leschenault also visited the Cape Verde Islands, Cape of Good Hope, Ceylon (Sri Lanka), Brazil and the Guianas. A mammal and four reptiles are named after him.

Le Souef

> Buff-banded Rail ssp. *Gallirallus philippensis lesouefi* **Mathews**, 1911

William Henry Dudley Le Souef (1856–1923) was an Australian zoologist. He became Assistant Secretary to the Zoological and Acclimatization Society of Victoria (1874). He collected in India, USA, Singapore, Sumatra, Europe, Japan and New Guinea (1880–1888). He was a founding member of the RAOU and Director of Melbourne Zoological Gardens (1902–1923). A former employee viciously mugged him (1919) for the staff wages, after which his health deteriorated and he suffered a stroke (1922).

Lesson, A.

> Lesson's Motmot *Momotus lessonii* **Lesson**, 1842
> [Alt. Blue-diademed Motmot; formerly *Momotus momota lessonii*]

Probably named after Pierre Adolphe Primevère Lesson (1805–1888). However, René Primevère Lesson, in Latin, dedicated the species to 'Adolpho Lessiano' – not making it clear whether this referred to his brother Pierre Adolphe or his father Adolph(e). As René travelled with his brother, and the latter often collected for him, it is most likely to be he who was honoured. Furthermore, he may well have been generally known as Adolph: Dumont d'Urville referred to him thus. Pierre was a botanist and plant collector in Europe, Africa, Australasia and the Pacific, notably aboard the *Astrolabe* (1826–1829), as ship's surgeon and botanist. He later became professor at a school of naval medicine. Both brothers were talented scientific graphic artists.

Lesson, R. P.

> Negrito genus *Lessonia* **Swainson**, 1832

> White-headed Petrel *Pterodroma lessonii* **Garnot**, 1826
> Lesson's Peacock Pheasant *Polyplectron chalcurum* Lesson, 1831
> [Alt. Bronze-tailed Peacock Pheasant]
> Lesson's Seedeater *Sporophila bouvronides* Lesson, 1831
> Lesson's Bird-of-Paradise *Diphyllodes seleucides* Lesson, 1834 NCR
> [Alt. Magnificent Bird-of-Paradise; JS *Diphyllodes magnificus*]
> Lesson's Euphonia *Euphonia affinis* Lesson, 1842
> [Alt. Scrub Euphonia, Black-throated Euphonia]
> Lesson's Oriole *Icterus chrysater* Lesson, 1844
> [Alt. Yellow-backed Oriole]
> Slaty Monarch *Mayrornis lessoni* **G. R. Gray**, 1846
> Lesson's Parrotlet *Forpus coelestis* Lesson, 1847
> [Alt: Pacific Parrotlet]
> Chat Tyrant sp. *Ochthoeca lessoni* **P. L. Sclater**, 1856 NCR
> [Alt. Rufous-breasted Chat Tyrant; JS *Ochthoeca rufipectoralis rufopectus*]
> Friarbird sp. *Tropidorhynchus lessoni* Gray, 1859 NCR
> [Alt. New Caledonian Friarbird; JS *Philemon diemenensis*]
> Honeyeater sp. *Xanthotis lessoni* **Sharpe**, 1909 NCR
> [Alt. Tawny-breasted Honeyeater; JS *Xanthotis flaviventer*]

> Azure Kingfisher ssp. *Ceyx azureus lessonii* **Cassin**, 1850
> Splendid Glossy Starling ssp. *Lamprotornis splendidus lessoni* **Pucheran**, 1858
> Blue-tailed Emerald ssp. *Chlorostilbon mellisugus lessoni* **E. L. Simon** & **Dalmas**, 1901 NCR
> [JS *Chlorostilbon mellisugus caribaeus*]
> Rufous-crested Coquette ssp. *Lophornis delattrei lessoni* E. L. Simon, 1921

René Primevère Lesson (1794–1849) was a French ornithologist and naturalist of enormous influence and importance. Whilst he was best known as a zoologist, he was also a skilled botanist and pharmacist. He was employed on the *Coquille* (1822) as botanist, and then on the *Astrolabe* (1826–1829) as naturalist and collector. The *Astrolabe*'s surgeon and botanist was René's brother, Pierre Adolphe Lesson (q.v.). Both voyages were in the Pacific and called at many islands, including New Zealand. Lesson named the Blue-throated Hummingbird, *Lampornis clemenciae* for his second wife Clémence (q.v.). He published a considerable number of ornithological texts including *Manuel d'Ornithologie* (1828), *Histoire Naturelle des Oiseaux-Mouches*

(1829), *Histoire Naturelle des Colibris* (1830), *Centurie Zoologique* (1830), *Traité d'Ornithologie* (1831), *Les Trochiides* (1831), *Illustrations d'Ornithologie* (1831) and *Histoire Naturelle des Oiseaux de Paradis et des Epimaques* (1834). He became Deputy Chief Pharmacist (1832), and later (1839) Chief Pharmacist, for the French Navy at Rochefort. He was awarded the Légion d'Honneur (1847). A mammal is named after him.

Lesueur

(See **Sueur**)

Letitia

> Letitia's Thorntail *Discosura letitiae* **Bourcier** & **Mulsant**, 1852
> [Alt. Coppery Thorntail]

Letizia del Gallo Roccagiovine (1848–1863) was the daughter of Princess Julie Charlotte Pauline Zénaïde Laetitia Désirée Bartholomée Bonaparte (1830–1900) and Marquese Alessandro Gaetano Carlo del Gallo Roccagiovine (1826–1892), and thus a granddaughter of the French ornithologist Prince Charles Bonaparte.

Letona

> Band-tailed Pigeon ssp. *Patagioenas fasciata letonai* **Dickey** & **van Rossem**, 1926

Dr Marcos Antonio Letona Espindola (1868–1948) was Secretary of Industry and Agriculture, El Salvador. The describers wrote that he 'spared no effort or courtesy' in assisting their field investigations.

Lett

> Maned Owl *Jubula lettii* **Büttikofer**, 1889

Mr Lett was a resident of Liberia (1911) and we know little more about him other than Büttikofer's (q.v.) words that the owl was named '… after its discoverer Mr Lett, our former landlord and huntsman at Schieffelinsville' and that he was once a servant of Schweitzer, an explorer and ornithological collector from Stettin (Szczecin, Poland). Major Hans Schomburgk, who was in Liberia (1911–1912), recorded in his *Zelte in Afrika. Eine Autobiographische Erzählung* (1931) that he regarded Lett as a skilful hunter and taxidermist but that 'his financial demands were very high'.

Leucippus

> Hummingbird genus *Leucippus* **Bonaparte**, 1850

Leucippus, in Greek mythology, was the son of King Oenomaus. He fell in love with the nymph Daphne and, in order to get close to her, disguised himself as a young woman. He was found out and killed by the nymph's attendants.

Levaillant, F.

> Levaillant's Parrot *Poicephalus robustus* **Gmelin**, 1788
> [Alt. Cape Parrot]
> Levaillant's Tchagra *Tchagra tchagra* **Vieillot**, 1816
> [Alt. Southern Tchagra]

> Levaillant's Barbet *Trachyphonus vaillantii* Ranzani, 1821
> [Alt. Crested Barbet]
> African Darter *Plotus levaillantii* **Lichtenstein**, 1823 NCR
> [JS *Anhinga rufa*]
> Red-winged Francolin *Scleroptila levaillantii* Valenciennes, 1825
> Levaillant's Cuckoo *Clamator levaillantii* **Swainson**, 1829
> Eastern Jungle Crow *Corvus levaillantii* **Lesson**, 1831
> Levaillant's Cisticola *Cisticola tinniens* Lichtenstein, 1842
> Levaillant's Amazon *Chrysotis levaillantii* **G. R. Gray**, 1859 NCR
> [NPRB *Amazona oratrix* (Ridgway, 1887)]
> Levaillant's Ostrich *Struthio bidactylus* (Hypothetical)

> Cape Penduline-tit ssp. *Anthoscopus minutus levaillanti* **Reichenow**, 1905 NCR
> [JS *Anthoscopus minutus minutus*]

François Levaillant (1753–1824) was a French traveller, explorer, collector and naturalist. He was born in Dutch Guiana (Suriname), son of the French Consul. Birds attracted his early interest and he spent much time collecting specimens, and he thus became acquainted with many of Europe's private collectors. He went to the Cape Province of South Africa (1781) in the employ of the Dutch East India Company, the first real ornithologist to visit the area. There he both explored and collected specimens, eventually publishing a 6-volume book *Historie Naturelle des Oiseaux d'Africa* (1801–1806), a classic of African ornithology. Published in Paris it contained 144 colour-printed engravings based on drawings by Barraband (q.v.). Over 2,000 skins were sent to Jacob Temminck (q.v.), who financed this expedition. His son Coenraad Temminck studied and categorised the specimens, incorporating them into the museum at Leiden. A large collection of his specimens was lost when a Dutch ship was attacked and sunk by the English. However, it is rumoured he may have invented a number of birds, even going so far as to create 'new' ones by putting together pieces from various specimens of other species! Levaillant was opposed to the systematic nomenclature of Linnaeus and only gave the new species he discovered French names. 'Levaillant's Ostrich' was named following reports of a dwarf form found in the African interior; it is now regarded as purely hypothetical.

Levaillant, J.

> Levaillant's Woodpecker *Picus vaillantii* **Malherbe**, 1847
> [Alt. Levaillant's Green Woodpecker]

General Jean-Jacques Rousseau Levaillant (1790–1876) was a soldier in Napoleon I's army who took part in the retreat from Moscow (1812) and battles including Leipzig (1814). He was captured at Dresden and kept prisoner until the Napoleonic Wars were over (1815). He was in the campaign in Spain (1823–1828), then served in Algeria (1838–1848) where he was in charge of the battalion responsible for scientific exploration there. He supervised the accumulation of collections for the MNHN. He took part in the siege of Rome and commanded the occupying troops (until 1853), then retired from the army (1854). He was the only one of the four sons of François Le Vaillant (q.v.) to show any interest in natural history and wrote *Histoire Naturelle. Introduction à l'Histoire des Mammifères et des Oiseaux du Nord de l'Afrique, ou Recherches sur les Lois de la Gravitation des Systèmes Naturels par la Reproduction des Germes dans les Milieux Variables* (1851).

Leventis

Camiguin Hawk Owl *Ninox leventisi* Rasmussen *et al.*, 2012

Dr Anastasios (Tasso) Paul Leventis is a Cypriot businessman, conservationist, photographer and Honorary Vice-President of BirdLife International. He is the Cyprus consul in Lagos, Nigeria, where he has worked for many years and where, as Chairman of the A. G. Leventis Foundation, he oversaw the construction and commissioning of the Natural History Museum of the Obafemi Awolowo University and the establishment of the A. P. Leventis Ornithological Research Institute (APLORI) at the University of Jos.

Lever

Magpie Tanager *Cissopis leveriana* **Gmelin**, 1788

Sir Ashton Lever (1729–1788) was an English naturalist and collector on a very big scale. In his youth he collected live birds, eventually amassing 4,000! He founded the Leverian Museum (1760), initially for a collection of shells and fossils. The live birds took up too much space they were given away to friends and replaced with stuffed ones instead. His collection included many curios, including a conch shell containing a swallow's nest and a swallow's nest built on the suspended corpse of an owl. He became acquainted and corresponded with Gilbert White (q.v.), although the relationship became strained as White regarded Lever as not having a proper background in natural history (having trained in law). Lever was also very interested in archery and was the first president of the Toxophilite Society.

Levraud

Rusty-flanked Crake *Laterallus levraudi* **P. L. Sclater** & **Salvin**, 1868

Léonce Levraud (DNF) was French consul at Guayaquil, Ecuador (1840), French Consul-General of France in Chile (1848) and French Consul in Venezuela (1856–1857). He collected birds in the vicinity of Caracas, sending specimens to the MNHN in Paris.

Levy

Veery ssp. *Catharus fuscescens levyi* **A. R. Phillips**, 1991
[May be a synonym of *Catharus fuscescens salicicola*]

Seymour H. Levy (b.1922) is an ornithologist and conservationist. He worked for the Idaho State Fish and Game Department (1949). He lived Arizona (1984–1992) and now lives in Indiana. He wrote *Summer Birds in Southern Idaho* (1950). Phillips stated that he had named the bird after Levy '… in recognition of his ill-rewarded contributions to ornithology and conservation, and of his help to Thomas D. Burleigh and myself'.

Lewin

Rail genus *Lewinia* **G. R. Gray**, 1855
Whistler genus *Lewinornis* **Mathews**, 1913 NCR
[Now in *Pachycephala*]

Lewin's Rail *Lewinia pectoralis* **Temminck**, 1831
Lewin's Honeyeater *Meliphaga lewinii* **Swainson**, 1837
Lesser Lewin Honeyeater *Meliphaga notata* **Gould**, 1867
[Alt. Yellow-spotted Honeyeater]

John William Lewin (1770–1819) was an English naturalist and engraver. He went to Sydney, Australia (1800) and collected widely there until his death. His father (some sources say his older brother), William Lewin, was the author of a 7-volume work, *Birds of Great Britain*. Lewin accompanied James Grant on his survey expeditions to the Bass Strait and then to the Hunter River. In Sydney he earned a meagre living as a portrait artist. Governor Macquarie appointed Lewin to the position of City Coroner (1814). He also accompanied Macquarie and made drawings during the construction of the road across the Blue Mountains. Macquarie commissioned Lewin to draw plants collected by the Surveyor-General, Henry Oxley, when exploring the country beyond Bathurst, the Liverpool Plains and New England District. As well as natural history, Lewin also painted landscapes and portraits of Aboriginals. He wrote *Prodromus Entomology – Natural History of Lepidopterous Insects of New South Wales* and *Birds of New Holland* (1808) the 1813 edition of which was the first illustrated book to be engraved and printed in Australia. He also wrote *A Natural History of the Birds of New South Wales* (1838).

Lewis, J.

Lewis's Silver Pheasant *Lophura nycthemera lewisi* **Delacour** & **Jabouille**, 1928

John Spedan Lewis (1885–1963) was a British businessman who took over his father's shops and turned them into the nationwide, employee-owned, John Lewis Partnership. He owned an important bird collection. He financed an expedition to Vietnam (1928) for his friend Delacour (q.v.) that sought to collect Edwards's Pheasants *Lophura edwardsi*. It was successful and (1929) Lewis bred chicks from birds obtained during this enterprise. A mammal is named after him.

Lewis, M.

Lewis's Woodpecker *Melanerpes lewis* **G. R. Gray**, 1849

Captain Meriwether Lewis (1774–1809) was one half of the 'Lewis and Clark' duo, whose famous expedition collected the woodpecker holotype (1805) near Helena, Montana. (It is now at Harvard and may be the only bird specimen left from that expedition.) Lewis was chosen to lead the expedition by President Thomas Jefferson; he was the latter's private secretary at the time. Jefferson wrote, 'It was impossible to find a character who, to a complete science in botany, natural history, mineralogy & astronomy, joined the firmness of constitution & character, prudence, habits adapted to the woods, & a familiarity with the Indian manners & character, requisite for this undertaking. All the latter qualifications Capt. Lewis has.' Before becoming Jefferson's secretary he grew up in the country, managed the family plantation, and spent time in the army as an ordinary soldier and then an officer. Lewis chose Clark (q.v.), a friend he had made whilst in the army, to accompany him. The famous journey of exploration took a year and a half and covered more than 4,000 miles to the Pacific Ocean. Lewis was fascinated with the Native Americans, plants, animals, fossils, geological formations, topography and other facets of the trip, all of which he recorded in his journal entries. As a reward for his success he was appointed to the Governorship of the Louisiana Territory. Lewis began a journey to Washington (1809) to clear his name, having been publicly accused of misusing public money, but in a Tennessee inn he met his death from two gunshot wounds to the head and chest, and it is still not known whether this was murder or suicide. As he was bi-polar, suicide appears more likely, although his family believed it was murder. A mammal is named after him.

Lewis, T. H.

Claret-breasted Fruit Dove ssp. *Ptilinopus viridis lewisii* **E. P. Ramsay**, 1882

T. H. Lewis (1854–1917) was a British naval surgeon and collector in the Pacific. He was on board HMS *Cormorant* in New Zealand waters (1882).

Lewis, W. S.

Rusty-fronted Barwing ssp. *Actinodura egertoni lewisi* **Ripley**, 1948

Wilmarth Sheldon 'Lefty' Lewis (1895–1979) was an American scholar and bibliophile. He was a keen collector of anything associated with Horace Walpole. He gave the collection to Yale of which he was a trustee. Ripley wrote that Lewis had 'encouraged and assisted' him on numerous occasions.

Leybold

Leybold's Firecrown *Sephanoides fernandensis leyboldi* **Gould**, 1870 EXTINCT
[Alt. Juan Fernandez Firecrown ssp.]

Friedrich Leybold (1827–1879) was a German botanist, pharmacist and collector. He arrived in Chile (c.1870) and visited the Juan Fernandez archipelago. Here he collected the hummingbird on Isla Alejandro Selkirk. The subspecies was last recorded 1908.

Leyland

Leyland's Quail *Colinus leucopogon leylandi* T. J. Moore, 1859
[Alt. Spot-bellied Bobwhite ssp.]

Joseph Leyland (b.1820) was an English natural history dealer in Liverpool who collected in South Africa (1848–1852), Honduras, Belize and Guatemala (1854–1859).

L'Herminier, F. J.

Forest Thrush genus *Cichlherminia* **Bonaparte**, 1854

Forest Thrush *Cichlherminia lherminieri* **Lafresnaye**, 1844
[Syn. *Turdus lherminieri*]

Dr Ferdinand Joseph L'Herminier (1802–1866) was a French osteologist, botanist, zoologist and ornithologist, and Director of the hospital on Guadeloupe. His father was Félix Louis L'Herminier (q.v.). All their notes and specimens were destroyed in an earthquake (1843). The genus name is made up of his surname combined with the Greek word for thrush: *kikhle*.

L'Herminier, F. L.

Guadeloupe Woodpecker *Melanerpes herminieri* **Lesson**, 1830
Audubon's Shearwater *Puffinus lherminieri* Lesson, 1839

Félix Louis L'Herminier (1779–1833) was a French pharmacist and naturalist who left France (1794) for America and then settled in Guadeloupe in the West Indies. He was at one time the Royal Pharmacist and Naturalist. He also sent specimens to Say (q.v.) at the Academy of Natural Sciences of Philadelphia. He was sent into exile from Guadeloupe (1815) as a result of disturbances on the island. He went to Charleston in South Carolina and eventually settled on the island of Saint-Barthélemy, US Virgin Islands, which, at that time belonged to Denmark. One of his ten children was Ferdinand Joseph L'Herminier (q.v.). Four reptiles are named after him.

L'Huys

Chinese Monal *Lophophorus lhuysii* **Geoffroy Saint-Hilaire & J. Verreaux**, 1866

White Eared Pheasant ssp. *Crossoptilon crossoptilon drouynii* **J. Verreaux**, 1868

Édouard Drouyn de L'Huys (1805–1881) was a French diplomat and statesman. He was Minister for Foreign Affairs several times (1848–1866) and President of La Société Nationale d'Acclimation (1861–1881). He fell out with Napoleon III for advocating an alliance with Austria against Prussia. (See also **Drouyn**).

Lichtenstein

Woodpecker genus *Lichtensteinipicus* **Bonaparte**, 1854 NCR
[Now in *Mulleripicus*]

Lichtenstein's Kingbird *Tyrannus melancholicus* **Vieillot**, 1819
[Alt. Tropical Kingbird]
Lichtenstein's (Desert) Finch *Rhodospiza obsoleta* Lichtenstein, 1823
[Alt. Desert Finch]
Lichtenstein's Sandgrouse *Pterocles lichtensteinii* **Temminck**, 1825
Lichtenstein's Oriole *Icterus gularis* **Wagler**, 1829
[Alt. Altamira Oriole]
Lichtenstein's Foliage-gleaner *Philydor lichtensteini* **Cabanis & Heine**, 1859
[Alt. Ochre-breasted Foliage-gleaner]
[NB. 'Lichtenstein's Foliage-gleaner' is sometimes used for *P. erythropterum*]
Guan sp. *Penelope lichtensteinii* **G. R. Gray**, 1860 NCR
[Alt. Band-tailed Guan; JS *Penelope argyrotis*]

Lichtenstein's Noble Macaw *Diopsittaca nobilis cumanensis*
Lichtenstein, 1823
[Alt. Red-shouldered Macaw ssp.]
Berylline Hummingbird ssp. *Amazilia beryllina lichtensteini*
R. T. Moore, 1950

Martin Heinrich Carl von Lichtenstein (1780–1857) was the German traveller and ornithologist head of the Berlin Zoological Museum (1813). He initially studied medicine, then travelled widely in South Africa (1802–1806), becoming personal physician to the Dutch Governor of the Cape of Good Hope. He wrote *Reisen in Sudlichen Africa* (1810) and became Professor of Zoology at the University of Berlin (1811) and Director of its Zoological Museum (1813). Along with Alexander von Humboldt (q.v.) he planned the creation of the Berlin Zoo (1815), but it took years for building to start (1841) and to open (1844). Lichtenstein studied many species sent to the museum by others and 'whilst he gave every species, or what he judged to be a species, a name, this was done without consulting the recent English and French literature. His only aim was to give the specimens in question a distinguishing mark for his personal needs. These names were used in Lichtenstein's registers and reappeared on the labels of the mounted specimens, but only exceptionally were they published by himself in connection with a scientific description'. This caused much unnecessary confusion and trouble to others! Two mammals and five reptiles are named after him.

Lidth

Lidth's Jay *Garrulus lidthi* **Bonaparte**, 1850
[Alt. Amami Jay, Purple Jay]

Theodoor Gerard van Lidth de Jeude (1788–1863) was a Dutch zoologist, and Director of the Rijks Veeartsenijschool (National Veterinary School) (1821–1826). He published extensively, mainly on veterinary matters, including an account of the bone structure of horses' heads.

Lighton

Black-headed Apalis ssp. *Apalis melanocephala lightoni*
J. A. Roberts, 1938

Norman Charles Kingsley Lighton (1904–1981) was a South African architect and bird artist. He was seconded from his architectural post at the Ministry of Public Works (1936) to the Transvaal Museum, Pretoria, to illustrate Roberts's (q.v.) *The Birds of South Africa* (1940).

Lila

Tawny Frogmouth ssp. *Podargus strigoides lilae* **Deignan**,
1951 NCR
[JS *Podargus strigoides phalaenoides*]

Mrs Lila Mary Mayo *née* Saxelby (fl.1870–at least 1931) married William McArthur Mayo, a commercial traveller, and emigrated from England to Brisbane, Queensland. She wrote an article called 'A Queensland sanctuary', which appeared in *Emu* (1931). Their daughter Daphne (1885–1982) was one of Australia's most famous sculptors.

Lilford

Lilford's Woodpecker *Dendrocopos leucotos lilfordi* **Sharpe**
& Dresser, 1871
[Alt. White-backed Woodpecker ssp.]
Lilford's Crane *Grus grus lilfordi* Sharpe, 1894
[Alt. Common Crane ssp.]

Lord Lilford or Thomas Littleton Powys, fourth Baron Lilford (1833–1896), had an early interest in animals. While at school (Harrow) he kept a small menagerie, and at university (Christ Church, Oxford) a larger one. He was plagued with gout throughout his life but, whenever his health allowed, he spent all his spare time travelling to study animals, especially birds. He went to the Isles of Scilly, Wales and Ireland (1853) and met Edward Clough Newcome, the best falconer of his day, and took up falconry. He joined his county militia (1854) and served in Dublin and Devonport (1855). He cruised the Mediterranean with Hercules Rowley (1856–1858) and visited Spain and the Mediterranean frequently (1864–1882) and rediscovered Audouin's Gull *Larus audouinii*. He was one of the founders of the BOU (1858) and its President (1867) and the first President of Northamptonshire Natural History Society. Here he successfully re-introduced the Little Owl *Athene noctua* into England (1888) already attempted in Kent by Meade-Waldo (q.v.). His aviaries at Lilford were noted for birds of prey. He wrote the 7-volume *Coloured Figures of the Birds of the British Islands* that was finished by Salvin posthumously. He wrote many papers on ornithology in various journals and the 2-volume *Notes on the Birds of Northamptonshire and Neighbourhood* (1895).

Lilian (Baldwin)

Lilian's Meadowlark *Sturnella magna lilianae* **Oberholser**,
1930
[Alt. Eastern Meadowlark ssp.]

Mrs Lilian Converse Hanna Baldwin (1852–1948) was the wife of Samuel Prentiss Baldwin (q.v.), an American patron of natural history. The collection of birds she presented to the Cleveland Museum contained a specimen of this form of meadowlark. The meadowlark has been regarded as a race of Eastern Meadowlark.

Lilian (Sclater)

Lilian's Lovebird *Agapornis lilianae* **G. E. Shelley**, 1894
[Alt. Nyasa Lovebird]

Lilian Elizabeth Lutley Sclater (1863–1957) was a British naturalist whose brother, Bertram Lutley Sclater (q.v), was on an expedition to Nyasaland (now Malawi) as Harry Johnston's (q.v.) right-hand man, when Alexander Whyte (q.v) collected the lovebird. Their father was a long-term associate of Shelley (q.v.). She married twice, to Harrington Morgan (1909) and Douglas Walter Campbell (1920).

Lilian (Witherby)

Middle Spotted Woodpecker ssp. *Dendrocopos medius*
lilianae **Witherby**, 1922 NCR
[NUI *Dendrocopos medius medius*]

Mrs Horatia Lilian Capper Witherby *née* Gillson (1883–1963) was the wife of the describer, Henry Forbes Witherby (q.v.).

Lilith

Lilith Owlet *Athene noctua lilith* **Hartert**, 1913
[Alt. Syrian Little Owl]

In Jewish mythology Lilith was a she-demon. In later folklore she became Adam's first wife. The only mention of Lilith in the Bible is found in *Isaiah* 34:14 – 'there too Lilith shall repose.' Apparently finding the term hard to translate, many Bibles rendered the passage as 'the screech owl also shall rest there'.

Lillie

Sapphire-bellied Hummingbird *Lepidopyga lilliae* **W. Stone**, 1917

Mrs Lillie Mary Stone *née* Lafferty (1872–1940) was a patron of the AOU and wife of the describer, ornithologist Dr. Witmer Stone (see **Stone, W.** and **Witmer**).

Lillo

Lillo's Canastero *Asthenes sclateri lilloi* **Oustalet**, 1904
[Alt. Puna Canastero ssp.]

Miguel Lillo (1862–1931) was an Argentinian naturalist. He was primarily a botanist, but had a wide, largely self-taught, knowledge of other sciences. He was an all-rounder interested in classical literature and languages, as well as science, and a life-long meteorological note-taker. His career was in teaching, research and administration of scientific institutions. He retired from teaching (1918), but remained as Honorary Director of the Museum of Natural History, University of Tucumán. Most of his natural history work was in his home province of Tucumán, although he undertook a study tour of Europe as a young man. He wrote a book on Tucumán's animals, concentrating on birds, which contained descriptions of new species (1905). He also wrote widely on the botany of Tucumán and published a work on Argentina's trees (1910). In December 1930, a few months before his death, Lillo donated all his property, including an extensive library, to the University of Tucumán. Lillo is also commemorated in the binomials of many plants.

Lima

Ochraceous Piculet *Picumnus limae* **Snethlage**, 1924

João Leonardo Lima (1874–1936) was a Brazilian zoologist. He started his working life as an assistant in the Ministry of the Interior but transferred to the museum of Pará (1895). He became preparator (1905), collector, and then collector-naturalist (1925). He only had two articles published, one on birds and the other on bats. An amphibian is named after him.

Limborg

Violet Cuckoo *Chrysococcyx limborgi* **Tweeddale**, 1877 NCR
[JS *Chrysococcyx xanthorhynchus*]

Gustaf Arthur Ossian Limborg (1848–1908) was a Swedish novelist, journalist, ship's captain and poet who collected in Burma (Myanmar) (1877). An amphibian is named after him.

Lina

Lina's Sunbird *Aethopyga linaraborae* Kennedy, Gonzales & Miranda, 1997

Philippine Trogon ssp. *Harpactes ardens linae* **Rand** & **Rabor**, 1959
Ruddy Kingfisher ssp. *Halcyon coromanda linae* Hubbard & **duPont**, 1974

Lina Rabor *née* Florendo (d.1997) was the wife of Dioscoro (Joe) Rabor (1911–1996) (q.v.), and accompanied him on all his research trips. Her husband first found the bird (1965) but seems to have misidentified it as the Apo Sunbird *Aethopyga boltoni*. Eventually the specimens, with mislabelled tags, made their way to the USNM and the Field Museum of Natural History in Chicago. Senior describer Kennedy noted that the 'new' sunbird had iridescent patches of blues, purples and emerald on its forehead, tail, wings and ears, which the Apo Sunbird lacks.

Linchi

Linchi Swiftlet *Collocalia linchi* **Horsfield** & F. Moore, 1854
[Alt. Cave Swiftlet]

Although the name *linchi* gives the impression that this bird is named after a person called Linch, it is in fact a Javanese word for a swiftlet.

Lincoln

Lincoln's Sparrow *Melospiza lincolnii* **Audubon**, 1834

Thomas Lincoln (1812–1883) was an American naturalist. He went on a trip with Audubon to Labrador, aged just 21, where he shot the bird that now bears his name. Audubon originally called it 'Tom's Finch', and it was the only new bird discovered on that expedition. Another expedition member described him as 'quiet, reserved, sensible, practical and reliable.' Back in his home state of Maine Lincoln became a farmer and noted local abolitionist. He later said of Audubon 'a nice man, but as Frenchy as thunder.'

Linden, C.

Pearly-breasted Cuckoo *Coccyzus lindeni* **J. A. Allen**, 1876 NCR
[JS *Coccyzus euleri*]

Professor Charles Linden (1832–1888) was a self-taught American naturalist born in Breslau, Germany (Wroclaw, Poland). He lived in Buffalo, New York (1866–1888), where he was Custodian of the Natural Science Museum (1866–1873) and taught science at the local high school (1866–1888). During summer holidays he collected in places as far apart as Florida, Brazil, Labrador and the West Indies. He wrote 'On the domestication of some of our wild ducks' (1882).

Linden, J. J.

Linden's Helmetcrest *Oxypogon lindenii* **Parzudaki**, 1845
[Alt. White-bearded Helmetcrest; Syn. *Oxypogon guerinii lindenii*]

Jean Jules Linden (1817–1898) was a Luxembourgian botanist who collected and dealt in exotic plants, and wrote *Catalogue du Plantes Exotiques*. He travelled extensively and collected plants in South America (1835–1837) at just 18, with a second expedition to Central America (1838–1841). He was appointed the first scientific Director of the Zoological Gardens of Brussels (1851), then gave up some of his duties and became Head of Botany (1852). When his replacement was dismissed Linden again became scientific director (1856–1861). He went into commercial horticulture and kept a nursery where he cultivated many exotics, which were sent to him. He is often regarded as the man who, in his day, introduced a passion for orchids into Belgium, France and beyond. The Oxford University Herbarium holds much of his collection. A number of Latin American rainforest plants were named in his honour. The 17-volume *Lindenia: Iconographie des Orchidées*, originally published as a periodical, was edited by Emile Rodigas and Linden's son Lucien (1853–1940) in his father's honour.

Lindermayer

Little Egret *Herodias lindermayeri* **C. L. Brehm**, 1855 NCR
[JS *Egretta garzetta garzetta*]

Dr Anton Ritter von Lindermayer (1808–1868) was a German physician and ornithologist who collected in Greece (1832–1868).

Lindo

Lindo's Ringneck Parrot *Barnardius barnardius lindoi* **S. A. White**, 1916 NCR
[Hybrid between *Barnardius zonarius zonarius* and *B. zonarius barnardi*]

John William Lindo (DNF) was an Australian stock farmer and collector in the Flinders Ranges.

Lindsay, H. H.

Spotted Kingfisher *Actenoides lindsayi* **Vigors**, 1831

Hugh Hamilton Lindsay (1802–1881) was an English naturalist and collector. He was an employee of the British East India Company and was sent by them (1832) from Macao, aboard the *Lady Amherst*, to report on the possibilities of opening trade with China via Shanghai. He reported (correctly) that Shanghai had great potential, but the Chinese authorities continued to insist that trade with foreigners could only take place in Canton (now Guangzhou). He published a number of works on China (1836–1840). He was a Member of Parliament at Westminster (1841–1847).

Lindsay, W. P.

Malagasy Paradise-flycatcher *Terpsiphone lindsayi* **Nicoll**, 1906 NCR
[JS *Terpsiphone mutata pretiosa*]

Walter Patrick Lindsay (1873–1936) was a civil engineer who was on Mayotte, Comoro Islands (1905), with the describer Michael Nicoll (q.v.). The latter wrote that Lindsay gave 'valuable assistance in collecting birds during the voyage.' He was a captain in the Middlesex Regiment and was wounded in WW1. He was living in London in 1928.

Linné

Woodpecker genus *Linneopicus* **Malherbe**, 1850 NCR
[Now in *Melanerpes*]

Eclectus Parrot *Eclectus linnaei* **Wagler**, 1832 NCR
[JS *Eclectus roratus*]
Linnaeus's Emerald *Thaumatias linnaei* **Bonaparte**, 1854 NCR
[Alt. Glittering-throated Emerald; JS *Amazilia fimbriata*]
Common Greenshank *Glottis linnei* Malm, 1877 NCR
[JS *Tringa nebularia*]

Clay-coloured Thrush ssp. *Turdus grayi linnaei* **A. R. Phillips**, 1966

Carl Linné (1707–1778) is now much better known by the latinised form of his name, Carolus Linnaeus (or just Linnaeus). Late in life (1761) he was ennobled and so could call himself Carl von Linné. In the natural sciences he was undoubtedly one of the great heavyweights of all time, ranking with Darwin (q.v.) and Wallace (q.v.). Primarily as a botanist, he nevertheless invented the binomial system, published in *Systema Naturae*, that is still in use today, albeit with modifications, for naming, ranking and classifying all living organisms. He entered Lund University (1727) to study medicine and transferred to Uppsala University (1728). At that time botanical study was part of medical training. His first expedition was to Lapland (1732); then he mounted an expedition to central Sweden (1734). He went to the Netherlands (1735) and finished his studies as a physician there before enrolling at Leiden University. He returned to Sweden (1738), lecturing and practising medicine in Stockholm. Appointed Professor at Uppsala (1742) he restored the university's botanical garden. He bought the manor estate of Hammarby outside Uppsala, where he built a small museum for his extensive personal collections (1758). This house and garden are now run by Uppsala University. His son, also Carl, succeeded to his professorship at Uppsala, but was never noteworthy as a botanist. When Carl the Younger died (1783) with no heirs, his mother and sisters sold the elder Linnaeus's library, manuscripts and natural history collections to the English natural historian Sir James Edward Smith (q.v.), who founded the Linnean Society of London to take care of them. Surprisingly few taxa have been named after Linnaeus, and even fewer which are still regarded as valid, but four reptiles, three mammals and two amphibians are among them.

Linton

Orange-banded Flycatcher *Nephelomyias lintoni* **Meyer de Schauensee**, 1951
[Syn. *Myiophobus lintoni*]

Morris Albert Linton (1887–1966) was an American actuary and businessman. He graduated with a master's degree at Haverford College (1910) and as an actuary at Zurich (1914). He worked for the Prudential Mutual Life Insurance Company of Philadelphia (1909–1966), having retired as Chairman (1956); he remained on the board until his death. Wildlife photography was one of his many interests.

Lipfert

Singing Honeyeater ssp. *Meliphaga virescens lipferti*
Mathews, 1942 NCR
[JS *Lichenostomus virescens forresti*]

Herman Franz Otto Lipfert (1864–1942) was a German immigrant to Western Australia (1892). He was employed as a collector and taxidermist by the Western Australian Museum, Perth (1894–1940). He wrote 'Notes on the birds of Crawley, Perth in the early 'nineties' (1937).

Lippens

Spotted Ground Thrush ssp. *Zoothera guttata lippensi*
Prigogine & Louette, 1984

Count Leon Lippens (1911–1986) was a Belgian ornithologist. After obtaining a law degree he went to the Belgian Congo (Zaire) with the intention of managing a family plantation, but became involved in the establishment of Albert National Park, becoming Warden. He returned to Belgium (1936) to marry and then published several books on his observations. Following his father into liberal politics he was Mayor of Knokke (1947–1966) and established the Zwin Nature Reserve there (1952), the first in Belgium. He was founder and president of the Belgian Society of Nature and Bird Sanctuaries. He co-wrote *Les Oiseaux du Zaïre* (1976).

Liriope

Mountain Velvetbreast ssp. *Lafresnaya lafresnayi liriope*
Bangs, 1910

In Greek mythology, Liriope was a naiad (water-nymph) and the mother of Narcissus.

Lisette

Purple-naped Sunbird ssp. *Hypogramma hypogrammicum lisettae* **Delacour**, 1926

Madame Elise 'Lisette' Pasquier *née* Pasquier (1888–1962) was both the niece and the wife of Pierre Marie Antoine Pasquier (q.v.).

Lister

Great Frigatebird ssp. *Fregata minor listeri* **Mathews**, 1914 NCR
[JS *Fregata minor minor*]

Joseph Jackson Lister (1857–1928) joined HMS *Egeria* as a volunteer naturalist on the ship's voyage to Christmas Island (1887–1888). He wrote 'On the natural history of Christmas Island in the Indian Ocean' (1888). Among other taxa a reptile is named after him.

Littaye

Melanesian Whistler ssp. *Pachycephala caledonica littayei*
Layard, 1878

Layard provided no etymology but probably had in mind Édouard William Marie Littaye (1833–1917) a French pioneer photographer and an administrator in the French colonial service. He worked in Cochinchina (Vietnam) (1866–1870), then New Caledonia, as Director of the Interior Department (1875–1879), and Senegal, and finally as direcrtor of the French naval base at Dunkirk until his retirement (1896). A number of photographs that he took in New Caledonia have survived.

Littler

Australian Robin genus *Littlera* **Mathews**, 1912 NCR
[Now in *Petroica*]

Striated Heron ssp. *Butorides striata littleri* Mathews, 1912
[SII *Butorides striata macrorhyncha*]
Southern Emu-wren ssp. *Stipiturus malachurus littleri*
Mathews, 1912

Frank Mervyn Littler (1880–1922) was an Australian accountant, entomologist, ornithologist and naturalist. He wrote *A Handbook of the Birds of Tasmania and its Dependencies* (1910).

Livingston, C. C.

Dusky Bush Tanager ssp. *Chlorospingus semifuscus livingstoni* **Bond** & **Meyer de Schauensee**, 1940

C. Carey Livingston (DNF) was an American ornithologist at the Academy of Natural Sciences, Philadelphia, where the describers worked.

Livingston, G. M.

Grey-sided Laughingthrush ssp. *Garrulax caerulatus livingstoni* **Ripley**, 1952

Gerald Moncrieffe Livingston (1883–1950) was a 'sportsman of New York' and father-in-law of the describer Ripley (q.v.). He seems to have been a keen breeder of pedigree dogs.

Livingstone

Livingstone's Turaco *Tauraco livingstonii* **G. R. Gray**, 1864
Livingstone's Flycatcher *Erythrocercus livingstonei*
G. R. Gray, 1870

Rev. Charles Livingstone (1821–1873) was the younger brother of David Livingstone, the great Scottish doctor, missionary and explorer. Charles accompanied David, as his mission's secretary, on the second Zambezi expedition, which sailed from Liverpool on HMS *Pearl* (March 1852) to the mouth of the river. The party, which included Dr (afterwards Sir) John Kirk (q.v.), ascended the river in a steam launch, the *Ma-Robert*, reaching Tete in September. Charles Livingstone and Kirk both became ill and returned to England (1863). The Livingstones were co-authors of *Narrative of an Expedition to the Zambesi and Its Tributaries* (1866).

Lizano

Lizano's Antpitta *Hylopezus perspicillatus lizanoi* **Cherrie**, 1891
[Alt. Streak-chested Antpitta ssp.]

Joaquín Lizano Gutiérrez (1829–1901) was a Costa Rican politician who was minister of state and Vice President (1890–1894).

Llane

Bougainville Thicketbird *Megalurulus llaneae* **Hadden**, 1983

Llane Hadden is a nurse. Llane and Don Hadden (q.v.) lived on the island of Bougainville (Papua New Guinea), where Don taught at Panguna International School (1976–1980 and 1999–2002). They now live in an Aboriginal community, south of Alice Springs, Northern Territory, Australia.

Lloyd, C. G. H.

Crested Bellbird ssp. *Oreoica gutturalis lloydi* **Mathews** 1917 NCR
[JS *Oreoica gutturalis gutturalis*]

Charles Griffiths Hamilton Lloyd (fl.1888–at least 1912) assisted Thomas Carter (q.v.) to make a collection of birds on Dirk Hartog Island, Shark Bay, Western Australia. He lived near New Norfolk, Tasmania, where he collected 121 species. He also discovered fossil remains in the area (1858). Lloyd acted as manager for J. Nicholas, who was the lessee of the island.

Lloyd, W.

Lloyd's Bushtit *Psaltriparus lloydi* **Sennett**, 1888 NCR
[Hybrid population of *Psaltriparus minimus plumbeus* x
  P. minimus dimorphicus]

William Lloyd (1854–1937) was born in Ireland but moved to Texas (1876). He was employed as a collector and obtained the bushtit holotype (1887) in Limpia Canyon near Fort Davis.

Loddiges

Hummingbird genus *Loddigesia* **Bonaparte**, 1850

Loddiges's Spatuletail *Loddigesia mirabilis* **Bourcier**, 1847
[Alt. Marvellous Spatuletail]

Loddiges's Plovercrest *Stephanoxis lalandi loddigesii*
  Gould, 1831
[Alt. Plovercrest ssp.]

George Loddiges (1786–1846) was a British naturalist, botanical artist, plant dealer, and taxidermist specialising in hummingbirds. He shared this obsession with Gould (q.v.), who became a lifelong friend and said of him: 'This gentleman and myself were imbued with a kindred spirit in the love we both entertained for this family of living gems. To describe the feeling which animated us with regard to them is impossible. It can, in fact, only be realised by those who have made natural history a study, and who pursue the investigation of its charming mysteries with ardour and delight'. The German traveller Richard Schomburgk (q.v.), on a visit to London (1840), said of a collection he saw: 'Amongst the many private collections which I had an opportunity of visiting, there was one that particularly engaged the whole of my attention: it was the beautiful, really fairy-like collection of hummingbirds, the property of Loddiges, the market-gardener, containing all the species of this interesting family at present known, and considerably richer in them than is the British Museum.' George Loddiges was the son of Joachim Conrad Loddiges. (See **Conrad (Loddiges)**)

Lodygin

Eurasian Sparrowhawk ssp. *Accipiter nisus lodygini*
  Bianchi, 1906 NCR
[JS *Accipiter nisus melaschistos*]

Alexander Nikolayevich Lodygin (1847–1923) was a Russian engineer and inventor, who made and patented an incandescent light bulb. He was an infantry officer in the Russian army (1865–1868). He became involved with a radical political group (the Narodniks) that assassinated Tsar Alexander II (1880), forcing Lodygin to emigrate for his own safety. He lived in France and USA (1884–1907). In the USA he developed a lamp with a tungsten filament, selling the invention to General Electric (1906). Back in Russia he taught at the Institute of Electrical Engineering (1907–1917). After the Russian Revolution (1917) he moved back to the USA and died in Brooklyn, New York.

Loe

European Turtle Dove ssp. *Streptopelia turtur loei* **von**
  Jordans, 1923 NCR
[NUI *Streptopelia turtur arenicola*]

Dr Felix Graf von Loë (1896–1944) was a German collector in the Balearic Islands (1922). He was killed in action at Schwaneburg, Latvia. His widow died in 2009 aged 105.

Loehken

Bronze-tailed Barbthroat ssp. *Threnetes niger loehkeni*
  Grantsau, 1969

W. Löhken (DNF) was a collector in Brazil in the 1960s.

Loetscher

Olive-backed Euphonia ssp. *Euphonia gouldi loetscheri*
  A. R. Phillips, 1966 NCR
[JS *Euphonia gouldi gouldi*]

Dr Frederick William Loetscher Jr (1913–2006) was an American botanist and ornithologist. Yale awarded his bachelor's degree (1935); his doctorate (1941) was the first awarded by Cornell in ornithology. He served in the US Army Medical Corps in WW2. His doctoral thesis, *Ornithology of the Mexican State of Veracruz*, was published much later (1978). He became Professor of Biology at Centre College, Danville, Kentucky (1963), retiring as Emeritus Professor.

Loftin

Large-billed Seed Finch ssp. *Oryzoborus crassirostris loftini*
  Wetmore, 1970 NCR
[JS *Oryzoborus nuttingi*]

Dr Horace Greeley Loftin (b.1927) was an American ornithologist and Professor, Biology Department, Florida State University, which has a research station in Panama where he ringed birds. He wrote 'Notes on autumn bird migrants in Panama' (1963).

Longstaff

> Temminck's Babbler ssp. *Pellorneum pyrrogenys longstaffi*
> **Harrisson** & **Hartley**, 1934

Captain Thomas George Longstaff (1875–1964) was an English physician, explorer and mountaineer. Before WW1 he travelled in Tibet (1905) and the Himalayas (1907–1900) and was the first person to scale a peak in excess of 7,000 metres (1907). Between the World Wars, in which he served in the British Army, he travelled and explored in the Himalayas, Spitsbergen, Greenland and Baffin Island. The holotype was collected by the Oxford University Exploration Club and the dedication mentions that Longstaff helped them.

Longuemare

> Longuemare's Hermit *Phaethornis longuemareus* **Lesson**,
> 1832
> [Alt. Little Hermit]
> Longuemare's Sunangel *Heliangelus clarisse* Longuemare,
> 1841
> [Alt. Amethyst-throated Sunangel; Syn. *Heliangelus amethysticollis clarisse*]

Lesson wrote that he had seen the hermit in the collection of 'M[onsieur] de Longuemare'. This may refer to Alphonse Pierre Françoise Le Touzé de Longuemare (1803-1881), a French antiquary and army officer. The original description states that the sunangel was described by M. Gouye de Longuemare; some sources say that this was Henri Victor Goüye de Longuemare (1823-1890) which seems unlikely as he was only 18 when the bird was described.

Lönnberg

> Bulbul sp. *Criniger lonnbergi* **Gyldenstolpe**, 1913 NCR
> [Alt. Grey-eyed Bulbul; JS *Iole propinqua propinqua*]

> Lönnberg's Skua *Stercorarius antarcticus lonnbergi*
> **Mathews**, 1912
> [Alt. Subantarctic Skua ssp.]
> Grey-olive Greenbul ssp. *Phyllastrephus cerviniventris
> lonnbergi* **Mearns**, 1914 NCR
> [JS *Phyllastrephus cerviniventris cerviniventris*]
> Willow Tit ssp. *Poecile montanus lonnbergi* **Zedlitz**, 1925
> NCR
> [JS *Poecile montanus borealis*]
> Eurasian Skylark ssp. *Alauda arvensis lonnbergi* **Hachisuka**,
> 1926
> Lesser Spotted Woodpecker ssp. *Dendrocopos minor
> lonnbergi* **Domaniewski**, 1927 NCR
> [JS *Dendrocopos minor minor*]
> Eurasian Wren ssp. *Troglodytes troglodytes lonnbergi*
> **Momiyama**, 1927 NCR
> [JS *Troglodytes troglodytes fumigatus*]
> Hawfinch ssp. *Coccothraustes coccothraustes lonnbergi*
> **Bergman**, 1931 NCR
> [JS *Coccothraustes coccothraustes japonicas*]

Professor Axel Johan Einar Lönnberg (1865–1942) was a Swedish zoologist who mainly worked at the Vertebrate Department of the Swedish Museum of Natural History in Stockholm. He travelled in the Caspian Sea region (1899) and

British East Africa (1910–1911), and was the last prefect of the Kristineberg Marine Zoological Station (1925–1942). An early conservationist, he worked for laws protecting reindeer and waterbirds. He founded the biological journal *Fauna och Flora*. Three amphibians are named after him.

Loomis

> Storm-petrel genus *Loomelania* **Mathews**, 1934 NCR
> [Now in *Oceanodroma*]

Leverett Mills Loomis (1857–1928) was an American ornithologist who was Director of the Museum of the California Academy of Sciences (1902–1912). The (1906) San Francisco earthquake destroyed the Academy building, and Loomis risked his life to rescue some of the most valuable books and bird-skins. He was particularly interested in the 'Tubinares' (now called the Procellariiformes). Mathews combined the first part of his surname with the binomial of *Oceanodroma melania* (Black Storm-petrel), which he proposed to move into the new genus *Loomelania*.

Lopez

> Lopez's Warbler *Bradypterus lopezi* **B. Alexander**, 1903
> [Alt. Evergreen Forest Warbler]
> White-tailed Warbler *Poliolais lopezi* B. Alexander, 1903

> Chocolate-backed Kingfisher ssp. *Halcyon badia lopezi*
> B. Alexander, 1903 NCR
> [*H. badia* NRM]
> Red-chested Goshawk ssp. *Accipiter toussenelii lopezi*
> B. Alexander, 1903
> Red-faced Pytilia ssp. *Pytilia hypogrammica lopezi* B.
> Alexander, 1906
> [Alt. Yellow-winged Pytilia. May be a colour-morph of (a
> then monotypic) *P. hypogrammica*]
> Lowland Akalat ssp. *Sheppardia cyornithopsis lopezi*
> B. Alexander, 1907

José Lopez (sometimes Lopes) (fl.1890–at least 1910) was a Portuguese collector in Africa, collecting for Boyd Alexander (1897) as a boy and returning to England with him. Boyd then took him on all his trips to Africa, including to the island of Bioko (Equatorial Guinea) (1902). He was still with him when local people killed Alexander (1910).

Løppenthin

> Eurasian Nuthatch ssp. *Sitta europaea loppenthini*
> Dunajewski, 1934 NCR
> [JS *Sitta europaea caesia*]

Dr Bernt Harvig Ove Fabricius Løppenthin (1904–1994) was a Danish zoologist who qualified as a physician (1929) and took a degree in zoology (1934). He practised medicine (1933–1938) and then became Head Librarian at the Copenhagen University Library. He took part in expeditions to Greenland (1930) and the Persian Gulf (1937–1938).

Lord Byron

> Lord Byron's Conure *Cyanoliseus patagonus byroni* **J. E.**
> **Gray**, 1831 NCR
> [Alt. Burrowing Parrot; now *Cyanoliseus patagonus bloxami*]

Admiral George Anson Byron, 7th Baron (1789–1868) succeeded to the peerage on the death (1824) of his cousin, the famous poet. He spent his life in the Royal Navy, being promoted Captain (1814) and commanding HMS *Blonde* (1824–1826) in South American and Hawaiian waters. The voyage's objective was to return the bodies of the King and Queen of the Sandwich Islands (Hawaii) for burial, after they had succumbed to measles whilst on a visit to London. He published *Voyage of HMS Blonde to the Sandwich Islands* (1826).

Lord Derby

Lord Derby's Toucanet *Aulacorhynchus derbianus* **Gould**, 1835
[Alt. Chestnut-tipped Toucanet]
Lord Derby's Mountain Pheasant *Oreophasis derbianus* **G. R. Gray**, 1844
[Alt. Horned Guan, Derby's Guan]
Lord Derby's Parakeet *Psittacula derbiana* **Fraser**, 1852
[Alt. Derbyan Parakeet]

Lord Derby's Flycatcher *Pitangus sulphuratus derbianus* **Kaup**, 1852
[Alt. Great Kiskadee ssp.]

The Hon Edward Smith Stanley (1775–1851) was 13th Earl of Derby. (See **Stanley** and **Derby**)

Lord Howe

Lord Howe's Rail *Gallirallus sylvestris* **P. L. Sclater**, 1869
[Alt. Lord Howe Woodhen]

The rail is named after Lord Howe Island. While sailing from Sydney to Norfolk Island (1788), Lieutenant Ball RN of HMS *Supply*, named the island after the, then, First Lord of the Admiralty, Richard Howe (1726–1799). A mammal is also named after this island.

Lorentz

Lorentz's Whistler *Pachycephala lorentzi* **Mayr**, 1931

Black-backed Honeyeater ssp. *Ptiloprora perstriata lorentzi* **Van Oort**, 1909 NCR
[NUI *Ptiloprora perstriata perstriata*]
Friendly Fantail ssp. *Rhipidura albolimbata lorentzi* Van Oort, 1909
White-shouldered Fairy-wren ssp. *Malurus alboscapulatus lorentzi* Van Oort, 1909
Lorentz's Painted Parrot *Psittacella picta lorentzi* Van Oort, 1910
[Alt. Painted Tiger Parrot ssp.]

Hendrikus Albertus Lorentz (1871–1944) was a Dutch explorer who studied law and biology. He participated in the expedition (1901) of Professor Wichmann to northern (Dutch) New Guinea and he himself led expeditions (1905–1906 and 1909–1910) in southern New Guinea, leading to the discovery of Wilhelmina Peak (named after the Dutch Queen Wilhelmina) in the Snow Mountains. Upon his return he entered the Dutch consular services, becoming Ambassador in Pretoria, South Africa (1929). The Lorentz River in New Guinea is named after him, as is a fish. Three mammals are named after him.

Lorenz, F.

Mountain Chiffchaff ssp. *Phylloscopus sindianus lorenzii* **T. K. Lorenz**, 1887

Professor Dr Friedrich Lorenz (1803–1861) was a director of all the German grammar schools in St Petersburg and editor of the *Deutsches Petersburger Zeitung*. He took a PhD at Berlin University, then became Professor at Halle University before moving to Russia (1832). His son, Theodore Lorenz (q.v.), was a zoologist who collected in Russia and named the bird in honour of his father.

Lorenz, L. R.

Lorenz's Seedeater *Sporophila lorenzi* **Hellmayr**, 1904 NCR
[Known from one specimen, believed to be an artifact]
Lorenz's Bulbul *Phyllastrephus lorenzi* **Sassi**, 1914
[Alt. Sassi's Olive Greenbul; Syn. *Phyllastrephus icterinus lorenzi*]

Dr Ludwig Ritter Lorenz von Liburnau (1856–1943) was an Austrian zoologist. Vienna University awarded his PhD (1879) and his Habilitation (1898). He was associated with the Vienna Museum's Bird Collection (1880–1922). Although Hellmayr obviously thought that the seedeater was a genuine taxon, it is now treated as 'a probable artifact' according to Meyer de Schauensee (1966). Presumably this means a deliberate forgery. A mammal is named after him.

Lorenz, T. K.

Altai Falcon *Falco lorenzi* **Sushkin**, 1938 NCR
[Taxonomy uncertain: probably a dark morph of Saker Falcon *Falco cherrug*]

Common Pheasant ssp. *Phasianus colchicus lorenzi* **Buturlin**, 1904 NCR
[NUI *Phasianus colchicus colchicus*]

Theodore Karlovic Lorenz (1842–1909) was a German zoologist and ornithologist who did most of his work in Russia. He was a taxidermist at the Zoological Gardens, Imperial University, and explored in the Caucasus. He wrote *Die Vögel der Moskauer Gouvernements* (1894). A reptile is named after him.

Loria

Loria's Satinbird *Cnemophilus loriae* **Salvadori**, 1894
[Alt. Loria's Bird-of-Paradise; Syn. *Loria loriae*]
Cassowary sp. *Casuarius loriae* **Rothschild**, 1898 NCR
[JS, or perhaps a subspecies of, Dwarf Cassowary *Casuarius bennetti*]

Papuan Pitta ssp. *Erythropitta macklotii loriae* Salvadori, 1890
Spotted Jewel-babbler ssp. *Ptilorrhoa leucosticta loriae* Salvadori, 1896

Dr Lamberto Loria (1855–1913) was an Italian ethnologist and explorer who collected in New Guinea (1889–1890). He founded the first Italian Museum of Ethnography (1906) in Florence; it was subsequently transferred to Rome, after he organised the first ethnography exhibition there (1911). When the 5,000 objects in the Florence museum were transferred to the capital, over 30,000 objects collected by Loria and his assistants from throughout the country were added to the collections. An amphibian, two mammals and three reptiles are named after him.

Loring

Red-faced Crombec ssp. *Sylvietta whytii loringi* **Mearns**, 1911

Blue-breasted Bee-eater ssp. *Merops variegatus loringi* Mearns, 1915

Buff-spotted Flufftail ssp. *Sarothrura elegans loringi* Mearns, 1915 NCR

[NUI *Sarothrura elegans elegans*]

John Alden Loring (1871–1947) was an American field collector. He was a field biologist with the Bureau of Biological Survey, US Department of Agriculture (1892–1897). He joined Theodore Roosevelt (q.v.) and his son Kermit (q.v.) on their lengthy safari (1909–1910) as a professional collector. He wrote and lectured about it under the title *Through Africa with Roosevelt*. The object of the expedition was to collect large game mammal specimens for the USNM in Washington, DC. They observed and collected more than 160 species of mammals in Kenya, Uganda and Sudan. A mammal is named after him.

Lort

Lilac-breasted Roller ssp. *Coracias caudatus lorti* **G. E. Shelley**, 1885

Orange River Francolin ssp. *Scleroptila levalliantoides lorti* **Sharpe**, 1897

[Alt. Archer's Francolin *Scleroptila gutturalis lorti*]

Ethelbert E Lort Phillips (See **Phillips, E. E. L.**)

Loten

Loten's Sunbird *Cinnyris lotenius* **Linnaeus**, 1766
[Alt. Long-billed Sunbird]

Johan (sometimes Joan) Gideon Loten (1710–1789) was the Dutch Governor of Makassar (Celebes, Indonesia) (1743–1749) and of Ceylon (Sri Lanka) (1752–1757). As an amateur naturalist he attempted the first systematic documentation of Ceylon's birdlife. He collected many specimens, especially of birds, but since they proved difficult to preserve, a young artist of mixed Singhalese-European descent named Pieter Cornelis de Bevere made coloured drawings of them. Loten's notes reveal that de Bevere drew from living or freshly dead specimens. The bulk of the illustrations, including 101 plates depicting birds, are in the British Museum, London. Loten himself lived in England for many years. He did not use binomial scientific names and this meant that the birds illustrated in his work remained unnamed (scientifically) until others identified them after his death.

Loudon

Shrike sp. *Lanius loudoni* **Buturlin**, 1907 NCR
[Hybrid: *Lanius collurio* x *Lanius phoenicuroides*?]

Eurasian Hoopoe ssp. *Upupa epops loudoni* **Tschusi**, 1902 NCR

[JS *Upupa epops epops*]

Common Starling ssp. *Sturnus vulgaris loudoni* Buturlin, 1904 NCR

[JS *Sturnus vulgaris porphyronotus*]

European Goldfinch ssp. *Carduelis carduelis loudoni* **Zarudny**, 1906

Eurasian Penduline-tit ssp. *Remiz macronyx loudoni* Zarudny, 1914 NCR

[Hybrid population of *Remiz macronyx* x *Remiz pendulinus*]

Barn Swallow ssp. *Hirundo rustica loudoni* Zarudny, 1923 NCR

[JS *Hirundo rustica rustica*]

Baron Harald von Loudon (1876–1959) lived in Latvia (1876–1914). He collected in Latvia, Turkmenistan and eastern Azerbaijan (1895–1914). His collection is mainly in Amsterdam as part of the Sillem-Van Marle Collection.

Louis Philippe

Comte de Paris's Starfrontlet *Coeligena lutetiae* **de Lattre** & **Bourcier**, 1846
[Alt. Buff-winged Starfrontlet]

Louis Philippe, Comte de Paris (1838–1894), was head of the House of Bourbon (1883–1894). The binomial *lutetiae* derives from 'Lutetia', the Latin name for Paris. (See **Comte de Paris**)

Louisa

Somali Thrush *Turdus ludoviciae* **E. L. Phillips**, 1895
Somali Grosbeak *Rhynchostruthus louisae* E. L. Philips, 1897
[Alt. Somali Golden-winged Grosbeak]

(See **Ludovicia**)

Louise

Green-fronted Lancebill *Doryfera ludovicae* **Bourcier** & **Mulsant**, 1847

(See **Ludovica**)

Lovat

Black-bellied Bustard *Lissotis lovati* **Ogilvie-Grant**, 1900 NCR

[JS *Lissotis melanogaster*]

Honeyguide sp. *Indicator lovati* Ogilvie-Grant, 1900 NCR
[Alt. Lesser Honeyguie; JS *Indicator minor diadematus*]

Weaver sp. *Othyphantes lovati* **G. E. Shelley**, 1905 NCR
[Alt. Baglafecht Weaver; JS *Ploceus baglafecht baglafecht*]

Simon Joseph Fraser, 14th Lord Lovat (1871–1933) (legally the 14th Lord, but he was referred to as the 16th), was an aristocrat and soldier who was commissioned into the Queen's Own Cameron Highlanders. He raised the Lovat Scouts (1899) who fought in the South African War. Lovat was awarded the DSO. He led an expedition through Abyssinia (Ethiopia) from Berbera to the Blue Nile (1899), collecting specimens of mammals and birds. He was the first Chairman of the Forestry Commission (1919). A mammal is named after him. (See also **Simon (Fraser)**)

Love

Desert Chat *Ashbyia lovensis* **Ashby**, 1911
[Alt. Gibberbird]

Rev. James Robert Beattie Love (1889–1947) was a missionary in the Kimberley area of north-west Australia. He wrote a book on the grammatical structure of the local language (Worrorra), into which language he translated the gospels according to St Luke and St Mark. He also wrote *Stone-age Bushmen of Today: Life and Adventure among a Tribe of Savages in North-Western Australia* (1936).

Lovejoy

Scaled Piculet ssp. *Picumnus squamulatus lovejoyi*
W. H. Phelps Jr & **Aveledo**, 1987

Professor Thomas Eugene Lovejoy (b.1941) is a tropical biologist and conservation biologist currently (2012) the first Biodiversity Chair of the H. John Heinz III Center for Science, Economics and the Environment. He is also Professor of Environmental Science and Policy at George Mason University. He introduced the term 'biological diversity' to the scientific community (1980). He is Chair of the Scientific Technical Advisory Panel (STAP) for the Global Environment Facility (GEF), the multibillion-dollar funding mechanism for developing countries to support their obligations under international environmental conventions. Yale University awarded his BSc (1964) and PhD (1971) in biology. He worked in the Brazilian Amazon as a Research Assistant for the Belem Project at the USNM (1965–1972). He was Director of Conservation at World Wildlife Fund (US) (1973–1987) and was Assisitant Secretary for Environmental and External Affairs at the USNM (1987–1998), then Chief Biodiversity Advisor to the President of the World Bank (1999–2002). He has advised US Presidents (Reagan, Bush and Clinton) and many institutions as well as the TV programme 'Nature'. He has edited and written articles and books including co-writing *Climate Change and Biodiversity* (2006).

Loven

Red-cheeked Cordon-bleu ssp. *Uraeginthus bengalus loveni*
Granvik, 1923 NCR
[JS *Uraeginthus bengalus littoralis*; or of *U. bengalus brunneigularis* if *littoralis* merged with latter]

Sven August Lovén (1879–1961) was Swedish Consul-General in Kenya, a zoologist, explorer and collector. He wrote *Kring Mount Elgon med vita vänner och svarta : anteckningar om Mount Elgonexpeditionens öden och äventyr (Around Mount Elgon with white and black friends: Notes about the Mount Elgon Expedition's various adventures)* (1921).

Loveridge

Loveridge's Sunbird *Cinnyris loveridgei* **Hartert**, 1922

Green-backed Woodpecker ssp. *Campethera cailliautii loveridgei* Hartert, 1920
Scaly Chatterer ssp. *Turdoides aylmeri loveridgei* Hartert, 1923 NCR
[JS *Turdoides aylmeri keniana*]

Black-faced Sandgrouse ssp. *Pterocles decoratus loveridgei* **Friedmann**, 1928

Arthur Loveridge (1891–1980) was a British herpetologist and ornithologist who became Curator of Nairobi Museum (1914). He worked for the Museum of Comparative Zoology, Harvard (1924–1957) and made several field trips to East Africa. His writings include *Many Happy Days I've Squandered* (1949) and *I Drank the Zambesi* (1954). He retired to the island of St Helena. Seven amphibians and thirteen reptiles are named after him.

Low, H. B.

Sulu Bald Starling *Sarcops calvus lowii* **Sharpe**, 1877
[Alt. Coleto ssp.]
Low's Swiftlet *Aerodramus maximus lowi* Sharpe, 1879
[Alt. Black-nest Swiftlet ssp.]

Sir Hugh Brooke Low (1824–1905) was a civil servant in the British Administration in Malaya and amateur collector in the Malay Archipelago. He was the first successful British Administrator of Perak (1877–1889). Subsequently his methods became models for British colonial operations in Malaya. Previously he had been an unremarkable Colonial Secretary of Labuan, an island off Borneo (1848–1877). There is a 'Historical Trail' to the summit of Mount Kinabalu, Sabah, named after him as he was first to climb it (1851) and used to collect specimens from the summit. He was instrumental in encouraging the planting of rubber trees throughout Malaya. He knew well Sir Charles Brooke (q.v.), the second 'White Rajah' of Sarawak, having travelled out with him when first appointed by the East India Company. He worked as Brooke's secretary for a few months before returning to England, where he published *Sarawak its Inhabitants and Productions,* an account that contained descriptions of two new beetles he had himself discovered. He is best remembered for many new orchids he found that are named after him. One reptile and two mammals are named after him.

Lowe, P. R.

Bananaquit ssp. *Coereba flaveola lowii* **Cory**, 1909
Black-throated Laughingthrush ssp. *Garrulax chinensis lowei* **La Touche**, 1922 NCR
[JS *Garrulax chinensis chinensis*]
Lowe's Green Pigeon *Treron apicauda lowei* **Delacour** & **Jabouille**, 1924
[Alt. Pin-tailed Green Pigeon ssp.]
Lowe's Grey Peacock Pheasant *Polyplectron bicalcaratum bailyi* P. R. Lowe, 1925 NCR
[NUI *Polyplectron bicalcaratum bicalcaratum*]
Calliope Hummingbird ssp. *Stellula calliope lowei* **Griscom**, 1934 NCR; NRM
Olive Sunbird ssp. *Cyanomitra olivacea lowei* **Vincent**, 1934
[Syn. *Cyanomitra obscura lowei*; or merged with *C. obscura ragazzii*]

Percy Roycroft Lowe (1870–1948) was an English ornithologist. He studied medicine at Cambridge and served as a civil surgeon during the Second Boer War (1899–1902). He became interested in ornithology whilst in South Africa. He succeeded Ogilvie-Grant (q.v.) as Curator of Birds at the Natural History Museum, London (1919). He was editor of the *Bulletin of the British Ornithologists' Club* (1920–1925), and President of the BOU (1938–1943). Lowe is credited with coining the term 'Darwin's Finches' in his paper in *Ibis* entitled 'The finches of the Galapagos in relation to Darwin's conception of species' (1936).

Lowe, W. P.

Iringa Akalat *Sheppardia lowei* **C. H. B. Grant** & **Mackworth-Praed**, 1941

Lowe's Four-banded Sandgrouse *Pterocles quadricinctus lowei* C. H. B. Grant, 1914 NCR; NRM

Mottled Swift ssp. *Tachymarptis aequatorialis lowei* **Bannerman**, 1920

Willoughby Prescott Lowe (1872–1949) was a British naturalist, explorer and collector for the British Museum. He was in the USA in the late 19th century, and in many parts of Africa and the Far East (1907–1935). He is reputed to have sent over 10,000 specimens to the Bird Room at the BMNH. He is also notorious for having shot eight specimens of Miss Waldron's Red Colobus in Ghana (1933): the monkey was already rare, and is now thought extinct. Five mammals are named after him.

Lowery

Long-whiskered Owlet *Xenoglaux loweryi* **O'Neill** & **Graves**, 1977

Great-tailed Grackle ssp. *Quiscalus mexicanus loweryi* **Dickerman** & **A. R. Phillips**, 1966

Professor George Hines Lowery Jr (1913–1978) was an American ornithologist who spent his professional career at the Louisiana State University (LSU), where he was Boyd Professor of Zoology (1955). Largely due to his efforts the LSU Museum of Zoology became the Museum of Natural Science. He played an important role in creating public awareness and support for a conservation ethic in Louisiana, where the term 'wildlife' was a synonym for hunting and fishing. His *Louisiana Birds* (1955) won the Louisiana Literary Award of that year. A reptile is named after him.

Loye Miller

Audubon's Shearwater ssp. *Puffinus lherminieri loyemilleri* **Wetmore**, 1959

Bronzed Cowbird ssp. *Molothrus aeneus loyei* **Parkes** & **E. R. Blake**, 1965

Dr Loye Holmes Miller (1874–1970) was an American ornithologist and palaeontologist. He was Professor and Chairman of the Biology Department, University of California, Los Angeles. He was interested in fossil birds and published many scientific papers on them. He wrote an autobiography, *Lifelong Boyhood* (1950).

Lübbert

Lübbert's Shrike *Lanius luebberti* **Reichenow**, 1902 NCR
[Alt. Lesser Grey Shrike; JS *Lanius minor*]

Masked Weaver sp. *Ploceus luebberti* Reichenow, 1902 NCR
[Alt. Lesser Masked Weaver; JS *Ploceus intermedius cabanisi*]

Familiar Chat ssp. *Cercomela familiaris luebberti* Reichenow, 1902 NCR
[JS *Cercomela familiaris galtoni*]

Dr Anton Lübbert (b.1860) was a German botanist and collector in Damaraland (Namibia) (1902–1903).

Lubomirski

Black-chested Fruiteater *Pipreola lubomirskii* **Taczanowski**, 1879

Prince Wladyslaw Jan Emanuel Lubomirski (1824–1882) was a Polish conchologist.

Lucas

Northern Golden Bulbul ssp. *Thapsinillas longirostris lucasi* **Hartert**, 1903

W. Lucas (d.1909) was a collector from Brussels who was in the Moluccas (Indonesia).

Luchs

Cliff Parakeet *Myiopsitta luchsi* **Finsch**, 1868

Dr Johann Nepomuk Karl Ernst Luchs (1812–1886) lived in Warmbrunn in Silesia, Germany (Cieplice Slaskie Zdroja, Poland). He qualified as a physician in Breslau, returning to Warmbrunn to practise. He sold (1878) his extensive ornithological collection to Ludwig, Count Schaffgotschow.

Lucian (Bonaparte)

Bonaparte's Parakeet *Pyrrhura lucianii* **Deville**, 1851

While the bird is named in the binomial after Charles Lucien Bonaparte, Deville's 'latinisation' renders the name as *lucianii* rather than *lucienii*. (See **Bonaparte**)

Lucian (Buquet)

Sapphire-vented Puffleg *Eriocnemis luciani* **Bourcier**, 1847

Jean Baptiste Lucien Buquet (1807–1889) was a French entomologist, hence the beetle family name: *Lucanidae*. His entomological collection was purchased, as part of a wider holding, by BMNH. He wrote 'Lettre de M. G. de Mniszech' in the *Annales de la Société Entomologique de France* (1854). He may have published other people's works on entomology in Paris. Gould painted the hummingbird shortly after it was described.

Lucie

Collared Babbler ssp. *Gampsorhynchus torquatus luciae*
Delacour, 1926

Madame Lucie Varenne was in Indochina (1925–1927) where her husband, Alexandre Claude Varenne, was Governor-General.

Lucifer

Hummingbird genus *Lucifer* **Reichenbach**, 1849 NCR
[Now in *Calothorax*]

Lucifer Hummingbird *Calothorax lucifer* **Swainson**, 1827

Streaked Rosefinch ssp. *Carpodacus rubicilloides lucifer*
R. Meinertzhagen & **A. Meinertzhagen**, 1926
Cinnamon-browed Honeyeater ssp. *Melidectes ochromelas lucifer* **Mayr**, 1931

The fallen angel Lucifer, of the Judeo-Christian tradition, is referred to in the fourteenth chapter of the Biblical book of *Isaiah*: 'How art thou fallen from heaven, O Lucifer, son of the morning! How art thou cut down to the ground, which didst weaken the nations!' The meaning of the name Lucifer is 'Light-bearer', and the name was given to the Morning Star (Venus). Presumably the birds were named without reference to any 'demonic' qualities. Two mammals also have 'Lucifer' in their names.

Lucille

Eyebrowed Wren-babbler ssp. *Napothera exsul lucilleae*
Meyer de Schauensee & **Ripley**, 1940 NCR
[JS *Napothera epilepidota diluta*]

Mrs Lucille Vanderbilt *née* Parsons (b.1912) was the first wife of George Washington Vanderbilt III and was with him in Sumatra (whence comes the wren-babbler) (1939) and perhaps earlier (1937). She was well known as an expert golfer, horsewoman and rifle shot. They divorced (1946) and she married Ronald Balcom (1949).

Lucy

Mountain Scops Owl ssp. *Otus spilocephalus luciae*
Sharpe, 1888

Sharpe gave no etymology in his original description. As John Whitehead (q.v.) collected the owl holotype it may be that *luciae* refers to a relative of his.

Lucy (Baird)

Lucy's Warbler *Leiothlypis luciae* **J. G. Cooper**, 1861
[Syn. *Vermivora luciae*]

Lucy Hunter Baird (1848–1913) was the only child of Spencer Baird (q.v.); she was 13 when the warbler was named after her.

Lucy (Brewer)

Lucy's Emerald *Amazilia luciae* **G. E. Lawrence**, 1868
[Alt. Honduran Emerald]

Lucy Stone Brewer (1854–1921) was the only daughter of the American ornithologist Dr Thomas Mayo Brewer

(1814–1880). Audubon (q.v.) and Brewer were friends, and she was named after Audubon's wife. In turn her father's friend, George Lawrence (q.v.), named this hummingbird after Lucy.

Lucy (Stoneham)

Black-rumped Buttonquail ssp. *Turnix nana luciana*
Stoneham, 1931 NCR; NRM

Mrs Alice Lucy Stoneham *née* Deacon (DNF) was the mother of the describer, amateur naturalist Hugh Frederic Stoneham.

Ludlow

Ludlow's Fulvetta *Fulvetta ludlowi* **Kinnear**, 1935
[Alt. Brown-throated Fulvetta]

Little Owl ssp. *Athene noctua ludlowi* **E. C. S. Baker**, 1926
Greenish Warbler ssp. *Phylloscopus trochiloides ludlowi*
Whistler, 1931
Crimson-breasted Woodpecker ssp. *Dendrocopos cathpharius ludlowi* **Vaurie**, 1959

Frank Ludlow (1885–1972) was a British educationist, botanist, ornithologist and explorer. During WW1 he was commissioned into the Ninety-Seventh Indian Infantry and thereafter went into the Indian Education Service. He collected c.7,000 bird specimens, but made his biggest impact as a botanist, being associated with the discovery of new species of peonies, primulas and rhododendrons. He collected widely in the Himalayas and Tibet, often with other collectors such as George Sheriff (q.v.).

Ludovic (Clarke)

Ludovic's Nightjar *Caprimulgus ludovicianus* **S. Clarke**, 1913
[JS (or possibly valid subspecies) of *Caprimulgus inornatus*]

Louis Colville Gray Clarke (1881–1960) was a member of the same family as Goland Clarke (q.v.) and Stephenson Clarke (q.v.). He was a naturalist who collected in Somalia and was Curator of the Fitzwilliam Museum of Anthropology and Archaeology at Cambridge, England (1937–1946). Stephenson Clarke collected the nightjar (1912) in south-west Ethiopia, and used *ludovici* as a latinised form of Louis.

Ludovic (Prince)

Puffleg sp. *Ornismya ludovicii* Da Silva Maia, 1852 NCR
[Alt. Glowing Puffleg; JS *Eriocnemis vestita*]

Louis Charles Marie Joseph (or Luigi Carlo Maria Giuseppe), Prince de Bourbon e Duas Sicílias, Conde d'Áquila (1824–1897), was the son of Francis I (King of the Two Sicilies). His wife was Januária Maria of Braganza, the Princess Imperial of Brazil. (Here *ludovici* is used as a latinised form of Louis/Luigi.)

Ludovica

Green-fronted Lancebill *Doryfera ludovicae* **Bourcier** & **Mulsant**, 1847

Louise Geoffroy Saint-Hilaire *née* Blacque-Belair (1810–1855) was the wife of the French zoologist Isidore Saint-Hilaire (see **Isidore**). (*Ludovicae* is a latinised version of Louise.)

Ludovicia

Somali Thrush *Turdus ludoviciae* **E. L. Phillips**, 1895
Somali Grosbeak *Rhynchostruthus louisae* E. L. Philips, 1897
[Alt. Somali Golden-winged Grosbeak]

Mrs Louisa Jane Forbes Lort-Phillips *née* Gunnis (d.1946) was the wife of E. L. Phillips. (*Ludoviciae* is a latinised version of Louisa.) A mammal is named after her.

Ludwig

Square-tailed Drongo *Dicrurus ludwigii* **A. Smith**, 1834
Ludwig's Bustard *Neotis ludwigii* **Rüppell**, 1837
Ludwig's Double-collared Sunbird *Cinnyris ludovicensis*
   Bocage, 1868
[Alt. Montane Double-collared Sunbird]

Orange River Francolin ssp. *Francolinus gariepensis ludwigi*
   Neumann, 1920 NCR
[JS *Scleroptila levalliantoides*]

Carl Ferdinand Heinrich von Ludwig, or Baron von Ludwig (1784–1847), was a German pharmacist. He responded to an advertisement in an Amsterdam newspaper (1805) and applied for a post as Pharmacy Assistant to Dr Liesching of Cape Town. He married well (1816), giving him enough time and money to indulge his interests in collecting natural history specimens. Ludwig shipped a collection of plants and insects to the Stuttgart Royal Museum, in recognition of which he was awarded a knighthood, allowing him the use of 'Baron'. He took a collection of plants, insects, birds and mammals with him on a visit to Germany (1828) that won him an honorary Ph.D. Bocage did not provide an etymology when naming the sunbird. Despite the species acquiring the vernacular name of 'Ludwig's' it may well be that *ludovicensis* was chosen to honour King Luis I of Portugal (1838–1889; reigned 1861–1889), a patron of the sciences, especially oceanography.

Luguier

Melanesian Flycatcher *Myiagra luguieri* **Tristram**, 1879 NCR
[JS *Myiagra caledonica viridinitens*]

M. Luguier (DNF) was French Resident on the Loyalty Islands. Layard (q.v.) reported that Luguier had introduced European edible snails there, as he adored 'escargots'.

Lühder

Lühder's Bush-shrike *Laniarius luehderi* **Reichenow**, 1874

Olive-bellied Sunbird ssp. *Cinnyris chloropygius luehderi*
   Reichenow, 1899 NCR
[NUI *Cinnyris chloropygius chloropygius*]

Dr Wilhelm Lühder (1847–1873) was a German naturalist who collected in Cameroon with Reichenow (q.v.) (1872). He wrote an article 'Notizen über den Bock' (1871).

Luiza

Cipo Canastero *Asthenes luizae* **Vielliard**, 1990

Luiza Lencioni is the wife of Brazilian ornithologist, collector and artist Frederico Lencioni.

Lulu

Lulu's Tody Tyrant *Poecilotriccus luluae* **Johnson** & Jones, 2001
[Alt. Lulu's Tody Flycatcher, Johnson's Tody Tyrant]

Lulu May Lloyd Von Hagen (1912–1998) was a Californian patroness of science. The bird was named in recognition of 'her generous and dedicated support of research in avian genetics'.

Lumholtz

Crested Shrike-tit ssp. *Falcunculus frontatus lumholtzi*
   Mathews, 1912 NCR
[JS *Falcunculus frontatus frontatus*]
Varied Sittella ssp. *Daphoenositta chrysoptera lumholtzi*
   Mathews, 1916 NCR
[JS *Daphoenositta chrysoptera leucocephala*]

Dr Carl Sophus Lumholtz (1851–1922) was a Norwegian naturalist, ethnologist, humanist and explorer. Immediately after graduating with a natural science degree he set off for north-eastern Australia, where he spent time living with Aboriginal peoples (1880–1884). He organised a number of expeditions, including one to explore the Sierra Madre, Mexico (1890), for the AMNH. He visited Borneo (1914), but a planned trip to New Guinea was prevented by the outbreak of WW1. Lumholtz National Park in Queensland is named after him, as is a mammal and a reptile.

Lumsden

Giant Babax ssp. *Babax waddelli lumsdeni* **Kinnear**, 1938
   NCR
[JS *Babax waddelli waddelli*]

Dr Kenneth Lumsden (1908–1995) was a British physician. After qualifying (1934) he became a consultant radiologist at Oxford (1946–1974). He was on an expedition that explored and collected in Tibet (1936). He served as a British Army Medical Officer in WW2.

Lupton

Lupton's Bird-of-Paradise *Paradisaea apoda luptoni* **Lowe**, 1923 NCR
[Hybrid of *Paradisaea raggiana* x *Paradisaea apoda novaeguineae*]

Arthur Sinclair Lupton (1877–1949) was a British barrister interested in birds. He became a civil servant (1900) working in Customs & Excise (1920–1937). As such he intercepted a consignment of bird specimens, which contained an unknown bird-of-paradise.

Lutley

Blue-browed Tanager ssp. *Tangara cyanotis lutleyi*
   Hellmayr, 1917

(See Philip Lutley **Sclater**)

Lyall

Lyall's Wren *Xenicus lyalli* **Rothschild**, 1894 EXTINCT
[Alt. Stephens Island Wren, Travers's Wren; Syn. *Traversia lyalli*]

David Lyall (b.1822) was the son of a lighthouse keeper who became lighthouse keeper on Stephens Island, Cook Strait, New Zealand (1894). A domestic cat on the island began to bring carcasses of a small bird to the keepers' dwellings. Lyall, who was interested in natural history, arranged to send a specimen to the New Zealand naturalist Walter Lawry Buller (q.v.). Lyall sold other specimens of this flightless 'wren' to Henry Travers (q.v.), who in turn sold the birds to Walter Rothschild (q.v.). Lyall himself was assigned to another lighthouse (1896), by which time the 'wren' was probably extinct. The oft-repeated story that Lyall's cat was solely responsible for wiping out the entire population of *Xenicus lyalli* is apparently apocryphal; there were other feral cats on the island until exterminated in 1925.

Lyard

Little Tinamou ssp. *Crypturellus soui lyardi* **Miranda-Ribeiro**, 1938 NCR
[JS *Crypturellus soui albigularis*]

This is almost certainly a misspelling of Layard; there was no etymology in the description. However, Layard described *Crypturellus pileatus* in *Ibis* (1873). It is also a syn. of *albigularis* and we think that Miranda-Ribeiro may have had that in mind. (See Edgar Leopold **Layard**)

Lyle

Greater Yellownape ssp. *Chrysophlegma flavinucha lylei* **Kloss**, 1918

Sir Thomas Harold Lyle (1878–1927) was a British consular official who served in Siam (Thailand). He was Vice Consul at Nan (1896–1907) and Consul at Chiang Mai (1907–1913). He was one of the founding members of Chiang Mai Gymkhana Club (1898). He was only 157 cm (5 feet 2 inches) tall and suffered from extremely bad health, but insisted in joining in all activities. He was undoubtedly very brave, as he played a prominent part in the Shan Rebellion by going alone to confront the insurgents. He persuaded many of them to lay down their arms and go home. When he left Chiang Mai he was knighted. At the end of WW1 he returned to Bangkok as Consul General for the rest of his life. Two mammals are named after him.

Lyman

Merlin ssp. *Falco columbarius lymani* **Bangs**, 1913
Black-necked Woodpecker ssp. *Chrysoptilus atricollis lymani* Bangs & Noble, 1918 NCR
[JS *Colaptes atricollis peruvianus*]

Dr Theodore Lyman (1874–1954) was a physicist who graduated from Harvard (1897), which also awarded his doctorate

(1900). He joined the faculty at there as an Assistant Professor and became Director, Jefferson Physical Laboratory (1908–1917). He financed and took part in an expedition to the Altai Mountains (1912) for the purpose of collecting fauna from a region then under-represented in American museums. He went to France as a Captain in the Signal Corps (1917), returning to Harvard after WW1. He was wealthy and generous with his money, as evidenced by the etymology for the woodpecker, which says it was '… named in honor of Professor Theodore Lyman of Harvard University … whose generosity enabled the Museum of Comparative Zoology to cooperate with the School of Tropical Medicine of Harvard University during its expedition to Peru (July–October, 1916)'. The crater Lyman on the far side of the Moon is also named after him, as is an amphibian.

Lynes

Lynes's Pygmy Bustard *Lophotis savilei* Lynes, 1920
[Alt. Savile's Bustard]
Moorchat sp. *Myrmecocichla lynesi* **Bannerman**, 1927 NCR
[Alt. Congo Moorchat; JS *Myrmecocichla tholloni*]
Lynes's Cisticola *Cisticola distinctus* Lynes, 1930
Namuli Apalis *Apalis lynesi* **Vincent**, 1933

African Pipit ssp. *Anthus cinnamomeus lynesi* Bannerman & **Bates**, 1926
Great Tit ssp. *Parus major lynesi* **Hartert**, 1926 NCR
[JS *Parus major excelsus*]
Crested Lark ssp. *Galerida cristata lynesi* **Whistler**, 1928
Arabian Bustard ssp. *Ardeotis arabs lynesi* Bannerman, 1930 EXTINCT
Coqui Francolin ssp. *Peliperdix coqui lynesi* **W. L. Sclater**, 1932 NCR
[JS *Peliperdix coqui coqui*]
Rusty Lark ssp. *Mirafra rufa lynesi* **C. H. B. Grant** & **Mackworth-Praed**, 1933
Lynes's Flufftail *Sarothrura lugens lynesi* C. H. B. Grant & Mackworth-Praed, 1934
[Alt. Chestnut-headed Flufftail ssp.]
White-headed Barbet ssp. *Lybius leucocephalus lynesi* C. H. B. Grant & Mackworth-Praed, 1938
Grey-headed Batis ssp. *Batis orientalis lynesi* C. H. B. Grant & Mackworth-Praed, 1940
Blue Swallow ssp. *Hirundo atrocaerulea lynesi* C. H. B. Grant & Mackworth-Praed, 1942 NCR; NRM
Great Spotted Woodpecker ssp. *Dendrocopos major lynesi* **J. M. Harrison**, 1944 NCR
[NUI *Dendrocopos major mauritanus*]

Rear-Admiral Hubert Lynes (1874–1942) was a naval officer and amateur ornithologist who collected in China (1910–1913) and, after his retirement from the Navy (1919), in Africa (1926–1927). During WW1 his ship *Penelope* was torpedoed and many of his natural history notes were lost. He wrote a *Review of the Genus Cisticola* (1930), which played a key part in giving order to some confusingly similar species. He was awarded the Godman-Salvin Gold Medal by the BOU (1936).

M

Maack

Black-browed Reed Warbler *Salicaria maackii* **Schrenck**, 1860 NCR
[JS *Acrocephalus bistrigiceps*]

Richard Karlovich Maack (1825–1886) was a naturalist and ethnographer, Professor of Natural History at Irkutsk (1852). He explored the Amur River (1855) and was Inspector of Schools in Eastern Siberia (1868–1879). A reptile is named after him.

Mab

Lewin's Honeyeater ssp. *Meliphaga lewinii mab* **Mathews**, 1912
Orange-bellied Parrot ssp. *Neophema chrysogaster mab* Mathews, 1912 NCR; NRM

In English folklore Mab was queen of the fairies. In Shakespeare's *Romeo and Juliet* she is described as a 'midwife' who helps sleepers 'give birth' to their dreams.

Mabbott

Dunnock ssp. *Prunella modularis mabbotti* Harper, 1919

Douglas Clifford Mabbott (1893–1918) was an American ornithologist with the US Biological Survey with which Harper also worked. He was killed in action with the US Marine Corps in France in WW1. The subspecies was named as a memorial to him as he was killed only a few weeks before the Armistice of 11 November 1918 and his obituary appeared only in *The Auk* in January 1919. The etymology reads Mabbott '... fell in the cause of human liberty.'

MacArthur

Shelley's Francolin ssp. *Scleroptila shelleyi macarthuri* **Someren**, 1938
White-starred Robin ssp. *Pogonocichla stellata macarthuri* Someren, 1939

C. G. MacArthur (fl.1915–1945) collected entomological specimens at Merifano (1932) on the Tana River in Kenya. He was employed by the Kenya Game Department (1939–1945). A mammal is named after him.

Macartney

Pheasant genus *Macartneya* **Lesson**, 1831 NCR
[Now in *Lophura*]

Delacour's Crested Fireback *Lophura ignita macartneyi* **Temminck**, 1813

George 1st Earl Macartney (1737–1806) was an English diplomat and colonial governor. He was Envoy to Russia (1764–1767), Governor of Madras (Chennai) (1780–1786), Ambassador to China (1792–1794) and Governor of the Cape of Good Hope (1796–1798). He is remembered as the person who described the British Empire as one 'on which the sun never sets'. Macartney was given a specimen of this pheasant during a visit to Batavia (Jakarta).

MacClounie

MacClounie's Barbet *Lybius minor macclounii* **G. E. Shelley**, 1899
[Alt. Black-backed Barbet ssp.]
MacClounie's Robin Chat *Cossypha anomala macclounii* G. E. Shelley, 1903
[Alt. Olive-flanked Robin Chat ssp.]

John M. MacClounie (sometimes McClounie) (b.1870) was a British naturalist who collected in Nyasaland (Malawi) (1895–1906), where he was head of research at the Zomba scientific department (1909). He wrote a report (1902) recommending that the Nyika Plateau, where he collected plant type specimens, should be declared a protected area. His recommendation was not accepted then, although much of the plateau is now a national park.

MacConnell

(See **McConnell**, also **Vavasour, F.** and **Frederick (McConnell)**)

MacCormick

(See **McCormick**)

MacCoy

(See **McCoy**)

Macdonald, J. D.

Cinnamon Bracken Warbler ssp. *Bradypterus cinnamomeus macdonaldi* **C. H. B. Grant** & **Mackworth-Praed**, 1941 NCR
[JS *Bradypterus cinnamomeus cinnamomeus*]
Abyssinian Lark ssp. *Mirafra africanoides macdonaldi* **C. M. N. White**, 1953 NCR
[Alt. Foxy Lark; JS *Calendulauda alopex intercedens*]

James 'Jim' David Macdonald (1908–2002) worked in the Bird Department of the British Museum Natural History (1935–1968), and retired (1968) as Senior Scientific Officer in charge of the Bird Room and Deputy Keeper, Zoology Department. He ran collecting expeditions to Sudan (1938) and wrote *Birds of the Sudan* (1955); to South West Africa (Namibia) (1950–1951); and to Australia, in a series of expeditions sponsored by Major Harold Hall (q.v.). After retirement he and his wife settled in Brisbane where he became the first President of the Queensland Ornithological Society (1969). He continued to write well into his seventies, including *The Illustrated Dictionary of Australian Birds by Common Names* (1987).

Macdonald, K. C.

Kalij Pheasant *Gennæus macdonaldi* **Oates**, 1906 NCR
[JS *Lophura leucomelanos williamsi*]

Kenneth Campbell Macdonald (b.1868) was a British officer who was a District Superintendent with the Burma Police (1886–1915).

Macdonald, M.

Hood Mockingbird *Mimus macdonaldi* **Ridgway**, 1890
[Alt. Española Mockingbird]

(See **McDonald**)

Macé

Fulvous-breasted Woodpecker *Dendrocopos macei*
Vieillot, 1818
Large Cuckooshrike *Coracina macei* **Lesson**, 1830

Marc Joseph Marion Dufresne (1724–1772) was also known as Macé. He was the son of a wealthy French shipowner. He explored in the Indian and Pacific Oceans before being killed by Maoris in New Zealand. Macé's journal has been lost, but that of his second-in-command, Julien Crozet, was published as *Nouveau Voyage à la Mer du Sud* (1783).

MacFarlane

MacFarlane's Screech Owl *Megascops kennicottii*
macfarlanei **Brewster**, 1891
[Alt. Western Screech Owl ssp.]

Roderick Ross MacFarlane (1833–1920) was a trader, collector and naturalist. He is usually remembered in relation to collecting specimens of Eskimo Curlew *Numenius borealis* (probably extinct). MacFarlane joined the Hudson's Bay Company, in the Northwest Territories, as an Assistant Clerk (1852). He was sent to explore trade possibilities with the Inuit (1857) at the same time as Robert Kennicott (q.v.), sponsored by the USNM, visited several of the Hudson's Bay Company posts. After three years of Kennicott's training and enthusiasm, MacFarlane became the most ardent Arctic collector for the USNM. His yearly egg-collecting trips (1862–1865) usually started at the peak of the nesting (and mosquito) season (mid-June to mid-July). Because of the unpredictable behaviour of the Inuit, his route varied. His party consisted of c.21 Native American packers, hunters, collectors and canoe

carriers, and pack dogs. He spent the long winters at Fort Anderson packing his material and writing notes by candlelight for each of his numbered specimens. In late winter or spring the collections were shipped out for forwarding to Spencer Baird (q.v.) at the USNM. The process of dispatch was not without its vicissitudes, however, as recorded in MacFarlane's own writings: 'For the reasons mentioned in my letters (1866) to Professor Baird, but a small portion of the "Anderson Collection" 1865, was forwarded by the boats of 1866. Several cases and parcels were carried overland to Fort Good Hope during the Summer, while the remainder thereof, together with the whole of the Small Collection of 1866, and certain notes and memoranda connected with Collections '65 and preceding years, was secured near "Lockhart" [Carnwath] river. Before these could, however, be sent from Good Hope, Wolverines and other animals broke into the cache, and destroyed all the bird skins, besides the rarest and finest eggs – in fact, everything that had been left there was destroyed irreparably, except one box containing a lot of Geese, Duck & other Eggs which shall be forwarded to Washington in 1869. The last of the cases of 1865 were sent off by the Boats of 1867, and have doubtless ere this reached their destination'. Scarlet fever literally wiped out the Native American hunters at the fort (1865) and measles was equally devastating to the Inuit. Much of what was collected thereafter was lost. Nevertheless, his success is evident by the fact that one collection (1862) consisted of 550 cases of specimens. He also wrote the *Birds and Mammals of Northern Canada*.

MacGillivray, A. S.

MacGillivray's Parakeet *Barnardius zonarius macgillivrayi*
North, 1900
[Alt. Australian Ringneck ssp., 'Cloncurry Ringneck']

Alexander Sykes MacGillivray (1853–1907) was the brother of W. D. MacGillivray (q.v.), for whom he collected. He was a Queensland pastoralist who collected the parrot holotype near his station at Cloncurry. He made field notes for Alfred J. North (q.v.) that were incorporated into North's works on Australian birds' nests and eggs.

MacGillivray, J.

MacGillivray's Petrel *Pseudobulweria macgillivrayi*
G. R. Gray, 1860
[Alt. Fiji Petrel]

Orange-footed Scrubfowl ssp. *Megapodius reinwardt*
macgillivrayi G. R. Gray, 1862
Yellow-fronted White-eye ssp. *Zosterops flavifrons*
macgillivrayi **Sharpe**, 1900

John MacGillivray (1821–1867) was the eldest son of William MacGillivray (q.v.). He was a botanist and natural history collector for, among others, Lord Derby (q.v.), e.g. on board HMS *Fly* as a medical student, during its surveying expedition of Australia (1842–1845). He also surveyed as naturalist on board HMS *Rattlesnake* (1847–1850) and HMS *Herald* (1853–1854). He was one of the earliest naturalists to visit Tristan da Cunha. He emigrated from Great Britain to Australia, living in

Sydney, where he became one of that country's most prominent naturalists. He wrote *Narrative of the Voyage of H.M.S. Rattlesnake, Commanded by the Late Captain Owen Stanley during the Years 1846–1850, including Discoveries and Surveys in New Guinea, the Louisiade Archipelago, etc., to Which is Added the Account of Mr. E. B. Kennedy's Expedition for the Exploration of the Cape York Peninsula [by William Carron]* (1852). Sharpe gave no etymology when describing the white-eye subspecies, but of the possible contenders John MacGillivray seems the most probable.

MacGillivray, W.

MacGillivray's Warbler *Oporornis tolmiei* **J. K. Townsend**, 1839

MacGillivray's Seaside Sparrow *Ammodramus maritimus macgillivraii* **Audubon**, 1834

William MacGillivray (1796–1852) was a Scottish ornithologist and artist. He also became conservator of the Museum of the Royal College of Surgeons. He grew up on Harris, Outer Hebrides, and built extensive knowledge of the fauna and flora through field trips, often walking hundreds of miles. He walked to London (1819) just to see the bird collections at the British Museum! He wrote (1819) 'If I were to be a painter by profession my aim would be to copy nature with the scrupulous, yea, even servile attention; instead of displaying a genius that scorned control by a masterly dash, which would produce the likeness of nothing on earth'. MacGillivray lived in Edinburgh (1830s), where he wrote 27 scientific papers and 13 books, and assisted Audubon (q.v.) with his *Ornithological Biography*. He wrote a 5-volume *History of British Birds*, which he intended to illustrate with coloured plates based on his watercolours, but the great expense involved prevented him. However, Yarrell (q.v.) published his identically titled *History of British Birds* at the same time. This volume was given better reviews than MacGillivray's work, becoming the preferred text for the next 50 years. Audubon named the warbler as *Sylvia macgillivrayi*, but the species was named by Townsend earlier so the latter has priority althought the vernacular name stuck.

MacGillivray, W. D. K.

Honeyeater genus *Macgillivrayornis* **Mathews**, 1914 NCR
[Now in *Glycichaera*]

MacGillivray's Prion *Pachyptila salvini macgillivrayi* Mathews, 1912
[Alt. Salvin's Prion ssp.]
Palm Cockatoo ssp. *Probosciger aterrimus macgillivrayi* Mathews, 1912
Eclectus Parrot ssp. *Eclectus roratus macgillivrayi* Mathews, 1913
MacGillivray's Fairy-wren *Malurus coronatus macgillivrayi* Mathews, 1913
[Alt. Purple-crowned Fairy-wren ssp.]
Southern Boobook ssp. *Ninox boobook macgillivrayi* Mathews, 1913 NCR
[JS *Ninox boobook ocellata*]
Blue-faced Parrotfinch ssp. *Erythrura trichroa macgillivrayi* Mathews, 1914

Red-browed Finch ssp. *Aegintha temporalis macgillivrayi* Mathews, 1914 NCR
[JS *Neochmia temporalis minor*]

William David Kerr MacGillivray (1867–1933) was a naturalist and noted surgeon in Broken Hill, New South Wales, Australia (1901–1933). A keen amateur ornithologist he wrote a number of articles and papers on birds. He was President of the RAOU (1916–1918). At the time of his death he was working on a school textbook on Australian birds. His son Ian was also an ornithologist. McLennan (q.v.), a professional collector, collected most of the birds he studied.

MacGregor, M.

MacGregor's Bowerbird *Amblyornis macgregoriae* **De Vis**, 1890
[Alt. Lady MacGregor's Bowerbird, Mocha-breasted Bowerbird]

Lady Mary MacGregor *née* Mary Jane Cocks (c.1839–1919) was the second wife of Sir William MacGregor (1846–1919), naturalist, collector and diplomat. She was the daughter of Captain Cocks, a trader and harbour-master at Suva, Fiji. The Bowerbird is a New Guinean species, which was first described when Sir William MacGregor (q.v.) was governor of its country of origin. Lady MacGregor also had a locomotive and a rhododendron named after her. (See also **Maria (MacGregor)**)

MacGregor, W.

Crested Satinbird *Cnemophilus macgregorii* **De Vis**, 1890
[Alt. Antenna Satinbird; formerly Sickle-crested Bird-of-Paradise]
MacGregor's Honeyeater *Macgregoria pulchra* De Vis, 1897
[Alt. Ochre-winged Honeyeater]

Sir William MacGregor (1846–1919) was a British diplomat, physician and amateur naturalist. He was in the diplomatic corps and served first as Assistant Medical Officer in the Seychelles and later was Fiji's Chief Medical Officer (1875–1888). Then as Administrator of British New Guinea, he explored the interior. He then became Governor of Lagos Colony (1899–1904) (Nigeria). His final appointment until retirement to Aberdeen was Governor of Queensland, Australia (1909–1914). A reptile is also named after him.

Macgrigor

(See **McGrigor**)

Machik

Fawn-breasted Thrush *Zoothera machiki* **H. O. Forbes**, 1883

Surgeon Captain Jozsef Machik (DNF) was a Hungarian conchologist and collector in the Moluccas.

Machris

Firewood-gatherer ssp. *Anumbius annumbi machrisi* **Stager**, 1959 NCR; NRM

Maurice Alfred Machris (1905–1980) was an American businessman in the oil industry, collector, trophy hunter and sponsor; he sponsored and went on many expeditions, including to Brazil (1956). He was President of the Shikari Safari Club.

MacIlvaine

(See **McIlvaine**)

MacIlwraith

Fig Parrot sp. *Cyclopsittacus macilwraithi* **Rothschild**, 1897 NCR
[Alt. Orange-breasted Fig Parrot; JS *Cyclopsitta gulielmitertii amabilis*]

The etymology given by Rothschild says the parrot is named 'in honour of Mr MacIlwraith, from whom I received the specimen' (… from Papua New Guinea). It seems likely that the man intended was named McIlwraith (rather than *MacIl-wraith*). Sir Thomas McIlwraith (1835–1900) was Prime Minister of Queensland. He came from the family that co-founded the Anglo-Australian shipping company 'McIl-wraith, McEacharn & Co.', which included New Guinea in its trading area. Sir Thomas was knighted (1882) so the 'Mr' [rather than 'Sir'] MacIlwraith mentioned by Rothschild was probably Andrew (1844–1932), who was based in London and ran the company's European operations. He in turn may have been sent the parrot specimen by his brother John (1828–1902), who worked at the company's head office in Melbourne. Ultimately we can only conjecture which, if any, of the brothers Rothschild was referring to.

MacIntyre

Gilded Barbet ssp. *Capito auratus macintyrei* **Brodkorb**, 1939 NCR
[JS *Capito auratus punctatus*]

William Clarke-MacIntyre (d.1952) was an American zoologist, entomologist and professional collector in Ecuador (1922–1952). He was Professor of Entomology, Head of the Department of Zoology and Director of the Museum in Quito (1938–1940). Asthma and a heart condition forced him to move to a farm at sea level (1944) where he was able to collect locally, helped by his family until he died.

Mackenzie, E. K.

White-bellied Bustard ssp. *Eupodotis senegalensis mackenziei* **C. M. N. White**, 1945

E. K. Mackenzie (b.1904) worked for the Game and Tsetse Department, Rhodesia (Zimbabwe). While attached to the Angola Border Cattle Cordon (1945) he kept the describer advised of ornithological possibilities.

Mackenzie, F. S.

Common Reed Bunting ssp. *Emberiza schoeniclus mackenziei* Bird, 1936 NCR
[JS *Emberiza schoeniclus schoeniclus*]

Major Finlay Simon Mackenzie (d.1963) was Laird of South Uist, Scotland. He had served in a Highland Regiment and the Royal Canadian Mounted Police before inheriting a hotel,

which he ran (1928–1960). On the outbreak of WW2 he raised his clansmen and was furious to be told that, at over 60 years old, he was not young enough to lead them into battle!

Mackenzie, G. S.

Brown Woodland Warbler ssp. *Phylloscopus umbrovirens mackenzianus* **Sharpe**, 1892

Sir George Sutherland Mackenzie (1844–1910) was a businessman and explorer who was the Administrator of the Imperial British East Africa Company (1888–1891).

Mackenzie, H.

Antshrike genus *Mackenziaena* **Chubb**, 1918

Helen Mackenzie McConnell (1871–1954) was the wife of collector Frederick McConnell (q.v.). (See **Helen (McConnell)**)

Mackinder

Mackinder's Eagle Owl *Bubo capensis mackinderi* **Sharpe**, 1899
[Alt. Cape Eagle Owl ssp.]

Sir Halford John Mackinder (1861–1947) was an English geographer and traveller, and first person recorded climbing to the summit of Mount Kenya, having trained in the Alps and hired two Italian guides to help him climb the Kenyan peak. Although he was not a mountaineer, he thought scaling Mount Kenya would establish him as an adventurer and explorer. They managed to reach the mountain (1899) despite local hostilities, and cut a way through the forests to the alpine zone. Several attempts on the summit failed, but eventually they managed to find a way to the top. In later years Mackinder became Director of the London School of Economics (1903–1908), and was elected to Parliament (1910) for a Glasgow constituency.

Mackinlay

Mackinlay's Cuckoo Dove *Macropygia mackinlayi* **E. P. Ramsay**, 1878
[Alt. Spot-breasted Cuckoo Dove]

Dr Archibald Mackinlay (1850–1924) was a surgeon on HMS *Nymphe*, and an amateur ornithologist who collected in the South-West Pacific.

Mackinnon

Mackinnon's Shrike *Lanius mackinnoni* **Sharpe**, 1891
[Alt. Mackinnon's Fiscal]

Dr Archibald Donald Mackinnon (1864–1937) graduated from Aberdeen University (1892) with a degree in medicine. He had been employed by the Imperial British East African Company (1888–1891) and returned to Uganda as Medical Officer (1894–1897), then Director of Transport (1898–1899).

Macklot

Macklot's Pitta *Erythropitta macklotii* **Temminck**, 1834
[Alt. Papuan Pitta; Syn. *Erythropitta erythrogaster macklotii*]
Macklot's Trogon *Apalharpactes mackloti* **S. Müller**, 1836
[Alt. Sumatran Trogon]

Macklot's Sunbird *Leptocoma calcostetha* **Jardine**, 1843
[Alt. Copper-throated Sunbird]
Flowerpecker sp. *Dicaeum mackloti* S. Müller, 1843 NCR
[Alt. Blue-cheeked Flowerpecker; JS *Dicaeum maugei*]

Dr Heinrich Christian Macklot (1799–1832) was a German naturalist and taxidermist appointed to assist members of the Dutch Natural Science Commission. He went on an expedition to New Guinea and Timor on the *Triton* (1828–1830). He apparently became so infuriated at seeing his collections burned by a mob in Java that he attacked the mob and was speared to death. A mammal and a reptile are also named after him.

Mackworth-Praed

(see **Praed**)

Maclatchy

African Black Duck ssp *Anas sparsa maclatchyi* **Berlioz**, 1947 NCR
[NUI *Anas sparsa leucostigma*]

Alain R. Maclatchy (fl.1949–1965) was a French colonial administrator in tropical Africa, a collector, big-game hunter and zoologist. He was in Gabon and the Congo for over 20 years (1920s–1940s). He retired to Rodez in southern France. Among other works he wrote, with René Malbrant (q.v.), *Faune de l'Equateur Africain Français* (1949).

Maclaud

African Spotted Creeper *Salpornis maclaudi* **Oustalet**, 1906 NCR
[JS *Salpornis salvadori emini*]

Dr Charles Maclaud (1866–1933) was the French Resident at Timbo in French Guinea (Republic of Guinea) (1898). He published 'Notes anthropologiques sur les Diolas de la Casamance' (1907) and *Gouvernement Général de l'Afrique Occidentale Française – Notes sur les Mammifères et les Oiseaux de l'Afrique Occidentale: Casamance, Fonta-Dialon, Guinée Française et Portugaise* (1906) (Fonta-Dialon, or Fouta-Djallon, is a highland region in west-central Guinea.). A mammal is named after him.

Maclean, G.

Sandgrouse genus *Macleanornis* **Wolters**, 1974 NCR
[Now in *Pterocles*]

Dr Gordon Lindsay Maclean (1937–2008) was a South African zoologist and ornithologist. He worked for De Beers Diamond Mines in Namibia before taking a bachelor's degree (1963) and his doctorate (1968) at Rhodes University. He was Emeritus Professor at the University of KwaZulu-Natal, which he joined as a lecturer (1968).

Maclean, J.

(See **McLean**).

Macleay, A.

Macleay's Kingfisher *Todiramphus macleayii* **Jardine** & **Selby**, 1830
[Alt. Forest Kingfisher]

Alexander Macleay (1767–1848) was born in Scotland but rose to prominence in Australia. He was appointed Colonial Secretary for New South Wales (1825). His chief interest was entomology and he had what was believed to be the most extensive collection of any private individual in Britain, possibly the world, at that time. When he left to take up his post in Sydney he took his family, library and collection with him. He was granted land at Elizabeth Bay where he laid out a botanic garden. His entomological collections became the basis of the Macleay Museum at the University of Sydney. He was uncle to William John MacLeay (q.v.).

Macleay, S.

Sicklebill sp. *Epimachus macleayanae* **E. P. Ramsay**, 1887 NCR
[Alt. Brown Sicklebill; JS *Epimachus meyeri meyeri*]

Lady Susan Emmeline Deas Macleay *née* Thompson (1838–1903) was the wife of Sir William John Macleay (q.v.).

Macleay, W. J.

Macleay's Honeyeater *Xanthotis macleayanus*
E. P. Ramsay, 1875
[Alt. Buff-striped Honeyeater]

Macleay's Fig Parrot *Cyclopsitta diophthalma macleayana*
Ramsay, 1874
[Alt. Double-eyed Fig Parrot ssp.]

Sir William John Macleay (1820–1891) was a Scottish medical student who followed his uncle Alexander (q.v.) to Sidney (1838), where he became an all-round naturalist. He wrote widely on entomology, ichthyology and zoology and took part in several collecting expeditions. He published a two-volume *Descriptive Catalogue of Australian Fishes* (1881). He was knighted two years before he died (1889). The whole of the Macleay family were avid naturalists and collectors, so prolifically that the Macleay Museum University of Sydney was built (1887) to house their vast collection. Alexander Macleay's insect collection (see above) was added to by his son, William Sharp Macleay (1792–1865), and expanded to include all aspects of natural history by William John Macleay, after whom a mammal and a reptile are named.

Maclennan

Songlark genus *Maclennania* **Mathews**, 1917 NCR
[Now in *Cincloramphus*]

Red-cheeked Parrot ssp. *Geoffroyus geoffroyi maclennani*
W. D. K. MacGillivray, 1913
Black-throated Finch ssp. *Poephila cincta maclennani*
Mathews, 1918 NCR
[JS *Poephila cincta atropygialis*]

(See **McLennan**)

Macmillan, L.

Silvereye ssp. *Zosterops lateralis macmillani* **Mayr**, 1937
Sacred Kingfisher ssp. *Todiramphus sanctus macmillani*
Mayr, 1940

L. Macmillan (DNF) was an explorer, adventurer and collector for the AMNH, New York, in the New Hebrides (1932 and 1936–1937), the Loyalty Islands (1937) and New Caledonia (1939).

Macmillan, W. N.

Black Chat sp. *Pentholaea macmillani* **Sharpe**, 1906 NCR
[Alt White-fronted Black Chat; JS *Pentholaea albifrons pachyrhyncha*]

Macmillan's Waxbill *Estrilda astrild macmillani* **Ogilvie-Grant**, 1907
[Alt. Common Waxbill ssp.]
Parasitic Weaver ssp. *Anomalospiza imberbis macmillani*
Bannerman, 1911 NCR; NRM
[Alt. Cuckoo-finch]
White-winged Flufftail ssp. *Sarothrura ayresi macmillani*
Bannerman, 1911 NCR; NRM

Sir William Northrop Macmillan (1872–1925) was a huge (reputedly seven feet tall – his sword belt measured 64 inches), rich American, from St Louis, Missouri. After ranching (1890–1893) he became a businessman working in St Louis (1893–1898) and England (1899–1901). He arrived in East Africa (1901) to go big-game hunting (1901–1905), fell in love with an area of Kenya, bought half a mountain (Ol Donyo Sabuk, 2,146 m) and developed a ranch at its base. Theodore Roosevelt (q.v.) stayed at this ranch (1911) while on his famous safari – Juja Farm is now a popular location for film crews. Macmillan and his wife Lady Lucie were philanthropists who founded the MacMillan Library (Nairobi) and bequeathed their ranch 'to the nation'; it is now a national park and they are both buried there. During WW1 Macmillan joined the 25th Royal Fusiliers – a rag, tag and bobtail outfit known officially as the 'Legion of Frontiersman', but was known as 'The Old and Bold' by its members. Philip Zaphiro (q.v.), who was employed by MacMillan, collected the type specimen of the waxbill at Ibago, Sudan, during his Sudan expedition (1903–1904). Two mammals are named after him.

Macneill

Macneill's Lophophorus *Lophophorus nigelli* **Jardine &
Selby**, 1829 NCR
[Alt. Caspian Snowcock; JS *Tetraogallus caspius*]

Sir John Macneill (1795–1883) studied medicine at Edinburgh, qualified (1814) and joined (1816) the British East India Company in Bombay as a surgeon. He was attached to the East India Company's legation in Tehran, Persia, as medical officer and later as political assistant (1824–1835). He served as Secretary to a special British embassy in Tehran (1835–1836), retired from the East India Company's service and returned to England (1836). He was afterwards (1838–1844) leader of the official British Government mission to Persia. He was appointed first Chairman of the Board (1845–1878) of the newly enacted Scottish Poor Law. He was married three

times: to Innes Robinson (1814), Elizabeth Wilson (1823), and Lady Emma Campbell, a daughter of the 7th Duke of Argyll, (1870). His only child was a daughter by his second wife, born in Persia and named Ferooza. He purchased Colonsay and Oronsay from his brother, Duncan in 1870. He was the first to bring the bird (a female) to Europe. He died in Cannes. *Nigelli* is a latinisation of Neill.

Macpherson, D. W. K.

Macpherson's White-winged Apalis *Apalis chariessa
macphersoni* **J. Vincent**, 1934

David William Kinloch Macpherson (1900–1982) was born in India where his father was Chief Secretary to the Government of Bengal. Age four he was sent back 'home' to Scotland to be educated. He was a keen ornithologist from early childhood. After training at the Royal Military College at Sandhurst he was commissioned Lieutenant in the Highland Light Infantry. He served in Africa with the King's African Rifles before being posted to India. He left the army to become a tobacco grower and manager of tea and coffee plantations, before deciding to farm for himself, leasing land in Nyasaland (Malawi) to grow tobacco. He raised the money by hunting elephants and collecting birds and their eggs (1928–1929); his youngest daughter Isabel Macpherson published his diaries for the period as *Little Birds and Elephants* (2005). During that time he suffered many tribulations and hardships, long treks and recurrent malaria. He called his plantation 'Kanongo', (place of the porcupine) but needed to suppliment his income by managing other estates. During one such period (1933) he continued collecting around Mount Cholo and donated the specimens to the BMNH. He raised three children including one he named Prinia.

Macpherson (Range)

Olive Whistler ssp. *Pachycephala olivacea macphersoniana*
H. L. White, 1920

Named after the Macpherson (or McPherson) Range of mountains in eastern Australia.

MacQueen

MacQueen's Bustard *Chlamydotis macqueenii* **J. E. Gray**,
1832

General Thomas Richard MacQueen (1792–1840) collected in the Himalayas and north-west India and presented the bustard to the BMNH. At the time when he collected it, he was a major in the 45th Bengal Native Infantry – a regiment in the Bengal Army of the Honourable East India Company.

Madame Verreaux

Madame Verreaux's Sunbird *Cinnyris johannae*
E & J. Verreaux, 1851
[Alt. Johanna's Sunbird]

Johanna Verreaux (DNF) was the wife of Jean Baptiste Edouard Verreaux. He and Jules Verreaux named the sunbird in her honour. (See **Verreaux**)

Madarász

Madarász's Parrot *Psittacella madaraszi* **A. B. Meyer**, 1886
[Alt. Madarász's Tiger Parrot]
Swift sp. *Chaetura madaraszi* Keve, 1959 NCR
[Alt. Böhm's Spinetail; JS *Neafrapus boehmi sheppardi*]

Chestnut-backed Sparrow Lark ssp. *Eremopterix leucotis madaraszi* **Zarudny**, 1902
Crested Lark ssp. *Galerida cristata madaraszi* Herman, 1903 NCR
[JS *Galerida cristata meridionalis*]
Little Shrike-thrush ssp. *Colluricincla megarhyncha madaraszi* **Rothschild** & **Hartert**, 1903
Tawny-breasted Honeyeater ssp. *Xanthotis flaviventer madaraszi* Rothschild & Hartert, 1903
European Greenfinch ssp. *Carduelis chloris madaraszi* **Tschusi**, 1911
Green Oriole ssp. *Oriolus flavocinctus madaraszi* **Mathews**, 1912 NCR
[JS *Oriolus flavocinctus kingi*]

Gyula von Madarász (Gyula Madarász Gyulara) (1858–1931) was a Hungarian ornithologist and painter, whose landscapes and birds were exhibited in the National Salon. He illustrated his own work *Magyarorszag Madarai* (Birds of Hungary), which was published in Hungarian and German. He was also a contributor to *Zeitschrift für Ornithologie* and made a special study of manakins (Pipridae). He visited Ceylon (Sri Lanka) (1896) and Sudan (1910). Remarkably, 'Madarász' means 'birder'. His papers were at the Royal Hungarian Museum, Budapest, where he was Curator of Birds (1881–1915), but it was destroyed (1956) during the fighting in the anti-Communist uprising.

Madoc

Blue Rock Thrush ssp. *Monticola solitarius madoci* **Chasen**, 1940

Guy Charles Madoc (1911–1999) was an ornithologist who joined the Federated Malay States Police (1930). He was captured on the fall of Singapore and was a prisoner-of-war in Changi Jail (1942–1945). He was attached to the British Embassy in Thailand (1948) and became Head of Special Branch (1952) and Director of Intelligence in Malaya (Malaysia) (1954–1957). He stayed on in Malaysia after its independence until 1959, when he retired to the Isle of Man, serving on the local Police Authority (1963–1988).

Madzoed

Asian Paradise-flycatcher ssp. *Terpsiphone paradisi madzoedi* **Chasen**, 1939 NCR
[JS *Terpsiphone paradise affinis*]

Madzoed or Madzud (DNF) was a Javanese collector and preparator at Zoologisch Museum Buitenzorg (Bogor), East Indies (Indonesia).

Maës

Maes's Laughingthrush *Garrulax maesi* **Oustalet**, 1890
[Alt. Grey Laughingthrush]

Albert Maës (1845–1914) was a French aviculturist who also made a collection of birds now held by the Natural History Museum of Bourges, France. He donated the holotype to the MNHN.

Magda

Crested Lark ssp. *Galerida cristata magdae* **Loudon** & **Zarudny**, 1903 NCR
[JS *Galerida cristata caucasica*]

Baroness Magda zur Mühlen von Loudon (b.1876) was the wife of Baron Harald von Loudon (q.v.).

Magellan

Magellan's Eagle Owl *Bubo magellanicus* **Gmelin**, 1788
[Alt. Magellanic Horned Owl, Lesser Horned Owl]
Magellanic Diving-petrel *Pelecanoides magellani* **Mathews**, 1912

These birds are named after the Straits of Magellan, where they occur, rather than after the discoverer of the straits, explorer Ferdinand Magellan (Magalhães) (c.1480–1521). The same is true of other 'Magellanic' species such as the Magellanic Penguin *Spheniscus magellanicus*. Only in the case of the owl has Magellan become used as a vernacular name. A reptile also has 'Magellan' in its name.

Magnolia

Magnolia Warbler *Dendroica magnolia* **A. Wilson**, 1811
[Syn. *Setophaga magnolia*]

Pierre Magnol (1638–1715) was a French physician and botanist who came from a long line of apothecaries. *Magnolia* trees and subsequently the warbler are named after him – the bird indirectly in relation to its supposed habitat, although it actually frequents conifers. Magnol renounced his Calvinist faith, as this was preventing him from gaining public office. In his *Prodromus Historiae Generalis Plantarum* (1689) he was among the first to classify plants in tables, making quick identification possible. He was also the first person to use the term 'family' in the sense of a natural group. In another work, *Botanicum Monspeliense* (1676), he described 1,354 species growing around the area where he lived. He gave their habitats and added notes on their medicinal or other uses. This work was the basis for the Linnaean dissertation on the Montpellier region (1756). His other works included *Hortus Regius Monspeliensis* (1697), which was a catalogue of his private garden, and *Novus Caracter Plantarum*, published posthumously (1720).

Magrath

Bulbul sp. *Molpastes magrathi* **C. H. Whitehead**, 1908 NCR
[Hybrid: *Pycnonotus leucogenys* x *Pycnonotus cafer intermedius*]

Magrath's Wren *Troglodytes troglodytes magrathi* C. H. Whitehead, 1907
[Alt. Eurasian Wren ssp.]

Lieutenant-Colonel Henry Augustus Frederick Magrath (1867–1940) was an officer in the British Army in India

(1888–1929), serving with Sikh regiments and commanding the 51st Sikhs in Mesopotamia (Iraq) (1916–1918). An amateur ornithologist, he co-wrote *Birds of Kohat* (1909).

Maguire

Black-fronted Tyrannulet ssp. *Phylloscartes nigrifrons maguirei* **W. H. Phelps** & **W. H. Phelps Jr**, 1951 NCR; NRM
Tepui Whitestart ssp. *Myioborus castaneocapilla maguirei* W. H. Phelps & W. H. PhelpsJr, 1961

Dr Bassett Maguire (1904–1991) was an American botanist, explorer and collector in many tropical American countries. The University of Georgia awarded his bachelor's degree (1926) and Cornell his doctorate (1938). He worked as Assistant Professor of Botany, Utah State Agricultural College (1931–1942), and Curator and Director, New York Botanical Gardens (1943–1975), retiring as Senior Scientist Emeritus.

Maharao

Coqui Francolin ssp. *Peliperdix coqui maharao* **W. L. Sclater**, 1927

Rao Kengarji III, Maharao of Kutch (1866–1942), was in Abyssinia (Ethiopia) (c.1926). He is noted for modernising his state greatly over the sixty-six years of his reign.

Maillard

Réunion Harrier *Circus maillardi* **J. Verreaux**, 1862

Louis Maillard (1814–1865) was a French engineer and botanist who wrote *Notes sur l'Ile de la Réunion* (1862), on which island (1854) he discovered the remains of a now-extinct species of giant tortoise.

Mailliard

Song Sparrow ssp. *Melospiza melodia mailliardi* **Grinnell**, 1911 NCR
[NUI *Melospiza melodia heermanni*]
Red-winged Blackbird ssp. *Agelaius phoeniceus mailliardorum* **Van Rossem**, 1926

Joseph Mailliard (1857–1945) was an ornithologist, collector and explorer in Alaska (1896) and Chile (1902). The Red-winged Blackbird subspecies is named after both him and his brother John Ward Mailliard (1862–1935), also an ornithologist. Both of them were involved in the California Academy of Sciences, to which they donated their large collection, and Joseph was its Curator (1919–1927).

Maingay

Brown Wood Owl ssp. *Strix leptogrammica maingayi* **Hume**, 1878

Dr Alexander Carroll Maingay (1836–1869) was a Scottish botanist and collector in Burma (Myanmar), Malaya (Malaysia), China and Japan. He graduated as a physician at Edinburgh (1858) and was Assistant Residency Surgeon at Malacca (1865). He became Superintendent of Rangoon Central Prison, where he was killed during a riot. Hume wrote

that he named the owl 'in memory of a gentleman whose long labours in the cause of botany and natural history appear to have met with very scant recognition'.

Major Mitchell

Major Mitchell's Cockatoo *Cacatua leadbeateri* **Vigors**, 1831 [Alt. Leadbeater's Cockatoo; Syn. *Lophochroa leadbeateri*]

(See **Mitchell, T. L.**)

Makawa

White-chested Tinkerbird *Pogoniulus makawai* **C. W. Benson** & **Irwin**, 1965

Olive Bush-shrike ssp. *Chlorophoneus olivaceus makawa* C. W. Benson, 1945

Jali Makawa (fl.1968) was an African hunter, taxidermist and collector for C. W. Benson (q.v.) in Northern Rhodesia (Zambia) (1932–1963) and accompanied him in WW2 in the Abyssinian campaign (1941), during which he made an outstanding collection of birds. Makawa, who came from Nyasaland (Malawi), stayed in Benson's employ until Benson retired (1965), after which he was recruited by others to collect in Madagascar and other Indian Ocean localities. He was based at the National Museum of Zambia, Livingstone (1968). Michael Irwin (q.v.) said to us in a letter 'Jali knew his birds better than anyone in Africa, white or black'. The tinkerbird is known only from the type specimen (1964) taken at Mayau, north-western Zambia.

Makayi

Red-capped Crombec ssp. *Sylvietta ruficapilla makayii* **C. M. N. White**, 1953

M. G. Makayi (DNF) was an African hunter and collector for the describer in Northern Rhodesia (Zambia).

Maki

Japanese Tit ssp. *Parus major makii* **Momiyama**, 1927 NCR [JS *Parus minor commixtus*]

S. Maki (DNF) collected the holotype (1920) on Formosa (Taiwan).

Malbrant

Malbrant's Lark *Mirafra africana malbranti* **Chapin**, 1946 [Alt. Rufous-naped Lark ssp.]

René Malbrant (1903–1961) was a French veterinarian, an expert in pests and diseases of cattle, in Chad (1927) who was elected as a Deputy in the National Constituent Assembly (Oubangui-Chari) (1946–1958). He was awarded the Légion d'Honneur, Croix de Guerre for his work for the resistance during WW2. Malbrant wrote, with Alain R. Maclatchy (q.v.), *Faune du Centre Africain Français (Mammifères et Oiseaux)* (1952). He died in Paris having spent much of his life alternating between Central Africa and France.

Malcolm

Malcolm's Babbler *Turdoides malcolmi* **Sykes**, 1832
[Alt. Large Grey Babbler]

Major-General Sir John Malcolm (1769–1833) was the Governor of Bombay (1827) and an amateur historian. He was a close friend and correspondent of the First Duke of Wellington. He was selected by Lord Wellesley to undertake a diplomatic mission to Persia, where he concluded two important treaties, one political and the other commercial (1800–1801). Malcolm is also credited with introducing the potato to Persia, having noticed on his first visit that the climate was suitable for this crop. He must have been enormously wealthy (or foolish), since he once lost £30,000 at a single sitting of cards at White's Club in London. He wrote a number of historical texts, including *Sketch of the Sikh* (1812), *A History of Persia* (1815), *Memoir of Central India* (1823), and *Political History of India from 1784 to 1823* (1826).

Malcolm Smith

Warbler sp. *Cryptolopha malcolmsmithi* **H. C. Robinson &
    Kloss**, 1919 NCR
[Alt. Ashy-throated Warbler; JS *Phylloscopus maculipennis*]

Dr Malcolm Arthur Smith (1875–1958) was an English herpetologist who was in Bangkok (1902–1924). He practised medicine there including five years as Court Physician, publishing his memoirs under the title *A Physician at the Court of Siam* (1947). He visited French Indo-China (1918). He was President of the British Herpetological Society (1949–1954). An amphibian and three reptiles are named acfter him.

Malherbe

Flicker genus *Malherbipicus* **Bonaparte**, 1854 NCR
[Now in *Colaptes*]

Malherbe's Golden-backed Woodpecker *Indopicus
    delesserti* Malherbe, 1849 NCR
[JS *Chrysocolaptes lucidus guttacristatus*]
Island Bronze-naped Pigeon *Columba malherbii*
    J. & E. Verreaux, 1851
[Alt. São Tomé Pigeon]
Malherbe's Flicker *Colaptes chrysoides* Malherbe, 1852
[Alt. Gilded Flicker]
Malherbe's Parakeet *Cyanoramphus malherbi* **Souancé**,
    1857
Green-backed Woodpecker *Chrysopicus malherbei* **Cassin**,
    1863 NCR
[JS *Campethera cailliautii*]

Crimson-crested Woodpecker ssp. *Campephilus
    melanoleucos malherbii* **G. R. Gray**, 1845

Alfred Malherbe (1804–1866) was a French magistrate, judge and passionate botanist, zoologist and ornithologist. His interest in nature was inspired by his Mauritius childhood. He produced the 4-volume *Monographie des Picidées* (1859–1862), a major study of woodpeckers describing every known species, with hand-colored lithographs accompanying the detailed reviews. He also wrote extensively on the birds of Algeria and Sicily. At one time he was Administrator of the Museum of Metz where he lived and where all his works

(1839 onwards) were published. He wrote *Revue des Collections Composant, en 1857, le Muséum d'Histoire Naturelle de la Ville de Metz* (1857).

Malzac

Ghazal Puffback *Dryoscopus gambensis malzacii* **Heuglin**,
    1870
[Alt. Northern Puffback ssp.]

Alphonse de Malzac (1800–1860) started out as a diplomat, being an attaché at the French embassy in Athens, but became an ivory hunter and slave-trader in the Sudan. He was said to reward his retainers by giving them slaves instead of wages. He was very knowledgeable about the White Nile and the fauna of that river and environs.

Mandel

Rusty Flowerpiercer ssp. *Diglossa sittoides mandeli*
    E. R. Blake, 1940

Leon Mandel II (1902–1974) sponsored the Field Museum's Mandel Venezuela Expedition (1931–1932). His family owned Chicago's major department store, Mandel Brothers, which traded for over a century (1855–1960). He also sponsored expeditions to Guatemala (1936) and the Galapagos Islands (1941), in which he and his wife Carola participated using his yacht *Carola* for transport and a base. He was friendly with the defendants in the infamous Leopold and Loeb case (1925) and gave evidence – they both went to prison for life (Leopold was eventually released, went to live in Puerto Rico and wrote *Checklist of Birds of Puerto Rico and the Virgin Islands*). An amphibian is named jointly after Leon and his brother, Fred.

Mandelli

Chestnut-breasted Partridge *Arborophila mandellii* **Hume**,
    1874
Russet Bush Warbler *Bradypterus mandelli* **W. E. Brooks**,
    1875
Mandelli's Snowfinch *Onychostruthus taczanowskii*
    Przevalsky, 1876
[Alt. White-rumped Snowfinch; Syn. *Pyrgilauda
    taczanowskii*]

Puff-throated Babbler ssp. *Pellorneum ruficeps mandellii*
    Blanford, 1871
Hodgson's Treecreeper ssp. *Certhia hodgsoni mandellii*
    W. E. Brooks, 1874
Rusty-capped Fulvetta ssp. *Alcippe dubia mandellii* **Austen**,
    1876
[Syn. *Schoeniparus dubius mandellii*]
Mandelli's Leaf Warbler *Phylloscopus humei mandellii*
    W. E. Brooks, 1879
[Hume's Leaf Warbler ssp.]

Louis H. Mandelli (1833–1880) was a tea plantation manager in Assam, India, and an avid ornithologist. He was the subject of a book by Fred Pinn, *Louis H. Mandelli, Darjeeling Tea Planter and Ornithologist* (1985), which contains several pages of birds identified and named by Mandelli, and refers to his excursions into 'the snow' seeking birds. Mandelli's collections are now in the BMNH, Tring, England. A mammal is named after him.

Mandt

Mandt's Guillemot *Cepphus grylle mandtii* Mandt, 1822
[Alt. Black Guillemot ssp.]

Martin Wilhelm Mandt (1800–1858) was a German zoologist and physician at the Russian imperial court of Tsar Nicholas I. He wrote *Observationes in Historiam Naturalem et Anatomiam Comparatam in Itinere Groenlandico Factæ* (1822). He also wrote several treatises on the treatment of cholera.

Mangin

Crimson Sunbird ssp. *Aethopyga siparaja mangini* **Delacour & Jabouille**, 1924

Professor Louis Mangin (1852–1937) was a French botanist, Professor of Botany (1904–1931) and Director (1919–1931) of the MNHN, Paris.

Manilia

Hummingbird genus *Manilia* **Mulsant & E. & J. Verreaux**, 1866 NCR
[Now in *Calothorax*]

Manilia was a Roman courtesan, mentioned in a work by the jurist Gaius Ateius Capito.

Mann

White-spotted Owlet-nightjar ssp. *Aegotheles wallacii manni* **Diamond**, 1969

Sir Alan Harbury Mann (1914–1970) was an Australian jurist, naturalist and collector and Chief Justice of Papua New Guinea. He served in the Royal Australian Air Force in WW2.

Manning

Black-fronted Bush-shrike ssp. *Chlorophoneus nigrifrons manningi* **G. E. Shelley**, 1899

Brigadier-General Sir William Henry Manning (1863–1932) was a British army officer in India (1888–1893). He became Deputy Commissioner for British Central Africa (1897), then first Inspector-General of the King's African Rifles (1901–1907). He became Governor of Nyasaland (Malawi) (1910–1913), Governor of Jamaica (1913–1918), and Governor of Ceylon (Sri Lanka) (1918–1925).

Mantell

Takahe genus *Mantellornis* **Mathews**, 1911 NCR
[Now in *Porphyrio*]

Mantell's Moa *Pachyornis geranoides* **Owen**, 1848 EXTINCT
Mantell's Rail *Porphyrio mantelli* Owen, 1848 EXTINCT
[Alt. North Island Takahe]
North Island Brown Kiwi *Apteryx mantelli* **A. D. Bartlett**, 1852

Walter Baldock Durant Mantell (1820–1895) was a British amateur naturalist and geologist, originally trained as a physician, who moved to New Zealand (1840) and settled in Wellington, where he became Deputy Postmaster. He learnt the Maori language and (1848) became a government land purchase agent in the South Island. Growing disillusioned with trying to get people to leave their land, he instead became the champion of Maori rights. He was in England (1855–1859) fighting on their behalf (without success). After his return to New Zealand he became a Member of Parliament and was Minister of Native Affairs in three governments. He resigned on every occasion when promises made to the Maoris were not kept. He searched for live moas but had to be content with sending his father, the geologist Gideon Mantell, some of the first subfossil remains to reach England.

Manteufel

Siberian Jay ssp. *Perisoreus infaustus manteufeli* **Stachanov**, 1928 NCR
[JS *Perisoreus infaustus infaustus*]

Dr P. A. Manteufel (b.1879) was a German physician and zoologist who became Professor at the Medical Health Department, Berlin. He investigated animal behaviour changes during eclipses of the sun. He wrote *The Life of Fur-bearing Animals* (1947).

Mantou

Mantou's Riflebird *Craspedophora mantoui* **Oustalet**, 1891 NCR
[Presumed hybrid: *Seleucidis melanoleucus* x *Ptiloris magnificus*]

M. Mantou (DNF) was a French plumassier (feather dealer) in Paris who presented the first known example of this bird to the Paris Museum.

Manuel, C. G.

Purple-throated Sunbird ssp. *Leptocoma sperata manueli* **Salomonsen**, 1952 NCR
[JS *Leptocoma sperata sperata*]
White-bellied Munia ssp. *Lonchura leucogastra manueli* **Parkes**, 1958

Dr Canuto Guevarra Manuel (b.1902) was a Filipino ornithologist, Curator of Zoological Collections (1946) and Chief Zoologist (1958) at the Bureau of Science, Manila. He wrote 'A review of Philippine pigeons' and *Fifty Philippine Birds* (1936), and 'Undescribed and newly recorded Philippine birds' (1952) among other works.

Manuel (Fereira)

Red-necked Spurfowl ssp. *Pternistis afer manueli* **C. M. N. White**, 1945 NCR
[JS *Pternistis afer cranchii*]

Manuel Fereira (DNF) was a Portuguese trader in Lumbala, Angola.

Manumeten

Spangled Drongo ssp. *Dicrurus hottentottus manumeten*
Stresemann, 1914 NCR
[JS *Dicrurus bracteatus amboinensis*]

Manumeten (fl.1680) was a priest and leader of the Alifuru people of Seram, Moluccas.

Mappin

Mappin's Moa *Pachyornis mappini* Archey, 1941 NCR
EXTINCT
[JS *Pachyornis geranoides* (Owen, 1848)]

Sir Frank Crossley Mappin Bt (1884–1975) was born in England but, even after he inherited a considerable estate in England (1920), decided that New Zealand was where he wanted to live. He bought an orchard there, then a 12-acre property near Auckland, which he and his wife turned into fine gardens with rare plants. He was also a noted philanthropist. Mappin accompanied Gilbert Archey, the Director of the Auckland Institute and Museum, on several trips to search caves for moa remains.

Marais

Olive Bush-shrike ssp. *Chlorophoneus olivaceus maraisi*
W. L. Sclater, 1901 NCR
[JS *Chlorophoneus olivaceus olivaceus*]
Grey Hornbill ssp. *Tockus nasutus maraisi* **J. A. Roberts**,
1914 NCR
[JS *Tockus nasutus epirhinus*]

J. van O. Marais (1871–1904) was a South African forestry officer at Knysna, Cape Province.

Marcgrave

Blue-crowned Motmot ssp. *Momotus momota*
marcgravianus **O. M. Pinto** & **Camargo**, 1961

Georg Marcgrave (also Marcgraf) (1610–1644) was a German naturalist and astronomer. He studied botany, astronomy, mathematics and medicine in Germany and Switzerland until 1633. He then practised medicine in the Netherlands (1633–1637). The Dutch West India Company offered him the position of Personal Physician to Count Johan Maurits van Nassau-Siegen (1637), who was at that time in South America in the capacity of Governor of the colony of Dutch Brazil. When Marcgrave arrived (1638) he undertook the first zoölogical, botanical and astronomical expedition in Brazil, exploring various parts of the colony to study natural history and geography. Cuvier (q.v.) praised him as one of the best and most diligent observers and recorders of the era. He was sent to Angola by the West India Company, but died shortly after his arrival. He was co-author, with Willem Piso, of *Historia Naturalis Brasiliae,* an 8-volume work on the botany and zoology of Brazil published posthumously (1648). A mammal is named after him.

March

Black-faced Grassquit ssp. *Tiaris bicolor marchii* **Baird**, 1863

William Thomas March (1795–1872) was a lawyer and collector in Jamaica. He was Clerk to the Supreme Court of Jamaica (1843) and Secretary to the Governor of Jamaica (1868). He wrote 'Notes on the birds of Jamaica' (1864). He was a regular correspondent with Hooker at Kew and was an expert on Jamaican ferns.

Marchant

Blackcap Illadopsis ssp. *Illadopsis cleaveri marchanti*
Serle, 1956

Stephen Marchant (1912–2003) was a British petroleum geologist and ornithologist. He worked in Egypt (1938–1939), Nigeria (1939–1940 and 1946–1950), Ecuador (1954–1958), and Iraq (1959–1962). He moved to Australia (1963), edited *Emu* (1969–1981) and co-founded the *Handbook of Australian, New Zealand and Antarctic Birds* (1990–2006).

Marche

Marche's Fruit Dove *Ptilinopus marchei* **Oustalet**, 1880
[Alt. Flame-breasted Fruit Dove]
Palawan Hornbill *Anthracoceros marchei* Oustalet, 1885
Golden White-eye *Cleptornis marchei* Oustalet, 1889

Rock Pratincole ssp. *Glareola nuchalis marchei* Oustalet,
1877 NCR
[JS *Glareola nuchalis nuchalis*]
Helmeted Guineafowl ssp. *Numida meleagris marchei*
Oustalet, 1882 NCR
[NUI *Numida meleagris galeata*]

Antoine-Alfred Marche (1844–1898) was a French explorer and naturalist. He wrote *Trois Voyages dans l'Afrique Occidentale* (1879), and *Luçon et Palaouan* (1887). After his visit to the Philippines (1879–1883) he wrote *Voyage aux Philippines*, which included local recipes for cooking wild boar and venison. He also conducted archaeological explorations and removed a vast number of artefacts, sending them to the Musée de l'Homme, in Paris. A mammal is named after him.

Marchesa

Scarlet Minivet ssp. *Pericrocotus speciosus marchesae*
Guillemard, 1885

Named after the vessel *Marchesa* used by Francis Guillemard (q.v.) in his expedition to the Pacific (1882–1884).

Mardi

Hill Blue Flycatcher ssp. *Cyornis banyumas mardii*
Hoogerwerf, 1962 NCR
[JS *Cyornis banyumas ligus*]

Mardi (fl.1951) was a taxidermist at Bogor Museum, Java.

Marescot

Cuckooshrike sp. *Edolisoma marescoti* **Pucheran**, 1854 NCR
[Alt. Black Cicadabird; JS *Coracina melas*]

Lieutenant Jacques Marie Eugène Marescot du Thilleul (1810–1839) was a French naval officer and explorer. He was on board *Astrolabe* in the Pacific (1837–1839) alongside Pucheran. Marescot Point in the Antarctic is named after him.

Margaret (Boulton)

Margaret's Batis *Batis margaritae* **Boulton**, 1934
[Alt. Boulton's Batis, Boulton's Puffback Flycatcher]

Margaret Hall Boulton *née* Lander (DNF) was the first wife of the American ornithologist Wolfrid Rudyerd Boulton. She had previously been married to ornithologist Ernest G. Holt (q.v.).

Margaret (Conover)

White-throated Quail Dove ssp. *Geotrygon frenata margaritae* **E. R. Blake, G. Hoy & Contino**, 1961

Margaret Boardman Conover (b.1888) was the sister of Henry Boardman Conover (q.v.).

Margaret (Gilliard)

Blue Bird-of-Paradise ssp. *Paradisaea rudolphi margaritae* **Mayr & Gilliard**, 1951
Forest Honeyeater ssp. *Meliphaga montana margaretae* **Greenway**, 1966

Mrs Margaret Fitzell Gilliard *née* Tifft (1918–2008) was the wife of Ernest Thomas Gilliard (q.v.).

Margaret (Heine)

Common Paradise Kingfisher ssp. *Tanysiptera galatea margarethae* **Heine**, 1860
Long-tailed Sylph ssp. *Aglaiocercus kingii margarethae* Heine, 1863

Margaret Heine (fl.1840) was the wife of German ornithologist Ferdinand Heine Sr (q.v.).

Margaret (Holt)

Pearly-breasted Conebill *Conirostrum margaritae* **E. G. Holt**, 1931

Margaret Hall Holt *née* Lander (b.1894) was the wife (1922–1933) of the American ornithologist Ernest G. Holt. The National Museum of Rio de Janeiro organised two expeditions (1921 and 1922) in the Itatiaia National Park, Brazil, where Holt, who worked for the AMNH, New York, identified 187 species. Holt named the conebill after his wife 'as a very paltry though peculiarly fitting tribute.' (See also **Margaret (Boulton)**)

Margaret (Moore)

Margaret's Hummingbird *Atthis heloisa margarethae* **R. T. Moore**, 1937
[Alt. Bumblebee Hummingbird ssp.]

Mrs Margaret Moore (DNF) was the describer's second wife and the first to observe the hummingbird. (See **Cleaves**)

Margaret (Smith)

Margaret's Hummingbird *Lampornis amethystinus margaritae* **Salvin & Godman**, 1889
[Alt. Amethyst-throated Hummingbird ssp.]
Burnished-buff Tanager ssp. *Tangara cayana margaritae* **J. A. Allen**, 1891

Blue-capped Tanager ssp. *Thraupis cyanocephala margaritae* **Chapman**, 1912

Mrs Amelia 'Daisy' Smith *née* Woolwirth (DNF) was the wife of Herbert Huntington Smith (1851–1919), an American naturalist and collector. After marrying Smith (1880) she travelled extensively with him, notably in Paraguay, Brazil (1881–1886), Mexico (1889), Colombia and Venezuela (1891). In naming the hummingbird Salvin and Godman said: '… her energy has added much to our knowledge of the distribution of Mexican birds.' *Margarita* is the Spanish word for her nickname 'Daisy', but the etymologies of the birds do not make this clear.

Margaret (Zimmer)

Collared Inca ssp. *Coeligena torquata margaretae* **Zimmer**, 1948

Margaret Zimmer *née* Thompson (d.1945) married the describer in 1917. Although there is nothing in Zimmer's etymology we believe this to be named after his (late) wife.

Margareta (Johansen)

Whinchat ssp. *Pratincola rubetra margaretae* **Johansen**, 1903 NCR
[Now *Saxicola rubetra*; NRM]

Margareta Johansen (DNF) was the describer's mother.

Margareta (Koenig)

Dupont's Lark ssp. *Chersophilus duponti margaritae* **Koenig**, 1888

Mrs Margareta Koenig *née* Westfal (1865–1943) was the wife of Professor Alexander Ferdinand Koenig.

Margareta (le Roi)

Northern White-faced Owl ssp. *Otus leucotis margarethae* **von Jordans** & Neubaur, 1932 NCR
[Now *Ptilopsis leucotis*; NRM]

Mrs Margareta le Roi (DNF) was the wife of Dr Otto le Roi (see **Roi**).

Margarete (Mayr)

Makira Thrush *Zoothera margaretae* **Mayr**, 1935
[Alt. San Cristobal Thrush, White-bellied Thrush]

Margarete 'Gretel' Mayr *née* Simon (1912–1990) was the wife of Ernst Mayr (q.v.).

Margarete (Platen)

Variable Dwarf Kingfisher ssp. *Ceyx lepidus margarethae* **Blasius**, 1890

(See **Platen, M.**)

Margaretha

Margaretha's Lorikeet *Charmosyna margarethae* **Tristram**, 1879
[Alt. Duchess Lorikeet]

Princess Louise Margaretha Alexandra Victoria Agnes (1860–1917) was the daughter of Prince Friedrich Karl von Preussen. She married Prince Arthur, Duke of Connaught and Strathearn (1879), the third son of Queen Victoria.

Margarethe

Black-naped Fruit Dove ssp. *Ptilinopus melanospilus margaretha* **Meise**, 1930 NCR
[NUI *Ptilinopus melanospilus melanauchen*]

Mrs Margarethe Stevensen (DNF) of Berlin. We can find nothing more about her.

Margaretta

Margaretta's Hermit *Phaethornis malaris margarettae* **Ruschi**, 1972
[Alt. Great-billed Hermit ssp.]

Margaretta Lammot Du Pont Greenwalt (1902–1991) was married to Crawford Hallock Greenwalt (q.v.), who financed many studies undertaken by his friend, the Brazilian ornithologist Dr Augusto Ruschi (1916–1986), who in turn named the bird for his friend's wife.

Margarita (Island)

Clapper Rail ssp. *Rallus longirostris margaritae* **Zimmer** & W. H. Phelps, 1944
Red-legged Tinamou ssp. *Crypturellus erythropus margaritae* **W. H. Phelps** & **W. H. Phelps Jr**, 1948
Cocoa Woodcreeper ssp. *Xiphorhynchus susurrans margaritae* W. H. Phelps & W. H. Phelps Jr, 1949
Ferruginous Pygmy Owl ssp. *Glaucidium brasilianum margaritae* W. H. Phelps & W. H. Phelps Jr, 1951

Named after Margarita Island, Venezuela, to which these subspecies are endemic. 'Margarita' also appears in the names of a mammal and a reptile.

Margarita (Vassal)

Common Green Magpie ssp. *Cissa chinensis margaritae* **H. C. Robinson** & **Kloss**, 1919

Gabrielle Maud Vassal *née* Candler (See **Gabrielle**). Although Robinson and Kloss wrote that the magpie was named after 'Mrs G. M. Vassal', her middle name was Maud not Margaret/Margarita, and it may be that a simple error was made.

Marguerite (Delacour)

Black-headed Parrotbill *Psittiparus margaritae* **Delacour**, 1927

Mrs Marguerite Delacour (1860–1954) was the mother of Jean Delacour (q.v.). She was in Indochina (1931–1932), New York (1942) and Los Angeles (1952–1954).

Marguerite (Dorst)

Margarit's Parakeet *Psilopsiagon aurifrons margaritae* **Berlioz** & **Dorst**, 1956
[Alt. Mountain Parakeet ssp.]

Mrs Marguerite J. Dorst (DNF) was the wife of Jean Dorst (q.v.) and was in Peru (1946).

Maria

Brown-mandibled Aracari *Pteroglossus mariae* **Gould**, 1854

Grand Duchess Maria (1819–1876) was the wife of Maximilian, Prince of Eichstadt. Her father was Tsar Nicholas I of Russia.

Maria (Deignan)

Purple-naped Sunbird ssp. *Hypogramma hypogrammicum mariae* **Deignan**, 1943

Mrs Stella Maria Aglaé Leche Deignan (1901–1993) was an American anthropologist and the wife of the describer, Herbert Deignan (q.v.).

Maria (de Korte)

Elegant Pitta ssp. *Pitta elegans maria* **Hartert**, 1896

Mrs Marie de Korte (DNF) gave the pitta holotype to William Doherty (q.v.). She was the daughter of a Dutch official on Sumba.

Maria (Island)

Tres Marias Cardinal *Cardinalis cardinalis mariae* E. W. Nelson, 1898
[Alt. Northern Cardinal ssp.]

Named after the type locality, Maria Madre Island, one of the Tres Marias Islands, Mexico. This island also gives its name to a reptile.

Maria (Jacquinot)

Many-coloured Fruit Dove ssp. *Ptilinopus perousii mariae* **Pucheran**, 1853

Marie Jacquinot (DNF) was the wife of Étienne Stephan Jacquinot (see **Stephan**) and mother of Honoré and Charles Hector Jacquinot (see **Jacquinot**).

Maria Koepcke

Maria Koepcke's Screech Owl *Megascops koepckeae* Hekstra, 1982
[Alt. Koepcke's Screech Owl]

(See **Koepcke**)

Maria (MacGregor)

Honeyeater genus *Maria* **Giglioli**, 1897 NCR
[Now in *Macgregoria*]

Satinbird sp. *Cnemophilus mariae* **De Vis**, 1894 NCR
[Alt. Loria's Satinbird; JS *Cnemophilus loriae*]

(See **MacGregor, M.**)

Maria (Martin)

Maria's Woodpecker *Picus martinae* **Audubon**, 1839 NCR
[Alt. Hairy Woodpecker; JS *Picoides villosus*]

(See **Martin, Maria**)

Maria (Reichenow)

Frau Reichenow's Bird-of-Paradise *Paradisaea maria*
Reichenow, 1894 NCR
[Hybrid: *Paradisaea raggiana* x *Paradisaea guilielmi*]

Maria Reichenow *née* Cabanis (DNF) was the wife of the describer, Anton Reichenow (q.v.), and daughter of Jean Cabanis (q.v.).

Maria (von Madarász)

Evergreen Forest Warbler ssp. *Bradypterus lopezi mariae*
Madarász, 1905

Maria von Madarász (fl.1905) was the wife of the describer, Gyula von Madarász.

Marian (Mathews)

Buttonquail genus *Marianornis* **Mathews**, 1917 NCR
[Now in *Turnix*]

Australian Raven *Corvus marianae* Mathews, 1911 NCR
[JS *Corvus coronoides*]
Thornbill sp. *Acanthiza marianae* **S. A. White**, 1915 NCR
[Alt. Slaty-backed Thornbill; JS *Acanthiza robustirostris*]

Mrs Cecil Marian Mathews (1865–1938) was the wife of Gregory M. Mathews (q.v.; see also **Cecilia**).

Marian (Scott)

Marian's Marsh Wren *Cistothorus palustris marianae*
W. E. D. Scott, 1888

Mrs Marian J. Scott (DNF) was the wife of the describer, W. E. D. Scott (q.v.).

Marie, E. A.

New Caledonian Thicketbird *Megalurulus mariei* **J. Verreaux**, 1869

Edouard Auguste Marie (1835–1889) was a French collector and naturalist. He was in New Caledonia (1869), Guadeloupe (1874) and Madagascar (1878). A reptile is named after him.

Marie (von Plessen)

Red-chested Flowerpecker ssp. *Dicaeum maugei mariae*
Neumann, 1942 NCR
[JS *Dicaeum maugei neglectum*]

Baronin Marie-Izabel von Plessen *née* von Jenisch (1906–1971) was the wife of the German ornithologist Victor Baron von Plessen (q.v.). (See also **Hutz**)

Marion (Alexander)

Bee-eater sp. *Merops marionis* **B. Alexander**, 1903 NCR
[Alt. Blue-headed Bee-eater; JS *Merops muelleri mentalis*]

Marion Alexander (DNF) was the sister of British explorer and ornithologist Boyd Alexander (q.v.).

Marion (Williamson)

Indochinese Bushlark *Mirafra assamica marionae*
E. C. S. Baker, 1915 NCR
[JS *Mirafra erythrocephala*]

Lady Marion Williamson (d.1945) was the wife of Sir Walter James Williamson (q.v.).

Marion (Vernay)

Blood Pheasant ssp. *Ithaginis cruentus marionae* **Mayr**, 1941

Mrs Marion Woodruff Vernay *née* Kelly (b.1883) was the wife of Arthur Stannard Vernay (q.v.). (See **Mrs Vernay**)

Marjory (Bradfield)

Alpine Swift ssp. *Tachymarptis melba marjoriae* **Bradfield**, 1935
Monteiro's Hornbill ssp. *Tockus monteiri marjoriae* Bradfield, 1935 NCR; NRM

Mrs Marjory Bradfield (DNF) was the wife of the describer, Rupert Bradfield (q.v.).

Marjory (Winterbottom)

Cape Clapper Lark ssp. *Mirafra apiata marjoriae*
Winterbottom, 1956

Mrs Marjory Winterbottom (b.1904) was the wife of the describer, Dr J. M. Winterbottom (q.v.).

Markham

Markham's Storm-petrel *Oceanodroma markhami* **Salvin**, 1883
[Alt. Sooty Storm-petrel]

Admiral of the Fleet Sir Albert Hastings Markham (1841–1918) was an officer in the Royal Navy who combined his naval duties with ornithology. He is perhaps most famous for being in command of HMS *Camperdown* (1893) when she rammed and sank her flagship HMS *Victoria*, a disaster in which 358 men including Admiral Sir George Tryon were lost. Despite that tragedy, Markham still made the highest rank in the Navy. He served on HMS *Blanche* in Australasian waters (1868–1871) and designed the national flag of New Zealand (1869). He was on the HMS *Alert* expedition – an attempt to reach the North Pole (1875–1876), and published *The Great Frozen Sea: A Personal Narrative of the Voyage of the 'Alert' During the Arctic Expedition of 1875–6* (1878). He commanded HMS *Triumph* in the Pacific (1880–1883) and cruised extensively off North, Central and South America, including making a number of visits to the Galapagos Islands. Although he collected many birds he had a particular interest in gulls, and Howard Saunders (q.v.) published his list of gulls (1882). Salvin (q.v.) wrote 'A list of the birds collected by Captain A. H. Markham on the west coast of America' (1883).

Markl

Rufous-collared Sparrow ssp. *Zonotrichia capensis markli*
  Koepcke, 1971

Dr Walter Otto Markl Prechter (1926–1966) was a Swiss entomologist and amateur ornithologist who lived and collected in Piura in northern Peru (1956–1959 and 1960–1966) discovering the bird. He graduated in zoology at the University of Basle, which houses his collection. Whenever he had cause to vist Lima he stayed at Maria Koepcke's house and she described the bird after his death.

Marley

Southern Brown-throated Weaver ssp. *Ploceus
  xanthopterus marleyi* **J. A. Roberts**, 1929
Grey-backed Camaroptera ssp. *Camaroptera brevicaudata
  marleyi* Roberts, 1932 NCR
[JS *Camaroptera brevicaudata beirensis*]
Cloud Cisticola ssp. *Cisticola textrix marleyi* Roberts, 1932

Harold Walter Bell-Marley (1872–1945) was a fisheries officer at Durban, South Africa, and a naturalist with a particular interest in entomology. He was born in England and went to South Africa at the time of the Boer War, then decided to stay on. He collected continuously for c.50 years nearly everywhere south of the Zambesi River. Museums in many parts of the world have specimens that he sent. He also collected birds' eggs, and his collection, now in the Pretoria Museum, is regarded as one of the most complete ever assembled. He contracted blackwater fever (1944) while collecting in northern Zululand and died soon after his return to Durban. A mammal is also named after him.

Marmora

Marmora's Warbler *Sylvia sarda* **Temminck**, 1820

Alberto Ferrero della Marmora (1789–1863) was an Italian general and naturalist who had a distinguished career in the Napoleonic Wars, being personally decorated by Napoleon I with the Légion d'Honneur. The King of Sardinia later employed him. He wrote *Voyage en Sardaigne de 1819 à 1825 ou Description Statistique, Physique et Politique de cette Île* (1826). He also has a mountain in Sardinia named after him. Marmora stands accused of having massacred those who revolted against the reign of the Savoys in Genoa by using his newly formed cadre of uniformed soldiers, all carrying a 'carbine', which he invented.

Marriner

Tomtit ssp. *Petroica macrocephala marrineri* **Mathews** &
  Iredale, 1913

George Reginald Marriner (1880–1910) was Curator of the Wanganui Museum at the time of his early death from a short painful illness. He was an Assistant in the Department of Biology, Canterbury College (New Zealand), and took part in the 1907 expedition to the Auckland and Campbell Islands during which the Tomtit holotype was collected. He wrote: *The Kea: a New Zealand problem. Including a full description of this very interesting bird, its habitat and ways, together* with a discussion of the theories advanced to explain its sheep-killing propensities (1908). He also had an interest in meteorites.

Marshall, A. J.

Marshall's Fig Parrot *Cyclopsitta diophthalma marshalli*
  Iredale, 1946
[Alt. Double-eyed Fig Parrot ssp.]

Professor Alan John 'Jock' Marshall (1911–1967) was an Australian collector and zoologist. He was an assistant in the Harvard University collecting expedition which visited Australia (1930–1933). He met John Randal Baker (q.v.) in Sydney (1933) and joined Baker's expedition to Espiritu Santu, the largest of the islands in the New Hebrides (Vanuatu). 'Jock' became Honorary Ornithologist of the Australian Museum, Sydney (1934), where Tom Iredale (q.v.), the Fig Parrot describer, had been conchologist (1924–1944). Marshall was Reader in Zoology and Comparative Anatomy at St Bartholomew's Medical College, London (1949–1959) and went on to become Professor of Zoology and Dean of Sciences of the newly created Monash University, Australia (1960–1977). Despite losing his left arm in a shooting accident (1927) he had a distinguished army war record (1941–1945). He took part in a number of other expeditions, several of which he led, and explored in Spitsbergen and New Guinea. He wrote a number of successful books including *Australia Limited* (1941), *Testament of Doubt* (1945), (co-written) *Journey Among Men* (1962) and *The Great Extermination* (1966).

Marshall, G. A. K.

Yellow Canary ssp. *Serinus flaviventris marshalli*
  G. E. Shelley, 1902
[Syn. *Crithagra flaviventris marshalli*]

Sir Guy Anstruther Knox Marshall (1871–1959), an entomologist, was an expert on African and oriental weevils. He was born in India and sent to England to be educated. He was responsible for founding the *Bulletin of Entomological Research* (1909) and the *Review of Applied Entomology* (1913). He was the first Director of the Commonwealth Institute of Entomology. He was in East Africa for some time, having collected specimens near Salisbury, Southern Rhodesia (Harare, Zimbabwe) (1901–1902). A reptile is named after him.

Marshall, G. F. L. & C. H. T.

Marshall's Iora *Aegithina nigrolutea* G. F. L. Marshall, 1876
[Alt. White-tailed Iora]

Great Barbet ssp. *Megalaima virens marshallorum*
  Swinhoe, 1870
Rufous-bellied Woodpecker ssp. *Hypopicus hyperythrus
  marshalli* **Hartert**, 1912

Major-General George Frederick Leycester Marshall (1843–1934) and his brother Colonel Charles Henry Tilson Marshall (1841–1927) wrote on ornithology together. G. F. L. Marshall, who was Chief Engineer in the Punjab (1892–1897), also wrote with L. De Niceville the 3-volume *Butterflies of India, Burmah & Ceylon* (1882–1890). C. H. T. Marshall served chiefly in the Punjab, where he collected, as well as in the Himalayas. He sent specimens to Hume (q.v.) and oversaw the plate production of Hume's *The Game Birds of India, Burmah and Ceylon* (1879–1881). The iora and woodpecker are both named after G. F. L. Marshall; the barbet after both brothers.

Marshall, J. T.

Cloud-forest Screech Owl *Megascops marshalli* **Weske** & **Terborgh**, 1981

White-throated Towhee ssp. *Melozone albicollis marshalli* **Parkes**, 1974

White-eyed Vireo ssp. *Vireo griseus marshalli* **A. R. Phillips**, 1991

Dr Joseph 'Joe' Truesdell Marshall Jr (b.1918) is an American field ornithologist. He was a parasitologist in the US Army, stationed in Micronesia during WW2, and Professor of Zoology and Curator of Birds at the University of Arizona, Tucson. He was a medical biologist (ornithology) at the SEATO Medical Research laboratory, Bangkok, and was a research zoologist with the US Fish & Wildlife Service at the USNM. He wrote *Birds of Pine-Oak Woodland in Southern Arizona and adjacent Mexico* (1957) and part (genus *Otus*) of a paper 'Hawks and owls of the world: a distributional and taxonomic list' (1988) and *The Gray-cheeked Thrush, Catharus minimus, and its New England Subspecies, Bicknell's Thrush, Catharus minimus bicknelli* (2001). A mammal is also named after him.

Marsyas

Hummingbird genus *Marsyas* **Mulsant**, 1876 NCR
[Now in *Chlorostilbon*]

In Greek mythology, Marsyas was a satyr and an expert player of the *aulos* (double-piped reed instrument). He made the mistake of entering a competition against the god Apollo, was defeated, and flayed alive for daring to challenge a god.

Martens, J.

Martens's Warbler *Seicercus omeiensis* Martens *et al.*, 1999
[Alt. Omei Spectacled Warbler]

Coal Tit ssp. *Periparus ater martensi* **Eck**, 1998

Dr Jochen Martens (b.1941) is a German zoologist, particularly interested in arachnology and ornithology. He is now Profesor Emeritus and researches at Johannes Gutenberg University, Mainz, where he started teaching (1976). He spent 16 months in Nepal (1969–1970) collecting birds and recording vocalisations. He has also spent time undertaking fieldwork in China, India, Russia, the Caucasus, Iran and the Philippines. In 2006 he was appointed visiting professor at the Hainan Normal University in China although he has officially 'retired'. He has written many scientific papers such as co-writing: 'The Great Tit (*Parus major*) – a misclassified ring species' (2005).

Martens, K. E.

Helmet-shrike sp. *Prionops martensi* **Reichenow**, 1901 NCR
[Alt. White-crested Helmet-shrike; JS *Prionops plumatus concinnatus*]

Dr Karl Eduard von Martens (1831–1904) was a German conchologist and zoologist. He was in the East Indies (Indonesia) (1860–1863). His doctorate was awarded by the University of Tübingen (1855) after which he moved to Berlin where he worked at the university and the Berlin Zoological Museum for the rest of his life, becoming Professor and Director there (1883–1887) as well as Curator of Malacology and other invertebrates (1864–1904). He was a member of the *Thetis* expedition to the Far East (1860–1862) and stayed on in South-East Asia (1863–1864) before returning to Berlin.

Martin

Sooty-headed Bulbul ssp. *Pycnonotus aurigaster martini* **Parrot**, 1910 NCR
[JS *Pycnonotus aurigaster aurigaster*]

Dr Johann Karl Ludwig Martin (1850–1942) was a German physician, geologist, ethnologist and zoologist. The University of Göttingen awarded his doctorate (1874). He was the first Professor of Geology at Leiden (1877–1922) and Director of the Geological Museum of Leiden (1880–1922). He was in the Netherlands Antilles (1884), the Moluccas and Sumatra (1892), the Celebes (1906 and 1912) and Java (1910). A reptile is named after him.

Martin, Maria

Maria's Woodpecker *Picus martinae* **Audubon**, 1839 NCR
[Alt. Hairy Woodpecker; JS *Picoides villosus*]

Miss Maria Martin (1796–1863) was Bachman's (q.v.) sister-in-law, and after her sister's death became Bachman's second wife. She was assistant to both Bachman and Audubon (q.v.) in their studies and was an accomplished artist. Audubon in his *Ornithological Biography* wrote 'In honouring this species with the name of Miss Maria Martin, I cannot refrain from intimating the respect, admiration, and sincere friendship which I feel towards her, and stating that, independently of her other accomplishments, and our mutual goodwill, I feel bound to make some ornithological acknowledgment for the aid she has on several occasions afforded me in embellishing my drawings of birds, by adding to them beautiful and correct representations of plants and flowers'.

Martini

Purple-throated Cuckooshrike ssp. *Campephaga quiscalina martini* **F. Jackson**, 1912

Antonio Martini (also known as James Martin) (1830–1924) was a Maltese mariner and sailmaker, adventurer and explorer. He was on board an American ship that was wrecked on Zanzibar, where he became a general handyman at a Mission in Freetown and joined Joseph Thomson's (q.v.) exploration of Masai territory. He then became second-in-command of the Sultan of Zanzibar's troops. He joined the Imperial British East Africa Company and went on a number of their expeditions with Frederick Jackson (q.v.), and later ran a rubber plantation in Kenya. In WW1 he worked in Intelligence and after the war lived in the Mombasa Club for the rest of his life. He was illiterate, and Jackson, who spent many hours trying to teach him to read and write, eventually gave up on it.

Martins, O.

Antpitta sp. *Grallaria martinsi* **Snethlage**, 1924 NCR
[Alt. White-browed Antpitta; JS *Hylopezus ochroleucus*]

Oscar Rodrigues Martins (d.1924) was a Brazilian who was a technical assistant and taxidermist at the Emilio Goeldi Museum, Pará (1908–1910). He explored and collected as an assistant to Maria Emilia Snethlage (q.v.) (1910). A mammal is also named after him.

Martins, V.

African Citril ssp. *Serinus citrinelloides martinsi* da Rosa
    Pinto, 1962
[Syn. *Crithagra citrinelloides martinsi*]

Professor V. C. Martins (fl.1960) was a Portuguese zoologist involved in the Institute for Investment in Angola.

Martius

Rufous Motmot *Baryphthengus martii* **Spix**, 1824

Carl Friedrich Phillip von Martius (1794–1868) was a German botanist and ethnographer. He was a member, with Spix, on Wied-Neuwied's expedition in Brazil (1817–1820). He continued with Spix's work after the latter's death. They had written three volumes of an unfinished work on the expedition. Martius founded *Flora Brasiliensis*. A reptile is named after him.

Marwitz

Uhehe Fiscal *Lanius marwitzi* **Reichenow**, 1901

Little Greenbul ssp. *Andropadus virens marwitzi*
    Reichenow, 1895 NCR
[NUI *Andropadus* (or *Eurillas*) *virens zombensis*]
Fawn-breasted Waxbill ssp. *Estrilda paludicola marwitzi*
    Reichenow, 1900
Pearl-breasted Swallow ssp. *Hirundo dimidiata marwitzi*
    Reichenow, 1903
Common Swift ssp. *Apus apus marwitzi* Reichenow, 1906
    NCR
[NUI *Apus apus pekinensis*]
Green Wood-hoopoe ssp. *Phoeniculus purpureus marwitzi*
    Reichenow, 1906
Mottled Spinetail ssp. *Telacanthura ussheri marwitzi*
    Reichenow, 1906 NCR
[JS *Telacanthura ussheri stictilaema*]
Southern Red Bishop ssp. *Euplectes orix marwitzi*
    Reichenow, 1906 NCR
[JS *Euplectes orix nigrifrons* (or species regarded as
    monotypic, with clinal size differences)]

Captain Adalbert von der Marwitz (1856–1929) was a German army officer who was in German East Africa (Tanzania) (1895–1906) and was later (1913) in the Cameroons.

Mary (Alexander)

Green Longtail ssp. *Urolais epichlorus mariae* **B. Alexander**,
    1903

Mrs Mary Alexander *née* Wilson (1843–1905) was the wife of Lieutenant Colonel Boyd Francis Alexander (q.v.) and the mother of Captain Boyd Alexander (q.v.).

Mary (Baker)

Rufous Fantail ssp. *Rhipidura rufifrons mariae* R. H. Baker,
    1946

Mrs Mary Elizabeth Waddell Baker (1913–2000) was the wife of the describer, American mammalogist Rollin Harold Baker (1916–2007).

Mary (Benson)

Moheli Brush Warbler *Nesillas mariae* **C. W. Benson**, 1960
[Alt. Mrs Benson's Warbler]

Churring Cisticola ssp. *Cisticola njombe mariae*
    C. W. Benson, 1945

Florence Mary Benson (1909–1993) was the wife of the describer (See **Mrs Benson**).

Mary (Davidson)

Crested Bobwhite ssp. *Colinus cristatus mariae* **Wetmore**,
    1962

Mrs Mary E. Davidson (DNF) was Curator of Birds at the Museum of Academic Sciences in San Francisco (1927–1938). She re-married (1938) to Robert A. Terry, a geologist.

Mary (Gartshore)

Jos Plateau Indigobird *Vidua maryae* Payne, 1982

Mary Eleanor Gartshore (b.1950) (formerly Dyer) is a conservation biologist, ornithologist, and former Bird Studies Canada board member. She graduated BSc Zoology (1973), University of Guelph, Canada, then worked out of Ahmadu Bello University and travelled extensively in Nigeria (1974–1977, 1979–1981). Payne's original etymology '… acknowledges the field work of Mary Dyer in Nigeria, in particular her help with the indigobirds.' For three decades she worked as a consultant or manager for a wide variety of conservation initiatives in Canada as well as field trips, consultancy work or project management in Mexico, Ghana, Cameroon and Cote d'Ivoire. She took part in two expeditions to Cross River National Park, Nigeria (2003 and 2004). She was advisor and manager of ecological restoration of Great Lakes sand plains for the Nature Conservancy of Canada (2006–2011). She has written more than 40 scientific papers, such as 'An avifaunal survey of Tai National Park' (1989). She continues to work in conservation mostly on ecological restoration projects through St Williams Nursery and Ecology

Centre where she is (2010) Senior Restoration Specialist and Vice President.

Mary (La Touche)

> Red-tailed Minla ssp. *Minla ignotincta mariae* **La Touche**, 1921

Mrs Elizabeth Mary Dillon *née* La Touche (DNF) was a daughter of J. D. D. La Touche (q.v.).

Mary (Ramsay)

> Scimitar Babbler sp. *Pomatorhinus mariae* **Walden**, 1875 NCR
> [Alt. Coral-billed Scimitar Babbler; JS *Pomatorhinus ferruginosus albogularis*]

Mrs Mary Alice Ramsay née Hogg (d.1951) was in India (1875–1882). She was the wife of Colonel Robert George Wardlaw Ramsay (q.v.).

Mary (Ripley)

> Long-tailed Minivet ssp. *Pericrocotus ethologus mariae* **Ripley**, 1952
> Black Sunbird ssp. *Leptocoma sericea mariae* Ripley, 1959
> Black-breasted Mannikin ssp. *Lonchura teerinki mariae* Ripley, 1964
> [Alt. Grand Valley Munia ssp.]

Mrs Mary Ripley *née* Livingston (d.1996) was an amateur entomologist and botanist. She was the wife of S. Dillon Ripley (q.v.) and daughter of Gerald M. Livingston (q.v.). A garden at the USNM is named in her honour.

Masaaki

> Small Buttonquail ssp. *Turnix sylvaticus masaaki* **Hachisuka**, 1931 NCR
> [NUI *Turnix sylvaticus celestinoi*]
> Varied Tit ssp. *Parus rubidus masaakii* **Momiyama**, 1940 NCR
> [JS *Poecile varius owstoni*]

Masaaki Marquis Hachisuka (1871–1932) was Vice-President of the House of Peers of Japan. His son was the ornithologist Masauji Hachisuka (q.v.), who named the buttonquail after him.

Masauji

> Little Kingfisher ssp. *Ceyx pusillus masauji* **Mathews**, 1927

Masauji Hachisuka (See **Hachisuka**)

Mason, G.

> Hooded Oriole ssp. *Icterus cucullatus masoni* **Griscom**, 1926 NCR
> [JS *Icterus cucullatus igneus*]

Gregory Mason (1889–1968) was an American anthropologist. He helped organise the Mason-Spinden Expedition to Yucatan (1926) searching for lost Mayan cities.

Mason, O.

> Red-ruffed Fruitcrow ssp. *Pyroderus scutatus masoni* **Ridgway**, 1886

Dr Otis Tufton Mason (1838–1908) was an American ethnologist and anthropologist. He graduated from Columbian University (George Washington University) (1861) and then worked there (1861–1884), becoming Professor of Anthropology. He was Curator of Anthropology, Smithsonian Institution (1884–1910), and a founder of the Anthropological Society of Washington.

Masséna

> Massena's Parrot *Pionus senilis* **Spix**, 1824
> [Alt. White-crowned Parrot]
> Massena's Partridge *Ortyx massena* **Lesson**, 1832 NCR
> [Alt. Montezuma Quail; JS *Cyrtonyx montezumae*]
> Massena Trogon *Trogon massena* **Gould**, 1838
> [Alt. Slaty-tailed Trogon]
>
> Massena's Lorikeet *Trichoglossus haematodus massena* **Bonaparte**, 1854
> [Alt. Coconut Lorikeet ssp.]

Francois Victor Masséna Prince D'Essling, Duc de Rivoli (1795–1863). (See **Rivoli** & **Prince of Essling**)

Master

> Chocó Vireo *Vireo masteri* Salaman & **Stiles**, 1996

Dr Bernard Master (b.1941) is a physician but also an internationally renowned ornithologist, having established the first ProAves bird preserve in Colombia. He is a leading light in the Ohio Ornithological Society. He co-authored the guidebook, *Annotated Checklist of the Birds of Ohio*. The scientific name was decided by an auction to raise money for conservation of the vireo's habitat: Master paid $70,000 for the honour! A trip was organised (2009) so that Dr Master could see the bird he helped preserve.

Masters

> Monarch Flycatcher genus *Mastersornis* **Mathews**, 1917 NCR
> [Now in *Myiagra*]
>
> Masters's Rosella *Platycercus mastersianus* **E. P. Ramsay**, 1877 NCR
> [Hybrid: *Platycercus elegans* x *Platycercus adscitus palliceps*]
> Gerygone sp. *Pseudogerygone mastersi* **Sharpe**, 1879 NCR
> [Alt. Mangrove Gerygone; JS *Gerygone levigaster*]
>
> Masters's Thornbill *Acanthiza inornata mastersi* **North**, 1901
> [Alt. Western Thornbill ssp.; species often regarded as monotypic]
> Purple-backed Fairy-wren ssp. *Malurus lamberti mastersi* Mathews, 1912 NCR
> [JS *Malurus lamberti assimilis*]
> Western Bristlebird ssp. *Dasyornis longirostris mastersi* Mathews, 1923 NCR; NRM

George Masters (1837–1912) was an entomologist who emigrated from England to Australia. He was Assistant Curator of the Australian Museum (1864), apparently on condition that he sold his private collection and made no new one, an agreement he ignored. He became Curator of the Macleay collection in Sydney (1874–1912) until his death in a carriage accident. He certainly collected widely as there are skeletons of Thylacines ('Tasmanian Tigers'), which he collected (1870s). Masters collected with Macleay (q.v.) when he was the latter's personal curator, including accompanying him to New Guinea (1875). He helped to plan and provision the expedition aboard *Chevert*, a barque, which Macleay paid for. In his diary for that year Macleay notes 'Masters continues still getting things ready for the expedition. It ought to be well found in everything from the bills coming in'. In addition to the birds, Masters' great knowledge of Australian fauna went largely unrecorded, as he disliked writing. A reptile is named after him.

Mathews, C. M.

Norfolk Island Gerygone *Gerygone mathewsae* **Mathews**, 1912 NCR
[JS *Gerygone modesta*]

Great-billed Heron ssp. *Ardea sumatrana mathewsae* Mathews, 1912
[Species often regarded as monotypic]

(See **Marian (Mathews)**)

Mathews, G. M.

Crane genus *Mathewsena* **Iredale**, 1914 NCR
[Now in *Grus*]
Riflebird genus *Mathewsiella* Iredale, 1922 NCR
[Now in *Ptiloris*]

Rufous Songlark *Cincloramphus mathewsi* Iredale, 1911

Yellow Thornbill ssp. *Acanthiza nana mathewsi* **Hartert**, 1910 NCR
[JS *Acanthiza nana modesta*]
Mathews's Cockatoo *Cacatua galerita fitzroyi* Mathews, 1912
[Alt. FitzRoy's Cockatoo, Sulphur-crested Cockatoo ssp.]
Mathews's Pink Cockatoo *Cacatua leadbeateri mollis* Mathews, 1912
[Alt. Major Mitchell's Cockatoo ssp.; Syn. *Lophochroa leadbeateri mollis*]
Black-naped Tern ssp. *Sterna sumatrana mathewsi* **Stresemann**, 1914
Weebill ssp. *Smicrornis brevirostris mathewsi* **S. A. White**, 1915 NCR
[Intergrade between *Smicrornis brevirostris flavescens* and *S. brevirostris ochrogaster*]
Mathews's Red-tailed Black Cockatoo *Calyptorhynchus banksii samueli* Mathews, 1917
Mathews's Little Corella *Cacatua sanguinea normantoni* Mathews, 1917
Grey Teal ssp. *Anas gibberifrons mathewsi* **J. C. Phillips**, 1923 NCR
[JS *Anas gracilis*]
Green Kingfisher ssp. *Chloroceryle americana mathewsii* **Laubmann**, 1927

Australasian Bittern ssp. *Botaurus poiciloptilus mathewsi* **Hachisuka**, 1931 NCR; NRM

Gregory Macalister Mathews (1876–1949) was an Australian ornithologist. He started working life droving cattle for six years, then became an orchardist. He moved to England (1902) after making a fortune in mining shares and marrying an Englishwoman. He developed a passion for hunting and horseracing, leading the life of a country squire in a Hampshire village (1910–1949), where he settled apart from returning to Australia (1940–1945) during WW2. Spurred by his observation of birds as a drover, and encouraged by R. B. Sharpe (q.v.), he decided to write what became the 12-volume *A Manual of the Birds of Australia* (1910–1927) assisted by Tom Iredale (q.v.). He donated his collection of books, pamphlets, reprints, manuscripts and pictorial material to the Australian National Library (1939), comprising what is now known as the Mathews Ornithological Library. It is considered to be the finest collection of material relating to the study of Australian birds. Among it are works by John Gould, J. W. Lewin and Lilian Medland. Mathews's main interest was in taxonomy: initially a conservative, he later gained the reputation of an arch 'splitter'. His *Reference List to the Birds of Australia* raised the number of described forms there from 800 to 1,500!

Mathilda

Pitta sp. *Pitta mathilda* **J. Verreaux**, 1857 NCR
[Alt. Elegant Pitta; JS *Pitta elegans concinna*]

Madame Mathilde Texier (DNF) was the wife of Professor Charles Texier (1802–1871), a French archaeologist and explorer.

Maton

Maton's Lorikeet *Trichoglossus chlorolepidotus* **Kuhl**, 1820
[Alt. Scaly-breasted Lorikeet; *Trichoglossus matoni* (**Vigors & Horsfield**, 1827) is a JS]

Dr William George Maton (1774–1835) was a British physician and botanist. He became a Fellow of the Linnean Society (1794), later its Vice-President. He wrote *Observations Relative Chiefly to the Natural History, Picturesque Scenery, and Antiquities of the Western Counties of England, made chiefly in the Years 1794 and 1796* (1797). He was also a Fellow of the London College of Physicians and was appointed Physician-Extraordinary to Queen Charlotte.

Matschie

Matschie's Brown Parrot *Poicephalus meyeri matschiei* **Neumann**, 1898
[Alt. Meyer's Parrot ssp.]
Brown-eared Bulbul ssp. *Microscelis amaurotis matchiae* **Momiyama**, 1923

George Friedrich Paul Matschie (1861–1926) was a German zoologist at the ornithological section of the Zoological Museum, Humboldt University, Berlin (1884) under Jean Louis Cabanis (q.v.). He became a Professor of Zoology (1902) and its second Director (1924). Two reptiles and four mammals are named after him. Momiyama misspelt Matschie's name in the bulbul's trinomial, forgetting to include an 's'.

Matsudaira

Matsudaira's Storm-petrel *Oceanodroma matsudairae*
Kuroda, 1922
Skua sp. *Catharacta matsudairae* **Taka-Tsukasa**, 1922 NCR
[Probably based on a specimen of *Stercorarius
maccormicki*]

Japanese Pygmy Woodpecker ssp. *Dendrocopos kizuki
matsudairai* Kuroda, 1921

Viscount Yorikatsu Matsudaira (1876–1945) was a Japanese
ornithologist. He wrote *A Hand-list of the Japanese Birds*
(1922).

Matthews, A.

Matthews's Panoplites *Boissonneaua matthewsii* **Bourcier**,
1847
[Alt. Chestnut-breasted Coronet]

Andrew Matthews (d.1841) was an English botanist who
collected in Peru and Chile (1830–1841). He was asked by his
friend George Loddiges (q.v.) to collect hummingbirds whilst
on his botanical explorations. Bourcier (q.v.) described this
species using the manuscript name that Loddiges had
suggested (Chestnut-breasted Coronet) but Gould's illustration used the vernacular eponym.

Matthias

Matthias Fantail *Rhipidura matthiae* **Heinroth**, 1902
[Alt. Mussau Fantail]

Named after the island of St Matthias, Bismarck Archipelago,
Papua New Guinea (Saint Matthias was the person chosen
by the disciples of Jesus to replace Judas).

Mattingley

Whistler genus *Mattingleya* **Mathews**, 1912 NCR
[Now in *Pachycephala*]

Antarctic Prion ssp. *Heteroprion desolatus mattingleyi*
Mathews, 1912 NCR
[JS *Pachyptila desolata*]
Pied Oystercatcher ssp. *Haematopus longirostris mattingleyi*
Mathews, 1912 NCR; NRM

Arthur Herbert Evelyn Mattingley (1870–1950) was an
Australian ornithologist and bird photographer. He was President, RAOU (1913–1914). He worked as a Customs Officer in
Melbourne (1891–1934) and, after retiring, led an expedition
to central and north-eastern Australia (1934).

Matuda

Streak-headed Woodcreeper ssp. *Lepidocolaptes souleyetii
matudae* **Brodkorb**, 1938 NCR
[JS *Lepidocolaptes souleyetii compressus*]
Red-winged Blackbird ssp. *Agelaius phoeniceus matudae*
Brodkorb, 1940 NCR
[JS *Agelaius phoeniceus richmondi*]
Grey-collared Becard ssp. *Pachyramphus major matudai*
A. R. Phillips, 1966

Dr Eizi Matuda (1894–1978) was a Japanese botanist who
moved to Mexico (1922), becoming a citizen (1928). All his
university education was in Formosa (Taiwan), then
controlled by Japan. The University of Tokyo awarded his
doctorate (1962). He taught in Japan and travelled to study
flora in mainland Asia (1914–1921). Back in Mexico he
became head of the Department of Botany, National Institute
of Forestry. Three reptiles and two amphibians are named
after him.

Maude

Rufous Whistler ssp. *Pachycephala rufiventris maudeae*
S. A. White, 1915 NCR
[NUI *Pachycephala rufiventris rufiventris*]

Mrs Louisa Maude White *née* Ebsworth (1866–1928) was the
wife of Henry Luke White (q.v.).

Maugé

Puerto Rican Emerald *Chlorostilbon maugaeus* **Audebert** &
Vieillot, 1801
Barred Dove *Geopelia maugei* **Temminck**, 1811
Maugé's Flowerpecker *Dicaeum maugei* **Lesson**, 1830
[Alt. Red-chested Flowerpecker, Blue-cheeked
Flowerpecker]

Maugé's Conure *Aratinga chloroptera maugei* **Souancé**,
1856 EXTINCT
[Alt. Hispaniolan Parakeet ssp. Syn. *Psittacara chloropterus
maugei*]

René Maugé de Cely (d.1802) was a French zoologist who
accompanied his friend Baudin (q.v.) on his great scientific
voyage (1800–1803). He collected in the West Indies (1796–
1798) and the Pacific (1800–1802). He fell ill in Timor and died
when the expedition arrived in Tasmania.

Maui

New Zealand Scaup ssp. *Aythya novaeseelandiae maui*
Mathews, 1937 NCR; NRM

In Maori tradition, Maui was the culture hero who fished the
North Island of New Zealand from the waters of the Pacific.

Mauri

Western Sandpiper *Calidris mauri* **Cabanis**, 1857

Ernesto Mauri (1791–1836) was an Italian botanist. With
Antonio Sebastiani, he wrote *Floræ Romanæ Prodromus*
(1818). Cabanis (q.v.) formally described the sandpiper (1858),
but Bonaparte (q.v.), who was a friend of Mauri, had first
called it *Calidris mauri* (1838). When Bonaparte heard that
Cabanis had subsequently described the species, he
persuaded him to change the name so that it still honoured
Mauri. Bonaparte also named a fish, *Smaris maurii*, after his
friend but this is a JS.

Maurice

Somali Fiscal *Lanius antinorii mauritii* **Neumann**, 1907 NCR
[JS *Lanius somalicus*; NRM]

Baron Maurice de Rothschild (see **Rothschild, M.**).

Mavors

Mavors's Sunangel *Heliangelus mavors* **Gould**, 1848
[Alt. Orange-throated Sunangel]

In Roman mythology Mavors was another name for Mars, God of war.

Max (Bartels)

Lemon-bellied White-eye ssp. *Zosterops chloris maxi*
Finsch, 1907

(See **Bartels**)

Max (Nikolaus)

Spotted Ground Thrush ssp. *Zoothera guttata maxis*
Nikolaus, 1982

Max Nikolaus is the son of the describer, Gerhard Nikolaus.

Maximilian

Parrot genus *Maximilicus* **G. R. Gray**, 1840 NCR
[Now in *Triclaria*]

Maximilian's Parrot *Pionus maximiliani* **Kuhl**, 1820
[Alt. Scaly-headed Parrot]
Hyacinth Macaw *Anodorhynchus maximiliani* **Spix**, 1824
NCR
[JS *Anodorhynchus hyacinthinus*]
Olive-crowned Crescentchest *Melanopareia maximiliani*
d'Orbigny, 1835
Maximilian's Jay *Gymnorhinus cyanocephalus* Maximilian,
1841
[Alt. Pinyon Jay]
Great-billed Seed Finch *Oryzoborus maximiliani* **Cabanis**,
1851

Maximilian's Aracari *Pteroglossus aracari wiedii* **Sturm**,
1847
[Alt. Black-necked Aracari ssp.]
Great Kiskadee ssp. *Pitangus sulphuratus maximiliani*
Cabanis & **Heine**, 1859
Sand Martin ssp. *Riparia riparia maximiliani* **Stejneger**, 1885
NCR
[NUI *Riparia riparia riparia*]

Alexander Philipp Maximilian zu Wied-Neuwied (1782–1867) was an aristocratic German explorer. He studied natural history at the University of Göttingen, then joined the Prussian army (1802) fighting Napoleon, ultimately becoming General-Major. He collected in Brazil (1815–1817), Guyana (1821) and North America (1832–1834). He made his famous journey of c.5,000 miles, principally up the Missouri River, with European artists and scientists, on the second voyage of the steamer *Yellowstone* (1833). They collected and painted the landscapes. He wrote *Reise nach Brasilien in den Jahren 1815 bis 1817* (1820), *Beiträge zur Naturgeschichte von Brasilien* (1825) and *Reise in das Innere Nord-Amerika in den Jahren 1832 bis 1834* (1840) when back in Europe. A mammal and two reptiles are named after him. (See also **Wied**)

Maxwell, H. E.

Maxwell's Black Weaver *Ploceus albinucha* **Bocage**, 1876

Maxwell's Black Weaver ssp. *Ploceus albinucha maxwelli*
B. Alexander, 1903

Sir Herbert Eustace Maxwell (1845–1937), 7th Baronet of Monreith, hailed from one of Scotland's oldest families. He became a Knight of the Thistle and MP for Wigtownshire, Lord of the Treasury, Secretary of State for Scotland, and Lord Lieutenant of the County of Wigtownshire. His garden at Monreith House became renowned for its many trees from around the world. He wrote both novels and non-fiction, the latter including *The Making of Scotland* (1911). His grandson was the naturalist and author Gavin Maxwell, famous for *Ring of Bright Water*.

Maxwell, M. A.

Maxwell's (Screech) Owl *Megascops asio maxwelliae*
Ridgway, 1877
[Alt. Eastern Screech Owl ssp.]

Mrs Martha Ann Maxwell (1831–1881) was an American field naturalist, taxidermist and collector, who travelled in Colorado Territory (1860) during the Pike's Peak Gold Rush.

Maxwell, M. H.

Rusty-winged Starling ssp. *Aplonis zelandica maxwellii*
H. O. Forbes, 1900

Maxwell Hyslop Maxwell (1862–1937) was a British businessman from a firm of tobacco importers. He was a Director of Cunard and a Committee Member of Liverpool Museum.

Maxwell, P. H.

Barred Rail ssp. *Hypotaenidia torquata maxwelli* **W. Lowe**,
1944 NCR
[JS *Gallirallus torquatus torquatus*]

Patrick 'Pat' Hall Maxwell (1912–1991) worked at Paignton Zoo and in the London Zoo (1944) and at Whipsnade Zoo (1946–1966), where he was keeper of the Parrot House, which contained a number of specimens that were his private property as well as birds that belonged to the zoo. He travelled (1940s through 1970s) to Africa, Samoa and the Solomon Islands. He donated over 300 birds to the Royal Albert Memorial Museum, Exeter (1939–1955), including the Barred Rail, which had died at London Zoo (1944). Lowe says 'The Exeter Museum has recently received from P. H. Maxwell, Esq., this new Rail, which died in the London Zoo. . . . . . It gives me pleasure to name this bird after Mr. Maxwell, who has generously presented to the Exeter Museum so many rare and valuable specimens.'

Maxwell, S.

Maxwell's Sapphire *Hylocharis ruficollis maxwelli* **Hartert**,
1898 NCR
[Alt. Gilded Sapphire; JS *Hylocharis chrysura*]

Arthur Maxwell Stuart (see **Stuart, A. M.**).

May

Fantail sp. *Rhipidura mayi* **Ashby**, 1911 NCR
[Alt. Arafura Fantail; JS *Rhipidura dryas*]

Tawny Grassbird ssp. *Dulciornis alisteri mayi* Ashby, 1914 NCR
[JS *Megalurus timoriensis alisteri*]
White-winged Triller ssp. *Karua leucomela mayi* Ashby, 1914 NCR
[JS *Lalage tricolor*]

C. E. May (d.1914) was an Australian collector at Anson Bay, Northern Territory. Ashby (q.v.) refers to him as 'my friend' and wrote that May 'has done so much good work in collecting the birds of the Northern Territory.'

Mayaud

Three-banded Courser ssp. *Rhinoptilus cinctus mayaudi* **Érard**, Hémery & Pasquet, 1993

Noël Mayaud (1899–1989) was a French ornithologist and co-founder of the journal *Alauda* (1929). He co-wrote *Les Oiseaux du Nord-Ouest de l'Afrique* (1962) with Heim de Balsac (q.v.).

Mayer, A. F. J.

Pink Pigeon *Nesoenas mayeri* **Prevost**, 1843
[Syn. *Columba mayeri*]

August Franz Joseph Carl Mayer (1787–1865) was a German anatomist and collector. He was Professor of Anatomy, Pathology and Physiology at Bern (1815) and at Bonn until retirement (1819–1856). He examined the skull of a Neanderthal, but could not identify it and thought it a Russian Cossack from the Napoleonic War!

Mayer, F. W. S.

Ribbon-tailed Astrapia *Astrapia mayeri* Stonor, 1939
[Alt. Ribbon-tailed Bird-of-Paradise]

Frederick William Shaw Mayer (See **Shaw Mayer**)

Maynard

Hairy Woodpecker ssp. *Picoides villosus maynardi* **Ridgway**, 1887
Maynard's Cuckoo *Coccyzus minor maynardi* Ridgway, 1887 NCR; NRM
[Alt. Mangrove Cuckoo ssp.]
White-eyed Vireo ssp. *Vireo griseus maynardi* **Brewster**, 1887

Professor Charles Johnson Maynard (1845–1929) was an American ornithologist and all-round naturalist described as 'Newtonville's [Massachusetts] enigmatic naturalist'. He was a well-known observer of birds, particularly in Florida and the Bahamas. For example (1896), he wrote 'The eagle hovers over a bunch of coots and endeavors by diving down towards the flock to make them scatter. The eagle will never attack a coot when surrounded by its fellows, but the instant one is separated from the flock his life is in jeopardy … Among coots, their safety lies in numbers, even if all be cowards, but the wonder is, not that the eagles know this, but that the coots themselves do.' Charles Batchelder (q.v.) wrote a catalogue of Maynard's publications. Maynard wrote books and articles including: *The Naturalist's Guide in Collecting and Preserving Objects of Natural History* (1870), *The Birds of Eastern North America* (1881), and *Birds of Washington and Vicinity* (1898). His published and competently illustrated many of them himself. A mammal and two reptiles are named after him.

Mayr

Monarch genus *Mayrornis* **Wetmore**, 1932
Mannikin genus *Mayrimunia* **Wolters**, 1949 NCR
[Now in *Lonchura*]

Mayr's Forest Rail *Rallina mayri* **Hartert**, 1930
[Alt. Black-tailed Forest Rail]
Mayr's Honeyeater *Ptiloprora mayri* Hartert, 1930
[Alt. Mayr's Streaked Honeyeater]
Mayr's Swiftlet *Aerodramus orientalis* Mayr, 1935
[Alt. Guadalcanal Swiftlet]

Lewin's Rail ssp. *Lewinia pectoralis mayri* Hartert, 1930
MacGregor's Bowerbird ssp. *Amblyornis macgregoriae mayri* Hartert, 1930
Mayr's Munia *Lonchura spectabilis mayri* Hartert, 1930
[Alt. Hooded Mannikin/Munia ssp.]
Papuan Grassbird ssp. *Megalurus macrurus mayri* Hartert, 1930
Spotted Jewel-babbler ssp. *Ptilorrhoa leucosticta mayri* Hartert, 1930
White-faced Robin ssp. *Tregellasia leucops mayri* Hartert, 1930
Mayr's Parrot *Psittacella brehmii harterti* Mayr, 1931
[Alt. Brehm's Tiger Parrot ssp.]
Thick-billed Ground Pigeon ssp. *Trugon terrestris mayri* **Rothschild**, 1931
Mayr's Pygmy Parrot *Micropsitta pusio harterti* Mayr, 1940
[Alt. Buff-faced Pygmy Parrot ssp.]
Mayr's Red-flanked Lorikeet *Charmosyna placentis ornata* Mayr, 1940
Spotted Owlet ssp. *Athene brama mayri* **Deignan**, 1941
Trumpet Manucode ssp. *Phonygammus keraudrenii mayri* Greenway, 1942
[SII *Phonygammus keraudrenii purpureoviolaceus*]
Yellow-fronted Tinkerbird ssp. *Pogoniulus chrysoconus mayri* **C. M. N. White**, 1946 NCR
[NUI *Pogoniulus chrysoconus extoni*]
Citrine Canary-flycatcher ssp. *Culicicapa helianthea mayri* Deignan, 1947
Great Barbet ssp. *Megalaima virens mayri* **Ripley**, 1948
[SII *Megalaima virens clamator*]
Mayr's Painted Parrot *Psittacella picta excelsa* Mayr & **Gilliard**, 1951
[Alt. Painted Tiger Parrot ssp.]
Pin-striped Tit Babbler ssp. *Macronus gularis mayri* Koelz, 1951 NCR
[JS *Macronus gularis rubicapilla*]
Baillon's Crake ssp. *Porzana pusilla mayri* **Junge**, 1952
Little Pied Flycatcher ssp. *Ficedula westermanni mayri* Ripley, 1952

Striated Swallow ssp. *Cecropis striolata mayri* **P. Hall**, 1953
Red-backed Buttonquail ssp. *Turnix maculosus mayri*
    Sutter, 1955
Chestnut Quail-thrush ssp. *Cinclosoma castanotum mayri*
    Condon, 1962 NCR
[NUI *Cinclosoma castanotum castanotum*]
Roraiman Barbtail ssp. *Roraimia adusta mayri* **W. H. Phelps
    Jr**, 1977

Dr Ernst Mayr (1904–2005) was a German ornithologist and zoologist who began serious bird studies in the South Pacific. He is best known as an eminent writer on evolution; it even being said: 'he is, without a doubt the most influential evolution theoretician of the twentieth century'. As a ten-year-old boy he could recognise all the local birds on sight and by song. He led ornithological expeditions to New Guinea (1928–1931), an experience that he said 'fulfilled the greatest ambition of my youth.' He collected 7,000 skins in 2.5 years. He later joined an expedition to the Solomon Islands, before returning to his academic career at the Berlin Museum. Mayr was employed by the Department of Ornithology of the AMNH (1931), at first as a visiting Curator, to catalogue the Whitney Expedition's collection of South Sea birds, in which Mayr had participated. He wrote 12 papers, describing 12 new species and 68 new subspecies, during his first year there. He went on to become the Alexander Agassiz Emeritus Professor of Zoology at Harvard University. He was the originator of the 'founder effect' idea of speciation and was a leading proponent of the Biological Species Concept. His many important books include *Systematics and the Origin of Species* (1942), *Animal Species and Evolution* (1963) and, with J. M. Diamond (q.v.), *The Birds of Melanesia: Speciation, Ecology and Biogeography* (2001). A mammal is named after him. (See **Ernst Mayr**)

Mazepa

Hermit sp. *Trochilus mazeppa* **Lesson**, 1831 NCR
[Alt. Rufous-breasted Hermit; JS *Glaucis hirsutus*]

Ivan Stepanovych Mazepa (1639–1709) was a patron of the arts and Hetman (commander) of the Zaporozhian Cossacks (1687–1708). He died in Turkey, having disastrously supported King Charles XII of Sweden at the Battle of Poltava (1709). Only 3,000 Cossacks followed Mazepa, with the rest remaining loyal to the Russian Tsar. His story has been a constant source of inspiration to poets and composers, including Lord Byron, Pushkin, Victor Hugo, Liszt and Tchaikovsky.

McAllan

Long-billed Corella ssp. *Cacatua tenuirostris mcallani* Wells
    & Wellington, 1992 NCR; NRM

Ian A. W. McAllan is an Australian ornithologist at Macquarie University, Sydney. He co-wrote *The Birds of New South Wales: a Working List* (1988).

McCall

Chestnut-capped Flycatcher *Erythrocercus mccallii* **Cassin**,
    1855

McCall's Screech Owl *Megascops asio mccallii* Cassin,
    1854
[Alt. Eastern Screech Owl ssp.]
Plain Chachalaca ssp. *Ortalis vetula mccalli* **Baird**, 1858

Brigadier-General George Archibald McCall (1802–1868) was a military man but also an amateur naturalist and collector. He graduated from West Point Military Academy (1822) and fought in the Second Seminole War and the Mexican-American War (1846–1848). He was Brigadier-General of Volunteers on the Union side in the American Civil War (1861), where he commanded the Pennsylvania Reserves Division. He was captured (June 1862) at Fraser's Farm and exchanged two months later for the Confederate General Simon Bolivar Buckner. Previous illness was aggravated by his imprisonment, and he was on sick leave until his resignation (March 1863). He wrote a brief paper (1851) entitled 'Some remarks on the habits &c., of birds met with in western Texas, between San Antonio and the Rio Grande and in New Mexico', which he sent to Audubon (q.v.). A reptile is named after him.

McChesney

Masked Lark ssp. *Spizocorys personata mcchesneyi*
    J. G. Williams, 1957

Donald S. McChesney (b.1895) was an American ornithologist. He led the Cornell University East African Expedition (1956).

McClelland

McClelland's Bulbul *Ixos mcclellandii* **Horsfield**, 1840
[Alt. Mountain Bulbul, Rufous-bellied Bulbul]
McClelland's Laughingthrush *Garrulax gularis* McClelland,
    1840
[Alt. Rufous-vented Laughingthrush]
Spot-breasted Scimitar Babbler *Pomatorhinus mcclellandi*
    Godwin-Austen, 1870

John McClelland (1805–1875) was a British doctor with interests in geology and natural history. He was sent on a mission (1835) to ascertain if tea could grow in north-eastern India. He was appointed (1836) Secretary of the 'Coal Committee', the forerunner of the Geological Survey of India, to advise the government how best to exploit Indian coal reserves. He was editor of the *Calcutta Journal of Natural History* (1841–1847).

McConnell

Tinamou sp. *Crypturus macconnelli* **Brabourne** & **Chubb**,
    1914 NCR
[Alt, Cinereous Tinamou; JS *Crypturellus cinereus*]
McConnell's Flycatcher *Mionectes macconnelli* Chubb,
    1919
McConnell's Spinetail *Synallaxis macconnelli* Chubb, 1919

McConnell's Rufous-collared Sparrow *Zonotrichia capensis
    macconnelli* **Sharpe**, 1900
White-barred Piculet ssp. *Picumnus cirratus macconnelli*
    Sharpe, 1901
McConnell's Mottled Owl *Strix virgata macconnelli* Chubb,
    1916

Rufous-sided Crake ssp. *Creciscus melanophæus macconnelli* Chubb, 1916 NCR
[JS *Laterallus melanophaius melanophaius*]

White-tipped Dove ssp. *Leptotila verreauxi macconnelli* Chubb, 1917 NCR
[JS *Leptotila verreauxi brasiliensis*]

Tawny-throated Leaftosser ssp. *Sclerurus mexicanus macconnelli* Chubb, 1919

Lowland Hepatic Tanager ssp. *Piranga flava macconnelli* Chubb, 1921

Frederick Vavasour McConnell (1868–1914) was an English traveller and collector. He made some of his collections (1894–1898) with a J. J. Quelch, who wrote *Animal Life in British Guiana* (1901). His collections inspired a book by C. H. Chubb (q.v.) – *The Birds of British Guiana* (1916–1921), to which McConnell's wife (see **Helen McConnell** and **Mackenzie, H.**) wrote the foreword. He presented his collections of mammals, spiders and birds, all obtained in British Guiana (Guyana) to the BMNH. He is commemorated in the names of four mammals and an amphibian. (See also **Frederick (McConnell)** and **Vavasour, F.**)

McCormick

McCormick's Skua *Stercorarius maccormicki* **H. Saunders**, 1893
[Alt. South Polar Skua]

Royal Albatross ssp. *Diomedea epomophora mccormicki* **Mathews**, 1912 NCR
[Alt. Southern Royal Albatross; NRM]

Robert M. McCormick (1800–1890) was a British naval surgeon, explorer and naturalist, described as a 'conscientious physician and zoological technician'. He was Assistant Surgeon on HMS *Hecla* (1827) during Parry's Arctic expedition, and was Ship's Surgeon on what he described as 'a small surveying ten gun brig' – HMS *Beagle* (1832). McCormick wrote (1832): 'We anchored off Porta Praya, in the island of St. Jago, Cape de Verde Islands, and I landed there … I paid a visit to the remarkable old baobab-tree … growing in an open space to the westward of the town … as a memento I cut my initials, with the date of the year, high up the main stem … and on measuring the baobab-tree, I found it 361/2 feet in circumference.' It was the tradition that Ship's Surgeons collected natural history specimens on such trips, so he was irritated Darwin (q.v.) did it instead, and he resigned once the ship returned from Salvador. He sailed back from Brazil to England on the *Tyne*. He was Ship's Surgeon on Sir James Clark Ross's Antarctic expedition (1839–1843) on HMS *Erebus* and *Terror*. On the Ross voyage (1839) he visited the Cape Verde Islands again and recorded in his journal: '… I made a visit to my old friend the baobab-tree, in the middle of the valley, and a mile to the eastward of the town … On reaching the baobab-tree, I ascended it, and looked for my own initials, which I cut, with the year 1832, in the main stem, about two-thirds up the tree, when here last in that year. Time had impressed them deeper, and they appeared larger, more marked and distinct from the contraction of the bark around. I now added the present year, 1839, beneath the former one …' When it was realised that

Franklin's (q.v.) Northwest Passage expedition had vanished, McCormick led an unsuccessful search party. However, he did chart the Wellington Channel from his ship, appropriately named the 'Forlorn Hope'. He wrote *Voyages of Discovery in the Arctic and Antarctic Seas, and Round the World*.

McCown

McCown's Longspur *Rhynchophanes mccownii* **G. N. Lawrence**, 1851
[Syn. *Calcarius mccownii*]

John Porter McCown (1815–1879) was an American soldier who attained the rank of Major-General. During the American Civil War he fought for the Confederacy and afterwards worked first as a teacher, then as a farmer. He collected the longspur when he fired on a group of Horned Larks *Eremophila alpestris*. His friend George Newbold Lawrence (q.v.) named the bird after him whilst working on the collections at the USNM.

McCoy

Bristlebird genus *Maccoyornis* **Mathews**, 1912 NCR
[Now in *Dasyornis*]

McCoy's Perroquet [archaic] *Cyclopsitta maccoyi* **Gould**, 1875 NCR
[Alt. Double-eyed Fig Parrot; JS *Cyclopsitta diophthalma macleayana*]

Sir Frederick McCoy (1817–1899) was an Irish palaeontologist and naturalist. He was educated at Cambridge and worked at the Woodwardian Museum on its fossil collection (1846–1850). He became Professor of Geology at Queen's College, Belfast (1850), then first Professor of Natural Science at the University of Melbourne (1854). He taught many different subjects over three decades. He was founding Director of the National Museum of Natural History and Geology, Melbourne. His last publication was 'Note on a New Australian *Pterygotus*' (1899). A reptile is named after him.

McDonald

Hood Mockingbird *Mimus macdonaldi* **Ridgway**, 1890
[Alt. Española Mockingbird]

Colonel Marshall 'Marsh' McDonald (1836–1895) was an American naturalist and ichthyologist. He served in the Confederacy Army in the American Civil War, including being Captain of Artillery at the siege of Vicksburg (1863), and was a professor at the Virginia Military Institute. He became Commissioner of Fish and Fisheries for the State of Virginia (1875). He devised a system of automatic fish-hatching jars (1881). He was appointed head of the US Fish Commission (1887) upon the death of Spencer Fullerton Baird (q.v.).

McGill

Brown Thornbill ssp. *Acanthiza pusilla mcgilli* Boles, 1983 NCR
[NUI *Acanthiza pusilla dawsonensis*]

Arnold Robert McGill (1905–1988) was an Australian businessman and ornithologist. He was Assistant Editor of *Emu*

(1948–1969) and President, RAOU (1958–1959). He wrote *Australian Warblers* (1970).

McGregor , J.

Yellow Chat ssp. *Epthianura crocea macgregori* **Keast**, 1958

J. McGregor (DNF), a naturalist, collected the Capricorn Yellow Chat (1859). In that same year near the same locality (Rockhampton, Queensland) he shot and stuffed a very large crocodile which he took to Sydney where it was exhibited.

McGregor, R. C.

McGregor's Cuckooshrike *Coracina mcgregori* **Mearns**, 1907
[Alt. Sharp-tailed Cuckooshrike]

McGregor's House Finch *Carpodacus mexicanus mcgregori* **H. E. Anthony**, 1897 EXTINCT
Philippine Cockatoo ssp. *Kakatoe haematuropygia mcgregori* **Hachisuka**, 1930 NCR; NRM
[JS *Cacatua haematuropygia*]

Richard Crittenden McGregor (1871–1936) was born in Australia, but moved to the USA with his American mother following the death of his father. He studied zoology at Stanford University, but his academic career was interrupted by a fish-collecting expedition to Panama delaying his bachelor's degree (1898) – in philosophy! He was Ornithologist to the Manila Bureau of Science in the Philippines. Among his many publications are several articles on the birds of Santa Cruz County, California. He also wrote *A Manual of Philippine Birds* (1909–1910), the first full treatment of the country's avifauna. An amphibian and a reptile that he collected (1907) are named after him.

McGrigor

Small Niltava *Niltava macgrigoriae* **E. Burton**, 1836

Jane Grant McGrigor (d.1902) was the daughter of Sir James McGrigor, Director-General of the Army Medical Department.

McIlhenny, A. E.

Short-eared Owl ssp. *Asio accipitrinus mcilhennyi* **W. Stone**, 1899 NCR
[JS *Asio flammeus flammeus*]

Edward Avery McIlhenny (1872–1949) was an American businessman, naturalist, conservationist, explorer and the son of Tabasco brand sauce inventor Edmund McIlhenny. He financed his own Arctic expedition to Alaska (1897) and used his 170-acre personal estate, known as Jungle Gardens, to propagate both native and imported plant varieties of plant. He was uncle to J. S. McIlhenny (q.v.).

McIlhenny, J. S.

Black-faced Cotinga *Conioptilon mcilhennyi* **Lowery & O'Neill**, 1966

John Stauffer McIlhenny (1910–1997) sponsored the Louisiana State University Peruvian Expedition (1964–1965), during which this cotinga was discovered. He was extremely wealthy, his family having invented the original recipe for Tabasco sauce in the 1860s (and still being in the business of producing it). A mammal is named after him.

McIlvaine

Forest Elaenia ssp. *Myiopagis gaimardii macilvainii* **G. N. Lawrence**, 1871

Professor Dr Joshua Hall McIlvaine (1815–1897), after schooling at Lafayette College, Easton, Pennsylvania, graduated from Princeton (1837) and attended a seminary, qualifying as a presbyterian preacher (1839). He lectured on philology and ethnology and wrote pamphlets such as *An Examination of the Malthusian Theory* (1867). He was awarded his doctor of divinity (1854) and was Professor of 'Belles Lettres' at Princeton (1860–1870). He delivered a series of lectures in Philadelphia (1869) on social science, at the close of which he was offered the chair of Social Science at the University of Pennsylvania, but declined. He was pastor of a New Jersey church (1870–1887), then left to establish Evelyn College for women at Princeton (1887). His other works included *The Life and Works of Lewis Henry Morgan* (1882). Lawrence's etymology says the bird was named after 'ethnologist and ornithologist J. H. McIlvaine of Philadelphia' – which we believe to be the above.

McKay

McKay's Bunting *Plectrophenax hyperboreus* **Ridgway**, 1884

Charles McKay (1855–1883) was an amateur naturalist in the United States signal corps. He was sent to Alaska, where he discovered the bunting that now bears his name. McKay collected c.400 birds for the USNM. He disappeared (April 1883) when on a collecting trip in a canoe. His body was never recovered.

McLean

Gerygone sp. *Pseudogerygone macleani* **Ogilvie-Grant**, 1907 NCR
[Alt. Grey Gerygone; JS *Gerygone igata*]

Yellow-crowned Parakeet ssp. *Cyanoramphus auriceps macleani* **Mathews & Iredale**, 1913 NCR; NRM

John C. McLean (1872–1918) was a New Zealand farmer, naturalist and oologist.

McLeannan

Ocellated Antbird *Phaenostictus mcleannani* **G. N. Lawrence**, 1860

James McLeannan (DNF) was an American engineer on the Panama Railway (described as 'trackmaster' of the Lion Hill station – an area now submerged under Lake Gatun). He made a collection with John R. Galbraith, a taxidermist (q.v.), of the 389 birds he found along the route during the 1850s, which were sent to various British and American ornithologists, but catalogued by Lawrence (q.v.) (1860–1863). Salvin (q.v.) visited McLeannan (1863), who made a second collection for him of 272 species.

McLennan

Songlark genus *Maclennania* **Mathews**, 1917 NCR
[Now in *Cincloramphus*]

Laughing Kookaburra ssp. *Dacelo novaeguineae mclennani*
   North, 1911 NCR
[JS *Dacelo novaeguineae minor*]
Red-cheeked Parrot ssp. *Geoffroyus geoffroyi maclennani*
   W. D. K. MacGillivray, 1913
Black-throated Finch ssp. *Poephila cincta maclennani*
   Mathews, 1918 NCR
[JS *Poephila cincta atropygialis*]

William Rae McLennan (1882–1935) was an Australian orni-
thologist and natural history collector. He learnt to skin birds
in a blacksmith's shop in Victoria before moving to Broken
Hill, New South Wales, where he did local fieldwork for W. D.
K. and Ian MacGillivray (q.v.). He collected for MacGillivray in
Cape York Peninsula (1909–1915) and in the Gulf of Carpen-
taria for H. L. White (q.v.) after his war service (1916–1918).
He took to gold prospecting (1922) until his health deterio-
rated, the long-term result of being gassed during WW1.
During all his collecting trips he made detailed notes.
Mathews (q.v.) wrote in his etymology for the songlark genus:
'Named in honour of William Rae McLennan, 1882, who, for
Dr. Macgillivray and others, has done such good work as a
collector'.

McLeod

Eared Poorwill *Nyctiphrynus mcleodii* **Brewster**, 1888

Rusty Sparrow ssp. *Aimophila rufescens mcleodii* Brewster,
   1888

Richard Randall McLeod (DNF) from Maine was a collector in
Mexico (1883–1888) and Arizona (1898).

McNicoll

Ribbon-tailed Astrapia *Taeniaparadisea macnicolli*
   Kinghorn, 1939 NCR
[JS *Astrapia mayeri*]

Brigadier-General Sir Walter Ramsay McNicoll (1877–1947)
was a career soldier. In WW1 he was wounded at Gallipoli
(1915) and later commanded a brigade in France (1916–1918).
He was Administrator of the Mandated Territory of New
Guinea (1934–1943).

Meade-Waldo

Meade-Waldo's Chat *Saxicola dacotiae* Meade-Waldo,
   1889
[Alt. Canary Islands Stonechat, Fuerteventura Stonechat]
Meade-Waldo's Oystercatcher *Haematopus meadewaldoi*
   Bannerman, 1913 EXTINCT
[Alt. Canarian Oystercatcher, Canary Islands Oystercatcher]

Common Linnet ssp. *Carduelis cannabina meadewaldoi*
   Hartert, 1901

Edmund Gustavus Bloomfield Meade-Waldo (1855–1934)
was an English explorer, ornithologist and aviculturist. He
collected widely, including the Canary Islands, where he

wrote a definitive checklist. He was the first (1896) to record
seeing a Pin-tailed Sandgrouse *Pterocles alchata* taking up
water in its breast feathers to take to its chicks. This reported
behaviour was ridiculed right up until the 1960s when it was
filmed a number of times in TV documentaries. He owned
Hever Castle in Kent (1896–1905), the same county where
(1874) he started an introduction scheme for the Little Owl,
later taken up by Lord Lilford (q.v.). He is also famed for
having seen an unidentified 'sea serpent' (1903). He discovered (1888) and described (1889) the chat.

Mearns

Swift genus *Mearnsia* **Ridgway**, 1911

Mearns's Flicker *Colaptes auratus* **Linnaeus**, 1758 [see note
   below]
Philippine Swiftlet *Aerodramus mearnsi* **Oberholser**, 1912

Mearns's Thrasher *Toxostoma cinereum mearnsi*
   A. W. Anthony, 1895
[Alt. Grey Thrasher ssp.]
Dark-eyed Junco ssp. *Junco hyemalis mearnsi* Ridgway,
   1897
Mearns's Quail *Cyrtonyx montezumae mearnsi* **Nelson**, 1900
[Alt. Montezuma Quail ssp.]
Loggerhead Shrike ssp. *Lanius ludovicianus mearnsi*
   Ridgway, 1903
Rufous-fronted Tailorbird ssp. *Orthotomus frontalis mearnsi*
   McGregor, 1907
Mearns's Gilded Flicker *Colaptes chrysoides mearnsi*
   Ridgway, 1911
White-winged Dove ssp. *Zenaida asiatica mearnsi* Ridgway,
   1915
Blue-spotted Wood Dove ssp. *Turtur afer mearnsi*
   W. L. Sclater & Mackworth-**Praed**, 1920 NCR; NRM
Red-winged Blackbird ssp. *Agelaius phoeniceus mearnsi*
   A. H. Howell & Van Rossem, 1928
Mearns's Woodpecker *Melanerpes formicivorus mearnsi*
   Van Rossem, 1934 NCR
[Alt. Acorn Woodpecker ssp.; JS *Melanerpes formicivorus
   formicivorus*]
Brown Tit Babbler ssp. *Macronus striaticeps mearnsi*
   Deignan, 1951 NCR
[NUI *Macronus striaticeps mindanensis*]

Lieutenant-Colonel Edgar Alexander Mearns (1856–1916)
was a surgeon in the United States Army. He was stationed
first in Mexico (1892–1894), then the Philippines (1903–1907),
and later Africa (1909–1911). He published widely on natural
history during the last decade of his life, including many
descriptions of African birds. Among his Philippines finds
was the rare Bagobo Babbler *Leonardina woodi*. Mearns
was a friend of Theodore Roosevelt (q.v.) and accompanied
him on his trip to East Africa (1909). Childs Frick (q.v.)
approached the USNM (1911) looking for a scientist to
accompany him on his collecting trip to Africa; Mearns was
chosen. Frick agreed to pay Mearns's salary and expenses
and to donate all bird collections to the USNM. This was
Mearns's last expedition. His life was beset with illness,
including a 'nervous breakdown complicated by malaria' and
various parasitic conditions. Eventually he developed

diabetes and, before insulin treatment was available, nothing could be done for him. Seven mammals and a reptile are named after him. [The name 'Mearns's Flicker' may be encountered attached to the scientific name *Colaptes auratus*. This probably arises as a result of sometime taxonomic opinion that *C. chrysoides* – as seen in Mearn's Gilded Flicker *Colaptes chrysoides mearnsi* – was conspecific with *C. auratus*.]

Mears

Kalij Pheasant *Gennæus mearsi* **Oates**, 1910 NCR
[JS *Lophura leucomelanos lathami*]

White-browed Scimitar Babbler ssp. *Pomatorhinus schisticeps mearsi* **Ogilvie-Grant**, 1905

Lieutenant-Colonel A. Mears (1869–1941) was a British army engineer and surveyor in India (1897–1921).

Mechow

Widowbird sp. *Urobrachia mechowi* **Cabanis**, 1881 NCR
[Alt. Fan-tailed Widowbird; JS *Euplectes axillaris bocagei*]
Mechow's Long-tailed Cuckoo *Cercococcyx mechowi* Cabanis, 1882
[Alt. Dusky Long-tailed Cuckoo]

Mechow's Chanting Goshawk *Melierax metabates mechowi* Cabanis, 1882
[Alt. Dark Chanting Goshawk ssp.]
Mechow's White-fronted Plover *Charadrius marginatus mechowi* Cabanis, 1884

Alexander von Mechow (1831–1890) was a Prussian officer, explorer and naturalist. He was born in Silesia, Austria, and at a young age joined the Prussian Army where he fought in the German War (1866) and Franco-Prussian War (1870–1871), being badly wounded in the battle of Wörth. He officially retired from the army (1874) with the rank of major. He led several collecting expeditions to Angola, including the first German Loango expedition (1873–1875), and a second longer expedition (1879–1882) that included exploring the middle course of the Kwango River (1880) collecting reptiles and amphibians. After this little was heard of him, and even the year of his death seems uncertain. A butterfly, a reptile and a mammal are named after him.

Medland

New Zealand Fantail ssp. *Rhipidura flabellifera medlandae* **Mathews**, 1926 NCR
[JS *Rhipidura fuliginosa placabilis*; trinomial originally spelled *melandae*]

Lilian Marguerite Iredale *née* Medland (1880–1955) was an English nurse, bird artist, and wife of ornithologist Tom Iredale (q.v.).

Meek

Louisiade Pitta *Erythropitta meeki* **Rothschild**, 1898
Meek's White-eye *Zosterops meeki* **Hartert**, 1898
[Alt. White-throated White-eye, Tagula White-eye]

Meek's Lorikeet *Charmosyna meeki* Rothschild & Hartert, 1901
Bougainville Crow *Corvus meeki* Rothschild, 1904
Meek's Pigeon *Microgoura meeki* Rothschild 1904 EXTINCT
[Alt. Choiseul Pigeon, Solomon Crowned Pigeon]
Meek's Streaked Honeyeater *Ptiloprora meekiana* Rothschild & Hartert, 1907
[Alt. Olive-streaked Honeyeater, Yellowish-streaked Honeyeater]
Meek's Hawk Owl *Ninox meeki* Rothschild & Hartert, 1914
[Alt. Manus Boobook, Manus Hawk Owl]
Meek's Pygmy Parrot *Micropsitta meeki* Rothschild & Hartert, 1914
[Alt. Yellow-breasted Pygmy Parrot]

Meek's Hanging Parrot *Loriculus aurantiifrons meeki* Hartert, 1895
[Alt. Orange-fronted Hanging Parrot ssp.]
Pygmy Longbill ssp. *Oedistoma pygmaeum meeki* Hartert, 1896
Marbled Frogmouth ssp. *Podargus ocellatus meeki* Hartert, 1898
White-bellied Whistler ssp. *Pachycephala leucogastra meeki* Hartert, 1898
Variable Kingfisher ssp. *Ceyx lepidus meeki* Rothschild, 1901
Yellow-billed Kingfisher ssp. *Syma torotoro meeki* Rothschild & Hartert, 1901
[SII *Syma torotoro torotoro*]
Fan-tailed Cuckoo ssp. *Cacomantis flabelliformis meeki* Rothschild & Hartert, 1902
[SII *Cacomantis flabelliformis pyrrophanus*]
New Guinea Cuckooshrike ssp. *Coracina melas meeki* Rothschild & Hartert, 1903
Spangled Drongo ssp. *Dicrurus bracteatus meeki* Rothschild & Hartert, 1903
Kolombangara Monarch ssp. *Monarcha browni meeki* Rothschild & Hartert, 1905
[Syn. *Symposiachrus browni meeki*]
Dusky Myzomela ssp. *Myzomela obscura meeki* Rothschild & Hartert, 1907 NCR
[JS *Myzomela obscura fumata*]
Eastern Barn Owl ssp. *Tyto delicatula meeki* Rothschild & Hartert, 1907
Queen Carola's Parotia ssp. *Parotia carolae meeki* Rothschild, 1910
Fan-tailed Berrypecker ssp. *Melanocharis versteri meeki* Rothschild & Hartert, 1911
Large Fig Parrot ssp. *Cyclopsitta blythi meeki* Rothschild, 1911 NCR
[JS *Psittaculirostris desmarestii godmani*]
Black Pitohui ssp. *Pitohui nigrescens meeki* Rothschild & Hartert, 1913
Papuan Scrubwren ssp. *Sericornis papuensis meeki* Rothschild & Hartert, 1913
Slender-billed Cuckoo Dove ssp. *Macropygia amboinensis meeki* Rothschild & Hartert, 1915
Large-tailed Nightjar ssp. *Caprimulgus macrurus meeki* Rothschild & Hartert, 1918 NCR
[NUI *Caprimulgus macrurus schlegelii*]

White-browed Crake ssp. *Porzana cinerea meeki* Hartert, 1924 NCR; NRM

Albert Stewart Meek (1871–1943) was an English explorer who collected in New Guinea, Australia and the Solomon Islands. He began collecting bird and insect specimens for Lord Rothschild (q.v.) (1894). He collected six specimens of a unique pigeon on the island of Choiseul (Solomon Islands) (1904), which Rothschild named *Microgoura meeki*. No European ever saw a living specimen again, as subsequent expeditions failed to find any. Meek wrote *A Naturalist in Cannibal Land* (1913).

Meena

Oriental Turtle Dove ssp. *Streptopelia orientalis meena* **Sykes**, 1832

Sykes (q.v.) did not explain his choice of scientific name. Meena is an Indian (female) name usually said to mean 'precious blue stone'. The word can also refer to the Meena community, a people found mainly in Rajasthan, India.

Mees

Mees's Monarch *Monarcha sacerdotum* Mees, 1973
[Alt. Flores Monarch; Syn. *Symposiachrus sacerdotum*]
Mees's Greenlet *Hylophilus puella* Mees, 1974 NCR
[Description based on a female *Terenura callinota*]
Mees's Nightjar *Caprimulgus meesi* Sangster & Rozendaal, 2004

Rufous Owl ssp. *Ninox rufa meesi* Mason & **Schodde**, 1980

Dr Gerlof Fokko Mees (1926–2013) was born in Holland, lived for some years in the East Indies, and retired to Australia. He studied biology at the University of Leiden. As a student, he collected birds on Trinidad and Tobago (1953–1954) and during his first year made a detailed study of white-eyes *Zosterops*, which had attracted his attention during his service as a volunteer in the Dutch army during the Indonesian independence war (1946–1949). This study ultimately resulted in a 762-page monograph: *A Systematic Review of the Indo-Australian Zosteropidae* (1957–1969). Volume I was his PhD thesis. He became Curator of Vertebrate Animals at the Western Australian Museum in Perth (1958), dividing his attention between ichthylogy and ornithology. He returned to Holland (1963) to become Curator of Birds at the University of Leiden after the death of Junge (q.v.), who had been his promoter during his study there. He made expeditions to Suriname (1965–1990), then returned to Australia (1991). He had a particular interest in *Zosterops* and visited Norfolk Island twice to study them there. He and his wife Veronica contributed to the revised reprint of François Haverschmidt's *Birds of Suriname* (1995).

Mehler

Squirrel Cuckoo ssp. *Piaya cayana mehleri* **Bonaparte**, 1850

Dr Eugen Mehler (1826–1896) was a German classical scholar and philologist who published dictionaries and editions of the Greek classics. He took Dutch nationality (1854).

Meiffren

Quail-plover *Ortyxelos meiffrenii* **Vieillot**, 1819
[Alt. Lark Buttonquail]

Guillaume Michel Jérôme Meiffren, Baron Laugier de Chartrouse (1772–1843), was a French ornithologist and collector. He wrote *Nouveau Recueil des Planches Coloriées d'Oiseaux* (1820) with Temminck (q.v.) as co-author.

Meinertzhagen, A.

Subantarctic Snipe ssp. *Coenocorypha aucklandica meinertzhagenae* **Rothschild**, 1927

(See also **Anne (Meinertzhagen)** and **Meinertzhagen, R.**)

Meinertzhagen, R.

Meinertzhagen's Snowfinch *Pyrgilauda theresae* Meinertzhagen, 1937
[Alt. Afghan Snowfinch, Theresa's Snowfinch]

African Broadbill ssp. *Smithornis capensis meinertzhageni* **Someren**, 1919
Rock Pipit ssp. *Anthus spinoletta meinertzhageni* E. G. Bird, 1936 NCR
[JS *Anthus petrosus petrosus*]
Dunnock ssp. *Prunella modularis meinertzhageni* **J. M. Harrison** & **Pateff**, 1937
Rock Bush Quail ssp. *Perdicula argoondah meinertzhageni* **Whistler**, 1937
Meinertzhagen's Warbler *Sylvia deserticola ticehursti* Meinertzhagen, 1939
[Alt. Tristram's Warbler ssp.]
Great Tit ssp. *Parus major meinertzhageni* **Koelz**, 1939 NCR
[Alt. Turkestan Tit; JS *Parus major bokharensis*]
Eurasian Treecreeper *Certhia familiaris meinertzhageni* **Clancey**, 1942 NCR
[JS *Certhia familiaris brittanica*]
Eurasian Wren ssp. *Troglodytes troglodytes meinertzhageni* Clancey, 1942 NCR
[JS *Troglodytes troglodytes troglodytes*]
Black-crowned Tchagra ssp. *Tchagra senegalus meinertzhageni* **Payn**, 1945 NCR
[JS *Tchagra senegalus cucullatus*]
Bimaculated Lark ssp. *Melanocorypha bimaculata meinertzhageni* **Wolters**, 1953 NCR
[JS *Melanocorypha bimaculata rufescens*]
Spike-heeled Lark ssp. *Chersomanes albofasciata meinertzhageni* **J. Macdonald**, 1953 NCR
[JS *Chersomanes albofasciata garrula*]

Richard Meinertzhagen (1878–1967) was a soldier, hunter, ornithologist, writer, spy, advocate of Zionism and, according to the preface of his *Kenya Diary*, a killer. He was an empire builder in the 'Boys Own' mould, although he lived in an era when the British Empire was in its final decline. He is truly one of the most remarkable characters to appear here. He was born to wealth and position. Although he was one of ten children he was still pampered, even to the extent of being bought an elephant by an eccentric uncle as a christening gift. The family had several homes and was well connected; family friends included Florence Nightingale. As a soldier,

Meinertzhagen served in India, East Africa, and Palestine. His primary role was intelligence gathering, but he was also known as a hardened killer. Meinertzhagen retired from the army (1925), after serving as Britain's chief political officer in Palestine and Syria where he indulged his twin hobbies of promoting a Jewish state and ornithology. He travelled widely, gathering material for books, which became standard references, but also frequently acting in support of British intelligence. His books include *Nicoll's Birds of Egypt* (1930), *The Birds of Arabia* (1954), *Kenya Diary* (1957), *Pirates and Predators* (1959), *Middle East Diary* (1959), *Army Diary* (1960) and *The Diary of a Black Sheep* (1964). He bequeathed his collection of c.25,000 bird specimens, among the best in the world, to the BMNH. However, in the past 20 years it has been conclusively shown that a very large proportion of these skins were, in fact, stolen from the BMNH in the first place with many others stolen from various collectors including Hugh Whistler (q.v.). The falsified data on their labels render his collection highly unreliable. Further revelations came in Brian Garfield's book *The Meinertzhagen Mystery* (2007), which was subtitled '*The Life and Legend of a Colossal Fraud*'. A mammal is also named after him.

Meise

Mountain Tailorbird ssp. *Phyllergates cuculatus meisei* **Stresemann**, 1931
Green-backed Gerygone ssp. *Gerygone chloronota meisei* Stresemann & Paludan 1932 NCR
[JS *Gerygone chloronota cinereiceps*]

Dr Wilhelm Meise (1901–2002) studied zoology, botany, chemistry, geography and mathematics under Stresemann (q.v.) at the University of Berlin, and worked for the Zoological Museum in Hamburg. He was Curator of Vertebrates at the Dresden Natural History Museum until WW2. Having survived the war and three years as a prisoner in Siberia, he returned to the Berlin Zoological Museum (1948). He became Curator of Ornithology at the Hamburg Natural History Museum (1951) and an Assistant Professor at the University of Hamburg. He took part in an expedition to Angola (1955), and in subsequent years published a number of papers on geographical variation, speciation and the evolution of African birds. Overall he produced c.170 publications, mostly on birds, but also taxonomy of molluscs, arachnids and reptiles. An exact contemporary of Ernst Mayr (q.v.), at his death he was the oldest and longest serving Corresponding Fellow of the AOU (64 years).

Meissner

Crested Tit Warbler ssp. *Leptopoecile elegans meissneri* **Schäfer**, 1937

Paul Meissner (DNF) was a German banker and financial expert in Shanghai who assisted Brooke Dolan (q.v.) in his Sichuan expedition.

Melanie

Dusky Coquette *Lophornis melaniae* Floericke, 1920 NCR
[Probably an aberrant individual or faded skin of *Lophornis delattrei*]

Melanie Floericke *née* Reiss (1881–1971) was the wife of the describer, Dr Kurt Floericke. The 'species' is known from a single specimen from Colombia that is now thought to be an atypical specimen of Rufous-crested Coquette.

Melchtal

Woodnymph sp. *Rhamphomicron melchtalianur* **W. Bertoni**, 1901 NCR
[Alt. Violet-capped Woodnymph; JS *Thalurania glaucopis*]

Arnold von Melchtal is one of the three legendary founding fathers of Switzerland who threw off the Hapsburg tyranny. The story is more folklore than fact, although still taught in Swiss schools as a historical truth.

Mélisande

Kelp Gull ssp. *Larus dominicanus melisandae* Jiguet, 2002

Mélisande Jiguet (b.2000) is the daughter of the describer, French ornithologist Frédéric Jiguet. (See **Judith**)

Melisseus

Drongo genus *Melisseus* **Hodgson**, 1841 NCR
[Now in *Dicrurus*]

In Greek mythology, Melisseus was father to the nymphs Adrasteia and Ide, who nursed the infant Zeus.

Mell

Mell's Oriole *Oriolus mellianus* **Stresemann**, 1922
[Alt. Silver Oriole, Stresemann's Oriole]

Black Baza ssp. *Baza lophotes melli* Stresemann, 1923 NCR
[JS *Aviceda leuphotes syama*]
Fairy Pittta ssp. *Pitta nympha melli* Stresemann, 1923 NCR
[*P. nympha* NRM]
Lesser Necklaced Laughingthrush ssp. *Garrulax monileger melli* Stresemann, 1923
Orange-bellied Leafbird ssp. *Chloropsis hardwickii melliana* Stresemann, 1923
Orange-headed Thrush ssp. *Zoothera citrina melli* Stresemann, 1923
Eastern Grass Owl ssp. *Tyto longimembris melli* **Yen**, 1933 NCR
[JS *Tyto longimembris chinensis*]

Dr Rudolph Emil Mell (1878–1970) was a German self-taught naturalist who taught biology until 1908, when he emigrated from Germany to China and founded the German-Chinese Secondary School in Canton (Guangzhou). Originally interested in entomology and herpetology, he developed interests in birds and mammals. He hired local collectors and undertook expeditions himself, such as to Ding Wu (north of Guangzhou) (1922). He discovered 76 new subspecies among the 431 bird species to be found in Kwantung province. He wrote on Lepidoptera including the 2-volume *Biologie und Systematik der Süd-chinesischen Sphingiden. Zugleich ein Versuch einer Biologie tropischer Lepidopteren überhaupt* (1922).

Meller

Meller's Duck *Anas melleri* **P. L. Sclater**, 1865

Charles James Meller (1836–1869) was a botanist who worked in Nyasaland (Malawi) (1861) and Mauritius (1865). A mammal and a reptile are named after him.

Mellon

Aztec Parakeet ssp. *Aratinga astec melloni* **Twomey**, 1950 NCR
[JS *Aratinga astec astec*; Syn. *Aratinga nana astec*]

Dr Matthew Taylor Mellon (1897–1992) was an American historian and philanthropist. He served in the US Navy in WW1. Princeton awarded his first degree (1922), Harvard his master's and the University of Freiburg, Germany, where he taught American Studies (1928–1939), his doctorate. The etymology states that: 'This extensive field research by the Section of Birds of Carnegie Museum has been made possible by the generosity and continued interest of Dr. Matthew T. Mellon, who has encouraged the author in every phase of this work.'

Mellor

Butcherbird genus *Melloria* **Mathews**, 1912 NCR
[Now in *Cracticus*]

Little Raven *Corvus mellori* Mathews, 1912

Australian Reed Warbler ssp. *Acrocephalus australis mellori* Mathews, 1912 NCR
[JS *Acrocephalus australis australis*]
Buff-banded Rail ssp. *Gallirallus philippensis mellori* Mathews, 1912
Chestnut-rumped Thornbill ssp. *Acanthiza uropygialis mellori* Mathews, 1912 NCR; NRM
Horsfield's Bronze Cuckoo ssp. *Chrysococcyx basalis mellori* Mathews, 1912 NCR; NRM
Pied Butcherbird ssp. *Cracticus nigrogularis mellori* Mathews, 1912 NCR
[JS *Cracticus nigrogularis nigrogularis*]
Varied Lorikeet ssp. *Psitteuteles versicolor mellori* Mathews, 1912 NCR; NRM
White-browed Scrubwren ssp. *Sericornis frontalis mellori* Mathews, 1912
White-plumed Honeyeater ssp. *Lichenostomus penicillatus mellori* Mathews, 1912 NCR
[NUI *Lichenostomus penicillatus penicillatus*]

John White Mellor (1868–1931) was an Australian ornithologist, a founder member, and President (1911–1912), of the RAOU. His brother-in-law was Samuel Albert White (q.v.).

Melpomene

Orange-billed Nightingale Thrush ssp. *Catharus aurantiirostris melpomene* **Cabanis**, 1850

In Greek mythology Melpomene was one of the Muses; initially the Muse of Singing, but later the Muse of Tragedy.

Memnon

Grey Nightjar ssp. *Caprimulgus indicus memnon* **Koelz**, 1954 NCR
[JS *Caprimulgus jotaka hazarae*]

In Greek mythology Memnon was a King of Ethiopia, and son of Eos (goddess of the dawn). He was one of the Trojan allies slain by Achilles at the siege of Troy. His followers were transformed into birds, which stayed close to Memnon's tomb.

Menage

Sulu Bleeding-heart *Gallicolumba menagei* **Bourns & Worcester**, 1894
Tablas Drongo *Dicrurus menagei* Bourns & Worcester, 1894

Philippine Frogmouth ssp. *Batrachostomus septimus menagei* Bourns & Worcester, 1894
Philippine Pygmy Woodpecker ssp. *Dendrocopos maculatus menagei* Bourns & Worcester, 1894

Louis F. Menage (1859–1924) was a wealthy American property developer and philanthropist. He sponsored the Menage Scientific Expedition to the Philippine Islands and Borneo (1890–1893) undertaken by the Minnesota Academy of Natural Sciences. By the time (1896) the press reported on the outcome of this expedition, Menage himself was described as 'a fugitive from justice in South America', evading a warrant for embezzlement.

Menbek

Menbek's Coucal *Centropus menbeki* **Lesson & Garnot**, 1828
[Alt. Greater Black Coucal, Ivory-billed Coucal]

'Menebiki' is a local New Guinean name for the bird, not an eponym.

Mencke

Mencke's Monarch *Monarcha menckei* **Heinroth**, 1902
[Alt. Mussau Monarch, White-breasted Monarch; Syn. *Symposiachrus menckei*]

Bruno Mencke (1876–1901) was a German naturalist and adventurer. When he inherited his father's confectionery and sugar factories, he bought a luxury yacht from Count Albert I of Monaco and re-named it *Eberhard* (probably after his late father). With a team of hired scientists, including Heinroth (q.v.), he led the first German South Seas Expedition (1900–1901). Mencke was reputed to be a rather unpleasant young man who destroyed other peoples' property – viz. coconut plantations belonging to the natives of St Matthias Island in Neuhannover (Papua New Guinea). This action resulted in him being speared to death by the disgruntled locals. Unfortunately several other members of his party were also killed. Shortly after his death the *Eberhard* was wrecked on a coral reef. The deaths of Mencke and his party resulted in the German colonial police carrying out reprisals and massacring 81 of the islanders. Five hundred 'specimens' supposedly from his expedition were folklore exhibits and had mostly been bought from other peoples' collections.

Menden

Red-and-black Thrush *Zoothera mendeni* **Neumann**, 1939
[Alt. Peleng Thrush]

Crescent-chested Babbler ssp. *Stachyris melanothorax mendeni* Neumann, 1935
Eyebrowed Wren-babbler ssp. *Napothera epilepidota mendeni* Neumann, 1937
[SiN *Napothera epilepidota diluta*]
Sulawesi Scops Owl ssp. *Otus manadensis mendeni* Neumann, 1939

J. J. Menden (DNF) was a Dutch collector and naturalist who was in the East Indies (1935–1939). He collected the holotype of the Taliabu Masked Owl *Tyto nigrobrunnea*.

Mendoza

Mendoza Monarch *Pomarea mendozae* **Hartlaub**, 1854
[Alt. Marquesan Monarch]

Named after the Marquesas Islands in the Pacific called 'Islas de la Marquesa de Mendoza' by the Spanish explorer Álvaro de Mendaña de Neira, which honoured the wife of the then Viceroy of Peru, García Hurtado de Mendoza y Manrique.

Menelik

Turaco genus *Menelikornis* **Boetticher**, 1947 NCR
[Now in *Turaco*]

Black-headed Oriole ssp. *Oriolus monacha meneliki* **Blundell** & **Lovat**, 1899

Menelik II (1844–1913) became King of Shewa (aged 11) and, having been imprisoned by usurpers, succeeded in becoming Emperor of Abyssinia (Ethiopia) (1889). Menelik II claimed to be directly descended from King Solomon and the Queen of Sheba. A mammal is named after him.

Ménétriés

Ménétriés's Wheatear *Oenanthe isabellina* **Temminck**, 1829
[Alt. Isabelline Wheatear]
Ménétriés's Warbler *Sylvia mystacea* Ménétriés, 1832
Ménétriés's Antwren *Myrmotherula menetriesii* **d'Orbigny**, 1837
[Alt. Grey Antwren]

Common Buzzard ssp. *Buteo buteo menetriesi* **Bogdanov**, 1879

Edouard P. Ménétriés (1801–1861) was a French zoologist who collected in Brazil and Russia, where he settled. He studied in Paris under Cuvier (q.v.), among others, on whose recommendation he participated in an expedition in Brazil under Langsdorff (q.v.) (1821–1825). He was invited to become Conservator of Collections of the Russian Academy of Sciences in St Petersburg (1826). He explored and collected in the Caucasus (1829–1830) and wrote *Catalogue Raisonée des Objects de Zoologie Recueillis dans un Voyage au Caucase et Jusqu'aux Frontiers Actuelles de la Perse* (1832). When the Zoological Museum of the Academy of Sciences was officially opened (1832) Ménétriés was

designated Curator of its entomological collections, a position he held for life. He studied the fauna of Siberia and also wrote one of the first works on the fauna of Kazakhstan, as well as many scientific papers. A reptile is named after him.

Mengel

Silky-tailed Nightjar ssp. *Caprimulgus sericocaudatus mengeli* **Dickerman**, 1975
[Syn. *Antrostomus sericocaudatus mengeli*]

Professor Robert Morrow Mengel (1921–1990) was an American ornithologist, artist and bibliographer. He wrote *The Birds of Kentucky* (1965).

Mennell

Mennell's Seedeater *Serinus mennelli* **Chubb**, 1908
[Alt. Black-eared Seedeater; Syn. *Crithagra mennelli*]

Frederic Philip Mennell (b.1880) was a British geologist and mineralogist. He was recruited as a Curator of the Bulawayo Museum in Southern Rhodesia (Zimbabwe) (1905) in order to establish a Department of Mineralogy there. He had an interest in an iron mine in the Sabi Valley where he also noted asbestos deposits. He wrote *The Zimbabwe Ruins* (1903).

Menzbier

Eurasian Penduline-tit ssp. *Remiz pendulinus menzbieri* **Zarudny**, 1913
European Greenfinch ssp. *Carduelis chloris menzbieri* **Moltshanov**, 1916 NCR
[JS *Carduelis chloris bilkevitchi*]
Lesser Spotted Woodpecker ssp. *Dendrocopos minor menzbieri* **Domaniewski**, 1927 NCR
[JS *Dendrocopos minor minor*]
Menzbier's Pipit *Anthus gustavi menzbieri* Shulpin, 1928
[Alt. Pechora Pipit ssp.]
Black-throated Accentor ssp. *Prunella atrogularis menzbieri* **Portenko**, 1929 NCR
[JS *Prunella atrogularis huttoni*]
Bar-tailed Godwit ssp. *Limosa lapponica menzbieri* Portenko, 1936
Common Chiffchaff ssp. *Phylloscopus collybita menzbieri* **Shestoperov**, 1937
Rough-legged Buzzard ssp. *Buteo lagopus menzbieri* **Dement'ev**, 1951

Mikhail Aleksandrovich Menzbier (1855–1935) was a Russian zoologist from Moscow. He described species across Russia, the eastern Soviet Union and northern China. He was a founding member of Russia's first ornithological society and has a number of such societies named after him. He was among the first to posit that birds are related to reptiles. His major work was on the taxonomy of birds of prey and he was responsible for much of their modern classification. He wrote *Ornitologicheskaya Geografiia Evropeiskoi Rossii* (1882) and *Ornithologie du Turkestan et des Pays Adjacents* (1888). He also wrote *Ptitsy Rossii* (Birds of Russia) (1893–1895) – the first critical review on systematics and biology of Russia's birds, as well as works on the zoogeography of the Palaearctic, comparative anatomy and Darwinism. He was a Professor at Moscow University (1886), but left (1911) in protest at the oppression of students. He became Rector of the University (1917). A mammal is named after him.

Mercedes Foster

MacConnell's Flycatcher ssp. *Mionectes macconnelli mercedesfosterae* **Dickerman** & **W. H. Phelps Jr**, 1987
[SII *Mionectes macconnelli roraimae*]

Mercedes S. McDiarmid *née* Foster (b.1942) is a research zoologist with the USNM, at the Patuxent Research Center. The University of California awarded both her bachelor's and master's degrees and the University of South Florida, Tampa her doctorate (1974). After teaching at the University of South Florida and the University of California, Berkeley, she joined the Museum Section of the US Fish and Wildlife Service, at the USNM, Washington, DC, as a research zoologist and Curator of Birds (now Patuxent Research Center). She has published many papers and longer works including (with others): *Neotropical Ornithology* (1985). She and close collaborator Roy McDiarmid co-wrote 'Additions to the reptile fauna of Paraguay with notes on a small herpetological collection from Amambay' (1987). An amphibian is also named after her.

Mercier

Red-moustached Fruit Dove *Ptilinopus mercierii* **Des Murs** & **Prevost**, 1849 EXTINCT

The describers honoured 'M[onsieur] Mercier' when naming this dove. The type specimen was collected by Mercier on the island of Nuku Hiva (French Polynesia) and brought by him to the Paris museum (1848). Nothing more seems to be known about him. He may have been a relative of Philippe Mercier (1781–1831), a botanist who collected in the Americas.

Merian

Turaco sp. *Corythaix meriani* **Rüppell**, 1851 NCR
[Alt. Yellow-billed Turaco. Name replaced by *Tauraco macrorhynchus verreauxii* (Schlegel, 1854)]

Professor Peter Merian (1795–1883) was a Swiss politician, naturalist and geologist at the University of Basel, who helped to develop the Basel Museum of Natural History.

Merion

Hummingbird genus *Merion* **Mulsant** & **E. Verreaux**, 1875 NCR
[Now in *Chlorostilbon*]

Merion, or Meriones, was the name of two characters in Greek mythology. The better known was a Cretan warrior at the siege of Troy, the other, famous for his opulence and avarice, was said to be a brother of the heroic Jason.

Merlin, M. H.

Annam Hill Partridge *Arborophila merlini* **Delacour** & **Jabouille**, 1924
[Alt. Annamese Hill Partridge]

Martial Henri Merlin (1860–1935) was the French Governor-General of Senegal (1907–1908), Madagascar (1917–1918),

Senegal again (1919–1923), and then Indochina (1923–1925). An amphibian is named after him.

Merlin, M. J.

Great Lizard Cuckoo *Coccyzus merlini* **d'Orbigny**, 1839
[Syn. *Saurothera merlini*]

Condesa Mercedes Jaruco de Merlin (Maria de las Mercedes Santa Cruz y Montalvo) (1789–1852) was a Cuban musician and writer who married a French general, only returning to Cuba upon his death (1840). During her time in Paris she was known as a 'Belle Dame'. She wrote *Viaje a la Habana* (1840), an ambitious 3-volume account of the political, social and economic life of Cuba.

Merriam

Merriam's Montezuma Quail *Cyrtonyx montezumae merriami* **Nelson**, 1897
Merriam's Turkey *Meleagris gallopavo merriamii* Nelson, 1900
[Alt. Wild Turkey ssp.]

Clinton Hart Merriam (1855–1942) was an American naturalist and physician, brother to Florence Augusta Bailey (q.v.). His father was a Congressman through whom he met Baird (q.v.) of the USNM (1871). This led to an invitation to work as a naturalist in Yellowstone, Wyoming, as a member of the Hayden Geological Survey. This experience sparked his interest and guided his further education; he studied biology and anatomy at Yale, finally graduating as a physician (1879). His natural history passtime continued while he practised medicine, until he opted for scientific work full-time (1883). When he became Chairman of the Bird Migration Committee of the AOU they successfully applied to Congress for funds to study birds on the grounds that such work would benefit farmers. Merriam became the first chief of the United States Biological Survey's (USBS) Division of Economic Ornithology and Mammalogy. He is most famed for his 'life zone' theory, which hypothesised that 'temperature extremes were the principal desiderata in determining the geographic distribution of organisms.' One of his longer ornithological works is *A Review of the Birds of Connecticut, with Remarks on their Habits* (1877). Later in life, Merriam's focus shifted to studying the Native American tribes of the western USA. Ten mammals and a reptile are named after him.

Merrill, E.

Merrill's Fruit Dove *Ptilinopus merrilli* **McGregor**, 1916
[Alt. Cream-bellied Fruit Dove]

Elmer Drew Merrill (1876–1956) was an American botanist who collected in the Philippines (1902–1929). He worked in the Philippines for 22 years, during which he became Director of the Bureau of Sciences, as well as Professor of Botany at the University of the Philippines. He wrote *Flora of Manila* (1912) and *Enumeration of Philippine Flowering Plants* (1922–1926). His expertise on the Philippines was put into service during WW2 when he compiled a *Handbook of Emergency Food Plants and Poisonous Plants of the Islands of the Pacific*. He became Dean of the California College of Agriculture (1923) and during his time there he added over 110,000 mounted specimens to the university herbarium and also published work on China, Borneo and the Philippines. He was the Director of the New York Botanical Garden (1929–1935) and Administrator of Botanical Collections for Harvard University. Merrill described c.3,000 new species of plants from the Philippines, Polynesia, China, the Moluccas and Borneo. At least seven plant genera are dedicated to him and some 220 other taxa.

Merrill, J.

Merrill's Pauraque *Nyctidromus albicollis merrilli* **Sennett**, 1888
[Alt. Common Pauraque ssp.]
Merrill's Horned Lark *Eremophila alpestris merrilli* **Dwight**, 1890
Merrill's Song Sparrow *Melospiza melodia merrilli* **Brewster**, 1896

James Cushing Merrill (1853–1902) was a Surgeon in the US army and naturalist. His second posting was to Fort Brown, Texas, where he made observations and collected the local birds (1876–1878). He published 'Notes on the ornithology of southern Texas' (1878), logging 252 species. He recorded 12 species and subspecies which he believed were new to the US. He gave many of the bird skins he collected to his friend William Brewster (q.v.) at the Museum of Comparative Zoology, Cambridge, Massachusetts. His various postings led to further study and publications including 'Notes on the birds of Fort Klamath, Oregon' (1883) and 'Notes on the birds of Fort Sherman, Idaho' (1897).

Merritt

Violet-headed Hummingbird ssp. *Klais guimeti merrittii* **G. N. Lawrence**, 1860

Dr Joseph King Merritt (1824–1882) was a physician and ethnologist who was the director of a gold mine at Veragua, New Grenada (Panama). He sent the holotype to Lawrence (q.v.). He wrote *Report on the Region of Mineral de Veraguas and its Gold Mines* (1854).

Merrotsy

Short-tailed Grasswren *Amytornis merrotsyi* **Mellor**, 1913

Arthur Leslie Merrotsy (b.1887) was an Australian collector, schoolteacher and soldier. He was just 17 when appointed as a teacher at Angaston School, South Australia (1904–1908). He was a sapper in the Australian Army throughout WW1, and received the Military Medal (1917). He collected botanical specimens in the Atherton Tablelands of Queensland (1923). Mellor (q.v.) wrote 'Specimens of a new species of Grasswren (*Amytornis*) have been forwarded to me by Mr. A. L. Merrotsy, who collected them in the spinifex or porcupine-grass country to the north-east of Lake Torrens'.

Mers

Smith's Longspur ssp. *Calcarius pictus mersi* **Kemsies**, 1961 NCR; NRM

William H. Mers (DNF) was an American ornithologist in Cincinatti, Ohio, where the describer was Curator of Ornithology at the University of Cincinatti.

Mertens

Hill Myna ssp. *Gracula religiosa mertensi* **Rensch**, 1928 NCR
[NUI *Gracula religiosa venerata*]

Robert Friedrich Wilhelm Mertens (1894–1975) was a German zoologist and herpetologist born in St Petersburg. He left Russia (1912) to study medicine and natural history, obtaining his doctorate from the University of Leipzig (1915). After serving in the German army during WW1, he started work as an assistant at the Senckenberg Museum in Frankfurt (1919–1920), eventually becoming Director and retiring (1960) as Director Emeritus. Mertens was a man of prodigious energy; his wife was his sole assistant (1920–1943). He lectured at Goethe-University, Frankfurt am Main (1932–1939) and became Professor (1939). In spite of these responsibilities and huge workload, he found time to publish c.800 scientific papers and 13 books. His first collecting trip was to Tunisia (1913), and he subsequently collected in 30 countries. During WW2 he evacuated most of the Senckenberg collection to small towns, where the specimens were set up in such locations as dance halls for use and study. He encouraged German soldiers fighting outside Germany to collect specimens for him, and a regular supply of reptiles and other fauna reached him via the German Field Post Office system. He died after a bite from a specimen of the Twig Snake *Thelotornis kirtlandi*, a South African snake that was his long-time pet. No antivenin existed then for this species. It took 18 very painful days for him to die, and he kept a diary of each day's events, remarking in it, with true gallows humour, 'für einen Herpetologen einzig angemessene Ende' (a singularly appropriate end for a herpetologist). Thirteen amphibians, seven reptiles and a mammal are named after him.

Merton

Cuckooshrike sp. *Graucalus mertoni* **Berlepsch**, 1911 NCR
[Alt. White-bellied Cuckooshrike; JS *Coracina papuensis hypoleuca*]
Kai Islands Swamphen *Porphyrio mertoni* Berlepsch, 1913 NCR
[Alt. Purple Swamphen; JS *Porphyrio porphyrio melanopterus*]

Dr Hugo Merton (1879–1939) was a German zoologist. He visited the Kai and Aru Islands in eastern Indonesia with Roux (1907–1908). He took a teaching post (1912) at the University of Heidelberg, becoming Professor of Zoology (1920), but was dismissed (1935) when the Nazis 'cleansed' the universities under the provisions of the Nuremberg Citizenship Laws. He moved to Britain (1937), to work at the Institute of Animal Genetics, University of Edinburgh. Late (1938) he tried to visit Heidelberg but was arrested and sent to Dachau Concentration Camp, and although he was later released and returned to Edinburgh his health was broken and he died soon after. A reptile and an amphibian are named after him.

Merzbacher

Barred Warbler ssp. *Sylvia nisoria merzbacheri* **Schalow**, 1907
Common Linnet ssp. *Acanthis cannabina merzbacheri* Schalow, 1907 NCR
[JS *Carduelis cannabina bella*]

Prof. Dr Gottfried Merzbacher (1843–1926) was a furrier in Munich (1868–1888) who sold his business to become a mountaineer, naturalist and explorer in the Caucasus (1890–1892) and Tien Shan (1902–1903 and 1907–1908). The University of Munich awarded his honorary doctorate (1902) and made him an honorary professor (1907). Merzbacher Glacier Lake is named after him.

Meston

Golden Bowerbird *Corymbicola mestoni* **De Vis**, 1889 NCR
[JS *Prionodura newtoniana*]
Whistler sp. *Pachycephala mestoni* De Vis, 1905 NCR
[Alt. Australian Golden Whistler; JS *Pachycephala pectoralis pectoralis*]

Archibald Meston (1851–1924) was a Scottish immigrant to Australia (1859). He was a journalist (1876–1881), a member of the Queensland Legislative Assembly (1878–1882), a farmer (1874 and 1885–1889), Protector of Aborigines for South Queensland (1898–1903), an explorer and naturalist. He led the Bellender-Ker Range Expedition (1883) and wrote *Geographic History of Queensland* (1895).

Metcalfe

Metcalfe's White-eye *Zosterops metcalfii* **Tristram**, 1894
[Alt. Yellow-throated White-eye]

Dr Percy Herbert Metcalfe (1853–1913) was a British physician and naturalist. He was resident medical officer on Norfolk Island (1882–1913) and collected in the Solomon Islands. He was a friend of the describer, Henry Baker Tristram (q.v.), who named the white-eye in recognition of Metcalfe's assistance.

Meves

Meves's Long-tailed Starling *Lamprotornis mevesii*
Wahlberg, 1856
[Alt. Meves's Glossy Starling]

Professor Friedrich Wilhelm Meves (1814–1891) was a German ornithologist, zoologist and teacher. He worked as an assistant in the Anatomical/Zoological Museum at Kiel (1840), where he met the ornithologist Friedrich Boie who had Swedish connections. Through Boie Meves became a Curator in the Zoological Department, Natural History Museum, Stockholm (1841–1877). From there he took part in many expeditions, both in Sweden and abroad, and made many valuable additions to the museum's collections. He wrote several ornithological works, including *Contribution to Swedish Ornithology* (1868). Wahlberg (q.v.) was his pupil.

Meyen

Chilean Swallow *Tachycineta meyeni* **Cabanis**, 1850
[Syn. *Tachycineta leucopyga* (Meyen, 1834)]
Lowland White-eye *Zosterops meyeni* **Bonaparte**, 1850

Franz Julius Ferdinand Meyen (1804–1840) was a German surgeon, botanist and collector. He suggested that new cells were created through cell division rather than by the creation of new free cells. He published this theory in his work on plant anatomy, *Phytotomie* (1830). He took part in an expedition to South America (1830–1832). There is disagreement over the valid scientific name of the swallow, depending on whether Meyen's name *leucopyga* is deemed to be preoccupied or not. A mammal is named after him.

Meyer, A. B.

Meyer's Bronze Cuckoo *Chrysococcyx meyeri* **Salvadori**, 1874
[Alt. White-eared Bronze Cuckoo]
Meyer's Friarbird *Philemon meyeri* Salvadori, 1878
Meyer's Goshawk *Accipiter meyerianus* **Sharpe**, 1878
[Alt. Papuan Goshawk]
Meyer's Sicklebill *Epimachus meyeri* **Finsch** & Meyer, 1885
[Alt. Brown Sicklebill]
Meyer's Whistler *Pachycephala meyeri* Salvadori, 1889
[Alt. Vogelkop Whistler, Grey-crowned Whistler]
Collared Kingfisher *Halcyon meyeri* Sharpe, 1892 NCR
[JS *Todiramphus chloris chloris*]
Meyer's Bowerbird *Chlamydodera recondita* Meyer, 1895 NCR
[Alt. Fawn-breasted Bowerbird; JS *Chlamydera cerviniventris*]

Meyer's Lorikeet *Trichoglossus flavoviridis meyeri* **Walden**, 1871
[Alt. Citrine Lorikeet ssp. Yellow-and-green Lorikeet ssp.]
Tawny-breasted Honeyeater ssp. *Xanthotis flaviventer meyerii* Salvadori, 1876
Common Cicadabird ssp. *Coracina tenuirostris meyerii* Salvadori, 1878
Black Myzomela ssp. *Myzomela nigrita meyeri* Salvadori, 1881
Meyer's Koklass (Pheasant) *Pucrasia macrolopha meyeri* **Madarász**, 1886
Common Paradise Kingfisher ssp. *Tanysiptera galatea meyeri* Salvadori, 1889
Meyer's Black-capped Lory *Lorius lory salvadorii* Meyer, 1891
Olive Straightbill ssp. *Timeliopsis fulvigula meyeri* Salvadori, 1896
Olive-backed Sunbird ssp. *Cinnyris jugularis meyeri* **Hartert**, 1897 NCR
[NUI *Cinnyris jugularis plateni*]
Variable Pitohui ssp. *Pitohui kirhocephalus meyeri* **Rothschild** & Hartert, 1903

Dr Adolf Bernard Meyer (1840–1911) was a German anthropologist, entomologist and ornithologist who collected in the East Indies and New Guinea (late 19th century). He was Professor at the Anthropological and Ethnographic Museum in Dresden. He co-wrote with Wiglesworth (q.v.) *The Birds of the Celebes and Neighbouring Islands* (1898), and made the first descriptions of a number of bird species from the East Indies. It was he who first recognised that the red male and green female King Parrot *Alisterus scapularis* constituted one species. He was very interested in the evolution debate and corresponded with Wallace (q.v.). He described the bowerbird *Chlamydodera recondita* on the basis of an egg discovered in New Guinea, with no knowledge of the bird that laid it. Even in the 1890s this caused many a raised ornithological eyebrow! One mammal, one amphibian and two reptiles are named after him.

Meyer, B.

> Meyer's Parrot *Poicephalus meyeri* **Cretzschmar**, 1827
> [Alt. Brown Parrot]

Dr Bernhard Meyer (1767–1836) was a German physician, ornithologist and botanist. He practised as a doctor in Hanau before moving to Offenbach. He made ornithological expeditions to Holland, Sweden and Denmark (1805–1827) and wrote *Naturgeschichte der Vögel Deutschlands* (1805) and *Tashenbuch der Deutschen Vogelkunde* (1810). He sold his ornithological collection to the Senckenberg Society for Natural Sciences, of which he was a founding member. Meyer's Parrot is native to tropical Africa, where Meyer never travelled; it was Rüppell (q.v.) who discovered the bird. Cretzschmar (q.v.) dedicated it to Meyer because of his outstanding contribution to the natural history of European birds.

Meyer, O.

> Clamorous Reed Warbler ssp. *Acrocephalus stentoreus*
> *meyeri* **Stresemann**, 1924 NCR
> [NUI *Acrocephalus stentoreus sumbae*]
> Yellow-bibbed Fruit Dove ssp. *Ptilinopus solomonensis*
> *meyeri* **Hartert**, 1926
> Buff-banded Rail ssp. *Gallirallus philippensis meyeri*
> Hartert, 1930

Father Otto Meyer (1877–1937) was a German missionary in the Bismarck Archipelago, Papua New Guinea (1902–1937). He is credited with being the first westerner to discover the distinctive 'Lapita' style of prehistoric pottery, on Watom Island (1908). He wrote *Rare Birds of New Britain* (1935). (See **Otto Meyer**)

Meyer de Schauensee

> Green-capped Tanager *Tangara meyerdeschauenseei*
> **Schulenberg** & Binford, 1985
> _____
> Ecuadorian Rail ssp. *Rallus aequatorialis*
> *meyerdeschauenseei* Fjeldså, 1990

Rodolphe Meyer de Schauensee (1901–1984). (See **De Schauensee**, and **Rodolphe**)

Meyler

> Golden-green White-eye ssp. *Zosterops nigrorum meyleri*
> **McGregor**, 1907

John Meyler (DNF) was the only American resident on Camiguin Island, near Mindanao, Philippines, at the time of the white-eye's discovery.

Michaelis

> Bianchi's Blood Pheasant *Ithaginis cruentus michaelis*
> **Bianchi**, 1903

J. Michaelis (DNF) was a collector in Gansu, western China.

Michahelles

> Yellow-legged Gull *Larus michahellis* **Naumann**, 1840

Dr Georg Wilhelm Christian Karl Michahelles (1807–1834) was a physician and an ichthyologist, ornithologist and herpetologist. He collected for the Munich and Wurzburg Museums in Dalmatia and Croatia (1829–1832) and went to Greece (1834) as an army doctor attached to a division of Bavarian volunteers. He wrote 'Neue südeuropäische Amphibien' (1830). He was a close friend of Louis Agassiz and sent him fish to study.

Michalowski

> Coal Tit ssp. *Periparus ater michalowskii* **Bogdanov**, 1879

(See **Mikhailovski, N.**)

Michie

> Michie's Cuckoo *Cuculus michieanus* **Swinhoe**, 1870 NCR
> [Alt. Indian Cuckoo; JS *Cuculus micropterus*]

Alexander Michie (1833–1902) was a Scottish businessman, Fellow of the Royal Geographical Society, and writer on China. He went to China and stayed for four decades (1854–1895), with occasional visits to England where he retired. He was Chairman of the Chamber of Commerce in Shanghai in which capacity – accompanied by the describer Swinhoe (q.v.), who was British Consul – he conducted a mission into western China, particularly Sichuan. He was the editor of the *Chinese Times*, an English-language newspaper (1880s) and wrote *The Siberian Overland Route* (1864), an account of his adventures while tracing an old route from Peking (Beijing) to St Petersburg, and much else, including *Missionaries in China* (1891). A mammal is named after him.

Michler

> Black-crowned Antpitta *Pittasoma michleri* **Cassin**, 1860

Brigadier-General Nathaniel Michler (1827–1881) was an American army officer and geographer in Panama. He had a distinguished military career in the American Civil War and held a number of appointments in the Army afterwards, including being Superintendent of Public Buildings and Grounds in the District of Columbia (1866–1871). He died of Bright's disease (nephritis).

Midas

> Striped Owl ssp. *Pseudoscops clamator midas* **Schlegel**,
> 1862
> [Syn. *Asio clamator midas*]

King Midas of Phrygia is a mythological figure, but the legends may have been based on a real king of that area (western Turkey) who lived in the eighth century BC. Midas was reputed to be very fond of gold, and was granted the gift of turning everything he touched into gold – often with unfortunate results. Two mammals and an amphibian are named after him.

Middendorff

Middendorff's Grasshopper Warbler *Locustella ochotensis*
  Middendorff, 1853
Middendorff's Stint *Calidris subminuta* Middendorff, 1853
[Alt. Long-toed Stint; Syn. *Ereunetes subminutus*]
Two-barred Warbler *Phyllopneuste middendorffii* **Meves,**
  1871 NCR
[JS *Phylloscopus plumbeitarsus*]

Middendorff's Bean Goose *Anser fabalis middendorffii*
  Severtzov, 1873

Alexander Theodor (Aleksander Fedorovich) von Middendorf
(1815–1894) was an Estonian of German extraction, and a
traveller and naturalist. He qualified as a physician (1837).
Together with Karl Maximilian von Baer (q.v.) he travelled
through the Kola Peninsula (1840). After journeying
throughout Siberia and the surrounding regions on an Imperial Academy expedition (1842–1845), he was made a member
of the Imperial Academy of Sciences at St Petersburg (1845).
His accounts of the Amur River and other remote regions
were the fullest by a naturalist and anthropologist of these
little-explored areas. As a naturalist he wrote on the spread
of permafrost and how this affected the distribution of plants
and animals. He explored Barabinsk forest-steppe (1870),
wrote *A journey to the North and East of Siberia* (1860 and
1877) and explored the Fergana valley (1878). Amongst his
observations of the people of the area he wrote that in
Northern Siberia no one would remove anything from a
sleigh left unattended, even if it contained much needed
food: 'It is well known that the inhabitants of the far North are
frequently on the verge of starvation, but to use any of the
supplies left – behind would be what we call a crime, and
such a crime might bring all sorts of evil upon the tribe'. He
was also famed for his selection work on horse breeding and
cattle-farming. Apart from the birds, two mammals and other
taxa, there is a cape on the island of Novaya Zemlya and a
bay on the Taimyr Peninsula named after him.

Mikado

Mikado Pheasant *Syrmaticus mikado* **Ogilvie-Grant,** 1906

Small Buttonquail ssp. *Turnix sylvatica mikado* **Hachisuka,**
  1931 NCR
[JS *Turnix sylvaticus davidi*]

Named after the title of the Japanese emperors in general,
but in each case a particular emperor is intended. The
pheasant was named after Meiji or Mutsuhito (1852–1912;
reigned 1867–1912) whose 'revolution' brought the shogu-
nate to an end and opened Japan up to the rest of the world
after centuries of isolation. The buttonquail was named after
Showa or Hirohito (1901–1989; reigned 1926–1989).

Miketta

Miketta's Greenlet *Vireolanius leucotis mikettae* **Hartert,**
  1900
[Alt. Slaty-capped Shrike Vireo ssp.]

R. Miketta (DNF) was a collector in Ecuador (1898–1899) for
the Tring Museum and for W. F. H. Rosenberg (q.v.), the

natural history dealer, who at that time had his premises in
Tring.

Mikhailovski, E. K.

Common Pheasant ssp. *Phasianus colchicus michailowski*
  Zarudny, 1909 NCR
[NUI *Phasianus colchicus bianchii*]

E. K. Mikhailovski (DNF) was a Russian publisher.

Mikhailovski, N.

Coal Tit ssp. *Periparus ater michalowskii* **Bogdanov,** 1879

Nikolai Georgievich Garin-Mikhailovsky (1852–1906) was a
Russian engineer, explorer and travel writer. After graduating
from the Railroad Engineers Institute he was in Bulgaria for
four years, and then worked on the construction of a port at
Batumi, Georgia, and the railway link from there to the
oil-field at Baku on the Caspian Sea, where the tit holotype
was collected in the Suram Pass (1878), probably by him.
Later he was involved in the construction of highways and
the Trans-Siberian Railway. He wrote *Around Korea,
Manchuria and Liaodong Peninsula* (1899).

Miki

Black-throated Mango ssp. *Anthracothorax nigricollis miki*
  Dunajewski, 1938 NCR; NRM

Mrs Maria Dunajewska (d.1944) was the wife of the
describer, Dr A. Dunajewski; we assume Miki to be his name
for her.

Mildbread

Cinnamon Bracken Warbler ssp. *Bradypterus cinnamomeus
  mildbreadi* **Reichenow,** 1908

Dr G. W. Johannes Mildbread (1879–1954) was a German
botanist who collected in German East Africa (Tanzania)
(1907–1908), German Cameroons and Fernando Po (1910–
1911). He was Professor of Botany at the Berlin Museum
(1903–1953).

Miles

Spotted Eagle Owl ssp. *Bubo africanus milesi* **Sharpe,** 1886

Colonel Samuel Barrett Miles (1838–1914) was a scholar,
Arabist antiquarian, and traveller in the Middle East. He was
amongst the first Europeans to travel widely in Oman, undertaking an expedition to Wadi Dayqah, Oman (1884) where he
was British Political Agent five times (1872–1887). He wrote
Countries and Tribes of the Persian Gulf, published posthumously and incomplete, only covering Oman (1919).

Millais

Rock Ptarmigan ssp *Lagopus muta millaisi* **Hartert,** 1923

John Guille Millais (1865–1931) was an English naturalist,
traveller, big-game hunter and artist whose father was the
Pre-Raphaelite painter, Sir John Everett Millais. He travelled
all over the world collecting anything and everything, and

created his own private museum at Horsham, Sussex. In WW1, although in his 50s, he served in naval intelligence as a spy in Iceland and Norway. He wrote on many subjects, including *The Natural History of British Game Birds* (1909).

Millard, J. F.

Red-capped Lark ssp. *Calandrella cinerea millardi*
Paterson, 1958

John Forster Millard (1911–2004) was a British civil servant and administrator. He was in various junior posts in Africa after taking a degree in Archaeology at Cambridge (1933). He was an artillery officer in WW2 serving in Ethiopia, Italy and Western Europe – including being present at the German surrender to Montgomery (1945) at Luneberg Heath. He served as District Commissioner in Tanganyika (Tanzania) before moving to Bechuanaland (Botswana) (1952), having been sent there to mediate between factions of the people at odds and on the point of civil war over the marriage (1948) of the hereditary chief, Sir Seretse Khama, to Ruth Williams. He retired to farm in Tanzania but finally moved to Kenya, where he died. He wrote his memoirs *Never a Dull Moment: The Autobiography of John Millard, Administrator, Soldier and Farmer* (1996).

Millard, W. S.

Common Hill Partridge ssp. *Arborophila torqueola millardi* **E. C. S. Baker**, 1921

Dr Walter Samuel Millard (1864–1952) was a British naturalist who was the Honorary Secretary of the Bombay Natural History Society, spending 35 years there. He was also a keen gardener and botanist. With E. Blatter he wrote *Some Beautiful Indian Trees* (1937). He showed (c.1906) the young Salim Ali (q.v.), India's most famed ornithologist, the Bombay Natural History Society's bird collection, and often helped him with identifications. Two mammals are named after him.

Miller

Miller's Emerald *Amazilia versicolor millerii* **Bourcier**, 1847
[Alt. Versicolored Emerald ssp.; Syn. *Agyrtria versicolor millerii*]

The notes of the collector Loddiges (q.v.) refer to a bird which he meant to carry the name *millerii*. Bourcier used Loddiges's manuscript name, but gave no indication as to the person being thus honoured. We think that William Miller (1796-1882), an engraver and water-colourist who illustrated many plants from Loddiges' (q.v. Conrad (Loddiges)) nursery (1820) may be the person intended.

Miller, A. Y.

Syrian Woodpecker ssp. *Dendrocopos syriacus milleri*
Zarudny, 1909

Aleksandr Yacovlevich Miller (1868–1940) became Russian Vice-Consul in Seistan, Persian Baluchistan (1900), and later Consul in Kerman. He was Russian Consul-General in Mongolia (1913–1916). He was described as 'effective and enterprising'.

Miller, J. F.

Miller's Rail *Porzana nigra* J. F. Miller, 1784 EXTINCT
[Alt. Tahiti Sooty Rail]

John Frederick Miller (1759–1796) was an English botanical illustrator. The son of an artist, John and his brother James made paintings using the sketches made by Sydney Parkinson (q.v.) during James Cook's (q.v.) first voyage of discovery. He also went with Joseph Banks (q.v.) on his voyage to Iceland (1772). Miller copied Georg Forster's (q.v.) painting of this rail done during the second Cook voyage, and published it with some remarks in *Cimelia Physica* (1784).

Miller, L. E.

Brown-banded Antpitta *Grallaria milleri* **Chapman**, 1912
Rufous-tailed Xenops *Xenops milleri* Chapman, 1914

Golden-collared Manakin ssp. *Manacus vitellinus milleri* Chapman, 1915

Leo Edward Miller (1887–1952) collected in tropical America (1910–1917) for the AMNH. He wrote *In the Wilds of South America: Six Years of Exploration in Colombia, Venezuela, British Guiana, Peru, Bolivia, Argentina, Paraguay, and Brazil* (1918). A mammal is named after him.

Miller, Waldron de Witt

Tepui Goldenthroat *Polytmus milleri* **Chapman**, 1929

Waldron de Witt Miller (1879–1929) was at the AMNH, New York, (1903) becoming Assistant Curator (1911) and then Associate Curator (1917). He was a co-founder of the New Jersey Audubon Society. He undertook a number of collecting trips such as to Nicaragua (1917) with Ludlow Griscom (q.v.). He co-wrote with Wetmore (q.v.) the fourth edition of the AOU *Check-list* (1926). Miller was injured in a motorcycle accident when he hit a bus, and died three days later. (See **Waldron**)

Millet

Millet's Laughingthrush *Garrulax milleti* **H. C. Robinson** & **Kloss**, 1919
[Alt. Black-hooded Laughingthrush]

Fernand Victor Millet (b.1878) was the superintendent of forests in Annam, in French colonial Vietnam. He was a professional hunter (1902) for many years. He wrote *Les Grands Animaux Sauvages de l'Annam, leurs Moeurs, leur Chasse, leur Tir* (1930). He appears to have been responsible for quite considerable slaughter. He acted as a professional guide for important visitors, including French journalist Albert Londres, who wrote that by the time they met in Indochina, Millet had already shot 47 tigers. Robinson (q.v.) and Kloss (q.v.) also named a mammal after him; Kloss wrote 'I am greatly indebted to Monsieur Millet for assistance and hospitality during my visit to the Langbian Plateau.' A mammal and an amphibian are named after him.

Milligan

Thornbill genus *Milligania* **Mathews**, 1912 NCR
[Now in *Acanthiza*]

Australian Bush Lark sp. *Mirafra milligani* Mathews, 1908
  NCR
[Alt. Horsfield's Bush Lark; JS *Mirafra javanica halli*]

Black-capped Sittella *Neositta pileata milligani* Mathews,
  1912 NCR
[Alt. Varied Sittella ssp.; JS *Daphoenositta chrysoptera
  pileata*]
Little Grassbird ssp. *Poodytes gramineus milligani*
  Mathews, 1921 NCR
[JS *Megalurus gramineus thomasi*]
Redthroat ssp. *Pyrrholaemus brunneus milligani* Mathews,
  1922 NCR; NRM
Brown Honeyeater ssp. *Lichmera indistincta milligani*
  Mathews, 1923 NCR
[JS *Lichmera indistincta indistincta*]

Alexander William Milligan (1858–1921) was an Australian
ornithologist and founding member of the RAOU. He worked
as an accountant and clerk in various legal firms in Western
Australia (1897–1908) and in Victoria (1908–1921). He was
Honorary Consulting Ornithologist to the Western Australian
Museum (1901–1908).

Mills

Hawaiian Rail *Pennula millsi* **Dole**, 1878 EXTINCT
[Syn. *Porzana sandwichensis*; binomial originally spelled
  millei in error]

James Dawkins Mills (1816–1887) was an English shopkeeper and amateur naturalist at Hilo, Hawaii (1851–1887). He
made very important collections of Hawaiian birds and drew
the comment from Munro in 1960 that '... we are indebted for
the only specimens of the Hawaiian rail in existence'.

Milne-Edwards

Red-tailed Laughingthrush *Trochalopteron milnei* **David**,
  1874
Milne-Edwards's Warbler *Phylloscopus armandii* A. Milne-
  Edwards, 1865
[Alt. Yellow-streaked Warbler]
Edwards's Fig Parrot *Psittaculirostris edwardsii* **Oustalet**,
  1885

Alphonse Milne-Edwards (1835–1900) was a French zoologist
and palaeontologist. He was Professor of Zoology at the
MNHN, Paris, (1876–1892), becoming Director (1892). His
interest in fossil birds led to the publication of *Recherches
Anatomiques et Paleontologiques pour servir a l'Histoire des
Oiseaux Fossiles de la France* (1867 and 1872). He had a close
working relationship with Prince Albert I of Monaco and may
have been influential in the prince establishing the Oceanographic Museum in Monte Carlo. The 'Prix Alphonse
Milne-Edwards' was created (1903) in his memory. Seven
mammals and a reptile are named after him. (See **Edwards**)
Alphonse's father, Henri Milne-Edwards (1800–1885) was
born in Belgium, the 27th son of a reproductively prolific

Englishman, and went on to become a renowned French
naturalist. He became Professor of Hygiene and Natural
History at the Collège Central des Arts et Manufactures
(1832). He succeeded Audouin (q.v.) (1841) as chair of entomology at the MNHN. He wrote works on crustaceans,
molluscs and corals. Many marine organisms were named
after him, but the birds were all named after Alphonse.

Milo

Buff-headed Coucal *Centropus milo* **Gould**, 1856

Milo of Croton (6th century BC) was a famous wrestler in
ancient Greece. Like other successful athletes of the time,
Milo was the subject of fantastic tales of strength and power,
including the story that he once carried a bull on his shoul-
ders. It was also said that he saved the life of the philosopher
Pythagoras, when a pillar collapsed in a banquet hall and
Milo supported the roof until Pythagoras could reach safety.

Minlos

Double-banded Greytail *Xenerpestes minlosi* **Berlepsch**,
  1886

Rufous-and-white Wren ssp. *Thryophilus rufalbus minlosi*
  Berlepsch, 1884

Emilio Minlos (DNF) was a German (Prussian) resident of
Bucaramanga, Colombia. He sent a large collection of bird
skins to the Lübeck Museum. Seemingly he was a trader who
paid others to collect for him.

Minton

Cryptic Forest Falcon *Micrastur mintoni* Whittaker, 2003

Dr Clive Dudley Thomas Minton (b.1934) is a British ornithologist, a friend and birding mentor of the describer. He is also a
metallurgist, administrator and management consultant. He
completed his PhD in metallurgy at Cambridge University.
Although involved in studies of various species of birds, his
main focus became migratory waders. He became the
founding Chairman of the Wash Wader Ringing Group and
was associated with the development of cannon-netting. He
moved to Australia (1978) as Managing Director of Imperial
Metal Industries Australia, Melbourne, Victoria. There he
introduced cannon-netting to the Victorian Wader Study
Group (VWSG), which became one of the most active
banding groups in the world. He was also instrumental in the
formation of the Australasian Wader Studies Group (AWSG)
of which he was founding Chair, as well as in the establishment of Broome Bird Observatory. Since the early 1980s he
has led regular wader study expeditions to north-west
Australia. Minton has also served the RAOU on its Research
Committee (1980–1988) and as Vice President (1989–1995).

Miquel

White-bibbed Fruit Dove ssp. *Ptilinopus rivoli miquelii*
  Schlegel, 1871

Dr Frederik Anton Willem Miquel (1811–1871) was a Dutch
physician, botanist, explorer and collector. He qualified

(1833) in medicine at the University of Groningen. He was Professor of Botany at the University of Amsterdam (1846) and at Utrecht (1858), and Director of the National Herbarium at Leiden (1862).

Miranda, A.

> Buff-breasted Tody Tyrant *Hemitriccus mirandae*
> **Snethlage**, 1925
> Goias Foliage-gleaner *Syndactyla mirandae* Snethlage,
> 1928 NCR
> [Alt. Planalto Foliage-gleaner; JS *Syndactyla dimidiata*]

Dr Alipio de Miranda Ribeiro (1874–1939) was one of the foremost Brazilian naturalists and zoologists of his era. He initially studied medicine but joined the National Museum (1894) before completing the course. He was Assistant Naturalist for the Museum (1897) and Secretary, Department of Zoology (1899), Deputy Head and Professor, Zoology Department (1910–1929). He explored the Amazon and created and directed the Inspectorate of Fisheries (1911), which was the first oceanographical service in South America. Ribeiro started to describe new fish species but he had competition from Goeldi (q.v.), who would go to a market and buy the fish from the same ship, describe them, and send them to the British Museum. Ribeiro became Professor of the Department of Zoology (1929), a post which he held until his death. He wrote *Fauna Brasiliensis* (1907–1915). Many taxa, including seven amphibians, are named in his honour. (See **Ribeiro**)

Miranda, S.

> Brown-capped Vireo ssp. *Vireo leucophrys mirandae*
> **Hartert**, 1917

Sebastián Francisco de Miranda y Rodríguez (1750–1816) was a Venezuelan revolutionary and national hero. He led a revolution against the Spanish rulers of South America but, after some initial success, was captured (1812) and died in a Spanish prison. He is regarded as the forerunner of Simon Bolivar.

Mirandolle

> Mirandolle's Forest Falcon *Micrastur mirandollei* **Schlegel**,
> 1862
> [Alt. Slaty-backed Forest Falcon]

Charles François Mirandolle (1789–1841) was the Dutch Resident of Suriname and President of the Court of Justice there (1815–1831). His son, Charles Jean François Mirandolle (1827–1884), a lawyer and politician, was born in Suriname. Schlegel's (q.v.) original entry for the type specimen reads: 'Adulte, Suriname, présenté par Mr Mirandolle: individu type, figure dans l'ouvrage cité'. It is not clear which Mirandolle is honoured, but it is most likely to be the son who gave the holotype to Schlegel.

Miré

> Desert Lark ssp. *Ammomanes deserti mirei* **Berlioz**, 1950
> NCR
> [Has been treated variously as a JS of *Ammomanes deserti
> deserti* or *A. deserti whitakeri*]

Philippe Bruneau de Miré (b.1921) was a French entomologist who was in French West Africa, including the Sahara. The Faculty of Sciences in Caen awarded his bachelor's degree in botany and zoology (1938). He appears to have annoyed the German occupation authorities in France during WW2, and was forbidden entrance to the Sorbonne, so he acquired a new identity card in the name of Vauquelin and eked out a living as a freelance journalist in Paris until the Liberation. During this time he became friendly with Baron Breunig, an officer in the Wehrmacht and a fellow entomologist, and together they prevailed upon the German Military Governor of Paris, General von Chölitz, not to allow the destruction of the best trees in the Forest of Fontainbleu. He became an official of the National Institute for Agriculture (1945). He went to Africa (1946–1956) investigating the desert locust, conducting expeditions and surveys in Morocco, Mauretania, Chad, Niger and the Sahel and Sahara. He took a year's sabbatical at the MNHN in Paris and became a Correspondent for the museum, involving further travels in Africa, visiting Chad and Cameroon (1958–1961). He was entomologist at the French Institute of Coffee and Cocoa Research Centre Agricultural, in Cameroon (1964–1974), then started a laboratory at Montpellier (1974) devoted to the identification of pests on tropical crops. He retired (1984) but continued his involvement in Africa and his concerns for the forests of France. An amphibian is named after him.

Mitchell, D. W.

> Mitchell's Plover *Phegornis mitchellii* **L. Fraser**, 1845
> [Alt. Diademed Sandpiper-Plover]
> Purple-throated Woodstar *Calliphlox mitchellii* **Bourcier**,
> 1847
>
> ---
>
> Grey-capped Woodpecker ssp. *Dendrocopos canicapillus
> mitchellii* **Malherbe**, 1849
> Mitchell's Lorikeet *Trichoglossus forsteni mitchellii*
> **G. R. Gray**, 1859
> [Alt. Sunset Lorikeet ssp.]

David William Mitchell (1814–1859) was an English zoologist who was Secretary of the Zoological Society of London (1847–1859). He was also a fine bird painter and was one of the illustrators of *The Genera of Birds* by G. R. Gray (q.v.). John Gould (q.v.) was a great friend and introduced him to Joseph Wolf (q.v.), who was acknowledged to be the best bird illustrator in Europe; it was he who finished illustrating *Genera*, rather leaving Mitchell in the shade.

Mitchell, T. L.

> Major Mitchell's Cockatoo *Cacatua leadbeateri* **Vigors**, 1831
> [Alt. Leadbeater's Cockatoo; Syn. *Lophochroa leadbeateri*]

Lieutenant-Colonel Sir Thomas Livingstone Mitchell (1792–1855) was a Scottish army surveyor and explorer. He was the Surveyor-General of New South Wales (1828–1855) and led various expeditions into eastern Australia (1831–1836) and to tropical Australia (1845–1846). A very life-like coloured plate of 'his' cockatoo appears in Mitchell's *Three Expeditions into the Interior of Eastern Australia* (1838). He was hot-headed and was the last person in Australia to challenge an opponent, in this case a politician, to a duel. Fortunately he only shot a hole in the man's hat, but honour was satisfied. He was heavily criticised for killing aboriginals, not only hostile tribesmen, but also harmless bystanders such as old women and children. His rash actions cost the lives of members of his exploration party, such as his botanist Richard Cunningham. A town in Queensland is also named after him, as is a mammal.

Mjoberg

Snowy-browed Flycatcher ssp. *Ficedula hyperythra mjobergi* **Hartert**, 1925
Brown-throated Sunbird ssp. *Anthreptes malacensis mjobergi* **Bangs** & **J. L. Peters**, 1927

Dr Eric Georg Mjöberg (1882–1938) was a naturalist, entomologist, ethnographer and explorer. He took his initial degree at Stockholm University (1908) and his master's at Lund University (1912). He held various jobs in Sweden, including working at the National Natural History Museum, Stockholm, and teaching in high schools (1903–1909). He led Swedish scientific expeditions in north-western Australia (1910–1911) and Queensland (1912–1913). He worked in Sumatra at an experimental station (1919–1922), combining those duties with being Swedish Consul. He was Curator of the Sarawak Museum (1922–1924) and led a scientific expedition in Borneo (1925–1926). Two reptiles and three amphibians are named after him.

Mademoiselle Thura

Mademoiselle Thura's Rosefinch *Carpodacus thura* **Bonaparte** & **Schlegel**, 1850
[Alt. White-browed Rosefinch, Thura's Rosefinch]

(See **Thura**)

Mlokosiewicz

Caucasian Black Grouse *Lyrurus mlokosiewiczi* **Taczanowski**, 1875

Ludwik Franciszek Mlokosiewicz (1831–1909) was a Polish botanist and naturalist. He served in the Russian army (1853–1861), resigning to travel to Persia. When he returned he was arrested and charged with inciting Poles to revolt in the Caucasus, where he had been stationed. He was sentenced to six years exile and his botanical collection was confiscated and never returned. He was freed and explored the mountains of Dagestan (1876). He again travelled to Persia (1878) getting as far as Baluchistan. When he returned to the Caucasus he was made Inspector of Forests and he settled in the town of Lagodekhi for the rest of his life. A peony is named after him.

Möbius

Malagasy Green Sunbird *Cinnyris moebii* **Reichenow**, 1887
[Syn. *Cinnyris notatus moebii*]

Professor Dr Karl August Möbius (1825–1908) was a German zoologist and pioneer in the field of ecology. He qualified as a teacher aged 19 but later (1849) studied natural science and philosophy at Humboldt University, Berlin. After graduating, he taught sciences in high school. He opened Germany's first seawater aquarium (1863). After obtaining his doctorate from the University of Halle (1868) he was appointed Professor of Zoology at the University of Kiel and Director of its Zoological Museum. His *Die Auster und die Austernwirtschaft* (1870) broke new ground in describing the ecosystem interactions between organisms in oyster beds. He became Director of the Zoology Museum in Berlin (1888) and Professor of Systematics and Geographical Zoology at the Kaiser Wilhelm University, teaching there until 1905 retiring aged 80!

Mocinno

Resplendent Quetzal *Pharomachrus mocinno* La Llave, 1832

José Mariano Mociño (1757–1820) was a Mexican physician, naturalist and botanist. The Spanish Government in Madrid sent a botanical expedition to New Spain (1787–1803) and Mociño was assistant to its commander, Martin de Sessé y Lacasta.

Mocquerys

Crested Bobwhite ssp. *Colinus cristatus mocquerysi* **Hartert**, 1894

Albert Mocquerys (DNF) was a French dental surgeon who collected in tropical Africa, Venezuela (1893–1894), and Madagascar. He was lame in one leg and explained this to friends with the following (probably apocryphal) story. He was one of five people captured by cannibals in the Belgian Congo. The canny anthropophagi broke their prisoners' legs to prevent them escaping the cooking pot but, fortunately for Mocquerys, some Belgians arrived in time to save them.

Modigliani

Gerygone sp. *Gerygone modiglianii* **Salvadori**, 1892 NCR
[Alt. Golden-bellied Gerygone; JS *Gerygone sulphurea sulphurea*]

Ruddy Cuckoo Dove ssp. *Macropygia emiliana modiglianii* Salvadori, 1887
Scarlet Minivet ssp. *Pericrocotus speciosus modiglianii* Salvadori, 1892

Emilio Modigliani (1860–1932) was an Italian zoologist and anthropologist who collected in Sumatra (1886–1894). Four reptiles and an amphibian are named after him.

Moebi

Malagasy Green Sunbird *Cinnyris moebii* **Reichenow**, 1887
[Syn. *Cinnyris notatus moebii*]

(See **Möbius**)

Moffitt

Moffitt's Kalij Pheasant *Lophura leucomelanos moffitti* **Hachisuka**, 1938
Moffitt's Canada Goose *Branta canadensis moffitti* **Aldrich**, 1946

James Kennedy Moffitt (1865–1955) was an American businessman and banker closely associated with the University of California, Berkeley, where he was Regent for many years

(1911–1940 and 1941–1948). In addition to ornithology and mammalogy he was greatly interested in the classics. He gave (1956) his entire book collection, including the complete works of the Roman poet Horace, to the Bancroft Library there and established an endowment of $100,000 in memory of his wife, to maintain and grow its collection.

Mogensen

Squirrel Cuckoo ssp. *Piaya cayana mogenseni* **J. L. Peters**, 1926
Striped Owl ssp. *Pseudoscops clamator mogenseni*
    L. & E. H. Kelso, 1935 NCR
[JS *Pseudoscops clamator midas*]

Juan Mogensen (b.1898) was a Danish collector resident in Argentina (1917–1932).

Mohun

Great Slaty Woodpecker ssp. *Mulleripicus pulverulentus mohun* **Ripley**, 1950
[SII *Mulleripicus pulverulentus harterti*]

Field-Marshal Maharaja Mohun Shamsher Jang Bahadur Rana (1885–1967) was Prime Minister of Nepal (1948–1951).

Mokrzecki

Rock Bunting ssp. *Emberiza cia mokrzeckyi* **Moltchanov**, 1917 NCR
[JS *Emberiza cia prageri*]

Dr Zygmunt Atanazy Mokrzecki (1865–1936) was a Polish entomologist. He was Professor of Entomology at University Taurydzkiego in Simferopol, and a Professor at Warsaw Agricultural University. He conducted studies of plant pests in Russia and Crimea (Ukraine).

Molesworth

Molesworth's Tragopan *Tragopan blythii molesworthi*
    E. C. S. Baker, 1914
[Alt. Blyth's Tragopan ssp.]

Brigadier Alec Lindsay Mortimer Molesworth (1881–1939) was an officer in the 8th Gurkha Rifles in the Indian Army (1904–1937), a naturalist and collector.

Molina

Penguin genus *Molinaea* Jarocki, 1821 NCR
[Now in *Spheniscus*]

Molina's Pelican *Pelecanus thagus* Molina, 1782
[Alt. Peruvian Pelican]
Rufous-tailed Plantcutter *Phytotoma molina* **Lesson**, 1834 NCR
[JS *Phytotoma rara*]
Molina's Conure *Pyrrhura molinae* **Massena** & **Souancé**, 1854
[Alt. Green-cheeked Parakeet]

Southern Lapwing ssp. *Belonopterus cayennensis molina* **Lowe**, 1921 NCR
[JS *Vanellus chilensis chilensis*]

Abbot Giovanni Ignazio (Juan Ignacio) Molina (1740–1829) was a Chilean Jesuit priest, as well as a naturalist, historian and geographer. He was forced to leave Chile (1768) when the Jesuits were expelled. He settled in Bologna, and wrote *Saggio sulla Storia Naturale del Chili* (1782), the first account of the natural history of Chile.

Möller

São Tomé Prinia *Prinia molleri* **Bocage,** 1887

Adolfo Frederico Möller (1842–1920) was a Portuguese botanist who collected on São Tomé. He studied sylviculture in Germany (1857–1860) and worked for the Forestry Department (1862–1874), then became Inspector of the Botanical Gardens at the University of Coimbra. He collected in Portugal and Africa, leading an expedition to São Tomé (1885). He returned to Portugal with 249 animal species and 735 samples for the herbarium as well as a valuable mineralogical and ethnographic collection. He published widely, including plant catalogues and articles on agriculture, sylviculture, floriculture, horticulture, pharmaceutical and medicinal botany. His most notable work was the *Catálogo das Plantas Medicinais que Habitam o Continente Português* (1878). Two amphibians are named after him.

Moloney

Moloney's Illadopsis *Illadopsis fulvescens moloneyana*
    Sharpe, 1892
[Alt. Brown Illadopsis ssp.]

Sir Cornelius Alfred Moloney (1848–1913) was a British civil servant. He was Administrator of the Gambia (1884–1886) and Governor of Lagos, Nigeria (1886–1889 and 1890–1891). He was Lieutenant-Governor of British Honduras (Belize) (1891–1897), and Governor of Trinidad (1900–1904). He wrote *Forest in West Africa*, which initiated a forestry policy in Nigeria. A mammal is named after him.

Moltchanov

Coal Tit ssp. *Periparus ater moltchanovi* **Menzbier**, 1903
Crested Lark ssp. *Galerida cristata moltschanowi*
    Gavrilenko, 1926 NCR
[JS *Galerida cristata tenuirostris*]

Professor Dr Lev Aleksandrovich Molchanov (1878–>1960) was a Russian ornithologist, meteorologist, geographer, teacher and author. He was Professor of Geography at Central Asia University (1921–1933), then became Chair of Physics and Agricultural Meteorology at Tashkent Agricultural Institute (1933–>1960). He particularly studied the climate of Central Asia. He was awarded the Order of Lenin. A reptile is named after him.

Moltoni

Moltoni's Warbler *Sylvia subalpina* **Temminck**, 1820
[*Sylvia cantillans moltonii* (**Orlando**, 1937) is a JS]

Professor Dr Edgardo Moltoni (1896–1980), the most eminent of Italian ornithologists, was Director of the Museo Civico di Storia Naturale in Milan (1951–1964), to which he bequeathed his personal ornithological library of about 6,000 books and pamplets. Previously he had been Curator of the ornithological collection of Count Ercole Turati (q.v.) (1922–1951). He travelled widely and wrote many papers, including a checklist of Italian birds. Orlando described the bird as *Sylvia cantillans moltonii* in 1937, but Temminck's earlier name has priority; the bird has recently been elevated to full species status and Moltoni's name is preserved in its English name.

Momiyama

Micronesian Imperial Pigeon ssp. *Ducula oceanica momiyamai* **Kuroda**, 1922 NCR
[JS *Ducula oceanica monacha*]
Ural Owl ssp. *Strix uralensis momiyamae* **Taka-Tsukasa**, 1931

Toku Taro Momiyama (1895–1962) was a Japanese ornithologist, author and collector. He wrote: *The Geographical Distribution of the birds in Botel Tobago* (1932).

Momus

Sardinian Warbler ssp. *Sylvia melanocephala momus* Hemprich & **Ehrenberg**, 1833

In Greek mythology, Momus was the god of satire and mockery.

Mondetour

Mondetour's Dove *Claravis mondetoura* **Bonaparte**, 1856
[Alt. Maroon-chested Ground Dove]

Angélique Jeanne Louise Pauline Brière de Mondetour (1780–1856) was married (1804) to French zoologist Étienne Geoffroy Saint-Hilaire (q.v.) (1772–1844) and was his widow when the dove was described.

Monguillot

Vietnamese Greenfinch *Carduelis monguilloti* **Delacour**, 1926

Maurice Antoine-François Monguillot (1874–1926) was a French colonial civil servant, Governor General of French Indochina (1919–1920 and 1925–1928).

Monica

Grey Heron ssp. *Ardea cinerea monicae* **Jouanin** & Roux, 1963

Madame Monica Avery de la Salle (DNF) was an assistant of French ornithologist, Fred Roux.

Monnard

European Robin ssp. *Erithacus rubecula monnardi* **Kleinschmidt**, 1916 NCR
[JS *Erithacus rubecula rubecula*]

Dr H. L. Monnard (b.1881) was an American army veterinary surgeon in Germany.

Monod

Desert Lark ssp. *Ammomanes deserti monodi* Dekeyser & Villiers, 1950 NCR
[Has been treated variously as a JS of *Ammomanes deserti geyri* or of *A. deserti payni*]

Théodore André Monod (1902–2000) was a French naturalist, humanist, philosopher and desert explorer. He was a professor at the MNHN, Paris, and the founder of the Institut Français d'Afrique Noire, Dakar, Senegal. Monod was an activist in pacifist and anti-nuclear campaigns.

Monson

Great-tailed Grackle ssp. *Quiscalus mexicanus monsoni* **A. R. Phillips**, 1950

Gale Wendell Monson (1912–2012) was an American naturalist and ornithologist. He worked for the US Bureau of Indian Affairs and the Soil Conservation Service in Arizona and New Mexico (1934–1940). He was then with the US Fish and Wildlife Service (1940–1969), except for service during WW2 in the US Army Medical Corps in the Far East. He was actve in helping run the Arizona-Sonora Desert Museum (1971–1977). He co-wrote *The Birds of Arizona* (1964).

Montagne

Andean Guan *Penelope montagnii* **Bonaparte**, 1856

Jean François Camille Montagne (1784–1866) was a French botanist. He entered the French navy (1798) and took part in Napoleon Bonaparte's expedition to Egypt. He returned to France (1802) and qualified as a doctor, becoming a surgeon (1804) at a military hospital at Boulogne. He was Chief Surgeon in Murat's army (1815 and 1819). He became head of the military hospital at Sedan (1830) and retired (1832) to devote his life to the study of cryptograms.

Montagu

Montagu's Harrier *Circus pygargus* **Linnaeus**, 1758

Colonel George Montagu (1751–1815) was a soldier and natural history writer. He attained the rank of Captain in the British army very young, and later served in the American Revolution as a Lieutenant-Colonel in the English militia. His military career was curtailed when he was court-martialled and cashiered for causing trouble among his brother officers by what was described as 'provocative marital skirmishing'. Montagu then devoted himself to science, particularly biology. In his own words; 'I have delighted in being an ornithologist from infancy, and, was I not bound by conjugal attachment, should like to ride my hobby to distant parts.' Montagu was among the first members of the Linnean Society. He was also an expert on shells and wrote *Testacea Britannica, a History of British Marine, Land and Freshwater Shells* (1803). He wrote many papers on the birds of southern England, but his greatest work was the *Ornithological Dictionary or Alphabetical Synopsis of British Birds* (1802). He was renowned for his meticulous work and observations bordering on the clinical. Such observations led to a better understanding of phenomena which had previously been romanticised – for example, he said of bird song: '… males of song-birds and of many others do not in general search for the female, but, on the contrary, their business in the spring is to perch on some conspicuous spot, breathing out their full and amorous notes, which, by instinct, the female knows, and repairs to the spot to choose her mate.' He died of lockjaw (tetanus) after stepping on a rusty nail. After his death, two continental ornithologists named the bird which Montagu had described in meticulous detail (1803) as different from the Hen Harrier *Circus cyaneus*, calling it *Le Busard Montagu* (although Linnaeus had already given it a name); William MacGillivray first used the term Montagu's Harrier (1836).

Montague

Fluttering Shearwater ssp. *Reinholdia reinholdi montaguei* **Mathews**, 1922 NCR; NRM
[JS *Puffinus gavia*]

Lieutenant Paul Denys Montague (1890–1917) of the Royal Flying Corps was, before WW1, a zoologist, ethnologist and explorer in north-west Australia (1912). He was killed when his plane was shot down during the Salonika campaign.

Montano

Montano's Hornbill *Anthracoceros montani* **Oustalet**, 1880
[Alt. Sulu Hornbill]

Dr Joseph Montano (b.1844) was a French anthropologist who spent some time undertaking a scientific survey in the Philippines (1879–1881). He accompanied Don Joaquin Rajal, the Spanish governor of Davao, on an expedition to Mount Apo on Mindanao (1880), the first ascent of the peak. Montano wrote *Voyages aux Philippines et en Malaisie* (1886).

Monteiro, J.

Monteiro's Twinspot *Clytospiza monteiri* **Hartlaub**, 1860
[Alt. Brown Twinspot]
Monteiro's Golden Weaver *Ploceus xanthops* Hartlaub, 1862
[Alt. Holub's Golden Weaver]
Monteiro's Hornbill *Tockus monteiri* Hartlaub, 1865
Monteiro's Bush-shrike *Malaconotus monteiri* **Sharpe**, 1870

Mosque Swallow ssp. *Cecropis senegalensis monteiri* Hartlaub, 1862

Joachim John (João José) Monteiro (1833–1878) was a Portuguese mining engineer who had English ancestors – hence 'John' and alternatives to it. He collected natural history specimens in Angola (1860–1875). He wrote *Angola and the River Congo* (1875).

Monteiro, L.

Monteiro's Storm-petrel *Oceanodroma monteiroi* Bolton *et al.*, 2008

Professor Dr Luis R Monteiro (1962–1999) was an environmentalist at the Department of Oceanography and Fisheries, University of the Azores. He was given a grant by the Junta Nacional de Investigação Científica and the International Fishmeal and Oil Manufacturers Association to study nesting petrels and co-ordinate the research, but he died in a plane accident on the island of San Jorge. He researched petrels in the Azores and co-wrote 'Speciation through temporal segregation of Madeiran storm petrel (*Oceanodroma castro*) populations in the Azores?' (1998). His final publication, on Fea's Petrel *Pterodroma feae*, co-written with Alec Zino (q.v.), appeared posthumously (2000).

Montezuma

Montezuma Oropendola *Psarocolius montezuma* **Lesson**, 1830
[Alt. Great Oropendola; Syn. *Gymnostinops montezuma*]
Montezuma Quail *Cyrtonyx montezumae* **Vigors**, 1830

Montezuma (1480–1520), also known as Moctezuma, was Emperor of the Aztecs, at the time of the Spanish conquest of Mexico. He also has an amphibian and a reptile named after him.

Montpellier

Grey-chinned Minivet ssp. *Pericrocotus solaris montpellieri* **La Touche**, 1922

Charles Baron de Montpellier de Vedrin (1896–1983) was a French administrator with the Chinese Maritime Customs (1922).

Montrosier

Long-tailed Triller ssp. *Lalage leucopyga montrosieri* **J. Verreaux** & **Des Murs**, 1860

Père Jean Hyacinthe Xavier Montrouzier (1820–1897) was a French botanist, zoologist, entomologist, explorer and Marist missionary. He was ordained (1843) and went as a missionary to the Solomon Islands (1845), where he was speared by an islander (1846), and then at Woodlark Island, Papua New Guinea (1847–1852), about which he wrote 'Essai sur la faune de l'île de Woodlark ou Mouiou' (1855). He spent the last three decades of his life in New Caledonia (1854–1897).

Moor

Grey-chested Illadopsis *Alethe moori* **Alexander**, 1903 NCR
[Alt. Grey-chested Babbler; JS *Kakamega poliothorax*]

Sir Ralph Denham Rayment Moor (1860–1909) was a British colonial administrator. He served in the Royal Irish Constabulary (1882–1891) and was Commandant of Constabulary in Oil Rivers Protectorate (Nigeria) (1891), rising to become Vice-Consul, Consul and Consul-General of the Niger Coast Protectorate (1892–1900). The protectorate passed from the Foreign Office to the Colonial Office (1900), and Moor became High Commissioner of Southern Nigeria. He caught malaria and blackwater fever and his health failed, which may have led to his suicide by poison.

Moore, G. E.

Pernambuco Pygmy Owl *Glaucidium mooreorum* Silva, Coelho & Gonzaga, 2002

Dr Gordon E. Moore (b.1929) and his wife, Betty, are Americans who have made significant contributions to conservation. They also made a generous donation to the

University of Cambridge for the construction of a new library for the physical sciences, technology and mathematics. Moore was Director of Development at Fairchild Semiconductor (1960s) and made very accurate predictions concerning the growth of computing power, which have become known as Moore's Law. This, combined with Intel Corporation's microprocessor, is the basis for today's microcomputer revolution. Moore is Chairman Emeritus of Intel Corporation, which he co-founded (1968). He and his wife set up (2000) the Gordon E. and Betty I. Moore Foundation to fund scientific, educational and environmental ventures. An amphibian and a mammal are also named after them.

Moore, G. H.

Haast's Eagle *Harpagornis moorei* **Haast**, 1872 EXTINCT

George Henry Moore (1812–1905) was a Manxman who went to Australia before settling in New Zealand where he managed (1854) then bought (1873) the Glenmark Station, building his home there (1888). He was sometimes known as 'Scabby Moore' as the station was notorious for its diseased sheep. Here many subfossil bird bones, including complete moa skeletons and the eagle, were excavated (1866–1871). Moore's reputation was furthered when, on a stormy night, he refused shelter to a swagman who then shot himself. Moore was described in the *Lyttelton Times* as a 'Mean, hard-hearted, barbarous, blasphemous man' (See **Haast**).

Moore, H. F.

Yellow Bittern ssp. *Ixobrychus sinensis moorei* **Wetmore**, 1919 NCR; NRM

Dr Henry Frank Moore (1867–1948) was an American zoologist who was the chief naturalist on board the *Albatross* (1896–1903) and with the US Bureau of Fisheries (1903–1923).

Moore, R. T.

Singing Quail ssp. *Dactylortyx thoracicus moorei* **Warner** & Harrell, 1957

Robert Thomas Moore (1882–1958) was an American ornithologist, poet, businessman and philanthropist. He graduated from University of Pennsylvania (1904) and received his master's degree from Harvard (1905). He married twice (1903 and 1922), his second wife being M. F. Cleaves (q.v.). While his love was ornithology he was a businessman and fur farmer. He was in Ecuador several times (1927–1929) and in Mexico nine times (1933–1948). He founded a laboratory (1950) at Los Angeles Occidental College which is now named in his honour the Moore Laboratory of Zoology and houses his collection, including 81 holotypes that he described. Moore bequeathed the land he owned on Borestone Mountain, Maine, to the Audubon Society. He co-wrote *Checklist of the Birds of Mexico* (1950).

Moore, W. E.

Moore's Hermit *Phaethornis malaris moorei* **G. N. Lawrence**, 1858
[Alt. Great-billed Hermit ssp.]

William E. Moore (DNF) was an explorer in Amazonia, who crossed South America (1857) following the line of the Equator.

Moorhouse

Island Leaf Warbler ssp. *Phylloscopus poliocephalus moorhousei* **Gilliard** & **LeCroy**, 1967

David Bruce Moorhouse (1936–2003) was an Australian Patrol Officer and Land Consultant in Papua New Guinea (1956–1974), becoming Deputy District Commissioner (1969–1974). He also worked as a consultant specialising in Papua New Guinea land issues (1989–1992). He led the National Geographic's (1960) expedition to the Whiteman Range, New Britain.

Moquin

Chat-shrike genus *Moquinus* **Bonaparte**, 1856 NCR
[Now in *Lanioturdus*]

African Oystercatcher *Haematopus moquini* **Bonaparte**, 1856

Christian Horace Bénédict Alfred Moquin-Tandon (1804–1863) was a French zoologist and botanist. He co-wrote of *L'Histoire Naturelle des Iles Canaries* (1835–1844). (See **Tandon**)

Morcom

Morcom's Hummingbird *Atthis morcomi* **Ridgway**, 1898 NCR
[Alt. Bumblebee Hummingbird; JS *Atthis heloisa*]

Morcom's Yellow Warbler *Dendroica aestiva morcomi* Coale, 1887

George Frean Morcom (1845–1932) was a British-born American amateur ornithologist, who settled in Chicago and later collected and recorded in California and Arizona (late 19th century). He took work as a bookkeeper and progressed in the company, becoming a partner and, eventually, sole proprietor. His free time was spent hunting and shooting, but also in collecting natural history specimens and observing wildlife, particularly birds. His first love was palaeontology, but his interest in ornithology was inspired by a friendship in England with John Gatcombe, who was a first-class painter of birds. Morcom became very involved in the Ridgway Ornithological Club in Chicago (1873). In addition to his own collecting, he occasionally commissioned others. He corresponded with Ridgway (q.v.) and, because he liked him, sent a number of his prize specimens to the USNM. These included the type specimen of the hummingbird, which Ridgway described as *Atthis morcomi*, amended to *Atthis heloisa morcomi* before his further research (1927) established that it was not a valid race. The California Academy of Sciences has Morcom's letters, together with the G. F. Morcom Collection. Henry Coale, the warbler subspecies describer, prepared many specimens for Morcom and wrote with him.

Morden

Morden's Scops Owl *Otus ireneae* **Ripley**, 1966
[Alt. Sokoke Scops Owl]

Myrtle Irene Morden (DNF) was an American sponsor of expeditions and a collector. William and Irene Morden undertook several collecting expeditions in Kenya (1953, 1956, 1965) on behalf of the AMNH, New York. They co-wrote *Our African Adventure* (1954) and several articles including 'The last four Strandlopers' (1955).

Moreau

Moreau's Tailorbird *Artisornis moreaui* **W. L. Sclater**, 1931
[Alt. Long-billed Tailorbird]
Moreau's Sunbird *Cinnyris moreaui* W. L. Sclater, 1933

Little Rush Warbler ssp. *Bradypterus brachypterus moreaui*
W. L. Sclater, 1931 NCR
[JS *Bradypterus baboecala tongensis*]

Reginald Ernest Moreau (1897–1970) was a British amateur ornithologist and collector. He wrote *The Bird Fauna of Africa and its Islands* (1966), *The Palaearctic-African Bird Migration Systems* (1972) and, with B. P. Hall, *An Atlas of Speciation in African Passerine Birds* (1970) as well as numerous papers, especially on African fauna and its evolution, distribution and migration.

Moreira

Itatiaia Thistletail *Asthenes moreirae* **Ribeiro**, 1905
[Syn. *Schizoeaca moreirae*]

Professor Carlos Moreira (1869–1946) was a Brazilian zoologist who was a specialist in crustaceans. He became Director, Museu Nacional, Rio de Janeiro (1916). Two amphibians are named after him.

Morel

Goshawk sp. *Nisuoides morelii* **Pollen**, 1866 NCR
[Alt. Frances's Sparrowhawk; JS *Accipiter francesiae*]

Louis Morel (DNF) was a lawyer and ichthyologist. He took over (1858) as Director of the Natural History Museum on Réunion after the sudden death of Joseph Bernier (q.v.). He was founding vice-president of the Société des Sciences et des Arts de la Réunion (1855), and its President (1859–1861). Auguste Lantz (q.v.), who collected the holotype in Madagascar, was originally employed at the museum as a taxidermist (1862) and then as Curator (1863–1866) and as Director (1866–1893). Dr Anthony Cheke theorised to us that sometime around 1865 or 1866, Morel must have caused enormous offence in some way and was ostracized, as he disappeared from both the museum and the Société des Sciences et des Arts around that time. Recent histories of the museum have even airbrushed him out, skipping straight from Bernier to Lantz.

Morel, G. & M.

Grey-headed Bristlebill ssp. *Bleda canicapillus morelorum*
Erard, 1992 NRM

Dr Gérard J. Morel (1925–2011) and his wife Dr Marie-Yvonne Morel were French ornithologists resident in Senegal (1953–1992). He worked at the ORSTOM Ornithological Station at Richard-Toll, Senegal (retired 1992) and made in-depth studies of the avifauna of the Sahel, extended later to the whole of the Senegambia region, and the ecology and dynamics of grain-eating passerines, particularly the Red-billed Quelea *Quelea quelea* and the Sudan Golden Sparrow *Passer luteus*. He was co-founder (1979) and President (1994–2011) of the West African Ornithological Society. They co-wrote together and separately, including *Les Oiseaux de Senegambie* (1990) and *Birds of West Africa* (1999).

Morelet

Morelet's Seedeater *Sporophila torqueola morelleti*
Bonaparte, 1850
[Alt. White-collared Seedeater ssp.]
Common Chaffinch ssp. *Fringilla coelebs moreletti*
Pucheran, 1859

Pierre Marie Arthur Morelet (1809–1892) was a French zoologist and malacologist. He was sent to collect specimens in Central America (1846) by the French Academy of Sciences. He was a member of the Commission to Algeria at the start of the French occupation. An amphibian and two reptiles are named after him. Oddly, neither Bonaparte nor Pucheran used the normal spelling of 'Morelet' in their scientific names, even though Pucheran refers to 'M. Morelet' two lines further down his text!

Moreno, A.

Cuban Pewee ssp. *Contopus caribaeus morenoi* **Burleigh**
& Duvall, 1948

Dr Abelardo Moreno Bonilla (1913–1992) was a Cuban zoologist, arachnologist, malacologist and conservationist. He was Professor of Zoology, University of Havana (1947–1962), Director of Havana Zoo, and founder of the Academy of Sciences of Cuba.

Moreno, F.

Moreno's Ground Dove *Metriopelia morenoi* **Sharpe**, 1902
[Alt. Bare-eyed Ground Dove]

Elegant Crested Tinamou ssp. *Calopezus elegans morenoi*
Chubb, 1917 NCR
[JS *Eudromia elegans elegans*]

Francisco Josué Pascasio Moreno (1852–1919) was an Argentine naturalist, geographer, anthropologist, explorer and founder and first director of La Plata Museum (1884). His collecting as a child was so prolific that his parents set aside a room for his specimens. He started exploring in Patagonia (1873), and achieved a long-held ambition to cross it from coast to coast (1875). He wrote *Viaje a la Patagonia Austral* about this journey (1876), which explored and mapped unknown territory. He donated c.15,000 specimens to the Anthropological and Ethnographic Museum of Buenos Aires (1879), which created a whole new department. He donated books and drew in many overseas scientists, as well as raising money for buildings; all of which culminated in his being appointed Director for life. As an anthropologist he was among the first to plead for attempts to 'civilise' native peoples to cease. He was briefly held prisoner (1879) by one of the tribes he went to study, and after escaping decided to go to Europe to study further. On his return he found those same people imprisoned and did everything he could to have them released and rehabilitated. He was instrumental (1880s) in establishing the border with Chile. Moreno's last collecting trip (1912) was made with American ex-President Theodore Roosevelt (q.v.). He also wrote the definitive work on his favourite area, *Voyage to Patagonia* (1879). He was granted territory, which he had explored, by the government, but he donated it back to create the Nahuel Huapi National Park. He is also honoured in the name of a large glacier and in the names of a mammal and a reptile.

Moreton

White-shouldered Fairy-wren ssp. *Malurus alboscapulatus moretoni* **De Vis**, 1892

Lieutenant Matthew Henry Reynolds-Moreton (1847–1909) served in the 78th Regiment of Foot (1866–1870). He became a colonial administrator and was Resident Magistrate in British New Guinea (1891–1907). He retired to Australia and died in Brisbane.

Morgan, A. M.

Australian Babbler genus *Morganornis* **Mathews**, 1912 NCR
[Now in *Pomatostomus*]

Thornbill sp. *Acanthiza morgani* Mathews, 1911 NCR
[Alt. Slender-billed Thornbill; JS *Acanthiza iredalei iredalei*]

Purple-backed Fairy-wren ssp. *Malurus lamberti morgani* **S. A. White**, 1912 NCR
[JS *Malurus lamberti assimilis*]
Thick-billed Grasswren ssp. *Diaphorillas textilis morgani* Mathews, 1912 NCR
[JS *Amytornis textilis textilis*]
Chestnut Quail-thrush ssp. *Cinclosoma castanotum morgani* **Condon**, 1951 NCR
[JS *Cinclosoma castanotum castanotum*]

Dr Alexander Matheson Morgan (1867–1935) was President of the South Australian Ornithologists' Association. He published regularly (1914–1935). Mostly he seems to have written alone, but in collaboration with A. Chenery he wrote 'The birds of rivers Murray and Darling and the district of Wentworth' (1920).

Morgan, J. J.

Lesser Spotted Woodpecker ssp. *Dendrocopos minor morgani* **Zarudny** & **Loudon**, 1904

Jacques Jean-Marie de Morgan (1857–1924) was a French archaeologist, geologist and explorer who prospected for tin in Malaya (1884) and excavated in Persia (1897–1907).

Mori

Black Woodpecker ssp. *Dryocopus martius morii* **Kuroda**, 1921 NCR
[JS *Dryocopus martius martius*]
Ural Owl ssp. *Strix uralensis morii* **Momiyama**, 1927 NCR
[JS *Strix uralensis nikolskii*]

Dr Tamezo Mori (1884–1962) was a Japanese entomologist, ichthyologist and ornithologist. He taught at Seoul Higher Common School for Keiji Imperial University, Seoul (1909–1945). He was expelled from Korea (1945) by the American authorities and became Director of Zoology at the Agricultural University Hyogo, retiring as Professor Emeritus (1961). A fish is named after him.

Mormon

Puffin genus *Mormon* **Illiger**, 1811 NCR
[Now in *Fratercula*]

In Greek myth, Mormo or Mormon was an ogress, useful for frightening children.

Morphoeus

White-fronted Nunbird *Monasa morphoeus* **Hahn** & Kuster, 1823

In Greek mythology, Morpheus was the god of dreams.

Morrell

Black-capped Babbler ssp. *Pellorneum capistratum morrelli* **Chasen** & **Kloss**, 1929

Edward William Morrell (1887–1936) was a British colonial administrator in British North Borneo (Sabah, Malaysia) (1909–1936).

Morrison, A. R. G.

White-tufted Grebe ssp. *Rollandia rolland morrisoni* Simmons, 1962

Alastair Robin Gwynne Morrison (1915–2009) was a colonial administrator, ornithologist and writer. He was born in Beijing, moved to England with his parents (1919), and after graduating travelled in South America, supporting himself by selling birds and other specimens to museums. He was back in Beijing in 1940 and volunteered to join a British military intelligence unit. After WW2 he entered the British Colonial Service and was sent to Sarawak. There he wrote frequently for the *Sarawak Gazette*. He and his wife moved to Canberra, Australia (1967). A keen birdwatcher, Morrison was also a bibliophile and connoisseur of Chinese porcelain.

Morrison, J. P. E.

Grey-necked Wood Rail ssp. *Aramides cajanea morrisoni* **Wetmore**, 1946

Dr Joseph Paul Eldred Morrison (1906–1983) was an American malacologist, biologist and collector. The University of Chicago awarded his bachelor's degree (1926), University of Wisconsin his master's (1929) and doctorate (1931). He worked at a number of institutions before joining the USNM (1934–1975), retiring as Associate Curator. He was a member of the Bikini Scientific Resurvey that looked at atomic bomb test sites in the Marshall Island (1946–1947).

Mortier

Mortier's Tribonyx *Tribonyx mortierii* **Du Bus de Gisignies**, 1840
[Alt. Tasmanian Native-hen; Syn. *Gallinula mortierii*]

Count Barthélemy Charles Joseph Dumortier (1797–1878) was a Belgian politician, botanist and zoologist. He published *Analyse des Familles des Plantes* (1829) and *Recherches sur la Structure Comparée et le Développement des Animaux et des Végétaux* (1832).

Morton, A.

White-winged Sittella sp. *Neositta mortoni* **North**, 1912 NCR
[Alt. Varied Sittella; JS *Daphoenositta chrysoptera leucoptera*]

Stephan's Emerald Dove ssp. *Chalcophaps stephani mortoni* **E. P. Ramsay**, 1882

Alexander Morton (1854–1907) was born in the USA, but moved to Australia as a child. He was Curator's Assistant at the Australian Museum, Sydney (1877–1882) and took part in expeditions to New Guinea (1877), the Solomon Islands (1881) and Queensland and Lord Howe Island (1882). He was Curator of the Royal Society of Tasmania's museum in Hobart (1884) and its library. He was made Director of both the Tasmanian Museum and Botanical Gardens (1904).

Morton, S. G.

Morton's Finch *Fringilla mortonii* **Audubon**, 1839 NCR
[Alt. Rufous-collared Sparrow; JS *Zonotrichia capensis*]

Samuel George Morton (1799–1851) was a Physician and Corresponding Secretary of the Academy of Natural Sciences of Philadelphia. He was very interested in craniology (study of skulls) and thought there was a correlation between intelligence and skull size. He studied 1,849 American skulls! He adopted the theory of diverse origins of the human race, carrying on an open and controversial discourse with Reverend John Bachman (q.v.). Audubon (q.v.), in his *Ornithological Biography*, wrote 'A single specimen of this pretty little bird, apparently an adult male, has been sent to me by Dr Townsend, who procured it in Upper California. Supposing it to be undescribed, I have named it after my excellent and much esteemed friend Dr Morton of Philadelphia …' He was wrong about it being undescribed, as Statius Muller had already named the species (1776).

Morton, W.

Stork sp. *Dissura mortoni* **Ogilvie-Grant**, 1902 NCR
[Alt. Storm's Stork; JS *Ciconia stormi*]

William Morton (DNF) was a British civil servant in Sarawak. In his etymology Ogilvie-Grant wrote: 'A Stork belonging to the genus *Dissoura* was recently shot by Mr William Morton on the Simunjan River, a tributary of the Sadong, in Southern Sarawak, and subsequently brought to me for identification.

An examination of the series of *Dissura episcopus* in the British Museum confirmed the suspicion that this bird belonged to a perfectly distinct species, but disclosed the fact that a similar specimen to that shot by Mr Morton had been sent home by Sir Hugh Low in 1876. Probably from lack of material the differences between this bird and *D. episcopus* had not been recognised, and the specimen had been catalogued under the latter title'. Ogilvie-Grant was right in saying the stork was distinct from *Ciconia episcopus* (Woolly-necked Stork), but Blasius (q.v.) had already named the taxon (1896).

Moseley, E. L.

Spotted Kingfisher ssp. *Actenoides lindsayi moseleyi* **Steere**, 1890

Dr Edwin Lincoln Moseley (1865–1948) was a US naturalist, botanist, teacher, meteorologist and collector in the Philippines, China and Japan. He taught at Bowling Green State University, Ohio (1914–1936), as Professor of Science until becoming Professor Emeritus (1936–1948).

Moseley, H. N.

Moseley's Rockhopper Penguin *Eudyptes moseleyi* **Mathews & Iredale**, 1921
[Alt. Northern Rockhopper Penguin, Tristan Penguin]

Professor Henry Nottidge Moseley (1844–1891) was Head of the Department of Zoology and Comparative Anatomy of the Pitt Rivers Museum, University of Oxford. He wrote *Notes by a Naturalist on HMS Challenger* (1879) and collected many specimens on that circumnavigation (1872–1876) which provided study material for many years. He also took part in expeditions to Ceylon (Sri Lanka) (1871) and to California and Oregon (1877). He has many species named after him, particularly of marine animals such as starfish and corals.

Mosilikatze

Mosilikatze's Roller *Coracias caudatus* **Linnaeus**, 1766
[Alt. Lilac-breasted Roller]

Mosilikatze was a General in the Zulu army (early 19th century). He decided to keep back some booty that Chaka, King of the Zulus, reckoned should have been handed in, and to avoid the King's wrath he fled (1820) with a large number of followers across the Drakensberg and ravaged the land between the Vaal and Limpopo rivers. The invaders were driven out of the Transvaal by the Boers (1837) and crossed the Limpopo and settled an area that they called Matabeleland (Zimbabwe) after their name for themselves. Matabeleland became a *de facto* independent state with Mosilikatze as their King. The bird was named thus, as it is said that he reserved its feathers for his exclusive use.

Mosquera

Golden-breasted Puffleg *Eriocnemis mosquera* **de Lattre & Bourcier**, 1846

General Tomás Cipriano de Mosquera (1798–1878) was Dictator of New Grenada (1845–1849) (a state in northern South America, which broke up into several present-day countries). He was President of the Grenadine Confederation (1862–1863) and President of Colombia (1863–1864 and 1866–1867).

Mossa

Purple-tailed Comet *Cometes mossia* **Gould**, 1853 NCR
[Hybrid of *Lesbia victoriae* x *Chalcostigma herrani*]

Monsieur Mossa (DNF) was a collector near Popayán, Colombia.

Moszkowski

Moszkowski's Green-winged King Parrot *Alisterus chloropterus moszkowskii* **Reichenow**, 1911
[Alt. Papuan King Parrot ssp.]
Grey-headed Cuckooshrike ssp. *Coracina schisticeps moszkowskii* **Neumann**, 1917 NCR
[JS *Coracina schisticeps reichenowi*]

Dr Max Moszkowski (1873–c.1950) was a German physician, botanist and ethnologist. He travelled to Ceylon (Sri Lanka) and Sumatra (1907), then undertook an expedition to New Guinea (1910–1913) in the area of the Van Rees Mountains and the Mamberano River. He tried to reach the Snow Mountains in Central New Guinea but failed because of food shortages. During the expedition he collected the eponymous parrot. He wrote the articles 'Expedition zur Erforschung des Mamberamo in Hollandish Neu-Guinea' and 'Wirtschaftsleben der primitiven Völker' (1911). Very interested in tribal customs and languages, he wrote 'Wörterverzeichnisse der Sprachen vom Zentralgebirge, vom Südfluß, des Tori, des Sidjuai, des Borumesu, des Pauwoi' (1913). He wrote a book *Inst Unerforschte Neuguinea, Erlebnisse mit Kopfjägern und Kannibalen* (1928).

Mother Carey

Mother Carey's Chickens Hydrobatidae
[Alt. Storm-petrels]
Mother Carey's Geese *Macronectes*
[Alt. Giant Petrels]

This name is probably a corruption of the Latin 'Mater cara', which means 'Dear Mother' and is a reference to the Virgin Mary. 'Mother Carey's Chickens' is a traditional seafarers' term for storm-petrels in general. The term petrel may itself derive from St Peter and relate to these birds' habit of 'walking on the water'. An old name used by sailors for the giant petrels was Mother Carey's Geese.

Motley

Oriental Dwarf Kingfisher ssp. *Ceyx erithaca motleyi* **Chasen** & **Kloss**, 1929

James Motley (1822–1859) was an English mining engineer and surveyor, botanist and naturalist in Borneo, Singapore and Sumatra (1849–1859). He and his wife and children were among a number of Europeans killed by Dyaks in Borneo at the start of the Banjarmasin War.

Mouhot

Ashy Drongo ssp. *Dicrurus leucophaeus mouhoti* **Walden**, 1870
Lesser Necklaced Laughingthrush ssp. *Garrulax monileger mouhoti* **Sharpe**, 1883

Alexandre Henri Mouhot (1826–1861) was a French traveller who is best known for having 'rediscovered' Angkor Wat in Cambodia (c.1859). In fact it had never been lost, and other Europeans had visited it before Mouhot, but he did much to popularise Angkor in the West with his evocative writings and detailed sketches. Earlier in life Mouhot had taught languages in Russia (1844–1854) and started studying natural science (1856). He married a descendant of the Scottish explorer Mungo Park – a connection that helped him when he decided to go to Indochina to collect botanical specimens (1857). The French authorities rejected his proposals, but the Royal Geographical Society and the Zoological Society of London supported the trip. He died in Laos and his book *Travels in Indo-China* appeared posthumously (1864). Two reptiles and an amphibian are named after him.

Moulton

Grey-breasted Babbler ssp. *Malacopteron albogulare moultoni* **H. C. Robinson** & Kloss, 1919
Mountain Black-eye ssp. *Chlorocharis emiliae moultoni* **Chasen** & **Kloss**, 1927

Major John Coney Moulton (1886–1926) was a British army officer, ornithologist and entomologist. He was Curator of Sarawak Museum (1908–1915), leaving to serve with the Wiltshire regiment in India and Singapore (1915–1919). He was Director of Raffles Museum, Singapore (1919–1923) but returned to Sarawak (1923) as Chief Secretary to Charles Vyner Brooke (q.v.), the second 'White Rajah'.

Mountford

Grey-crowned Babbler ssp. *Pomatostomus temporalis mountfordae* **Deignan**, 1950 NCR
[JS *Pomatostomus temporalis rubeculus*]

Bessie Ilma Mountford *née* Johnstone (1890–1996) was an Australian civil servant, and second wife of anthropologist Charles Pearcy Mountford.

Moussier

Moussier's Redstart *Phoenicurus moussieri* Olphe-Galliard, 1852

Jean Moussier (1795–1850) was a surgeon in the French army during the Napoleonic Wars and an amateur naturalist.

Mrs Bailey

Mrs Bailey's Chickadee *Poecile gambeli baileyae* **Grinnell**, 1908
[Alt. Mountain Chickadee ssp.]

(See **Bailey**)

Mrs Benson

Mrs Benson's Warbler *Nesillas mariae* **C. W. Benson**, 1960
[Alt. Moheli Brush Warbler]

Florence Mary Benson (1909–1993) was the wife of Constantine 'Con' Walter Benson (1909–1982) (q.v.) who named the warbler after her. (See **Mary (Benson)** & **Benson, C. W.**)

Mrs Boulton

Mrs Boulton's Woodland Warbler *Phylloscopus laurae*
Boulton, 1931
[Alt. Laura's Woodland Warbler]

Mrs Laura Theresa Boulton *née* Crayton (1899–1980) was the second wife of the American ornithologist Wolfrid Rudyerd Boulton (1901–1983), who named the warbler for her after she collected it. She was also an ornithologist, as well as a famous musicologist, or musical anthropologist, who in her lifetime collected c.30,000 examples of music from around the world.

Mrs Forbes

Mrs. Forbes's Honeyeater *Myzomela boiei annabellae*
P. L. Sclater, 1883
[Alt. Banda Myzomela ssp.]

Mrs Annabelle Forbes *née* Keith (d.1922) was the wife of Henry Ogg Forbes (q.v.). She wrote *Insulinde: Experiences of a Naturalist's Wife in the Eastern Archipelago* (1887).

Mrs Forbes-Watson

Mrs Forbes-Watson's Black Flycatcher *Melaenornis annamarulae* **Forbes-Watson**, 1970
[Alt. Nimba Flycatcher, Liberian Black Flycatcher]

(See **Anna Marula**)

Mrs Gould

Mrs Gould's Sunbird *Aethopyga gouldiae* **Vigors**, 1831
[Alt. Gould's Sunbird, Blue-throated Sunbird]

Elizabeth Gould (1804–1841) was the artist wife of John Gould (1804–1881). Vigors (q.v.), who described the sunbird, said he named it after 'Mrs Gould, who executed the plates of these Himalayan Birds'. (See also **Gould, E.**)

Mrs Hall

Mrs Hall's Greenbul *Andropadus hallae* **Prigogine**, 1972
NCR
[Alt. Little Greenbul; Syn. *Eurillas virens*]

(See **Hall, P.**)

Mrs Hume

Mrs Hume's Pheasant *Syrmaticus humiae* **Hume**, 1881
[Alt. Hume's Pheasant, Bar-tailed Pheasant]

Mary Ann Hume *née* Grindall (d.1890) was the wife of the describer, Allan Hume (q.v.).

Mrs McConnell

Mrs McConnell's Manakin *Neopipo cinnamomea helenae*
McConnell, 1911
[Alt. Cinnamon Tyrant Manakin, Cinnamon Neopipo ssp.]

Mrs Helen Mackenzie McConnell (1871–1954) was the wife of the describer, Frederick McConnell (q.v.).

Mrs Moreau

Mrs Moreau's Warbler *Scepomycter winifredae* **Moreau**,
1938
[Alt. Winifred's Warbler; Syn. *Bathmocercus winifredae*]

(See **Winifred**)

Mrs Sage

Mrs Sage's Blood Pheasant *Ithaginis cruentus annae* **Mayr**
& Birckhead, 1937 NCR
[NUI *Ithaginis cruentus berezowskii*]

Mrs Anne Sage *née* Tilney (1910–1996) was the first wife of wealthy American lawyer Dean Sage Jr (1909–1963) (After they divorced, she married Lieutenant Lanman T. Holmes.) She was in Western China (1934–1935) as part of the Sage West China Expedition, which mainly collected mammal specimens (they shot and skinned at least two Giant Pandas *Ailuropoda melanoleuca*) but also collected 426 bird specimens. Mrs Sage 'collected and prepared a large proportion of the bird-skins'.

Mrs Swinhoe

Mrs Swinhoe's Sunbird *Aethopyga christinae* **Swinhoe**,
1869
[Alt. Fork-tailed Sunbird]

(See **Christina**)

Mrs Vernay

Mrs Vernay's Blood Pheasant *Ithaginis cruentus marionae*
Mayr, 1941

(See **Marion (Vernay)**)

Mrs Wilson

Mrs Wilson's Tanager *Tangara larvata franciscae*
P. L. Sclater, 1856
[Alt. Golden-hooded Tanager ssp.]

(See **Fanny**)

Msiri

Little Rush Warbler ssp. *Bradypterus baboecala msiri*
Neave, 1909

Mwenda Msiri Ngelengwa Shitambi (c.1830–1891) was the founder and ruler of the Yeke or Garanganze Kingdom in Katanga, Congo. He was killed by Belgian troops, sent by King Leopold II to annex Katanga.

Mu

Timor Coucal *Centropus phasianinus mui* Mason & McKean, 1984

Peter Mu (DNF) lived in East Timor. The describers named the coucal '… in honour of the late Peter Mu, to whom we owe much for his assistance during our trips to East Timor'.

Muhamed-ben-Abdullah

Yellow-necked Spurfowl ssp. *Pternistis leucoscepus muhamed-ben-abdullah* **Erlanger**, 1904 NCR; NRM

Muhammed bin Abdullah (Sayyid Muhammad 'Abd Allah al-Hasan) (1856–1920) was a Somali religious leader. The British knew him as 'the Mad Mullah'. His Dervish army inflicted heavy losses on their various enemies (1901–1904) – the Ethiopians, British and Italians.

Mühle

European Greenfinch ssp. *Carduelis chloris muehlei* **Parrot**, 1905
Crested Lark ssp. *Galerida cristata muehlei* **Stresemann**, 1920 NCR
[JS *Galerida cristata meridionalis*]

Heinrich Karl Leopold Graf von der Mühle (1810–1855) was a German ornithologist who collected in Greece (1834–1838) and co-wrote *Beitrage zur Ornithologie Griechenlands* (1844).

Muir

Muir's Corella *Cacatua pastinator pastinator* **Gould**, 1841
[Alt. Western Corella ssp.]

Named after Lake Muir in southwest Western Australia, which was named after the Muir family of farming pioneers. Thomas Muir located the lake after listening to local Aboriginal advice.

Muir, J.

Muir's Winter Wren *Troglodytes pacificus muiri* **Rea**, 1986
[Alt. Pacific Wren; Syn. *Troglodytes troglodytes muiri*]

John Muir (1838–1914) was a Scottish-born American author, naturalist and pioneer conservationist. He undertook a walk of about 1,000 miles (1,600 km) from Indiana to Florida (1867), with no specific route planned except to go by the '… wildest, leafiest, and least trodden way I could find'. He was co-founder and first president of the Sierra Club (1892). He is regarded as the 'Father of the National Parks'.

Müller

Piping Cisticola ssp. *Cisticola fulvicapilla muelleri* **B. Alexander**, 1899

Major Müller (d.1898) was in command of the rear guard of Major Alfred St Hill Gibbon's expedition (1898–1900) to discover the sources of the rivers Congo and Zambezi (Boyd Alexander was a member). Müller died at Tete, Mozambique. An amphibian is named after him.

Müller, A.

Eurasian Wren ssp. *Troglodytes troglodytes muelleri* **von Jordans**, 1928 NCR
[JS *Troglodytes troglodytes kabylorum*]

Alfred Müller (d.1931) was the German Consul in Palma, Majorca.

Müller, C.

African Quailfinch ssp. *Ortygospiza fuscocrissa muelleri* **Zedlitz**, 1911
Blue-capped Cordon-bleu ssp. *Uraeginthus cyanocephalus muelleri* Zedlitz, 1912 NCR; NRM
Spotted Palm-thrush ssp. *Cichladusa guttata muelleri* Zedlitz, 1916 NCR
[JS *Cichladusa guttata rufipennis*]

C. Müller (fl.1910) was a skinner for Graf von Zedlitz (q.v.) in tropical Africa (1908–1910).

Müller, D.

Yemen Sunbird sp. *Nectarinia muelleri* **Lorenz** & **Hellmayr**, 1901 NCR
[Alt. Nile Valley Sunbird; JS *Hedydipna metallica*]

Prof. David Heinrich Müller (1846–1912) was an Austrian orientalist, linguist and explorer in Arabia (1897–1898). He was Professor of Semitic Philology at the University of Vienna (1881). A reptile is named after him.

Müller, F. J. H.

Auckland Islands Rail *Lewinia muelleri* **Rothschild**, 1893

Baron Sir Ferdinand Jakob Heinrich von Müller (1825–1896) was a German-born Australian botanist and explorer. He studied pharmacy at Kiel and emigrated (1847) for health reasons. He was the government botanist for Victoria (1857–1896) and Director of the Zoological and Botanical Gardens in Melbourne (1857–1873).

Müller, J. W.

Blue-headed Bee-eater *Merops muelleri* **Cassin**, 1857

Baron Johann Wilhelm von Müller (1824–1866) was a German who collected in North America, Mexico and tropical Africa. His African expedition (1847–1852) established his reputation as a zoologist. This expedition started in Egypt, travelled up the Nile into the Sudan, then crossed the Red Sea finishing in the Sinai Peninsula. He took Alfred Brehm (q.v.) with him as his Secretary and Assistant. The Kingdom of Belgium opened a museum and botanical and zoological gardens near Brussels and J. J. Linden (q.v.) was appointed Scientific Director (1851). Von Müller, on the basis of his successful African expedition, was appointed Zoological Director (1852), Linden remaining as Director of Botany. The zoo had 570 animals, but in his first year von Müller succeeded in losing 140 of them; he was sacked (1856), and Linden again became overall Scientific Director. A picture we have seen of von Müller shows him dressed in the uniform of an Austrian consul, but we do not know where or when he was appointed to act as an Austrian official.

Müller, L.

Scaled Spinetail *Cranioleuca muelleri* **Hellmayr**, 1911

Long-tailed Hermit ssp. *Phaethornis superciliosus muelleri* Hellmayr, 1911

Lorenz Müller (1868–1953) was a German herpetologist who trained as an artist in Paris and the Low Countries and worked as a scientific illustrator at State Zoological Collection in Munich. He was particularly interested in amphibians and reptiles, and as that department in Munich had no curator (1903) he took on those additional duties. He collected in Brazil as a member of the museum's expedition to the Lower Amazon (1909–1910). He served in the German Army in WW1 but was able to get posted to the Balkans and spent most of his time collecting specimens! After the war he returned to Munich and he became the Chief Curator of Zoology (1928). During WW2 both his own private and the museum's collections were largely destroyed in air raids, but Müller doggedly re-built them. During his life he published over 100 articles, monographs and papers on herpetology. A reptile and an amphibian are named after him.

Müller, S.

Woodpecker genus *Mulleripicus* **Bonaparte**, 1854
Thicketbird genus *Mulleria* **Büttikofer**, 1895 NCR
[Now in *Buettikoferella*]

Müller's Parrot *Tanygnathus sumatranus* **Raffles**, 1822
[Alt. Blue-backed Parrot; *Tanygnathus muelleri* is a JS]
Müller's Fruit Pigeon *Ducula mullerii* **Temminck**, 1835
[Alt. Collared Imperial Pigeon]
Müller's Barbet *Megalaima oorti* Müller, 1836
[Alt. Black-browed Barbet]
Müller's Greybird *Coracina morio* Müller, 1843
[Alt. Sulawesi Cicadabird]
Spot-breasted White-eye *Heleia muelleri* **Hartlaub**, 1865
[Alt. Spot-breasted Heleia, Timor White-eye]
Moluccan Goshawk *Accipiter muelleri* **Wallace**, 1865 NCR
[JS *Accipiter henicogrammus*]
Müller's Bush Warbler *Cettia vulcania* **Blyth**, 1870
[Alt. Sunda Bush Warbler]
Torrent Flyrobin *Monachella muelleriana* **Schlegel**, 1871
Müller's Wren-babbler *Napothera marmorata* **R. G. W. Ramsay**, 1880
[Alt. Marbled Wren-babbler]

Green Oriole ssp. *Oriolus flavocinctus muelleri* Bonaparte, 1850
Hooded Pitta ssp. *Pitta sordida mulleri* Bonaparte, 1850
Chestnut-bellied Fantail ssp. *Rhipidura hyperythra muelleri* **A. B. Meyer**, 1874
Slender-billed Cicadabird ssp. *Coracina tenuirostris muellerii* **Salvadori**, 1876
Rufous-chested Flycatcher ssp. *Ficedula dumetoria muelleri* **Sharpe**, 1879
Chestnut-crowned Warbler ssp. *Seicercus castaniceps muelleri* **Robinson** & **Kloss**, 1916

Dr Salomon Müller (1804–1864) was a German naturalist who collected in Indonesia (1826) where he assisted members of the Netherlands Natural Sciences Commission as a taxidermist. He went to New Guinea and Timor (1828), collected in Java (1831), and explored in western Sumatra (1833–1835). Three mammals and three reptiles are named after him.

Mulsant

Hummingbird subgenus *Mulsantia* **Reichenbach**, 1854 NCR
[Now in *Loddigesia*]

Mulsant's Woodstar *Chaetocercus mulsant* **Bourcier**, 1842
[Alt. White-bellied Woodstar]

Martial Etienne Mulsant (1797–1880) was a French collector, general naturalist and most famously an entomologist. The best known of his works was *Histoire Naturelle des Coléoptères de France*. He also wrote *Lettres à Julie sur l'Ornithologie* (1868), and *Histoire Naturelle des Oiseaux-Mouches ou Colibris* (1874).

Mundy

Ascension Crake genus *Mundia* **W. R. P. Bourne**, Ashmole & Simmons, 2003 EXTINCT

Sir Peter D. Mundy (1596–1667) was a British traveller in India and the Far East. He witnessed the construction of the Taj Mahal, writing: 'the building is begun and goes on with excessive labour and cost, gold and silver being used like common metal'. When visiting Ascension Island (1656), he made a crude drawing of this extinct crake, which is otherwise known from subfossil bones.

Munoz Tebar

Rufous Spinetail ssp. *Synallaxis unirufa munoztebari* **W. H. Phelps** & **W. H. Phelps Jr**, 1953

Dr Ricardo Muñoz Tébar (DNF) was Assistant Curator of the Phelps Collection Caracas, Venezuela.

Munro, G. C.

Lana'i Hookbill *Dysmorodrepanis munroi* **R. C. L. Perkins**, 1919 EXTINCT
Akiapolaau *Hemignathus munroi* **H. D. Pratt**, 1979
[JS *Hemignathus wilsoni*]

George Campbell Munro (1866–1963) was a New Zealand ornithologist and collector who wrote *Birds of Hawaii* (1960). He went to Hawaii (1910) and set about sowing the seeds of Norfolk pines. The George C. Munro Award for Environmental Law was established in his honour.

Munro, J. A.

Sooty Grouse ssp. *Dendragapus obscurus munroi* **Griscom**, 1923 NCR
[JS *Dendragapus fuliginosus sitkensis*]
Hermit Thrush ssp. *Catharus guttatus munroi* **A. R. Phillips**, 1962

James Alexander Munro (1884–1958) was a Canadian ornithologist and conservationist. He was Chief Federal Migratory Birds Officer for various Canadian provinces (1920–1949) and was Dominion Wildlife Officer (1947).

Münzner

Bush-shrike sp. *Chlorophoneus muenzneri* **Reichenow**, 1915 NCR
[Alt. Black-fronted Bush-shrike; JS *Chlorophoneus nigrifrons nigrifrons*]

Common Waxbill ssp. *Estrilda astrild muenzneri* Kothe, 1911 NCR
[JS *Estrilda astrild cavendishi*]
Purple-throated Cuckooshrike ssp. *Campephaga quiscalina muenzneri* Reichenow, 1915
Square-tailed Drongo ssp. *Dicrurus ludwigii muenzneri* Reichenow, 1915
Placid Greenbul ssp. *Phyllostrephus placidus muenzneri* Reichenow 1916 NCR
[JS *Phyllastrephus fischeri* (monotypic)]
Singing Cisticola ssp. *Cisticola cantans muenzneri* Reichenow, 1916

Max Münzner (fl.1911) was a sergeant in the German army in German East Africa (Tanzania) (1905–1907). He was on Fromm's (q.v.) expedition to Lake Tanganyika (1908–1909).

Murat

Common Rosefinch ssp. *Carpodacus erythrinus murati* **Delacour**, 1926 NCR
[JS *Carpodacus erythrinus roseatus*]

Paul Jérome Michel Joachim Napoléon Prince Murat (1893–1964) was a French aviculturist and collector who was Président de la Ligue Française pour la Protection des Oiseaux.

Murie

Willow Ptarmigan ssp. *Lagopus lagopus muriei* **Gabrielson** & Lincoln, 1949

Dr Olaus Johan Murie (1889–1963) was an American zoologist, biologist, conservationist and ecologist. Pacific University, Oregon, awarded his bachelor's degree (1912). He was an Oregon State conservation official (1912–1920) and took part in expeditions to Hudson Bay and Labrador (1914–1917). He worked as a wildlife biologist for the US Bureau of Biological Survey (1920–1945). He was Director of the Wilderness Society (1945–1963). He wrote *The Elk of North America* (1951).

Muriel (Bannerman)

Muriel's Chat *Saxicola dacotiae murielae* **Bannerman**, 1913 EXTINCT
[Alt. Canary Islands Stonechat ssp.]

Muriel Bannerman *née* Morgan (d.1945) was the first wife of Dr David Armitage Bannerman (1886–1979). He named the chat after her when she accompanied him on an expedition to the Canaries (1913).

Muriel (Ezra)

White-bellied Green Pigeon *Treron sieboldii murielae* **Delacour**, 1927

Mrs Muriel Helene Ezra *née* Sassoon (b.1897) was the wife of Alfred Aaron D. Ezra (q.v.).

Murphy, C. B. G.

Black-collared Apalis ssp. *Oreolais pulcher murphyi* **Chapin**, 1932

Charles B. G. Murphy (1906–1977) was an American philanthropist who graduated from Yale (1928), then explored in Tanganyika (Tanzania) and the Belgian Congo (Democratic Republic of Congo) with J. Sterling Rockefeller (1928–1929 and 1933). He co-wrote 'The Rediscovery of *Pseudocalyptomena*' (1933).

Murphy, R. C.

Peruvian Martin *Progne murphyi* **Chapman**, 1925
Murphy's White-eye *Zosterops murphyi* **Hartert**, 1929
[Alt. Hermit White-eye, Kolombangara Mountain White-eye]
Murphy's Petrel *Pterodroma ultima* Murphy, 1949

Blue Petrel ssp. *Halobaena caerulea murphyi* **W. S. Brooks**, 1917 NCR; NRM
Brown Pelican ssp. *Pelecanus occidentalis murphyi* **Wetmore**, 1945

Robert Cushman Murphy (1887–1973) was an American naturalist who worked at the AMNH and became a world authority on marine birds. He is also famed for persuading Rachel Carson to write *Silent Spring*, after he was unable to persuade the US government to stop spraying DDT. His *Oceanic Birds of South America* (1936) was awarded the John Burroughs Medal for excellence in natural history and the Brewster Medal of the AOU. Murphy also spent time excavating the remains of moas in New Zealand. On Bermuda (1951) he 'slipped a noose onto a pole, slid it down a tunnel between some ocean side rocks, and pulled out a sea bird called a cahow' – the first living member of that species (Bermuda Petrel *Pterodroma cahow*) seen since the early 17th century. A country park and a school in New York State are named after him. According to his obituary in the *New York Times*, there are also two mountains, a fish, a spider, a lizard and a louse named in his honour. 'As a scientist,' he once said, 'I'd as soon have a louse named for me as a mountain.'

Murray

Henderson Island Crake *Porzana murrayi* **Ogilvie-Grant**, 1913 NCR
[JS *Porzana atra*]

Yellow-bibbed Fruit Dove ssp. *Ptilinopus solomonensis johannis* **P. L. Sclater**, 1877

Sir John Murray (1841–1914) was a Canadian-born Scottish oceanographer and marine biologist. He was naturalist in charge of collections on the *Challenger* expedition (1872–1876), and edited the *Report on the Scientific Results of the*

Voyage of HMS Challenger, which appeared in instalments (1880–1895). He wrote *On the Origin and Structure of Coral Reefs and Islands* (1880). Murray was killed when his car overturned near his home. (See **Johannis (John M.)**)

Musschenbroek

Fruit dove sp. *Ptilinopus musschenbroekii* **Schlegel**, 1871 NCR
[Alt. Claret-breasted Fruit Dove; JS *Ptilinopus viridis geelvinkianus*]
Musschenbroek's Lorikeet *Neopsittacus musschenbroekii* Schlegel, 1871
[Alt. Yellow-billed Lorikeet]
Moluccan Drongo Cuckoo *Surniculus musschenbroeki* **A. B. Meyer**, 1878

White-breasted Woodswallow ssp. *Artamus leucorynchus musschenbroeki* A. B. Meyer, 1884

Samuel Cornelis Jan Willem van Musschenbroek (1827–1883) graduated as a lawyer and trained as a seaman, passing his first mate's examination. He quickly rose to become a Dutch colonial administrator in the East Indies (1855–1876). He was Resident of Ternate (1873) and of Menado (1875). He travelled extensively in the Moluccas with Beccari (q.v.). He was in Leiden (1879) and was first Director of the Museum der Koloniale Vereniging (Colonial Museum), but died shortly after his appointment. Many bird skins, which he had collected in Indonesia were exhibited there until they were transferred by Voous (q.v.) to the Zoological Museum of Amsterdam. Musschenbroek was regarded as an expert on large parts of the Dutch East Indies, as is shown by his pioneering maps of Minahassa and the northern parts of the Moluccas. The Artis Library, University of Amsterdam, advised us that he wrote the article 'Iets over de fauna van Noord-Celebes' (1867) and created the map *Kaart van de Bocht van Tomini* (1880). He has two mammals named after him.

Muttu

Brown-breasted Flycatcher *Muscicapa muttui* **Layard**, 1854
[Alt. Layard's Flycatcher]

Muttu was a Tamil cook who worked for the describer, Edgar Leopold Layard, when he was in Ceylon (Sri Lanka).

Mwaki

Mountain Thrush ssp. *Turdus abyssinicus mwaki* **Someren & Schifter**, 1981 NCR
[NUI *Turdus abyssinicus abyssinicus*]

Joseph Mwaki (DNF) was a Kenyan collector and skinner for A. D. Forbes-Watson (q.v.).

Myrtha

Brown Wood Owl ssp. *Strix leptogrammica myrtha* **Bonaparte**, 1850

Myrtha, Queen of the Wilis, is a character from the ballet *Giselle,* which premiered in Paris (1841). The ghost-like wilis seek out their male prey between twilight and dawn, and magically force them to dance until their hearts give out.

Myrtle (Ashmole)

Newell's Shearwater ssp. *Puffinus newelli myrtae* **Bourne**, 1959

Mrs Myrtle J. Ashmole *née* Goodacre worked at the Edward Grey Institute (1959–1962) and is the wife of Dr N. Philip Ashmole (b.1934), who worked on Ascension Island (1957–1959) and at the Edward Grey Institute (1959–1963). Both she and her husband are dedicated ecologists with a particular interest in southern Atlantic oceanic islands.

Myrtle (Morgan)

Australian Ringneck ssp. *Barnardius zonarius myrtae* **S. A. White**, 1915 NCR
[JS *Barnardius zonarius zonarius*]

Mrs Myrtle Ellen Morgan *née* Green (b.1879) was the wife of Australian oculist and ornithologist Dr Alexander M. Morgan (q.v.).

Myrtis

Hummingbird genus *Myrtis* **Reichenbach**, 1854

Myrtis was a Greek poetess in the 4th century BC.

N

Nadezda

Rock Ptarmigan ssp. *Lagopus muta nadezdae* **Serebrovski**, 1926

Professor Nadezda Nikolaevne Sushkina *née* Popova (1889–1975) was an ornithologist who worked at the Zoological Museum of the Academy of Sciences in Leningrad (St Petersburg), as did her husband Professor Petr Petrovich Sushkin (q.v.). She was the first female Professor of Microbiology at Moscow State University.

Nadia

Brown Accentor ssp. *Prunella fulvescens nadiae* **Bangs** & **J. L. Peters**, 1928 NCR
[JS *Prunella fulvescens nanshanica*]

Nadia is a shortened form of **Nadezda** (see above).

Naenia

Inca Tern genus *Naenia* **Boie**, 1844 NCR
[Now in *Larosterna*]

In Roman mythology, Naenia was the goddess of funerals.

Nagamichi

Grey-capped Woodpecker ssp. *Dendrocopos canicapillus nagamichii* **La Touche**, 1932
Japanese Pygmy Woodpecker ssp. *Dendrocopos kizuki nagamichi* **Bergman**, 1935 NCR
[JS *Dendrocopos kizuki ijimae* or *D. kizuki seebohmi*]
Chestnut-eared Bulbul ssp. *Microscelis amaurotis nagamichii* **Mayr** & **Greenway**, 1960 NCR
[JS *Microscelis amaurotis harterti*]

Dr Nagamichi Kuroda (See **Kuroda**)

Nahan

Nahan's Partridge *Ptilopachus nahani* **A. Dubois**, 1905
[Alt. Nahan's Francolin]

Commandant Paul François Joseph Nahan (1867–1930) was a Belgian traveller who explored in tropical Africa. He served in the Belgian Army (1883–1894) and first arrived in the Congo (1891). After leaving the army he stayed on in the Congo exploring and collecting for the Tervuren Museum (1895–1899 & 1904). He returned to Belgium (1904).

Naim

White-shouldered Fairy-wren ssp. *Malurus alboscapulatus naimii* **Albertis**, 1875

Naim (fl.1875) was a Papuan servant to d'Albertis in New Guinea.

Nakamura

Whiskered Treeswift ssp. *Hemiprocne comata nakamurai* **Hachisuka**, 1930 NCR
[JS *Hemiprocne comata major*]
Comb-crested Jacana ssp. *Irediparra gallinacea nakamurai* Hachisuka, 1932 NCR
[JS *Irediparra gallinacea gallinacea* or species may be monotypic]

Yukio Nakamura (b.1889) was a Japanese field ornithologist who collected in the Philippines (1929–1930).

Nakaoka

Eurasian Jay ssp. *Garrulus glandarius nakaokae* **Momiyama**, 1927 NCR
[JS *Garrulus glandarius japonicus*]
Eurasian Nuthatch ssp *Sitta europaea nakaokae* Momiyama, 1931 NCR
[JS *Sitta europaea hondoensis*]

Mrs Koma Nakaoka (DNF) was an ornithologist on Shikoku, Japan.

Namiye

Ryukyu Robin ssp. *Erithacus komadori namiyei* **Stejneger**, 1886
White-backed Woodpecker ssp. *Dendrocopos leucotos namiyei* Stejneger, 1886
Pacific Swallow ssp. *Hirundo tahitica namiyei* Stejneger, 1887
Varied Tit ssp. *Poecile varius namiyei* **Kuroda**, 1918
Eurasian Jay ssp. *Garrulus glandarius namiyei* Kuroda, 1922 NCR
[JS *Garrulus glandarius japonicus*]
Meadow Bunting ssp. *Emberiza cioides namiyei* **Momiyama**, 1923 NCR
[JS *Emberiza cioides ciopsis*]

Motoyoshi Namiye (1854–1918) was a Japanese naturalist and herpetologist. He was a member of the Faculty of Zoology at the Tokyo Educational Museum, and a Corresponding Fellow of the AOU. Stejneger wrote a report on a collection of birds Namiye had made in the Riu Kiu Islands (1886). He wrote 'Oviposition of a blind snake from Okinawa' (1912). He is remembered in the names of other taxa, including a fish and an amphibian.

Nana

Puffleg genus *Nania* **Mulsant**, 1876 NCR
[Now in *Eriocnemis*]

In Greek mythology, Nana was a daughter of the river-god Sangarius. She was the mother of Atys, a beautiful shepherd loved by the goddess Cybele.

Napoleon

Napoleon Weaver *Euplectes afer* **Gmelin**, 1789
[Alt. Yellow-crowned Bishop]
Napoleon's Peacock Pheasant *Polyplectron napoleonis* **Lesson**, 1831
[Alt. Palawan Peacock Pheasant; Syn. *Polyplectron emphanum*]

Napoleon Bonaparte (1769–1821) was Emperor of France (1804–1815) and is so well known no biography is necessary. Although Lesson described the pheasant ten years after Napoleon's death, he based it on an earlier manuscript of Massena's in which he coined the name.

Narina

Narina Trogon *Apaloderma narina* **J. F. Stephens**, 1815

Narina was a beautiful Khoi Khoi (Hottentot) girl. Le Vaillant (q.v.), whose mistress she may have been according to Lesson (q.v.), first applied this name to the bird. Le Vaillant said: 'I found her name difficult to be pronounced, disagreeable to the ear, and very insignificant according to my ideas; I therefore gave her a new one and called her Narina, which in the Hottentot language signifies a flower.'

Narosky

Narosky's Seedeater *Sporophila zelichi* Narosky, 1977 NCR
[Alt. Entre Ríos Seedeater. Now regarded as a colour morph of *Sporophila palustris*]

Ash-breasted Sierra Finch ssp. *Phrygilus plebejus naroskyi* Nores & **Yzurieta**, 1983

Samuel 'Tito' Narosky (b.1932) is an expert on Argentine birds and molluscs. He wrote *Guia de Aves de Patagonia y Tierra del Fuego* (2001), *Manual del Observador de Aves* (1996) and, with the late Dario Yzurieta (q.v.), *Birds of Argentina and Uruguay: A Field Guide*.

Natalia

Song Thrush ssp. *Turdus philomelos nataliae* **Buturlin**, 1929

Countess Natalie Obolensky-Neledinsky-Meletzky *née* Mesentzev (1820–1895) was a lady-in-waiting at the Imperial Russian court and a relation by marriage to the describer Buturlin (q.v.), himself a member of the Russian aristocracy.

Natalie

Natalie's Sapsucker *Sphyrapicus thyroideus nataliae* **Malherbe**, 1854
[Alt. Williamson's Sapsucker ssp.]

Miss Natalie Kaup (DNF) was the daughter of German ornithologist Johann Jakob von Kaup (q.v.).

Nation

Rufescent Flycatcher *Myiobius nationi* **P. L. Sclater**, 1866 NCR
[JS *Myiophobus (fasciatus) rufescens*]
Southern Pochard *Fuligula nationi* P. L. Sclater & **Salvin**, 1877 NCR
[JS *Netta erythrophthalma*]
Rusty-bellied Brush Finch *Atlapetes nationi* P. L. Sclater, 1881

Professor William Nation (1826–1907) was a British botanist who lived and taught in Lima, Peru (1862–1880). He sent all his specimens to Philip Sclater (q.v.).

Natorp

Willow Tit ssp. *Poecile montanus natorpi* **Kleinschmidt**, 1917 NCR
[JS *Poecile montanus salicarius*]

Dr Otto Natorp (1876–1956) was a German surgeon and field ornithologist in Bavaria.

Natterer

Natterer's Vizorbearer *Augastes scutatus* **Temminck**, 1824
[Alt. Hyacinth Visorbearer]
Tawny-tufted Toucanet *Selenidera nattereri* **Gould**, 1836
Natterer's Cotinga *Cotinga nattererii* **A Boissonneau**, 1840
[Alt. Blue Cotinga]
Natterer's Emerald *Ptochoptera iolaima* **Reichenbach**, 1854 NCR
[Probable hybrid: *Chlorostilbon alice* x *Chlorostilbon poortmani*?]
Snow-capped Manakin *Lepidothrix nattereri* **P. L. Sclater**, 1865
Natterer's Slaty Antshrike *Thamnophilus stictocephalus* **Pelzeln**, 1868
Natterer's Piculet *Picumnus fuscus* Pelzeln, 1870
[Alt. Rusty-necked Piculet]
Ochre-breasted Pipit *Anthus nattereri* P. L. Sclater, 1878
Cinnamon-throated Hermit *Phaethornis nattereri* **Berlepsch**, 1887
Natterer's Tody Tyrant *Poecilotriccus nattereri* **Hellmayr**, 1903 NCR
[JS *Poecilotriccus latirostris ochropterus*]
Speckle-breasted Antpitta *Hylopezus nattereri* **Pinto**, 1937

Natterer's Nighthawk *Lurocalis semitorquatus nattereri* Temminck, 1822
[Alt. Short-tailed Nighthawk ssp.]
Green-barred Woodpecker ssp. *Colaptes melanochloros nattereri* **Malherbe**, 1845
Blue-crowned Motmot ssp. *Momotus momota nattereri* Sclater, 1858
Natterer's Piping Guan *Pipile cujubi nattereri* Reichenbach, 1861
[Alt. Red-throated Piping Guan ssp.]
Natterer's Amazon *Amazona ochrocephala nattereri* **Finsch**, 1865
[Alt. Yellow-crowned Amazon ssp.]
Roadside Hawk ssp. *Rupornis magnirostris nattereri* Sclater & **Salvin**, 1869

Natterer's Tanager *Tachyphonus cristatus nattereri* Pelzeln,
1870
[Alt. Flame-crested Tanager ssp.]
Natterer's Curassow *Crax fasciolata pinima* Pelzeln, 1870
[Alt. Bare-faced Curassow ssp.]
White-crested Spadebill ssp. *Platyrinchus platyrhynchos
nattereri* **Hartert** & Hellmayr, 1902
White-eyed Attila ssp. *Attila bolivianus nattereri* Hellmayr,
1902
Rufous-capped Nunlet ssp. *Nonnula ruficapilla nattereri*
Hellmayr, 1921
Solitary Tinamou ssp. *Tinamus solitarius nattereri* **Miranda-
Ribeiro**, 1938 NCR
[JS *Tinamus solitarius solitarius*]

Dr Johann Natterer (1787–1843) was an Austrian naturalist
and collector. He studied botany, zoology, mineralogy, chemistry and anatomy and was appointed as a taxidermist to
what became the Natural History Museum in Vienna. As a
zoologist, he took part with Spix (q.v.) and others in the expedition to Brazil (1817), which started on the occasion of
Archduchess Leopoldina's wedding to Dom Pedro, the
Brazilian Crown Prince. The entire suite travelled in two
Austrian frigates, *Austria* and *Principessa Augusta*. Natterer
explored a potential river route to Paraguay (1818–1819), and
subsequently undertook (1821–1835) five expeditions,
exploring the Mato Grosso and the Amazon Basin before
returning to Vienna. He accumulated a huge collection
numbering 12,293 birds and c.24,000 insects, which can still
be seen today in the Vienna Museum. He lost the majority of
his possessions in the Civil War then being waged in Brazil,
so his total collection must have been staggering. He ended
his career at the Austrian Imperial Museum of Natural
History, dying of a lung ailment. He did not publish an account
of his travels and, unfortunately for posterity, his notebooks
and diary were destroyed by fire (1848). Fortunately, August
von Pelzeln (q.v.) was able to reconstruct some of Natterer's
itinerary and information and wrote up the collection (1868–
1871) Natterer's specimens are beautifully prepared and his
tiny labels always meticulous and clear – something that is a
real rarity for the period. The overall result was that he never
received the credit in Austria that he should have, but abroad
he was held in high esteem, being, for instance, made an
honorary Doctor of Philosophy at Heidelberg University.
Three reptiles, two amphibians and two mammals are named
after him.

Naumann, J. A.

Naumann's Thrush *Turdus naumanni* **Temminck**, 1820

Johann Andreas Naumann (1744–1826) was a German farmer
and amateur naturalist. He travelled widely from the North
Sea coast to Hungary to study better the birds rarely found in
his own region. A definitive book on Germany's birds, *Naturgeschichte der Vögel Deutschlands* was published under his
name (1822–1844), although his son, Johann Friederich
Naumann (q.v.) was reputed to have written it. It was
completely revised and re-issued at the end of the 19th
century by a group of ornithologists, which included Blasius
(q.v.).

Naumann, J. F.

Lesser Kestrel *Falco naumanni* Fleischer, 1818

Atlantic Puffin ssp. *Fratercula arctica naumanni*
A. H. Norton, 1901

Johann Friedrich Naumann (1780–1857) was the eldest son of
Johann Andreas Naumann (q.v.). He is regarded by some as
the founder of scientific ornithology in Europe. Duke August
Christian Friedrich von Anhalt-Koethen was Naumann's
patron and bought his collection (1821). The Duke made him
Curator of his Natural History Cabinet, and the collection can
still be seen at the Naumann-Museum Köthen. Naumann
was one of the founders of the Deutsche Ornithologen-
Gesellschaft (German Ornithologists' Society), whose journal
Naumannia was named after him.

Naumburg

Brown Jacamar ssp. *Brachygalba lugubris naumburgi*
Chapman, 1931
Blue-throated Piping Guan ssp. *Pipile cumanensis
naumburgae* **W. E. C. Todd**, 1932 NCR
[JS *Pipile cumanensis cumanensis*]
White-bellied Tody Tyrant ssp. *Hemitriccus griseipectus
naumburgae* **Zimmer**, 1945
Brown-breasted Bamboo Tyrant ssp. *Hemitriccus obsoletus
naumburgae* Zimmer, 1953 NCR
[NPRB *Hemitriccus obsoletus zimmeri*]

Mrs Elsie Margaret Binger Naumburg (formerly Reichenberger) (1880–1953) was an American ornithologist at the
AMNH, New York (1918–1953). She studied the birds
collected on the Theodore Roosevelt/Colonel Rondon expedition to Brazil, the 'River of Doubt' expedition, which led to
the publication of the 400-page paper 'The birds of Matto
Grosso, Brazil' (1930).

Nava, J.

Nava's Wren *Hylorchilus navai* **Crossin** & **Ely**, 1973

Yucatan Flycatcher ssp. *Myiarchus yucatanensis navai*
Parkes, 1982

Juan Nava Solorio (DNF) was a Mexican ornithologist who
worked with Crossin (q.v.), and together they collected the
type specimen of the wren (1971) in Chiapas, Mexico, originally describing it as a subspecies of *Hylorchilus sumichrasti*
although when its voice was recorded it was elevated to a
full species. The etymology states: '… named for Juan Nava
Solario, who, through his devotion to learning the birds of his
native Mexico, has earned the admiration and respect of
numerous American ornithologists'. He was in Quintana Roo
(1986–1992). He also collected a mammal holotype (1964).

Nava, M.

Black-fronted Wood Quail ssp. *Odontophorus atrifrons navai*
Aveledo & **Pons**, 1952

Moisés C. Nava (d.1954) was a Venezuelan collector who
was based in Maracaibo.

Navarro

Tawny-bellied Babbler ssp. *Dumetia hyperythra navarroi*
  Abdulali, 1959 NCR
[JS *Dumetia hyperythra abuensis*; or *D. hyperythra albogularis* if *abuensis* not recognised as valid]

Reverend Brother Antonio Navarro (1903–1987) was a Jesuit priest and ornithologist who taught at St Xavier's High School, Bombay (Mumbai).

Navas

Cordilleran Canastero ssp. *Asthenes modesta navasi*
  Contreras, 1979
Great Grebe ssp. *Podicephorus major navasi* Manghi, 1984
[Syn. *Podiceps major navasi*]

Dr Jorge Rafael Navas (1921–2009) was an Argentine zoologist, ornithologist and biologist well-versed in botany, history, geography and archaeology. He graduated from the National College Bernardino Rivadavia (1945) and taught at the School of Mineralogy, Buenos Aires (1945–1947). He took a degree in biology at the Faculty of Natural Sciences, Museum of La Plata (1950) and a doctorate (1955). He worked at the Argentine Museum of Natural Sciences 'Bernadino Rivadavia' (1947–1997) becoming a Professor there and at the Universidad Nacional del Sur and the Museum of La Plata. He became an honorary researcher (1992) and retired as emeritus head of ornithology (1997). He was honorary curator of the ornithological collection (2003–2009).

Neave

Meyer's Parrot ssp. *Poicephalus meyeri neavei*
  C. H. B. Grant, 1914 NCR
[JS *Poicephalus meyeri matschiei*]
Natal Spurfowl ssp. *Pternistis natalensis neavei*
  Mackworth-Praed, 1920

Dr Sheffield Airey Neave (1879–1961) was a British entomologist who wrote *The History of the Entomological Society of London, 1833–1933*. He was Secretary of the Zoological Society of London, for which he edited *Nomenclator Zoologicus* (1939), and was employed by the BMNH. He travelled extensively in central Africa and collected invertebrates, fish, amphibians and reptiles in Northern Rhodesia (Zambia) (1911). A mammal is named after him.

Néboux

Blue-footed Booby *Sula nebouxii* **Milne-Edwards**, 1882

Blue Noddy ssp. *Procelsterna cerulea nebouxi* **Mathews**, 1912

Adolphe-Simon Néboux (1806–at least 1842) was a French surgeon, explorer and naturalist. He took part in the circumnavigation by the frigate *Vénus* (1836–1839), during which he visited the Galapagos Islands. Among the species he described was the Swallow-tailed Gull *Creagrus furcatus*.

Neergaard

Neergaard's Sunbird *Cinnyris neergaardi* **C. H. B. Grant**, 1908
[Alt. Coguno Double-collared Sunbird]

Paul Neergaard (DNF) was a recruiting official in southern Mozambique for the Witwatersrand mines (1907–1927) and travelled widely in Nyasaland (Malawi) to recruit extra labour (1928). He assisted Claude Grant (q.v.) on his expedition.

Negret

Munchique Wood Wren *Henicorhina negreti* Salaman *et al.*, 2003

Álvaro José Negret (1949–1998) was a Colombian naturalist. As his obituary in *Cotinga* stated, 'his untimely death robbed Colombia of one of its greatest naturalists and conservationists'. He started collecting birds for the Natural History Museum of Cauca University in Popayán as a child. As an undergraduate, he co-founded the Natural History Museum at Caldas University (Manizales). After his Masters in ecology and management of natural resources, University of Brasilia, he continued his research throughout Brazil. He later returned to Colombia as Professor at Popayán University, and became Director of the Natural History Museum at Cauca University, a position he held for the rest of his life. He travelled widely throughout Colombia and was heavily involved in conservation, especially in the Chocó region, where he founded and managed Tambito Nature Reserve. He also initiated and directed the Naya Corridor Program, aiming to unify two national p\arks. A dedicated conservationist and naturalist, he always remained an enthusiastic ornithologist. At the time of his death a manuscript entitled *Aves Colombianas Amenazadas de Extinción* remained unfinished.

Nehrkorn

Nehrkorn's Flowerpecker *Dicaeum nehrkorni* **Blasius**, 1886
[Alt. Crimson-crowned Flowerpecker, Red-headed Flowerpecker]
Nehrkorn's Sylph *Neolesbia nehrkorni* **Berlepsch**, 1887 NCR
[Hybrid: perhaps *Aglaiocerus kingi* x *Ramphomicron microrhynchum*, or *Aglaiocerus kingi* x *Thalurania* sp.]
Sangihe White-eye *Zosterops nehrkorni* Blasius, 1888

Common Cicadabird ssp. *Coracina tenuirostris nehrkorni*
  Salvadori, 1889
White-fronted Tit ssp. *Parus semilarvatus nehrkorni*
  Blasius, 1890
Violet-backed Hyliota ssp. *Hyliota violacea nehrkorni*
  Hartlaub, 1892

Adolph Nehrkorn (1841–1916) was a German oologist and ornithologist. He wrote *Katalog der Eiersammlung, nebst Beschreibungen der Aussereuropäischen Eier* (1910). His large collection of birds' eggs was bequeathed to the Berlin Zoological Museum.

Neiff

Short-billed Canastero ssp. *Asthenes baeri neiffi* **Contreras**, 1980
[Originally described as *Tripophaga steinbachi neiffi*]

Dr Juan José Neiff (b.1947) is an Argentine ecologist and biologist. His bachelor's degree (1969) and master's were awarded by Universidad Nacional del Litoral-Santa Fé, while the University of the Northeast awarded his doctorate. He is presently Director of Centre de Ecologia Aplicada del Litoral, Corrientes, Argentina, and Professor and Chairman, Aquatic Ecology in the Universidad del Litoral.

Neisna

Waxbill genus *Neisna* **Sharpe**, 1890 NCR
[Now in *Amandava*]

There are several contestants if this is an eponym. Neis was a wife of the youthful Endymion. Nais was (1) an Oceanid, mother of Chiron, (2) a nymph, mother of Aesepus and Pedasus, and (3) a nymph of the Red Sea who transformed her lovers into fishes! More likely, however, is Nais or Neis, one of a number of deities, the Naiades or Naides, who presided over rivers, springs, wells and fountains. Just to confuse things, there is a Greek word *neis* meaning feeble. *Estrelda neisna* was first used as a specific name by Lichtenstein (c.1842) and was quoted by Bonaparte (1850) when he created the genus *Neisna*. *Estrelda neisna* is a synonym of *Coccopygia melanotis*, the Swee Waxbill of South Africa. These birds common name refers to their feeble sibilant call of '*swee swee*' when they feed, so it could also account for the scientific name.

Neiva

Olive Oropendola ssp. *Psarocolius bifasciatus neivae*
Snethlage, 1925
[Syn. *Psarocolius yuracares neivae*]

Dr. Arthur Neiva (1880–1943) was an epidemiologist and biologist. He qualified in medicine in Rio de Janeiro (1903), and did entomological research at the Institute of Manguinhos. He organised the Medical Section of Zoology and Parasitology, Instituto Bacteriológico, Buenos Aires, for the Argentine government (1915). He returned to Brazil (1916), becoming Director of Public Health for São Paulo State, a member of the staff at Instituto Butantan, São Paulo, and then Director of the National Museum, Rio de Janeiro (1923). He was first Director, Institute Bacteriológico (1928–1932). After the revolution (1930) he held a number of appointments, including being Director-General of Research, Ministry of Agriculture. He entered politics (1933–1937) and then gave it up to resume his original research at the Institute of Manguinhos. A reptile is also named after him.

Nelicourvi

Nelicourvi Weaver *Ploceus nelicourvi* **Scopoli**, 1786
Nelicourvi Weaver *Ploceus sakalava* **Hartlaub**, 1861
[Alt. Sakalava Weaver]

This is an 'eponym' that originates from a mis-transcription, as it is not named after anyone or anything. The name comes from a corruption of a Sri Lankan word, 'nellukuruvi', which is a general name for finches/weavers. Despite the origin of the word, both the species are from Madagascar. The name may be found attached to the Sakalava Weaver because it has at times been considered conspecific with the Nelicourvi Weaver.

Nelson, E. W.

Nelson's Sparrow *Ammodramus nelsoni* **J. A. Allen**, 1875
Nelson's Gull *Larus nelsoni* **Henshaw**, 1884 NCR
[Hybrid: *Larus hyperboreus* x *Larus argentatus/ smithsonianus*]
Hooded Yellowthroat *Geothlypis nelsoni* **Richmond**, 1900
Yucatan Nightjar *Antrostomus nelsoni* **Ridgway**, 1912 NCR
[JS *Antrostomus badius*]
Nelson's Vireo *Vireo nelsoni* **Bond**, 1936
[Alt. Dwarf Vireo]

Nelson's Ptarmigan *Lagopus muta nelsoni* **Stejneger**, 1884
[Alt. Rock Ptarmigan ssp.]
Nelson's Oriole *Icterus cucullatus nelsoni* Ridgway, 1885
[Alt. Hooded Oriole ssp.]
Nelson's Downy Woodpecker *Picoides pubescens nelsoni*
Oberholser, 1896 NCR
[NUI *Picoides pubescens medianus*]
Great-tailed Grackle ssp. *Quiscalus mexicanus nelsoni*
Ridgway, 1901
Red-crowned Ant Tanager ssp. *Habia rubica nelsoni*
Ridgway, 1902
White-breasted Nuthatch ssp. *Sitta carolinensis nelsoni*
Mearns, 1902
Blue-grey Gnatcatcher ssp. *Polioptila caerulea nelsoni*
Ridgway, 1903
Grey-barred Wren ssp. *Campylorhynchus megalopterus
nelsoni* Ridgway, 1903
Brown-crested Flycatcher ssp. *Myiarchus magister nelsoni*
Ridgway, 1907 NCR
[JS *Myiarchus tyrannulus cooperi*]
Stripe-tailed Hummingbird ssp. *Eupherusa eximia nelsoni*
Ridgway, 1910
Bridled Tern ssp. *Onychoprion anaethetus nelsoni* Ridgway,
1911
Pale-billed Woodpecker ssp. *Campephilus guatemalensis
nelsoni* Ridgway, 1911
Common Pauraque ssp. *Nyctidromus albicollis nelsoni*
Ridgway, 1912 NCR
[JS *Nyctidromus albicollis yucatanensis*]
Nelson's Loggerhead Shrike *Lanius ludovicianus nelsoni*
Oberholser, 1918 NCR
[NUI *Lanius ludovicianus mexicanus*]
San Blas Jay ssp. *Cyanocorax sanblasianus nelsoni* **Bangs
& T. E. Penard**, 1919
Stripe-throated Hermit ssp. *Phaethornis adolphi nelsoni*
Bangs & **Barbour**, 1922 NCR
[JS *Phaethornis striigularis saturatus*]
Tawny-crowned Greenlet ssp. *Hylophilus ochraceiceps
nelsoni* **Todd**, 1929
Sharp-tailed Streamcreeper ssp. *Lochmias nematura
nelsoni* **Aldrich**, 1945
Red-winged Blackbird ssp. *Agelaius phoeniceus nelsoni*
Dickerman, 1965
Ruddy-capped Nightingale Thrush ssp. *Catharus frantzii
nelsoni* **A. R. Phillips**, 1969
Western Bluebird ssp. *Sialia mexicana nelsoni* A. R. Phillips,
1991

Edward William Nelson (1855–1934) was an outstanding American naturalist. As a young man he was sent as a weather observer to Alaska (1877). Although his major objective was to make meteorological observations, he was also tasked to '… obtain all the information possible on the geography, ethnology, and zoology of the surrounding region'. With the help of Inuits, dog sleds, and kayaks, he explored (1877–1881) areas where no Europeans had been before. He was the naturalist on board the Corwin during its search for the missing Arctic exploration vessel Jeanette (1881) – this expedition was the first to reach and explore Wrangel Island. He was a member of the Death Valley Expedition (1890) and then conducted a field survey of Mexico (1891–1905) before returning to the Bureau of Biological Survey (1906–1929), including being Chief of the Bureau (1916–1927), and President of both the American Ornithologists' Union and the American Society of Mammalogists. His greatest lasting contribution was the Migratory Bird Treaty, which is still in force today. Fifteen mammals, an amphibian and five reptiles are named after him.

Nelson, G.

Vitelline Warbler ssp. *Dendroica vitellina nelsoni* **Bangs**, 1919

George Nelson (b.1873) was a botanist, lecturer, zoologist and photographer who became Chief Taxidermist at the Museum of Comparative Zoology, Harvard (1901–1946). He specialised in the fauna of Florida, spending his winters studying Brown Pelicans *Pelecanus occidentalis*; he acquired land and built a house there in 1910. He wrote a study of Pelican Island (1911) and the changes to its ecology after the 1910 hurricane. A turtle is named after him.

Neser

Capped Wheatear ssp. *Oenanthe pileata neseri* **J. Macdonald**, 1952

John Neser (DNF) was Civil Secretary to the South West Africa Government (Namibia) at the time of the British Museum Expedition (1949–1950). Macdonald stated that Neser '… showed great kindness to the members of the expedition'.

Nerly

Cuban Pewee ssp. *Contopus caribaeus nerlyi* Garrido, 1978

Dr Nerly Lorenzo Hernández (DNF) was a Cuban ornithologist and parasitologist at the Instituto de Zoología (Academia de Ciencias de Cuba). He wrote or co-wrote many articles and books including *Eficacia de algunos cesticidas en aves* (1969). He collected the holotype of the pewee.

Nestor

New Zealand parrot genus *Nestor* **Lesson**, 1830

Hoary-headed Grebe *Podiceps nestor* **Gould**, 1837 NCR
[JS *Poliocephalus poliocephalus*]

In Greek mythology, Nestor (c.1265–c.1185 BC) was King of Pylos and was present at the siege of Troy. He was noted for his wisdom and great age. In the Iliad he is about 70 years old – in times when men usually died in their thirties on the battlefield or from diseases, a very respectable age. Ovid

parodied his antique garrulity by making him 200 years old. Because of this Nestor became the symbol of longevity and age.

Neuhaus

Black-tailed Whistler ssp. *Pachycephala pectoralis neuhausi* **Stresemann**, 1934 NCR
[JS *Pachycephala melanura dahli*]

Father Karl Neuhaus (<1885–1942) was a German priest in charge of the Sacred Heart Mission at Namatanai, New Ireland (1911–1929). He was the first resident priest on the island of Lihir (1931), where the whistler holotype was taken. He 'disappeared' in mid-1942 and his fate remains uncertain: he was probably executed by the occupying Japanese forces. His ethnography *Beiträge zur Ethnographie der Pala, Mittel Neu Irland* was published posthumously (1962).

Neumann

Neumann's Waxbill *Estrilda thomensis* **Sousa**, 1888
[Alt. Cinderella Waxbill]
Woodpecker sp. *Dendromus neumanni* **Reichenow**, 1896 NCR
[Alt. Nubian Woodpecker; JS *Campethera nubica*]
Neumann's Starling *Onychognathus neumanni*
    B. Alexander, 1908
Neumann's Warbler *Hemitesia neumanni* **Rothschild**, 1908
[Alt. Short-tailed Warbler]
Uluguru Greenbul *Arizelocichla neumanni* **Hartert**, 1922

Pale-billed Hornbill ssp. *Tockus pallidirostris neumanni*
    Reichenow, 1894
Rufous-chested Swallow ssp. *Cecropis semirufa neumanni*
    Reichenow, 1901 NCR
[NUI *Cecropis semirufa gordoni*]
Crested Lark ssp. *Galerida cristata neumanni* Hilgert, 1907
African Grey Flycatcher ssp. *Bradornis microrhynchus
    neumanni* Hilgert, 1908
Village Indigobird ssp. *Vidua chalybeata neumanni*
    Alexander, 1908
Grey-headed Sparrow ssp. *Passer griseus neumanni*
    Zedlitz, 1908 NCR
[JS *Passer swainsonii*]
Neumann's Coucal *Centropus leucogaster neumanni*
    Alexander, 1908
[Alt Black-throated Coucal ssp.]
Dark Chanting Goshawk ssp. *Melierax metabates neumanni*
    Hartert, 1914
Neumann's Duyvenbode's Lory *Chalcopsitta duivenbodei
    syringanuchalis* Neumann, 1915
[Alt. Brown Lory ssp.]
Neumann's Orange-breasted Fig Parrot *Cyclopsitta
    gulielmitertii ramuensis* Neumann, 1915
Neumann's Ring-necked Parakeet *Psittacula krameri
    borealis* Neumann, 1915
Retz's Helmet-shrike ssp. *Prionops retzii neumanni* Zedlitz,
    1915 NCR
[JS *Prionops retzii graculinus*]
Eastern Violet-backed Sunbird ssp. *Anthreptes orientalis
    neumanni* Zedlitz, 1916 NCR; NRM
Yellow-breasted Apalis ssp. *Apalis flavida neumanni* Zedlitz,
    1916 NCR

[JS *Apalis flavida flavocincta*]

Trumpet Manucode ssp. *Phonygammus keraudrenii neumanni* Reichenow, 1918

Buffy Pipit ssp. *Anthus vaalensis neumanni* **Meinertzhagen**, 1920

Neumann's Red-cheeked Parrot *Geoffroyus geoffroyi minor* Neumann, 1922

Baglafecht Weaver ssp. *Ploceus baglafecht neumanni* **Bannerman**, 1923

African Dusky Flycatcher ssp. *Muscicapa minima neumanniana* **Grote**, 1924 NCR

[JS *Muscicapa adusta minima*]

Red-bellied Paradise-flycatcher ssp. *Terpsiphone rufiventer neumanni* **Stresemann**, 1924

White-bellied Kingfisher ssp *Corythornis leucogaster neumanni* Laubmann, 1926 NCR

[JS *Corythornis leucogaster bowdleri*]

Yellow-bibbed Fruit Dove ssp. *Ptilinopus solomonensis neumanni* Hartert, 1926

Neumann's Pearly Conure *Pyrrhura lepida anerythra* Neumann, 1927

[Alt. Pearly Parakeet]

Black-bellied Seedcracker ssp. *Pyrenestes ostrinus neumanni* Neunzig, 1928 NCR; NRM

Fan-tailed Widowbird ssp. *Euplectes axillaris neumanni* Neunzig, 1928 NCR

[JS *Euplectes axillaris phoeniceus*]

Dusky Moorhen ssp. *Gallinula tenebrosa neumanni* Hartert, 1930

Neumann's Macaw *Diopsittaca nobilis longipennis* Neumann, 1931

[Alt. Red-shouldered Macaw, Long-winged Macaw]

Blue Jewel-babbler ssp. *Ptilorrhoa caerulescens neumanni* **Mayr** & **de Schauensee**, 1939

Blue-crowned Parakeet ssp. *Aratinga acuticaudata neumanni* **E. R. Blake** & **Traylor**, 1947

[Syn. *Thectocercus acuticaudatus*]

Blackstart ssp. *Cercomela melanura neumanni* Ripley, 1952

Professor Oskar Rudolph Neumann (1867–1946) was a German ornithologist who collected widely in east and north-east Africa (1892–1894). He travelled through Somaliland and southern Ethiopia with Erlanger (q.v.) (1900–1901). He was bankrupt (1908), and worked for a few months that year at the Rothschild Museum in Tring, England, but due to Walter Rothschild's (q.v.) own financial difficulties he had to leave, after which he became a stockbroker in Berlin. Although he received the Iron Cross (WW1) when an officer, he had to flee from the Nazi regime (1941). Via Switzerland and Cuba he reached the USA, where he worked the last few years of his life for the Field Museum of Natural History in Chicago. The results of his expeditions to Africa were published in the *Journal für Ornithologie*. Neumann's name became attached to the waxbill when he described it as *E. cinderella* (1908), but Sousa's description takes priority. Neumann's Duyvenbode's Lory is the only example, which we know of where a vernacular name simultaneously honours two persons. Three mammals, two reptiles and an amphibian are named after him.

Neumayer

Neumayer's Rock Nuthatch *Sitta neumayer* **Michahelles**, 1830

[Alt. Western Rock Nuthatch]

Franz Neumayer (1791–1842) was an Austrian botanist, entomologist, ornithologist and natural history dealer who lived in Ragusa (Dubrovnik) and collected in Dalmatia (Croatia).

Neven

Bar-backed Partridge ssp. *Arborophila brunneopectus neveni* **Delacour**, 1926 NCR

[JS *Arborophila brunneopectus henrici*]

A. Neven (DNF) was the French Director of Saigon Zoological Gardens.

Nevermann

Grey-crowned Munia *Lonchura nevermanni* **Stresemann**, 1934

Dr Hans Nevermann (1902–1982) was a German ethnologist. He was employed at the Berlin Museum of Ethnology (1926–1928) and by museums in Hamburg and Dresden (1928–1932). He was on an expedition to New Guinea, New Caledonia, the Loyalty Islands and the New Hebrides (1933–1934). He was head of the Indian Department of the Berlin Ethnology Museum (1945–1953), but ill health forced his early retirement (1957). He was the author of a number of books on anthropology including, with Margaret Trowell, *African and Oceanic Art* & *Bibliography of Nigerian Sculpture* (1968). He also wrote (1938) about an expedition to the Marshall Islands in 1910. He wrote numerous articles and books including, with many others, *The Religion of the Pacific and Australia* (1968).

Newell

Newell's Shearwater *Puffinus newelli* **Henshaw**, 1900

Brother Matthias Newell (1854–1939) was a missionary to Hawaii (1886–1924). He obtained the type specimen of the shearwater (1894).

Newman

Rusty Sparrow ssp. *Aimophila rufescens newmani* **A. R. Phillips**, 1966 NCR

[JS *Aimophila rufescens pyrgitoides*]

Dr Robert James Newman (1907–1988) was a zoologist and ornithologist. University of Pennsylvania awarded his BA in English (1927). He was first Curator of Birds at the Zoological Museum, Louisiana State University, which also awarded his MS (1951) and doctorate (1956) and where he worked until retirement (1976). He made a number of field trips to Mexico with his wife, but ill health curtailed this (1960s). He had a life-long collaboration with G. H. Lowery Jr (q.v.), particularly studying migration. He was Treasurer of AOU (1964). He spent his last two decades birding whenever he could and a friend and colleague, H. D. Pratt, described him as 'a birder's ornithologist and an ornithologist's birder'. Two reptiles are named after him and his wife, Marcella.

Newton, A.

Newton's Parakeet *Psittacula exsul* Newton, 1872 EXTINCT
Newton's Golden Bowerbird *Prionodura newtoniana*
   De Vis, 1883
[Alt. Golden Bowerbird]
Leaf Warbler sp. *Phylloscopus newtoni* **Gätke**, 1889 NCR
[Alt. Lemon-rumped Warbler; JS *Phylloscopus chloronotus*]
Newton's Scrub Warbler *Bradypterus alfredi* **Hartlaub**, 1890
[Alt. Bamboo Warbler]
Newton's Owl *Strix newtoni* **Rothschild**, 1907 NCR
   (EXTINCT)
[Alt. Mauritius Owl; JS *Mascarenotus sauzieri*]

Puerto Rican Screech Owl ssp. *Megascops nudipes*
   newtoni **Lawrence**, 1860
Bananaquit ssp. *Coereba flaveola newtoni* **Baird**, 1873
Maui Alauahio ssp. *Paroreomyza montana newtoni*
   Rothschild, 1893
[Sometimes regarded as monotypic]
Great Tit ssp. *Parus major newtoni* Prazák, 1894

Professor Alfred Newton FRS (1829–1907) was a Geneva-born British zoologist and co-founder of the BOU (1858). He studied ornithology in Lapland, Iceland, the West Indies, and North America (1854–1865), and became Professor of Zoology and Comparative Anatomy at Cambridge (1866–1907). He was a winner of both the Royal Medal of the Royal Society and the Gold Medal of the Linnean Society. He edited *Ibis* (1865–1870). He had a very low opinion of John Gould's *Birds of Great Britain,* which he was asked by Gould to look over, and wrote to his brother: '… the utter ignorance they sometimes betray is amazing. He has no personal knowledge of any English birds, except those between Eton and Maidenhead, and about those species he fancies no one else knows anything'. Although it is seldom acknowledged, he was instrumental in launching the bird protection movement in England and the rest of the world. For example, he said 'Fair and innocent as the snowy plumes may appear on a lady's hat, I must tell the wearer the truth – she bears the murderer's brand on her forehead' (1886). He studied the vanishing birds of the Mascarene Islands – a task made all the easier by the fact that his brother was appointed Assistant Colonial Secretary on Mauritius (1859).He was generally interested in extinct birds and spent a considerable amount of time and energy over a thirty year period in searching for signs that the Great Auk *Pinguinus impennis* was not in fact extinct. His travels came to an end after a fall on Heligoland left him unable to walk without the aid of two sticks. He wrote *Zoology of Ancient Europe* (1862) and *A Dictionary of Birds* (1893) with Hans Gadow. The bananaquit subspecies, despite its binomial being in the singular, appears to be named after Alfred and his brother, Sir Edward Newton (below). A mammal is named after him.

Newton, E.

Newtonia genus *Newtonia* **Schlegel**, 1867

Madagascar Swamp Warbler *Acrocephalus newtoni*
   Hartlaub, 1863
Newton's Kestrel *Falco newtoni* **Gurney**, 1863
[Alt. Madagascar Kestrel, Malagasy Spotted Kestrel]

Réunion Cuckooshrike *Coracina newtoni* **Pollen**, 1866
Mascarene Coot *Fulica newtoni* **Milne-Edwards**, 1867
   EXTINCT
Mascarene White-eye sp. *Zosterops enewtoni* **Hartlaub**,
   1877 NCR
[Alt. Reunion Grey White-eye; JS *Zosterops borbonicus*]

Sir Edward Newton (1832–1897) was, according to Hartlaub (1862) '… a gentleman who has recently visited Madagascar, and whose zealous efforts have very materially forwarded our knowledge of the ornithology of the east African archipelago'. Edward's brother Alfred (q.v.) was Professor of Zoology and Comparative Anatomy at Cambridge (1866–1907) whilst Edward was a colonial administrator in Mauritius (1859–1877). Edward sent many specimens to his brother, including remains of the Dodo *Raphus cucullatus*, although he sent the kestrel to the raptorphile Gurney. Edward wrote 'On a collection of birds from the Island of Anjuan',and the brothers jointly published 'On the osteology of the Solitaire' (1869).

Newton, F.

Newton's Sunbird *Anabathmis newtonii* **Bocage**, 1887
[Alt. Yellow-breasted Sunbird]
Newton's Fiscal Shrike *Lanius newtoni* Bocage, 1891
[Alt. São Tomé Fiscal]

Grey-backed Cisticola ssp. *Cisticola subruficapilla newtoni*
   da Rosa Pinto, 1967

Colonel Francisco Xavier Aguilar O'Kelly Azeredo Newton (1864–1909) was a Portuguese explorer and naturalist who collected on São Tomé and other islands in the Gulf of Guinea (1885–1895), and in Timor (1896). He wrote accounts of his travels and findings, and was ahead of his time in recording especially meticulous detailed information on the localities and ecology of the specimens he collected. Two amphibians and two reptiles are named after him.

Newton, R.

Plain Martin ssp. *Riparia paludicola newtoni* **Bannerman**,
   1937

Robert Newton (1908–1983) was a British colonial administrator in the Cameroons (1931–1937) and Mauritius (1955–1961).

Nice

Bewick's Wren ssp. *Thryomanes bewickii niceae* **G. Sutton**,
   1934 NCR
[JS *Thryomanes bewickii cryptus*]
Song Sparrow ssp. *Melospiza melodia niceae* **Dickerman**,
   1963 NCR
[JS *Melospiza melodia mexicana*]

Dr Margaret Morse Nice (1883–1974) was an American ornithologist, behaviourist and child psychologist. Mount Holyoke College awarded her bachelor's degree (1906) and Clark University, Worcester, Massachusetts, her master's (1915). She was the first female President of the Wilson Ornithological Society (1937–1939) and the leading authority on the Song Sparrow, writing *Studies in the Life History of the Song Sparrow* (1937).

Nicéforo

Nicéforo's Wren *Thryophilus nicefori* **Meyer de Schauensee**, 1946

Nicéforo's Pintail *Anas georgica niceforoi* **Wetmore** & **Borrero**, 1946 EXTINCT
[Alt. Yellow-billed Pintail ssp.]
White-eyed Parakeet ssp. *Aratinga leucophthalma nicefori* Meyer de Schauensee, 1946
[Syn. *Psittacara leucophthalma nicefori*]
Spot-billed Ground Tyrant ssp. *Muscisaxicola maculirostris niceforoi* **Zimmer**, 1947

Brother Nicéforo Maria (1888–1980) was a Frenchman originally named Antoine Rouhaire, who became a missionary in Colombia under his monastic name. He went to Medellin (1908) and was given the task of forming a natural history museum (1913). He was primarily a herpetologist and an excellent taxidermist. He is also remembered in the names of a mammal, ten amphibians, and seven reptiles.

Nicholson

Nicholson's [Rock] Pipit *Anthus similis nicholsoni* **Sharpe**, 1884
[Alt. Long-billed Pipit ssp.]

Francis Nicholson (1843–1925) was a cotton merchant and active member of the Zoological Society of London. He wrote on African species, as well as those of the English Lake District, but never visited Africa.

Nicole

Tropical Shearwater ssp. *Puffinus bailloni nicolae* **Jouanin**, 1971

Nicole Jouanin is the wife of the describer, Christian Jouanin (q.v.).

Nicoll

Nicoll's Weaver *Ploceus nicolli* **W. L. Sclater**, 1931
[Alt. Usambara Weaver, Tanzanian Mountain Weaver]
Booby sp. *Sula nicolli* **C. H. B. Grant** & Mackworth-Praed, 1933 NCR
[Alt. Red-footed Booby; JS *Sula sula rubripes*]

Lesser Short-toed Lark ssp. *Calandrella rufescens nicolli* **Hartert**, 1909
Great Frigatebird ssp. *Fregata minor nicolli* **Mathews**, 1914
Sand Partridge ssp. *Ammoperdix heyi nicolli* Hartert, 1919

Michael John Nicoll (1880–1925) was an assistant to R. E. Moreau (q.v.) when the latter was working in Egypt. He wrote 'The Water Pipit as a visitor to England' in *The Zoologist* (1906), and *Three Voyages of a Naturalist* (1908), which covered 1902–1906 and 72,000 miles – all on board the Earl of Crawford's yacht *Valhalla*. He was assistant to Major S. S. Flower (q.v.) at the Egyptian Government's zoological gardens at Giza (1906) and acted as Flower's successor (1923–1924) but resigned as his health forced him to retire to England. He

published *Handlist of the Birds of Egypt* (1919) and started work on *Birds of Egypt*, which was completed by Meinertzhagen and published in 1930. Nicoll is also famed for a description of a 'sea serpent' while (1905), off the coast of Brazil aboard *Valhalla*. A fellow voyager, Meade-Waldo (q.v.), noticed a large, six-foot-long 'fin or frill' in the water about a hundred yards from the boat. Looking more closely, he could see a large body beneath the surface. Just as he got out his binoculars, a huge head and neck rose up seven to eight feet out of the water and as thick as 'a slight man's body'; the head was about the same thickness and resembled a turtle's, as did the eye. Both head and neck were dark brown on top, whitish underneath. Nicoll's account of the beast was similar to Meade-Waldo's except his impression was of a mammal, not a reptile, although he was not certain.

Niedieck

European Goldfinch ssp. *Carduelis carduelis niediecki* **Reichenow**, 1907
Common Waxbill ssp. *Estrilda astrild niediecki* Reichenow, 1916

Paul Niedieck (DNF) was a German big-game hunter and explorer. He travelled very widely, as evidenced by the title of his book: *With Rifle in Five Continents* (1908) and *Cruises in The Bering Sea: Being Records of Further Sport and Travel* (1909). He was wounded in the Sudan (1902) when an elephant trod on his left foot and crushed it.

Niethammer

Spotted Great Rosefinch *Carpodacus rubicilla niethammeri* Keve, 1943 NCR
[JS *Carpodacus severtzovi*]
Asian Short-toed Lark ssp. *Calandrella cheleensis niethammeri* **Kumerloeve**, 1963
Andean Tinamou ssp. *Nothoprocta pentlandii niethammeri* **Koepcke**, 1968
Zebra Waxbill ssp. *Amandava subflava niethammeri* **Wolters**, 1969
Great Tit ssp. *Parus major niethammeri* **von Jordans**, 1970

Dr Günther T. Niethammer (1908–1974) was a German zoologist and ornithologist who after graduating (University of Leipzig, 1933) worked in museums in Berlin, Vienna and Bonn. We infer that he may have travelled in China (1930s) from his co-authorship with Adolf von Jordans of *Eine Vogelausbeute aus Fukien* (1940). He joined both the Nazi Party and the SS (1937) and then the Waffen SS (1940). He was at Auschwitz, firstly as a guard and then appointed to be the camp ornithologist and published 'Observations on the avifauna of Auschwitz' (1942). He became (1943) a Research Fellow at the Sven Hedin Institute and of an institute concerned with ancestral heritage. After WW2 he was arrested by the Allies (1946) and imprisoned before being extradited to Poland. Here (1948) he was sentenced to 3 to 8 years imprisonment, but was released and returned to Germany (1950) and to his old job at Museum Alexander Koenig, Bonn, where he became Head of the Department of Ornithology (1950) and was appointed Professor (1957). He was joint editor with Stresemann of *Journal für Ornithologie*

(1956–1961), then sole editor (1961–1970). He was President of the German Ornithological Society (1968–1973). (See also **Günther (Niethammer)**)

Nieuwenhuis

> Bronze Cuckoo sp. *Chalcococcyx nieuwenhuisi* **Vorderman**, 1898 NCR
> [Alt. Little Bronze Cuckoo; JS *Chrysococcyx minutillus*]
> Nieuwenhuis's Bulbul *Pycnonotus nieuwenhuisii* **Finsch**, 1901
> [Alt. Blue-wattled Bulbul; possibly of hybrid origin]

Dr Anton Willem Nieuwenhuis (1864–1953) was a physician in the Dutch East Indian Army (1889–1901) and in Borneo (1893–1900). He led an expeditions to central Borneo (1894 & 1896–1897 & 1898–1900). He was also an ethnologist and specialised in collecting tribal items and artefacts. He became Professor of Geography and Ethnolgy at Leiden University (1904–1934). He wrote *In Central Borneo* (1900). A reptile is named after him.

Nigell

> Nigell's Pheasant Grouse *Tetraogallus nigelli* **Jardine &** Selby, 1829 NCR
> [Alt. Caspian Snowcock; JS *Tetraogallus caspius*]

The original description states: 'It has been named Loph. Nigelli in remembrance of the individual by whom it was first transmitted to Europe'. This was Sir John Macneill (q.v.) and *nigelli* is a latinised form of Neill (originally *Niall* in its Gaelic form) – a later ornithologist no doubt misconstrued this as referring to a surname 'Nigell'. (See **Macneill**)

Nijo

> Nightingale Reed Warbler ssp. *Acrocephalus luscinius nijoi* **Yamashina**, 1940

Baron T. Nijo (fl.1940) was a Japanese collector and naturalist.

Nikerson

> Grey-headed Sparrow *Passer nikersoni* **Madarász**, 1911 NCR
> [JS *Passer griseus*]

Colonel George Snyder Nickerson (1873–1911) was a physician who qualified at Manchester (1896) and served in the Royal Army Medical Corps (1897–1909). He took part in the Nile Expedition (1899) and was at the battle of Omdurman. He was Governor of Sennar Province in the Sudan (1909–1911). He died after falling off his horse.

Nikolsky

> Eurasian Eagle Owl ssp. *Bubo bubo nikolskii* **Zarudny**, 1905
> Ural Owl ssp. *Strix uralensis nikolskii* **Buturlin**, 1907
> European Goldfinch ssp. *Carduelis carduelis nikolskii* **Moltchanov**, 1917 NCR
> [JS *Carduelis carduelis colchica*]

Dr Alexander Mikhailovich Nikolsky (1858–1942) was a herpetologist and zoologist. He studied at St Petersburg University (1877–1881), taking his doctorate in the year (1887) he became Assistant Professor and a curator of the zoological collection there. He was Director of the Department of Herpetology, Natural History Museum, Russian Academy of Sciences (1895–1903), then Professor, Kharkov University, Ukraine (1903). He made a number of expeditions to the Caucasus Mountains, Iran, Siberia and Japan (1881–1891). Today in Russia the A. M. Nikolsky Herpetological Society commemorates him. Nine reptiles are named after him.

Niles

> Grey-backed Tailorbird ssp. *Orthotomus derbianus nilesi* **Parkes**, 1988

Dr David M. Niles was head of the Ornithology Department, Delaware Museum of Natural History, Greenville, USA, when he wrote, jointly with Parkes, 'Notes on Philippine birds' (1988). He was a graduate student at the Museum of Southwestern Biology, New Mexico (1960s). The University of Kansas awarded his doctorate (1972).

Niobe

> Widowbird genus *Niobe* **Reichenbach**, 1862 NCR
> [Now in *Euplectes*]

In Greek mythology, Niobe was the daughter of Tantalus and wife of Amphion, King of Thebes, by whom she had 14 children. She made the mistake of teasing the goddess Latona for only having two. Latona's children were Artemis and Apollo, gods who did not take kindly to anyone insulting their mother. They killed all of Niobe's sons and daughters; it is unwise to mock the gods! Niobe, who is the personification of maternal grief, wept continuously until she died and was transformed into a stone, from which water ran. William Shakespeare put it rather well: 'Like Niobe, all tears' (*Hamlet* 1.2.149). Three mammals are named after her.

Nisbett

> Silver Pheasant sp. *Gennæus nisbetti* **Oates**, 1903 NCR
> [Hybrid of *Lophura nycthemera* subspecies]

Lieutenant-Colonel W. G. Nisbett (1865–1920) was a British army officer in India (1889–1919). He collected the pheasant holotype in the Kachin Hills, Burma (Myanmar), whilst in the Military Police there.

Niven, C. & J.

> Red-capped Lark ssp. *Calandrella cinerea niveni* **J. Macdonald**, 1952
> Cinnamon-breasted Bunting ssp. *Emberiza tahapisi nivenorum* **Winterbottom**, 1965

Cecily Kathleen Niven *née* FitzPatrick (1899–1992) was a South African ornithologist who founded the Percy FitzPatrick Institute of African Ornithology (1960), named after her father, Sir James Percy FitzPatrick. Her husband (1923) Jack P. Mackie Niven (d.1985) was a South African businessman. The Niven library is named after the whole family (1981), including their sons Patrick, Dan and Desmond in recognition of the contribution the family made to African ornithology through the FitzPatrick Institute initiative and their financial and moral support. The lark is named after her (despite the masculine trinomial); the bunting is named after both her and her husband.

Noceda

> Antthrush sp. *Chamaeza nocedae* **M. Bertoni**, 1901 NCR
> [Alt. Short-tailed Antthrush; JS *Chamaeza campanisona*]

Father Pedro Bias Noseda (DNF) was the Curate of a small Paraguayan village near San Ignacio Guazu in southern Paraguay (1781–1800) and was a friend of the naturalist Félix Manuel de Azara (q.v.), acting as his sometime assistant. He was regarded as the Gilbert White (q.v.) of his country and Azara had the benefit of his observations.

Noguchi

> Noguchi's Woodpecker *Sapheopipo noguchii* **Seebohm**, 1887
> [Alt. Okinawa Woodpecker, Pryer's Woodpecker]

T. Noguchi (DNF) was a Japanese collector. When naming the species, Seebohm (q.v.) wrote that he did so 'according to Mr. Pryer's instructions' (see **Pryer, H. J. S.**), but gave no information about Noguchi.

Noomé

> Parasitic Weaver *Heliospiza noomeae* **Gunning**, 1907 NCR
> [JS *Anomalospiza imberbis*]

> Grey-backed Camaroptera ssp. *Camaroptera griseoviridis noomei* Gunning & **J. A. Roberts**, 1911 NCR
> [JS *Camaroptera brevicaudata sharpei*]
> Sombre Greenbul ssp. *Andropadus importunus noomei* J. A. Roberts, 1917 NCR
> [NUI *Andropadus importunus importunus*]
> Cardinal Woodpecker ssp. *Dendropicos fuscescens noomei* J. A. Roberts, 1924 NCR
> [NUI *Dendropicos fuscescens intermedius*]
> Cut-throat Finch ssp. *Amadina fasciata noomei* J. A. Roberts, 1932 NCR
> [JS *Amadina fasciata meridionalis*]

F. O. Noomé (DNF) was a South African taxidermist and collector, often with Roberts (q.v.), for the Transvaal Museum. The feminine form of binomial used for the weaver (*noomeae*) is no error, as Gunning named it for 'Mrs. Noomé, who with her husband has been fortunate and kind enough to procure many a rare and interesting specimen for our Museum'.

Nordmann

> Nordmann's Greenshank *Tringa guttifer* Nordmann, 1835
> [Alt. Spotted Greenshank]
> Black-winged Pratincole *Glareola nordmanni* **J. G. Fischer**, 1842

Alexander von Nordmann (1803–1866) was a Finnish-born biologist who was interested in everything from palaeontology to botany and birds to molluscs. He collected extensively in southern Russia, which is where the majority

of his travelling was done, except for a trip (1830) to the Harz Mountains in Germany, where he met Blumenbach (q.v.). He went to Berlin (1827) and became a Professor at Odessa (1832), finally becoming Professor of Zoology at Helsinki University (1849). He travelled very extensively over more than 40 years – 26 trips, some perhaps for pleasure, such as Paris, Brussels and London (1850). The Nordmann Fir *Abies nordmanniana* is also named after him.

Norman

> Sunda Cuckooshrike ssp. *Coracina larvata normani* **Sharpe**, 1887

George Cameron Norman (1861–1910) was a British ornithologist. He was elected to the BOU (1887) and wrote a number of papers in its journal *Ibis*, such as 'Note on the geographical distribution of the Crested Cuckoos' (1887).

Norman (River)

> Little Shrike-thrush ssp. *Colluricincla megarhyncha normani* **Mathews**, 1914
> Zitting Cisticola ssp. *Cisticola juncidis normani* Mathews, 1914

Named after the Norman River, Queensland, Australia, which was named after master mariner Captain William Henry Norman (1812–1869).

Norris

> Sardinian Warbler ssp. *Sylvia melanocephala norrisae* **Nicoll**, 1917 EXTINCT

Mrs Norris Nicoll *née* Lyon (fl.1917) was the wife of the describer, Michael John Nicoll (q.v.). This warbler subspecies, from Faiyum, Egypt, is believed extinct (c.1940).

North

> Australian Parrot genus *Northipsitta* **Mathews**, 1912 NCR
> [Now in *Polytelis*]
> Bluebonnet genus *Northiella* Mathews, 1912

> North's Crake *Porzana atra* North, 1908
> [Alt. Henderson Crake, Henderson Island Crake]

> North's Parakeet *Barnardius zonarius macgillivrayi* North, 1900
> [Alt. Australian Ringneck ssp.]
> Red-tailed Black Cockatoo ssp. *Calyptorhynchus banksii northi* Mathews, 1912 NCR
> [JS *Calyptorhynchus banksii banksii*]
> Broad-tailed Thornbill ssp. *Acanthiza pusilla northi* Mathews, 1922 NCR
> [Intergrade between *Acanthiza apicalis apicalis* and *A. apicalis whitlocki*]
> Chestnut-breasted Munia ssp. *Donacola castaneothorax northi* Mathews, 1923 NCR
> [JS *Lonchura castaneothorax castaneothorax*]
> Golden-backed Honeyeater ssp. *Melithreptus laetior northi* Mathews, 1923 NCR
> [JS *Melithreptus gularis laetior*]

White-throated Treecreeper ssp. *Cormobates minor northi*
Mathews, 1923 NCR
[JS *Cormobates leucophaea minor*]

Alfred John North (1855–1917) was an Australian ornithologist and oologist at the Australian Museum, Sydney, having been temporarily employed to arrange an egg collection (1886), then becoming Assistant in Ornithology (1891–1917). He started working life as a jeweller. He wrote *Descriptive Catalogue of the Nests and Eggs of Birds Found Breeding in Australia and Tasmania* (1889) and *List of the Insectivorous Birds of New South Wales* (1897). He was a founding member of the Field Naturalists' Club of Victoria.

Northcott

Bee-eater sp. *Merops northcotti* **Sharpe**, 1900 NCR
[Alt. Blue-headed Bee-eater; JS *Merops (muelleri) mentalis*]

Lieutenant-Colonel Henry Ponting Northcott (1856–1899) was a British army officer (1876–1899). He served in Zululand (1883) and the Gold Coast (Ghana) (1895–1899), and was killed in action at the Battle of Modder River, South Africa (1899) during the Boer War.

Northrop

Northrop's Oriole *Icterus northropi* **J. A. Allen**, 1890
[Alt. Bahama Oriole]

John Isaiah Northrop (1861–1891) taught botany and zoology at Columbia University. He was the husband (1889) of Alice Rich Northrop (1864–1922). They spent six months in the Bahamas collecting animal, plant and mineral specimens (1890), then the most extensive natural history survey undertaken there. When she finished her analysis of the botanical material, ten years later, Alice found she had discovered 18 new species. *A Naturalist in the Bahamas* (1910) was a collection of John's and Alice's papers, edited by Henry Fairfield Osborn, and published posthumously under the names of Northrop and Osborn as co-authors, John was killed in a laboratory explosion (1891) a week before the birth of their only child, a son, John Howard Northrop (1891–1987) who won a Nobel Prize for Chemistry (1946). She travelled widely in the Americas and became a Professor at Hunter College. She was killed when her car stalled on a level crossing and was hit by a train.

Norton

Spectacled Prickletail ssp. *Siptornis striaticollis nortoni*
Graves & **Robbins**, 1987

Dr David William Norton (b.1944) is an ecologist, biologist and ornithologist who has spent most of his working life in the Arctic region. He graduated from Harvard, after which he collected in Ecuador and wrote 'Notes on some non-passerine birds from eastern Ecuador' (1965). He went to Alaska for graduate study in animal physiology at the University of Alaska, Fairbanks. He joined the Alaska Department of Fish and Game (1973), monitored the construction of the Trans-Alaska Pipeline (1974–1977), and was an environmental research manager (1978–1985) while the US federal government scrambled to lease Outer Continental Shelf (OCS) submerged lands to the petroleum industry for exploration drilling and production. The grounding of the 'Exxon Valdez' and the resulting oil spill meant he was invited to Barrow Community College to develop a Natural Sciences department. He retired (1999) but now, inter alia, operates river craft to support teams investigating Cretaceous dinosaurs of Alaska.

Noska

Whinchat ssp. *Saxicola rubetra noskae* **Tschusi**, 1902 NCR;
NRM

Max Noska (DNF) was a Russian forestry officer in the Caucasus. He wrote several monographs with Tschusi (q.v.) such as *Das Kaukasische Birkhuhn* (1895).

Nouhuys

Large Scrubwren *Sericornis nouhuysi* **Oort**, 1909
Short-bearded Honeyeater *Melidectes nouhuysi* Oort, 1910

Captain Jan Willem van Nouhuys (1869–1963) was a Dutch naval officer (1888–1915) commanding the ships on expeditions in New Guinea (1903, 1907 and 1909). He carried out independent biological and geological research on the islands of Sula, Indonesia, and was one of the first to reach the eternal snows in the mountains of tropical New Guinea. He returned to Holland and was Director of the Museum of Asian and Caribbean Studies and the Maritime Museum Prins Hendrik (1915–1934). A mammal is named after him.

Novaes

Novaes's Foliage-gleaner *Philydor novaesi* Teixeira &
Gonzaga, 1983
[Alt. Alagoas Foliage-gleaner]

Rufous-collared Sparrow ssp. *Zonotrichia capensis novaesi*
Oren, 1985
Least Nighthawk ssp. *Chordeiles pusillus novaesi*
Dickerman, 1988

Dr Fernando da Costa Novaes (1927–2004) was a Brazilian ornithologist based at the Museu Paraense Emílio Goeldi, Belém, where he assembled a huge collection of bird skins and skeletons. Novaes wrote several books and many papers on the avifauna of various areas of Brazil.

Nuna

Green-tailed Trainbearer *Lesbia nuna* **Lesson**, 1832

Lesson's (q.v.) original description explains that he is naming this hummingbird after Nouna-Koali, a graceful young female character in the romantic novel *Ismael Ben Kaïzar ou la Découverte du Nouveau-Monde* (1829). The author of this work, Ferdinand Denis, was a friend of Lesson's.

Nuñez

Nuñez's Saucerottia *Saucerottia nunezi* **Boucard**, 1892 NCR
[Based on a melanistic specimen of *Amazilia cyanifrons*]

Rafael Wenceslao Núñez Moledo (1825–1894) was President of Colombia (1880–1882, 1884–1888 and 1892–1894).

Nuttall

Tyrant flycatcher genus *Nuttallornis* **Ridgway**, 1887 NCR
[Now in *Contopus*]

Nuttall's Lesser Marsh Wren *Cistothorus platensis* **Latham**, 1790
[Alt. Sedge Wren]
Yellow-billed Magpie *Pica nuttalli* **Audubon**, 1837
Nuttall's Woodpecker *Picoides nuttallii* **Gambel**, 1843
Nuttall's Poorwill *Phalaenoptilus nuttallii* Audubon, 1844
[Alt. Common Poorwill]

Nuttall's White-crowned Sparrow *Zonotrichia leucophrys nuttalli* Ridgway, 1899

Thomas Nuttall (1786–1859), an English botanist and zoologist, collected for Pennsylvania University. He went to the USA (1808), but when war between Britain and America seemed imminent he returned to England. He returned to America (1815), published *The Genera of North American Plants* (1818) and became curator of the botanical gardens at Harvard University (1825). He was the basis for the character 'Old Curious', the naturalist in Richard Henry Dana's *Two Years Before the Mast*. Nuttall was the first to publish a small, inexpensive field guide to North American birds, and also wrote *Manual of the Ornithology of the United States and Canada* (1832). Audubon (q.v.) in his *Ornithological Biography* (1831) specifically named the wren for Nuttall, using the following words: 'I hope, kind reader, you will approve of the liberty which I have taken in prefixing the name of my friend NUTTALL to the present species, which was discovered by his indefatigable and enthusiastic devotion to science'. Two mammals are named after him.

Nutting

Nutting's Flycatcher *Myiarchus nuttingi* **Ridgway**, 1882
[Alt. Pale-throated Flycatcher]
Nicaraguan Seed Finch *Oryzoborus nuttingi* Ridgway, 1884

White-tipped Dove ssp. *Leptotila verreauxi nuttingi* Ridgway, 1915

Charles Cleveland Nutting (1858–1927) was an American naturalist and collector. He became Curator of the Museum of Natural History of the University of Iowa (1886), and worked to build up the museum's collection further when he became Professor of Systematic Zoology (1888). He generated public and private support to finance several expeditions. His journeys to the Bay of Fundy, the Bahamas, Nicaragua and Costa Rica added many more specimens, including seabirds, to the collection. On these trips he was said to be an 'energetic, forceful character, his organizing abilities and his enthusiasm for collecting helped to ensure the success of his trips.' As well as birds, Nutting was particularly interested in hydrozoans (relatives of jellyfish and corals).

Nyctimene

Owl genus *Nyctimene* **Heine** & **Reichenow**, 1890 NCR
[Now in *Strix*]

In Greek mythology Nyctimene was the daughter of King Epopeus of Lesbos. Raped by her own father, she fled into the woods filled with shame and despair. Here the goddess Athena turned her into an owl (apparently an act of pity, though it is not exactly clear how this metamorphosis helped the situation).

Nye

West Indian Woodpecker ssp. *Melanerpes superciliaris nyeanus* **Ridgway**, 1886

Willard Nye Jr (DNF) was an ornithologist, naturalist and dealer in real estate in Massachusetts who sailed as a volunteer on the US Fisheries research vessel *Albatross*. He also collected in the Bahamas. He wrote 'A Bahamian bird (*Centurus nyeanus*) apparently extinct' (1899).

Nyman

Red-throated Myzomela ssp. *Myzomela eques nymani* **Rothschild** & **Hartert**, 1903

Dr Erik Olof August Nyman (1866–1900) was a Swedish botanist whose bachelor's degree (1889), master's (1895) and doctorate (1896) were all awarded by Uppsala University. He collected in Java (1897–1898) and New Guinea (1898–1899). He died in Munich on his way home to Sweden.

O

Oates

Rusty-naped Pitta *Hydrornis oatesi* **Hume**, 1873
Oates's Leaf Warbler *Phylloscopus davisoni* Oates, 1889
[Alt. Davison's Leaf Warbler, White-tailed Leaf Warbler]
Chin Hills Wren-babbler *Spelaeornis oatesi* **Rippon**, 1904

Vivid Niltava ssp. *Niltava vivida oatesi* **Salvadori**, 1887
Oates's Kalij Pheasant *Lophura leucomelanos oatesi*
  Ogilvie-Grant, 1893
Eurasian Jay ssp. *Garrulus glandarius oatesi* **Sharpe**, 1896
Pale-billed Parrotbill ssp. *Chleuasicus atrosuperciliaris
  oatesi* Sharpe, 1903
[Alt. Black-browed Parrotbill ssp.; Syn. *Paradoxornis
  atrosuperciliaris oatesi*]

Eugene William Oates (1845–1911) worked in the Public Works Department in British Colonial India and Burma (Myanmar), and was an amateur naturalist. When he moved back to England he became Secretary of the BOU (1898–1901). He wrote *The Fauna of British India* (1889). (See also **Eugene**). An amphibian and a reptile are named after him.

Oatley

Brown Scrub Robin ssp. *Erythropygia signata oatleyi*
  Clancey, 1956 NCR
[NUI *Erythropygia signata signata*]

Terence Barry Oatley (b.1934) is a South African ornithologist, ecologist and photographer who was with the Natal Parks, Game and Fish Preservation Board, Pietermaritzburg, and is now associated with the Avian Demography Unit, Statistical Sciences Department, University of Cape Town. He wrote 'Competition and local migration in some African Turdidae' (1966).

Obama

Western Striolated Puffbird *Nystalus obamai* Whitney *et al.*, 2013

Barack Hussein Obama II (b.1961), the 44th President of the United States of America (2009–2017) needs no biography here.

Ober

Lesser Antillean Flycatcher *Myiarchus oberi*
  G. N. Lawrence, 1877
Montserrat Oriole *Icterus oberi* Lawrence, 1880

Plain Antvireo ssp. *Dysithamnus mentalis oberi* **Ridgway**, 1908

Striped Owl ssp. *Pseudoscops clamator oberi* **Kelso**, 1936
[Syn. *Asio clamator oberi*]

Frederick Albion Ober (1849–1913) was an American writer and naturalist. Aged 13 he worked as a shoemaker (1862–1867) but then attended Massachusetts Agricultural College. He had to leave due to lack of funds, then worked in a drugstore and in shoemaking again (1867–1870). He abandoned his business pursuits (1872) to hunt in Florida, and was then in the Lesser Antilles (1876–1878) undertaking ornithological surveys before further travelling in Mexico (1881). He wrote more than 40 books over 30 years, mostly about travel, including *Camps in the Caribbees: The Adventures of a Naturalist in the Lesser Antilles*. He was also a founder member (1904) of the Explorers' Club.

Oberholser

Towee genus *Oberholseria* **Richmond**, 1915 NCR
[Now in *Pipilo*]

Oberholser's Tyrannulet *Phylloscartes pammictus*
  Oberholser, 1902 NCR
[Alt. Mottle-cheeked Tyrannulet; JS *Phylloscartes ventralis*]
Oberholser's Piculet *Picumnus arileucus* Oberholser, 1931
  NCR
[Alt. White-wedged Piculet; JS *Picumnus albosquamatus*]
Dusky Flycatcher *Empidonax oberholseri* **A. R. Phillips**,
  1939

Glossy Swiftlet ssp. *Collocalia esculenta oberholseri*
  Stresemann, 1912
Black-naped Monarch ssp. *Hypothymis azurea oberholseri*
  Stresemann, 1913
Large-tailed Nightjar ssp. *Caprimulgus macrurus oberholseri*
  Rothschild & **Hartert**, 1918 NCR
[NUI *Caprimulgus macrurus schlegelii*]
Oberholser's Fruit Dove *Ptilinopus subgularis epius*
  Oberholser, 1918
[Alt. Maroon-chinned Fruit Dove ssp.]
Brown-backed Solitaire ssp. *Myadestes occidentalis
  oberholseri* **Dickey** & **van Rossem**, 1925
Curve-billed Thrasher ssp. *Toxostoma curvirostre
  oberholseri* Law, 1928
Cinnamon-headed Green Pigeon ssp. *Treron fulvicollis
  oberholseri* **Chasen**, 1935
Dollarbird ssp. *Eurystomus orientalis oberholseri* **Junge**,
  1936
Colima Pygmy Owl ssp. *Glaucidium palmarum oberholseri*
  R. T. Moore, 1937

White-breasted Nuthatch ssp. *Sitta carolinensis oberholseri*
  H. W. Brandt, 1938
  [SII *Sitta carolinensis nelsoni*]
Carolina Wren ssp. *Thryothorus ludovicianus oberholseri*
  Lowery, 1940
Purple Martin ssp. *Progne subis oberholseri* Brandt, 1951
  NCR
  [JS *Progne subis hesperia*]

Dr Harry Church Oberholser (1870–1963) was an American ornithologist who worked for the United States Bureau of Biological Survey (US Fish and Wildlife Service) (1895–1941). Thereafter he worked as Curator of Ornithology at the Cleveland Museum of Natural History. He made extensive ornithological explorations in the USA and Canada. He wrote *Birds of Mt Kilimanjaro* (1905), 'Birds of the Anamba Islands' (1917) and *The Bird Life of Louisiana* (1938). He also wrote the massive two-volume work *The Bird Life of Texas*, published posthumously (1974). Louis Agassiz Fuertes (q.v.), who went with him on his collecting trip to Texas, illustrated the book.

Oberländer

Oberländer's Ground Thrush *Zoothera oberlaenderi* **Sassi**, 1914
  [Alt. Forest Ground Thrush]

Philipp (Filip) von Oberländer (1875–1911) was an industrialist who sponsored an expedition to the Congo. He was also a traveller in Africa and America, a buffalo hunter and sportsman who used to send 'trophies' to Czech castles and museums, including the Natural History Museum in Vienna where Sassi (q.v.) was Curator. He was born in Bohemia (Czech Republic) and died in the Sudan. He wrote: *Jagdfahrten in Nordamerika* (1911).

Oberon

Mallee Whipbird ssp. *Psophodes leucogaster oberon*
  Schodde & Mason, 1991

In Mediaeval mythology, Oberon was king of the fairies and Titania's husband. A reptile is also named after him.

O'Brien

Kaempfer's Woodpecker *Celeus obrieni* Short, 1973
  [Alt. Piauí Woodpecker]

Dr Charles E. O'Brien (1905–1987) was an American ornithologist who worked at the AMNH. where he was the ornithological collection manager in charge of the move of the 185 large packing cases containing the Rothschild collection of 280,000 bird skins that the AMNH acquired (1932). He became Assistant Curator (1954). He retired in 1972 after nearly 50 years at the museum. He was honoured because he had pointed out to Short an unidentified specimen of Brazilian woodpecker in the museum, collected earlier by Emil Kaempfer (q.v.) (1926).

Obst

Sulawesi Scops Owl ssp. *Otus manadensis obsti* **Eck**, 1973
  NCR
  [JS *Otus manadensis manadensis*]

Professor Dr Fritz Jürgen Obst (b.1939) is a German herpetologist. He studied psychology and biology at Heidelberg Univeristy and Hohenheim University, the latter awarding his doctorate (1996). He was Curator of Lower Vertebrates and Insects at the Zoological Gardens, Stuttgart (1990–1996) and became herpetologist at the State Natural History Museum, Dresden, becoming Deputy Director (1997) then Director (2001). He also taught at Leipzig University (1997). He wrote *Turtles, Tortoises and Terrapins* (1988). Three reptiles and an amphibian are named after him.

Oca

D'Oca's Amazilia *Amazilia ocai* **Gould**, 1859 NCR
  [Hybrid of *Amazilia cyanocephala* x *Amazilia beryllina*]
  Collared Towhee *Pipilo ocai* **G. N. Lawrence**, 1865

Rafael Montes de Oca (d.1880) was a Mexican naturalist and artist. He created a series of plates showing Mexican hummingbirds and orchids published many years later (1963). He was employed as a naturalist on the Mexican-Guatemalan Boundary Commission.

Ochoterena

Maroon-chested Ground Dove ssp. *Claravis mondetoura ochoterena* **van Rossem**, 1934

Dr Isaac Ochoterena Mendieta (1885–1950) was Professor of Histology and Embryology at, and Director of, Instituto de Biologia, Universidad Nacional de Mexico, Mexico City. He was a Lieutenant Colonel in the Mexican army, having been Professor of Histology at the Mexican Army Medical School. Three reptiles are named after him.

Ockenden

Great Thrush ssp. *Turdus fuscater ockendeni* **Hellmayr**, 1906

George Richard Ockenden (1868–1906) was a highly regarded professional collector, especially of lepidoptera. He undertook an expedition (1900–1906) to the Peruvian Andes to collect zoological specimens of all kinds for W. F. H. Rosenberg (q.v.). He was taken ill with typhoid while in the mountains, a long way from any hospital, and died before any help could reach him. An amphibian is named after him.

Octavia

Blue-throated Brown Sunbird ssp. *Cyanomitra cyanolaema octaviae* **Amadon**, 1953

Mrs Octavia Amadon *née* Gardella (DNF) was Dean Amadon's (q.v.) wife. The original description contains no etymology so this attribution is our assumption.

Odom

Neblina Metaltail *Metallura odomae* **Graves**, 1980

Mrs Babette Odom *née* Moore (1911–1984) was a wealthy American birdwatcher and one of the people after whom the Moore-Odom Wildlife Foundation was named (2004). She owned a number of farms in Orange, Texas, and sponsored field programmes in Peru.

Oenone

Golden-tailed Sapphire *Chrysuronia oenone* **Lesson**, 1832

In Greek mythology Oenone was a nymph who was loved by Paris, but was deserted by him in favour of Helen of Troy.

Ogawa

Brown-eared Bulbul ssp. *Microscelis amaurotis ogawae* **Hartert**, 1907
Eurasian Wren ssp. *Troglodytes troglodytes ogawae* Hartert, 1910
[Syn. *Nannus troglodytes ogawae*]
Great Tit ssp. *Parus major ogawai* **Momiyama**, 1923 NCR
[Alt. Japanese Tit ssp; JS *Parus minor minor*]

Dr Minori Ogawa (1876–1908) was a Japanese physician and ornithologist. He wrote *A Hand List of the Birds of Japan* (1908).

Ogilvie-Grant

Grant's Blue Flycatcher *Cyornis hainanus* Ogilvie-Grant, 1900
[Alt. Hainan Blue Flycatcher]
Ogilvie-Grant's Warbler *Phylloscopus subaffinis* Ogilvie-Grant, 1900
[Alt. Buff-throated Warbler]
Grant's Wood-hoopoe *Phoeniculus granti* **Neumann**, 1903
[Syn. *Phoeniculus damarensis granti*]
Grant's Starling *Aplonis mystacea* Ogilvie-Grant, 1911
[Alt. Yellow-eyed Starling]
Sunbird sp. *Anthreptes ogilviegranti* **Bannerman**, 1921 NCR
[Alt. Olive-bellied Sunbird; JS *Cinnyris chloropygius*]
Kloss's Leaf Warbler *Phylloscopus ogilviegranti* **La Touche**, 1922

Grey-headed Kingfisher ssp. *Halcyon leucocephala ogilviei* **C. H. B. Grant**, 1914 NCR
[JS *Halcyon leucocephala pallidiventris*]
Ruddy Pigeon ssp. *Patagioenas subvinacea ogilviegranti* **Chubb**, 1917
Double-spurred Francolin ssp. *Pternistis bicalcaratus ogilviegranti* Bannerman, 1922
Red-billed Brush-turkey ssp. *Talegalla cuvieri granti* **Roselaar**, 1994

William Robert Ogilvie-Grant (1863–1924) was a Scottish ornithologist. He was Curator of Birds at the BMNH (1909–1918), having started work there aged 19. He enlisted with the First Battalion of the County of London Regiment at the beginning of WW1 and suffered a stroke whilst helping to build fortifications near London (1916). He is famed for describing a number of well-known species, such as the huge Philippine Eagle *Pithecophaga jefferyi*. He wrote *A Hand-book to the Game Birds* (1895) and is also remembered in the name of a mammal. (Also see **Grant, Ogilvie**)

Ogle

Ogle's Laughingthrush *Garrulax nuchalis* **Godwin-Austen**, 1876
[Alt. Chestnut-backed Laughingthrush]

Ogle's Spotted Babbler *Stachyris oglei* Godwin-Austen, 1877
[Alt. Snowy-throated Babbler, Austen's Spotted Babbler]

M. J. Ogle (1842–1892) was a British surveyor in Assam, India. He collected natural history specimens whilst working for the Topographical Survey, and sent the birds to Godwin-Austen.

Ognev

Wallcreeper ssp. *Tichodroma muraria ognewi* **Portenko**, 1954 NCR
[JS *Tichodroma muraria nepalensis*]

Professor Sergei Ivanovich Ognev (1886–1951) was a Russian zoologist who specialised in mammals and their taxonomy. Moscow University awarded his doctorate (1910) and he stayed on as a member of the staff, becoming a professor (1928). Under the Soviet regime he was awarded the degree of Doctor of Science (1935). He was ranked as a Scientist of Merit, was twice awarded the Stalin Prize, and given the Order of Lenin. Based on his own expeditions and field studies he wrote a number of books, including a 7-volume work, *The Mammals of the USSR and Adjacent Countries* (1928–1950). Two mammals are named after him.

Okada

Willow Ptarmigan ssp. *Lagopus lagopus okadai* **Momiyama**, 1928
Whydah ssp. *Prosteganura haagneri okadai* **Yamashina**, 1930 NCR
[Hybrid: *Vidua paradisaea* x *Vidua chalybeata* or *Vidua funerea*]

Professor Dr Yaichiro Okada (1892–1976) was a herpetologist and ichthyologist who wrote over 400 books and papers dealing with nearly all branches of zoology, including *Ecology and Evolution of Reptiles* (1932). He was Curator of Marquis Yamashina's (q.v.) museum (1930).

Oken

Oken's Plover *Charadrius okeni* **Wagler**, 1827 NCR
[Alt. Piping Plover; JS *Charadrius melodus*]

Dr Lorenz Oken (*né* Okenfuss) (1779–1851) was a German physician, zoologist and natural philosopher, who graduated from Freiburg University (1804). He was Professor of Medicine at Jena (1807–1819), and Professor of Physiology at Munich University (1827–1832). Oken then became Professor of Natural History, Natural Philosophy and Physiology at the newly founded University of Zurich (1832) and soon after a citizen of Zurich (1835). He was editor of the journal *Isis* (1817–1848) and founded (1822) Gesellschaft Deutscher Naturforscher und Ärzte (Society of German Natural Scientists and Physicians), which still exists.

Olalla

Olalla's Wren *Odontorchilus olallae* **O. Pinto**, 1937 NCR
[Alt. Grey Wren; JS *Cantorchilus griseus*]
Maroon-banded Aracari *Pteroglossus olallae* **Gyldenstolpe**, 1942 NCR

[Hybrid: *Pteroglossus mariae* x *Pteroglossus inscriptus humboldti*]
Foothill Elaenia *Myiopagis olallai* Coopmans & Krabbe, 2000

Olalla's Blue-winged Parrotlet *Forpus xanthopterygius olallae* Gyldenstolpe, 1941 NCR
[NUI *Forpus xanthopterygius crassirostris*]
Red-necked Woodpecker ssp. *Campephilus rubricollis olallae* Gyldenstolpe, 1945

Alfonso Maria Olalla (d.1971) was an Ecuadorian professional collector, who lived in Brazil (mid-1930s) and took Brazilian citizenship. He was with Spillman (q.v.) collecting birds in Ecuador (1930). He sent many of his specimens, particularly of New World monkeys and squirrels, to the Museum of Zoology in Rio di Janeiro,.but also to United States and European institutions. He discovered the first specimen of the Masked Antpitta *Hylopezus auricularis* (1937). Collecting ran in the Olalla family: his father Carlos and his brothers Manuel, Rosalino and Ramón also collected, as did their sons. Two mammals are named after him and the other members of the Ollala family.

Oldean

Oldean's Thrush *Turdus abyssinicus oldeani* **W. L. Sclater** & **Moreau**, 1935
[Alt. Abyssinian/Mountain Thrush ssp.]

The type specimen was described from the Oldeani Forest in Tanganyika (Tanzania). We assume that someone mistook *oldeani* as an honorific for a person called Oldean, and mis-transcribed it as Oldean's Thrush rather than the Oldeani Thrush.

Olga

Calandra Lark ssp. *Melanocorypha calandra olgae* Gavrilenko, 1928 NCR
[JS *Melanocorypha calandra calandra*]

Olga Antonova Kapeller (1892–1975) was a botanist who collected and explored in the Caucasus region (c.1925).

Olivares

Chiribiquete Emerald *Chlorostilbon olivaresi* **Stiles**, 1996

Whooping Motmot ssp. *Momotus subrufescens olivaresi* Hernandez-Camacho & Romero-Zambrano, 1978

Fr. Antonio Olivares (1917–1975) was a Colombian ornithologist. He was professor of Natural Sciences at the Franciscan seminary in Bogota (1944–1947), then visited the USNM in Washington and met Alexander Wetmore (q.v.). He made a very important collection of birds of Colombia, held in the Institute of Natural Sciences at the National University, and undertook several expeditions for collecting purposes. These included a trip (1959) to the area of the Sierra de la Macarena, taking 551 specimens of 204 species and subspecies. The original etymology for the emerald reads: 'I take pleasure in naming this species for Fr. Antonio Olivares in honor of his many pioneering contributions to Colombian ornithology and his indefatigable labor in building the bird collection of the Instituto de Ciencias Naturales.'

Olive, E. A. C.

Olive's Buttonquail *Turnix olivii* **H. C. Robinson**, 1900
[Alt. Buff-breasted Buttonquail, Robinson's Buttonquail]

Olive-backed Sunbird ssp. *Cinnyris frenata olivei* **Mathews**, 1912 NCR
[JS *Cinnyris jugularis frenatus*]
Australian Owlet-nightjar ssp. *Aegotheles cristata olivei* Mathews, 1918 NCR
[JS *Aegotheles cristatus cristatus*]

Edmund Abraham Cumberbatch Olive (1844–1921) was born in England but emigrated to Australia (1870s). He arrived in Cooktown, north Queensland (c.1875), at the peak of the Palmer gold rush. He became an auctioneer and commission agent (1875), remaining thus employed until his death. He developed a natural history interest from his home, Mount Olive, outside Cooktown, with the help of an Aboriginal man known as Billy Olive. He accumulated impressive collections of various native fauna at his home and sent many specimens to Australian, European and American taxonomists, collectors and museums. He collected in Cooktown but also in Bellenden Ker and in New Guinea. Several other taxa were named in his honour including a reptile. His children Edmund Olive (1874–1923), Carlton Chaloner Olive (1881–1958), Herbert Lawrence Olive (b.1885) and John Henry Olive (b.1890) all pursued interests in natural history and sent specimens to Australian museums and to private collectors. A reptile is named after him.

Olive (Archer)

Somali Pigeon *Columba oliviae* **S. Clarke**, 1919

Lady Olive Archer (DNF) was the wife of Sir Geoffrey Francis Archer (q.v.).

Olive (MacLeod)

Crombec sp. *Sylviella oliviae* **B. Alexander**, 1908 NCR
[Alt. Northern Crombec; JS *Sylvietta brachyura carnapi*]

Olive MacLeod (1886–1936) was a traveller and explorer. She was engaged to Boyd Alexander (q.v.), who described this bird but included no etymology – so we cannot be completely sure that he was honouring his fiancée. She travelled to West Africa (1910) to discover more about the death of her fiancé and to see his grave. She later married Charles Lindsay Temple (1871–1929), a colonial administrator, traveller and author (1912). She wrote *Chiefs and Cities of Central Africa, across Lake Chad by the way of British, French and German Territories* (1912).

Oliver

Caspian Tern ssp. *Hydroprogne tschegrava oliveri* **Mathews** & **Iredale**, 1913 NCR
[Now *Hydroprogne caspia*; NRM]
Spotless Crake ssp. *Porzana tabuensis oliveri* Mathews & Iredale, 1914 NCR
[NUI *Porzana tabuensis tabuensis*]
Spotted Shag ssp. *Phalacrocorax punctatus oliveri* Mathews, 1930

Fairy Prion ssp. *Pseudoprion turtur oliveri* Mathews, 1932
NCR
[JS *Pachyptila turtur turtur*]

Dr Walter Reginald Brook Oliver (1883–1957) was a New Zealand ornithologist and avian palaeontologist. He wrote the definitive *New Zealand Birds* (1930) and revised it several times. He also wrote *The Moas of New Zealand and Australia* (1949). His research notes for *New Zealand Birds* (1920–1960) and his collected works (1910–1957) are in the archives of the Te Papa Tongarewa, the Museum of New Zealand at Wellington, where he was Director (1928) until his retirement (1947). He was described a: 'our last true biologist equally authoritative about animals or plants ...' by Professor G. T. S. Baylis. A reptile is named after him.

Olivier

Olivier's Rail *Amaurornis olivieri* **Grandidier** & **Berlioz**, 1929
[Alt. Sakalava Rail]

Guillaume C. Olivier (DNF) was a French zoologist who collected in Madagascar, although the rail might have been named after Marcel Achille Olivier (1879–1945), who was the Governor-General of Madagascar (1924–1929) in which latter year Grandidier and Berlioz described the bird.

Olmec

Scrub Euphonia ssp. *Euphonia affinis olmecorum*
Dickerman, 1981

Dickerman named this subspecies after the Olmec Indians of the Caribbean lowlands of Veracruz province in Mexico.

Olrog

Olrog's Gull *Larus atlanticus* Olrog, 1958
Olrog's Tyrannulet *Tyranniscus australis* Olrog & **Contino**,
1966 NCR
[Alt. Sclater's Tyrannulet; JS *Phyllomyias sclateri*]
Olrog's Cinclodes *Cinclodes olrogi* Nores & **Yzurieta**, 1979

Barred Forest Falcon ssp. *Micrastur ruficollis olrogi*
Amadon, 1964
Pale-crested Woodpecker ssp. *Celeus lugubris olrogi* **Fraga**
& **E. C. Dickinson**, 2008
[Replacement name for *Celeus lugubris castaneus* (Olrog,
1963)]

Claes Christian Olrog (1912–1985) was a Swedish ornithologist who lived in Argentina and wrote widely on South American birds. He published *Las Aves Argentinas* (1959) and an annotated Peruvian checklist, plus many articles on distribution and other topics. He also wrote *Destination Eldslandet* (*Destination Tierra del Fuego*) (1943). The Institute for the Administration of Protected Areas in Buenos Aires is named after him, as are two mammals.

Olson

Olson's Petrel *Bulweria bifax* Olson, 1975 EXTINCT
Olson's Shearwater *Puffinus olsoni* McMinn, Jaume &
Alcover, 1990 EXTINCT
[Alt. Lava Shearwater]

Ascension Night Heron *Nycticorax olsoni* Ashmole,
Simmons & **W. R. P. Bourne**, 2003 EXTINCT

Mangrove Vireo ssp. *Vireo pallens olsoni* **A. R. Phillips**,
1991 NCR
[NUI *Vireo pallens salvini*]
Common Bush Tanager ssp. *Chlorospingus flavopectus
olsoni* Avendaño, **Stiles** & Cadena, 2013

Dr Storrs Lovejoy Olson (b.1944) was Curator of the Division of Birds in the USNM (1975–2009), which houses the third-largest collection of bird skins in the world. Olson's main specialisation is fossil birds and he was also *de facto* curator of the fossil bird collection in the Department of Paleobiology, which is the largest in the world by far. He graduated (1966) from the Florida State University and went on to achieve his doctorate (1972) at Johns Hopkins University. He is the author of well over 300 publications in a variety of scientific journals. His interests are primarily in avian palaeontology and systematics, avifaunas of oceanic islands prior to human-caused extinction events, and biogeography and systematics of Neotropical birds, especially those of the Panamanian isthmus. Olson has received many honours from ornithological institutions. The extinct shearwater was described from bones found in Fuerteventura, Canary Islands, the bird itself having apparently been eaten by the indigenous people.

Ommaney

Kalij Pheasant ssp. *Gennæus macdonaldi ommaneyi* **E. C. S.
Baker**, 1915 NCR
[JS *Lophura leucomelanos williamsi*]

J. L. Ommaney (1869–1945) was a British officer in the Indian Police. He was Deputy Inspector General (1920).

O'Neill

Pardusco *Nephelornis oneilli* **Lowery** & **Tallman**, 1976

John Patton O'Neill (b.1942) is an American field ornithologist and artist whose career has been closely associated with the Louisiana State University Museum of Zoology and with that institution's explorations of Peru. He has co-written several books such as *Great Texas Birds* (1999) and *Birds of Peru* (2007). A reptile is named after him.

Onslow

Chatham [Island] Shag *Leucocarbo onslowi* **H. O. Forbes**,
1893
[Syn. *Phalacrocorax onslowi*]

The Right Honourable Sir William Hillier Onslow, 4th Earl of Onslow (1853–1911), was Parliamentary Secretary to the Board of Trade, Governor of New Zealand (1889–1892), Under Secretary of State for India (1895–1900), for Colonies (1900–1903) and President of the Board of Agriculture with Cabinet rank (1903–1905). His last posts were as Chairman of Committees and Deputy Speaker of the House of Lords (1906–1911).

Oort, E. van

Orange-crowned Fairy-wren ssp. *Clytomyias insignis oorti* **Rothschild** & **Hartert**, 1907
Large Scrubwren ssp. *Sericornis nouhuysi oorti* Rothschild & Hartert, 1913
White-eared Catbird ssp. *Ailuroedus buccoides oorti* Rothschild & Hartert, 1913 NCR
[NUI *Ailuroedus buccoides buccoides*]

(See **van Oort, E.**)

Oort, P.

Oort's Barbet *Megalaima oorti* **S. Muller**, 1836
[Alt. Black-browed Barbet; Syn. *Psilopogon oorti*]

(See **van Oort, P.**)

Oppenheim

Common Starling ssp. *Sturnus vulgaris oppenheimi* **Neumann**, 1915

Max Freiherr von Oppenheim (1860–1946) was a German diplomat in Egypt (1896–1903) who resigned to become a private archaeologist in the Middle East, financing his own digs at Tell Halaf (Syria) (1911–1913 and 1929). In WW1 he was in the German Intelligence Bureau, attempting to raise rebellions in Egypt and India. He opened his own museum in Berlin (1930) but a bombing raid obliterated his collection of neo-Hittite material (1943).

Orbigny

Grey-breasted Seedsnipe *Thinocorus orbignyianus* **Saint-Hilaire** & **Lesson**, 1831
Andean Parakeet *Bolborhynchus orbygnesius* **Souancé**, 1856

(See under **D'Orbigny**)

Orces

El Oro Parakeet *Pyrrhura orcesi* **Ridgely** & **Robbins**, 1988

Professor Gustavo Orces (1902–1999) was an Ecuadorian zoologist who worked at the Polytechnic in Quito; his principal interest was herpetology. The Fundación Herpetológica Gustavo Orces (Museum of Natural History) holds his collection. One visiting zoologist, Dr Janis Roze, made a point of noting that he '… made available to me his large coral snake collection and provided other valuable assistance'. Ridgely & Robbins wrote in their etymology 'We are pleased to name this species in honor of Dr. Gustavo Orces V., in recognition of his many contributions to Ecuadorian ornithology and his continuing encouragement of younger generations of field biologists. The proposed English name, El Oro Parakeet, refers to the province in Ecuador where this species was discovered.' He is also remembered in the names of two mammals, two amphibians and five reptiles.

Ord

Brown-banded Puffbird *Notharcus ordii* **Cassin**, 1851

George Ord (1781–1866) was an American philologist, collector and naturalist. He was originally a ships' chandler, but became one of the earliest members of the active Philadelphia natural history community. This brought him to the attention of President Thomas Jefferson, who sent him many specimens from the Lewis (q.v.) and Clarke (q.v.) Expedition, and Ord named many of western North America's familiar birds. When Bonaparte (q.v.) tried to get Audubon (q.v.) accepted by the Academy of Natural Sciences (1824) he was opposed by Ord, who detested Audubon and regarded him with contempt. He also thought that Thomas Nuttall (q.v.) was a 'presumptuous ass' and said that he deserved 'the lash'.He read a memoir of Thomas Say (q.v.) before the American Philosophical Society (1834). A mammal is named after him.

Oren

Bamboo Antwren *Myrmotherula oreni* Miranda *et al.*, 2013

Dr David C. Oren is an ornithologist with the Emilio Goeldi Museum, Belém. Harvard awarded his doctorate. He is a specialist in Brazilian Amazonian biodiversity and also a cryptozoologist who is searching for mythical beasts such as Mapinguari – a kind of South American Giant Sloth or Bigfoot.

Orii

Varied Tit ssp. *Poecile varius orii* **Kuroda**, 1923
Eurasian Jay ssp. *Garrulus glandarius orii* Kuroda, 1923
Japanese Pygmy Woodpecker ssp. *Dendrocopos kizuki orii* Kuroda, 1923
Light-vented Bulbul ssp. *Pycnonotus sinensis orii* Kuroda, 1923
Brown-headed Thrush ssp. *Turdus chrysolaus orii* **Yamashina**, 1929 NCR; NRM
Collared Kingfisher ssp. *Todiramphus chloris orii* **Taka-Tsukasa** & **Momiyama**, 1931
Micronesian Starling ssp. *Aplonis opaca orii* Taka-Tsukasa & Yamashina, 1931
Oriental Turtle Dove ssp. *Streptopelia orientalis orii* Yamashina, 1932
Eurasian Wren ssp. *Troglodytes troglodytes orii* Yamashina, 1938 EXTINCT?
[Alt. Daito Wren; SII *Troglodytes troglodytes mosukei*]

Hyojiro Orii (1883–1948) was a Japanese collector. His main task was to collect ornithological specimens for Nagamichi Kuroda (q.v.). He also worked for Marquis Yamashina (q.v.), who published a small work called 'On Korean birds collected by Mr H. Orii', and about whom Orii wrote an appreciation (1948). Two mammals are named after him.

Orlando

Lanner Falcon ssp. *Falco biarmicus orlandoi* Trischitta, 1939 NCR
[JS *Falco biarmicus feldeggii*]
Rock Partridge ssp. *Alectoris graeca orlandoi* Priolo, 1984
[SII *Alectoris graeca saxatilis*]

Carlo Orlando (1898–1976) was an Italian ornithologist. He founded the museum at Terrasini, Sicily (c. 1929).

Orleans

Silverbird ssp. *Empidornis semipartitus orleansi* **Rothschild**, 1922 NCR; NRM

Louis Philippe Robert Duc d'Orléans and Duc de Montpensier (1869–1926) claimed (1894–1926) to be King Philippe VIII of France. His claim was not accepted! The holotype examined by Rothschild was in Orleans' private collection.

Orpheus

Mockingbird genus *Orpheus* **Swainson**, 1827 NCR
[Now in *Mimus*]
Bulbul genus *Orpheus* **Temminck** & **Schlegel**, 1848 NCR
[Now in *Microscelis*]

Orphean Warbler *Sylvia hortensis* **J. F. Gmelin**, 1789
Fawn-breasted Whistler *Pachycephala orpheus* **Jardine**, 1849

Northern Mockingbird ssp. *Mimus polyglottos orpheus* **Linnaeus**, 1758

Orpheus was the son of the Muse Calliope. His father was either the god Apollo or Oeagrus, king of Thrace, depending on which version of the myth you prefer. He was presented with a lyre and taught to play upon it to such perfection that nothing could withstand the charm of his music.

Ortiz

Grey-winged Inca Finch *Incaspiza ortizi* **Zimmer**, 1952

Javier Ortiz de la Puente Denegri (1928–1952) was a Peruvian zoologist and ornithologist. The Museo de Historia Natural de la Universidad Nacional Mayor de San Marcos awarded his bachelor's degree (1948). Whilst still an undergraduate he visited both the department of Pasco and Lake Titicaca. He was invited to visit the Field Museum, Chicago (1950), and afterwards studied at the AMNH, New York for 9 months. He was killed when he was run over by a motorcar. He wrote 'Estudio monográfico de los quirópteros de Lima' (1951).

Ortlepp

Black-chested Prinia ssp. *Prinia flavicans ortleppi* **Tristram**, 1869

Albert Frederick Ortlepp (1840–1891) was a land surveyor and naturalist in South Africa. He took part in the Kimberley Diamond Rush (1869) and must have done quite well as he is recorded presenting the Natal Museum with 'fourteen small diamonds'.

Orton, C.

Dusky Miner ssp *Myzantha obscura ortoni* **Ashby**, 1922 NCR
[JS *Manorina flavigula obscura*]

Clarence Leonard Egremont Orton (1883–1965) emigrated from England to Australia (1904). He was a land agent, farmer in Western Australia, ornithologist, oologist and collector. His egg collection is now held by the Western Australian Museum, Perth.

Orton, J.

Orton's Comet *Lesbia ortoni* **G. N. Lawrence**, 1869 NCR
[Hybrid: *Lesbia victoriae* x *Ramphomicron microrhynchum*]
Orton's Guan *Penelope ortoni* **Salvin**, 1874
[Alt. Baudó Guan]

Professor James Orton (1830–1877) was an American zoologist who collected in South America (1860s). He taught at Vassar College, New York (1866–1877). He wrote *The Andes and the Amazon* (1876) and numerous articles, such as 'Notes on some birds in the Museum of Vassar College' (1871). Two reptiles are named after him.

Osa

Large-billed Crow ssp. *Corvus macrorhynchos osai* **Ogawa**, 1905

Osa (DNF) was a Japanese collector employed by Alan Owston (q.v.). He collected in the Riu Kiu Islands (1904).

Osbert

Fork-tailed Emerald ssp. *Chlorostilbon canivetii osberti* **Gould**, 1860

(See **Salvin**)

Osburn

Blue Mountain Vireo *Vireo osburni* **P. L. Sclater**, 1861

William Osburn (d.1860) was a collector in Jamaica (1858–1860). He conducted early studies on bats at Oxford Cave and other sites (1858–1860).

Osceola

Osceola Turkey *Meleagris gallopavo osceola* **W. E. D. Scott**, 1890
[Alt. Wild Turkey ssp.]

Osceola (also known as Billy Powell) (1804–1838) was a leader of the Seminole Native American people of Florida, although born in Tallassee, Alabama, a town populated by people of mixed White and Black Americans, British, Irish and Native American ancestry. He led a small band of warriors in the Seminole resistance during the second Seminole War when the US tried to remove them from their ancestral lands. His life was made into a feature film *Naked in the Sun* (1957).

Osculati

Quail Dove genus *Osculatia* **Bonaparte**, 1855 NCR
[Now in *Geotrygon*]

Gaetano Osculati (1808–1894) was an Italian entomologist, ethnographer, cartographer, botanist, explorer and collector. He was in the Near East and Middle East (1831–1833), in Persia (Iran) (1841) and South America (1834–1836 and 1846–1848) and Hindustan and China (1857).

Osery

Casqued Oropendola *Clypicterus oseryi* **Deville**, 1849
d'Osery's Hermit *Phaethornis oseryi* **Bourcier** & **Mulsant**,
1852 NCR
[Alt. White-bearded Hermit; JS *Phaethornis hispidus*]

(See **d'Osery**)

Osgood

Black Tinamou *Tinamus osgoodi* **Conover**, 1949

Spruce Grouse ssp. *Falcipennis canadensis osgoodi*
Bishop, 1900
Osgood's Jay *Cyanocitta stelleri carlottae* Osgood, 1901
[Alt. Queen Charlotte's Jay, Steller's Jay ssp.]
Whooping Motmot ssp. *Momotus subrufescens osgoodi*
Cory, 1913
Chestnut-tipped Toucanet ssp. *Aulacorhynchus derbianus*
osgoodi **E. R. Blake**, 1941
Hermit Thrush ssp. *Catharus guttatus osgoodi* **A. R. Phillips**,
1991 NCR
[NUI *Catharus guttatus nanus*]

Wilfred Hudson Osgood (1875–1947) was an American zoologist who collected in North America, Chile and elsewhere.
The United States Biological Survey employed him in Alaska.
He was Curator of Zoology at The Field Museum, Chicago
(1909–1941). He gave the museum his personal collection of
mammalogy and ornithology books containing many important works, including numerous editions of the works of
Linnaeus and other 18th century authors. Whilst at the
museum he encouraged Conover (q.v.) to develop 'from a
wealthy young sportsman to a scientific ornithologist'. He
wrote *The Mammals of Chile* (1923). He made an expedition
with Louis Agassiz Fuertes (q.v.) to Ethiopia (1920s) and, with
him, co-wrote *Artist and Naturalist in Ethiopia* (1936). He is
remembered in the names of an amphibian and seven
mammals.

Oshiro

Eastern Buzzard ssp *Buteo japonicus oshiroi* **Kuroda**, 1971
[Syn. *Buteo buteo oshiroi*]

Masao Oshiro (fl.1968) was a Japanese aviculturist.

Osiris

Marico Sunbird ssp. *Cinnyris mariquensis osiris* **Finsch**,
1870

In Egyptian mythology, Osiris was the god of fertility and the
dead, ruler of the underworld, husband and brother of Isis
(q.v.), and father to Horus (q.v.).

Osmaston

Pale-footed Bush Warbler ssp. *Cettia pallidipes osmastoni*
Hartert, 1908
Stork-billed Kingfisher ssp. *Pelargopsis capensis osmastoni*
E. C. S. Baker, 1934
Small Minivet ssp. *Pericrocotus cinnamomeus osmastoni*
Roselaar & Prins, 2000
[SII *Pericrocotus cinnamomeus vividus*]

Bertram Beresford Osmaston (1867–1961) was a British
forestry officer in India (1888–1919) and amateur naturalist
who made and published notes of his observations during
postings to various parts of India such as Ladakh and
Kashmir.

Otero

Violet-fronted Brilliant ssp. *Heliodoxa leadbeateri otero*
Tschudi, 1844

José de Otero (fl.1838) was a resident of Peru. Unfortunately
Tschudi says nothing about him in his description of this
hummingbird.

Othello

Tufted Antshrike *Thamnophilus othello* **Lesson**, 1831 NCR
[JS *Mackenziaena severa*]

Othello, a Moorish general in the Venetian army, is one of
Shakespeare's best-known characters. Presumably his
name was attached to the antshrike because of the male
bird's blackish plumage.

Othmar

Common Reed Bunting ssp. *Emberiza schoeniclus othmari*
Hartert, 1904 NCR
[JS *Emberiza schoeniclus tschusii*]

(See **Reiser**)

Otto (Finsch)

Black-backed Fruit Dove ssp. *Ptilinopus cinctus ottonis*
Hartert, 1904

(See **Finsch**)

Otto (Garlepp)

Rusty-fronted Canastero *Asthenes ottonis* **Berlepsch**, 1901

Marble-faced Bristle Tyrant ssp. *Pogonotriccus*
ophthalmicus ottonis Berlepsch, 1901

Otto Garlepp (1864–1959) was the younger brother of Gustav
Garlepp (q.v.). The brothers collected together in Bolivia, and
Otto made a collection of over 200 skins in south-eastern
Peru. These collections are in the Senckenberg Museum,
Frankfurt-am-Main, Germany.

Ottolander

Chestnut-backed Scimitar Babbler ssp. *Pomatorhinus*
montanus ottolanderi **H. C. Robinson**, 1918

Teun Ottolander (1854–1935) was a Dutch planter in Java
(1879–1935), President of the Netherlands Indies Agricultural
Syndicate. He was a botanist and collector of botanical
specimens in both Java and Sumatra (Indonesia).

Otto Meyer

Louisiade White-eye ssp. *Zosterops longirostris ottomeyeri*
Stresemann, 1930 NCR
[JS *Zosterops griseotinctus eichhorni*]

Bismarck Whistler ssp. *Pachycephala citreogaster ottomeyeri* Stresemann, 1933
Varied Triller ssp. *Lalage leucomela ottomeyeri* Stresemann, 1933

(See **Meyer, O.**)

Ottow

Rufous-banded Miner ssp. *Geositta rufipennis ottowi* **G. A. Hoy**, 1968

Dr Johann Ottow (DNF) was a German ornithologist who was in South America (1930s and 1940s). He was interned in Brazil as an enemy alien (1943–1944) after Brazil declared war on Germany and Italy (1942). He later moved to Argentina where he met and became a friend of Hoy (q.v.). They co-wrote 'Biological and oological studies of the molothrine cowbirds (Icteridae) of Argentina' (1964), which represented the culmination of eight years' fieldwork in Salta and Jujuy (observations by Hoy and oology by Ottow). Ottow was in South Africa (1960s) and he co-wrote 'Zur Kenntnis der Fortpflanzung von *Chrysococcyx caprius* und *Cuculus canorus gularis* in Süd-Africa' (1965), by which time he had settled in Skellefteå, Sweden.

Oustalet

Oustalet's Swiftlet *Aerodramus germani* Oustalet, 1876
[Alt. German's Swiftlet]
Oustalet's Sunbird *Cinnyris oustaleti* **Bocage**, 1878
[Alt. Angola White-bellied Sunbird]
Grass Owl sp. *Strix oustaleti* **Hartlaub**, 1879 NCR
[Alt. Eastern Grass Owl; JS *Tyto longimembris*]
Oustalet's Bustard *Lophotis gindiana* Oustalet, 1881
[Alt. Buff-crested Bustard]
Fishing Owl sp. *Scotopelia oustaleti* Rochebrune, 1883 NCR
[Alt. Pel's Fishing Owl; JS *Scotopelia peli*]
Oustalet's Tyrannulet *Phylloscartes oustaleti* **P. L. Sclater**, 1887
Oustalet's Duck *Anas oustaleti* **Salvadori**, 1894 EXTINCT
[Alt. Mariana Mallard; Syn. *Anas platyrhynchos oustaleti*]
Oustalet's Greenfinch *Carduelis ambigua* Oustalet, 1896
[Alt. Black-headed Greenfinch, Tibetan Greenfinch]
Grey-flanked Cinclodes *Cinclodes oustaleti* **W. E. D. Scott**, 1900

Andean Tinamou ssp. *Nothoprocta pentlandii oustaleti* **Berlepsch & Stolzmann**, 1901
Black-faced Laughingthrush ssp. *Trochalopteron affine oustaleti* **Hartert**, 1909
Dusky Scrubfowl ssp. *Megapodius freycinet oustaleti* **Roselaar**, 1994

Dr Jean-Frédéric Émile Oustalet (1844–1905) was a French zoologist who worked in the Far East. He wrote *Les Oiseaux du Cambodge* (1899) and, with Père Armand David, *Les Oiseaux de la Chine* (1877). He succeeded Jules Verreaux at the Paris Natural History Museum (1873) and succeeded Alphonse Milne-Edwards (with whom he co-authored a number of works) when he died (1900). A reptile and a mammal are also named after him.

Oviedo

Palmchat ssp. *Dulus dominicus oviedo* **Wetmore**, 1929 NCR; NRM

Gonzalo Fernández de Oviedo y Valdés (1478–1557) was a Spanish historian and chronicler. He wrote *La Historia General y Natural de las Indias*.

Owen, Richard

Owen's Kiwi *Apteryx owenii* **Gould**, 1847
[Alt. Little Spotted Kiwi]

Professor Sir Richard Owen (1804–1892) was a British anatomist and biologist who studied medicine at the University of Edinburgh, and became Assistant Curator of the Hunterian Collection at the Royal College of Surgeons. His fame as a scientist led to his appointment to teach natural history to Queen Victoria's children. He was largely responsible for the creation of the Natural History Museum in London, having successfully separated it from the British Museum. He was the first person to conceptualise what a moa looked like (especially its size), based on his study of a few bones sent from New Zealand, and he named the moa genus *Dinornis*, meaning 'terrible bird'. He also named the *Dinosauria* ('terrible lizards'), so creating the present-day dinosaur industry. Despite his fame, Owen had many critics. The palaeontologist Gideon Mantell (father of Walter Mantell, q.v.) said of him that it was 'a pity a man so talented should be so dastardly and envious'. When Mantell suffered an accident that left him permanently crippled, Owen exploited the opportunity by renaming several dinosaurs which had already been named by Mantell, even claiming credit for their discovery himself. Owen's unwillingness to come off the fence concerning the debate over evolutionary theory also became increasingly damaging to his reputation. Two mammals and a reptile are named after him.

Owen (Wynne), Richard

Black-eared Cuckoo genus *Owenavis* **Mathews**, 1912 NCR
[Now in *Chrysococcyx*]

Striated Grasswren ssp. *Amytornis striatus oweni* Mathews, 1911 NCR
[NUI *Amytornis striatus striatus*]
Great Bowerbird ssp. *Chlamydera nuchalis oweni* Mathews, 1912 NCR
[NUI *Chlamydera nuchalis nuchalis*]
Swinhoe's Snipe ssp. *Gallinago australis oweni* Mathews, 1912 NCR
[JS *Gallinago megala* (monotypic)]
Tawny Grassbird ssp. *Megalurus alisteri oweni* Mathews, 1912 NCR
[JS *Megalurus timoriensis alisteri*]

Colonel Richard Owen Wynne (1892–1969) was a British army officer. The describer was his stepfather, although because Mathews almost never supplied etymologies, it cannot be said with certainty whether he was honouring his stepson in every instance, or whether he sometimes intended to honour Richard Owen the anatomist (above).

Owen, Robert

> Owen's Sabrewing *Phaeochroa cuvierii roberti* **Salvin**, 1861
> [Alt. Robert's Hummingbird, Scaly-breasted Hummingbird ssp.]

Robert Owen (fl.1860) was a resident of Vera Paz in Guatemala where Salvin (q.v.) and Godman (q.v.) both made use of his assistance in their collecting. Owen was the first person to find and collect the nest and egg of the Resplendent Quetzal *Pharomachrus mocinno*. (See also **Robert**)

Owston

> Guam Rail *Gallirallus owstoni* **Rothschild**, 1895
> Storm-petrel sp. *Cymochorea owstoni* **Mathews** & **Iredale**, 1915 NCR
> [Alt. Tristram's Storm-petrel; JS *Oceanodroma tristrami*]
>
> ---
>
> Owston's Varied Tit *Poecile varius owstoni* **Ijima**, 1893
> Citrine White-eye ssp. *Zosterops semperi owstoni* **Hartert**, 1900
> Narcissus Flycatcher ssp. *Ficedula narcissina owstoni* **Bangs**, 1901

> Chinese Hwamei ssp. *Garrulax canorus owstoni* Rothschild, 1903
> Collared Kingfisher ssp. *Todiramphus chloris owstoni* Rothschild, 1904
> Owston's Woodpecker *Dendrocopos leucotos owstoni* **Ogawa**, 1905
> [Alt. White-backed Woodpecker ssp.]
> Grey-headed Bullfinch ssp. *Pyrrhula erythaca owstoni* Hartert & Rothschild, 1907
> Crimson Sunbird ssp. *Aethopyga siparaja owstoni* Rothschild, 1910
> Spotted Nutcracker ssp. *Nucifraga caryocatactes owstoni* **Ingram**, 1910

Alan Owston (1853–1915) was an English collector of Asian wildlife, as well as a businessman and yachtsman. He left for the Orient when still quite young. He married Shimada Rei Jkao (c.1880) in Japan and they had one child, Susie. He later married Kame (Edith) Miyahara (c.1893), having eight children by that marriage. Owston's most active collecting period was in the early 20th century. He died of lung cancer in Yokohama, Japan. He is also commemorated in the name of a mammal, an amphibian and various other taxa.

P

Pabst

Pabst's Cinclodes *Cinclodes pabsti* **Sick**, 1969
[Alt. Long-tailed Cinclodes]

Dr Guido Frederico João Pabst (1914–1980) was a Brazilian botanist and taxonomist famed for his work on orchids. He founded the Herbarium Bradeanum, in Rio de Janeiro (1958) and the orchid genus *Pabstia* is named after him. He wrote the two-volume *Orchidaceae Brasiliensis* (1975) with F. Dungs.

Pacheco

Planalto Tapaculo *Scytalopus pachecoi* Mauricio, 2005

José Fernando Pacheco is a Brazilian biologist and ornithologist who studied for his master's degree in animal biology (1999–2001) at Universidade Federal Rural do Rio de Janeiro. He additionally worked (1990–1998) in various capacities in the university's Institute of Biology, Department of Zoology. He is Director of the Brazilian Committee of Ornithological Records in São Paulo and specialises in recording bird song. He has published some 200 articles and papers, normally as a co-author, such as, with R. Parrini, 'Brief history of knowledge of the avifauna of the region of Alto Juruá' (2002).

Pagel

Buffy Fish Owl ssp. *Ketupa ketupu pageli* **Neumann**, 1935

Dr Pagel (fl.1901) was a German physician in North Borneo (1891–1901).

Painter

Starling genus *Painterius* **Oberholser**, 1930 NCR
[Now in *Lamprotornis*]

Kenyon Vickers Painter (1867–1940) was an American bank director and hunter in East Africa (1912–1913) and aviculturist in Cleveland, Ohio. Convicted of embezzling $2 million (1935) he spent time in prison, but was pardoned by Governor Davey. His aviary, containing 500 birds, was given to Cleveland Zoo.

Palamedes

Horned Screamer genus *Palamedea* **Linnaeus**, 1766 NCR
[Now in *Anhima*]

In Greek mythology, Palamedes was Prince of Euboea. He died at the siege of Troy, having incurred the hatred of his fellow Greek, Odysseus. When Palamedes advised the Greek troops to give up the siege and return home, Odysseus hid gold in his tent along with a fake letter purportedly from Priam, the king of Troy. The letter was found, the Greeks accused Palamedes of being a traitor, and stoned him to death.

Pallas

Lark genus *Pallasia* **Homeyer**, 1873 NCR
[Now in *Melanocorypha*]
Cormorant genus *Pallasicarbo* **Coues**, 1903 NCR
[Now in *Phalacrocorax*]

Pallas's Fish Eagle *Haliaeetus leucoryphus* Pallas, 1771
[Alt. Pallas's Sea Eagle]
Pallas's Gull *Ichthyaetus ichthyaetus* Pallas, 1773
[Alt. Great Black-headed Gull; note *Ichthyaetus pallasi*
    (Bonaparte, 1856) is a junior synonym]
Pallas's Sandgrouse *Syrrhaptes paradoxus* Pallas, 1773
Pallas's Crake *Porzana pusilla* Pallas, 1776
[Alt. Baillon's Crake]
Pallas's Rosefinch *Carpodacus roseus* Pallas, 1776
[Alt. Siberian Rosefinch]
Pallas's Cormorant *Phalacrocorax perspicillatus* Pallas,
    1811 EXTINCT
[Alt. Spectacled Cormorant]
Pallas's Eared Pheasant *Crossoptilon auritum* Pallas, 1811
[Alt. Blue Eared Pheasant]
Pallas's Flycatcher *Musciapa dauurica* Pallas, 1811
[Alt. Asian Brown Flycatcher]
Pallas's Grasshopper Warbler *Locustella certhiola* Pallas,
    1811
Pallas's Leaf Warbler *Phylloscopus proregulus* Pallas, 1811
Pallas's Dipper *Cinclus pallasii* **Temminck**, 1820
[Alt. Brown Dipper]
Hermit Thrush *Turdus pallasii* **Cabanis**, 1847 NCR
[JS *Catharus guttatus*]
Pallas's Bunting *Emberiza pallasi* Cabanis, 1851
[Alt. Pallas's Reed Bunting]

Pallas's Murre *Uria lomvia arra* Pallas, 1811
[Alt. Brünnich's Guillemot/Thick-billed Murre ssp.]
Common Kingfisher ssp. *Alcedo atthis pallasii* **Reichenbach**,
    1851 NCR
[NUI *Alcedo atthis atthis*]
Pallas's Ring-necked Pheasant *Phasianus colchicus pallasi*
    Rothschild, 1903
[Alt. Common Pheasant ssp.]

Northern Hawk Owl ssp. *Surnia ulula pallasi* **Buturlin**, 1907 NCR

[JS *Surnia ulula ulula*]

Crested Guineafowl ssp. *Guttera edouardi pallasi* Stone, 1912 NCR

[JS *Guttera pucherani verreauxi*]

Peter Simon Pallas (1741–1811) was a German-born Russian (1767) zoologist and one of greatest 18th century naturalists. Being very bright he earned his doctorate from the University of Leiden aged 19! He went to London (1761) to study the English hospital system and was enchanted by the Sussex coast and the countryside in Oxfordshire. The Empress Catherine II summoned him to Russia (1767) to become the Professor of Natural History at the St Petersburg Academy of Sciences and to investigate Russia's natural environment. He was also a geographer and traveller and explored widely in lesser known areas of Russia. He headed an Academy of Sciences expedition (1768–1774) which studied many regions of Russia, including southern Siberia (Altai, Lake Baikal and the region to the east of Baikal). He described many new species of mammals, birds, fish, insects and fossils. His works include *A Journey through Various Provinces of the Russian State* (1771), *Flora of Russia* (1774), and *A History of the Mongolian People and Asian-Russian Fauna* (1811), as well as works on zoology, palaeontology, botany, ethnography, etc. He found a mass of iron weighing 700 kilograms (1772) that was a new kind meteorite, which was named pallasite after him. Seven mammals and three reptiles are named after him. A volcano on the Kurile Islands and a reef off New Guinea were also named in his honour.

Palliser

Palliser's Warbler *Elaphrornis palliseri* **Blyth**, 1851

[Alt. Sri Lanka Bush Warbler]

Captain Edward Mathew Palliser (1828–1907) and his brother Frederick Hugh Palliser (1826–1883) were travellers who collected together in Sri Lanka (1845–1851). (Samuel White Baker mentions Palliser in *Eight Years' Wanderings in Ceylon*, describing himself and Palliser shooting rogue elephants together.)

Palmer, E.

Palmer's Thrasher *Toxostoma curvirostre palmeri* **Coues**, 1872

[Alt. Curve-billed Thrasher ssp.]

Edward Palmer (1829–1911) was a British-born, self-taught botanist employed as such by the US Department of Agriculture. He led an expedition exploring the flora and fauna of California (1891), particularly Death Valley. Later he collected in Mexico and other parts of the Neotropics. He is remembered in the names of c.200 plants. He also worked as a field assistant for the Bureau of American Ethnology (1882–1884) in their Mound Exploration Division, contributing to the understanding that these were left by American Indians, not a mythical 'mound builders' race. From 1894 he was based in Washington DC working on natural history in general and botany in particular.

Palmer, H. C.

Hawaiian Honeycreeper genus *Palmeria* **Rothschild**, 1893

Greater Koa Finch *Rhodacanthis palmeri* Rothschild, 1892 EXTINCT

Laysan Rail *Porzana palmeri* Frohawk, 1892 EXTINCT

Palmer's Thrush *Myadestes palmeri* Rothschild, 1893

[Alt. Small Kauai Thrush, Puaiohi]

Henry C. Palmer (d. 1894/1895) was Australian and probably a sailor, but very little is known about him. Rothschild (q.v.) employed him as a collector and sent him to the Chatham Islands late in 1889. A year later he collected in Hawaii (1890–1893) and sent back a considerable number of skins from which Rothschild named 18 new species and subspecies. He worked in Hawaii with a couple of New Zealanders as assistants – George Munro (q.v.) (until March 1892) then Ted Wolstenholme (q.v.). Having left Hawaii he made a short visit to England (presumably to Tring among other places) and then returned to Australia. Here he went to the goldfields, where he was reportedly murdered.

Palmer, M. G.

Palmer's Tanager *Tangara palmeri* **Hellmayr**, 1909

[Alt. Grey-and-gold Tanager]

Mervyn George Palmer (1882–1954) was an indefatigable English traveller and collector. After graduating he became an analytical chemist in the cement industry before deciding on a career as a freelance collector and naturalist. He collected mammals, birds, reptiles, fish, butterflies and shells for the British Museum and other institutions, in Colombia, Ecuador and Nicaragua (1904–1910). Here he learnt to speak Spanish and two South American Indian languages, married a South American woman, undertook archaeological digs, and explored and mapped the Río Segovia between Nicaragua and Honduras. He became a Fellow of the Royal Geographical Society by virtue of his explorations in Central and South America, which were all on foot or by canoe, and was made a Freeman of the City of London. At one time he edited the Natural Science Gazette. He worked for commercial concerns in Ecuador (1910–1918) before moving to London. He was declared unfit for overseas army service during WW1, having had yellow fever and malaria, but he used his commercial skills for his regiment's benefit. He then worked in Venezuela (1919–1921). Later, until retirement, he was based in England but travelled widely in South and Central America and the Caribbean, visiting every South American country except Paraguay. He wrote *Through Unknown Nicaragua – The Adventures of a Naturalist on a Wild-Goose Chase* (1945). He retired to Ilfracombe in Devon, England, where he established a library, museum and field club for the town (1932), as he 'wanted something to do'. He was curator there until his death in 1954. He wrote *The Homeland Guide to North Devon*, which was published posthumously in 1960. Two amphibians and a reptile are also named after him.

Palmer, R. S.

> Brown-capped Vireo ssp. *Vireo leucophrys palmeri*
> **A. R. Phillips**, 1991

Ralph Simon Palmer (1914–2003) was an American ornithologist and mammalogist. The University of Maine awarded his first degree (1937) and Cornell his doctorate (1940). He worked for the New York Conservation Department before becoming Assistant Professor at Vassar College (1942–1949). During WW2 he was a naval ensign. He was Senior Scientist at New York State Museum (1949–1977). He retired but became a Research Associate at the USNM (1978–1983) and (1981) a Faculty Associate at the University of Maine, which also awarded him an honorary doctorate (1994). He wrote and illustrated the *Mammal Guide* (1954) and edited the never-completed *Handbook of North American Birds* (started 1962). He collected books from early childhood and donated several collections to museums as well as leaving 1,700 mammal specimens to the Museum of Comparative Zoology at Harvard.

Palmer, T. S.

> Mountain Quail ssp. *Oreortyx pictus palmeri* **Oberholser**,
> 1923 NCR
> [NUI *Oreortyx pictus pictus*]

Dr Theodore Sherman Palmer (1868–1958) was an American naturalist. He worked for the US Biological Survey as Assistant Chief (1889–1933) and was also the Law Enforcement Officer of the US Fish and Wildlife Service (1900–1916). He led an expedition (1891) to study the flora and fauna of Death Valley. Following a broken hip he was confined to his house for his last two years. He wrote *Index Generum Mammalium* (1904). Two reptiles and a mammal are named after him.

Palmer, W.

> Barn Swallow ssp. *Hirundo rustica palmeri* **Grinnell**, 1902
> NCR
> [JS *Hirundo rustica erythrogaster*]
> Blue-breasted Quail ssp. *Excalfactoria chinensis palmeri*
> **Riley**, 1919
> Collared Kingfisher ssp. *Todiramphus chloris palmeri*
> **Oberholser**, 1919

William Palmer (1856–1921) was a US taxidermist at the USNM (1874–1921). He collected in Cuba (1900) and Java (1909–1910).

Palmer (River)

> Little Shrike-thrush ssp. *Colluricincla megarhyncha palmeri*
> **Rand**, 1938

Named after the Palmer River, southern New Guinea, where the holotype was collected.

Palmquist

> African Nightjar sp. *Caprimulgus palmquisti* **Sjöstedt**, 1908
> NCR
> [Alt. Montane Nightjar; JS *Caprimulgus poliocephalus*]

Per Palmquist (d.1887) was a Swedish publisher who belonged to the revivalist movement and published many religious tracts. His company published Sjöstedt's report of the Swedish 1905–1906 Zoological Expedition to German East Africa (1908), wherein was printed the original description of the nightjar. It *may* be that the bird was named after the company rather than the man.

Paluka

> Yellowhammer ssp. *Emberiza citrinella palukae* **Parrot**, 1905
> NCR
> [JS *Emberiza citrinella citrinella*]

Adolph Paluka (DNF) collected the type specimen in Constantinople (Istanbul).

Pamela (Lovibond)

> Arabian Scops Owl *Otus pamelae* **G. L. Bates**, 1937
> [Syn. *Otus senegalensis pamelae*]

Miss Pamela Lovibond (DNF) was the librarian at the Athenaeum Club in London. A member of the club, Harry St John Bridger Philby (q.v.), suggested the name to Bates (q.v.). She married Major David Francis Pawson (1946) and thereafter lived in Greece.

Pamela (Martipon)

> Pamela's Sunbeam *Aglaeactis pamela* **d'Orbigny**, 1838
> [Alt. Black-hooded Sunbeam]

Pamela Martipon (1810–1842) was Alcide d'Orbigny's (q.v.) first wife.

Pan

> Toucan Barbet genus *Pan* **Richmond**, 1899 NCR
> [Now in *Semnornis*]

Pan was the Greek god of nature. He had the hindquarters and legs of a goat, thus mixing human and animal characteristics. The name is not used because it already had a zoological usage as the genus for chimpanzees.

Pander

> Pander's Ground Jay *Podoces panderi* **J. Fischer**, 1821
> [Alt. Turkestan Ground Jay]

Christian Heinrich Pander (1794–1865) was an affluent Latvian of German descent (Latvia then being part of the Romanov Empire). He was a palaeontologist, biologist and evolutionist who joined the Academy of Sciences in St Petersburg. He was the first to describe conodont fossils – extinct aquatic eel-like creatures (early 1830s). He was a friend and colleague of Baer (q.v.). He wrote on general and comparative morphology of animals. One of his most notable scientific achievements was to describe the three germ layers (ectoderm, mesoderm and endoderm) in chick embryos (1817).

Pandion

Osprey genus *Pandion* **Savigny**, 1809

In Greek mythology Pandion was a King of Athens whose son, Nisus, was metamorphosed into a hawk or sea-eagle.

Panin

Visayan Tarictic Hornbill *Penelopides panini* Boddaert, 1783

Count Nikita Ivanovitz Panin (1718–1783) was a statesman at the Russian court and oversaw the education of the heir to the throne. He was also an ambassador to Stockholm. Pieter Boddaert, the describer of the hornbill, was a contemporary and it may be presumed they were friends.

Pantchenko

Pantchenko's Conure *Pyrrhura picta pantchenkoi*
  W. H. Phelps Jr, 1977
[Alt. Painted Parakeet; Synonymised with *Pyrrhura
  caeruleiceps* by some authorities]

Georges Pantchenko (d.2003) was a Venezuelan member of the joint Venezuelan/Brazilian team responsible for establishing the border between the two countries. They did this by descending from helicopters into the jungle and literally hacking 'landmarks' to delimit the boundary. He was working for the Commission (1957) when he was lost in the jungle for a time. He was asked at a conference in Brasília about his surname and is reported to have replied in humorous vein 'My parents are Russian, I was born in Bulgaria, grew up in France, have Venezuelan nationality but my heart is Brazilian!'

Papantzin

Magnificent Hummingbird *Trochilus papantzin* de la Llave,
  1833 NCR
[JS *Eugenes fulgens*]

Papantzin (fl.1525) was the sister-in-law of the Aztec emperor, Montezuma, She was said to have risen from the grave to warn her brother of the impending ruin of his empire. She was one of the first Aztecs to convert to Christianity (1525).

Paphos

Hummingbird genus *Paphosia* **Mulsant & E & J. Verreaux**,
  1866 NCR
[Now in *Lophornis*]

In Greek mythology, Paphos was the son of Pygmalion by the statue he had carved, which had been turned into a living woman by the goddess Aphrodite. He founded the eponymous city in Cyprus.

Parelius

Grey-throated Tit Flycatcher ssp. *Myioparus griseigularis
  parelii* **Traylor**, 1970

Daniel Arthur Parelius (b.1949) was brought up in the Ivory Coast where his parents were American missionaries. He collected over 3,000 specimens for the Field Museum, Chicago. Later he was a rural delivery carrier for the US Postal Service for 29 years. He co-wrote with Traylor (q.v.) 'A collection of birds from the Ivory Coast' (1967).

Paris Brothers

Scott's Oriole *Icterus parisorum* **Bonaparte**, 1838

The Paris brothers were French natural history dealers who were on business in Mexico (1837) and organised and financed the collection and transport to France of a large number of natural history specimens. Bonaparte (q.v.) gave no indication of their names and some commentators cast doubt about their surname, as Bonaparte may have thought of them simply as 'The Paris-based Brothers', as that is where their main place of business was.

Paris, P.

Coal Tit ssp. *Parus ater parisi* **Jouard**, 1928 NCR
[JS *Periparus ater ater*]
Short-toed Treecreeper ssp. *Certhia brachydactyla parisi*
  Jouard, 1929 NCR
[JS *Certhia brachydactyla brachydactyla*]

Paul Paris (1875–1938) was a French ornithologist. He was one of the founders of the journal *Alauda*, and its editor (1928–1938).

Parish

Greater Antillean Bullfinch ssp. *Loxigilla violacea parishi*
  Wetmore, 1931 NCR
[NUI *Loxigilla violacea affinis*]

Lee Hamilton Parish (1903–1931) was leader of the 1930 Parish-Smithsonian Expedition to Haiti, which was financed by his parents Semmes Wilder and Katherine Parish. A reptile is named after him.

Parker, M. S.

Black Francolin ssp. *Francolinus francolinus parkerae*
  Van Tyne & **Koelz**, 1936 NCR
[JS *Francolinus francolinus asiae*]

Mrs Margaret Selkirk Parker *née* Watson (1867–1936) was an American art collector and sponsor, who was the wife of ophthalmologist Dr Walter R. Parker. She bequeathed her art collection to the University of Michigan. The original etymology names 'Mrs Walter Parker of Detroit, in grateful recognition of her active interest in the junior author's researches in India.'

Parker, S. A.

Spiny-cheeked Honeyeater ssp. *Acanthagenys rufogularis
  parkeri* **Parkes**, 1980 NCR
[*A. rufogularis* NRM]
Chestnut-rumped Heathwren ssp. *Calamanthus
  pyrrhopygius parkeri* **Schodde** & Mason, 1999

Shane Alwyne Parker (1943–1992) worked at the BMNH in London (1959–1966), taking part in the second Harold Hall expedition to northern Queensland (1964). He decided to move to Australia (1967) and worked as a scientific assistant in Alice Springs and Darwin (1967–1970). He then moved to Adelaide, returned to school to matriculate (1971), and receiving a bachelor's degree (1975) from the University of Adelaide. He was Curator of Birds, South Australian Museum (1976–1991), combined with being Curator of Lower Marine Invertebrates (1985–1991). He was an archetypical English eccentric whose interests ranged from playing the harp, learning Gaelic, singing bawdy English folk songs and collecting stamps.

Parker, T. A.

Yellow-shouldered Grosbeak genus *Parkerthraustes* **Remsen**, 1997

Parker's Spinetail *Cranioleuca vulpecula* **P. L. Sclater** & **Salvin**, 1866

Ash-throated Antwren *Herpsilochmus parkeri* T. J. Davis & **O'Neill**, 1986

Parker's Antbird *Cercomacra parkeri* **Graves**, 1997

Subtropical Pygmy Owl *Glaucidium parkeri* **Robbins** & Howell, 1995

Cinnamon-faced Tyrannulet *Phylloscartes parkeri* **Fitzpatrick** & Stotz, 1997

Chusquea Tapaculo *Scytalopus parkeri* Krabbe & **Schulenberg**, 1997

Coppery Metaltail ssp. *Metallura theresiae parkeri* Graves, 1981

Natterer's Slaty Antshrike ssp. *Thamnophilus stictocephalus parkeri* **Isler**, Isler & **Whitney**, 1997

Black-and-yellow Silky Flycatcher ssp. *Phainoptila melanoxantha parkeri* Barrantes & Sánchez, 2000

Theodore (Ted) Albert Parker III (1953–1993) was a professional American ornithologist and a former member of the Cornell Laboratory's Administrative Board. He has been described as a 'great Neotropical conservationist and recordist' and also as the 'world's leading authority on Neotropical bird identification and distribution'. He was the single largest contributor to the Library of Natural Sounds at Cornell University, providing more than 10,000 bird recordings. He could recognise 4,000 species of bird by call alone. He revolutionised bird-surveying methods by employing a tape recorder in the field. The book *A Parrot Without a Name: The Search for the Last Unknown Birds on Earth* by Don Stap, was written about Parker's discoveries, and those of John O'Neill (q.v.), of new bird species in Peru. There was a less serious side to Ted; at the age of 18 (1971), he set a then record of 626 species seen in one year in the USA. A friend, Kenn Kaufman, said of him that he was destined never to slow down: 'he was a runaway train, except that he was running on tracks that he had planned out for himself, and he knew exactly where he was going'. Parker died tragically in a plane crash in Ecuador (1993).

Parkes

Waterthrush genus *Parkesia* Sangster, 2008

Chestnut-capped Brush Finch ssp. *Arremon brunneinucha parkesi* **A. R. Phillips**, 1966 NCR
[JS *Arremon brunneinucha brunneinucha*]

Mountain White-eye ssp. *Zosterops montanus parkesi* **duPont**, 1971

Philippine Bulbul ssp. *Hypsipetes philippinus parkesi* duPont, 1980

Golden-billed Saltator ssp. *Saltator aurantiirostris parkesi* Cardoso da Silva, 1990

Hutton's Vireo ssp. *Vireo huttoni parkesi* **Rea**, 1991

Yellow Warbler ssp. *Dendroica aestiva parkesi* **Browning**, 1994

Yellow-throated Cuckoo ssp. *Chrysococcyx flavigularis parkesi* **Dickerman**, 1994 NCR, NRM

Kenneth Carroll Parkes (1922–2007) was Senior Curator of Birds, Carnegie Museum of Natural History, Pittsburgh (retired 1996). He graduated from Cornell University (1943) going on to take his MSc there (1948) and being awarded his PhD (1952). He served in the army during WW2 (1943–1946). He trained as Curator of Birds (1947–1952) under A. A. Allen, his doctoral supervisor. He became Assistant Curator of Birds at Carnegie Museum (1953). His first overseas collecting trip (1956) was to Luzon, Philippines. He concentrated on plugging the collection's gaps in Neotropical birds, first visiting Argentina (1961) then 35 different countries (1963–1981). His research interests were wide, spanning systematics, distribution, hybridisation etc. He published over 400 papers. He was an active member of the Wilson Ornithological Club for over 40 years.

Parkin

House Sparrow ssp. *Passer domesticus parkini* **Whistler**, 1920

Thomas Parkin (1845–1932) was a British naturalist, oologist and collector. He took a master's degree at Cambridge (1871) and was called to the Bar (1874). He travelled widely including to Australia and New Zealand via the Cape of Good Hope, returning home via Cape Horn, making a collection of seabirds that he donated to the British Museum. He also collected in the southern states, USA and southern Spain. He made a number of trips to North Africa after which he stopped off in Paris, always making sure he returned to England in time for the cricket season!

Parkinson, J.

Lyrebird genus *Parkinsonius* **Bechstein**, 1811 NCR
[Now in *Menura*]

John Parkinson (1775–1847) was an amateur mineralogist whose father, James Parkinson, won a lottery (1786) and found he was the new owner and proprietor of Sir Ashton Lever's famous Leicester House museum in London ('Leverian Museum'). John helped his father dispose of the museum, eventually by auction of 8,000 lots over 65 days (1806). After his father's death (1813) he became a diplomat (1819) in Europe and British Consul in Pernambuco, Brazil, and finally in Mexico. He died in Paris.

Parkinson, S.

Parkinson's Petrel *Procellaria parkinsoni* **G. R. Gray**, 1862
[Alt. Black Petrel]

Sydney Parkinson (1745–1771) was an English explorer and collector who was a natural history artist on Cook's (q.v.)

voyage on HMS *Endeavour*, in the company of Banks (q.v.) and Solander (q.v.). He drew the large collection of plants and seabirds obtained along the Australian east coast. The young artist suffered seasickness and other tribulations in the cramped quarters. Joseph Banks noted '... at one point a swarm of flies was eating the paint off his (Parkinson's) paper as fast as he could lay it on'. Parkinson died at sea from malaria and dysentery contracted in Jakarta, whilst enroute to Cape Town. His brother Stanfield published Sydney's *Journal of a Voyage to the South Seas*.

Parkman

Parkman's Wren *Troglodytes aedon parkmanii* **Audubon**, 1839
[Alt. House Wren ssp.]

Dr George Parkman (1790–1849) was a physician in Boston, Massachusetts. He was very wealthy and very thrifty. Parkman was a great friend of Audubon (q.v.) and wrote the introduction to Audubon's *Quadrupeds of North America*. Audubon in his *Ornithological Biography* states he '... named it after my most kind, generous, and highly talented friend George Parkman, Esq., M. D., of Boston, as an indication of the esteem in which I hold him, and of the gratitude which I ever cherish towards him'. Whilst walking home he was murdered; one John Webster was tried, convicted and hanged (1850) for the offence.

Parodi

Parodi's Hemispingus *Hemispingus parodii* **Weske** & **Terborgh**, 1974
[Alt. Parodi's Tanager]

José 'Pepe' Guillermo Parodi Vargas (b.1929 or 1930 – his parents lived in a very remote area and never knew the exact date – and d.2012) was a Peruvian landowner and a congressman during the first Belaunde government. Pepe and his father were pioneers who constructed a small airport on his property, which the government used to survey the area, and eventually raised money to construct a road from Ayacucho to their Hacienda Luisiana on the Río Apurímac. The Sendero Luminoso (Shining Path) terrorist group attacked and destroyed most of the farm (1983). Parodi also liked to travel to unexplored areas of Peru such as the Cutivineri Falls. When he died he was running an agricultural property south of Lima.

Parrot

Great Spotted Woodpecker ssp. *Dendrocopos major parroti* **Hartert**, 1911
[SII *Dendrocopos major harterti*]
Greater Coucal ssp. *Centropus sinensis parroti* **Stresemann**, 1913
Corn Bunting ssp. *Emberiza calandra parroti* Gornitz, 1921 NCR
[JS *Emberiza calandra calandra*]

Dr Carl P. A. Parrot (1867–1911) was a German ornithologist. He was the Founder and President of the Bavarian Ornithological Association (1897–1911).

Parsons, F.

Kakapo ssp. *Strigops habroptilus parsonsi* **Mathews** & **Iredale**, 1913 NCR; NRM
Chestnut-crowned Babbler ssp. *Pomatostomus ruficeps parsonsi* Mathews, 1918 NCR; NRM
White-browed Treecreeper ssp. *Climacteris erythrops parsonsi* Mellor, 1919 NCR
[JS *Climacteris affinis affinis*]
Welcome Swallow ssp. *Hirundo neoxena parsonsi* White, 1936 NCR
[NUI *Hirundo neoxena neoxena*]
Striated Fieldwren ssp. *Calamanthus fuliginosus parsonsi* **Condon**, 1951 NCR
[JS *Calamanthus campestris winiam*]

Frank Elliot Parsons (1882–1968) was an Australian land surveyor, ornithologist, collector and taxidermist.

Parsons, W.

Olive-backed Flowerpecker ssp. *Prionochilus olivaceus parsonsi* **McGregor**, 1927

William Parsons (DNF) was an American aviculturist who was in Manila, Philippines (1927).

Partridge

White-throated Spadebill ssp. *Platyrinchus mystaceus partridgei* Short, 1969
Burrowing Owl ssp. *Athene cunicularia partridgei* **Olrog**, 1976

William Henry Partridge (1924–1966) was an Argentine zoologist, ornithologist, entomologist and botanist at Argentine National Science Museum, Buenos Aires, and a specialist in the birds of Misiones Province.

Parvex

Horned Lark *Otocorys parvexi* **Taczanowski**, 1876 NCR
[JS *Eremophila alpestris brandti*]

Alfons Xavier (Ksawery) Parvex (1833–at least 1890) was a Polish-born Swiss amateur ornithologist who took part in the Polish uprising while working as a clerk in a paper factory. He was exiled with Godlewski (q.v.) to Siberia (1863–1868) where he studied and collected with Dybowski (q.v.) (1865–1868). He worked in a pet shop in Paris (1879), then collected in Canada and Brazil (1881–1882). The Verreaux Brothers (q.v.) in Paris employed him as a taxidermist (1883). There is no record of him after 1890. He is also commemorated in the name of a sponge.

Parzudaki

Hummingbird genus *Parzudakia* **Reichenbach**, 1855 NCR
[Now in *Heliangelus*]

Emerald sp. *Ornismya parzudaki* **Lesson**, 1838 NCR
[Alt. Cuban Emerald; JS *Chlorostilbon ricordii*]
Flame-faced Tanager *Tangara parzudakii* **Lafresnaye**, 1843
Parzudaki's Starfrontlet *Heliotrypha parzudakii* **Bonaparte**, 1854 NCR
[Alt. Tourmaline Sunangel; JS *Heliangelus exortis*]

Charles Parzudaki (d.1847) was a French traveller and Paris-based dealer in natural history specimens. He collected in Colombia (1841–1845). He was almost certainly related to Émile Parzudaki (see **Emile**), who may have taken over the family business after Charles's death.

Pasiphae

Sardinian Warbler ssp. *Sylvia melanocephala pasiphae*
Stresemann & **Schiebel**, 1925 NCR
[NUI *Sylvia melanocephala melanocephala*]

In Greek mythology, Pasiphaë was the wife of King Minos of Crete and mother of the Minotaur.

Pasquier

White-throated Wren-babbler *Rimator pasquieri*
Delacour & **Jabouille**, 1930

Lesser Necklaced Laughingthrush ssp. *Garrulax monileger pasquieri* Delacour & Jabouille, 1924

Pierre Marie Antoine Pasquier (1877–1934) was the Governor General of French Indochina (1927–1934); he died in an air crash.

Passek

Long-tailed Tit ssp. *Aegithalos caudatus passekii* **Zarudny**, 1904

Nikolai P. Passek (b.1850) was a Russian diplomat who was Consul-General in Bender-Bushire, Persia (Iran) (1876–1900). He was educated at Kings College, London, and took a law degree at Moscow University. He was the Russian Consul-General in Melbourne (1900) for Australia and New Zealand, and held a similar position in Canada (1913).

Passerini

Passerini's Tanager *Ramphocelus passerinii* **Bonaparte**, 1831
[Alt. Scarlet-rumped Tanager]

Professor Carlo(s) Passerini (1793–1857) was an Italian entomologist whose collection and papers are now at the Natural History Museum of the University of Pisa. He was an early enthusiast of scientific photography.

Pateff

Garden Warbler ssp *Sylvia borin pateffi* **von Jordans**, 1940 NCR
[JS *Sylvia borin woodwardi*]

Dr Pavel Pateff or Patev (1889–1950) was a Bulgarian ornithologist who was Curator of Birds at the Bulgarian Academy of Sciences (1928–1948), and Manager and Director of the Zoological Gardens, Sofia (1930–1948). He wrote *The Birds of Bulgaria* (1950).

Pat(ricia)

Barlow's Lark ssp. *Calendulauda barlowi patae*
J. Macdonald, 1953
Chestnut-capped Babbler ssp. *Timalia pileata patriciae*
Deignan, 1955

Jackson's Francolin ssp. *Pternistis jacksoni patriciae* **Ripley** & G. Bond, 1971 NCR; NRM
[Syn *Francolinus jacksoni patriciae*]

(See **Hall, P.**)

Paterson

Collared Sunbird ssp. *Anthodiaeta collaris patersonae*
Irwin, 1960
[SII *Anthodiaeta collaris zuluensis*]

Mrs Mary Lorajo Ball *née* Paterson (b. 1934) was a secretary at the National Museum, Bulawayo, Southern Rhodesia (Zimbabwe). She became very involved with the bird section until she resigned (1959) to marry. She now lives in Harare. She co-wrote *A Check List of the Birds of Southern Rhodesia* (1957).

Patricia

Black-throated Parrotbill ssp. *Suthora nipalensis patriciae*
Koelz,1954
Grey Wagtail ssp. *Motacilla cinerea patriciae* **Vaurie**, 1957

Mrs Patricia Vaurie *née* Wilson (1909–1982) was an American entomologist and collector and wife of the describer, Charles Vaurie (q.v.).

Patrick

Great Cormorant ssp. *Phalacrocorax carbo patricki* **J. G. Williams**, 1966 NCR
[Alt. White-breasted Cormorant; JS *Phalacrocorax lucidus*]

Patrick Curry (DNF) was the son of a friend of the describer. The describer says that '… whilst visiting the Queen Elizabeth National Park, Western Uganda, with Mrs Peter Curry, her son Mr Patrick Curry and Mr Bill Ryan, Patrick noticed some unusual-looking cormorants among some typical-looking White-breasted Cormorants'. (They now seem to be regarded just as a colour morph.)

Patrizi

Chestnut-naped Francolin ssp. *Francolinus castaneicollis patrizii* Toschi, 1958 NCR
[JS *Pternistis castaneicollis kaffanus*]

Marquis Don Saverio Patrizi Naro Montoro (1902–1957) was an Italian explorer, zoologist, collector and speleologist. He was the Italian signatory of a 1930s League of Nations treaty on African mammal preservation. During the Italian occupation of Ethiopia, Patrizi collected and prepared almost all the initial mammal and bird specimens currently displayed at Ethiopia's Zoological Natural History Museum. They were donated to the University College of Addis Ababa following their discovery in storage at Akaki (1955). Patrizi collected in central Africa (1920s) as well as in Ethiopia (1930s and 1940s). He collected in Kenya (1946–1947), having been interned there (WW2). A mammal is also named after him.

Paulian

Comoro Scops Owl *Otus pauliani* **C. W. Benson**, 1960

Renaud Paulian (1913–2003) was a French zoologist who was in Madagascar (1947–1961). He then served as Director of the Institut Scientifique de Congo-Brazzaville and head of the local university (1961–1966), moving to become Head of the Université d'Abidjan, Ivory Coast (1966–1969). He returned to France (1969) to become Rector of the Academy of Amiens, then Rector of the Academy of Bordeaux. He was elected correspondent of the Academy of Sciences (1975). He wrote a number of works including *Butterflies of Madagascar* (1951) and *Observations Ecologiques en Forêt de Basse Cote d'Ivoire* (1997). Two amphibians, two reptiles and a mammal are named after him.

Pauline (Barthod)

Metaltail sp. *Ornismya paulinae* **Boissonneau**, 1839 NCR
[Alt. Tyrian Metaltail; JS *Metallura tyrianthina*]

Pauline Barthod (fl.1839) is described as being the 'fiancée de l' intrépide voyageur à qui la science doit la découverte de beaucoup d'Oiseaux' – without giving his name!

Pauline (Knip)

Green Imperial Pigeon ssp. *Ducula aenea paulina*
    Bonaparte, 1854

Antoinette Pauline Jacqueline Knip *née* de Courcelles (1781–1851) was a French bird artist who wrote *Les Pigeons* (1811), and was employed as a painter on porcelain at the Sèvres factory, Paris.

Pauline (Rockefeller)

Yellow-breasted Warbler ssp. *Seicercus montis paulinae*
    Mayr, 1944

Mrs Pauline Watjen Rockefeller (1905–2000) was the wife of John Sterling Rockefeller (q.v.).

Paulucci

Blackcap ssp. *Sylvia atricapilla pauluccii* **Arrigoni**, 1902

Marianna Panciatichi Ximenes d'Aragona Marchesa Paulucci (1835–1919) was an Italian botanist, malacologist, ornithologist and collector. Her husband, also a botanist, cultivated a wide variety of plants at their villa near Parma.

Pauwels

Northern Double-collared Sunbird *Cinnyris chloropygius*
    pauwelsi **A. Dubois**, 1911 NCR
[JS *Cinnyris reichenowi*]

Captain Henry Pauwels (DNF) was a Belgian army officer in the Belgian Congo (Democratic Republic of Congo). His collection is in the Royal Museum for Central Africa, Tervuren, Belgium.

Payes

Little Bittern ssp. *Ixobrychus minutus payesii* **Hartlaub**, 1858

M. Payes (DNF) was a collector in Senegal.

Paykull

Band-bellied Crake *Porzana paykullii* Ljungh, 1813

Friherre (Baron) Gustav von Paykull (1757–1826) was a Swedish naturalist and collector specialising in entomology. He was 'hovmarskalk' (Lord Chamberlain) to the Royal Court. He presented his entire collection of 8,600 insects to the state (1819) in order to establish a National Natural History Museum in Stockholm. The museum possesses a specimen of the Great Auk *Pinguinis impennis*, which also came from him. His name is connected to a large number of beetles, and he described and named the Crab Plover *Dromas ardeola*.

Payn

Desert Lark ssp. *Ammomanes deserti payni* **Hartert**, 1924

Lieutenant-Colonel William Arthur Payn (1871–1955) was a British army officer who served in both the Boer War and WW1, in which he was badly wounded and afflicted by 'shell shock' for most of the rest of his life. He visited Kenya (1920–1921) and much of Mediterranean Europe and North Africa (1920s and 1930s), including Corsica and Cyprus (1938). He was a friend of Richard Meinertzhagen (q.v.), who wrote his obituary. He wrote 'Further notes on the birds of Corsica' (1931). He should not be confuse with Billy Payn (q.v.) – both of whom were known to Meinertzhagen and apparently in North Africa in the 1930s.

Paynter

Paynter's Brush Finch *Atlapetes leucopterus paynteri*
    Fitzpatrick, 1980
[Alt. White-winged Brush Finch ssp.]

Dr Raymond Andrew Paynter Jr (1925–2003) was an American ornithologist who was Assistant Curator (1953), then Curator (1961) of the Museum of Comparative Zoology, Harvard University. He edited *Avian Energetics* (1974) and wrote *Nearctic Passerine Migrants in South America* (1995), as well as a whole series of *Ornithological Gazetteers*. He received the Elliott Coues Award (2001). He revised Ernest Choate's *The Dictionary of American Bird Names*. His son is a wildlife artist and photographer.

Peale

Storm-petrel genus *Pealea* **Ridgway**, 1886 NCR
[Now in *Fregetta*]
Storm-petrel genus *Pealeornis* **Mathews**, 1932 NCR
[Now in *Fregetta*]

Peale's Egret *Ardea pealii* **Bonaparte**, 1827 NCR
[White morph of Reddish Egret, *Egretta rufescens*]
Peale's Petrel *Pterodroma inexpectata* **J. R. Forster**, 1844
[Alt. Mottled Petrel, Scaled Petrel]
Peale's Imperial Pigeon *Ducula latrans* Peale, 1848
[Alt. Barking Imperial Pigeon]
Peale's Parrotfinch *Erythrura pealii* **Hartlaub**, 1852
[Alt. Fiji Parrotfinch]

Collared Kingfisher ssp. *Todiramphus chloris pealei* **Finsch**
    & Hartlaub, 1867
Peale's Falcon *Falco peregrinus pealei* Ridgway, 1873
[Alt. Peregrine Falcon ssp.]

Titian Ramsay Peale (1799–1885) was an American naturalist and artist who collected in the Pacific – Tahiti, Fiji and Samoa (1838–1842) on the Wilkes (q.v.) expedition, aboard the *Peacock*. He was also noted for his pen and ink sketches of North American Indians and bison, made when assistant naturalist during an extensive expedition (1820) under Major Long, which followed the Lewis (q.v.) and Clark (q.v.) route. He accompanied an expedition (1818) to Florida, where he recorded the flora and fauna. Many of the birds which he obtained from around the world contributed to the original collections of the USNM. Peale Islet, near Wake Island in the Pacific Ocean, is named after him, as are three mammals. (Peale's Egret is the white morph of the Reddish Egret *Egretta rufescens*, regarded as a separate species for many years. Audubon wrote: 'While sailing towards the Florida Keys, my mind was agitated with anticipations of the delight I should experience in exploring a region whose productions were very imperfectly known. Often did I think of the Heron named after TITIAN PEALE, by my learned friend the Prince of MUSIGNANO [i.e. Bonaparte]. Mr. PEALE had procured only a single specimen, and in the winter season, but whether or not the species was abundant on the Keys of Florida remained to be discovered'.

Pearson

> Pearson's Cisticola *Dryodromas pearsoni* **Neave**, 1909 NCR
> [Alt. Black-tailed Cisticola; JS *Cisticola melanurus*]

Dr Arthur Pearson (1870–1947) was Principal Medical Officer to the Tanganyika Concession Ltd. (a British company created by mining engineer and explorer Sir Robert Williams).

Pease

> Brown-backed Honeybird ssp. *Prodotiscus regulus peasei* **Ogilvie-Grant**, 1901 NCR
> [JS *Prodotiscus regulus regulus*]
> Common Waxbill ssp. *Estrilda astrild peasei* **G. E. Shelley**, 1903
> Common Bulbul ssp. *Pycnonotus layardi peasei* **Mearns**, 1911 NCR
> [JS *Pycnonotus (barbatus) dodsoni*]

Sir Alfred Edward Pease (1857–1939) was an English MP (1885–1902), banker and magistrate, and a pioneer settler in Kenya. He was a big-game hunter and explorer in Somaliland (Somalia) (1900), also exploring in Sudan and the northern Sahara.

Pecile

> Pale Flycatcher sp. *Cossypha pecilei* **Oustalet**, 1886 NCR
> [JS *Bradornis pallidus murinus*]

Cavaliere Attilio Pecile (1856–1931) was an Italian explorer, naturalist and ethnographer in the Congo (1883–1886) with de Brazza (q.v.). He dropped out of university, secretly leaving for Africa to prevent his father stopping him. He was a great admirer of de Brazza, was deeply saddened by his death, and never again returned to Africa.

Pecquet

> Pesquet's Parrot *Psittrichas fulgidus* **Lesson**, 1830

M. Pecquet presented the holotype to Lesson (q.v.) with the information that he had obtained it from a quay in Le Havre, France. Lesson noted it was part of a consignment of birds from Patagonia and the River Plate region. However, he wrote later (correctly) that the specimen probably came from New Guinea. The bird was collected during the (1826–1829) circumnavigation by the French vessel *L'Astrolabe*. During this voyage Lesson's brother, Pierre Adolphe Lesson, was aboard as ship's surgeon and naturalist. This bird's taxonomic history is an excellent example of the problems of namings in the 19th century. René Primevère Lesson had first named the species *Banksianus fulgidus* (1830) (so that binomial has priority), but he re-named the same species *Psittacus pecquetii* (1831) and then used *Psittrichas pesqueti* (1838). Selby (1836) had used the name *Dasyptilus pequeti*; and various other combinations can be found in 19th-century ornithological works.

Pedler

> Chestnut-rumped Heathwren ssp. *Calamanthus pyrrhopygius pedleri* **Schodde** & Mason, 1999
> Short-tailed Grasswren ssp. *Amytornis merrotsyi pedleri* Christidis, Horton & Norman, 2008

Lynn P. Pedler is an ornithologist and biologist working for the Government of South Australia on Kanagroo Island, combining this with farming on the mainland.

Pedro Lima

> Band-winged Nightjar ssp. *Caprimulgus longirostris pedrolimai* **Grantsau**, 2008

Pedro Lima is a Brazilian ornithologist and photographer. He is based Bahia province and was in charge of the Cetrel Company's wildlife preservation programme at its Camaçari Petrochemical Complex (1988). He was still working for Cetrel (2002) and he was involved (2010) in projects on the coast of Bahia province in relation to migratory birds.

Peet

> American Bittern ssp. *Botaurus lentiginosus peeti* **Brodkorb**, 1936 NCR; NRM
> Black-headed Saltator ssp. *Saltator atriceps peeti* Brodkorb, 1940

Dr Max Minor Peet (1885–1949) was a neurosurgeon, Professor of Surgery at the University of Michigan, and keen ornithologist. His ornithological interests pre-dated his medical training (1910). His collection comprised over 32,000 specimens, which he bequeathed to the university where he graduated and worked. He was elected Associate of the AOU (1933). Brodkorb (q.v.), who was also at the University of Michigan, wrote: 'I take pleasure in naming this bird for Dr Max M. Peet in appreciation of his generous support of our work in Mexico.'

Pel

Pel's Fishing Owl *Scotopelia peli* **Bonaparte**, 1850
Bristle-nosed Barbet *Gymnobucco peli* **Hartlaub**, 1857

Hendrik Severinus Pel (1818–1876) was the Dutch Governor of the Gold Coast (Ghana) (c.1840–1850). He was also an amateur naturalist and trained taxidermist, and acted as such for the Leiden Museum, to which he spent shipments of animal specimens. He published a paper (1851) entitled 'Over de jagt aan de Goudkust, volgens eene tienjarige eigene ondervinding'. He is remembered in the names of other taxa, including two mammals.

Pel'tzam

Great Spotted Woodpecker ssp. *Dendrocopos major poelzami* **Bogdanov**, 1879

E. D. Pel'tzam (b.1837) was in the area of the Caspian Sea (1870) and a few years later was in northern Persia (now Iran), but we have failed to find any further information about him.

Pelzeln

Pelzeln's Grebe *Tachybaptus pelzelnii* **Hartlaub**, 1861
[Alt. Madagascar Grebe]
Rose-breasted Chat *Granatellus pelzelni* **P. L. Sclater**, 1865
Pelzeln's Tody Tyrant *Hemitriccus inornatus* Pelzeln, 1868
Pohnpei Starling *Aplonis pelzelni* **Finsch**, 1876
Bronze-olive Pygmy Tyrant *Pseudotriccus pelzelni*
Taczanowski & **Berlepsch**, 1885
Slender-billed Weaver *Ploceus pelzelni* Hartlaub, 1887
Grey-bellied Antbird *Ammonastes pelzelni* Sclater, 1890
[Syn. *Myrmeciza pelzelni*]
Brownish Elaenia *Elaenia pelzelni* Berlepsch, 1907
Planalto Slaty Antshrike *Thamnophilus pelzelni* **Hellmayr**, 1924

Grey Fantail ssp. *Rhipidura albiscapa pelzelni* **G. R. Gray**, 1862
Pelzeln's Finch *Sicalis flaveola pelzelni* Sclater, 1872
[Alt. Saffron Finch ssp.]
Barred Forest Falcon ssp. *Micrastur ruficollis pelzelni*
Ridgway, 1876
Pelzeln's Flycatcher *Myiarchus swainsoni pelzelni*
Berlepsch, 1883
[Alt. Swainson's Flycatcher ssp.]
Black-crowned Tityra ssp. *Tityra inquisitor pelzelni* **Salvin** &
Godman, 1890
Red-throated Caracara ssp. *Daptrius americanus pelzelni*
Pinto, 1948 NCR; NRM
[Syn. *Ibycter americanus*]

August Pelzel von Pelzeln (1825–1891) was an Austrian ornithologist. He was in charge of the mammal and bird collections at the Imperial Museum of Vienna for 40 years, where he worked on the 343 species of birds that Natterer (q.v.), a fellow Austrian, had collected in Brazil. He wrote *Ornithologie Brasileiras* (1871), and *Beitrage zur Ornithologie Sud Afrikas* (1882). A mammal is named after him.

Pemberton, C.

Sand Martin sp. *Cotile pembertoni* **Hartert**, 1902 NCR
[Alt. Brown-throated Martin; JS *Riparia paludicola*]

C. Hubert Pemberton (DNF) was a collector In Angola for the Tring Museum.

Pemberton, J. R.

Austral Thrush ssp. *Turdus falcklandii pembertoni* **Wetmore**, 1923
Gambel's Quail ssp. *Callipepla gambelii pembertoni* **van Rossem**, 1932

John Roy Pemberton (1884–1968) was known from childhood, at his own instigation, as 'Bill'. His father first aroused his interest in natural history by taking him on frequent outings. By age 11 he was a keen ornithologist. Before his degree he worked in the petroleum industry. He was a renowned student athlete and could have become a professional boxer. Torn between ornithology or geology, he decided on the latter on the grounds that there were more career opportunities, and being a geologist was likely to mean he would travel – so could study and collect local fauna in his spare time. Until paralysed by a stroke (1960), he collected bird skins, nests and eggs wherever he travelled. After graduation (1909) he worked as a geologist in Argentina (1910) before returning to the USA (1915). He worked for several petroleum companies in the USA, while continuing to collect and publish papers on ornithology. Among other things, he taught himself to play the piano and invented a new type of boomerang. In his spare time he used his own yacht to explore the islands off the Californian and Mexican coasts. He set up in private practice as a geologist (1940–1960), but continued to publish and take volunteer positions with the Cooper Bird Club. He was the leading expert on the California Condor *Gymnogyps californianus*. A mammal is also named after him.

Penard

Piculet sp. *Picumnus penardi* F. & A. Penard, 1908 NCR
[Alt. Golden-spanged Piculet; JS *Picumnus exilis buffoni*]

Smoky-fronted Tody Flycatcher ssp. *Poecilotriccus fumifrons penardi* **Hellmayr**, 1905
Great Jacamar ssp. *Jacamerops aureus penardi* **Bangs** &
Barbour, 1922

Thomas Edward Penard (1878–1936) was a Dutch-born US engineer and ornithologist. Born in Suriname he went to the USA (1891) for his education, graduating in electrical engineering (1900) from Massachusetts Institute of Technology. He started work (1901) with the Edison Electric Illuminating Company of Boston and was associated with them for the rest of his life. He returned to Suriname from time to time to add to his collections, which he eventually sold to the Museum of Comparative Zoology (1930). (See also **Arthur** & **Frederick** his brothers)

Penelope

Guan genus *Penelope* Merrem, 1786

Eurasian Wigeon *Anas penelope* **Linnaeus**, 1758

In Greek mythology, Penelope was the wife of Odysseus – see Homer's *Odyssey*. Her mother, Periboea, hid her infant daughter as soon as she was born, knowing that her husband wanted a son. When he discovered the truth, he threw the baby into the sea to drown. However, Penelope was rescued by a family of ducks. In contrast to the Homeric, Penelope was considered by some classical writers to be a lascivious adultress and she was depicted as being covered in contrasting colours of paints which might easily bring Neotropical guans to mind.

Pennant

Pennant's Parakeet *Platycercus elegans* **J. F. Gmelin**, 1788
[Alt. Crimson Rosella; *Psittacus pennantii* (Latham, 1790) is a JS]
King Penguin *Aptenodytes pennantii* **G. R. Gray**, 1844 NCR
[JS *Aptenodytes patagonicus*]

Thomas Pennant (1726–1798) was a highly regarded British naturalist, antiquary and traveller around the British Isles. His early work, *Tour in Scotland* (1771), was instrumental in encouraging tourism in the Highlands. Gilbert White (q.v.) published his *Natural History of Selbourne* in the form of letters to Thomas Pennant and Daines Barrington. Pennant published on the Arctic, Britain and India, and wrote about quadrupeds as well as birds. Among his other publications were *Genera of Birds* (1773), and *Arctic Zoology* (1784, 1785 and 1787). He was said to make 'dry and technical material interesting'. Three mammals are named after him.

Pentland

Puna Tinamou *Tinamotis pentlandii* **Vigors**, 1837
Andean Tinamou *Nothoprocta pentlandii* **G. R. Gray**, 1867

Joseph Barclay Pentland (1796–1873) was an Irish explorer and diplomat (1836–1839) in Bolivia. He worked for a time in Paris, studying with Cuvier (q.v.). He was particularly interested in geology, and the mineral pentlandite (a nickel iron sulphide) is named after him.

Penton

Erckel's Francolin ssp. *Francolinus erckelii pentoni* **Mackworth-Praed**, 1920 NCR; NRM

Colonel Richard Hugh Penton (1863–1934) was a British surgeon who joined the army (1887) and was seconded to the Egyptian Army (1892–1905), becoming Principal Medical Officer (1898–1905). He took part in Kitchener's campaigns to conquer Sudan: he was with the Dongala Expeditionary Force (1896), served in both Nile Expeditions (1898–1899) and was present at the Battles of Atbara and Khartoum. He was Assistant Director of Medical Services of 1st Indian Cavalry Division in Flanders (1915), later fulfilling the same function at Dunkirk. An amphibian is named after him.

Percival

Arabian Golden-winged Grosbeak *Rhynchostruthus percivali* **Ogilvie-Grant**, 1900
Percival's Oriole *Oriolus percivali* Ogilvie-Grant, 1903
[Alt. Montane Oriole, Black-tailed Oriole]

Black-crowned Tchagra ssp. *Tchagra senegalus percivali* Ogilvie-Grant, 1900
Slender-billed Greenbul ssp. *Stelgidillas gracilirostris percivali* **Neumann**, 1903
[Syn. *Andropadus gracilirostris percivali*]
Green-winged Pytilia ssp. *Pytilia melba percivali* Someren, 1919

Arthur Blayney Percival (1874–1940) was a British game warden in East Africa (1901–1928) who retired there, dying in Kenya. With the taxidermist W. Dodson he took part in a Royal Society expedition to Arabia (1899). He was one of the founders of the East Africa and Uganda Natural History Society (1909) and was known as one of the most knowledgeable wildlife experts and hunters in East Africa. He wrote *A Game Ranger's Notebook* (1924) and *A Game Ranger on Safari* (1928). Three mammals and a reptile are named after him. (See also **Blayney**)

Père David

(See under **David**)

Pereyra

Pereyra's Yellow Finch *Sicalis striata* Pereyra, 1937 NCR
[Alt. Saffron Finch; JS *Sicalis flaveola pelzelni*]

José A. Pereyra (d.1940) was an Argentine ornithologist, who published extensively, notably the 300-page paper 'Aves de la Región Ribereña Nordeste de la Provincia de Buenos Aires' (1938).

Pérez Chinchilla

Rufous-collared Sparrow ssp. *Zonotrichia capensis perezchinchillorum* **W. H. Phelps** Jr & **Aveledo**, 1984

Luis A. Pérez Chinchilla often worked with the junior author at the Phelps Ornithological Collection and they co-wrote 'Descripción de nueve subespecies nuevas y comentarios sobre dos especies de aves de Venezuela' (1994), while with Aveledo and others he wrote 'An annotated list of the birds of the Cerro Tamacuari region, Serranía de Tapirapecó, Federal Territory of Amazonas, Venezuela' (1995). The trinomial also includes Gilberto Pérez Chinchilla, whom Phelps employed as a collector. The two of them are described in the etymology (in translation) as '… who over 20 years have collaborated with great dedication to the study of our avifauna.' We have not been able to establish how the two men were related.

Péringuey

Antarctic Prion ssp. *Heteroprion desolatus peringueyi* **Mathews**, 1912 NCR
[JS *Pachyptila desolata banksi* (or species sometimes regarded as monotypic)]

Dr Louis Albert Péringuey (1855–1924) was a French entomologist and naturalist. He left France (1879) for South Africa, where he became a Scientific Assistant, South African Museum, Cape Town (1884), was in charge of the Invertebrates Collection (1885), and became the museum's director (1906–1924). He collapsed and died while walking home from the museum. Two reptiles are named after him.

Perkeo

Pygmy Batis *Batis perkeo* **Neumann**, 1907

Perkeo of Heidelberg (born Pankert Clemens or Giovanni Clementi) (1702–at least 1782) was a Tyrolean dwarf who was court-jester (1720) to the Elector Palatine Karl Philipp (reigned 1716–1743). He was notorious for a massive capacity for wine and is celebrated in a Heidelberg students' drinking-song. 'Perkeo' is supposedly derived from 'perché no?' (meaning 'why not?' in Italian), his reply whenever asked if he wanted another glass of wine. He fell ill one day and instead of wine drank water – which it is claimed killed him!

Perkins

Perkins's Mamo *Drepanis funerea* **A. Newton**, 1894
    EXTINCT
[Alt. Black Mamo]
Perkins's Creeper *Oreomystis perkinsi* **Rothschild**, 1900 NCR
[Hybrid? – *Manucerthia mana* x *Hemignathus virens*]

Dr Robert Cyril Layton Perkins (1866–1955) was an English naturalist and collector, primarily an entomologist, who moved to Hawaii (1892) to become Director of the Entomology Division at the Hawaiian Sugar Planters' Experimental Station. He discovered the Black Mamo (1893) in the Pelekunu Valley on Molokai; it was last sighted 1907. Perkins virtually lived in the field, only occasionally returning to Honolulu for provisions. Much of his collecting was sponsored by the Bishop Museum, where he became Assistant Superintendent of Entomology. He wrote dozens of scientific papers (1896–1938). He returned to England (1912), settling in Devon and working mainly on taxonomy. He was elected a Fellow of the Royal Society (1920) and was awarded the Linnean Society's top award – the Linnean Medal.

Perks

Yellow-rumped Thornbill ssp. *Acanthiza chrysorrhoa perksi*
    Mathews, 1912 NCR
[JS *Acanthiza chrysorrhoa leighi*]

Dr Robert Howell Perks (1854–1929) was an English physician, naturalist, photographer and collector in South Australia (1890–1897) as Medical Superintendent of the Adelaide Hospital. He was greatly opposed to the use of vivisection as a way to advance science and wrote *Why I Condemn Vivisection* (1905).

Pernetty

Plover genus *Pernettyva* **Mathews**, 1913 NCR
[Now in *Charadrius*]

Dom Antoine-Joseph Pernetty (1716–1801) was a Benedictine monk who was one of the original settlers to Les Îles Malouines (Falkland Islands). He was resident priest but doubled as botanist and chronicler of the expedition, which was led by Antoine Louis de Bougainville. He published his account (1770) as *Journal Historique d'un Voyage Fait aux Îles Malouines en 1763 et 1764 pour Reconnaître, et y Former un Établissement; et Deux Voyages au Détroit de Magellan avec une Relation sur les Patagons*. Pernetty wrote about the Warrah (Falklands 'Wolf') *Dusicyon australis*, which was the only endemic Falklands land mammal (extirpated 1876). Pernetty was a member of the Academy of Sciences in Paris and was Librarian to Frederick the Great at Berlin for several years. He is said to have founded a Rosicrucian order called the Illuminati of Avignon (c.1770) with Cagliostro and Mesmer as members; by this time he had been unfrocked for being a Cabalist and alchemist. A mammal and a botanical genus, *Pernettya*, are also named after him.

Perny

Crimson-breasted Woodpecker ssp. *Dendrocopos*
    cathpharius pernyii **J. Verreaux**, 1867

Abbé Paul Hubert Perny (1818–1907) was a parish priest at Besançon before becoming a missionary in Guizhou Province, China (1850–1860), and at Chungking, Sikian Province (1862–1868). He returned to France (1869) after a short period in Singapore. He was more interested in entomology than botany but collected specimens of each. He was particularly interested in silkworms and sent 500 silkmoth cocoons to the natural history musem in Lyon where the species was named *Bombyx pernyi*. He was imprisoned under the Paris Commune until his friend Jean-Pierre Guillaume Pauthier persuaded the authorities that he was harmless. He published a 459-page French/Latin/Mandarin Chinese dictionary (1869), also containing notes on the geography and administration of China and its plants. Perny later published translations of Chinese proverbs and other material. He is also commemorated in the names of other taxa including a mammal.

Péron

Emu genus *Peronista* **Mathews**, 1912 NCR
[Now in *Dromaius*]

Peron's Ground Thrush *Zoothera peronii* **Vieillot**, 1818
[Alt. Orange-banded Thrush]
Malaysian Plover *Charadrius peronii* **Schlegel**, 1865
Kangaroo Island Emu *Dromaius peroni* **Rothschild**, 1907
    NCR EXTINCT
[Now known as *Dromaius baudinianus*]

Whistling Duck ssp. *Dendrocygna javanica peroni*
    Mathews, 1912 NCR
[Alt. Wandering Whistling Duck ssp.; JS *Dendrocygna arcuata australis*]
Rufous Fieldwren ssp. *Calamanthus campestris peroni*
    Mathews, 1917 NCR
[JS *Calamanthus campestris rubiginosus*]

François Péron (1775–1810) was a French voyager and naturalist. He was a member of the Nicolas Baudin (q.v.) scientific expedition with the ships *Geographe* and *Naturaliste* to southern and western Australia (1800–1804). This visited

New Holland, Maria Island and Van Diemen's Land (Tasmania), as well as Timor in Indonesia. The Peron Peninsula, Western Australia, is named after him, while the fieldwren subspecies appears to be named after the peninsula (the type locality) rather than the man. Peron died of tuberculosis before he could properly study the vast collection of zoological specimens he made in Australia. Three reptiles, two amphibians and two mammals are named after him.

Perous

Many-coloured Fruit Dove *Ptilinopus perousii* **Peale**, 1848

(See **Laperouse**)

Perrein

Grey Waxbill *Estrilda perreini* **Vieillot**, 1817
[Alt. Black-tailed Waxbill]
Perrin's Bush-shrike *Telophorus viridis* Vieillot, 1817
[Alt. Gorgeous Bush-shrike]

Jean Perrein (1750–1805) was a French naturalist who travelled and collected in northern Africa, Cabinda, Arabia, Persia and India. He presented his specimens to the Academy of Sciences of Bordeaux, where he was elected an associate member. He went to North America (1794) and visited Hudson Bay, the Rockies, Quebec, Ontario and all the New England states. He died of malaria on his return journey to France. Buffon (q.v.) used many of his manuscripts in his own works. Perrein's *Voyage chez les Indiens de l'Amerique du Nord, avec un Aperçu des Usages et du Caractère de ces Peuples* was published posthumously (1809). The use of 'Perrin's Bush-shrike' as a common name for *Telophorus viridis* is something of a mystery, but Perrin is an anglicised version of Perrein, and the German name for this species is 'Perreinwürger'.

Perrett

Scarlet Sunbird ssp. *Aethopyga mystacalis perretti* **Harrisson** & **Hartley**, 1934 NCR
[Alt. Temminck's Sunbird; JS *Aethopyga temminckii*]

William Hedley Perrett (b.1903) was a British ornithologist and technical assistant in the Department of Zoology, BMNH (1917–1948).

Perrot

Red-billed Woodcreeper *Hylexetastes perrotii* **Lafresnaye**, 1844

Jean Perrot (1790–1858) was a French taxidermist.

Perrygo

Rufous-browed Peppershrike ssp. *Cyclarhis gujanensis perrygoi* **Wetmore**, 1950

Watson Mondell Perrygo (1906–1984) was an American taxidermist who collected (1927–1958) for and worked at the USNM (1927–1964). He was in Haiti (1929–1930) and Panama (1946–1953). He co-wrote with Alexander Wetmore (q.v.) 'The cruise of the "Esperanza" to Haiti' (1931).

Pesquet

Pesquet's Parrot *Psittrichas fulgidus* **Lesson**, 1830

(See **Pecquet**)

Peters, E.

Venezuelan Tyrannulet ssp. *Zimmerius improbus petersi* **Berlepsch**, 1907

Ernst Peters (DNF) was a theology student from Hamburg who made a collection during a short visit to Curaçao (1890).

Peters, J. L.

Hispaniolan Lizard Cuckoo ssp. *Coccyzus longirostris petersi* **Richmond** & **Swales**, 1924
Peters's Conure *Aratinga wagleri transilis* J. L. Peters, 1927
[Alt. Scarlet-fronted Parakeet ssp.]
Japanese Pygmy Woodpecker ssp. *Dryobates kizuki petersi* **Kuroda**, 1929 NCR
[JS *Dendrocopos kizuki matsudairai*; and latter SII *Dendrocopos kizuki kizuki*]
Spot-breasted Wren ssp. *Pheugopedius maculipectus petersi* **Griscom**, 1930
Blue-rumped Pitta ssp. *Hydrornis soror petersi* **Delacour**, 1934
Lineated Woodpecker ssp. *Dryocopus lineatus petersi* **van Rossem**, 1934 NCR
[NUI *Dryocopus lineatus similis*]
Yellow-winged Blackbird ssp. *Agelasticus thilius petersii* **Laubmann**, 1934
Slaty Cuckooshrike ssp. *Coracina schistacea petersi* **Neumann**, 1939 NCR; NRM
Peters's Dacnis *Xenodacnis parina petersi* **Bond** & **Meyer de Schauensee**, 1939
[Alt. Tit-like Dacnis ssp.]
Peters's Apalis *Apalis argentea eidos* J. L. Peters & **Loveridge**, 1942
[Alt. Kungwe Apalis ssp.]
Black-throated Sunbird ssp. *Aethopyga saturata petersi* **Deignan**, 1948
Chotoy Spinetail ssp. *Schoeniophylax phryganophilus petersi* **O. Pinto**, 1949

James Lee Peters (1889–1952) was an American ornithologist who was Curator of Birds at the Museum of Comparative Zoology, Harvard University. He is the renowned author of the authoritative *Check-list of Birds of the World*, commonly referred to simply as the *Peters*, published by Harvard University Press. He died before finishing the work, and the later volumes were completed by others. Peters was president of the AOU (1942–1945).

Peters, W. K. H.

Peters's Twinspot *Hypargos niveoguttatus* W. K. H. Peters, 1868
[Alt. Red-throated Twinspot]

Peters's Finfoot *Podica senegalensis petersii* **Hartlaub**, 1852
[Alt. African Finfoot ssp.]

Wilhelm Karl Hartwig Peters (1815–1883) was a German zoologist and traveller who made some very important collections in Mozambique. Most of his published works are on herpetology. For many years he was the Head of the Berlin Zoology Museum. He was elected a corresponding member of the Russian Academy of Sciences (1876). As many as 23 mammals, 18 amphibians and 39 reptiles are named after him.

Petersen

Sharp-tailed Cuckooshrike ssp. *Coracina mcgregori
    peterseni* **Salomonsen**, 1953 NCR; NRM
Mountain Leaf Warbler ssp. *Phylloscopus trivirgatus
    peterseni* Salomonsen, 1962

Erik Petersen (d.1961) was a Danish zoologist, Curator of Ornithology, Zoological Museum, Copenhagen. He was on the Danish Philippines Expedition (1951–1952).

Peterson

Cinnamon Screech Owl *Megascops petersoni* **Fitzpatrick &
    O'Neill**, 1986

Roger Tory Peterson (1908–1996) was an American ornithologist and a prolific author whose works include *A Field Guide to the Birds of North America* (1934), *A Field Guide to Western Birds* (1941), *A Field Guide to the Birds of Britain and Europe* (1954), and *A Field Guide to Mexican Birds* (1973). He is, of course, the man who invented the 'Peterson Identification System', which some birders have described as '… the greatest invention since binoculars'. The Roger Peterson Institute of Natural History in Jamestown, New York State, is named after him.

Petit

Petit's Cuckooshrike *Campephaga petiti* **Oustalet**, 1884

Petit's Saw-wing *Psalidoprocne* (*pristoptera*) *petiti* **Sharpe
    & Bouvier**, 1876
[Alt. Black Saw-wing ssp.]
Nkulengu Rail ssp. *Himantornis haematopus petiti* Oustalet,
    1884 NRM

Louis Petit Jr (1856–1943) was a French naturalist who collected in Angola (1876–1884) but not to be confused with his father, who had the same names, and was also an ornithologist who collected in Africa. Three reptiles are named after him.

Pettingill

Singing Quail ssp. *Dactylortyx thoracicus pettingilli* **Warner
    & Harrell**, 1957
Guatemalan Screech Owl ssp. *Megascops guatemalae
    pettingilli* Hekstra, 1982 NCR
[JS *Megascops guatemalae hastatus*]

Dr Olin Sewall Pettingill Jr (1907–2001) was an American ornithologist, collector and filmmaker. Cornell University awarded his doctorate (1933); he was later (1960–1973) Director of the Laboratory of Ornithology there. He was President, Wilson Ornithological Club (1948–1950). He wrote *Ornithology in Laboratory and Field* (1939).

Petz

Petz's Conure *Aratinga canicularis* **Linneaus**, 1758
[Alt. Orange-fronted Parakeet; *Sittace petzii* (Wagler, 1832)
    is a JS]

Dr Petz (DNF) is a rather enigmatic figure. It seems he may have collected the conure holotype in Mexico and deposited it in the museum in Würzburg, where Wagler (q.v.) found it and gave it his name.

Pewzow

Common Pheasant ssp. *Phasianus gmelini pewzowi*
    Alphéraky & **Bianchi**, 1907 NCR
[JS *Phasianus colchicus kiangsuensis*]

General Michail Wassiljewitsch Pewzow (1843–1902) was a geographer, cartographer and explorer who led an expedition to the Kuen Lun Mountains and Yarkand (1889–1890). A amphibian is named after him.

Pfrimer

Pfrimer's Parakeet *Pyrrhura pfrimeri* Ribeiro, 1920
[Syn. *Pyrrhura leucotis pfrimeri*]

Rudolph Pfrimer (d.c.1933) was a geologist and professional bird collector operating in Brazil (c.1910). He acquired a farm (mid-1920s) – Fazenda São Bento – to support his rock crystal exploration; this later became a centre for crystal extraction for several decades, operated by his nephew Walter Pfrimer. There are records of frogs he collected in the same area.

Pharaoh

Pharaoh Eagle Owl *Bubo ascalaphus* **Savigny**, 1809
[Alt. Savigny's Eagle Owl]

This owl occurs in Egypt, and across North Africa and parts of the Middle East, and is illustrated on ancient glyphs. There is no evidence of it being named after any particular pharaoh.

Phayre

Ashy-headed Green Pigeon *Treron phayrei* **Blyth**, 1862
[Syn. *Treron pompadora phayrei*]
Phayre's Pitta *Hydrornis phayrei* Blyth, 1862
[Alt. Eared Pitta]

Phayre's Francolin *Francolinus pintadeanus phayrei* Blyth,
    1843
[Alt. Chinese Francolin ssp.]
Brown-cheeked Fulvetta ssp. *Alcippe poioicephala phayrei*
    Blyth, 1845
Coral-billed Scimitar Babbler ssp. *Pomatorhinus
    ferruginosus phayrei* Blyth, 1847

Lieutenant-General Sir Arthur Purves Phayre (1812–1885) was Commissioner in Burma (Myanmar) (1862–1867), and Governor of Mauritius (1871–1878). He wrote a *History of Burmah* (1883). He also has a reptile and three mammals named after him.

Phelps

White-bearded Flycatcher genus *Phelpsia* **Lanyon**, 1984

Phelps's Brush Finch *Arremon perijanus* W. H. Phelps & **Gilliard**, 1940
[Alt. Perija Brush Finch; Syn. *Arremon torquatus perijanus*]
Phelps's Swift *Streptoprocne phelpsi* C. T. Collins, 1972
[Alt. Tepui Swift]

Great Antpitta ssp. *Grallaria excelsa phelpsi* Gilliard, 1939
Sharpbill ssp. *Oxyruncus cristatus phelpsi* **Chapman**, 1939
Clapper Rail ssp. *Rallus longirostris phelpsi* **Wetmore**, 1941
Yellow-bellied Tanager ssp. *Tangara xanthogastra phelpsi* **Zimmer**, 1943
Streaked Xenops ssp. *Xenops rutilans phelpsi* **Meyer de Schauensee**, 1959
Yellow-breasted Brush Finch ssp. *Atlapetes latinuchus phelpsi* **Paynter**, 1970

William Henry Phelps (1875–1965) was an American-born Venezuelan ornithologist. He first visited Venezuela as a Harvard student (1896). He wrote *Lista de las Aves de Venezuela* (1958). His son William 'Billy' Henry Phelps Jr (1902–1988) wrote *A Guide to the Birds of Venezuela* (1978) with Rodolphe Meyer de Schauensee (q.v.). The 'Colección Ornitológica Phelps' in Caracas consists of over 75,000 skins, mostly of Venezuelan origin. William, Billy and Billy's wife Kathleen (q.v.) built up the collection. The Phelps family had their own specially equipped yacht, *Ornis*, in which they made 49 trips to the Caribbean islands and to the hinterland of Venezuela. Phelps Sr. married the daughter of wealthy British settlers called Tucker in San Antonio de Maturin where he started selling coffee – the first of many successful business ventures. The flycatcher genus was named after William, Billy and Kathleen 'in recognition of their monumental contribution to our understanding of the ornithology of Venezuela'.

Phemonoë

Hummingbird genus *Phemonoe* **Reichenbach**, 1855 NCR
[Now in *Eriocnemis*]

In Greek mythology Phemonoë was a priestess and oracle of Apollo, and said to have been Apollo's daughter. Some traditions claim she invented hexameter verse.

Philby

Philby's Partridge *Alectoris philbyi* **Lowe**, 1934

Arabian Serin ssp. *Serinus angolensis philbyi* **Bates**, 1935 NCR
[JS *Serinus* (*Crithagra*) *rothschildi*]

Harry St John Bridger Philby (1885–1960) was a noted Arabist and a British explorer who was among the first Europeans to travel in the southern Arabian provinces. He became dissatisfied with British policy towards the Middle East and resigned from the Foreign Service (1925). He was an advisor to King ibn Saud of Saudi Arabia for 30 years. Philby became a Muslim, re-naming himself Hajj Abdullah. He wrote several works including *Heart of Arabia* (1923), *Saudi Arabia* (1955), and *Forty Years in the Wilderness* (1957). His son was Kim Philby, who spied for the Soviet Union in Britain. His wife is also remembered in a bird name (see **Dora**). He has a reptile named after him.

Philip

Common Iora ssp. *Aegithina tiphia philipi* **Oustalet**, 1885

Dr Philip (DNF) in Hue, Annam (Vietnam) sent specimens to Oustalet (1874 onwards) and provided the holotype of this bird (1883).

Philip (Sclater)

Cassowary sp. *Casuarius philipi* **Rothschild**, 1897 NCR
[Alt. Northern Cassowary; JS *Casuarius unappendiculatus*]

(See **Sclater**).

Philippa (Bruce)

Yellow-legged Pigeon *Ianthenas philippanae* **E. P. Ramsay**, 1882 NCR
[JS *Columba pallidiceps*]

Lady Philippa Bruce (1847–1924) was the wife of Admiral Sir James Andrew Bruce (1846–1921).

Philippa (Williams)

Philippa's Crombec *Sylvietta philippae* **J. G. Williams**, 1955
[Alt. Short-billed Crombec]

Dr Philippa Williams *née* Gaffikin (d.1993) was an anthropologist, archaeologist and physician in the Royal Air Force and married to the describer John Williams (q.v.).

Philippi

Philippi's Hermit *Phaethornis philippii* **Bourcier**, 1847
[Alt. Needle-billed Hermit]

Filippo de' Filippi (See **De Filippi**)

Philippina

Rufous-breasted Hermit *Heteroglaucis philippinae* **T. E. Penard**, 1922 NCR
[JS *Glaucis hirsutus*]

Mrs Philippina Penard *née* Salomons (fl.1922) was the mother of the describer, Dutch ornithologist Thomas Penard (q.v.).

Phillip

Norfolk Island Petrel *Pterodroma phillipii* **G. R. Gray**, 1862 NCR
[Alt. Providence Petrel; JS *Pterodroma solandri*]

Captain Arthur Phillip (1738–1814) was a British naval officer with a long and distinguished career, whose greatest task and success was leading the expedition to found the settlement at Botany Bay, Australia (1788). He was the first Governor of New South Wales and his five-year term of office greatly influenced the future development of Australia.

Phillips, A. R.

Sira Tanager *Tangara phillipsi* **Graves** & **Weske**, 1987

Bridled Titmouse ssp. *Baeolophus wollweberi phillipsi* **Van Rossem**, 1947
Orchard Oriole ssp. *Icterus spurius phillipsi* **Dickerman** & Warner, 1962
Boat-billed Heron ssp. *Cochlearius cochlearius phillipsi* Dickerman, 1973
Rufous-crowned Sparrow ssp. *Aimophila ruficeps phillipsi* Hubbard & **Crossin**, 1974
Lesser Greenlet ssp. *Hylophilus decurtatus phillipsi* **Parkes**, 1991
Swainson's Thrush ssp. *Catharus ustulatus phillipsi* **M. A. Ramos**, 1991
Mangrove Warbler ssp. *Dendroica petechia phillipsi* **Browning**, 1994
American Treecreeper ssp. *Certhia americana phillipsi* **Unitt** & **Rea**, 1997
Northern Cardinal ssp. *Cardinalis cardinalis phillipsi* Parkes, 1997

Allan Robert Phillips (1914–1996) was an American ornithologist and taxonomist whose speciality was the avifauna of south-west USA and Mexico. He co-authored *Birds of Arizona* (1964) and wrote *The Known Birds of North and Middle America* (1986 and 1991). He did not mince his words: an obituary said of him that his writings 'left many an ornithological sacred cow wounded on the arena floor'. A mammal is named after him.

Phillips, B.

Phillips's Woodpecker *Picus phillipsii* **Audubon** 1839 NCR
[Alt. Hairy Woodpecker; JS *Picoides villosus*]

Dr Benjamin Phillips (1805–1861) was an English physician and Audubon's doctor in London. Audubon, in his *Ornithological Biography*, wrote 'In naming it after my friend BENJAMIN PHILLIPS, Esq., F.R.S., I have the pleasure of testifying my esteem and gratitude towards one whose kindness and generosity has often been experienced by me and every member of my family'. A reptile is named after him.

Phillips, E. E. L.

Phillips's Wheatear *Oenanthe phillipsi* **G. E. Shelley**, 1885
[Alt. Somali Wheatear]

Ethelbert Edward Lort Phillips (1857–1926) was a British traveller and hunter who shot big game around the world. He was also a collector of natural history specimens, particularly mammals and birds. He was in East Africa (1884–1895) and, with a party of friends, explored parts of Somaliland (Somalia) (1895). This must have been a grand Victorian odyssey, as the party included both Mrs Phillips (see **Louisa**) and a Miss Gillett, with her brother Frederick Alfred Gillett (q.v.). Phillips is also remembered in Norway as the man who developed an estate called Vangshaugen on Lake Storvatnet, where he planted a rhododendron garden – virtually unknown in Norway at that time. He became Vice-President of the Zoological Society of London. Three reptiles and two mammals are also named after him. (See also **Lort**)

Phillips, J. C.

American Robin ssp. *Turdus migratorius phillipsi* **Bangs**, 1915
Correndera Pipit ssp. *Anthus correndera phillipsi* **W. S. Brooks**, 1916 NCR
[JS *Anthus correndera grayi*]

Dr John Charles Phillips (1876–1938) was an American physician and traveller. He commanded a field hospital during WW1 before and after which he travelled widely, making zoological collections for Harvard's Museum of Comparative Zoology. He often represented the USA in international conservation meetings and was also largely responsible for the Migratory Bird Treaty, which involved the USA, Canada, and Mexico. He co-authored *Extinct and Vanishing Mammals of the Old and New World* (in sections 1942–1945). An mammal and four reptiles are named after him.

Phillips, J. G.

Collared Sunbird ssp. *Anthodiaeta collaris phillipsi* **C. M. N. White**, 1950 NCR
[JS *Anthodiaeta collaris garguensis*]

J. G. Phillips (DNF) was a British academic who was at Oxford with the describer Charles White (q.v.).

Phillips, W. W. A.

Blue-eared Kingfisher ssp. *Alcedo meninting phillipsi* **E. C. S. Baker**, 1927
Tawny-bellied Babbler ssp. *Dumetia hyperythra phillipsi* **Whistler**, 1941
Yellow-footed Green Pigeon ssp. *Treron phoenicopterus phillipsi* **Ripley**, 1949

Major William Watt Addison Phillips (1892–1981) was a tea and rubber planter and naturalist in Ceylon (Sri Lanka). While a WW1 prisoner-of-war in Turkey he developed an interest in zoology. He was Secretary and later Chairman of the Ceylon Bird Club. He also collected herpetofauna. He wrote *Check List of the Birds of Ceylon* (1952). He returned to England in 1956. A reptile is also named after him.

Phipps

Kakariki (parakeet) genus *Phippspsittacus* **McAllan** & M. D. Bruce, 1989 NCR
[Now in *Cyanoramphus*]

Graeme R. Phipps is an Australian aviculturist and conservationist. He was a Senior Curator at Taronga Zoo, Sydney.

Phoebe

Eastern Phoebe *Sayornis phoebe* **Latham**, 1790
Black Metaltail *Metallura phoebe* **Lesson** & **de Lattre**, 1839

In Greek mythology Phoebe was one of the Titans, the children of Uranus and Gaia. She was traditionally associated with the moon. Sometimes Phoebe is used as another name for the goddess Artemis. The name is now used as a common name for New World flycatchers of the genus *Sayornis*.

Piaggi

Abyssinian Ground Thrush *Zoothera piaggiae* **Bouvier**, 1877

Carlo Piaggi (1827–1882) was an Italian explorer who collected in tropical Africa (1876–1882).

Pickering

Pickering's Imperial Pigeon *Ducula pickeringii* **Cassin**, 1854
[Alt. Grey Imperial Pigeon]

Dr Charles F. Pickering (1805–1878) was an American ornithologist and friend of Audubon (q.v.). He explored the White Mountains in New Hampshire (1825), qualified as a physician (1826) at Harvard and set up in practice (1827) in Philadelphia. He was associated with the Philadelphia Academy of Natural Sciences as Librarian (1828–1833) and thereafter as Curator. He took part in the United States Exploring Expedition (1838–1842) as naturalist, sailing on the *Vincennes*. He helped Holbrook with the publication of *North American Herpetology* (1842) and travelled (1844) to study ethnology from Egypt to Zanzibar, and then to India, writing *The Races of Man: and Their Geographical Distribution* (1847). He visited Thomas Nuttall (q.v.) (1854) in England and is remembered in the name of a reptile and an amphibian.

Pielou

Red-bellied Woodpecker ssp. *Melanerpes carolinus harpaceus* **Koelz**, 1954 NCR; NRM

William Pielou (DNF) was a Profesor of Biology at Alma College, Michigan (1952–1953) (see **Harpaceus**).

Pierre

Greater Yellownape ssp. *Chrysophlegma flavinucha pierrei* **Oustalet**, 1889

M. E. Pierre (DNF) was a French colonial administrator in Cochinchina (Vietnam) (1888).

Piersma

Red Knot ssp. *Calidris canutus piersmai* **Tomkovich**, 2001

Dr Theunis Piersma (b.1958) is a Dutch biologist, ecologist and ornithologist who became (2003) Professor of Animal Ecology, University of Groningen, Netherlands, where he gained his bachelor's degree (1980), master's (1984) and doctorate (1994). He is most closely associated with the study of shorebirds.

Pifano

Variegated Bristle Tyrant ssp. *Pogonotriccus poecilotis pifanoi* **Aveledo** & **Pons**, 1952

Professor Félix Pifano Capdevielle (1912–2003) was a Venezuelan physician, parasitologist and founder (1941) of the Faculty of Tropical Medicine, Central University, Caracas.

Pilette

Crossley's Ground Thrush ssp. *Zoothera crossleyi pilettei* **Schouteden**, 1918
Dusky Crimsonwing ssp. *Cryptospiza jacksoni pilettei* Schouteden, 1918 NCR; NRM

André Pilette (DNF) was a Belgian hunter and collector in the Belgian Congo (Democratic Republic of Congo). He wrote *À Travers l'Afrique Équatoriale* (1914).

Pilumnus

Sapsucker genus *Pilumnus* **Bonaparte**, 1854 NCR
[Now in *Sphyrapicus*]

In Roman mythology, Pilumnus was brother to Picumnus, the personification of the woodpecker.

Pimento

Nicobar Hawk Owl ssp. *Ninox affinis rexpimenti* **Abdulali**, 1979
[SII *Ninox affinis isolata*]

(See under **Rex Pimento**)

Pinchot

Green-breasted Mango ssp. *Anthracothorax prevostii pinchoti* **Wetmore**, 1930
[NUI *Anthracothorax prevostii hendersoni*]

Gifford Pinchot (1865–1946) went to Yale (1885) but found no suitable course there, and after taking an arts degree, went to Nancy, France, to study forestry. After returning to the USA he worked as resident forester for the Vanderbilt family. He was in government service (1898–1910) as Chief of the Division of Forestry, Department of the Interior, transferring as first head of a new Forestry Service (1905). During his administration the number of national forests in the USA grew enormously, but President Taft sacked him (1910) in a controversy over coal claims in Alaska. He entered politics and was Governor of Pennsylvania (1923–1927 and 1931–1935). He was in the Caribbean (1929) on an epic voyage with his son, aboard the *Mary Pinchot*, from New York to the Society Islands via Key West, Colombia, the Galapagos, and the Marquesas. A reptile is also named after him.

Ping

Greater Necklaced Laughingthrush ssp. *Garrulax pectoralis pingi* **Cheng**, 1963

Dr Chih Ping (1886–1965), whose doctorate was awarded (1918) by Cornell, became a zoology investigator at Wistar Institute, Philadelphia, then Professor of Zoology, College of Agriculture, National Southeastern University, Nanking (1921). He was director of the biological laboratory at Nanking (Nanjing) (1935). Two amphibians are named after him.

Pinon

Pinon's Imperial Pigeon *Ducula pinon* **Quoy** & **Gaimard**, 1824
[Alt. Bare-eyed Imperial Pigeon]

Rose Pinon de Freycinet (1796–1832) was the wife of the French explorer Captain Louis Freycinet (q.v.). Freycinet led a round-the-world expedition (1817) aboard the 350-ton corvette, *L'Uranie*; his wife disguised herself as a young man and boarded the ship as a stowaway in order to accompany him. Her journal was published (1996) under the title *A Woman of Courage: The Journal of Rose de Freycinet on her Voyage around the World, 1817–1820*. She died of cholera in Paris after nursing Louis through the same illness.

Pinsker

Pinsker's Hawk Eagle *Nisaetus pinskeri* Preleuthner & Gamauf, 1998
[Syn. *Nisaetus philippensis pinskeri*]

Professsor Dr Wilhelm Pinsker (b.1945) studied zoology and botany at the University of Vienna (1968–1973) and became Associate Professor at the Institute of General Biology there (1988–2003). He was co-author of the paper (1998) 'Distribution and field identification of Philippine birds of prey'. Dr Pinsker told us 'A former doctoral student of mine (Monika Preleuthner) went for a field trip to the Philippines and decided to name the first new taxon she might discover after me. It happened that she described a new subspecies (*Spizaetus philippensis pinskeri*). Later another former student of mine (Elisabeth Haring) published a molecular phylogeny of the genus *Spizaetus* where my subspecies was raised to species rank and moved into the genus *Nisaetus*.'

Pintado

Pintado Petrel *Daption capense* **Linnaeus**, 1758
[Alt. Cape Petrel]

This is another example where someone has assumed a person is involved in the name when they are not. 'Pintado' means 'painted' in Spanish and is a description of how the bird looks.

Pinto, A. A.

Flappet Lark ssp. *Mirafra rufocinnamomea pintoi* **C. M. N. White**, 1956
Stierling's Wren Warbler ssp. *Calamonastes stierlingi pintoi* **M. P. S. Irwin**, 1960
Pinto's (Double-collared) Sunbird *Cinnyris manoensis pintoi* **Wolters**, 1965
[Alt. Miombo Double-collared Sunbird ssp.]
Chin-spot Batis ssp. *Batis molitor pintoi* **W. J. Lawson**, 1966

Dr Antonio Augusto da Rosa Pinto (1904–1986) was a Portuguese zoologist and ornithologist. He wrote on several national parks and their avifaunas, and a standard work on the ornithology of Angola, *Ornitologia de Angola*. He was Director of the Museum in Lourenço Marques (Maputo), Mozambique (1953–1960).

Pinto, O. M.

Pinto's Spinetail *Synallaxis infuscata* Pinto, 1950
[Alt. Plain Spinetail]
Sulphur-breasted Parakeet *Aratinga pintoi* Silveira *et al.*, 2005 NCR
[JS *Aratinga maculata* (Müller, 1776)]

Tropical Screech Owl ssp. *Otus choliba pintoi* **Kelso**, 1937 NCR

[Regarded as a JS of *Megascops atricapilla* or *Megascops sanctaecatarinae*]
Channel-billed Toucan ssp. *Ramphastos vitellinus pintoi* **J. L. Peters**, 1945 NCR
[Hybrid population: *R. vitellinus culminatus* x *R. vitellinus ariel*]
Lesser Woodcreeper ssp. *Xiphorhynchus fuscus pintoi* Longmore & Silveira, 2005

Olivério Mario de Oliveira Pinto (1896–1981) was a Brazilian ornithologist. He was the Head of the Zoological Museum of the University of São Paulo (1940s), and led expeditions under its auspices. He wrote *Catalogo sas Aves do Brasil* (1938). An amphibian is also named after him.

Pinwill

White-browed Scimitar Babbler ssp. *Pomatorhinus schisticeps pinwilli* **Sharpe**, 1883 NCR
[JS *Pomatorhinus schisticeps leucogaster*]

Captain William Stackhouse Church Pinwill (1831–1926) was an ornithologist and a British army officer in the 27th Regiment of Foot (1852–1868). He served in India (1861–1868) but resigned his commission on inheriting an estate and garden in Cornwall, England. He let the farmland and concentrated on turning the garden into a show place.

Pittier

White-breasted Wood Wren ssp. *Henicorhina leucosticta pittieri* **Cherrie**, 1893

Henri François Pittier (1857–1950) was a Swiss botanist, geographer, civil engineer and all-round naturalist. He published over 300 papers, monographs and books in various languages covering a wide variety of subjects including geography, botany, forestry, archaeology, ethnography, linguistics, geology and climatology. He moved to Costa Rica (1887), but travelled over much of Latin America. The oldest nature reserve in Venezuela (created 1937) was designated the 'Henri Pittier National Park' (1953). It was well known to Pittier himself, who visited often over the course of 30 years. A mammal and an amphibian are also named after him.

Planas

Grey-fronted Honeyeater ssp. *Lichenostomus plumulus planasi* **A. J. Campbell**, 1910

Father Planas (DNF) was a Spanish Benedictine missionary. He was head of Drysdale River Mission, Kimberley Division, Western Australia.

Plate

Buller's Albatross ssp. *Thalassarche bulleri platei* **Reichenow**, 1898

Dr Ludwig Hermann Plate (1862–1937) was a German zoologist and geneticist. He studied under Haeckel at Friedrich Schiller University, Jena, which awarded his doctorate (1886). He qualified in zoology at Philipps University, Marburg (1888), then taught at the Veterinary High School in Berlin (1898–1905) and at the Agricultural College (1905–1908) as

Professor of Zoology. He was Haeckel's successor as Professor at Jena (1909–1935) but became embroiled in an unpleasant case when he accused Haeckel and his circle of slandering him. He was a convinced Darwinist but was also a virulent anti-Semite. A reptile is also named after him.

Platen, C. C.

Platen's Rail *Aramidopsis plateni* **W. H. Blasius**, 1886
[Alt. Snoring Rail]
Platen's Flowerpecker *Prionochilus plateni* W. H. Blasius, 1888
[Alt. Palawan Flowerpecker]
Platen's Babbler *Sterrhoptilus plateni* W. H. Blasius, 1890
[Alt. Mindanao Pygmy Babbler; Syn *Dasycrotapha plateni*]
Mindanao Hawk Owl *Ninox plateni* **Hartlaub**, 1899 NCR
[JS *Ninox mindorensis*]

Olive-backed Sunbird ssp. *Cinnyris jugularis plateni* W. H. Blasius, 1885
Mangrove Whistler ssp. *Pachycephala cinerea plateni* W. H. Blasius, 1888

Carl Constantin Platen (1843–1899) was a German doctor who collected in the East Indies and southern Philippines (1878–1894). He has butterflies named after him and also seems to have imported live birds (1879). There are also a number of birds whose scientific names honour his wife (q.v.). Platen was reputedly nearly blind and deaf by the time he returned to Germany (1894).

Platen, M.

Blue-headed Racquet-tail *Prioniturus platenae* **W. H. Blasius**, 1888
Palawan Flycatcher *Ficedula platenae* W. H. Blasius, 1888
Philippine Dwarf Kingfisher *Ceyx platenae* W. H. Blasius, 1890 NCR
[JS *Ceyx melanurus mindanensis*]
Mindoro Bleeding-heart *Gallicolumba platenae* **Salvadori**, 1893

Northern Golden Bulbul ssp. *Thapsinillas longirostris platenae* W. H. Blasius, 1888

Margarete Platen *née* Geisler (DNF) was the wife of the German collector Carl Platen (above). Together they visited the Philippines during several summers of 1887, 1889 and 1890. Blasius published the ornithological results, not in an ornithological journal, but in an ordinary Braunschweig newspaper, thus greatly upsetting the scientific world. (See also **Margarete (Platen)**)

Plenge

Plenge's Thistletail *Asthenes fuliginosa plengei* **O'Neill** & **T. A. Parker**, 1976
[Alt. White-chinned Thistletail ssp.]
Unstreaked Tit Tyrant ssp. *Anairetes agraphia plengei* Schulenberg & Graham, 1981
[Syn. *Uromyias agraphia plengei*]

Manuel Alberto Plenge (b.1937) is a Peruvian ornithologist and conservationist who in his youth emigrated from Peru to the USA. He served in the US Army and was stationed in Korea for 17 months. He eventually returned to Peru and worked for a copper mining company (1962–2008), retiring as Director of Logistics. He is a member of many different ornithological and conservation organisations as well as author of scientific papers. He co-wrote *An Annotated Checklist of Peruvian Birds* (1982). He also prepared a *List of the Birds of Peru* (2002) and *Bibliography of the Birds of Peru* (2011) and uploaded both to the webpage of the 'Unión de Ornitólogos del Perú', updating them periodically.

Pleschanka

Pleschanka's Pied Chat *Oenanthe pleschanka* Lepechin, 1770
[Alt. Pied Wheatear]

This is not named after a person, but is another example of transcription error: 'pleschanka' is simply the Russian name for the Pied Wheatear.

Pleske

Pleske's Warbler *Locustella pleskei* **Taczanowski**, 1890
[Alt. Styan's Grasshopper Warbler]
Pleske's Ground Jay *Podoces pleskei* **Zarudny**, 1896
[Alt. Iranian Ground Jay]

Rock Ptarmigan ssp. *Lagopus muta pleskei* Serebrovski, 1926
Western Capercaillie ssp. *Tetrao urogallus pleskei* **B. Stegmann**, 1926
Peregrine Falcon ssp. *Falco peregrinus pleskei* Dementiev, 1934 NCR
[JS *Falco peregrinus japonensis*]

Fedor Dimitrievich Pleske (1858–1932) was a Russian zoologist, geographer and ethnographer. From childhood he gathered a collection of birds and insects in the different provinces of European Russia. He also analysed ornithological collections of other travellers in central Asia, including Przewalski (q.v.), describing several new species. He became Scientific Secretary of the St Petersburg Natural History Society (1881) and joined their expedition to the Kola Peninsula. He graduated from St Petersburg University (1882) and became a Fellow of the Russian Imperial Academy of Science, St Petersburg (1886), then Scientific Keeper (Curator) at their Zoological Museum, and finally Director (1892–1896) before retiring through ill health. He was an active member of the USSR Zoological Museum (1918). He wrote *Ornithological Fauna of Imperial Russia* (1891) and *Ornithographia Rossica* (1898) as well as on the systematics of Arctic birds, and other taxa such as Diptera (gadflies, horse-flies etc.) His final monograph *The Birds of Eurasian Tundra* (1928) was written in English and edited in the USA. It was devoted largely to the heroic role of Admiral Aleksandr Vasiliyevich Kolchak (1874–1920): since Kolchak later (1918) commanded the White Russian forces trying to overthrow the communist regime in Russia, Pleske would have risked at least a long spell in the gulag or even a bullet, for mentioning Kolchak in any positive light was dangerous even as late as 1970s. A reptile and an amphibian are named after him.

Plessen

Southern Boobook ssp. *Ninox boobook plesseni*
Stresemann, 1929

Golden-bellied Gerygone ssp. *Gerygone sulphurea plesseni*
Stresemann, 1926 NCR

[JS *Gerygone sulphurea sulphurea*]

Elegant Pitta ssp. *Pitta elegans plesseni* **Meise**, 1929 NCR

[JS *Pitta elegans virginalis*]

Helmeted Friarbird ssp. *Philemon timoriensis plesseni*
Rensch, 1929 NCR

[NUI *Philemon buceroides neglectus*]

Cave Swiftlet ssp. *Collocalia esculenta plesseni* Meise,
1941 NCR

[JS *Collocalia linchi linchi*]

Baron Viktor von Plessen (1900–1980) was a German explorer and ornithologist in South-East Asia. After military service he studied art (1917–1918). His first expeditions were to the Malay Peninsula (1924–1925 and 1927–1928). He was in the Dutch East Indies (1930s) acting as ethnological adviser for the production of a film called *Legong: Dance of the Virgins: A Story of the South Seas* (1935), and later in Borneo (1934–1935 and 1937–1938) for the film *Head Hunters of Borneo* (1938). He also made scientific films always accompanied by his wife Marie-Isabel. (See also **Hutz** and **Mary**)

Plowes

Cape Bunting ssp. *Emberiza capensis plowesi* **J. Vincent**,
1950

Darrel Charles Herbert Plowes (b.1925) is a citizen of Zimbabwe with over 50 years of field experience in southern Africa. He was an agricultural director and is now retired and lives in Mutare, Zimbabwe, but he remains active in photography of the natural world and is still collecting plants. A reptile is named after him.

Pluto

Black Swan *Anas plutonia* Shaw & Nodder, 1791 NCR

[JS *Cygnus atratus*]

Pohnpei Flycatcher *Myiagra pluto* **Finsch**, 1876

Oriental Magpie Robin ssp. *Copsychus saularis pluto*
Bonaparte, 1850

Black Myzomela ssp. *Myzomela nigrita pluto* Forbes, 1879

In Greek mythology, Pluto was the god of the underworld. His name is usually applied to birds (and other animals) of a very dark colour. Three mammals are named after Pluto.

Poelzam

Great Spotted Woodpecker ssp. *Dendrocopos major
poelzami* **Bogdanov**, 1879

(See under **Pel'tzam**)

Poeppig

Curl-crested Aracari *Pteroglossus poeppigii* **Wagler**, 1932
NCR

[JS *Pteroglossus beauharnaesii*]

Professor Eduard Friedrich Poeppig (1798–1868) was a German naturalist, botanist and collector. He studied medicine and natural science at the University of Leipzig, leaving to undertake an expedition to Cuba and the USA. He was on an expedition to Brazil, Chile and Peru (1826–1832). This involved sailing from Baltimore, rounding Cape Horn, travelling through Chile and Peru to cross the Andes and by canoe to join the Amazon at Nauta and thence to Pará, Brazil, to take ship for to Germany. Here he became Professor of Zoology back at the University of Leipzig. He wrote *Reise nach Chili, Peru, und auf dem Amazonen-Flusse* (1835). A mammal, reptile and amphibian are named after him.

Pogge

Little Grebe ssp. *Tachybaptus ruficollis poggei* **Reichenow**,
1902

Lieutenant Karl Pogge (1873–1959) was a German army officer who was in China at the time of the Boxer Rebellion (1900).

Poilane

Pale-headed Woodpecker ssp. *Gecinulus grantia poilanei*
Deignan, 1950

Eugène Poilane (1887–1964) was a French botanist who was in Cochinchina (Vietnam) and neighbouring areas of French Indochina (1909–1964). He started a coffee plantation (1918). He fathered 10 children, five of them after the age of 60. He was assassinated by Viet Cong troops. Two reptiles are named after him.

Polatzek

Lesser Short-toed Lark ssp. *Calandrella rufescens polatzeki*
Hartert, 1904

Blue Chaffinch ssp. *Fringilla teydea polatzeki* Hartert, 1905

Thekla Lark ssp. *Galerida theklae polatzeki* Hartert, 1912
NCR

[JS *Galerida theklae theklae*]

Captain R. Johann Polatzek (1839–1927) was an Austrian ornithologist and taxidermist who collected in the Canary Islands (1902–1905) and Balearic Islands (1910). He published an important series of papers under the title *Die Vögel der Kanaren* (1908–1909).

Poliakov

European Goldfinch ssp. *Carduelis caniceps poliakovi*
Sushkin, 1925 NCR

[JS *Carduelis carduelis subulata*]

Grigoriy Ivanovich Polyakov (1876–1939) was a Russian taxidermist and collector. He graduated from evening classes at Shanyavsky People's University of Moscow. He was imprisoned (1927–1932) in Solovki, Solovetsky Islands in the White Sea. He served there as head of the biological research station of the society of local naturalists. He later lived near Moscow (1932–1939).

Polivanov

Polivanov's Parrotbill *Paradoxornis polivanovi* **Stepanyan**, 1974

[Alt. Reed Parrotbill, Chinese Parrotbill; Syn. *Paradoxornis heudei polivanovi*]

Dr.Vladimir Mikhailovich Polivanov (DNF) worked for many years in the Russian Far East (on Lake Khanka) and in the northern Caucasus, mainly in the Teberda Nature Reserve.

Pollen, F. P. L.

Comoros Olive Pigeon *Columba pollenii* **Schlegel**, 1865
Pollen's Vanga *Xenopirostris polleni* Schlegel, 1868

François Paul Louis Pollen (1842–1886) was a Dutch naturalist who collected in Madagascar (1862–1866) for the Leiden Museum. He must have been wealthy, as he paid for his own collecting trips and sponsored other research in Madagascar. He wrote *Récherches sur la Faune de Madagascar et de ses Dépendances – d'Après les Découvertes de François P. L. Pollen et D. C. van Dam* (1868). Despite his name, the French colony where he collected, and the language in which he published, he was *not* French. Two reptiles are also named after him.

Pollen, W. M. & R. F.

Jackson's Francolin ssp. *Pternistis jacksoni pollenorum* **Meinertzhagen**, 1937 NCR; NRM

Captain Sir Walter Michael Hungerford Pollen (1894–1968) served in the British army in WW1 and became a colonial administrator in the Sudan (1918–1928). He and his wife Lady Rosalind Frances Pollen *née* Benson (b.1899) are both honoured in the name of the francolin. Meinertzhagen (q.v.) was a friend, as he and Theresa Clay (q.v.) wrote in conjunction with Captain Pollen in regard to bird fleas that they had collected on the island of Ushant, Brittany.

Poltaratsky

Poltaratsky's Starling *Sturnus vulgaris poltaratskyi* **Finsch**, 1878

[Alt. Common Starling ssp.]

Major-General W. A. Poltaratsky (DNF) was the Russian Governor of Semipalatinsk. He explored in Central Asia (1867–1868) and Siberia.

Polymnia

Hummingbird genus *Polymnia* **Mulsant & E & J. Verreaux**, 1865 NCR

[Now in *Chaetocercus*]

In Greek mythology, Polymnia (or Polyhymnia) was the Muse of lyric poetry and the mother of Orpheus. An amphibian is also named after her.

Polyxemus

Hummingbird genus *Polyxemus* **Mulsant & E. Verreaux**, 1877 NCR

[Now in *Chaetocercus*]

In Greek mythology, Polyxemus was a son of the sorceress Medea. He is also known as Medus. It is possible that the describers (who were fond of taking names from mythology) were thinking of Polyxena, a beautiful young Trojan princess.

Pomare

Monarch flycatcher genus *Pomarea* **Bonaparte**, 1854

Maupiti Monarch *Pomarea pomarea* Lesson & **Garnot**, 1828 EXTINCT
Gull sp. *Gelastes pomare* Bonaparte, 1856 NCR
[Alt. Silver Gull; JS *Chroicocephalus novaehollandiae*]

Tu Tunuieaiteatua Pomare (1774–1821) was King Pomare II of Tahiti (1803–1808 and 1815–1821) succeeded by his son, Pomare III (reigned 1821–1827).

Pompadour

Pompadour Cotinga *Xipholena punicea* **Pallas**, 1764
Pompadour Green Pigeon *Treron pompadora* **J. F. Gmelin**, 1789

Jeanne Antoine Poisson, Marquise de Pompadour (1721–1764), was the mistress of King Louis XV of France (from 1745) and known as 'Little Queen' to her friends. She is better known as Madame de Pompadour. Her favourite colour was widely reported to be purple, allegedly the colour of her underwear. The scientific name *punicea* means purple.

Pons

Red-billed Parrot ssp. *Pionus sordidus ponsi* **Aveledo & Ginés**, 1950
Common Bush Tanager ssp. *Chlorospingus flavopectus ponsi* **W. H. Phelps & W. H. Phelps Jr**, 1952

Dr Adolfo R. Pons (1914–1982) was a Venezuelan pathologist and microbiologist who graduated as a physician (1936) and then did post-graduate studies in London (1937) and Rio de Janeiro (1938). He was Chief Medical Officer, Division of Yellow Fever, Ministry of Health, Venezuela (1939–1942) and afterwards a Professor at Zulia University (1949–1978).

Ponty

Blue-breasted Kingfisher ssp. *Halcyon torquatus pontyi* Millet-Horsin, 1921 NCR

[JS *Halcyon malimbica forbesi*]

Amédée William Merlaud-Ponty (1866–1915) was a French colonial administrator who was Governor-General, French West Africa (1908–1915).

Poortman

Poortman's Emerald *Chlorostilbon poortmani* **Bourcier**, 1843
[Alt. Short-tailed Emerald]

Willem Poortman (1819–1891) was a Dutch ornithologist who collected in Colombia (1840s).

Popelaire

Popelaire's Coquette *Discosura popelairii* **Du Bus de Gisignies**, 1846
[Alt. Wire-crested Thorntail; Syn. *Popelairia popelairii*]

Baron Jean Baptiste Joseph Louis Popelaire de Terloo (1810–1870) was a Belgian naturalist who collected in Peru (1841–1843), where he acquired a number of Inca mummies, now in Belgium. He may have also had a role as an unofficial diplomat for Belgium to South America as his journey to Peru took place shortly after Belgium became an independent nation. Popelaire sent all his specimens to the describer Bernard-Aimé Leonard, Vicomte Du Bus de Gisignies (1808–1874). The coquette is sometimes placed in the eponymous genus *Popelairia* (Reichenbach, 1854). Popelaire's name is also given to a giant snail *Strophocheilus popelarianus* which can grow up to six inches!

Portenko

> Pallas's Rosefinch ssp. *Carpodacus roseus portenkoi*
> **Browning**, 1988

Dr Leonid Aleksandrowitsch Portenko (1896–1972) was a Russian zoologist, primarily an ornithologist. He completed his PhD at the St Petersburg Zoological Institute (1929). He was at the Arctic Institute 1929–1939 but returned to St Petersburg (1940) to spend the rest of his working life there. He made a number of expeditions to little-studied and remote areas of the USSR, including the northern Urals, Novaya Zemlya, Wrangel Island, Kamchatka, and the Kuril Islands. He wrote a great many papers and books, and edited and contributed several chapters to the multivolume work *The Birds of the USSR* (published 1951–1954). He is also remembered for having described an important zoogeographical concept on the unity of the fauna of circumpolar tundra, and was an advocate of the single Holarctic zoogeographic region. Two mammals are named after him.

Porter

> Wandering Tattler ssp. *Heteroscelus incanus porteri*
> **Mathews**, 1916 NCR
> [Syn. *Tringa incana* NRM]

John Porter Rogers (see **Rogers, J. P.**)

Potanin

> Pere David's Snowfinch ssp. *Pyrgilauda davidiana potanini*
> **Sushkin**, 1925
> Chukar Partridge ssp. *Alectoris chukar potanini* Sushkin, 1927

Grigorii Nikolaevich Potanin (1835–1920) was a Russian botanist, ethnologist and explorer in Central Asia (1861–1892). He graduated from Tomsk military academy (1952) and joined the Russian army (1953–1958), then studied at the University of St Petersburg (1859–1862). He became involved in the Society for Siberian Independence and in student riots (1861) leading to his imprisonment (1865–1874), and he was later exiled and sentenced to hard labour. He went on a number of expeditions, to Zaisan and the Tarbagatai Range (1863–1864), Mongolia and Tuva (1876–1877 and 1879–1880), China, Tibet and Mongolia (1884–1886 and 1892–1893) and the Greater Khingan Mountains (1899). Two reptiles are named after him.

Potts

> New Zealand Falcon *Nesierax pottsi* **Mathews** & **Iredale**, 1913 NCR
> [JS *Falco novaeseelandiae*]

Thomas Henry Potts (1824–1888) was an English businessman, botanist, ornithologist, entomologist, naturalist and conservationist who settled in New Zealand (1854) and described a number of new species, notably the Great Spotted Kiwi *Apteryx haastii*. He was a Member of the New Zealand Parliament (1866–1870).

Poty

> Long-tailed Tit ssp. *Aegithalos caudatus potyi* **Jouard**, 1929 NCR
> [JS *Aegithalos caudatus europaeus*]

Dr Paul Poty (1889–1961) was a French ornithologist. He was one of the founders of the French periodical *Alauda* in 1929.

Pouchet

> Fairy sp. *Ornismya pouchetii* **Lesson**, 1840 NCR
> [Alt. Black-eared Fairy; JS *Heliothryx auritus auriculatus*]
> Swallow sp. *Hirundo poucheti* **Petit**, 1884 NCR
> [Alt. Grey-rumped Swallow; JS *Pseudhirundo griseopyga melbina*]

Félix-Archimède Pouchet (1800–1872) was a French zoologist and naturalist who was a proponent of the theory of spontaneous generation. He was Director of the Rouen Natural History Museum (1828–1872) and Professor at the Rouen School of Medicine from 1838. He wrote *Hétérogénie* (1859).

Powell, H. L.

> Eurasian Scops Owl *Otus scops powelli* **Meinertzhagen**, 1920 NCR
> [JS *Otus scops cycladum*]

H. L. Powell (d.1927) was Meinertzhagen's (q.v.) taxidermist during expeditions to Crete (1920) and Iraq (1922–1923). They appear to have met in Palestine (1919) when Powell was serving as a trooper in the Gloucestershire Yeomanry and helped Meinertzhagen with his collecting. Powell was also on the Oxford University Spitsbergen Expedition (1921).

Powell, R.

> Powell's Buttonquail *Turnix suscitator powelli* **Guillemard** 1885
> [Alt. Barred Buttonquail ssp.]

Commander Richmond ffolliott Powell (1851–1938) was on board the schooner yacht *Marchesa* with the buttonquail's describer, Dr Francis Guillemard (q.v.), on voyages to Kamchatka, Japan, the East Indies and New Guinea (1882–1884). Powell retired from the Navy (1891) and was appointed Governor of Osborne House, Isle of Wight (1903).

Powell, T.

Lesser Shrikebill ssp. *Clytorhynchus vitiensis powelli* **Salvin**, 1879

Reverend Thomas Powell (1817–1887) was a missionary in Samoa (1845–1885). He sent a small collection of bird specimens to Philip Lutley Sclater (q.v.).

Praed

Red-headed Malimbe ssp. *Malimbus rubricollis praedi* **Bannerman**, 1921
Broad-billed Roller ssp. *Eurystomus afer praedi* Bannerman, 1921 NCR
[JS *Eurystomus glaucurus aethiopicus*]

Cyril Winthrop Mackworth-Praed (1891–1974) was a British stockbroker and ornithologist who was in Kenya (1914). He represented Great Britain at two Olympic Games (1924 and 1952), winning one gold and two silver medals for shooting. He served in the Scots Guards in WW1 and commanded the Commando Training School in Scotland in WW2. He worked very closely with Claude Grant (q.v.), with whom he produced the 6-volume *African Handbook of Birds*.

Prager

Coal Tit ssp. *Parus ater prageri* **Hellmayr**, 1915 NCR
[JS *Periparus ater michalowskii*]
Rock Bunting ssp. *Emberiza cia prageri* **Laubmann**, 1915
[SII *Emberiza cia par*]

Max Prager (DNF) was a German collector in the Caucasus (1914).

Prater

Indian Nuthatch ssp. *Sitta castanea prateri* **Whistler** & **Kinnear**, 1932

Stanley Henry Prater (1890–1960) was Curator of the Bombay Natural History Society (1923–1947). He wrote *The Book of Indian Animals* (1948), since updated and republished several times. A subspecies of Jungle Cat *Felis chaus* is also named after him.

Pratt

American Oystercatcher ssp. *Haematopus palliatus prattii* **C. J. Maynard**, 1899 NCR
[NUI *Haematopus palliatus palliatus*]

Marland L. Pratt (DNF) was an oologist and friend of the describer Charles Maynard (q.v.). He owned a yacht called *Cleopatra* in which he sailed to the Bahamas and West Indies.

Praxilla

Hummingbird genus *Praxilla* **Reichenbach**, 1854 NCR
[Now in *Colibri*]

Praxilla (fl.492 BC) was a lyric poetess who lived in Sicyon, Greece. Very little of her poetry has survived to the modern era, but she was highly regarded in ancient times.

Preiss

Preiss's Fantail *Rhipidura albiscapa preissi* **Cabanis**, 1850
[Alt. Grey Fantail ssp.]

Dr Johann August Ludwig Preiss (1811–1883) was a German naturalist who explored and collected birds around the Swan River in Western Australia (1838–1841). He was also a well-known plant collector, having been apprenticed as a gardener at the botanical gardens in Göttingen. He later returned to the town of his birth – Herzberg – and bought land there.

Prendergast

Kalij Pheasant *Gennæus prendergasti* **Oates**, 1906 NCR
[JS *Lophura leucomelanos lathami*]

C. M. Prendergast (1866–1917) was a British colonial administrator in the Burma Police Force and Deputy Commissioner in the Arakan, where he made a collection of birds (1905).

Prentice

Blue-breasted Kingfisher ssp. *Halcyon malimbica prenticei* **Mearns**, 1915 NCR
[NUI *Halcyon malimbica malimbica*]

Father Arthur Prentice (1872–1964) was converted to Catholicism whilst in Canada (1890–1896), became an English missionary (ordained 1903) to Uganda (1904–1921 and 1927–1950) and returned to be a parish priest at Heston, England (1951).

Prentiss Gray

Brown-hooded Kingfisher ssp. *Halcyon albiventris prentissgrayi* **Bowen**, 1930

Prentiss Nathaniel Gray (1884–1935), amateur hunter and collector, was a financier and a director of many important businesses, and was closely associated with the US Government when Herbert Hoover was US President. He ran the Gray Equatorial Africa Expedition (1929), which covered British East Africa and Angola, on behalf of the Academy of Natural Sciences in Philadelphia. The describer W. W. Bowen (q.v.) was a member of that expedition representing the Academy. Gray also wrote books on hunting and the like, such as *African Game Lands* and *From the Peace to the Fraser*.

Prêtre

Prêtre's Parrot *Amazona pretrei* **Temminck**, 1830
[Alt. Red-spectacled Amazon]
Prêtre's Hermit *Phaethornis pretrei* **Lesson** & **de Lattre**, 1839
[Alt. Planalto Hermit]

Prêtre's Tanager *Spindalis zena pretrei* Lesson, 1831
[Alt. Western Spindalis ssp.]

Jean Gabriel Prêtre (c.1800–1845) was a French artist employed by the Natural History Museum in Paris. He illustrated Louis Vieillot's classic work *Histoire Naturelle des plus Beaux Oiseaux de la Zone Torride*, published in 12 parts (1805–1809). He also illustrated *Animal Kingdom*, published posthumously (1850). A reptile is named after him. (See also **Gabriel**)

Preuss

Preuss's Weaver *Ploceus preussi* **Reichenow**, 1892
Preuss's Cliff Swallow *Petrochelidon preussi* Reichenow, 1898
African Cuckooshrike sp. *Campephaga preussi* Reichenow, 1899 NCR
[Alt. Purple-throated Cuckooshrike; JS *Campephaga quiscalina*]

Grey Cuckooshrike ssp. *Coracina caesia preussi* Reichenow, 1892 NCR
[NUI *Coracina caesia pura*]
Preuss's Double-collared Sunbird *Cinnyris reichenowi preussi* Reichenow, 1892
[Alt. Northern Double-collared Sunbird ssp.]
Waller's Starling ssp. *Onychognathus walleri preussi* Reichenow, 1892

Professor Paul Preuss (1861–1926) was a Poland-born German naturalist, botanist and horticulturalist. He collected in West Africa (1886–1898), New Guinea (c.1903) and again in West Africa (1910). He was a member of Zintgraff's (1888–1891) military expedition to explore the hinterland of Cameroon, then a German colony. Whilst storming a native village, the troop commander was killed and the second-in-command severely wounded; Preuss took over command and led the remaining troops back to the coast. He constructed the botanical gardens of Victoria (Limbe), Cameroon (1901), being employed by the colonial government. Three mammals, an amphibian and a reptile are named after him.

Prévost

Helmet Vanga *Euryceros prevostii* **Lesson**, 1831
Prévost's Mango *Anthracothorax prevostii* Lesson, 1832
[Alt. Green-breasted Mango]
Prévost's Ground Sparrow *Melozone biarcuata* Prévost & **Des Murs**, 1846
[Alt. White-faced Ground Sparrow]

Florent Prévost [or Prevot] (1794–1870) was a French artist and writer who worked on museum collections. He wrote *Iconographie Ornithologique* (1845) and *Histoire Naturelle des Oiseaux d'Europe* (1864). He also illustrated works by Temminck, Bonaparte and Buffon (all q.v.). A mammal and a reptile are named after him.

Pridi

Golden-fronted Leafbird ssp. *Chloropsis aurifrons pridii* **Deignan**, 1946

Pridi Banomyong (also known as Pridi Phanomyong) (1900–1983) was a Thai politician and statesman. He was Prime Minister of Thailand (1946–1947). In November 1947 a coup by army troops forced hm to flee the country but he secretly returned in 1949 to stage a pro-democracy coup, but when it failed he left for China, never to return. From China he travelled to France, where he spent the remainder of his life.

Priest

Barratt's Warbler ssp. *Bradypterus barratti priesti* **C. W. Benson**, 1946

Captain Cecil Damer Priest (1887–1955) was a well-travelled British Naval Captain who led an interesting life – we found reference to him giving a lecture on 'Queer Birds' of Rhodesia, in Australia (1937) and collecting a bird skin from an island off Chile (1932). He wrote the 4-volume *Birds of Southern Rhodesia* (1929–1936), and *Eggs of Birds of Southern Africa* (1948).

Prigogine

Prigogine's Bay Owl *Philodus prigoginei* **Schouteden**, 1952
[Alt. Congo Bay Owl, Itombwe Owl]
Prigogine's Apalis *Apalis kaboboensis* Prigogine, 1955
[Alt. Kabobo Apalis]
Prigogine's Alseonax *Muscicapa itombwensis* Prigogine, 1957
[Alt. Itombwe Flycatcher; Syn. *Muscicapa lendu itombwensis*]
Prigogine's Double-collared Sunbird *Cinnyris prigoginei* **J. Macdonald**, 1958
[Alt. Marungu Sunbird]
Prigogine's Greenbul *Chlorocichla prigoginei* De Roo, 1967
Prigogine's Ground Thrush *Zoothera kibalensis* Prigogine, 1978
[Alt. Kibale Ground Thrush; Syn. *Geokichla camaronensis kibalensis*]
Prigogine's Owlet *Glaucidium albertinum* Prigogine, 1983
[Alt. Albertine Owlet]
Prigogine's Nightjar *Caprimulgus prigoginei* Louett, 1990
[Alt. Itombwe Nightjar]

Oriole Finch ssp. *Linurgus olivaceus prigoginei* Schouteden, 1950
Brown Parisoma ssp. *Sylvia lugens prigoginei* Schouteden, 1952

Alexandre Prigogine (1913–1991) was a Russian-born Belgian ornithologist and explorer, whose brother Ilya (q.v.) won the Nobel Prize for Chemistry (1977). The family left Moscow after the Revolution, settling in Belgium. Ilya mentioned Alexandre in his Nobel Prize acceptance speech as also having studied chemistry. Alexandre wrote extensively on birds and sponsored a number of expeditions to Central Africa, especially to the former Belgian Congo. He wrote *Les Oiseaux de l'Itombwe et de son Hinterland*, in three volumes (1971–1984). A mammal is named after him.

Prillwitz

Grey-cheeked Tit Babbler ssp. *Macronus flavicollis prillwitzi* **Hartert**, 1901
Little Spiderhunter ssp. *Arachnothera longirostra prillwitzi* Hartert, 1901
Cream-vented Bulbul ssp. *Pycnonotus simplex prillwitzi* Hartert, 1902

Conrad Ernst August Prillwitz (1856–1922) was a German collector who lived in Java (1897–1922).

Primoli, J. N.

Primoli's Hummingbird *Metallura williami primolina*
Bourcier, 1853
[Alt. Viridian Metaltail ssp.]

Count Joseph Napoleon Primoli (1851–1927) was a scholar and photographer who founded the Napoleonic Museum in Rome. His mother was Charlotte Primoli di Foglia (q.v.), daughter of Prince Charles Lucien Bonaparte (q.v.). Bourcier wrote in his original etymology: 'Nous dédions cette espèce au petit-fils du prince Charles Bonaparte et fils de la comtesse de Primoli, avec l'espoir de voir se perpétuer dans cette illustre famille le goût et l'étude des sciences naturelles.'

Primoli, P.

Macaw genus *Primolius* **Bonaparte**, 1857

Pietro, Count Primoli di Foglia (1820–1883) was married to Charlotte Honorine Joséphine Pauline Primoli di Foglia (q.v.), a daughter of Charles Bonaparte (q.v.), the author.

Primrose

Red-vented Bulbul ssp. *Pycnonotus cafer primrosei*
Deignan, 1949 NCR
[JS *Pycnonotus cafer bengalensis*]

Archibald M. Primrose (1872–1922) was a British tea-planter and ornithologist at Goalpara, Assam, and the Nilgiri Hills, India, in the early 20th century.

Prince

Prince's Hawk *Leucopternis princeps* **P. L. Sclater**, 1865
[Alt. Barred Hawk]

This is not named after a specific person but rather as 'the prince of birds'.

Prince Albert

Prince Albert's Lyrebird *Menura alberti* **Bonaparte**, 1850
[Alt. Albert's Lyrebird, Northern Lyrebird]
Prince Albert's Curassow *Crax alberti* **L. Fraser**, 1852
[Alt. Blue-billed Curassow, Blue-knobbed Curassow]

(See under **Albert**)

Prince Henri

Prince Henri's Laughingthrush *Trochalopteron henrici*
Oustalet, 1892
[Alt. Brown-cheeked Laughingthrush]

Henri Prince d'Orléans (1867–1901) was born in Richmond, England, but was a French Bourbon prince of the Royal House. He was an explorer and geographer who collected in China and Tibet (1889–1890), south-east Africa (1892), and Indochina (1895). He also discovered the source of the Irrawaddy River. He wrote *Around Tonkin and Siam* (1894) and *From Tonkin to India* (1897). He died in Saigon (Ho Chi Minh city), Vietnam.

Prince Lucian

Prince Lucian's Conure *Pyrrhura lucianii* **Deville**, 1851
[Alt. Bonaparte's Parakeet, Deville's Conure]

Prince Lucian Bonaparte (See **Bonaparte**)

Prince of Essling

Prince of Essling's Parrot *Nestor esslingii* **Souancé**, 1856
NCR
[Alt. New Zealand Kaka; JS *Nestor meridionalis*; Based on an unusually coloured specimen of the species.]

(See **Rivoli**, **Massena** and **Victor**)

Prince of Wales

Prince of Wales's Pheasant *Phasianus colchicus principalis*
P. L. Sclater, 1855
[Alt. Common Pheasant ssp.]

The Prince of Wales, later King Edward VII (1841–1910), succeeded to the British throne on the death of Queen Victoria (1901), but he was Prince of Wales when Sclater described the pheasant.

Prince of Wales (Island)

Prince of Wales Spruce Grouse *Falcipennis canadensis isleibi* **Dickerman** & Gustafson, 1996

This subspecies is named after Prince of Wales Island, south-eastern Alaska.

Prince Rudolf/Rudolph

Prince Rudolf's Bird-of-Paradise *Paradisaea rudolphi*
Finsch, 1885
[Alt. Blue Bird-of-Paradise]

Archduke Rudolf, Crown Prince of Austria-Hungary (1858–1889), was tutored by Ferdinand von Hochstetter (q.v.) and became very interested in the natural sciences. However, in an incident that would have major consequences for the Habsburg Empire and the course of European history, he and his mistress were found dead in mysterious circumstances and have been the subject of many studies and films. It is generally supposed that they were involved in a suicide pact, but this has never been proven and there are advocates of the theory that they were murdered. (See **Rudolf**)

Prince Ruspoli

Prince Ruspoli's Turaco *Tauraco ruspolii* **Salvadori**, 1896

Prince Eugenio Ruspoli (1866–1893) was an explorer who came from a family of eminent Roman aristocrats with relations to the Bonaparte dynasty. An ancestor had been both Scarlatti's and Handel's patron before the latter moved to England. Ruspoli explored in Ethiopia (1891–1893), where he was killed in an encounter with an elephant that he had wounded. Prince Ruspoli had collected the turaco in 1892 or 1893 but had left no note of the precise locality or date before meeting his fate. Two reptiles are named after him.

Prince, E.

Grey Ground Thrush *Zoothera princei* **Sharpe**, 1874

Edwin C. Prince (d.1873) was an English naturalist and John Gould's (q.v.) assistant (1830).

Princess Alexandra/Princess of Wales

Princess Alexandra's Parrot *Polytelis alexandrae* **Gould**, 1863
[Alt. Princess Parrot, Princess of Wales Parakeet]

(See **Alexandra**)

Princess Helene

Princess Helene's Coquette *Lophornis helenae* **de Lattre**, 1843
[Alt. Black-crested Coquette]

(See **Helene**)

Princess Sirindhorn

Princess Sirindhorn's Bird *Pseudochelidon sirintarae* Thonglongya, 1968
[Alt. White-eyed River Martin]

Princess Maha Chakri Sirindhorn of Thailand (b.1955) is a member of the Thai Royal Family. In Thailand, she is often referred to as the 'Princess of Information Technology' due to her interest in applying science for her country's development.

Princess Stephanie

Princess Stephanie's Astrapia *Astrapia stephaniae* **Finsch**, 1885
[Alt. Princess Stephanie's Bird-of-Paradise]

(See **Stephanie**)

Princess Therese

Princess Therese's Parrot *Hapalopsittaca amazonina theresae* **Hellmayr**, 1915
[Alt. Rusty-faced Parrot ssp.]

Princess Therese of Bavaria (1850–1925) was very interested in ornithology and had a small collection of South American birds. The collection was later moved to the Munich Museum.

Pringle, C. G.

Pringle's Hummer *Delattria pringlei* **Nelson**, 1897 NCR
[Alt. Amethyst-throated Mountaingem; JS *Lampornis amethystinus*]

Cyrus Guernsey Pringle (1838–1911) was an American horti-culturist and botanist. He started his first nursery (1858) growing fruit and potatoes and was successful in crossing and producing new varieties of potato. He was drafted into the Union army in the American Civil War but, as a Quaker, he refused to take part and was subjected to torture by being staked to the ground, racked (1863) and threatened with death – but he still declined to bear arms. He was only released from this treatment and from the army on Abraham Lincoln's personal intervention. He finally recovered his health (1868) and started on a career of collecting in the USA, Canada and especially Mexico whch he surveyed (1885–1910). Over his career he collected c.20,000 species of which c.2,500 were new to science.

Pringle, J. W.

Pringle's Puffback *Dryoscopus pringlii* **F. J. Jackson**, 1893

Colonel Sir John Wallace Pringle (1863–1938) took part in the Burmese Expedition (1885–1886) and was the Chief Inspector of Railways in Uganda (1891–1892). He subsequently became Chief Inspector of Railways for the Board of Trade in England (1916–1929) until his retirement. The etymology does not mention why he was honoured by Jackson.

Pritchard

Pritchard's Scrubfowl *Megapodius pritchardii* **G. R. Gray**, 1864
[Alt. Polynesian/Tongan Megapode]

William Thomas Pritchard (1829–1909) was a British Consular Agent in Samoa (1856), and then the British Consul in Fiji (1857–1862), having been instrumental in the addition of Fiji to the Empire. Ratu Seru Cakobau, leader of the most powerful clan, virtually sold Fiji to Queen Victoria (q.v.). Cakobau had run up huge debts to an American, John B. Williams, so he made an offer to Pritchard to cede Fiji to Britain if Britain would cover his debts to Williams. The deal did not go through at first, but (1874) Britain took over. Pritchard wrote *Polynesian Reminiscences* (1866).

Pritzbuer

Pritzbuer's Thrush *Turdus poliocephalus pritzbueri* **Layard**, 1878
[Alt. Island Thrush ssp.]

Admiral Léopold Eberhard Ludovic de Pritzbuer (1824–1889) was Governor of New Caledonia (1875–1878).

Prjevalsky

(see **Przewalski**)

Procne/Progne

Bellbird genus *Procnias* **Illiger**, 1811
Martin genus *Progne* **F. Boie**, 1826

In Greek mythology, Procne was a daughter of King Pandion of Athens. She was metamorphosed into a swallow.

Prometheus

Ground Woodpecker ssp. *Geocolaptes olivaceus prometheus* **Clancey**, 1952

In Greek mythology, Prometheus was a Titan who stole fire from Zeus and gave it to mortals for their use. Zeus was not pleased at this, so he had Prometheus chained to a rock. Every day an eagle came to devour his liver. Prometheus was, of course, immortal, so his liver grew back every night. Thus he was forced to suffer agony every day until the hero Heracles eventually freed him. A mammal is also named after him.

Proserpina

> Black Sunbird ssp. *Leptocoma sericea proserpina* **Wallace**, 1863

In Roman mythology, Proserpine was the equivalent of the Greek Persephone: a daughter of the goddess Ceres who was abducted by Pluto (q.v.) and carried off to the Underworld. There is a wonderful statue called 'The Rape of Proserpina' by Bernini in the Villa Borghese in Rome.

Proteus

> Variable Pitohui ssp. *Pitohui kirhocephalus proteus* **Hartert**, 1932 NCR
> [JS *Pitohui kirhocephalus meyeri*]

In Greek mythology, Proteus was a sea-god who could change his shape.

Prunelle

> Prunelle's Coeligene *Coeligena prunellei* **Bourcier**, 1843
> [Alt. Black Inca]

Clement François Victor Gabriel Prunelle (1777–1853) was a French physician and professor of medical history.

Pryer, H. J. S.

> Marsh Grassbird *Locustella pryeri* **Seebohm**, 1884
> [Alt. Japanese Swamp Warbler; Syn. *Megalurus pryeri*]
> Pryer's Woodpecker *Sapheopipo noguchii* Seebohm, 1887
> [Alt. Okinawa Woodpecker]
>
> Brown-eared Bulbul ssp. *Microscelis amaurotis pryeri* **Stejneger**, 1887
> Japanese Scops Owl ssp. *Otus semitorques pryeri* **Gurney**, 1889

Henry James Stovin Pryer (1850–1888) was a British lepidopterist who went to China (1871) and shortly after settled in Japan. He devoted his spare time to collecting natural history specimens, later donating them to the British Museum. He died of pneumonia in Yokohama. He also has a reptile named after him.

Pryer, W. B.

> Pryer's Flowerpecker *Dicaeum pryeri* **Sharpe**, 1881 NCR
> [Alt. Scarlet-backed Flowerpecker; JS *Dicaeum cruentatum nigrimentum*]

William Burges Pryer (1843–1899) was a British civil servant. He founded the town of Sandakan (1879), which became the headquarters of the British North Borneo Chartered Company of which Pryer was the first Resident in Sabah (Malaysia) when the company was officially formed (1882).

He wrote 'Notes on north-eastern Borneo and the Sulu Islands' (1883).

Przewalski

> Przewalski's Finch *Urocynchramus pylzowi* Przewalski, 1876
> [Alt. Pink-tailed Bunting/Rosefinch]
> Przewalski's Redstart *Phoenicurus alaschanicus* Przewalski, 1876
> [Alt. Ala Shan Redstart]
> Przewalski's Rock Partridge *Alectoris magna* Przewalski, 1876
> [Alt. Rusty-necklaced Partridge]
> Przewalski's Thrush *Turdus kessleri* Przewalski, 1876
> [Alt. Kessler's Thrush, White-backed Thrush]
> Przewalski's Antpitta *Grallaria przewalskii* **Taczanowski**, 1882
> [Alt. Rusty-tinged Antpitta]
> Przewalski's Nuthatch *Sitta przewalskii* **Berezowski** & **Bianchi**, 1891
> Przewalski's Parrotbill *Sinosuthora przewalskii* Berezowski & Bianchi, 1891
> [Alt. Rusty-throated Parrotbill]
>
> ---
>
> Elliot's Laughingthrush ssp. *Trochalopteron elliotii prjevalskii* **Menzbier**, 1887
> [Species sometimes regarded as monotypic]
> Siberian Stonechat ssp. *Saxicola maurus przewalskii* **Pleske**, 1889
> Przewalski's Horned Lark *Eremophila alpestris przewalskii* Bianchi, 1904
> White-throated Dipper ssp. *Cinclus cinclus przewalskii* Bianchi, 1905
> Tibetan Snowcock ssp. *Tetraogallus tibetanus przewalskii* Bianchi, 1907
> Spotted Bush Warbler ssp. *Bradypterus thoracicus przevalskii* Sushkin, 1925
> Daurian Partridge ssp. *Perdix barbata przewalskii* Sushkin, 1926 NCR
> [NUI *Perdix dauurica suschkini*]
> Bluethroat ssp. *Luscinia svecica przevalskii* Tugarinov, 1929
> Common Crossbill ssp. *Loxia curvirostra przewalskii* Dementiev, 1932 NCR
> [JS *Loxia curvirostra tianschanica*]

General Nikolai Mikhailovitch Przewalski (1839–1888) was a Russian Cossack naturalist who explored Central Asia. He was undoubtedly one of the greatest explorers the world has ever seen, making five major expeditions: one to the Russian Far East and the others to Mongolia. He wrote *Mongolia, and the Tangut Country* (1875) and *From Kulja, across the Tian Shan to Lob-Nor* (1879). He is best known for having the wild horse *Equus przewalskii*, which he discovered, named after him. The Russian Academy of Sciences instituted the Przewalski Gold Medal (1946). There are at least half-a-dozen different spellings of his name including Przewalski and Prjevalsky (pronounced 'Shev-al-ski'). He died of typhus aged 49 whilst preparing for another expedition. Tsar Alexander II decreed that the town where he died, Karakol, should immediately have its name changed to Przhevalsk.

Psamathe

Bush Warbler genus *Psamathia* **Hartlaub** & **Finsch**, 1868 NCR
[Now in *Cettia*]

In Greek mythology, Psamathe was a Nereid (sea-nymph) and the wife of Proteus (q.v.).

Pucheran

Whistler genus *Pucherania* **Bonaparte**, 1854 NCR
[Now in *Pachycephala*]

Greater Scythebill *Drymotoxeres pucherani* **Des Murs**, 1849
Pucheran's Woodpecker *Melanerpes pucherani* **Malherbe**, 1849
[Alt. Black-cheeked Woodpecker]
Red-billed Ground Cuckoo *Neomorphus pucheranii* **Deville**, 1851
Crested Guineafowl *Guttera pucherani* **Hartlaub**, 1860

Pucheran's Emerald *Chlorostilbon lucidus pucherani* **Bourcier** & **Mulsant**, 1848
[Alt. Glittering-throated Emerald ssp.; Syn. *Chlorostilbon aureoventris pucherani*]
Roadside Hawk ssp. *Rupornis magnirostris pucherani* **J** & **E Verreaux**, 1855
Pucheran's Red-cheeked Parrot *Geoffroyus geoffroyi pucherani* **Souancé**, 1856

Jacques Pucheran (1817–1894) was a French zoologist who went on the expedition of the *Astrolabe* with Dumont d'Urville, Gaimard (q.v.) and Jacquinot (q.v.). He later worked as a zoologist at naturalist at Muséum Nationale d'Histoire Naturelle, Paris. He wrote extensively on ornithology, mammalogy and anthropology as well as *Voyage au Pole Sud et dans l'Oceanie sur les Corvettes l'Astrolabe et la Zélée* (1842). A mammal is named after him.

Pui

Greater Short-toed Lark ssp. *Calandrella cinerea puii* **Yamashina**, 1939 NCR
[JS *Calandrella brachydactyla orientalis*; latter itself SII *Calandrella brachydactyla longipennis*]

Pu'i or Xüan-Tong (1906–1967) was the last Qing Emperor of China (1908–1912) and (as K'ang-de) Emperor of Manchukuo – a Japanese puppet state (1934–1945). He ended his days as a gardener.

Pujol

Pujoli Sunbird *Anthreptes pujoli* **Berlioz**, 1958 NCR
[Based on a juvenile *Anthreptes rectirostris*]

Raymond Pujol (b.1927) is an agronomist and ethnologist. He is a Professor Emeritus of Ethno-Zoology at MNHN, Paris, where he was originally employed in the entomological department before working in West Africa in the second half of the 20th century. He led an expedition to Sérédou (1957–1958). He was put in charge of the ethno-zoological department in 1966. He co-wrote *Dictionnaire Raisonné de Biologie* (2003). An amphibian is also named after him.

Pulich, W. & A.

Bewick's Wren ssp. *Thryomanes bewickii pulichi* **A. R. Phillips**, 1986
Veery ssp. *Catharus fuscescens pulichorum* A. R. Phillips, 1991 NCR
[NUI *Catharus fuscescens fuscescens*]

Dr Warren Mark Pulich (1919–2010) was an American ornithologist. His wife Anne Marie Pulich *née* Doles (d. 2003), an artist, illustrated his books. He served in the US Army Air Force in WW2 and finished his bachelor's degree after the war (1948) at the University of Arizona-Tucson. He took his master's (1958) at Southern Methodist University, Dallas, and his doctorate (1963) at the University of Oklahoma. After working for the US Fish and Wildlife Service, he was a founding member of the Biology Department of the newly established University of Dallas, Irving (1956), and retired as Associate Professor Emeritus (2006). He was the first resident professional ornithologist in his part of Texas. He wrote *Birds of Tarrant County* (1979). Whereas the wren subspecies is named after him, the Veery is named after both Warren and Anne Marie.

Pulitzer

Pulitzer's Longbill *Macrosphenus pulitzeri* **Boulton**, 1931

Ralph Pulitzer (1879–1939) was a publisher and a sponsor of speed flying, among other enterprises. He was the son of Joseph Pulitzer, who is commemorated by the 'Pulitzer Prize', a prestigious award for American journalism, literature or music.

Purnell

Honeyeater genus *Purnellornis* **Mathews**, 1914 NCR
[Now in *Phylidonyris*]

Dusky Grasswren *Amytornis purnelli* Mathews, 1914

Herbert A. Purnell (1884–1962) was an Australian collector and naturalist.

Puvel

Puvel's Illadopsis *Illadopsis puveli* **Salvadori**, 1901
[Alt. Puvel's Thrush-babbler]

Pierre Puvel (DNF) was a French settler in Guinea-Bissau.

Pycraft

Red-chested Owlet ssp. *Glaucidium tephronotum pycrafti* **G. L. Bates**, 1911

William Plane Pycraft (1868–1942) was an English zoologist and osteologist. He was a museum assistant at the Town Museum, Leicester (1891–1898), moved to the BMNH (1898) and was put in charge of osteology (1907). He wrote many books and articles including *Birds of Great Britain and their Natural History* (1934).

Pycroft

Pycroft's Petrel *Pterodroma pycrofti* **Falla**, 1933

Arthur Thomas Pycroft (1875–1971) was a New Zealand naturalist, historian and conservationist. He published an article entitled 'Santa Cruz Red Feather Money – its manufacture and use' in the *Journal of the Polynesian Society*.

Pylzov

Przewalski's Finch *Urocynchramus pylzowi* **Przewalski**, 1876

Lieutentant Mikhail Pylzov (DNF) was a Russian explorer in Central Asia c.1870. A reptile is named after him.

Pym

Pym's Coucal *Centropus pymi* **J. A. Roberts**, 1914 NCR [Alt. Burchell's Coucal; JS *Centropus burchellii*]

Frank A. O. Pym (1879–1920) was a South African zoologist who was an Assistant Curator at the Albany Museum, Grahamstown, and then Curator, Kingwilliamstown Museum, South Africa (1898–1920). He fought in both the Boer War and the southern African campaign in WW1. He was sent to England for training in artillery, but the damp and cold northern winter broke his health and probably fostered the tuberculosis from which he died. He was a direct descendant of the famous John Pym, the Parliamentarian who opposed King Charles I.

Q

Quartin

Yellow-bellied Waxbill *Coccopygia quartinia* **Bonaparte**, 1850
[Alt. Yellow-bellied Swee; Syn. *Estrilda quartinia*]

Dr Leon Richard Quartin-Dillon (d.1840) was a physician and botanist. Another naturalist and physician, Dr Petit, accompanied him on a French government-sponsored scientific expedition to Abyssinia (now Ethiopia). They travelled widely there (1839–1840), but Quartin-Dillon died and Petit became gravely ill. Despite this and further deaths (including Petit 1841) the expedition as a whole continued before returning to France (1843).

Queen Alexandra

Queen Alexandra's Parrot *Polytelis alexandrae* **Gould**, 1863
[Alt. Princess Parrot, Princess of Wales Parakeet]

(See **Alexandra (Princess)**)

Queen Carola

Queen Carola's Parotia *Parotia carolae* **A. B. Meyer**, 1894
[Alt. Carola's Parotia, Queen Carola's Six-wired Bird-of-Paradise]

Queen Carola of Saxony (1833–1907) was the wife of Albert King of Saxony (q.v.). Her social conscience led her to found the Albert Society (1867) caring for sick and wounded, and several schools and homes in Dresden.

Queen Charlotte

Queen Charlotte Jay *Cyanocitta stelleri carlottae* **Osgood**, 1901
[Alt. Steller's Jay spp.]
Queen Charlotte Woodpecker *Picoides villosus picoideus* Osgood, 1901
[Alt. Hairy Woodpecker ssp.]
Queen Charlotte Owl *Aegolius acadicus brooksi* **J. H. Fleming**, 1916
[Alt. Northern Saw-whet Owl ssp.]

These birds are named after the Queen Charlotte Islands (now Haida Gwaii) off British Columbia, Canada, which were named after Sophia Charlotte, Princess of Mecklenburg-Strelitz (1744–1818), the Queen and Consort of George III, King of Great Britain, Ireland and Hanover. A mammal is also named after these islands.

Queen of Bavaria

Queen of Bavaria's Conure *Guaruba guarouba* **J. F. Gmelin**, 1788
[Alt. Golden Parakeet]

Marie Wilhelmine Auguste of Hessen-Darmstadt (1765–1796) was the first wife (1785) of Maximilian I, King of Bavaria (1756–1825).

Queen Victoria

Queen Victoria's Egret *Erodius victoriae* W. **MacGillivray**, 1842 NCR
[Alt. Great Egret; JS *Ardea alba*]
Queen Victoria Riflebird *Ptiloris victoriae* **Gould**, 1850
[Alt. Victoria's Riflebird]

Queen Victoria Lyrebird *Menura novaehollandiae victoriae* Gould, 1865
[Alt. Superb Lyrebird ssp.]

(See **Victoria, Queen**)

Queen Wilhelmina

Queen Wilhelmina's Parakeet *Alisterus chloropterus wilhelminae* **Ogilvie-Grant**, 1911 NCR
[Alt. Papuan King Parrot; NUI *Alisterus chloropterus callopterus*]

Wilhemina Helena Pauline Maria (1880–1962) was Queen of the Netherlands. She succeeded her father William III (1890); her mother Queen Emma acted as regent (1890–1898). She abdicated (1948) in favour of her eldest daughter, the late Queen Juliana of the Netherlands. A mammal is named after her. (See **Wilhelmina (Queen)**)

Queseda

Plain-capped Ground Tyrant ssp. *Muscisaxicola alpinus quesadae* **Meyer de Schauensee**, 1942

Captain-General Gonzalo Jiménez de Quesada (1495–1579) was a Spanish explorer, conquistador, Chief Justice of Santa Marta and founder of the settlement of New Granada (Santa Fé de Bogotá, Colombia). He explored extensively in the northern part of South America, whilst looking for gold and jewels. It has been suggested he was the model for Cervantes's Don Quixote. He died of leprosy.

Quillin

Lark Sparrow ssp. *Chondestes grammacus quillini*
Oberholser, 1974 NCR
[NUI *Chondestes grammacus strigatus*]

Roy William Quillin (1894–1974) was an American oil company executive and amateur ornithologist and oologist at San Antonio, Texas. He co-wrote 'The breeding birds of Bexar County, Texas' (1918).

Quinton

Yellow Canary ssp. *Serinus flaviventris quintoni*
Winterbottom, 1959 NCR
[NUI *Serinus (Crithagra) flaviventris flaviventris*]

William F. Quinton (b.1906) was a South African naturalist, ornithologist and farmer. He co-wrote 'A list of the birds of the Beaufort West District' (1968).

Quisumbing

Barred Rail ssp. *Gallirallus torquatus quisumbingi* **Gilliard**,
1949 NCR
[JS *Gallirallus torquatus torquatus*]

Dr Eduardo A. Quisumbing y Argüelles (1895–1986) was a Filipino botanist, specialising in orchids, and an expert on medicinal plants of the Philippines. The University of Philippines, Los Bāņos, awarded his bachelor's degree (1918) and master's (1921). The University of Chicago awarded his doctorate (1923). He was attached to the University of Philippines (1920–1926) and the University of California (1926–1928). He became Systematic Botanist (1928), Acting Chief (1934), and Chief (1940–1945) of the Natural History Museum of the Bureau of Science, Manila. He was Director, National Museum, Manila (1947–1962) and, after retirement, he was attached to Araneta University.

Quoy

Yellow Robin genus *Quoyornis* **Mathews**, 1912 NCR
[Now in *Eopsaltria*]

Quoy Butcherbird *Cracticus quoyi* **Lesson**, 1827
[Alt. Black Butcherbird]
Oystercatcher sp. *Haematopus quoyi* Brabourne & **Chubb**,
1912 NCR
[Alt. Blackish Oystercatcher; JS *Haematopus ater*]

Red-capped Robin ssp. *Petroica goodenovii quoyi*
Mathews, 1912 NCR; NRM
Western Yellow Robin ssp. *Eopsaltria griseogularis quoyi*
Mathews, 1920 NCR
[JS *Eopsaltria griseogularis griseogularis*]

Jean René Constant Quoy (1790–1869) was a French naval surgeon and zoologist who named and described birds, often with Joseph Paul Gaimard (q.v.). He is commemorated in the names of other taxa, including a parrotfish *Scarus quoyi* and a skink *Sphenomorphus quoyi*. He took part in a number of voyages of discovery, including a circumnavigation aboard the *Astrolabe* (1826–1829) with Jules Dumont d'Urville. He became chief medical officer of the naval hospital at Toulon (1835). Two amphibians and a reptile are named after him.

Quy

Black-winged Cuckooshrike ssp. *Coracina melaschistos
quyi* Dao Van Tien, 1961 NCR
[JS *Coracina melaschistos saturata*]

(See **Vo Quy**)

R

Raalten

Brown Quail ssp. *Synoicus ypsilophorus raaltenii* **S. Müller**, 1842
[Syn. *Coturnix ypsilophora raaltenii*]

Gerrit van Raalten (1797–1829) was a Dutch taxidermist and collector in the East Indies (Indonesia) (1820–1829). He survived an attack by a Javan Rhinoceros *Rhinoceros sondaicus* and died of a tropical disease some years later.

Rabbitts

Boreal Chickadee ssp. *Parus hudsonicus rabbittsi* **J. W. Aldrich** & Nutt, 1939 NCR
[JS *Poecile hudsonicus hudsonicus*]

Gower E. Rabbitts (1883–1946) was the Clerk of Game & Inland Fisheries, later Secretary of the Department of Natural Resources, St John's, Newfoundland.

Rabier

Rabier's Woodpecker *Picus rabieri* **Oustalet**, 1898
[Alt. Red-collared Woodpecker]

Lieutenant-Colonel Paul Rabier (DNF) was a French army officer who served in Indochina (1897–1904). He may have been the same person as Gustave Marie Paul Stanislas Rabier (1878–1915), a French officer from Indochina who was killed in WW1 and who, with others described as French Shanghai patriots, has a street named after him there.

Rabor

Rabor's Wren-babbler *Robsonius rabori* **Rand**, 1960
[Alt. Rusty-faced Babbler, Cordillera Ground Babbler; Syn. *Napothera rabori*]

Stripe-breasted Rhabdornis ssp. *Rhabdornis inornatus rabori* Rand, 1950
Little Pied Flycatcher ssp. *Ficedula westermanni rabori* **Ripley**, 1952
Philippine Tailorbird ssp. *Orthotomus castaneiceps rabori* **Parkes**, 1961
Celestial Monarch ssp. *Hypothymis coelestis rabori* Rand, 1970

Dr Dioscoro Siarot 'Joe' Rabor (1911–1996) was the pre-eminent Filipino zoologist, ornithologist, conservationist and collector of his day. He graduated from the University of the Philippines (1934) and read for a Doctorate under Dillon Ripley (q.v.) at Yale. He worked in a variety of other fields including ichthyology, fisheries, mammalogy, herpetology and ecology. He filled 30 positions during his academic and research career. He founded a museum in Los Baños, Laguna, which houses his collections and from which eight full species and 61 subspecies of birds have been described. The museum concentrated on the flora and fauna of Mindanao, Sulu and Palawan. He led over 50 expeditions (1935–1977), more often than not accompanied by his wife and six children, of whom four became physicians. His passion for nature is reflected in the names he gave his four daughters – Iole Irena, Nectarinia Julia, and Ardea Ardeola were all named after birds, and Alectis Cyrene was named after a fish! He wrote *Philippine Birds and Mammals* and, with Austin Rand (q.v.), who described the wren-babbler, 'Birds of the Philippine Islands: Siquijor, Mount Malindang, Bohol and Samar'. He was a Fellow of the John Simon Guggenheim Memorial Foundation. The sunbird *Aethopyga linaraborae* is named for his wife, Lina (q.v.). Four mammals, two reptiles and an amphibian are named after him.

Rachel

Rachel's Malimbe *Malimbus racheliae* **Cassin**, 1857

Rachel Cassin (b.1844) was the daughter of the describer John Cassin (q.v.). She married Thomas Campbell Davis (1869).

Radama

Peregrine Falcon ssp. *Falco peregrinus radama* **Hartlaub**, 1861

Radama II Rakotosehenondradama (1829–1863) reigned as King of Madagascar (1861–1863).

Radcliffe, D. J. C.

Spot-flanked Barbet ssp. *Tricholaema lacrymosa radcliffei* **Ogilvie-Grant**, 1904

Brigadier-General Sir Denis John Charles Delmé-Radcliffe (1864–1937) was in the British army in India (1884–1890) and Uganda (1898–1904). He was Military Attaché at Rome (1906–1911) and wrote an eyewitness account of Mount Vesuvius's eruption (1906). He wrote *Part of the Nile Province* (1903). He was the son of Charles Delmé-Radcliffe (q.v.) and brother to Henry Delmé-Radcliffe (q.v.)

Radcliffe, E. C.

Yellow-rumped Honeyguide ssp. *Indicator xanthonotus radcliffi* **Hume**, 1870

Lieutenant-Colonel Emilius Charles Delmé-Radcliffe (1832–1907) was a British army officer in 88th Regiment of Foot in India. He was the father of both Henry Delmé-Radcliffe (q.v.) and Denis Delmé-Radcliffe (q.v.) He was a falconer and wrote *Falconry: Notes on the Falconidae used in India in Falconry* (1871).

Radcliffe, H.

Spectacled Barwing ssp. *Actinodura ramsayi radcliffei* **H. H. Harington**, 1910
Dark-backed Sibia ssp. *Heterophasia melanoleuca radcliffei* **E. C. S. Baker**, 1922

Lieutenant-Colonel Henry Delmé-Radcliffe (1866–1947) was a British army officer in Burma (Myanmar) (1909). He served in France in WW1 and commanded a reserve battalion of Royal Welsh Fusiliers (1915), but became embroiled in a scandal and appears to have been the scapegoat. He was removed from his command (1917), never re-employed, and put on half pay (1918); he retired in 1921. He was the son of Charles Delmé-Radcliffe (q.v.) and brother to Denis Delmé-Radcliffe (q.v.).

Radde

Radde's Warbler *Phylloscopus schwarzi* Radde, 1863
Radde's Accentor *Prunella ocularis* Radde, 1884
Shrike sp. *Lanius raddei* **Dresser**, 1888 NCR
[Hybrid: *Lanius collurio* x *Lanius isabellinus*]

Calandra Lark ssp. *Melanocorypha calandra raddei* **Zarudny** & **Loudon**, 1904 NCR
[JS *Melanocorypha calandra psammochroa*]
Blue Tit ssp. *Cyanistes caeruleus raddei* Zarudny, 1908
Yellow Wagtail ssp. *Motacilla flava raddei* **Härms**, 1909 NCR
[JS *Motacilla flava feldegg*]
Common Snipe ssp. *Gallinago gallinago raddei* **Buturlin**, 1912 NCR
[JS *Gallinago gallinago gallinago*]

Gustav Ferdinand Richard Radde (1831–1903) was originally trained as an apothecary. Born in Danzig in Prussia (Gdansk, Poland), he settled in Russia (1852). He participated in numerous expeditions through eastern Siberia, the Crimea, Caucasus, Trans-Caucasus and other regions of Russia, and also through Iran and Turkey, during which he gathered an extensive zoological, botanical and ethnographic collection. He came across a 'skulking warbler in a kitchen garden in the heart of Central Asia' (1856). He collected it and found it to be a hitherto undiscovered species. He settled in Georgia (1863) and founded the Caucasian museum in Tbilisi (1867). He wrote the 2-volume *Reisen im Süden von Ost-Sibirien in den Jahren 1855–1859* (1862–1863) and *Die Vogelwelt des Kaukasus* (1884). He was the first person to give a detailed description of the flora of the Caucasus. He made two further journeys, both as part of the suite of members of the Russian Imperial family; to India and Japan (1895) with the Grand Duke Michael, and with other members of the family to North Africa (1897). Three reptiles, two mammals and an amphibian are named after him.

Rafferty

Double-banded Courser ssp. *Rhinoptilus africanus raffertyi* **Mearns**, 1915
[Syn. *Smutsornis africanus raffertyi*]

Dr Donald George Rafferty (1882–1954) was an American physician, qualifying at Harvard (1906) and practising in Pittsburgh. He was also a zoologist, explorer and collector in Abyssinia (Ethiopia) with the Childs Frick Expedition (1911–1912), of which Mearns (q.v.) was a member.

Raffles

Raffles's Malkoha *Rhinortha chlorophaea* Raffles, 1822
Olive-backed Woodpecker *Dinopium rafflesii* **Vigors**, 1830
Red-crowned Barbet *Megalaima rafflesii* **Lesson**, 1839

Yellow-bellied Prinia ssp. *Prinia flaviventris rafflesi* **Tweeddale**, 1877

Sir Thomas Stamford Bingley Raffles (1781–1826) was a colonial officer, Lieutenant-Governor of Java (1811–1815) and Governor-General of Bencoolen (Sumatra) (1818–1924) and founded the city-state of Singapore (1819). He was noted for his 'liberal attitude toward peoples under colonial rule, his rigorous suppression of the slave trade, and his zeal in collecting historical and scientific information'. He was also the first President of the Zoological Society of London, and wrote a *History of Java* (1817). He employed zoologists and botanists to collect specimens, paying them out of his own pocket. On his return journey to England (1824) on HMS *Fame*, he lost a huge collection of specimens, notes and drawings to a fire. As the local vicar of the parish where Raffles died was a man whose family had made money out of the Jamaica slave trade and disliked abolitionists, he refused to allow Raffles to be buried in the local church. Two mammals are named after him as are the plant genus *Rafflesia* and many other taxa. The Raffles Museum has long been re-named the National Museum of Singapore.

Ragazzi

Olive Sunbird ssp. *Cyanomitra olivacea ragazzii* **Salvadori**, 1888

Dr Vincenzo Ragazzi (1856–1929) of the Modena Natural History Society explored and collected in Ethiopia. He was a physician posted (1884) to the Italian research station Let Marefia, Ethiopia, later being Director. He was on good terms with the Emperor Menelik and accompanied his military expeditions to Harrar (1886–1887) to make geographical surveys. Menelik selected (1887) Ragazzi as his emissary to Italy. Two reptiles and an amphibian are named after him.

Raggi

Raggiana Bird-of-Paradise *Paradisaea raggiana* **P. L. Sclater**, 1873
[Alt. Count Raggi's Bird-of-Paradise]

(See **Count Raggi**)

Ragionieri

Common Quail ssp. *Coturnix coturnix ragionierii* Trischitta, 1939 NCR
[NUI *Coturnix coturnix coturnix*]

Dr Renzo Ragionieri (DNF) was an Italian naturalist and taxidermist. He co-wrote 'Altre nuove forme di uccelli italiani' (1939). (See **Renzo**)

Ragless

Thick-billed Grasswren ssp. *Amytornis modestus raglessi* Black, 2011

Gordon Ragless (1909–2002) was an Australian oologist and amateur ornithologist who was President of the South Australian Ornithological Association (1957–1960)). He collected eggs from the same birds over a number of seasons demonstrating that, for example, Wedge-tailed Eagles *Aquila audax* are capable of a breeding life of thirty years. He donated his collection of over five thousand sets of eggs (1999) to the Australian National Wildlife Collection.

Raimondi

Raimondi's Yellow Finch *Sicalis raimondii* **Taczanowski**, 1874
Peruvian Plantcutter *Phytotoma raimondii* Taczanowski, 1883

Antonio Raimondi (1822–1890) was an Italian explorer and naturalist who collected in Peru (1850–1890), where he made numerous journeys of investigation. He was dedicated to the study of Peru's geography, mineralogy, botany, zoology and ethnography. His motto was 'lose not an instant'. He is perhaps most famed for the bromeliad *Puya raimondii*, which is named after him, which can have 8,000 individual florets on a spike 35 feet tall, the tallest inflorescence in the world, and which also holds the record for flowering intervals, which can be as much as 150 years apart! Raimondi wrote about encountering this 'living fossil' relative of the pineapple thus: 'The travelling botanist who has the thrill to surprise these strange and admirable plants while they are in flower can do nothing but stop and contemplate ecstatically for some time such a beautiful spectacle.' What is more, the spiky rosette offers protection to a variety of birds, which may nest within the leaves of one plant. The Museo Raimondi in Lima is named after him.

Rainey

Sunbird sp. *Helionympha raineyi* **Mearns**, 1911 NCR
[Alt. Marico Sunbird; JS *Cinnyris mariquensis suahelicus*]

Orange Ground Thrush ssp. *Zoothera gurneyi raineyi* Mearns, 1913

Paul James Rainey (1877–1923) was an American multimillionaire, hunter and playboy. His family money came from coal and coke production. He was reckless and wild, the odd man out in a family that was sober and conservative. He bought land in an area where other families had neither running water nor electricity, building himself an enormous estate that had an indoor heated swimming pool and a room the size of a small house to accommodate his worldwide hunting trophies. He also owned a large plantation near Nairobi, racing stables in England and America, and a 23,000-acre duck preserve in Vermilion, Louisiana, that was given to the National Audubon Society after his death. He was a member of the American Geographical Society, the AMNH, the Zoological Society of New York, the USNM, and the National Institute of Social Sciences. He was the first person to take a cameraman on his safari, producing the first films of African big game. He went to England (1923) to buy a new pack of hounds, which he intended to take to India to hunt tigers from horseback, like foxhunting. Sadly for him, but fortunately for the tigers, he died at sea before reaching Cape Town. Two mammals are named after him.

Raisuli

Short-toed Treecreeper ssp. *Certhia brachydactyla raisulii* **Bannerman**, 1926 NCR
[JS *Certhia brachydactyla mauritanica*]

Mulai Ahmed ibn-Muhammad er Raisuli (1871–1925) was a Rif chieftain in Morocco. Depending on your point of view, he is either a folk hero or a criminal. He certainly went in for extortion, torture, kidnapping and armed rebellion, being known as 'The last of the Barbary Pirates'. Hollywood made a highly fictionalised film about his life: *The Wind and the Lion*. Bannerman was asked to name the bird after Raisuli by Admiral Lynes, who collected the holotype (1919).

Rajah

Rajah Scops Owl *Otus brookii* **Sharpe**, 1892
[Alt. Brooke's Scops Owl]
Serpent Eagle sp. *Spilornis raja* Sharpe, 1893 NCR
[Alt. Crested Serpent Eagle; JS *Spilornis cheela pallidus*]

Rajah's Lory *Chalcopsitta atra insignis* **Oustalet**, 1878
[Alt. Black Lory ssp.]

(See **Brooke**)

Ralph, C. J. & C. P.

Great O'ahu Crake *Porzana ralphorum* **Olson** & James, 1991 EXTINCT

Dr C. John Ralph and Dr Carol Pearson Ralph helped the describers, Storrs Olson (q.v.) and Helen James, when they were researching on O'ahu. John Ralph is a wildlife biologist with the Forest Service of the US Department of Agriculture. The University of California, Berkeley, awarded his bachelor's degree; his master's was awarded by San Jose State University, California, and his doctorate by Johns Hopkins University, Baltimore, Maryland. He worked in Hawaii (1976–1981) and researched in New Zealand (1980) on the re-introduction of native species to an island cleared of non-native fauna. He moved to the Redwood Sciences Laboratory, Arcata, California (1981) and has since directed research (1994) at a bird monitoring station in Costa Rica. Carol is a botanist and ornithologist. This bird became extinct shortly after the first Polynesian settlers arrived in Hawaii.

Ralph, W. L.

Grey-crowned Yellowthroat ssp. *Geothlypis poliocephala ralphi* **Ridgway**, 1894

Dr William Legrange Ralph (1851–1907) was an American physician, ornithologist, collector and oologist. He qualified as a physician (1879) and practised in Utica, New York, but his health was not good and he gave up medicine for natural history. He was chosen as Custodian (1897) in succession to Bendire (q.v.) and his title was changed (1904) to Curator of Birds' Eggs at the USNM.

Rama

Sykes's Warbler *Iduna rama* **Sykes**, 1832
[Syn. *Hippolais rama*]

In Hinduism, Rama is the seventh incarnation of the supreme god Vishnu.

Ramon

Amazonian Trogon *Trogon ramonianus* **Deville** & **Des Murs**, 1849

Brother Ramon Busquet (1772–c.1845) was a missionary in Peru. He was the first to explore parts of the Santa Ana, Cocabambilla and Urumamba Rivers (1799–1807). He died whilst accompanying Castelnau's (q.v.) expedition to the Urumamba River.

Ramos, M. A.

Green Shrike Vireo ssp. *Vireolanius pulchellus ramosi* **A. R. Phillips**, 1991

Dr Mario Alberto Ramos Olmos (1949–2006) was a Mexican ornithologist. He worked for the World Bank, Washington DC, Environmental Assessment (1991–2006). He was part of the team that studied the Los Tuxtlas area in Mexico and collected the holotype there, spending 36 months in the field (1973–1987).

Ramos, N. V.

Rufous Paradise-flycatcher ssp. *Terpsiphone unirufa ramosi* Manuel, 1957 NCR
[JS *Terpsiphone cinnamomea unirufa*]

Dr N. V. de Ramos (DNF) was a Filipino biologist.

Ramsay, E. P.

Honeyeater genus *Ramsayornis* **Mathews**, 1912

Ramsay's Diamond Bird *Pardalotus assimilis* Ramsay, 1878 NCR
[Alt. Spotted Pardalote; JS *Pardalotus striatus ornatus*]
Red-capped Robin *Petroeca ramsayi* **Sharpe**, 1879 NCR
[JS *Petroica goodenovii*]

Silvereye ssp. *Zosterops lateralis ramsayi* **G. Masters**, 1876 NCR
[This population now divided between *Zosterops lateralis vegetus* and *Z. lateralis cornwalli*]

Bush Stone-curlew ssp. *Burhinus magnirostris ramsayi* Mathews, 1912 NCR; NRM
[Syn *Esacus magnirostris*]
Little Kingfisher ssp. *Ceyx pusillus ramsayi* **North**, 1912
Ebony Myzomela *ssp. Myzomela pammelaena ramsayi* **Finsch**, 1886
Brown-backed Honeyeater ssp. *Gliciphila modesta ramsayi* Mathews, 1912 NCR
[Syn *Ramsayornis modestus* NRM]
Sacred Kingfisher ssp. *Halcyon sanctus ramsayi* Mathews, 1912 NCR
[JS *Todiramphus sanctus sanctus*]
White-headed Sittella ssp. *Neositta albata ramsayi* Mathews, 1923 NCR
[JS *Daphoenositta chrysoptera leucocephala*]
Chestnut-breasted Munia ssp. *Lonchura castaneothorax ramsayi* **Delacour**, 1943

Edward Pearson Ramsay (1842–1916) was an Australian naturalist, oologist, ornithologist and particularly marine zoologist. He corresponded with and sent specimens to John Gould (q.v.), who persuaded him not to continue publishing on his oological studies, which he had done (1882 and 1883), as Gould anticipated producing a work on this subject himself – a project that never materialised. Ramsay became Curator of the Australian Museum (1874). Two reptiles are named after him.

Ramsay, R. G. W.

Ramsay's Barwing *Actinodura ramsayi* **Walden**, 1875
[Alt. Spectacled Barwing]
Ramsay's Woodpecker *Dendrocopos ramsayi* **Hargitt**, 1881
[Alt. Sulu (Pygmy) Woodpecker]
Ramsay's Blue Flycatcher *Cyornis lemprieri* **Sharpe**, 1884
[Alt. Palawan Blue Flycatcher]

Golden-throated Barbet ssp. *Megalaima franklinii ramsayi* Walden, 1875
Silver-eared Laughingthrush ssp. *Trochalopteron melanostigma ramsayi* **Ogilvie-Grant**, 1904

Colonel Robert George Wardlaw Ramsay (1852–1921) was a British ornithologist who spent time in India and Burma (Myanmar) (1872–1882). He was a President of the BOU (1913–1918). He was also a nephew of Arthur Hay (q.v.), Lord Walden, Marquis of Tweeddale and sent him the specimens he collected.

Ramsden

King Rail ssp. *Rallus elegans ramsdeni* **Riley**, 1913
Red-legged Honeycreeper ssp. *Cyanerpes cyaneus ramsdeni* **Bangs**, 1913 NCR
[JS *Cyanerpes cyaneus cyaneus*]

Dr Charles Theodore Ramsden (1876–1951) was an entomologist, herpetologist, and naturalist who received his doctorate from the University of La Habana (1917). He collected mainly in eastern Cuba and co-wrote *The Herpetology of Cuba* (1919). The museum at Oriente University, Santiago de Cuba, is named after him, as is a reptile. (The Cuban population of Red-legged Honeycreeper is now believed to be an introduced population rather than a valid subspecies.)

Rana

Brown-headed Barbet ssp. *Megalaima zeylanica rana* **Ripley**, 1950 NCR
[NUI Lineated Barbet *Megalaima lineata hodgsoni*]

Named after the Rana family of Nepal. Ripley described them as 'the family of the Prime Ministers of Nepal, who have made and shaped the destiny of that Country for over one hundred years.'

Rand

Rand's Warbler genus *Randia* **Delacour**, 1931

Rand's Warbler *Randia pseudozosterops* Delacour, 1931
Rand's Owlet-nightjar *Aegotheles tatei* Rand, 1941
[Alt. Spangled Owlet-nightjar]
Chocolate Boobook *Ninox randi* **Deignan**, 1951
Rand's Robin Chat *Cossypha heinrichi* Rand, 1955
[Alt. White-headed Robin Chat]
Rand's Red-billed Helmet-shrike *Prionops gabela* Rand, 1957
[Alt. Gabela Helmet-shrike]
Ashy-breasted Flycatcher *Muscicapa randi* **Amadon** & **duPont**, 1970

Tropical Scrubwren ssp. *Sericornis beccarii randi* **Mayr**, 1937
Rand's Modest Parrot *Psittacella modesta subcollaris* Rand, 1941
[Alt. Modest Tiger Parrot ssp.]
Buff-banded Rail ssp. *Gallirallus philippensis randi* Mayr & **Gilliard**, 1951
White-shouldered Fairy-wren ssp. *Malurus alboscapulatus randi* **Junge**, 1952 NCR
[NUI *Malurus alboscapulatus aida*]
Little Spiderhunter ssp. *Arachnothera longirostra randi* **Salomonsen**, 1955
Pied Bushchat ssp. *Saxicola caprata randi* **Parkes**, 1960
Goldenface ssp. *Pachycare flavogriseum randi* Gilliard, 1961

Austin Loomer Rand (1905–1982) was a Canadian ornithologist who collected in several countries. His first degree was awarded by Acadia University, which also awared him an honorary DSc (1961). He was on a bird-collecting expedition to Madagascar (1929) which included Richard Archbold (q.v.), writing it up as *The Distribution and Habits of Madagascar Birds* (1936). Archbold subsequently financed and led a series of biological expeditions to New Guinea (1930s), which Rand co-led and which ultimately resulted in his *Handbook of New Guinea Birds* (1967). He named a number of birds after Archbold, and became Assistant Zoologist at the National Museum of Canada (1942–1947), then Curator of Birds at the Field Museum in Chicago (1947–1955), and finally Chief Curator of Zoology (1955–1970). He was President of the American Ornithologists' Union (1962–1964).

Randon

Maghreb Lark ssp. *Galerida macrorhyncha randonii* Loche, 1860
[Alt. Long-billed Crested Lark]

Jacques Louis César Alexandre Randon, 1st Comte Randon (1795–1871), was a French military and political leader, also Marshal of France (1856) and Governor of Algeria (1851–1858). He enlisted aged 16, being promoted to Sergeant at 17. He took part in the French campaign that took Moscow and then had to retreat. He also fought for Napoleon in Germany and France. Loche's etymology reads: 'C'est à M. le maréchal, comte Randon, gouverneur général de l'Algérie, à la bienveillance duquel nous avons du de pouvoir explorer fructueusement le sud de l'Algérie, que nous avons dédié cette belle espèce.'

Randrianasolo

Cryptic Warbler *Cryptosylvicola randrianasoloi* Goodman, Langrand & **Whitney**, 1996

Georges Randrianasolo (1930–1989) was a Malagasy zoologist with an interest in both ornithology and mammology. He was an expert on lemur distribution and worked at the fauna department of the Parc Botanique et Zoologique de Tsimbazaza, latterly as Director. He wrote at least one book, with others, *Faune de Madagascar* (1973). He has a mammal named after him.

Ranfurly

Bounty Islands Shag *Leucocarbo ranfurlyi* **Ogilvie-Grant**, 1901
[Syn. *Phalacrocorax ranfurlyi*]

Uchter John Mark Knox, 5th Earl of Ranfurly (1865–1933), was Governor of New Zealand (1897–1904). He was requested to obtain bird specimens for the BMNH (1897). Canadian National Railways named a siding and town site 'Ranfurly' in his honour (1905), presumably as he crossed Canada by rail to embark in British Columbia for New Zealand as his lordship appears never to have set foot in Canada again.

Rankine

Grey Penduline-tit ssp. *Anthoscopus caroli rankinei* Irwin, 1963

Ronald William Rankine (b.1912) was a wealthy tobacco farmer and politician who was very interested in natural history. He collected the holotype in Rhodesia (Zimbabwe). Rankine collected together with Irwin (q.v.) in Tanganyika (Tanzania) (1951) and in Mozambique (1952). He lost a leg to a crocodile when the Kariba dam was being constructed, and unsuccessfully stood for the Rhodesian parliament (1962). Irwin in a letter to us described him as '… but like so many, never got down to putting pen to paper.' But he did write a booklet, illustrated with photographs, called *Birds of Rhodesia* (1968).

Raper

Pigeon genus *Raperia* **Mathews**, 1915 NCR
[Now in *Columba*]

Lord Howe Swamphen *Porphyrio raperi* Mathews, 1928 NCR
[JS *Porphyrio albus* EXTINCT]

George Raper (1769–1797) was a British explorer and artist. He entered the Royal Navy (1783) as a captain's servant and, as an able seaman, joined HMS *Sirius* (1786). *Sirius* was flagship of the First Fleet and Raper was a midshipman by the time they arrived at Botany Bay (1788). He was still on board when *Sirius* was wrecked on Norfolk Island (1790). He returned to Sydney (1791) then back to England (1792). He was promoted to lieutenant (1793) and served on board HMS *Cumberland*. The Admiralty record of his death states he was 'Late Commander HMS Cutter *Expedition*'. His watercolours, mainly to do with Australian subjects, are highly prized. One shows the now-extinct Lord Howe Swamphen *Porphyrio albus*.

Rapine

Coal Tit ssp. *Parus ater rapinensis* **Jouard**, 1928 NCR
[JS *Periparus ater ater*]

Joseph Rapine (1884–1958) was an ornithologist and President of the French Ornithological Society (1925–1931).

Rathbone

Rathbone's Wood Warbler *Sylvicola rathbonii* **Audubon**, 1839 NCR
[Alt. Mangrove Warbler; JS *Dendroica petechia*]

Named after the whole Rathbone family. Richard Rathbone (1788–1860) was a Liverpool businessman and banker who befriended Audubon (q.v.) when he arrived from the USA (1826). Audubon became a friend of all the family as is evidenced by his attribution in *Ornithological Biography* as follows: 'Kind reader, you are now presented with a new and beautiful little species of Warbler, which I have honoured with the name of a family that must ever be dear to me. Were I at liberty here to express the gratitude, which swells my heart, when the remembrance of all the unmerited kindness and unlooked-for friendship, which I have received from the Rathbones of Liverpool comes to my mind, I might produce a volume of thanks. But I must content myself with informing you, that the small tribute of gratitude which alone it is in my power to pay, I now joyfully accord, by naming after them one of those birds …' The Rathbone family continued to prosper and was a generous supporter and one of the founders of the National Trust in England. They were the only people to get a free copy of Audubon's great work as it was published.

Rathbun

Grey Jay ssp. *Perisoreus canadensis rathbuni* **Oberholser**, 1917 NCR
[JS *Perisoreus canadensis obscurus*]

Samuel Frederick Rathbun (1858–1946) was an American ornithologist, taxidermist and banker. He moved to Seattle (1890) and served as City Treasurer (1900–1902). He left his collection of birds to the Washington State Museum, where he had been Honorary Curator of Ornithology.

Raven

Raven's Whistler *Coracornis raveni* **Riley**, 1918
[Alt. Maroon-backed Whistler; Syn. *Pachycephala raveni*]

Common Flameback ssp. *Dinopium javanense raveni* Riley, 1927

Henry Cushier Raven (1889–1944) was an American explorer who collected in Borneo and Sulawesi (1912–1917), when he led an expedition on behalf of the USNM. William Abbott (q.v.) sponsored the trip although he was too ill to go himself. Raven's 178-page field journal is at the USNM.

Rawnsley

Rawnsley's Bowerbird *Ptilonorhynchus rawnsleyi* **Diggles,** 1867 NCR
[Hybrid: *Ptilonorhynchus violaceus* x *Sericulus chrysocephalus*]

Henry Charles Rawnsley (1818–1873) was a surveyor, or so he claimed when he arrived in Australia from England, and a naturalist. He surveyed the Mount Remarkable area for the Colonial Government for three months before being recalled. He certainly did undertake survey work (1850s and 1860s) and bought land at Witton Creek near Brisbane. It was in this area that he shot the bird that bears his name (1867). No other similar specimen was seen until photographs were taken of a living bird (2003). A landmark, and now a field station, is named 'Rawnsley's Bluff' after him.

Ray, J.

Grey Goshawk *Astur raii* **Vigors** & **Hosfield**, 1827 NCR
[JS *Accipiter novaehollandiae*]
Yellow Wagtail *Budytes rayi* **Bonaparte**, 1838 NCR
[JS *Motacilla flava flavissima*]

Rev. John Ray (1627–1705) was the son of a village blacksmith, from a very humble, deeply religious family. The local parish priest recognised that he had potential, and arranged for him to attend a grammar school and then Cambridge. He could not afford the fees so, as was the case with Isaac Newton subsequently, he was exempted by becoming a servant of the staff of his college (Trinity) as well as a student. He obtained a master's degree (1651) and stayed on as a lecturer. Botany and zoology were not then on the Cambridge curriculum, but Ray invited his students to join him in the study of local flora and fauna, and published his first book on botany (1660). He lived in difficult times, the period of the English Civil War and the Puritan Commonwealth. He disagreed with ritual in the State religion, re-established after the Restoration (1660), the year in which he was ordained as a priest in the Church of England so he was forced to leave the University (1662). At this time one of his former students, Francis Willughby, suggested they undertake scientific research together at Willughby's expense. Over the next decade they travelled all over the British Isles and Europe, observing and collecting, Willughby concentrating on animals and Ray on plants. Ray published a catalogue of plants of the British Isles (1670) and was elected a Fellow of the Royal Society in recognition of his achievements. When Willughby died (1672) he willed Ray enough money to continue with their work. Ray published Willughby's work on birds in Latin (1676) and in an English translation, *The Ornithology of Francis Willughby* (1678). He published on fish (1685) and on mammals and reptiles (1693). In this last book, he stated that fossils were petrified remains of extinct creatures – another example of how far ahead of his time he was, as this was not accepted for a century. He started the documentation and classification of European plants and specifically defined what was meant by a species. Linnaeus later used Ray's work as the building blocks of his classification system.

Ray, M.

Eared Poorwill ssp. *Nyctiphrynus mcleodii rayi* A. H. Miller, 1948

Mexican Chickadee ssp. *Poecile sclateri rayi* A. H. Miller & **Storer**, 1950

Milton Smith Ray (1881–1946) was a San Francisco field ornithologist and collector, as well as a poet and wealthy industrialist. He wrote *The Literary and Other Principles in Ornithological Writing* (1911).

Raymond

Slender-billed Starling ssp. *Onychognathus tenuirostris raymondi* **Meinertzhagen**, 1936 NCR

[JS *Onychognathus tenuirostris theresae*; or species sometimes regarded as monotypic]

Raymond Hook (1892–1973) was a farmer, hunter, guide and naturalist in Kenya (1912). He had a farm which straddled the equator, and could thus be said to be in both hemispheres! He became involved (1937) in an abortive scheme to introduce cheetah racing in England. He made a lot of money from selling Bongo antelopes to zoos in the USA and Europe. During the Mau-Mau emergency in Kenya (1950s), Hook was told by Army Intelligence that he was number two on the hit list of white settlers – the number one man on the list had recently been murdered by being buried alive – but he let nothing worry him and although he carried a gun, he never bothered to lock his front door.

Rea

Common/Red Crossbill ssp. *Loxia curvirostra reai* **A. R. Phillips**, 1981

Chihuahuan Raven ssp. *Corvus cryptoleucus reai* A. R. Phillips, 1986 NCR; NRM

Dr Amadeo M. Rea is an American ornithologist and ethnobiologist. He is an Adjunct Professor, San Diego State University, and a Research Associate, San Diego Natural History Museum. The University of Arizona awarded his doctorate (1977). He wrote: *Wings in the Desert* (2007).

Reeve

Reeve's Thrush *Turdus reevei* **G. N. Lawrence**, 1869
[Alt. Plumbeous-backed Thrush]

J. F. Reeve (DNF) was an American resident in Guayaquil, Ecuador, who collected birds – including the thrush, first taken on Puna Island. Lawrence wrote: 'I have conferred upon this species the name of J. F. Reeve, Esq., of Guayaquil, who (as I am informed by Prof. Jas. Orton) is a gentleman of great energy of character and courage, which latter quality is of importance in any explorations on Puna Island, where collections are made at great personal risk, from the ferocious nature of the wild animals with which it abounds.'

Reeves, J.

Reeves's Pheasant *Syrmaticus reevesii* **J. E. Gray**, 1829

John Reeves (1774–1856) was an English amateur naturalist and collector who served in China, chiefly Canton and Macao, as a civil servant (1812–1831). The East India Company employed him as an 'Inspector of Tea'. He sent the first specimens of the Reeves's Muntjac *Muntiacus reevesi* back to England, where escapees from collections have established it as a feral species. Reeves also commissioned local artists to paint accurate pictures of Chinese flora and fauna; this collection of over 2,000 paintings is now in the BMNH. He is also commemorated in the names of three reptiles and several fish species.

Reeves, T.

Reeves's Woodnymph *Eucephala caeruleolavata* **Gould**, 1860 NCR
[Hybrid involving *Thalurania furcata*?]

Thomas Reeves (DNF) was a lawyer and ornithologist. He edited Descourtilz's *Ornithologie Brésilienne ou Histoire des Oiseaux du Brésil, Remarquables par leur Plumage, leur Chant ou leurs Habitudes*, publishing it in Brazil (1854–1856). He supplied Gould (q.v.) with the type – and only known – specimen of the woodnymph, which was collected in south-eastern Brazil.

Regent

Regent Bowerbird *Sericulus chrysocephalus* **Lewin**, 1808
Regent Parrot *Polytelis anthopeplus* **Lear**, 1831

Named after George, the Prince Regent (1762–1830), who acted as head of state when his father King George III was deemed insane (he actually probably suffered from porphyria). The Prince Regent became King George IV (1820). His mother was Queen Charlotte (q.v.). Some references say that the Prince Regent's colours were black and yellow, like the male bowerbird.

Regina

Spangled Coquette *Lophornis reginae* **Gould**, 1847 NCR
[Invalid name; species now known as *Lophornis stictolophus*]

(See **Queen Victoria** and **Victoria**)

Rehn

Rufous-browed Wren ssp. *Troglodytes rufociliatus rehni* **W. Stone**, 1932

James Abram Garfield Rehn (1881–1965) was an American mammalogist, entomologist and collector. He joined the staff of the Philadelphia Academy of Sciences (1900) and travelled widely, spending over 30 seasons in the field, including

Africa, Colombia, Costa Rica, Honduras (1930) and Brazil. He never went back to his desk at the Academy after his wife died (1964). He wrote mainly on orthoptera but also on mammals, such as the paper 'Three new American bats' (1902).

Rehse

Nauru Reed Warbler *Acrocephalus rehsei* **Finsch**, 1883

Ernst Rehse (DNF) was a German ornithologist who collected in the Pacific region (c.1870).

Reichard

Reichard's Seedeater *Serinus reichardi* **Reichenow**, 1882
[Alt. Stripe-breasted Seedeater; Syn. *Crithagra reichardi*]
Tanzanian Masked Weaver *Ploceus reichardi* Reichenow, 1886

Paul Reichard (1845–1938) was a German geographer and engineer. He collected in East Africa (1880–1884), sending ethnographical specimens to the museum in Berlin. On this expedition he travelled with Böhm (q.v.). They discovered Lake Upemba (Tanzania) and Böhm died there of fever. Reichard continued on alone to Katanga (Belgian Congo) and discovered copper deposits there, which led eventually to a huge mining operation. He wrote a number of papers including 'Deutsch-Ostafrika, Das Land und seine Bewohner' (1882), and 'Das Afrikanische Eisenbahn und sein Handel' (1899).

Reichenbach

Pigeon sp. *Crossophthalmus reichenbachi* **Bonaparte**, 1855 NCR
[Alt. Picazuro Pigeon; JS *Patagioenas picazuro*]
Reichenbach's Whitethroat *Leucochloris malvina* Reichenbach, 1855 NCR
[Hybrid: *Leucochloris albicollis* x *Chlorostilbon aureoventris*]
Reichenbach's Sunbird *Anabathmis reichenbachii* **Hartlaub**, 1857
Chestnut Woodpecker *Celeopicus reichenbachi* **Malherbe**, 1862 NCR
[JS *Celeus elegans*]

Micronesian Kingfisher ssp. *Todiramphus cinnamominus reichenbachii* Hartlaub, 1852
Smoky-brown Woodpecker ssp. *Picoides fumigatus reichenbachi* **Cabanis** & **Heine**, 1863
Grey-fronted Dove ssp. *Leptotila rufaxilla reichenbachii* **Pelzeln**, 1870

Dr Heinrich Gottlieb Ludwig Reichenbach (1793–1879) was a German zoologist and botanist. He studied medicine and natural science at Leipzig (1810) and practised there after graduating. He then also qualified as a private tutor in medicine and natural science, but was appointed Professor of Natural History (1820) at the Surgical/Medical Academy in Dresden, beoming responsible for the Natural History Museum. He founded the Dresden Botanical Gardens. He researched very extensively and his legacy included over 6,000 drawings, most of them his own work. Reichenbach devised a unique method of botanical classification, under which there are eight categories depending upon the plants' organs. He wrote *Praktische Naturgeschichte der Vögel* (1845) and *Avium Systema Naturale, das Naturliche System der Vögel* (1849). He retired in 1862.

Reichenow

Parrotfinch genus *Reichenowia* Poche, 1904
[Usually included in *Erythrura*, or treated as a subgenus of latter]

Reichenow's Crimsonwing *Cryptospiza reichenovii* **Hartlaub**, 1874
[Alt. Red-faced Crimsonwing]
Golden-winged Sunbird *Drepanorhynchus reichenowi* **Fischer**, 1884
Reichenow's Seedeater *Serinus reichenowi* **Salvadori**, 1888
[Alt. Kenya Yellow-rumped Seedeater; Syn. *Crithagra reichenowi*]
Northern Double-collared Sunbird *Cinnyris reichenowi* **Sharpe**, 1891
Reichenow's Nightjar *Caprimulgus clarus* Reichenow, 1892
[Alt. Slender-tailed Nightjar]
Reichenow's Woodpecker *Campethera scriptoricauda* Reichenow, 1896
[Alt. Speckle-throated Woodpecker]
Reichenow's Wattle-eye *Platysteira chalybea* Reichenow, 1897
[Alt. Black-necked Wattle-eye]
Bulbul sp. *Pycnonotus reichenowi* Lorenz & **Hellmayr**, 1901 NCR
[Alt. White-spectacled Bulbul; JS *Pycnonotus xanthopygos*]
Green-breasted Pitta *Pitta reichenowi* **Madarász**, 1901
Reichenow's Dove *Streptopelia reichenowi* **Erlanger**, 1901
[Alt. African White-winged Dove]
Red-fronted Warbler *Spiloptila reichenowi* Madarász, 1904 NCR
[JS *Urorhipis rufifrons*]
Reichenow's Swift *Apus reichenowi* **Neumann**, 1908 NCR
[Alt. Mottled Swift; JS *Tachymarptis aequatorialis*]
Plain-backed Sunbird *Anthreptes reichenowi* **Gunning**, 1909
Reichenow's Firefinch *Lagonosticta umbrinodorsalis* Reichenow, 1910
[Alt. Chad Firefinch, Pink-backed Firefinch]
Reichenow's Batis *Batis reichenowi* **Grote**, 1911
Reichenow's Melidectes *Melidectes rufocrissalis* Reichenow, 1915
[Alt. Yellow-browed Honeyeater]

Reichenow's Turaco *Tauraco livingstonii reichenowi* Fischer, 1880
[Alt. Livingstone's Turaco ssp.]
Reichenow's Blue-headed Parrot *Pionus menstruus reichenowi* **Heine**, 1884
Reichenow's Weaver *Ploceus baglafecht reichenowi* Fischer, 1884
[Alt. Baglafecht Weaver ssp.]
White-chinned Prinia ssp. *Schistolais leucopogon reichenowi* Hartlaub, 1890
Reichenow's Orange-breasted Fig Parrot *Cyclopsitta gulielmitertii amabilis* Reichenow, 1891
Gabon Woodpecker ssp. *Dendropicos gabonensis reichenowi* **Sjostedt**, 1893

Buff-spotted Flufftail ssp. *Sarothrura elegans reichenovi* Sharpe, 1894

Reichenow's Guineafowl *Numida meleagris reichenowi* **Ogilvie-Grant**, 1894

[Alt. Helmeted Guineafowl ssp.]

Brown Parrot ssp. *Pocephalus meyeri reichenowi* Neumann, 1898

Reichenow's Grey-headed Parrot *Poicephalus fuscicollis suahelicus* Reichenow, 1898

Crested Lark ssp. *Galerida cristata reichenowi* Erlanger, 1899 NCR

[JS *Galerida cristata arenicola*]

Blackcap Bush-shrike ssp. *Bocagia minuta reichenowi* Neumann, 1900

[Alt. Marsh Tchagra ssp.; Syn. *Tchagra minutus reichenowi*]

Arafura Fantail ssp. *Rhipidura dryas reichenowi* **Finsch**, 1901

Abyssinian Slaty Flycatcher ssp. *Dioptrornis chocolatinus reichenowi* Neumann, 1902

Black Saw-wing ssp. *Psalidoprocne pristoptera reichenowi* Neumann, 1904

Green-backed Honeyguide ssp. *Prodotiscus zambesiae reichenowi* Madarász, 1904 NCR

[JS *Prodotiscus zambesiae ellenbecki*]

Black Woodpecker ssp. *Dryocopus martius reichenowi* Kothe, 1906 NCR

[NUI *Dryocopus martius martius*]

Forest Scrub Robin ssp. *Erythropygia leucosticta reichenowi* **Hartert**, 1907

Red-necked Aracari ssp. *Pteroglossus bitorquatus reichenowi* **Snethlage**, 1907

Reichenow's Reed Warbler *Bradypterus cinnamomeus mildbreadi* Reichenow, 1908

[Alt. Cinnamon Bracken Warbler ssp.]

Little Green Bee-eater ssp. *Merops viridis reichenowi* Parrot, 1910 NCR

[JS *Merops orientalis viridissimus*]

Reichenow's Orange-bellied Parrot *Poicephalus senegalus mesotypus* Reichenow, 1910

[SII *Poicephalus senegalus senegalus*]

African Yellow White-eye ssp. *Zosterops senegalensis reichenowi* Dubois, 1911

Rusty Whistler ssp. *Pachycephala hyperythra reichenowi* **Rothschild** & Hartert, 1911

Short-winged Cisticola ssp. *Cisticola brachypterus reichenowi* **Mearns**, 1911

White-browed Piculet ssp. *Sasia ochracea reichenowi* **Hesse**, 1911

White-rumped Swiftlet ssp. *Aerodramus spodiopygius reichenowi* **Stresemann**, 1912

Black-headed Oriole ssp. *Oriolus larvatus reichenowi* **Zedlitz**, 1916

Grey-headed Cuckooshrike ssp. *Coracina schisticeps reichenowi* Neumann, 1917

Dr Anton Reichenow (1847–1941) was the German son-in-law of Cabanis (q.v.), whom he succeeded as editor of *Journal für Ornithologie* (1894–1921), and dominated German ornithology for many years. He became an assistant in the Zoological Museum of the Humboldt University of Berlin (1874), becoming Curator of Birds (1888–1906). He was promoted to Vice-Director in 1906, succeeding Cabanis. He also succeeded Cabanis as Secretary General of the German Ornithologists' Society (1894). He established the first German bird observatories (1875), but his reputation was built around his expert knowledge of African birds. He only visited Africa once, on a collecting expedition to West Africa, including the Gold Coast (Ghana), Gabon and Cameroon (1872–1873). He wrote *Die Vogelwelt von Kamerun* (1890 and 1892), *Die Vögel Deutsch-Ostafrikas* (1894) and a 3-volume handbook on the birds of Africa, *Die Vögel Afrikas* (1900–1905), but also produced *Die Vögel der Bismarckinseln* (1899). On his retirement (1921) he moved to Hamburg, where he was actively engaged in the local Natural History Museum. Reichenow described over 1,000 species and subspecies of birds. Two reptiles are named after him.

Reichert

Golden Weaver ssp. *Ploceus aureoflavus reicherti* **Meise**, 1934 NCR

[JS *Ploceus subaureus aureoflavus*]

Robert Reichert (1897–1959) was a German zoologist, taxidermist, and explorer who collected in Tanganyika (Tanzania) (1931–1932). He was associated with the Dresden Museum for many years, including being in charge of finding safe temporary homes for the exhibits during WW2 – two-thirds survived the war, so he was able to stage Dresden's first post-war exhibition (1949). He became Acting Director of the museum (1950–1957) and then Director (1957–1959).

Reid

Cape Bunting ssp. *Emberiza capensis reidi* **G. E. Shelley**, 1902

Captain Philip Savile Grey Reid (1845–1915) was a British army engineer (commissioned 1865) who served in Gibraltar (1871–1873), Bermuda (1875) and Natal (1881), before retiring (1884). He wrote 'Winter notes from Morocco' in *Ibis* (1885).

Rein

Purple-chested Hummingbird ssp. *Polyerata rosenbergi reini* **Berlepsch**, 1897 NCR; NRM

[Syn. *Amazilia rosenbergi*]

Dr Johannes Justus Rein (1835–1918) was a German geographer, traveller and Japanologist. He was on an expedition to Morocco (1872–1873) and also travelled in the Americas, Great Britain and Spain, and was in Japan (1873–1875). He was appointed Professor of Geography at the University of Marburg (1876) and to the same position at the University of Bonn (1883).

Reinard

Palm Swift genus *Reinarda* **Hartert**, 1915 NCR

[Now in *Tachornis*]

Claudia Reinard (DNF) was the wife of the describer (See **Claudia**)

Reinhardt, J. C. H.

Reinhardt's Ptarmigan *Lagopus muta reinhardi* **C. L. Brehm**, 1824
[Alt. Rock Ptarmigan ssp.; SII *Lagopus muta rupestris*]

Dr Johannes Christopher Hagemann Reinhardt (1776–1845) was Professor of Zoology (1814) at the University of Copenhagen. Born in Norway, Reinhardt studied theology, zoology, botany, mineralogy and anatomy in Copenhagen, Freiberg, Göttingen and Paris. His son was J. T. Reinhardt (q.v.).

Reinhardt, J. T.

Yellow-scarfed Tanager *Iridosornis reinhardti* **P. L. Sclater**, 1865

Dr Johannes Theodor Reinhardt (1816–1882) was a Danish zoologist who was Director of the National Natural History Museum, Copenhagen. He wrote a 'List of birds hitherto observed in Greenland' in *Ibis* (1861), which included the Eskimo Curlew *Numenius borealis* and the Great Auk *Pinguinus impennis*. His father was J. C. H. Reinhardt (q.v.). An amphibian and five reptiles are named after him.

Reinhold

Shearwater genus *Reinholdia* **Mathews**, 1912 NCR
[Now in *Puffinus*]

Fluttering Shearwater *Puffinus reinholdi* Mathews, 1912 NCR
[JS *Puffinus gavia*]

(See **Forster, J.**)

Reinwardt

Cuckoo Dove genus *Reinwardtoena* **Bonaparte**, 1854
Woodpecker genus *Reinwardtipicus* Bonaparte, 1854

Reinwardt's Blue-tailed Trogon *Apalharpactes reinwardtii* **Temminck**, 1822
[Alt. Javan Trogon]
Reinwardt's Scrubfowl *Megapodius reinwardt* **C. H. F. Dumont**, 1823
[Alt. Orange-footed Scrubfowl/Megapode]
Reinwardt's Long-tailed Pigeon *Reinwardtoena reinwardtii* Temminck, 1824
[Alt. Great Cuckoo Dove; binomial was originally spelt *reinwardtsi* and retained thus by some authorities]
Golden-collared Toucanet *Selenidera reinwardtii* **Wagler**, 1827
Reinwardt's Babbler *Turdoides reinwardtii* **Swainson**, 1831
[Alt. Blackcap Babbler]

Pacific Baza ssp. *Aviceda subcristata reinwardtii* **Schlegel & Muller**, 1841

Caspar Georg Carl Reinwardt (1773–1854) was a German-born Dutch ornithologist who collected in Java (1817–1822). At 14 he went to study botany and chemistry in Amsterdam, where he also studied mathematics, classical and modern languages, and history. He gained a doctorate in Natural Philosophy and Medical Science (1801). That year he also became Professor of Chemistry and Natural History at the University of Harderwijk, where he was responsible for the botanical garden. Louis Napoleon made him Director of his menagerie in Amsterdam, and he became Professor of Natural History there. He was made responsible for all matters concerning agriculture, arts and sciences in Java (1816–1821) and contributed greatly to education and public health there. He founded the Botanical Gardens of Buitenzorg (Bogor) and travelled extensively. He laid the foundation for two important museum collections in Leiden: the National Museum of Natural History and the National Museum of Ethnography. While collecting he was meticulous and detailed in the documentation of his specimens. He was very concerned with the systematic development of collections and their documentation, and was always pleased to extend opportunities for research to others. It was through this that many fellow naturalists honoured him in the names of various plants. An amphibian is named after him.

Reischek

Reischek's Parakeet *Cyanoramphus hochstetteri* Reischek, 1889
Variable Oystercatcher *Haematopus reischeki* **Rothschild**, 1899 NCR
[JS *Haematopus unicolor*]

New Zealand Pipit ssp. *Anthus novaeseelandiae reischeki* Lorenz von Liburnau, 1902
Weka ssp. *Gallirallus hectori reischeki* **Iredale**, 1913 NCR
[JS *Gallirallus australis hectori*]

Anton (or Andreas) Reischek (1845–1902) was an Austrian taxidermist, mountaineer, collector and hunter. He started his career in the service of the 'Alpenjägern' (1866–1875), a section of the Austrian army, after which he became a dealer in natural history objects in Vienna. At the request of the Vienna Museum of Natural History he accepted a post in Christchurch, New Zealand, as a taxidermist (1877–1889). From the Maori King Tawhiao he got permission to travel into a territory in central North Island that was usually forbidden to foreigners. An early conservationist, he warned of the destructive European influence on the New Zealand environment and on Maori culture. He disinterred two mummified Maori corpses and took them to the Ethnographical Museum in Vienna. The surviving mummy, of a Maori chief, was returned to New Zealand (1985) and re-interred in a private ceremony. Upon his return to Vienna, Reischek worked in the Natural History Museum, but did not get tenure so (1896) became curator of the newly established Museum of Natural History in Linz.

Reiser

Reiser's Recurvebill *Megaxenops parnaguae* Reiser, 1905
[Alt. Great Xenops]
Reiser's Spinetail *Gyalophylax hellmayri* Reiser, 1905
[Alt. Red-shouldered Spinetail]
Reiser's Tyrannulet *Phyllomyias reiseri* **Hellmayr**, 1905

Alpine Accentor ssp. *Accentor collaris reiseri* Tschusi, 1901 NCR
[JS *Prunella collaris subalpina*]

Reed Bunting ssp. *Emberiza schoeniclus reiseri* **Hartert**, 1904

Eastern Olivaceous Warbler ssp. *Iduna pallida reiseri* Hilgert, 1908

Mistle Thrush ssp. *Turdus viscivorus reiseri* Schiebel, 1911 NCR

[JS *Turdus viscivorus deichleri*]

Rock Partridge ssp. *Alectoris saxatilis reiseri* **Reichenow**, 1911 NCR

[JS *Alectoris graeca graeca*]

Reiser's Woodcreeper *Sittasomus griseicapillus reiseri* Hellmayr, 1917

[Alt. Olivaceous Woodcreeper ssp.]

Rusty-backed Spinetail ssp. *Cranioleuca vulpina reiseri* Reichenberger, 1922

Dr C. H. Othmar Reiser (1861–1936) was an Austrian oologist and collector in north-eastern Brazil with the expedition of the Austrian Academy of Sciences (1903). He wrote an account of this enterprise, *Die Ergebnisse der Zool. Expedition der Akad. der Wissenschaften nach nordostbrasilien im Jahre 1903*. Reiser was curator of the museum in Sarajevo (1887–1920), Bosnia, and wrote a 4-volume series; *Materialen zu einer Ornis Balcanica*, the second volume of which was devoted entirely to the birds of Bulgaria. (See **Othmar**)

Remsen

Chestnut-bellied Cotinga *Doliornis remseni* **Robbins**, G. H. Rosenberg & Molina, 1994

Dr James Van Remsen Jr is Curator of Birds, McIlhenny Distinguished Professor of Natural Science, and Adjunct Professor of Biological Sciences at the Louisiana State University, USA. He became interested in wildlife when just five years old, and by the time he was 11 he was an avid birdwatcher. Berkeley awarded his PhD (1978), his dissertation being 'Geographical ecology of Neotropical kingfishers'. Currently his area of interest is the ecology, evolution and biogeography of Neotropical birds, particularly those of the Andes and the Amazon basin. He has numerous published papers such as (1984) 'High incidence of "leap-frog" pattern of geographic variation in Andean birds: implications for the speciation process'. Robbins *et al.* wrote in their etymology: 'We take great pleasure in naming this new cotinga after our friend and colleague J. Van Remsen Jr, in recognition of his many contributions to Neotropical ornithology, and his special interest in Andean birds'.

Renate

Mountain Serin ssp. *Serinus estherae renatae* Schuchmann & **Wolters**, 1982

Dr Renate van den Elzen recently retired as Curator of Ornithology at the Alexander Koenig Museum in Bonn, Germany. She graduated with a doctorate in biology at the University of Vienna (1973). She was Secretary, German Ornithologists' Society (1990–1996) and started working with the Zambian Biodiversity Project (2006).

Renauld

Renauld's Ground Cuckoo *Carpococcyx renauldi* **Oustalet**, 1896

[Alt. Coral-billed Ground Cuckoo]

Father J. N. Renauld (1839–1898) was a French missionary in Vietnam. His many specimens of Vietnamese fauna and flora collected in Quang Tri province (1869–1896) were used to found the Vietnamese Natural History Museum (1896). Oustalet (q.v.) described Edwards's Pheasant *Lophura edwardsi* (1896) from skins which Renauld sent to the MNHN.

Rendall

Parasitic Weaver *Crithagra rendalli* **Tristram**, 1895 NCR

[JS *Anomalospiza imberbis*]

Red-billed Firefinch ssp. *Lagonosticta senegala rendalli* **Hartert**, 1898

Dr Percy Rendall (1861–1948) was an itinerant zoologist who collected over much of Africa and in Trinidad and other Caribbean locations in the late 19th century. He made a collection of new fish species from the Upper Shiré River, British Central Africa, which was presented to the British Museum by Sir Harry Johnston (q.v.). He published 'Notes on the ornithology of the Gambia' (1892). A fish and a mammal are named after him.

Rengger

Tyrannulet genus *Renggerornis* **A. Bertoni**, 1901 NCR

[Now in *Camptostoma*]

Dr Johann Rudolph Rengger (1795–1832) was a German-born Swiss physician and naturalist who studied at University of Lausanne and the Eberhard Karls University at Tübingen. He travelled extensively and lived in Paraguay (1819–1825). He wrote *Natural History of the Mammals of Paraguay* (1830). A reptile is also named after him.

Rens

Burnt-necked Eremomela ssp. *Eremomela usticollis rensi* **C. W. Benson**, 1943

J. H. Rens (1886–1957) was a Dutch Reformed Church missionary to Nyasaland (Malawi) (1915–1949). Benson (q.v.) was a district officer in Nyasaland and expressed his gratitude for all the collecting Rens did on his behalf.

Rensch

Great Spotted Woodpecker ssp. *Dryobates major renschi* **Kuroda**, 1929 NCR

[JS *Dendrocopos major cabanisi*]

Spotted Kestrel ssp. *Falco moluccensis renschi* Siebers, 1930 NCR

[NUI *Falco moluccensis microbalius*]

Bonelli's Eagle ssp. *Hieraeetus fasciatus renschi* **Stresemann**, 1932

[Syn. *Aquila fasciata renschi*]

Wallacean Drongo ssp. *Dicrurus hottentottus renschi* **Vaurie**, 1949 NCR

[NUI *Dicrurus densus bimaensis*]

Professor Bernhard Rensch (1900–1990) was a zoologist, biologist, philosopher and artist. He was at the Zoological Museum of Berlin University (1925–1936) and was Director of the Zoological Institute of the University of Munster (1947–1968). He visited the Lesser Sunda Islands with his wife, the botanist Ilse Rensch-Maier (1927). A reptile is named after him.

Renzo

Eurasian Hoopoe ssp. *Upupa epops renzoi* Trischitta, 1939 NCR
[JS *Upupa epops epops*]

Dr Renzo Ragionieri (See **Ragionieri**)

Retz

Retz's Helmet-shrike *Prionops retzii* **Wahlberg**, 1856
[Alt. Retz's Red-billed Helmet-shrike]

Anders Adolph Retzius (1796–1860) was Professor in Anatomy and Physiology at the Karolinska Institute, Stockholm. He was also Wahlberg's (q.v.) brother-in-law and helped finance his expeditions.

Reuben

Black-hooded Oriole ssp. *Oriolus xanthornus reubeni*
Abdulali, 1977

Dr Rachel Reuben (1934–2010) was an Indian scientist, medical entomologist and conservationist. She was a member of one of India's smallest minorities – Bene Israel Jews of Maharashtra – of whom only about 5,000 are left out of some 20,000, the majority having emigrated to Israel. Both Abdulali (q.v.) and Salim Ali (q.v.) were long-standing friends of her family.

Révoil

Somali Bee-eater *Merops revoilii* **Oustalet**, 1882

Georges Emmanuel Joseph Révoil (1852–1894) was a French naturalist who collected in Somaliland (1878–1880) and wrote *La Vallée du Darro: Voyage aux Pays Somalis* (1882). A mammal and a reptile are named after him.

Rex

Anchieta's Barbet ssp. *Stactolaema anchietae rex*
Neumann, 1908

Carlos Fernando Luís Maria Vítor Miguel Rafael Gabriel Gonzaga Xavier Francisco de Assis José Simão de Bragança Sabóia Bourbon Saxe-Coburgo-Gotha (1863–1908) was King of Portugal, reigning simply as Carlos I (1889–1908) until assassinated. He was a talented watercolour painter of birds and was an oceanographer of international repute. *Rex* is, of course, Latin for King.

Rex Pimento

Nicobar Hawk Owl ssp. *Ninox affinis rexpimenti* **Abdulali**, 1979
[SII *Ninox affinis isolata*]

Rex Pimento (DNF) was a Field assistant at the Bombay Natural History Society. He collected in the Andaman and Nicobar Islands.

Rey, J. G.

Grey-capped Hemispingus *Hemispingus reyi* **Berlepsch**, 1885

Dr J. G. C. Eugene Rey (1838–1909) was a German oologist, who was born in Berlin and died in Leipzig.

Rey, S.

Sulu Hawk Owl *Ninox reyi* **Oustalet**, 1880

Dr S. Rey (DNF) was a German naturalist in Leipzig.

Reynaud

Kalij Pheasant sp. *Phasianus reynaudii* **Lesson**, 1831 NCR
[Probably a JS of *Lophura leucomelanos lineata*]
Red-fronted Coua *Coua reynaudii* **Pucheran**, 1845

Auguste Adolphe Marc Reynaud (1804–1872) was a French naval surgeon and explorer on board *La Chevrette* (1827–1828). He was later Chief Inspector of Medical Services (1858–1872). He collected in Burma (Myanmar) and Madagascar.

Reynolds, B.

Eastern Clapper Lark ssp. *Mirafra fasciolata reynoldsi*
C. W. Benson & **Irwin**, 1965

Dr Barrie Reynolds is an anthropologist who was Keeper of Ethnography, National Museums of Zambia (which includes Barotseland). He wrote his PhD on 'An ethnographic study of the Kwandu People, south-western Barotseland, with particular reference to the role of craftsmen in the society' (1968).

Reynolds, P. W.

Austral Blackbird ssp. *Curaeus curaeus reynoldsi*
W. L. Sclater, 1939

Percival William Reynolds (1904–1940) was a British ornithologist and explorer in Tierra del Fuego (1922–1940). He was the grandson of the first Anglican missionary there whose descendants stayed on and became sheep farmers. He wrote 'Notes on the birds of Cape Horn' (1935).

Rham

Garnet-throated Hummingbird *Lamprolaima rhami* **Lesson**, 1839

Henri Casimir de Rham (1785–1873) (See **De Rham**)

Rhea

Rhea genus *Rhea* Brisson, 1760

In Greek mythology, Rhea was one of the Titans, a daughter of Uranus and Gaia. She was known as 'The Mother of Gods'. Though the word has now also become the common name for these ratites, the reasons behind Brisson's choice of nomenclature remain obscure.

Rheinard

Rheinard's (Argus) Pheasant *Rheinardia ocellata*
D. G. Elliot, 1871
[Alt. Crested Argus]

Lieutenant-Colonel Pierre-Paul Rheinart (1840–1902) was an officer in the French army and an administrator in Vietnam. He was Chargé d'Affaires in Annam (1879–1889) and Résident-Générale in Annam-Tonkin (1889). He also explored in Laos (1859) and sent the first specimen of the pheasant to Paris. The pheasant should more correctly be called Rheinart's Argus, but the spelling of the genus has led to his name usually being corrupted to 'Rheinard'.

Rhodes

Common Kestrel ssp. *Falco tinnunculus rhodesi* Finch-Davies, 1920 NCR
[Alt. Rock Kestrel; JS *Falco* (*tinnunculus*) *rupicolus*]

Cecil John Rhodes (1853–1902) was an English businessman, mining magnate and politician who was Prime Minister of Cape Colony (South Africa) (1890–1896) and founder of Rhodesia (Zambia and Zimbabwe) (1895). He is the epitome of the 19th century empire-builder!

Rhodopis

Oasis Hummingbird genus *Rhodopis* **Reichenbach**, 1854

Rhodopis was the heroine of the original 'Cinderella' story, first recorded in the 1st century BC by the Greek historian Strabo. She was a slave in Egypt who lost one of her slippers. Naturally it came into the hands of the Pharaoh, who sought out its owner and married her!

Ribeiro

Ribeiro's Scaly-headed Parrot *Pionus maximiliani melanoblepharus* Ribeiro, 1920

(See **Miranda, A.**)

Rich

Yellow Rail ssp. *Coturnicops noveboracensis richii*
H. H. Bailey, 1935 NCR
[JS *Coturnicops noveboracensis noveboracensis*]

Walter Herbert Rich (1866–1948) was a watercolourist who worked for the US Fisheries Commission (23 years) and became the Curator of the Portland Society of Natural History's Museum. Among other works he wrote *Feathered Game of the Northeast* (1907).

Richard

Richard's Pipit *Anthus richardi* **Vieillot**, 1818

Monsieur Richard (fl.1815) of Lunéville was a French naturalist and collector. It is possible that Vieillot's 'Monsieur Richard' was Achilles Richard (1794–1852), who was a botanist at Abbéville, or his father Louis Claude Marie Richard

(1754–1821), who collected in Central America and the West Indies (1780–1789) and became Professor of Botany at the School of Medicine (1790).

Richard (Liversidge)

Familiar Chat ssp. *Cercomela familiaris richardi*
J. D. Macdonald, 1953 NCR
[JS *Cercomela familiaris galtoni*]

Dr Richard Liversidge (1926–2003) was a South African ornithologist with the Port Elizabeth Museum (1956–1966) before becoming Director, McGregor Museum, Kimberley (1966–1986).

Richards, G.

Bougainville Whistler *Pachycephala richardsi* **Mayr**, 1932
[Syn. *Pachycephala implicata richardsi*]

Guy Richards (b.1905) was an Associate at the AMNH. He was educated at Yale, graduating PhD (1927). He was a member of the Whitney South Sea Expedition to New Guinea and the Solomon Islands (1927) and later worked as a journalist for several New York newspapers, becoming City Editor of the *New York Journal*. He was also a freelance author, writing a number of papers including: *Trials and Tribulations of Bougainville – Bird Collecting Adventures on the Mountain Slopes of a South Sea Island* (1931). Mayr (q.v.) wrote that he named the bird after 'Mr Guy Richards, who was a member of the party that so successfully explored the mountains of Bougainville Island'.

Richards, G. E.

Richards's Monarch *Monarcha richardsii* **E. P. Ramsay**, 1881
[Alt. White-capped Monarch]
Richards's Fruit Dove *Ptilinopus richardsii* E. P. Ramsay, 1882
[Alt. Silver-capped Fruit Dove]

White-bellied Woodpecker ssp. *Dryocopus javensis richardsi* **Tristram**, 1879
Little Kingfisher ssp. *Ceyx pusillus richardsi* Tristram, 1882

Rear-Admiral George Edward Richards (1852–1927) was an English geographer who collected in the East Indies.

Richardson, J.

Richardson's Canada Goose *Branta hutchinsii* Richardson, 1831
[Alt. Cackling Goose]
Eastern Phoebe *Tyrannula richardsoni* **Swainson**, 1832 NCR
[JS *Sayornis phoebe*]
Richardson's Skua *Lestris richardsonii* Swainson, 1832 NCR
[Alt. Arctic Skua; JS *Stercorarius parasiticus*]
Lemon-throated Barbet *Eubucco richardsoni* **G. R. Gray**, 1846
Western Wood Pewee *Contopus richardsonii* **Baird**, 1858 NCR
[Now known as *Contopus sordidulus veliei*]

Richardson's Grouse *Dendragapus obscurus richardsonii* **Douglas**, 1829
[Alt. Dusky Grouse ssp.]

Richardson's Owl *Aegolius funereus richardsoni* **Bonaparte**, 1838
[Alt. Boreal Owl ssp.]
Richardson's Merlin *Falco columbarius richardsonii* **Ridgway**, 1871

Sir John Richardson (1787–1865) was a Scottish naval surgeon and Arctic explorer, knighted (1846), who assisted Swainson (q.v.). He was a friend of Sir John Franklin (q.v.), to whom he was also related by marriage, and took part in Franklin's expeditions (1819–1822 and 1825–1827). He also participated (1847) in the vain search for Franklin and his colleagues. The Richardson Mountains in Canada are also named after him as are five mammals and four reptiles.

Richardson, W. B.

Richardson's Warbler *Basileuterus luteoviridis richardsoni* **Chapman**, 1912
[Alt. Citrine Warbler ssp.]
Greyish Saltator ssp. *Saltator coerulescens richardsoni* **Van Rossem**, 1938 NCR
[NUI *Saltator coerulescens plumbiceps*]
Audubon's Oriole ssp. *Icterus graduacauda richardsoni* **W. L. Sclater**, 1939 NCR
[JS *Icterus graduacauda dickeyae*]

William Blaney Richardson (1868–1927) was an American ornithologist who collected in Mexico (1887–1890), Guatemala (1897), Colombia (1900) and Guatemala again (1907–1917). He was the first person (1889) to collect the Colima Warbler *Vermivora crissalis*.

Richardson, W. B. Jr

Spotless Crake ssp. *Porzana tabuensis richardsoni* **Rand**, 1940

William Blaine Richardson Jr (1891–1972) was the son of William Blaney Richardson (q.v.), the American zoologist who settled in Nicaragua (1891). He was born on a boat en route to Nicaragua, where he spent his childhood but was taken to Boston for his education after his mother's death (1907). His main occupation was as a banker in Mexico, but he clearly inherited his father's interests in natural history as the original citation for the crake recordss that it was named after 'W. B. Richardson, mammalogist of the 1938–1939 Archbold New Guinea expedition.' Richardson kept a detailed journal of that trip.

Richardson, W. E.

St Lucia Black Finch *Melanospiza richardsoni* **Cory**, 1886

William Everett Richardson (1825–1902) was the father and grandfather of W. B. Richardson (q.v.) and W. B. Richardson Jr (q.v.). Cory (q.v.) obtained the type specimen of the finch from Richardson who acquired it, alongside various other specimens, 'from a native living in the interior of St Lucia'.

Richmond

Cardinal genus *Richmondena* **Mathews** & **Iredale**, 1918 NCR
[Now in *Cardinalis*]

Black-naped Fruit Dove *Haemataena richmondena* Mathews, 1926 NCR
[JS *Ptilinopus melanospilus melanauchen*]

Red-winged Blackbird ssp. *Agelaius phoeniceus richmondi* **Nelson**, 1897
Black-striped Sparrow ssp. *Arremonops conirostris richmondi* **Ridgway**, 1898
Grey-chested Jungle Flycatcher ssp. *Rhinomyias umbratilis richmondi* Stone, 1902 NCR; NRM
Yellowish White-eye ssp. *Zosterops nigrorum richmondi* **McGregor**, 1904
Richmond's Swift *Chaetura vauxi richmondi* Ridgway, 1910
[Alt. Dusky-backed Swift, Vaux's Swift]
Black-naped Monarch ssp. *Hypothymis azurea richmondi* **Oberholser**, 1911
Black-browed Albatross ssp. *Thalassarche melanophris richmondi* Mathews, 1912 NCR
[JS *Thalassarche melanophris melanophris* (or species is monotypic if *impavida* raised to full species)]
Black-naped Oriole ssp. *Oriolus chinensis richmondi* Oberholser, 1912
Buffy Hummingbird ssp. *Leucippus fallax richmondi* **Cory**, 1915 NCR
[*L. fallax* NRM]
Crested Serpent Eagle ssp. *Spilornis cheela richmondi* Swann, 1922
Rose-throated Becard ssp. *Platypsaris aglaiae richmondi* **Van Rossem**, 1930 NCR
[JS *Pachyramphus aglaiae albiventris*]

Charles Wallace Richmond (1868–1932) was an American ornithologist who specialised in being a nomenclaturist and bibliographer. He was a colleague of Ridgway (q.v.) at the USNM and he collected all Ridgway's notes after he died, filing them for future use. He created 'The Richmond Index to the Genera and Species of Birds', his life's work lasting 40 years (started 1889) and culminating in 70,000 file cards. A mammal and an amphibian are named after him.

Richmond (River)

Brown Gerygone ssp. *Gerygone mouki richmondi* **Mathews**, 1915

Named after the Richmond River, northern New South Wales, Australia, where the holotype was collected. The river was named after the 5th Duke of Richmond (1791–1860).

Rickett

Rickett's Hill Partridge *Arborophila gingica* **Gmelin**, 1789
[Alt. White-necklaced Partridge; *Arboricola ricketti* is a JS]
Rickett's Leaf Warbler *Phylloscopus ricketti* **Slater**, 1897
[Alt. Sulphur-breasted Warbler, Slater's Leaf Warbler]

Greater Yellownape ssp. *Chrysophlegma flavinucha ricketti* **Styan**, 1898
[Syn. *Picus flavinucha ricketti*]
Rickett's Shrike Babbler *Pteruthius aeralatus ricketti* **Ogilvie-Grant**, 1904
[Alt. Blyth's Shrike Babbler ssp.]
Brown Bullfinch ssp. *Pyrrhula nipalensis ricketti* **La Touche**, 1905

Crested Serpent Eagle ssp. *Spilornis cheela ricketti*
W. L. Sclater, 1919

Rickett's Parrotbill *Sinosuthora brunnea ricketti* **Rothschild**, 1922

[Alt. Brown-winged Parrotbill ssp.]

Silver-eared Mesia ssp. *Leiothrix argentauris ricketti*
La Touche, 1923

[Syn. *Mesia argentauris ricketti*]

Yellow-browed Tit ssp. *Sylviparus modestus ricketti*
La Touche, 1923 NCR

[JS *Sylviparus modestus modestus*]

Charles Boughey Rickett (1851–1943) was a British banker and an amateur ornithologist. He collected in China, Japan, the Straits Settlements (Malaysia), Java and India. Slater said of him: 'I have named this species after Mr C. B. Rickett, who is doing so much at present for Central Chinese ornithology'. He worked for the Hong Kong & Shanghai Banking Corporation in Asia (23 years). He left records of his observations in Scotland and Cornwall (1901–1907). A mammal and an amphibian are named after him.

Ricord

Ricord's Hummingbird *Chlorostilbon ricordii* **Gervais**, 1835
[Alt. Cuban Emerald]

Alexandre Ricord (1798–1876) was a French naval surgeon who qualified as a physician in Paris (1824) and became a Corresponding Member of the French Academy of Medicine (1838). He collected in Latin America (1826–1834). An amphibian and two reptiles are named after him.

Ridgely

Ridgely's Antpitta *Grallaria ridgelyi* Krabbe *et al.*, 1999
[Alt. Jocotoco Antpitta]

Dr Robert 'Bob' Sterling Ridgely (b.1946) is an American ornithologist specialising in Neotropical birds, who worked at the Academy of Natural Sciences at Philadelphia for many years and is now working for World Land Trust US. He is President and Founder of the Fundación Jocotoco, which owns and manages eight nature reserves in Ecuador. He has co-written three books, including *The Birds of Ecuador* (2001). He was the co-discoverer of this species of antpitta (1997).

Ridgway

Aztec Thrush genus *Ridgwayia* **Stejneger**, 1883

Tyrannulet genus *Ridgwayornis* **A. Bertoni**, 1925 NCR
[Now in *Serpophaga*]

Ridgway's Rail *Rallus obsoletus* Ridgway, 1874
[Syn. *Rallus longirostris obsoletus*]

Puna Ibis *Plegadis ridgwayi* **J. A. Allen**, 1876

Ridgway's Hawk *Buteo ridgwayi* **Cory**, 1883
[Alt. Hispaniolan Hawk]

Ridgway's Cotinga *Cotinga ridgwayi* Ridgway, 1887
[Alt. Turquoise Cotinga]

Cocos Flycatcher *Nesotriccus ridgwayi* **C. H. Townsend**, 1895

Ridgway's Whip-poor-will *Caprimulgus ridgwayi* **Nelson**, 1897
[Alt. Buff-collared Nightjar]

Mexican Woodnymph *Thalurania ridgwayi* Nelson, 1900

Ridgway's Titmouse *Baeolophus ridgwayi* **Richmond**, 1902
[Alt. Juniper Titmouse]

Unspotted Saw-whet Owl *Aegolius ridgwayi* **Alfaro**, 1905

Ridgway's Junco *Junco hyemalis aikeni* Ridgway, 1873
[Alt. Dark-eyed Junco ssp.]

Ridgway's Pygmy Owl *Glaucidium brasilianum ridgwayi*
Sharpe, 1875
[Alt. Ferruginous Pygmy Owl ssp.]

Northern Beardless Tyrannulet ssp. *Camptostoma imberbe ridgwayi* **Brewster**, 1882

Great Black Hawk ssp. *Buteogallus urubitinga ridgwayi*
Gurney, 1884

Rock Ptarmigan ssp. *Lagopus muta ridgwayi* Stejneger, 1884

Northern Bobwhite ssp. *Colinus virginianus ridgwayi*
Brewster, 1885

Hawaii Elepaio ssp. *Chasiempis sandwichensis ridgwayi*
Stejneger, 1887

Osprey ssp. *Pandion haliaetus ridgwayi* **Maynard**, 1887

Black-tailed Flycatcher ssp. *Myiobius atricaudus ridgwayi*
Berlepsch, 1888

Ridgway's Parrotlet *Forpus coelestis lucida* Ridgway, 1888
[Alt. Pacific Parrotlet ssp.; species often regarded as monotypic]

Coraya Wren ssp. *Pheugopedius coraya ridgwayi*
Berlepsch, 1889

Lesser Antillean Bullfinch ssp. *Loxigilla noctis ridgwayi*
Cory, 1892

Scaled Dove ssp. *Columbina squammata ridgwayi*
Richmond, 1896

Brown Noddy ssp. *Anous stolidus ridgwayi* **A. W. Anthony**, 1898

Grey Warbler Finch ssp. *Certhidea fusca ridgwayi*
Rothschild & **Hartert**, 1899

Ridgway's Rough-winged Swallow *Stelgidopteryx serripennis ridgwayi* Nelson, 1901
[Alt. Northern Rough-winged Swallow ssp.]

Venezuelan Troupial ssp. *Icterus icterus ridgwayi* Hartert, 1902

Plain-brown Woodcreeper ssp. *Dendrocincla fuliginosa ridgwayi* **Oberholser**, 1904

Amazonian Barred Woodcreeper ssp. *Dendrocolaptes certhia ridgwayi* **Hellmayr**, 1905 NCR
[Intergrade between *Dendrocolaptes certhia concolor* and *D. certhia medius*]

Ladder-backed Woodpecker ssp. *Picoides scalaris ridgwayi*
Oberholser, 1911 NCR
[JS *Picoides scalaris scalaris*]

Great Frigatebird ssp. *Fregata minor ridgwayi* **Mathews**, 1914

Plain Xenops ssp. *Xenops minutus ridgwayi* Hartert & **Goodson**, 1917

Nashville Warbler ssp. *Leiothlypis ruficapilla ridgwayi*
Van Rossem, 1929

Grey-crowned Yellowthroat ssp. *Geothlypis poliocephala ridgwayi* **Griscom**, 1930

Chestnut-headed Oropendola ssp. *Psarocolius wagleri ridgwayi* Van Rossem, 1934

Olive Sparrow ssp. *Arremonops rufivirgatus ridgwayi*
Sutton & Burleigh, 1941

Great Jacamar ssp. *Jacamerops aureus ridgwayi* **Todd**, 1943

Boat-billed Heron ssp. *Cochlearius cochlearius ridgwayi* **Dickerman**, 1973

Robert Ridgway (1850–1929) was a zoologist. At just seventeen he was appointed zoologist on a geological survey of the 40th parallel. He was Curator of Birds at the United States National Museum (USNM) (1880–1929). He was Founder President of the American Ornithologists' Union. Ridgway co-wrote *A History of North American Birds* with Baird (q.v.) & Brewer (q.v.) (1901), and the 11-volume *The Birds of Middle and North America* with Herbert Friedmann (q.v.) (1901–1950). He was also a fine illustrator, famed for having sketched and collected birds around his home in Richland County, Illinois. Because he encountered an almost infinite number of colours and needed accuracy for a scientific description he realised that this would only be possible through some form of standardisation. He therefore proposed a colour system, which was published (1912) under the title Colour Standards and Nomenclature. Ridgway's system exploits the possibilities of additive colour mixing. The basis for the required systematic order of colours is a circle subdivided into 36 pure, solid colours (full colours). Ridgway's method (through a series of other steps) gives 1,115 colour standards intended for use in the identification of the colours of birds. He developed an 18-acre area near his home as a bird sanctuary, which he called Bird Haven. He also established an experimental plot for the cultivation of trees and plants native to his area, and it remains as a memorial. Ridgway seems at first glance to have named *Cotinga ridgwayi* after himself, but he wrote the description using a manuscript name of Zeledon's (q.v.), which had not been officially published. (See **Robert (Ridgway)**)

Ridley

Noronha Elaenia *Elaenia ridleyana* **Sharpe**, 1888

Henry Nicholas Ridley (1855–1956) was a British botanist and collector on the island of Fernando de Noronha (Brazil) (1887). He was known as 'Mad Ridley' or 'Rubber Ridley', as he was keen to ensure that the rubber tree was transplanted to British territory, and so free Britain from dependency on supplies of latex from Brazil. He was Superintendent of the Tropical Gardens in Singapore (1888–1912), where early experiments in growing rubber trees outside Brazil took place, the first successful growth having been achieved at Kew and plants then shipped to Singapore. He wrote 'The natural history of the island of Fernando de Noronha based on the collections made by the British Museum Expedition' (1887) and 'The habits of Malay reptiles' (1889). Three reptiles and two mammals are named after him.

Riedel

Biak Paradise Kingfisher *Tanysiptera riedelii* **J. Verreaux**, 1866

Triller sp. *Lalage riedelii* **A. B. Meyer**, 1884 NCR
[Alt. White-shouldered Triller; JS *Lalage sueurii*]

Tanimbar Flycatcher *Ficedula riedeli* **Büttikofer**, 1886

Riedel's Eclectus Parrot *Eclectus roratus riedeli* A. B. Meyer, 1882

Mountain Tailorbird ssp. *Phyllergates cuculatus riedeli* A. B. Meyer & Wiglesworth, 1895

Johan Gerard Friedrich Riedel (1832–1911) was a Dutch administrator and amateur naturalist in the colonial Dutch East Indies (Indonesia) (1853–1883). It was said of the parrot: 'This interesting bird was sent first from Timor Laut by Mr Riedel, recently Dutch resident at Amboina, to Dr Meyer at Dresden, by whom it was named after the discoverer. Like all the green and red parrots, the usual differences of the sexes are observed, the male being green and the female red'.

Rieffer

Rieffer's Hummingbird *Amazilia tzacatl* La Llave, 1833
[Alt. Rufous-tailed Hummingbird]

Rieffer's Cotinga *Pipreola riefferii* **Boissonneau**, 1840
[Alt. Green-and-black Fruiteater]

Rieffer's Tanager *Chlorornis riefferii* Boissonneau, 1840
[Alt. Grass-green Tanager]

Gabriel Rieffer (DNF) was a collector in South America (1830s), but little seems to be known about him. Bourcier (q.v.) and Mulsant (q.v.) described *Trochilus riefferi* (1843), a junior synonym of *Amazilia tzacatl*, and said simply that Rieffer was a: 'voyageur dans cette partie de L'Amérique méridional'.

Riggenbach

Shikra *Astur riggenbachi* **Neumann**, 1908 NCR
[JS *Accipiter badius sphenurus*]

Crested Lark ssp. *Galerida cristata riggenbachi* **Hartert**, 1902

White-rumped Seedeater ssp. *Serinus leucopygius riggenbachi* Neumann, 1908
[Syn. *Crithagra leucopygia riggenbachi*]

Black Wheatear ssp. *Oenanthe leucura riggenbachi* Hartert, 1909 NCR
[JS *Oenanthe leucura syenitica*]

African Spotted Creeper ssp. *Salpornis salvadori riggenbachi* **Reichenow**, 1909 NCR
[JS *Salpornis salvadori emini*]

Lesser Honeyguide ssp. *Indicator minor riggenbachi* **Zedlitz**, 1915

Fine-spotted Woodpecker ssp. *Campethera punctuligera riggenbachi* Grote, 1923 NCR
[JS *Campethera punctuligera punctuligera*]

European Shag ssp. *Phalacrocorax aristotelis riggenbachi* Hartert, 1923

Fritz Wilhelm Riggenbach (1864–1944) was a Swiss zoologist and collector. He travelled to Morocco a number of times (1888–1893 and 1894–1909), to New Guinea (1910), Senegal and Cameroon while employed on several of the German Central African Expeditions (1902–1911), which were led by Adolf Friedrich Duke of Mecklenburg (q.v. as Ducis). He then worked for a US insurance company (1911–1933) and became an art dealer. A mammal, a reptile, an amphibian and a number of fish are named after him.

Riis

Blue Flycatcher sp. *Muscicapa riisii* **Hartlaub**, 1857
SPURIOUS SPECIES

Andreas Riis (1804–1854) was a Danish missionary of the Basel Mission to the Gold Coast (Ghana) (1832–1845). This flycatcher, said to come from the Gold Coast, was apparently based on a partial artefact of the Asian species *Cyornis banyumas*.

Riise

Caribbean Elaenia ssp. *Elaenia martinica riisii* **P. L. Sclater**, 1860

Albert Heinrich Riise (1810–1882) was a Danish pharmacist, botanist and collector who supplied much material to the Zoological Museum, Copenhagen. He was an apprentice pharmacist (1824–1830), then moved to Copenhagen where he graduated (1832) and worked (1832–1838). The Danish King appointed him a pharmacist in the Danish West Indies (US Virgin Islands) with an exclusive licence to open a retail shop: A. H. Riise in St Thomas, which is still in business. Epidemics of cholera, smallpox, and yellow fever broke out in St Thomas (1868), so Riise returned to Denmark with his family and never went back. A reptile is named after him.

Riker

Point-tailed Palmcreeper *Berlepschia rikeri* **Ridgway**, 1887

White-fronted Nunbird ssp. *Monasa morphoeus rikeri* Ridgway, 1912

Clarence Barrington Riker (1863–1947) was an American collector in Latin America. He was in the Amazon region of Brazil (1884–1887) with his wife, Jessie (q.v. as Jessie (Riker)), and sons Bowman (q.v.) and Herbert (q.v.). Ridgway (q.v.) reported that Riker had collected near Santarem no fewer than three new genera and 15 new species in June–July 1887.

Riley

Blue Flycatcher genus *Rileyornis* **Mathews**, 1927 NCR
[Now in *Cyornis*]

Riley's Mangrove Cuckoo *Coccyzus minor rileyi* **Ridgway**, 1915 NCR; NRM
Black-naped Oriole ssp. *Broderipus chinensis rileyi* Mathews, 1925 NCR
[JS *Oriolus chinensis celebensis*]
Cerulean Cuckooshrike ssp. *Coracina temminckii rileyi* **Meise**, 1931
Grey-throated Babbler ssp. *Stachyris nigriceps rileyi* **Chasen**, 1936
Brown Wood Owl ssp. *Strix indranee rileyi* **Kelso**, 1937 NCR
[JS *Strix leptogrammica maingayi*]
Blue Whistling Thrush ssp. *Myophonus caeruleus rileyi* **Deignan**, 1938 NCR
[JS *Myophonus caeruleus temminckii*]
Riley's Eastern Towhee *Pipilo erythrophthalmus rileyi* **Koelz**, 1939
Blue-hooded Euphonia ssp. *Euphonia elegantissima rileyi* **Van Rossem**, 1942

Joseph Harvey Riley (1873–1941) was a biologist and ornithologist at the USNM (1896–1941) who became Associate Curator of Birds (1932). He travelled in Cuba (1900) and the Bahamas (1905). A reptile is named after him.

Rinchen

Mountain Bamboo Partridge ssp. *Bambusicola fytchii rincheni* **Koelz**, 1954 NCR
[JS *Bambusicola fytchii hopkinsoni*]

Rinchen Gailtsen (DNF) was a Tibetan friend of Koelz and one of his companions during his travels in India, Afghanistan and Iran.

Riney

New Zealand Rockwren ssp. *Xenicus gilviventris rineyi* **Falla**, 1953 NCR; NRM

Thane Albert Riney (1918–at least 1992) was an American zoologist, ecologist and wildlife biologist. He worked for the New Zealand Department of Internal Affairs and then the New Zealand Forest Service. He resigned (1958) and moved to Southern Rhodesia (Zimbabwe) to work with the Food and Agriculture Organization. He wrote: *Birds aboard Ship* (1946).

Riocour

African Swallow-tailed Kite *Chelictinia riocourii* **Vieillot**, 1822
[Alt. Scissor-tailed Kite]

Antoine Nicolas Francois Du Boys Comte de Riocour (1761–1841) was a French collector and naturalist.

Riordan

Australian Masked Owl ssp. *Tyto novaehollandiae riordani* **Mathews**, 1912 NCR
[JS *Tyto novaehollandiae novaehollandiae*]
Brown Whistler ssp. *Pachycephala simplex riordani* Mathews, 1912 NCR
[JS *Pachycephala simplex simplex*]
Common Bronzewing ssp. *Phaps chalcoptera riordani* Mathews, 1912 NCR; NRM
Splendid Fairy-wren ssp. *Malurus splendens riordani* Mathews, 1912 NCR
[JS *Malurus splendens splendens*]

Hugh H. Riordan (d.1950) was an Australian bird photographer who illustrated *Birds of Geelong* (1914). As the name McKnight was in danger of disappearing, he took his mother's maiden name as his own surname.

Riotte

White-tipped Dove ssp. *Leptotila verreauxi riottei* **Lawrence**, 1868
[SII *Leptotila verreauxi verreauxi*]

Charles N. Riotte (d.1873) was a French immigrant, living in Texas, who became the US Minister to Costa Rica (1861–1867) and Nicaragua (1869–1873).

Ripley

Ripley's Fruit Dove *Ptilinopus arcanus* Ripley & **Rabor**, 1955
[Alt. Negros Fruit Dove]
Ripley's Shrike Babbler *Pteruthius ripleyi* **Biswas**, 1960
[Alt. Himalayan Shrike Babbler]

Mountain Mouse-warbler ssp. *Crateroscelis robusta ripleyi*
Mayr & **Meyer de Schauensee**, 1939
Mountain Serin ssp. *Serinus estherae ripleyi* **Chasen**, 1939
NCR
[JS *Serinus estherae vanderbilti*]
Malayan Banded Pitta ssp. *Hydrornis irena ripleyi* **Deignan**,
1946
[Syn. *Pitta guajana ripleyi*]
Puff-throated Babbler ssp. *Pellorneum ruficeps ripleyi*
Deignan, 1947
Chestnut-headed Tesia ssp. *Tesia castaneocoronata ripleyi*
Deignan, 1951
[Syn. *Oligura castaneocoronata ripleyi*]
Green-tailed Sunbird ssp. *Aethopyga nipalensis ripleyi*
Koelz, 1952 NCR
[JS *Aethopyga nipalensis koelzi*]
Colombian Crake ssp. *Neocrex colombiana ripleyi*
Wetmore, 1967
Black-bibbed Cicadabird ssp. *Coracina mindanensis ripleyi*
Parkes, 1971
Sri Lanka Bay Owl ssp. *Phodilus assimilis ripleyi* Hussain &
Reza Khan, 1978
Cave Swiftlet ssp. *Collocalia linchi ripleyi* Somadikarta, 1986

Dr Sidney Dillon Ripley II (1913–2001) was an eminent American ornithologist who was awarded the Presidential Medal of Freedom, the highest civilian honour of the United States. He was privately educated and travelled widely in both Europe and India. He graduated in history at Yale, but then studied zoology at Columbia University where he specialised in ornithology. He joined an expedition to New Guinea (1936), where he spent 18 months collecting bird specimens. Always a bird lover, at 17 he built a pond to attract waterfowl. His service during WW2 was with the Office of Strategic Services, coordinating United States and British intelligence efforts in South-East Asia. After the war he taught at Yale, and was then Director of the Peabody Museum of Natural History, until becoming Secretary the USNM (1964–1984). On retirement he was reported to have said: 'I shall enjoy my freedom from the tyranny of the In and Out boxes'. He travelled 'around the world from Patagonia to Pakistan'. Ripley was instrumental in the founding of the Charles Darwin Foundation for the Galápagos Islands (1959). He is recognised as one of the giants of Indian ornithology. Among his many publications, he co-wrote with Salim Ali (q.v.) the 10-volume *Handbook of the Birds of India* (1968–1975) and (as sole author) *Rails of the World*. (See also **Dillon, R.**)

Rippon

Scarlet-faced Liocichla *Liocichla ripponi* **Oates**, 1900

Black-faced Warbler ssp. *Abroscopus schisticeps ripponi*
Sharpe, 1902
Rippon's Silver Pheasant *Lophura nycthemera ripponi*
Sharpe, 1902
Black-throated Parrotbill ssp. *Suthora nipalensis ripponi*
Sharpe, 1905
Rusty-fronted Barwing ssp. *Actinodura egertoni ripponi*
Ogilvie-Grant & **La Touche**, 1907
Alpine Accentor ssp. *Prunella collaris ripponi* **Hartert**, 1910
NCR
[JS *Prunella collaris nipalensis*]
White-browed Scimitar Babbler ssp. *Pomatorhinus
schisticeps ripponi* **Harington**, 1910
White-browed Fulvetta ssp. *Fulvetta vinipectus ripponi*
Harington, 1913
Long-tailed Minivet ssp. *Pericrocotus ethologus ripponi*
E. C. S. Baker, 1924
Bar-tailed Treecreeper ssp. *Certhia himalayana ripponi*
Kinnear, 1929

Lieutenant-Colonel George Rippon (1861–1927) was commissioned into the 29th Madras Infantry as 2nd Lieutenant (1880), advancing to Captain (1881). After serving in Burma (Myanmar) and India, he was placed on the Retired List (1912), but was recalled on the outbreak of WW1 (1914) and served in the UK as Commanding Officer of the 8th Battalion, Liverpool Regiment. He was a keen amateur ornithologist, and published articles such as 'On the birds of the southern Shan States, Burma' (1901).

Rivera

Oriental Bay Owl ssp. *Phodilus badius riverae* **McGregor**,
1927
[Validity uncertain; known from 1 specimen, now destroyed]

Francisco Rivera (DNF) was a Filipino collector and expedition assistant.

Rivet

Buff-fronted Foliage-gleaner ssp. *Philydor rufum riveti*
Menegaux & **Hellmayr**, 1906

Dr Paul Rivet (1876–1958) originally trained as a physician. He was on the second French Geodesic mission (1901) to Ecuador, staying until 1906. After returning to Paris he worked at the MNHN and was one of the founders of the Institut d'Ethnologie (1926). He became Director of the MNHN (1928) and founded Musée de l'Homme (1937). Later he founded the Anthropological Institute in Colombia (1942–1945). A reptile and an amphibian are named after him.

Riviere

Riviere's Hawk *Buteo platypterus rivierei* **A. H. Verrill**, 1905
[Alt. Broad-winged Hawk ssp.]

Dr Riviere (DNF) was a collector on Dominica, West Indies.

Rivoli

Rivoli's Hummingbird *Eugenes fulgens* **Swainson**, 1827
[Alt. Magnificent Hummingbird]
Crimson-mantled Woodpecker *Colaptes rivolii*
Boissonneau, 1840
White-bibbed Fruit Dove *Ptilinopus rivoli* **Prévost**, 1843

François Victor Masséna Prince d'Essling, Duc de Rivoli (1798–1863), was the son of one of Napoleon's marshals and an amateur ornithologist. He amassed a huge collection of hummingbirds. His wife Anna has a hummingbird named after her – they may be the only husband and wife to have birds of the same family named after them. (See also **Anna (d'Essling)**, **Masséna**, **Prince of Essling** and **Victor**)

Robbins

Ecuadorian Tapaculo *Scytalopus robbinsi* Krabbe & Schulenberg, 1997
[Alt. El Oro Tapaculo]

Mark Blair Robbins (b.1954) is manager of the ornithology collection at the University of Kansas Natural History Museum, having previously fulfilled the same role at the Academy of Natural Sciences in Philadelphia (1982–1993). During a quarter of a century he has made more than 40 foreign expeditions, as leader or co-leader, for the explicit purpose of enhancing those collections, but also making thousands of sound recordings. His work also led to the discovery of several species new to science. The original etymology reads: 'We take the pleasure of naming this bird after Mark B. Robbins, who was the first to tape-record and collect it … ; his tape-recordings greatly facilitated the collecting of further specimens and sound material. We also take the opportunity to acknowledge his substantial contribution to Neotropical ornithology.'

Robert, A.

Hooded Gnateater *Conopophaga roberti* **Hellmayr**, 1905

Alphonse Robert (DNF) was a French collector, particularly in South America but also in other parts of the world. He took part in C. I. Forsyth Major's expedition to Madagascar (1894–1896) as his assistant, and collected in Brazil (1901 and 1903) for the BMNH. Five mammals are named after him.

Robert, F. W.

Robert's Wedge-billed Babbler *Sphenocichla roberti* **Godwin-Austen** & **Walden**, 1875
[Alt. Chevron-breasted Babbler, Cachar Wedge-billed Babbler]

Eyebrowed Wren-babbler ssp. *Napothera epilepidota roberti* Godwin-Austen & Walden, 1875

Frederick 'Fred' William Robert (DNF) was an Australian who was employed by the Indian Survey Department (1875–1895). He was an expert mountaineer and was in the Himalayas (1893) and referred to by Godwin-Austen as 'mapping in the Cachar Hills'.

Robert (Alexander)

White-bellied Robin Chat *Cossyphicula roberti* **B. Alexander**, 1903

Major Robert Alexander (1873–1928) was the twin brother of the describer Boyd Alexander (q.v.).

Robert (Mathews)

Black Swan ssp. *Chenopis atrata roberti* **Mathews**, 1912 NCR; NRM
[Syn *Cygnus atratus*]
Spotless Crake ssp. *Porzana plumbea roberti* Mathews, 1912 NCR
[JS *Porzana tabuensis tabuensis*]

Robert Hamilton Mathews (1841–1918) was a licensed surveyor in New South Wales (1870–1890). He moved to Paramatta (1899) where he became Deputy Coroner and wrote *Handbook to Magisterial Inquiries and Coroners' Inquests*. He became extremely interested in Aboriginal life and customs, and travelled extensively in all parts of Australia. He published more than 150 articles in anthropological magazines and scientific journals. His sons were Hamilton Robert Mathews (q.v.) and Gregory Macalister Mathews (q.v.).

Robert (Owen)

Robert's Hummingbird *Phaeochroa cuvierii roberti* **Salvin**, 1861
[Alt. Scaly-breasted Hummingbird ssp.; Syn. *Campylopterus cuvierii roberti*]

Robert Owen (DNF) was a resident of Vera Paz in Guatemala where he collected for both Salvin (q.v.) and Godman (q.v.). Owen was the first person to find and collect the nest and egg of the Resplendent Quetzal (See **Owen, Robert**)

Robert (Ridgway)

Plain Wren ssp. *Thryothorus modestus roberti* **A. R. Phillips**, 1986

(See **Ridgway**)

Roberts, H. R.

Roberts's Parakeet *Bolborhynchus aurifrons robertsi* **Carriker**, 1933
[Alt. Mountain Parakeet ssp.]

Dr Howard Radclyffe Roberts (1906–1982) was an entomologist. He served as part of the Medical Entomological Department of the US Army and worked in the Malaria Survey Unit in the Philippines and New Guinea. During WW2, he collected extensively in the Philippines and the Academy of Natural Sciences, Philadelphia, now houses many thousands of his specimens of *Aedes* (mosquito) and other genera of medical importance. He became Managing Director of the Academy (1947–1972). Roberts co-wrote the *Mosquito Atlas* (1940s), a standard reference work for medical servicemen, describing mosquitoes that transmit malaria. The parakeet was collected during Roberts's expedition to Peru with Carriker (q.v.), as were other taxa. Two amphibians are named after him.

Roberts, J. A.

Roberts's Warbler *Oreophilais robertsi* **C. W. Benson**, 1946
[Alt. Roberts's Prinia, Briar Warbler; Syn. *Prinia robertsi*]

Grey Penduline-tit ssp. *Anthoscopus caroli robertsi*
Haagner, 1909

African Barred Owlet ssp. *Glaucidium capense robertsi*
J. L. Peters, 1940 NCR

[NUI *Glaucidium capense ngamiense*]

Malachite Kingfisher ssp. *Corythornis cristatus robertsi*
J. L. Peters, 1945 NCR

[JS *Corythornis cristatus cristatus*]

Spike-heeled Lark ssp. *Chersomanes albofasciata robertsi*
Macdonald, 1953 NCR

[JS *Chersomanes albofasciata alticola*]

J. Austin Roberts (1883–1948) was a South African zoologist. During the first half of the 20th century he was the most prominent ornithologist in southern Africa. He worked at the Transvaal Museum for nearly four decades studying birds (1910–1946). He amassed 30,000 bird skins and 9,000 mammal specimens there. Although he did not have formal academic training, he received several high academic awards and an honorary doctorate. Roberts is best remembered for his *Birds of South Africa* (1940), a landmark publication in African ornithology which has developed in size and authority with repeated posthumous editions. He died in a traffic accident. The Austin Roberts Bird Sanctuary was established in his hometown, Pretoria (1958). Two mammals and two reptiles are named after him.

Robertson

Golden Myna ssp. *Mino anais robertsoni* **D'Albertis**, 1877

Sir John Robertson (1816–1891) was an Australian statesman, Premier of New South Wales five times (1860–1886).

Robin, E. J.

Greater Necklaced Laughingthrush ssp. *Garrulax pectoralis robini* **Delacour**, 1927

Yellow-billed Blue Magpie ssp. *Urocissa flavirostris robini*
Delacour & **Jabouille**, 1930

Eugène Jean Louis René Robin (1872–1954) was the French Resident Superior in Tonkin (Vietnam) (1925–1930) and Governor-General of Indochina (1934–1936).

Robin (Kemp)

Lesser Crested Tern ssp. *Thalasseus bengalensis robini*
Mathews, 1916 NCR

[JS *Thalasseus bengalensis torresii*]

Large-billed Gerygone ssp. *Ethelornis cairnsensis robini*
Mathews, 1920 NCR

[JS *Gerygone magnirostris cairnsensis*]

(See **Kemp**)

Robinson, H. C.

Robinson's Buttonquail *Turnix olivii* Robinson, 1900

[Alt. Buff-breasted Buttonquail, Olive's Buttonquail]

Robinson's Whistling Thrush *Myophonus robinsoni* **Ogilvie-Grant**, 1905

[Alt. Malaysian Whistling Thrush]

Streak-eared Bulbul ssp. *Pycnonotus blanfordi robinsoni*
Ogilvie-Grant, 1905 NCR

[JS *Pycnonotus blanfordi conradi*]

Common Green Magpie ssp. *Cissa chinensis robinsoni*
Ogilvie-Grant, 1906

Grey-headed Woodpecker ssp. *Picus canus robinsoni*
Ogilvie-Grant, 1906

Wedge-tailed Green Pigeon ssp. *Treron sphenurus robinsoni*
Ogilvie-Grant, 1906

Red-necked Crake ssp. *Rallina tricolor robinsoni* **Mathews**,
1911 NCR; NRM

Australian Brush-turkey ssp. *Alectura lathami robinsoni*
Mathews, 1912 NCR

[JS *Alectura lathami lathami*]

Brown Cuckoo Dove ssp. *Macropygia phasianella robinsoni*
Mathews, 1912

Pied Currawong ssp. *Strepera graculina robinsoni*
Mathews, 1912

Bamboo Woodpecker ssp. *Gecinulus viridis robinsoni*
Kloss, 1918

[*G. viridis* sometimes regarded as monotypic]

Blue-eared Barbet ssp. *Cyanops duvauceli robinsoni*
E. C. S. Baker, 1918 NCR

[JS *Megalaima australis duvaucelii*]

Black-headed Sibia ssp. *Heterophasia desgodinsi robinsoni*
Rothschild, 1921

Spotted Forktail ssp. *Enicurus maculatus robinsoni*
E. C. S. Baker, 1922

Maroon Oriole ssp. *Oriolus traillii robinsoni* **Delacour**, 1927

Emerald Dove ssp. *Chalcophaps indica robinsoni*
E. C. S. Baker, 1928

Black-nest Swiftlet ssp. *Collocalia lowi robinsoni*
Stresemann, 1931

[JS *Aerodramus maximus maximus*]

Robinson's Shrike Babbler *Pteruthius aeralatus robinsoni*
Chasen & Kloss, 1931

[Alt. Blyth's Shrike Babbler ssp.]

Puff-throated Bulbul ssp. *Alophoixus pallidus robinsoni*
Ticehurst, 1932

White-throated Fantail ssp. *Rhipidura albicollis robinsoni*
Chasen, 1941 NCR

[JS *Rhipidura albicollis atrata*]

Herbert Christopher Robinson (1874–1929) was a British zoologist and ornithologist. After education at Marlborough College (1894), he went to Switzerland because of ill health. He made a trip to Queensland (1896) and on his return worked as assistant to Dr H. O. Forbes (q.v.) at the Liverpool Museum (1897–1900). He then spent 30 years in the tropics, initially with Dr Annandale in the Malay Peninsula. He became Curator of the Federated Malay States Museum at Selangor (1903–1926) and joined Cecil Boden Kloss (q.v.) (1908) exploring the Indo-Malay region. He sent many of the specimens that he collected to Liverpool and the BMNH. He co-authored *The Birds of the Malay Peninsula* (4 vols, 1927–1939) with Frederick N. Chasen (q.v.), and many articles. Four mammals, two amphibians and two reptiles are named after him.

Robinson, W.

Striated Heron *Butorides robinsoni* **Richmond**, 1896 NCR
[JS *Butorides striata striata*]

Eared Dove ssp. *Zenaida ruficauda robinsoni* **Ridgway**, 1915 NCR
[JS *Zenaida auriculata rubripes*]

Wirt Robinson (1864–1929) was a US Army General whose major area of expertise was the chemistry of explosives. He taught chemistry at West Point Military Academy (1906–1928) and among his pupils was Eisenhower, the future US President. However, Robinson was an enthusiastic naturalist on his vacations, and collected in Venezuela for the USNM (1895) and then the Department of Mammalogy (1899) with Marcus Lyon, an Assistant Curator. He wrote a number of articles, notably 'Some rare Rhode Island birds' and 'Some rare Virginia birds' (1889) and a series of articles on collections made in South America. A mammal is named after him.

Roborovski

Roborovski's Rosefinch *Kozlowia roborowskii* **Przewalski**, 1887
[Alt. Tibetan Rosefinch; Syn. *Carpodacus roborowskii*]

Captain Vladimir Ivanovich Roborovski (1856–1910) was a Russian explorer of parts of China and Tibet. He accompanied Przewalski (q.v.) on his third and fourth expeditions (1879–1880 and 1883–1885). He then accompanied M. V. Pevtsov on his expedition to Chinese Turkestan (1889–1890). He was the leader of his own expedition in Eastern Tien-Shan, Nanshan and Northern Tibet (1893–1895) during which he gathered zoological, botanical and geological material. On this expedition he was hit by paralysis, despite which he continued the handling of field material and then published the expedition results (1899–1901). He wrote *Ekspeditisii v storonu ot pugey Tibetskoy ekspeditsii* (1896) and *Otchet nachalnika ekspeditskii* (1900). He has a mammal and two reptiles named after him.

Robson

Ground Warbler genus *Robsonius* (Collar, 2006)

Golden-breasted Fulvetta ssp. *Lioparus chrysotis robsoni* Eames, 2002

Craig R. Robson (b.1959) is a professional ornithologist, naturalist and artist, a leading authority on Asian birds, and a founding member of the Oriental Bird Club, who has visited Asia many times over 35 years. Initially he funded his birding travel by working in Scotland for the oil industry but has been leading tours for 'Birdquest' since 1991, and also carries out ecological survey work in the UK. He has undertaken pioneering surveys of Western Tragopans *Tragopan melanocephalus* in Pakistan, Gurney's Pittas *Pitta gurneyi* in Thailand and rare pheasants and other forest birds in Vietnam, and has spent some time working on the distribution of Asia's restricted range birds for BirdLife International. During his travels and pioneering adventures, he has made many major discoveries and rediscoveries. Among his many papers and longer written works are: *A Field Guide to the Birds of South-East Asia* (2000) and *A Field Guide to the Birds of Thailand* (2002), and major contributions to the *Handbook of the Birds of the World* (babblers, parrotbills).

Roccati

Grey Penduline-tit ssp. *Anthoscopus caroli roccatii* **Salvadori**, 1906

Dr Alessandro Roccati (1872–1928) was an Italian geologist, collector and explorer. He was director of the geomineralogical laboratory of Turin Polytechnic and was a member of the Duke of Abruzzi's (q.v. under Aloysius) expedition to Uganda (1906). He was also on a geological expedition to Brazil (1926).

Roch

Madagascar Cuckoo *Cuculus rochii* **Hartlaub**, 1863

Sampson Roch (1829–1906) was an army surgeon. He studied medicine at Trinity College, Dublin, and became MRCS (Eng.) in 1854. On qualifying, he joined the army medical corps and was at the siege of Sebastopol (1854). He was 'mentioned in dispatches' for his bravery as a field surgeon (1855). He served in Bengal during the 'Mutiny' (1857–1859) before being posted to Mauritius (1860–1865). He was in Madagascar (1861) as part of the embassy of Colonel Middleton from the Governor of Mauritius to the new king, Radama I. He volunteered for Abyssinia (Ethiopia) (1867) and was put in charge of the hospital ship *Golden Fleece*. He left the army (1882) to take a diploma and eventually retired to his native Ireland (1892). His collection of bird specimens is now in the National Museum of Ireland.

Roché

Somali Lark ssp. *Mirafra somalica rochei* **Colston**, 1982

Dr Jean-Claude Roché (b.1931) is a French ornithologist who has made many recordings of birds' songs.

Rochussen

Moluccan Woodcock *Scopolax rochussenii* **Schlegel**, 1866

Jan Jacob Rochussen (1797–1871) was a Dutch politician who was Governor-General of the Dutch East Indies (1845–1851).

Rock

Rock's Blood Pheasant *Ithaginis cruentus rocki* **Riley**, 1925
Spectacled Parrotbill ssp. *Sinosuthora conspicillata rocki* **Bangs** & **J. L. Peters**, 1928
Bar-winged Wren-babbler ssp. *Spelaeornis troglodytoides rocki* Riley, 1929
Ashy Drongo ssp. *Dicrurus leucophaeus rocki* Riley, 1940 NCR
[JS *Dicrurus leucophaeus bondi*]
Brown Prinia ssp. *Prinia polychroa rocki* **Deignan**, 1957

Dr Joseph Francis Charles Rock (1884–1962) was an Austrian-born naturalised American. He was a self-taught botanist in Hawaii and (1920–1949) collected plant specimens for a number of American institutions including Harvard

University, the National Geographic Society, and the USNM. He was also an anthropologist and ornithologist, and led the National Geographic Society's Expedition to Yunnan-Sichuan, China (1927–1930), and spent 27 years living in Yunnan, becoming the leading authority on the Naxi people who live there. Bangs & Peters (1928) and J. H. Riley (1931) wrote papers on his collections. He was the author of a number of books and articles, including *Seeking the Mountains of Mystery*, *Life among the Lamas of Choni* and *Through the Great River Trenches of Asia*. In his career in China he collected 60,000 plant specimens and 1,600 birds, which were sent to American ornithologists.

Rockefeller

Rockefeller's Sunbird *Cinnyris rockefelleri* **Chapin**, 1932

John Sterling Rockefeller (1904–1944) was an American who collected in the Congo (1928–1929) and sponsored other collecting expeditions. He was an enthusiastic conservationist; for example, he purchased Kent Island in Maine, USA (1930) and made it a bird sanctuary. Earlier, Allan Moses, who accompanied Rockefeller on the African expedition, had described the problems in the Grand Manan Archipelago, which led to his decision to purchase this island. (See **Sterling**)

Rodger

Lesser Yellownape ssp. *Picus chlorolophus rodgeri* **Hartert & Butler**, 1898

Sir John Pickersgill Rodger (1851–1910) was a British lawyer (1877), diplomat and colonial administrator in Malaya (Malaysia), being Resident of Selangor (1884–1885 and 1896–1902), Pahang (1888–1896) and Perak (1902–1904). He was Governor of the Gold Coast (Ghana) (1904–1910). He played first-class cricket for Kent.

Rodgers, J.

Rodgers's Fulmar *Fulmarus glacialis rodgersii* **Cassin**, 1862
[Alt. Northern Fulmar ssp.]

Commander (later Rear-Admiral) John Rodgers (1812–1882) of the United States Navy led various expeditions to Kamchatka (1853–1856), and elsewhere, commanding the *Vincennes*, which had been the flagship of the Wilkes (q.v.) expedition.

Rodgers, W. A.

Swynnerton's Robin ssp. *Swynnertonia swynnertoni rodgersi* Jensen & Stuart, 1982

Dr William Alan Rodgers (1944–2009) was a British zoologist and conservationist in Tanzania (1965–1984 and 1992–2009), and India (1984–1991). The University of Aberdeen awarded his master's degree and the University of Nairobi his doctorate.

Rodolphe

Deignan's Babbler *Stachyris rodolphei* **Deignan**, 1939
[Now often regarded as a synonym of *Stachyridopsis rufifrons rufifrons*]

Chestnut-bellied Malkoha ssp. *Rhopodytes sumatranus rodolphi* **Ripley**, 1942 NCR; NRM
[Syn *Phaenicophaeus sumatranus*]

Rodolphe Meyer de Schauensee (see **De Schauensee**)

Rodrigues

Rodrigues Warbler *Acrocephalus rodericanus* **A. Newton**, 1865
Rodrigues Starling *Necrospar rodericanus* **Slater**, 1879 EXTINCT

Named after the island of Rodrigues, rather than directly after the Portuguese navigator, Diego Rodrigues, who discovered the island (1528). A reptile is also named after this island.

Rodríguez, J. J.

Spot-backed Antshrike ssp. *Thamnophilus guttatus rodriguezianus* **Bertoni**, 1901 NCR; NRM
[Syn. *Hypoedaleus guttatus*]

Juan J. Rodríguez Luna (1840–1916) was a Guatemalan lawyer, zoologist and naturalist who founded and was first Director of Museo Nacional de Historia Natural, Guatemala (1896–1916). He wrote *Memoria sobre la Fauna de Guatemala* (1894). Two reptiles are named after him.

Rodríguez, J. V.

Upper Magdalena Tapaculo *Scytalopus rodriguezi* **Krabbe** *et al.*, 2005

José Vicente Rodríguez-Mahecha (b.1949) is a Colombian biologist, herpetologist and ornithologist who works for the Species Conservation Unit of the Andes Biological University and Conservation International. He is the committee co-ordinator for the group dealing with conservation of mammals in Colombia. He was editor for R. T. Defler's 2005 book *Primates of Colombia*.

Rodway

Violaceous Euphonia ssp. *Euphonia violacea rodwayi* **T. E. Penard**, 1919

James Rodway (1848–1926) was a British botanist and author who was Curator of Georgetown Museum, British Guiana (Guyana) (1900–1925). He wrote *Guiana: British, Dutch, and French* (1912).

Roehl, K.

Roehl's Thrush *Turdus roehli* **Reichenow**, 1905
[Alt. Usambara Thrush]
Roehl's Swamp Warbler *Bradypterus roehli* **Grote**, 1920 NCR
[Alt. Evergreen Forest Warbler; JS *Bradypterus lopezi usambarae*]

Shelley's Greenbul ssp. *Arizelocichla masukuensis roehli* Reichenow, 1905
African Black Swift ssp. *Apus barbatus roehli* Reichenow, 1906

African Dusky Flycatcher ssp. *Muscicapa adusta roehli*
Grote, 1919
[NUI *Muscicapa adusta murina*]

Pastor Karl Roehl (1870–1951) was a missionary at Usambara, Tanganyika Territory (1896–1908), and in Rwanda (1908–1916), when he was taken prisoner. He was conscripted to serve in the German Colonial Army formation in East Africa (1914). The Belgians captured him. He returned to Tanganyika (1926–1934). Roehl is most famous for having translated the Bible into Swahili, a language he was the leading authority on. This effort was recognised by the award of the Leibniz Citizens Medal of the Berlin Academy of Sciences and by an honorary doctorate from the University of Hamburg. He was an amateur naturalist and collected the type specimens of the birds named after him. He wrote *Versuch einer Systematischen Grammatik der Schambala-Sprache* (1911), *Ostafrikas Heldenkampf* (1918) and 'The linguistic situation in East Africa' (1930).

Rogacheva

Whimbrel ssp. *Numenius phaeopus rogachevae* Tomkovich, 2008

Professor Helena V. Rogacheva and her husband, Evgeni Syroechkovski, started ringing birds and generally researching in Siberia when they started out from Taymyr (1956), being later joined by their son and daughter-in-law, who are both also ornithologists. The Institute of Animal Morphology and Ecology, Russian Academy of Sciences, Moscow employed them. She wrote *The Birds of Central Siberia* (1992).

Rogers, C. G.

Blue-throated Flycatcher ssp. *Cyornis rubeculoides rogersi*
H. C. Robinson & **Kinnear**, 1928

Charles Gilbert Rogers (1864–1937) was a forester with the Imperial Forestry Commission in India and Burma (Myanmar) (1888–1919), retiring as Chief Conservator of Forests. He collected botanical, entomological and ornithological specimens for several organisations including the BMNH and Lord Rothschild (q.v.) at Tring. He wrote *A Manual of Forest Engineering for India* (1900).

Rogers, C. H.

Whiskered Yuhina ssp. *Yuhina flavicollis rogersi* **Deignan**, 1937
Himalayan Swiftlet ssp. *Aerodramus brevirostris rogersi*
Deignan, 1955

Charles Henry Rogers (1888–1977) was an American ornithologist and conservationist. He graduated from Princeton (1909) and went to work on Wall Street (1909–1912). He joined the bird department, AMNH, New York (1912), leaving to become an infantryman in WW1. After demobilisation he returned to Princeton (1920) to teach biology and ornithology and to be curator of the bird collection at Princeton Museum of Zoology. The Charles H. Rogers Wildlife Refuge, named after him, is in Princeton.

Rogers, H. E.

Cassowary sp. *Casuarius rogersi* **Rothschild**, 1928 NCR
[Alt. Dwarf Cassowary; JS *Casuarius bennetti*]

H. E. Rogers (DNF) was a collector who set up on his own as a dealer (1919) in Liverpool. The type specimen of the cassowary was purchased (alive) from Rogers and lived at the London Zoo – the mounted specimen later went to Rothschild (q.v.) at Tring.

Rogers, H. M. C.

Inaccessible Island Rail *Atlantisia rogersi* **P. R. Lowe**, 1923

The Reverend Henry Martyn C. Rogers (1879–1926) was a missionary to Tristan da Cunha (1922–1925). He wrote *The Lonely Island* (1926).

Rogers, J. P.

Bowerbird genus *Rogersornis* **Mathews**, 1912 NCR
[Now in *Chlamydera*]
Lapwing genus *Rogibyx* Mathews, 1913 NCR
[Now in *Vanellus*]

Black-eared Cuckoo ssp. *Owenavis osculans rogersi*
Mathews, 1912 NCR
[Syn *Chrysococcyx osculans* NRM]
Brown Quail ssp. *Coturnix australis rogersi* Mathews, 1912 NCR
[JS *Synoicus ypsilophorus australis*]
Brown Songlark ssp. *Cincloramphus cruralis rogersi*
Mathews, 1912 NCR; NRM
Grey Teal ssp. *Nettion castaneum rogersi* Mathews, 1912 NCR
[JS *Anas gracilis*]
Pacific Black Duck ssp. *Anas superciliosa rogersi*
Mathews, 1912
Pacific Emerald Dove ssp. *Chalcophaps longirostris rogersi*
Mathews, 1912
Papuan Frogmouth ssp. *Podargus papuensis rogersi*
Mathews, 1912
Rogers's Fairy-wren *Malurus lamberti rogersi* Mathews, 1912
[Alt. Variegated Fairy-wren ssp.]
Singing Honeyeater ssp. *Ptilotis sonora rogersi* Mathews, 1912 NCR
[JS *Lichenostomus virescens forresti*]
Spotted Harrier ssp. *Circus assimilis rogersi* Mathews, 1912 NCR; NRM
Striated Pardalote ssp. *Pardalotus striatus rogersi*
Mathews, 1912 NCR
[JS *Pardalotus striatus substriatus*]
Tree Martin ssp. *Petrochelidon nigricans rogersi* Mathews 1912 NCR
[JS *Petrochelidon nigricans neglecta*]
Weebill ssp. *Smicrornis brevirostris rogersi* Mathews, 1912 NCR
[JS *Smicrornis brevirostris flavescens*]
Varied Sittella ssp. *Neositta leucoptera rogersi* Mathews 1912 NCR
[JS *Daphoenositta chrysoptera leucoptera*]

Australian Pipit ssp. *Anthus australis rogersi* Mathews, 1913

Brown Booby ssp. *Sula leucogaster rogersi* Mathews, 1913 NCR

[JS *Sula leucogaster plotus*]

Red Knot ssp. *Calidris canutus rogersi* Mathews, 1913

Spinifexbird ssp. *Eremiornis carteri rogersi* Mathews, 1913 NCR; NRM

Bridled Tern ssp. *Melanosterna anaethetus rogersi* Mathews, 1915 NCR

[JS *Onychoprion anaethetus anaethetus*]

Green Pygmy-goose ssp. *Cheniscus pulchellus rogersi* Mathews, 1916 NCR

[Syn *Nettapus pulchellus* NRM]

Princess Parrot ssp. *Spathopterus alexandrae rogersi* Mathews, 1916 NCR

[Syn *Polytelis alexandrae* NRM]

Broad-billed Sandpiper ssp. *Limicola falcinellus rogersi* Mathews, 1917 NCR

[JS *Limicola falcinellus sibirica*]

Large-tailed Nightjar ssp. *Rossornis macrurus rogersi* Mathews, 1918 NCR

[JS *Caprimulgus macrurus schlegelii*]

Paperbark Flycatcher ssp. *Seisura inquieta rogersi* Mathews, 1921 NCR

[JS *Myiagra nana*]

Purple-crowned Fairy-wren ssp. *Rosina coronata rogersiana* Mathews, 1922 NCR

[JS *Malurus coronatus coronatus*]

John Porter Rogers (1873–1941) was an English collector and gold prospector. He collected for Mathews (q.v.) in northern Australia. (See also **Porter**)

Rogosov

Siberian Jay ssp. *Perisoreus infaustus rogosowi* **Sushkin & B. Stegmann**, 1929 NCR

[NUI *Perisoreus infaustus sibericus*]

V. V. Rogosov (DNF) was a Russian zoologist and explorer in Siberia.

Rohde

Antshrike sp. *Thamnophilus rohdei* **Berlepsch**, 1887 NCR

[Alt. Great Antshrike; JS *Taraba major*]

Ricardo Rohde (DNF) was a collector who operated in Paraguay (1885–1886). He later joined the German New Guinea Company (1889). An amphibian and a reptile are named after him.

Röhl, E.

Scaled Piculet ssp. *Picumnus squamulatus roehli* **Zimmer & W. H. Phelps**, 1944

Dr Eduardo Röhl Arriens (1891–1959) was a Venezuelan scientist, zoologist and humanist, noted for research in climatology, meteorology, ornithology, astronomy, history and geography. He was President, Venezuelan Society for Natural Sciences (1940), and Director of the Astronomical Observatory, Caracas, when this bird was described with the

words that Röhl 'initiated the study of modern ornithology in Venezuela.' He also initiated the creation of the National Observatory of Venezuela (1952). He wrote *Fauna Descriptiva de Venezuela* (1942).

Rohu

Wandering Albatross ssp. *Diomedea exulans rohui* **Mathews**, 1915 NCR

[Indeterminate: could be either *Diomedea exulans gibsoni* or *D. exulans antipodensis*]

Shy Albatross ssp. *Diomedea cauta rohui* Mathews, 1916 NCR

[JS *Thalassarche cauta cauta*]

Sylvester Edwin Rohu (1882–1945) was an Australian gunsmith, yachtsman, naturalist and collector. He served in the Australian Imperial Force artillery in France (WW1) (1916–1918), then returned to Australia (1919).

Rojas

Rufous-breasted Hermit *Glaucis rojasi* **Boucard**, 1895 NCR

[JS *Glaucis hirsutus*]

Dr C. E. Rojas (DNF) was a Venezuelan zoologist and entomologist in Caracas (1870–1895).

Roll

Roll's Hill Partridge *Arborophila rolli* **Rothschild**, 1909

[Alt. Roll's Partridge]

U van Roll (DNF) was a Dutch coffee planter in Sumatra who collected the holotype and helped Gustav Schneider (q.v.) during his expedition to Sumatra's Battak Mountains.

Rolland

Grebe genus *Rollandia* **Bonaparte**, 1856

Rolland's Grebe *Rollandia rolland* **Quoy & Gaimard**, 1824

[Alt. White-tufted Grebe]

Thomas Pierre Rolland (1776–1847) was a master gunner in the French Navy. He was on board the corvette *Uranie* during her circumnavigation of the globe (1817–1820), under the command of Louis Freycinet (q.v.), whose account of the voyage was later published. The vessel suffered severe weather damage when rounding Cape Horn and had to be beached in the Falkland Islands (where the grebe is found). The vessel was abandoned as a shipwreck, but the expedition and its notes and collections were saved and continued on their journey in the *Physicienne*. Rolland then joined Dumont d'Urville's circumnavigation of the globe (1822–1825) on the *Coquille*. He left a manuscript account of this voyage.

Rollet

Black-breasted Barbet *Lybius rolleti* **Filippi**, 1853

Black-headed Oriole ssp. *Oriolus larvatus rolleti* **Salvadori**, 1864

Antoine Brun-Rollet (1810–1857) was a Savoyard (Savoy was culturally, but not politically, part of France) trader, explorer and elephant hunter who originally intended to be a priest. He changed his mind and left France (1831) for Alexandria, where he worked as a cook calling himself Ya'cub. He then went to Sudan where he made a fortune trading. He founded and operated a trading post at Bilinian on the White Nile and explored in Abyssinia (Ethiopia). He wrote a book *Le Nil Blanc et le Soudan: Études sur l'Afrique Centrale, Moeurs et Coutumes des Sauvages* (1855) which advocated the development of the country by using railways and steamships. He became the Sardinian consul in Khartoum (1856), the city where he died.

Romaine

Spot-tailed Nightjar ssp. *Antiurus maculicaudatus romainei* **M. A. Carriker**, 1935 NCR; NRM
[Syn. *Caprimulgus maculicaudus*]

Dr Melbourne Romaine Carriker (1915–2007) was the son of the describer, Melbourne Armstrong Carriker Jr (q.v.).

Romero

Bicoloured Antpitta ssp. *Grallaria rufocinerea romeroana* Hernández & **J. V. Rodríguez**, 1979

Hernando Romero-Zambrano (1943–1983) was a Colombian ornithologist. He joined the Faculty of Science, National University of Colombia, Bogota (1973), then became Director (1975–1983).

Rookmaker

Silver-eared Mesia ssp. *Leiothrix argentauris rookmakeri* **Junge**, 1948
[Syn. *Mesia argentauris rookmakeri*]

Hendrik Roelof Rookmaaker (1887–1945) was a Dutch colonial administrator (1911–1937) who was Assistant Resident on Flores. He supplied Mertens (q.v.) with two Komodo Dragons *Varanus komodoensis* for the Frankfurt Zoo. He captured 12 Komodo Dragons (1927) and sent living specimens to zoos in London, Rotterdam, Amsterdam, Berlin and Surabaya. An amphibian is named after him.

Roonwal

Sri Lanka Frogmouth ssp. *Batrachostomus moniliger roonwali* Dutta, 2009

Professor Dr Mithan Lal Roonwal (1908–1990) was Director (1956–1965) of the Zoological Survey of India (ZSI), Calcutta. He was at University of Lucknow where he studied for his first degree (1930) and where he started his career as a Research Fellow (1930). His PhD (1935) was awarded by Cambridge University, which also conferred on him a DSc (1962). He worked on locust research at Punjab Agricultural University (1931–1935) and the Pasni Field Station (1935–1939). He was in charge of the Bird & Mammal Section of ZSI (1939–1949) (with a break during the war as a Major). He then began work in the forestry section of ZSI (1949–1956). He became Professor of Zoology at Jodhpur University (1966–1969), then Professor Emeritus (1969–1971). He published

widely, particularly on insects and mammals, co-authoring volume two of *The Fauna of India, including Pakistan, Burma and Ceylon: Mammalia. Rodentia* (1961).

Roosevelt

Woodpecker sp. *Celeus roosevelti* **Cherrie**, 1916 NCR
[Probable hybrid: *Celeus elegans* x *Celeus lugubris*]
Roosevelt Stipple-throated Antwren *Epinecrophylla dentei* Whitney *et al.*, 2013

Purple Grenadier ssp. *Uraeginthus ianthinogaster roosevelti* **Mearns**, 1913 NCR; NRM

Theodore Roosevelt (1858–1919) was the 26th President of the United States of America (1901–1909). His long and distinguished career and achievements need no re-iteration here. It is however, worth noting that he belonged to a generation that believed that there was nothing reprehensible in the wholesale slaughter of wildlife: 'On the 21 April 1909, Teddy Roosevelt's safari set off from Mombasa, Kenya. By the time the entourage arrived in Khartoum 8 months later, they had slaughtered 5,013 mammals, 4,453 birds, 2,322 reptiles and amphibians and similar numbers of fish, invertebrates, shells, and plants. The skins, etc. were sent to the Smithsonian; among these were Roosevelt's gazelle and Roosevelt's sable.' He was joint-leader of the Roosevelt-Rondon Scientific Expedition to Brazil (1913–1914), during which he became so ill as to be near death when a wounded leg became infected. His health never fully recovered. Five mammals are named after him. The Roosevelt referred to in the common name of the antwren is the Rio Roosevelt in Brazil. This river was formerly known as the River of Doubt and its modern name is after Theodore Roosevelt to commemorate his Roosevelt-Rondon Expedition in 1913–14.

Roquette

Minas Gerais Tyrannulet *Phylloscartes roquettei* **Snethlage**, 1928

Professor Edgard Roquette-Pinto (1884–1954) was a Brazilian physician who was also an ornithologist, anthropologist, ethnologist, archaeologist and writer. He became an assistant Professor of Anthropology at the Brazilian National Museum (1906). After further study in Europe he accompanied official expeditions to lesser known parts of the Mato Grosso, studying tribal medicine and collecting natural history and anthropological items. He then became fascinated with filmmaking. He died in his flat after a fall.

Roscoe

Roscoe's Yellowthroat *Sylvia roscoe* **Audubon**, 1831 NCR
[Presumed to be a JS of *Geothlypis trichas*]

William Roscoe (1753–1831) was an English attorney, abolitionist and historian. Among other works, he wrote *The Life and Pontificate of Leo the Tenth* (1805). Audubon collected the bird (1821) and, as he describes in his *Ornithological Biography*, he named the yellowthroat after 'the author of the Life of Leo the Tenth'. Audubon believed the bird to be different from the Common Yellowthroat *Geothlypis trichas*, but no type specimen exists and it seems that it was based on the Mississippi Valley population of *G. trichas*.

Rose

Rufous-shafted Woodstar ssp. *Chaetocercus jourdanii rosae* **Bourcier** & **Mulsant**, 1846

Madame Rose Duquaire (DNF) was the wife of a notary in Lyon. Her sister was Madame E. Mulsant, the wife of Martial Etienne Mulsant (q.v.).

Roselaar

Red Knot ssp. *Calidris canutus roselaari* **Tomkovich**, 1990

Dr Cornelis 'Cees' Simon Roselaar (b.1947) is a Dutch biologist and ornithologist who was at the Zoological Museum, University of Amsterdam, until that museum was incorporated into the Naturalis Biodiversity Centre in Leiden, where he now works. He wrote *Geographic Variation in Waders* (1998).

Rosenberg, C. B. H.

Rosenberg's Owl *Tyto rosenbergii* **Schlegel**, 1866
[Alt. Sulawesi Owl, Celebes Masked Owl]
Rosenberg's Rail *Gymnocrex rosenbergii* Schlegel, 1866
[Alt. Bald-faced Rail, Blue-faced Rail]
Rosenberg's Lorikeet *Trichoglossus rosenbergii* Schlegel, 1871
[Alt. Biak Lorikeet; Syn. *Trichoglossus haematodus rosenbergii*]
Rosenberg's Myzomela *Myzomela rosenbergii* Schlegel, 1871
[Alt. Red-collared Myzomela]
Rosenberg's Woodcock *Scolopax rosenbergii* Schlegel, 1871
[Alt. New Guinea Woodcock; Syn. *Scolopax saturata rosenbergii*]
Fantail sp. *Rhipidura rosenbergi* **Büttikofer**, 1892 NCR
[Alt. Sooty Thicket Fantail; JS *Rhipidura threnothorax*]

Rosenberg's Pitta *Pitta sordida rosenbergii* Schlegel, 1871
[Alt. Hooded Pitta ssp.]

Baron Carl (originally Karl) Benjamin Hermann von Rosenberg (1817–1888) was a German naturalist and geographer who collected in the East Indies and later took Dutch nationality (1865). He enlisted in the Dutch colonial army and was in the Malay Archipelago three decades, as topographic draughtsman on the island of Sumatra (1839–1855), then as a civil servant in the Moluccas and around New Guinea (travelling there in the Dutch warship *Etna*, during which he met Wallace (q.v.). He went to Europe (1866–1868) but returned to the East Indies, finally returning to Europe (1871) in poor health. All the while he pursued his interest in ornithology, including writing a series of articles. He wrote *Reistochten naar de Geelvinkbaai op Nieuw-Guinea in de Jaren 1869 en 1870*, an important zoological and ethnographical study on New Guinea. He also wrote *Der Malayische Archipel, Land und Leute* (1878). He died in The Hague, but was buried in the family vault in Darmstadt. A mammal and a reptile re named after him.

Rosenberg, W. F. H.

Chocó Poorwill *Nyctiphrynus rosenbergi* **Hartert**, 1895
Rosenberg's Hummingbird *Amazilia rosenbergi* **Boucard**, 1895
[Alt. Purple-chested Hummingbird; Syn. *Polyerata rosenbergi*]
Scarlet-and-white Tanager *Nemosia rosenbergi* **Rothschild**, 1897 NCR
[JS *Chrysothlypis salmoni*]
Esmeraldas Antbird *Sipia rosenbergi* Hartert, 1898 NCR
[JS *Myrmeciza nigricauda*]

Northern Schiffornis ssp. *Schiffornis veraepacis rosenbergi* Hartert, 1898
Rufous Mourner ssp. *Rhytipterna holerythra rosenbergi* Hartert, 1905
Cocoa Woodcreeper ssp. *Xiphorhynchus susurrans rosenbergi* **Bangs**, 1910
Rufous-crowned Antpitta ssp. *Pittasoma rufopileatum rosenbergi* **Hellmayr**, 1911

William Frederik Henry Rosenberg (1868–1957) was a traveller, naturalist, entomologist and natural history dealer and importer, based in London's Charing Cross Road (1930–1948). He had earlier had premises in Tring. He collected in South America (1896–1898), but also employed other collectors and bought in skins. Rothschild (q.v.), among others, would send Rosenberg specimens not wanted for his own collections. An amphibian is named after him.

Rosina

Fairy-wren genus *Rosina* **Mathews**, 1912 NCR
[Now in *Malurus*]

Malleefowl ssp. *Leipoa ocellata rosinae* Mathews, 1912 NCR; NRM
Mulga Parrot ssp. *Psephotus varius rosinae* Mathews, 1912 NCR; NRM
Sulphur-crested Cockatoo ssp. *Cacatua galerita rosinae* Mathews, 1912 NCR
[JS *Cacatua galerita galerita*]
Western Yellow Robin ssp. *Eopsaltria griseogularis rosinae* Mathews, 1912
White-browed Scrubwren ssp. *Sericornis frontalis rosinae* Mathews, 1912
White-plumed Honeyeater ssp. *Ptilotis penicillata rosinae* Mathews, 1912 NCR
[Intergrade between *Lichenostomus penicillatus leilavalensis* and *L. p. penicillatus*]
Rosina's Slender-billed Thornbill *Acanthiza iredalei rosinae* Mathews, 1913
Beautiful Firetail ssp. *Zonaeginthus bellus rosinae* Mathews, 1923 NCR
[JS *Stagonopleura bella bella*]

(See **White, S. A. & E. R.** and **Ethel**)

Rosita

Rosita's Bunting *Passerina rositae* **G. N. Lawrence**, 1874
[Alt. Rose-bellied Bunting]

Rosita Sumichrast was the wife of Adrien de Sumichrast (q.v.).

Ross, B. R.

Ross's Goose *Anser rossii* **Cassin**, 1861

Bernard Rogan Ross (1827–1874) was an Irish trader and administrator who was a chief factor in the Hudson's Bay Company. He may have obtained his position largely as a result of marrying his boss's daughter, but nonetheless proved to be a very able administrator. Cassin (q.v.) named the goose after Ross with encouragement from Kennicott (q.v.), in appreciation for the cooperation that Ross showed Kennicott in arranging transportation of the latter's specimens to the USNM (of which Ross was himself an associate).

Ross, E.

Ross's Turaco *Musophaga rossae* **Gould**, 1852
[Alt. Lady Ross's Turaco, Ross's Lourie]

Lady Eliza Ross *née* Pritchard, formerly Bennett (1818–1890) was the second wife of Major-General Sir Patrick Ross (1778–1850), Governor of St. Helena (1846–1850). The description was based on a sketch and some moulted feathers which she sent to Gould (q.v.). Her first husband, George Brooks Bennett attended Napoleon's funeral at the age of five; his father lived near where Napoleon was exiled on St Helena and, when he died, decided to provide a mahogany coffin for him, but as no mahogany could be found he used his large mahogany table! She was born and died on St Helena.

Ross, E. C. & E. D.

Pied Bushchat ssp. *Saxicola caprata rossorum* **Hartert**, 1910

We have to conjecture, as Hartert gave no etymology for his use of *rossorum* but the following two gentlemen (since *rossorum* is a plural) could be the honorands. Sir Edward Charles Ross (1836–1913) entered the Indian Civil Service (1855) and was Resident and Consul-General in the Persian Gulf (1872–1891). Sir Edward Denison Ross (1871–1940) was an orientalist who explored in Persia (Iran) and Turkestan. He was multilingual, speaking 30 oriental languages and reading a further 19. He was Professor of Persian at University College, London (1896–1901) then became Principal of the Calcutta Madrasah for the education of Muslims (1901), combining that post (1906) with Curator of Records for the Government of India. He was the first Director of the School of Oriental and African Studies, London (retiring 1937), and became head of the British Information Bureau, Istanbul (1939–1940).

Ross, F. W.

Violaceous Trogon *Trogon rossi* **W. P. Lowe**, 1939 NCR
[Alt. Guianan Trogon; JS *Trogon violaceus*]

Francis William Locke Ross (1793–1860) was an English naturalist and collector. He was an officer in the Royal Navy, retiring (1830) to Topsham, Devon, where he created his own museum with exhibits in the fields of archaeology, ethnography, conchology and geology as well as ornithology. Most of his collection is housed at the Royal Albert Memorial Museum and Art Gallery in Exeter.

Ross, J. A.

Nightjar genus *Rossornis* **Mathews**, 1918 NCR
[Now in *Caprimulgus*]

Tawny Frogmouth ssp. *Podargus strigoides rossi* Mathews, 1912 NCR
[JS *Podargus strigoides brachypterus*]

John Alexander Ross (1868–1957) was an Australian solicitor and ornithologist. He worked for the Victorian Crown Solicitor's Office (1886–1930) and in private practice (1930–1952). He was President of the RAOU (1926). He was also a noted shot, winning trophies at Bisley, England (1898).

Ross, J. C.

Ross's Gull *Rhodostethia rosea* **W. MacGillivray**, 1824
Auckland Islands Shore Plover *Thinornis rossii* **G. R. Gray**, 1845 EXTINCT
[Status uncertain; known only from one specimen]

Rear-Admiral Sir James Clark Ross (1802–1862) discovered the Ross Sea and the Ross Ice Shelf and an island (1841) which Scott, on his first expedition, named Ross Island in his honour. Ross joined the Royal Navy aged 12! He was also a member of several important expeditions to the Arctic. He commanded *Erebus* and *Terror* during the Antarctic expedition (1839–1843). It was while close to the Magnetic South Pole that he broke through a wide expanse of pack ice and into a large and clear sea that later bore his name. A mammal is named after him.

Rossikow

Eurasian Bullfinch ssp. *Pyrrhula pyrrhula rossikowi* **Derjugin** & **Bianchi**, 1900

Konstantin Nikolaevich Rossikow (often Rossikov) (1854–1910) was a zoologist, ichthyologist, arachnologist and 'zoographer'. He graduated from the University of St Petersburg and travelled in the Northern Caucasus region (1890), where he mentioned seeing 'multi-coloured' vipers. He wrote, in Russian, a long paper 'In the mountains of the north-western Caucasus: a trip to Zaakdan and the Bolshaya Laba riverhead for the purpose of zoogeographic research' (1890). A reptile is named after him.

Roth

Rufous-rumped Seedeater ssp. *Sporophila hypochroma rothi* Singh, 1960 NCR
[Probably a hybrid: *Sporophila castaneiventris* x *Sporophila minuta*]

Dr Walter Edmund Roth (1860–1933) was a British surgeon, anthropologist and anatomist in Australia (1898–1906), where he was Chief Protector of Aborigines, and in British Guiana (Guyana) as Protector of Indians (1906–1928). He was Curator, British Guiana Museum, Georgetown (1928–1933).

Rothschild, L. W.

Hawaiian Honeycreeper genus *Rothschildia* **S. B. Wilson** &
Evans, 1899 NCR
[Now in *Magumma*]

Rothschild's Grosbeak *Cyanocompsa rothschildii* **E. Bartlett**,
1890
[Alt. Blue-black Grosbeak; Syn. *Cyanocompsa cyanoides
rothschildii*]
Rothschild's Sunangel *Heliangelus rothschildi* **Boucard**,
1892 NCR
[Presumed hybrid: *Eriocnemis* sp. x *Heliangelus* sp. or
Heliangelus sp. x *Ramphomicron microrhynchum*]
Rothschild's Parakeet *Psittacula intermedia* Rothschild,
1895
[Alt. Intermediate Parakeet; hybrid *Psittacula cyanocephala*
x *Psittacula himalayana*]
Golden-chested Tanager *Bangsia rothschildi* **Berlepsch**,
1897
Rothschild's Lobe-billed Bird-of-Paradise *Loborhamphus
nobilis* Rothschild, 1901 NCR
[Presumed hybrid: *Paradigalla carunculata* x *Lophorina
superba?*]
Rothschild's Canary *Serinus rothschildi* **Ogilvie-Grant**, 1902
[Alt. Arabian Serin; Syn. *Crithagra rothschildi*]
Rothschild's Peacock Pheasant *Polyplectron inopinatum*
Rothschild, 1903
[Alt. Mountain Peacock Pheasant]
Rothschild's Astrapia *Astrapia rothschildi* **F. Förster**, 1906
[Alt. Huon Astrapia]
Rothschild's Tody Tyrant *Idioptilon rothschildi* Berlepsch,
1907 NCR
[Alt. White-eyed Tody Tyrant; JS *Hemitriccus zosterops
zosterops*]
Rhea sp. *Rhea rothschildi* **Brabourne** & Chubb, 1911 NCR
[Alt. Greater Rhea; JS *Rhea americana albescens*]
Rothschild's Fody *Foudia omissa* Rothschild 1912
[Alt. Forest Fody]
Rothschild's Myna *Leucopsar rothschildi* **Stresemann**, 1912
[Alt. Bali Starling, Bali Myna]
Rothschild's Bird-of-Paradise *Paradisea mixta* Rothschild,
1921 NCR
[Hybrid: *Paradisaea raggiana* x *Paradisaea minor finschi*]
Cochoa sp. *Cochoa rothschildi* **E. C. S. Baker**, 1924 NCR
[Alt. Green Cochoa; JS *Cochoa viridis*]
Rothschild's Swift *Cypseloides rothschildi* **Zimmer**, 1945
[Alt. Giant Swift]

Rothschild's Amazon *Amazona barbadensis rothschildi*
Hartert, 1892 NCR; NRM
[Alt. Yellow-shouldered Amazon ssp.]
Silver-breasted Broadbill ssp. *Serilophus lunatus rothschildi*
Hartert & **A. L. Butler**, 1898
Common Cactus Finch ssp. *Geospiza scandens rothschildi*
Heller & Snodgrass, 1901
Northern Cassowary ssp. *Casuarius unappendiculatus
rothschildi* **Matschie**, 1901 NCR; NRM
Spotted Nutcracker ssp. *Nucifraga caryocatactes
rothschildi* Hartert, 1903
Hooded Pitta ssp. *Pitta atricapilla rothschildi* **Parrot**, 1907
NCR
[JS *Pitta sordida sordida*]

Little Spiderhunter ssp. *Arachnothera longirostra rothschildi*
van Oort, 1910
Yellow-browed Camaroptera ssp. *Camaroptera
brevicaudata rothschildi* **Zedlitz**, 1911 NCR
[JS *Camaroptera superciliaris*]
Comb-crested Jacana ssp. *Irediparra gallinacea rothschildi*
Mathews, 1912 NCR
[JS *Irediparra gallinacea novaehollandiae* – or species
regarded as monotypic]
Emu ssp. *Dromaius novaehollandiae rothschildi* Mathews,
1912
Regent Bowerbird ssp. *Sericulus chrysocephalus rothschildi*
Mathews, 1912 NCR; NRM
Southern Emu-wren ssp. *Stipiturus malachurus rothschildi*
Mathews, 1912 NCR
[JS *Stipiturus malachurus westernensis*]
Striated Sittella ssp. *Neositta striata rothschildi* Mathews,
1912 NCR
[JS *Daphoenositta chrysoptera striata*]
Rothschild's Red Lory *Eos bornea rothschildi* Stresemann,
1912 NCR
[NUI *Eos bornea bornea*]
Wandering Albatross ssp. *Diomedea exulans rothschildi*
Mathews, 1912 NCR
[Indeterminate: could be either *Diomedea exulans gibsoni* or
D. exulans antipodensis]
Magnificent Frigatebird ssp. *Fregata magnificens rothschildi*
Mathews, 1915 NCR; NRM
Red-tailed Tropicbird ssp. *Phaethon rubricauda rothschildi*
Mathews, 1915 NCR
[JS *Phaethon rubricauda melanorhynchos*]
Common Buzzard ssp. *Buteo buteo rothschildi* Swann, 1919
Olive Ibis ssp. *Bostrychia olivacea rothschildi* **Bannerman**,
1919
Purple Grenadier ssp. *Uraeginthus ianthinogaster
rothschildi* van Someren, 1919 NCR; NRM
Buff-faced Pygmy Parrot ssp. *Micropsitta pusio rothschildi*
Stresemann, 1922 NCR
[JS *Micropsitta pusio pusio*]
Rothschild's Pheasant *Phasianus colchicus rothschildi*
La Touche, 1922
[Alt. Common Pheasant ssp.]
Dark-sided Flycatcher ssp. *Muscicapa sibirica rothschildi*
E. C. S. Baker, 1923
Red-breasted Paradise Kingfisher ssp. *Tanysiptera nympha
rothschildi* Laubmann, 1924 NCR; NRM
Sapphire Quail Dove ssp. *Geotrygon saphirina rothschildi*
Stolzmann, 1926
Red-vented Barbet ssp. *Megalaima lagrandieri rothschildi*
Delacour, 1927
White Tern ssp. *Gygis alba rothschildi* Hartert, 1927
[NUI *Gygis alba candida*]
Ashy Robin ssp. *Heteromyias albispecularis rothschildi*
Hartert, 1930
Rothschild's Fairy Lorikeet *Charmosyna pulchella rothschildi*
Hartert, 1930
Slender-billed Scimitar Babbler ssp. *Pomatorhinus
superciliaris rothschildi* Delacour & **Jabouille**, 1930
Black-fronted White-eye ssp. *Zosterops minor rothschildi*
Stresemann & Paludan, 1934
White-bellied Cuckooshrike ssp. *Coracina papuensis
rothschildi* Kok, 2008

Lord Lionel Walter Rothschild (1868–1937) was the Founder of the Tring Museum. It is now known as The Walter Rothschild Zoological Museum and comprises the BMNH ornithology section. According to the history of the museum: 'As a child Walter Rothschild knew exactly what he was going to do when he grew up, announcing at the age of seven, 'Mama, Papa, I am going to make a museum'. He had already started collecting insects and stuffed animals by then, and a year later started setting his own collection of butterflies. By the time he was 10, Walter had enough natural history objects to start his first museum – in a garden shed! Before long Walter's insect and bird collections were so large that they had to be stored in rented rooms and sheds around Tring. Then in 1889, when Walter Rothschild was 21, his father gave him some land on the outskirts of Tring Park. Two small cottages were built, one to house his books and insect collection, the other for a caretaker. Behind these was a much larger building, which would contain Lord Rothschild's collection of mounted specimens. This was the beginning of the zoological museum which opened to the public in 1892 and the beginning of Lord Rothschild's life long passion for natural history.' He amassed the largest bird collection in the world; 300,000 bird skins, 200,000 birds' eggs and 30,000 books. Ernst Hartert (q.v.) was one of his curators. Rothschild lived in Tring and was Member of Parliament for Aylesbury, a major in the Buckinghamshire Yeomanry, a Justice of the Peace, and a Deputy Lieutenant for the County of Buckinghamshire. Six mammals are named after him. (See **Walter**)

Rothschild, M.

Cuckooshrike sp. *Campephaga rothschildi* **Neumann**, 1907 NCR
[Alt. Red-shouldered Cuckooshrike; JS *Campephaga phoenicea*]

Red-chested Swallow ssp. *Hirundo lucida rothschildi* Neumann, 1904
Black-winged Bishop ssp. *Euplectes hordeaceus rothschildi* Neumann, 1907 NCR
[JS *Euplectes hordeaceus craspedopterus* – or species regarded as monotypic]
Buff-bellied Penduline-tit ssp. *Anthoscopus caroli rothschildi* Neumann, 1907 NCR
[JS *Anthoscopus caroli sylviella*]
Slate-coloured Boubou ssp. *Laniarius funebris rothschildi* Neumann, 1907 NCR
[JS *Laniarius funebris funebris*]
Western Bronze-naped Pigeon ssp. *Columba iriditorques rothschildi* Neumann, 1908 NCR; NRM
Rothschild's Seedcracker *Pyrenestes ostrinus rothschildi* Neumann, 1910 NCR; NRM
[Alt. Black-bellied Seedcracker]

Baron Maurice de Rothschild (1881–1957) was a member of the French branch of the famous banking family. In his youth he was a well-known playboy and quarreled with his relations over an investment they regarded as risky – but which turned out to be highly profitable, as did many of his subsequent ventures. He became a politician and was a member of the Chamber of Deputies (1919–1929) and a Senator (1929–1945), one of the few to vote against giving Marshall Pétain

full powers (1940). He was instrumental in helping de Gaulle become the leader of the Free French in exile in England during WW2, but he later upset him and was virtually banished from France to the Bahamas. He travelled in East Africa (1904–1905). He also collected art and was a patron, philanthropist and big-game hunter. A reptile is named after him. (See **Maurice**)

Rougeot

Red-throated Wryneck ssp. *Jynx ruficollis rougeoti* **Berlioz**, 1953 NCR
[JS *Jynx ruficollis ruficollis*]

Pierre Claude Rougeot (1920–2002) was a naturalist, entomologist, collector and colonial administrator in French Equatorial Africa (1950). He wrote *Les Attacides (Saturnides) de l'Equateur Africain Français* (1955).

Rouget

Rouget's Rail genus *Rougetius* **Bonaparte**, 1856

Rouget's Rail *Rougetius rougetii* **Guérin-Méneville**, 1843

J. Rouget (d.1840) was a French explorer who collected in Ethiopia (1839–1840).

Round

Rufous-backed Sibia ssp. *Heterophasia annectans roundi* Eames, 2002

Professor Philip David Round (b.1953) is a British ornithologist and conservationist. He began work for Association for the Conservation of Wildlife in Thailand (1980) after graduating from the University of Aberdeen. He is Assistant Professor in the Department of Biology, Faculty of Science, Mahidol University, Bangkok, Thailand (where he lives with his wife Thiyapa and daughter Mim) as well as regional representative for the UK NGO, The Wetland Trust. He has a strong interest in promoting bird-banding, having been a ringer since 1969, and has research interests in ecology, life history and seasonality of resident and migrant birds, taxonomy and biogeography. He has a special affinity for hard-to-identify warblers, being the first to discover Manchurian Reed Warbler *Acrocephalus tangorum* anywhere in its wintering areas (1981) and rediscovering the Large-billed Reed Warbler *A. orinus* (2006), formerly known from one specimen, collected 139 years earlier in NW India. He also rediscovered the globally threatened Gurney's Pitta *Pitta gurneyi* (1986), spending over a decade fighting to protect its lowland forest habitat from encroachment. He has written or co-written five books on Thai birds including *A Guide to the Birds of Thailand* (1991) and *Birds of the Bangkok Area* (2008), and over 80 scientific papers. He is generally recognised as the authority on Thai birds and it was for that and his struggle on behalf of Gurney's Pitta that Eames named the sibia *roundi*.

Roure

Cherry-throated Tanager *Nemosia rourei* **Cabanis**, 1870

Jean de Roure (DNF) was a French traveller who had been collecting in Brazil for over 30 years when he discovered the tanager holotype (1870).

Rousseau

Noddy sp. *Anous rousseaui* **Hartlaub**, 1861 NCR
[Alt. Brown Noddy; JS *Anous stolidus pileatus*]

Louis Rousseau (1811–1874) was a French malacologist and collector at MNHN, Paris, and a pioneer photographer. He collected mainly in Madagascar but took part in expeditions to Zanzibar, the Seychelles, and Russia. His brother, Théodore, was an important landscape painter.

Roux, E.

Whiskered Yuhina ssp. *Yuhina flavicollis rouxi* **Oustalet**, 1896

Lieutenant Emile François Louis Roux (1868–1917) was a French naval officer who explored in Yunnan (1894–1895). He was with Prince Henri d'Orléans (q.v. under Prince Henry) on an expedition to find the sources of the rivers Mekong and Irrawaddy. He wrote *Expédition du Prince Henri d'Orléans, de MM. Roux et Briffaut, du Tonkin aux Indes* (1895).

Roux, J.

Fan-tailed Gerygone ssp. *Gerygone flavolateralis rouxi* **F. Sarasin**, 1913

Dr Jean Roux (1876–1939) was a zoologist. Geneva University awarded his doctorate (1899). He studied protozoa in Berlin (1899–1902), then was Curator, Natural History Museum, Basel, Switzerland (1902–1930). He was in New Guinea and Australia (1907–1908) and in New Caledonia and the Loyalty Islands (1911–1912) with Fritz Sarasin (q.v.). He wrote *Les Reptiles de la Nouvelle-Calédonie et des Îles Loyalty* (1913). Six reptiles and an amphibian are named after him.

Rowan

Sandhill Crane ssp. *Grus canadensis rowani* Walkinshaw, 1965

Professor William Rowan (1891–1957) was a Swiss-born, French-educated, British/Canadian zoologist, conservationist and wildlife artist who went to Canada and worked as a ranch hand (1908–1911). He studied at University College London (1911–1914), only graduating (1917) after service in WW1 with the London Scottish Regiment. He returned to Canada as a lecturer at the University of Manitoba (1919) but then founded, and was Director of, the Department of Zoology, Albert University (1920–1956).

Rowe, E.

Abyssinian Ground Thrush ssp. *Zoothera piaggiae rowei* **C. H. B. Grant** & **Mackworth-Praed**, 1937

Eric George Rowe (1904–1987) was an ornithologist and British colonial administrator in Tanganyika (Tanzania) (1928–1958). He worked for the Institute of Colonial Studies, Oxford (1959). The BMNH at Tring holds his manuscript notes on birds and fishing in Tanganyika and on birds of Kashmir.

Rowe, S. & S.

Smith's Longspur ssp. *Calcarius pictus roweorum* Kemsies, 1961 NCR; NRM

Stanley M. Rowe Sr (1890–1987) was an American businessman and arborealist, and his son Stanley M. Rowe Jr (1918–2009) was a businessman and naturalist in Cincinnati. They founded and supported the Cincinnati Nature Center.

Rowett

Gough Island Finch genus *Rowettia* **P. R. Lowe**, 1923

John Quiller Rowett (1876–1924) was an English businessman who financed Shackleton's (1921–1922) Antarctic Expedition. He was responsible for founding the Rowett Institute of Research in Animal Nutrition at Aberdeen University. Rowett Island in the Weddell Sea is named after him.

Rowley, G. D.

Rowley's Flycatcher *Eutrichomyias rowleyi* **A. B. Meyer**, 1878
[Alt. Cerulean Paradise-flycatcher]

George Dawson Rowley (1822–1878) was an English amateur ornithologist. He wrote *Ornithological Miscellany* (1875). He was also a skilled artist and his pictures of extinct birds have been used in posters.

Rowley, I. C.

Striated Grasswren ssp. *Amytornis striatus rowleyi* **Schodde** & Mason, 1999

Ian Cecil Robert Rowley (1926–2009) was a Scottish-born Australian ornithologist, who was editor of *Emu* (1990–2000). He served in the Royal Navy in WW2, moved to Australia (1949), and took a degree in Agricultural Science at Melbourne University. He worked for CSIRO from 1952.

Rowley, J. S.

Montezuma Quail ssp. *Cyrtonyx montezumae rowleyi* **A. R. Phillips**, 1966
Red Warbler ssp. *Ergaticus ruber rowleyi* Orr & Webster, 1968

John Stuart Rowley (1907–1968) was an ornithologist who was a Research Associate, Department of Ornithology and Mammalogy, California Academy of Sciences. He took his degree at Berkeley and went to work (1933) in the family business (weighing machines), which he sold (1957) to devote the rest of his life to birds. He was killed in a fall from a cliff. A reptile is named after him.

Roy

Gerygone genus *Royigerygone* **Mathews**, 1912 NCR
[Now in *Gerygone*]

Storm-petrel sp. *Fregettornis royanus* Mathews, 1914 NCR
[Alt. White-bellied Storm-petrel; JS *Fregetta grallaria grallaria*]

Norfolk Island Boobook *Ninox boobook royana* Mathews, 1912 NCR

[Alt. Morepork ssp.; JS *Ninox novaeseelandiae undulata*]

Wedge-tailed Shearwater ssp. *Puffinus pacificus royanus* Mathews, 1912 NCR; NRM

White Tern ssp. *Gygis alba royana* Mathews, 1912 NCR

[NUI *Gygis alba candida*]

Raoul 'Roy' Sunday Bell (1882–1966) was a naturalist, ornithologist and photographer. Born on Sunday Island (one of the Kermadecs), his parents gave him a name to remember it by! He collected for Gregory Mathews (q.v.) (1908) and settled on Norfolk Island (1910). He served in WW1 as an aerial photographer for the Australian Flying Corps.

Rück

Rück's Blue Flycatcher *Cyornis ruckii* **Oustalet**, 1881
[Alt. Rueck's Blue Flycatcher]

M. Rück (DNF) was a French traveller who collected in Malacca (Malaysia) (1880).

Rucker

Rucker's Hermit *Threnetes ruckeri* **Bourcier**, 1847
[Alt. Band-tailed Barbthroat]

Sigismund Rucker (1809–1876) was an English naturalist, collector of fine art, hummingbirds and orchids. Bourcier bestowed the name 'having come across new hummingbirds in Mr Rucker's collection'. Rucker was one of the pioneers of glasshouse cultivation and his garden in Wandsworth, south-west London, held plants from Ecuador – where he collected in 1846 – that had never before been cultivated in Europe.

Rudd

Rudd's Apalis *Apalis ruddi* **C. H. B. Grant**, 1908
Rudd's Lark *Heteromirafra ruddi* C. H. B. Grant, 1908

Charles Dunnel Rudd (1844–1916) was an associate of Cecil Rhodes (q.v.) and attended to their mining business while Rhodes got himself into politics, obtained the concession (1883) to go into Mashonaland to establish mining, and founded Rhodesia. Rudd financed Captain C. H. B. Grant (q.v.), who collected and described both birds. Two mammals are named after him.

Rudebeck

Fairy Flycatcher ssp. *Stenostira scita rudebecki* **Clancey**, 1955

Dr Gustaf E. Rudebeck (1913–2005) was a zoologist, ornithologist and entomologist at Lund University, Sweden, where he became Professor Emeritus in Ecology. He was attached to the Transvaal Museum, South Africa (1950–1951 and 1954–1956), and was on the Swedish Lund University expedition to South Africa (1956). A reptile is named after him.

Rüdiger

Common Blackbird *Turdus ruedigeri* **Kleinschmidt**, 1919 NCR
[JS *Turdus merula merula*]

W. Rüdiger (1875–1957) was a German forester and oologist.

Rudolf (Grauer)

Many-coloured Bush-shrike ssp. *Laniarius rubiginosus rudolfi* **Hartert**, 1908 NCR
[JS *Chlorophoneus multicolor graueri*]

Dusky Twinspot ssp. *Estrilda cinereovinacea rudolfi* Hartert, 1919 NCR
[Unnecessary replacement name for *Euschistospiza cinereovinacea graueri*]

(See **Grauer**)

Rudolf (Kmunke)

Moorland Chat ssp. *Pinarochroa sordida rudolfi* **Madarász**, 1912 NCR
[NUI *Pinarochroa sordida ernesti*]

Rudolf Kmunke (1866–1918) trained and qualified as an architect in Vienna, Austria. His first expedition was to eastern Greenland (1909). He led expeditions to East Africa (1911–1912) and Morocco (1913). He died from Spanish Flu in the pandemic. He wrote *Quer durch Uganda. Eine Forschungsreise in Zentralafrika 1911/1912* (1913).

Rudolf/Rudolph, Prince

Sumba Boobook *Ninox rudolfi* **Meyer**, 1882
Blue Bird-of-Paradise *Paradisaea rudolphi* **Finsch** & Meyer, 1885

(See **Prince Rudolf/Rudolph**)

Rueck

Rueck's Blue Flycatcher *Cyornis ruckii* **Oustalet**, 1881
[Binomial sometimes amended (wrongly) to *ruecki*]

(See **Rück**)

Rumsey

Cebu Hawk Owl *Ninox rumseyi* Rasmussen *et al.* 2012

Stephen John Raymond Rumsey (b.1950) is a British businessman, ornithologist and conservationist. He was an investment banker and stockbroker, but now devotes his energies mainly into conservation. He founded the Wetland Trust, and owns and runs a reserve at Icklesham in Sussex where more birds are ringed each year than anywhere else in the UK. He served as Treasurer of BirdLife International (2004–2009).

Rup Chand

Pied Bushchat ssp. *Saxicola caprata rupchandi* **Koelz**, 1939 NCR
[JS *Saxicola caprata bicolor*]

Rup Chand's Needletail *Hirundapus cochinchinensis rupchandi* **Biswas**, 1951
[Alt. Silver-backed Needletail ssp.]

White-cheeked Partridge ssp. *Arborophila atrogularis rupchandi* Koelz, 1953 NCR; NRM

Blyth's Tragopan ssp. *Tragopan blythii rupchandi* Koelz, 1954 NCR
[JS *Tragopan blythii blythii*]

Broad-billed Warbler ssp. *Tickellia hodgsoni rupchandi* Koelz, 1954 NCR
[JS *Tickellia hodgsoni hodgsoni*]

Dark-rumped Swift ssp. *Apus acuticauda rupchandi* Koelz, 1954 NCR; NRM

Hodgson's Frogmouth ssp. *Batrachostomus hodgsoni rupchandi* Koelz, 1954 NCR
[JS *Batrachostomus hodgsoni hodgsoni*]

Thakur Rup Chand (DNF) was a Tibetan assistant to Walter Koelz (q.v.). They collected together in the Indian subcontinent (1931–1953). 'Thakur' is an Indian feudal title, rather than a first name.

Rüppell

Rüppell's (Swift) Tern *Thalasseus bergii* **Lichtenstein**, 1823
[Alt. Greater Crested Tern]

Rüppell's Warbler *Sylvia rueppelli* **Temminck**, 1823

Rüppell's Robin Chat *Cossypha semirufa* Rüppell, 1837

Rüppell's Weaver *Ploceus galbula* Rüppell, 1840

Rüppell's Glossy Starling *Lamprotornis purpuroptera* Rüppell, 1845
[Alt. Rüppell's Long-tailed Starling]

Rüppell's Parrot *Poicephalus rueppellii* **G. R. Gray**, 1849

Rüppell's Griffon (Vulture) *Gyps rueppellii* **A. E. Brehm**, 1852

Rüppell's White-crowned Shrike *Eurocephalus ruppelli* **Bonaparte**, 1853
[Alt. White-rumped Shrike, Northern White-crowned Shrike]

Rüppell's Bustard *Eupodotis rueppelii* **Wahlberg**, 1856
[Alt. Rüppell's Korhaan]

Yellow-billed Duck ssp. *Anas undulata ruppelli* **Blyth**, 1855

Red-winged Starling ssp. *Onychognathus morio rueppellii* **J. Verreaux**, 1856

Wilhelm Peter Eduard Simon Rüppell (1794–1884) was a German collector. He went to Egypt and ascended the Nile as far as Aswan (1817), and later made two extended expeditions to northern and eastern Africa, Sudan (1821–1827) and Ethiopia (1830–1834). Abdim Bey (q.v.) helped him in Egypt. Although he brought back large zoological and ethnographical collections, his expeditions impoverished him. He wrote *Reisen in Nubien, Kordofan und dem Petraischen Arabien* (1829), *Reise in Abyssinien* (1838–1840) and *Systematische Uebersicht der Vögel Nord-Ost-Afrikas* (1845). He also collected in the broadest sense, and presented his collection of coins and rare manuscripts to the Historical Museum in Frankfurt (his home town). Five mammals, an amphibian and two reptiles are named after him.

Rush

Tit-hylia *Pholidornis rushiae* **Cassin**, 1855

Phoebe Ann Rush *née* Ridgway (1799–1857) was the wife of the American pioneer psychologist, Dr James Rush (1786–1869). She was a member of a wealthy Philadelphia merchant family and lived with her husband in a house big enough to accommodate 800 guests – they gave memorable parties! Cassin (q.v.) honoured her because of her 'enlightened encouragement of men devoted to the sciences and to the arts'.

Ruspoli, Prince

Ruspoli's Turaco *Tauraco ruspolii* **Salvadori**, 1896
[Alt. Prince Ruspoli's Turaco]

White-headed Buffalo Weaver ssp. *Dinemellia dinemelli ruspolii* Salvadori, 1894 NCR
[JS *Dinemellia dinemelli dinemelli*]

(See **Prince Ruspoli**)

Russ

Russ's Weaver-bird *Ploceus russi* **Finsch**, 1877 NCR
[Alt. Red-billed Quelea; JS *Quelea quelea*]

Dr Karl Friedrich Otto Russ (1833–1899) was a German aviculturist. He wrote *Die Sprechenden Papageien* (1882).

Russell, M.

Yellow-throated Euphonia ssp. *Euphonia hirundinacea russelli* **A. R. Phillips**, 1966 NCR
[JS *Euphonia hirundinacea hirundinacea*]

Sedge Wren ssp. *Cistothorus platensis russelli* **Dickerman**, 1975

Dr Stephen M. Russell (b.1931) is an American zoologist and ornithologist. He was at the Department of Biology, Louisiana State University (LSU), and collected in British Honduras (Belize) (1956). LSU awarded his doctorate and he taught there (1958–1964) before becoming Professor of Zoology, University of Arizona (1964) where he retired (1996) as Emeritus Professor and Curator Emeritus of the university's bird collection. He co-wrote *The Birds of Sonora* (1998).

Russell, W. C.

Mountain Quail ssp. *Oreortyx pictus russelli* A. H. Miller, 1946

Ward C. Russell (b.1907) was a 'skilled and veteran collector of birds and mammals' at the Museum of Vertebrate Zoology, University of California, Berkeley. He had been one of Grinnell's (q.v.) students and was the museum preparatory for 40 years; he collected the holotype (1945).

Rutenberg

Striated Heron ssp. *Butorides striata rutenbergi* **Hartlaub**, 1880

Diedrich Christian Rutenberg (1851–1878) was an explorer and traveller who studied medicine and natural history at Jena, Germany. He was in Dalmatia and Montenegro (1872). He went to South Africa (1877), collecting botanical specimens in Natal and Transvaal, and thence to Mauritius and Madagascar, where the local inhabitants killed him. An amphibian is named after him.

Ruth

> Fawn-breasted Waxbill ssp. *Estrilda paludicola ruthae* **Chapin**, 1950
> Forest Penduline-tit ssp. *Anthoscopus flavifrons ruthae* Chapin, 1958
> White-browed Forest Flycatcher ssp. *Fraseria cinerascens ruthae* **Dickerman**, 1994

Mrs Ruth Chapin (née Trimble) (d.1994) was an ornithologist who was an assistant curator of birds at the Carnegie Museum until her marriage to Dr James Chapin (q.v.) in 1940. She wrote 'The behaviour of the Olive Weaver-finch *Necocharis ansorgei*' (1959).

Rutherford

> Alcock's Snake Eagle *Spilornis cheela rutherfordi* **Swinhoe**, 1870
> [Alt. Crested Serpent Eagle ssp.]

Dr Rutherford Alcock (See **Alcock**)

Rutledge

> Finn's Weaver *Ploceus rutledgii* **Finn**, 1899 NCR
> [JS *Ploceus megarhynchus*]

William Rutledge (d.1908) was a British animal dealer in Calcutta.

Ruwet

> Ruwet's Masked Weaver *Ploceus ruweti* Louette & **C. W. Benson**, 1982
> [Alt. Lufira Masked Weaver]

Professor Emeritus Jean-Claude Ruwet (d. 2007) was a Belgian zoologist at the University of Liège. He published extensively, mostly in the *Belgian Journal of Zoology*. He worked on the Black Grouse *Tetrao tetrix* and presented a paper (2000) with others entitled: 'Compared modelling of the climate's influence on the dynamics of six European Black Grouse populations'.

Ruys

> Ruys's Bird-of-Paradise *Neoparadisea ruysi* **van Oort**, 1906 NCR
> [Presumed hybrid: *Paradisaea minor* x *Diphyllodes magnificus*]

Theodor H. Ruys (DNF) was a Dutch traveller who spent four years in New Guinea 'for commercial purposes' in the early years of the 20th century. He obtained the type (and unique) specimen of this bird-of-paradise and presented it to the Leiden Museum, The Netherlands. The 'species' is now regarded as a hybrid between the Lesser Bird-of-Paradise and the Magnificent Bird-of-Paradise.

Ryan

> Fairy-wren genus *Ryania* **Mathews**, 1912 NCR
> [Now in *Malurus*]
> Honeyeater genus *Ryanornis* **A. J. Campbell**, 1919 NCR
> [Now in *Ramsayornis*]

Sir Charles Snodgrass Ryan (1853–1926) was an Australian surgeon and ornithologist. He studied medicine at the University of Melbourne (1870–1872) and qualified at Edinburgh University (1875). He did postgraduate study in Bonn and Vienna and, whilst in Rome, saw an advertisement that led him to join the Turkish army as a military surgeon. He saw action in the Turko-Serbian War (1876) and in the Russo-Turkish War (1877–1878). The Russians captured him when they took Erzerum in Turkish Armenia. He returned to Melbourne (1878) and worked as a surgeon at the Royal Melbourne Hospital (1879–1913). In WW1 he was Assistant Director 1st Division, Australian Imperial Force, and with them landed at Gallipoli (1915) to fight an army in which he had served nearly 40 years earlier. He served in London as a consulting surgeon at the Australian Imperial Force's headquarters. He finally returned to Australia (1919) with the honorary rank of Major-General. He was co-founder of the RAOU and its President (1905–1907). He wrote of his experiences in the Turkish army in *Under the Red Crescent* (1897) and kept his contact with Turkey, being Turkish Consul in Victoria. He died at sea.

Rymill

> Wedge Island Scrubwren *Sericornis maculatus rymilli* **S. A. White**, 1916 NCR
> [Alt. White-browed Scrubwren ssp.; JS *Sericornis frontalis mellori*]

Ernest S. Rymill and A. G. Rymill (fl.1930) were father-and-son South Australian yachtsmen. They invited the describer, Captain Samuel White (q.v.) to take part in a cruise aboard their yacht *Avocet*. White makes it clear in his original etymology that, despite the singular trinomial, the scrubwren is named after both men.

S

Saän

White-bellied Erpornis ssp. *Erpornis zantholeuca saani*
Chasen, 1939
[Alt: White-bellied Yuhina ssp.]
Black-naped Oriole ssp. *Oriolus chinensis saani* Jany, 1955
NCR
[NUI *Oriolus chinensis formosus*]

Mantri Saän (DNF) was an Indonesian collector and taxidermist at Buitenzorg (Bogor) Museum, Java (1934–1953).

Sabine, E.

Sabine's Gull *Xema sabini* **J. Sabine**, 1819
Sabine's Spinetail *Rhaphidura sabini* **J. E. Gray**, 1829
[Alt. Sabine's Spine-tailed Swift]
Sabine's Puffback *Dryoscopus sabini* J. E. Gray, 1831
[Alt. Large-billed Puffback]

General Sir Edward Sabine (1788–1883) combined a successful military career in the Royal Artillery for an incredible 74 years (1803–1877), with notable achievements in physics and astronomy, and being an explorer. He was elected a Fellow of the Royal Society (1818) and was its Treasurer (1850–1861) and President (1862–1871). His elder brother Joseph (q.v.) named the gull after Edward, who saw the bird on John Ross's expedition searching for a Northwest Passage (1818). (John Ross was the uncle of James Clark Ross q.v.). Sir Edward was the expedition's astronomer and geologist, and being a keen ornithologist he was also expected to write up the expedition's zoological findings. Ross and Sabine quarrelled, however, and the task fell to the ship's surgeon and assistant surgeon. It was then edited and corrected by William Leach (q.v.). Following the dispute, Joseph Sabine was quick to publish 'A memoir of the birds of Greenland' (1819) in the *Transactions of the Linnean Society*, before the official report could appear, and also wrote the official description of the gull himself.

Sabine, J.

Sabine's Snipe *Scolopax sabini* **Vigors**, 1825 NCR
[Based on a melanistic example of *Gallinago gallinago*]

Ruffed Grouse ssp. *Bonasa umbellus sabini* **D. Douglas**,
1829

Joseph Sabine (1770–1837) was an English lawyer and elder brother to Sir Edward Sabine (q.v.). He practised law, but then became Inspector General of Taxes (1808–1835). He had a life-long interest in natural history and was an original Fellow of the Linnean Society, Secretary of the Royal Horticultural Society (1810–1830) and vice-chairman of the Zoological Society of London. A Fellow of the Royal Society, he was both a gifted botanist and authority on British birds. He sold a collection of 2,000 birds to the museum in Glasgow. Douglas named the grouse as 'a tribute to the merits of my friend Joseph Sabine, Esq.' A tree is also named after him.

Sabino

Willis's Antbird ssp. *Cercomacra laeta sabinoi* **O. Pinto**, 1939

Dr F. Sabino (DNF) was a Brazilian biologist in São Bento, Pernambuco.

Saby

Saby's Helmeted Guineafowl *Numida meleagris sabyi*
Hartert, 1919
[Alt. Helmeted Guineafowl ssp.]

Lieutenant Paul Saby (DNF) was a French Inspector of Forests in Algeria (1905). He became a professional collector, taking the guineafowl holotype in Morocco for Rothschild (q.v.).

Sacerdotum

Flores Monarch *Monarcha sacerdotum* **Mees**, 1973
[Syn. *Symposiachrus sacerdotum*]

'*Sacerdotum*' means 'of the priests, in this case two Catholic missionaries to the island of Flores, Indonesia. Both men were botanists, naturalists, ornithologists and collectors. Father Jilis Antonius Josephus Verheijen was on Flores (1935–1993) except for a period of internment by the Japanese on Sulawesi in WW2. He retired to a monastery in Holland. Father Erwin Schmutz (b.1932) is still a missionary. He trained as a pharmacist and, having studied theology (1956–1962), went to Flores (1963). He co-wrote 'Living space of *Varanus (Odatria) t. timorensis* (Gray, 1931) (Sauria: Varanidae)' (1986) and has written on the birds of Indonesia. A reptile is named after him.

Sada

Bewick's Wren ssp. *Thryomanes bewickii sadai*
A. R. Phillips, 1986

Andrés Marcelo Sada Zambrano (b.1930) is a Mexican mechanical engineer, ornithologist and sound recordist who has supplied Mexican bird vocalisations to Cornell's Macaulay Library. He was one of the founders and a former President of Pronatura, established as Mexican Association Pro Conservation of Nature (1981). He is currently Chairman of the Board of the Mexican Fund for the Conservation of Nature. He co-wrote *Nombres en Castellano para las Aves Mexicanas* (1987) with Phillips (q.v.) and Ramos (q.v.).

Sadie Coats

Roraiman Flycatcher ssp. *Myiophobus roraimae sadiecoatsae* **Dickerman & W. H. Phelps Jr**, 1987

Sadie L. Coats (fl.1930–c.1990) was an American ornithologist. She was at the University of California, Berkeley (1979). She was an Associate in the Department of Ornithology, AMNH, New York. She co-wrote *The birds of Cerro Neblina, Territorio Federal Amazonas, Venezuela* (1991).

Safford

Micronesian Myzomela ssp. *Myzomela rubratra saffordi* **Wetmore**, 1917

Lieutenant Dr William Edwin Safford (1859–1926) was an anthropologist and botanist who served in the US Navy (1880–1900), during which time he was Lieutenant-Governor of Guam (1889–1900). His doctorate was awarded by George Washington University (1920). He was Assistant Botanist, US Department of Agriculture (1900–1924). He wrote 'Useful plants of the island of Guam' (1905).

Sagra

La Sagra's Flycatcher *Myiarchus sagrae* **Gundlach**, 1852

(See **La Sagra**)

Saint-Hilaire

Streamcreeper sp. *Furnarius sanctihilarii* **Lesson**, 1830 NCR
[Alt. Sharp-tailed Streamcreeper; JS *Lochmias nematura*]

Auguste François César Provençal de Saint-Hilaire (1779–1853) was a French naturalist and botanist who explored in Brazil (1816–1822). He was President of the French Academy of Sciences (1835).

Saisset

New Caledonian Parakeet *Cyanoramphus saisetti* **J. Verreaux & Des Murs**, 1860

Vice-Admiral Jean-Marie Joseph Théodore de Saisset (1810–1879) was Commandant of French Oceania in New Caledonia (1858–1860). He was a member of l'Assemblé Nationale for the Seine Départment (1870–1876) and was Commander-in-Chief of the National Guard in the Seine (1871).

Salim Ali

Salim Ali's Swift *Apus salimalii* **Lack**, 1958

Oriental White-eye ssp. *Zosterops palpebrosus salimalii* **Whistler**, 1933

Zitting Cisticola ssp. *Cisticola juncidis salimalii* **Whistler**, 1936

Sinai Rosefinch ssp. *Carpodacus synoicus salimalii* **Meinertzhagen**, 1938

Rock Bush Quail ssp. *Perdicula argoondah salimalii* **Whistler**, 1943

White-browed Scimitar Babbler ssp. *Pomatorhinus schisticeps salimalii* **Ripley**, 1948

Goldcrest ssp. *Regulus regulus salimalii* **Deignan**, 1954 NCR
[JS *Regulus regulus himalayensis*]

Emerald Dove ssp. *Chalcophaps indica salimalii* **Mukherjee**, 1960 NCR

[JS *Chalcophaps indica indica*]

Finn's Weaver ssp. *Ploceus megarhynchus salimalii* **Abdulali**, 1961

Dr Salim Moizuddin Abdul Ali (1896–1987), was an Indian ornithologist and conservationist. At the age of 10 he shot a sparrow that had a yellow streak below its neck. Plucking up his courage, he took it to the Bombay Natural History Society's offices and was seen by W. S. Millard (q.v.), who told him it was a Yellow-throated Sparrow *Petronia xanthocollis* and showed the young Salim Ali the society's collection of stuffed birds. That experience persuaded him to become an ornithologist. He moved to Burma (1919) to work in his family's timber business. When he returned he could not get a job at the Zoological Survey of India because he was only a Bachelor of Science and had no Doctorate. He went to Germany, where he studied under Stresemann (q.v.). He then spent 20 years wandering over India and becoming the foremost expert on the birds of the subcontinent. Despite Richard Meinertzhagen's (q.v.) well-known disdain for 'colonials', he and Salim Ali went on a number of expeditions together. Another of his collaborators was Sidney Dillon Ripley (q.v.) with whom he co-wrote *Handbook of the Birds of India and Pakistan* (10 volumes 1968–1975). His involvement with Ripley caused some extraordinary ideas that the CIA were involved in bird-ringing in India because Ripley had been an agent in the Office of Strategic Services (now the CIA) during WW2. At Indian independence, Salim Ali took over the Bombay Natural History Society and made sure that funds were found to prevent it from closing down. Founded in 1883, it had acquired the Bombay records of the Honourable East India Company from 1750s onward. His intervention was responsible for saving the Bharatpur Bird Sanctuary and the Silent Valley National Park. The Salim Ali Centre for Ornithology and Natural History at Coimbatore, (Tamil Nadu), was established in his name. A mammal is also named after him.

Salinas

Black Rail ssp. *Laterallus jamaicensis salinasi* **Philippi**, 1857

Eulogio Salinas Rengifo (1822–1878) was a Chilean businessman, ornithologist and collector.

Sallé

Sallé's Amazon *Amazona ventralis* P. L. S. Müller, 1776
[Alt. Hispaniolan Amazon; *Chrysotis sallaei* (Sclater, 1858) is a JS]

Sallé's Hermit *Phaethornis augusti* **Bourcier**, 1847
[Alt. Sooty-capped Hermit]

Grey-throated Chat *Granatellus sallaei* **Bonaparte**, 1856
Jacobin sp. *Florisuga sallei* **Hartert**, 1891 NCR
[Alt. White-necked Jacobin; JS *Florisuga mellivora*]

Gartered Trogon ssp. *Trogon caligatus sallaei* Bonaparte, 1856
Thicket Tinamou ssp. *Crypturellus cinnamomeus sallaei* Bonaparte, 1856
Sallé's Quail *Cyrtonyx montezumae sallei* **J. Verreaux**, 1859
[Alt. Montezuma Quail ssp.]

Auguste Sallé (1820–1896) was a French taxonomist and entomologist who collected in the US, West Indies, Venezuela and Central America including Mexico (1846–1856). His mother accompanied him on his expeditions. He became a very successful natural history dealer in Paris, specialising in insects. Many other taxa including two reptiles are named after him.

Salmon

Salmon's Jacamar *Brachygalba salmoni* **P. L. Sclater** & **Salvin**, 1879
[Alt. Dusky-backed Jacamar]
Scarlet-and-white Tanager *Chrysothlypis salmoni* P. L. Sclater, 1886

Salmon's Tiger Heron *Tigrisoma fasciatum salmoni* P. L. Sclater & Salvin, 1875
[Alt. Fasciated Tiger Heron ssp.]
Russet-backed Oropendola ssp. *Psarocolius angustifrons salmoni* Sclater, 1883
Checker-throated Antwren ssp. *Epinecrophylla fulviventris salmoni* **Chubb**, 1918
Grey-necked Wood Rail ssp. *Aramides cajanea salmoni* Chubb, 1918 NCR
[JS *Aramides cajanea cajanea*]

Colonel Thomas Knight Salmon (1840–1878) was a British mechanical engineer employed by a railway company in Guildford, England. He suffered from lung disease, forcing him to retire from the railways and opened a naturalist's shop. However, his health deteriorated further (1870) and he was advised to move to a better climate. He went to Medellin in Colombia, where he spent the next seven years working for the Republic of Colombia as an engineer and collecting zoological specimens. He was in England when he died, and his collection of about 3,500 bird skins formed the subject of a memoir by P. L. Sclater (q.v.) and Salvin (q.v.) that was read to a meeting of the Zoological Society of London (1878). Salmon also wrote vividly about his experiences.

Salomonsen

Tree Pipit ssp. *Anthus trivialis salomonseni* **Clancey**, 1950 NCR
[JS *Anthus trivialis trivialis*]
Salomonsen's Blue-naped Parrot *Tanygnathus lucionensis hybridus* Salomonsen, 1952
Salomonsen's Racket-tailed Parrot *Prioniturus discurus whiteheadi* Salomonsen, 1953
[Alt. Blue-crowned Racquet-tail ssp.]
Pygmy Flowerpecker ssp. *Dicaeum pygmaeum salomonseni* Parkes, 1962

Forest Rock Thrush ssp. *Monticola sharpei salomonseni* **Farkas**, 1973

Finn Salomonsen (1909–1983) was a Danish ornithologist and artist. He led a natural history expedition to north-west Greenland (1936) and assembled much of the ornithology collection of the Zoological Museum at the University of Copenhagen covering Denmark and the North Atlantic Dependencies. He was in the Philippines (1951–1952) and in the Bismarck Archipelago, Papua New Guinea (1962). He co-wrote and illustrated *Birds of Greenland* (1952) and *Øversigt øver Danmarks Fugle* (1963). A mammal is named after him.

Salt

Barbet sp. *Bucco saltii* **E. S. Stanley**, 1816 NCR
[Alt. Black-billed Barbet; JS *Lybius guifsobalito*]

Sir Henry Salt (1780–1827) was an explorer and diplomat who had originally been trained as a painter. He visited Egypt and India (1802–1806) and returned to Africa (1809) to attempt to establish contact with the King of Abyssinia on behalf of the British government, which took two years. Here he collected animals, including the barbet holotype, which on his return were sent to Stanley and so to the Liverpool Museum. He was British Consul General in Alexandria (1815–1827), during which time he accumulated antiquities that he bequeathed to the British Museum. He carried out major excavations at Giza investigating the Great Pyramid, and employed Caviglia to excavate the Sphinx, but the two men fell out because Caviglia kept looking for mummy pits. A mammal is also named after him.

Salvadori

Salvadori's Antwren *Myrmotherula minor* Salvadori, 1864
African Spotted Creeper *Salpornis salvadori* **Bocage**, 1878
[Syn. *Salpornis spilonotus salvadori*]
Salvadori's Nightjar *Caprimulgus pulchellus* Salvadori, 1879
Salvadori's Pheasant *Lophura inornata* Salvadori, 1879
Salvadori's Fig Parrot *Psittaculirostris salvadorii* **Oustalet**, 1880
Bristle-crowned Starling *Onychognathus salvadorii* **Sharpe**, 1891
Salvadori's Conure *Pyrrhura emma* Salvadori, 1891
[Alt. Venezuelan Parakeet; Syn. *Pyrrhura leucotis emma*]
Salvadori's Eremomela *Eremomela salvadorii* **Reichenow**, 1891
Salvadori's Crimsonwing *Cryptospiza salvadorii* Reichenow, 1892
[Alt. Abyssinian Crimsonwing]
Parakeet sp. *Palaeornis salvadorii* Oustalet, 1893 NCR
[Alt. Derbyan Parakeet; JS *Psittacula derbiana*]
Enggano White-eye *Zosterops salvadorii* **A. B. Meyer** & **Wiglesworth**, 1894
Palm Cockatoo *Microglossus salvadorii* A. B. Meyer, 1894 NCR
[JS *Probosciger aterrimus goliath*]
Salvadori's Duck *Salvadorina waigiuensis* **Rothschild** & **Hartert**, 1894
[Alt. Salvadori's Teal]

Nias Serpent Eagle *Spilornis salvadorii* **Berlepsch**, 1895
NCR
[JS *Spilornis cheela asturinus*]
Salvadori's Seedeater *Serinus xantholaemus* Salvadori,
1896
[Alt. Salvadori's Serin; Syn. *Crithagra xantholaema*]
Salvadori's Weaver *Ploceus dichrocephalus* Salvadori, 1896
[Alt. Juba Weaver]
Bracken Warbler sp. *Bradypterus salvadorii* **Neumann**,
1900 NCR
[Alt. Cinnamon Bracken Warbler; JS *Bradypterus
cinnamomeus*]
Mouse-warbler sp. *Sericornis salvadorii* Reichenow, 1901
NCR
[Alt. Mountain Mouse-warbler; JS *Crateroscelis robusta*]
Salvadori's Hanging Parrot *Loriculus salvadorii* **Hachisuka**,
1930 NCR
[Alt. Philippine Hanging Parrot; JS *Loriculus philippensis
apicalis*]

Afghan Babbler ssp. *Turdoides huttoni salvadorii* **De Filippi**,
1865
Large-tailed Nightjar ssp. *Caprimulgus macrurus salvadorii*
Sharpe, 1875
Salvadori's Orange-breasted Fig Parrot *Cyclopsitta
gulielmitertii fuscifrons* Salvadori, 1876
Australasian Figbird ssp. *Sphecotheres vieilloti salvadorii*
Sharpe, 1877
Black Sunbird ssp. *Leptocoma sericea salvadorii*
G. E. Shelley, 1877
Salvadori's Black-capped Lory *Lorius lory erythrothorax*
Salvadori, 1877
Sulawesi Cicadabird ssp. *Coracina morio salvadorii* Sharpe,
1878
Buff-breasted Paradise Kingfisher ssp. *Tanysiptera sylvia
salvadoriana* **E. P. Ramsay**, 1879
Salvadori's King Parrot *Alisterus chloropterus callopterus*
D'Albertis & Salvadori, 1879
[Alt. Papuan King Parrot ssp.]
Pinon's Imperial Pigeon ssp. *Ducula pinon salvadorii*
Tristram, 1882
Spectacled Bulbul ssp. *Pycnonotus erythropthalmos
salvadorii* Sharpe, 1882 NCR; NRM
Scaly-breasted Honeyeater ssp. *Lichmera squamata
salvadorii* Meyer, 1884 NCR; NRM
Blue-tailed Bee-eater ssp. *Merops philippinus salvadorii*
Meyer, 1891
Salvadori's Blue-winged Parrotlet *Forpus xanthopterygius
flavescens* Salvadori, 1891
Salvadori's (Black-capped) Lory *Lorius lory salvadorii*
A. B. Meyer, 1891
Claret-breasted Fruit Dove ssp. *Ptilinopus viridis salvadorii*
Rothschild, 1892
Mountain Owlet-nightjar ssp. *Aegotheles albertisi salvadorii*
Hartert, 1892
Black-throated Honeyeater ssp. *Lichenostomus subfrenatus
salvadorii* Hartert, 1896
Salvadori's Blue-naped Parrot *Tanygnathus lucionensis
salvadorii* **Ogilvie-Grant**, 1896
African Green Pigeon ssp. *Treron calvus salvadorii*
A. Dubois, 1897

Rusty Whistler ssp. *Pachycephala hyperythra salvadorii*
Rothschild, 1897
Pacific Koel ssp. *Eudynamys orientalis salvadorii* Hartert,
1900
Slaty Robin ssp. *Poecilodryas cyanus salvadorii* Rothschild
& Hartert, 1900 NCR
[JS *Peneothello cyanus subcyanea*]
Salvadori's (Buff-faced) Pygmy Parrot *Micropsitta pusio
salvadorii* Rothschild & Hartert, 1901 NCR
[NUI *Micropsitta pusio beccarii*]
Banded Barbet ssp. *Lybius undatus salvadorii* Neumann,
1903
Darwin's Nothura ssp. *Nothura darwinii salvadorii* Hartert,
1909
Zenaida Dove ssp. *Zenaida aurita salvadorii* **Ridgway**, 1916
Little Bronze Cuckoo ssp. *Chrysococcyx minutillus salvadorii*
Hartert & Stresemann, 1925
Variable Pitohui ssp. *Pitohui kirhocephalus salvadorii*
Meise, 1929 NCR
[Intergrade between *Pitohui kirhocephalus kirhocephalus*
and *P. kirhocephalus dohertyi*]
Salvadori's Bird-of-Paradise *Paradisaea raggiana salvadorii*
Mayr & **Rand**, 1935
[Alt. Raggiana Bird-of-Paradise ssp.]

Conte Adelardo Tommaso Paleotti Salvadori (1835–1923) was
an eminent Italian physician, author, educator and ornithologist. He was Vice-Director of the Museum of Zoology at the
University of Turin (1879–1923). He was also Medical Officer
in Garibaldi's battalion during his second 'expedition' in
Sicily. His collection was donated to the Natural Science
Museum of Villa Vitali. He wrote *Catalogo Sistimatico degli
Uccelli di Borneo* (1874) and *Ornitologia della Papuasia e
delle Molucche* (1880). A reptile is also named after him. (See
also **Tommaso**)

Salvago-Raggi

Pel's Fishing Owl ssp. *Scotopelia peli salvago-raggii* **Zelditz**,
1908 NCR; NRM

Giuseppe Marchese Salvago-Raggi (1866–1946) was an
Italian diplomat who was Ambassador to China (1897–1901)
during the Boxer Rebellion and Governor-General of Italian
Somaliland (Somalia) (1906–1907) and of Eritrea (1907–1915).
He was Ambassador to France (1916–1917). He wrote of his
travels in *Lettere dall'Oriente*, which is still in print.

Salvin

Prion genus *Salviprion* **Mathews** & **Hallstrom**, 1943 NCR
[Now in *Pachyptila*]

Salvin's Emerald *Chlorostilbon salvini* **Cabanis** & **Heine**,
1860
[Alt. Canivet's Emerald; Syn. *Chlorostilbon canivetii salvini*]
Salvin's Antbird *Sipia laemosticta* Salvin, 1865
[Alt. Dull-mantled Antbird; Syn. *Myrmeciza laemosticta*]
Salvin's Negrito *Lessonia oreas* **P. L. Sclater** & Salvin, 1869
[Alt. Andean Negrito]
Salvin's Wren *Thryothorus semibadius* Salvin, 1870
[Alt. Riverside Wren]

Salvin's Rusty-faced Parrot *Hapalopsittaca pyrrhops* Salvin, 1876
[Alt. Red-faced Parrot]
Salvin's Silky Flycatcher *Phainoptila melanoxantha* Salvin, 1877
[Alt. Black-and-yellow Silky Flycatcher]
Tumbes Tyrant *Tumbezia salvini* **Taczanowski**, 1877
Salvin's (Razor-billed) Curassow *Mitu salvini* **Reinhardt**, 1879
Brilliant sp. *Xanthogenyx salvini* D'Hamonville, 1883 NCR
[Alt. Velvet-browed Brilliant; JS *Heliodoxa xanthogonys*]
Salvin's Petrel *Pterodroma hypoleuca* Salvin, 1888
[Alt. Bonin Petrel]
Salvin's Chuck-will *Caprimulgus salvini* **Hartert**, 1892
[Alt. Tawny-collared Nightjar]
Salvin's Albatross *Thalassarche salvini* **Rothschild**, 1893
[Alt. Salvin's Mollymawk]
Salvin's Piculet *Picumnus salvini* **Hargitt**, 1893 NCR
[Now considered to be a synonym of *Picumnus squamulatus obsoletus*]
Salvin's Pigeon *Patagioenas oenops* Salvin, 1895
[Alt. Maranon Pigeon, Peruvian Pigeon]
Salvin's Screech Owl *Megascops ingens* Salvin, 1897
[Alt. Rufescent Screech Owl]
White-throated Antbird *Gymnopithys salvini* **Berlepsch**, 1901
Salvin's Prion *Pachyptila salvini* Mathews, 1912
[Alt. Medium-billed Prion]

Rufous-vented Ground Cuckoo ssp. *Neomorphus geoffroyi salvini* P. L. Sclater, 1866
Yellow-tailed Oriole ssp. *Icterus mesomelas salvinii* **Cassin**, 1867
White-tipped Sicklebill ssp. *Eutoxeres aquila salvini* **Gould**, 1868
Salvin's Ant Tanager *Habia fuscicauda salvini* Berlepsch, 1883
[Alt. Red-throated Ant Tanager ssp.]
Thick-billed Seed Finch ssp. *Oryzoborus funereus salvini* **Ridgway**, 1884
Salvin's Spindalis *Spindalis zena salvini* **Cory**, 1886
[Alt. Western Spindalis ssp.]
Rufous-tailed Tyrant ssp. *Knipolegus poecilurus salvini* P. L. Sclater, 1888
Salvin's Amazon *Amazona autumnalis salvini* **Salvadori**, 1891
[Alt. Red-lored Amazon ssp.]
Salvin's Barbet *Eubucco bourcierii salvini* **G. E. Shelley**, 1891
[Alt. Red-headed Barbet ssp.]
Salvin's Warbler *Basileuterus rufifrons salvini* **Cherrie**, 1891
[Alt. Rufous-capped Warbler ssp.]
Small Tree-Finch ssp. *Camarhynchus parvulus salvini* Ridgway, 1894
Variegated Tinamou ssp. *Crypturellus variegatus salvini* Salvadori, 1895 NCR; NRM
Northern Bobwhite ssp. *Colinus virginianus salvini* **E. W. Nelson**, 1897
Black-and-white Becard ssp. *Pachyramphus albogriseus salvini* **Richmond**, 1899

Scaly-throated Leaftosser ssp. *Sclerurus guatemalensis salvini* Salvadori & **Festa**, 1899
Amethyst-throated Hummingbird ssp. *Lampornis amethystinus salvini* Ridgway, 1908
Green Thorntail ssp. *Popelairia conversii salvini* **Zeledon**, 1911 NCR
[Syn *Discosura conversii* NRM]
Greenish Yellow Finch ssp. *Sicalis olivascens salvini* **Chubb**, 1919
Maroon-chested Ground Dove ssp. *Claravis mondetoura salvini* **Griscom**, 1930
Mangrove Vireo ssp. *Vireo pallens salvini* **Van Rossem**, 1934

Osbert Salvin (1835–1898) was an English naturalist who became a Fellow of the Royal Society. He was a life-long friend of Godman (q.v.) from their Cambridge University days, where Salvin studied mathematics. He wrote (1861) that he was 'determined, rain or no rain, to be off to the mountain forests in search of quetzals, to see and shoot which has been a daydream for me ever since I set foot in Central America.' He was the first European to record observing a Resplendent Quetzal *Pharomacrus mocinno*, pronouncing it 'unequalled for splendour among the birds of the New World' – and promptly shot it. During the course of the next three decades, thousands of quetzal plumes crossed the Atlantic to fill the specimen cabinets of European collectors and adorn the fashionable milliners' shops of Paris, Amsterdam and London. Salvin redeemed himself by co-authoring with Godman the incredible 40-volume *Biologia Centrali Americana* (1879), which provided a near-complete catalogue of Middle American species. Earlier he had written *Exotic Ornithology* (1866) and *Nomenclatur Avium Neotropicalium* (1873). Salvin and Godman combined their bird collections and presented them over a 15-year period (1885–1900) to the BMNH. The Godman-Salvin Medal, a prestigious award of the British Ornithologists' Union, is named after them. A amphibian, two mammals and four reptiles are also named after him. (Also see **Osbert**)

Samuel

Quail-thrush genus *Samuela* **Mathews**, 1912 NCR
[Now in *Cinclosoma*]

Beautiful Firetail ssp. *Stagonopleura bella samueli* Mathews, 1912
[*S. bella* sometimes regarded as monotypic]
Purple-gaped Honeyeater ssp. *Ptilotis cratitia samueli* Mathews, 1912 NCR
[JS *Lichenostomus cratitius occidentalis*]
Scarlet Robin ssp. *Petroica multicolor samueli* Mathews, 1912 NCR
[JS *Petroica boodang boodang*]
Superb Fairy-wren ssp. *Malurus cyaneus samueli* Mathews, 1912
Yellow-faced Honeyeater ssp. *Lichenostomus chrysops samueli* Mathews, 1912
Brown Thornbill ssp. *Acanthiza pusilla samueli* Mathews, 1913 NCR
[JS *Acanthiza pusilla pusilla*]
Cinnamon Quail-thrush ssp. *Cinclosoma cinnamomeum samueli* Mathews, 1916 NCR
[NUI *Cinclosoma cinnamomeum cinnamomeum*]

Red-tailed Black Cockatoo ssp. *Calyptorhynchus banksii samueli* Mathews, 1917

(See **White, S. A.**)

Samuels

Song Sparrow ssp. *Melospiza melodia samuelis* **Baird**, 1858

Emanuel Samuels (1816–1886) was an American collector. He went to California (1855) to collect for the USNM, the Boston Society of Natural History, and the Academy of Natural Sciences, Philadelphia.

Sam Veasna

Mekong Wagtail *Motacilla samveasnae* Duckworth *et al.*, 2001

Sam Veasna (1966–1999) was one of Cambodia's leading ornithologists and conservationists. He was head of Siem Reap Provincial Wildlife Office where the Sam Veasna Center for Wildlife Conservation was established in his honour. He discovered populations of Sarus Crane *Grus antigone* and Bengal Florican *Houbaropsis bengalensis* in Cambodia (where they were thought extirpated). He died of malaria contracted while undertaking fieldwork, working closely with village people promoting conservation.

San Andrés

San Andrés Vireo *Vireo caribaeus* **Bond** & **Meyer de Schauensee**, 1942

San Andrés is a Colombian island in the western Caribbean, where the bird is endemic and after which it was named.

Sanborn

Chilean Tinamou ssp. *Nothoprocta perdicaria sanborni* **Conover**, 1924

Rufous-collared Sparrow ssp. *Zonotrichia capensis sanborni* **Hellmayr**, 1932

Rufous-legged Owl ssp. *Strix rufipes sanborni* Wheeler, 1938

Dr Colin Campbell Sanborn (1897–1962) was a biologist interested in both birds and mammals. He was Assistant Curator, Ornithology Section, Field Museum (Chicago) when Wilfred Osgood (q.v.) was overall Curator of Zoology. Sanborn subsequently became Curator, Mammals. He wrote *Birds of the Chicago Region* (1934). Six mammals and two amphibians are named after him.

Sanchez, C.

Tamaulipas Pygmy Owl *Glaucidium sanchezi* **Lowery** & **R. J. Newman**, 1949

Carlos Sanchez Mejurado (DNF) was a Mexican naturalist from San Luis Potosi.

Sanchez, M. & M.

Varzea Thrush *Turdus sanchezorum* **O'Neill**, Lane & Naka, 2011

Manuel Sánchez S. and his wife, Marta Chávez de Sánchez, are long-term friends and field companions of O'Neill (q.v.)

and have worked with many researchers in Peru, Bolivia and Venezuela since the 1960s.

Sander

Buff-throated Apalis ssp. *Apalis rufogularis sanderi* **Serle**, 1951

Frederick Sander (b.1906) was a British civil servant employed overseas as chief accountant with Nigerian Railways.

Sandground

Black-fronted Bush-shrike ssp. *Chlorophoneus nigrifrons sandgroundi* **Bangs**, 1931

Dr John 'Jack' Henry Sandground (1899–1976) was an American parasitologist and helminthologist. He worked as a Curator at the Department of Tropical Medicine, Harvard Medical School (1925–1928 and 1930s). He researched in Southern Rhodesia (Zimbabwe) (1929–1930).

Sanford

White-eye genus *Sanfordia* **R. C. Murphy** & **Mathews**, 1929 NCR
[NUI *Woodfordia*]

Northern Royal Albatross *Diomedea sanfordi* R. C. Murphy, 1917

Sanford's White-eye *Woodfordia lacertosa* R. C. Murphy & Mathews, 1929

Long-billed White-eye *Rhamphozosterops sanfordi* **Mayr**, 1931 NCR
[JS *Rukia longirostra*]

Sanford's Niltava *Cyornis sanfordi* **Stresemann**, 1931
[Alt. Matinan Blue Flycatcher]

Sanford's Sea Eagle *Haliaeetus sanfordi* Mayr, 1935
[Alt. Solomons Sea Eagle]

Sanford's Ptarmigan *Lagopus muta sanfordi* **Bent**, 1912
[Alt. Rock Ptarmigan ssp.]

Sanford's Elf Owl *Micrathene whitneyi sanfordi* **Ridgway**, 1914

Short-eared Owl ssp. *Asio flammeus sanfordi* **Bangs**, 1919

Long-tailed Myna ssp. *Mino kreffti sanfordi* **Hartert**, 1929

Mountain Mouse-warbler ssp. *Crateroscelis robusta sanfordi* Hartert, 1930

Blue-faced Parrotfinch ssp. *Erythrura trichroa sanfordi* Stresemann, 1931

Cardinal Honeyeater ssp. *Myzomela cardinalis sanfordi* Mayr, 1931

Oriole Whistler ssp. *Pachycephala orioloides sanfordi* Mayr, 1931

Long-billed Cuckoo ssp. *Rhamphomantis megarhynchus sanfordi* Stresemann & Paludan, 1932
[Syn. *Chrysococcyx megarhynchus sanfordi*; validity of subspecies questionable]

Sulawesi Hornbill ssp. *Penelopides exarhatus sanfordi* Stresemann, 1932

Sanford's Bowerbird *Archboldia papuensis sanfordi* Mayr & **Gilliard**, 1950
[Alt. Archbold's Bowerbird ssp.]

Dr Leonard Cutler Sanford (1868–1950) was an American surgeon and amateur zoologist. He was a trustee of the AMNH, and acquired many bird specimens for its collection. He wrote *The Water-fowl Family* (1924) with L. B. Bishop and T. S. van Dyke. A mammal and a reptile are named after him.

Santiago

White-tipped Dove ssp. *Leptotila verreauxi santiago*
van Rossem & **Hachisuka**, 1937 NCR
[JS *Leptotila verreauxi angelica*]

James 'Don Santiago' McCarty (d.1937) was the owner of Guirocoba (Buzzard Head) Ranch in Sonora, Mexico. He was described as being the 'friend of every naturalist who has worked in southeastern Sonora in recent years'.

Saphiro

Plain-backed Pipit ssp. *Anthus leucophrys saphiroi*
Neumann, 1906
[Sometimes placed with Buffy Pipit, as *Anthus vaalensis saphiroi*]

Philip Photious Constantine Zaphiro (1877–1933) was a collector employed by W. N. Macmillan (q.v.) during his Sudan expedition (1903–1904). He was born in Constantinople (Istanbul). He went to Ethiopia as a medical dispenser, before working (1904–1911) for the British East Africa Protectorate on the Kenyan/Ethiopian border. He was employed at the British Legation (1909–1919) as interpreter for the minister, Wilfred Thesiger, whose son, also Wilfred, was a famous explorer and travel writer. Zaphiro became Vice-Consul (1915), then Oriental Secretary (1921), working at the British Embassy in Addis Ababa until his death. A mammal is named after him. (See also **Zaphiro**)

Saposhnikoff

Black-headed Penduline-tit ssp. *Remiz macronyx ssaposhnikowi* **Johansen**, 1907

Professor V. V. Saposhnikoff (1861–1924) of Tomsk University was in Turkestan (1902).

Sappho

Red-tailed Comet genus *Sappho* **Reichenbach**, 1849

Sappho (d.c.570 BC) was a Greek lyric poetess who lived on the island of Lesbos. Little is known for certain about her life, and most of her poetry, much admired throughout antiquity, has been lost.

Sapsworth

White-throated Dipper ssp. *Cinclus cinclus sapsworthi*
Arrigoni, 1902 NCR
[JS *Cinclus cinclus cinclus*]

Arnold Duer Sapsworth (1872–1957) was a British businessman, landowner, ornithologist and collector. He was in Corsica in the early years of the 20th century.

Sarasin

White-eye sp. *Zosterops sarasinorum* **A. B. Meyer** & **Wiglesworth**, 1894 NCR
[Alt. Mountain White-eye; JS *Zosterops montanus montanus*]
Sarasins' Myza *Myza sarasinorum* Meyer & Wiglesworth, 1895
[Alt. Greater Streaked Honeyeater, White-eared Myza]
Sulawesi Leaf Warbler *Phylloscopus sarasinorum* Meyer & Wiglesworth, 1896

Lesser Coucal ssp. *Centropus bengalensis sarasinorum* **Stresemann**, 1912
Whistling Kite ssp. *Haliastur sphenurus sarasini* **Mathews**, 1916 NCR; NRM

Paul Benedikt Sarasin (1856–1929) and Karl Friedrich 'Fritz' Sarasin (1859–1942) were cousins. They were Swiss zoologists, explorers and collectors and they also wrote together: *Reisen in Celebes* (1905) and *Die Vögel Neu-Caledoniens und der Loyalty Inseln* (1913). Among other taxa five reptiles, an amphibian and a mammal are named after them.

Sardar Patel

Baya Weaver ssp. *Ploceus philippinus sardarpateli* **Koelz**, 1952 NCR
[JS *Ploceus philippinus philippinus*]

Sardar Vallabhbhai Jhaverbhai Patel (1875–1950) was one of the leaders of the Indian National Congress, and Deputy Prime Minister of India (1948–1950). ('Sardar' is a title, meaning 'chief', not a forename.)

Sartori

Mexican Barred Owl *Strix sartorii* **Ridgway**, 1874
[Syn. *Ciccaba sartorii, Strix varia sartorii*]

Dr Carl Christian Wilhelm Sartorius (1796–1872) was a German writer and naturalist. Educated at the University of Giessen, Germany, he faced persecution because of his liberal political views. He emigrated to Mexico (1824), where he managed a silver mine. Later he established a hacienda called El Mirador, which became a magnet for visiting German scientists, especially botanists. Sartorius himself seems to have collected everything and anything and is mentioned in the literature in connection with, among other topics, botany, herpetology and ornithology. A mammal and a reptile are named after him.

Sarudny

Black Francolin ssp. *Francolinus orientalis sarudnyi*
Buturlin, 1907 NCR
[JS *Francolinus francolinus francolinus*]
European Nightjar ssp. *Caprimulgus europaeus sarudnyi*
Hartert, 1912
Eurasian Wryneck ssp. *Jynx torquilla sarudnyi* **Loudon**, 1912
Mourning Wheatear ssp. *Oenanthe lugens sarudnyi* **Harms**, 1926 NCR
[JS *Oenanthe lugens persica*]
Spotted Flycatcher ssp. *Muscicapa striata sarudnyi*
Snigirewski, 1928

(See **Zarudny**)

Sassi

Sassi's Olive Greenbul *Phyllastrephus lorenzi* Sassi, 1914
[Alt. Lorenz's Bulbul/Greenbul]

Yellow-crowned Canary ssp. *Serinus canicollis sassii*
Neumann, 1922
[Syn. *Serinus flavivertex sassii*]
Little Sparrowhawk ssp. *Accipiter minullus sassii*
Stresemann, 1924 NCR
[JS *Accipiter erythropus zenkeri*]
White-winged Widowbird ssp. *Euplectes albonotatus sassii*
Neunzig, 1928 NCR
[NUI *Euplectes albonotatus eques*]
Spotted Shag ssp. *Sticticarbo punctatus sassi* **Mathews**,
1929 NCR
[JS *Phalacrocorax punctatus punctatus*]
Spotted Nutcracker ssp. *Nucifraga caryocatactes sassii*
Keve, 1943 NCR
[JS *Nucifraga caryocatactes macrorhynchos*]

Dr Moriz Sassi (1880–1967) was an Austrian zoologist, ornithologist and curator of the bird collection at the Natural History Museum in Vienna (1915–1940 and 1946–1949). He wrote *Eine Sudanreise* (1911) about an expedition he took part in.

Satunin

Common Starling ssp. *Sturnus poltaratskyi satunini*
Buturlin, 1904 NCR
[JS *Sturnus vulgaris caucasicus*]
Eurasian Blue Tit ssp. *Cyanistes caeruleus satunini*
Zarudny, 1908

Konstantin Alexeevitsch Satunin (1863–1915) was an eminent Russian zoologist who studied the fauna of the Caucasus region. Like his predecessor Radde (q.v.), he was initially most interested in birds, e.g. writing *A Systematic Catalogue of the Birds of the Caucasian Region* (1912). However, he was an all-rounder, collecting fish, insects and mammals and writing on an extensive range of topics. He wrote 'New mammals from Transcaucasia' (1914) and 'Mammals of the Caucasian Region' (1915). A mammal is named after him.

Saucerotte

Hummingbird genus *Saucerottia* **Bonaparte**, 1850
[Often merged into *Amazilia*, or regarded as a subgenus of latter]

Steely-vented Hummingbird *Amazilia saucerrottei* **de Lattre**
& Bourcier, 1846

Dr Antoine Constant Saucerotte (1805–1884) was a French physician, naturalist and collector. He wrote *Histoire Critique de La Doctrine Physiologique* (1847).

Saul, G.

Blue-headed Fantail ssp. *Rhipidura cyaniceps sauli* **Bourns**
& Worcester, 1894

George Medhurst Saul (b.1851) was a British-born American resident in the Philippines who worked for Hoskyn and Company (the first department store in the Philippines) and

who befriended the describers when they became stranded in Ilo Ilo. He lent them a house and use of servants in Guimaras, where they stayed for a month.

Saul, J.

Mountain Velvetbreast ssp. *Lafresnaya lafresnayi saul* **de**
Lattre & Bourcier, 1846

Miss Jane Saul (1807–1895) was a British conchologist and collector who lived in Limehouse, London (1846).

Saunders, G.

Eastern Meadowlark ssp. *Sturnella magna saundersi*
Dickerman & A. R. Phillips, 1970

Dr George B. Saunders (1907–2001) was an American ornithologist. He took his first degree at Oklahoma University, then his PhD at Cornell (1932). His doctoral dissertation was on the taxonomy of *Sturnella* and in their etymology the authors thank Saunders '... who has contributed much to our knowledge of *Sturnella*, and to whom we are indebted for helpful information on the more northern subspecies.' He went on an expedition to South Africa (1930) as a Research Fellow of the National Research Council and as Staff Ornithologist at the Michigan Department of Conservation. He joined the US Fish & Wildlife Service (1937) with which he spent the rest of his working life. He undertook several other fieldwork assignments, notably to Guatemala (1947–1948) as well as a number of trips to Mexico and throughout the US and Canada. He once said 'I believe that wildlife biologists are among the happiest people to be found, especially if they have a wife who is also a biologist and shares their adventures in the field'.

Saunders, H.

Gull genus *Saundersia* **Dwight**, 1925 NCR
[Name unavailable, replaced by *Saundersilarus*]
Gull genus *Saundersilarus* Dwight, 1926
[SII *Chroicocephalus*]

Saunders's Gull *Chroicocephalus saunderi* **Swinhoe**, 1871
[Alt. Chinese Black-headed Gull; Syn. *Saundersilarus saundersi*]
Saunders's Tern *Sternula saundersi* **Hume**, 1877
[Alt. Black-shafted Tern]

European Green Woodpecker ssp. *Picus viridis saundersi*
Taczanowski, 1878 NCR
[JS *Picus viridis karelini*]

Howard Saunders (1835–1907) was a British ornithologist who worked at the BMNH. He was the 19th century's foremost expert on gulls and terns, writing *Sternæ* and *An Illustrated Manual of British Birds* (1889). He co-wrote *Catalogue of the Gaviae and Tubinares in the Collection of the British Museum* (1896) with Salvin (q.v.). His contemporary, Swinhoe (q.v.), named the gull when Saunders was completing a study of that family. Saunders applied strict standards to sight records, at a time when field identification was in its infancy and optical aids were inferior to modern equipment. He rejected his own sighting of a Masked Shrike

Lanius nubicus near Gibraltar (1863) writing 'as I am frequently sceptical of other people's identifications … I do not want anyone to accept mine until the bird can be produced as proof.'

Sauzier

Sauzier's Owl *Mascarenotus sauzieri* **E. Newton & Gadow**, 1893 EXTINCT
[Alt. Mauritius Owl, Commerson's Scops Owl]
Sauzier's Teal *Anas theodori* E. Newton & Gadow, 1893 EXTINCT
[Alt. Mauritius Duck, Mascarene Teal]

Théodore Sauzier (1829–1904) was a notary public, palaeontologist and naturalist who was born on Réunion and lived and practised in Paris, where he was a close associate of Alphonse Milne-Edwards (q.v.). He visited Mauritius several times and dug for bones in Mare aux Songes including when Sir Edward Newton (q.v.) was Colonial Secretary of Mauritius, to whom he presented a number of bones (1889). He co-wrote *Un Projet de Republique A l'ile D'Eden, L'Ile Bourbon, en 1689* (1887).

Savés

New Caledonian Owlet-nightjar *Aegotheles savesi*
E. L. Layard & E. L. C. Layard, 1881

Théodore Savés (1855–1918) was a French naturalist who travelled in New Caledonia. He collected botanical specimens on that island (1870s and 1880s).

Savi

Savi's Warbler *Locustella luscinioides* Savi, 1824

Paolo Savi (1798–1871) was an Italian naturalist, zoologist, palaeontologist and geologist. He studied physics and natural science at Pisa University, becoming Professor of Natural History there, and also Director of the Museum. He became an Italian senator (1862). His greatest work, *Ornitologia Italiana* was published posthumously (1873–1876). Three mammals are named after him.

Savigny

Savigny's Eagle Owl *Bubo ascalaphus* Savigny, 1809
[Alt. Desert/Pharaoh Eagle Owl]
Black-capped Bee-eater *Merops savignii* Audouin, 1825 NCR
[Alt. White-throated Bee-eater; JS *Merops albicollis*]

Barn Swallow ssp. *Hirundo rustica savignii* **J. F. Stephens**, 1817

Marie Jules César Lelorgne de Savigny (1777–1851) was a French zoologist and artist. He studied medicine but under Geoffroy Saint-Hilaire's (q.v.) influence turned to zoology. He was in Egypt (1798–1800) and undertook several expeditions, including one studying the avifauna of Lake Manzala, sending specimens to Saint-Hilaire. He wrote *Description d'Egypte; ou Recueil des Observations et des Recherches qui ont été Faites en Egypte Pendant l'Expédition de l'Armée Française* (1798–1801), *Histoire Naturelle et Mythologique de*

l'Ibis (1805) and *Système des Oiseaux de l'Égypte et de la Syrie* (1810), derived from research conducted during Napoleon's occupation of Egypt (1790s). He illustrated most of the birds in his works, although some are by Barraband (q.v.). Savigny fell out with Saint-Hilaire, who blocked him from becoming Professor at the Natural History Museum. He became ill (1823) and spent all his latter years in poor health. Two reptiles and an amphibian are named after him.

Savile

Savile's Bustard *Lophotis savilei* **Lynes**, 1920
[Alt. Lynes's Pygmy Bustard]

Lieutenant-Colonel Robert Vesey Savile (1873–1947) was a British soldier and diplomat who served in Sudan (1901), becoming a provincial governor (1917–1923). He was known as 'Savile Pasha'.

Savorgnan

Coucal sp. *Centropus savorgnani* **Oustalet**, 1886 NCR
[Alt. Blue-headed Coucal; JS *Centropus monachus*]

Count Pierre Paul François Camille Savorgnan de Brazza (1852–1905) was the elder brother of Jacques (or Giacomo) C. Savorgnan de Brazza (q.v.). The brothers were born in Rio de Janeiro of Italian descent. Count Pierre was a distinguished explorer who entered the French navy (1870) and served in Gabon. The French government gave him 100,000 francs for exploring the country north of the Congo (1878), where he secured vast grants of land for France and founded stations including that of Brazzaville on the north shore of Stanley Pool. He returned (1883) largely unsubsidised by the French government, and established 26 stations (1886) and continued to explore (1886–1897). He was Commissioner-General of the French Congo (1896–1898).

Sawitzky

Barn Swallow ssp. *Hirundo rustica sawitzkii* **Loudon**, 1904 NCR
[JS *Hirundo rustica rustica*]

William (Vaseli) Sawitzky (1879–1947) was a Latvian-born Polish ornithologist who collected in the Baltic States and Turkestan (1903). He went to the US as a newspaper correspondent (1911), stayed on and entered the field of art scholarship (1913) as a librarian. He wrote 'Die Vogelwelt der Stadt Riga und Umgegend' (1899).

Sawtell

Sawtell's Swiftlet *Aerodramus sawtelli* Holyoak, 1974
[Alt. Atiu Swiftlet, Cook Islands Swiftlet]

Gordon Henry Sawtell (1929–2010) was a British-born civil servant educated in Britain and New Zealand. Following military service he worked as a civil servant in New Zealand and then Cook Islands, becoming Secretary to the Prime Minister. He retired (1982) on medical grounds and stayed in the Cook Islands. He was instrumental in the swiftlet's discovery (1973) in Annatake-take cave. The cave is also inhabited by many land crabs, which despatch any nestlings that fall from their nests. Sawtell told us (August 2002) 'I use the name Kopeka on my personalised motor vehicle registration plate, Kopeka being the Polynesian name for "my" bird'. Sawtell made arrangements for Holyoak to visit all of the islands and to view the caves (1973), as he knew of the swiftlets. He was surprised when he found out that Holyoak had named the swiftlet after him.

Say

Phoebe genus *Sayornis* **Bonaparte**, 1854

Say's Phoebe *Sayornis saya* Bonaparte, 1825

Thomas Say (1787–1834) was a self-taught American naturalist whose primary interest was entomology. He described over 1,000 new species of beetles and over 400 new insects of other orders. He became a charter member and founder of the Academy of Natural Sciences of Philadelphia (1812) and was appointed chief zoologist with Major Stephen H. Long's expeditions, which explored the Rocky Mountains (1819–1820). He lived at the utopian village of 'New Harmony' in Indiana (1826–1834). Say wrote *American Entomology, or Descriptions of the Insects of North America* (1824–1828) and *American Conchology* (1830–1834). A mammal and a reptile are named after him.

Scarlett

Scarlett's Duck *Malacorhynchus scarletti* **Olson**, 1977 EXTINCT
Scarlett's Shearwater *Puffinus spelaeus* Holdaway & Worthy, 1994 EXTINCT

Ronald James Scarlett (1911–2002) was a New Zealand palaeozoologist, as well as a collector of stamps, coins and postcards. His nomination for the Meritorious Service Award of the Ornithological Society of New Zealand read 'Ron Scarlett has made a major contribution to New Zealand ornithology, maintaining single-handedly the field of palaeornithology here for 30 years. He collected and catalogued bird fossils from throughout New Zealand, and described several species, including Eyles's Harrier, Hodgens's Waterhen, and the New Zealand Owlet-nightjar. Many papers on fossil and living birds, and a book on bone identification, brought the extinct avifauna into the lives of others. The large, carefully catalogued, skeleton collection he built up in Canterbury Museum is now the basis of much research on the history of the New Zealand avifauna.' He published widely including a paper he co-wrote on 'The extinct Auckland Islands Merganser *Mergus australis*' (1970).

Schaefer (Schäfer)

Lesser Sand Plover ssp. *Charadrius mongolus schaeferi* **Meyer de Schauensee**, 1937
Russet Sparrow ssp. *Passer rutilans schaeferi* **Stresemann**, 1939 NCR
[JS *Passer rutilans cinnamomeus*]
Yellow-billed Blue Magpie ssp. *Urocissa flavirostris schaeferi* **Sick**, 1939

Short-tailed Nighthawk ssp. *Lurocalis semitorquatus schaeferi* **W. H. Phelps** & W. H. Phelps Jr, 1952
Tawny-breasted Flycatcher ssp. *Myiobius villosus schaeferi* Aveledo & **Pons**, 1952

Ernst Schäfer (1910–1992) was a German hunter, biologist and ornithologist who led the German expedition to Tibet (1938). He and all the other scientists on the expedition were members of the SS and confirmed Nazis. He had joined the SS (1933) and was a member of Himmler's inner circle. Schäfer had first visited Tibet in 1930 and returned there in 1931–1932, and again in 1934–1936 as a member of the American Brooke-Dolan expeditions, which also visited China and Siberia. He spent some time (1932–1933) studying the collections of the BMNH in London. He appears to have had different aims than those approved by Himmler, who later lost patience with him and had him posted to fight on the Eastern Front (1943). He was arrested (1945) and charged with war crimes, but was acquitted and released (1947). He lectured on zoology and biology in Caracas and researched the Venezuelan fauna, with particular reference to birds (1949–1954). He met King Baudouin of Belgium, who invited him to Brussels to advise on an intended expedition to the Belgian Congo (1955), and accompanied Baudouin and his companions on their travels there (1956–1959). He was head of the Zoological Department, Lower Saxony State Museum, Hanover (1960–1970) and visited Alaska, Tanzania, Kenya, Uganda, Namibia, Zimbabwe and Mozambique, as well as Venezuela again (1970–1984). He suffered badly from arthritis (1986–1992). His collection is in the Natural History Museum in Berlin. A mammal is named after him.

Schaeffer

David's Fulvetta ssp. *Alcippe davidi schaefferi* **La Touche**, 1923
[Syn. *Alcippe morrisonia schaefferi*]

Monsieur Schaeffer (DNF) was an engineer for the railways in French Indochina. La Touche wrote that he was much indebted to Schaeffer for his kind hospitality and assistance, but neglected to give his first name(s).

Schaldach

Scaly-throated Foliage-gleaner ssp. *Anabacerthia variegaticeps schaldachi* Winker, 1997

Dr William 'Willy' Joseph Schaldach Jr (1924–2005), the 'Dean of Mexican Ornithology', was an American ornithologist resident in Mexico. His bachelor's degree was awarded by Dartmouth College, New Hampshire (1947), and his doctorate by the University of Arizona. He left an unfinished checklist of the birds of the State of Vera Cruz, Mexico, when he died of neglect, totally impoverished.

Schalow

Schalow's Turaco *Tauraco schalowi* **Reichenow**, 1891

African Green Pigeon ssp. *Treron calvus schalowi* Reichenow, 1880
Schalow's Wheatear *Oenanthe lugubris schalowi* **G. A. Fischer** & Reichenow, 1884
[Alt. Mourning Wheatear ssp.]

Common Scimitarbill ssp. *Rhinopomastus cyanomelas*
   schalowi **Neumann**, 1900
Blue-capped Ifrit ssp. *Ifrita kowaldi schalowiana*
   Stresemann, 1922 NCR
[JS *Ifrita kowaldi kowaldi*]

Herman Schalow (1852–1925) was a German banker in Berlin
and an amateur ornithologist. He worked with both Cabanis
(q.v.) and Reichenow (q.v.). He wrote *Die Musophagidae*
(1886) and 'Beiträge zur Vogelfauna der Mark Brandenburg'
(1919). He gave his library to the German Ornithological
Society (1922), of which he was President (1907–1921). After
his death it was given to the Zoological Museum Library in
Berlin and re-named the Schalow Library in his honour.

Schauensee

Lesser Necklaced Laughingthrush ssp. *Garrulax monileger*
   schauenseei **Delacour** & **J. Greenway**, 1939
Sooty-headed Bulbul ssp. *Pycnonotus aurigaster*
   schauenseei Delacour, 1943
Blyth's Shrike Babbler ssp. *Pteruthius aeralatus*
   schauenseei **Deignan**, 1946

(See **De Schauensee**; also **Meyer de Schauensee** and
Rodolphe)

Scheepmaker

Scheepmaker's Crowned Pigeon *Goura scheepmakeri*
   Finsch, 1876
[Alt. Southern Crowned Pigeon]

Cornelius Scheepmaker (DNF) was a Dutch civil servant and
collector active (c.1875) at Surabaya, Java, in the Dutch East
Indies, who traded in live animals from New Guinea. These
he sent to Europe, e.g. to Artis Zoo, Amsterdam, where Otto
Finsch (q.v.) saw a specimen of this pigeon, although his type
specimen was a skin obtained from another Dutch dealer, G.
A. Frank (q.v.).

Scheffler

Scheffler's Owlet *Glaucidium capense scheffleri* **Neumann**,
   1911
[Alt. African Barred Owlet ssp.]

Georg Scheffler (d.1910) was a German collector in East
Africa, particularly north-east Tanzania (1899–1910). A
number of African shrubs, trees, a reptile and an amphibian
are named after him.

Schiebel

Common Chaffinch ssp. *Fringilla coelebs schiebeli*
   Stresemann, 1925
Cetti's Warbler ssp. *Cettia cetti schiebeli* Rokitansky, 1934
   NCR
[JS *Cettia cetti cetti*]
Italian Sparrow ssp. *Passer italiae schiebeli* Rokitansky,
   1934 NCR
[Formerly regarded as a stable hybrid of *P. domesticus* x
   P. hispaniolensis]
Willow Tit ssp. *Parus salicarius schiebeli* **Kleinschmidt**,
   1937 NCR
[JS *Poecile montanus montanus*]

European Goldfinch ssp. *Carduelis carduelis schiebeli*
   von Jordans & **Steinbacher**, 1943 NCR
[JS *Carduelis carduelis balcanica*]

Dr Guido Schiebel (1881–1956) was an Austrian zoologist and
ornithologist. He collected in Corsica (1910), Crete (1925),
Sicily (1934 and 1935) and the former Yugoslavia. The University of Vienna awarded his doctorate. He lived in Innsbruck
(1906–1907) and taught in high schools in Freistadt and
Klagenfurt, and subsequently at the University of Graz, where
he became a professor.

Schierbrand

Lesser Cuckooshrike ssp. *Coracina fimbriata schierbrandii*
   Pelzeln, 1865

Lieutenant-General Wolf Curt van Schierbrand (1807–1888)
was a German army officer, an engineer and topographer in
the service of the Dutch East India Company (1825–1867). He
was on Java (1825) and Borneo (1850), and became
Commander-in-Chief, East Indies Army (1862), when he took
Dutch nationality. He was interested in ethnography and
zoology and sent much material, including the cuckooshrike
holotype, to the Dresden Museum in the city he retired to.

Schiff

Schiffornis genus *Schiffornis* **Bonaparte**, 1854

Moritz Schiff (1823–1896) was a German zoologist, anatomist
and amateur ornithologist. He was awarded his MD at
Göttingen (1844). He studied brain physiology in Paris and
spent much time among the animal specimens at the Jardin
des Plantes, publishing over 200 scientific papers. Back in
Germany he became head of the ornithological collection of
the Senckenberg Institute, responsible for cataloguing and
determining taxonomy. After working in the Frankfurt Zoological Museum he moved to Bern (1854) becoming Professor of
Comparative Anatomy, then became Professor of Physiology
in Florence (1863) where his brother held the chemistry chair.
Throughout he was continuing to work on his catalogue of
birds. A campaign against vivisection forced Schiff to leave
Florence (1876), and he became Professor of Physiology at
Geneva.

Schillings

Golden Weaver sp. *Ploceus schillingsi* **Reichenow**, 1902
   NCR
[Alt. Taveta Weaver; JS *Ploceus castaneiceps*]
Brownbul sp. *Calamocichla schillingsi* Reichenow, 1904
   NCR
[Alt. Northern Brownbul; JS *Phyllastrephus strepitans*]
Bush Lark sp. *Mirafra schillingsi* Reichenow, 1916 NCR
[Alt. Singing Bush Lark; JS *Mirafra cantillans marginata*]
Sunbird sp. *Cinnyris schillingsi* Reichenow, 1916 NCR
[Alt. Purple-banded Sunbird; JS *Cinnyris bifasciatus*
   microrhynchus]

White-backed Vulture ssp. *Pseudogyps africanus schillingsi*
   Erlanger, 1903 NCR; NRM
[Syn. *Gyps africanus*]
Ashy Cisticola ssp. *Cisticola cinereolus schillingsi*
   Reichenow, 1905

Karl Georg Joseph Schillings (1866–1921) was a German wildlife photographer and explorer in German East Africa (Tanzania) (1896–1904). He wrote *Mit Blitzlicht und Büchse* (1905).

Schillmöller

Large-tailed Nightjar ssp. *Caprimulgus macrurus schillmollerii* **Stresemann**, 1931 NCR
[NUI *Caprimulgus macrurus schlegelii*]

Captain Schillmöller (DNF) was a Dutch administrator on Halmahera, Dutch East Indies (Indonesia). He may be the same person as Major B. F. A. Schillmoeller who commanded the Dutch forces at the Battle of Menado (1942).

Schimper

Gull sp. *Larus schimperi* **Schlegel**, 1863 NCR
[Alt. Saunders's Gull; Syn. *Chroicocephalus saundersi*]

Rock Pigeon ssp. *Columba livia schimperi* **Bonaparte**, 1854

Georg Heinrich Wilhelm Schimper (1804–1878) was a German botanist, explorer and collector in North Africa (1830s), the Middle East and Abyssinia (Ethiopia), where he settled (1836). He married a local and became a governor of the district of Enticho. The Emperor Tewodros II had him imprisoned in Magdala as a hostage (1868). He wrote *Reise nach Algier 1831–1832* (1834). Although Schlegel's name (1863) is an older one than Swinhoe's (1871) for Saunders's Gull it is, in taxonomic parlance, a *nomen oblitum* ('forgotten name'): i.e. a more recent name for the same taxon is in common usage.

Schinz

Dunlin ssp. *Calidris alpina schinzii* **C. L. Brehm**, 1822

Heinrich Rudolph Schinz (1777–1861) was a Swiss zoologist. He wrote *Die Vögel der Schweiz* (1815) with F. Meisner. He also wrote *Naturgeschichte und Abbildungen der Saugethiere* (1824). A reptile is named after him.

Schiøler

Northern Wheatear ssp. *Oenanthe oenanthe schioleri* **Salomonsen**, 1927 NCR
[JS *Oenanthe oenanthe leucorhoa*]
Red-necked Grebe ssp. *Podiceps grisegena schioleri* Hortling, 1929 NCR
[JS *Podiceps grisegena holbollii*]
Red-breasted Merganser ssp. *Mergus serrator schioleri* Salomonsen, 1949 NCR; NRM

Eiler Lauritz Theodor Lehn Schiøler (1874–1929) was a very wealthy Danish banker and keen ornithologist. He was in Greenland (1925). He constructed a large museum for his collection of over 25,000 specimens. He was Chairman, Danish Ornithologists' Union.

Schleep

Schleep's Rapacious Gull [archaic] *Lestris schleepii* **C. L. Brehm**, 1824 NCR
[Alt. Arctic Skua; JS *Stercorarius parasiticus*]

Dr Bernhard Christian Schleep (1768–1838) was a German naturalist who lived at Gottorp, Schleswig-Holstein.

Schlegel

Bird-of-Paradise genus *Schlegelia* **Bernstein**, 1864 NCR
[Now in *Diphyllodes*]

Golden-winged Sparrow *Arremon schlegeli* **Bonaparte**, 1850
Schlegel's Chat *Emarginata schlegelii* **Wahlberg**, 1855
[Alt. Karoo Chat; Syn. *Cercomela schlegelii*]
Schlegel's Francolin *Peliperdrix schlegelii* **Heuglin**, 1863
Schlegel's Fruit Dove *Ptilinopus insolitus* Schlegel, 1863
[Alt. Knob-billed Fruit Dove]
Schlegel's Petrel *Pterodroma incerta* Schlegel, 1863
[Alt. Atlantic Petrel]
Schlegel's Dove *Turtur brehmeri* **Hartlaub**, 1865
[Alt. Blue-headed Wood Dove]
Schlegel's Myna *Streptocitta albertinae* Schlegel, 1866
[Alt. Albertina's Myna, Bare-eyed Myna]
Schlegel's Asity *Philepitta schlegeli* Schlegel, 1867
Schlegel's Whistler *Pachycephala schlegelii* Schlegel, 1871
[Alt. Regent Whistler]
Royal Penguin *Eudyptes schlegeli* **Finsch**, 1876
Southern Ground-hornbill *Bucorvus schlegeli* **Roberts**, 1926 NCR
[JS *Bucorvus leadbeateri*]

Island Thrush ssp. *Turdus poliocephalus schlegelii* **P. L. Sclater**, 1861
Schlegel's Parrotlet *Forpus passerinus cyanochlorus* Schlegel, 1864
[Alt. Green-rumped Parrotlet ssp.]
New Guinea Bronzewing ssp. *Henicophaps albifrons schlegeli* **C. B. H. Rosenberg**, 1866
Schlegel's Twinspot *Mandingoa nitidula schlegeli* **Sharpe**, 1870
[Alt. Green-backed Twinspot ssp.]

Hermann Schlegel (1804–1884) was a German-born zoologist who spent much of his life in the Netherlands. He was the first person (1844) to use trinomials to describe separate races. Schlegel made a trip on foot through large parts of Germany and Austria (1824–1825). While he was in Vienna (1825) he received a letter through Johan Natterer (q.v.) from Jacob Coenraad Temminck (q.v.) who was looking for a researcher to explore parts of Indonesia. Schlegel went to Leiden, and worked so hard and well for Temminck that the latter decided that Schlegel had to stay at Leiden and was too valuable to risk on an overseas assignment. Schlegel had many publications, some co-written with Temminck, including *Fauna Japonica–Aves* and *Kritische Uebersicht der Europäischen Vögel*, and he succeeded Temminck as Director at the Museum (1858). Schlegel married twice. He honoured his second wife, Albertina Pfeiffer, in the scientific name of the Bare-eyed Myna. Two amphibians, two mammals and ten reptiles are named after him.

Schleiermacher

Bornean Peacock Pheasant *Polyplectron schleiermacheri*
Bruggemann, 1877

The Schleiermacher family were hereditary holders of the position of Curator and Director of the Grossherzoglich Hessischen Museen, Darmstadt, occupying that position for over 100 years. Ernst Christian Schleiermacher (1755–1844), who was interested in zoology and palaeontology, was Confidential Secretary of the Grand-Duke of Hesse, who made him first Curator and Director (1779–1844). His son, Ludwig Johann Schleiermacher (1785–1844) followed, and was succeeded (1844–1854) by his brother, Andreas August Schleiermacher (1787–1858). He in turn was succeeded by his son, Heinrich August Schleiermacher (1816–1892). Unfortunately the etymology just mentions Herr Schleiermacher, without initials, but we think it is named after Heinrich August, who was Director when the bird was described.

Schlüter

Schluter's Hermit *Phaethornis fumosus* Schlüter, 1901 NCR
[Based on melanistic specimens of *Phaethornis augusti*]
Lark sp. *Galerida schlueteri* **Kleinschmidt**, 1904 NCR
[Alt. Thekla Lark; JS *Galerida theklae ruficolor*]

Tree Pipit ssp. *Anthus trivialis schlueteri* Kleinschmidt, 1920
Western Jackdaw ssp. *Corvus monedula schluteri*
Kleinschmidt, 1935 NCR
[JS *Coloeus monedula soemmerringii*]

Wilhelm Schlüter (1828–1919) founded the eponymous firm of natural history dealers in Halle, Germany (1853). His father, Friedrich, was an entomologist and malacologist. His brother, Julius, emigrated to Brazil and was one of a network of collectors, including the Geisler brothers (q.v.) in Australia and New Guinea, which kept the company supplied with specimens. The company was pre-eminent in the trade, and dealt with many major museums and private collectors. Wilhelm's sons, both of whom had studied natural sciences, later ran the company: Wilhelm 'Willy' Schlüter Jr (1866–1938) and Curt (1881–1944). Two reptiles and an amphibian are named after Wilhelm senior as are three of the birds; the Jackdaw is named after Wilhelm junior.

Schmacker

Mindoro Bulbul *Jole schmackeri* **Hartert**, 1890 NCR
[JS *Hypsipetes mindorensis*]
Tarictic Hornbill sp. *Penelopides schmackeri* Hartert, 1891 NCR
[Alt. Mindoro Hornbill; JS *Penelopides mindorensis*]

Lesser Necklaced Laughingthrush ssp. *Garrulax monileger schmackeri* **Hartlaub**, 1898

Philipp Bernhard Schmacker (1852–1896) came from a trading family. He went to Hong Kong (1872) and worked in Canton and Shanghai. He collected molluscs in China and other Asian regions. He and Oscar Boettger jointly presented a paper to the Malacological Society (1894). He died in Yokohama and left his collection to the city of Bremen. Two amphibians are named after him.

Schmitz

Barn Owl ssp. *Tyto alba schmitzi* **Hartert**, 1900
Grey Wagtail ssp. *Motacilla cinerea schmitzi* **Tschusi**, 1900

Father Ernst Schmitz (1845–1922) was an ornithologist, naturalist and Lazarite priest (1869). He was in Portugal (1875–1879) but was then transferred to Madeira, where he remained till 1898. He moved (1898–1902) to Belgium, but then returned to Madeira where he stayed until being transferred to Jerusalem (1908). He spent the rest of his life in the Middle East. In Jerusalem he ran the Lazarite seminary, which included a small natural history museum. He entered enthusiastically into fieldwork and accumulated many varieties of ant, of which 10 were new to science. A mammal is named after him.

Schneider

Schneider's Pitta *Hydrornis schneideri* **Hartert**, 1909

Gustav Schneider (1867–1958) was a Swiss zoologist who collected in Sumatra (1897) for the museum in Basel. F. Werner wrote an article on 'Reptilien und Batrachier aus Sumatra gesammelt von Herrn Gustav Schneider im Jahre 1897/1898' (1900). There is some confusion as Gustav Schneider Snr. (1834–1900) had also been conservator at Basel (1859–1875), but given the timing of the description we are confident that the eponym is for the younger man. An amphibian is named after him.

Schnitzer

Rufous Chatterer ssp. *Turdoides rubiginosa schnitzeri* **Deignan**, 1964

Isaak Eduard Schnitzer was the given name of Emin Pasha (q.v.).

Schodde

Schodde's Bird-of-Paradise – informal common name suggested by Frith & Frith, 1998 (no scientific name)
[Hybrid: *Parotia lawesii* x *Paradisaea rudolphi*]

Channel-billed Cuckoo ssp. *Scythrops novaehollandiae schoddei* Mason & Forrester, 1996

Dr Richard Schodde (b.1936) is an Australian botanist and ornithologist. The University of Adelaide awarded his bachelor's degree (1960) and his doctorate (1970). He worked for CSIRO (1960s–1998), first in Papua New Guinea and (1970–1998) as founder Curator and Director of the Australian National Wildlife Collection. He is one of those who can claim credit for helping establish the Kakadu National Park. He co-wrote *The Directory of Australian Birds: Passerines. A Taxonomic and Zoogeographic Atlas of the Biodiversity of Birds of Australia and its Territories* (1999).

Schoede

Rainbow Lorikeet ssp. *Trichoglossus cyanogrammus schoedei* **Reichenow**, 1910 NCR
[JS *Trichoglossus haematodus flavicans*]

Hermann Schoede (DNF) was a wealthy German amateur collector in the Solomon Islands and New Guinea. He leased a schooner and explored the islands of German New Guinea (1909–1910). A reptile is also named after him.

Schomburgk

> Schomburgk's Crake *Micropygia schomburgkii* **Cabanis**, 1848
> [Alt. Ocellated Crake, Dotted Crake]
> Schomburgk's Parrotlet *Forpus modestus* Cabanis, 1848
> [Alt. Dusky-billed Parrotlet; taxon formerly known as *Forpus sclateri eidos*]

> Ladder-tailed Nightjar ssp. *Hydropsalis climacocerca schomburgki* **P. L. Sclater**, 1866

Sir Robert Hermann Schomburgk (1804–1865) was a German businessman who was asked to supervise transporting Saxon sheep to Virginia, USA (1828), where he stayed. He lost his fortune partly because he failed as a tobacco grower, partly because he lost all his belongings in a fire on the Caribbean island of St Thomas. Afterwards he went to the British Virgin Islands to map, at his own expense, the coast of the island of Anega, notorious for its shipwrecks. The results published by him drew the attention of the Royal Geographical Society, who asked him to explore British Guiana (1835–1839), during which he discovered the famous giant water lily *Victoria regina*. The British Government asked him to undertake a second expedition (1841–1844) to Guiana to investigate its southern borders with Venezuela and the Dutch colony of Suriname. He also urged the British government to establish the borders with Brazil, because of the repeated enslavement of local Indian tribes by Portuguese Brazilian slave drivers. Upon his return Queen Victoria knighted him and he became Consul at Barbados (1846), Consul at the Dominican Republic (1848–1857), and finally Consul-General in Bangkok (Thailand) (1857–1864). Hampered by health problems, he retired to Germany but died the next year. His brother Richard Moritz (1811–1891), a gardener and botanist, accompanied him during his second expedition to Guiana and Venezuela. He was also a collector, who emigrated, with a third brother, Otto (1849), to southern Australia to escape the political turmoil in Europe (1848). Richard became Director of the Botanical Gardens, Adelaide (1866–1891), until his death from a heart attack. Otto (1810–1857) edited Richard's travelogue *R. H. Schomburgk's Reisen in Guiana und am Orinoco während 1835–1839* (1841). A mammal is named after Robert and a reptile after Richard.

Schott

> Bay Wren ssp. *Cantorchilus nigricapillus schottii* **Baird**, 1864

Arthur Carl Victor Schott (1814–1875) was born in Stuttgart, Germany, where he was apprenticed at the Royal Gardens. He studied at the Institute of Agriculture, Hohenheim. He spent 10 years in Hungary, managing a mining property and studying geology, botany and zoology. He travelled in Europe and the Near East, then went to the USA (1850) where the Corps of Topographical Engineers in Washington employed him. He was a member of the U.S.-Mexican border survey (1853–1855) and collected animals, fossils and minerals in the Rio Grande valley. He was naturalist and geologist on

Michler's (q.v.) survey of the Isthmus of Darien (1857), and surveyed in the Yucatan Peninsula (1864–1866). Two reptiles are named after him.

Schouteden

> Swift genus *Schoutedenapus* **De Roo**, 1968

> Schouteden's Apalis *Apalis schoutedeni* **Chapin**, 1937 NCR
> [Alt. Gosling's Apalis; JS *Apalis goslingi*]
> Schouteden's Swift *Schoutedenapus schoutedeni* **Prigogine**, 1960

> Schouteden's Crested Guineafowl *Guttera edouardi schoutedeni* Chapin, 1923 NCR
> [JS *Guttera pucherani verreauxi*]
> Black-bellied Seedcracker ssp. *Pyrenestes ostrinus schoutedeni* Neunzig, 1928 NCR; NRM
> Wood Pipit ssp. *Anthus nyassae schoutedeni* Chapin, 1937
> Bocage's Akalat ssp. *Sheppardia bocagei schoutedeni* Prigogine, 1952
> Red-capped Crombec ssp. *Sylvietta ruficapilla schoutedeni* **C. M. N. White**, 1953
> Joyful Greenbul ssp. *Chlorocichla laetissima schoutedeni* Prigogine, 1954
> Luapula Cisticola ssp. *Cisticola luapula schoutedeni* White, 1954 NCR; NRM
> Red-faced Woodland Warbler ssp. *Phylloscopus laetus schoutedeni* Prigogine, 1955
> Flappet Lark ssp. *Mirafra rufocinnamomea schoutedeni* White, 1956
> Common Waxbill ssp. *Estrilda astrild schoutedeni* **Wolters**, 1962
> Grey-olive Greenbul ssp. *Phyllastrephus cerviniventris schoutedeni* Prigogine, 1969
> [*P. cerviniventris* sometimes regarded as monotypic]

Henri Eugene Alphonse Hubert Schouteden (1881–1972) was a Belgian zoologist who undertook many expeditions to the Congo. He was an expert on swifts and published on both ornithology and entomology. He wrote *De Vogels van Belgisch-Congo en van Ruanda-Urundi*. Three reptiles, two amphibians and two mammals are named after him.

Schrader

> Stout Cisticola ssp. *Cisticola robustus schraderi* **Neumann**, 1906
> Lesser Blue-eared Starling ssp. *Lamprotornis chloropterus schraderi* Neumann, 1908 NCR
> [JS *Lamprotornis chloropterus chloropterus*]

Gustav Schrader (1852–1942) was a German natural history dealer in Port Said who collected in Egypt, Eritrea, Somaliland (Somalia), Abyssinia (Ethiopia) and Asia Minor (Turkey) (1875–1891). Rothschild (q.v.) at Tring and Schlüter (q.v.) at Halle were among his regular customers.

Schrank

> Schrank's Tanager *Tangara schrankii* **Spix**, 1825
> [Alt. Green-and-gold Tanager]

Dr Father Franz von Paula von Schrank (1747–1835) was a German Jesuit priest as well as being a botanist, entomologist and author. He was Professor of Botany and Agronomy at Ingolstadt (1784–1809) and was founder and first Director (1809–1832) of the Botanical Gardens in Munich.

Schreibers

Schreibers's Hummingbird *Heliodoxa schreibersii* **Bourcier,** 1847
[Alt. Black-throated Brilliant]

Carl Franz Anton Ritter von Schreibers (1775–1852) was an Austrian naturalist who collected in Brazil (1817). He qualified as a physician and studied botany and mineralogy. He became Director of the Viennese Natural History Collections (1806) and worked for decades to overhaul them, including documenting the expeditions and collecting activities of Natterer, Spix and Martius (all q.v.). His main interest was meteorites, on which he wrote many papers, most of which were burned (October 1848) when much of the museum was destroyed by fire and his life's work literally went up in smoke. Schreibers could not bear the loss and retired a broken man (1851), dying the following year. A mammal and a reptile are named after him.

Schreiner

White-bellied Nothura *Nothura schreineri* Miranda-Ribeiro, 1938 NCR
[JS *Nothura boraquira*]

Karl (or Carl/Carlos) Schreiner (1849–1896) was a German-born Brazilian ornithologist and collector. He was an Assistant Naturalist at the Museu Nacional, Rio de Janeiro (1872), a travelling naturalist (1889) and sub-Director of the Zoological Section (1895). Miranda Ribeiro (q.v. under Miranda, A.) collected with him and wrote his obituary.

Schrenck

Schrenck's Reed Warbler *Acrocephalus bistrigiceps* **Swinhoe,** 1860
[Alt. Black-browed Reed Warbler]
Schrenck's Little Bittern *Ixobrychus eurhythmus* Swinhoe, 1873
[Alt. Von Schrenck's Bittern]

Leopold Ivanovich von Schrenck (sometimes Schenk or Shrenk) (1826–1894) was a Russo-German zoologist, geographer and ethnographer who was a Fellow and a Member of the Council and Director (1879) of the Imperial Academy of Sciences at St Petersburg. He explored the Amur River and Sakhalin Island on an Academy of Science expedition (1854–1856), the results of which were published in the 4-volume *Reisen und Forschungen im Amur-Lande in den Jahren 1854–1856* (1860–1900). He also wrote the 3-volume work on Russian ethnography *On non-Russians of Amurian Territory* (1883–1903). He coined the term 'Paleoasiatic nations'. His main work was on the native nations of Amurian Krai (territory). He believed that the mammoths, which he found, preserved in the permafrost, must have perished only recently – and thought it was a creature that lived underground on soil! He was associated with several branches of natural science, and other taxa bear his name including a fish, an amphibian and a reptile.

Schubotz

Greenbul sp. *Phyllastrephus schubotzi* **Reichenow,** 1908 NCR
[Alt. Olive-breasted Greenbul; JS *Arizelocichla kikuyuensis*]
Weaver sp. *Ploceus schubotzi* Reichenow, 1908 NCR
[Alt. Strange Weaver; JS *Ploceus alienus*]
White-eye sp. *Zosterops schubotzi* Reichenow, 1908 NCR
[Alt. African Yellow White-eye; JS *Zosterops senegalensis stuhlmanni*]
Yellow Warbler sp. *Chloropeta schubotzi* Reichenow, 1908 NCR
[Alt. Mountain Yellow Warbler; JS *Iduna similis* (Syn. *Chloropeta similis*)]

Mottled Swift ssp. *Tachymarptis aequatorialis schubotzi* Reichenow, 1908 NCR
[NUI *Tachymarptis aequatorialis aequatorialis*]
Stuhlmann's Double-collared Sunbird ssp. *Cinnyris stuhlmanni schubotzi* Reichenow, 1908
Red-bellied Paradise-flycatcher ssp. *Terpsiphone rufiventer schubotzi* Reichenow, 1911
Schubotz's Forest Francolin *Peliperdix lathami schubotzi* Reichenow, 1912
[Alt. Latham's Francolin ssp.]
Schubotz's Plumed Guineafowl *Guttera plumifera schubotzi* Reichenow, 1912
Yellow-fronted Tinkerbird ssp. *Pogoniulus chrysoconus schubotzi* Reichenow, 1912 NCR
[NUI *Pogoniulus chrysoconus chrysoconus*]

Johann G. Hermann Schubotz (1881–1955) was a German naturalist who was an Assistant in Zoology at the Humboldt University, Berlin, and participated in the two expeditions that Duke Adolf Friedrich zu Mecklenburg (q.v. under Adolf) undertook to Equatorial Africa (1907–1908 and 1910–1911). He was appointed Professor of Zoology (1916), then Cultural Attaché at the German Embassy, Stockholm (1919–1926). He emigrated from Germany to South West Africa (Namibia) (1935) where he bred sheep. He returned to Germany (1952) and worked at the Broadcoasting Corporation ('Rundfunk'). A reptile and two amphibians are named after him.

Schuchov

Icterine Warbler ssp. *Hippolais icterina schuchovi* Snigirewski, 1931 NCR; NRM

I. N. Shukhov (DNF) worked at the Zoological Museum of the Academy of Sciences in St Petersburg (1902). He was later reported in Siberia.

Schuett

Black-billed Turaco *Tauraco schuetti* **Cabanis,** 1879

(See under **Schütt**)

Schuhmacher

Rufous Hornero ssp. *Furnarius rufus schuhmacheri* **Laubmann,** 1933
[SII *Furnarius rufus commersoni*]

Eugen Josef Robert Schuhmacher (1906–1973) was a German zoologist, wildlife filmmaker, and explorer who was on the Gran Chaco Expedition (1931–1932).

Schulenberg

Diademed Tacapulo *Scytalopus schulenbergi*
B. M. Whitney, 1994

Plain-tailed Wren ssp. *Pheugopedius euophrys schulenbergi*
T. A. Parker & **O'Neill**, 1985

Dr Thomas S. Schulenberg is an American zoologist and ornithologist. Louisiana State University awarded his master's degree and the University of Chicago awarded his doctorate. He worked as Field Biologist, Birds and Mammals, for the Field Museum in Chicago and is now is a Research Associate at the Cornell Laboratory of Ornithology. He is a member of the editorial board of *Bird Conservation International* and co-wrote *Birds of Peru* (2007).

Schulpin

Hawfinch ssp. *Coccothraustes coccothraustes schulpini*
H. Johansen, 1944

(See **Shulpin**)

Schultze

Barbet sp. *Tricholaema schultzei* **Reichenow**, 1911 NCR
[Alt. Hairy-breasted Barbet; JS *Tricholaema hirsuta flavipunctata*]
Woodpecker sp. *Mesopicos schultzei* Reichenow, 1912 NCR
[Alt. Elliot's Woodpecker; JS *Dendropicos elliotii johnstoni*]

Dr Arnold Schultze (1875–1948) was a German entomologist and specialist in Lepidoptera. He collected in tropical Africa including the Congo (1907–1910 and 1929–1932), then Colombia (1926–1928), and Ecuador (1935–1939). His collections are in the museums of Stuttgart and Berlin.

Schulz

Rufous-throated Dipper *Cinclus schulzii* **Cabanis**, 1882
Black-bodied Woodpecker *Dryocopus schulzi* Cabanis, 1883

Slaty-headed Tody Flycatcher ssp. *Poecilotriccus sylvia schulzi* **Berlepsch**, 1907

Fritz (Federico) W. Schulz (<1850–1933) was a German zoologist who collected in Argentina (1866–1933). He became a preparator at the Museum of Zoology, University of Cordoba, and collected in the mountains (1881) where he discovered several new species which Cabanis (q.v.) described. He drew up a list of bird species in Cordoba province with Stempelmann: *Enumeración de las Aves de la Provincia de Córdoba* (1887). He later became became curator at the museum (1896).

Schulze

Red-crested Cardinal ssp. *Paroaria coronata schulzei*
Brodkorb, 1937 NCR; NRM
Blue-and-yellow Tanager ssp. *Thraupis bonariensis schulzei*
Brodkorb, 1938

Alberto Schulze (DNF) was a Paraguayan ornithologist and collector (1930–1941). Details of him are sparse and he is sometimes referred to as 'the mysterious Alberto Schulze'.

He had a team of collectors who were only known by their surnames and the chronology and geographical spread of the labels on the specimens, always bearing his name, suggest he took the credit of his co-workers' efforts.

Schummer

Calandra Lark ssp. *Melanocorypha calandra schummeri*
Charlemagne, 1927 NCR
[JS *Melanocorypha calandra calandra*]

A. A. Schummer (DNF) was a Russian ornithologist. He co-wrote: *Short List of Birds of the Vicinities of Kiew* (1909).

Schuster

Cisticola sp. *Cisticola schusteri* **Reichenow**, 1913 NCR
[Alt. Trilling Cisticola; JS *Cisticola woosnami woosnami*]

Yellow-throated Greenbul ssp. *Phyllastrephus chlorigula schusteri* Reichenow, 1913 NCR; NRM
[Syn *Arizelocichla chlorigula*]

Dr Ludwig Schuster (1883–1954) was a German settler and ornithologist in German East Africa (Tanzania) (1909–1918). He was Vice-President of the German Ornithologists' Society and was a corresponding fellow of the AOU.

Schütt

Black-billed Turaco *Tauraco schuetti* **Cabanis**, 1879
Thrush sp. *Peliocichla schuetti* Cabanis, 1882 NCR
[Alt. Kurrichane Thrush; JS *Turdus libonyana verrauxi*]

Scaly Francolin ssp. *Pternistis squamatus schuetti* Cabanis, 1880

Otto Schütt (1843–1888) was a German railway engineer who was in Mesopotamia (Iraq) (1875), Northern Angola (1877–1879), and Constantinople (Istanbul) (c.1888). He wrote *Reisen im Südwestlichen Becken des Kongo* (1881).

Schvedow

Northern Goshawk ssp. *Accipiter gentilis schvedowi*
Menzbier, 1882

Grigorij Schvedow (DNF) was a Russian collector in Siberia.

Schwaner

Flycatcher genus *Schwaneria* **Bonaparte**, 1857 NCR
[Now in *Cyornis*]

Bornean Banded Pitta *Hydrornis schwaneri* Bonaparte, 1850
[Syn. *Pitta guajana schwaneri*]

Yellow-bellied Warbler ssp. *Abroscopus superciliaris schwaneri* **Blyth**, 1870

Dr Carl Anton Ludwig Maria Schwaner (1817–1851) was a geologist and explorer. He graduated in mineralogy and geology (1842) from Heidelberg. That year he was appointed to the scientific commission in Indonesia on the recommendation of Temminck (q.v.). He explored in Borneo (1843–1847) looking for coal and other minerals. He was to join a second planned expedition (1851) but died before he could take part.

Schwartz, A.

Scaly-breasted Thrasher ssp. *Allenia fusca schwartzi* Buden, 1993

Dr Albert Schwartz (1923–1992) was a biologist and entomologist. He was Professor Emeritus of Biology, Miami-Dade Community College, and was associated with the Florida Museum of Natural History, the USNM, and the Museo Nacional de Historia Natural, Santo Domingo. He was a specialist in Caribbean fauna, writing extensively on amphibians, reptiles and lepidoptera, including *The Butterflies of Hispaniola*. Seven reptiles, two amphibians, and two mammals are named after him.

Schwartz, P. A.

Schwartz's Antthrush *Chamaeza turdina* **Cabanis** & **Heine**, 1859
[Alt. Scalloped Antthrush]

Paul A. Schwartz (1917–1979) was a mechanical engineer. He worked at Instituto Neotropical, Caracas, Venezuela and was a Research Associate, Rancho Grande Biological Research Station, Venezuela. He was a pioneer in the recording of Neotropical birdsong, making over 5,000 recordings. He tape-recorded (1971) two different song-types from what was thought to be a single taxon, *Chamaeza ruficauda*, thus proving them separate species. The common name was adopted (2004) as it was said of Schwartz that he 'arguably knew *chionogaster* better than anyone, probably made the first tape recordings of the species, and his observations regarding the different voices of antthrushes in eastern Brazil led Willis down the path of discovery that resolved the species-limits that we recognize today'. He was made a life member of the Wilson Ornithological Society (1960). He wrote a life history of the Rusty-breasted Antpitta *Grallaricula ferrugineipectus* and contributed photographs to Gilliard's (q.v.) *Living Birds of the World*.

Schwarz

Radde's Warbler *Phylloscopus schwarzi* **Radde**, 1863

Professor Doctor Peter Carl Ludwig Schwarz (1822–1894) was a German astronomer who studied mathematics at Tartu University, Estonia, and became an assistant at the observatory there (1846). He took part in an expedition to the Amur region (1849), and was appointed leader of the Imperial Russian Geographical Society's expedition (1854–1862) to eastern Siberia and Northern China. He moved his family from St Petersburg to Berlin to undertake further study there and at Gotha (1863). Subsequently he returned to Dorpat in Estonia as Professor of Astronomy (1872).

Schwebisch

Blue Flycatcher sp. *Elminia schwebischi* **Oustalet**, 1892 NCR
[Alt. African Blue Flycatcher; JS *Elminia longicauda teresita*]

Dr D. Schwebisch (DNF) was a French naval physician. As part of the Brazza (q.v.) mission he explored in Gabon (1883) and the Congo. He co-wrote *Étude sur les Mammifères du Congo Français* (1884).

Sclater

Antbird genus *Sclateria* **Oberholser**, 1899
Greenbul genus *Sclaterillas* **J. A. Roberts**, 1922 NCR
[Now in *Phyllastrephus*]
Sunbird genus *Sclaterornis* J. A. Roberts, 1922 NCR
[Now in *Nectarinia*]

Fiery-throated Fruiteater *Euchlornis sclateri* Cornalia, 1852 NCR
[JS *Pipreola chlorolepidota*]
Black-banded Crake *Micropygia sclateri* **Bonaparte**, 1856 NCR
[Name replaced by *Laterallus fasciatus*]
Sclater's Wren *Campylorhynchus humilis* P. L. Sclater, 1856
Sclater's Hanging Parrot *Loriculus sclateri* **Wallace**, 1863
[Alt. Sula Hanging Parrot]
Sclater's Pygmy Parrot *Micropsitta pusio* P. L. Sclater, 1865
[Alt. Buff-faced Pygmy Parrot]
Sclater's Curassow *Crax sclateri* **G. R. Gray**, 1867 NCR
[Alt. Bare-faced Curassow; JS *Crax fasciolata*]
Makira Honeyeater *Meliarchus sclateri* G. R. Gray, 1870
[Alt. San Cristobal Honeyeater]
Sclater's Monal *Lophophorus sclateri* **Jerdon**, 1870
Bay-vented Cotinga *Doliornis sclateri* **Taczanowski**, 1874
Sclater's Whistler *Pachycephala soror* P. L. Sclater, 1874
Ecuadorian Piculet *Picumnus sclateri* Taczanowski, 1877
Sclater's Woodswallow *Artamus insignis* P. L. Sclater, 1877
[Alt. Bismarck Woodswallow, White-backed Woodswallow]
Puna Canastero *Asthenes sclateri* **Cabanis**, 1878
Sclater's Myzomela *Myzomela cineracea* P. L. Sclater, 1879
[Alt. Ashy Myzomela, Bismarck Honeyeater]
Sclater's Myzomela *Myzomela sclateri* **Forbes**, 1879
[Alt. Scarlet-bibbed Myzomela/Honeyeater]
Speckle-breasted Wren *Pheugopedius sclateri* Taczanowski, 1879
Sclater's Mannikin *Lonchura melaena* P. L. Sclater, 1880
[Alt. New Britain Mannikin, Thick-billed Mannikin]
Sclater's Rail *Gallirallus insignis* P. L. Sclater, 1880
[Alt. New Britain Rail, Pink-legged Rail]
Kauai Elepaio *Chasiempis sclateri* **Ridgway**, 1882
Tepui Greenlet *Hylophilus sclateri* **Salvin** & **Godman**, 1883
Ecuadorian Cacique *Cacicus sclateri* **A. Dubois**, 1887
Sclater's Penguin *Eudyptes sclateri* **Buller**, 1888
[Alt. Erect-crested Penguin]
Sclater's Aracari *Pteroglossus didymus* P. L. Sclater, 1890 NCR
[Alt. Lettered Aracari; JS *Pteroglossus inscriptus*]
Crested Doradito *Pseudocolopteryx sclateri* **Oustalet**, 1892
Greater Ground Robin *Amalocichla sclateriana* **De Vis**, 1892
Mexican Chickadee *Poecile sclateri* **Kleinschmidt**, 1897
Sclater's Tyrannulet *Phyllomyias sclateri* **Berlepsch**, 1901
Sclater's Lark *Spizocorys sclateri* **G. E. Shelley**, 1902
Black Riverside Tyrant *Knipolegus sclateri* **Hellmayr**, 1906
[Syn. *Knipolegus orenocensis sclateri*]
Fulvous-chinned Nunlet *Nonnula sclateri* Hellmayr, 1907
Sclater's Antwren *Myrmotherula sclateri* **Snethlage**, 1912
Sclater's Forest Falcon *Micrastur plumbeus* W. L. Sclater, 1918
[Alt. Plumbeous Forest Falcon]

Antillean Euphonia ssp. *Euphonia musica sclateri*
P. L. Sclater, 1854

Sclater's Tanager *Tangara arthus sclateri* **Lafresnaye**, 1854
[Alt. Golden Tanager ssp.]

Common Tody Flycatcher ssp. *Todirostrum cinereum sclateri*
Cabanis & Heine, 1859

Sclater's Parrotlet *Forpus modestus sclateri* G. R. Gray, 1859
[Alt. Dusky-billed Parrotlet ssp.]

Andean Guan ssp. *Penelope montagnii sclateri* G. R. Gray,
1860

Bright-rumped Attila ssp. *Attila spadiceus sclateri*
Lawrence, 1862

Streak-backed Oriole ssp. *Icterus pustulatus sclateri*
Cassin, 1867

Grey-rumped Swift ssp. *Chaetura cinereiventris sclateri*
Pelzeln, 1868

Cliff Flycatcher ssp. *Hirundinea ferruginea sclateri*
J. T. Reinhardt, 1870

Sclater's Bare-eyed Cockatoo *Cacatua sanguinea gymnopis*
P. L. Sclater 1871
[Alt. Little Corella ssp.]

Sclater's Crowned Pigeon *Goura scheepmakeri sclaterii*
Salvadori, 1876
[Alt. Southern Crowned Pigeon ssp.]

Sclater's Pleasing Lorikeet *Charmosyna placentis*
subplacens P. L. Sclater, 1876
[Alt. Red-flanked Lorikeet ssp.]

Sclater's Cassowary *Casuarius casuarius sclaterii*
Salvadori, 1878 NCR; NRM
[Alt. Southern Cassowary ssp.]

White-bellied Cuckooshrike ssp. *Coracina papuensis*
sclaterii Salvadori, 1878

Lesser Antillean Flycatcher ssp. *Myiarchus oberi sclateri*
Lawrence, 1879

Lesser Antillean Bullfinch ssp. *Loxigilla noctis sclateri*
J. A. Allen, 1880

Southern Beardless Tyrannulet ssp. *Camptostoma*
obsoletum sclateri Berlepsch & Taczanowski, 1883

Golden Swallow ssp. *Tachycineta euchrysea sclateri* **Cory**,
1884

Strong-billed Woodcreeper ssp. *Xiphocolaptes*
promeropirhynchus sclateri Ridgway, 1890

Grey Apalis ssp. *Apalis cinerea sclateri* **B. Alexander**, 1903

Grey-crowned Flatbill ssp. *Tolmomyias poliocephalus*
sclateri Hellmayr, 1903

Sclater's Antbird *Cercomacra cinerascens sclateri*
Hellmayr, 1905
[Alt. Grey Antbird ssp.]

Streak-capped Treehunter ssp. *Thripadectes virgaticeps*
sclateri Berlepsch, 1907

White-fronted Nunbird ssp. *Monasa morphoeus sclateri*
Ridgway, 1912

Blue-spotted Wood Dove ssp. *Turtur afer sclateri*
Rothschild, 1917 NCR; NRM

Little Rock Thrush ssp. *Monticola rufocinereus sclateri*
Hartert, 1917

Whiskered Tern ssp. *Chlidonias hybrida sclateri* **Mathews**
& **Iredale**, 1921 NCR
[JS *Chlidonias hybrida delalandii*]

Rufous Treepie ssp. *Dendrocitta vagabunda sclateri*
E. C. S. Baker, 1922

Amazonian Antshrike ssp. *Thamnophilus punctatus sclateri*
Stoltzmann, 1926 NCR
[JS *Thamnophilus amazonicus amazonicus*]

Long-tailed Nightjar ssp. *Caprimulgus climacurus sclateri*
G. L. Bates, 1927

White-browed Scrub Robin ssp. *Erythropygia leucophrys*
sclateri Grote, 1930

Dark-backed Weaver ssp. *Ploceus bicolor sclateri* Roberts,
1931

Stripe-throated Jery ssp. *Neomixis striatigula sclateri*
Delacour, 1931

Olive Sunbird ssp. *Cyanomitra olivacea sclateri* **J. Vincent**,
1934

These birds are named after father and son, but it is not always clear which species was named after whom. Between them they described no fewer than 913 species; 189 non-passerines and 724 passerines! The father (Philip Lutley) wrote more descriptions, and certainly the majority of the birds are named after him. The antbird and sunbird genera were named after the father; the greenbul genus after the son. Some of the later descriptions of African birds, such as *Ploceus bicolor sclateri*, were definitely named after the son.

P. L. Sclater

Dr Philip Lutley Sclater (1829–1913) was a graduate of Oxford University and practised law for many years. He was the founder and first Editor of *Ibis* (1858–1865) and again (1877–1912), the journal of the BOU. He was also Secretary of the Zoological Society of London (1860–1902). Sclater's study of bird distribution resulted in the classification of the biogeographical regions of the world into six major categories. He later adapted his scheme for mammals, and it is still the basis for work in biogeography. He wrote *A Monograph of the Birds Forming the Tanagrine Genus Calliste* (1857), *Exotic Ornithology* (1866), *The Curassows* (1875), *A Monograph of the Jacamars and Puffbirds* (1879), *Birds of the Challenger Expedition* (1881) and *Argentine Ornithology* (1888). Six mammals and a reptile are named after him.

W. L. Sclater

William Lutley Sclater (1863–1944) was the son of Philip Lutley Sclater and the brother of Lilian Sclater (q.v.). Like his father, he was educated at Oxford, obtaining a first-class degree in Natural Science (1885). He was President of the BOU (1928–1933). For a few years he was Deputy-Superintendent of the Indian Museum in Calcutta, the first Director of the South African Museum, Cape Town (1896) and President of the South African Ornithologists' Union. Sclater resigned from the South African Museum (1906) and worked at the BMNH (1906–1936). He succeeded his father as Editor of *Ibis* (1913–1930). In July 1944 he was killed by a V1 flying bomb in London. He wrote *Systema Avium Aethiopicarum* (1924). Two mammals are named after him.

Scopoli

Scopoli's Shearwater *Calonectris diomedea* Scopoli, 1769

Johannes Antonius Scopoli (1723–1788) was an Italian-Austrian who studied medicine and became Professor of

Mineralogy and Metallurgy in Schemnitz (now Branska Stiavnica, Slovakia), and then Professor of Chemistry and Botany at the University of Pavia. His most famous publication was *Flora Carniolica* (1760). He corresponded with Linnaeus (q.v.) and adopted his system of classification. The drug scopalamine, an alkaloid of henbane which is a sedative and one of the so-called 'truth drugs', is named after him.

Scoresby

Scoresby's Gull *Leucophaeus scoresbii* **Traill**, 1823
[Alt. Dolphin Gull]

William Scoresby (1789–1857) was an English captain of a whaling ship and an arctic explorer. His father was taken out of school at the age of nine (by his father) to work on the family farm. When hired out to another farm he was ill-treated and decided not to follow the family trade but to seek a life at sea. He signed up at the age of ten, and by the 1790s had worked his way up to second officer, mostly aboard whalers. He invented the 'crow's nest' (c.1789). William Scoresby was apprenticed aboard his father's ship *Resolution* (1803), a Greenland whaler in which his father owned a one-eighth share. At 17 he became mate, and by 21 captain. He left the sea to study at Edinburgh University (1806), which sparked his interest in science. Nevertheless, Scoresby returned to sea in the Royal Navy without completing his studies, although he later met Sir Joseph Banks (q.v.) who persuaded him to return. He commanded the *Esk* (1813) which undertook scientific studies at sea, and he developed scientific instruments with Banks. He commanded the *Baffin* (1819), which was built to his specifications, and began his exploration of Arctic waters. He wrote *The Arctic Regions* (1820). Scoresby continued to explore and collect over the next few years and went on inventing scientific instruments. Many of his original instruments and much of his written work may be seen at the Whitby Museum. Later in life he took holy orders.

Scortecci

Blue Rock Thrush ssp. *Monticola solitarius scorteccii*
Moltoni, 1934 NCR
[JS *Monticola solitarius solitarius*]

Dr Giuseppe Scortecci (1898–1973) was an Italian zoologist and herpetologist. He took his doctorate at the University of Florence (1921) and joined the staff of the Institute of Comparative Anatomy there. He became Professor of Zoology, University of Genoa (1942). Before WW2 he explored in the Sahara, Italian Somaliland (Somalia) and Ethiopia. He produced around 50 publications on herpetology, particularly about the fauna of desert regions. Twelve reptiles and an amphibian are named after him.

Scott, A. E.

Chukar Partridge ssp. *Alectoris graeca scotti*
C. M. N. White, 1937 NCR
[JS *Alectoris chukar cypriotes*]

Algot Erling Scott-Skovso (1899–1943) was a Danish zoologist and ornithologist who led an expedition in Thailand and Malaya (Malaysia) on behalf the Tring Museum (1938–1940). He served as a lieutenant in the British Army (WW2) and was killed whilst serving with Special Operations Executive in Malaya. There is a reference in the BMNH archives to a list of parasites found on birds and mammals that he collected. White and Scott collected the holotype on Crete (1936).

Scott, G. F.

African Yellow White-eye sp. *Zosterops scotti* **Neumann**, 1899 NCR
[JS *Zosterops senegalensis stuhlmanni*]

Captain George Francis Scott Elliot (1862–1934) was a botanist, traveller and author. He took bachelors' degrees in mathematics at Cambridge (1882) and science at Edinburgh (1885). He went to South Africa (1885) and from there to Mauritius via Madagascar. His next expeditions were to Tripoli and Egypt; and, as botanist, with the French/English Commission to define the Sierra Leone boundary. From West Africa he set out for Uganda, where he collected reptiles. He was Professor of Botany, Glasgow Veterinary College (1896–1904). As he had been a soldier, he re-enlisted on the outbreak of WW1 (1914) at the age of 52. Posted to Egypt (1915), he fought at the Battle of Romani during which nearly all the men in his command were killed or wounded, and later fought at Gaza. He was ordered to return (1917) to the UK, but his ship was torpedoed. He survived, arriving home wearing an Italian officer's uniform and a pair of white slippers. He wrote *A Naturalist in Mid-Africa: Being an Account of a Journey to the Mountains of the Moon and Tanganyika* (1896). A reptile is named after him.

Scott, J. H.

Weka ssp. *Gallirallus australis scotti* **Ogilvie-Grant**, 1905

Professor John Halliday Scott (1851–1914) was a Scottish surgeon, anthropologist and watercolourist who settled in New Zealand (1877). He became Professor of Anatomy and Physiology and Dean of the Medical School, University of Otago. He qualified in medicine at the University of Edinburgh (1874) and became a Member of the Royal College of Surgeons (1877).

Scott, W.

Scott's Oriole *Icterus parisorum* **Bonaparte**, 1838

Winfield Scott (1786–1866) was a commander in the United States forces during the Mexican War. He was known (probably not to his face) as 'Old Fuss and Feathers'. Scott stood as a Whig for the Presidency (1852) and was defeated by Franklin Pierce. Although a Southerner, he remained loyal to the Union throughout the American Civil War and died at West Point, New York.

Scott, W. E. D.

Clapper Rail ssp. *Rallus longirostris scottii* **Sennett**, 1888
[Syn. *Rallus crepitans scottii*]
Scott's Sparrow *Aimophila ruficeps scottii* Sennett, 1888
[Alt. Rufous-crowned Sparrow ssp.]

William Earl Dodge Scott (1852–1910) was Curator of the Department of Ornithology at the Museum of Biology, Princeton University (1875–1910).

Scouler

Little Forktail *Enicurus scouleri* **Vigors**, 1832

Dr John Scouler (1804–1871) was a Scottish physician and naturalist who travelled extensively in the American Northwest with David Douglas (q.v.). He was appointed (1834) Professor of Mineralogy – and subsequently of geology, zoology, and botany – to the Royal Dublin Society, Ireland.

Scratchley

Grey-headed Munia ssp. *Lonchura caniceps scratchleyana* **Sharpe**, 1898

The original description contains no etymology. A likely person is Major-General Sir Peter Henry Scratchley (1835–1885), who was in the British Army (1854–1882). He was commissioned into the Royal Engineers (1854) and served in both the Crimean War and the Indian Mutiny. He was in Australia (1860–1863) in charge of building defences for Melbourne and Geelong. He served in England (1864–1877). He was commissioner of defences for all six Australian colonies and for New Zealand (1877–1882). After retiring he was special commissioner in New Guinea (1884–1885). He died from malaria. His brother-in-law was Thomas Alexander Browne, better known under his pen name of Rolf Boldrewood whose most famous Australian classic novel was *Robbery under Arms*.

Scripps, E. B.

Hairy Woodpecker ssp. *Dryobates villosus scrippsae* **L. M. Huey**, 1927 NCR
[NUI *Picoides villosus hyloscopus*]

Ellen Browning Scripps (1836–1932) was an English-born philanthropist, newspaper owner and columnist who emigrated to the USA with her father (1844). Among her many foundations is the Scripps Institute of Oceanography, California (1903). She was the aunt of R. P. Scripps (q.v.).

Scripps, R. P.

Scripps's Murrelet *Synthliboramphus scrippsi* Green & Arnold, 1939
[Syn. *Synthliboramphus hypoleucus scrippsi*]

Robert Paine Scripps (1895–1938) was an American newspaper baron who died mysteriously aged 42. He joined the company which his father had set up when he was 16 and became editorial director of the chain (1917). He became President and chief stockholder in the 1920s. He made many donations to the Scripps Institution of Oceanography (formerly the San Diego Marine Biological Institution) set up by his aunt Ellen Scripps (q.v.). He died unexpectedly aboard his yacht off Baja California, the cause of death being recorded as a 'throat hemorrhage'. To deepen the mystery, 12 years previously, his father had also died whilst aboard *his* yacht off the African coast.

Scully

Scully's Red-rumped Swallow *Hirundo rufula scullii* **Seebohm**, 1883 NCR
[JS *Cecropis daurica rufula*]

Dr John Scully (1846–1912) was Assistant Surgeon in the Indian Army (1872–1899), eventually being promoted to the rank of Lieutenant-Colonel. He was Nepal's Resident Surgeon (1876–1877). He made a collection of c.300 birds and was the first person to describe the status of birds in the Kathmandu valley. A mammal is named after him.

Seba

Lory sp. *Psittacus sebanus* Shaw, 1812 NCR
[Alt. Black-capped Lory; JS *Lorius lory*]

Albert Seba (1665–1736) was an extremely wealthy Dutch collector and apothecary who lived in Amsterdam and created the richest museum of his time. He sold a huge collection (1717) to the Russian Tsar, Peter the Great, and then re-started collecting. Linnaeus (q.v.) visited him (1735) and Seba's broad collection-cataloguing systems influenced him in the shaping of his own system, and many of Seba's animals became Linnaeus's holotypes. A mammal and three reptiles are named after him.

Seebohm

Seebohm's Feather-tailed Warbler *Amphilais seebohmi* **Sharpe**, 1879
[Alt. Grey Emu-tail; Syn. *Dromaeocercus seebohmi*]
Bay-crowned Brush Finch *Atlapetes seebohmi* **Taczanowski**, 1883
Marsh Tit *Parus seebohmi* **Stejneger**, 1892 NCR
[JS *Poecile palustris hensoni*]
Philippine Bush Warbler *Cettia seebohmi* **Ogilvie-Grant**, 1894
Benguet Bush Warbler *Bradypterus seebohmi* Ogilvie-Grant, 1895

Seebohm's Wheatear *Oenanthe oenanthe seebohmi* **Dixon**, 1882
[Alt. Northern Wheatear ssp.]
Japanese Pygmy Woodpecker ssp. *Dendrocopos kizuki seebohmi* **Hargitt**, 1884
Seebohm's Thrush *Turdus poliocephalus seebohmi* Sharpe, 1888
[Alt. Island Thrush ssp.]
Asian Short-toed Lark ssp. *Calandrella cheleensis seebohmi* Sharpe, 1890
Seebohm's Courser *Rhinoptilus cinctus seebohmi* Sharpe, 1893
[Alt. Heuglin's/Three-banded Courser]
Kentish Plover ssp. *Charadrius alexandrinus seebohmi* **Hartert** & Jackson, 1915

Henry Seebohm (1832–1895) was a British businessman and amateur ornithologist, oologist and traveller, who explored the Yenisey tundra, Siberia. Blakiston (q.v.) sent skins to England, which were the subject of a series of papers by Seebohm with Robert Swinhoe (q.v.). He employed Charles Dixon (q.v.) as a collector. Meinhertzhagen (q.v.) recorded in his diary that he used to visit Seebohm to examine his collection. Among other works he wrote *A History of British Birds* (1883) and *The Birds of Siberia* (1901). He died of influenza.

Sefton

Northern Cardinal ssp. *Cardinalis cardinalis seftoni*
L. M. Huey, 1940

Joseph Weller Sefton Jr (1881–1966) was an American banker, amateur ornithologist and philanthropist in San Diego. Stanford awarded his bachelor's degree (1904). He was president of the local natural history society and formed a foundation that bought and equipped a vessel for oceanic research (1948).

Segeth

White-bellied Storm-petrel ssp. *Fregetta grallaria segethi*
R.A.Philippi & **Landbeck**, 1860

Dr Karl Segeth (Seghet) (d.1890) was a German surgeon and naturalist. He was the surgeon-naturalist on board the Prussian vessel *Princess Louise* collecting on behalf of the Berlin Museum. A seaman on board the same vessel, Bernhard 'Bernardo' Philippi, worked as Segeth's assistant (1837). The two decided to go into partnership and jumped ship to become professional collectors in Chile. Bernardo Philippi was killed in Tierra del Fuego (1852). His brother Rodolpho Amando Philippi (1808–1904), a zoologist and palaeontologist, described the petrel.

Seilern

Black-banded Woodcreeper ssp. *Dendrocolaptes picumnus seilerni* **Hartert** & **Goodson**, 1917
Eurasian Wren ssp. *Troglodytes troglodytes seilerni* **Sassi**, 1937 NCR
[JS *Troglodytes troglodytes cypriotes*]

Josef Karl Franz Maria Johann Nepomuk Graf von Seilern und Aspang (1883–1939) was an Austrian ornithologist and collector, who founded the Lesná Museum in the South Moravian region of the Czech Republic.

Seimund

Lemon Dove *Haplopelia seimundi* **Sharpe**, 1904 NCR
[JS *Columba larvata inornata*]
Seimund's Sunbird *Anthreptes seimundi* **Ogilvie-Grant**, 1908
[Alt. Little Green Sunbird]
Seimund's Green Pigeon *Treron seimundi* **H. C. Robinson**, 1910
[Alt. Yellow-vented Green Pigeon]

Chat Flycatcher ssp. *Bradornis infuscatus seimundi* Ogilvie-Grant, 1913

Eibert Carl Henry Seimund (1878–1942) was a British taxidermist employed in the Zoology Department, BMNH (1897–1906). He collected in South Africa (1899–1903) whilst fighting in the South African War (1899–1902), in Fernando Po (1904) on a British Museum expedition, Thailand (1913) and Malaya (1916), having become Assistant Curator, Selangor State Museum, Kuala Lumpur (1906). The Raffles Museum in Singapore holds specimens from his last expedition. A mammal is named after him.

Selby

Selby's Flycatcher *Muscicapa selbii* **Audubon**, 1831 NCR
[Alt. Hooded Warbler; Syn. *Setophaga citrina, Wilsonia citrina*]

Prideaux John Selby (1788–1867) was a British landowner and squire with the time and money to devote to studying the fauna and flora on his estate, Twizell House in Northumberland. As a boy he had learned how to preserve and set up his specimens. He was an extremely skilled painter and engraver, producing the 19-part *Illustrations of British Ornithology* (1821–1834). Audubon (q.v.) knew and admired Selby, whom he mentions specifically in connection with this 'flycatcher' in his *Ornithological Biography*. Selby believed that character and mental capacity could be measured by feeling the bumps on the skull, as he became a corresponding member of the Phrenological Society (1821).

Sellow

Caatinga Antwren *Herpsilochmus sellowi* **B. M. Whitney** et al., 2000

Friedrich Sellow (1789–1831) was a German explorer and naturalist who first worked in the Botanical Gardens in Berlin. He then studied in Paris (1810) and London (1811–1814) before sailing for Rio de Janeiro. He travelled (1814–1831) in Brazil and Uruguay, sending many specimens to BMNH, which also now holds his journals. He collected with Wied (q.v.) and corresponded with Darwin (q.v.). He drowned in the Rio Doce in Minas Gerais, Brazil.

Selys

Thrush genus *Cichloselys* **Bonaparte**, 1854 NCR
[Now in *Turdus*]

Mountain Tanager sp. *Dubusia selysia* **Bonaparte**, 1851 NCR
[Alt. Buff-breasted Mountain Tanager; JS *Dubusia taeniata*]

Michel Edmond Baron de Selys-Longchamps (1813–1900) was a Belgian zoologist and politician who was a member of the Belgian parliament (1846) and President of the Senate (1880–1884). He was an entomologist, and a world expert on Odonata (dragonflies and damselflies). His private collection included a skin and an egg of the Great Auk *Pinguinus impennis*. The thrush genus is made by combining his name with the Greek word for thrush, *kikhle*.

Semenov

Common Redstart *Ruticilla semenowi* **Zarudny**, 1904 NCR
[JS *Phoenicurus phoenicurus samamisicus*]

European Roller ssp. *Coracias garrulus semenowi* **Loudon** & **Tschusi**, 1902
Cinereous Bunting ssp. *Emberiza cineracea semenowi* Zarudny, 1904
Brown Fish Owl ssp. *Ketupa zeylonensis semenowi* Zarudny, 1905
Caspian Snowcock ssp. *Tetraogallus caspius semenowtianschanskii* Zarudny, 1908

Pyotr Petrovich Semenov-Tian-Shansky (1827–1914) was a Russian statistician, geographer and entomologist who explored in Central Asia (1856–1857). He graduated in natural science at the University of St Petersburg (1848) and studied geography and geology in Switzerland, Germany, Italy and France (1853–1855). His surname was Semenov but he added 'Tian-Shansky' to his name (1906) to commemorate the 50th anniversary of his exploration and scientific study of the Tian-Shan mountains, after which Zarudny (q.v.) used the 'expanded' surname in the snowcock's trinomial. Semenov was Vice-Chairman of the Imperial Russian Geographical Society (1873–1914). As a statistician he conducted the first population census in Russia (1897) and created the system of European Russia's economic regions. His work was admired and used by Karl Marx and Lenin. He explored in Transcaspia and Turkistan (1888–1889). His son was Andrei Pyotrovich Semenov-Tian-Shansky (1866–1942), a Russian entomologist and taxonomist at the Museum of Zoology, Academy of Sciences, Leningrad (St Petersburg).

Semper, C. G.

> Caroline Islands White-eye *Zosterops semperi* **Hartlaub**, 1868
> [Alt. Citrine White-eye]

Carl Gottfried Semper (1832–1893) was a German ethnologist and zoologist. He was in the Philippines and Palau (1858–1865), where he was noted for his humane and unbiased attitude towards the indigenous cultures. Three reptiles are named after him.

Semper, J. E.

> Semper's Warbler *Leucopeza semperi* **P. L. Sclater**, 1877
> EXTINCT?

Reverend John E. Semper (DNF) was a parish priest and amateur ornithologist who was resident in St Lucia.

Semple

> Semple's Blue Jay *Cyanocitta cristata semplei* **Todd**, 1928

John Bonner Semple (1869–1947) was an American businessman, ornithologist and collector in Florida. He sponsored expeditions in Hudson Bay (1926), Michigan (1936) and Mexico (1939). He was a Trustee of the Carnegie Museum, Pittsburgh.

Sennett

> Sennett's (Olive-backed) Warbler *Parula pitiayumi* **Vieillot**, 1817
> [Alt. Tropical Parula; Syn. *Setophaga pitiayumi*]

> Sennett's White-tailed Hawk *Buteo albicaudatus hypospodius* **Gurney**, 1876
> Sennett's Nighthawk *Chordeiles minor sennetti* **Coues**, 1888
> [Alt. Common Nighthawk ssp.]
> Sennett's Thrasher *Toxostoma longirostre sennetti* **Ridgway**, 1888
> [Alt. Long-billed Thrasher ssp.]
> Seaside Sparrow ssp. *Ammodramus maritimus sennetti* **J. A. Allen**, 1888
> Sennett's Oriole *Icterus cucullatus sennetti* Ridgway, 1901
> [Alt. Hooded Oriole ssp.]

> Sennett's Titmouse *Baeolophus atricristatus sennetti* Ridgway, 1904
> [Alt. Black-crested Titmouse ssp.]

George Burritt Sennett (1840–1900) was a businessman and naturalist from New York. After schooling, he spent four years travelling in Europe and then managed the family iron foundry. He is best known for his studies of Texan birds, particularly those of the Lower Rio Grande Valley. He made collecting trips to Texas (1877, 1878 and 1882) and then paid collectors, including Frazar (q.v.), to send specimens from Texas and Mexico to him. His collection is now in the AMNH, New York. Sennett collected at Galveston, Corpus Christi, Brownsville and Hidalgo (1877). The account of that last expedition was an annotated list of 150 species. His trip along the coast (1878) recorded 168 species, including Sennett's White-tailed Hawk. He was one of the original members of the AOU and was Chairman of the Committee on the Protection of North American Birds (1886–1893). A mammal is named after him.

Serebrowsky

> Willow Ptarmigan ssp. *Lagopus lagopus sserebrowsky* **Domaniewski**, 1933
> Rock Bunting ssp. *Emberiza cia serebrowskii* **Johansen**, 1944 NCR
> [JS *Emberiza cia par*]

Pavel V. Serebrovsky (1893–1942) was a Russian ornithologist who worked at the Department of Ornithology, Zoological Institute, Leningrad (St Petersburg) (1924–1942), becoming Director (1928–1940). He wrote 'The influence of climate on the evolution of birds' (1925).

Serle

> Serle's Bush Shrike *Chlorophoneus kupeensis* **Serle**, 1951
> [Alt. Mount Kupe Bush-shrike]

> Xavier's Greenbul ssp. *Phyllastrephus xavieri serlei* **Chapin**, 1949
> [*P. xavieri* sometimes regarded as monotypic]
> Flappet Lark ssp. *Mirafra rufocinnamomea serlei* **C. M. N. White**, 1960
> African Black Swift ssp. *Apus barbatus serlei* **De Roo**, 1970

The Reverend Dr William Serle (1912–1992) was a Scottish physician, minister of religion, and ornithologist. He graduated as a doctor (1936) and sailed for Lagos (1937) to work for the Colonial Medical Service. He spent 20 years in West Africa, interrupted by service in India and Burma (WW2). He started to train for the Church (1956), becoming a minister in Kincardineshire, Scotland (1959), where he stayed until retirement (1987). He gave most of his collection of bird skins to BMNH, Tring, and his egg collection to the National Museums of Scotland. He co-wrote *A Field Guide to the Birds of West Africa*.

Serna

> Antioquia Wren *Thryophilus sernai* Lara et al., 2012

Brother Marco Antonio Serna Diaz (1936–1991) was a biologist and naturalist who was Professor of Ornithological

Biology at La Salle University, Colombia, and in charge of its museum (1970–1991). Two amphibians are named after him.

Serre

Rockhopper Penguin *Eudyptula serresiana* **Oustalet**, 1879 NCR
[Alt. Southern Rockhopper Penguin; JS *Eudyptes chrysocome*]

Rear-Admiral Paul Serre (1818–1890) was a French naval officer (1834–1880). He commanded a French squadron in the Pacific (1877). He wrote *Le Siège de Pylos, par le Contre-Amiral Serre* (1891).

Serrès

Imperial Pigeon genus *Serresius* **Bonaparte**, 1855 NCR
[Now in *Ducula*]

Red-breasted Coua *Coua serriana* **Pucheran**, 1845

Professor Antoine Etienne Renaud Augustin Serrès (1786–1868) was a French anatomist. He qualified as a physician in Paris (1810) and worked in various hospitals, then taught comparative anatomy at the Jardin des Plantes (1839). He became President, French Academy of Sciences (1841).

Serventy

Slender-billed Prion ssp. *Heteroprion belcheri serventyi* **Mathews**, 1935 NCR
[Now *Pachyptila belcheri* and regarded as monotypic]

Dr Dominic Louis Serventy (1904–1988) was an outstanding ornithologist, interested in all aspects of ornithology from biogeography and speciation to breeding seasons, and had a long-term influence on conservation and government policies. After a bachelor's degree from the University of Western Australia he took a doctorate at Cambridge (1933). He was Assistant Lecturer in Zoology, University of Western Australia (1934–1937), then worked for the Fisheries Division of CSIRO (1937–1951) and the Wildlife Survey Section in Perth (1951). He was a major contributor to scientific journals and *Western Australian Naturalist* (1947–1980), and was President, RAOU (1947–1949). He helped his sister Lucy, and brother Vincent, a well-known naturalist, to revive the Western Australian Naturalists' Club after WW2. He co-wrote *Birds of Western Australia*. A reptile is named after him.

Seth-Smith, D.

Buff-banded Rail ssp. *Gallirallus philippensis sethsmithi* **Mathews**, 1911

David William Seth-Smith (1875–1963) was an English ornithologist, aviculturist, bird artist and broadcaster. He was Curator of Mammals and Birds, Zoological Society of London (1909–1939). He presented nature programmes on the BBC as 'The Zoo Man'. He was the 14th of 22 children and elder brother of the subject of the next entry.

Seth-Smith, L. M.

Yellow-footed Flycatcher *Muscicapa sethsmithi* **Someren**, 1922

Seth-Smith's Crested Guineafowl *Guttera cristata sethsmithi* **Neumann**, 1908 NCR
[JS *Guttera pucherani verreauxi*]
Xavier's Bulbul ssp. *Phyllastrephus xavieri sethsmithi* **Hartert** & Neumann, 1910 NCR
[JS *Phyllastrephus xavieri xavieri*]

Leslie Moffat Seth-Smith (1879–1955) was a government employee, a surveyor in Uganda where he collected. He served in WW1 and was awarded a Military Cross. He was the 15th of 22 children and younger brother of the subject of the previous entry.

Setzer

Spotted Dove ssp. *Streptopelia chinensis setzeri* **Deignan**, 1955 NCR
[JS *Spilopelia chinensis chinensis*]

Dr Henry W. Setzer (1916–1992) was an American zoologist whose bachelor's and master's degrees were awarded by the University of Utah, and his doctorate by the University of Kansas. He served in the US Army (WW2) and moved to Washington DC (1947) where he was Assistant Curator (1948–1969), then Curator, at the USNM. He ran their African Mammal Project, and his teams of field collectors acquired c.63,000 specimens from across Africa. He wrote the *National Geographic Book of Mammals* (1998). Three mammals are named after him.

Severns

Severns's Crake *Porzana severnsi* **Olson** & James, 1991 EXTINCT
[Alt. Great Maui Crake]

R. Michael Severns (DNF) knew most of the important fossil locations on Maui, Hawaiian Islands, and willingly shared this with the crake's describers, Storrs Olson (q.v.) and Helen James.

Severtzov

Severtzov's Tit Warbler *Leptopoecile sophiae* **Severtzov**, 1873
[Alt. White-browed Tit Warbler]
Severtzov's Grouse *Tetrastes sewerzowi* **Przewalski**, 1876
[Alt. Chinese Grouse, Black-breasted Hazel Grouse]
Severtzov's Rosefinch *Carpodacus severtzovi* **Sharpe**, 1886
[Alt. Spotted Great Rosefinch]

Eurasian Jay ssp. *Garrulus glandarius sewerzowii* **Bogdanov**, 1871 NCR
[NUI *Garrulus glandarius brandtii*]
European Nightjar ssp. *Caprimulgus europaeus severzowi* **Zarudny**, 1907 NCR
[JS *Caprimulgus europaeus unwini*]
Himalayan Snowcock ssp. *Tetraogallus himalayensis sewerzowi* Zarudny, 1910

Nikolai Alekseevich Severtsov (1827–1885) [sometimes Severtsov, Severtzow, or Severzow] was a Russian zoologist and zoographer who explored in Central Asia. He is considered to be one of the Russian pioneers of ecology and evolutionary science. He wrote works on the zoogeographical division of the regions of the Palaearctic, and on the birds of Russia and Turkestan, including mapping migration routes. After becoming acquainted with Darwin's (q.v.) theory of natural selection he tested it against his own observations and became an eager supporter and propagandist for Darwinism. He made extensive collections on his travels, including 12,000 bird skins. He wrote several full-length works including: *Ornithology and Ornithological Geography of European and Asian Russia* (1867). A mountain peak in Pamiro-Alai and several glaciers in Pamir and Zailijaskoe are named after him, as are a number of fish and four mammals.

Severzov/Severzow

(See **Severtzov** above)

Sganzin

Comoro Blue Pigeon *Alectroenas sganzini* **Bonaparte**, 1854

Captain Victor Sganzin (d.1841) was a French naval gunnery officer and entomologist who was Governor of Île Sainte-Marie. He travelled in Madagascar (1831–1832) where he claimed to have seen the gigantic egg of the extinct 'elephant bird', and perhaps acquired it. It was rumoured that an egg was sold to the natural history dealers Verreaux (q.v.), but that the ship carrying it to France ran aground on the rocks of La Rochelle and all specimens lost. He wrote 'Notes sur les mammifères et sur l'ornithologie de l'île de Madagascar (1831 et 1832)'.

Sharpe, A.

Sharpe's Greenbul *Phyllastrephus alfredi* **G. E. Shelley**, 1903
[Alt. Malawi Greenbul; Syn. *Phyllastrephus flavostriatus alfredi*]

Baglafecht Weaver ssp. *Ploceus baglafecht sharpii* Shelley, 1898

Sir Alfred Sharpe (1853–1935) was the Crown's Commissioner and Consul-General for the British Central Africa Protectorate (1896–1910) and Nyasaland (Malawi) (1907). Sharpe started his career in Fiji but transferred to Africa as a professional hunter, and worked for Cecil Rhodes (1890–1896). He was an amateur naturalist. Two mammals are named after him.

Sharpe, R. B.

Weaver genus *Sharpia* **Bocage**, 1878 NCR
[Now in *Ploceus*]

Bornean Kingfisher sp. *Ceyx sharpei* **Salvadori**, 1869 NCR
[Alt. Oriental Dwarf Kingfisher; JS *Ceyx rufidorsa*]
Sharpe's Giant Kingfisher *Ceryle sharpii* **Gould**, 1869 NCR
[JS *Megaceryle maxima gigantea*]
Forest Rock Thrush *Monticola sharpei* **G. R. Gray**, 1871
Sharpe's Woodpecker *Picus sharpei* **Saunders**, 1872
[Alt. Iberian Green Woodpecker; Syn. *Picus viridis sharpei*]

Sharpe's Rosefinch *Carpodacus verreauxii* **David** & **Oustalet**, 1877
Sharpe's Wren *Cinnycerthia olivascens* Sharpe, 1882
Sharpe's Apalis *Apalis sharpii* **G. E. Shelley**, 1884
Sharpe's Pied Babbler *Turdoides sharpei* **Reichenow**, 1891
[Alt. Black-lored Babbler]
Sharpe's Frogmouth *Batrachostomus mixtus* Sharpe, 1892
[Alt. Bornean Frogmouth]
Sharpe's Rail *Gallirallus sharpei* **Büttikofer**, 1893
[Unique specimen – probably a morph of the Buff-banded Rail]
Sharpe's Starling *Pholia sharpii* **F. J. Jackson**, 1898
[Syn. *Poeoptera sharpii*]
Bee-eater sp. *Melittophagus sharpei* **Hartert**, 1899 NCR
[Alt. Little Bee-eater; JS *Merops pusillus cyanostictus*]
Galapagos Crake *Creciscus sharpei* **Rothschild** & Hartert, 1899 NCR
[JS *Laterallus spilonota*]
Samoan Triller *Lalage sharpei* Rothschild, 1900
Black-crowned White-eye *Zosterops sharpei* **Finsch**, 1901 NCR
[JS *Zosterops atrifrons*]
Tawny-browed Owl *Pulsatrix sharpei* **Berlepsch**, 1901 NCR
[JS *Pulsatrix koeniswaldiana*]
Yellow-rumped Antwren *Euchrepornis sharpei* Berlepsch, 1901
[Syn. *Terenura sharpei*]
Grey-headed Broadbill *Smithornis sharpei* **B. Alexander**, 1903
Sharpe's Akalat *Sheppardia sharpei* G. E. Shelley, 1903
Burmese Bushtit *Aegithalos sharpei* **Rippon**, 1904
Sharpe's Longclaw *Macronyx sharpei* F. J. Jackson, 1904
Sharpe's Lobe-billed Riflebird *Loborhamphus ptilorhis* Sharpe, 1908 NCR
[Presumed hybrid: *Parotia sefilata* x *Paradigalla carunculata*]

Piping Hornbill ssp. *Bycanistes fistulator sharpii* **D. G. Elliot**, 1873
Pied Goshawk ssp. *Accipiter albogularis sharpei* Oustalet, 1875
Cardinal Woodpecker ssp. *Dendropicos fuscescens sharpii* Oustalet, 1879
Sharpe's Drongo *Dicrurus ludwigii sharpei* Oustalet, 1879
[Alt. Square-tailed Drongo ssp.]
Yellow-throated Whistler ssp. *Pachycephala macrorhyncha sharpei* **A. B. Meyer**, 1884
Bananaquit ssp. *Coereba flaveola sharpei* **Cory**, 1886
Hooded Crow ssp. *Corvus cornix sharpii* **Oates**, 1889
Sharpe's Seedeater *Sporophila torqueola sharpei* **G. N. Lawrence**, 1889
[Alt. White-collared Seedeater ssp.]
Sharpe's Francolin *Pternistis clappertoni sharpii* **Ogilvie-Grant**, 1892
[Alt. Clapperton's Francolin ssp.; sometimes regarded as monotypic]
Black-faced Grassquit ssp. *Tiaris bicolor sharpei* Hartert, 1893
Eastern Bronze-naped Pigeon ssp. *Columba delegorguei sharpei* Salvadori, 1893
Bright-rumped Yellow Finch ssp. *Sicalis uropygialis sharpei* Berlepsch & Stolzmann, 1894
Chestnut-breasted Munia ssp. *Lonchura castaneothorax sharpii* **Madarász**, 1894

Sarus Crane ssp. *Grus antigone sharpii* **Blanford**, 1895

Sharpe's Lark *Mirafra africana sharpii* D. G. Elliot, 1897
[Alt. Rufous-naped Lark ssp.]

Black-billed Turaco ssp. *Tauraco schuettii sharpei*
Reichenow, 1898 NCR
[NUI *Tauraco schuettii emini*]

Kalij Pheasant spp. *Gennaeus lineatus sharpei* Oates, 1898
NCR
[JS *Lophura leucomelanos lineata*]

Rufous-crowned Roller ssp. *Coracias naevius sharpei*
Reichenow, 1899 NCR
[Alt. Purple Roller ssp.; JS *Coracias naevius naevius*]

Brimstone Canary ssp. *Serinus sulphuratus sharpii*
Neumann, 1900
[Syn. *Crithagra sulphurata sharpii*]

Red-tailed Laughingthrush ssp. *Trochalopteron milnei*
sharpei Rippon, 1901

Freckled Nightjar ssp. *Caprimulgus tristigma sharpei*
B. Alexander, 1901

African Green Pigeon ssp. *Treron calvus sharpei*
Reichenow, 1902

Guttate Piculet ssp. *Picumnus sagittatus sharpei* **Ihering**,
1902 NCR
[Alt. White-wedged Piculet; JS *Picumnus albosquamatus*
guttifer]

Puff-backed Honeyeater ssp. *Meliphaga aruensis sharpei*
Rothschild & Hartert, 1903

Singing Quail ssp. *Dactylortyx thoracicus sharpei*
E. W. Nelson, 1903

Grey Penduline-tit ssp. *Anthoscopus caroli sharpei* Hartert,
1905

Yellow-bellied Eremomela ssp. *Eremomela flaviventris*
sharpei Reichenow, 1905 NCR
[JS *Eremomela icteropygialis saturatior*]

Yellow-billed Blue Magpie ssp. *Urocissa flavirostris sharpii*
Parrot, 1907 NCR
[JS *Urocissa flavirostris cucullata*]

Yellow-rumped Tinkerbird ssp. *Pogoniulus bilineatus sharpei*
Ogilvie-Grant, 1907 NCR
[NUI *Pogoniulus bilineatus leucolaimus*]

Mottled Spinetail ssp. *Telacanthura ussheri sharpei*
Neumann, 1908

Grey-backed Camaroptera ssp. *Camaroptera brevicaudata*
sharpei **Zedlitz**, 1911

African Emerald Cuckoo ssp. *Chrysococcyx cupreus sharpei*
van Someren, 1922 NCR; NRM

Richard Bowdler Sharpe (1847–1909) was a British zoologist
who was Assistant Keeper of the Vertebrate Section of the
British Museum's Zoology Department (1895–1909) until his
death from pneumonia. His interest was in classification and
phylogeny and its relation to evolution. He wrote descriptions
of over 200 bird species. He was also Librarian to the Royal
Society of London (1867–1872) and co-author of the series
Birds of Europe. When he joined the BMNH (1872) as a Senior
Assistant in the Department of Zoology, he began his
27-volume *Catalogue of the Birds in the British Museum*
(1874–1898). Sharpe founded the British Ornithologists' Club
(1892) and edited its *Bulletin* for many years. He and his wife
Emily had many daughters, as evidenced by C. E. Jackson's

biography *Richard Bowdler Sharpe and His Ten Daughters*
(1994). (See also **Bowdler**)

Shattuck

Shattuck's Bunting *Emberiza shattuckii* **Audubon**, 1843 NCR
[JS Clay-coloured Sparrow *Spizella pallida* Swainson, 1832]

George Cheyne Shattuck (1783–1854) was a physician in
Boston. He contributed funds for the foundation of the Shattuck School there, which is named after him. Audubon (q.v.)
wrote: 'I have great pleasure in naming this species after my
worthy young friend George C. Shattuck, Esq., M. D., of
Boston, one of the amiable gentlemen who accompanied me
on my voyage to the coast of Labrador.'

Shaw

Common Pheasant ssp. *Phasianus colchicus shawii*
D. G. Elliot, 1870

Robert Barkley Shaw (1839–1879) was a British tea planter in
the Himalayas (1859–1868) and traveller in central Asia
(1868–1869). He became British Joint Commissioner, Ladakh
(1874), and later British Political Resident at the court of the
King of Burma (Myanmar) at Mandalay (1878–1879). He was
the first European to get to Yarkand and return alive to tell of
it, in *A Visit to High Tartary, Yarkand and Kashgar* (1871). He
died from rheumatic fever.

Shaw Mayer

Shaw Mayer's Bird-of-Paradise *Astrapia mayeri* Stonor,
1939
[Alt. Ribbon-tailed Astrapia/Bird-of-Paradise]

Dwarf Cassowary ssp. *Casuarius bennetti shawmayeri*
Rothschild, 1937 NCR; NRM
Papuan White-eye ssp. *Zosterops novaeguineae*
shawmeyeri **Mayr** & **Gilliard**, 1951 NCR
[JS *Zosterops novaeguineae wahgiensis*]

Frederick 'Fred' William Shaw Mayer (1899–1989) was an
Australian collector who worked for Hallstrom (q.v.) in Papua
New Guinea, and collected for Rothschild's (q.v.) Tring
Museum, also capturing live birds for the collections of Dela-
cour, Ezra, Spedan Lewis and Whitney (all q.v.). Seven
mammals are named after him.

Sheba

Guadalcanal Goshawk sp. *Astur shebae* **R. Sharpe**, 1888
NCR
[Alt. Variable Goshawk; JS *Accipiter hiogaster pulchellus*]

Etymology not explained by Sharpe. Possibly a pun: the bird
came from the *Solomon* Islands, and the Queen of Sheba
famously visited King Solomon.

Sheffler

Military Macaw ssp. *Ara militaris sheffleri* **Van Rossem** &
Hachisuka, 1939 NCR
[JS *Ara militaris mexicanus*]
Black Solitary Eagle ssp. *Harpyhaliaetus solitarius sheffleri*
Van Rossem, 1948

Northern Barred Woodcreeper ssp. *Dendrocolaptes sanctithomae sheffleri* **Binford**, 1965

William J. Sheffler (1893–1967) was an American businessman, ornithologist and aviculturist. He sponsored and led the Mexico expedition (1950) of Los Angeles County Museum where was Secretary (1951).

Shekar

Stork-billed Kingfisher ssp. *Pelargopsis capensis shekarii* **Abdulali**, 1964 NCR
[JS *Pelargopsis capensis osmastoni*]

P. B. Shekar (DNF) prepared skins for the describer.

Shelley, E.

Shelley's Wheatear *Saxicola shelleyi* **Sharpe**, 1877 NCR
[Alt. White-headed Black Chat, Arnot's Chat; Syn. *Myrmecocichla arnotti*; JS *Pentholaea arnotti*]

Sir Edward Shelley (1827–1890) was an English baronet and traveller in tropical Africa. He was a cousin of G. E. Shelley (q.v.) and both were related to the poet, Percy Bysshe Shelley.

Shelley, G. E.

Sunbird genus *Shelleyia* **J. A. Roberts**, 1922 NCR
[Now in *Cinnyris*]

Shelley's Eagle Owl *Bubo shelleyi* **Sharpe & Ussher**, 1872
Lovely Sunbird *Aethopyga shelleyi* Sharpe, 1876
Shelley's Red-throated Sunbird *Anthreptes rhodolaemus* Shelley, 1878
Nyanza Swift *Cypselus shelleyi* **Salvadori**, 1888 NCR
[JS *Apus niansae*]
Shelley's Francolin *Scleroptila shelleyi* **Ogilvie-Grant**, 1890
Shelley's Starling *Lamprotornis shelleyi* Sharpe, 1890
Shelley's Sparrow *Passer shelleyi* Sharpe, 1891
[Alt. Rufous Sparrow ssp; Syn *Passer motitensis shelleyi*]
Shelley's Greenbul *Andropadus masukuensis* Shelley, 1897
Shelley's Sunbird *Cinnyris shelleyi* **B. Alexander**, 1899
[Alt. Shelley's Double-collared Sunbird]
Barbet sp. *Capito shelleyi* **Dalmas**, 1900 NCR
[Alt. Red-headed Barbet; JS *Eubucco bourcierii aequatorialis*]
Shelley's Crimsonwing *Cryptospiza shelleyi* Sharpe, 1902
Shelley's Oliveback *Nesocharis shelleyi* Alexander, 1903
[Alt. Fernando Po Oliveback]

Sand Martin ssp. *Riparia riparia shelleyi* Sharpe, 1885
Red-and-yellow Barbet ssp. *Trachyphonus erythrocephalus shelleyi* **Hartlaub**, 1886
African Mourning Dove ssp. *Streptopelia decipiens shelleyi* Salvadori, 1893
Brimstone Canary ssp. *Serinus sulphuratus shelleyi* **Neumann**, 1903 NCR
[JS *Serinus sulphuratus sharpii*]
Mangrove Cuckoo ssp. *Coccyzus minor shelleyi* **Riley**, 1904 NCR; NRM

Captain George Ernest Shelley (1840–1910), nephew of the famous poet, was a geologist and amateur ornithologist. He was educated in England and the Lycée de Versailles, after which he joined the Grenadier Guards (1863), retiring a few years later with the rank of captain. The government of South Africa sent him on a geological survey. He wrote books on the birds of Egypt and a review of sunbirds, *A Monograph of the Nectariniidae* (1880). He collected in Africa, Australia and Burma but suffered a paralysing stroke (1906), which prevented him travelling.

Shelly

Golden-winged Sunbird ssp. *Drepanorhynchus reichenowi shellyae* **Prigogine**, 1952

Madame Shelly Prigogine (DNF) was the wife of the describer, Alexandre Prigogine (q.v.).

Shemley

White-browed Shama ssp. *Copsychus luzoniensis shemleyi* **duPont**, 1976

Lawrence P. Shemley is an aviator who was described by duPont (q.v.) as his 'pilot, friend and able assistant'. He landed his helicopter to go to the aid of two people whose plane had crashed in Pennsylvania, and flew them to hospital (1989).

Sheppard

Akalat genus *Sheppardia* **Haagner**, 1909

Violet-breasted Sunbird *Cinnyris sheppardi* **F. J. Jackson**, 1910 NCR
[JS *Cinnyris chalcomelas*]

Woodward's Batis ssp. *Batis fratrum sheppardi* Haagner, 1909 NCR; NRM
Böhm's Spinetail ssp. *Neafrapus boehmi sheppardi* **Roberts**, 1922

Peter A. Sheppard (1875–1958) was a British farmer, collector and oologist who lived in Rhodesia (Zimbabwe) and Mozambique (1898–1916).

Sherriff

Bar-winged Wren-babbler ssp. *Spelaeornis troglodytoides sherriffi* **Kinnear**, 1934

Major George Sherriff (1898–1967) was a British army officer, botanist, explorer and plant-hunter. He was Vice-Consul/Consul-General at Kashgar (1929–1934) and Resident in Lhasa (1943–1945). He explored in the Himalayas, Bhutan and Tibet (1933–1938, 1946–1947 and 1949). He retired to Scotland (1950) and created a Himalayan garden with every known species of primula.

Shestoperov

Chukar Partridge ssp. *Alectoris chukar shestoperovi* **Sushkin**, 1927 NCR
[NUI *Alectoris chukar koroviakovi*]

E. L. Shestoperov (1885–1940) was a Russian entomologist, explorer and collector in Turkmenistan.

Shimoizumi

Eurasian Jay ssp. *Garrulus japonicus shimoizumii*
Momiyama, 1939 NCR
[JS *Garrulus glandarius hiugaensis*]

Dr Jukichi Shimoizumi (DNF) was a Japanese zoologist who was President, Tsuru Bunka University, and Professor Emeritus, Tokyo University of Education. He introduced the concept of 'conservation education', especially into teacher training, based on ecology and sensibility to nature.

Shimokoriyama

Long-tailed Tit ssp. *Aegithalos caudatus shimokoriyamae*
Kuroda, 1923 NCR
[JS *Aegithalos caudatus magnus*]

Seiichi Shimokoriyama (b.1883) was a Japanese ornithologist, botanist, and collector who went to Korea (1911) and wrote *Hand-List of the Birds of Korea* (1914).

Shipton

White-shouldered Woodpecker *Dryocopus shiptoni*
Dabbene, 1915 NCR
[Colour-variant of Black-bodied Woodpecker, *Dryocopus schulzi*]

Paramo Pipit ssp. *Anthus bogotensis shiptoni* **Chubb**, 1923

Stewart Shipton (1869–1939) was an Englishman who spent most of his life in Argentina. He was employed as an accountant and then as general manager of a sugar mill at Concepcion. He was an amateur naturalist and ornithologist, and there is a Shipton Bird Collection at the Zoology Department of Fundación Miguel Lillo, Tucumán, Argentina. A mammal is named after him.

Shirai

Pacific Swift ssp. *Apus pacificus shiraii* Mishima, 1960 NCR
[JS *Apus pacificus pacificus*]

Kunihko Shirai (DNF) was a Japanese ornithologist and collector. He was Curator of the Nagasaki Aquarium (1967).

Sho

Narcissus Flycatcher ssp. *Zanthopygia narcissina shonis*
Kuroda, 1923 NCR
[JS *Ficedula narcissina owstoni*]

Kei Sho of Okinawa (1889–1922) was a younger brother of Marquis Sho and a member of the Ornithological Society of Japan.

Shore

Himalayan Flameback *Dinopium shorii* **Vigors**, 1832

Hon. Frederick John Shore (1799–1837) was a member of the East India Company's civil service. He wrote the 2-volume *Notes on Indian Affairs* (1837).

Shortridge

Australian Dove sp. *Geopelia shortridgei* **Ogilvie-Grant**,
1909 NCR
[Hybrid: *Geopelia cuneata* x *Geopelia placida*]
Australian White-eye sp. *Zosterops shortridgii* Ogilvie-Grant, 1909 NCR
[Alt. Silvereye; JS *Zosterops lateralis chloronotus*]

Rainbow Bee-eater ssp. *Merops ornatus shortridgei*
Mathews, 1912 NCR; NRM

Captain Guy Chester Shortridge (1880–1949) was originally trained as a geologist, but worked as a taxidermist for the South African Museum (1902–1903). He was an indefatigable zoological and entomological collector for the BMNH and AMNH. He worked for these clients in South-East Asia when he was attached to the Raffles Museum in Singapore, in Australia (1904–1907), and in Africa. On the effect of Europeans on Australia he wrote in dismay: 'Animals are dying out here as fast as Aboriginals – both are fading before the products of a tougher civilisation.' He was very far-sighted, and appealed to Western Australia to set aside reserves in which the fauna of the state could be preserved. He was the Director of the Kaffrarian Museum in King William's Town (1921–1949), which holds the Shortridge Mammal Collection. He led at least five Percy Sladen (q.v.) and Kaffrarian Museum expeditions. He wrote *The Mammals of South West Africa* (1934). Eleven mammals are named after him.

Shufeldt

Shufeldt's Junco *Junco hyemalis shufeldti* Coale, 1887
[Alt. Dark-eyed Junco ssp.]

Professor Dr Robert Wilson Shufeldt (1850–1934) was an expert working on extant and fossil bird bones (1885–1925). He made a number of expeditions to Africa but was profoundly racist and published *America's Greatest Problem: The Negro* (1915). Shufeldt served as a major in the medical corps in WW1.

Shulpin

Hawfinch ssp. *Coccothraustes coccothraustes schulpini*
Johansen, 1944
Willow Tit ssp. *Penthestes montanus shulpini* **Portenko**,
1954 NCR
[JS *Poecile montanus baicalensis*]

L. M. Shulpin (sometimes Schulpin) (1904–1942) was a Russian ornithologist, systematist, and geographer at Leningrad University. He researched on the Amur River (1926–1927) and in Kazakhstan (1933–1935). He wrote: *Game Birds and Birds of Prey in Primorye* (1936).

Siberg

Olive-winged Bulbul ssp. *Pycnonotus plumosus sibergi*
Hoogerwerf, 1965 NCR
[NUI *Pycnonotus plumosus plumosus*]

Johannes Alting Siberg (1800–1857) was a Dutch magistrate, colonial administrator, ornithologist and collector at Bawean,

Dutch East Indies (Indonesia). He wrote *Account of the Island Bawean* (1851).

Sick

Sick's Manakin *Lepidothrix vilasboasi* Sick, 1959
[Alt. Golden-crowned Manakin]
Orange-bellied Antwren *Terenura sicki* Teixeira & Gonzaga, 1983

Cinnamon Tanager ssp. *Schistochlamys ruficapillus sicki*
O. Pinto & Camargo, 1952

Helmut Sick (1910–1991) was a German-born Brazilian ornithologist who worked for the National Museum, Rio de Janeiro, for many years. He rediscovered (1978) the critically endangered Lear's Macaw *Anodorhynchus leari*. His particular interest was macaws and he was also responsible for the re-ordering of some genera in taxonomic sequence. It has been said that he 'was one of the most dedicated and active ornithologists ever to work in South America'. He wrote *Ornitologia Brasileira, uma Introdução* (1984). This work was translated into English (1993) as *Birds in Brazil – A Natural History.*

Sida

Japanese Paradise-flycatcher ssp. *Tchitrea atrocaudata sidai* **Momiyama**, 1932 NCR
[JS *Terpsiphone atrocaudata atrocaudata*]

Minoru Sida (DNF) collected the holotype (1928).

Sieber

Jay genus *Sieberocitta* **Coues**, 1903 NCR
[Now in *Aphelocoma*]

Sieber's Jay *Aphelocoma ultramarina* **Bonaparte**, 1825
[Alt. Transvolcanic Jay; *Pica sieberii* (Wagler, 1827) is a JS]

Franz Wilhelm Sieber (1789–1844) was a botanist and collector from Prague (Czech Republic), then in the Austro-Hungarian Empire. He travelled in Greece and Egypt, and circumnavigated the world (1822–1825), making an important collection of botanical specimens in Australia. His behaviour became progressively erratic, and he quarreled constantly with the Prague authorities. Eventually he was placed in the Prague insane asylum (1830–1844).

Siebers

Cinnamon-chested Flycatcher ssp. *Ficedula buruensis siebersi* **Hartert**, 1924
Clamorous Reed Warbler ssp. *Acrocephalus stentoreus siebersi* **Salomonsen**, 1928
Scarlet Minivet ssp. *Pericrocotus speciosus siebersi* **Rensch**, 1928

Hendrik Cornelis Siebers (1890–1949) was a Dutch zoologist and ornithologist in the East Indies (Indonesia). He worked for the Amsterdam University Zoological Museum (1920–1947) and took part in their central-east Borneo expedition (1925). He wrote *Fauna Buruana, Aves* (1930). A mammal and four reptiles are named after him.

Siebold

Waxwing genus *Sieboldornis* **Momiyama**, 1928 NCR
[Now in *Bombycilla*]

Siebold's Green Pigeon *Treron sieboldii* **Temminck**, 1835
[Alt. White-bellied Green Pigeon]
Siebold's Bunting *Emberiza sulphurata* Temminck & **Schlegel**, 1848
[Alt. Japanese Yellow Bunting]

Dr Philipp Franz Balthasar von Siebold (1796–1866) was a German physician, biologist, traveller and medical officer to the Dutch East Indian Army in Batavia and at the Dutch Trading Post, Dejima Island, Nagasaki, Japan. He taught Western medicine and treated Japanese patients, accepting ethnographic and art objects as payment, and established a boarding school in the outskirts of Nagasaki. Using local Japanese agents he collected in the interior (1823–1829). With the connivance of the Imperial librarian and astronomer, he copied a map of the northern regions of Japan, so upsetting the government that all his known Japanese contacts were imprisoned, his house was searched and many possessions confiscated. He packed all of his manuscripts, maps and books in a large lead-lined chest, which was then hidden. He was banished from Japan (1829) and was forced to leave behind his young Japanese mistress and a two-year-old daughter (shades *of Madame Butterfly*), returned to Holland, prepared his Japanese materials for publication, and was appointed by the King to advise on Japanese affairs (1831). The Japanese ban was lifted (1859) and he became (1861) chief negotiator for all European nations who were trying to establish trade links. However, his mission was a failure and he was pensioned off (1863). His eldest son served the Japanese government as an interpreter and adviser for external and financial affairs (1870–1911). He wrote much on Japan including *Fauna Japonica – Aves* (1844). In Nagasaki the Siebold Memorial Museum was founded to honour his contributions to the modernisation of Japan. Two reptiles and an amphibian are named after him.

Siemiradzki

Siemeradski's Siskin *Carduelis siemiradzkii* **Berlepsch** & **Taczanowski**, 1884
[Alt. Saffron Siskin]

Olive Spinetail ssp. *Cranioleuca obsoleta siemiradskii* **Stolzmann**, 1926 NCR; NRM

Dr Jozef Siemiradski (1858–1933) was a Polish geologist who was Professor of Geology at the University of Lvov (1901–1933). He was in Ecuador (1883) and Brazil (1882–1883, 1892 and 1895), where he did the first geographical survey of the state of Para. He wrote *Die Geognostischen Verhaltnisse der Insel Martinique* (1884).

Siemssen

Slaty Bunting *Emberiza siemsseni* G. H. Martens, 1906
[Syn. *Latoucheornis siemsseni*]

Gustav Theodor Siemssen (1857–1915) was a natural history collector, merchant, and German Consul-General in Fukien (China) (1903).

Sigman

Zapata Sparrow ssp. *Torreornis inexpectata sigmani*
M. J. Spence & B. L. Smith, 1961

Arthur Tucker Sigman (d.1959) was an American conservationist and amateur ornithologist in Pennsylvania.

Silberbauer

Neddicky ssp. *Cisticola fulvicapilla silberbaueri*
J. A. Roberts, 1919
[Alt. Piping Cisticola ssp.]

Conrad Christian Silberbauer (1863–1944) was a South African lawyer (admitted to the Bar 1885) and a prominent Freemason in the Transvaal.

Sillem

Sillem's Mountain Finch *Leucosticte sillemi* **Roselaar**, 1992
[Alt. Sillem's Rosy Finch]

Jérome Alexander 'Lex' Sillem (1902–1986) was a Dutch banker who became the Director of the Bank Mees & Hope in The Hague. He was also an amateur ornithologist. Together with two of his brothers, and with his nephews (who were brothers) of the Van Marle family, he founded the Sillem-Van Marle Society for the advancement of knowledge of ornithological science. The society attempted to assemble a large collection of study skins to investigate geographical variation in Palearctic birds, mostly by buying skins from dealers, but occasionally by funding collecting expeditions, e.g. to Romania and Portugal. The collection of c.10,000 bird skins was acquired by the Bird Department of the Zoological Museum, University of Amsterdam (1979). Of the four brothers, only Lex Sillem and Johan Gottleib van Marle (q.v.) were active field birdwatchers; the other two had a more marginal interest, offering financial support. Shortly after obtaining his BA at the Economic University of Rotterdam, Lex Sillem was zoological collector in the Third Netherlands Karakoram Expedition (1929–1930). It traversed the remote western Tibetan Plateau and overwintered in western Xinjiang. In Kashgar, where the expedition wintered as guests at the British Consulate-General, Sillem met Frank Ludlow (q.v.), a professional bird collector working for the BMNH and they travelled during the winter to explore the western fringes of the Takla Maklan desert, during which Sillem's bird-skinning skills improved. A spider genus is named after him.

Silliman

White-browed Shortwing ssp. *Brachypteryx montana
sillimani* **Ripley** & **Rabor**, 1962

Dr Robert Benton Silliman (1902–1988) was an American philanthropist who lived on Negros Island in the Philippines, where Silliman University is located. In spite of the locality and name, he was *not* related to Horace B. Silliman after whom the university is named. He wrote: *Pocket of Resistance – Guerrilla Warfare in Negros Island, the Philippines* (1980). The etymology says that he was honoured for: '… his interest in and active stimulation of biological research in the university.'

Silvestri

Antvireo genus *Silvestrius* **M. Bertoni**, 1901 NCR
[Now in *Dysithamnus*]

Streamcreeper sp. *Hydrolegus silvestrianus* Bertoni, 1901 NCR
[Alt. Sharp-tailed Streamcreeper; JS *Lochmias nematura*]

Professor Dr Filippo Silvestri (1873–1949) was a zoologist and entomologist. He attended Università degli Studi di Roma 'La Sapienza' (1892), later moving to Università degli Studi di Palermo, where he graduated (1896). He worked at the Institute of Comparative Anatomy, Rome (1896–1902), then went to Laboratorio di Zoologia Generale e Agraria della R. Scuola Superiore d'Agricoltura, Portico, becoming Director (1904–1949). He visited South America (c.1900). A reptile is named after him.

Simeon

Simeon's Gull *Larus belcheri* **Vigors**, 1829
[Alt. Belcher's Gull, Band-tailed Gull]
Simeon's Gull *Larus atlanticus* **Olrog**, 1958
[Alt. Olrog's Gull]

Our opinion is that these are not named after a person but are a transcription error. We believe that the original name was 'Simeon Gull', possibly referring to an area where it occurred rather than to a person who found it. The fact that the name has been used for two closely related species is due to Olrog's Gull having originally been treated as a subspecies of Belcher's Gull.

Simon, E. L.

Hummingbird genus *Simonula* **Chubb**, 1916 NCR
[Now in *Anthocephala*]

Sunangel sp. *Heliotrypha simoni* **Boucard**, 1892 NCR
[Hybrid? – *Heliangelus* sp. x *Eriocnemis cupreoventris*]
Simon's Emerald *Chlorostilbon vitticeps* Simon, 1910 NCR
[Alt. Blue-tailed Emerald; JS *Chlorostilbon mellisugus
    phoeopygus*]

Simon's Woodnymph *Thalurania furcata simoni* **Hellmayr**,
    1906
[Alt. Fork-tailed Woodnymph ssp.]
White-lored Spinetail ssp. *Synallaxis albilora simoni*
    Hellmayr, 1907
Cerise-throated Hummingbird ssp. *Selasphorus flammula
    simoni* **Carriker**, 1910
Swallow-tailed Hummingbird ssp. *Eupetomena macroura
    simoni* Hellmayr, 1929

Eugene Louis Simon (1848–1924) was a French arachnologist, ornithologist and expert on hummingbirds. He travelled and collected in France, Italy (1864), Spain (1865–1868), Corsica, Sicily and Morocco (1869), Tunisia and Algeria (1875), Egypt, Suez and Aden (1889–1890), the Philippines (1890–1891), Venezuela (1887–1888), South Africa (1893), and Ceylon (Sri Lanka) and southern India (1892). He wrote *Histoire Naturelle des Trochilidae* (1921). A mammal is named after him.

Simon (Fraser)

Woodpecker sp. *Dendropicos simoni* **Ogilvie-Grant**, 1900 NCR
[Alt. Cardinal Woodpecker ssp.; JS *Dendropicos fuscescens lepidus*]

Simon Fraser, 16th Lord Lovat and 3rd Baron Lovat (1871–1933), was an aristocrat and soldier who served in the 1st Life Guards. He raised the Lovat Scouts (1899), who fought in the South African War. In WW1 he commanded the Highland Mounted Brigade and was awarded the DSO. He led an expedition through Abyssinia (Ethiopia) from Berbera to the Blue Nile (1899), which collected widely. A mammal is named after him. (See also **Lovat**)

Simons

Puna Tapaculo *Scytalopus simonsi* **Chubb**, 1917

Simons's Brush Finch *Atlapetes seebohmi simonsi* **Sharpe**, 1900
[Alt. Bay-crowned Brush Finch ssp.]
Plumbeous Rail ssp. *Pardirallus sanguinolentus simonsi* Chubb, 1917
Rufous-bellied Seedsnipe ssp. *Attagis gayi simonsi* Chubb, 1918
Marbled Wood Quail ssp. *Odontophorus gujanensis simonsi* Chubb, 1919
Tawny-throated Dotterel ssp. *Oreopholus ruficollis simonsi* Chubb, 1919 NCR
[JS *Oreopholus ruficollis ruficollis*]
Red-backed Hawk ssp. *Buteo erythronotus simonsi* Swann, 1922 NCR
[JS *Buteo polyosoma polyosoma*]

Perry Oveitt Simons (1869–1901) was an American citizen who collected in South America, taking reptiles and amphibians in Peru (c.1900) and birds in Bolivia (1901). When crossing the Andes his lone guide murdered him. Chubb (q.v.) studied Simons' collections extensively in the second decade of the 20th century. There are more than a dozen holotypes in the BMNH which he collected. Four reptiles, two amphibians and a mammal are named after him.

Sin

White-necklaced Partridge ssp. *Arborophila ricketti sini* **Delacour**, 1930 NCR
[JS *Arborophila gingica*]

S. S. Sin (b.1891) was a botanist who was Professor and Head of the Biology Department, Sun Yat-sen University, Canton (China).

Sindel

Superb Parrot subgenus *Sindelia* Wells & Wellington, 1992 NCR
[Now in genus *Polytelis*]

Stanley 'Stan' Raymond Sindel (b.1935) is an Australian aviculturist who has been keeping birds since he was eight years old. He wrote *Mutations of Australian Parrots* (1986).

Sirindhorn/Sirintara

Princess Sirindhorn's Bird *Pseudochelidon sirintarae* Thonglongya, 1968
[Alt. White-eyed River Martin]

(See **Princess Sirindhorn**)

Sisson

Socorro Wren *Troglodytes sissonii* **Grayson**, 1868

Isaac Sisson (1828–1906) was the US Consul at Mazatlan, Sinaloa, Mexico. He was Commercial Agent in Mazatlan (1866), later becoming Consul (c.1870). He settled and married but became ill (1874) and returned to the USA, never to return. His in-laws raised his two children in Mexico although he had another child after his return to the USA, who lived with him and his wife in New York.

Sjöstedt

Sjöstedt's Barred Owlet *Glaucidium sjostedti* **Reichenow**, 1893
Sjöstedt's Greenbul *Baeopogon clamans* Sjöstedt, 1893
[Alt. White-tailed Greenbul]
Sjöstedt's Pigeon *Columba sjostedti* Reichenow, 1901
[Alt. Cameroon Olive Pigeon]

African Dusky Flycatcher ssp. *Muscicapa adusta sjostedti* **Grote**, 1936 NCR
[JS *Muscicapa adusta poensis*]

Bror Yngve Sjöstedt (1866–1948) was a Swedish entomologist and ornithologist. Between 1890 and 1891 he was in Cameroon collecting for the Uppsala University Zoological Department and for the State Natural History Museum. He joined the entomology section of the National Natural History Museum (1897) and went on an expedition to the USA and Canada to visit entomological stations and to study their methods (1898). He was part of the Swedish Zoological Expedition to Mount Kilimanjaro (1905–1906). He published extensively, including *Zur ornithologie Kameruns* (1895), and he edited the *Wissenschaftliche Ergenbisse der Schwedischen Expedition nach dem Kilimanjaro* (1905–1906). An amphibian and a reptile are named after him.

Skottsberg

Grey Noddy ssp. *Procelsterna albivitta skottsbergii* **Lönnberg**, 1921

Dr Carl Johan Fredrik Skottsberg (1880–1963) was a Swedish botanist and explorer in Antarctica (1901–1903) who led the Swedish Magellanic Expedition to Patagonia (1907–1909). His doctorate was awarded by Uppsala University (1907). He was a curator at Uppsala University Botanical Museum (1909–1914) and was in charge of the development of the botanical gardens in Gothenburg (1915), then Director and Professor (1919).

Skutch

Antbird genus *Skutchia* **Willis**, 1968
[Genus often merged with *Phlegopsis*]

Dr Alexander Frank Skutch (1904–2004) was an American ornithologist, collector, field naturalist and author who died aged 99. After his doctorate (1928) he left almost immediately for Panama to study birds. He established a farm, 'Los Cusingos', as his home in Costa Rica. He wrote hundreds of scientific papers and more than 40 books including *Life Histories of Central American Birds* (1954), *A Guide to the Birds of Costa Rica* (1990) and *Trogons, Laughing Falcons, and Other Neotropical Birds* (1999). He once wrote: 'For a large and growing number of people, birds are the strongest bond with the living world of nature. They charm us with lovely plumage and melodious songs; our quest of them takes us to the fairest places; to find them and uncover some of their well-guarded secrets we exert ourselves greatly and live intensely. In the measure that we appreciate and understand them and are grateful for our coexistence with them, we help to bring to fruition the age of long travail that made them and us. This, I am convinced, is the highest significance of our relationship with birds.'

Sladen, C.

Fernando Po Swift *Apus sladeniae* **Ogilvie-Grant**, 1904

Mrs Constance Sladen *née* Anderson (1848–1906) was the wife (1890) of Walter Percy Sladen, the British ornithologist (q.v.). They met (1870) but did not marry for 20 years. A contemporary described the union as '... a union of heart and mind, yielding a bright and tender sympathy, which strengthened and stimulated him in his life's work.' She donated his collection of echinoderms (starfish, etc.) to the Royal Albert Memorial Museum in Exeter (1903) but did not live to see the exhibit opened (1910). She also established the Percy Sladen Memorial Trust with an endowment of £20,000. She was an exhibiting artist of some repute and an authority on the archaeology of Yorkshire. She was also a natural historian, joining the Linnean Society when it first allowed women to become members (1904).

Sladen, W. P.

Sladen's Barbet *Gymnobucco sladeni* **Ogilvie-Grant**, 1907

Island Thrush ssp. *Turdus poliocephalus sladeni* Cain & Galbraith, 1955

Walter Percy Sladen (1849–1900) was a self-taught British academic biologist whose main interest was ichthyology. He is best known for the work he did on the specimens brought back by the *Challenger* expedition, which took years and broke his health. His travelling was restricted to visiting collections of material in European museums and working in Naples with Dr Anton Dohrn (see **Dohrn, H. W. L.**), the founder of the Stazione Zoologica di Messina 'Anton Dohrn'. He gained a reputation as a taxonomic 'splitter', as he declared many specimens to belong to new genera or species. Exeter Museum hold his huge collection of echinoderms. He was a member of the Zoological Society of London, as was Ogilvie-Grant (q.v.), which probably explains why a bird was named after an ichthyologist. He resigned as Secretary of the Linnean Society after 10 years (1895) because of illness and in order to manage his uncle's country estate. He collapsed and died in Florence when walking back to his hotel at the end of a six-week holiday in Italy.

Slater

Slater's Leaf Warbler *Phylloscopus ricketti* Slater, 1897
[Alt. Rickett's Leaf Warbler, Sulphur-breasted Warbler]

The Reverend Henry Horrocks Slater (1851–1934) was a Northamptonshire clergyman and zoologist. He was naturalist on the British Transit of Venus expedition aboard HMS *Challenger* (1872–1875), becoming the naturalist stationed on Rodrigues. He made a collection there which BMNH holds. He wrote *Manual of the Birds of Iceland* (1901).

Slatin

Green-winged Pytilia *Pytelia slatini* **Madarász**, 1914 NCR
[JS *Pytilia melba soudanensis*]

Southern Hyliota ssp. *Hyliota australis slatini* **Sassi**, 1914

Major-General Sir Rudolf Anton Carl Freiherr von Slatin, known as Slatin Pasha (1857–1932), was an Austrian soldier-of-fortune in the British and Egyptian service as Governor of Dara (1879), Governor-General of Darfur (1881–1883) – after which he was imprisoned by the Mahdists (1883–1895) – and Inspector-General of Sudan (1900–1914). He wrote *Fire and Sword in the Sudan* (1896). Madarász wrote of the pytilia: 'Therefore I think the Sudanese birds ... must get a new name, for which I propose to call *P. Slatini* in honour of His Exellency Baron K. Slatin – Pasha'.

Slevin

Hermit Thrush ssp. *Catharus guttatus slevini* **J. Grinnell**, 1901

Thomas Edwards Slevin (1871–1902) was an American ornithologist and collector.

Sloet

Golden Cuckooshrike *Campochaera sloetii* **Schlegel**, 1866

Baron Ludolf Anne Jan Wuilt Sloet van de Beele (1806–1890) was Governor of the Dutch East Indies (1860–1866). He was also a jurist and historian.

Sloggett

Lark-like Bunting ssp. *Emberiza impetuani sloggetti* **J. D. Macdonald**, 1957

Lieutenant General Sir Arthur Thomas Sloggett (1857–1929) was a Colonel in the Royal Army Medical Corps and in charge of No. 21 General Hospital at Deelfontein, Cape Colony, during the Boer War. He was on duty at the hospital (1900) until the cessation of hostilities. He eventually became Surgeon General and Director General of Medical Services in the British Army, and in that capacity received a report on the introduction of steel helmets (1916). He must have had an interest in natural history, as he presented a collection of mammals from Deelfontein to the BMNH. A mammal is named after him.

Smith, A.

Broadbill genus *Smithornis* **Bonaparte**, 1850
Owl genus *Smithiglaux* Bonaparte, 1854
[Often included within *Glaucidium*]

Snake Eagle genus *Smithaetus* **J. A. Roberts**, 1922 NCR
[Now in *Circaetus*]

Cape Penduline-tit *Aegithalus smithii* **Jardine**, 1831 NCR
[JS *Anthoscopus minutus*]
Smith's Chestnut-vented Sandgrouse *Pterocles gutturalis*
A. Smith, 1836
[Alt. Yellow-throated Sandgrouse]
Annobón Paradise Flycatcher *Terpsiphone smithii* **Fraser**,
1843
Karoo Thrush *Turdus smithi* Bonaparte, 1850
Smith's Shoveler *Anas smithii* **Hartert**, 1891
[Alt. Cape/South African Shoveler]

Smith's Helmet-shrike *Prionops plumatus talacoma*
A. Smith, 1836 NCR
[Alt. White-crested Helmet-shrike ssp.; JS *Prionops*
plumatus poliocephalus]
Common Fiscal ssp. *Lanius collaris smithii* Fraser, 1843
Golden-tailed Woodpecker ssp. *Campethera abingoni*
smithii **Malherbe**, 1845 NCR
[JS *Campethera abingoni abingoni*]
Chestnut-backed Sparrow Lark ssp. *Eremopterix leucotis*
smithi Bonaparte, 1850

Sir Andrew Smith (1797–1872) was a Scotsman who was a
surgeon in the Army Medical Service (1819) after graduating
from the University of Edinburgh. He was a scrupulously
accurate zoologist whose first love was reptiles. He was
posted to the Cape Colony, South Africa (1820–1837), and was
the first Superintendent of the South African Museum of
Natural History, Cape Town (1825). He travelled to
Namaqualand to discover more about its inhabitants (1828)
and later published a paper on the history and lives of
'Bushmen' (1831). He also led the first scientific expedition
into the South African interior (1834–1836). He wrote the
5-volume *Illustrations of the Zoology of South Africa* (1838–
1850). However, he returned both to Britain and to medicine,
becoming Principal Medical Officer at Fort Pitt, Chatham
(1841), and later Director General of the Army Medical
Services (1853), a post that included organising medical
services in the Crimean War (an enquiry cleared him of
charges of inefficiency and incompetence instigated by the
Times newspaper). Much of his private collection was given
to Edinburgh University and is now in the Royal Museum of
Scotland. Nine reptiles, two amphibians and four mammals
are named after him.

Smith, A. D.

White-eye sp. *Zosterops smithi* **Neumann**, 1902 NCR
[Alt. Abyssinian White-eye; JS *Zosterops abyssinicus*
jubaensis]

White-rumped Babbler ssp. *Turdoides leucopygia smithii*
Sharpe, 1895
Red-fronted Warbler ssp. *Urorhipis rufifrons smithi* Sharpe,
1895

(See **Donaldson Smith**)

Smith, A. P.

Black-breasted Wood Quail *Odontophorus smithianus*
Oberholser, 1932 NCR
[JS *Odontophorus leucolaemus*]

Smith's Nightingale Thrush *Catharus mexicanus smithi*
Nelson, 1909 NCR
[Alt. Black-headed Nightingale Thrush ssp.; JS *Catharus*
mexicanus mexicanus]

Austin Paul Smith (1881–1948) was an American bird and
plant collector in Mexico (1909), Guatemala (1919) and Costa
Rica (1920–1948).

Smith, C.

Wire-tailed Swallow *Hirundo smithii* **Leach**, 1818

Professor Dr Christen Smith (1785–1816) was a Norwegian
naturalist and physician who qualified at the University of
Copenhagen. He became Professor of Botany and Land
Economy at the newly founded University, Christiania (Oslo)
(1814). (He is sometimes described as Danish because
Norway was then Danish-owned.) He visited England (1814)
and met Sir Joseph Banks (q.v.). He studied all aspects of
natural history and science in the Canary Islands (1815). That
year Banks persuaded him to be the botanist and geologist
on a scientific expedition under Captain James Tuckey to
discover if the Congo River had any connection to the Niger
basin. Smith died on the expedition. The type specimen of the
swallow was taken during it, and described in 'Narrative of
an expedition to explore the river Zaire, usually called the
Congo, in South Africa, in 1816, under the direction of Captain
J. K. Tuckey, R.N. To which is added, The journal of Professor
Smith.' A reptile is named after him.

Smith, G.

Smith's Lark-bunting *Plectrophanes smithii* **Audubon**, 1844
[JS Smith's Longspur *Calcarius pictus* **Swainson**, 1832]

Dr Gideon B. Smith (1793–1867) was physician, editor and
friend of Audubon (q.v.), who named the bird for him although
Swainson had already described the species. A man of the
same name (and it may be our man) patented the first 'fake
gold coin' detector (1853).

Smith, Harry M.

Chestnut-capped Babbler ssp. *Timalia pileata smithi*
Deignan, 1955

Dr Harry Madison Smith (b.1918) was an American zoologist
at Columbia University, New York. He collected in Burma
(Myanmar) (1951–1952) and joined Springfield College,
Massachusetts (1962).

Smith, Hugh M.

Niltava sp. *Niltava smithii* **Riley**, 1929 NCR
[Alt. Vivid Niltava; JS *Niltava vivida oatesi*]

Stork-billed Kingfisher ssp. *Pelargopsis capensis smithi*
Mearns, 1909
Puff-throated Babbler ssp. *Pellorneum ruficeps smithi* Riley,
1924

Dr Hugh McCormick Smith (1865–1941) was an American physician (1888) and became an ichthyologist. He led an expedition that explored in the Philippines (1907–1910). He worked for the US Bureau of Fisheries (1886–1922), was Fisheries Advisor in Siam (Thailand) (1923–1934), and Curator of Zoology at the USNM (1935–1941).

Smith, J. E.

> Catbird sp. *Ptilonorhynchus smithii* **Vigors** & **Horsfield**, 1827 NCR
> [Alt. Green Catbird; JS *Ailuroedus crassirostris*]
> Smith's Bronzewing *Geophaps smithii* **Jardine** & **Selby**, 1830
> [Alt. Partridge Pigeon, Bare-eyed Partridge Bronzewing]

Sir James Edward Smith (1759–1828) was a Scottish botanist. He was founder and became the first President of the Linnaean Society of London (1788–1828). He was a friend of Sir Joseph Banks (q.v.) and wrote many botanical papers as well as *Flora Brittanica*, and the 4-volume *English Flora*.

Smith, M. A.

(See **Malcolm Smith**)

Smithe, F. B.

> White-breasted Wood Wren ssp. *Henicorhina leucosticta smithei* **Dickerman**, 1973

Frank Bertram Smithe (*né* Schmidt) (1892–1989) was an American engineer, ornithologist, archaeologist and conservationist. His degree in mechanical engineering was awarded by Columbia University, New York (1914), and he then worked in his family business. He wrote *Birds of Tikal* (1966).

Smithe, G.

> European Greenfinch ssp. *Carduelis chloris smithae* **Koelz**, 1939 NCR
> [JS *Carduelis chloris turkestanica*]

Geneva Josephine Smithe (1891–1968) was an American field naturalist. She was a leading member of the Michigan Audubon Society, and a friend of the describer Walter Koelz (q.v.).

Smithers

> Cape Bunting ssp. *Emberiza capensis smithersii* **Plowes**, 1951
> Quailfinch ssp. *Ortygospiza atricollis smithersi* **Benson**, 1955
> [Syn. *Ortygospiza fuscocrissa smithersi*]
> Flappet Lark ssp. *Mirafra rufocinnamomea smithersi* **C. M. N. White**, 1956
> Rattling Cisticola ssp. *Cisticola chiniana smithersi* **Hall**, 1956

Dr Reay Henry Noble Smithers (1907–1987) was a South African zoologist, ornithologist, botanist and collector who was educated in Scotland and England. He returned to South African and worked in his family's business (1928–1930) but decided that he wanted to work in natural history and joined an expedition to Angola (1930). He worked for the South

African Museum, Cape Town (1934–1939), and served in the South African forces (WW2) in the manufacture of explosives. He then worked in the chemical industry, but became Director of Museums, Rhodesia (Zimbabwe) (1948–1976). After retirement he continued to work as Curator of Vertebrates (1976–1978).

Smithson

> American Herring Gull *Larus smithsonianus* **Coues**, 1862
> Fruit Dove sp. *Ptilopus smithsonianus* **Salvadori**, 1893 NCR
> [Alt. Atoll Fruit Dove; JS *Ptilinopus coralensis*]

James Smithson (born James Lewis Macie) (c.1765–1829) was an illegitimate son of Hugh Smithson Percy, 1st Duke of Northumberland. At twenty-two he changed his surname to Smithson. He was a mineralogist, chemist and philosopher, and in 1787 became the youngest person ever to be elected to the Royal Society (1787). Although he never visited the USA, he left the huge sum of over £100,000 in his will '… to the United States of America, to found at Washington, under the name of the Smithsonian, an Establishment for the increase & diffusion of knowledge among men.' Thus the Smithsonian Institution was started (1846).

Smuts

> Courser genus *Smutsornis* **Roberts**, 1922
> [Often included in *Rhinoptilus*]

Field-Marshall Jan Christiaan Smuts (1870–1950) was a South African writer, philosopher and statesman who fought against the British in the Boer War but was reconciled and became Prime Minister of South Africa (1919–1924 and 1939–1948).

Smythies

> Yellow-bellied Warbler ssp. *Abroscopus superciliaris smythiesi* **Deignan**, 1947
> White-bellied Munia ssp. *Lonchura leucogastra smythiesi* Parkes, 1958

Bertram Evelyn 'Bill' Smythies (1912–1999) was an ornithologist, botanist and a British Forestry Officer in Burma (1934–1940) and Sarawak (1949–1964). He wrote *Birds of Borneo* (fourth edition 1999) and *Birds of Burma* (third edition 1986), and bequeathed a significant sum to various charities including the BTO and Oriental Bird Club (which now operates a Bertram Smythies Memorial Fund for conservation work).

Sneidern

> Montane Woodcreeper ssp. *Lepidocolaptes lacrymiger sneiderni* **Meyer de Schauensee**, 1945

Kjell Eriksson von Sneidern Johansson (1910–2000) was a Swedish naturalist and taxidermist who settled in South America and worked as a collector in Colombia for the Philadelphia Academy of Natural Sciences. He became Deputy Director (1946), and later Director, of the Natural History Museum at Universidad del Cauca (1946–1983). His son, Erik, runs shooting lodges in Colombia and Paraguay. A reptile is named after him.

Snethlage, Emil

Black-tailed Flycatcher ssp. *Myiobius atricaudus snethlagei*
Hellmayr, 1927

Dr Emil Heinrich Snethlage (1897–1939) was a German botanist, zoologist and ethnographer. He served in the German
Navy WW1 (1917–1919), receiving injuries which eventually
led to his death. He studied at Freiburg, Kiel and Berlin
Universities, from which he received his doctorate (1923). He
was in Brazil (1923–1926 and 1933–1935). On his first visit he
explored with his aunt, Emilie Snethlage (q.v.) of the Goeldi
Museum (see below), and later on his own in north-eastern
Brazil on behalf of the Field Museum, Chicago. He worked for
the Museum für Völkerkunde, Berlin (1927–1933) until leaving
for Brazil again to make ethnology collections for the
Baessler Foundation, Berlin.

Snethlage, Emilie

Tody Tyrant genus *Snethlagea* **Berlepsch**, 1909
[SII *Hemitriccus*]

Snethlage's Tody Tyrant *Snethlagea minor* Snethlage, 1907
[Syn. *Hemitriccus minor*]
Snethlage's Antpitta *Hylopezus paraensis* Snethlage, 1910
[Syn. *Hylopezus macularius paraensis*]

Snethlage's Gnateater *Conopophaga aurita snethlageae*
Berlepsch, 1912
[Alt. Chestnut-belted Gnateater ssp.]
Snethlage's Woodcreeper *Xiphocolaptes falcirostris
franciscanus* Snethlage, 1927
[Alt. Moustached Woodcreeper ssp.]
Red-billed Scythebill ssp. *Campylorhamphus trochilirostris
snethlageae* **Zimmer**, 1934
Snethlage's Tody Tyrant ssp. *Snethlagea minor snethlageae*
E. H. Snethlage, 1937
Santarem Parakeet ssp. *Pyrrhura amazonum snethlageae*
Joseph & Bates, 2002

Dr Henriette Mathilde Maria Elisabeth Emilie Snethlage
(1868–1929) was a German ornithologist (former assistant in
zoology at the Berlin Museum specialising in ornithology)
who collected in the Amazon rainforests (1905–1929), having
been recommended by Reichenow (q.v.) to the Goeldi
Museum. She succeeded Goeldi (q.v.) as head of the zoological section (1914) but was suspended (1917) when Brazil
entered WW1 against Germany, being reinstated after the
Armistice (1918). She wrote *Catalogo das Aves Amazonicas*
(1914). She also wrote on local languages. She was the first
woman scientist to direct a museum in Brazil and to work in
Amazonia. Six mammals, two reptiles and an amphibian are
named after her. (See also **Emilia**)

Snigirewski

Snigirewski's Small Whitethroat *Sylvia curruca snigirewskii*
Stachanow, 1929 NCR
[Alt. Desert Whitethroat ssp.; JS *Sylvia minula minula*]

Sergey Ivanovich Snigirewski (b.1901) was a Russian ornithologist and collector at the Zoological Institute, Leningrad
(St Petersburg).

Snouckaert

Orange-spotted Bulbul ssp. *Pycnonotus bimaculatus
snouckaerti* **Siebers**, 1928

Dr René Charles Edouard George Jean Baron Snouckaert
van Schauburg (1857–1936) was a Dutch ornithologist and
President of the Dutch Ornithological Society. There were
differences of opinion over the issue of shooting, and he set
up the Club of Dutch Ornithologists (1911), which was in
favour of shooting. The two societies later merged as the
Dutch Ornithological Union.

Snow, D. W.

Cotinga genus *Snowornis* Prum, 2001

Snow's Cotinga *Tijuca condita* Snow, 1980
[Alt. Grey-winged Cotinga]
Alagoas Antwren *Myrmotherula snowi* Teixeira & Gonzaga,
1985

White-fronted Tit ssp. *Parus semilarvatus snowi* **Parkes**,
1971

Dr David William Snow (1924–2009) was a Demonstrator at
the Edward Grey Institute of Field Ornithology at Oxford
University (1949–1956). He then served as Resident Naturalist
for the New York Zoological Society at its tropical field station
in Trinidad (1957–1961), and as Director of the Charles Darwin
Research Station in the Galápagos (1963–1964). Thereafter,
he was Director of Research for the British Trust for Ornithology (1964–1968) before becoming Senior Principal
Scientific Officer at the BMNH (1968) until his retirement
(1984). He wrote several books including *A Study of Black-
birds* (1958), *The Web of Adaptation* (1976) and *The Cotingas*
(1982). He was also an editor of the *Handbook of the Birds of
the Western Palearctic*, and wrote many scientific papers.

Snow, H. J.

Pigeon Guillemot ssp. *Cepphus columba snowi* **Stejneger**,
1897

Captain Henry James Snow (1848–1915) was an English
seaman, hunter, author and geographer who charted the
Kurile Islands (1873–1896). He wrote *Notes on the Kuril
Islands* (1897). The Snow Strait in the Kuriles is named after
him.

Soares

Hyacinth Visorbearer ssp. *Augastes scutatus soaresi*
Ruschi, 1963

Dr Júlio Soares (DNF) was a Brazilian zoologist.

Söderberg

Australasian Bush Lark ssp. *Mirafra javanica soderbergi*
Mathews, 1921
[Alt. Horsfield's Bush Lark ssp.]

Dr Rudolf Söderberg (1881–1958) was a Swedish ornithologist, entomologist and collector in East Africa (1908) and
Australia (1910–1912) with the Swedish Australia expedition.

After returning he became a teacher in Swedish secondary schools. His bachelor's degree was in biology awarded by Uppsala University and his (honorary) doctorate was awarded by Stockholm University shortly before his death.

Söderstrom

Söderström's Puffleg *Eriocnemis soderstromi* Butler, 1926 NCR
[Hybrid: *Eriocnemis luciani* x *Eriocnemis nigrivestis*]

Tourmaline Sunangel ssp. *Heliangelus exortis soderstromi* **Oberholser**, 1902 NCR; NRM
Ecuadorian Hillstar ssp. *Oreotrochilus chimborazo soderstromi* Lönnberg & Rendahl, 1922
Rufous-breasted Wood Quail ssp. *Odontophorus speciosus soderstromii* Lönnberg, 1922
Band-tailed Seedeater ssp. *Catamenia analis soderstromi* **Chapman**, 1924
Stripe-chested Antwren ssp. *Myrmotherula longicauda soderstromi* **Gyldenstolpe**, 1930

Ludovic Söderström (1843–1927) was the Swedish Consul-General in Quito, Ecuador (1912–1927), having earlier worked in Ecuador as a businessman. He collected natural history specimens for various museums. A mammal is named after him.

Soemmerring

Soemmerring's Pheasant *Syrmaticus soemmerringii* **Temminck**, 1830
[Alt. Copper Pheasant]

Western Jackdaw ssp. *Coloeus monedula soemmerringii* **Fischer**, 1811

Samuel Thomas von Soemmerring (1755–1830) was a German anatomist, physician, palaeontologist and notable freemason. He was author of a large body of work on anatomy. He was interested in many scientific and philosophical fields and corresponded with Goethe, Kant, Blumenbach (q.v.), J. G. A. Forster (q.v.) and von Humboldt (q.v.). He was Professor of Anatomy at Kassel and prepared an elephant for display. He also developed the first German hot air balloon. He was opposed to the use of the guillotine and the wearing of corsets! A number of parts of the human body are named after him; e.g. Soemmerring's ganglion. A mammal is also named after him.

Sokolnikov

Siberian Jay ssp. *Perisoreus infaustus sokolnikowi* Dementiev, 1935 NCR
[JS *Perisoreus infaustus yakutensis*]

Nikolai P. Sokol'nikov (fl.1914) was a Russian collector, naturalist and archaeologist who was on Bering Island (1911–1914). He had collected lichens (1903–1907) in Anadyrland and became Governor of the area (1927).

Sokolov

Fork-tailed Sunbird ssp. *Aethopyga christinae sokolovi* Stepanyan, 1985

Dr Leonid V. Sokolov (b.1949) is a Russian ornithologist at the Russian academy of Sciences' biological station at Kaliningrad (2001).

Solander

Solander's Cockatoo *Calyptorhynchus solandri* **Temminck**, 1821 NCR
[Alt. Glossy Black Cockatoo; JS *Calyptorhynchus lathami*]
Solander's Petrel *Pterodroma solandri* **Gould**, 1844
[Alt. Providence Petrel]

Daniel Carl [Karl] Solander (1733–1782) was a Swedish naturalist and explorer who was one of Linnaeus's (q.v. under Linné) pupils at Uppsala. He undertook an expedition to the extreme north of the Scandinavian Peninsula (1756). On Linnaeus's recommendation, he went to England to continue his natural history studies (1760). In London he associated with J. Ellis and P. Collinson and through them met Sir Joseph Banks (q.v.). It was Banks's influence which resulted in Solander sailing on Cook's (q.v.) first expedition on HMS Endeavour to the Southern Ocean, together with Banks and Sydney Parkinson (q.v.) (1768–1771). The botanical observations of Banks and Solander were published under the aegis of the British Museum (1900–1905) as *Illustrations of the Botany of Captain Cook's Voyage Round the World*. Solander also accompanied Banks on an expedition to Iceland (1772). He is credited with an unpublished manuscript 'Descriptions of plants from various parts of the world' (1767). He was the official Curator of the Duchess of Portland's considerable collection at Bulstrode in Buckinghamshire (1779). A monument was erected at Botany Bay, New South Wales (1914), to mark the spot where Cook, Banks and Solander landed in Australia (1770). He died in London and was buried in the Swedish Church there, although his remains were removed (1913) to the Swedish Cemetery in Woking.

Solange

Yellow-billed Nuthatch *Sitta solangiae* **Delacour** & **Jabouille**, 1930

Solange de la Rochefoucauld Princesse Murat (1894–1955) was the wife of the French naturalist and collector Prince Paul Murat (q.v.).

Solomko

Common Chaffinch ssp. *Fringilla coelebs solomkoi* **Menzbier** & **Sushkin**, 1913

Colonel J. Solomko (DNF) was a Russian soldier in the Crimea who assisted the chaffinch's describers during their expedition.

Somadikarta

Togian White-eye *Zosterops somadikartai* Indrawan *et al.*, 2008

Dr Soekarja Somadikarta (b.1930) is Indonesia's leading avian taxonomist and was mentor to the senior describer

Mochamad Indrawan. He is Emeritus Professor of Biology, Universitas Indonesia. The Free University of Berlin awarded his doctorate (1959). He was appointed Honorary President of the XXV International Ornithological Congress. Indrawan first spotted the white-eye on the remote Togian Islands, Indonesia in 1996. He and Somadikarta collaborated with well-known Asian bird specialist Pamela Rasmussen of Michigan State University to identify it as a new species.

Someren, N.

Kikuyu White-eye *Zosterops virens somereni* **Hartert**, 1928 NCR
[JS *Zosterops kikuyuensis*]

(See **Van Someren, N.**)

Someren, R. A. L.

Yellow-rumped Seedeater ssp. *Serinus atrogularis somereni* **Hartert**, 1912
[Syn. *Crithagra atrogularis somereni*]

(See **Van Someren, R. A. L.**)

Someren, V. G. L.

Streaky-breasted Flufftail *Sarothrura somereni* **Bannerman**, 1919 NCR
[JS *Sarothrura boehmi*]

Purple Grenadier ssp. *Uraeginthus ianthinogaster somereni* **Delacour**, 1943 NCR; NRM
Red-bellied Paradise-flycatcher ssp. *Terpsiphone rufiventer somereni* **Chapin**, 1948
Collared Sunbird ssp. *Anthodiaeta collaris somereni* Chapin, 1949
African Finfoot ssp. *Podica senegalensis somereni* Chapin, 1954
Klaas's Cuckoo ssp. *Chrysococcyx klaasi somereni* Chapin, 1954 NCR; NRM
[Note *klaasi* is an incorrect spelling of binomial *klaas*]

(See **Van Someren, V. G. L.**)

Somerville

Jungle Babbler ssp. *Turdoides striata somervillei* **Sykes**, 1832

Dr William Somerville (1771–1860) was a Scottish army surgeon. He was inspector-general of Canadian hospitals (1807–1811), becoming head of the Army Medical Department in Scotland (1813–1816). He explored in South Africa, where he served as a civil commissioner. He married (1812) his cousin Mary Greig, who became better known as the scientist and mathematician Mary Somerville.

Sommerfeld

Lyre-tailed Honeyguide *Melichneutes sommerfeldi* **Reichenow**, 1910 NCR
[JS *Melichneutes robustus*]

Lieutenant W. von Sommerfeld (DNF) was an officer in the German army in the Cameroons where he made a collection of ants.

Somov

Crested Tit ssp. *Parus cristatus somovi* Fediuschin, 1927 NCR
[JS *Lophophanes cristatus cristatus*]

Nikolaus Nikolaevich Somov (fl.1927) was a Russian ornithologist at the University of Kharkov (1897). He started gathering information on the avifauna of Kharkov province in the early 1870s.

Sonnerat

Banded Bay Cuckoo *Cacomantis sonneratii* **Latham**, 1790
[Syn. *Penthoceryx sonneratii*]
Sonnerat's Junglefowl *Gallus sonneratii* **Temminck**, 1813
[Alt. Grey Junglefowl]
White-browed Hawk Owl *Strix sonnerati* Temminck, 1820 NCR
[JS *Ninox superciliaris*]
Greater Green Leafbird *Chloropsis sonnerati* **Jardine** & **Selby**, 1827

Pierre Sonnerat (1748–1814) was a French explorer, naturalist and collector. He wrote *Voyage à la Nouvelle Guinée* (1776), although he never set foot on the island (only landing from a ship called *Isle de France* on nearby islands). He also wrote *Voyage aux Indes Orientales et à la Chine* (1782), both books being illustrated with engravings taken from his own drawings. He brought back many natural history specimens and other curiosities, which were exhibited by the Crown. He also recognised that India and China were the seats of ancient civilisations, writing about the former: 'We find among the Indians the vestiges of the most remote antiquity … We know that all peoples came there to draw the elements of their knowledge … India, in her splendour, gave religions and laws to all the other peoples; Egypt and Greece owed to her both their fables and their wisdom …' Sonnerat discovered the junglefowl on his voyage to the Far East (1774–1781).

Sonnini

Crested Eagle *Falco sonnini* **Shaw**, 1809 NCR
[JS *Morphnus guianensis*]

Crested Bobwhite ssp. *Colinus cristatus sonnini* **Temminck**, 1815

Charles Nicolas Sigisbert Sonnini de Manoncourt (1751–1812) was a French naturalist and explorer, who travelled in Egypt, the Greek islands and Asia Minor (Turkey), and lived for several years in Cayenne (French Guiana). He wrote the 127-volume *Histoire Naturelle* (1799–1808).

Sophia

White-browed Tit Warbler *Leptopoecile sophiae* **Severtzov**, 1873

Tsarina Sophia Maria Alexandrovna (1824–1880), formerly Princess Maximiliane Wilhelmine Auguste Sophie Marie, was the wife of Tsar Alexander II.

Sophia/Sophie

> Western Jackdaw ssp. *Coloeus monedula sophiae*
> Dunajewski, 1938 NCR
> [JS *Coloeus monedula soemmerringii*]

The original description contains no etymology and we have been unable to identify her. Andrzej (Andreas) Dunajewski (1908–1944) was the Curator of Ornithology at Museum of the Polish Academy of Science in Warsaw. He collected the holotype himself (1935) at Dolsk.

Sophie (Gairal)

> Sophia's Erythronote *Trochilus sophiae* **Bourcier** &
> **Mulsant**, 1846 NCR
> [Alt. Steely-vented Hummingbird; JS *Amazilia saucerrottei*]

Philiberte Sophie Gairal de Serezin *née* Tuffet (b.1815) was related to the junior author. Her mother's maiden name was Mulsant. Gould coined the common name 'Sophia's Erythronote' when depicting this hummingbird (1861).

Sophie (Garlepp)

> Golden Tanager ssp. *Tangara arthus sophiae* **Berlepsch**,
> 1901

Mrs Sophie Garlepp *née* Pölysius (DNF) was the mother of Otto and Gustav Garlepp (q.v.).

Sordahl

> White-throated Canary ssp. *Serinus albogularis sordahlae*
> **Friedmann**, 1932
> [Syn. *Crithagra albogularis sordahlae*]

Mrs Margaret Sordahl *née* Froiland (1906–1995) was a biologist who, with her brother and her husband Louis, was stationed at the USNM's astrophysical observatory in Namaqualand, South West Africa (Namibia). She made a collection of birds from the area (1929–1932) and collected the canary holotype.

Souancé

> Souancé's Maroon-tailed Parakeet *Pyrrhura melanura*
> *souancei* **J. Verreaux**, 1858

Charles Jacques Gabriel Guillier Baron de Souancé (1823–1896) was a French ornithologist with a particular interest in parrots. He began his career as a purser in the French navy. He was a nephew of Victor Masséna, Duc de Rivoli (see under **Massena** and **Rivoli**).

Soulé

> Ladder-backed Woodpecker ssp. *Picoides scalaris soulei* **R.**
> **Banks**, 1963

Dr Michael E. Soulé (b.1936) graduated (1959) BA from California State University and went on to study for his MA (1962) in Biology and then his PhD (1964) at Stanford University. Soulé and Banks were members of the Belvedere Scientific Expedition to the Gulf of California (1962) collecting, for example, rattlesnakes together and several of the eponymous woodpeckers. He went on to become a lecturer in

Zoology at the University of Malawi 1965–1967). He was then at the University of California San Diego (1967–1979). He was Director of a Buddhist Institute (1978–1983) and Associate Director of the Rocky Mountain Biological Laboratory (1981–1983). After being Associate Professor at the University of Michigan (1984–1989) he joined the University of California Santa Cruz, taking the chair of Environmental Studies (1989) and becoming Professor Emeritus after retirement (1999), and is now in private consultancy. Banks says in his etymology: 'This bird is named for my friend and field companion, Michael Soulé.' He has written a great many articles and papers and edited, written or co-written ten books including: *Conservation Biology: The Science of Scarcity and Diversity* (1986).

Souleyet

> Souleyet's Woodcreeper *Lepidocolaptes souleyetii* **Des**
> **Murs**, 1849
> [Alt. Streak-headed Woodcreeper]

Louis François Auguste Souleyet (1811–1852) was a French naval surgeon and naturalist in the Pacific (1836–1837). He was aboard the *Bonite*, which undertook a circumnavigation of the globe. He died of yellow fever in Martinique.

Soulie

> Soulie's Barwing *Actinodura souliei* **Oustalet**, 1897
> [Alt. Streaked Barwing]

> Brown Dipper ssp. *Cinclus pallasii soulei* Oustalet, 1892
> NCR
> [JS *Cinclus pallasii pallasii*]
> Bar-winged Wren-babbler ssp. *Spelaeornis troglodytoides*
> *souliei* Oustalet, 1898

Père (Father) Jean André Soulie (1858–1905) was a French missionary to China, and a botanist. He discovered a shrub, which is now very widespread in Europe, the buddleia or butterfly bush *Buddleia davidiii* and named it after Père David (q.v,) He was killed during a revolt by Tibetans (1905).

Soumagne

> Soumagne's Owl *Tyto soumagnei* **Grandidier**, 1878
> [Alt. Madagascar Grass Owl, Red Owl]

M. Soumagne (DNF) was a trader, oologist and honorary French Consul at Tamatave, Madagascar (1863–1877). He sent many ornithological specimens to Grandidier (q.v.).

Sousa

> Long-billed Tailorbird ssp. *Artisornis moreaui sousae* **C. W.**
> **Benson**, 1945

A. Baptista de Sousa (DNF) was the Portuguese Provincial Commissioner at Vila Cabral, Mozambique.

Souza

> Souza's Shrike *Lanius souzae* **Bocage**, 1878

> Souza's Tchagra *Tchagra australis souzae* Bocage, 1892
> [Alt. Brown-crowned Tchagra ssp.]

Common Waxbill ssp. *Estrilda astrild sousae* **Reichenow,**
1904
[SII *Estrilda astrild jagoensis*]

José Augusto de Souza (1837–1889) was a Portuguese ornithologist, Director of Ornithology at the Museum of Lisbon. He wrote numerous articles on African birds although he never visited Africa.

Sowerby, A.

Yellow-billed Grosbeak ssp. *Eophona migratoria sowerbyi*
Riley, 1915
Common Blackbird ssp. *Turdus merula sowerbyi* **Deignan,**
1951
Black-streaked Scimitar Babbler ssp. *Pomatorhinus gravivox decarlei* Deignan, 1952

Arthur de Carle Sowerby (1885–1954) was a naturalist, explorer and artist who was born in China, where his father was a Baptist missionary. He went to Bristol University but only stayed a short time before returning to China, where he began to collect specimens for the BMNH in Tai-yuan Fu. He collected mammals (1907) for the BMNH during an expedition to the Ordos Desert in Mongolia, and (1908) was part of the Clark Expedition to Shansi (Shanxi) and Kansu (Gansu) provinces. He wrote *Through Shên Kan, the Account of the Clark Expedition in North China 1908–09* (1912). R. S. Clark was a very wealthy man (he was heir to the Singer Sewing Machine fortune) and he financed a number of collecting trips for Sowerby. There was a revolution in China (1911) and Sowerby led an expedition to evacuate foreign missionaries from Shanxi and Shaanxi provinces. During WW1, Sowerby was a technical officer in the Chinese Labour Corps and saw service in France. After the war he settled in Shanghai and established *The China Journal of Science and Arts*, which he edited until the Japanese occupied Shanghai during WW2. The Japanese Army in Shanghai interned him for the duration, but despite this he appears to have been able to go on writing and publishing, as evidenced by 'Birds recorded from or known to inhabit the Shanghai area' (1943). He emigrated to the USA (1949) and lived the rest of his life in Washington DC, spending his time in genealogical research, which resulted in a family history, *The Sowerby Saga*. An amphibian and a reptile are named after him.

Sowerby, J. L.

Sowerby's Barbet *Stactolaema whytii sowerbyi* **Sharpe,**
1898
[Alt. Whyte's Barbet ssp.]

John Lawrence Sowerby (c.1875–1957) was in charge of a detachment of the British South African Mounted Police as well as being a natural history collector. He collected this race of barbet near Fort Chiquaqua along with 65 other birds (1897), while he was serving during the Matabele Rebellion (1896) in Southern Rhodesia (Zimbabwe). All were shot with his issued rifle – Sharpe said of this 'It is not given to every young ornithologist to shoot a Hoopoe with a bullet, and then make a good skin of it'. He and Sharpe wrote an article about the collection (1898). He wrote *Through the Mashonaland War with the BSAP* (1901).

Spalding

Spalding's Logrunner *Orthonyx spaldingii* **R. G. W. Ramsay,**
1868
[Alt. Chowchilla, Northern Logrunner]

Spalding's Butcherbird *Cracticus quoyi spaldingi* **Masters,**
1878
[Alt. Black Butcherbird ssp.]

Edward Spalding (1836–1900) was an Australian entomologist, collector and taxidermist. He also collected aboriginal artifacts, many of which are now part of the Macleay Museum Ethnographic Collection. He was employed by Macleay (q.v.) as a collector (1870s) and went with him (1875) on his expedition to Papua New Guinea. He was a taxidermist at the Queensland Museum (1880–1894). A reptile is named after him.

Spangenberg

White-winged Woodpecker ssp. *Dendrocopos leucopterus spangenbergi* **Gladkov,** 1951 NCR; NRM

Evgeniy Pavlovich Spangenberg (1898–1968) was a Russian ornithologist, naturalist and oologist who collected in Turkestan (1946). He served in the Red Army in post-revolutionary Russia until demobilised (1921). He graduated from Moscow University (1930) and then worked for the Research Institute for Game and Fur Farming and took part in many expeditions for them. He joined the permanent staff of Moscow University (1946) and became Senior Researcher, Section of Ornithology (1950). He was co-author of *Birds of the Soviet Union* (published in sections 1951–1954).

Sparrman

Greater Honeyguide *Indicator sparrmanii* **J. F. Stephens,**
1815 NCR
[JS *Indicator indicator*]

Dr Anders Erikson Sparrman (1748–1820) was a Swedish zoologist, explorer and collector in South Africa (1772–1776). He enrolled as a student at Uppsala University at the age of nine and studied medicine (at 14) under Linnaeus (q.v.). He went as ship's doctor to China (1765–1767) and later to Cape Town (1772) to become a tutor, but joined Captain Cook's (q.v.) second voyage (1773–1775) as an assistant naturalist to Johann and Georg Forster (q.v.). He returned to Cape Town (1775) to practise medicine and explore the interior. He returned to Sweden (1776) and was appointed Keeper of the natural history collections of the Academy of Sciences (1780) and Professor of Natural History and Pharmacology (1781). He took part in an expedition to West Africa (1787). He wrote (in English) *A Voyage to the Cape of Good Hope, Towards the Antarctic Polar Circle, and Round the World: But chiefly into the Country of the Hottentots and Caffres, from the Year 1772 to 1776* (1789). The Asteroid 16646 Sparrman is named after him.

Spatz

Barbary Partridge ssp. *Alectoris barbara spatzi* **Reichenow**, 1895

Common Kingfisher ssp. *Alcedo ispida spatzii* **Erlanger**, 1900 NCR

[JS *Alcedo atthis atthis*]

Whinchat ssp. *Saxicola rubetra spatzi* Erlanger, 1900 NCR; NRM

Pale Crag Martin ssp. *Ptyonoprogne obsoleta spatzi* Geyr von Schweppenburg, 1916

Common Ostrich ssp. *Struthio camelus spatzi* **Stresemann**, 1926 NCR

[JS *Struthio camelus camelus*]

Short-toed Treecreeper ssp. *Certhia brachydactyla spatzi* Stresemann, 1926 NCR

[JS *Certhia brachydactyla dorotheae*]

Kentish Plover ssp. *Charadrius alexandrinus spatzi* **Neumann**, 1929 NCR

[JS *Charadrius alexandrinus alexandrinus*]

Paul W. H. Spatz (1865–1942) was an explorer, collector and dealer in natural history specimens in north-west Africa. He lived in Gabès (Tunisia) (1893–1928) and covered the countries from Tunisia to Gambia. After returning he set up a company in Diemitz (Germany). An amphibian is named after him.

Speke

Speke's Weaver *Ploceus spekei* **Heuglin**, 1861

Captain John Hanning Speke (1827–1864) was a British explorer. He was the first European to see Lake Victoria (Lake Nyanza) and it was he who proved it to be the source of the Nile. Speke joined Richard Burton's (q.v.) expedition to discover the source of the Nile, not because he was particularly interested in finding it, but more because he wanted the chance to hunt big game. By the time he parted from Burton, who went on to Lake Tanganyika, Speke too had caught the source-location obsession. Speke hunted to supply the expedition, but he also observed the behaviour and ecology of birds. His own shotgun killed him when he stumbled over a stile whilst out shooting in England, although some believe it was suicide. Three mammals and two reptiles are named after him.

Spence, J. M.

Red-legged Tinamou ssp. *Crypturellus erythropus spencei* **Brabourne** & **Chubb**, 1914

James Mudie Spence (1836–1878) was an English traveller, naturalist and collector. He travelled in Norway, Arizona and California before visiting Venezuela (1871–1872). He wrote *The Land of Bolivar* (1878).

Spence, R. A.

Indian Blackbird ssp. *Turdus simillimus spencei* **Whistler** & **Kinnear**, 1932 NCR

[NUI *Turdus simillimus nigropileus*]

Sir Reginald Arthur Spence (1880–1961) was a British businessman and naturalist in India, where he first went (1901) to join Phipson & Co. He served in the Bombay Light Horse (1901–1922), was Honorary Secretary of the Bombay Natural History Society (1920–1933), and was an active Freemason and became a member of the Council of State for India (1930). In WW2 he was a Captain in the Sussex Home Guard.

Spence, W.

Spence's Sunangel *Heliangelus clarisse spencei* **Bourcier**, 1847

[Alt. Mérida Sunangel; Syn. *Heliangelus amethysticollis spencei*]

William Spence (1783–1860) was an English entomologist. He was a close friend of the Reverend W. Kirby, a fellow entomologist, with whom he wrote and published the 4-volume *Introduction to Entomology* (1815–1826). He was the founder of the Entomological Society (1833), later becoming President (1847). His son, William Blundell Spence (1814–1900), who became a noted artist, was also present at the founding and was given the task of maintaining relations with foreign entomologists (Bourcier was certainly one of these and had various papers read at meetings of the society in London). He later moved to Italy but remained a member of the society the rest of life.

Spengel

Spengel's Parrotlet *Forpus xanthopterygius spengeli* **Hartlaub**, 1885

[Alt. Turquoise-rumped Parrotlet]

Johann Wilhelm Spengel (1852–1921) was a German zoologist who was Director of the Bremen City Natural History and Ethnography Museum. He wrote *Die Fortschritte des Darwinismus* (1874) and *Die Enteropneusten des Golfes von Neapel und der Angrenzenden Meeres-Abschnitte* (1893). He founded the journal *Zoology* as *Zoologische Jahrbücher* (1886).

Spillmann

Spillmann's Tapaculo *Scytalopus spillmanni* **Stresemann**, 1937

Professor Dr Franz Spillmann was an Austrian zoologist who lived in Ecuador. Whilst there he discovered (1928) the complete skeleton of a mastodon. His collection of fossils and other specimens was donated to the National Polytechnic School in Quito, Ecuador (c.1946). He became a Curator at the Upper Austria Federal Museum (1946–1948).

Spix

Spix's Guan *Penelope jacquacu* Spix, 1825

Spix's Warbling Antbird *Hypocnemis striata* Spix, 1825

Spix's Woodcreeper *Xiphorhynchus spixii* **Lesson**, 1830

Spix's Macaw *Cyanopsitta spixii* **Wagler**, 1832

[Alt. Little Blue Macaw]

Spix's Spinetail *Synallaxis spixi* **P. L. Sclater**, 1856

[Alt. Chicli Spinetail]

Spix's Sawbill *Grypus spixi* **Gould**, 1860 NCR

[Alt. Hook-billed Hermit; JS *Glaucis dohrnii*]

Chachalaca sp. *Ortalis spixi* **Hellmayr**, 1906 NCR
[Alt. Buff-browed Chachalaca; JS *Ortalis superciliaris*]

White-winged Becard ssp. *Pachyramphus polychopterus spixii* **Swainson**, 1838

Johann Baptist Ritter von Spix (1781–1826) was a German naturalist working in Brazil (1817–1820). He gained his PhD aged 19! He studied theology for three years in Würzburg, then medicine and the natural sciences, qualifying as a medical doctor (1806). He was awarded a scholarship by the King of Bavaria (1808) and went to Paris to study zoology. At that time Paris was *the* centre for the natural sciences, with renowned scientists such as Cuvier (q.v.), Buffon (q.v.), Lamarck and Etienne Geoffroy de Saint-Hilaire (q.v.) at the height of their reputations. The King appointed him Assistant to the Bavarian Royal Academy of Sciences (1810) with special responsibility for the natural history exhibits. A group of academicians was invited to travel to Brazil (1816) and King Maximilian I agreed that two members of the Bavarian Academy of Sciences should accompany them. When Spix went to South America (1817), Natterer (q.v.) was also on board. Spix returned (1820) with specimens of 85 species of mammal, 350 birds, 130 amphibian, 116 fish and 2,700 insects as well as 6,500 botanical items. His party also brought back 57 species of living animals, mainly monkeys, parrots and curassows. This was to form the basis for the Natural History Museum in Munich. The King awarded him a knighthood and a pension for life. When he returned he catalogued and published his findings despite extremely poor health caused by his stay in Brazil. The 3-volume report on the expedition was published in 1823, 1828 and 1831. He wrote *Avium Brasiliensium Species Novae* (1824), which included a description of the Hyacinth Macaw he dedicated to his royal sponsor as *Anodorhynchus maximiliani* [now *A. hyacinthinus*]. Eight mammals, four amphibians and five reptiles are named after him.

Sprague

Sprague's Pipit *Anthus spragueii* **Audubon**, 1844

Isaac S. Sprague (1811–1895) was a self-taught artist who accompanied Audubon (q.v.) on his Missouri trip (1843), during which Edward Harris (q.v.) and John Bell (q.v.) shot a small brown bird near the mouth of the Yellowstone River. Audubon recognised it as a new species and named it 'Sprague's Missouri Lark'. Apparently, a few days later, whilst sitting drawing specimens, Sprague put down his drawing tools and found his namesake's nest and eggs in a mound of prairie grass. Sprague became America's best-known botanical illustrator, and was chosen to illustrate Asa Gray's classics, the *Botanical Textbook* and *Flora*. He also worked with the naturalist Henry Henshaw (q.v.).

Spurrell

Scops owl sp. *Scops spurrelli* **Ogilvie-Grant**, 1912 NCR
[Alt. Sandy Scops Owl; JS *Otus icterorhynchus*]

Professor Dr Herbert George Flaxman Spurrell (1877–1918) was a British physician and zoologist, a Fellow of the Zoological Society who collected in both Ghana and Colombia. He

wrote *Modern Man and His Forerunners: A Short Study of the Human Species Living and Extinct* (1917). He served in the Royal Army Medical Corps (WW1) as a Captain. He died in Egypt of pneumonia. Three mammals, two amphibians and three reptiles are named after him.

Ssaposhnikow

Black-headed Penduline-tit ssp. *Remiz macronyx ssaposhnikowi* **Johansen**, 1907

(See **Saposhnikoff**)

Sserebrowsky

Willow Ptarmigan ssp. *Lagopus lagopus sserebrowsky* **Domaniewski**, 1933

(See **Serebrowsky**)

St Cuthbert

St Cuthbert's Duck *Somateria mollissima* **Linnaeus**, 1758
[Alt. Common Eider, Cuddy's Duck]

St Cuthbert (c.635–687) was an English soldier, monk, hermit and Prior and Bishop at the monastery of Lindisfarne off the Northumbrian coast, where he actively provided protection for the eider. He was buried on Lindisfarne but the monks had to flee to the mainland to avoid Danish invaders (875). His body, which was greatly reverenced as a source of miracles, was moved from place to place to avoid the Danes and even as late as the 11th Century was restored to Lindisfarne to avoid the attentions of William the Conqueror. Finally (1104) he was buried in Durham Cathedral, where his shrine can be seen to this day. The use of the name Cuddy's Duck appears to be restricted to Northumberland.

St John

St John's Black Hawk *Buteo lagopus sanctijohannis* **J. F. Gmelin**, 1788
[Alt. Rough-legged Hawk ssp.]

This is an obsolete English name for an American bird. It is probably named after St John's in Newfoundland, Canada, where the species certainly occurs. According to tradition, the city earned its name when explorer John Cabot became the first European to sail into the harbour, on 24 June 1497, the feast day of Saint John the Baptist.

St John, O. B. C.

Middle Spotted Woodpecker ssp. *Dendrocopos medius sanctijohannis* **Blanford**, 1873

Colonel Sir Oliver Beauchamp Coventry St John (1837–1891) joined the Bengal Army of the Honourable East India Company (1856). He was on a special mission (1860) to Persia (Iran) to try to improve the speed of communication between London and India. He was Chief Commissioner in Baluchistan (1877–1887) and Chief Commissioner of Mysore (1889–1891). He returned to Baluchistan but died of pneumonia shortly after. A reptile is named after him.

St Thomas

St Thomas's Mango *Lampornis virginalis* **Gould**, 1861 NCR
[Alt. Antillean Mango; JS *Anthracothorax dominicus aurulentus*]

This bird is named after an island, rather than directly after a saint. St Thomas is a Caribbean island in the American Virgin Islands.

Stachanov

Pine Bunting ssp. *Emberiza leucocephalos stachanowi* **Boetticher**, 1935 NCR
[JS *Emberiza leucocephalos leucocephalos*]

W. S. Stachanov (DNF) was a Russian ornithologist.

Staebler

Least Tern ssp. *Sternula antillarum staebleri* Brodkorb, 1940 NCR
[NUI *Sternula antillarum browni*]

Dr Arthur Eugene Staebler (1915–2007) was an American palaeontologist, zoologist, biologist and ornithologist. University of Michigan awarded all three of his degrees. He was Director of the W. W. Kellogg Biological Station, Michigan State University (1948–1954). He taught at California State University, Fresno (1955–1980), retiring as Professor Emeritus of Biology.

Stager

Stager's Piculet *Picumnus subtilis* Stager, 1968
[Alt. Fine-barred Piculet]

Puff-throated Babbler ssp. *Pellorneum ruficeps stageri* **Deignan**, 1947

Dr Kenneth Earl Stager (1915–2009) was a mammalogist and ornithologist who joined Los Angeles County Natural History Museum as a volunteer (1930s) and became a full-time employee (1941) after the University of California, Los Angeles, awarded his bachelor's degree in zoology (1940). He served in the US Army (WW2) in Asia, including action with a special forces (Merrill's Marauders) for which he was decorated. He was in charge of the ornithology and mammal laboratory of the USA Typhus Commission Burma-India Field Party. He was discharged from the army as a Captain (1946) and went to the University of Southern California, which awarded his master's (1953) and doctorate (1962). He retired from Los Angeles County Natural History Museum (1976) as Emeritus Senior Curator of Ornithology and Mammalogy. He was a Fellow of the AOU and an honorary life member of the Cooper Ornithological Society. He, with others, wrote a number of booklets on the results of various collecting expeditions in East Africa. He demonstrated that the Turkey Vulture *Cathartes aura* relies on a keen sense of smell to find carrion, and that some other birds such as honeyguides (Indicatoridae) have well-developed olfactory lobes of the brain.

Stair

Tongan Ground Dove *Gallicolumba stairi* **G. R. Gray**, 1856
[Alt. Friendly Ground Dove]

Rev. John Bettridge Stair (1815–1898) was a missionary in Samoa. He wrote *Old Samoa; or, Flotsam and Jetsam from the Pacific Ocean* (1897).

Stalker

Figbird sp. *Sphecotheres stalkeri* **Ingram**, 1908 NCR
[Alt. Australasian Figbird; JS *Sphecotheres vieilloti* – perhaps a subspecific hybrid]
Bicoloured White-eye *Tephrozosterops stalkeri* **Ogilvie-Grant**, 1910
[Alt. Rufescent Dark-eye]
Seram White-eye *Zosterops stalkeri* Ogilvie-Grant, 1910
Seram Mountain Pigeon *Gymnophaps stalkeri* Ogilvie-Grant, 1911

Crested Pigeon ssp. *Ocyphaps lophotes stalkeri* **Mathews**, 1912 NCR
[JS *Ocyphaps lophotes lophotes*]
Red-necked Avocet ssp. *Recurvirostra novaehollandiae stalkeri* Mathews, 1912 NCR; NRM
White-bellied Cuckooshrike ssp. *Coracina papuensis stalkeri* Mathews, 1912 NCR
[Regarded as referring to an intergrade population; closest to *C. papuensis oriomo*]
Spectacled Monarch ssp. *Symposiachrus trivirgatus stalkeri* Mathews, 1916 NCR
[JS *Symposiachrus trivirgatus gouldii*]

Wilfred Stalker (1879–1910) was an Australian collector and ornithologist in Australia, the Moluccas and New Guinea, where he drowned during the BOU expedition. A mammal is named after him.

Stampfli

Crombec sp. *Sylvietta stampflii* **Büttikofer**, 1886 NCR
[Alt. Green Crombec; JS *Sylvietta virens flaviventris*]

Franz Xavier Stampfli (1847–1903) was a German (possibly Swiss) naturalist who was working in Liberia (1879–1887). Büttikofer (q.v.), who was Stampfli's companion on at least one expedition, wrote a paper (1886) entitled 'Zoological researches in Liberia: a list of birds, collected by Mr. F. X. Stampfli near Monrovia, on the Messurado River, and on the Junk River with Its tributaries'. A mammal is named after him.

Stanford

Minivet sp. *Pericrocotus stanfordi* **R. E. Vaughan** & K. H. Jones, 1913 NCR
[A hybrid population of *Pericrocotus cantonensis* x *P. roseus*]

Nepal Fulvetta ssp. *Alcippe nipalensis stanfordi* **Ticehurst**, 1930
Coral-billed Scimitar Babbler ssp. *Pomatorhinus ferruginosus stanfordi* Ticehurst, 1935
Striated Swallow ssp. *Cecropis striolata stanfordi* **Mayr**, 1941

Red-vented Bulbul ssp. *Pycnonotus cafer stanfordi*
Deignan, 1949

Lieutenant-Colonel John Keith Stanford (1892–1971) was a British army officer, writer, sportsman and colonial administrator in India and Burma (1919–1938). He fought in the Tank Corps in France (WW1). Most of his time between the wars was spent in Burma undertaking ornithological work (1927–1939), and after his retirement (1938) he took part in the Vernay-Cutting Expedition to the north-east Burma Hills. He wrote *The Birds of Northern Burma* (1938).

Stanger

Green-throated Sunbird ssp. *Chalcomitra rubescens stangerii* **Jardine**, 1842

Dr William Stanger (1811–1854) was a British geologist and explorer. He took part in the Niger expedition (1841), which used three ships to sail up the Niger River. He suffered from fever (presumably malaria) intermittently on his return to England. He was the author of the geological report of the expedition. He was the first Surveyor General of Natal (1845–1854), where a town was named after him. He was instrumental in establishing the Durban Botanical Gardens (1848). A Durban writer reported, 'Stanger who had come to Natal fresh from exploring the Niger River and was well pickled with tropical diseases died in Durban on the 14th March 1854.' He was buried in England in 1857! A mammal and a reptile are named after him.

Stanley, E.

Stanley Crane *Grus paradisea* **Lichtenstein**, 1793
[Alt. Blue Crane; Syn. *Anthropoides paradisea*]
Stanley Rosella *Platycercus icterotis* **Kuhl**, 1820
[Alt. Western Rosella]
Stanley's Hawk *Falco stanleii* **Audubon**, 1828 NCR
[Alt. Cooper's Hawk; JS *Accipiter cooperii*]
Stanley Goldfinch *Carduelis stanleyi* Audubon, 1839 NCR
[Alt. Black-chinned Siskin; JS *Carduelis barbata*]
Blue-mantled Thornbill *Chalcostigma stanleyi* **Bourcier**, 1851
New Zealand Gallinule *Porphyrio stanleyi* **G. D. Rowley**, 1875 NCR
[Alt. Purple Swamphen; JS *Porphyrio (porphyrio) melanotus*]

Stanley's Bustard *Neotis denhami stanleyi* **J. E. Gray**, 1831
[Alt. Denham's Bustard ssp.]

The Hon Edward Smith-Stanley (1775–1851), 13th Earl of Derby, was an English politician, landowner, zoologist and collector. He founded the Derby Museum, which was formed from his specimen collection and material derived from the animals, including 318 bird species, which he kept at Knowsley Park near Liverpool. He was President of the Linnean Society and of the Zoological Society of London for c.20 years. He was first elected Member of Parliament for Preston, Lancashire (1796–1812), when he was just 21. In 1826 he met Audubon (q.v.) who gave him a few Passenger Pigeons *Ectopistes migratorius*. These started breeding (1832), but they quickly became a nuisance-sized flock of 70

and Stanley allowed them to fly free – if only he had built up the flock he might have saved the species from extinction. He employed (1832–1837) Edward Lear (q.v.) to draw the plates for the *Knowsley Menagerie* (1846). Lear also invented *A Book of Nonsense* (1846) for Stanley's grandchildren. Stanley purchased Bartram's (q.v.) original work, some of which is still in the present Earl of Derby's library. His father gave his name to the world-famous horse race and his son, the 14th Earl, was three times Prime Minister of the United Kingdom. (See **Derby**, **Earl of Derby** and **Lord Derby**)

The Blue-mantled Thornbill is sometimes said to have been named after Edward Henry Stanley, 15th Earl of Derby (1826–1893). Bourcier's original text carries no etymology so so the issue remains unresolved.

Stanley (Cramp)

Fan-tailed Raven ssp. *Corvus rhipidurus stanleyi* **Roselaar**, 1993

Stanley Cramp (1913–1987) was a British civil servant and ornithologist. Manchester University awarded his BA (1934) whereafter he joined the Customs & Excise in Manchester, later transferring to London (1938) where he stayed – apart from his war service in the RAF (1944–1946) – until taking early retirement (1970). He was active in RSPB, BTO and BOU, serving in various posts in all three. He began on the board of *British Birds* (1960) before becoming Senior Editor (1963) for the rest of his life. He started the 9-volume *Birds of the Western Palearctic* and was Chief Editor, working virtually full-time on it for 17 years (1970–1987). He was still working on volume 5 when he had a stroke and subsequently died of pneumonia. He had a particularly strong interest in crows and the Middle East.

Stanley (Kemp)

White-throated Fantail ssp. *Rhipidura albicollis stanleyi* **E. C. S. Baker**, 1916

Dr Stanley Wells Kemp (1882–1945) was a zoologist and anthropologist. He joined the Fisheries Research Section of the Department of Agriculture in Dublin (1903) as Assistant Naturalist. He joined the Indian Museum in Calcutta as Superintendent of the zoological section (1911). There he worked very closely with, and became a great friend of, the Director Nelson Annandale. He took part in the Abor Punitive Expedition (1911–1912). He joined the Colonial Office (1924) as Director of Research of the *Discovery* Committee, and led the second *Discovery* Expedition to the Antarctic (1924) in relation to Whale Fisheries. He became Director of the Plymouth Marine Laboratory (1936–1945) but lost all his personal possessions, his library, and his unpublished works as the result of a German air raid (1941).

Stanton

Acacia Francolin ssp. *Francolinus africanus stantoni* **Cave**, 1940 NCR
[Alt. Orange River Francolin; JS *Scleroptila levalliantoides lorti*]

Lieutenant-Colonel Guy Manning Stanton (b.1906) was a British army officer who was seconded to the Sudan Defence

Force (1936) and was based in Khartoum (1945–1955) as Director, Stores and Ordnance Department. He was also an honorary Game Warden and a member of the Zoological Gardens Advisory Committee. He co-wrote *Dangerous Game* (1950).

Stark

Stark's Lark *Spizocorys starki* **G. E. Shelley**, 1902
[Syn. *Eremalauda starki*]

Arthur Cowell Stark (1846–1899) was a British physician and naturalist who travelled widely to collect birds. He was co-author, with W. L. Sclater, of *Fauna of South Africa*. He was killed by shellfire at the siege of Ladysmith (Boer War).

Stavorinus

Lory genus *Stavorinius* **Bonaparte**, 1850 NCR
[Now in *Chalcopsitta*]

Lory sp. *Psittacus stavorini* **Lesson**, 1828 NCR
[Indeterminate: perhaps refers to *Chalcopsitta atra insignis*]

Captain Johan Splinter Stavorinus (1739–1788) of the Dutch East India Company traded and explored in Suriname, the East Indies (Indonesia), and South Africa (1768–1788). He held the rank of Rear-Admiral in the Navy of the States General in the Netherlands. He wrote *Voyages to the East Indies* (English edition 1798).

Stead

White-chinned Petrel ssp. *Procellaria aequinoctialis steadi* **Mathews**, 1912 NCR; NRM
Great Cormorant ssp. *Carbo carbo steadi* Mathews & **Iredale**, 1913 NCR
[JS *Phalacrocorax carbo novaehollandiae*]
Fairy Prion ssp. *Pseudoprion turtur steadi* Mathews, 1932 NCR
[JS *Pachyptila turtur turtur*]
White-capped Albatross *Thalassarche cauta steadi* **Falla**, 1933
[Alt. Shy Albatross ssp.]
Stead's Bush Wren *Xenicus longipes variabilis* Stead, 1936 EXTINCT

Edgar Fraser Stead (1882–1949) was a New Zealand electrical engineer, plant breeder, ornithologist and marksman. He began birdwatching at his parents' home at Strowan, now the site of St Andrew's College. He trained as an electrical engineer and worked for a time in America. His father was a grain merchant who left him a fortune (1908). The inscription on his gravestone says that when he inherited this fortune he 'angled in summer, shot animals and birds in winter and was a world-class marksman'. When he returned to New Zealand he bought a property at Ilam, which is now the University of Canterbury staff club. There he developed an extensive collection, now known as the Edgar Stead Hall of New Zealand Birds. He wrote *Life Histories of New Zealand Birds* (1932).

Stechow

White-backed Woodpecker ssp. *Dryobates leucotos stechowi* Sachtleben, 1919 NCR
[JS *Dendrocopos leucotos leucotos*]
Three-toed Woodpecker ssp. *Picoides tridactylus stechowi* Sachtleben, 1920 NCR
[JS *Picoides tridactylus tridactylus*]

Dr Reinhard Theodor Walther Eberhard Stechow (1883–1959) was a German zoologist at Munich Museum (1905–1948).

Steere

Steere's Broadbill *Sarcophanops steerii* **Sharpe**, 1876
[Alt. Mindanao Wattled Broadbill]
Steere's Pitta *Pitta steerii* Sharpe, 1876
[Alt. Azure-breasted Pitta]
Steere's Liocichla *Liocichla steerii* **Swinhoe**, 1877
Philippine Oriole *Oriolus steerii* Sharpe, 1877
Indigo-banded Kingfisher *Ceyx steerii* Sharpe, 1892 NCR
[JS *Ceyx cyanopectus*]
Steere's Coucal *Centropus steerii* **Bourns** & **Worcester**, 1894
[Alt. Black-hooded Coucal]
Steere's Honey-buzzard *Pernis steerei* **W. L. Sclater**, 1919
[Alt. Philippine Honey-buzzard]

Versicoloured Barbet ssp. *Eubucco versicolor steerii* **P. L. Sclater** & **Salvin**, 1878
Mantanani Scops Owl ssp. *Otus mantananensis steerei* **Mearns**, 1909 NCR
[JS *Otus mantananensis sibutuensis*]
White-cheeked Bullfinch ssp. *Pyrrhula leucogenis steerei* Mearns, 1909

Professor Dr Joseph Beal Steere (1842–1940) was an American ornithologist who took a trip round the world (1870–1875). During this, he first went to Brazil and up the Amazon as far as he could by boat, crossed the Andes to Peru, and took ship for China. He collected in the Philippines (1874–1875 and 1887–1888) and published *A List of the Birds and Mammals Collected by the Steere Expedition to the Philippines* (1890). He was Curator of the USNM (1876–1894) and expanded the collections by c.2,500 specimens, c.50 previously undescribed, which he collected during his explorations in the Amazon, Peru, Formosa (Taiwan), Celebes (Sulawesi) and especially in the Philippines. Two mammals and a reptile are named after him.

Stegmann, B.

Azure-winged Magpie ssp. *Cyanopica cyanus stegmanni* **Meise**, 1932
Long-tailed Rosefinch ssp. *Uragus sibiricus stegmanni* **Hartert** & **Steinbacher**, 1932 NCR
[JS *Uragus sibiricus sibiricus*]
Eurasian Tree Sparrow ssp. *Passer montanus stegmanni* **Dementiev**, 1933 NCR
[JS *Passer montanus montanus*]
Asian Short-toed Lark ssp. *Calandrella rufescens stegmanni* Meise, 1937 NCR
[JS *Calandrella cheleensis beicki*]

Lesser Sand Plover ssp. *Charadrius mongolus stegmanni*
Portenko, 1939
Thick-billed Warbler ssp. *Acrocephalus aedon stegmanni*
G. E. Watson, 1985 NCR
[JS *Iduna aedon rufescens*]

Dr Boris Karlovich Stegmann (or Shtegman) (1897–1975) was a Russian ornithologist, systematist and zoogeographer at the Department of Ornithology, Zoological Museum, Leningrad (St Petersburg) (1921–1938). He was arrested and imprisoned on false charges (1938–1940). As he was an ethnic German, he was exiled from Leningrad (WW2) and lived in the delta of the Ili River, Kazakhstan (1941–1946), where he studied the biology and trapping of muskrats. When he was allowed back to Leningrad, he was not employed again in the museum but allowed to visit and continue his research. He wrote *In the Reedbeds of Pribalkhashie. Life and Adventures of an Exiled Naturalist 1941–1946* (2004).

Stegmann, K.

Crowned Hornbill ssp. *Tockus alboterminatus stegmanni*
Neumann, 1923 NCR; NRM

Lieutenant Kurt von Stegmann und Stein (DNF) was the German Resident at Usumbura, Burundi, German East Africa (1911–1912) and was living near Lake Kivu (1920).

Stein

Black Myzomela ssp. *Myzomela nigrita steini* **Stresemann**
& Paludan, 1932
Black-sided Robin ssp. *Poecilodryas hypoleuca steini*
Stresemann & Paludan, 1932
Forest Honeyeater ssp. *Meliphaga montana steini*
Stresemann & Paludan, 1932
Island Monarch ssp. *Monarcha cinerascens steini*
Stresemann & Paludan, 1932
Uniform Swiftlet ssp. *Aerodramus vanikorensis steini*
Stresemann & Paludan, 1932
Mountain Mouse-warbler ssp. *Crateroscelis robusta steini*
Stresemann & Paludan, 1934 NCR
[JS *Crateroscelis robusta sanfordi*]
Olive-yellow Robin ssp. *Poecilodryas placens steini* **Hartert**
& Paludan, 1936 NCR; NRM
Papuan Treecreeper ssp. *Cormobates placens steini* **Mayr**,
1936
Striated Heron ssp. *Butorides striata steini* Mayr, 1943
Mountain White-eye ssp. *Zosterops montanus steini* Mayr,
1944 NCR
[NUI *Zosterops montanus montanus*]

Georg Hermann Wilhelm Stein (1897–1976) was a German collector and zoologist. Sterling Rockefeller (q.v.) sent him on a collecting expedition to New Guinea (1931–1932) for the Museum of Comparative Zoology, Harvard. The prime focus was the collection of thousands of bird skins. He made similar expeditions to other parts of Indonesia. Whilst in New Guinea he visited Waigeu and the Weyland Mountains. He also wrote formal descriptions of several New Guinea species, such as the bandicoot *Echymipera clara*, named after his wife who accompanied him on this expedition. Three mammals are named after him.

Steinbach

Steinbach's Canastero *Asthenes steinbachi* **Hartert**, 1909
[Alt. Chestnut Canastero]

Brown-winged Schiffornis ssp. *Schiffornis turdina*
steinbachi **Todd**, 1928
[Alt. Thrush-like Schiffornis ssp.]
Variable Antshrike ssp. *Thamnophilus aspersiventer*
steinbachi **Carriker**, 1932 NCR
[JS *Thamnophilus caerulescens aspersiventer*]

Dr José (Joseph) Steinbach Kemmerich (1875–1930) was born in Germany and first went to Bolivia (1904) where he stayed for the rest of his life and shortened his name by dropping the Kemmerich. He became a Bolivian citizen, married and his descendants are still a prominent family there. He was a collector in Argentina and Bolivia for the Field Museum of Natural History, Chicago, but sold various natural history collections to museums and universities all over the world. The Carnegie Museum of Natural History holds his collection and many of the plants he collected are in the Darwin Institute at San Isidro in Argentina. His grandson Roy F. Steinbach was a professional collector in the 1960s. Two mammals and a reptile are named after him.

Steinbacher

Common Reed Bunting ssp. *Emberiza schoeniclus*
steinbacheri **Dementiev**, 1937 NCR
[JS *Emberiza schoeniclus schoeniclus*]

Dr Friedrich Steinbacher (1877–1938) was a German ornithologist who studied mathematics, coming to ornithology as an amateur (c.1920). He was a Professor of Mathematics and Biology at a high school through his working life. His nephew Joachim Steinbacher (1911–2005) was an ornithologist at the Senckenberg Museum, Frankfurt-am-Main.

Steindachner

Speckle-chested Piculet *Picumnus steindachneri*
Taczanowski, 1882

New Zealand Pipit ssp. *Anthus novaeseelandiae*
steindachneri Reischek, 1889

Franz Steindachner (1834–1919) was an Austrian zoologist who specialised in herpetology and ichthyology. He originally planned to become a lawyer, but became interested in fossil fish and (1860) joined the Natural History Museum in Vienna, becoming a Curator (1861) and Head of the Zoology Department (1874) and finally Director of the entire museum (1898–1919). Unlike many museum curators he also travelled actively and collected in the Americas including the Galapagos Islands, Africa and the Middle East. His major work was to write up the amphibian and reptile sections of the results of the circumnavigation by the Austrian frigate *Novara*. Seven amphibians and ten reptiles are named after him.

Stejneger

Kauai Amakihi *Chlorodrepanis stejnegeri* **S. B. Wilson**, 1890
[Syn. *Hemignathus kauaiensis*]
Stejneger's Petrel *Pterodroma longirostris* Stejneger, 1893

Oriental Tit sp. *Parus stejnegeri* **Bangs**, 1901 NCR
[Alt. Japanese Tit; JS *Parus minor nigriloris*]
Stejneger's Stonechat *Saxicola stejnegeri* **Parrot**, 1908
[Alt. Siberian Stonechat; Syn. *Saxicola maurus stejnegeri*]

Stejneger's Scoter *Melanitta deglandi stejnegeri* **Ridgway**, 1887
[Alt. White-winged Scoter ssp.]
Yellow-bellied Siskin ssp. *Carduelis xanthogastra stejnegeri* **Sharpe**, 1888
Greater Akialoa ssp. *Akialoa ellisiana stejnegeri* S. B. Wilson, 1889 EXTINCT
[Syn. *Hemignathus ellisianus stejnegeri*]
Japanese White-eye ssp. *Zosterops japonicus stejnegeri* **Seebohm**, 1891
Brown-eared Bulbul ssp. *Microscelis amaurotis stejnegeri* **Hartert**, 1907
White-backed Woodpecker ssp. *Dendrocopos leucotos stejnegeri* **Kuroda**, 1921
Japanese Wood Pigeon ssp. *Columba janthina stejnegeri* **Kuroda**, 1923

Dr Leonhard Hess Stejneger (1851–1943) was a Norwegian ornithologist and herpetologist who settled in the USA, where he became the USNM's vertebrate expert. He was the first full-time Curator of the Herpetology Division and held the position of Curator of the Department of Reptiles and Batrachians (1889–1943). He wrote the *Aves* volume in the series *Standard Natural History* (1885) and *Birds of the Commander Islands and Kamtschatka*. He had a lifelong fascination with Steller (q.v.), writing a biography of him (1936) and retracing many of his journeys, discovering the petrel during one of them. Two mammals, sixteen reptiles and ten amphibians are named after him.

Stella (Cherrie)

Stella's Piculet *Picumnus stellae* **Berlepsch** & **Hartert**, 1902 NCR
[Alt. Orinoco Piculet; JS *Picumnus pumilus*]

Mrs Stella M. Cherrie (1876–1967) was the wife of George Kruck Cherrie (q.v.). She travelled with him (1897–1899) in the Orinoco basin.

Stella (Deignan)

Orange-breasted Trogon ssp. *Harpactes oreskios stellae* H. G. **Deignan**, 1941
Ruby-cheeked Sunbird ssp. *Chalcoparia singalensis stellae* H. G. **Deignan**, 1950 NCR
[JS *Chalcoparia singalensis koratensis*]

Dr Stella Maria Aglaé Leche Deignan (1901–1993), an anthropologist, was the wife of the describer, Herbert Deignan (q.v.).

Stella (Erggelet)

Stella's Lorikeet *Charmosyna papou stellae* **A. B. Meyer**, 1886
[Alt. Papuan Lorikeet ssp.]

Stella Baroness von Erggelet (DNF) was the wife of Alex Freiherr von Salzburg. She was a patroness of the sciences, so

perhaps Meyer was thanking her for support or hoping to get money from her for research.

Steller

Eider genus *Stelleria* **Bonaparte**, 1842 NCR
[Now in *Polysticta*]
Jay genus *Stellerocitta* **Coues**, 1903 NCR
[Now in *Cyanocitta*]

Steller's Albatross *Phoebastria albatrus* **Pallas**, 1769
[Alt. Short-tailed Albatross]
Steller's Eider *Polysticta stelleri* Pallas, 1769
Steller's Jay *Cyanocitta stelleri* **J. F. Gmelin**, 1788
Steller's Flightless Cormorant *Phalacrocorax perspicillatus* Pallas, 1811 EXTINCT
[Alt. Pallas's Cormorant, Spectacled Cormorant]
Steller's Sea Eagle *Haliaeetus pelagicus* Pallas, 1811

Georg Wilhelm Steller (originally Stöhler) (1709–1746) was a German naturalist and explorer in the Russian service. He studied medicine at Halle and went to Russia (1731–1734) as a physician in the Russian Army. He became an Assistant at the Academy of Sciences in St Petersburg (1734) and left for Kamchatka (1737) accompanying Vitus Bering (q.v.) on his second expedition (1738–1742) to Alaska on board *St Peter*, which was accompanied by the *St Paul*. This expedition ended when the *St Peter* was wrecked on a desolate island, now called Bering Island, where Bering died and the surviving crew had to spend the winter in crude huts. Steller and the Danish first lieutenant Waxell proved effective in ensuring their survival. After nine months a boat was constructed from the wreckage of the *St Peter*, enabling the survivors to leave the island for Kamchatka (1742). Steller worked in Petropavlovsk (1742–1744) but died on his return journey from there to St Petersburg. He published *Journal of a Voyage with Bering 1741–1742* (1743) in which he informally described the marine mammal now known as Steller's Sea Cow. Soon afterwards the animal was hunted to extinction, so Steller's expedition members were the only scientists to see it alive. A second mammal is also named after him.

Stepanyan

Snowfinch genus *Stepaniania* Kasin, 1982 NCR
[Now in *Pyrgilauda*]

Stepanyan's Warbler *Locustella amnicola* Stepanyan, 1972
[Alt. Sakhalin Grasshopper Warbler; Syn. *Locustella fasciolata amnicola*]

Rufous-winged Fulvetta ssp. *Alcippe castaneceps stepanyani* Eames, 2002
[Syn. *Pseudominla castaneceps stepanyani*]

Leo Surenovich Stepanyan (1931–2002) was a Moscow-based Armenian ornithologist, taxonomist and member of the Severtzov Institute of Ecology and Evolution, Russian Academy of Sciences. He wrote *Composition and Distribution of the Fauna of Birds of the USSR* (1990) and *Birds of Vietnam* (1995). The former was based on his investigations (1978–1990) and on extensive published material in the *Russian Journal of Zoology*. Much of his work was on taxonomy, based on studies of existing collections of skins, and he described many bird subspecies.

Stephan

Stephan's Dove *Chalcophaps stephani* **Pucheran**, 1853
[Alt. Stephan's Emerald Dove]

Étienne Stephan Jacquinot (1776–1840) was the father of the French explorer Vice-Admiral Charles Hector Jacquinot (q.v.), who was ensign on the *Coquille* (1822–1825) and second-in-command to Dumont d'Urville on the *Astrolabe* (1837–1840). His brother Honoré Jacquinot (q.v.) was a ship's surgeon on the same expeditions. 'Honoré Jacquinot, the captain of the Zélée, relates that when he brought a bird down with a shot from his musket, the initial fright of his native companions turned to shocked amazement as they saw the creature lying lifeless on the ground'. Honoré collaborated closely with Pucheran (q.v.) in preparing descriptions of specimens. We believe that Honoré asked Pucheran to name it after his father. (See also **Jacquinot** and **Maria (Jacquinot)**)

Stephanie

Stephanie's Astrapia *Astrapia stephaniae* **Finsch** &
A. B. Meyer, 1885
[Alt. Princess Stephanie's Bird-of-Paradise]

Princess Stephanie of Belgium (1864–1945) was the wife of Prince Rudolph (q.v.) of Austria-Hungary – an arranged marriage when Stephanie was 16. After his death she married Count Lonyay.

Stephen

Stephen's Lorikeet *Vini stepheni* **A. J. North**, 1908
[Alt. Henderson Lorikeet]

Alfred Ernest Stephen (1879–1961) was an Australian civil servant, businessman and collector. He made a small collection of birds on Henderson Island, Pitcairn Islands (1907). The *Records of the Australian Museum* include a (1903) paper entitled 'Notes on the zoology of Paanopa or Ocean Island and Nauru or Pleasant Island, Gilbert group', by A. J. North, F. Danvers Power and A. E. Stephen.

Stephens, F.

Stephens's Whip-poor-will *Antrostomus arizonae* **Brewster**, 1881
[Alt. Mexican Whip-poor-will]

Stephens's Vireo *Vireo huttoni stephensi* Brewster, 1882
[Alt. Hutton's Vireo ssp.]
Stephens's Fox Sparrow *Passerella iliaca stephensi*
A. W. Anthony, 1895
[Syn. *Passerella megarhyncha stephensi*]

Frank Stephens (1849–1937) was an ornithologist and mammalogist. He was Curator Emeritus of the San Diego Society of Natural History and a member of the Death Valley expedition (1891). In his early years he collected birds and their nests and eggs for Aiken (q.v.). His wife, Kate, who lived to be over 100 years old, was a conchologist of note. As her husband had a reputation for being careless as well as deaf, she insisted on travelling with him on all his trips. For example, they were both members of the Alexander

Expedition to south-eastern Alaska (1907), and they accompanied Grinnell (q.v.) on the Colorado River (1910). In that same year Stephens gave 2,000 bird and mammal specimens to the San Diego Society. He was knocked down by a tram and died 10 days later; we speculate that he might not have heard the tram coming. Two mammals and a reptile are named after him.

Stephens, S. E.

Scarlet Myzomela ssp. *Myzomela sanguinolenta stephensi*
Mathews, 1912 NCR; NRM

Stephen Ernest 'Ern' Stephens (d.1958) was an Australian horticulurist and local historian. He was President, North Queensland Naturalists' Club, Cairns (1948), and was the first President of the Cairns Historical Society, which offers an annual S. E. Stephens History Award.

Stephenson

Great Spotted Woodpecker ssp. *Dryobates cabanisi stephensoni* **E. C. S. Baker**, 1926 NCR
[JS *Dendrocopos major stresemanni*]

(See **Clarke, S. R.**)

Sterling

Sterling's Thrush *Turdus poliocephalus sterlingi* **Mayr**, 1944
[Alt. Island Thrush ssp.]

(See **Rockefeller**)

Stevens

Stevens's Hornbill *Aceros nipalensis* **Hodgson**, 1829
[Alt. Rufous-necked Hornbill]

Ashy Drongo ssp. *Dicrurus leucophaeus stevensi*
E. C. S. Baker, 1918 NCR
[JS *Dicrurus leucophaeus hopwoodi*]
Blunt-winged Warbler ssp. *Acrocephalus concinens stevensi* E. C. S. Baker, 1922
Graceful Prinia ssp. *Prinia gracilis stevensi* Hartert, 1923
Rufous-throated Fulvetta ssp. *Alcippe rufogularis stevensi*
Kinnear, 1924
Streaked Wren-babbler ssp. *Napothera brevicaudata stevensi* **Kinnear**, 1925
Papuan Treecreeper ssp. *Climacteris placens stevensi*
Greenway, 1934 NCR
[JS *Cormobates placens meridionalis*]
Stevens's Honeyeater *Meliphaga cinereifrons stevensi*
Rand, 1936
[Alt. Elegant Honeyeater ssp.; Syn. *Meliphaga analoga stevensi* – taxonomic status uncertain]

Herbert Stevens (1877–1964) was a tea planter in Sikkim (India) and an ornithologist and zoologist. He took part in a number of expeditions, normally as a museum collector. He was in Tonkin (1923–1924 and 1929) and was ornithologist on the Kelley-Roosevelt expedition of the Field Museum, Chicago, to Indochina. He was in New Guinea (1932–1933), collecting for the Museum of Comparative Zoology, Harvard. He wrote 'Notes on the birds of the Sikkim Himalaya' and his autobiography *Through Deep Defiles to Tibetan Uplands* (1934). He was a great benefactor of the BOC and bequeathed to them his house in Tring (Hertfordshire).

Stevenson, D. H.

Stevenson's Pacific Wren *Troglodytes pacificus stevensoni* **Oberholser**, 1930
[Syn. *Nannus pacificus stevensoni*]

Donald H. Stevenson (d.1926) worked for the U.S. Biological Survey. He was the Reservation Warden for the Aleutian Islands (1920–1925). He collected specimens of lemmings on Umnak Island in the Aleutians (1920–1924), and one is named after him.

Stevenson, H.

Sparrowhawk sp. *Accipiter stevensoni* **Gurney**, 1863 NCR
[Alt. Japanese Sparrowhawk; JS *Accipiter gularis*]

Henry Stevenson (1833–1888) was an English ornithologist who was a newspaper proprietor (*Norfolk Chronicle*), Secretary to the Norfolk & Norwich Museum (1855–1888) and author of *The Birds of Norfolk*. Gurney (q.v.) described this sparrowhawk from specimens held in the Norwich Museum.

Stewart, J. S.

Square-tailed Drongo Cuckoo ssp. *Surniculus lugubris stewarti* **E. C. S. Baker**, 1919
Red Spurfowl ssp. *Galloperdix spadicea stewarti* E. C. S. Baker, 1919

Dr J. S. Stewart (d.1942) was in Travancore, India (1907–1913).

Stewart, L. C.

Stewart's Bunting *Emberiza stewarti* **Blyth**, 1854
[Alt. White-capped Bunting]

Ashy Prinia ssp. *Prinia socialis stewarti* Blyth, 1847

Surgeon-General Ludovick Charles Stewart (1819–1888) was an officer in the British Army in India. He was an ardent amateur botanist and ornithologist and collected many bird skins. His correspondence and papers (1834–1887) are in Cambridge University Library. He was commissioned (1841) and served with the 29th Regiment throughout the Punjab campaign (1848–1849), including the passage of the Chenab and the battles of Chillianwallah and Goojerat (Gujarat). He was a member of the Zoological Society of London (1885–1888) and wrote *Natural History and Sport in the Himalayas* (1886).

Stewart, R. E.

Indian Scops Owl ssp. *Otus bakkamoena stewarti* **Koelz**, 1939 NCR
[JS *Otus bakkamoena gangeticus*]

Robert E. Stewart (1913–1993) was an American wildlife biologist and ornithologist whose entire professional career was with the US Government. The University of Iowa awarded his bachelor's degree (1936) and the University of Michigan his master's (1937). He was stationed at the Patuxent Wildlife Research Center, Maryland (1940–1960), interrupted by service with the US Navy Medical Corps in WW2. He transferred (1960) to the Northern Prairie Wildlife Research Center, Jamestown, North Dakota, and worked there until his retirement from the US Fish and Wildlife Service (1976). He wrote *Breeding Birds of North Dakota* (1975).

Stewart, R. M.

Brown Creeper ssp. *Certhia americana stewarti* J. D. Webster, 1986

Ronald McDonald Stewart (1881–1958) was an English-born ornithologist and naturalist. He went to Australia (1897) and then to the Queen Charlotte Islands, British Columbia (1907). The Government of British Columbia in the Forestry Service employed him (1919–1938) as a Game Warden. In retirement he was Dominion Wildfowl Officer (1946) based in the Queen Charlotte Islands and collected much of the material on which the description of this subspecies, long after his death, was based.

Stieber

Arabian Bustard ssp. *Ardeotis arabs stieberi* **Neumann**, 1907

Major Sylvester Stieber (1867–1914) was an officer in the Prussian army (1888–1900), retiring to join the German colonial army in West Africa (1901–1910) including acting as the German Resident at Kusseri, Cameroon. After leaving the German colonial army he was part of the Grand Duke of Mecklenburg's Rifle Battalion.

Stierling

Stierling's Woodpecker *Dendropicos stierlingi* **Reichenow**, 1901
Stierling's Wren Warbler *Calamonastes stierlingi* Reichenow, 1901

Stierling's Hill Babbler *Pseudoalcippe abyssinica stierlingi* Reichenow, 1898
[Alt African Hill Babbler ssp.]
African Yellow White-eye ssp. *Zosterops senegalensis stierlingi* Reichenow, 1899
Groundscraper Thrush ssp. *Psophocichla litsitsirupa stierlingi* Reichenow, 1900

Dr N. Stierling (fl.1887–at least 1901) was a German naturalist who collected in Nyasaland (Malawi) and Tanganyika (Tanzania) (1887–1901). He was a doctor with the German colonial forces in German East Africa and had to help deal with a cholera outbreak in Zanzibar during the Maji-Maji rebellion.

Stiles

Stiles's Tapaculo *Scytalopus stilesi* Cuervo, Cadena, Krabbe & Renjifo, 2005

Yucatan Vireo ssp. *Vireo magister stilesi* **A. R. Phillips**, 1991

Dr F. Gary Stiles is Curator of the bird collection at the National University of Colombia in Bogotá. His career in

Neotropical ornithology includes both fieldwork in Costa Rica and Colombia, and museum research concentrating on ecology and taxonomy – most notably of hummingbirds. Since arriving in Colombia (1988) he has had considerable influence on the development of scientific ornithology there, and contributed to the growth of the Asociacion Bogotana de Ornitologia and the Asociacion Colombiana de Ornitologia. He co-wrote *The Field Guide to the Birds of Costa Rica* (1989).

Stimpson

Oriental Turtle Dove ssp. *Streptopelia orientalis stimpsoni* **Stejneger**, 1887

William Stimpson (1832–1872) was an American engineer, conchologist and zoologist. He was a member of the North Pacific Exploring Expedition (1853–1856). He was Curator and Director, Chicago Academy of Sciences (1864–1871). He died of tuberculosis.

Stirton

Squirrel Cuckoo ssp. *Piaya cayana stirtoni* **van Rossem**, 1930 NCR
[NUI *Piaya cayana thermophila*]

Ruben Arthur Stirton (1901–1966) was an American palaeontologist who specialised in mammals, and a Professor at the University of California, Berkeley. His earlier work was on extant North American mammals. He conducted fieldwork in South America (1940s), then on Neogene mammalian fauna from Australia (1950s). A mammal is named after him.

Stoddard

Black Rail ssp. *Laterallus jamaicensis stoddardi* Coale, 1923 NCR
[JS *Laterallus jamaicensis jamaicensis*]
Stoddard's Yellow-throated Warbler *Dendroica dominica stoddardi* **G. M. Sutton**, 1951

Herbert Lee Stoddard (1889–1970) was an American taxidermist, ecologist and conservationist. He was at the Milwaukee Museum (1910–1913 and 1920–1924), spending the interval at the Field Museum, Chicago (1913–1920). He joined the US Biological Survey (1924). He wrote *The Bobwhite Quail* (1931).

Stoehr

Stoehr's Black Chat *Myrmecocichla stoehri* **Sclater**, 1941 NCR
[Alt. Sooty Chat; JS *Myrmecocichla nigra*]

Dr Frederick Otto Stoehr (1871–1946) was a British physician and surgeon in Northern Rhodesia (Zambia) (1905–1946). He changed his name by deed poll (1911) to Stohr.

Stötzner

Willow Tit ssp. *Poecile montanus stoetzneri* **Kleinschmidt**, 1921
[Alt. Songar Tit; Syn. *Poecile songarus stoetzneri*]
Brown-breasted Flycatcher ssp. *Muscicapa muttui stoetzneri* **Weigold**, 1922 NCR; NRM
Sand Martin ssp. *Riparia riparia stoetzneriana* **Meise**, 1934 NCR
[JS *Riparia riparia taczanowskii*]

Walter Stötzner (1882–1965) was a German traveller and ethnologist who explored in China and Tibet (1914), where he was one of the first Europeans to see a live wild Giant Panda *Ailuropoda melanoleuca*. He avoided being interned in Hong Kong at the outbreak of WW1, later (1917) managing to get back to Germany. He explored in Mongolia and Manchuria (1927–1929).

Stokes, J. L.

North Island Bushwren *Xenicus longipes stokesii* **G. R. Gray**, 1862 EXTINCT

Admiral John Lort Stokes (1812–1885) was an officer in the Royal Navy who served on HMS *Beagle*. He served under Commander John Wickham for a survey of Australasian waters, and when Wickham was invalided (1841) he took command of the Beagle. On returning to England (1843), he wrote the 2-volume *Discoveries in Australia, with an Account of the Coasts and Rivers Explored and Surveyed during the Voyage of the Beagle, 1837–1843* (1846). He was (1846) promoted to captain and commanded the steam ship HMS *Acheron* surveying in New Zealand waters. He retired (1863) yet was promoted to the rank of rear admiral, vice-admiral (1871) and admiral (1877). Two reptiles are named after him.

Stokes, P.

Hummingbird genus *Stokoesiella* **Bonaparte**, 1854 NCR
[Now in *Sephanoides*]

Stokes's Hummingbird *Trochilus stokesii* **King**, 1831 NCR
[Alt. Juan Fernandez Firecrown; JS *Sephanoides fernandensis fernandensis*]

We believe this hummingbird is named after Captain Pringle Stokes (d.1828), although the describer – Captain Phillip Parker King – gave no etymology. HMS *Beagle* set sail on her first voyage (May 1826) under the command of Captain Stokes, accompanying the larger ship HMS *Adventure* on a survey of Patagonia and Tierra del Fuego under the overall command of Captain Phillip Parker King. Stokes fell into a deep depression and shot himself, dying 10 days after leaving port. We think King might have honoured Stokes by naming the hummingbird after him; perhaps the stigma attached to suicide at that time prevented King from openly stating his etymology.

Stoliczka

Tit Warbler genus *Stoliczkana* **Hume**, 1874 NCR
[Now in *Leptopoecile*]

Stoliczka's Treecreeper *Certhia nipalensis* **Blyth**, 1845
[Alt. Rusty-flanked Treecreeper; *Certhia stoliczkae* (Brooks, 1874) is a JS]
Stoliczka's Bushchat *Saxicola macrorhynchus* Stoliczka 1872
[Alt. White-browed Bushchat]

Eurasian Collared Dove ssp. *Streptopelia decaocto stoliczkae* Hume, 1874 NCR
[Now known to be based on a feral population]
Sinai Rosefinch ssp. *Carpodacus synoicus stoliczkae* Hume, 1874
Saxaul Sparrow ssp. *Passer ammodendri stoliczkae* Hume, 1874

Stoliczka's Tit Warbler *Leptopoecile sophiae stoliczkae*
Hume, 1874
[Alt. White-browed Tit Warbler ssp.]
White-crowned Penduline-tit ssp. *Remiz coronatus*
stoliczkae Hume, 1874

Dr Ferdinand Stoliczka (1838–1874) was an Austrian palaeontologist and zoologist who was born in Moravia (Czech Republic). He was educated at Prague and at the University of Vienna, where he obtained his doctorate. He collected during travels throughout India (1864–1874) as an Assistant Superintendent of the Geological Survey there. He wrote some scientific papers, such as 'Contribution towards the knowledge of Indian Arachnoidea', in the *Journal of the Asiatic Society*. During the last ten years of his life he published geological memoirs on the western Himalayas and Tibet, and many papers on Indian zoology, from mammals to insects and corals. He took part in the Second Yarkand Mission (1873–1874) but collapsed and died of spinal meningitis when 'returning loaded with the spoils and notes of nearly a year's research in one of the least-known parts of Central Asia', according to his obituary in *Nature*. He is commemorated in species from many phyla including nine reptiles, three mammals and an amphibian.

Stolzmann

Tumbes Sparrow *Rhynchospiza stolzmanni* **Taczanowski,**
1877
[Syn. *Aimophila stolzmanni*]
Ochre-breasted Tanager *Chlorothraupis stolzmanni*
Berlepsch & Taczanowski, 1884
Black-backed Bush Tanager *Urothraupis stolzmanni*
Berlepsch & Taczanowski, 1885
Tumbes Swallow *Tachycineta stolzmanni* **Philippi**, 1902
Dwarf Tyrant Manakin *Tyranneutes stolzmanni* **Hellmayr,**
1906

Stolzmann's Hillstar *Oreotrochilus estella stolzmanni* **Salvin,**
1895
[Alt. Green-headed Hillstar ssp.]
Highland Elaenia ssp. *Elaenia frantzii stolzmanni* **Ridgway,**
1906 NCR
[JS *Elaenia obscura obscura*]
Eurasian Nuthatch ssp. *Sitta europaea sztolcmani*
Domaniewski 1915 NCR
[Transitional form: *Sitta europaea europaea* x *S. europaea*
caesia]
Fawn-breasted Tanager ssp. *Pipraeidea melanonota*
sztolcmani Dunajewski 1939 NCR
[JS *Pipraeidea melanonota venezuelensis*]

Jean Stanislas Stolzmann (or Jan Sztolcman) (1854–1928) was a Polish naturalist who went to Peru (1875), a trip initiated by Taczanowski (q.v.), where he collected with Jelski (q.v.) for the Branicki brothers (q.v.). After a brief return to Poland (1881–1882) he went back to South America where he collected in Ecuador (1882–1884). He succeeded Taczanowski as Curator of the Zoological Cabinet of the University of Warsaw (1887–1928). He also became Director of the Branicki Museum, until these merged (1919) and became the Zoological Museum of Warsaw. The scientific

name *stolzmanni* really ought to be *stolzmani*, as the name in Polish is written as Stolzman and not Stolzmann (which is German). One mammal, one amphibian and one reptile are named after him.

Stone, A. C.

White-backed Swallow ssp. *Cheramoeca leucosterna*
stonei **Mathews**, 1912 NCR; NRM

A. Charles Stone (d.1920) was an English immigrant to Australia. He was a businessman, a field ornithologist and Honorary Secretary of the RAOU (1913).

Stone, J. E.

Black-streaked Scimitar Babbler ssp. *Pomatorhinus*
erythrogenys stoneae **Deignan**, 1952 NCR
[JS *Pomatorhinus gravivox dedekensi*]

Mrs Joan Evelyn Stone *née* Sowerby (fl.1952) was the wife of F. A. Stone, and a relative of Arthur de Carle Sowerby (q.v.).

Stone, J. J.

Stone's Pheasant *Phasianus colchicus elegans* **D. G. Elliot,**
1870
[Alt. Common Pheasant ssp.]

John J. Stone (DNF) was an aviculturist involved in collecting various pheasant taxa, and trying to establish them in Britain. Elliot wrote 'It is to the exertions of Mr. Stone, who has succeeded in bringing to Europe many of the rarer species of this family, that we are indebted for the opportunity of being able to describe this new form'. Stone is also mentioned – as 'the late Mr. John J. Stone' – in William Tegetmeier's work *Pheasants: their Natural History and Practical Management* (1881), but little seems to have been recorded about his life.

Stone, O. C.

Stone's Catbird *Ailuroedus buccoides stonii* **Sharpe**, 1876
[Alt. White-eared Catbird ssp.]

Octavius C. Stone (d.1933) was a collector who supplied Sharpe (q.v.) with a number of specimens from New Guinea (1875–1878), but we can find no record of him having supplied anyone else. He wrote for the Royal Geographical Society's *Journal* an article entitled 'Description of the country and natives of Port Moresby and neighbourhood, New Guinea' (1876), noting of certain indigenous people 'They paint the face with streaks by means of a rose-coloured lime'. This account may well have been upstaged by Stanley's report of his meeting with Livingstone, which appeared in the same issue.

Stone, W.

Whistling Thrush sp. *Myiophoneus stonei* **Meyer de**
Schauensee, 1929 NCR
[Alt. Blue Whistling Thrush; JS *Myophonus caeruleus*
eugenei]

Short-tailed Nighthawk ssp. *Lurocalis semitorquatus stonei*
Huber, 1923

African Stonechat ssp. *Saxicola torquatus stonei* **Bowen**, 1931

Common Grackle ssp. *Quiscalus quiscula stonei* **Chapman**, 1935

Dr Witmer Stone (1866–1939) worked for over 50 years in the ornithology department of the Academy of Natural Sciences, Philadelphia (1888–1939), and followed in the tradition of Audubon (q.v.) and Wilson (q.v.) in the study of the *Birds of Pennsylvania*. He produced numerous other writings in addition to those in *The Auk,* which he edited (1912–1936). His important works include *Birds of Eastern Pennsylvania and New Jersey* (1894) and *Bird Studies at Old Cape May* (1937). (See also **Witmer**)

Stoney

Boreal Chickadee ssp. *Poecile hudsonicus stoneyi* **Ridgway**, 1887

Commander George M. Stoney (1853–1905) was an officer in the US Navy who led two expeditions to Alaska (1883–1886).

Stonowa

Storm-petrel genus *Stonowa* **Mathews & Hallstrom**, 1943 NCR
[Now in *Cymochorea* (formerly *Oceanodroma*)]

This genus is an anagram of the initial and surname of Alan **Owston** (q.v.)

Storer

White-fronted Swift *Cypseloides storeri* Navarro, A. T. Peterson, Escalante & Benitez, 1992

Professor Dr Robert Winthrop Storer (1914–2008) was an American ornithologist. His first degree (1936) from Princeton was in chemistry, but his MA (1942) and doctorate (1949) from the University of California were in zoology. He started work as an Assistant Curator of Birds at the University of Michigan's Museum of Zoology and spent his entire career there until retiring as Acting Director (1949–1982). In addition he was a President of the AOU (1970–1972) and editor of *Auk* (1953–1957). He published more than 230 scientific papers and articles. He was particularly known for his work on evolution and systematics, and for his interest in the grebe family. The original etymology stated 'We take pleasure in naming this species for Dr Robert W. Storer in recognition of his many contributions to the knowledge of the birds of Guerrero and Michoacan.'

Storey

African Yellow Warbler *Chloropeta storeyi* **Ogilvie-Grant**, 1906 NCR
[JS *Iduna natalensis massaica*]

Charles B. C. Storey (b.1868) was a collector in British East Africa (c.1903). A mammal is named after him. We have been unable to find more about him.

Storm

Storm's Stork *Ciconia stormi* **Blasius**, 1896

Captain Hugo Storm (DNF) was a German seaman from Lübeck who was captain of the steamship *Lübeck* (1887 to mid-1890s). He collected in those parts of the Far East where his vessel traded – Sumatra, Singapore, Java, Borneo and the Moluccas – and visited areas where few Europeans had been before. He resigned his command (1896) and settled in a town called Tower in North America. Storm sent zoological specimens, including Orang-utans *Pongo pygmaeus*, from the East Indies to the museum in Lübeck, and he continued sending specimens there from North America. The museum's magazine (1901) named him as a donor and a life-sized portrait (now lost) was placed in the collection room. A reptile is also named after him.

Storms

Madagascar Cuckoo *Cuculus stormsi* **Dubois**, 1887 NCR
[JS *Cuculus rochii*]

African Thrush ssp. *Turdus pelios stormsi* **Hartlaub**, 1886

Lieutenant-General Émile Pierre Joseph Storms (1846–1918) was a soldier in the Belgian army and explorer in the Congo (1882–1891). He was violently opposed to slavery in Africa.

Stotz

Aripuana Antwren *Herpsilochmus stotzi* Whitney *et al.*, 2013

Dr Douglas F. Stotz is an ornithologist and conservation ecologist at the Field Museum, Chicago, which he joined in 1994. The University of Arizona awarded his bachelor's degree and the University of Chicago his doctorate. He co-wrote *Birds of Peru* (2007).

Strachey

Rock Bunting ssp. *Emberiza cia stracheyi* F. Moore, 1856

Lieutenant-General Sir Richard Strachey (1817–1908) went to India (1836), returned to England (1850) and went back to India (1855). He served in the Bengal Engineers, rising from Lieutenant (1841) to Lieutenant-General (1875). Frequent attacks of fever compelled him to go to Nani Tal in the Kumaon Himalayas for his health (1847). There he met Major E. Madden, under whose guidance he studied botany and geology, making expeditions into the western Himalayas for scientific purposes. He served as an administrator in several capacities (1862–1871) and was a member of the Council of India (1875–1889). He was President of the Royal Geographical Society (1887–1889). Jointly with his brother, Sir John Strachey, who was also a colonial administrator, he wrote *The Finances and Public Works of India* (1882). He wrote many monographs and papers, such as 'On the physical geography of the provinces of Kumaon and Garhwal, in the Himalaya Mountains, and of the adjoining parts of Tibet' (1851). A mammal is named after him.

Straneck

Straneck's Tyrannulet *Serpophaga griseicapilla* Straneck,
2007
[Alt. Grey-crowned Tyrannulet]

Ferruginous Pygmy Owl ssp. *Glaucidium brasilianum
stranecki* C. Koenig & Wink, 1995

Roberto Juan Straneck is an Argentinian ornithologist and
research biologist at the Laboratorio de Sonidos Naturales,
División Ornitología Museo Argentino de Ciencias Naturales
'Bernardino Rivadavia', Buenos Aires. He has written a
number of papers such as one describing two new owls, and
co-wrote *Field Check-list to the Birds of Argentina* (1999). He
is particularly interested in animal vocalisations.

Strange

Croaking Cisticola ssp. *Cisticola natalensis strangei*
L. Fraser, 1843

Admiral James N. Strange (1812–1895) served in the Royal
Navy (1827–1876). He was on the Niger River Expedition
(1840–1841).

Strassen

Strassen's Helmeted Guineafowl *Numida meleagris
strasseni* **Reichenow**, 1911 NCR
[JS *Numida meleagris galeata*]

Dr Otto Karl Ladislaus zur Strassen (1869–1961) was a
German zoologist. His doctorate was awarded by the Univer-
sity of Leipzig (1892). He was a member of the *Valdivia*
expedition (1898–1899) and was appointed Associate
Professor of Zoology at Leipzig (1901). He was Director of the
Senckenberg Natural History Museum (1909–1934) and
Professor of Zoology at the University of Frankfurt
(1914–1937).

Strauch

Strauch's Pheasant *Phasianus colchicus strauchi*
Przewalski, 1876
[Alt. Common Pheasant ssp.]

Professor Dr Alexander Alexandrovich Strauch (1832–1893)
was a Russian-German zoologist. He finished his training as
a physician in Estonia (1859), but he was also a naturalist,
mainly interested in herpetology, and his doctoral dissertation was on zoology. He was sent to Algeria (1859–1860). He
became Director of the Zoological Museum in St Petersburg
(1879) and (1890) Permanent Secretary of the library of the
Academy of Science. He wrote *Essai d'une Erpétologie de
L'Algérie* (1862). He mainly wrote on herpetological zoogeography. He died in St Petersburg and is buried in the Lutheran
cemetery there. Seven reptiles and an amphibian are named
after him.

Straus

Straus's Apalis *Apalis chapini strausae* **Boulton**, 1931
[Alt. Chapin's Apalis ssp.]

Mrs Oscar Straus (DNF) was the wife of the American
collector and explorer Oscar Straus. Together and with
others they organised and undertook the Straus Central
African Expedition, which collected botanical and ornithological specimens under the auspices of the AMNH, New
York (1929–1931).

Streich

Eurasian Hobby ssp. *Falco subbuteo streichi* **Hartert** &
Neumann, 1907

Ivo Streich (DNF) was the German Consul in Swatow, China
(1906). The European residents of Swatow regarded him as a
'tower of strength'.

Stresemann

Bougainville Honeyeater genus *Stresemannia* **Meise**, 1950

Stresemann's Oriole *Oriolus mellianus* Stresemann, 1922
[Alt. Silver Oriole, Mell's Oriole]
Bar-bellied Woodcreeper *Hylexetastes stresemanni*
Snethlage, 1925
Stresemann's Scops Owl *Otus stresemanni* **H. C. Robinson**,
1927
[Alt. Sumatran Scops Owl; taxonomy uncertain]
Malaita White-eye *Zosterops stresemanni* **Mayr**, 1931
Stresemann's Lory *Lorius amabilis* Stresemann, 1931 NCR
[Based on an abberant individual of *Lorius hypoinochrous*]
Stresemann's Elaenia *Elaenia aenigma* Stresemann, 1937
NCR
[Alt. Small-billed Elaenia; JS *Elaenia parvirostris*]
Stresemann's Bush-crow *Zavattariornis stresemanni*
Moltoni, 1938
[Alt. Ethiopian Bush-crow]
White-cheeked Cotinga *Zaratornis stresemanni* **Koepcke**,
1954
Stresemann's Bristlefront *Merulaxis stresemanni* **Sick**, 1960

Stresemann's Orange-fronted Hanging Parrot *Loriculus
aurantiifrons batavorum* Stresemann, 1913
Glossy Swiftlet ssp. *Collocalia esculenta stresemanni*
Rothschild & **Hartert**, 1914
Cardinal Woodpecker ssp. *Dendropicos fuscescens
stresemanni* **Grote**, 1922 NCR
[JS *Dendropicos fuscescens centralis*]
Collared Kingfisher ssp. *Todiramphus chloris stresemanni*
Laubmann, 1923
Long-tailed Shrike ssp. *Lanius schach stresemanni*
Mertens, 1923
Rufous-naped Lark ssp. *Mirafra africana stresemanni*
Bannerman 1923
Great Spotted Woodpecker ssp. *Dendrocopos major
stresemanni* **Rensch**, 1924
Rusty-breasted Cuckoo ssp. *Cacomantis variolosus
stresemanni* Hartert, 1925 NCR
[JS *Cacomantis sepulcralis aeruginosus*]
Stresemann's Pygmy Parrot *Micropsitta pusio stresemanni*
Hartert, 1926
[Alt. Buff-faced Pygmy Parrot ssp.]
Eurasian Wren ssp. *Troglodytes troglodytes stresemanni*
Schiebel, 1926 NCR
[JS *Troglodytes troglodytes cypriotes*]

Papuan King Parrot ssp. *Alisterus amboinensis stresemanni*
Neumann, 1927 NCR

[JS *Alisterus chloropterus calloptercus*]

Scarlet-headed Flowerpecker ssp. *Dicaeum trochileum stresemanni* Rensch, 1928

Russet-backed Jungle Flycatcher ssp. *Rhinomyias oscillans stresemanni* **Siebers**, 1928

Stresemann's Lorikeet *Trichoglossus forsteni stresemanni* **Meise**, 1929

[Alt. Coconut Lorikeet; Syn. *Trichoglossus haematodus stresemanni*]

Black Sicklebill ssp. *Epimachus fastuosus stresemanni* Hartert, 1930 NCR

[JS *Epimachus fastuosus atratus*]

Large Scrubwren ssp. *Sericornis nouhuysi stresemanni* Mayr, 1930

Long-billed Honeyeater ssp. *Melilestes megarhynchus stresemanni* Hartert, 1930

Pacific Baza ssp. *Aviceda subcristata stresemanni* Siebers, 1930

Papuan Grassbird ssp. *Megalurus macrurus stresemanni* Hartert, 1930

Stresemann's Rosefinch *Carpodacus waltoni eos* Stresemann, 1930

[Alt. Pink-rumped Rosefinch ssp.]

Belford's Honeyeater ssp. *Melidectes belfordi stresemanni* Mayr, 1931 NCR

[Hybrid population: *Melidectes belfordi* x *M. rufocrissalis*]

Brown Goshawk ssp. *Accipiter fasciatus stresemanni* Rensch, 1931

Malia ssp. *Malia grata stresemanni* Meise, 1931

Chestnut-crowned Warbler ssp. *Seicercus castaniceps stresemanni* **Delacour**, 1932

Grey-headed Woodpecker ssp. *Picus canus stresemanni* **Yen**, 1933 NCR

[JS *Picus canus kogo*]

Ashy-throated Parrotbill ssp. *Sinosuthora alphonsiana stresemanni* Yen, 1934

Short-toed Treecreeper ssp. *Certhia brachydactyla stresemanni* Kummerlowe & Niethammer, 1934

Stresemann's Bird-of-Paradise *Lophorina superba pseudoparotia* Stresemann, 1934 NCR

[Hybrid: *Lophorina superba* x *Parotia carolae*]

Whyte's Barbet ssp. *Stactolaema whytii stresemanni* Grote, 1934

Stresemann's Red-cheeked Parrot *Geoffroyus geoffroyi stresemanni* **Salomonsen**, 1937 NCR

[NUI *Geoffroyus geoffroyi rhodops*]

Whiskered Treeswift ssp. *Hemiprocne comata stresemanni* Neumann, 1937 NCR

[NUI *Hemiprocne comata comata*]

Island Thrush ssp. *Turdus poliocephalus stresemanni* Bartels, 1938

Black-naped Oriole ssp. *Oriolus chinensis stresemanni* Neumann, 1939

Streak-headed White-eye ssp. *Lophozosterops squamiceps stresemanni* van Marle, 1940

Black Bulbul ssp. *Hypsipetes leucocephalus stresemanni* Mayr, 1942

Wallace's Hawk Eagle ssp. *Nisaetus nanus stresemanni* **Amadon**, 1953

Aberrant Bush Warbler ssp. *Cettia flavolivacea stresemanni* **Koelz**, 1954

Island Whistler ssp. *Pachycephala phaionota stresemanni* Jany, 1955 NCR; NRM

Erwin Friedrich Theodor Stresemann (1889–1972) was a German ornithologist and collector in the Far East. He was President of the German Ornithological Society and was chairman of the Standing Committee on Ornithological Nomenclature of the International Ornithological Congress (1954), as well as Curator of Birds at the Berlin Natural History Museum. He was both the patron and teacher of Ernst Mayr (q.v.) and persuaded him to study natural sciences. He was the editor of the highly authoritative *Journal für Ornithologie* for many years. He wrote *Aves* (1927) and *Die Entwicklung der Ornithologie von Aristoteles biz zur Gegenwart* (1951). An odd fact is that he is buried in Berlin in the same grave as Ernst Hartert (q.v.). A reptile is also name after him.

Streubel

White-rumped Swift ssp. *Apus caffer streubelii* **Hartlaub**, 1861 NCR; NRM

August Vollrath Streubel (DNF) was a German ornithologist, with a special interest in swifts. He named the swift genus *Cypseloides*.

Strickland

Strickland's Snipe *Chubbia stricklandii* **G. R. Gray**, 1845

[Alt. Fuegian Snipe; Syn. *Gallinago stricklandii*]

Strickland's Woodpecker *Picoides stricklandi* **Malherbe**, 1845

Strickland's Babbler *Trichastoma celebense* Strickland, 1849

[Alt. Sulawesi Babbler]

Strickland's Jay *Cyanolyca pumilo* Strickland, 1849

[Alt. Black-throated Jay]

Crimson-backed Flameback *Chrysocolaptes stricklandi* **Layard**, 1854

Strickland's Bunting *Emberiza cinerea* **C. L. Brehm**, 1855

[Alt. Cinereous Bunting]

Strickland's Shama *Copsychus stricklandii* **Motley** & **Dillwyn**, 1855

[Alt. White-crowned Shama]

Crested Owl ssp. *Lophostrix cristata stricklandi* **P. L. Sclater** & **Salvin**, 1859

Common Crossbill ssp. *Loxia curvirostra stricklandi* **Ridgway**, 1885

Hugh Edwin Strickland (1811–1853) was a geologist and naturalist who travelled after graduating from Oxford. He married (1845) Catherine, daughter of Sir William Jardine (q.v.), taking her on a European tour through Holland, Copenhagen, Berlin, Frankfurt and Brussels, visiting most of the museums on the way. Strickland was a well-liked and respected naturalist who corresponded with all the giants of his day, including Darwin (q.v.). He eventually became the Deputy Reader of Geology at Oxford. Strickland collected both fossils and birds; his home was described as 'quite a museum of ornithology'. Most of his fossil collection is now in the Sedgwick Museum, Cambridge. The birds also formed the original nucleus of the skin collection of the University Museum of Zoology, Cambridge, which houses many of the specimens collected by Darwin. After his death his wife and Sir William Jardine edited his book *Ornithological Synonyms* (1855). He is best remembered for creating the 'Strickland Code' (1842), which tried (in vain) to standardise zoological nomenclature for all time. He was killed in an accident in a railway cutting near Hull when making a pencil sketch of the rock strata in pursuit of his geological interests.

Stronach

Karamoja Apalis ssp. *Apalis karamojae stronachi* Stuart & **Collar**, 1985

Brian William Hemsworth Stronach (1928–1984) was born in Kenya and educated in Ireland. He was a game warden, wildlife college lecturer, and Tsetse Officer in Tanganyika (Tanzania). He stayed on after independence, but eventually returned to Ireland where he died. He was one of the collectors of the holotype.

Strümpell

White-eye sp. *Zosterops struempelli* **Reichenow**, 1910 NCR
[Alt. African Yellow White-eye; JS *Zosterops senegalensis*]

Sun Lark ssp. *Galerida modesta struempelli* Reichenow, 1910
Emin's Shrike ssp. *Lanius gubernator struempelli* Reichenow, 1910 NCR; NRM

Lieutenant-Colonel Kurt F. Strümpell (1872–1923) was an officer in the German colonial army in the Cameroons (1900–1910). He was a geographer, ethnographer, and linguist and was in charge of the Colonial Section, the Foreign Office, Berlin (1917–1919).

Struthers

Ibisbill *Ibidorhyncha struthersii* Vigors, 1832

Dr John Struthers (DNF) was described by Vigors as 'a zealous naturalist' who had collected birds in the Himalayas. Very little seems to be known of him; a Scottish physician of that name is known to have practised in Glasgow (1831).

Stuart, A. M.

White-browed Hermit *Phaethornis stuarti* **Hartert**, 1897

Arthur Maxwell Stuart (DNF) was a collector in Bolivia towards the end of the 19th century.

Stuart, E. C.

Blue-eared Barbet ssp. *Megalaima australis stuarti* **H. C. Robinson** & **Kloss**, 1919
Lesser Necklaced Laughingthrush ssp. *Garrulax monileger stuarti* **Meyer de Schauensee**, 1955

(See **Stuart Baker** & **Baker, E. C. S.**)

Stuart, L. C.

Northern Rough-winged Swallow ssp. *Stelgidopteryx serripennis stuarti* **Brodkorb**, 1942

Dr Laurence Cooper Stuart (1907–1983) of the University of Michigan's Museum of Zoology was an expert on the herpetofauna of Guatemala, where he collected the holotype of the swallow (1938). Eight reptiles and four amphibians are named after him.

Stuart, M.

Long-billed Starthroat ssp. *Heliomaster longirostris stuartae* **G. N. Lawrence**, 1860 NCR
[NUI *Heliomaster longirostris longirostris*]

Mrs Mary Stuart *née* McCrea (1810–1891) was an American collector of books and *objets d'art*, and the wife of sugar magnate and philanthropist Robert Leighton Stuart. The type specimen of this race of hummingbird was in 'a small collection of skins from Bogota', in the possession of Robert Stuart, who was a friend of the describer.

Stuart Baker

Indian Robin ssp. *Saxicoloides fulicatus stuartbakeri* **Koelz**, 1939 NCR
[JS *Saxicoloides fulicatus erythrurus*]

(See **Baker, E. C. S.**)

Stuart Irwin

Yellow-bellied Waxbill ssp. *Coccopygia quartinia stuartirwini* **Clancey**, 1969

Michael Patrick Stuart Irwin (b.1925) is a British ornithologist who spent 63 years in Africa, mainly in Harare, Zimbabwe, before moving back (2012) to Norfolk, England. He wrote *The Birds of Zimbabwe* (1981) and was editor of *Honeyguide* for many years. (See also **Irwin**)

Stuart Keith

Malachite Kingfisher ssp. *Corythornis cristatus stuartkeithi* **Dickerman**, 1989

(See **Keith**)

Stübel

Buffy Helmetcrest *Oxypogon stuebelii* **A. B. Meyer**, 1884
[Syn. *Oxypogon guerinii stuebelii*]

Dr Moritz Alphons Stübel (1835–1904) was a German geologist, vulcanologist, archaeologist, explorer, ethnologist and collector who was educated at the University of Leipzig. He visited Santorini, Greece (1866) and was in Colombia and Ecuador (1868–1874), and thereafter in Peru, Brazil, Argentina, Uruguay, Chile and Bolivia (1874–1877).

Stuhlmann

Dusky-blue Flycatcher *Pedilorhynchus stuhlmanni* **Reichenow**, 1892 NCR
[JS *Muscicapa comitata*]

Stuhlmann's Double-collared Sunbird *Cinnyris stuhlmanni* Reichenow, 1893
[Alt. Rwenzori Double-collared Sunbird]
Stuhlmann's Starling *Poeoptera stuhlmanni* Reichenow, 1893

African Yellow White-eye ssp. *Zosterops senegalensis stuhlmanni* Reichenow, 1892
Stuhlmann's Weaver *Ploceus baglafecht stuhlmanni* Reichenow, 1893
[Alt. Baglafecht Weaver ssp.]

Professor Dr Franz Ludwig Stuhlmann (1863–1928) was a German zoologist and naturalist who collected in East Africa (1888–1900). He made his career in the German Colonial Forces and Civil Service. He did not confine himself to zoological specimens, as a number of the artefacts he collected in Africa are in anthropological exhibits. Stuhlmann travelled with Emin Pasha (q.v.) and after Emin's murder, he and others who had survived an outbreak of smallpox went back from the area of Lake Albert with a large collection and a lot of cartographic material from which the first comprehensive map of German East Africa (Tanzania) was made. The German Government published a monograph by Stuhlmann, *Dr. Franz Stuhlmann: Mit Emin Pasha ins Herz von Africa* (1894). Two mammals, a reptile and an amphibian are named after him.

Sturm

Dwarf Bittern *Ixobrychus sturmii* **Wagler**, 1827

Johann Heinrich Christian Friedrich Sturm (1805–1862) was a German collector and bird artist.

Styan

Styan's Grasshopper Warbler *Locustella pleskei* **Taczanowski**, 1890
[Alt. Pleske's (Grasshopper) Warbler]
Styan's Bulbul *Pycnonotus taivanus* Styan, 1893
[Alt. Taiwan Bulbul]
Styan's Laughingthrush *Trochalopteron styani* **Oustalet**, 1898 NCR
[Alt. Moustached Laughingthrush; JS *Garrulax cineraceus cinereiceps*]
Koklass Pheasant *Pucrasia styani* **Ogilvie-Grant**, 1908 NCR
[JS *Pucrasia macrolopha darwini*]

Streak-breasted Scimitar Babbler ssp. *Pomatorhinus ruficollis styani* **Seebohm**, 1884
Black-naped Monarch ssp. *Hypothymis azurea styani* **Hartlaub**, 1899
Greater Yellownape ssp. *Chrysophlegma flavinucha styani* Ogilvie-Grant, 1899
Brown-winged Parrotbill ssp. *Sinosuthora brunnea styani* Rippon, 1903
Long-tailed Minivet ssp. *Pericrocotus brevirostris styani* **E. C. S. Baker**, 1920 NCR
[JS *Pericrocotus ethologus ethologus*]
Common Iora ssp. *Aegithina tiphia styani* **La Touche**, 1923 NCR
[JS *Aegithina tiphia philipi*]

Rock Bunting ssp. *Emberiza cia styani* La Touche, 1923 NCR
[Alt. Godlewski's Bunting ssp.; JS *Emberiza godlewskii omissa*]

Frederick William Styan (1838–1934) was a tea trader and collector in China for 27 years who corresponded from Kiukiang. He was a Fellow of the Zoological Society of London and was elected as a Member of the BOU (1887). A mammal and three reptiles are named after him.

Such

Such's Antthrush *Chamaeza meruloides* **Vigors**, 1825
[Alt. Cryptic Antthrush]
Yellow-rumped Marshbird *Leistes suchii* Vigors, 1825 NCR
[JS *Pseudoleistes guirahuro*]
Black-goggled Tanager *Tachyphonus suchii* **Swainson**, 1826 NCR
[JS *Trichothraupis melanops*]

Dr George Such (1798–1879) was a physician who practised in London, and a Fellow of the Linnean Society. Vigors stated: 'Dr Such obtained the antthrush in the "Brazils".'

Suchitra

Scarlet Minivet ssp. *Pericrocotus flammeus suchitrae* **Deignan**, 1948 NCR
[JS *Pericrocotus speciosus semiruber*]

Dr Suchitra Punyaratabandhu (DNF) from Bangkok, Thailand, also worked in Washington. She was Dean, School of Public Administration, National Institute of Public Administration, Bangkok.

Suckley

Suckley's Gull *Larus suckleyi* **G. N. Lawrence**, 1858 NCR
[Alt. Mew Gull; JS *Larus canus brachyrhynchus*]
Rhinoceros Auklet *Sagmatorrhina suckleyi* **Coues**, 1868 NCR
[JS *Cerorhinca monocerata*]

Suckley's Pigeon-Hawk *Falco columbarius suckleyi* **Ridgway**, 1873
[Alt. Merlin ssp.]

George Suckley (1830–1869) was an American army surgeon and naturalist. He was appointed as assistant surgeon and naturalist of the Pacific Railway Survey between Minnesota and the Puget Sound (1853). Later, he explored the Oregon and Washington territories, which had not yet been admitted as States of the Union. He resigned from the army (1856) to concentrate on natural history, then rejoined the Union Army and served as a surgeon throughout the Civil War. He co-wrote *Natural History of Washington Territory* (1859). A fish is named after him.

Sueur

Sueur's Triller *Lalage sueurii* **Vieillot**, 1818
[Alt. White-shouldered Triller]

Charles Alexandre Lesueur (Le Sueur) (1778–1846) was a French naturalist, artist and explorer. At 23 he set sail for

Australia and Tasmania aboard *Le Géographe* as an assistant gunner. Baudin (q.v.) appointed him as an official expedition artist when the original artists jumped ship in Mauritius. During the next four years he and fellow naturalist François Péron (q.v.) collected more than 100,000 zoological specimens representing 2,500 new species, and Lesueur had made 1,500 drawings. From these drawings he produced a series of watercolours on vellum, which were published (1807–1816) in the expedition's official report, *Voyage de Découvertes aux Terres Australes*. Lesueur lived in the USA (1815–1837) and undertook some local travels and collecting. In 1824 he met Audubon (q.v.) and so admired his work that he suggested he should try again to get it published in France. Lesueur was appointed Curator of the Natural History Museum in Le Havre (1845), which was created to house his drawings and paintings.

Sukatschev

Sukatschev's Laughingthrush *Garrulax sukatschewi*
Berezowski & **Bianchi**, 1891
[Alt. Snowy-cheeked Laughingthrush]

Vladimir Platonovich Sukatschev (1849–1919) was a Russian trader, philanthropist, passionate art collector, and explorer who founded the Art Museum (Gallery) in Irkutsk (starting with many paintings from his own home), the town where he was born – and also where a great many Polish scientists were exiled. His training was originally in the law but he had wide interests as well as devoting much of his time to the Eastern Siberian Branch of the Russian Geographical Society. He used some of his inherited wealth to educate local people and to finance a trip to Mongolia and China (1884–1887); whatever he collected went into his local museum. He used his own money to publish *Irkutsk – Its Place and Significance in the History and Culture of East Siberia* (1892) and *The First Century of Irkutsk* (1902), the revenue from which went to pay for local scholars. He also published a magazine in St Petersburg called *Siberian Questions*.

Sumichrast

Sumichrast's Blackbird *Dives dives* **Deppe**, 1830
[Alt. Melodious Blackbird]
Sumichrast's Sparrow *Peucaea sumichrasti*
G. N. Lawrence, 1871
[Alt. Cinnamon-tailed Sparrow]
Sumichrast's Wren *Hylorchilus sumichrasti* G. N. Lawrence, 1871
[Alt. Slender-billed Wren]

Sumichrast's Scrub Jay *Aphelocoma californica sumichrasti*
Ridgway, 1874
[Alt. Western Scrub Jay ssp.]
Olive Sparrow ssp. *Arremonops rufivirgatus sumichrasti*
Sharpe, 1888
Berylline Hummingbird ssp. *Amazilia beryllina sumichrasti*
Salvin, 1891
Rose-throated Becard ssp. *Pachyramphus aglaiae sumichrasti* **Nelson**, 1897
Common Pauraque ssp. *Nyctidromus albicollis sumichrasti* Ridgway, 1912 NCR

[JS *Nyctidromus albicollis yucatanensis*]
Citreoline Trogon ssp. *Trogon citreolus sumichrasti*
Brodkorb, 1942

Adrien Jean Louis François de Sumichrast (1828–1882) was a Swiss naturalist who accompanied Saussure on his travels (1854–1856) in the West Indies, USA and Mexico. They made considerable collections of specimens, which Saussure took back to Geneva (1856). Sumichrast stayed in Mexico for the rest of his life. He may have taken Mexican nationality, although sources in Mexico record him as a Swiss naturalist called A. L. François Sumichrast. The Smithsonian Institution knew him as Professor François Sumichrast and thought him a French naturalist, under which guise he undertook an expedition in Mexico for them. He applied to Baird (q.v.) at the USNM for a second expedition (1870) but funding was refused. However, he did undertake several other collecting expeditions under their auspices. Three mammals, three reptiles and an amphibian are named after him.

Sundevall

Sundevall's Waxbill *Estrilda rhodopyga* Sundevall, 1850
[Alt. Crimson-rumped Waxbill, Rosy-rumped Waxbill]
Lava Heron *Butorides sundevalli* **Reichenow**, 1877
Camaroptera sp. *Camaroptera sundevalli* **Sharpe**, 1882 NCR
[Alt. Green-backed Camaroptera; JS *Camaroptera brachyura*]

Southern Red Bishop ssp. *Euplectes orix sundevalli*
Bonaparte, 1850
[*E. orix* sometimes regarded as monotypic]
Pale White-eye ssp. *Zosterops pallidus sundevalli* **Hartlaub**, 1865
[Alt. Orange River White-eye ssp.]

Carl Jakob Sundevall (sometimes written Sundewall) (1801–1875) was a Swedish zoologist and ornithologist who had a strong interest in spiders as well as birds. He wrote *Svenska Fåglarna* (1856) and *Methodi Naturalis Avium Disponendarum Tentamen* (1889). Two mammals and four reptiles are named after him.

Susan

Bird-of-Paradise sp. *Paradisea susannae* **E. Ramsay**, 1883 NCR
[Alt. Goldie's Bird-of-Paradise; JS *Paradisaea decora*]

Lady Susan Emmeline Macleay (1838–1903) was the wife of Sir William John Macleay (q.v.).

Sushkin/Suschkin

Black-throated Diver ssp. *Gavia arctica suschkini* **Zarudny**, 1912 NCR
[JS *Gavia arctica arctica*]
Willow Tit ssp. *Poecile baicalensis suschkini* Hachlor, 1912 NCR
[JS *Poecile montanus baicalensis*]
Daurian Partridge ssp. *Perdix dauurica suschkini* **Poliakov**, 1915
Brown Accentor ssp. *Prunella fulvescens sushkini* **Collin** & **Hartert**, 1927 NCR
[JS *Prunella fulvescens khamensis*]

Crested Barbet ssp. *Trachyphonus vaillantii suschkini*
  Grote, 1929 NCR
[JS *Trachyphonus vaillantii suahelicus*]
David's Bush Warbler ssp. *Bradypterus davidi suschkini* **B.
  Stegmann**, 1929
Eurasian Curlew ssp. *Numenius arquata sushkini* **Neumann**,
  1929
[SII *Numenius arquata orientalis*]
Yellow-breasted Bunting ssp. *Emberiza aureola suschkini*
  Stanchinsky, 1929 NCR
[JS *Emberiza aureola aureola*]
Pallas's Reed Bunting ssp. *Emberiza pallasi suschkiniana*
  Grote, 1931 NCR
[JS *Emberiza pallasi pallasi*]
Asian Rosy Finch ssp. *Leucosticte arctoa suschkini*
  Stegmann, 1932
Siberian Jay ssp. *Perisoreus infaustus suschkini* **Dementiev**
  1932 NCR
[JS *Perisoreus infaustus sibericus*]
Common Kingfisher ssp. *Alcedo atthis suschkini* Pusanov,
  1933 NCR
[JS *Alcedo atthis atthis*]
Eurasian Skylark ssp. *Alauda arvensis sushkini*
  Domaniewski, 1933 NCR
[JS *Alauda arvensis kiborti*]
Twite ssp. *Carduelis flavirostris sushkini* Sudilovskaya, 1938
  NCR
[JS *Carduelis flavirostris miniakensis*]
Northern Goshawk ssp. *Accipiter gentilis suschkini*
  Dementiev 1940 NCR
[JS *Accipiter gentilis schvedowi*]
Great Rosefinch ssp. *Carpodacus rubicilla sushkini* Keve,
  1943 NCR
[JS *Carpodacus rubicilla kobdensis*]

Professor Petr Petrovich Sushkin (1868–1928) was a palaeontologist, anatomist, ornithologist and lepidopterist working
at the Zoological Museum of the Academy of Sciences in
Leningrad (St Petersburg), becoming Director (1921–1928).
He wrote *Birds of Soviet Altai* (1925). He travelled widely in
the former Soviet Union. Among his students was Dr Elizaveta Vladimirovna Kozlova (q.v.). Many subspecies of
lepidoptera are also named after him

Sutton, G. M.

Sutton's Warbler *Dendroica potomac* Haller, 1940 NCR
[Hybrid: *Parula americana* x *Dendroica dominica*]

Sutton's Sharp-shinned Hawk *Accipiter striatus suttoni*
  Van Rossem, 1939
Sutton's Screech Owl *Megascops kennicottii suttoni*
  R. T. Moore, 1941
[Alt. Western Screech Owl ssp.]
Woodhouse's Scrub Jay ssp. *Aphelocoma woodhouseii*
  suttoni **A. R. Phillips**, 1964 NCR
[JS *Aphelocoma woodhouseii nevadae*]
Yellow-throated Euphonia ssp. *Euphonia hirundinacea*
  suttoni A. R. Phillips, 1966
[SII *Euphonia hirundinacea hirundinacea*]
Rufous-crowned Sparrow ssp. *Aimophila ruficeps suttoni*
  J. P. Hubbard, 1975

White-throated Thrush ssp. *Turdus assimilis suttoni*
  A. R. Phillips, 1991
Red-tailed Hawk ssp. *Buteo jamaicensis suttoni* **Dickerman**,
  1993

George Miksch 'Doc' Sutton (1898–1982) was an American
ornithologist and bird artist. At 17 he spent some weeks with
Louis Agassiz Fuertes (q.v.), with whom he had corresponded
for a year. He began his professional life as Keeper of the egg
collection at the Carnegie Museum. He became State Ornithologist for Pennsylvania (1926) and spent his time trying to
persuade hunters that raptors should not be shot. He
completed his doctorate at Cornell, where he remained for
some years as Curator of Birds, enabling him to undertake a
number of trips and expeditions in the USA, Canada and
Mexico. He was Curator of Birds at the Stovall Museum, at
the University of Oklahoma (1952) and stayed on after retiring
(1968) as George Lynn Cross Research Professor Emeritus of
Zoology. The George Miksch Sutton Avian Research Center
(GMSARC) of Bartlesville, Oklahoma, is named after him.
According to the center 'A recently-published bibliography
of his works lists 13 books, 18 monographs and museum
publications, 201 journal articles, 12 book reviews, four obituaries, 18 popular articles and eight essays. He also illustrated
at least another 18 books'. His books included *Iceland
Summer: Adventures of a Bird Painter*, *Portraits of Mexican
Birds*, *Fifty Selected Paintings* and *Bird Student* (1980), his
autobiography.

Sutton, J.

Rufous Fieldwren ssp. *Calamanthus fuliginosus suttoni*
  Condon, 1951 NCR
[JS *Calamanthus campestris campestris*]

John Sutton (1866–1938) was an Australian banker and ornithologist who was Assistant Honorary Curator (1923–1934)
and Honorary Curator of Birds at the South Australian
Museum, Adelaide (1934–1938). His business career was
mainly in Melbourne but when he retired from banking (1917)
he moved to Adelaide, where he lectured on banking at the
University of Adelaide.

Suzuki

Crake sp. *Rallina suzuki* **Momiyama**, 1930 NCR
[Alt. Red-legged Crake; JS *Rallina fasciata*]

Mrs Z. Suzuki (DNF) was a Japanese collector on Taiwan.
She made a collection on the small island of Botel Tobago
(Orchid Island) (1929).

Swainson

Swainson's Lorikeet *Trichoglossus moluccanus* **F. J. Gmelin**,
  1788
[Alt. Coconut Lorikeet; Syn. *Trichoglossus haematodus
  moluccanus*]
Swainson's Hummingbird *Avocettula recurvirostris*
  Swainson, 1822
[Alt. Fiery-tailed Awlbill]
Pearl Kite *Gampsonyx swainsonii* **Vigors**, 1825
Swainson's Antbird *Myrmeciza longipes* Swainson, 1825
[Alt. White-bellied Antbird]

Swainson's Cuckooshrike *Coracina lineata* Swainson, 1825
[Alt. Barred Cuckooshrike]
Swainson's Fire-eye *Pyriglena atra* Swainson, 1825
[Alt. Fringe-backed Fire-eye]
Swainson's Fruit Dove *Ptilinopus regina* Swainson, 1825
[Alt. Rose-crowned Fruit Dove]
Superb Parrot *Polytelis swainsonii* **Desmarest**, 1826
[Alt. Barraband's Parakeet]
Swainson's Tanager *Piranga bidentata* Swainson, 1827
[Alt. Flame-coloured Tanager]
Hispaniolan Emerald *Chlorostilbon swainsonii* **Lesson**, 1829
Seedsnipe sp. *Tinochorus swainsonii* Lesson, 1831 NCR
[Indeterminate: perhaps a JS of *Thinocorus rumicivorus*]
Swainson's Toucan *Ramphastos swainsonii* **Gould**, 1833
[Alt. Chestnut-mandibled Toucan; Syn. *Ramphastos ambiguus swainsonii*]
Swainson's Warbler *Limnothlypis swainsonii* **Audubon**, 1834
Swainson's Francolin *Pternistis swainsonii* **A. Smith**, 1836
[Alt. Swainson's Spurfowl]
Swainson's Glossy Starling *Lamprotornis chloropterus* Swainson, 1838
[Alt. Lesser Blue-eared Starling]
Swainson's Hawk *Buteo swainsoni* **Bonaparte**, 1838
Swainson's Sparrow *Passer swainsonii* **Rüppell**, 1840
Swainson's Thrush *Catharus ustulatus* **Nuttall**, 1840
[Alt. Olive-backed Thrush]
Buff-bellied Puffbird *Notharchus swainsoni* **G. R. Gray**, 1846
Cliff Swallow sp. *Petrochelidon swainsoni* **Sclater**, 1858 NCR
[Alt. American Cliff Swallow; JS *Petrochelidon pyrrhonota melanogaster*]
Swainson's Royal Flycatcher *Onychorhynchus swainsoni* **Pelzeln**, 1858
[Alt. Atlantic Royal Flycatcher]
Swainson's Flycatcher *Myiarchus swainsoni* **Cabanis & Heine** 1859

Grey-headed Kingfisher ssp. *Halcyon leucocephala swainsoni* Smith, 1834 NCR
[NPRB *Halcyon leucocephala pallidiventris*]
Swainson's Weaver *Ploceus nigricollis brachypterus* Swainson, 1837
[Alt. Black-necked Weaver ssp.]
Swainson's Thrush ssp. *Catharus ustulatus swainsoni* **Tschudi**, 1845
Swainson's Vireo *Vireo gilvus swainsoni* **Baird**, 1858
[Alt. Warbling Vireo ssp.]
Apostlebird ssp. *Struthidea cinerea swainsoni* **Mathews**, 1912 NCR
[Imprecise type locality, and regarded as invalid: possibly an intergrade between *S. cinerea cinerea* and *S. cinerea dalyi*]

William Swainson (1789–1855) was a naturalist and bird illustrator. He was born in Liverpool, the son of a collector of customs duty. After elementary education, he worked as a junior clerk and then in the army commissariat in Malta and Sicily. Before going abroad he drew up, at the request of the Liverpool Museum, the *Instructions for Collecting and Preserving Subjects of Natural History* (1808). He served (1807–1815) with the army commissariat and amassed a

collection of zoological specimens. At the end of the Napoleonic wars he retired on half-pay. He left for Brazil and travelled, collecting specimens, through Pernambuco to the Rio São Francisco and then on to Rio de Janeiro (1816–1818). On his return he published a sketch (1819), very briefly describing the voyage without any scientific detail. He then endeavoured to sort his zoological specimens. He learned the new technique of lithography and produced the 3-volume *Zoological Illustration* (1820–1823), the *Naturalist's Guide* (1822) and *Exotic Conchology*. In 1828 he visited museums in Paris under the guidance of Cuvier (q.v.) and St Hilaire (q.v.), meeting great French naturalists, and in the same year moved to the English countryside and worked as a full-time artist and author. However, in 1840 he became New Zealand's first Attorney General, losing most of his specimen collection on the voyage, and lived out his life there. He published many papers, as well as the 5-volume *Birds of Brazil* (1834–1835). He wrote the bird section of Sir John Richardson's (q.v.) *Fauna Boreali-Americana* and contributed to the 11-volume *Cabinet Encyclopaedia* (1834–1840) and the 3-volume *Naturalist's Library* (1833–1846). A mammal is also named after him.

Swales

La Selle Thrush *Turdus swalesi* **Wetmore**, 1927

Bradshaw Swales (1875–1928) was an American ornithologist. He qualified and practised as a lawyer in Detroit and appears to have lived most of his life in Michigan. He was interested in oology and anthropology, especially in relation to Native Americans. With Alexander Wetmore (q.v.) he wrote *Birds of Haiti and the Dominican Republic* (1931). He was one of the founders of the Baird Ornithological Club, and was President when he died.

Swanzy

Singing Cisticola ssp. *Cisticola cantans swanzii* **Sharpe**, 1870

Alfred Swanzy (1818–1879) was an entomologist and lepidopterist who sponsored a number of expeditions to the Gold Coast (Ghana), where his company, F. & A. Swanzy, had strong links and a physical presence in its trading stations. He was a Fellow of the Linnean Society (1868–1879).

Swarth

Swarth's Ground Finch *Camarhynchus conjunctus* Swarth, 1929 NCR
[Hybrid: *Camarhynchus parvulus* x *Certhidea olivacea*]

Northern Pygmy Owl ssp. *Glaucidium californicum swarthi* **Grinnell**, 1913
Sulphur-bellied Flycatcher ssp. *Myiodynastes luteiventris swarthi* **Van Rossem**, 1927 NCR; NRM
Fox Sparrow ssp. *Passerella iliaca swarthi* Behle & Selander, 1951
[Syn. *Passerella schistacea swarthi*]

Dr Harry Schelwaldt Swarth (1878–1935) was an American zoologist. He began collecting birds (1894), then (1896) he made a trip to the Huachuca Mountains, Arizona, developing such an interest in the area that he took five more trips to different regions of the state and became the leading authority on Arizona ornithology. He became Assistant in the Department of Zoology at the Field Museum, Chicago (1904), and Assistant in Ornithology at the Museum of Vertebrate Zoology, University of California, Berkeley (1908) and Curator (1910). He joined the California Academy of Sciences (1912) and was appointed Curator of the Department of Ornithology and Mammalogy there (1927), having left (1913–1916). He wrote *The Avifauna of the Galapagos Islands* (1931) and 'A new bird family (Geospizidae) from the Galapagos Islands' (1929). He led the Templeton Crocker expedition of the California Academy of Sciences to the Galápagos (1932). A mammal is named after him.

Sweet

Black-hooded Laughingthrush ssp. *Garrulax milleti sweeti* Eames, 2002

Paul R. Sweet is a Bristol-born ornithologist who is Collections Manager in the Bird Department, AMNH, New York. He was a member of the Mount Tay Con Linh, Vietnam expedition (2000).

Swenk

Eastern Screech Owl ssp. *Otus asio swenki* **Oberholser**, 1937 NCR
[NUI *Megascops asio maxwelliae*]

Professor Myron Harmon Swenk (1883–1941) was an American ornithologist and entomologist at the University of Nebraska, which awarded his bachelor's (1907) and master's (1908) degrees, after he joined the staff as a Laboratory Assistant (1904). He retired because of ill health (1941) as Full Professor and Chairman of the Department of Entomology.

Swierstra

Swierstra's Francolin *Pternistis swierstrai* **J. A. Roberts**, 1929

Cornelis Jacobus Swierstra (1874–1952) was a Dutch-South African entomologist who was born in Groningen and died in Pretoria. He started his career in the Transvaal Museum (1897) as a General Assistant, becoming Deputy Director under J. W. B. Gunning (q.v.) and eventually succeeding him as Director (1922–1946). He was an accomplished collector of butterflies, who described many holotypes. He managed the ethnology collection, having conducted some field studies among the Hananwa people (1912). A fish is named after him.

Swindells

Buff-banded Rail ssp. *Gallirallus philippensis swindellsi* **Mathews**, 1911

Arthur William Swindells (1877–1960) was an Australian accountant, horticulturist and oologist.

Swinderen

Swinderen's Lovebird *Agapornis swindernianus* **Kuhl**, 1820
[Alt. Black-collared Lovebird]

Professor Dr Theodorus van Swinderen (1784–1851) was a Dutch naturalist and Doctor of Linguistics and Law who was a school inspector for 40 years. He was born in Groningen, where he became a student at the University (1799) and established the Royal Natural History Society (Koninklijk Natuurkundig Genootschap) (1801). He was appointed Professor of Natural History at the University of Groningen (1814) where he taught both Kuhl (q.v.) and van Hasselt (q.v.). He initiated the erection of various historical monuments in Groningen, including one to commemorate the battle of Heiligerlee, the first battle in the 80-year war to secure independence from Spain (1568–1648). A mammal is named after him.

Swinhoe

Swinhoe's Pitta *Pitta nympha* **Temminck** & **Schlegel**, 1850
[Alt. Fairy Pitta, Chinese Pitta]
Swinhoe's Egret *Egretta eulophotes* Swinhoe, 1860
[Alt. Chinese Egret]
Swinhoe's Finch-billed Bulbul *Spizixos semitorques* Swinhoe, 1861
[Alt. Collared Finchbill]
Swinhoe's Minivet *Pericrocotus cantonensis* Swinhoe, 1861
[Alt. Brown-rumped Minivet]
Swinhoe's Snipe *Gallinago megala* Swinhoe, 1861
Swinhoe's Pheasant *Lophura swinhoii* **Gould**, 1863
Swinhoe's (Red-tailed) Robin *Luscinia sibilans* Swinhoe, 1863
[Alt. Rufous-tailed Robin]
Swinhoe's Rock Thrush *Monticola gularis* Swinhoe, 1863
[Alt. White-throated Rock Thrush]
Swinhoe's Storm-petrel *Cymochorea monorhis* Swinhoe, 1867
[Syn. *Oceanodroma monorhis*]
Bee-eater sp. *Melittophagus swinhoei* **Hume**, 1873 NCR
[Alt. Chestnut-headed Bee-eater; JS *Merops leschenaulti*]
Swinhoe's Rail *Coturnicops exquisitus* Swinhoe, 1873
Grey-sided Scimitar Babbler *Pomatorhinus swinhoei* **David**, 1874

Swinhoe's Wagtail *Motacilla alba ocularis* Swinhoe, 1860
[Alt. White Wagtail ssp.]
Swinhoe's Bush Warbler *Cettia fortipes robustipes* Swinhoe, 1866
[Alt. Brownish-flanked Bush Warbler ssp.]
Golden-breasted Fulvetta ssp. *Lioparus chrysotis swinhoii* **J. Verreaux**, 1870
Swinhoe's Wagtail *Motacilla alba baicalensis* Swinhoe, 1871
[Alt. White Wagtail ssp.; different taxon from above]
White-rumped Munia ssp. *Lonchura striata swinhoei* **Cabanis**, 1882
Azure-winged Magpie ssp. *Cyanopica cyanus swinhoei* **Hartert**, 1903
Spot-necked Babbler ssp. *Stachyris strialata swinhoei* **Rothschild**, 1903
Grey-capped Woodpecker ssp. *Dendrocopos canicapillus swinhoei* Hartert, 1910

Whiskered Tern ssp. *Chlidonias hybrida swinhoei* **Mathews,** 1912

[SII *Chlidonias hybrida hybrida*]

Eurasian Eagle Owl ssp. *Bubo bubo swinhoei* Hartert, 1913

[SII *Bubo bubo kiautschensis*]

Black-naped Oriole ssp. *Oriolus chinensis swinhoii* **Momiyama** & Isii, 1928 NCR

[JS *Oriolus chinensis diffusus*]

Robert Swinhoe (1836–1877) was born in Calcutta (Kolkata), India, but was sent to London (1852) to be educated. While at the University of London (1854) he was recruited into the China Consular Corps by the Foreign Office. Before he left for Hong Kong he deposited a small collection of British birds' nests and eggs with the British Museum. His time in China as a diplomat gave him great opportunities as a naturalist; he explored a vast area which had not been open previously to any other collector. As a result he discovered new species at the rate of about one per month throughout the nearly two decades he was there. He was primarily an ornithologist, but his name is also associated with various Chinese mammals, fish and insects. He returned to London in 1862 and brought part of his vast collection of specimens to meetings of the Zoological Society, as well as to their counterparts in France and Holland. He was somewhat taken aback by having to allow someone else to name the 200-plus new bird species which he had discovered, as he himself related; 'I have been blamed by some naturalists for allowing Mr. Gould to reap the fruits of my labours, in having the privilege of describing most of my novelties. I must briefly state, in explanation, that I returned to England elated with the fine new species I had discovered, and was particularly anxious that they should comprise one entire part of Mr. Gould's fine work on the Birds of Asia, still in progress. On an interview with Mr. Gould, I found that the only way to achieve this was to consent to his describing the entire series to be figured, as he would include none in the part but novelties, which he should himself name and describe. I somewhat reluctantly complied; but as he has done me the honour to name the most important species after me, I suppose I have no right to complain …' He began suffering from partial paralysis (c.1871) and his ill health forced him to leave China (1875). An amphibian, four mammals and four reptiles are named after him.

Swynnerton

Swynnerton's Robin *Swynnertonia swynnertoni* **G. E. Shelley,** 1906

White-tailed Crested Flycatcher ssp. *Elminia albonotata swynnertoni* **Neumann,** 1908

Swynnerton's Thrush *Turdus olivaceus swynnertoni* **Bannerman,** 1913

[Alt. Olive Thrush ssp.]

Red-necked Spurfowl ssp. *Pternistis afer swynnertoni* **W. L. Sclater,** 1921

Charles Francis Massy Swynnerton (1877–1938) was principally an entomologist. He was born in Suffolk, spent his early years in India, and worked in Africa. He became manager of Gungunyana Farm close to the Chirinda Forest, Southern Rhodesia (Zimbabwe), in 1900. He used it as a base and

worked on comprehensive collections of local plants, birds and insects. He was appointed (1919) the first game warden of Tanganyika (Tanzania). He later spent ten years (1929–1938) as head of tsetse research in East Africa. He published papers on many aspects of natural history, including 'On the birds of Gazaland, Southern Rhodesia' (1907). He was killed in an air-crash on his way to Dar-es-Salaam, where he was to have been invested with the order of St Michael and St George. An amphibian, a mammal and a reptile are named after him.

Sybil

Green-breasted Mountaingem *Lampornis sybillae* **Salvin & Godman,** 1892

The original description has no etymology or any hint as to who Sybil was. Perhaps this hummingbird is named after one of the legendary Sybils (female prophets) of the ancient world, such as the Cumaean Sybil of Roman mythology.

Sykes

Sykes's Lark *Galerida deva* Sykes, 1832

[Alt. Tawny Lark]

Sykes's Nightjar *Caprimulgus mahrattensis* Sykes, 1832

Sykes's Warbler *Iduna rama* Sykes, 1832

Sykes's Yellow Wagtail *Motacilla flava beema* Sykes, 1832

Black-headed Cuckooshrike ssp. *Coracina melanoptera sykesi* **Strickland,** 1844

Black-naped Monarch ssp. *Hypothymis azurea sykesi* **E. C. S. Baker,** 1920 NCR

[JS *Hypothymis azurea styani*]

Colonel William Henry Sykes (1790–1872) was an English ornithologist and army officer. He saw plenty of action after he joined the Bombay Army aged 14 (1804–1824), a part of the armed forces of the Honourable East India Company. He was then appointed (1824) as a statistical reporter to the Bombay government, and later his statistical researches involved him in natural history. He wrote *Catalogue of Birds of the Rapotorial and Incessorial Orders Observed in the Dukkan* (1832). He retired from active service (1833) and later became a Director of the East India Company, Rector of Aberdeen University, and Member of Parliament for Aberdeen. A mammal is named after him.

Symons

Symons's Siskin *Serinus symonsi* **J. A. Roberts,** 1916

[Alt. Drakensberg Siskin; Syn. *Crithagra symonsi*]

Crested Guineafowl ssp. *Guttera edouardi symonsi* Roberts, 1917 NCR

[JS *Guttera pucherani edouardi*]

Roden E. Symons (1884–1974) was a South African game warden and collector.

Szalay

Brown Oriole *Oriolus szalayi* **Madarász,** 1900

Freiherr Imre von Szalay (1846–1918) was a Hungarian author, and explorer in Central Asia. He became Director of the Hungarian National Museum, Budapest (1894–1916).

Széchenyi

Széchenyi's Monal Partridge *Tetraophasis szechenyii*
Madárasz, 1885
[Alt. Széchenyi's Pheasant Partridge]

Graf Bela Széchenyi (1837–1918) was a Hungarian explorer of Central Asia, India, China and Japan on a three-year expedition (1877–1880) he financed, The results were published by Kreitner (1881) in Vienna and attracted great attention among scientists throughout Europe. His father employed Xantus's (q.v.) father on his estate.

T

Taczanowski

Tyrannulet genus *Taczanowskia* **Stolzmann** 1926 NCR
[Now in *Serpophaga*]

Asian Dowitcher *Micropalama tacksanowskia* **J. Verreaux**, 1860 NCR
[JS *Limnodromus semipalmatus*]
Chinese Bush Warbler *Bradypterus tacsanowskius* **Swinhoe**, 1871
Taczanowski's Tinamou *Nothoprocta taczanowskii* **P. L. Sclater** & **Salvin**, 1875
Taczanowski's Finch *Onychostruthus taczanowskii* **Przewalski**, 1876
[Alt. White-rumped Snowfinch; Syn. *Pyrgilauda taczanowskii*]
Taczanowski's White-throat *Leucippus taczanowskii* P. L. Sclater, 1879
[Alt. Spot-throated Hummingbird]
Taczanowski's Nighthawk *Lurocalis rufiventris* Taczanowski, 1884
[Alt. Rufous-bellied Nighthawk]
Sulphur-throated Finch *Sicalis taczanowskii* **Sharpe**, 1888
Taczanowski's Cinclodes *Cinclodes taczanowskii* **Berlepsch** & Stolzmann, 1892
[Alt. Peruvian Seaside Cinclodes, Surf Cinclodes]
Hill Pigeon *Columba taczanowskii* **Stejneger**, 1893 NCR
[JS *Columba rupestris rupestris*]
Taczanowski's Grebe *Podiceps taczanowskii* Berlepsch & Stolzmann, 1894
[Alt. Junín Grebe, Puna Grebe]
Inca Flycatcher *Leptopogon taczanowskii* **Hellmayr**, 1917

Taczanowski's Brush Finch *Atlapetes schistaceus taczanowskii* P. L. Sclater & Salvin, 1875
[Alt. Slaty Brush Finch ssp.]
Western Capercaillie ssp. *Tetrao urogallus taczanowskii* Stejneger, 1885
Purple-throated Euphonia ssp. *Euphonia chlorotica taczanowskii* P. L. Sclater, 1886
Many-striped Canastero ssp. *Asthenes flammulata taczanowskii* Belepsch & Stolzmann 1894
Yellow-tailed Oriole ssp. *Icterus mesomelas taczanowskii* **Ridgway**, 1901
Eurasian Jay ssp. *Garrulus glandarius taczanowskii* **Lönnberg**, 1908 NCR
[JS *Garrulus glandarius brandtii*]
Lafresnaye's Piculet ssp. *Picumnus lafresnayi taczanowskii* Domaniewski, 1925
Sand Martin ssp. *Riparia riparia taczanowskii* **B. Stegmann**, 1925

Fork-tailed Woodnymph ssp. *Thalurania furcata taczanowskii* Dunajewski, 1938 NCR
[NUI *Thalurania furcata viridipectus*]

Władysław Ladislaus Taczanowski (1819–1890) was Conservator, then Curator, of the Royal University of Warsaw Zoological Cabinet (1862–1887), which he transformed from a teaching institution into a scientific centre. An outstanding zoologist and ornithologist describing many new species, he promoted the protection of birds of prey and (1866–1867) was a member of the Branicki expedition to Algeria. He wrote the 4-volume *Ornithologie du Pérou* (1884–1886), based on the collections of Kalinowski, Jelski and Stolzman (q.v. all). As the first birds handbook of a Neotropical country it was an important benchmark for South American ornithology. He also wrote *Faune Ornithologique de la Sibérie Orientale* (1891–1893), published posthumously in two parts, based on collections made by Dybowski, Godlewski and Jankowski (q.v. all), who were banished to Siberia by the Russians who occupied Poland. He was also an arachnologist, writing a list of spiders of Warsaw and describing spiders from French Guiana and Peru. Two mammals, a reptile and a fish are named after him.

Tada

Japanese Paradise Flycatcher ssp. *Terpsiphone atrocaudata tadai* **Momiyama**, 1931 NCR
[JS *Terpsiphone atrocaudata periophthalmica*]

Tsunesuke Tada (fl.1897) was a Japanese naturalist and collector on Taiwan on behalf of Tokyo Imperial University.

Taimur

Desert Lark ssp. *Ammomanes deserti taimuri* **Meyer de Schauensee** & **Ripley**, 1953

Sultan al-Wasik Billah al-Majid Sayyed Said bin Taimur (1910–1972) became Sultan of Muscat and Oman (1932), but was deposed (1970).

Taissia

Northern Harrier sp. *Circus taissiae* **Buturlin**, 1908 NCR
[Alt. Hen Harrier; JS *Circus cyaneus cyaneus*]

Dr Taissia Mikhailovna Akimova (DNF) was a Russian physician and collector at Kolyma, Siberia. She sent the harrier holotype to the describer.

Tait, D.

Henderson Reed Warbler *Acrocephalus taiti* **Ogilvie-Grant**, 1913

David R. Tait (DNF) collected birds on Henderson Island, Pitcairns (1912).

Tait, W.

Long-tailed Tit ssp. *Aegithalos caudatus taiti* **C. Ingram**, 1913
Eurasian Skylark ssp. *Alauda arvensis taiti* **Weigold**, 1913 NCR
[JS *Alauda arvensis sierrae*]

William Chester Tait (1844–1928) was a British businessman, ornithologist, botanist and sportsman resident in Portugal. He worked in the family shipping agency and trading business (founded 1834) in Oporto, where the same family still own and run the business. He wrote *Birds of Portugal* (1924). (See also **Guillelm**)

Tajan

Black Guillemot ssp. *Cepphus grylle tajani* **Portenko**, 1944 NCR
[JS *Cepphus grylle mandtii*]

Tayan, an Inuit ('Eskimo'), was a former chief of Wrangel Island. He was an important guide and mentor to Portenko during his expedition there (1939) for the USSR Academy of Sciences (Russian Federation).

Takahashi

White-backed Woodpecker ssp. *Dendrocopos leucotos takahashii* **Kuroda** & Mori, 1920
Coal Tit ssp. *Periparus ater takahashii* **Momiyama**, 1927 NCR
[JS *Periparus ater insularis*]
Japanese Bush Warbler ssp. *Cettia diphone takahashii* Momiyama, 1927 NCR
[JS *Cettia diphone cantans*]
Great Tit ssp. *Parus major takahashii* Momiyama, 1927 NCR
[Alt. Japanese Tit ssp.; JS *Parus minor minor*]

Eizo Takahashi (1872–1956) was Japanese and collected in Korea (1920–1930) for Mori (q.v.) and other Japanese ornithologists. He taught at Seoul Higher School, Korea (during the Japanese occupation).

Taka-Tsukasa

Flycatcher genus *Takatsukasaia* **Hachisuka**, 1935 NCR
[Now in *Ficedula*]

Tinian Monarch *Monarcha takatsukasae* **Yamashina**, 1931

Japanese Green Woodpecker ssp. *Picus awokera takatsukasae* **Kuroda**, 1921
Citrine White-eye ssp. *Zosterops semperi takatsukasai* **Momiyama**, 1922
Common Pheasant ssp. *Phasianus colchicus takatsukasae* **Delacour**, 1927
Coal Tit ssp. *Periparus ater takatsukasae* **Bergman**, 1931 NCR
[JS *Periparus ater insularis*]

Eurasian Nuthatch ssp. *Sitta europaea takatsukasae* Momiyama, 1931
[SI *Sitta europaea asiatica*]
Takatsukasa's Parakeet *Brotogeris sanctithomae takatsukasae* **Neumann**, 1931
[Alt. Tui Parakeet ssp.]
Takatsukasa's Reed Warbler *Acrocephalus yamashinae* Taka-Tsukasa, 1931 EXTINCT
[Alt. Pagan Reed Warbler; Syn. *Acrocephalus luscinius yamashinae*]
Lesser Coucal ssp. *Centropus bengalensis takatsukasai* Momiyama, 1932 NCR
[JS *Centropus bengalensis lignator*]

Dr Nobusuke Taka-Tsukasa (1889–1959) was a Japanese zoologist. He was a member of the aristocracy in Japan, heir to Prince Hiromichi, whom he succeeded (1918). He wrote the 8-part *Birds of Nippon* (1932–1943).

Talbot

Rufous Warbler sp. *Bathmedonia talboti* **B. Alexander**, 1907 NCR
[Alt. Black-faced Rufous Warbler; JS *Bathmocercus rufus vulpinus*]

Dr Percy Amaury Talbot (1877–1945) was a British Civil Servant in the Nigerian Political Service in Nigeria and the Cameroons (1902–1931). He was an anthropologist, explorer, naturalist and member of the Niger-Nile Expedition (1904–1905). He wrote *From the Gulf of Guinea to the Central Sudan* (1912).

Tallman

Tallmans' Fruiteater *Pipreola riefferii tallmanorum* **O'Neill** & **T. A. Parker**, 1981
[Alt. Black-headed Fruiteater]

Ornithologists Dr Dan Allen Tallman (b.1947) and Dr Erika J. Tallman (b.1949) are married, and live in Minnesota. He was a Professor of Biology, Northern State University, Aberdeen, South Dakota (1979–2006). Louisiana State University (LSU), Baton Rouge, awarded his master's and his doctorate (1979). He is associated with the Patuxent Wildlife Research Center's Bird Banding Program and was editor of *South Dakota Bird Notes*. He also co-wrote *The Birds of South Dakota* (1991). She also gained her master's and doctorate at LSU, and was Chief Information Officer for the Northern State University, Aberdeen, South Dakota (1981–2007). They have published a great many articles, jointly and separately, in national and international journals. They discovered a new bird taxon, the Pardusco *Nephelornis oneillii*, whilst Dan was undertaking fieldwork in Peru for his master's thesis.

Talpacoti

Talpacoti's Dove *Columbina talpacoti* **Temminck**, 1810
[Alt. Ruddy Ground Dove]

This is not an eponym but a Suriname Amerindian name for the dove – another over-enthusiastic use of an apostrophe creating a false eponym.

Tamemoto

Meadow Bunting ssp. *Emberiza cioides tamemoto*
Momiyama, 1923 NCR
[JS *Emberiza cioides ciopsis*]

Minamoto no Tamemoto (1139–1170) was a Japanese samurai exiled to the Seven Islands of Izu.

Tanaka

Black-hooded Oriole ssp. *Oriolus xanthornus tanakae*
Kuroda, 1925

Seikichi Tanaka (DNF) collected the holotype in British North Borneo (1925).

Tandon

White-tailed Shrike *Moquinus tandonus* **Bonaparte**, 1856 NCR
[Alt. Chat-shrike; JS *Lanioturdus torquatus*]

(See under **Moquin**)

Tang, J. C.

White-backed Woodpecker ssp. *Dendrocopos leucotos tangi* Cheng, 1956

Tang Jui Chang (b.1913) was a Chinese taxidermist at the Biology Department, Wuhan University.

Tang, W. W.

Manchurian Reed Warbler *Acrocephalus tangorum*
La Touche, 1912
[Syn. *Notiocichla tangorum*]

Tang Wang Wang and his brother (fl.1912) were Chinese collectors employed by La Touche.

Tanner

Clarion Wren *Troglodytes tanneri* **C. H. Townsend**, 1890

Bahama Yellowthroat ssp. *Geothlypis rostrata tanneri*
Ridgway, 1886

Lieutenant-Commander Zera Luther Tanner (1835–1906) was a US naval officer. He was involved in the design and construction of the Fish Commission's steamship *Albatross*, becoming its first Commanding Officer. He was an oceanographer and deep-sea explorer who developed an improved method of depth-sounding using instruments of his own design.

Tantalus

Stork genus *Tantalus* **Linnaeus**, 1758 NCR
[Now in *Mycteria*]

In Greek mythology, Tantalus was King of Sipylos and a son of Zeus. He was invited to eat and drink with the gods on Olympus, but stole nectar and ambrosia to take back to his people. He also sacrificed his son, Pelops, to the gods, boiled his body, and served pieces of it as food to them. The gods were neither deceived nor forgiving. Tantalus was sentenced to eternal punishment in Hades, condemned to stand forever in a pool of water beneath a fruit tree. Whenever he tried to pick a fruit, the branch raised itself beyond his grasp, and whenever he bent to drink, the water receded. It is the proverbial punishment of temptation without satisfaction. A mammal is also named after him.

Tappenbeck

White-shouldered Fairy-wren ssp. *Malurus alboscapulatus tappenbecki* **Reichenow**, 1897 NCR
[NUI *Malurus alboscapulatus naimii*]
Little Shrike-thrush ssp. *Colluricincla megarhyncha tappenbecki* Reichenow, 1898

Ernst Tappenbeck (fl.1898) was a German agriculturist and explorer in the Cameroons (1889) and New Guinea, working for the German New Guinea Company (1891–1898). He wrote *Deutsch-Neuguinea* (1901).

Tariho

Long-tailed Tit ssp. *Aegithalos caudatus tarihoae*
Momiyama, 1927 NCR
[Variously treated as a JS of either *Aegithalos caudatus magnus* or *A. caudatus trivirgatus*]

Tariho Takahashi (DNF) was the daughter of Eizo Takahashi, who collected the holotype (1926).

Tarn

Black-throated Huet-huet *Pteroptochos tarnii* **P. P. King**, 1831

John Tarn (1794–1877) was an English naval surgeon. When he was on board HMS *Adventure* in South American waters (1825–1830) King (q.v.) was his commanding officer.

Tarral

Pacific Imperial Pigeon ssp. *Ducula pacifica tarrali*
Bonaparte, 1854 NCR
[JS *Ducula pacifica pacifica*]

Dr Claude Tarral (DNF) was a French surgeon and physician, and Bonaparte's (q.v.) friend.

Tate

White-throated Barbtail *Premnoplex tatei* **Chapman**, 1925
Spangled Owlet-nightjar *Aegotheles tatei* **Rand**, 1941
[Alt. Rand's Owlet-nightjar, Starry Owlet-nightjar]

White-tipped Swift ssp. *Aeronautes montivagus tatei*
Chapman, 1929

George Henry Hamilton Tate (1894–1953) was an American mammalogist and author. He collected in Ecuador (1921–1924) and Venezuela (1925–1928), as well as Australasia. He wrote *Mammals of Eastern Asia*. Five mammals and a reptile are named after him.

Tatibana

Great Tit ssp. *Parus major tatibanai* **Momiyama**, 1927 NCR
[Alt. Japanese Tit ssp.; JS *Parus minor minor*]
Ural Owl ssp. *Strix uralensis tatibanai* Momiyama, 1927 NCR
[JS *Strix uralensis nikolskii*]

Matikiti Tatibana (DNF) was a Japanese collector on Sakhalin (1926–1927)

Taunay

Taunay's Woodcreeper *Dendroconcla turdina taunayi*
O. Pinto, 1939
[Alt. Plain-winged Woodcreeper ssp.; Syn. *Dendrocincla fuliginosa taunayi*]

Afonso d'Escragnolle Taunay (1876–1958) was the son of a Brazilian viscount when Brazil was still a monarchy. He graduated in Rio de Janeiro as a civil engineer (1900), and taught engineering in São Paulo (1904–1910). He was Director, Museum Paulista (1917–1939), and Professor, Faculty of Philosophy, Science, and Art, University of São Paulo (1934–1937). He was more interested in history than zoology, writing an 11-volume account of the coffee industry in Brazil (1929–1941). A reptile is named after him.

Taverner

Timberline Sparrow *Spizella taverneri* **Swarth** &
A. C. Brooks, 1925
[Syn. *Spizella breweri taverneri*]

Purple Finch ssp. *Carpodacus purpureus taverneri* **Rand**, 1946 NCR
[JS *Carpodacus purpureus purpureus*]
Taverner's Cackling Goose *Branta hutchinsii taverneri*
Delacour, 1951

Sir Percy Algernon Taverner (1875–1947) was a Canadian ornithologist at the National Museum of Natural Sciences, Ottawa (1911–1942). He was a leading advocate of conservation and bird protection and his research-based recommendations played a major part in designating Point Pelee National Park, and establishing other bird sanctuaries along the north shore of the Gulf of St Lawrence and Bonaventure Island. He was an active member of the Council of the Ottawa Field-Naturalists' Club and associate editor and frequent contributor to *Canadian Field-Naturalist*. He wrote *Birds of Eastern Canada, Birds of Western Canada* and *Birds of Canada*. A special issue of *Canadian Field-Naturalist*, written by John L. Cranmer-Byng, was devoted to his life and work. The Taverner Cup, a 24-hour bird race in Canada, is named after him.

Tavistock

Firetail genus *Tavistocka* **Mathews**, 1919 NCR
[Now in *Stagonopleura*]

Hastings William Sackville Russell, 12th Duke of Bedford (1888–1953), was an aviculturist. The Marquess of Tavistock is a courtesy title used by the heirs to the Dukedom of Bedford.

Taylor, A.

Golden-naped Tanager ssp. *Tangara ruficervix taylori*
Taczanowski & **Berlepsch**, 1885

Anthony Taylor (DNF) was a collector in Ecuador, where he accompanied Louis Fraser (q.v.) and collected for Clarence Buckley (q.v.)

Taylor, C. B.

Doctor Bird *Aithurus taylori* **Rothschild**, 1894
[Alt. Red-billed Streamertail; JS *Trochilus polytmus*]

Cuban Bullfinch ssp. *Melopyrrha nigra taylori* **Hartert**, 1896

Charles B. Taylor (fl.1900) was a lepidopterist and collector in Jamaica and the Cayman Islands, where he made a large collection of birds (1896) for Rothschild's (q.v.) Tring Museum. He was acting Curator for the Department of Zoology, Jamaica Institute, Kingston (1890). Two amphibians are named after him.

Taylor, C. H.

Olive Bush-shrike ssp. *Chlorophoneus olivaceus taylori*
J. A. Roberts, 1914 NCR
[JS *Chlorophoneus olivaceus olivaceus*]

Claude H. Taylor (DNF) was a South African naturalist. He wrote an article entitled 'Notes on a collection of birds made in the Amsterdam District upon the Transvaal-Swazieland Border' (1907).

Taylor, E. C.

Puerto Rican Kingbird *Tyrannus taylori* **P. L. Sclater**, 1864
[Syn. *Tyrannus caudifasciatus taylori*]

Edward Cavendish Taylor (1831–1905) was a clergyman, traveller, ornithologist and collector in Egypt (1853), Tunisia and Algeria (1859), and the West Indies and northern Venezuela (1862–1863). He resigned from the priesthood (1870) to concentrate on ornithology. He was a founder member of the BOU (1858), and in its journal *Ibis* he published 'Five months in the West Indies' (1864).

Taylor, F. M.

Northern Bobwhite ssp. *Colinus virginianus taylori*
F. C. Lincoln, 1915

Frank Mansfield Taylor (1850–1930) was one of the founding trustees of the Colorado Museum of Natural History, Denver (1901–1930), and President (1916–1930). He graduated as a mining engineer from Amherst College (1871) and moved to Colorado (1875).

Taylor, F. W.

Singing Quail ssp. *Dactylortyx thoracicus taylori*
Van Rossem, 1932

Frederic William Taylor (1876–1944) was an American agronomist and botanist and Director General of Agriculture for El Salvador (1923–1927).

Taylor, G.

Black-headed Greenfinch ssp. *Carduelis ambigua taylori* **Kinnear**, 1939

Dr Sir George Taylor (1904–1993) was a Scottish botanist and collector. He was in southern Tibet and Bhutan (1938) and served in the Air Ministry in WW2 (1940–1945), then worked at the BMNH, London (1945–1956), latterly as Keeper of Botany. He became Director of the Royal Botanic Gardens, Kew (1956–1971).

Teerink

Teerink's Munia *Lonchura teerinki* **Rand**, 1940
[Alt. Black-breasted Mannikin]

Carel Gerrit Jan Teerink (1897–1942) was a major in the Dutch Army in the East Indies. As a captain (1938) he commanded an army contingent (56 officers and men), which accompanied the third Archbold Expedition to New Guinea. He commanded the Fifth Battalion of Regular Infantry (early 1942) in heavy fighting in Java against Japanese invaders, and died in action.

Tegima

Ryukyu Minivet *Pericrocotus tegimae* **Stejneger**, 1887

Seiichi Tegima (DNF) was a Japanese naturalist and late 19th century Director of the Educational Museum in Tokyo. He was the official Japanese Commissioner at the St Louis World Fair (1904).

Tehmina

Black-rumped Flameback ssp. *Dinopium benghalense tehminae* **Whistler** & **Kinnear**, 1934

Tehmina (1902–1939) was the wife of the Indian ornithologist Salim Ali (q.v.).

Teijsmann

Sumba Green Pigeon *Treron teysmannii* **Schlegel**, 1879
Teijsmann's Fantail *Rhipidura teysmanni* **Büttikofer**, 1892
[Alt. Rusty-bellied Fantail, Rusty-flanked Fantail]

Olive-backed Sunbird ssp. *Cinnyris jugularis teysmanni* Büttikofer, 1893
Rusty-breasted Whistler ssp. *Pachycephala fulvotincta teysmanni* Büttikofer, 1893

Johannes Elias Teijsmann (or Teysmann) (1808–1882) was a Dutch botanist and plant collector. He was Curator of the Buitenzorg Botanic Gardens in Java (1831–1869), which were subsequently renamed in his honour. He took part in numerous botanical expeditions in the Dutch East Indies (Indonesia).

Telamon

Hummingbird genus *Telamon* **Mulsant** & **E.** & **J. Verreaux**, 1865 NCR
[Now in *Lophornis*]

In Greek mythology, Telamon was one of Jason's Argonauts on the quest for the Golden Fleece. He was the father of Ajax (q.v.), one of the most famous Greek warriors in the Trojan War.

Telasco

Chestnut-throated Seedeater *Sporophila telasco* Lesson, 1828

Telasco is an Inca character in a novel by Jean François Marmontel – *Les Incas ou la Destruction de L'Empire du Pérou* (1777).

Teleschov

Horned Lark ssp. *Eremophila alpestris teleschowi* **Przewalski**, 1887

Lieutenant P. P. Teleschov (DNF) was a Russian explorer in the Trans-Baikal region. He was an officer in the Trans-Baikalian Cossacks and accompanied Przewalski (q.v.) on his 4th expedition (1883–1884).

Temminck

Temminck's Stint *Calidris temminckii* Leisler, 1812
[Syn. *Ereunetes temminckii*]
Temminck's Gull *Larus crassirostris* **Vieillot**, 1818
[Alt. Black-tailed Gull]
Temminck's Roller *Coracias temminckii* Vieillot, 1819
[Alt. Purple-winged Roller]
Temminck's Seedeater *Sporophila falcirostris* Temminck, 1820
Australian Logrunner *Orthonyx temminckii* Ranzani, 1822
Temminck's Courser *Cursorius temminckii* **Swainson**, 1822
Temminck's Hornbill *Penelopides exarhatus* Temminck, 1823
[Alt. Sulawesi Hornbill]
Temminck's Lark *Eremophila bilopha* Temminck, 1823
[Alt. Temminck's Horned Lark]
Temminck's Kingfisher *Cittura cyanotis* Temminck, 1824
[Alt. Lilac-cheeked Kingfisher, Lilac Kingfisher]
Temminck's White-eye *Zosterops palpebrosus* Temminck, 1824
[Alt. Oriental White-eye]
Temminck's Babbler *Pellorneum pyrrogenys* Temminck, 1827
Starthroat sp. *Ornismya temminckii* **Lesson**, 1829 NCR
[Alt. Stripe-breasted Starthroat; JS *Heliomaster squamosus*]
Temminck's Tragopan *Tragopan temminckii* **J. E. Gray**, 1831
Temminck's Robin *Erithacus komadori* Temminck, 1835
[Alt. Ryukyu Robin]
Temminck's Murrelet *Synthliboramphus wumizusume* Temminck, 1836
[Alt. Japanese Murrelet]
Malaysian Eared Nightjar *Lyncornis temminckii* **Gould**, 1838
Temminck's Cuckooshrike *Coracina temminckii* **S. Müller**, 1843
[Alt. Cerulean Cuckooshrike]
Temminck's Sunbird *Aethopyga temminckii* S. Müller, 1843
Curassow sp. *Crax temminckii* **Tschudi**, 1844 NCR
[Alt. Great Curassow; JS *Crax rubra*]
Ochre-collared Piculet *Picumnus temminckii* **Lafresnaye**, 1845
Temminck's Crowned Warbler *Phylloscopus coronatus* Temminck & **Schlegel**, 1847
[Alt. Eastern Crowned Warbler]

Temminck's Pygmy Woodpecker *Dendrocopos temminckii*
Malherbe, 1849
[Alt. Sulawesi Pygmy Woodpecker]
Temminck's Cormorant *Phalacrocorax capillatus* Temminck
& Schlegel, 1850
[Alt. Japanese Cormorant]
Wattled Cuckooshrike sp. *Lobotos temminckii* **Reichenbach**,
1850 NCR
[Alt. Western Wattled Cuckooshrike; JS *Lobotos lobatus*]

Blue Whistling Thrush ssp. *Myophonus caeruleus*
temminckii **Vigors**, 1832
Superb Fruit Dove ssp. *Ptilinopus superbus temminckii*
Des Murs & **Prevost**, 1849
Black Sparrowhawk ssp. *Accipiter melanoleucus*
temminckii **Hartlaub**, 1855

Coenraad Jacob Temminck (1778–1858) was a Dutch ornithologist, illustrator and collector. He was the first Director of
the National Museum of Natural History, in Leiden (1820–
1858). He was a wealthy man who had a very large collection
of specimens and live birds. His first ornithological task was
cataloguing his father's (Jacob) extensive collection, after
whom he may have named some birds. Le Vaillant (q.v.)
collected specimens for Jacob. He issued his *Manuel d'Ornithologie, ou Tableau Systematique des Oiseaux qui se
Trouvent en Europe* (1815) and wrote *Nouveau Recueil de
Planches Coloriées d'Oiseaux* (1820). Thirteen mammals, two
reptiles, and several fish are named after him.

Tempest

Island Thrush ssp. *Turdus poliocephalus tempesti* **Layard**,
1876

A. Tempest (DNF) was a collector in Vanua Levu, Fiji, for
Layard (q.v.) when the latter was the colony's administrator.
Tempest collected the holotype.

Templeton

Templeton's Mynah *Gracula ptilogenys* **Blyth**, 1846
[Alt. Sri Lanka Hill Myna]

Dr Robert Templeton (1802–1892) was an Irish naturalist and
entomologist. His father was the botanist John Templeton. He
studied medicine at Edinburgh and (1833) became Assistant
Surgeon in an artillery regiment at Woolwich. He was posted
to Mauritius (1834), visited Brazil (1835) and then went to
Colombo, Ceylon (Sri Lanka), and became a corresponding
member of the Zoological Society of London. He was then
posted to Malta (1836), followed by a long stay back in Ceylon
(1839–1851). He was Assistant Surgeon (1837–1851), then
Surgeon (1847) and served during the Crimean War. He
retired (1860) with the honorary rank of Deputy Inspector
General of Hospitals. A reptile is named after him.

Tengmalm

Tengmalm's Owl *Aegolius funereus* **Linnaeus**, 1758
[Alt. Boreal Owl; *Strix tengmalmi* (Gmelin, 1788) is a JS]

Dr Peter Gustaf Tengmalm (1754–1803) was a Swedish physician and naturalist. He qualified at Uppsala University and

was a general practitioner in Uppsala, Västerås and Eskil-
stuna. He spent his spare time studying birds and taxidermy.
Particularly interested in owls, he wrote a paper improving
upon Linnaeus's classification of them. He left comprehensive medical and ornithological notes.

Ten Kate

Gerygone sp. *Acanthiza tenkatei* **Büttikofer**, 1892 NCR
[Alt. Golden-bellied Gerygone; JS *Gerygone sulphurea*
sulphurea]

Northern Fantail ssp. *Rhipidura rufiventris tenkatei*
Büttikofer, 1892

Dr Herman Frederik Carel ten Kate Jr (1858–1931) was a
Dutch physician, anthropologist, zoologist, and collector in
the Americas, Scandinavia, Polynesia, the East Indies and
Japan. He became Curator of Museo de la Plata, Argentina
(1893). He lived in Japan (1898–1909 and 1913–1919). He
finally settled in North Africa and died at Carthage.

Tennent

Red-capped Robin Chat ssp. *Cossypha natalensis tennenti*
J. G. Williams, 1962 NCR
[NUI *Cossypha natalensis intensa*]

John R. M. Tennent (b.1926) was a British diplomat, colonial
administrator, ornithologist and collector in Kenya (1962) and
Cairo (1965). He wrote 'Notes on the birds of Kakamega
forest' (1965).

Teraoka

Collared Kingfisher ssp. *Todiramphus chloris teraokai*
Kuroda, 1915
Coal Tit ssp. *Periparus ater teraokai* Kuroda, 1922 NCR
[JS *Periparus ater insularis*]
Micronesian Imperial Pigeon ssp. *Ducula oceanica teraokai*
Momiyama, 1922

N. Teraoka (1885–1955) was a Japanese collector.

Terborgh

Vilcabamba Brush Finch *Atlapetes terborghi* **Remsen**, 1993

Dr John Whittle Terborgh (b.1936) is an American ecologist,
conservation biologist and ornithologist. His bachelor's
(1958), master's (1960) and doctorate (1963) are all from
Harvard. He was a member of the faculty at Princeton (1971–
1989) and then moved to Duke University as a faculty member
of the Nicholas School of the Environment, and founded the
Duke University Center for Tropical Conservation. He is now
Research Professor Emeritus and Director there. He wrote
Requiem for Nature (1999).

Termeulen, J.

Hair-crested Drongo ssp. *Dicrurus hottentottus termeuleni*
Finsch, 1907
[Often included in *Dicrurus hottentottus jentincki*]

Jan ter Meulen Jr (1846–1916) was an Amsterdam businessman with interests in marine insurance. He attended the

International Maritime Conference in Paris (1900). He had a house built in Amsterdam using pigeons as a major decorative motif (1895).

Termeulen, P.

See-see Partridge ssp. *Ammoperdix bonhami termeuleni*
Zarudny & **Loudon**, 1904 NCR
[JS *Ammoperdix griseogularis* NRM]

P. P. ter Meulen (DNF) was a successful businessman who was both Dutch and Russian Consul in Ahwaz, Persia (1904–1928).

Terris

Maroon-fronted Parrot *Rhynchopsitta terrisi* **R. T. Moore**, 1947

Dr Terris Moore (1908–1993) was an American mountaineer and explorer. He was also the author's son. He was the 2nd President of the University of Alaska (1949–1953), known as the 'Flying President' as he was an expert high-altitude pilot. During WW2 Moore served as consultant on arctic and mountain conditions to the US military. Later, he was president of the New England Society of Natural History. He wrote *Mount Mckinley: Pioneer Climbs* (1981).

Tessmann

Saw-wing sp. *Psalidoprocne tessmanni* **Reichenow**, 1907 NCR
[Alt. Black Saw-wing; JS *Psalidoprocne pristoptera petiti*]
Tessman's Flycatcher *Muscicapa tessmanni* Reichenow, 1907
Lemon Dove *Aplopelia tessmanni* Reichenow, 1909 NCR
[JS *Columba larvata inornata*]
African Thrush *Turdus tessmanni* Reichenow, 1921 NCR
[JS *Turdus pelios saturatus*]
Shrike sp. *Lanius tessmanni* Reichenow, 1921 NCR
[Alt. Emin's Shrike; JS *Lanius gubernator*]

Pale Flycatcher ssp. *Bradornis pallidus tessmanni*
Reichenow, 1915 NCR
[JS *Bradornis pallidus modestus*]
Golden-tailed Woodpecker ssp. *Campethera abingoni tessmanni* Reichenow, 1921 NCR
[NUI *Campethera abingoni chrysura*]
Green-backed Eremomela ssp. *Eremomela pusilla tessmanni*
Grote, 1921 NCR
[JS *Eremomela canescens canescens*]
Grey-winged Robin Chat ssp. *Cossypha polioptera tessmanni* Reichenow, 1921
Vieillot's Barbet ssp. *Lybius vieilloti tessmanni* Grote, 1923 NCR
[JS *Lybius vieilloti rubescens*]

Günther Tessmann (1884–1969) was a German botanist and ethnologist who collected in Guinea, Gabon and Cameroon (1904–1914) and Peru (1923). He reported on homosexuality amongst different ethnic groups in Cameroon (1921). He studied the Ngi cult, which was based around the gorilla, and represented fire and positive power, whereas the chimpanzee represented evil. He was later based in Curitiba in Parana province, Brazil, working at the Museu Paranaense. He wrote *Die Indianer Nordost-perus* (1930) and *Die Völker und Sprachen Kamerun* (1932). He contributed to the translation and understanding of at least one Andean language.

Tethys

Storm-petrel genus *Tethysia* **Mathews**, 1933 NCR
[Now in *Halocyptena*]

Wedge-rumped Storm-petrel *Halocyptena tethys*
Bonaparte, 1852
[Syn. *Oceanodroma tethys*]

In Greek mythology, Tethys, a daughter of Uranus and Gaia and wife of Oceanus, was the supreme sea goddess and mother of the Oceanides.

Teysmann

(See under **Teijsmann**)

Thanner

Corn Bunting ssp. *Emberiza calandra thanneri* Tschusi zu Schmidhoffen, 1903 NCR
[NUI *Emberiza calandra calandra*]
Great Spotted Woodpecker ssp. *Dendrocopos major thanneri* Le Roi, 1911

Rudolf von Thanner (born von Tschusi zu Schmidhoffen) (1872–1922) was an Austrian ornithologist, taxidermist and professional collector in the Azores and Canary Islands (1902–1919) for Alexander Koenig (q.v.) in Bonn and Rothschild (q.v.) at Tring. He returned to Austria (1919), but his relatives still live in Tenerife. His change of name (1902) may have been brought about by circumstances caused by the fact that he was illegitimate, but he was clearly acknowledged by his father who, when he described the Corn Bunting subspecies named it after him.

Thayer

Thayer's Gull *Larus thayeri* **W. S. Brooks**, 1915

Ruffed Grouse ssp. *Bonasa umbellus thayeri* **Bangs**, 1912
[SII *Bonasa umbellus togata*]
Thayer's Bobwhite *Colinus virginianus thayeri* Bangs & **J. L. Peters**, 1928
[Alt. Northern Bobwhite ssp.]

Colonel John Eliot Thayer (1862–1933) was an American ornithologist and collector who amassed a huge collection and ornithological library. His father and brother were merchant bankers. He used his wealth to sponsor natural history expeditions. When Thayer became ill (1928) he donated his collection of 28,000 skins and 15,000 eggs and nests to Harvard University.

Thekla

Thekla Lark *Galerida theklae* **C. L. Brehm**, 1857

Thekla Brehm (1832–1857) was the daughter of the describer, Christian Ludwig Brehm (q.v.) – she died of heart disease.

Her brother, Alfred Brehm (q.v.), also named his eldest daughter Thekla.

Thélie

Cuckooshrike sp. *Campephaga theliei* **Schouteden**, 1914 NCR
[Alt. Purple-throated Cuckooshrike; JS *Campephaga quiscalina martini*]

Many-coloured Bush-shrike ssp. *Chlorophoneus multicolor theliei* Schouteden, 1914 NCR
[JS *Chlorophoneus multicolor batesi*]

Monsieur Thélie (DNF) was in the Ituri District, Belgian Congo (Democratic Republic of Congo). He made a collection of birds that was sent to the Royal Museum for Central Africa, Tervuren, Belgium.

Théodore

Sauzier's Teal *Anas theodori* E. **Newton & Gadow**, 1893 EXTINCT
[Alt. Mauritius Duck, Mascarene Teal]

Théodore Sauzier (See **Sauzier**)

Theresa (Baer)

Coppery Metaltail *Metallura theresiae* **E. L. Simon**, 1902

Theresa Baer (DNF) was the wife of Gustave-Adolphe Baer (q.v.).

Theresa (Clay)

Theresa's Snowfinch *Pyrgilauda theresae* **Meinertzhagen**, 1937
[Alt. Afghan Snowfinch]

Afghan Babbler ssp. *Turdoides caudatus theresae* Meinertzhagen, 1930 NCR
[JS *Turdoides huttoni salvadorii*]
European Stonechat ssp. *Saxicola torquata theresae* Meinertzhagen, 1934 NCR
[JS *Saxicola rubicola hibernans*]
Asian Desert Warbler ssp. *Sylvia nana theresae* Meinertzhagen, 1937 NCR
[*S. nana* NRM]
Moorland Francolin ssp. *Francolinus shelleyi theresae* Meinertzhagen, 1937 NCR
[JS *Scleroptila psilolaema elgonensis*]
Slender-billed Starling ssp. *Onychognathus tenuirostris theresae* Meintertzhagen, 1937
Barbary Partridge ssp. *Alectoris barbara theresae* Meinertzhagen, 1939 NCR
[JS *Alectoris barbara barbara*]
Eurasian Crag Martin ssp. *Ptyonoprogne rupestris theresae* Meinertzhagen, 1939 NCR; NRM
Eurasian Jay ssp. *Garrulus glandarius theresae* Meinertzhagen, 1939 NCR
[JS *Garrulus glandarius minor*]
Hawfinch ssp. *Coccothraustes coccothraustes theresae* Meinertzhagen, 1939 NCR
[JS *Coccothraustes coccothraustes buvryi*]

House Bunting ssp. *Emberiza striolata theresae* Meinertzhagen, 1939 NCR
[JS *Emberiza sahari sahari* (species may be monotypic)]
Mistle Thrush ssp. *Turdus viscivorus theresae* Meinertzhagen, 1939 NCR
[JS *Turdus viscivorus deichleri*]
Dark Chanting Goshawk ssp. *Melierax metabates theresae* Meinertzhagen, 1939
Red-rumped Wheatear ssp. *Oenanthe moesta theresae* Meinertzhagen, 1939 NCR
[JS *Oenanthe moesta moesta*]
Streaked Scrub Warbler ssp. *Scotocerca inquieta theresae* Meinertzhagen, 1939
Thekla Lark ssp. *Galerida theklae theresae* Meinertzhagen, 1939
Trumpeter Finch ssp. *Erythrospiza githaginea theresae* Meinertzhagen, 1939 NCR
[JS *Bucanetes githagineus zedlitzi*]
Eurasian Skylark ssp. *Alauda arvensis theresae* Meinertzhagen, 1947 NCR
[JS *Alauda arvensis scotica*]
Sclater's Lark ssp. *Spizocorys sclateri theresae* Meinertzhagen, 1949 NCR; NRM
White-throated Canary ssp. *Serinus albogularis theresae* Meinertzhagen, 1949 NCR
[JS *Serinus albogularis sordahlae*]
Meadow Pipit ssp. *Anthus pratensis theresae* Meinertzhagen, 1953 NCR
[JS *Anthus pratensis whistleri*]
Southern Grey Shrike ssp. *Lanius meridionalis theresae* Meinertzhagen, 1953

Theresa Rachel Clay (1911–1995) was a British parasitologist and ornithologist, and a companion to Richard Meinertzhagen (q.v.) who described all the bird taxa named after her. She co-wrote *Fleas, Flukes and Cuckoos – A Study of Bird Parasites* with Miriam Rothschild (1952), the year she was appointed Senior Scientific Officer at the BMNH.

Theresa (Doria)

Sunbird sp. *Hermotimia theresia* **Salvadori**, 1874 NCR
[Alt. Black Sunbird; JS *Leptocoma sericea chlorolaema*]

Teresa Durazzo Marchesa Doria (DNF) was the mother of the Italian explorer, Giacomo Marchese Doria.

Theresa (Empress)

Green-tailed Goldenthroat *Polytmus theresiae* Da Silva Maia, 1843

Spot-backed Antbird ssp. *Hylophylax naevius theresae* Des Murs, 1856

Empress Theresa (1822–1889) was the wife of Pedro II, Emperor of Brazil. However, this eponym is uncertain, due to the lack of etymology.

Theresa (Princess)

Indian Roller ssp. *Coracias affinis theresiae* **Parrot**, 1908 NCR
[JS *Coracias benghalensis affinis*]
Rusty-faced Parrot ssp. *Hapalopsittaca amazonina theresae* **Hellmayr**, 1915

Therese Charlotte Marianne Auguste von Bayern (Princess Theresa of Bavaria) (1850–1925) was a zoologist, botanist, anthropologist and explorer in Tunisia, Russia, the Arctic, Mexico, Brazil and western South America. Her father was Luitpold, Prince Regent of Bavaria. She was the first woman to receive an honorary degree from Ludwig Maximilian University, Munich (1897). She wrote books under the pseudonym 'Th. v. Bayer', such as *Ausflug nach Tunis* (1880). Her South American anthropological collection is held by the State Museum of Ethnology, Munich. A reptile is named after her.

Theresa (Reiser)

Toucan sp. *Ramphastos theresae* **O Reiser**, 1905 NCR
[Hybrid: *Ramphastos vitellinus culminatus* x *R. vitellinus ariel*]

Theresa Reiser (DNF) was the wife of Austrian ornithologist, Dr Othmar Reiser (q.v.).

Therese (Waelchli)

Purple-throated Sunbird ssp. *Leptocoma sperata thereseae* **ET Gilliard**, 1950 NCR
[Hybrid population: *Leptocoma sperata sperata* x *L. sperata henkei*]

Marie Therese Waelchli (1879–1948) was a Swiss-born schoolteacher in Ruxton, Maryland, who arrived via Ellis Island (1903). She married Ernest Gilliard and was aunt to the describer Thomas Gilliard (q.v.) and helped inspire and encourage his bird studies.

Thierfelder

Pheasant Coucal ssp. *Centropus phasianinus thierfelderi* **Stresemann**, 1927

Professor Dr Max Thierfelder (1885–1957) was a German medical researcher in Dutch New Guinea.

Thierry

African Coucal sp. *Centropus thierryi* **Reichenow**, 1899 NCR
[Alt. Black Coucal; JS *Centropus grillii*]

Grey-headed Sparrow ssp. *Passer diffusus thierryi* Reichenow, 1899 NCR
[JS *Passer griseus griseus*]

Gaston Thierry (1866–1904) was, despite his French names, an Oberleutnant (First Lieutenant) in the Imperial German Army. He was sent to Togoland (1896) to establish a series of bases, after the French and German governments reached agreement on the border between German Togoland (partly in Ghana and partly Togo) and French Dahomey (Benin), to enforce German control over the country. The government in Berlin expected its officers to collect examples of the local fauna, and Thierry was one who did. He left Togo (1899) and was killed in Cameroon (a German colony) by a poisoned arrow (1904). A mammal and a reptile are named after him.

Thilenius

South Melanesian Cuckooshrike ssp. *Coracina caledonica thilenii* **Neumann**, 1915

Dr Georg Christian Thilenius (1868–1937) was a German physician, anthropologist, ethnologist and anatomist. He was Professor of Anthropology and Ethnology, University of Breslau (Wrocław, Poland) (1900–1904). He was Director of the Museum of Ethnography, Hamburg (1904–1935), and was co-ordinator of the German South Sea Expedition (1908–1910).

Thilo Hoffmann

Serendib Scops Owl *Otus thilohoffmanni* Warakagoda & Rasmussen, 2004

Thilo W. Hoffmann is a Swiss conservationist, resident in Sri Lanka. He is Chairman of the Ceylon Bird Club. He revised and enlarged Henry's *A Guide to the Birds of Sri Lanka* with Warakagoda & Ekanayake (1998) and wrote *Threatened Birds of Sri Lanka National Red List* (1999). Deepal Warakagoda told us that he discovered this scops owl in 1995, but it was only six years later that he was sure it was the first new species to be found in Sri Lanka for 132 years!

Thollon

Thollon's Moorchat *Myrmecocichla tholloni* **Oustalet**, 1886
[Alt. Congo Moorchat]

François-Romain Thollon (1855–1896) was a French botanist and collector in the Congo, and a member of the de Brazza (q.v.) Mission in Gabon. Many taxa are named after him including a mammal, a reptile, several fish and the elephant tick *Amblyomma tholloni* that is a vector of the 'heartwater' cattle disease.

Thomas

Little Grassbird ssp. *Megalurus gramineus thomasi* **Mathews**, 1912
White-eared Honeyeater ssp. *Lichenostomus leucotis thomasi* Mathews, 1912

Mathews was notorious for not supplying etymologies for the taxa he described. They *may* be named after Thomas Carter (q.v.), but this remains a guess.

Thomas (Gilliard)

Yellow-browed Melidectes ssp. *Melidectes rufocrissalis thomasi* **Diamond**, 1969

(See **Gilliard**)

Thomasson

Thomasson's Thrush *Turdus poliocephalus thomassoni* **Seebohm**, 1894
[Alt. Island Thrush ssp.]

John Pennington Thomasson (1841–1904) was a businessman from a Lancashire Quaker family that made its money from cotton. He was also Liberal Member of

Parliament for Bolton (1880–1885). He was very generous in buying premises (1883) for the town to use as a museum and in buying natural history exhibits for it.

Thompson, E. H.

Middle American Screech Owl ssp. *Megascops guatemalae thompsoni* L. Cole, 1906

Edward Herbert Thompson (1857–1935) was an American archaeologist and diplomat who was US Consul, Yucatan, Mexico (1885). He took the post to subsidise his archaeological explorations, and spent most of the rest of his life in Mexico. He acquired a plantation, the Hacienda Chichén, which included the ruins of the Mayan city Chichén Itza.

Thompson, H. C.

Mottled Petrel ssp. *Pterodroma inexpectata thompsoni* **Mathews**, 1915 NCR; NRM

Hubert Charles Thompson (1869–1953) was an Australian oologist and photographer in Launceston, Tasmania. He was a founding member of the RAOU.

Thompson, H. N.

White-headed Bulbul *Cerasophila thompsoni* **Bingham**, 1900

Short-tailed Parrotbill ssp. *Neosuthora davidiana thompsoni* Bingham, 1903

Henry Nilus Thompson (d.1938) was a British forestry officer in Burma (Myanmar) and subsequently Director of Forestry in Nigeria. Bingham (q.v.) described him as 'a keen observer and field-naturalist'. Thompson collected botanical specimens in West Africa, and wrote *Gold Coast: Report on Forests* (1910).

Thompson, L. C.

Crested Francolin ssp. *Dendroperdix sephaena thompsoni* **J. A. Roberts**, 1924 NCR
[JS *Dendroperdix sephaena zambesiae*]

Dr Louis C. Thompson (b.1877) was an English collector in Damaraland, South Africa.

Thompson, M. C.

Sierra Madre Ground Warbler *Robsonius thompsoni* Hosner et al., 2013

Red-bellied Pitta ssp. *Erythropitta erythrogaster thompsoni* **Ripley** & **Rabor**, 1962
[SII *Erythropitta erythrogaster erythrogaster*]

Professor Dr Max C. Thompson is an American ornithologist who was for 33 years Professor of Biology at Southwestern College, in Winfield, Kansas, and is currently a Research Associate at the University of Kansas Museum of Natural History. His main interest is ornithology but his PhD thesis at Ohio State University was in Horticulture and he is an avid orchid collector. He collected birds in the Hawaiian Islands and Philippines in particular, but also in Africa, Latin America,

Asia and Australasia. His collections are deposited at the University of Kansas, Smithsonian Institution, Bernice P. Bishop Museum, and American Museum of Natural History (AMNH), providing an invaluable resource for the ornithological community. Among his written works are *Birds from North Borneo* (1964) and the co-authored *Birds of Kansas* (1998–2011).

Thompson, Mrs L. C.

Cape Canary ssp. *Serinus canicollis thompsonae* **J. A. Roberts**, 1924

Mrs L. C. Thompson (DNF) was the wife of Dr Louis Thompson (q.v.).

Thoms

Crested Lark ssp. *Galerida cristata thomsi* **Ripley**, 1951 NCR
[JS *Galerida cristata tardinata*]

Dr William Wells Thoms (1903–1971) was an American medical missionary to Iraq, Kuwait, Bahrain and Muscat (1931–1970).

Thomson, B. H.

Manucode sp. *Manucodia thomsoni* **Tristram**, 1889 NCR
[Alt. Trumpet Manucode; JS *Phonygammus keraudrenii hunsteini*]

Sir Basil Home Thomson (1861–1939) was a British intelligence officer, police officer, prison governor, colonial administrator and writer. Educated at Hendon and Eton he attended New College, Oxford, for just two terms but suffering bouts of depression left, spending time in Iowa, USA, as a farmer (1881–1882). Needing to be secure in order to marry he joined the Colonial Office (1883). He was posted to Fiji (1884), then New Guinea, but was invalided out due to malaria. He returned to London where he married (1890), then went back to Fiji with his wife, becoming Commissioner of Native Lands. He resigned from the Colonial Office (1893) and returned to England because of his wife's ill health. He started writing, producing *South Sea Yarns* (1894), *The Diversions of a Prime Minister* (1894) and *The Indiscretions of Lady Asenath* (1898). He read for the Bar (1890s) and was admitted (1896), but did not practice. Instead he became Deputy Governor of Liverpool Prison and later Governor of Northampton, Cardiff, Dartmoor and Wormwood Scrubs prisons, eventually becoming Secretary of the Prison Commission (1908–1913). He was appointed Assistant Commissioner of the Metropolitan Police (1913) as head of CID and the Secret Service Bureau when war broke out (1914), and became known as a 'spycatcher', which he documented in his memoirs *The Scene Changes*. Notably he interrogated Mata Hari. On a darker note he was foremost in the suppression of the Suffragettes, English Marxists and Irish republicans. He went on to become Director of Intelligence (1919–1921). In 1925 he was arrested in London's Hyde Park and charged with public indecency with a young woman, Miss Thelma de Lava. Thomson rejected the charges, insisting that he was engaged in conversation with the woman for the purposes of research for a book on London vice. Nonetheless, he was found guilty and fined £5. An amphibian is named after him.

Thomson, J.

Thomson's Flycatcher *Erythrocercus livingstonei thomsoni*
 G. E. Shelley, 1882
 [Alt. Livingstone's Flycatcher ssp.]

Joseph Thomson (1858–1895) was a Scot described as 'one of the most colourful and prudent of 19th century African explorers'. Aged just 20, he was placed second in charge on his first exploration for the Royal Geographic Society (1879). Very early in the expedition, the leader, Scottish cartographer Keith Johnston, succumbed to dysentery. Thomson buried him under a tree on which he carved Johnston's initials and the date of his death (28 June 1879), near the village of Behobeho, near the Rufiji River, southern Tanzania. He said that he then asked himself, 'should we simply turn back?' and answered, 'I feel I must go forward, whatever might be my destiny.' He carried on and successfully led the expedition to Lake Nyasa (Lake Malawi) and Lake Tanganyika. Thereafter, he made many more safe and successful explorations of routes through Kenya and Tanzania (1878–1884), Nigeria (1885) and Morocco (1888). He was first to explore Masai land successfully, simply because he was not confrontational, writing: 'in my opinion the traveller's strength would lie more in his manner towards and treatment of the natives than in his guns and revolvers.' Despite his relaxed 'fun' style of leadership he always put the safety of the party first and his motto was 'He who goes slowly, goes safely; he who goes safely, goes far.' He wrote *Through Masai Land: A Journey of Exploration among the Snowclad Volcanic Mountains and Strange Tribes of Eastern Equatorial Africa, being the Narrative of the Royal Geographical Society's Expedition to Mount Kenya and Lake Victoria Nyanza, 1883–1884* (1885). The diseases he caught in tropical Africa led to his early death from pneumonia. He is most famous for having Thomson's Gazelle *Eudorcas thomsonii*, which he discovered, named after him.

Thorbecke

Red-throated Wryneck ssp. *Jynx ruficollis thorbeckei*
 Reichenow, 1912 NCR
 [JS *Jynx ruficollis pulchricollis*]

Professor Franz Thorbecke (1875–1945) was a German geographer and ethnographer in the Cameroons (1907–1908 and 1911–1913). He became Professor at the Graduate School, Mannheim, and subsequently at the University of Cologne. He wrote the 3-volume *Im Hochland von Mittel-Kamerun* (1914–1919).

Thorne

Double-spurred Francolin ssp. *Pternistis bicalcaratus*
 thornei **Ogilvie-Grant**, 1902

Major Henry Albert Thorne (1866–1903) was in the British Army in West Africa (1899–1903). He wrote *Notes on Kit for West Africa* (1902). He collected the holotype in Sierra Leone (1899).

Thorpe

Munia sp. *Lonchura thorpei* **Mathews**, 1913 NCR
 [Alt. Chestbut-breasted Mannikin; JS *Lonchura castaneothorax castaneothorax*]

Star Finch ssp. *Aegintha ruficauda thorpei* Mathews, 1912 NCR
 [JS *Neochmia ruficauda subclarescens*]

T. Thorpe (DNF) was an English collector in Australia.

Thouars

Fruit Dove genus *Thouarsitreron* **Bonaparte**, 1854 NCR
 [Now in *Ptilinopus*]

(See **Dupetit-Thouars**)

Thrupp

Acacia Tit *Parus thruppi* **G. E. Shelley**, 1885
 [Alt. Somali Tit, Northern Grey Tit]

James Godfrey Thrupp (1848–1913) was a British surgeon in Somaliland (1884–1885). He was Assistant Surgeon at St George's Hospital, London, and gave medical evidence for the prosecution at the Old Bailey (1872) in the case of Regina v. Smith. He was a Civilian Surgeon attached to the 24th Regiment during the Anglo-Zulu War (1878–1879). He was fortunate to arrive at Isandlwana too late for the battle, but was one of the first to arrive at Rorke's Drift after that action was over. He collected botanical specimens in Somaliland (1888) and co-wrote *The Unknown Horn of Africa: An Exploration from Berbera to the Leopard River* (1888).

Thunberg

Thunberg's Swiftlet *Aerodramus fuciphagus* Thunberg, 1812
 [Alt. Edible-nest Swiftlet]

Thunberg's Wagtail *Motacilla flava thunbergi* Billberg, 1828
 [Alt. Grey-headed (Yellow) Wagtail]

Carl Peter Thunberg (1743–1828) was a Swedish student of Linnaeus (q.v.). He journeyed through the Low Countries to Paris (1770), where he stayed for six months to study medicine and natural history. He travelled to South Africa, Japan, Java for the Dutch East India Company (1772–1775), and later to Ceylon (1778) before returning to Europe (1779). His major publications include *Flora Japonica* (1784), *Flora Capensis* and *Prodromus Plantarum Capensium* (1794). He was subject to much ridicule during his life for his 'Thunbergisms'. For example: 'Water is the element which makes sea journeys both outside and inside of the Netherlands so nimble and comfortable'. Or, on first going to France: 'To me it could not but seem both strange and ridiculous to hear Burghers and Farmers all speak that, in other places so noble, language.'

Thura

Thura's Rosefinch *Carpodacus thura* **Bonaparte** & **Schlegel**, 1850
 [Alt. White-browed Rosefinch]

Thura Nilsson (DNF) was the daughter of the Swedish ornithologist, geologist, palaeontologist and ethnologist Svenn Nilsson (1787–1883).

Thurber

Thurber's Junco *Junco hyemalis thurberi* **A. W. Anthony**, 1890
[Alt. Dark-eyed Junco ssp.]

Dr George Thurber (1821–1890) was a self-educated chemist and botanist. Often called the most accomplished horticulturist in America, he first worked as a pharmacist. He served as botanist of the United States Boundary Commission, which surveyed the US/Mexico boundary (1850–1854). He took a master's degree in chemistry at Brown University and held a succession of jobs until he became editor of *American Agriculturist* for 22 years. He was an expert on grasses, some of which are named after him, leaving an uncompleted manuscript on American grasses when he died.

Tiboli

Apo Sunbird ssp. *Aethopyga boltoni tibolii* Kennedy,
Gonzales & Miranda, 1997

This bird is named after the T'Boli people, the original inhabitants of southern Mindanao.

Ticehurst

Yellow-throated Bunting ssp. *Emberiza elegans ticehursti* **Sushkin**, 1926 NCR
[NUI *Emberiza elegans elegans*]
Eurasian Skylark ssp. *Alauda arvensis ticehursti* **Whistler**, 1928 NCR
[JS *Alauda arvensis guillelmi*]
Brown Wood Owl ssp. *Strix leptogrammica ticehursti* **Delacour**, 1930
Common Blackbird ssp. *Turdus merula ticehursti* **Clancey**, 1938 NCR
[JS *Turdus merula merula*]
Blyth's Leaf Warbler ssp. *Phylloscopus reguloides ticehursti* Delacour & **J. C. Greenway**, 1939
Meinertzhagen's Warbler *Sylvia deserticola ticehursti* **Meinertzhagen**, 1939
[Alt. Tristram's Warbler ssp.]
Pin-striped Tit Babbler ssp. *Macronus gularis ticehursti* **Stresemann**, 1940
Wreathed Hornbill ssp. *Rhyticeros undulatus ticehursti* **Deignan**, 1941 NCR; NRM
Brown Bush Warbler ssp. *Bradypterus luteoventris ticehursti* Deignan, 1943 NCR; NRM

Dr Claud Buchanan Ticehurst (1881–1941) was a British physician and ornithologist. He served (WW1) in Mesopotamia (Iraq) and was in India (1917–1920). He edited *Ibis* (1931–1941), a period when he was working closely with Whistler (q.v.) on plans for a handbook of Indian birds – the manuscript material being used in due course by Salim Ali (q.v.) and Dillon Ripley (q.v.) in their own handbook. His collection of 10,000 bird skins was bequeathed to the BMNH.

Tichelman

Black-nest Swiftlet ssp. *Aerodramus maximus tichelmani* **Stresemann**, 1926

Gerardus Lowrens Tichelman (1892–1952) was a Dutch colonial administrator and anthropologist in the Dutch East Indies (Indonesia).

Tickell

Broad-billed Warbler genus *Tickellia* **Blyth**, 1861

Tickell's Flowerpecker *Dicaeum erythrorhynchos* **Latham**, 1790
[Alt. Pale-billed Flowerpecker]
Tickell's Leaf Warbler *Phylloscopus affinis* Tickell, 1833
Tickell's Thrush *Turdus unicolor* Tickell, 1833
Tickell's Blue Flycatcher *Cyornis tickelliae* Blyth, 1843
[Alt. Orange-breasted Blue Flycatcher]
Tickell's Brown Hornbill *Anorrhinus tickelli* Blyth, 1855
Tickell's Laughingthrush *Garrulax strepitans* Blyth, 1855
[Alt. White-necked Laughingthrush]
Tickell's Babbler *Pellorneum tickelli* Blyth, 1859
[Alt. Buff-breasted Babbler]
Tickell's Sibia *Heterophasia melanoleuca* Blyth, 1859
[Alt. Dark-backed Sibia; Syn. *Malacias melanoleucus*]

Mountain Bulbul ssp. *Ixos mcclellandii tickelli* Blyth, 1855
Large Scimitar Babbler ssp. *Pomatorhinus hypoleucos tickelli* **Hume**, 1877
Rufous-throated Partridge ssp. *Arborophila rufogularis tickelli* Hume, 1880

Colonel Samuel Richard Tickell (1811–1875) was a British army officer, artist and ornithologist in India and Burma (Myanmar). He was described by Kinnear (q.v.) as 'one of the best field naturalists India has known'. He made important early contributions to Indian ornithology and mammalogy through field observations and collecting specimens, while he was stationed in several localities (1830s and 1840s). He planned to publish a book on the birds and mammals of India, but never did. However, his manuscript notes and illustrations are preserved in the library of the Zoological Society of London. These notes contain many references to observations of birds in Bihar, Orissa, Darjeeling and Tenasserim. He also published on the structure and vocabulary of the Ho language. A mammal is named after him. The blue flycatcher was probably named after Tickell's wife, given the feminine form of *tickelliae*, but this is not made clear in the original description.

Timolia

Hummingbird genus *Timolia* **Mulsant**, 1875 NCR
[Applied to a hybrid: *Thalurania* x *Chrysuronia*]

Timolus (or Tmolus) was the name of two characters in Greek mythology: one was a mountain-god who acted as a judge in a musical competition between Apollo and Marsyas; the other was a king of Lydia who was gored to death by a bull. The two are sometimes conflated.

Timothy

Stub-tailed Spadebill ssp. *Platyrinchus cancrominus timothei* **Paynter**, 1954

Timothy H. Laughlin (d.c.1953) assisted the describer, Raymond Paynter (q.v.), during fieldwork (1950–1951) in the Yucatan Peninsula. Paynter wrote that Laughlin's 'untimely death came shortly after our return to this country' (US).

Tischler

Goshawk sp. *Accipiter tischleri* **Kleinschmidt**, 1938 NCR
[Alt. Northern Goshawk; JS *Accipiter gentilis gentilis*]

Willow Tit ssp. *Parus borealis tischleri* Kleinschmidt, 1917 NCR
[JS *Poecile montanus borealis*]
Western Jackdaw ssp. *Corvus coloeus tischleri* Kleinschmidt, 1935 NCR
[JS *Coloeus monedula soemmerringii*]

Dr Friedrich Tischler (1881–1945) was a German ornithologist and magistrate in East Prussia.

Titan

White-bellied Storm-petrel ssp. *Fregetta grallaria titan* **R. C. Murphy**, 1928

In Greek mythology, the Titans were a primeval race of immortal giants. Sometimes 'Titan' (singular) is used as a name for the sun-god Helios. The storm petrel was apparently given this name because of its larger size than the nominate race.

Titania

Black-chinned Yuhina ssp. *Yuhina nigrimenta titania* **Koelz**, 1954 NCR
[JS *Yuhina nigrimenta nigrimenta*]

Titania is Queen of the Fairies in Shakespeare's *A Midsummer Night's Dream*. In traditional folklore the queen of the fairies has no name, and Shakespeare took the name from Ovid's *Metamorphoses*, where it is used as a term for the daughters of the Titans. Another name for the Fairy Queen is Mab (q.v.). A mammal is named after her.

Tiwari

Andaman Cuckoo Dove ssp. *Macropygia rufipennis tiwarii* **Abdulali**, 1957 NCR; NRM

Dr Krishna Kant Tiwari is a zoologist and carcinologist. He was a member of the Zoological Survey of India (1951), becoming Joint Director (1980–1981). He is a retired Vice Chancellor, Jiwaji University, Gwalior. He co-wrote 'Two new reptiles from the Great Nicobar Islands' (1973). Two reptiles are named after him.

Tkachenko

Siberian Jay ssp. *Perisoreus infaustus tkachenkoi* **Sushkin & B. Stegmann**, 1929

Mikhail Elevferovich Tkachenko (b.1878) was a Russian forester who was Director of Forestry, Experimental Station, Leningrad (St Petersburg).

Todd

Todd's Nightjar *Caprimulgus heterurus* Todd, 1915
Todd's Scrub Flycatcher *Sublegatus obscurior* Todd, 1920
[Alt. Amazonian Scrub Flycatcher]
Todd's Sirystes *Sirystes subcanescens* Todd, 1920
[Syn. *Sirystes sibilator subcanescens*]

Todd's Flycatcher *Myiarchus toddi* **Chapman**, 1923 NCR
[Based on an aberrant *Myiarchus phaeocephalus*]
Todd's Antwren *Herpsilochmus stictocephalus* Todd, 1927
Todd's Parakeet *Pyrrhura caeruleiceps* Todd, 1947
[Syn. *Pyrrhura picta caeruleiceps*]

Bay-headed Tanager ssp. *Tangara gyrola toddi* **Bangs & T. E. Penard**, 1921
Silvered Antbird ssp. *Sclateria naevia toddi* **Hellmayr**, 1924
Todd's Canada Goose *Branta canadensis interior* Todd, 1938
Hooded Siskin ssp. *Carduelis magellanica toddornis* **Wolters**, 1949 NCR
[JS *Carduelis magellanica longirostris*]
Highland Hepatic Tanager ssp. *Piranga lutea toddi* **Parkes**, 1969
Blue-black Grosbeak ssp. *Cyanocompsa cyanoides toddi* **Paynter**, 1970 NCR
[JS *Cyanocompsa cyanoides caerulescens*]

Walter Edmond Clyde Todd (1874–1969) was an American ornithologist at the Carnegie Museum, Pittsburgh. He wrote *Birds of Western Pennsylvania* (1940) and *Birds of the Labrador Peninsula and Adjacent Areas* (1963). Todd also wrote on Neotropical avifauna, although he never did fieldwork there due to malaria he contracted in Washington, DC.

Tokumi

Grey-capped Greenfinch ssp. *Carduelis sinica tokumii* Mishima, 1961 NCR
[JS *Carduelis sinica minor*]

Y. Tokumi (DNF) was a Japanese collector.

Tokunaga

Eurasian Tree Sparrow ssp. *Passer montanus tokunagai* **Kuroda & Yamashina**, 1935 NCR
[JS *Passer montanus dilutus*]

Dr Masaaki Tokunaga (1903–1998) was a Japanese entomologist and collector at the Kyoto Prefectural University. He led the expedition (1933) to Manchoukuo (Manchuria) where the holotype was collected.

Tolmie

MacGillivray's Warbler *Oporornis tolmiei* J. K. Townsend, 1839

William Fraser Tolmie (1812–1886) was a Scottish naturalist and collector who lived in Canada (1833–1886). He studied medicine at the University of Glasgow, but never qualified. However, this did not prevent the Hudson's Bay Company employing him in the dual capacity of clerk and surgeon at Fort Vancouver (May 1833). He was keenly interested in natural history and local Native Americans. He sent at least two collections of birds, other fauna, and native artifacts to Scotland. A number of plants are named after him.

Tomirdus

Rail genus *Tomirdus* **Mathews**, 1912 NCR
[Now in *Rallina*]

(See under **Iredale**)

Tomkovich

Grey Plover ssp. *Pluvialis squatarola tomkovichi* Engelmoer
& **Roselaar**, 1998

Dr Pavel Stanislavovich Tomkovich is a Russian ornithologist
and conservationist who is head of the Division of Ornithology, Zoological Museum, Moscow State University. He is
an expert on Arctic-breeding waders, and co-wrote 'First
indications of a sharp population decline in the globally
threatened Spoon-billed Sandpiper, *Eurynorhynchus
pygmeus*' (2002).

Tomlin

Middle American Screech Owl ssp. *Megascops guatemalae
tomlini* **R. T. Moore**, 1937

Dr Francis Harry Tomlin (DNF) was an American dentist, traveller and ornithologist and friend of the describer Moore
(q.v.).

Tommaso

Black Cicadabird ssp. *Coracina melas tommasonis*
Rothschild & Hartert, 1903

(See **Salvadori**)

Tomyris

Eurasian Blue Tit ssp. *Parus caeruleus tomyris* Floericke,
1926 NCR
[JS *Cyanistes caeruleus orientalis*]

Tomyris, or Tahm-Rayis (fl.530 BC), was a semi-legendary
queen of the Massagetae, a people of the steppes east of the
Caspian Sea. They defeated and killed Cyrus the Great, King
of Persia. Herodotus, who lived c.100 years later, wrote of her
in his *Histories*.

Topiltzin

Blue-throated Hummingbird *Trochilus topiltzin* de la Llave,
1833 NCR
[Alt. Blue-throated Mountaingem; JS *Lampornis
clemenciae*]

Topiltzin was a name or title implying divinity, commonly
associated with the pre-Columbian Mexican deity,
Quetzalcoatl.

Toro

Broad-billed Hummingbird ssp. *Cynanthus latirostris toroi*
Berlioz, 1937 NCR
[Intergrade between *Cynanthus latirostris propinquus* and
C. doubledayi]
Canyon Towhee ssp. *Melozone fusca toroi* **R. T. Moore**, 1942

Mario del Toro Avilés (DNF) was a Mexican collector and
ornithologist who, according to Moore's original etymology,
'... has made expeditions into the more remote portions of
Mexico without regard for difficulties or dangers.' He lived at
La Estancita, Guerrero, Mexico, where the type specimen of
the Broad-billed Hummingbird ssp. was taken.

Torre

Zapata Sparrow genus *Torreornis* **Barbour** & **J. L. Peters**,
1927

Professor Carlos de la Torre y la Huerta (1858–1950) of
Havana University was regarded as the foremost Cuban
naturalist of his generation. He was closely associated with
the USNM in Washington, DC before Castro took power. He
was a leading figure in the Academia de Ciencias Medicas,
Fisicas y Naturales de la Habana. An extinct Cuban mammal
and a reptile are named after him.

Torrey

Boat-tailed Grackle ssp. *Quiscalus major torreyi* Harper,
1934

Bradford Torrey (1843–1912) was an American journalist and
naturalist living in Weymouth, Massachusetts, where a bird
sanctuary is named after him. He travelled quite widely in
North America, including visiting Yosemite and Florida, and
wrote *A Florida Sketch-Book* (1894).

Torrington

Sri Lanka Wood Pigeon *Columba torringtoniae* **Kelaart**, 1853

Viscountess Mary Anne Byng (*née* Astley), Lady Torrington
(1805–1885), was the wife (1833) of George Byng, 7th Viscount
Torrington (1812–1884), who was Governor of Ceylon (Sri
Lanka) (1847–1851). It has sometimes been assumed that the
bird was named after the Viscount, as it was long known as
Columba torringtonii.

Touchen

Chinese Hwamei *Trochalopteron touchena* **Mathews**, 1934
NCR
[JS *Garrulax canorus canorus*]

(See **La Touche**)

Toulson

Toulson's Swift *Apus toulsoni* **Bocage**, 1877 NCR
[Alt. Loanda Swift; status uncertain – probably a dark morph
of *Apus horus*]

A. Toulson (DNF) was a collector in Angola (c.1870). Bocage
described him as 'an amateur ... former trader with Loanda',
but little is known about him. An amphibian is named after
him.

Toussenel

Red-chested Goshawk *Accipter tousseneli*
J. & E. Verreaux, 1855

Alphonse Toussenel (1803–1885) was a French journalist and
amateur naturalist. He wrote *L'Esprit des Bêtes: Le Monde
des Oiseaux Ornithologie Passionnelle* (1853).

Townsend, C. H.

Townsend's Shearwater *Puffinus auricularis*
C. H. Townsend, 1890
Black Storm-petrel *Oceanodroma townsendi* **Ridgway**,
1893 NCR
[JS *Oceanodroma melania*]

Snow Bunting ssp. *Plectrophenax nivalis townsendi*
Ridgway, 1887
Crowned Woodnymph ssp. *Thalurania colombica townsendi*
Ridgway, 1888
[Alt. Violet-crowned Woodnymph]
Townsend's Junco *Junco hyemalis townsendi*
A. W. Anthony, 1889
[Alt. Dark-eyed Junco ssp.]
Townsend's Ptarmigan *Lagopus muta townsendi* **D. G. Elliot**,
1896
[Alt. Rock Ptarmigan ssp.]
White-rumped Swiftlet ssp. *Aerodramus spodiopygius
townsendi* **Oberholser**, 1906
Micronesian Imperial Pigeon ssp. *Ducula oceanica
townsendi* **Wetmore**, 1919
Vanikoro Flycatcher ssp. *Myiagra vanikorensis townsendi*
Wetmore, 1919
Northern Cardinal ssp. *Cardinalis cardinalis townsendi*
Van Rossem, 1932

Charles Haskins Townsend (1859–1944) was an American
zoologist who worked for the US Fish Commission, and later
became Director of New York Aquarium. He explored
northern California (1883–1884) and the Kobuk River, Alaska
(1885). He wrote 'Field notes on the mammals, birds, and
reptiles of northern California' (1887). A mammal and two
reptiles are named after him. Two junco subspecies have
been given the vernacular name of 'Townsend's Junco' (q.v.
Townsend, J. K.).

Townsend, C. W.

Western Spindalis ssp. *Spindalis zena townsendi* **Ridgway**,
1887

Dr Charles Wendell Townsend (1859–1934) was an American
obstetrician, field ornithologist, worldwide traveller (1906–
1932) and collector. He qualified as a physician at Harvard
Medical School (1885) and worked in various hospitals
(1887–1909). He was particularly interested in following
Audubon's (q.v.) travels in the Americas, and in Labrador. He
co-wrote *Birds of Labrador* (1907) with Glover M. Allen (q.v.).

Townsend, J. K.

Townsend's Bunting *Spiza townsendii* **Audubon**, 1834 NCR
[Alt. Townsend's Dickcissel; now believed to be a colour
morph of Dickcissel (*Spiza americana*)]
Townsend's Warbler *Dendroica townsendi* Townsend, 1837
[Syn. *Setophaga townsendi*]
Townsend's Cormorant *Phalacrocorax townsendi* Audubon,
1838 NCR
[Alt. Brandt's Cormorant; JS *Phalacrocorax penicillatus*]
Townsend's Oystercatcher *Haematopus townsendi*
Audubon, 1838 NCR
[Alt. Blackish Oystercatcher; JS *Haematopus ater*]

Townsend's Solitaire *Myadestes townsendi* Audubon, 1838
Townsend's Surfbird *Aphriza townsendi* Audubon, 1839 NCR
[JS *Aphriza virgata*]

Townsend's Junco *Junco hyemalis oreganus* Townsend,
1837
[Alt. Dark-eyed Junco ssp.]
Townsend's Fox Sparrow *Passerella iliaca townsendi*
Audubon, 1838
[Syn. *Passerella unalaschcensis townsendi*]

John Kirk Townsend (1809–1851) was an American naturalist,
ornithologist and collector. He trained as a physician and
pharmacist, but developed an interest in natural history. He
was invited (1833) to join botanist Thomas Nuttall (q.v.) on an
expedition across the Rocky Mountains to the Pacific Ocean.
He later made two visits to the Hawaiian Islands (1835 and
1837). Audubon (q.v.), for his *Birds of America* and *Viviparous
Quadrupeds*, used Townsend's bird and mammal specimens.
Ironically Townsend died of arsenic poisoning, the 'secret'
ingredient of the powder he formulated to use in taxidermy.
Seven mammals are named after him. Two junco subspecies
have been given the vernacular name of 'Townsend's Junco'
(q.v. Townsend, C. H.).

Townson

Weka *Gallirallus townsoni* **Mathews** & **Iredale**, 1914 NCR
[JS *Gallirallus australis australis*]

William Lewis Townson (1855–1926) was an English-born
plant collector and pharmaceutical chemist who emigrated
to New Zealand (c.1876). He set up in business as a chemist
in Westport, South Island (1888). An orchid genus is named
after him.

Toxopeus

Blue-fronted Lorikeet *Charmosyna toxopei* **Siebers**, 1930

Clamorous Reed Warbler ssp. *Acrocephalus stentoreus
toxopei* **Hartert**, 1924 NCR
[JS *Acrocephalus stentoreus sumbae*]
Papuan Sittella ssp. *Daphoenositta papuensis toxopeusi*
Rand, 1940

Professor Lambertus Johannes Toxopeus (1894–1951) was a
Dutch zoologist and collector, born in Java. He lectured at an
Indonesian University. His main interest was Lepidoptera.

Toyoshima

Eastern Buzzard ssp. *Buteo japonicus toyoshimai*
Momiyama, 1927

K. Toyoshima (DNF) was a Japanese botanist on the Bonin
Islands.

Trai

Chestnut-tailed Minla ssp. *Minla strigula traii* Eames, 2002

Le Trong Trai is a Vietnamese ornithologist and collector who
worked for BirdLife International (2002). He was previously a
researcher and planner for the Forest Inventory and Planning
Institute, Ministry of Agriculture and Rural Development.

Traill

Traill's Flycatcher *Empidonax traillii* **Audubon**, 1828
[Alt. Willow Flycatcher]
Maroon Oriole *Oriolus traillii* **Vigors**, 1832

Dr Thomas Stewart Traill (1781–1862) was a Scottish zoologist, physician and philosopher. He was a founder of the Royal Institution of Liverpool and was appointed Professor of Medical Jurisprudence at Edinburgh University (1832). He edited *Encyclopaedia Britannica* (eighth edition). He was also influential in separating the Natural History Museum from the British Museum and in the former's reconstitution in South Kensington, London. Audubon (q.v.) visited Traill in Liverpool and in his description wrote: 'I have named this species after my learned friend, Doctor Thomas Stewart Traill, of Edinburgh, in evidence of the gratitude which I cherish toward that gentleman for all his kind attention to me'.

Trapnell

Fawn-coloured Lark ssp. *Calendulauda africanoides trapnelli* **C. M. N. White**, 1943

Colin Graham Trapnell (1907–2004) was a British conservationist, environmentalist, botanist and explorer, a government ecologist in Northern Rhodesia (Zambia) (1931–1950) and Kenya (1950–1960). He retired to England and was a founder of the Somerset Trust for Nature Conservation (Somerset Wildlife Trust) and an active member of the National Trust.

Travers

New Zealand Wren genus *Traversia* **Rothschild**, 1894
[Now in *Xenicus*]

Black Robin *Petroica traversi* **W. L. Buller**, 1872
[Alt. Chatham Island Robin]
Macquarie Shag *Phalacrocorax traversi* Rothschild, 1898 NCR
[JS *Leucocarbo purpurascens*]

Henry Hammersley Travers (1844–1928) was an English-born New Zealand lawyer and naturalist. He went to New Zealand with his parents (1850) and made two collecting trips to the Chatham Islands (1864 and 1871), returning with plant, bird and insect specimens, rock samples, and Moriori artefacts. He was Curator of the Newtown Museum (1913–1915), but was dismissed.

Traversi

Fan-tailed Widowbird ssp. *Euplectes axillaris traversii* **Salvadori**, 1888

Dr Raffaele Leopoldo Traversi (1856–1949) was an Italian physician and explorer in Abyssinia (Ethiopia) (1884–1887 and 1890–1893).

Traviès

Lilac-fronted Starfrontlet *Coeligena traviesii* **Mulsant** & **E. Verreaux**, 1866 NCR
[Probable hybrid: *Coeligena torquata* x *Coeligena lutetiae*]

Édouard Traviès (1807–1867) was a French artist and naturalist. He illustrated some of Alcide d'Orbigny's (q.v.) publications.

Traylor

Traylor's Forest Falcon *Micrastur buckleyi* Swann, 1919
[Alt. Buckley's Forest Falcon]
Orange-eyed Flatbill *Tolmomyias traylori* **Schulenberg** & **T. A. Parker**, 1997

Traylor's Tinamou *Crypturellus obsoletus traylori* **E. R. Blake**, 1961
[Alt. Brown Tinamou ssp.]
Double-banded Courser ssp. *Rhinoptilus africanus traylori* **Irwin**, 1963
Bush Pipit ssp. *Anthus caffer traylori* **Clancey**, 1964
Little Green Sunbird ssp. *Nectarinia seimundi traylori* **Wolters**, 1965 NCR
[JS *Anthreptes seimundi minor*]
Desert Cisticola ssp. *Cisticola aridulus traylori* **C. W. Benson** & Irwin, 1966

Major Melvin Alvah Traylor Jr (1915–2008) was an American ornithologist and collector, the son of an unsuccessful candidate for the Democratic nomination for President. As a Marine Corp officer, Traylor Jr was severely injured during the Battle of Tarawa (WW2) in the Pacific theatre, losing one eye and suffering arm and upper body wounds during the beach assault. He became Assistant Curator of Birds at the Field Museum, Chicago (1956), in recognition of his earlier contributions to the ornithology division as an expedition collector and unpaid associate. He wrote *Birds of Angola* (1963) and co-wrote *An Annotated List of the Birds of Bolivia* (1989). He also collaborated with Raymond Paynter (q.v.) in the production of a number of *Ornithological Gazetteers* on Peru (1983) and Brazil (1991). Traylor and Paynter were joint recipients of the Elliot Coues Award (2001).

Treacher

Chestnut-hooded Laughingthrush *Garrulax treacheri* **Sharpe**, 1879

Sir William Hood Treacher (1849–1919) was Governor of British North Borneo (1881–1887), British Resident at Perak (1896–1901), and Resident General at the Federated Malay States (Malaysia) (1902–1904). He gave his collection of mammals to Oxford University Museum of Natural History (1878) when he was Consul in Labuan and Acting Consul-General in Borneo. He was largely instrumental in obtaining the concessions in Borneo that led to the creation of the British North Borneo Company. He wrote *British Borneo, Sketches of Brunai, Sarawak, Labuan, and North Borneo* (1891).

Treganza

Treganza's Blue Heron *Ardea herodias treganzai* Court, 1908 NCR
[Alt. Great Blue Heron ssp.; JS *Ardea herodias herodias*]

Alberto Owen Treganza (b.1870) was an architect, able artist and amateur ornithologist. His homestead was purchased for

$5 (1906) by his parents, who arrived in San Diego County (1890) by wagon from Salt Lake City, Utah. His wife Antoinette Treganza (*née* Kaufman) was a writer and also an ornithologist. The Lemon Grove Chamber of Commerce commissioned Alberto to design a float for the Fourth of July parade in San Diego (1928). Alberto designed a seven-by-twelve-foot plaster lemon mounted on a wooden platform adorned with lemons, oranges and grapefruit from local orchards and borne on a truck from the Lemon Grove Fruit Packing House. Alberto passed on his interest in natural history to his son Professor Adaan Eduardo Treganza, known as 'Don' or 'Trig' (1916–1968), who was primarily an anthropologist and archaeologist. The Treganzas studied 'snowy herons' in the Tule marshes around Great Salt Lake (c.1914).

Tregellas

Australasian Robin genus *Tregellasia* **Mathews**, 1912

Woodswallow sp. *Artamus tregellasi* Mathews, 1911 NCR
[Alt. Black-faced Woodswallow; JS *Artamus cinereus cinereus*]

Grey-crowned Babbler ssp. *Pomatostomus temporalis tregellasi* Mathews, 1912 NCR
[JS *Pomatostomus temporalis temporalis*]
Hooded Plover ssp. *Charadrius cucullatus tregellasi* Mathews, 1912 NCR
[JS *Thinornis rubricollis* NRM]
Laughing Kookaburra ssp. *Dacelo gigas tregellasi* Mathews, 1912 NCR
[JS *Dacelo novaeguineae novaeguineae*]
Olive Whistler ssp. *Pachycephala olivacea tregellasi* Mathews, 1912 NCR
[JS *Pachycephala olivacea olivacea*]
Red-browed Finch ssp. *Aegintha temporalis tregellasi* Mathews, 1912 NCR
[JS *Neochmia temporalis temporalis*]
Red Wattlebird ssp. *Anthochaera carunculata tregellasi* Mathews, 1912 NCR
[JS *Anthochaera carunculata carunculata*]
Regent Honeyeater ssp. *Meliphaga phrygia tregellasi* Mathews, 1912 NCR; NRM
[Syn. *Anthochaera phrygia*]
Southern Emu-wren ssp. *Stipiturus malachurus tregellasi* Mathews, 1912 NCR
[JS *Stipiturus malachurus malachurus*]
Swift Parrot ssp. *Lathamus discolor tregellasi* Mathews, 1912 NCR; NRM
Southern Boobook ssp. *Spiloglaux boobook tregellasi* Mathews, 1913 NCR
[JS *Ninox boobook boobook*]
Tasmanian Scrubwren ssp. *Sericornis humilis tregellasi* Mathews, 1914

Thomas Henry Tregellas (1864–1938) was an Australian field ornithologist. He specialised in the study of lyrebirds and was one of the first to use radio to broadcast the song of the lyrebird.

Trense

Scaly-breasted Illadopsis ssp. *Illadopsis albipectus trensei* **Meise**, 1978

Werner Trense (b.1922) is a German explorer, conservationist, and big-game hunter. He was in Iran (1957) and was a farmer in Angola (1959–1964). He was Secretary-General of Conseil International de la Chasse et de la Conservation du Gibier for 35 years. He wrote *The Big Game of the World* (1989).

Trimen

Roller genus *Trimenornis* **J. A. Roberts**, 1922 NCR
[Now in *Coracias*]

Roland Trimen (1840–1916) was an English zoologist and entomologist in South Africa (1859), when he was in the Civil Service. He became Curator of the South African Museum, Cape Town (1872), retiring on grounds of ill health (1893). He co-wrote *South African Butterflies* (1887–1889).

Trischitta

Northern Goshawk ssp. *Accipiter gentilis trischittae* Ragioneri, 1946 NCR
[JS *Accipiter gentilis gentilis*]

Dr Antonio Trischitta (1882–1966) was an Italian ornithologist.

Tristram

Tristram's Wheatear *Oenanthe moesta* **Lichtenstein**, 1823
[Alt. Red-rumped Wheatear]
Tristram's Serin *Serinus syriacus* **Bonaparte**, 1850
[Alt. Syrian Serin]
Tristram's Starling *Onychognathus tristramii* **P. L. Sclater**, 1858
[Alt. Tristram's Grackle]
Tristram's Warbler *Sylvia deserticola* Tristram, 1859
Tristram's Bunting *Emberiza tristrami* **Swinhoe**, 1870
Tristram's Scrubfowl *Megapodius layardi* Tristram, 1879
[Alt. Vanuatu Megapode/Scrubfowl]
Tristram's Honeyeater *Myzomela tristrami* **E. P. Ramsay**, 1881
[Alt. Sooty Myzomela]
Tristram's Flowerpecker *Dicaeum tristrami* **Sharpe**, 1884
[Alt. Mottled Flowerpecker]
Subantarctic Snipe *Gallinago tristrami* **Rothschild**, 1893 NCR
[JS *Coenocorypha aucklandica aucklandica*]
Tristram's Storm-petrel *Cymochorea tristrami* **Salvin**, 1896
[Syn. *Oceanodroma tristrami*]

Tristram's Woodpecker *Dryocopus javensis richardsi* Tristram, 1879
[Alt. White-bellied Woodpecker ssp.]
Collared Kingfisher ssp. *Todiramphus chloris tristrami* **Layard**, 1880
Red-moustached Fruit Dove ssp. *Ptilinopus mercierii tristrami* **Salvadori**, 1892 EXTINCT
Tristram's Pygmy Parrot *Micropsitta finschii tristrami* Rothschild & **Hartert**, 1902
[Alt. Green Pygmy Parrot ssp.]

The Reverend Henry Baker Tristram (1822–1906) was Canon of Durham Cathedral and a traveller, archaeologist, naturalist and antiquarian, who assembled a large collection of birds. Despite being a churchman he was an early supporter of Darwin (q.v.). He wrote a number of accounts of his explorations, including *The Great Sahara: Wanderings South of the Atlas Mountains* (1860), which he undertook in the company of his friend Upcher (q.v.) and *A Journal of Travels in Palestine* (1865). In his Sahara book he describes how he penetrated far into the desert and made an ornithological collection in the course of gathering material for his work. He writes interestingly on the indigenous peoples and their customs, as well as on the natural history of the region. He originally went there because of ill health. Salvin (q.v.) was Tristram's cousin by marriage. Despite his early penchant for collecting with a gun, Tristram went on to be a vice-president of the RSPB (1904–1906). A mammal and a reptile are named after him.

Triton

Triton's Cockatoo *Cacatua galerita triton* **Temminck**, 1849
[Alt. Sulphur-crested Cockatoo ssp.]

More correctly known as the Triton Cockatoo, not 'Triton's'. In Greek mythology, Triton was a sea-god, the son of Poseidon. He blew on a conch-shell trumpet, producing a sound so terrible that it once put the race of giants to flight. Temminck does not comment on why he chose the name 'triton' but it might be after a Dutch corvette *Triton* that was operating off the Dutch New Guinea coastline in that period. It is also possible that the cockatoo's call sounded terrible to the author. Two mammals are also named after him.

Trizna

White-throated Dipper ssp. *Cinclus leucogaster triznae* **Zarudny**, 1909 NCR
[JS *Cinclus cinclus leucogaster*]

Captain Boris Petrovich Trizna (1867–1938) was a palaeontologist friend of the describer Zarudny (q.v.) who collected in Turkestan (Kazakhstan), was Director of the Chimkent Museum (1924) and had charge of the Aksu-Dzhabagly reserve in Kazakhstan (1927–1937).

Troilius

Guillemot sp. *Colymbus troille* Linnaeus, 1761 NCR
[Alt. Common Guillemot/Murre; JS *Uria aalge*]

Samuel Troilius (1706–1764) was a Swedish priest (1736). He was chaplain to the royal court in Stockholm (1740–1751), Bishop of Västerås (1751–1758), and Archbishop of Uppsala (1758–1764). His family was ennobled and used the name 'von Troil'.

Trotha

Slaty Flycatcher sp. *Dioptrornis trothae* **Reichenow**, 1900 NCR
[Alt. White-eyed Slaty Flycatcher; JS *Melaenornis fischeri nyikensis*]

Shelley's Francolin ssp. *Francolinus shelleyi trothae* Reichenow, 1901 NCR
[JS *Scleroptila shelleyi shelleyi*]
Chestnut Weaver ssp. *Ploceus rubiginosus trothae* Reichenow, 1905
Spotted Eagle Owl ssp. *Bubo ascalaphus trothae* Reichenow, 1906 NCR
[JS *Bubo africanus africanus*]

General Adrian Dietrich Lothar von Trotha (1848–1920) joined the Prussian army (1865) and fought in both the Austro-Prussian and Franco-Prussian wars. He became commander of the German colonial army in German East Africa (Tanzania) (1894–1897), China (Boxer Rebellion, 1901), and German South West Africa (Namibia) (1904–1906). He is generally blamed for the virtual extermination of the native Herero people. He died of typhoid.

Trouessart

Tahiti Petrel ssp. *Pseudobulweria rostrata trouessarti* Brasil, 1917

Édouard Louis Trouessart (1842–1927) was a French naturalist who was associated with the Museum of Natural History, Angers, and the MNHN, Paris. He described (1870s–1890s) a number of species endemic to Madagascar, evidenced by his paper 'Description de mammifères nouveaux d'Afrique et de Madagascar' (1906). He was elected an honorary member of the American Society of Mammalogists (1921). He published much on halacarids (aquatic mites) and other Acarina; some of them and a mammal are named after him.

Troughton

Australian Masked Owl ssp. *Tyto novaehollandiae troughtoni* **Cayley**, 1931
[Taxonomic status disputed; often included in *Tyto novaehollandiae novaehollandiae*]

Ellis Le Geyt Troughton (1893–1974) was an Australian zoologist who was Curator of Mammals at the Australian Museum, Sydney (1921–1958). He was involved in field investigations into scrub typhus in New Guinea (WW2). His best-known work is *Furred Animals of Australia* (1941). Other taxa, including three mammals, are named after him.

Trudeau

Trudeau's Tern *Sterna trudeaui* **Audubon**, 1838
[Alt. Snowy-crowned Tern]

James de Berty Trudeau (1817–1887) was at various times a physician, surgeon, painter, explorer, collector and General in the artillery of the Confederacy in the American Civil War. He was severely wounded at Shiloh and later captured by Union forces. He was a friend of Audubon's (q.v.) and may have accompanied him over the Rockies (1842).

Tschebaeiv

White-tailed Rubythroat ssp. *Luscinia pectoralis tschebaiewi* **Przewalski**, 1876

Pamfili Tschebaeiv (b.1852) was a Cossack who accompanied Przewalski (q.v.) on his explorations in Transbaikalia.

Tschersky

Great Spotted Woodpecker ssp. *Dendrocopos major tscherskii* **Buturlin**, 1910 NCR
[NUI *Dendrocopos major japonicus*]

A. J. Tschersky (DNF) was a Russian zoologist and collector in Ussuriland, Siberia (1909).

Tschitscherin

Western Rock Nuthatch ssp. *Sitta neumayer tschitscherini* **Zarudny**, 1904

Tikhon Sergeyevich Tschitscherine (1869–1904) was a Russian entomologist and collector at the Zoological Museum, Academy of Sciences, St Petersburg. He graduated in law (1889), then served as public prosecutor in Dagestan (1890–1892). He travelled in Germany, Belgium and France (1896), and Madagascar (1900), but on the voyage lost all his money playing cards with fellow passengers, so was forced to return home immediately. This passion proved fatal, as he gambled away everything and committed suicide.

Tschudi

Tschudi's Parrot *Pionus tumultuosus* Tschudi, 1844
[Alt. Plum-crowned Parrot, Speckle-faced Parrot]
Tschudi's Tapaculo *Scytalopus acutirostris* Tschudi, 1844
[Alt. Sharp-billed Tacapulo]
Tschudi's Tyrannulet *Zimmerius viridiflavus* Tschudi, 1844
[Alt. Peruvian Tyrannulet]
Tschudi's Woodcreeper *Xiphorhynchus chunchotambo* Tschudi, 1844
Scaled Fruiteater *Ampelioides tschudii* **G. R. Gray**, 1846
Tschudi's Woodnymph *Thalurania tschudii* **P. L. Sclater**, 1858 NCR
[Alt. Fork-tailed Woodnymph; JS *Thalurania furcata viridipectus*]

Blackish-grey Antshrike ssp. *Thamnophilus nigrocinereus tschudii* **Pelzeln**, 1868
Wing-barred Piprites ssp. *Piprites chloris tschudii* **Cabanis**, 1874
Sickle-winged Guan ssp. *Chamaepetes goudotii tschudii* **Taczanowski**, 1886
Plumbeous Rail ssp. *Pardirallus sanguinolentus tschudii* **Chubb**, 1919

Baron Dr Johann Jacob von Tschudi (1818–1889) was a Swiss explorer who travelled in Peru (1838–1842), Brazil, Argentina and Chile. He was also a physician, diplomat, naturalist, student of South America, hunter, anthropologist, cultural historian, language researcher and statesman. He wrote *Untersuchungen über die Fauna Peruana Ornithologie* (1844). Five mammals, six reptiles and five amphibians are named after him.

Tschusi

Black Grouse ssp. *Lyrurus tetrix tschusii* **H. Johansen**, 1898
[SII *Lyrurus tetrix viridanus*]
Common Reed Bunting ssp. *Emberiza schoeniclus tschusii* **Reiser** & **Almásy**, 1898
Marsh Tit ssp. *Parus communis tschusii* **Hellmayr** 1901 NCR
[JS *Poecile palustris italicus*]
European Goldfinch ssp. *Carduelis carduels tschusii* **Arrigoni**, 1902
Rook ssp. *Corvus frugilegus tschusii* **Hartert**, 1903 NCR
[JS *Corvus frugilegus frugilegus*]
Eurasian Wryneck ssp. *Jynx torquilla tschusii* **Kleinschmidt**, 1907
Lanner Falcon ssp. *Falco hierofalco tschusii* Kleinschmidt, 1907 NCR
[JS *Falco biarmicus erlangeri*]
Alpine Accentor ssp. *Prunella collaris tschusii* **Schiebel**, 1910 NCR
[JS *Prunella collaris collaris*]
European Scops Owl ssp. *Otus scops tschusii* Schiebel, 1910 NCR
[JS *Otus scops scops*]

Dr Victor Ritter von Tschusi zu Schmidhoffen (1847–1924) was an Austrian ornithologist and founding editor of *Ornithologisches Jahrbuch* (1890). He travelled widely in Europe, especially in Germany, Italy, Austria, Bohemia (Czech Republic) and Heligoland (1868–1870). The University of Innsbruck awarded his honorary doctorate (1921). He was the father of Rudolf von Thanner (q.v.).

Tucker

Bornean Treepie *Dendrocitta sinensis tuckeri* **Harrisson** & **Hartley**, 1934 NCR
[JS *Dendrocitta cinerascens*]

Bernard William Tucker (1901–1950) was an English ornithologist and a Lecturer in Zoology, Oxford (1926–1950). He was founding Secretary, BTO. He co-wrote *The Handbook of British Birds* (1938).

Tueros

Junin Rail *Laterallus tuerosi* Fjeldså, 1983
[Alt. Black Rail; Syn. *Laterallus jamaicensis tuerosi*]

Justo Tueros Aldana (b.1953) is a Peruvian miner, naturalist and amateur collector from a family that made its living from Lake Junin, where the holotype was collected. He learnt taxidermy as a pastime at school and now lives in the US. Justo and his brothers, Francisco and Máximo, were keen observers of Lake Junin's birds. Francisco, in particular, has been involved in conservation and in the study of the status of the Junin Giant Frog *Batrachophrynus macrostomus*.

Tufts

Long-eared Owl ssp. *Asio otus tuftsi* **W. E. Godfrey**, 1948

Dr Robie W. Tufts (1884–1982) was a Nova Scotian ornithologist and conservationist who worked as a banker until appointed Chief Federal Migratory Birds Officer for the Maritimes (1919). He retired (1947) from the organisation that developed into the Canadian Wildlife Service. He was awarded a number of honorary degrees, and wrote *The Birds of Nova Scotia* (1961).

Tullberg

Tullberg's Woodpecker *Campethera tullbergi* **Sjöstedt**, 1892
[Alt. Fine-banded Woodpecker]

Tycho Fredrik Hugo Tullberg (1842–1920) was a Swedish zoologist. He gained a doctorate in philosophy at Uppsala (1869), becoming a lecturer there (1871) and Professor of Zoology (1882–1907). He was instrumental in developing modern Swedish zoological veterinary medicine. He was Chairman of the Linnean Society at Hammarby (1902). As far as we can ascertain he was an academic rather than active in the field. He published in Swedish, English and German (the preferred learned language of his day). His works included '*Neomenia*, a new genus of invertebrate animals' (1875) (that odd title was published in English) and *Djurriket* (*The Animal Kingdom*) (1885). A mammal is named after him.

Tunney

Black Butcherbird ssp. *Cracticus quoyi tunneyi* **Hartert**, 1905 NCR
[JS *Cracticus quoyi spaldingi*]
Black-faced Cormorant ssp. *Carbo gouldi tunneyi* **Mathews**, 1912 NCR
[JS *Phalacrocorax fuscescens* NRM]
Little Shearwater ssp. *Puffinus assimilis tunneyi* Mathews, 1912
Yellow Chat ssp. *Epthianura crocea tunneyi* Mathews, 1912
Lesser Frigatebird ssp. *Fregata ariel tunnyi* Mathews, 1914 NCR
[JS *Fregata ariel ariel*]

John Thomas Tunney (1871–1929) was an Australian farmer and professional collector for the Western Australian Museum (1895–1906). He led the Tunney expedition to the Northern Territory (1903) for Lord Rothschild (q.v.) of the Tring Museum. Oldfield Thomas wrote a paper on the expedition's findings (1904), entitled 'On a collection of mammals made by Mr J. T. Tunney in Arnhem Land, Northern Territory of South Australia'. A mammal is also named after him. We assume the spelling *tunnyi* (not *tunneyi*) in the frigatebird's trinomial to be a *lapsus calami* on Mathews's part, but the notorious lack of etymologies in his descriptions makes it impossible to know for sure.

Tupinier

Thorn-tailed Rayadito *Synallaxis tupinieri* **Lesson**, 1828 NCR
[JS *Aphrastura spinicauda*]

Baron Jean Marguerite Tupinier (1779–1850) was a French politician and naval engineer. He was present at the re-conquest of Santo Domingo (1802). He held a number of posts under the First Empire, but on Napoleon's fall (1815) was disgraced. He was restored to favour (1818), and entered parliament (1834). He was a great supporter of exploration in general and of Dumont d'Urville in particular. He retired from public life after the 1848 Revolution.

Turati

Turati's Boubou *Laniarius turatii* **J. Verreaux**, 1858
[Alt. Turati's Bush-shrike]

Downy Woodpecker ssp. *Picoides pubescens turati* **Malherbe**, 1860

Conte Ercole Turati (1829–1881) was a Milanese banker, naturalist and collector. Although he accumulated an enormous collection (c.20,000 skins), he had no time to study them, so made them available for others. A palace that was constructed for him in Milan (1880) is now the Chamber of Commerce. Giacinto Martorelli wrote a book about him, *Commemorazione Scientifica del Conte Ercole Turati* (1898).

Turner, H. H.

Kalij Pheasant sp. *Gennæus turneri* **Finn**, 1900 NCR
[JS *Lophura leucomelanos williamsi*]

Colonel H. H. Turner (1867–1930) of the Royal Engineers was a surveyor in India.

Turner, H. J. A.

Turner's Eremomela *Eremomela turneri* **Van Someren**, 1920

White-headed Mousebird ssp. *Colius leucocephalus turneri* Someren, 1919

H. J. Allen Turner (1876–1952) was a British taxidermist who lived in Kenya (1908–1952) and worked at the Coryndon Museum, Nairobi (1909–1952). He collected birds in East Africa (mainly Kenya) (1915–1917). A reptile is named after him.

Turner, J.

Torrent Duck ssp. *Merganetta armata turneri* **P. L. Sclater** & **Salvin**, 1869

We cannot be certain of Turner's identity, as the original description only says that the holotype 'was shot and skinned by Mr. Turner, a friend of Mr. Whitely's, near Tinta. We have, therefore, acceded to Mr. Whitely's request to call it, if new, after his friend's name' (q.v. Whitely, H. Jr). We think it refers to a wealthy landowner, Dr J. Turner (d.1881), who married (1871) Clorinda Matto (1852–1909), a well-known Peruvian author of the period. They are known to have lived on his estate at Tinta (1871–1881).

Turner, L. M.

Turner's Ptarmigan *Lagopus muta atkhensis* Turner, 1882
[Alt. Rock Ptarmigan ssp.]
Black-capped Chickadee ssp. *Poecile atricapillus turneri* **Ridgway**, 1884

Lucien McShann Turner (1848–1909) was a member of the Army Signal Corps who collected natural history and ethnological specimens for the USNM. He was a meteorological observer for the Signal Service at St Michael, Alaska (1874–1877), and then trained voluntary observers in the Aleutians (1878–1881) before being sent to Fort Chimo, Labrador, as an observer (1882–1884). Turner made extensive collection and also had a rapport with the local people, the Innu and Inuit, spending his free time studying and recording their culture, routines, language and stories. His pictures of them and their camps are among the earliest photographs of the Arctic. He wrote a number of books including *Contributions to the Natural History of Alaska* (1886) and *Ethnology of the Ungava District, Hudson Bay Territory* (1894).

Tweeddale

Tweeddale's Babbler *Dasycrotapha speciosa* Tweeddale, 1878
[Alt. Flame-templed Babbler]
Tweeddale's Hawk Owl *Ninox spilocephala* Tweeddale, 1879
[Alt. Mindanao Hawk Owl]

Asian Fairy-bluebird ssp. *Irena puella tweeddalei* **Sharpe**, 1877
Tweeddale's Woodpecker *Mulleripicus funebris fuliginosus* Tweeddale, 1877
[Alt. Sooty Woodpecker ssp.]
Tawny Grassbird ssp. *Megalurus timoriensis tweeddalei* **McGregor**, 1908

Arthur Hay (1824–1878) became the ninth Marquess of Tweeddale, inheriting the title late (1876), formerly being known as Viscount Walden (1862–1876). He was President of the Zoological Society of London and an amateur ornithologist who had a large private collection of insects, birds and mammals. He employed Carl Bock (q.v.) to collect further specimens from the Malay Archipelago. Tweeddale described c.40 new species from Bock's efforts. (See **Walden**)

Twomey

Common Nighthawk ssp. *Chordeiles minor twomeyi* **W. Hawkins**, 1948 NCR
[JS *Chordeiles minor hesperis*]
Green-backed Sparrow ssp. *Arremonops chloronotus twomeyi* Monroe, 1963

Dr Arthur Cornelius Twomey (1908–1996) was an American zoologist and ornithologist. The University of Alberta awarded his bachelor's degree (1934), and the University of Illinois both his master's (1935) and doctorate (1937). He joined the Carnegie Museum, Pittsburgh (1936), as Assistant and Field Collector in Ornithology, and worked there until retiring (1973) as Director of Education. He collected in Chile, Peru and the Galapagos Islands (1939) and on four big-game hunting expeditions to Africa (1960–1964). He wrote *Needle to the North* (1942). He died of Alzheimer's disease.

Tyler

Duida Elaenia *Elaenia dayi tyleri* **Chapman**, 1929
[Alt. Great Elaenia ssp.]

Sidney F. Tyler Jr (1850–1937) was an American lawyer and banker in Philadelphia. He sponsored the Tyler-Duida Expedition of the American Museum of Natural History (1928–1929).

Chapman wrote up its results. A mammal and a reptile are named after him.

Tyro

Spangled Kookaburra *Dacelo tyro* **G. R. Gray**, 1858

In Greek mythology, Tyro was a Thessalian princess who fell in love with the river Enipeus.

Tyson

Yucatan Woodpecker ssp. *Melanerpes pygmaeus tysoni* **Bond**, 1936

Canby S. Tyson Jr (DNF) of Philadelphia was an American businessman and philanthropist.

Tytler

Tytler's Leaf Warbler *Phylloscopus tytleri* **W. E. Brooks**, 1872

Golden-headed Cisticola ssp. *Cisticola exilis tytleri* **Jerdon**, 1863
Barn Swallow ssp. *Hirundo rustica tytleri* Jerdon, 1864
Black-naped Monarch ssp. *Hypothymis azurea tytleri* **Beavan**, 1867
Asian Glossy Starling ssp. *Aplonis panayensis tytleri* **Hume**, 1873
Tytler's Parakeet *Psittacula longicauda tytleri* Hume, 1874
[Alt. Long-tailed Parakeet ssp.]

Colonel Robert Christopher Tytler (1818–1872) was a naturalist, photographer and collector. He served very actively in the British Army throughout India, and in Kabul and other parts of Afghanistan (1835–1864), after which he was placed at the disposal of the Home Department. This service was only broken by two trips back to England for two years (1850s) due to ill health, and eighteen months (1860s). He was the third superintendent of the convict settlement at Port Blair, part of the Andaman Islands administration, a position he was not suited for being heavily criticised by the Indian Government for his botched investigation of an alleged murder. This resulted in him being put on the retired list and put in charge of the museum at Simla for the last six months of his life. He and his wife Harriet Christina Eart (1827–1907) were very keen photographers and took c.500 large-format calotype negatives of scenes associated with the Indian Mutiny (1857). He amassed a collection of c.2,500 birds. An amphibian and a reptile are named after him.

Tzacatl

Rufous-tailed Hummingbird *Amazilia tzacatl* De la Llave, 1833

In Aztec mythology, Tzacatl was a warrior chief.

U

Uchida

Brown Bullfinch ssp. *Pyrrhula nipalensis uchidai* **Kuroda**, 1916

Plain Flowerpecker ssp. *Dicaeum minullum uchidai* Kuroda, 1920

Great Tit ssp. *Parus major uchidae* Kuroda, 1923 NCR
[Alt. Japanese Tit ssp.; JS *Parus minor amamiensis*]

White-backed Woodpecker ssp. *Dryobates leucotos uchidai* **Momiyama**, 1927 NCR
[JS *Dendrocopos leucotos namiyei*]

White-throated Needletail ssp. *Hirundapus caudacutus uchidai* Ishizawa, 1928 NCR
[JS *Hirundapus caudacutus caudacutus*]

Dr Seinosuke Uchida (1884–1975) was a Japanese ornithologist and entomologist, Forestry Department, Japanese Ministry of Agriculture, and President, Ornithology Congress of Japan. He wrote widely on Mallophaga (bird lice), such as 'Mallophaga from birds of Formosa' (1917).

Újhelyi

Spot-breasted Woodpecker ssp. *Colaptes punctigula ujhelyii* **Madarász**, 1912

József Újhelyi-Uhl (1879–1933) was a Hungarian entomologist and collector in Colombia (1912)

Ulysses

Hummingbird genus *Ulysses* **Mulsant & E. Verreaux**, 1875 NCR
[Now in *Hylocharis*]

Ulysses is the Latin name for Odysseus, King of Ithaca. In Greek mythology he was one of the leading commanders at the siege of Troy. His subsequent adventures formed the subject of Homer's *Odyssey*.

Underwood

Booted Racket-tail *Ocreatus underwoodii* **Lesson**, 1832

Mr Underwood (DNF) sent Lesson (q.v.) a drawing of this hummingbird on behalf of Charles Stokes, a collector in London. We have been unable to find out anything more about him.

Underwood, C. F.

Underwood's Hummingbird *Selasphorus underwoodi* **Salvin**, 1897 NCR

[Presumed hybrid: *Selasphorus flammula* x *Selasphorus scintilla*]

Collared Trogon ssp. *Trogon aurantiiventris underwoodi* **Bangs**, 1908

Cecil F. Underwood (1867–1943) left London for Costa Rica (1889) to collect natural history specimens for a living, staying for the rest of his life. He was an all-round naturalist who collected for a number of overseas museums, and was a taxidermist at Costa Rica's National Museum. He described many new mammals from Central America, often with George Goodwin. Four mammals and two amphibians are named after him.

Unger

Chaco Chachalaca ssp. *Ortalis canicollis ungeri* **Steinbacher**, 1962 NCR
[NUI *Ortalis canicollis canicollis*]

Jakob Unger (1894–1959) was a German collector for the Senckenberg Museum, Frankfurt-am-Main, Germany. He left Manchuria (1930) and became resident at the Mennonite (Anabaptist) colony in Paraguay (1932–1959), where the Museo Jakob Unger in Filadelfia is named after him.

Unitt

Hutton's Vireo ssp. *Vireo huttoni unitti* Rea, 1991

Philip Unitt is an American ornithologist, Curator, Department of Birds and Mammals, San Diego Natural History Museum. He wrote *The Birds of San Diego County* (1984) and edits *Western Birds*.

Unwin

European Nightjar ssp. *Caprimulgus europaeus unwini* **Hume**, 1871

Colonel William Heathcote Unwin (1840–1929) of the Bengal Staff Corps was a British army officer in India (1857–1884). He wrote 'On the breeding of the Golden Eagle (*Aquila chrysaëtos*) in north-western India' (1874).

Upcher

Upcher's Warbler *Hippolais languida* **Hemprich & Ehrenberg**, 1833

Henry Morris Upcher (1839–1921), Deputy-Lieutenant of Norfolk, was a close friend and travelling companion of

Henry B. Tristram (q.v.). Upcher was born at Sheringham Hall, near Cley, Norfolk, and became an important county figure, encouraging developments to the local economy, such as the construction of a golf course and continuing the family tradition of financing the local lifeboat. Hemprich (q.v.) and Ehrenberg's (q.v.) original description of the warbler (1833) was 'rather vague', and Tristram named '*Hippolais upcheri*' as a new species (1864), clearly unaware that it had been described 30 years earlier. Upcher had accompanied Tristram on a trip to Palestine. Although the original binomial *languida* takes precedence, the common name 'Upcher's Warbler' has endured.

Urbano

Musician Wren ssp. *Cyphorhinus arada urbanoi* **Zimmer** & **W. H. Phelps**, 1946

Ramón Urbano (b.1917) was a Venezuelan collector who worked for the Phelps family for 20 years at Estación Biológica de Rancho Grande. He collected the holotype (1944).

Urich

Urich's Tyrannulet *Phyllomyias urichi* **Chapman**, 1899

Friederich William Urich (1872–1936) was a Trinidadian naturalist. He was a founder member of the Trinidad Field Naturalists Club (c.1891), most famous for discovering an eyeless cave-dwelling catfish which was named after him, *Caecorhamdia urichi* (now believed to be a local, cave-adapted population of the widely distributed catfish species *Rhamdia quelen*). He published his finding in the *Field Naturalists' Club Journal* (1895) as 'A visit to the Guacharo Cave of Oropuche'. He was described as Adjutant (1915) with the rank of Captain to the Military Department of the Government of Trinidad; presumably a temporary WW1 post. An amphibian and two mammals are also named after him.

Ursch

Madagascar Ibis ssp. *Lophotibis cristata urschi* **Lavauden**, 1929

Eugène Ursch (1882–1962) was a French botanist and orchid collector on Madagascar, Director of the Zoological and Botanical Gardens at Tsimbazaza.

Ursula

Ursula's Sunbird *Cinnyris ursulae* **B. Alexander**, 1903

Ursula Davies (DNF) was the niece of the describer, Boyd Alexander (q.v.).

Usa

Red-naped Trogon ssp. *Pyrotrogon kasumba usa* **Harrisson** & Hartley, 1934 NCR
[JS *Harpactes kasumba impavidus*]

Uyan Usa (fl.1933) was a Bornean collector in Sarawak.

Usher

Double-banded Sandgrouse ssp. *Eremialector bicinctus usheri* **C. W. Benson**, 1947 NCR
[JS *Pterocles bicinctus multicolor*]
Pale-tailed Canastero ssp. *Asthenes huancavelicae usheri* **A. R. G. Morrison**, 1947
[Syn. *Asthenes dorbignyi usheri*]
Streak-breasted Scimitar Babbler ssp. *Pomatorhinus ruficollis usheri* **B. P. Hall**, 1954 NCR
[JS *Pomatorhinus ruficollis eidos*]

Harold Bench Usher (1893–1990) was a British ornithologist who worked in the Bird Room at the British Museum (Natural History) (1908–1953), having joined the staff as a Boy Attendant. Morrison thanked him as having 'worked very hard in helping me to identify certain species'.

Ussher

Ussher's Spinetail *Telacanthura ussheri* **Sharpe**, 1870
[Alt. Mottled Spinetail, Mottle-throated Spinetail]
Ussher's Flycatcher *Muscicapa ussheri* Sharpe, 1871
Ussher's Owl *Scotopelia ussheri* Sharpe, 1871
[Alt Rufous Fishing Owl]
Black-crowned Pitta *Erythropitta ussheri* Gould, 1877
[Syn. *Pitta granatina ussheri*]

Brown-crowned Tchagra ssp. *Tchagra australis ussheri* Sharpe, 1882
Thick-billed Honeyguide ssp. *Indicator conirostris ussheri* Sharpe, 1902
Tit-hylia ssp. *Pholidornis rushiae ussheri* **Reichenow**, 1905
Honeyguide Greenbul ssp. *Andropadus indicator ussheri* **Bannerman**, 1920 NCR
[JS *Baeopogon indicator leucurus*]

Herbert Taylor Ussher (1836–1880) was variously, the Governor of the Gold Coast (Ghana), the Governor of Tobago, and Consul-General in Borneo. He sent many specimens back to the British Museum. He was highly regarded by his contemporaries since he is honoured in the binomials of fish, butterflies and other taxa. He wrote 'Notes on the ornithology of the Gold Coast' (1874) and was a co-author with Sharpe (q.v.).

Utano

Eurasian Wren ssp. *Troglodytes troglodytes utanoi* **Kuroda**, 1922 NCR
[JS *Troglodytes troglodytes fumigatus*]

Yoshiho Utano (b.1891) was a Japanese teacher and naturalist on Tsushima Island.

Uthai

Stripe-throated Yuhina ssp. *Yuhina gularis uthaii* Eames, 2002

Uthai Treesucon is a Thai biologist, ornithologist and conservationist. He worked as a wildlife researcher at Mahidol University and has led birding tours since 1983. He is a former chairman of the Bird Conservation Society of Thailand.

V

Vaillant, F.

Crested Barbet *Trachyphonus vaillantii* Ranzani, 1821

(See **Levaillant, F.**)

Vaillant, J.

Levaillant's Woodpecker *Picus vaillantii* **Malherbe**, 1847

(See **Levaillant, J.**)

Valdizán

Yellow-vented Woodpecker ssp. *Veniliornis dignus valdizani*
Berlepsch & **Stolzmann**, 1894

Dario Valdizán (d.1927) was a Peruvian civil engineer who was involved in government road and railway projects (c.1888–1927). He befriended Jan Kalinowski (q.v.) when he was in Lima. A street in Lima is named after him.

Valencio Bueno

Variable Oriole ssp. *Icterus pyrrhopterus valenciobuenoi*
Ihering, 1902

Valencio Bueno de Toledo (DNF) was a resident of Piracicaba in São Paulo, Brazil. He donated the holotype to Museu Paulista, São Paulo.

Valentin

Bianchi's Warbler *Seicercus valentini* **Hartert**, 1907

(See **Bianchi**)

Valentine

White-browed Fulvetta ssp. *Fulvetta vinipectus valentinae*
Delacour & **Jabouille**, 1930

Madame Valentine Lécallier (DNF) was a French aviculturist who lived in the Seine-Inférieure department of France. Her great interest was breeding budgerigars, and she was among the first to succeed in breeding a mauve bird (1924).

Valéry

Black-shouldered Tanager *Tachyphonus valeryi*
J. & **E. Verreaux**, 1855 NCR
[Alt. Velvet-fronted Grackle; JS *Lampropsar tanagrinus*]

Valéry Louis Victor Potiez (1806–1870) was a French zoologist and malacologist, Director of Douai Museum (1855), who co-wrote *Catalogue Méthodique, Descriptif et Raisonné des Mollusques et Coquilles du Muséum de Douai* (1835).

Valisnera

Canvasback *Aythya valisineria* **A. Wilson**, 1814

Professor Antonio Vallisneri de Valisnera (1661–1730) was an Italian naturalist, physician and author. He taught Practical and Theoretical Medicine at the University of Padua and collected natural history and mineral specimens. Linnaeus (q.v.) named the eelgrass genus *Vallisneria* after him, and the binomial of the Canvasback is derived from it – or so claimed George Bird Grinnell (q.v.) in *American Duck Shooting* (1901): 'The food of the canvas-back, from which it gets its specific name, and to which it owes its delicious flavor, is the so called wild celery.'

Valverde

Sardinian Warbler ssp. *Sylvia melanocephala valverdei*
Cabot & Urdiales, 2005

Professor José Antonio Valverde (1926–2003) was a Spanish biologist, ecologist and environmentalist who helped to preserve Doñana National Park. Among his publications is *Birds of the Spanish Sahara* (1957). The University of Salamanca holds his collection of books, field notebooks, drawings, diagrams and photographs. A reptile is named after him.

Vámbéry

Crested Lark ssp. *Galerida cristata vamberyi* **Harms**, 1907
NCR
[JS *Galerida cristata iwanowi*]

Ármin Vámbéry (*né* Hermann Bamberger or Wamberger) (1832–1913) was a Hungarian orientalist, geographer, explorer and linguist who knew c.30 languages. He was in Constantinople (Istanbul) as a private tutor of languages (1853–1861) and then travelled, disguised as a dervish, through Anatolia (Turkey), Persia (Iran) and Turkistan (1861–1864) and visited both Bokhara and Samarkand. Files released (2005) by the UK National Archives showed that he had been a secret agent of the British Government, charged with combating Russian expansion in central Asia (threatening Britain's position in the Indian subcontinent). He was Professor of Oriental Languages, University of Budapest (1865–1905). He wrote *Travels in Central Asia* (1864).

van Bemmel

Besra ssp. *Accipiter virgatus vanbemmeli* **Voous**, 1950

Dr Adriaan Cornelis Valentin van Bemmel (1908–1990) was a Dutch zoologist and ornithologist. He was deputy director of Rotterdam Zoo (1980).

van Dam

van Dam's Vanga *Xenopirostris damii* **Schlegel**, 1865

Douwe Casparius van Dam (1827–1898) was a Dutch naturalist who collected in Reunion, Mayotte and Madagascar (1860s–1880s), initially in the company of François P. L. Pollen (q.v.) until the latter's marriage (1866). His work was the subject of a book by Bleeker, *Recherches sur la Faune de Madagascar et de ses Dépendances d'après les Découvertes de François P. L. Pollen et D. C. van Dam* (1875). All the specimens from Pollen and van Dam were distributed by Pollen, who sent most of them to Hermann Schlegel (q.v.). Van Dam is also remembered in the scientific names of other taxa including a fish.

van den Broecke

van den Broecke's Rail *Aphanapteryx bonasia* **Sélys**, 1848
EXTINCT
[Alt. Mauritius Red Rail]

Pieter van den Broecke (1585–1640) was a Dutch traveller who was a merchant commander in West Africa and subsequently served the Dutch East India Company in Java, Arabia, Persia and India. He wrote a journal of his voyages to Cape Verde, Guinea and Angola (1605–1612). He brought home a fleet from India (1630) and was rewarded by the company for 17 years of service with a gold chain worth 1,200 guilders; it is just visible in his portrait. Van den Broecke knew Frans Hals well, and was a witness at the baptism of Hals's daughter, Susanna (1634). He visited Mauritius (1617) and made a (rather poor) drawing, which shows a dodo, a red rail, and a one-horned sheep. The rail was only known from such 17th-century descriptions and illustrations until subfossil remains were described (1869).

van de Poll

Imperial Pigeon sp. *Carpophaga vandepolli* **Büttikofer**, 1896
NCR
[Alt. Green Imperial Pigeon; JS *Ducula aenea consobrina*]

Mangrove Whistler ssp. *Pachycephala cinerea vandepolli* **Finsch**, 1899 NCR
[NUI *Pachycephala cinerea cinerea*]

Jacob Rudolph Hendrick Neervoort van de Poll (1862–1925) was a Dutch entomologist and pioneer photographer who loaned equipment to Büttikofer. His speciality was the study of Coleoptera.

Vanderbilt

Vanderbilt's Babbler *Malacocincla vanderbilti* **Meyer de Schauensee** & **Ripley**, 1940 NCR
[JS *Malacocincla sepiaria barussana*]

Mountain Serin ssp. *Serinus estherae vanderbilti* Meyer de Schauensee, 1939
Glossy Swiftlet ssp. *Collocalia esculenta vanderbilti* Meyer de Schauensee & Ripley, 1940

George Washington Vanderbilt III (1914–1961) was a scientific explorer whose main interest was marine life. He owned several yachts, using them to combine exploration with the pleasure of sailing. He went on at least five major expeditions, including several to Africa. On board his schooner *Cressida* he sailed in the Indo-Pacific (1937–1939), most importantly to Sumatra to carry out a systematic study of more than 10,000 fish specimens. His 5th major expedition (1941) was on board his schooner *Pioneer* to the Bahamas, Caribbean, Panama, Galapagos and islands off the Pacific coast of Mexico. He established the George Vanderbilt Foundation for scientific research. He was found dead on the pavement in front of a skyscraper in San Francisco. The official verdict was that he committed suicide by jumping from his 101st floor apartment, but the truth remains a mystery.

Van Devender

Bridled Titmouse ssp. *Baeolophus wollweberi vandevenderi* Rea, 1986

Dr Thomas Roger Van Devender (b.1946) is an American zoologist, herpetologist, botanist and palaeo-ecologist at the Arizona-Sonora Desert Museum, Tucson. Lamar University, Texas, awarded his bachelor's degree (1968), and the University of Arizona his master's (1969) and doctorate (1973).

van de Water

Vandewater's Scops Owl *Otus spilocephalus vandewateri* **H. C. Robinson** & **Kloss**, 1916
[Alt. Mountain Scops Owl ssp.]

Lieutenant A. van de Water (DNF) was an officer in the Dutch East Indies Army. He was in New Guinea (1913), assisting Kloss (q.v.) and Wollaston (q.v.) in an expedition. He was also in Sumatra (1914), where the owl holotype was collected from Mt Korinchi. A glacier in New Guinea is named after him.

Vande weghe

Brown-chested Alethe ssp. *Pseudalethe poliocephala vandeweghei* **Prigogine**, 1984

Dr Jean Pierre Vande weghe (b.1940) is a Belgian physician, zoologist, conservationist and forest specialist. He lived and worked in Africa for over 40 years based mainly in Rwanda and Gabon (since 1999). He wrote *Forests of Central Africa* (2004) and *Birds in Rwanda: An Atlas and Handbook* (2011), the latter with his son, Gael Ruboneka Vande weghe.

Van Diemen

Van Diemen's Parrot *Platycercus caledonicus* **Gmelin**, 1788
[Alt. Green Rosella]

An archaic vernacular name, and probably after Van Diemen's Land (Tasmania), where the rosella is found, rather than directly after Anthony van Diemen (1593–1645) who sent Abel Tasman out on his voyage of discovery (1642).

van Hasselt

van Hasselt's Sunbird *Leptocoma brasiliana* **Gmelin**, 1788
[*Nectarinia hasseltii* (Temminck, 1825) is a JS]

Dr Johan Coenraad van Hasselt (1797–1823) came to fame as the first person to climb Mount Pangrango in Java. Heinrich Kuhl (q.v.) and van Hasselt, two young biologists working for The Netherlands Commission for Natural Sciences, made the first ascent (August 1821). Van Hasselt qualified as a physician, but was more interested in natural history, as was his fellow student and close friend Kuhl. They made various excursions in Europe, visiting natural history museums where they met Cuvier (q.v.) and other famous zoologists of the time. They were sent to Java (1820) to study natural history but started their work on the way, studying the pelagic fauna, as well as that of Madeira, the Cape of Good Hope and the Cocos Islands. When Kuhl died after less than a year in Java, van Hasselt spent his time working still harder until he himself died two years later. It is confusing that an A. W. M. van Hasselt was also a zoologist who worked in Java in the 1820s and for whom we believe a large number of arachnids, fish and other taxa are named, but a mammal is named after Johann. (See **Hasselt**)

van Heurn

Garnet Pitta ssp. *Pitta granatina vanheurni* **Kloss**, 1921 NCR
[JS *Erythropitta granatina coccinea*]
Great-billed Mannikin ssp. *Lonchura grandis heurni* **Hartert**, 1932
Rusty Pitohui ssp. *Pitohui ferrugineus heurni* Hartert, 1932 NCR
[JS *Pitohui ferrugineus holerythrus*]
White-faced Robin ssp. *Tregellasia leucops heurni* Hartert, 1932
Changeable Hawk Eagle ssp. *Nisaetus cirrhatus vanheurni* **Junge**, 1936

Willem Cornelis van Heurn (1887–1972) was a Dutch taxonomist, civil engineer, botanist, educationalist, collector and biologist who worked for a period at the Natural History Museum, Leiden. He came from a wealthy family but chose to work all his life. He went to Suriname (1911), to Simeulue (off Sumatra) (1913), and to Dutch New Guinea (1920–1921). He then lived in the Dutch East Indies (mostly Java) (1924–1939), where he ran a laboratory for sea research; studied rat control on Java, Timor and Flores; was a schoolteacher; and served as head of the Botany Department at the Netherlands Indies Medical School before returning to Holland. Wherever he travelled or settled he collected natural history specimens which he meticulously prepared and labelled. Most he sent to the Leiden Museum, where he himself worked as an Assistant Curator for Fossil Mammals (1941–1945). He was a prolific writer, publishing c.100 articles on a wide range of topics, including such gems as 'The safety instinct in chickens' (1927), 'Cannibalism in frogs' (1928), 'Do tits lay eggs together as the result of a housing shortage?' (1955) and 'Wrinkled eggs' (1958). It was said of him in a memorial booklet published by the museum, 'He made natural history collections wherever he went and gave his attention to almost all animal groups. He was an excellent shot, and a competent preparator; his mammal and bird skins are exemplary.' Many different taxa, including mammals, reptiles and fish are named after him.

van Heyst

Flowerpecker sp. *Dicaeum vanheysti* **H. C. Robinson** & **Kloss**, 1918 NCR
[Alt. Fire-breasted Flowerpecker; JS *Dicaeum ignipectus beccarii*]
Blue Flycatcher sp. *Cyornis vanheysti* H. C. Robinson & Kloss, 1919 NCR
[Alt. Rück's Blue Flycatcher; JS *Cyornis ruckii*]

Lesser Yellownape ssp. *Picus chlorolophus vanheysti* H. C. Robinson & Kloss, 1919

August Floris Charles André van Heyst (1887–1962) was a Dutch planter and collector on Sumatra, Indonesia.

Vanhoeffen

Crozet Shag *Phalacrocorax vanhoeffeni* **Reichenow**, 1904 NCR
[JS *Leucocarbo melanogenis*]

Dr Ernst Vanhöffen (1858–1918) was a German zoologist, geologist, explorer and collector. He studied at the Universities of Berlin and Königsberg. He was at the Naples Zoological Institute (1888–1889), and a member of Drygalski's (q.v.) expedition to Greenland (1892–1893). He took part in the German Deep Sea Expedition (1898–1899) aboard the steamship *Valdivia*, returned to Germany and lectured at the University of Kiel. He was aboard *Gauss* in the Antarctic (1901–1904) as part of the German South Pole Expedition.

van Marle

European Greenfinch ssp. *Carduelis chloris vanmarli* **Voous**, 1951

Johan Gottlieb van Marle (1901–1979) was a Dutch businessman and ornithologist. He was a relative of Sillem (q.v.) and together they formed the Sillem-van Marle Society, along with a very important collection. With Voous (q.v.) he co-wrote *The Birds of Sumatra* (1988).

van Oort, E.

van Oort's Fantail *Rhipidura dedemi* van Oort, 1911
[Alt. Streak-breasted Fantail, Seram Rufous Fantail]

Orange-crowned Fairy-wren ssp. *Clytomyias insignis oorti* **Rothschild** & **Hartert**, 1907
van Oort's Black Lory *Chalcopsitta atra spectabilis* van Oort, 1908 NCR
[Status uncertain; possible hybrid *Chalcopsitta atra insignis* x *C. sintillata*]
van Oort's Fig Parrot *Psittaculirostris desmarestii intermedius* van Oort, 1909
[Alt. Golden-headed Fig Parrot ssp.]
van Oort's Palm Cockatoo *Probosciger aterrimus stenolophus* van Oort, 1911
Large Scrubwren ssp. *Sericornis nouhuysi oorti* Rothschild & Hartert, 1913

White-eared Catbird ssp. *Ailuroedus buccoides oorti* Rothschild & Hartert, 1913 NCR
[NUI *Ailuroedus buccoides buccoides*]

Dr Eduard Daniel van Oort (1876–1933) was a Dutch zoologist, ornithologist and botanist who collected in the East Indies (Indonesia). He was Curator of the bird collections at the Leiden; later becoming Director (1915–1933). He was the author of *Ornithologia Neerlandica, de Vogels van Nederland* (1922–1935). A reptile is named after him.

van Oort, P.

Oort's Barbet *Megalaima oorti* **S. Muller**, 1836
[Alt. Black-browed Barbet; Syn. *Psilopogon oorti*]

Pieter van Oort (1804–1834) was a Dutch naturalist and collector in the East Indies (Indonesia).

Van Rossem

Turquoise-browed Motmot ssp. *Eumomota superciliosa vanrossemi* **Griscom**, 1929
Van Rossem's Gull-billed Tern *Gelochelidon nilotica vanrossemi* **Bancroft**, 1929
Van Rossem's Wood Rail *Aramides cajanea vanrossemi* **Dickey**, 1929
[Alt. Grey-necked Wood Rail ssp.]
Russet-crowned Motmot ssp. *Momotus mexicanus vanrossemi* **R. T. Moore**, 1932
Van Rossem's Yellow Grosbeak *Pheucticus chrysopeplus dilutus* Van Rossem, 1934
[Alt. Mexican Yellow Grosbeak ssp.]
Green Kingfisher ssp. *Chloroceryle americana vanrossemi* Brodkorb, 1940 NCR
[JS *Chloroceryle americana septentrionalis*]
White-lored Gnatcatcher ssp. *Polioptila albiloris vanrossemi* Brodkorb, 1944
Nutting's Flycatcher ssp. *Myiarchus nuttingi vanrossemi* **A. R. Phillips**, 1961 NCR
[JS *Myiarchus nuttingi inquietus*]
Plain Wren ssp. *Cantorchilus modestus vanrossemi* A. R. Phillips, 1986

Adriaan Joseph Van Rossem (1892–1949) was an American ornithologist who published most prolifically in the 1920s–1930s. He collected in the southern USA, northern Mexico and El Salvador. He had a long association with Donald R. Dickey (q.v.) and together they wrote *The Birds of El Salvador* (1938). Dickey's large collections were moved to the University of California at Los Angeles (1940) and Van Rossem was given the position of Curator of the Dickey Collection. The University appointed him Lecturer in Zoology (1946). Among his later works was *A Distributional Survey of the Birds of Sonora, Mexico* (1945).

van Someren, N.

Kikuyu White-eye *Zosterops virens somereni* **Hartert**, 1928 NCR
[JS *Zosterops kikuyuensis*]

Noël van Someren (d.1921), like his brothers (below), was born in Midlothian, Scotland. He was commissioned as a second lieutenant (1917) but was soon (1919) collecting in Africa. His brother Victor described how Noël shot a Mountain Buzzard *Buteo oreophilus* 6,000ft up Mount Kenya (1919), and there are records of skins sent to AMNH from him (1920–1921) from German East Africa (Tanzania). Hartert wrote that the white-eye was '… Named in remembrance of Noël van Someren, who was killed some months later by a buffalo. He sent us twelve skins from Mt Kenya …'

van Someren, R. A. L.

Yellow-rumped Seedeater ssp. *Serinus atrogularis somereni* **Hartert**, 1912
[Syn. *Crithagra atrogularis somereni*]

Dr Robert Abraham Logan van Someren (1880–1955) was a naturalist who lived in Uganda for many years (from 1905); he was brother to Noël (q.v.) and Victor (q.v.). He co-wrote a paper (1926) with Victor about the life-cycle of certain butterflies. They started (1906) a survey of the birds of Kenya and Uganda, and their collection ultimately exceeded 25,000 items.

van Someren, V. G. L.

Streaky-breasted Flufftail *Sarothrura somereni* **Bannerman**, 1919 NCR
[JS *Sarothrura boehmi*]
van Someren's Canary *Serinus koliensis* **C. H. B. Grant** & **Mackworth-Praed**, 1952
[Alt. Papyrus Canary; Syn. *Crithagra koliensis*]

White-browed Scrub Robin ssp. *Erythropygia leucophrys vansomereni* **W. L. Sclater**, 1929 NCR
[NUI *Cercotrichas leucophrys zambesiana*]
Purple Grenadier ssp. *Uraeginthus ianthinogaster somereni* **Delacour**, 1943 NCR; NRM
Red-bellied Paradise-flycatcher ssp. *Terpsiphone rufiventer somereni* **Chapin**, 1948
Collared Sunbird ssp. *Anthodiaeta collaris somereni* Chapin, 1949
African Finfoot ssp. *Podica senegalensis somereni* Chapin, 1954
Klaas's Cuckoo ssp. *Chrysococcyx klaasi somereni* Chapin, 1954 NCR; NRM
[Note *klaasi* is an incorrect spelling of binomial *klaas*]

Dr Victor Gurnet Logan van Someren (1886–1976) qualified at Edinburgh University in both medicine and dentistry. He was appointed medical officer in British East Africa (Kenya) and spent 40 years practising there, during which time he studied its natural history. He and his brother Robert (q.v.) started a survey of the birds of Kenya and Uganda (1906) and their collection ultimately exceeded 25,000 specimens. He was honorary Curator of the Natural History Museum in Nairobi (1914–1938), and a Fellow of both the Linnean Society of London and the Royal Entomological Society. His collection of butterflies and other insects is in the BMNH, and his collection of African birds is in the Field Museum, Chicago. He wrote, among other works, *Days with Birds – Studies of Habits of some East African Species* (1956), continuing to write into the 1960s.

Van Son

Yellow-fronted Canary ssp. *Serinus mozambicus vansoni*
J. A. Roberts, 1932
[Syn. *Crithagra mozambica vansoni*]

Dr Georges Van Son (1898–1967) was an entomologist and botanist. He was born in Russia to a Dutch diplomat father and a Russian countess. They had only French as a common language, which became Georges's mother tongue. He was an Imperial Russian Navy cadet and visited China and Japan (1915). During the Russian revolution a Bolshevik sniper killed his father, and the family was imprisoned until 1921 when the Dutch Embassy obtained their release. They emigrated to Holland to join the father's family. Georges worked at National Natural History Museum, Leiden, from where he was recruited by Dr A. J. T. Janse to work with his private entomological collection in Pretoria, South Africa (1923). He was later employed at the Transvaal Museum (1925–1967), mainly working on butterflies. He was a pioneer in cultivating South African orchids and succulents. He was the expedition botanist and entomologist on the Vernay-Lang Kalahari expedition (1932). A reptile is named after him.

Van Tyne

Little Spiderhunter ssp. *Arachnothera longirostra vantynei*
Koelz, 1939 NCR
[JS *Arachnothera longirostra longirostra*]
Black-crested Bulbul ssp. *Pycnonotus flaviventris vantynei*
Deignan, 1948
Sirkeer Malkoha ssp. *Taccocua leschenaultii vantynei*
Koelz, 1954 NCR
[JS *Taccocua leschenaultii sirkee*]
Botteri's Sparrow ssp. *Peucaea botterii vantynei* Webster,
1959

Dr Josselyn Van Tyne (1902–1957) was an American ornithologist, collector and Curator of Birds at the Museum of Zoology, University of Michigan (1931–1957). Harvard awarded his bachelor's degree (1925) and University of Michigan his doctorate (1928). He was on a number of expeditions, including the Kelley-Roosevelt Expedition to Indo-China (1928–1929). Among his more than 90 papers is the co-written *The Birds of Brewster County, Texas* (1937). He was also editor of *Wilson Bulletin* (1939–1948).

Van Wyck

Island Imperial Pigeon ssp. *Ducula pistrinaria vanwyckii*
Cassin, 1862

Lieutenant William Van Wyck Reilly (1824–1854) was an American naval officer. He obtained the type specimen of this pigeon in New Ireland (Bismarck Archipelago). He drowned when USS *Porpoise* was lost with all hands during a typhoon in the Formosa Straits. His only son, a cavalry officer, was killed at the Battle of Little Bighorn (1876).

Varenne

Rufous-cheeked Laughingthrush ssp. *Garrulax castanotis*
varennei **Delacour**, 1926

Alexandre Claude Varenne (1870–1947) was a French socialist journalist and politician. He was Governor-General of Indochina (1925–1927).

Vargas

Oriental Dwarf Kingfisher ssp. *Ceyx erithaca vargasi*
Manuel, 1939 NCR
[NUI *Ceyx erithaca motleyi*]

Jorge B. Vargas (1890–1980) was a Filipino lawyer and politician. He was the Presiding Officer of the Japanese-sponsored Philippine Executive Commission (1942–1943) and later (WW2) served as Philippine ambassador to Japan. He was a leader in the Filipino Boy Scouts movement and was the first Filipino member of the International Olympic Committee.

Varona

Zapata Sparrow ssp. *Torreornis inexpectata varonai*
Regalado Ruiz, 1981

Dr Luis Sánchez Varona y Calvo (1923–1989) was a Cuban palaeontologist, geologist and zoologist. He worked at the Zoology Department, Cuban Academy of Sciences (1962–1979). He wrote *Mamiferos de Cuba* (1980). A mammal is named after him.

Vassal

White-cheeked Laughingthrush *Garrulax vassali* **Ogilvie-Grant**, 1906

Doctor Joseph Marguerite Jean Vassal (1867–1957) was a French Army Doctor and collector who was stationed in Vietnam (1907–1910). He worked closely with Yersin (q.v.) and together they wrote 'Une maladie rappelant le typhus exanthématique observée en Indochine' (1908). Sometime after WW1 he was appointed Director of Public Health for the French Congo and we know he was there until c.1925, as in that year his wife, Gabrielle M. Vassal (q.v. as Gabrielle), published a book entitled *Life in French Congo*.

Vassori

Vassori's Tanager *Tangara vassorii* **Boissonneau**, 1840
[Alt. Blue-and-black Tanager]

Boissonneau's original description says that he is naming the tanager after one of his best friends, but gives no further details about Vassori.

Vasvári

Common Crossbill ssp. *Loxia curvirostra vasvarii* Keve, 1943
NCR
[JS *Loxia curvirostra guillemardi*]

Dr Miklós Vasvári (1898–1945) was a Hungarian ornithologist who collected in Turkey (1936–1937). He worked in the ornithology department of the Hungarian National Museum, Budapest (1928), and was Assistant-in-Chief (1940).

Vaughan, J. H.

Pemba White-eye *Zosterops vaughani* **Bannerman**, 1924

Black-bellied Starling ssp. *Notopholia corrusca vaughani*
Bannerman, 1926

John Henry Vaughan (1892–1965) was an English jurist and ornithologist who wrote *Birds of Zanzibar and Pemba* (1930). He was later in Fiji (1945–1952).

Vaughan, R. E.

Pitcairn Reed Warbler *Acrocephalus vaughani* **Sharpe**, 1900

Commander Robert E. Vaughan (1874–1937) was an English naval officer who was in the Pacific, including Pitcairn Island where he collected specimens of the reed warbler. He was in China (1900–1906), where he appears to have commanded a river gunboat suitably named for an ornithologist – HMS *Moorhen*.

Vaughan-Jones

Copper Sunbird ssp. *Cinnyris cupreus vaughanjonesi*
   C. M. N. White, 1944 NCR
[JS *Cinnyris cupreus chalceus*]
Dark-capped Bulbul ssp. *Pycnonotus tricolor vaughanjonesi*
   C. M. N. White, 1944 NCR
[Syn. *Pycnonotus barbatus tricolor*]

Thomas George Clayton Vaughan-Jones (1907–1986) was a British colonial administrator in Africa (1929–1967). He was Director of Game and Tsetse Control, Northern Rhodesia (Zambia) (1945).

Vaurie

Flycatcher genus *Vauriella* **Wouters**, 1980
[Formerly in *Rhinomyias*]

Vaurie's Flycatcher *Ficedula crypta* Vaurie, 1951
[Alt. Cryptic Flycatcher]
Vaurie's Nightjar *Caprimulgus centralasicus* Vaurie, 1960

Abyssinian Wheatear ssp. *Oenanthe lugubris vauriei*
   Meinertzhagen, 1949
Cinereous Tit ssp. *Parus cinereus vauriei* **Ripley**, 1950
[Syn. *Parus major vauriei*]
Sand Lark ssp. *Calandrella raytal vauriei* **Koelz**, 1954 NCR
[NUI *Calandrella raytal raytal*]

Dr Charles J. Vaurie (1906–1975) was a French-born ornithologist and systematist who emigrated to the USA whilst still a schoolboy. He qualified as a dental surgeon at the University of Pennsylvania (1928) and practised dentistry in New York, where he was a volunteer (1942) at the AMNH. He gave up dentistry when he was appointed Assistant Curator, Department of Ornithology, American Museum of Natural History (1956) and Curator (1967–1972). He is best known for his *Birds of the Palearctic Fauna* (1959–1965) and *Tibet and Its Birds* (1972). He was also a keen stamp collector.

Vaux

Vaux's Swift *Chaetura vauxi* **J. K. Townsend**, 1839

William Sansom Vaux (1811–1882) was an American mineralogist and archaeologist who was a long-time member of the Academy of Natural Sciences, Philadelphia. He was also a member of the Zoological Society of Philadelphia. During his life Vaux acquired an extensive mineralogical collection, which he bequeathed to the Academy of Natural Sciences, together with his library and a handsome endowment for their preservation. His friend John Kirk Townsend (q.v.) named the swift.

Vavasour

Bellbird genus *Vavasouria* **Chubb**, 1920 NCR
[Now in *Procnias*]

Antwren sp. *Myrmotherula vavasouri* **Chubb**, 1918 NCR
[Alt. Long-winged Antwren; JS *Myrmotherula longipennis longipennis*]

Frederick Vavasour McConnell (see **Frederick (McConnell)** and **McConnell**).

Vega

Vega Gull *Larus vegae* Palmén, 1887
[Syn. *Larus argentatus vegae*]

Vega was the name of the vessel used by Arctic explorer Nils Adolf Erik Baron Nordenskjøld. He led an expedition (1878–1879), which was the first complete crossing of the Northeast Passage: also known as *Vega* Expedition.

Velez

Velez's Rusty-faced Parrot *Hapalopsittaca amazonina velezi*
   Graves & Uribe Restrepo, 1989

Jesús Hernán Vélez Estrada (b.1949) founded (1975) and was first Director of the Museum, University of Caldas, Colombia, until retiring (2004). He is an active zoologist, collecting the paratypes of this parrot, preparing the type series, and assisting the junior describer (Uribe Restrepo) in his research in the Rio Branco watershed. He co-wrote 'Mamíferos del Departamento de Caldas – Colombia' (2003). He has a number of butterflies named after him.

Velie

Western Wood Pewee ssp. *Contopus sordidulus veliei*
   Coues, 1866

Dr Jacob Wilber Velie (b.1829) was a physician who practised pharmacy and dentistry rather than medicine. He was mainly interested in ornithology and conchology. He explored in Colorado (1864) and Florida (1870) He was Acting Curator of the Chicago Academy of Sciences Museum (1873–1876), then Curator (1879–1893), and was Secretary of the Academy (1879–1896).

Velizhanin

Red-backed Shrike ssp. *Lanius collurio velizhanini* **Buturlin**, 1909 NCR
[Hybrid: *Lanius collurio* x *Lanius phoenicuroides*]

Dr Andrei Petrovich Velizhanin (1875–1937) was a Russian ornithologist and collector in Siberia who was head of the Altai Department of the Russian Geographic Society in Barnaul (1913–1930). He was also a physician at the local hospital for tubercular children. He was arrested (1933) and given a five-year suspended sentence of hard labour, but was later re-arrested as 'an enemy of the state' and executed by the NKVD. He wrote *Sketches of the Altai Territory* (1925).

Venilia

Woodpecker genus *Venilia* **Bonaparte**, 1850 NCR
[Now in *Blythipicus*]
Woodpecker genus *Veniliornis* Bonaparte, 1854

In Roman mythology, Venilia was a beautiful nymph and the wife of Faunus, a son of Picus, who was metamorphosed into a woodpecker: thus there is an indirect link between Venilia and woodpeckers!

Venning

Streaked Wren-babbler ssp. *Napothera brevicaudata venningi* **H. H. Harington**, 1913

Brigadier Francis Esmond Wingate Venning (1882–1970) was born in Ceylon (Sri Lanka) and educated at Sandhurst, then served in the Indian army (1902–1933). He was an ornithologist and oologist who collected in Burma (Myanmar), north-west India (Pakistan), and Iraq. In retirement he was active in the Botany Section, Hampshire Field Club and Archaeological Society. A reptile is named after him.

Venturi

Band-tailed Sierra Finch ssp. *Phrygilus alaudinus venturii* **Hartert**, 1909
Picazuro Pigeon ssp. *Columba picazuro venturiana* Hartert, 1909 NCR
[JS *Patagioenas picazuro picazuro*]
Least Seedsnipe ssp. *Thinocorus rumicivorus venturii* **Rothschild**, 1921 NCR
[JS *Thinocorus rumicivorus rumicivorus*]

Santiago Venturi (d.1930) was an Argentinian naturalist, oologist, botanist and plant-collector. He wrote the manuscript 'List of birds' eggs in the Venturi Argentine collection, 1897–1906' (1906). Lord Rothschild (q.v.) bought the collection and it was described (1909) by Hartert (q.v.) and Venturi in *Novitates Zoologicae*.

Veragua

Veragua's Parakeet *Aratinga pertinax ocularis* **P. L. Sclater & Salvin**, 1864
[Alt. Brown-throated Parakeet ssp.; Syn *Eupsittula pertinax ocularis*]

This should more properly be called the Veraguas Parakeet since it is named after the province of Veraguas in Panama, not after a person.

Verheyen

Blue-headed Coucal ssp. *Centropus monachus verheyeni* Louette, 1987 NCR
[JS *Centropus monachus fischeri*]

Professor Dr René Karel Verheyen (1907–1961) was a Belgian ornithologist. He studied biology at the University of Ghent and worked at the Royal Belgian Institute Natural History Museum. He specialised in the ornithology of the Belgian Congo and visited the country several times after 1948. He was a corresponding Fellow of the AOU.

Vernay

Large Woodshrike ssp. *Tephrodornis virgatus vernayi* **Kinnear**, 1924
Striated Swallow ssp. *Cecropis striolata vernayi* Kinnear, 1924
Rufous Treepie ssp. *Dendrocitta vagabunda vernayi* Kinnear & **Whistler**, 1930 NCR
[JS *Dendrocitta vagabunda pallida*]
Naked-faced Barbet ssp. *Gymnobucco calvus vernayi* **Boulton**, 1931
Pale-chinned Blue Flycatcher ssp. *Cyornis poliogenys vernayi* Whistler, 1931
White-spotted Fantail ssp. *Rhipidura albogularis vernayi* Whistler, 1931
Coqui Francolin ssp. *Francolinus coqui vernayi* **J. A. Roberts**, 1932 NCR
[JS *Peliperdix coqui coqui*]
Black-throated Munia ssp. *Lonchura kelaarti vernayi* Whistler & Kinnear, 1933
Silver-eared Mesia ssp. *Leiothrix argentauris vernayi* **Mayr & Greenway**, 1938
Oriental Skylark ssp. *Alauda gulgula vernayi* Mayr, 1941

Arthur Stannard Vernay (1877–1960) was an English antiques dealer and philanthropist who lived in the USA. He had a deep interest in natural history, and was a Trustee of the AMNH. Jointly with Colonel John Faunthorpe he financed six expeditions to Burma, India and Thailand (1922–1928) and financed a BMNH collecting trip to Tunisia (1925). He was a friend of Theodore and Kermit Roosevelt (q.v.) and, especially, of an American millionaire named Suydam Cutting (q.v., with whom he travelled quite often; notably they journeyed to Lhasa (1935) and met the 13th Dalai Lama. Having sold his business and all his collections and antiques, he retired to the Bahamas (1940). A reptile and three mammals are named after him.

Verox

Mouse-coloured Sunbird *Cyanomitra veroxii* **A. Smith**, 1831

'Verox' is used here as a variant of Verreaux, and refers to J. B. E. Verreaux (see below).

Verreaux

African Piculet genus *Verreauxia* **Hartlaub**, 1856 NCR (JBE & JP)
[Now in *Sasia*]

Verreaux's Eagle Owl *Bubo lacteus* **Temminck**, 1820
[Alt. Milky Eagle Owl]
Verreaux's Eagle *Aquila verreauxii* **Lesson**, 1831 (JBE)
[Alt. Black Eagle]
Verreaux's Twinspot *Hypargos margaritatus* **Strickland**, 1844
[Alt. Pink-throated Twinspot]
Verreaux's Batis *Batis minima* J. & E. Verreaux, 1855
[Alt. Gabon Batis]
White-tipped Dove *Leptotila verreauxi* **Bonaparte**, 1855
Verreaux's Coua *Coua verreauxi* **Grandidier**, 1867
Verreaux's Monal Partridge *Tetraophasis obscurus* J. Verreaux, 1869
[Alt. Chestnut-throated Monal Partridge]

Streaked Fantail *Rhipidura verreauxi* **E. A. Marié**, 1870

Verreaux's Bush Warbler *Cettia acanthizoides* J. Verreaux, 1870

[Alt. Yellowish-bellied Bush Warbler]

Verreaux's Fulvetta *Fulvetta ruficapilla* J. Verreaux 1870

[Alt. Spectacled Fulvetta]

Chestnut-headed Crake *Ortygometra verreauxi* **G. R. Gray**, 1871 NCR

[JS *Anurolimnas castaneiceps*]

Sharpe's Rosefinch *Carpodacus verreauxii* **David & Oustalet**, 1877

Golden Parrotbill *Suthora verreauxi* **Sharpe**, 1883

[Syn. *Paradoxornis verreauxi*]

Verreaux's Song Thrush *Turdus mupinensis* **Laubmann**, 1920

[Alt. Chinese Thrush]

African Cuckoo Hawk ssp. *Aviceda cuculoides verreauxii* **Lafresnaye**, 1846

Blue-eared Kingfisher ssp. *Alcedo meninting verreauxii* De la Berge, 1851

Festive Coquette ssp. *Lophornis chalybeus verreauxii* **Bourcier**, 1853 (JBE)

Verreaux's Yellow-billed Turaco *Tauraco macrorhynchus verreauxii* **Schlegel**, 1854

Scaly-breasted Woodpecker ssp. *Celeus grammicus verreauxii* **Malherbe**, 1858 (JBE & JP)

Kurrichane Thrush ssp. *Turdus libonyana verrauxii* **Bocage**, 1869

Amethyst Starling ssp. *Cinnyricinclus leucogaster verreauxi* Bocage, 1870

Verreaux's Crested Guineafowl *Guttera pucherani verreauxi* **D. G. Elliot**, 1870

Red-tailed Greenbul ssp. *Criniger calurus verreauxi* Sharpe, 1871

Jean Baptiste Edouard Verreaux (JBE) (1810–1868) and Jules Pierre Verreaux (JP) (1807–1873) were brothers, naturalists, dealers in natural history specimens, collectors and taxidermists. The Verreaux family traded in Paris from a huge emporium for stuffed birds and feathers which they called 'Maison Verreaux'. They were clearly ambitious taxidermists and gained notoriety for once attending the funeral of a tribal chief whose body they then disinterred, took to Cape Town and stuffed! The Catalan veterinarian Francisco Darder, then Curator of the Barcelona Zoo, purchased the 'specimen' from one of the brothers' sons, Edouard Verreaux (1888). This controversial exhibit was on show in Barcelona until the end of the 20th century, when the man's descendants demanded that it should be returned for a decent burial. JBE was in South Africa to help JP pack and ship a huge collection that they had sold (1830 and 1832), and then took charge of the family business in Paris (1834). JP accompanied his uncle, Pierre Antoine Delalande (q.v.), to Cape Colony (1818–1822) and after studying in Paris was again in South Africa (1825–1832). Here he assisted Andrew Smith (q.v.) in founding the South African National Museum at Cape Town and was Curator (1829–1832). He travelled in Indochina, Indonesia and the Philippines (1832–1837), after which he worked in the family business and as an ornithologist and plant collector for the MNHN, Paris, which sent him to Australia (1842). He made collections of plants around Hobart, Tasmania (1844)

and around Sydney, New South Wales (1844–1846). His servant, Émile, collected plants for him in New South Wales and Queensland. JP returned to France (c.1851) with a natural history collection of 115,000 items! He fled from Paris at the start of the Franco-Prussian War (1870) and went to London, where Sharpe (q.v.) enthusiastically welcomed him. He died in England. (See also **Julius**)

The birds in the list are all named after Jules Pierre Verreaux (JP) except those marked with the initials of JBE. In two cases it appears that both brothers were intended. Between them they have three mammals, two reptiles and an amphibian named after them.

Verrill, A. H. & G. E.

Red-legged Thrush *Mimocichla verrillorum* **J. A. Allen**, 1891 NCR

[JS *Turdus plumbeus albiventris*]

Alpheus Hyatt Verrill (1871–1954), who was known as Hyatt, and his brother George Elliott Verrill (1866–1946), were the sons of Professor Addison Emery Verrill, who became the first Professor of Zoology at Yale (1864–1907). The two brothers collected bird specimens on the island of Dominica (1890). Hyatt went on to become natural history editor of *Webster's International Dictionary* (1896), and also wrote science fiction tales for the pulp magazine *Amazing Stories*.

Versteeg

Versteeg's Thrush *Turdus poliocephalus versteegi* **Junge**, 1939

[Alt. Island Thrush ssp.]

Dr Gerard Martinus Versteeg (1876–1943) qualified as a physician (1905) and went to the East Indies as army surgeon. He joined two expeditions to Dutch New Guinea (1907–1913). He was in charge of disease control in part of northern Java (1919–1923) and in the Health Department (1928–1931). He retired to Holland and was a medical administrator, Central Bureau for Statistics, The Hague (1931–1943). A reptile is named after him.

Verster

Verster's Berrypecker *Melanocharis versteri* **Finsch**, 1876

[Alt. Fan-tailed Berrypecker]

Verster's Senegal Parrot *Poicephalus senegalus versteri* Finsch, 1863

Florentius Abraham Verster heer van Wulverhorst (1826–1923), known as 'Frits', was a Dutch professional zoologist. He was Inspector of Game and Fishery at his father's office (1852–1857) and became (1860–1920) administrator of the National Museum of Natural History, Leiden, for half the salary of his predecessor. He stayed in post to the age of 93 when he reluctantly, against the wishes of the director, had to step down. Verster was also (1857–1901) Steward of the 'hoogheemraadschap' (a kind of district water board, an important body in watery Holland) of Rijnland. He held various social positions in the local community, including member of the city council of Leiden. His son Floris Hendrik Verster (1861–1926) became a very well known 'dark' impressionist painter.

Vesey-Fitzgerald

Sabota Lark ssp. *Calendulauda sabota veseyfitzgeraldi*
C. M. N. White, 1956 NCR
[JS *Calendulauda sabota waibeli*]

Leslie Desmond Edward Foster Vesey-Fitzgerald (1909–1974) was an Irish zoologist, herpetologist, and environmentalist. He was in Trinidad with Hampton Wildman Parker (1933–1934) collecting for the BMNH. He was in the Seychelles (1938) and led an expedition to the Comoro Islands (1940). He undertook several wildlife surveys in the Abu Dhabi desert and collected many insects, amphibians and reptiles. He was involved in locust control in Northern Rhodesia (Zambia) (1949–1964). The Tanganyika (Tanzania) Government employed him (1965) as National Parks Officer. His colleagues called him Vesey, and local Africans termed him 'Bwana Mungosi' (Mr Skins, referring to the boots that he always wore). He wrote *East African Grasslands* (1973). A reptile is named after him.

Victoire

Black-tailed Trainbearer *Lesbia victoriae* **Bourcier** & **Mulsant**, 1846

Victoire Mulsant (DNF) was the wife of one of the describers, Martial Etienne Mulsant.

Victor (Vittorio)

'Dodo' genus *Victoriornis* **Hachisuka**, 1937 NCR

Vittorio Emanuele III (1869–1947) was King of Italy (1900–1946). He abdicated after a referendum showed the Italians wanted a republic. Hachisuka based his species of 'Réunion White Dodo' *Victoriornis imperialis* on vague accounts which probably related to the extinct ibis *Threskiornis solitarius* rather than a dodo.

Victoria (Mount)

Victoria Nuthatch *Sitta victoriae* **Rippon**, 1904
[Alt. White-browed Nuthatch]
Treecreeper sp. *Certhia victoriae* Rippon, 1906 NCR
[Alt. Hume's Treecreeper; JS *Certhia manipurensis manipurensis*]

Green-tailed Sunbird ssp. *Aethopyga nipalensis victoriae* Rippon, 1904
Brown Bullfinch ssp. *Pyrrhula nipalensis victoriae* Rippon, 1906
Brown-capped Laughingthrush ssp. *Trochalopteron austeni victoriae* Rippon, 1906
Chestnut-tailed Minla ssp. *Siva strigula victoriae* **Meinertzhagen**, 1926 NCR
[JS *Minla strigula yunnanensis*]
Puff-throated Babbler ssp. *Pellorneum ruficeps victoriae* **Deignan**, 1947

These birds are named after Mount Victoria, also known as Nat Ma Taung, part of the Chin Hills range in Burma (Myanmar).

Victoria (Mountains)

Warbling Vireo ssp. *Vireo gilvus victoriae* Sibley, 1940

Named after the Victoria Mountains of Baja California, Mexico.

Victoria, Princess

Silktail *Lamprolia victoriae* **Finsch**, 1874

Victoria Adelaide Mary Louise of Prussia (1840–1901) was the eldest daughter of Queen Victoria of Great Britain. She was the Princess Royal of the United Kingdom and married Prince Frederick William of Prussia (1858).

Victoria, Queen

Queen Victoria's Egret *Erodius victoriae* **MacGillivray**, 1842 NCR
[Alt. Great Egret; JS *Ardea alba*]
Victoria Crowned Pigeon *Goura victoria* **L. Fraser**, 1844
[Alt. White-tipped Goura]
Victoria's Riflebird *Ptiloris victoriae* **Gould**, 1850

Queen Victoria Lyrebird *Menura novaehollandiae victoriae* Gould, 1865
[Alt. Superb Lyrebird ssp.]

Queen Victoria (1819–1901) was Queen of Great Britain and Ireland (1837) and later also Empress of India (1877). Whilst many royal personages were honoured in the names of birds, few perhaps deserved the accolade. However, Victoria played a role in bird conservation as she was opposed to the feather trade and ordered that her regiments should stop wearing plumes as part of their uniform. Two mammals are named after her. (See **Queen Victoria**)

Victorin, J. F.

Victorin's Warbler *Cryptillas victorini* **Sundevall**, 1860
[Alt. Victorin's Scrub Warbler]

Johan Frederik Victorin (1831–1855) was a Swedish traveller who visited South Africa's Cape Colony (1853–1855), where he died of tuberculosis. He wrote *Resa I Laplandet Åren 1853–1855* (Journey to the Cape Land in the Years 1853–1855) and *Jakt och Naturbilder* (Hunting and Nature Scenes) (1863). Johan Victorin sometimes feasted on the species he had come to study, some of which, including the Narina Trogon, he noted were both 'beautiful and good to eat'.

Victorin (Masséna)

Blue-winged Mountain Tanager ssp. *Anisognathus somptuosus victorini* **Lafresnaye**, 1842

Victor Masséna (1836–1910) was the son of François Victor Masséna Prince d'Essling and Duc de Rivoli (see **Masséna** & **Rivoli**). According to Lafresnaye's (q.v.) wording, at 6 the boy already had, like his father, a taste for ornithology. He succeeded to the titles of Prince d'Essling and Duc de Rivoli on his brother's death (1898). He served as a Deputy during the Second Empire.

Vidal, G. W.

Collared Kingfisher ssp. *Todiramphus chloris vidali* **Sharpe**, 1892

Jungle Bush Quail ssp. *Perdicula asiatica vidali* **Whistler** & **Kinnear**, 1936

George William Vidal (1845–1907) was a British civil servant who served in India (1867–1897). After being called to the Bar (1867) at Lincoln's Inn, he started his career in Bombay (Mumbai) as an assistant magistrate and forest settlement officer, and later for collection of salt tax. He was a member of the mixed Anglo-Portuguese commission at Goa (1880 and 1885). He became a member of the legislative council (1894) and Chief Secretary of the Political Department (1896). He was an early exponent of the game of badminton and was Honorary Secretary and Honorary Treasurer of the Badminton Association (1899–1906) and Vice-President when he died.

Vidal, I.

Little Owl ssp. *Athene noctua vidalii* **A. E. Brehm**, 1857

Professor I. Vidal (d.1859) was a Spanish zoologist. He was Professor of Zoology at Valencia University.

Vidgen

Cuckoo genus *Vidgenia* **Mathews**, 1918 NCR
[Now in *Cacomantis*]

Herbert Graham Vidgen (d.1948) was an Australian pearl-fisher and collector in the Pacific. He was owner and master of the pearling schooner *Olive* (1899). He wrote 'Birds visiting Cape York Peninsula and New Guinea' (1921).

Vieillot

Vieillot's Barbet *Lybius vieilloti* **Leach**, 1815
Vieillot's Storm-petrel *Fregetta grallaria* Vieillot, 1817
[Alt. White-bellied Storm-petrel]
Vieillot's Starling *Streptocitta albicollis* Vieillot, 1818
[Alt. White-necked Myna]
Vieillot's Black Weaver *Ploceus nigerrimus* Vieillot, 1819
Australasian Figbird *Sphecotheres vieilloti* **Vigors** & **Horsfield**, 1827
Puerto Rican Lizard Cuckoo *Coccyzus vieilloti* **Bonaparte**, 1850
[Syn. *Saurothera vieilloti*]

Vervain Hummingbird ssp. *Mellisuga minima vieilloti* Shaw, 1812
Vieillot's Crested Fireback *Lophura ignita rufa* **Raffles**, 1822
Chestnut Flycatcher ssp. *Pyrrhomyias cinnamomeus vieillotioides* **Lafresnaye**, 1848
Turquoise Tanager ssp. *Tangara mexicana vieilloti* **P. L. Sclater**, 1857
Grey Currawong ssp. *Strepera versicolor vieilloti* **Mathews**, 1912 NCR
[JS *Strepera versicolor versicolor*]

Louis Jean Pierre Vieillot (1748–1831) was a French ornithologist and businessman based in Haiti. He and his family eventually fled to the United States during the French Revolution but he later returned, only to die in poverty. His fellow Frenchmen Buffon (q.v.) and Cuvier (q.v.) have largely overshadowed him, but his contributions to ornithology are very significant and Lesson (q.v.) described him as a genius. In Vieillot's time a number of ornithologists mistook juveniles, females or moulting individuals of known birds for new species. Vieillot was the first to study these plumage changes, which was in itself a major development in accurate classification. He was also an early proponent of studying living birds and not just the specimens and skins in museum collections. He described a great many species. He wrote *Oiseaux Dorés ou à Reflets Metalliques* (1800), *Histoire Naturelle des plus Beaux Oiseaux Chanteurs de la Zone Torride* (1805), and *La Galerie des Oiseaux* (1820).

Vieira, D.

Gorgeous Bush-shrike ssp. *Telophorus viridis vieirae* **C. M. N. White**, 1946 NCR; NRM

D. D. de M. Vieira (fl.1946) was a Portuguese Benedictine missionary and collector at Cazombo, Angola.

Vieira, L.

Coal Tit ssp. *Periparus ater vieirae* **F. Nicholson**, 1906

Dr Lopez Vieira (fl.1906) was a Portuguese zoologist. He was Professor of Zoology at the University of Coimbra (1887).

Vielliard

Plain-tailed Nighthawk *Nyctiprogne vielliardi* Lencioni-Neto, 1994

Pygmy Nightjar ssp. *Caprimulgus hirundinaceus vielliardi* Ribon, 1995

Dr Jacques Marie Edme Vielliard (1944–2010) was a French ornithologist, zoologist and ecologist. The University of Paris awarded his bachelor's degree (1967), master's (1968) and doctorate (1971). Interested in birds from an early age, he made his first expedition aged 18 to Cota Donaña, Spain. He continued to travel widely, from Romania to the Himalayas (1967–1971). He went on to study birds in Africa before being invited by the Brazilian Academy of Science to help develop ecological study there (1976). After four years he took a post at Universidade Estadual de Campinas, Brazil, where he created the Laboratory of Bioacoustics and was a professor. His particular interest was in the migratory ecology of aquatic birds. He died following kidney surgery. An amphibian is named after him.

Vigors

Vasa Parrot genus *Vigorsia* **Swainson**, 1837 NCR
[Now in *Coracopsis*]

Vigors's Bustard *Eupodotis vigorsii* **A. Smith**, 1831
[Alt. Karoo Bustard, Karoo Korhaan]
Vigors's Crested Tit *Parus melanolophus* Vigors, 1831
[Alt. Black-crested Tit, Spot-winged Tit; Syn. *Periparus ater melanolophus*]
Vigors's Warbler *Sylvia vigorsii* **Audubon**, 1831 NCR
[Alt. Pine Warbler; JS *Setophaga pinus*]

Vigors's Grosbeak *Pheucticus chrysopeplus* Vigors, 1832
[Alt. Yellow Grosbeak]
Vigors's Sunbird *Aethopyga vigorsii* **Sykes**, 1832

Vigors's Pitta *Pitta elegans vigorsii* **Gould**, 1838
[Alt. Elegant Pitta ssp.]
Vigors's Wren *Thryomanes bewickii spilurus* Vigors, 1839
[Alt. Bewick's Wren ssp.]
Greyish Saltator ssp. *Saltator coerulescens vigorsii*
    G. R. Gray, 1844
Hooded Robin ssp. *Petroica cucullata vigorsi* **Mathews**,
    1912 NCR
[JS *Melanodryas cucullata cucullata*]

Nicholas Aylward Vigors (1785–1840) was an Irish zoologist and politician. He served in the Peninsular War (1809–1811), then returned to Oxford and took his MA (1817). He was elected a Fellow of the Royal Society (1826), and was the first Secretary of the Zoological Society of London (1826–1833). He was a Member of Parliament at Westminster (1828–1835 &1837–1840). He contributed a chapter on ornithology in *Zoology of Captain Beechey's Voyage* (1839). He named several species, including some collected by Collie (q.v.) in California. Audubon (q.v.) was his friend and named Vigors's Warbler in his honour, but the name disappeared when the type specimen proved to be an immature Pine Warbler.

Viguier

Viguier's Dacnis *Dacnis viguieri* **Oustalet**, 1883
[Alt. Viridian Dacnis]

Dr Camille Viguier (1850–1930) was a French zoologist who was both a Doctor of Medicine and a Doctor of Science. He studied at, and graduated from, the University at Nancy. His early career was in Algiers, where he started a maritime laboratory and became Professor of Science (1888). He collected in Panama (1881). He was a corresponding member of the French Academy of Sciences for the section dealing with Anatomy and Zoology (1920–1930).

Villa

Song Sparrow ssp. *Melospiza melodia villai* **A. R. Phillips** &
    Dickerman, 1957
Ocellated Thrasher ssp. *Toxostoma ocellatum villai*
    A. R. Phillips, 1986

Dr Bernardo Villa-Ramirez (1911–2006) obtained a doctorate in biology at the National Autonomous University in Mexico City. He was head of the Department of Zoology at the Institute of Biology, University in Mexico. He was also a leading member of the Association of Mexican Mammalogists, an honorary member of the Society for Marine Mammalogy, and an honorary trustee of the USNM. He wrote a number of papers. Two mammals are named after him.

Villada

Green-fronted Hummingbird ssp. *Amazilia viridifrons villadai*
    A. T. Peterson & Navarro-Siguenza, 2000

Manuel Maria Villada Peimbert (1841–1924) was a Mexican naturalist at Toluca, where he was Director of the institute and museum that is now named after him.

Villarejo

Mishana Tyrannulet *Zimmerius villarejoi* Alvarez &
    B. M. Whitney, 2001

Padre Avencio Jesús Carnero Villarejo (1910–2000) was an Augustinian priest who was born, educated and ordained (1934) in Spain. He spent his life in Peru, becoming a citizen (1959). He worked in Loreto department and explored widely in the Peruvian Amazon. He wrote *Asi es la Selva* (1943).

Villas Boas

Golden-crowned Manakin *Lepidothrix vilasboasi* Sick, 1959

This manakin is named after three brothers whose surname was Villas Boas: Claudio (1916–1998), Leonardo (1918–1961), and Orlando (1914–2002). They were Brazilian explorers, anthropologists and authors who wrote *Xingu; the Indians, their Myths* (1975). The brothers were among the few non-missionaries to live permanently with indigenous tribes, treating them as equals and friends. They persuaded tribes to end their internecine feuds and unite to confront encroaching settlement.

Villaviscensio

Villaviscensio Sabrewing *Campylopterus villaviscensio*
    Bourcier, 1851
[Alt. Napo Sabrewing]

Manuel Villaviscensio (1804–1871) was an Ecuadorian geographer who collected in his native country (1850–1865). He produced a street plan of the capital, Quito (1858). That same year he wrote the first description of a beverage widely known as *ayahuasca* and related its use in sorcery and divination among the peoples of the Upper Río Napo. He described his own self-intoxication when using it, an experience that made him feel that he was 'flying' to most marvellous places. He reported that 'natives using this drink were able to foresee and answer accurately in difficult cases, be it to reply opportunely to ambassadors from other tribes in a question of war; to decipher plans of the enemy through the medium of this magic drink and take proper steps for attack and defence; to ascertain, when a relative is sick, what sorcerer has put on the hex; to carry out a friendly visit to other tribes; to welcome foreign travellers or, at least to make sure of the love of their womenfolk'. He reported 'tar' on the surface of the Río Hollin (1888): this turned out to be a natural surface-seepage of crude oil, which is now commercially produced in Ecuador.

Vincent

Vincent's Bunting *Emberiza vincenti* **P. R. Lowe**, 1932
[Alt. Cape Bunting; Syn. *Emberiza capensis vincenti*]
Seedcracker sp. *Pirenestes vincenti* **C. W. Benson**, 1955
    NCR
[Alt. Lesser Seedcracker; JS *Pyrenestes minor*]

Fawn-coloured Lark ssp. *Calendulauda africanoides*
    vincenti **J. A. Roberts**, 1938
Yellow-streaked Greenbul ssp. *Phyllastrephus flavostriatus*
    vincenti **C. H. B. Grant** & **Mackworth-Praed**, 1940

Olive Sunbird ssp. *Cyanomitra olivacea vincenti* C. H. B. Grant & Mackworth-Praed, 1943
Whistling Cisticola ssp. *Cisticola lateralis vincenti* **Chapin,** 1953 NCR
[JS *Cisticola lateralis modestus*]

Colonel Jack Vincent (1904–1999) was a British bird collector and the first Director of the Natal Parks Board. He first visited South Africa aged 21 and worked on farms before returning to the UK (1920s), becoming a collector for the BMNH. He then accompanied Admiral Hubert Lynes (q.v.) on several ornithological expeditions to East, Central and Southern Africa (1920s-1930s). His most fruitful trip was to Mozambique (1932), where he discovered several new birds, including the Namuli Apalis *Apalis lynesi* named after the expedition leader. Vincent married in Cape Town (1934) and later bought a farm (1937) in Natal. He served in WW2 and then Palestine, after which he became Director of the Natal Park Service (1949). He also edited *Ostrich*. He took part in a number of conservation projects under the aegis of ICBP (forerunner of BirdLife International) including preparation of the first bird Red Data Book (1963–1967). He continued working for the Parks Board until retirement (1974). A mammal is named after him.

Violani

Rosella subgenus *Violania* Wells & Wellington, 1992
[Not widely accepted. NUI *Platycercus*]

Dr Carlo Violani (b.1946) is an Italian zoologist and ornithologist at the Department of Animal Biology, University of Pavia, where he is Professor at the Museum of Natural History.

Viosca

Viosca's Pigeon *Patagioenas fasciata vioscae* **Brewster,** 1888
[Alt. Band-tailed Pigeon ssp.]

James Viosca (d.1895) was appointed first as US Vice Consul (1877), then as US Consul at La Paz, Baja California, Mexico (1882–1895). The pigeon was named in Viosca's honour at the request of Marston Abbott Frazar (q.v.), who collected the type specimen, because Viosca 'has been most kind and helpful in furthering the success of Mr. Frazar's explorations.'

Virginia (Anderson)

Virginia's Warbler *Leiothlypis virginiae* **Baird,** 1860
[Syn. *Oreothlypis virginiae*]

Mary Virginia Childs Anderson (1833–1912) was the wife of American army surgeon Dr William Wallace Anderson, who discovered the species (c.1858).

Virginia (Correia)

Green-backed Twinspot ssp. *Mandingoa nitidula virginiae* **Amadon,** 1953
[SII *Mandingoa nitidula schlegeli*]

Mrs Virginia Correia (c.1888–at least 1928) was the wife of José Correia (q.v.). She was in the Azores (1921–1928) with her husband, who was a native of those islands.

Vis

Bird-of-Paradise genus *Visendavis* **Iredale,** 1948 NCR
[Applied to a hybrid form of *Paradisaea*]

Tawny-breasted Honeyeater ssp. *Xanthotis flaviventer visi* **Hartert,** 1896

(See **De Vis**)

Vitry

Red-tailed Laughingthrush ssp. *Trochalopteron milnei vitryi* **Delacour,** 1932

Paul Vitry (d.1944) was a French colonial administrator in Indochina (Vietnam) in the early 20th century (retired 1933). He wrote *Etude sur le Régime Financier de l'Empire d'Annam* (1905).

Vlangali

Common Pheasant ssp. *Phasianus colchicus vlangalii* **Przewalski,** 1876

Major General Aleksandr Georgiyevich Vlangali (1823–1908) was a traveller and diplomat who was the Russian Envoy to Peking (Beijing) c.1870. He was in Paris (1877) and was Russian Ambassador in Rome (1894). He wrote *Reise nach der Östlichen Kirgisen-Steppe* (1894). A reptile is named after him.

Vlasova

Snow Bunting ssp. *Plectrophenax nivalis vlasowae* **Portenko,** 1937

V. F. Vlasova (DNF) lived and collected on Wrangel Island in the Arctic Ocean for five years during the 1930s, and is mentioned by Portenko (q.v.) in his *Birds of the Chukchi Peninsula and Wrangel Island* (1972).

Voelcker

Yellow-throated Woodland Warbler ssp. *Phylloscopus ruficapilla voelckeri* **J. A. Roberts,** 1941

John Voelcker (b.1898) was a South African businessman and ornithologist. He was President, South African Ornithological Society (1935–1950).

Voeltzkow

Malagasy White-eye ssp. *Zosterops maderaspatanus voeltzkowi* **Reichenow,** 1905
Malagasy Green Sunbird ssp. *Cinnyris notatus voeltzkowi* Reichenow, 1905
Malagasy Paradise-flycatcher ssp. *Terpsiphone mutata voeltzkowiana* **Stresemann,** 1924
African Stonechat ssp. *Saxicola torquatus voeltzkowi* **Grote,** 1926

Professor Dr Alfred Voeltzkow (1860–1947) was a German palaeontologist, collector, traveller and zoologist who spent many years in East Africa. He travelled in Zanzibar and Pemba (1889–1895), was in the Comoro Islands (1906), and at some stage visited the island of Aldabra. His East African collection is in the Natural History Museum in Vienna. A mammal is named after him.

Vogt

Common House Martin ssp. *Hirundo urbica vogti*
O. Kleinschmidt, 1943 NCR
[JS *Delichon urbicum urbicum*]

Vogt (DNF) was a collector for Kleinschmidt and that is about all the etymology that there is. He collected the holotype (1916) in the Odenwald mountain range in Germany. We have been unable to discover anything more about him.

Volkmann

Emerald-spotted Wood Dove ssp. *Turtur chalcospilos volkmanni* **Reichenow**, 1902 NCR; NRM

Captain Richard Volkmann (b.1870) was a German army officer (1889–1906). He was in German South West Africa (Namibia) (1894–1906), where he became Director of the Lüderitz Bay Company (1906).

Volz

Grey-capped Woodpecker ssp. *Dendrocopos canicapillus volzi* **Stresemann**, 1920

Dr Walter Volz (1875–1907) was a Swiss zoologist, traveller and ichthyologist. He was in Indonesia and Thailand (1901–1902), visiting Japan and Hawaii en route to Bern. He published on the fishes of Sumatra (1903). He travelled in Sierra Leone and Liberia (1906–1907) and was killed ('collateral damage') by French troops conducting a punitive raid. An amphibian is named after him.

von Bloeker

Band-backed Wren ssp. *Campylorhynchus zonatus vonbloekeri* J. S. Rowley, 1968 NCR
[JS *Campylorhynchus zonatus zonatus*]

Jack Christian von Bloeker Jr (1909–1991) was a naturalist, ornithologist and collector. He was a member of the Cooper Ornithological Society (1927–1991). He wrote *The Mammals of Monterey County, California* (1938).

von der Decken

von der Decken's Hornbill *Tockus deckeni* **Cabanis**, 1869

Baron Karl Claus von der Decken (1833–1865) was a German explorer who died in Somalia. He explored in East Africa and was the first European to try to climb Mount Kilimanjaro. Decken explored the region of Lake Nyasa on his first expedition (1860). He ascended Kilimanjaro (1862) to 13,780 feet, seeing its permanent snowcap, and also establishing its height as about 20,000 feet. Another expedition (1863) took him to Madagascar, the Comoro Islands, and the Mascarene Islands. Then, in Somalia (1865), he sailed the Jubba River, where his ship *Welf* foundered in the rapids above Bardera, where Somalis killed him and three other Europeans. He sent a considerable quantity of specimens from Somalia to the Museum in Hamburg. His letters were edited and published (1869) in book form under the title *Reisen in Ost-Afrika*. The giant lobelia *Lobelia deckenii* and a mammal are named after him.

von Jordans

Lesser Spotted Woodpecker ssp. *Dendrocopos minor jordansi* **Götz**, 1925 NCR
[JS *Dendrocopos minor hortorum*]
Eurasian Stone-curlew *Burhinus oedicnemus jordansi* **Neumann**, 1932 NCR
[NUI *Burhinus oedicnemus oedicnemus*]
Common Whitethroat ssp. *Sylvia communis jordansi* **Clancey**, 1950 NCR
[JS *Sylvia communis communis*]
Common Raven ssp. *Corvus corax jordansi* **Niethammer**, 1953 NCR
[JS *Corvus corax canariensis*]
Crested Lark ssp. *Galerida cristata jordansi* Niethammer, 1955
Eurasian Jay ssp. *Garrulus glandarius jordansi* Keve, 1966 NCR
[NUI *Garrulus glandarius albipectus*]

Professor Dr Adolph von Jordans (1892–1974) was a German ornithologist and collector who was Director of Museum Alexander Koenig, Bonn. He made a particular study of the birds of the Balearic Islands.

Vönöczky

Pine Grosbeak ssp. *Pinicola enucleator vonoczkyi* Keve, 1943 NCR
[JS *Pinicola enucleator pacata*]

Dr Jákab Vönöczky Schenk (1876–1945) was a Hungarian ornithologist who was Director of the Hungarian Institute of Ornithology (1935).

von Plessen

(see **Plessen**)

Voous

Abbott's Babbler ssp. *Malacocincla abbotti voousi* Wynne, 1955 NCR
[JS *Malacocincla abbotti concreta*]
European Greenfinch ssp. *Carduelis chloris voousi* **Roselaar**, 1993

Professor Karel Hendrik Voous (1920–2002) was a Dutch ornithologist and systematist. He was Secretary-General (1966–1970) and Honorary President (1990–1994) of the International Ornithological Committee. One of his best known publications is his *List of Recent Holarctic Bird Species* (1977), published by the BOU.

Vo Quy

Vo Quy's Pheasant *Lophura edwardsi hatinhensis* Vo Quy, 1975 NCR
[Alt. Vietnamese Pheasant; now treated as variant *Lophura edwardsi*]

Professor Vo Quy (b.1929) is a Vietnamese ornithologist. During the Vietnamese uprising against French colonial rule (1951) he walked to China, where he and his colleagues set up a temporary university. Vo Quy, amongst others, founded the University of Hanoi (1956) and established the Department of Zoology, which he later headed. For years he was Dean of the Faculty of Biology. He has a PhD in ornithology from the University of Moscow (1966). Vo Quy has inspired and trained several generations of conservationists. He led the first team of environmental scientists south of the 17th parallel to investigate the environmental damage caused during the Vietnam War. He founded the country's first conservation and management training institution (1985), the Centre for Natural Resources and Environmental Studies (CRES), which he ran on a shoestring and undertook activities which the government could not afford as it recovered from the ravages of war. Vo Quy was awarded the World Wide Fund for Nature's highest honour, the Gold Medal (1996), in recognition of his pioneering environmental conservation and education in Indochina. He was the architect of an agreement between Laos, Cambodia and Vietnam for international cooperation in protecting rare and endangered migratory species, and in establishing trans-frontier reserves, or 'Peace Parks', on their shared borders. (See also **Quy**)

Vorderman

Rufous-chested Flycatcher *Siphia vordermani* **Sharpe**, 1890 NCR
[JS *Ficedula dumetoria dumetoria*]

Yellow-bellied Warbler ssp. *Abroscopus superciliaris vordermani* **Büttikofer**, 1893
Bar-bellied Cuckooshrike ssp. *Coracina striata vordermani* **Hartert**, 1901
Grey-cheeked Green Pigeon ssp. *Treron griseicauda vordermani* **Finsch**, 1901

Dr Adolphe Guillaume Vorderman (1844–1902) was a Dutch naval surgeon in the Dutch East Indies (1866). He worked (1890–1902) as a government physician of the Dutch health inspectorate in Indonesia, principally as an inspector of prisons. Clearly enlightened, he was concerned about disease among prisoners, and his observations helped link beriberi to vitamin deficiency. Vorderman found that in those prisons using mostly brown rice less than 1 prisoner in 10,000 had developed beriberi, while in those using mainly white rice the proportion was 1 in 39. He was also a keen naturalist and mounted several brief expeditions, notably to Belitung, an island west of Borneo (1891). Vorderman was the great-grandfather of the British television personality Carol Vorderman. Two mammals are named after him.

Vorhies

Brown-throated Wren ssp. *Troglodytes brunneicollis vorhiesi* H. Brandt, 1945 NCR
[Alt. House Wren ssp.; JS *Troglodytes aedon cahooni*]
Abert's Towhee ssp. *Pipilo aberti vorhiesi* **A. R. Phillips**, 1962 NCR
[JS *Melozone aberti dumeticola*]

Dr Charles Taylor Vorhies (1879–1949) was an American zoologist, ornithologist and entomologist. Iowa Wesleyan College awarded his bachelor's degree (1902) and the University of Wisconsin his doctorate in zoology (1908). He was Professor of Zoology and Botany, University of Utah (1908–1915), and Professor of Entomology and Economic Zoology, University of Arizona (1915–1949).

Vose

Honeyeater genus *Vosea* **Gilliard**, 1960

Vose's Melidectes *Vosea whitemanensis* Gilliard, 1960
[Alt. Bismarck Honeyeater, Gilliard's Honeyeater; Syn. *Melidectes whitemanensis*]

Charles Redfield Vose (1890–1957) was an American explorer, businessman and sponsor of expeditions.

Vosmaer

Vosmaer's Eclectus Parrot *Eclectus roratus vosmaeri* **Rothschild**, 1922

Arnout Vosmaer (1720–1799) was the Curator of the Menagerie and the Museum of the Stadtholder (Dutch head of State) in The Hague. He published extensively in Dutch and French, and among the birds he described was a 'Purpur-roode Loeri' he thought came from Ceylon but was an Eclectus Parrot from the Moluccas. Two reptiles are named after him.

Vosseler

Usambara Eagle Owl *Bubo vosseleri* Reichenow, 1908
[Alt. Nduk Eagle Owl]

Professor Julius Vosseler (1861–1933) was a German zoologist. He was in German East Africa (Tanzania) (1903–1908). He was Director of the Hamburg Zoo (1910). A reptile is named after him.

Voznesensky

White-backed Woodpecker ssp. *Dendrocopos leucotos voznesenskii* **Buturlin**, 1907 NCR
[NUI *Dendrocopos leucotos leucotos*]

(See **Woznesensky**)

Vuilleumier

Vuilleumier's Flowerpiercer *Diglossa brunneiventris vuilleumieri* **G. R. Graves**, 1980
[Alt. Black-throated Flowerpiercer ssp.]

Dr François Vuilleumier (b.1938) was awarded his PhD in biology by Harvard University (1967) under Professor Ernst Mayr (q.v.). He was born in Switzerland, but for much of his working life has been associated with the AMNH. A former Chairman of the Department of Ornithology at the museum, he is now Curator Emeritus. He has taught at the University of Paris, the University of Lausanne, Switzerland, the University of the Andes in Merida, Venezuela, and the College of the Atlantic in Bar Harbor, USA. He carried out research on the evolution and systematics of Andean birds (1964–1986). During that period he wrote a monograph on the genus *Diglossa*, as well as numerous other papers. Gary Graves (q.v.) named the flowerpiercer subspecies with the words 'I take pleasure in naming this new form for François Vuille-umier in recognition of his contributions to Andean evolutionary biology.'

Vulsin

African Darter ssp. *Anhinga rufa vulsini* **Bangs**, 1918

(See **Wulsin**)

Vylder

African Green Pigeon ssp. *Treron calvus vylderi* **Gyldenstolpe**, 1924

Gustaf de Vylder (1827–1908) was a Swedish entomologist, trader and explorer in Cape Colony and South West Africa (Namibia) (1871–1875 and 1879–1887).

W

Wache

Bearded Barbet *Apatelornis wachei* **Reichenow**, 1926 NCR
[JS *Lybius dubius*]

Ernst Wache (DNF) was the German traveller, agent and collector for Hagenbeck (q.v.) in Abyssinia (Ethiopia) (1909–1910) and West Africa (1925).

Waddell, E. G. R.

Rufous Whistler ssp. *Lewinornis rufiventris waddelli* **Mathews**, 1920 NCR
[JS *Pachycephala rufiventris rufiventris*]

Miss Elizabeth Gaston Ralston Peddie-Waddell (1865–1929) was a Scottish aviculturist.

Waddell, L. A.

Giant Babax *Babax waddelli* Dresser, 1905

Lieutenant-Colonel Professor Dr Laurence Austine Waddell (1854–1938) was a noted British physician, orientalist, archaeologist, collector and explorer in Tibet. He accompanied Younghusband's expedition to Lhasa (1903–1904) as a cultural expert, as he had been stationed for years in Darjeeling and had repeatedly, in disguise, crossed the border and visited the forbidden land of Tibet to study its language, customs and culture. He was Professor of Chemistry and Pathology at the Calcutta Medical College (1881–1886). Among many articles, monographs and books he wrote *The Birds of Sikkim* (1893) and *Lhasa and its Mysteries* (1905). He left his collection of c.700 books, notes and photographs etc. to Glasgow University.

Waga

Waga's Mouse-coloured Tyrannulet *Phaeomyias murina wagae* **Taczanowski**, 1884

Anton Waga (1799–1890) was a Polish teacher at the Gymnasium of Warsaw who accompanied the Branicki brothers (q.v.) as naturalist on their hunting expeditions to Egypt, Palestine and Nubia (northern Sudan), and south-western Europe. He convinced them to turn these trips into scientific expeditions. He was primarily an entomologist who discovered and described many new species of insects.

Wagler

Blue-headed Hummingbird *Ornismya wagleri* **Lesson**, 1829 NCR
[Wagler's Woodnymph *Thalurania wagleri*; JS *Cyanophaia bicolor*]

Wagler's Toucanet *Aulacorhynchus wagleri* **J. H. Sturm** & J. W. Sturm, 1841
[Syn. *Aulacorhynchus prasinus wagleri*]

Wagler's Oropendola *Psarocolius wagleri* **G. R. Gray**, 1845
[Alt. Chestnut-headed Oropendola]

Wagler's Conure *Aratinga wagleri* G. R. Gray, 1845
[Alt. Scarlet-fronted Parakeet; Syn. *Psittacara wagleri*]

Wagler's Oriole *Icterus wagleri* **P. L. Sclater**, 1857
[Alt. Black-vented Oriole]

Gentoo Penguin *Pygosceles wagleri* P. L. Sclater, 1860 NCR
[Invalid replacement name for *Pygoscelis papua papua*]

Wagler's Chachalaca *Ortalis wagleri* G. R. Gray, 1867
[Alt. Rufous-bellied Chachalaca]

Wagler's Woodpecker *Melanerpes wagleri* **Salvin** & **Godman**, 1895 NCR
[Alt. Red-crowned Woodpecker; JS *Melanerpes rubricapillus rubricapillus*]

Wagler's Macaw *Ara glaucogularis* **Dabbene**, 1921
[Alt. Blue-throated Macaw]

Wagler's Woodcreeper *Lepidocolaptes squamatus wagleri* **Spix**, 1824
[Alt. Scaled Woodcreeper ssp.]

Wagler's Pearly Conure *Pyrrhura lepida lepida* Wagler, 1832
[Alt. Pearly Parakeet ssp.]

Johann Georg Wagler (1800–1832) was a German herpetologist. He was Spix's (q.v.) assistant, upon whose death (1826) he became the Director of the Zoological Museum of the University of Munich, and worked there to continue the treatment of the extensive collections from Brazil. In addition he worked on systematics of amphibians and reptiles there. He wrote the highly regarded *Monographia Psittacorum*, which includes various descriptions of the blue macaws (1832). Wagler died accidentally from a self-inflicted gunshot wound whilst out collecting. Two amphibians and eight reptiles are named after him.

Wagner, A. J.

Eurasian Blue Tit ssp. *Parus caeruleus wagneri* Floericke, 1921 NCR
[JS *Cyanistes caeruleus ogliastrae*]

Black-and-chestnut Warbling Finch ssp. *Poospiza whitii wagneri* **Stolzmann**, 1926

Woodlark ssp. *Lullula arborea wagneri* Floericke, 1926 NCR
[JS *Lullula arborea arborea*]
Lesser Spotted Woodpecker ssp. *Dendrocopos minor
    wagneri* **Domaniewski**, 1927 NCR
[JS *Dendrocopos minor buturlini*]

Dr Anton J. Wagner (1860–1928) was Director of the Polish Museum of Zoology, Warsaw.

Wagner, H. O.

Cinnamon-sided Hummingbird *Amazilia wagneri*
    A. R. Phillips, 1966
[Syn. *Amazilia viridifrons wagneri*]

Dr Helmuth Otto Wagner (1897–1977) was a German zoologist and ornithologist. He studied (1925–1929) at Frankfurt and Göttingen, which awarded his doctorate, after which he became an assistant at Cologne Zoo and at the zoological institute of Georg-August University Göttingen. He was an animal dealer in Australia, Indonesia and Central America (1933–1940) and worked as a scientist and collector in Mexico (1940–1950). He was Director, Ubersee Museum, Bremen, Germany (1951–1962). He retired to Mexico City. He was an expert on Mexican hummingbirds and Phillips (q.v.) refers to him as a source in a number of his papers.

Wagstaffe

Rufous-winged Fulvetta ssp. *Alcippe castaneceps
    wagstaffei* Wynne, 1954 NCR
[JS *Alcippe castaneceps castaneceps*]

Reginald Wagstaffe (b.1907) was a British zoologist and taxidermist who worked at various English museums. He was Keeper of Vertebrate Zoology at Liverpool Museum, and wrote *Type Specimens of Birds in the Merseyside County Museums* (1978).

Wahlberg

Wahlberg's Eagle *Hieraaetus wahlbergi* **Sundevall**, 1850
[Syn. *Aquila wahlbergi*]
Wahlberg's Honeyguide *Prodotiscus regulus* Sundevall,
    1850
[Alt. Brown-backed Honeyguide, Sharp-billed Honeyguide]
Wahlberg's Cormorant *Phalacrocorax neglectus* Wahlberg,
    1855
[Alt. Bank Cormorant]

Black Coucal ssp. *Centropus grillii wahlbergi* **C. H. B. Grant**,
    1915 NCR; NRM

Johan August Wahlberg (1810–1856) was a Swedish naturalist and collector. He studied chemistry and pharmacy at Uppsala (1829) and worked in a chemist's shop in Stockholm while studying at the Forestry Institute. He travelled and collected widely in southern Africa (1838–1856), sending thousands of specimens home to Sweden. He returned briefly to Sweden (1853) but was soon back in Africa where he was in Walvis Bay (1854). He was exploring the headwaters of the Limpopo when a wounded elephant killed him. An amphibian, mammal and four reptiles are named after him.

Wahnes

Wahnes's Parotia *Parotia wahnesi* **Rothschild**, 1906
[Alt. Wahnes's Six-wired Bird-of-Paradise]

Fairy Gerygone ssp. *Gerygone palpebrosa wahnesi*
    A. B. Meyer, 1899
Wahnes's Lorikeet *Charmosyna papou wahnesi* Rothschild,
    1906
[Alt. Papuan Lorikeet ssp.]

Carl Wahnes (1835–1910) was a German naturalist who collected in New Guinea.

Waibel

Eurasian Hoopoe ssp. *Upupa epops waibeli* **Reichenow**,
    1913
Sabota Lark ssp. *Calendulauda sabota waibeli* **Grote**, 1922

Prof. Leo Heinrich Waibel (1888–1951) was a German geographer and zoologist. He took part in expeditions to the Cameroons (1911–1912) and South West Africa (1914–1919). He was Professor of Geography at the University of Bonn (1929–1937), but his career in Germany ended due to his open dislike of the Nazis. He moved to the USA (1939), returning to Germany shortly before his death.

Wait

Banded Bay Cuckoo ssp. *Cacomantis sonnerati waiti*
    E. C. S. Baker, 1919

Walter Ernest Wait (1878–1961) was a British civil administrator, ornithologist and oologist in Ceylon (Sri Lanka). He was Deputy Director of Customs, Colombo (1916), and Director of the Museum in Colombo (1925). He wrote *Manual of the Birds of Ceylon* (1925).

Waite, E. R.

Australian Treecreeper sp. *Climacteris waitei* **S. A. White**,
    1917 NCR
[Alt. Brown Treecreeper; JS *Climacteris picumnus
    picumnus*]

Edgar Ravenswood Waite (1866–1928) was an English-born Australian zoologist and ichthyologist. After studying at Manchester University he worked at the Leeds Museum (1888–1892). He went to work at the Australian Museum, Sydney (1892), where he was Curator of Ichthyology (1893–1905). He was in New Zealand (1906–1914) as Curator of the Canterbury Museum, Christchurch, but then General Director of the South Australian Public Library, Museum, and Art Gallery (1914–1928). He wrote 'Notes on Australian Typhlopidae' (1894). He had malaria, contracted in New Guinea, which compromised his health, and he died of enteric fever while attending a meeting in Hobart, Tasmania. A reptile is also named after him.

Waite, H. W.

Paddyfield Pipit ssp. *Anthus rufulus waitei* **Whistler**, 1936
    NCR
[JS *Anthus rufulus rufulus*]

Major Herbert William Waite (1887–1967) was an ornithologist and a British officer in the Indian Police in the Punjab (1907–1942), and Principal of the Pakistan Police Training School (1949–1954).

Wakefield

Wakefield's Green Pigeon *Treron calvus wakefieldii* **Sharpe**, 1874
[Alt. African Green Pigeon ssp]

Rev. Thomas Wakefield (1836–1901) was an English Methodist missionary to Kenya (1861–1888), an explorer, cartographer and botanical collector for Kew. Known to the Africans as 'Bwana Wakfili', he made four expeditions to previously unexplored areas of Kenya. The data he produced (1870) on native caravan routes from the interior to the coast led to the first comprehensive maps of East Africa. He was elected a Fellow of the Royal Geographical Society (1889).

Walden

Swallow genus *Waldenia* **Sharpe**, 1869 NCR
[Now in *Hirundo*]

Mayotte Drongo *Dicrurus waldenii* **Schlegel**, 1865
Walden's Barwing *Actinodura waldeni* **Godwin-Austen**, 1874
[Alt. Austen's Barwing, Streak-throated Barwing]
Walden's Hornbill *Aceros waldeni* Sharpe, 1877
[Alt. Rufous-headed Hornbill, Visayan Wrinkled Hornbill]

(See **Tweeddale**)

Waldron, F.

Forest Penduline-tit ssp. *Anthoscopus flavifrons waldroni*
Bannerman, 1935
[Trinomial often given as *waldronae*]

Miss Fanny Waldron (DNF) was an employee of the BMNH in London. She accompanied W. P. Lowe (q.v.) on his expedition to the Gold Coast (Ghana) in 1934–1935 when she was well over 60. She also has a subspecies of monkey named after her. Bannerman spelt the trinomial *waldroni*, and this is often corrected to the feminine *waldronae*.

Waldron (de Witt Miller)

Hummingbird genus *Waldronia* **Chapman**, 1929 NCR
[Now in *Polytmus*]

Ruddy-capped Nightingale Thrush ssp. *Catharus frantzii waldroni* **A. R. Phillips**, 1969

(See **Miller, W. de W.**)

Wallace, A. R.

Wallace's Fruit Dove *Ptilinopus wallacii* **G. R. Gray**, 1858
[Alt. Golden-shouldered Fruit Dove]
Wallace's Owlet-nightjar *Aegotheles wallacii* G. R. Gray, 1859
[Alt. White-spotted Owlet-nightjar]
Wallace's Standardwing *Semioptera wallacii* G. R. Gray, 1859
[Alt. Standardwing Bird-of-Paradise]

Wallace's Rail *Habroptila wallacii* G. R. Gray, 1860
[Alt. Invisible Rail]
Wallace's Scrubfowl *Eulipoa wallacei* G. R. Gray 1861
[Alt. Moluccan Scrubfowl/Megapode, Painted Megapode]
Wallace's Fairy-wren *Sipodotus wallacii* G. R. Gray, 1862
Wallace's Hanging Parrot *Loriculus flosculus* Wallace, 1864
[Alt. Flores Hanging Parrot]
Wallace's Scops Owl *Otus silvicola* Wallace, 1864
Wallace's Lory *Eos wallacei* **Finsch**, 1865 NCR
[Alt. Violet-necked Lory; JS *Eos squamata squamata*]
Wallace's Hawk Eagle *Nisaetus nanus* Wallace, 1868
Wallace's White-eye *Zosterops wallacei* Finsch, 1901
[Alt. Yellow-ringed White-eye]

Grey-rumped Treeswift ssp. *Hemiprocne longipennis wallacii* **Gould**, 1859
Wallace's Fig Parrot *Psittaculirostris desmarestii blythii* Wallace, 1864
[Alt. Large Fig Parrot ssp.]
Thrush-like Schiffornis ssp. *Schiffornis turdina wallacii* **P. L. Sclater** & **Salvin**, 1867
Variable Kingfisher ssp. *Ceyx lepidus wallacii* **Sharpe**, 1868
Australasian Goshawk ssp. *Accipiter fasciatus wallacii* Sharpe, 1874
Ashy Woodpecker ssp. *Mulleripicus fulvus wallacei* **Tweeddale**, 1877
Stephan's Emerald Dove ssp. *Chalcophaps stephani wallacei* Bruggemann, 1877
Bar-breasted Piculet ssp. *Picumnus aurifrons wallacii* **Hargitt**, 1889
Five-coloured Munia ssp. *Lonchura quinticolor wallacii* Sharpe, 1890 NCR; NRM
Plumbeous Pigeon ssp. *Patagioenas plumbea wallacei* **Chubb**, 1917
[JS *Patagioenas plumbea pallescens*]
Ochre-bellied Flycatcher ssp. *Mionectes oleagineus wallacei* Chubb, 1919
Greater Racket-tailed Drongo ssp. *Dicrurus paradiseus wallacei* **Hachisuka**, 1926 NCR
[JS *Dicrurus paradiseus formosus*]
White Bellbird ssp. *Procnias albus wallacei* Oren & **Novaes**, 1985

Alfred Russel Wallace (1823–1913) was an English naturalist, evolutionary scientist, geographer and anthropologist, regarded as the father of zoogeography. He was also a social critic and theorist, a follower of the utopian socialist Robert Owen. His interest in natural history began whilst working as an apprentice surveyor, at which time he also attended public lectures. He went to Brazil (1848) on a self-sustaining natural history collecting expedition. Even on this first expedition he was very interested in how geography limited or facilitated the extension of species' ranges. He not only collected but also mapped, using his surveying skills. His return to England (1852) was a near disaster: his ship, the brig *Helen*, caught fire and sank with all his specimens lost, and he was lucky to be rescued by a passing vessel. He spent the next two years writing *Palm Trees of the Amazon and Their Uses* and *A Narrative of Travels on the Amazon and Rio Negro*, and organising another collecting expedition to the Indonesian archipelago. He managed to get a grant to cover his passage to Singapore (1862) and had the benefit of letters of introduction prepared for him by representatives of the British and Dutch governments. He spent nearly eight years there, during which he undertook about 70 different expeditions involving a total of some 14,000 miles of travel. He visited every important island in the archipelago at least once, some many times. He collected a remarkable 125,660 specimens, including c.1,000 new taxa. He wrote *The Malay Archipelago* (1869), perhaps the most celebrated of all writings on Indonesia and ranking as one of the 19th-century's best scientific travel books. He also wrote *Contributions to the History of Natural Selection* (1870), and *Island Life* (1880). His essay 'On the law which has regulated the introduction of new species', which encapsulated his most profound theories on evolution, was sent to Darwin (q.v.). He later sent Darwin his essay 'On the tendency of varieties to depart indefinitely from the original type', presenting the theory of 'survival of the fittest'. Darwin and Lyell presented this essay, together with Darwin's own work, to the Linnean Society. Wallace's thinking spurred Darwin to encapsulate these ideas in *The Origin of Species*; the rest is history. Wallace developed the theory of natural selection, based on the differential survival of variable individuals, half-way through his stay in Indonesia. He remained for four more years, continuing his systematic exploration of the region's fauna, flora and people. For the rest of his life he was known as the greatest living authority on the region and its zoogeography, including his discovery and description of the faunal discontinuity that now bears his name – Wallace's Line. This natural boundary runs between the islands of Bali and Lombok in the south and Borneo and Sulawesi in the north, and separates the Oriental and Australasian faunal regions. Two mammals, one amphibian and a reptile are also named after him.

Wallace, G. J.

Red-bellied Woodpecker ssp. *Melanerpes carolinus harpaceus* **W. Koelz**, 1954 NRM

Professor George J. Wallace (1907–1986) earned his bachelor's degree, master's, and PhD from Michigan State University. He worked at Vermont Fish and Game Service, before becoming Director of Pleasant Valley Bird and Wildlife Sanctuary in Massachusetts. He returned to Michigan State University (1942–1972) where he was very active in the fight against use of DDT. He published numerous articles and textbooks on ornithology, several monographs, and other books including the autobiographical *My World of Birds* (1979). The word *harpaceus* is a compound of pieces of three person's names: J. W. Hardy (q.v.), E. C. Pielou (q.v.) and Wallace's.

Wallace, R. B.

Scarlet-banded Barbet *Capito wallacei* **O'Neill** et al., 2000

Robert B. Wallace (1918–2002) was a businessman and philanthropist. His father had been a member of Franklin D. Roosevelt's administration. He founded (1995) the Wallace Global Fund which is concerned, *inter alia*, with population stabilisation and environmental protection. Iowa State University awarded his bachelor's degree in biology (1940). He was a veteran of WW2. The original etymology reads: 'We

name this species in honor of Robert B. Wallace of Washington, D.C. in recognition of his intense interest in, and support of, ornithological exploration by the Louisiana State University Museum of Natural Science in Peru. His understanding of the need to study areas that are biologically unknown, before they are forever changed by development, is greatly appreciated.'

Waller, E.

Waller's Owl *Tyto longimembris walleri* **Diggles**, 1866
[Alt. Eastern Grass Owl ssp.; SII *Tyto longimembris longimembris*]

Eli Waller (d.1875) of Brisbane was an Australian dealer, collector, painter and taxidermist who regularly sent specimens to and corresponded with John Gould (q.v.). He had a large collection that he made freely available to the describer, Silvester Diggles (q.v.).

Waller, G.

Waller's Starling *Onychognathus walleri* **G. E. Shelley**, 1880
[Alt. Waller's Red-winged Starling]

Gerald Waller (DNF) was an English amateur naturalist who collected in East Africa whilst involved in diplomatic negotiations with the Sultan of Zanzibar (in hopes that the Sultan would cede to Britain his land on the African mainland). He is known to have taken rubber plants from Kew to Sir John Kirk (q.v.) in Zanzibar, and promised to collect Zanzibarian flora for Kew. Waller seems to have been employed as an agent of the Florida Land & Colonization Company, but details of his later life have proved elusive. A mammal is named after him.

Wallich

Wallich's Pheasant *Catreus wallichii* **Hardwicke**, 1827
[Alt. Cheer Pheasant]

Dr Nathaniel Wallich (1786–1854) was a Danish physician and botanist. He graduated from the Royal Academy of Surgeons in Copenhagen (1806), and the same year was appointed surgeon to the Danish settlement at Serampore, in Frederischnagor (Bengal) (1807). By the time he arrived, the British had annexed the area and they interned him. He was later paroled into the service of the East India Company. He was instrumental in establishing a museum in Calcutta (1814) with many of his own botanical specimens. He supervised the gardens belonging to the East India Company in Calcutta (1815–1846). He prepared a catalogue of c.20,000 specimens, in addition to publishing *Tentamen Flora Nepalensis Illustratae* (1824) and *Plantae Asiaticae Rariories* (1830). He also collected in Nepal (1820–1822). He retired to London (1847). A mammal and several plants are named after him, including *Ulmus wallichiana*, the Himalayan Elm.

Walter

Brandt's Mountain Finch ssp. *Leucosticte brandti walteri* **Hartert**, 1904

(See **Rothschild, L. W.**)

Walton

Pink-rumped Rosefinch *Carpodacus waltoni* **Sharpe**, 1905
[Formerly *Carpodacus pulcherrimus waltoni*]

Lieutenant-Colonel Herbert James Walton (1869–1938) was a physician who qualified in London (1893), then joined the Indian Medical Service (1896). He was part of the force that relieved Peking (Beijing) in the 1900 Boxer Rebellion, and part of the Lhasa Expedition (1903–1904). He was co-author of *The Opening of Tibet; an Account of Lhasa and the Country and People of Central Tibet and of the Progress of the Mission sent there by the English Government in the Year 1903–4* (1905). During this expedition he collected 500 bird skins, including the holotype of the rosefinch. He retired from the army after WW1.

Wangyel

Grey Junglefowl ssp. *Gallus sonneratii wangyeli* **Koelz**, 1954
NCR; NRM

Wangyel was a Tibetan friend of Koelz (q.v.) and one of his companions during his travels in India, Afghanistan and Iran.

Ward, C.

Ward's Shrike-flycatcher *Pseudobias wardi* **Sharpe**, 1870
[Alt. Ward's Flycatcher]

Christopher Ward (DNF) was an English naturalist whose chief interest was entomology. He was also a collector and sponsored Alfred Crossley's (q.v.) expedition to Madagascar. The type specimen of the shrike-flycatcher was acquired by Crossley. Ward wrote *African Lepidoptera, Being Descriptions of New Species* published in three parts (1873–1875), which contained descriptions of at least 55 new species.

Ward, C. W.

Ward's Heron *Ardea herodias wardi* **Ridgway**, 1882
[Alt. Great Blue Heron ssp.]

Charles Willis Ward (1856–1920) was a conservationist, businessman and amateur naturalist from Pontiac, Michigan. He grew carnations on a commercial scale on Long Island, New York. He was one of the businessmen who put up the funds to buy 54,000 acres (220 km²) of coastal marshland in Louisiana to create what is today a state wildlife refuge. Towards the end of the 19th century, he explored the Everglades on behalf of the Smithsonian. It was presumably then that he spent several weeks observing herons at their breeding grounds in Florida.

Ward, F. Kingdon

Ward's Trogon *Harpactes wardi* **Kinnear** 1927

Hoary-throated Barwing ssp. *Actinodura nipalensis wardi* Kinnear, 1932 NCR
[JS *Actinodura waldeni saturatior*]

(See **Kingdon-Ward**)

Ward, S.

Seychelles Parakeet *Psittacula wardi* **E. Newton**, 1867
EXTINCT

Swinburne Ward (1830–1897) was the British Civil Commissioner in the Seychelles (1862–1868). Later he stayed in Panama: a report of the voyage around the world of the Italian corvette *Vettor Pisani* (1882) mentions a remark made by Swinburne Ward on the length ('50 feet or more') of a gigantic shark, caught by the *Vettor Pisani* in the Gulf of Panama.

Ward, S. N.

Pied Thrush *Zoothera wardii* **Blyth**, 1842

Samuel Neville Ward (1813–1897) was a British colonial administrator with the Madras Civil Service of the Honourable East India Company (1832–1863).

Wardell

Shining Flycatcher ssp. *Myiagra alecto wardelli* **Mathews**, 1911

Mathews (q.v.) wrote that this bird was 'Named in honour of Dr Wardell, of Stotfold' without further explanation.

Wardlaw

Eurasian Bullfinch ssp. *Pyrrhula pyrrhula wardlawi* **Clancey**, 1947 NCR
[JS *Pyrrhula pyrrhula pileata*]

Wilfred James Plowden-Wardlaw (b.1905) was a Scottish lawyer, philanthropist and naturalist who collected the holotype in Scotland (1947). He travelled much of the world, the possessor of a large expense account courtesy of a Scottish brewery company. He lived in British Columbia and spent time in the West Indies, New South Wales, New Zealand and South Africa. He adopted the name 'The Campbell of Craigie' (1960s). He wrote *The Birds of Barbados and the Birds of St. Lucia* (1954).

Ware

Common Woodshrike ssp. *Tephrodornis pondiceriana warei* **Koelz**, 1939 NCR
[JS *Tephrodornis pondicerianus pondicerianus*]

S. J. Ware (DNF) was a friend of the describer, who gave no further details.

Warner

Vaux's Swift ssp. *Chaetura vauxi warneri* **A. R. Phillips**, 1966 NCR
[JS *Chaetura vauxi richmondi*]
Sedge Wren ssp. *Cistothorus platensis warneri* **Dickerman**, 1975
Grey-breasted Martin ssp. *Progne chalybea warneri* A. R. Phillips, 1986
Emerald Toucanet ssp. *Aulacorhynchus prasinus warneri* Winker, 2000

Dr Dwain Willard Warner (1917–2005) was an American naturalist. He graduated from Carleton College, Minnesota (1939), but was only awarded his doctorate by Cornell University (1947) after service in WW2 in the Pacific with the US Army (1942–1945). He wrote his dissertation on the avifauna of New Caledonia and the Loyalty Islands. He was Curator of Ornithology at the Museum of Natural History, University of Minnesota (1947–1987). In retirement he took up leading safaris in Kenya.

Warr

Basalt Wheatear *Oenanthe lugens warriae* Shirihai & Kirwan, 2011
[Alt. Mourning Wheatear ssp.]

Mrs Frances E. ('Effie') Warr was Librarian at BMNH (in Tring) where she still volunteers, and is a long-time stalwart of the Ornithological Society of the Middle East. She wrote *Manuscripts and Drawings in the Ornithology and Rothschild Libraries of The Natural History Museum at Tring* (1996).

Warszewicz

Hummingbird genus *Warszewiczia* **Boucard**, 1895 NCR
[Now in *Heliangelus*]

Scrub Blackbird *Dives warczewiczi* **Cabanis**, 1861
[Binomial sometimes 'corrected' to *warszewiczi*]

Warszewiscz's Rainbow *Coeligena iris aurora* **Gould**, 1854
[Alt. Rainbow Starfrontlet ssp.; *Coeligena warszewizii* is a JS]
Montane Woodcreeper ssp. *Lepidocolaptes lacrymiger warscewiczi* Cabanis & **Heine**, 1859
Steely-vented Hummingbird ssp. *Amazilia saucerrottei warscewiczi* Cabanis & Heine, 1860

Joseph Ritter von Rawicz Warszewicz (1812–1866) was a Lithuanian-Polish botanist. He was an assistant at the Berlin Botanical Garden (1840–1844), before moving to Guatemala and establishing himself as a plant collector. In 1850 he returned to Europe after an attack of yellow fever, but was soon back collecting in Ecuador and Peru (1851–1852). He is regarded as one of the greatest orchid collectors of all time. An amphibian is named after him.

Washington

Washington's Sea Eagle *Haliaeetus leucocephalus washingtoniensis* **Audubon**, 1827 NCR
[Alt. Bald Eagle ssp.]

George Washington (1732–1799) was the first President of the United States of America. Audubon (q.v.) described this bird from Kentucky, and at times used other variations on the binomial – *washingtoni* and *washingtoniana* – but the (incorrect) form *washingtoniensis* was used first, and must take precedence.

Waters

Waters's Flufftail *Sarothrura watersi* **Bartlett**, 1880
[Alt. Slender-billed Flufftail]

Thomas Waters (1840–1904) was in the British Consular Service in East Africa and collected specimens in central

and south-eastern Madagascar. Bartlett (q.v.) presented three papers to the Zoological Society of London on mammals and birds collected by Waters (1875–1879).

Waterstradt

Waterstradt's Racket-tailed Parrot *Prioniturus waterstradti* **Rothschild**, 1904
[Alt. Mindanao Racket-tail]

Brown Bullfinch ssp. *Pyrrhula nipalensis waterstradti* **Hartert**, 1902
Hill Prinia ssp. *Prinia superciliaris waterstradti* Hartert, 1902
Island Leaf Warbler ssp. *Phylloscopus poliocephalus waterstradti* Hartert, 1903

Johannes Waterstradt (1869–1944) was a Danish botanist, ornithologist and entomologist who conducted extensive expeditions in Ceylon (Sri Lanka), Borneo, the Malay Peninsula, the Moluccas and Philippines (1891–1912). In 1913 he returned to Denmark and took up horticulture, particularly of orchids. Several insects are also named after him. See also **John** (Waterstradt)

Waterton

Waterton's Woodnymph *Thalurania watertonii* **Bourcier**, 1847
[Alt. Long-tailed Woodnymph]

Charles Waterton (1782–1865) was a highly idiosyncratic English naturalist and collector. He has been called 'the type specimen of the British eccentric naturalist' and was known as 'the Squire of Walton Hall'. He was one of the first to convert land to the sole purpose of a wildlife sanctuary. 'He hated scientific names, John James Audubon (whom he called a charlatan), Protestants, Hanoverians, Hanoverian Protestants, rats and Charles Darwin; he loved the natural world, birds, taxidermy, climbing, and practical jokes'. He travelled and collected in South America, and wrote *Wanderings* (1825), which included an account of riding a caiman and trying deliberately to get bitten by a vampire bat (without success). He thought bloodletting was the best way to treat all ills. His most long-lasting practical jokes were his many taxidermy mounts of imaginary (fake) animals, which he enjoyed foisting on the public. He was extraordinary to the end; a Dr Hobson wrote 'When Mr. Waterton was seventy-seven years of age, I was witness to his scratching the back part of his head with the big toe of his right foot'. There is a nice ironic footnote to his hatred of Darwin (q.v.). Waterton kept a slave whom he taught how to help him collect and prepare skins – John Edmonstone – who was freed when he came to Edinburgh (all slaves who got to Scotland at that time were declared freemen). Edmonston[e] set up in Edinburgh as a taxidermist and taught Darwin how to prepare and preserve skins and the two men became friends. Waterton's greatest friend was Charles Edmonstone, a plantation owner in British Guiana (Guyana) who also tracked down runaway slaves – and one can presume, by virtue of the common practice of naming a slave after his master, that John was one of 'his' slaves. At the age of 48 Waterton married Charles's daughter by an Arawak princess. Waterton was present at the christening and apparently fell in love with the girl when she was an infant, then went to Guyana to collect her when she was 17 and made her his bride – she died about a year later in childbirth. Waterton was dismissive of many of the 'greats' of ornithology: Swainson (q.v.), Jameson (q.v.) and MacGillivray (q.v.) '… shall have their ignorance brought home to them'; Cuvier (q.v.) '… knew no more about the real habits of most birds than I did about his grandmother'; Audubon was an '… ornithological impostor [who] ought to be exposed'. Bonaparte (q.v.), too, was dismissed. However, in 1841 Waterton was leaving Italy with his family on a small steamer, the *Pollux*, when it collided with another vessel, the *Monjibellow*. The *Pollux* would have sunk but for the quick thinking of a passenger aboard the latter ship who took the tiller, keeping the vessels together long enough for all of *Pollux*'s passengers to move to the less damaged ship. That passenger was none other than the Prince of Canino, Lucien Bonaparte!

Watkins

Watkins's Antpitta *Grallaria watkinsi* **Chapman**, 1919
[Alt. Scrub Antpitta]
Little Inca Finch *Incaspiza watkinsi* Chapman, 1925

Ruddy Foliage-gleaner ssp. *Automolus rubiginosus watkinsi* **Hellmayr**, 1912

Henry George 'Harry' Watkins (DNF) was an English collector resident in Peru (1910–c.1926). He was employed by the AMNH to map the southern boundary of the Tumbesian endemic area (1919) and to collect in the Pomara region (1923–1924). He was known to still be alive in 1926 when the USNM credited him and his brother, Casimir, with whom he travelled and collected, with the holotype of an insect. He should not be confused with the Arctic explorer Henry George Watkins (1907–1932).

Watson

Tawny-bellied Screech Owl *Megascops watsonii* **Cassin**, 1848

Dr Gavin Watson (1796–1858) was an American physician, naturalist and author. He presented many specimens to the Philadelphia Academy of Natural Sciences. Cassin (q.v.) wrote that Watson was 'an especial admirer of the Owls'.

Watt

Orange River Francolin ssp. *Francolinus levaillantoides wattii* **J. D. Macdonald**, 1953 NCR
[JS *Scleroptila levalliantoides pallidior*]

Dr James Shaw Watt (1906–2002) was a Scottish-born veterinary surgeon and botanist who was brought up in South Africa. He was a Government vet at Walvis Bay, South West Africa (Namibia) (1928–1931) and then at Windhoek and Okahandja (1932–1940). He was in South Africa (1940–1946), returning to South West Africa as Director of Agriculture (1946–1969). After retiring he lived in Swakopmund.

Watters

Pygmy Woodpecker sp. *Dendrocopos wattersi* **Salvadori & Giglioli**, 1885 NCR
[Alt. Grey-capped Pygmy Woodpecker; JS *Dendrocopos canicapillus kaleensis*]

Watters's Skylark *Alauda gulgula wattersi* **Swinhoe**, 1871
[Alt. Oriental Skylark ssp.]

Thomas Watters (1840–1901) was in the British Consular Service (1863–1895). He was Robert Swinhoe's (q.v.) Consular Assistant at Taiwan-foo and Takow (1865), becoming Acting Consul (1866). He returned to Taiwan as Acting Consul (1876–1877) and again as Vice-Consul at Tamsui (1880–1883). After Swinhoe left Taiwan (1866) Watters kept him supplied with specimens. He was Acting Consul-General in Seoul, Korea (1887–1888), and Consul at Canton (1891–1893). He retired on the grounds of ill health. He was extremely interested in Chinese and Korean culture and published extensively on this. He donated the first collection of Korean art to the Victoria & Albert Museum in London.

Wayne

Wayne's Clapper Rail *Rallus longirostris waynei* **Brewster**, 1899
[Syn. *Rallus crepitans waynei*]
Wayne's Warbler *Dendroica virens waynei* **Bangs**, 1918
[Alt. Black-throated Green Warbler ssp.; *D. virens* sometimes regarded as monotypic]
Seaside Sparrow ssp. *Thryospiza maritima waynei* **Oberholser**, 1931 NCR
[JS *Ammodramus maritimus macgillivraii*]
Wayne's Marsh Wren *Cistothorus palustris waynei* Dingle & Sprunt, 1932

Arthur Trezevant Wayne (1863–1930) was an American ornithologist from South Carolina. An early friendship with the Director of the Charleston Museum led to him obtaining birds for the museum and learning the arts of skinning and preparing specimens. He made a collecting trip to Florida (1892–1893) during which he collected Ivory-billed Woodpeckers *Campephilus principalis* and Carolina Parakeets *Conuropsis carolinensis*, both now gone forever. He wrote *Birds of South Carolina* (1910), an annotated list of birds from the state.

Weatherill

Mangrove Gerygone ssp. *Ethelornis cantator weatherilli* **Mathews**, 1920 NCR
[JS *Gerygone levigaster cantator*]

William Edward Weatherill (b.1891) was an Australian naturalist at the Brisbane Museum (1908–1911). His publications include 'List of birds occurring within a 12–mile radius of Brisbane' (1910).

Webb

Webb's Parrotbill *Sinosuthora webbiana* **Gould**, 1852
[Alt. Vinous-throated Parrotbill]

The holotype of this bird was supplied by 'Mr. Webb, a gentleman who has been instrumental in introducing to our notice considerable collections of birds and quadrupeds from the neighbourhood of Shanghai in China'. There are also references to 'J. Webb Esq.' collecting birds in Shanghai (1850), but we have not been able to identify him further.

Webb, C.

Narrow-tailed Starling ssp. *Poeoptera lugubris webbi* **Keith**, 1968 NCR. NRM

Cecil Stanley 'Webbie' Webb (1895–c.1964) was a British natural history collector who was Curator-Collector for the Zoological Society of London and responsible for the restocking of both the London and Whipsnade Zoos after WW2. He started life collecting birds in South Africa for the feather trade, and went on to collect animals all over Africa for London Zoo before working at that zoo and others. He became Curator of Birds and Mammals at London Zoo and was appointed Superintendent of the Dublin Zoo (1952) but resigned that year to move to Nairobi to become the Superintendent of the Nairobi Snake Park (1961) and a Trustee of the Royal National Parks of Kenya (1963). He wrote a number of articles, such as 'Collecting waterfowl in Madagascar' (1935), and wrote *The Odyssey of an Animal Collector* (1954). His collections were sent to the BMNH. Keith wrote in his etymology: 'The bird is named in honour of the late Cecil Webb of Nairobi who guided me on my first visit to these forests [Kibale and Impenetrable Forests, Uganda] in 1961.' An amphibian and a mammal are named after him.

Weber, J.

Scaly-breasted Bulbul ssp. *Pycnonotus squamatus webberi* **Hume**, 1879
[Binomial often 'corrected' to *weberi*, but originally given as *webberi*]
Streak-breasted Woodpecker ssp. *Picus viridanus weberi* A. Müller, 1882

Captain Johann Weber (DNF) was a collector in Peninsular Siam (Thailand) (1879).

Weber, M. W. C.

Weber's Lorikeet *Trichoglossus weberi* **Büttikofer**, 1894
[Alt. Flores Lorikeet, Leaf Lorikeet; Syn. *Trichoglossus haematodus weberi*]

Max Wilhelm Carl Weber van Bosse (1852–1937) was a German-born Dutch physician and zoologist, and Director of the Zoological Museum in Amsterdam (1883) (when he became a naturalised Dutch citizen). He was educated in Germany at Bonn and Berlin. He did military service in the German army – half the time as a doctor and half as a hussar. He made a voyage in the small schooner *Willem Barents* (1881), appropriately to the Barents Sea. He combined the roles of watch-keeping officer, ship's doctor and naturalist. His wife was a skilled and learned botanist and after their marriage the Webers spent three summers in Norway where he could dissect whales and she could collect algae – her speciality. They made a number of other voyages to Sumatra, Java, Sulawesi and Flores (1888), and to South Africa (1894). He was co-author with De Beaufort (q.v.) of the authoritative *Fishes of the Indo-Australian Archipelago*. He also established 'Weber's Line', an important zoogeographical line between Sulawesi and the Moluccas, which is sometimes preferred over Wallace's Line (between Sulawesi and Borneo) as the dividing line between the Oriental and Australasian faunas. The *Siboga* expedition to Indonesian waters was carried out under Weber's personal leadership (1899–1900). Two amphibians, two mammals and three reptiles are named after him.

Weber, O. F.

Aberrant Bush Warbler ssp. *Cettia flavolivacea weberi* **Mayr**, 1941

Orlando Franklin Weber (1879–1945) was an American businessman and philanthropist who was a major supporter of AMNH.

Weber, W. H.

Chestnut-capped Piha *Lipaugus weberi* Cuervo *et al.*, 2001

Walter H. Weber is best described in the words of the original etymology: 'The species epithet is dedicated to Walter H. Weber of Medellín, Colombia, for his enormous and ongoing contribution to Sociedad Antioqueña de Ornitología (SAO) and for promoting Colombian ornithology and conservation. His commitment to Colombian bird conservation and his encouragement of young ornithologists are providing the country with renewed hope.' He was a long-term friend of Roger Tory Peterson (q.v.).

Webster, F. B.

Red-footed Booby ssp. *Sula sula websteri* **Rothschild**, 1898

Frank Blake Webster (1850–1921) was an American natural history dealer who arranged a collecting expedition to the Galapagos Islands (1897), funded by Lord Rothschild (q.v.). He was publisher of, and wrote or co-wrote a number of volumes of, *The Ornithologist and Oologist*. He was an amateur taxidermist before becoming a full-time naturalist.

Webster, H.

Webster's Kingfisher *Ceyx websteri* **Hartert**, 1898
[Alt. Bismarck Kingfisher; Syn. *Alcedo websteri*]

Brush Cuckoo ssp. *Cacomantis variolosus websteri* Hartert, 1898

Captain Herbert Cayley-Webster (DNF) was an English explorer. He wrote *Through New Guinea and the Cannibal Coutries* (1898). He collected and travelled quite widely in (quoting Hartert) 'the Papuan Islands'.

Weddell

Weddell's Conure *Aratinga weddellii* **Deville**, 1851
[Alt. Dusky-headed Parakeet]
Tinamou sp. *Tinamus weddelli* **Bonaparte**, 1856 NCR
[Alt. Grey Tinamou; JS *Tinamus tao kleei*]

Ocellated Woodcreeper ssp. *Xiphorhynchus ocellatus weddellii* **Des Murs**, 1855

Hugh Algernon Weddell (1819–1877) was a botanist, born in England but raised in France. He explored in South America (1843–1847) as a member of the Castelnau (q.v.) Expedition to Brazil. Before leaving Paris he had been particularly instructed to make a thorough investigation of the *Cinchona* plant, the source of quinine, in its native habitat. He wrote *Voyage dans le Nord de la Bolivie et dans les Parties Voisines du Pérou* (1853) and the 2-volume *Chloris Andina – Essai d'une Flore de la Région Alpine des Cordillières de l'Amerique du Sud* (1855–1861). He was not related to the sealer Captain James Weddell (1787–1834), after whom the Weddell Sea and the Weddell Seal are named.

Wedel

Crested Owl ssp. *Lophostrix cristata wedeli* **Griscom**, 1932

Dr Hans von Wedel (DNF) was a German collector in Panama (1926–1930 and 1940–1941).

Weeden

Yungas Tyrannulet *Phyllomyias weedeni* Herzog, Kessler & Balderrama, 2008

Alan Weeden (b.1924) is on the Board of Directors of the Weeden Foundation, which was founded by his brother Frank Weeden (1963) and acquires land for conservation. Alan, who was President of the foundation (1980–2001), was educated at Stanford University and served in the US Navy (WW2). He was CEO of Weeden & Company, a securities firm and chairman until retiring (1981). He serves on numerous boards of both corporations and non-profit environmental organisations.

Weid

Weid's Crested Flycatcher *Myiarchus tyrannulus* P. L. S. Müller, 1776
[Alt. Brown-crested Flycatcher]

This is a transcription error for 'Wied' (see **Wied**).

Weigall

Weigall's Roller *Coracias spatulatus weigalli* **Dresser**, 1890
[Alt. Racket-tailed Roller ssp.]

Reverend Spencer Weigall (1861–1925) was an English missionary in Central Africa (1884–1924).

Weigold

Sichuan Tit *Poecile weigoldicus* **Kleinschmidt**, 1921
[Syn. *Poecile songarus weigoldicus*]

Crested Lark ssp. *Galerida cristata weigoldi* **Kollibay**, 1912 NCR
[JS *Galerida cristata subtaurica*]
Eurasian Goldfinch ssp. *Carduelis carduelis weigoldi* **Reichenow**, 1913 NCR
[JS *Carduelis carduelis parva*]
Crested Tit ssp. *Lophophanes cristatus weigoldi* Tratz, 1914
Woodchat Shrike ssp. *Lanius senator weigoldi* Kleinschmidt, 1919 NCR
[JS *Lanius senator rutilans*]
Oriental Skylark ssp. *Alauda gulgula weigoldi* **Hartert**, 1922
Dusky Fulvetta ssp. *Alcippe brunnea weigoldi* **Stresemann**, 1923

Meadow Bunting ssp. *Emberiza cioides weigoldi* **Jacobi**, 1923
Smoky Warbler ssp. *Phylloscopus fuligiventer weigoldi* Stresemann, 1923
Eurasian Wren ssp. *Troglodytes troglodytes weigoldi* **von Jordans**, 1923 NCR
[JS *Troglodytes troglodytes troglodytes*]
Bluethroat ssp. *Luscinia svecica weigoldi* Kleinschmidt, 1924 NCR
[JS *Luscinia svecica svecica*]
Weigold's Citrine Wagtail *Motacilla citreola weigoldi* **Rensch**, 1924 NCR
[JS *Motacilla citreola calcarata*]

Dr Hugo Max Weigold (1886–1973) was a German zoologist and ornithologist, and probably the first Westerner to see a living Giant Panda *Ailuropoda melanoleuca*. He bought a cub (1916), which died shortly afterwards, while he was a member of the Walter Stotzner Expedition (1914–1916) on behalf of the Dresden Museum of Anthropology and Ethnography. He was back in China in Chengdu (1931) with the First Dolan Expedition from the Academy of Natural Sciences, Philadelphia. Dr Weigold is most well known for having established the Heligoland Bird Observatory, one of the world's first such observatories. During his six years in China he discovered a flycatcher which he named after his wife, Elisa (q.v.). Two amphibians are also named after him.

Weiler

Eastern Bearded Greenbul ssp. *Criniger chloronotus weileri* **Gyldenstolpe**, 1922 NCR; NRM

Major Max Weiler (DNF) was a Belgian army officer in the Congo. He was an aide to Prince Wilhelm of Sweden (q.v. under Wilhelm), who led the Swedish Central African Expedition (1920–1921).

Weiske

Australian Treecreeper sp. *Climacteris weiskei* **Reichenow**, 1900 NCR
[Alt. White-throated Treecreeper; JS *Cormobates leucophaea minor*]
New Guinea Kingfisher sp. *Syma weiskei* Reichenow, 1900 NCR
[Alt. Mountain Kingfisher; JS *Syma megarhyncha megarhyncha*]
Pygmy Eagle *Hieraaetus weiskei* Reichenow, 1900
[Alt. New Guinea Hawk Eagle]

Chestnut-breasted Cuckoo ssp. *Cacomantis castaneiventris weiskei* Reichenow, 1900

Emil Weiske (1867–1950) was a German traveller and specimen preparator who collected animals and ethnological objects and founded a private museum. He was in California (1890–1892), unsuccessfully prospecting for gold but successfully learning English. He went to Hawaii (1892) to collect birds and insects, then moved on to the Fiji Islands (1894). He moved to North Queensland (1895) for two years, then went to Papua (1897). He had the unwise habit of fishing with dynamite, and a stick of dynamite exploded shattering his right hand (1900). His companion, a local, saved him from bleeding to death. He returned to Germany (1900) and established a mobile museum (1904) which he took on tour. He went to Lake Baikal (1908) in Siberia to collect seals, etc., taking an assistant, Otto Taschmann, to make up for his one-handedness. His final trip abroad was to Argentina (1911–1913). Back again in Germany he bought a house and established his private museum.

Welch

Welch's Ptarmigan *Lagopus muta welchi* **Brewster**, 1885
[Alt. Rock Ptarmigan ssp.]

George O. Welch (DNF) of Lynn, Massachusetts, was an ornithologist who collected in North America. Brewster (q.v.) mentions his observations of movements of woodpeckers (1860). He collected the ptarmigan holotype (1883).

Welchman

North Melanesian Cuckooshrike *Coracina welchmani*
Tristram, 1892

Rev. Dr Henry Palmer Welchman (1850–1908) was a British medical missionary at Bogutu, Solomon Islands (1888–1908). He collected objects of anthropological interest for the Pitt Rivers Museum in Oxford.

Wells, J.

Wells's Dove *Leptotila wellsi* **G. N. Lawrence**, 1884
[Alt. Grenada Dove]
Grenada Hummingbird sp. *Saucerottia wellsi* **Boucard**, 1893
NCR
[Vagrant specimen of *Amazilia tobaci tobaci*?]

John Grant Wells (DNF) was an ornithologist who was active in the West Indies, particularly Grenada where he lived (1886–1902). He sent all his specimens to G. N. Lawrence of the USNM. He wrote 'A catalogue of the birds of Grenada, West Indies, with observations thereon' (1886), and 'Birds of the island of Carriacou' (1902).

Wells, T.

Horned Lark *Otocorys wellsi* **Babault**, 1920 NCR
[JS *Eremophila alpestris longirostris*]

Black-tailed Treecreeper ssp. *Climacteris melanurus wellsi*
Ogilvie-Grant, 1909
Spectacled Monarch ssp. *Monarcha trivirgatus wellsi*
Ogilvie-Grant, 1911 NCR
[JS *Symposiachrus trivirgatus nigrimentum*]
Wells's Wagtail *Motacilla capensis wellsi* Ogilvie-Grant,
1911
[Alt. Cape Wagtail ssp.]
Grey Crested Tit ssp. *Lophophanes dichrous wellsi*
E. C. S. Baker, 1917
Mountain Kingfisher ssp. *Syma megarhyncha wellsi*
Mathews, 1918
Dusky Long-tailed Cuckoo ssp. *Cercococcyx mechowi
wellsi* **Bannerman**, 1919 NCR; NRM
Scarlet-faced Liocichla ssp. *Liocichla ripponi wellsi*
La Touche, 1921

Lesser Yellownape ssp. *Picus chlorolophus wellsi*
Meinertzhagen, 1924
Chinese Francolin ssp. *Francolinus pintadeanus wellsi*
Delacour, 1926 NCR
[JS *Francolinus pintadeanus phayrei*]
Tullberg's Woodpecker ssp. *Campethera tullbergi wellsi*
G. L. Bates, 1926 NCR
[JS *Campethera tullbergi tullbergi*]

Thomas 'Jimmy' Wells (1868–1939) worked at the BMNH, London, where he started as a Boy Attendant (1883) and assisted in moving the cabinets containing the bird collection from the old building at Bloomsbury to South Kensington. By the time he retired (1930) after 47 years' service, he was a higher-grade clerk in the Bird Section. As an obituary put it: 'His official designation by no means indicated the wonderful knowledge of systematic ornithology which he possessed.' He is also mentioned in Bannerman's *Birds of Tropical West Africa* as having compiled the indexes. 'Much unobtrusive work was done by him in assisting authors with their systematic contributions and his name is perpetuated in ornithological literature by many species named in his honour; as an indexer and proof-reader he had few equals and was the possessor of beautifully clear handwriting as hundreds of bird labels can testify.' He is further mentioned in Christopher Frost's *A History of British Taxidermy*.

Welwitsch

Square-tailed Nightjar ssp. *Caprimulgus fossii welwitschii*
Bocage, 1867
[Alt. Mozambique Nightjar]

Dr Friedrich Martin Josef Welwitsch (1806–1872) was an Austrian botanist. He studied medicine and botany in Vienna before travelling to Portugal to escape the consequences of what was termed a 'youthful indiscretion'. He explored most of Portugal, forming a herbarium of 9,000 specimens. The Portuguese government sent him to explore and collect botanical specimens in Angola (1853–1861), where he accumulated c.5,000 specimens, many new to science. He caused an international quarrel by sending a large proportion of his collection to the BMNH, London, instead of to Lisbon. The Portuguese took the view that, as they had paid him, the collection belonged to them. The collection's duplicate specimens were split, so both museums got something out of it. Welwitsch moved to London (1863) and worked at Kew Gardens. Two mammals and a reptile are named after him, as is the plant genus *Welwitschia*.

Wenman

Galapagos Mockingbird ssp. *Mimus parvulus wenmani*
Swarth, 1931

This bird is named after Wenman Island in the Galapagos. Nowadays, the island is more commonly called Wolf. The island's name was probably given in honour of Philip, third viscount Wenman (1610–1686).

Wera (Buturlin)

Citrine Wagtail ssp. *Motacilla citreola werae* **Buturlin**, 1907

Wera V. Buturlin (DNF) was the wife of the describer Buturlin (q.v.).

Wera (Zarudny)

Chukar Partridge ssp. *Alectoris chukar werae* **Zarudny** & **Loudon**, 1904

Wera Zarudny (DNF) was the wife of the senior describer Zarudny (q.v.).

Werther

Red Bishop sp. *Pyromelana wertheri* **Reichenow**, 1897 NCR
[Alt. Southern Red Bishop; JS *Euplectes orix nigrifrons*]

Lieutenant C. Waldemar Werther (1867–1932) was a German army officer and geographer who explored in East Africa (1896–1897). He wrote *Zum Victoria Nyanza: eine Antisklaverei-Expedition und Forschungsreise* (1894).

Weske

Marcapata Spinetail ssp. *Cranioleuca marcapatae weskei* **Remsen**, 1984

Dr John S. Weske is an American ornithologist and collector. The University of Oklahoma awarded his doctorate (1972). He is based at National Fish and Wildlife Laboratory, USNM.

Westerman

Westerman's Flycatcher *Ficedula westermanni* **Sharpe**, 1888
[Alt. Little Pied Flycatcher]

Westerman's Eclectus Parrot *Eclectus roratus westermani* **Bonaparte**, 1850 NCR
[Based on aviary birds. Status uncertain; perhaps hybrids]
Fulvous-breasted Woodpecker ssp. *Dendrocopos macei westermani* **Blyth**, 1870

Gerardus Frederick Westerman (1807–1890) was a book dealer and an enthusiastic pigeon fancier who became a zoologist. He founded, with J. H. W. Werlemann and Jan Jacob Wijsmuller (the so called '3 Ws' – fellow members of the Amsterdam Free Mason Lodge 'La Bienne Aimée'), the Amsterdam Zoo 'Natura Artis Magistra', usually known as 'Artis' (1838). He was the first Director of the Zoo (1838–1890) and was married to Maria Eleonora van der Schroeff (q.v. under Eleonora). He wrote De Toerakos (The Turacos) (1860). Sharpe spelt his name wrongly by adding an additional 'n' to the end. A reptile is named after him.

Weston

Boat-tailed Grackle ssp. *Quiscalus major westoni* **Sprunt**, 1934

Francis Marion Weston (1887–1969) was an American ornithologist and collector. He was Student Assistant at the Charleston Museum, South Carolina (1906). The College of Charleston awarded his bachelor's degree (1907). He worked as a civil service draftsman in Charleston and Washington (1908–1916), then moved to Pensacola, Florida, to work at the Pensacola Naval Air Station (1916). He became a leading expert on Florida's birds. The Francis M. Weston Audubon Society of Pensacola is named after him.

Wetmore

Tanager genus *Wetmorethraupis* **Lowery** & **O'Neill**, 1964

Masked Mountain Tanager *Buthraupis wetmorei* **R. T. Moore**, 1934
Wetmore's Rail *Rallus wetmorei* **Zimmer** & **W. H. Phelps**, 1944
[Alt. Plain-flanked Rail]
Wetmore's Bush Tanager *Chlorospingus wetmorei* Lowery & **R. J. Newman**, 1949
[Alt. Common Bush Tanager; Syn. *Chlorospingus flavopectus wetmorei*]
Wetmore's Tanager *Wetmorethraupis sterrhopteron* Lowery & O'Neill, 1964
[Alt. Orange-throated Tanager]

Red-throated Ant Tanager ssp. *Habia salvini wetmorei* **Dickey** & **Van Rossem**, 1927 NCR
[JS *Habia fuscicauda salvini*]
Plain Pigeon ssp. *Patagioenas inornata wetmorei* **J. L. Peters**, 1937
Tropical Screech Owl ssp. *Megascops choliba wetmorei* **Brodkorb**, 1937
Savannah Sparrow ssp. *Passerculus sandwichensis wetmorei* Van Rossem, 1938
Everett's White-eye ssp. *Zosterops everetti wetmorei* **Deignan**, 1943
White-shouldered Antshrike ssp. *Thamnophilus aethiops wetmorei* **Meyer de Schauensee**, 1945
White-fronted Tyrannulet ssp. *Phyllomyias zeledoni wetmorei* Aveledo & **Pons**, 1953
Tepui Wren ssp. *Troglodytes rufulus wetmorei* Phelps & Phelps, 1955
Red-vented Bulbul ssp. *Pycnonotus cafer wetmorei* Deignan, 1960 NCR
[Unnecessary replacement name for *Pycnonotus cafer saturatus*]
Ruddy-capped Nightingale Thrush ssp. *Catharus frantzii wetmorei* **A. R. Phillips**, 1969
Common Tody Flycatcher ssp. *Todirostrum cinereum wetmorei* **Parkes**, 1976
Elegant Crested Tinamou ssp. *Eudromia elegans wetmorei* **R. C. Banks**, 1977
Mangrove Vireo ssp. *Vireo pallens wetmorei* A. R. Phillips, 1991

Frank 'Alexander' Wetmore (1886–1978) was an American ornithologist and avian palaeontologist who conducted extensive fieldwork in Latin America. His first job was as a bird taxidermist at the Denver Museum of Natural History, Colorado (1909). He spent time in Puerto Rico studying its avifauna (1911). He travelled throughout South America for two years, investigating bird migration between continents, whilst working for the US Bureau of Biological Survey. He was Assistant Secretary of the USNM (1925–1946). He was President of the AOU (1926–1929), then became the USNM's sixth Secretary (1945–1952). Wetmore made a number of short trips to Haiti, the Dominican Republic, Guatemala, Mexico, Costa Rica and Colombia, and conducted a research programme in Panama (1946–1966), during which he made an exhaustive survey of the birds of the isthmus. A canopy bridge in the Río Bayano Basin, in Panama, was named after him (1973). He wrote 'A systematic classification for the birds of the world' (1930), which he revised twice (1951 and 1960). Therein he devised the Wetmore Order, a sequence of bird classification, which had widespread acceptance until recently (although still in use). His other publications included *Birds of Haiti and the Dominican Republic* (1931) and *The Birds of the Republic of Panamá* (1965). Numerous taxa, comprising 56 new genera, species and subspecies of birds (both recent and fossil), mammals, amphibians, insects, molluscs and plants, are named after him. He wrote the first descriptions of 189 species and subspecies of living birds, mostly from Central and South America. There is a nice story behind the naming of the tanager genus. In the early 1960s Wetmore declared unequivocally that all the birds in the world had been discovered, aside perhaps from some slightly different forms that might show up in museum drawers. Not long afterwards, a Louisiana State University expedition to a previously unvisited area of Peru discovered a strikingly different tanager and ironically named the new genus after the man who had discounted the possibility of anything 'truly new' ever being discovered again. An amphibian, a mammal and five reptiles are named after him. (See also **Alexander (Wetmore)**)

Wettstein

Woodlark ssp. *Lullula arborea wettsteini* **Niethammer**, 1943 NCR
[JS *Lullula arborea pallida*]

Dr Otto von Wettstein Ritter von Westersheimb (1892–1967) was an Austrian zoologist who was Professor of Natural History, Natural History Museum, Vienna, where he was Curator of Herpetology (1920–1945). Originally interested in birds and mammals, he found that herpetology was the only department with a vacancy (1915). During WW2 he success-fully kept the collection of tens of thousands of specimens, preserved in alcohol, safe in bunkers or in mines. He took over many other duties, even succeeding in publishing the *Annals* of the museum (1941–1944). The Allies (1945) officially barred him from the museum, so he retired and worked for the Department of Forest Protection, studying insects and their parasites. He published 60 scientific papers on herpe-tology. An amphibian and a reptile are named after him.

Weyns

Weyns's Weaver *Ploceus weynsi* **A. J. C. Dubois**, 1900

Lieutenant Colonel Auguste F. G. Weyns (1854–1944) was a Belgian explorer who collected in Central Africa (1888–1903). He was also the governor (1900–1903) of the semi-autonomous state of Katanga within the Congo, as 'representative' of the Comité Spécial du Katanga. A mammal is named after him.

Wharton

Christmas (Island) Imperial Pigeon *Ducula whartoni* **Sharpe**, 1887

Rear-Admiral Sir William James Lloyd Wharton (1843–1905) was an English hydrographer. He joined the Royal Navy

(1857) and after several promotions was appointed (1884) as Hydrographer of the Navy. He was very interested in fossils and mineralogy, about which he corresponded at length with Sir Archibald Geikie (1835–1924), Professor of Geology at Edinburgh University. He wrote an 'Account of Christmas Island' in the *Proceedings of the Royal Geographical Society* (1888).

Whistler

Whistler's Warbler *Seicercus whistleri* **Ticehurst**, 1925

White-throated Laughingthrush ssp. *Garrulax albogularis whistleri* **E. C. S. Baker**, 1921
Golden Bush Robin ssp. *Tarsiger chrysaeus whistleri* Ticehurst, 1922
Grey-sided Bush Warbler ssp. *Cettia brunnifrons whistleri* Ticehurst, 1923
Rufous-bellied Niltava ssp. *Niltava sundara whistleri* Ticehurst, 1926
Rufous-vented Tit ssp. *Periparus rubidiventris whistleri* **Stresemann**, 1931
[SII *Periparus rubidiventris beavani*]
White-tailed Nuthatch ssp. *Sitta himalayensis whistleri* **Delacour**, 1932 NCR; NRM
Meadow Pipit ssp. *Anthus pratensis whistleri* **Clancey**, 1942
Purple-rumped Sunbird ssp. *Leptocoma zeylonica whistleri* **Ripley**, 1946 NCR
[JS *Leptocoma zeylonica flaviventris*]
Red-whiskered Bulbul ssp. *Pycnonotus jocosus whistleri* **Deignan**, 1948

Hugh Whistler (1889–1943) was a British policeman (1909–1924) in the Punjab (most of present-day Pakistan, Himachal Pradesh and Haryana). He was also an ornithologist attached to the BMNH and a Fellow of the Zoological Society of London. Suffering ill-health he had to retire prematurely to England, living at Battle, Sussex, where he was a Justice of the Peace. Sussex University now awards the Whistler Prize to the best essay on natural history or archaeology. He worked very closely with C. B. Ticehurst (q.v.) in Spain, Albania, India and Ceylon (Sri Lanka). He spent much time in India and undertook a number of surveys there, e.g. Rajasthan (1938), which listed 300 species. Whistler and Ticehurst published a number of articles, including 'On the avifauna of Galicia' (1928) and 'On the ornithology of Albania' (1932). They wrote a monograph, *Birds of India*, which exists only as a manuscript in the USNM archives where it formed the basis for much of Ripley's (q.v.) work on Indian birds with Salim Ali (q.v.). Whistler wrote a number of articles alone, such as 'Migration notes from a passenger steamer' (1916) and the *Popular Handbook of Indian Birds* (1928). The latter has been updated many times and is regarded as the book which popularised birdwatching as a hobby amongst the Indian elite. Salim Ali included Whistler's notes in *The Birds of Mysore* (1942).

Whitaker

Eurasian Jay ssp. *Garrulus glandarius whitakeri* **Hartert**, 1903
Desert Lark ssp. *Ammomanes deserti whitakeri* Hartert, 1911

Crested Lark ssp. *Galerida cristata whitakeri* **Bannerman**, 1922 NCR
[JS *Galerida cristata arenicola*]

Rock Partridge ssp. *Alectoris graeca whitakeri* **Schiebel**, 1934

Cetti's Warbler ssp. *Cettia cetti whitakeri* **C. Orlando**, 1937 NCR
[JS *Cettia cetti cetti*]

Joseph Isaac Spadafora Whitaker (1850–1936) came from a British family which settled in Palermo, Sicily (1806), and became influential and extremely wealthy as a leading producer of Marsala wine. Joseph was a grandson of the company founder. He had no interest in business and devoted himself to ornithology, archaeology and botany. He travelled in Tunisia and wrote *Birds of Tunisia* (1905). He bought the island of Mozia, off the Sicilian coast, and excavated there. His villa in Palermo was a centre for high society, and its gardens held exotic plants including a huge banyan tree. His collection of bird skins is housed in Edinburgh. A mammal is also named after him.

White, A. H. E.

White's Honeyeater *Conopophila whitei* **North**, 1910
[Alt. Grey Honeyeater]

Alfred Henry Ebsworth White (1901–1964) was the son of Australian ornithologist, oologist and grazier Henry Luke White (q.v.). North (q.v.) wrote: '… in response to a request from the owner of the specimens, who has done so much recently to advance Australian ornithology, I have associated with it the name of his son, Mr Alfred Henry Ebsworth White, who, although yet young in years, I am informed is worthily following in his father's footsteps.' He was a keen cricketer and played for Cambridge University (1922–1924) and for New South Wales (1925–1926).

White, C. M. N.

Bar-throated Apalis ssp. *Apalis thoracica whitei* **Grant** & **Mackworth-Praed**, 1937

Scaly Francolin ssp. *Francolinus squamatus whitei* **Schouteden**, 1954 NCR
[JS *Pternistis squamatus squamatus*]

Charles Matthew Newton White (1914–1978) was a British ornithologist who studied the biology and taxonomy of francolins. After graduating from Balliol, Oxford, he went into the Colonial Administration Service (1938) in Northern Rhodesia (Zambia) as a District Officer before transferring to national government service after independence (1964). Thereafter he was a member of the Zambian Land Law Commission (1965–1967) and of a special commission that reported on African Customary Law (1960–1970). Throughout his service he collected birds and published his studies, such as *Check List of the Birds of Northern Rhodesia* (1957) with C. W. Benson (q.v.), and many papers on systematics including one on the taxonomy of cassowaries. He also wrote papers on linguistics. After he retired to the UK he began co-writing *Birds of Wallacea*, later published by the BOU (1986).

White, E. W.

Black-and-chestnut Warbling Finch *Poospiza whitii*
P. L. Sclater, 1883
[Syn. *Poospiza nigrorufa whitii*]

Ochre-cheeked Spinetail ssp. *Synallaxis scutata whitii*
P. L. Sclater, 1881

Ernest William White (1858–1884) was a collector who travelled widely in Argentina. He was taken to South America when aged six. There he was forbidden books and roamed free, developing a love of all wildlife. At his own request he was sent to London to meet other naturalists and was taught how to prepare specimens. However, he developed tuberculosis and was called home, barely making it after haemorrhaging on the voyage. He was sent to Mendoza and was soon enjoying nature again. Here he spent five years wandering, travelling on mule-back, outdoors in all weathers, and regaining his strength. He returned to Buenos Aires where he married. Having studied medicine, he decided upon dentistry as a way of earning a good living whilst still being able to travel and study natural history. Unable to find a place in London he decided to go to Philadelphia and study dentistry there. He completed two years of the course and was about to qualify, but contracted typhus during the epidemic and died. He wrote the 2-volume *Cameos from the Silver-Land; or The Experiences of a Young Naturalist in the Argentine Republic* (1881). Throughout the book he makes detailed remarks on birdlife. A reptile is also named after him.

White, G.

White's Thrush *Zoothera (dauma) aurea* Holandre, 1825
[Alt. Scaly Thrush (when *aurea* is included in *Z. dauma*); *Turdus whitei* is a JS]

Reverend Gilbert White (1720–1793) was a country parson at Selbourne in Hampshire, England. He was a keen amateur naturalist famed for his meticulous observations; for example, he played a major part in recognising that the *Phylloscopus* warblers nesting in Britain belonged to three different species. He is also noted for his correspondence with other enthusiasts of his day. His letter to Daines Barrington (1772), at a time when there was much speculation regarding whether birds migrated or hibernated, provided his recognition of the former; he wrote 'We must not, I think, deny migration in general; because migration certainly does subsist in some places, as my brother [Reverend John White] in Andalusia has fully informed me. Of the motions of these birds he has ocular demonstration, for many weeks together, both spring and fall, during which periods myriads of the swallow kind traverse the Straits from north to south and from south to north, according to the season'. He published his observations and letters (1789) as *The Natural History and Antiquities of Selbourne*, one of the most delightful books ever written and still in print over 200 years later.

White, H. L.

Northern Shrike-tit *Falcunculus whitei* **A. J. Campbell**, 1910
[Syn. *Falcunculus frontatus whitei*]

Splendid Fairy-wren ssp. *Malurus splendens whitei*
A. J. Campbell, 1901 NCR
[JS *Malurus splendens melanotus*]
Striated Grasswren ssp. *Amytornis striatus whitei*
Mathews, 1910

Henry Luke White (1860–1927) was an Australian ornithologist, oologist and grazier. He donated his collection of c.8,500 bird skins to the Museum of Victoria (1917), along with 4,200 clutches of eggs from 800 species. He was also a benefactor of the RAOU. He qualified as a surveyor (1884) but went into many types of farming, animal husbandry and land management. He employed a number of collectors including S. W. Jackson (q.v.), who went on to curate the collection. His books were donated to the Mitchell Library in Sydney, and other collections were donated to Melbourne's National Museum. (See **Harry White**)

White, H. T.

Hill Blue Flycatcher ssp. *Cyornis banyumas whitei*
Harington, 1908

Sir Herbert Thirkell White (1855–1931) was a British colonial administrator in Burma (Myanmar) (1878–1910), becoming Lieutenant-Governor (1905–1910).

White, S.

Pale-eyed Honeyeater *Tephras whitei* **E. Ramsay**, 1882 NCR
[Alt. Green-backed Honeyeater; JS *Glycichaera fallax poliocephala*]

Samuel White (1835–1880) was an English-born Australian ornithologist, collector and explorer. He returned to England (1869–1870) specifically to see John Gould (q.v.). Using his own vessel, *Elsea* he cruised and collected in the Aru Islands (1880). He had trouble with the crew, became ill, and returned to Sydney where he died of pneumonia. Despite instructions in his will, the collection at his home at Fulham, Adelaide, was dispersed after his death.

White, S. A. & E. R.

Australian Robin genus *Whiteornis* **Mathews**, 1912 NCR
[Now in *Petroica*]

Australian Crake ssp. *Porzana fluminea whitei* Mathews, 1912 NCR; NRM
Black-tailed Native-hen ssp. *Tribonyx ventralis whitei* Mathews, 1912 NCR; NRM
Grey Shrike-thrush ssp. *Colluricincla harmonica whitei* Mathews, 1912 NCR
[JS *Colluricincla harmonica rufiventris*]
Australian Masked Owl ssp. *Tyto novaehollandiae whitei* Mathews, 1912 NCR
[JS *Tyto novaehollandiae novaehollandiae*]
Rufous Bristlebird ssp. *Dasyornis broadbenti whitei* Mathews, 1912 NCR
[JS *Dasyornis broadbenti broadbenti*]
Striated Thornbill ssp. *Acanthiza lineata whitei* Mathews, 1912
Varied Lorikeet ssp. *Psitteuteles versicolor whitei* Mathews, 1912 NCR; NRM

White's Parakeet *Barnardius barnardi whitei* Mathews, 1912
[Taxonomy unclear: sometimes regarded as a hybrid population between *Barnardius* (*zonarius*) *zonarius* and *B.* (*zonarius*) *barnardi*]
White-plumed Honeyeater ssp. *Ptilotis penicillata whitei* Mathews, 1912 NCR
[JS *Lichenostomus penicillatus penicillatus*]
White-winged Chough ssp. *Corcorax melanoramphos whiteae* Mathews, 1912
Yellow-tailed Black Cockatoo ssp. *Calyptorhynchus funereus whiteae* Mathews, 1912
Southern Whiteface ssp. *Aphelocephala leucopsis whitei* Mathews, 1914 NCR
[NUI *Aphelocephala leucopsis leucopsis*]
Night Parrot ssp. *Geopsittacus occidentalis whiteae* Mathews, 1915 NCR
[Syn. *Pezoporus occidentalis* NRM]
Superb Parrot ssp. *Polytelis swainsonii whitei* Mathews, 1916 NCR; NRM
Grey-headed Honeyeater ssp. *Sacramela keartlandi whiteorum* **Mathews**, 1924 NCR; NRM
[Syn. *Ptilotula keartlandi*]

Captain Samuel Albert White (1870–1954) was an independently wealthy hunter, explorer and noted racehorse owner who became an influential ornithologist. He made a number of ornithological trips in Australia (1887–1891). He fought in the Boer War in South Africa (1900–1903), then went big game hunting in East and Central Africa. He travelled extensively in the interior of Australia (1906–1922) and collaborated with Gregory Mathews (q.v.) in the production of *The Birds of Australia* that appeared in stages (1910–1927). He was President of the RAOU (1914–1916). In 1905 he married his first wife Ethel Rosina Toms (q.v. under Ethel and Rosina), who accompanied him on many of his collecting trips in Australia (1906–1916). Species named *whiteae* above are after her and *whiteorum* honours both. White and Mathews were very close friends – White used the name *marianae* to honour Mrs Mathews and Mathews reciprocated by naming birds after White.

White, W. T.

Social Weaver genus *Whitellus* **Oberholser**, 1945 NCR
[Now in *Pseudonigrita*]

Windsor Thomas White (1866–1958) was an American ornithologist and businessman who produced 'White Steamer Cars' (1901), and later steam-driven trucks, which were used in WW1. He was a trustee of the Cleveland Museum, and was leader of the White-Fuller expedition to Kenya.

Whitehead, C. H.

Plain-backed Thrush ssp. *Zoothera mollissima whiteheadi*
E. C. S. Baker, 1913

Major Charles Hughes Tempest Whitehead (1880–1915) was a British army officer in 56th Punjab Rifles, Indian Army, and a collector. He wrote 'On the birds of Kohat and Kurram, northern India' (1909). He was killed in action in France (WW1).

Whitehead, J.

Whitehead's Scops Owl *Otus megalotis* **Walden**, 1875
[Alt. Philippine Scops Owl]
Corsican Nuthatch *Sitta whiteheadi* **Sharpe**, 1884
Whitehead's Spiderhunter *Arachnothera juliae* Sharpe, 1887
Mangrove Whistler *Hyloterpe whiteheadi* Sharpe, 1888 NCR
[JS *Pachycephala cinerea plateni*]
Spotted Wood Owl *Syrnium whiteheadi* Sharpe, 1888 NCR
[JS *Strix seloputo wiepkeni*]
Whitehead's Broadbill *Calyptomena whiteheadi* Sharpe, 1888
Whitehead's Stubtail *Urosphena whiteheadi* Sharpe, 1888
[Alt. Bornean Stubtail]
Whitehead's Trogon *Harpactes whiteheadi* Sharpe, 1888
Whitehead's Tree Babbler *Zosterornis whiteheadi* **Ogilvie-Grant**, 1894
[Alt. Chestnut-faced Babbler; Syn. *Stachyris whiteheadi*]
Whitehead's Swiftlet *Aerodramus whiteheadi* Ogilvie-Grant, 1895
Whitehead's Magpie *Urocissa whiteheadi* Ogilvie-Grant, 1899
[Alt. White-winged Magpie]

Island Thrush ssp. *Turdus poliocephalus whiteheadi* **Seebohm**, 1893
Common Buttonquail ssp. *Turnix sylvaticus whiteheadi* Ogilvie-Grant, 1897
Silver Pheasant ssp. *Lophura nycthemera whiteheadi* Ogilvie-Grant, 1899
Mountain White-eye ssp. *Zosterops montanus whiteheadi* **Hartert**, 1903
Ferruginous Babbler ssp. *Erythrocichla bicolor whiteheadi* Hartert, 1915 NCR
[Syn *Trichastoma bicolor* NRM]
Black-belted Flowerpecker ssp. *Dicaeum haematostictum whiteheadi* **Hachisuka**, 1926 NCR; NRM
Mountain Hawk Eagle ssp. *Spizaetus nipalensis whiteheadi* Swann, 1933 NCR
[JS *Nisaetus nipalensis nipalensis*]
Blue-crowned Racket-tail ssp. *Prioniturus discurus whiteheadi* **Salomonsen**, 1953

John Whitehead (1860–1899) was a British explorer and naturalist who collected in Borneo (1885–1888), the Philippines (1893–1896) and Hainan (1899). He wrote *Explorations of Mount Kina Balu, North Borneo* (1893) and he may have been the first European to reach the summit of the mountain. He died of fever in Hainan, at the age of 38. An amphibian and five mammals are named after him. (See also **John (Whitehead)** and **Jeffery**)

Whitely, H. Jr

Red-banded Fruiteater *Pipreola whitelyi* **Salvin & Godman**, 1884
Whitely's Nightjar *Caprimulgus whitelyi* Salvin, 1885
[Alt. Roraiman Nightjar; Syn. *Setopagis whitelyi*]
Emerald sp. *Uranomitra whitelyi* **Boucard**, 1893 NCR
[Alt. White-chested Emerald; JS *Amazilia brevirostris brevirostris*]

Black-throated Brilliant ssp. *Heliodoxa schreibersii whitelyana* **Gould**, 1872
[Originally described as *Iolaema whitelyana*]

Chestnut-tipped Toucanet ssp. *Aulacorhynchus derbianus whitelianus* Salvin & Godman, 1882
Whitely's Flycatcher *Myiarchus swainsoni phaeonotus* Salvin & Godman, 1883
[Alt. Swainson's Flycatcher ssp.]
Whitely's Tanager *Tangara cyanoptera whitelyi* Salvin & Godman, 1884
[Alt. Black-headed Tanager ssp.]
Plumbeous Seedeater ssp. *Sporophila plumbea whiteleyana* **Sharpe**, 1888
Straight-billed Hermit ssp. *Phaethornis bourcieri whitelyi* Boucard, 1891 NCR
[JS *Phaethornis bourcieri bourcieri*]

Henry Whitely Jr (1844–1892), son of Henry Whitely Sr (q.v.), was an Englishman who collected in parts of Brazil, Peru (1867–1876) and British Guiana (Guyana). In Peru he explored very extensively, often in extreme discomfort, and travelled for a time with a Mr Gibson, who may have been collecting for Loddiges (q.v.). Among his welcome discoveries was a good pub in the Andes, kept by an Englishman whom, unfortunately, he does not name in his letters! He wrote 'Notes on humming-birds collected in high Peru' (1873). He died in British Guiana – according to his companions, he shot himself after losing his bird specimens in a boat mishap.

Whitely, H. Sr

House Martin sp. *Chelidon whiteleyi* **Swinhoe**, 1863 NCR
[Alt. Common House Martin; JS *Delichon urbicum lagopodum*]

Asian Barred Owlet ssp. *Glaucidium cuculoides whitelyi* **Blyth**, 1867

Henry Whitely Sr (1817–1898), father to Henry Whitely Jr (above), was a Curator at the Royal Artillery Institution Museum (1865) at Woolwich Arsenal and later owned a taxidermy shop in Woolwich (1866–1880). The martin holotype was part of a collection of birds made in China by 'Mr Fleming of the Royal Artillery', which came into Whitely's possession.

Whiteside

Nkulengu Rail ssp. *Himantornis haematopus whitesidei* **Sharpe**, 1909 NCR; NRM

Rev. H. M. Whiteside (DNF) was a Baptist missionary in the Lolanga district of what is now Democratic Republic of Congo. The American naturalist James Chapin (q.v.), who was in the Congo (1909–1915), records that Whiteside provided him with some ornithological specimens. Chapin also mentions that Whiteside was in the habit of sending specimens to the BMNH. Whiteside also has a subspecies of monkey named after him.

Whitley

Whitley's Parakeet *Cyanoliseus whitleyi* **Kinnear**, 1926
[Probably hybrid *Cyanoliseus* x *Aratinga*]

Herbert Whitley (1886–1955) was a very wealthy Englishman, often described as an eccentric, who devoted his life to natural history. He founded Paignton Zoo, Devon, initially as his private menagerie. The collection still flourishes today.

Whitlock

Australian Treecreeper genus *Whitlocka* **Mathews**, 1912 NCR
[Now in *Climacteris*]

White-naped Honeyeater sp. *Melithreptus whitlocki* Mathews, 1909 NCR
[Alt. Swan River Honeyeater; JS *Melithreptus chloropsis*]

Whitlock's Thornbill *Acanthiza apicalis whitlocki* **North**, 1909
[Alt. Inland Thornbill ssp.]

Black-capped Sittella ssp. *Neositta pileata whitlocki* Mathews, 1912 NCR
[JS *Daphoenositta chrysoptera pileata*]

Crested Pigeon ssp. *Ocyphaps lophotes whitlocki* Mathews, 1912

Inland Dotterel ssp. *Peltohyas australis whitlocki* Mathews, 1912 NCR; NRM

Purple-crowned Lorikeet ssp. *Glossopsitta porphyrocephala whitlocki* Mathews, 1912 NCR; NRM

Shy Heathwren ssp. *Calamanthus cautus whitlocki* Mathews, 1912
[Syn. *Hylacola cauta whitlocki*]

Spotted Pardalote ssp. *Pardalotus punctatus whitlocki* Mathews, 1912 NCR
[JS *Pardalotus punctatus punctatus*]

Western Rosella ssp. *Platycercus icterotis whitlocki* Mathews, 1912 NCR
[JS *Platycercus icterotis xanthogenys*]

Dusky Gerygone ssp. *Gerygone tenebrosa whitlocki* Mathews, 1915 NCR
[JS *Gerygone tenebrosa christophori*]

Frederick Bulstrode Lawson Whitlock (1860–1953) was a British ornithologist. He took all of the money from the safe of the bank he managed in Nottingham and disappeared (1897). A reward of £100 was offered for 'information leading to his arrest'. The notice included this description: '… he has a shifty expression when talking. He is a clever bicyclist; and a collector of birds and birds' eggs, upon which he is a considerable authority. He usually converses on this subject when in company …' He eluded the British authorities and reached Australia, where he was arrested (1898) and sent back to England to face trial; the money was never found. Any period of incarceration cannot have been very long, as he was certainly back in Australia in 1901. He collected more extensively than any other ornithologist in Western Australia, including for H. L. White (q.v.). He also visited the Hermannsburg Range, in central Australia, in search of the rare Night Parrot *Pezoporus occidentalis*. He was the first to collect the eggs and nests of many species. His bird skins are in the Western Australian Museum, Perth, and in the H. L. White Collection in the National Museum, Melbourne, along with most of the eggs that he collected. His research and travels were recorded in numerous articles in *Emu*. He was made an Honorary Life Member of the Royal Australian Ornithologists' Union and the Western Australian Naturalists' Club. He wrote *Birds of Derbyshire* (1893).

Whitmee

Polynesian Triller ssp. *Lalage maculosa whitmeei* **Sharpe**, 1878

Rev. Samuel James Whitmee (1838–1925) was a British missionary to Samoa (1863–1876). He became the incumbent at churches first in Dublin and subsequently in Bristol (1879–1910). He was asked to return to Samoa, which he did (1891), where he became a great friend of a recent arrival – Robert Louis Stevenson. He left again (1893), returning to England (1894) via Australia and New Zealand. He was an amateur naturalist and botanist and presented hundreds of specimens to the British Museum and to Kew Gardens.

Whitney (Family)

Golden Monarch ssp. *Carterornis chrysomela whitneyorum* **Mayr**, 1955

Mayr (q.v.) wrote that he was naming this bird after three generations of the Whitney family which 'have sponsored exploration and zoological research in the most generous manner.' The patriarch of this family was William Collins Whitney (1841–1904), an American financier, philanthropist, lawyer and politician who was Secretary of the Navy (1885–1889). His son was Harry P. Whitney (q.v.). Harry's son, Cornelius Vanderbilt Whitney (1899–1992), was interested in marine life and helped to found the oceanarium 'Marineland of Florida'.

Whitney, B. M.

Bahia Spinetail *Synallaxis whitneyi* **Pacheco** & Gonzaga, 1995

Bret Meyers Whitney (b.1955) is an American ornithologist, Research Associate at the Museum of Natural Science at Louisiana State University, and Associate of the Laboratory of Ornithology at Cornell. He specialises in sound recordings of Neotropical birds. He has produced many recordings including, with others, *Sounds of Neotropical Rainforest Mammals: An Audio Field Guide* (1998) and, solely, *Voices of New World Parrots* (2002). His truly remarkable skill at making recordings has meant that he has discovered several species new to science including, reportedly on his first day in the field in Madagascar, *Cryptosylvicola randrianasoloi*. He also leads bird tours, being one of the founders of the company Field Guides Inc. in Austin, Texas.

Whitney, H. P.

Fatuhiva Monarch *Pomarea whitneyi* **Murphy** & **Mathews**, 1928

Whitney's Thicket Warbler *Megalurulus whitneyi* **Mayr**, 1933
[Alt. Guadalcanal Thicketbird]

Wedge-tailed Shearwater ssp. *Puffinus pacificus whitneyi* **P. R. Lowe**, 1925 NCR; NRM
[Syn. *Ardenna pacifica*]

Island Thrush ssp. *Turdus poliocephalus whitneyi* Mayr, 1941

Harry Payne Whitney (1872–1930) was an American banker and heir to a tobacco and oil empire. He was a philanthropist who co-sponsored a number of expeditions to the Pacific, notably the Sanford-Whitney Expeditions (1921 and 1929). He married banker's daughter and artist Gertrude Vanderbilt (see **Gertrude (Whitney)**). Whitney was a passionate race-horse owner who captained the United States polo team (1910). He also competed with his yacht *Vanitie* in the America's Cup.

Whitney, J. D.

Elf Owl *Micrathene whitneyi* **J. G. Cooper**, 1861

Josiah Dwight Whitney (1819–1896) was a Californian state geologist who founded the Harvard School of Mining. Mount Whitney in California is also named for him. He wrote *The Yosemite Book – A Description of the Yosemite Valley and the Adjacent Region of the Sierra Nevada, and of the Big Trees of California* (1868). A member of his staff, James G. Cooper, the son of W. C. Cooper (q.v. both), found and described the owl.

Whittaker

Alta Floresta Antpitta *Hylopezus whittakeri* Carneiro *et al.*, 2012
[Syn. *Hylopezus macularius whittakeri*]

Andrew Whittaker (b.1960) is a British ornithologist who founded the birding tour company 'Birding Brazil Tours'. He moved to Brazil (1987) to work for the USNM, banding rain-forest birds in Manaus. His special interest is bird vocalisations and, along with colleagues, he published (2008) a collection of four CDs of *Voices of the Brazilian Amazon*. His exceptional knowledge of voices has enabled him to make several important ornithological discoveries, including describing several species new to science. He was the first to discover the antpitta (2007). He is currently working on a comprehensive field guide for Brazil. He is a member of the Brazilian Records Committee and is a very active conservationist, living in his preferred Amazonian biome in Manaus.

Whittell

Medium-billed Prion ssp. *Pachyptila gouldi whittelli* **Mathews**, 1938 NCR
[Alt. Salvin's Prion; JS *Pachyptila salvini salvini*]

Major Hubert Massey Whittell (1883–1954) intended to be a physician and studied at Edinburgh (1899–1902). He abandoned medicine to become a soldier and was commissioned (1904). He went to India (1905) and was appointed (1907) to the 56th Punjabi Rifles, Indian Army. He served in Belgium, France and Egypt (WW1). He settled in Western Australia (1925), developed a dairy farm and orchard, and started collecting. He was a very skilled taxidermist, often providing specimens for Mathews (q.v.). He wrote *The Literature of Australian Birds* (1954). Whittel [*sic*] Island, north-west of Perth, is named after him.

Whymper

Alpine Accentor ssp. *Prunella collaris whymperi* **E. C. S. Baker**, 1915

Samuel Leigh Whymper (1857–1941) was a British oologist, collector and big-game hunter. He was an analytical chemist

who went to India (1877) and managed a brewery at Jeolikote, India (1878–1912). He celebrated his retirement by taking a round-the-world voyage before settling in London. He did most of his collecting in the Garhwal region of the Himalayas at altitudes of 14,000–18,000 feet and wrote 'Birds-nesting in Garhwal' (1904). His eldest brother was Edward Whymper, a celebrated mountaineer.

Whyte

Whyte's Barbet *Stactolaema whytii* **G. E. Shelley**, 1893
Red-faced Crombec *Sylvietta whytii* G. E. Shelley, 1894
Yellow-browed Seedeater *Serinus whytii* G. E. Shelley, 1897
[Syn. *Crithagra whytii*]

Shelley's Francolin ssp. *Scleroptila shelleyi whytei* **Neumann**, 1908
Montane Double-collared Sunbird ssp. *Cinnyris ludovicensis whytei* **C. W. Benson**, 1948

Alexander Whyte (1834–1908) was a government naturalist in Nyasaland (Malawi), where he collected extensively under the patronage of Sir Harry Johnston (q.v.). Among several articles referring to his collecting activities, Britten wrote 'The plants of Milanji, Nyasa-land, collected by Mr Alexander Whyte' (1894); and Oldfield Thomas wrote 'On the mammals obtained by Mr A Whyte in Nyasaland, and presented to the British Museum by Sir H. H. Johnston' (1897). Three mammals and a reptile are named after Whyte.

Wickham

Wickham's Silver Pheasant *Gennaeus wickhami* **Oates**, 1899 NCR
[Hybrid of *Lophura* subspecies]

Indian Blue Robin ssp. *Luscinia brunnea wickhami* **E. C. S. Baker**, 1916

Percy Frederic Wickham (1869–1949) was a British civil servant and engineer in India and Burma (Myanmar) (1892–1922). He wrote 'Catalogue of P. F. Wickham Burmah egg collection' (1898).

Wied

Wied's Crested Flycatcher *Myiarchus tyrannulus* P. L. S. Müller, 1776
[Name applied also to the species below; Alt. Brown-crested Flycatcher]
Wied's Parrotlet *Touit melanonotus* Wied-Neuwied, 1820
[Alt. Brown-backed Parrotlet; *Urochroma wiedi* is a JS]
Wied's Tinamou *Crypturellus noctivagus* Wied-Neuwied, 1820
[Alt. Yellow-legged Tinamou]
Wied's Tyrant Manakin *Neopelma aurifrons* Wied-Neuwied, 1831
Wied's Crested Flycatcher *Myiarchus oberi* **G. N. Lawrence**, 1877
[Name applied also to the species above; Alt. Lesser Antillean Flycatcher]
Wied's Emerald *Chlorostilbon wiedi* **Boucard**, 1895 NCR
[Alt. Glittering-bellied Emerald; JS *Chlorostilbon lucidus pucherani*]

Black-necked Aracari ssp. *Pteroglossus aracari wiedii*
J. H. C. F. Sturm & J. W. Sturm, 1847

(See **Maximilian**)

Wiedenfeld

Barred Owlet-nightjar ssp. *Aegotheles bennettii wiedenfeldi*
Laubmann, 1914
Papuan King Parrot ssp. *Alisterus amboinensis wiedenfeldi*
Neumann, 1927 NCR
[JS *Alisterus chloropterus moszkowskii*]

Dr L. von Wiedenfeld (DNF) was a German collector in New
Guinea (1909–1910).

Wiepken

Spotted Wood Owl ssp. *Strix seloputo wiepkeni* Blasius,
1888

Dr Carl Friedrich Wiepken (1815–1897) was a German zoologist and ornithologist at the Oldenburg Natural History
Museum (1837–1895), becoming its Director (1879–1895).

Wiglesworth

Chattering Kingfisher *Todirhamphus wiglesworthi* **Sharpe**,
1906 NCR
[JS *Todiramphus tutus*]

Brown-throated Sunbird ssp. *Anthreptes malacensis*
wiglesworthi **Hartert**, 1902
Sulawesi Cicadabird ssp. *Coracina morio wiglesworthi*
van Oort, 1907 NCR
[NUI *Coracina morio morio*]
Fiji Shrikebill ssp. *Clytorhynchus vitiensis wiglesworthi*
Mayr, 1933
Black Sunbird ssp. *Nectarinia sericea wiglesworthi*
Delacour, 1944 NCR
[JS *Leptocoma sericea porphyrolaema*]

Lionel William Wiglesworth (1865–1901) was an English ornithologist. He studied in Braunschweig (1889–1891) and
worked for the Zoological Museum in Dresden (1891–1900).
He left Europe (1900), stopping at Fiji en route to New Zealand
and Australia, where he contracted dysentery and died. He
co-wrote, with A. B. Meyer (q.v.), *The Birds of Celebes and
the Neighbouring Islands* (1898).

Wilcox

Pale-eyed Pygmy Tyrant ssp. *Atalotriccus pilaris wilcoxi*
Griscom, 1924

A. L. Wilcox (DNF) was an American businessman. He was
President of Tropical Forest Products Company, Veraguas,
Panama (1924–1928). He was Griscom's (q.v.) host for several
weeks during his expedition to Panama (1924), when the
pygmy tyrant holotype was collected.

Wilder

Grey-headed Bullfinch ssp. *Pyrrhula erythaca wilderi* **Riley**,
1918
Brown Dipper ssp. *Cinclus pallasii wilderi* **La Touche**, 1925
NCR

[JS *Cinclus pallasii pallasii*]
Japanese Pygmy Woodpecker ssp. *Dendrocopos kizuki*
wilderi **Kuroda**, 1926
[SII *Dendrocopos kizuki kizuki*]

Rev. George Durand Wilder (1869–1946) was an American
missionary to China (1894–1938). He was ordained (1894) and
went to Tungcho, where he was stationed during the Boxer
Rebellion (1900). He taught theology at the University of
Peking (1910–1938). After retiring from the university he
helped co-ordinate emergency relief in Shatung province
during the Sino-Japanese War, but was detained in Japan
until repatriated to the USA (1943).

Wiley

Gundlach's Hawk ssp. *Accipiter gundlachi wileyi* Wotzkow,
1991

James W. Wiley is an American ornithologist who is a
Research Associate at the Western Foundation of Vertebrate
Zoology, Camarillo, California. He was at the US National
Biological Survey (1996) and the University of Maryland
(2004). He has a particular interest in the birds of the Carib-
bean, and wrote *Ornithology in Puerto Rico and the Virgin
Islands* (1996).

Wilhelm

Brown Woodland Warbler ssp. *Phylloscopus umbrovirens*
wilhelmi **Gyldenstolpe**, 1922

Carl Wilhelm Ludwig, Prince of Sweden and Duke of Söder-
manland (1884–1965), led the Swedish Central African
Expedition (1920–1921).

Wilhelmina (Meyer)

Sicklebill sp. *Epimachus wilhelminae* **A. B. Meyer**, 1873 NCR
[Alt. Black-billed Sicklebill; JS *Drepanornis albertisi*]
Wilhelmina's Lorikeet *Charmosyna wilhelminae* A. B. Meyer,
1874
[Alt. Pygmy Lorikeet]

Wilhelmina Meyer (DNF) was the wife of the describer,
Adolph Meyer (q.v.).

Wilhelmina, Queen

Wilhelmina's Bird-of-Paradise *Lamprothorax wilhelminae*
A. B. Meyer, 1894 NCR
[Presumed hybrid: *Lophorina superba* x *Cicinnurus*
magnificus]

Blood-breasted Flowerpecker ssp. *Dicaeum sanguinolentum*
wilhelminae **Büttikofer**, 1892
Papuan King Parrot ssp. *Alisterus chloropterus wilhelminae*
Ogilvie-Grant, 1911 NCR
[NUI *Alisterus chloropterus callopterus*]

(See **Queen Wilhelmina**)

Wilkes

Wilkes's Pigeon *Ducula aurorae wilkesii* **Peale**, 1848
[Alt. Polynesian Imperial Pigeon ssp.]

Charles Wilkes (1798–1877) was a pioneering navy captain who led the US Exploring Expedition (1838–1842), which was a milestone in American science. It comprised four naval vessels, the flagship *Vincennes*, the *Peacock*, the *Porpoise* and the store ship *Relief*. Two New York pilot boats, the *Sea Gull* and the *Flying Fish*, were used as survey vessels close to shore. They visited Brazil, Tierra del Fuego, Antarctica, Chile, Australia, New Zealand, the west coast of North America, the Philippines and the East Indies. The two penetrations into Antarctic waters sighted land and provided the first proof of an Antarctic continent. Wilkes was a strict disciplinarian who was disliked by many of the crew. He took with him 82 officers, nine naturalists, scientists and artists, and 342 sailors. Of the latter, only 223 returned. During the voyage, 62 were discharged as unsuitable, 42 deserted, and 15 died of disease, injury or drowning. However, he brought back a wealth of geological, botanical, zoological, anthropological and other material, which was to be the foundation for much of American science. Peale (q.v.) was one of the expedition scientists and official illustrator.

Wilkins

Wilkins's Bunting *Nesospiza wilkinsi* **P. R. Lowe**, 1923
[Alt. Wilkins's Finch, Grosbeak Bunting]

White-bellied Cuckooshrike ssp. *Coracina papuensis wilkinsi* **Kinnear**, 1924 NCR
[JS *Coracina papuensis robusta*]

Captain Sir George Hubert Wilkins (1888–1958) was an Australian polar explorer and ornithologist. He explored both the Arctic (1913–1917) and the Antarctic (1920–1922). He was the 13th child born to a South Australian sheep-farming family. An official biography lists his career as war correspondent, polar explorer, naturalist, geographer, climatologist, aviator, author, balloonist, war hero, reporter, secret agent, submariner and navigator! His twin loves were aviation (he always carried a miniature Australian flag in his cockpit) and cinematography. However, his exploits were legion and we recommend further reading on this extraordinary man. His adventures (1928–1930) included the purchase of a surplus WW1 submarine for $1 that he renamed *Nautilus* and attempted to cruise beneath the ice to the North Pole. Unfortunately, the old ship broke down and the expedition failed. Wilkins was the cameraman and naturalist on board Shackleton's ship *Quest* which called briefly at the Tristan da Cunha islands (1922) where the bunting was discovered. He was on board *Wyatt Earp* (1936), the support ship for Lincoln Ellsworth, the American pioneering aviator/explorer who was stranded at Admiral Byrd's abandoned camp in Antarctica. Wilkins wrote *No Foxes Seen, A Log of Arctic Flying Adventures* (1928). His last expedition was to Antarctica (aged 69) as a guest of Operation Deepfreeze. He died of a heart attack in the USA and his ashes were taken to the North Pole and scattered there. A reptile is also named after him.

Wilkinson

Buff-banded Rail ssp. *Gallirallus philippensis wilkinsoni* **Mathews**, 1911

Johnson Wilkinson (d.1928) was an English naturalist and ornithologist in Huddersfield, Yorkshire.

Will

Croaking Cisticola ssp. *Cisticola natalensis willi* **C. M. N. White**, 1945 NCR
[JS *Cisticola natalensis katanga*]

Dr J. W. O. Will (b.1909) was a physician in the British colonial medical service in Northern Rhodesia (Zambia) (1944).

Willard

Willard's Sooty Boubou *Laniarius willardi* Voelker *et al.*, 2010

Velvet-browed Brilliant ssp. *Heliodoxa xanthogonys willardi* Weller & Renner, 2001

Dr David E. Willard (b. 1946) is an American ornithologist. He was collection manager of the Bird Division, Field Museum, Chicago (1978–2012). His bachelor's degree was awarded by Carleton College, Minnesota (1968), and his doctorate by Princeton (1975). (See **David Willard**)

Willcocks

Willcocks's Honeyguide *Indicator willcocksi* **B. Alexander**, 1901

General Sir James Willcocks (1857–1926) commanded the Ashanti expedition (1900), intended to suppress the Ashanti Rebellion led by (female) leader Yaa Asantewaa. The then-Colonel Wilcocks led 1,400 soldiers to Kumasi in the Gold Coast (Ghana) from Nigeria. He wrote *From Kabul to Kumassi – Twenty-Four Years of Soldiering and Sport* (1904). He commanded (1914) the Indian Army Corps in the British Expeditionary Force in France (WW1). However, he resigned (1915) following differences with the Commander, Sir Douglas (later Earl) Haig. He became Governor of Bermuda (1917–1922). He died in India, at Bharatpur.

Willett

Leach's Storm-petrel ssp. *Oceanodroma leucorhoa willetti* **Van Rossem**, 1942 NCR
[JS *Oceanodroma leucorhoa leucorhoa*]

George Willett (1879–1945) was a Canadian malacologist and ornithologist whose family moved to California (1888). He was the first Curator of Birds at the Los Angeles County Museum of Natural History (1926–1945). He collected in California and Alaska. He wrote 'Mammals of Los Angeles County, California' (1944).

William (Chikundulo)

African Thrush ssp. *Turdus olivaceus williami* **C. M. N. White**, 1949 NCR
[JS *Turdus pelios stormsi*]

William Washa Chikundulo was an African collector for the describer in Northern Rhodesia (Zambia).

William (Doherty)

Dark-backed Imperial Pigeon ssp. *Ducula lacernulata williami* **Hartert**, 1896

(See **Doherty**)

William (King of Holland)

William's Fig Parrot *Cyclopsitta gulielmitertii* **Schlegel**, 1866 [Alt. Orange-breasted Fig Parrot, King of Holland's Fig Parrot]

(See **King of Holland**)

William (Wilson)

Viridian Metaltail *Metallura williami* **de Lattre** & **Bourcier**, 1846

William Savory Wilson (DNF) was an American cloth merchant and financier. He and his wife, Adda (q.v.), lived in Paris (1840s) but returned to Philadelphia (1853). He was a brother of E. Wilson (q.v.) and T. B. Wilson (q.v.).

Williamina

Williamina's Niltava *Niltava williaminae* **Meyer de Schauensee**, 1929 NCR [Alt. Vivid Niltava; JS *Niltava vivida oatesi*]

Bradfield's Hornbill ssp. *Tockus bradfieldi williaminae* Meyer de Schauensee, 1931 NCR; NRM White Hawk ssp. *Leucopternis albicollis williaminae* Meyer de Schauensee, 1950

Williamina Wemyss Meyer de Schauensee *née* Wentz (b.1905) was the wife of the describer Rodolphe Meyer de Schauensee (q.v.). She accompanied him on an expedition in southern and eastern Africa (1930) when the hornbill was collected.

Williams, F. T.

Williams's Kalij Pheasant *Lophura leucomelanos williamsi* **Oates**, 1898

Captain Frederic Thesiger Williams (1862–1902) was commissioned a British Army Officer (1881) seconded to 26th Madras Native Infantry (1885). He served in Burma (Myanmar) on the western frontier (1885–1887), and was in the Chin Hills (Burma) where he collected the pheasant holotype (1894–1897). He was Deputy Assistant Quarter Master General, 4th Infantry Brigade, China Expeditionary Force (1900). He was promoted to Major (1901). He died after being mauled by a wounded tiger.

Williams, J. G.

Williams's Lark *Mirafra williamsi* **J. D. Macdonald**, 1956

Abyssinian Ground Thrush ssp. *Zoothera piaggiae williamsi* J. D. Macdonald, 1948 NCR [JS *Zoothera piaggiae piaggiae*]

Red-capped Lark ssp. *Calandrella cinerea williamsi* **Clancey**, 1952 Brown Woodland Warbler ssp. *Phylloscopus umbrovirens williamsi* Clancey, 1956

John George Williams (1913–1997) was a British ornithologist, taxidermist and illustrator who was Curator (1946–1966) of the Coryndon Museum (National Museum), Nairobi, Kenya. He discovered several new taxa during his expeditions in eastern Africa, including his eponymous lark and Philippa's Crombec, named after his wife (q.v.). He wrote *A Field Guide to the Birds of East Africa* (1980), which was an expanded version of his pioneering earlier work (1963). He was also the author of *A Field Guide to the National Parks of East Africa* (1967). His obituary in *Ibis* (1999) notes that: 'Early on, he saw the need for conservation. The establishment of a National Park at Lake Nakuru, famous for its flamingos, was in large part due to his efforts.' He and his wife returned to England (1978) but he never fully retired. Just before his death he prepared bird specimens for the Royal Scottish Museum in Edinburgh.

Williamson, R. S.

Williamson's Sapsucker *Sphyrapicus thyroideus* **Cassin**, 1852

Lieutenant Robert Stockton Williamson (1825–1882) was an American soldier and engineer. He was assigned to conduct surveys for proposed routes for the transcontinental railroad in California and Oregon. John Strong Newberry (1822–1892), an army surgeon and geologist, collected the sapsucker and named it *Picus williamsonii*, believing it to be a new species (1857). Only in the 1870s was it realised that *williamsonii* was in fact the male plumage of a species already named by Cassin (*thyroideus*). Mount Williamson in California was perhaps also named after Robert S. Williamson (1864), or perhaps not: in an interview, Don McLain told John Robinson that he had named this peak for Will Williamson, a friend of his. When reminded of Lt. Robert Williamson, McLain added 'Well, yes, I named [it] for him too'. Historian Don Hedly reported that McLain's widow told him this same story.

Williamson, W. J. F.

Horsfield's Bush Lark ssp. *Mirafra javanica williamsoni* **E. C. S. Baker**, 1915 Rufous Woodpecker ssp. *Micropternus brachyurus williamsoni* **Kloss**, 1918 Oriental White-eye ssp. *Zosterops palpebrosus williamsoni* **H. C. Robinson** & Kloss, 1919 Abbott's Babbler ssp. *Malacocincla abbotti williamsoni* **Deignan**, 1948 Streaked Weaver ssp. *Ploceus manyar williamsoni* **B. P. Hall**, 1957 Williamson's Flycatcher *Muscicapa dauurica williamsoni* Deignan, 1957 [Alt. Asian Brown Flycatcher ssp.]

Sir Walter James Franklin Williamson (1867–1954) was the financial adviser to the government of Siam (Thailand) (1904–1925). He was co-editor of *Journal of the Natural History*

Society of Siam and wrote a number of articles in it, such as 'The birds of Bangkok'. A mammal is also named after him.

Willis

Antbird genus _Willisornis_ Agne & **Pacheco**, 2007

Willis's Antbird _Cercomacra laeta_ **Todd**, 1920

Red-throated Ant Tanager ssp. _Habia fuscicauda willisi_
     Parkes, 1969

Dr Edwin O'Neill Willis (b.1935), an American ornithologist, was Professor of Zoology at the Universidade Estadual Paulista in Rio Claro, São Paulo, Brazil. He retired in 2005. Willis has done a great deal of work on antbird ecology and systematics, chiefly in Brazil but also in Panama. He is a pioneer of field studies of Neotropical birds and the world's recognised expert on Neotropical ant-following birds. Many of his papers deal with 'applied avian biogeography'. His publications include 'The behaviour of Ocellated Antbirds' (1973) and 'The composition of avian communities in remanescent woodlots in southern Brazil' (1979). In reply to our inquiry about the naming of Willis's Antbird, he wrote: 'I was the first person to note that the bird had a different voice from the ordinary species, both at Belém and Manaus.' Todd first described the taxon as a subspecies of the Dusky Antbird _C. tyrannina_.

Willkonski

Tawny Owl ssp. _Strix aluco willkonskii_ **Menzbier**, 1896

Willkonski (DNF) was a Russian collector in Batum (now Batumi) on the Black Sea coast, Georgia. That is all the detail we have been able to find about him; even Menzbier was confused, originally giving his name as 'Willkousky'. He provided no first name(s) for Willkonski.

Willoughby

Yellow-browed Camaroptera ssp. _Camaroptera superciliaris_
     willoughbyi **Bannerman**, 1923 NCR; NRM
Blue Pitta ssp. _Hydrornis cyaneus willoughbyi_ **Delacour**,
     1926

(See **Lowe, W. P.**)

Wills

Alpine Swift ssp. _Tachymarptis melba willsi_ **Hartert**, 1896

Rev. James Wills (1836–1898) was an English ornithologist and missionary in Madagascar (1870–1898). He wrote 'Notes on some Malagasy birds rarely seen in the Interior' (1893) and made a collection of birds that was later sold to the USNM. A reptile is also named after him.

Wilson, A.

New World Warbler genus _Wilsonia_ **Bonaparte**, 1838

Wilson's Warbler _Cardellina pusilla_ Wilson, 1811
[Syn. _Wilsonia pusilla_]
Wilson's Plover _Charadrius wilsonia_ **Ord**, 1814
Wilson's Thrush _Catharus fuscescens_ J. F. Stephens, 1817
[Alt. Veery]

Wilson's Phalarope _Phalaropus tricolor_ **Vieillot**, 1819
Wilson's Storm-petrel _Oceanites oceanicus_ **Kuhl**, 1820
[_Procellaria wilsonii_ (Bonaparte, 1823) is a JS]
Hawk sp. _Falco wilsonii_ Bonaparte, 1824 NCR
[Alt. Broad-winged Hawk; JS _Buteo platypterus_]
Wilson's Snipe _Gallinago delicata_ Ord, 1825
Sandpiper sp. _Tringa wilsonii_ **Nuttall**, 1834 NCR
[Alt. Least Sandpiper; JS _Calidris minutilla_]

Long-eared Owl ssp. _Asio otus wilsonianus_ **Lesson**, 1830

Alexander Wilson (1766–1813) was a pioneering American ornithologist, the first to study American birds in their native habitats. He is often called the 'Father of American Ornithology'. Wilson was born in Paisley, Scotland, where he earned a meagre livelihood as an itinerant poet and peddler of muslin. His narrative poem _Watty and Meg_ was published anonymously (1792), attaining great popularity, but was ascribed to the Scottish poet Robert Burns. Subsequently, during a labour dispute in Paisley, Wilson wrote satiric verses lampooning the manufacturers and was imprisoned for libel. Following his release (1794) he emigrated from Scotland to the USA, where Bartram (q.v.) befriended him. While working as a village schoolmaster in Pennsylvania, Wilson began to collect material for a comprehensive work, illustrated with his own drawings, on the birds of America. Seven volumes of his _American Ornithology_ were published (1808–1813); an additional two volumes were edited and published after his death from dysentery. He briefly met Audubon (q.v.) (1810) when he entered the latter's shop looking for subscribers to his publication. Audubon offered him drawings but Wilson seems to have ignored this and after a subsequent encounter at the home of Rembrandt Peale, Audubon described him as exhibiting 'a strong feeling of discontent, or a decided melancholy. The Scotch airs which he played sweetly on his flute made me melancholy too, and I felt for him'. Wilson is noted for the accuracy of his descriptions and for his superior illustrations. A mammal is named after him.

Wilson, E.

Wilson's Bird-of-Paradise _Diphyllodes respublica_
     Bonaparte, 1850

Edward Wilson Jr (fl.1807–at least 1857) lived in Pembrokeshire (Wales), but was born in Philadelphia into a Welsh family that returned to Britain (1830). His father (Edward Wilson Sr), who made a large fortune out of the iron trade between Philadelphia and Liverpool, had several sons, some of whom remained in Philadelphia when their parents went home – one of these was Dr Thomas Bellerby Wilson (q.v.) and another William Savory Wilson [William (Wilson)] (q.v.). Edward Jr was a Corresponding Member of the Philadelphia Academy of Natural Sciences and presented its library with c.15,000 books. His wife was Mrs Frances 'Fanny' Wilson (see **Fanny**) and their grandson was Dr Edward Adrian Wilson (1872–1912) who died in the Antarctic on Scott's last expedition. (See also **Edward (Wilson)**)

Wilson, E. A.

Petrel sp. *Oestrelata wilsoni* **Sharpe**, 1902 NCR
[Alt. Trindade Petrel; JS *Pterodroma arminjoniana*]

Southern Giant Petrel ssp. *Macronectes giganteus wilsoni*
Mathews, 1912 NCR; NRM
South Polar Skua ssp. *Catharacta maccormicki wilsoni*
Mathews, 1913 NCR
[Syn *Stercorarius maccormicki* NRM]

Dr Edward Adrian Wilson (1872–1912) was a naturalist and Antarctic explorer. He was appointed as the Assistant Surgeon and Vertebrate Zoologist to the British National Antarctic Expedition (1901–1904) aboard *Discovery,* under Commander Robert Falcon Scott. He was also a gifted artist and was probably the last 'exploration artist', as after his time photography became the major medium for recording explorations. He returned to the Antarctic (1910) with Captain Scott aboard *Terra Nova* as Chief of Scientific Staff. He died with his comrades on the return from the South Pole (1912). A mammal is named after him.

Wilson, E. H.

Blood Pheasant *Ithagenes wilsoni* **Thayer** & **Bangs**, 1912 NCR
[JS *Ithaginis cruentus geoffroyi*]

Ernest Henry Wilson (1876–1930) was an English botanist and plant collector in China (1899–1910), Japan (1911–1916), Korea and Formosa (Taiwan) (1917–1918) and Australia, New Zealand, India, Central and South America, and East Africa (all 1922–1924). He worked for the Arnold Arboretum, Boston, Massachusetts (1905–1930) and eventually became Keeper of the Arboretum (1927–1930). When in China (1910) he suffered a broken leg and used his camera tripod as a splint, limping for three days to get treatment. He and his wife both died in a car crash in Worcester, Massachusetts. He wrote *Lilies of Eastern Asia; a Monograph* (1925). About 60 Asian plants are named after him.

Wilson, F. E.

Gerygone genus *Wilsonavis* **Mathews**, 1912 NCR
[Now in *Gerygone*]

Black-eared Miner ssp. *Myzantha flavigula wilsoni*
Mathews, 1912 NCR
[JS *Manorina melanotis* NRM]

Francis Erasmus Wilson (1888–1960) was an Australian analytical chemist, entomologist, ornithologist and collector. He was Secretary of the RAOU (1911–1912). He worked for the same firm (Moran and Cato) for over 45 years (c.1906–1953), eventually as manager of the manufacturing department.

Wilson, M.

Wilson's Indigobird *Vidua wilsoni* **Hartert**, 1901
[Alt. Pale-winged Indigobird]

Captain Malcolm Wilson (1869–1900) was with the British Army in West Africa and collected birds on the Niger River.

Wilson, R. A.

New Zealand Fernbird ssp. *Megalurus punctatus wilsoni*
Stead, 1936
[Syn. *Bowdleria punctata wilsoni*]

Major Robert A. Wilson (1882–1964) was a New Zealand farmer and ornithologist. He served in France during WW1 (1916–1918) in both English and New Zealand regiments, and was famed as a deerstalker and hunter. He had a lifelong association with Stead (q.v.) through field trips, and later recorded in some detail the bush birds of the offshore islands around Stewart Island (1930s), writing *Bird Islands of New Zealand* (1959).

Wilson, R. B.

Barratt's Warbler ssp. *Bradypterus barratti wilsoni*
J. A. Roberts, 1933 NCR
[Alt. African Scrub Warbler ssp.; JS *Bradypterus barratti godfreyi*]
Bearded Scrub Robin ssp. *Erythropygia quadrivirgata wilsoni* Roberts, 1936 NCR
[JS *Erythropygia quadrivirgata quadrivirgata*]
Brimstone Canary ssp. *Crithagra sulphurata wilsoni*
J. A. Roberts, 1936
[Syn. *Serinus sulphuratus wilsoni*]

Lieutenant-Commander R. B. Wilson (d.1955) collected the holotype of the scrub warbler (1932) and other birds in Natal, South Africa.

Wilson, S. B.

Akiapolaau *Hemignathus wilsoni* **Rothschild**, 1893
[Name now replaced by *Hemignathus munroi* (Pratt, 1979)]
Hawaiian Crake *Pennula wilsoni* **Finsch**, 1898 EXTINCT NCR
[Alt. Hawaiian Rail; JS *Porzana sandwichensis*]

Hawaii Amakihi ssp. *Hemignathus virens wilsoni* Rothschild, 1893
[Syn. *Chlorodrepanis virens wilsoni*]

Scott Burchard Wilson (1865–1923) was an English ornithologist and collector who wrote *Aves Hawaiienses: The Birds of the Sandwich Islands* (1890).

Wilson, T. B.

Wilson's Coeligene *Coeligena wilsoni* **de Lattre** & **Bourcier**, 1846
[Alt. Brown Inca]
Cuban Kite *Chondrohierax wilsonii* **Cassin**, 1847
North Island Kokako *Callaeas wilsoni* **Bonaparte**, 1850

Dr Thomas Bellerby Wilson (1807–1865) was an American amateur ornithologist, naturalist, philanthropist and physician who graduated from the University of Pennsylvania (1830). He was a Trustee of the Academy of Natural Sciences, Philadelphia, and bought the John Gould (q.v.) Collection of Australian Birds on their behalf (1848). He was largely responsible for accumulating their enormous collection of birds and his patronage was recognised when he was eventually elected their President. Cassin (q.v.) catalogued the collection and served as Vice-president.

Winchell, H. V.

White-vented Whistler ssp. *Pachycephala homeyeri winchelli* **Bourns** & **Worcester**, 1894

Horace Vaughn Winchell (1865–1923) was the son of Newton H. Winchell (q.v.). He was a mining geologist and was associated with a number of US railway companies. The describers remarked upon his interest in the Steere expedition to the Philippines.

Winchell, N. H.

Winchell's Kingfisher *Todiramphus winchelli* **Sharpe**, 1877
[Alt. Rufous-lored Kingfisher]

Newton Horace Winchell (1839–1914) was an American archaeologist and geologist who took part in the expedition which led to the discovery of gold in the Black Hills of Dakota. (The leader of the expedition, Lieutenant-Colonel George Armstrong Custer, later gained notoriety as General Custer when the Sioux killed him and his entire company at the Battle of the Little Bighorn, Montana.) Winchell's career included being a schoolmaster in New York State (1855–1857) and later (1866–1872) Superintendent of Schools in Michigan, before settling in Minnesota (1872) where he was appointed to direct 'The Geological and Natural History Survey of Minnesota'. He published a classic study of the post-glacial retreat of St Anthony Falls, which carved the Mississippi River gorge in Minneapolis. The Winchell Trail, a path through the gorge, commemorates him, as does the Newton Horace Winchell School of Earth Sciences at the University of Minnesota. He wrote *The Aborigines of Minnesota* (1910). Sharpe (q.v.) described the kingfisher as part of a collection made in the Philippines by Professor Joseph B. Steere (q.v.), and wrote: 'This species is named, by Dr Steere's request, after his friend and old tutor, Mr Winchell.'

Wingate

Blue-winged Minla ssp. *Minla cyanouroptera wingatei* **Ogilvie-Grant**, 1900

Colonel Alfred Woodrow Stanley Wingate (1861–1938) was a British army officer in India (1884–1913). He explored in Burma (Myanmar) and China (1898–1899) and was at the relief of Peking (Beijing) (1900 Boxer rebellion). He wrote: *A Cavalier in China* (1940).

Winifred

Winifred's Warbler *Scepomycter winifredae* **Moreau**, 1938
[Alt. Mrs Moreau's Warbler; Syn. *Bathmocercus winifredae*]

Montane White-eye ssp. *Zosterops poliogastrus winifredae* Moreau & **W. L. Sclater**, 1934

Winifred Muriel Moreau (1891–1981) was the wife of the describer Reginald Moreau (q.v.). (See also **Mrs Moreau**)

Winkler, A.

Micronesian Imperial Pigeon ssp. *Muscadivora oceanica winkleri* **Neumann**, 1922 NCR
[JS *Ducula oceanica monacha*]

Dr A. Winkler (c.1890–?1945) was an Austrian entomologist who was involved in collecting from caves in the Carpathian mountains, Romania. He was in the Palau Islands (1922).

Winkler, H.

Luzon Honey-buzzard *Pernis steerei winkleri* Gamauf & Preleuthner, 1998

Prof. Dr Hans Winkler (b.1945) is an Austrian ornithologist and eco-ethologist. He is director of the Konrad Lorenz-Institute for Comparative Ethology, Austrian Academy of Sciences. He co-wrote *Woodpeckers: a Guide to the Woodpeckers, Piculets and Wrynecks of the World* (1995).

Winterbottom

Grey Penduline-tit ssp. *Anthoscopus caroli winterbottomi* **C. M. N. White**, 1946
Lesser Swamp Warbler ssp. *Acrocephalus gracilirostris winterbottomi* C. M. N. White, 1947
Madagascar Magpie Robin ssp. *Copsychus albospecularis winterbottomi* **Farkas**, 1972
[SII *Copsychus albospecularis inexpectatus*]
African Pipit ssp. *Anthus cinnamomeus winterbottomi* **Clancey**, 1985

Professor Dr Jack M. Winterbottom (1903–1984) held an honorary Professorship at the University of Cape Town and was appointed Director of the Percy FitzPatrick Institute (1960). He joined the British Colonial Service in Gold Coast (Ghana) (1927) before transferring to Northern Rhodesia (Zambia) as Education Officer (1931). He earned a PhD in zoology (1932). He stayed with the service until taking early retirement (1950), when he moved to Cape Town, eventually retiring from his second career as an ornithologist (1971). He edited *Checklist of the Birds of the South Western Cape* (1955). It was said of him: 'Jack Winterbottom's observational skills were an eye opener especially on the (to us) rather dull Peninsula Park section with its picnic areas, public open space and waterside paths. Within ten minutes he would have a list of 30 species at least ten of which the rest of us had overlooked and he was always surprised that we needed to have them pointed out to us.'

Winton

Longclaw sp. *Macronyx wintoni* **Sharpe**, 1891 NCR
[Alt. Rosy-throated Longclaw; JS *Macronyx ameliae*]

William Edward de Winton (1856–1922) was Superintendent of the Zoological Gardens, London (London Zoo). He wrote 'On the moulting of the King Penguin' (1900). Most of his scientific papers were commentaries on collections made by others from all over the world. He completed *Zoology of Egypt – Mammalia* (1902) after Dr John Anderson's (q.v.) death (1900). Four mammals are named after him.

Wirth

Blue-winged Minla ssp. *Minla cyanouroptera wirthi* Collar, 2011

Roland Wirth (b.1954) is a German amateur zoologist who founded and for thirty years (1982–2012) was Chairman of the

Munich-based Zoologische Gesellschaft für Arten- und Populationsschutz (Zoological Society for the Conservation of Species and Populations) (ZGAP). He remains Senior Project Advisor and was chair of the IUCN/SSC Small Carnivore Specialist Group for 10 years. He is the author of many articles on mammal (particularly primate) and bird taxonomy, distribution and conservation. He was honoured in the name because of his contribution to the conservation of several birds and mammals in Indochina.

Witherby

Witherby's Lark *Alaemon hamertoni* Witherby, 1905
[Alt. Lesser Hoopoe Lark]

European Robin ssp. *Erithacus rubecula witherbyi* **Hartert**, 1910
Witherby's Tree Pipit *Anthus trivialis haringtoni* Witherby, 1917
Common Reed Bunting ssp. *Emberiza schoeniclus witherbyi* **von Jordans**, 1923

Henry Forbes 'Harry F.' Witherby (1873–1943) was an English ornithologist who was an expert on British and Spanish birds. His varied activities had an enduring influence in most fields of British ornithology. The family firm H. F. & G. Witherby & Co. began specialising in bird books in the early 20th century when Harry published his own works. He edited the 5-volume *Handbook of British Birds* widely referred to as 'Witherby' (1938–1941). He started the first bird-ringing scheme in Britain (1909), incorporating a second scheme at the outbreak of WW1. Witherby transferred the control of his scheme to the BTO (1937), which still administers it. He was an early Vice-Chairman of the BTO and his donations were vital to it. They included a capital endowment derived from the proceeds of the sale of his collection of skins to BMNH. He was Chairman of the BOC (1924–1927) and also of the Council of the BOU. He was awarded the Godman-Salvin Medal of the BOU (1938), a signal honour for distinguished ornithological work. He founded *British Birds* (1907), a monthly journal that has become an institution. A mammal is named after him.

Witmer

Babbler sp. *Aethostoma witmeri* **Sharpe**, 1903 NCR
[Alt. White-chested Babbler; JS *Trichastoma rostratum macropterum*]

(See **Stone, W.**)

Witt

Lesser Goldfinch ssp. *Carduelis psaltria witti* P. R. Grant, 1964

L. Witt (DNF) was a Canadian taxidermist at the University of British Columbia.

Wolf, H.

Spotted Nutcracker ssp. *Nucifraga caryocatactes wolfi* **von Jordans**, 1940 NCR
[JS *Nucifraga caryocatactes caryocatactes*]

Dr Heinrich Wolf (DNF) was a German zoologist, Director of the Zoological Museum, Bonn.

Wolf, J.

Harrier sp. *Circus wolfi* **Gurney**, 1865 NCR
[Alt. Swamp Harrier; JS *Circus approximans*]

Joseph Wolf (1820–1899) was a German wildlife artist whose observations of living birds helped him to produce illustrations in very accurate and lifelike stances. He moved to London (1848) where his work was greatly admired. Edwin Landseer, the landscape painter, described Wolf as '… without exception, the best all-round animal artist who ever lived.'

Wolf, W. F. T.

Brown Wood Rail *Aramides wolfi* **Berlepsch** & **Taczanowski**, 1884

Professor W. Franz Theodor Wolf (1841–1924) was a German natural historian. He became a novitiate in the Jesuit order (1856), which sent him to University at Bonn where he studied botany, zoology, geology and mineralogy. He became Professor of Natural History at the Jesuit College at Lake Leach, Luxembourg (1864). He made an extensive collection of volcanic minerals in that region and his collection is still at the University of Bonn. He was consecrated as a priest (1870) and sent to Quito, Ecuador, where he was appointed Professor of Mineralology and Geology. The Jesuits expected him to preach against Darwinism, but Wolf was unable to follow the party line and resigned from the order, which dismissed him (1874). He moved to Guayaquil, and the new government of Ecuador offered him a job (1875) as Geologist of the State of Ecuador, in which capacity he witnessed the eruption of Cotopaxi (1877). He visited the Galapagos Islands (1875 and 1878) and wrote *Ein Besuch der Galápagos-Inseln, Sammlung von Vorträgen für das deutsche Volk* (1879). He returned to Dresden, Germany (1891). During the rampant inflation (after WW1) his savings were quickly exhausted and he became destitute, but the Government of Ecuador granted him a generous pension. He wrote a highly regarded monograph on the botanical genus *Potentilla* and several plants are named after him.

Wolfe

Oriental Skylark ssp. *Alauda gulgula wolfei* **Hachisuka**, 1930
[Sometimes merged with *Alauda gulgula wattersi*]
Black Baza ssp. *Aviceda leuphotes wolfei* **Deignan**, 1948

Colonel Lloyd Raymond 'Lobo' Wolfe (1891–1989) served in the US Army in the Philippines and Korea. He wrote *Check-list of the Birds of Texas* (1956). He was an early opponent of the use of DDT.

Wollaston

Alpine Pipit ssp. *Anthus gutturalis wollastoni* **Ogilvie-Grant**, 1913

Dr Alexander 'Sandy' Frederick Richmond Wollaston (1875–1930) was a physician, botanist, ornithologist and explorer.

He travelled extensively, including taking part in an expedition to the Ruwenzori Mountains in Uganda (1905). He led an expedition to Dutch New Guinea (1912–1913), having been on an earlier expedition (1910–1911), which had been deliberately misdirected by Dutch authorities. He was naval surgeon in East Africa (WW1) where he met his future brother-in-law, Richard Meinertzhagen (q.v.). He was on an Everest expedition (1922) as physician and botanist. He wrote 'An expedition to Dutch New Guinea' (1914) and *Life of Alfred Newton* (1921). While teaching, he was shot dead in his Cambridge rooms by a deranged student, Douglas Potts, who then shot and killed a policeman who had come to arrest him and finally committed suicide (1930). Two mammals and an amphibian are named after him.

Wollweber

Wollweber's Titmouse *Baeolophus wollweberi* **Bonaparte**, 1850
[Alt. Bridled Titmouse]
Wollweber's Jay *Aphelocoma wollweberi* **Kaup**, 1855
[Alt. Mexican Jay; Syn. *Aphelocoma ultramarina wollweberi*]

Wollweber (DNF) was a traveller and collector in Mexico, particularly the western state of Zacatecas where he took bird skins in the 1840s. He sent specimens to the Darmstadt Museum, Germany. Almost nothing seems to have been recorded about his life.

Wolstenholme

Akepa ssp. *Loxops coccineus wolstenholmei* **Rothschild**, 1893 EXTINCT

Edward 'Ted' B. Wolstenholme (DNF) was a New Zealander who was hired by Rothschild (q.v.) to help Henry Palmer (q.v.) in collecting bird specimens in the Hawaiian Islands (1892).

Wolters

Saw-wing genus *Woltersia* **Boetticher**, 1943 NCR
[Now in *Psalidoprocne* – may be valid as a subgenus]

European Serin ssp. *Serinus canaria woltersi* **Laubmann**, 1954 NCR
[JS *Serinus serinus* NRM]
Black-and-white Mannikin ssp. *Lonchura bicolor woltersi* **Schouteden**, 1956
Grosbeak Weaver ssp. *Amblyospiza albifrons woltersi* **Clancey**, 1956
[SII *Amblyospiza albifrons albifrons*]
Shaft-tailed Whydah ssp. *Vidua regia woltersi* **Pinto**, 1961 NCR; NRM
Greater Short-toed Lark ssp. *Calandrella brachydactyla woltersi* Kumerloeve, 1969

Hans Edmund Wolters (1915–1991) was a German ornithologist, an associate member of Museum Alexander König, Bonn (1960–1973) and head of the museum's ornithology department thereafter. He was one of the first European ornithologists to use a cladistic classification.

Wolterstorff

Eurasian Sparrowhawk ssp. *Accipiter nisus wolterstorffi* **Kleinschmidt**, 1901

Dr Willy Georg Wolterstorff (1864–1943) was a geologist and herpetologist. An illness deprived him of his hearing and power of speech (1871) but he learned to lip-read. He was also very myopic and so had a lonely childhood, compensating by collecting and keeping amphibians; they remained his major lifelong interest. His death two years before spared him seeing the total destruction of the Magdeburg museum and all his work, including 12,000 specimens in glass jars, by the RAF (1945). A reptile and three amphibians are named after him.

Wombey

Brown-headed Honeyeater ssp. *Melithreptus brevirostris wombeyi* **Schodde** & Mason, 1999

John Charles Wombey (b.1945) is an Australian zoologist who worked for CSIRO Division of Wildlife Research (1970–2001). For some time he was Collection Manager, Australian National Wildlife Collection, and after retiring he remained as a fellow for several years. The curator at that time was Richard Schodde (q.v.) who, in collaboration with fellow staff member Ian Mason, and with Wombey's assistance, spent many years gathering data which culminated in the publication of *The Directory of Australian Birds – Passerines* (1999). It was for this assistance that they honoured him in the honeyeater's trinomial. He retains a life-long interest in herpetology and was fortunate enough occasionally to use this expertise in the CSIRO, but the bulk of his time there was spent in the study of birds and mammals. He co-wrote *List of Australian Vertebrates: A Reference with Conservation Status* (2006). A claim to fame is that he rediscovered the Inland Taipan *Oxyuranus microlepidotus*, the most venomous land snake in the world. A variety of taxa, including a gecko, are named after him.

Wood

Pin-striped Tit Babbler ssp. *Macronus gularis woodi* **Sharpe**, 1877

Sharpe provided no etymology, and we are unable to identify this person.

Wood, C. A.

Polynesian Triller ssp. *Lalage maculosa woodi* **Wetmore**, 1925
Yellow-breasted Crake ssp. *Porzana flaviventer woodi* **Van Rossem**, 1934

Dr Casey Albert Wood (1856–1942) was a Canadian-born ophthalmologist. A keen amateur ornithologist, he established the Emma Shearer Wood Library of Ornithology at McGill University, Montreal. He wrote *Birds of Fiji* (1925).

Wood, H.

Mount Victoria Babax *Babax woodi* **Finn**, 1902

Assam Laughingthrush ssp. *Trochalopteron chrysopterum woodi* **E. C. S. Baker**, 1914

Colonel Henry Wood (1872–1940) joined the Royal Engineers (1892) and served with the Survey of India. The Government of Nepal permitted him to survey (1903) and identify the correct locations of various peaks in the Himalaya, including Mount Everest, which he proved the world's highest mountain. He was responsible (WW1) for extensive, accurate mapping of the topography of several fronts in France and in Macedonia. He was a notable early photographer and his work is now part of the Fox Talbot collection at Lacock Abbey in Wiltshire.

Wood, L. Jr

Olive-backed Sunbird ssp. *Cinnyris jugularis woodi* **Mearns**, 1909

Leonard Wood Jr (1892–1931) collected the sunbird holotype (1906) in the Sulu Archipelago, southern Philippines, when his father, also Leonard Wood (q.v.), was Governor of Moro Province in the Philippines. He was studying at Cornell but left to join the British Army (1916) and transferred to the US Army after the USA entered WW1. In civilian life he was involved in a number of financial disasters in banking and oil, and was declared bankrupt (1925). He was finally reduced to polishing the brass rails of a houseboat. He died of pneumonia.

Wood, L. Sr

Bagobo Babbler *Leonardina woodi* **Mearns**, 1905

Major-General Dr Leonard Wood (1860–1927) was a noted American soldier who became American Governor-General of the Philippines. He graduated from Harvard as a physician (1883). He served in the campaign against the Native American leader Geronimo and was awarded the Congressional Medal of Honour for his gallantry as both a medical and line officer. He was personal physician to President William McKinley and his family. He took part in the war against Spain, commanding (1898) the 1st Volunteer Cavalry, the 'Rough Riders'. His second-in-command was the former Assistant Secretary of Navy, Theodore Roosevelt (q.v.). He was military Governor of Cuba (1900–1902). He fought the Moros in the Philippine Insurrection (1904), staying on as Governor of Moro Province (1906–1908). He became Chief of Staff of the Army (1910), the only medical officer to ever hold the position, until resigning (1914). He tried for the Republican nomination for the Presidency (1920) but Warren Harding was chosen instead. Wood retired from the Army (1921), taking the post of Governor-General of the Philippines (1921–1927). He died after unsuccessful surgery to remove a brain tumour. Mearns (q.v.) served under Wood in the Philippines. (See **Leonardina**)

Wood, N. A.

Sandhill Crane ssp. *Grus canadensis woodi* **H. H. Bailey**, 1930 NCR
[JS *Grus canadensis tabida*]
Black-rumped Flameback ssp. *Brachypternus benghalensis woodi* **Koelz**, 1939 NCR
[JS *Dinopium benghalense tehminae*]

Norman Asa Wood (1857–1943) was an American zoologist at the Museum of Zoology, University of Michigan (1895–1932). He wrote *A Preliminary Survey of the Bird Life of North Dakota* (1923).

Wood, T. W.

Wood's Argus *Argusianus bipunctatus* Wood, 1871
[Alt. Double-banded Argus; now considered invalid]

Thomas W. Wood (DNF) was an illustrator and naturalist who named this mysterious bird in a letter printed in *Annals and Magazine of Natural History*. To quote Beebe (q.v.): 'This species is known only from the portion of a primary feather of a male bird. In 1871 this was found among some loose Argus feathers in the British Museum, described and named by Mr T W Wood. It differs so decidedly from any corresponding feather in the known species that there is little doubt that it represents a new species, although we have no idea of the country in which it lives …' Nothing similar has come to scientific notice ever since. Wood found the anomalous feather 'amongst some loose feathers of *Argus giganteus*' (i.e. *Argusianus argus*); perhaps suggesting it came from an aberrant specimen of that species or was an anomalous product of a normal bird – now considered the most likely explanation.

Woodford, C.

White-eye genus *Woodfordia* **North**, 1906

White-billed Crow *Corvus woodfordi* **Ogilvie-Grant**, 1887
Woodford's Rail *Nesoclopeus woodfordi* Ogilvie-Grant, 1889
Woodford's White-eye *Woodfordia superciliosa* North, 1906
[Alt. Bare-eyed White-eye]

Black Bittern ssp. *Dupetor flavicollis woodfordi* Ogilvie-Grant, 1888
Pied Goshawk ssp. *Accipiter albogularis woodfordi* **Sharpe**, 1888
Moustached Treeswift ssp. *Hemiprocne mystacea woodfordiana* **Hartert**, 1896
Blue-faced Parrotfinch ssp. *Erythrura trichroa woodfordi* Hartert, 1900

Charles Morris Woodford (1852–1927) was Resident Commissioner in the Solomon Islands Protectorate (1896–1914). He was a British adventurer, naturalist and philatelist. He established the first postal service in the islands and issued their first stamps, personally franking the envelopes. He wrote *A Naturalist Among the Head-Hunters* (1890), which is referred to in a letter by his friend, the novelist Jack London. Two mammals and two reptiles are also named after him.

Woodford, E.

Woodford's Owl *Strix woodfordii* **A. Smith**, 1834
[Alt. African Wood Owl]

Smith (q.v.) gave no etymology when describing this owl, but perhaps named it after Colonel Emperor John Alexander Woodford (1761–1835) ('Emperor' being derived from his mother's maiden name, Mary Emperor). He fought with distinction at the Battle of Salamanca (1812) and was specifically mentioned by Wellington in his post-battle despatch. He fought at Waterloo (1815) and commanded a battalion of the Coldstream Guards as part of the garrison at Cambrai (c.1816). Woodford was also a natural history collector who dealt in bird art in London. He built up a fine library of books concerning natural history, English antiquities and English and foreign history.

Woodhouse

Woodhouse's Scrub Jay *Aphelocoma woodhouseii* **Baird**, 1858
[Syn. *Aphelocoma californica woodhouseii*]
Woodhouse's Antpecker *Parmoptila woodhousei* **Cassin**, 1859
[Alt. Flowerpecker Finch]

Samuel Washington Woodhouse (1821–1904) was an American surgeon, explorer and naturalist who collected in the USA. He was on the Sitgreaves exploration of the Colorado and Zuni Rivers (1852) with the joint roles of doctor and naturalist. This expedition was the first to find Arizona's Petrified Forest. He published his journals under the title *A Naturalist in Indian Territory: The Journal of S. W. Woodhouse, 1849–50*. An amphibian and a mammal are also named after him.

Woodward, B. H.

White-throated Grasswren *Amytornis woodwardi* **Hartert**, 1905
Woodward's Shrike-thrush *Colluricincla woodwardi* Hartert, 1905
[Alt. Sandstone Shrike-thrush]

Woodward's Bush Lark *Mirafra javanica woodwardi* **Milligan**, 1901
[Alt. Horsfield's Bush Lark ssp., Australasian Bush Lark ssp.]
Australian Swamphen ssp. *Porphyrio melanotus woodwardi* **Mathews**, 1912 NCR
[JS *Porphyrio melanotus bellus*; Syn. *Porphyrio porphyrio bellus*]
Emu ssp. *Dromaius novaehollandiae woodwardi* Mathews, 1912
[Often included in *Dromaius novaehollandiae novaehollandiae*]
Red Wattlebird ssp. *Anthochaera carunculata woodwardi* Mathews, 1912

Bernard Henry Woodward (1846–1916) was a wine merchant in London (1875). He suffered from bronchial disease, and so moved to Western Australia for his health (1889). Here he worked as government analyst (1889–1895), responsible for assaying and examining mineral oils. He also became Curator of the Geological Museum, Perth (1889). This expanded to become the Western Australian Museum and Art Gallery, and Woodward became Director (1897). He founded the Western Australia Natural History Society (1890). He wrote an article on 'National parks and the fauna and flora reserves in Australia' (1907). Two mammals are named after him.

Woodward, R. B. & J. D. S.

Woodwards' Batis *Batis fratrum* **G. E. Shelley**, 1900

Garden Warbler ssp. *Sylvia borin woodwardi* **Sharpe**, 1877
Woodward's Barbet *Stactolaema olivacea woodwardi* Shelley, 1895
[Alt. Green Barbet ssp.]

Reverend Robert Blake Woodward (b.1848) and his brother John Deverell Stewart Woodward (b.1849) were Anglican missionaries in Natal, South Africa. They first went to the Transvaal in the early 1870s to farm sheep, and later moved to Natal to run a plantation. They were both ordained into the Anglican Church (1881). They are believed to have returned to England (1905), although one brother (it is not clear which) is reported to have drowned in the Tugela River. They were both deeply interested in ornithology and sent many specimens to Sharpe (q.v.) at the British Museum. They were co-authors of *Natal Birds* (1899). The batis's binomial *fratrum* means 'of the brothers'.

Woosnam

Trilling Cisticola *Cisticola woosnami* **Ogilvie-Grant**, 1908

Fire-crested Alethe ssp. *Alethe castanea woosnami* Ogilvie-Grant, 1906
Red-tailed Bristlebill ssp. *Bleda syndactylus woosnami* Ogilvie-Grant, 1907

Major Richard Bowen Woosnam (1880–1915) was a British army officer who served in the Boer War in South Africa and the Dardenelles campaign (WW1). He explored and collected in the Middle East and in tropical Africa, including being a game ranger in Kenya. Two mammals and a reptile are also named after him.

Worcester

Worcester's Buttonquail *Turnix worcesteri* **McGregor**, 1904
[Alt. Luzon Buttonquail]
Hair-crested Drongo *Chibia worcesteri* McGregor, 1905 NCR
[JS *Dicrurus hottentottus cuyensis*]

Worcester's Hanging Parrot *Loriculus philippensis worcesteri* **Steere**, 1890
[Alt. Philippine Hanging Parrot ssp.]
Black Noddy ssp. *Anous minutus worcesteri* McGregor, 1911

Dean Conant Worcester (1866–1924) was an American ornithologist, collector and Professor of Zoology at the University of Michigan who became a colonial administrator and businessman. He first went to the Philippines with the Steere Expedition (1887–1889) along with Frank Swift Bourns (q.v.). He wrote *The Philippine Islands and Their People* (1898). He served as a member of the Philippine Commission (1899–1913) as Secretary of the Interior. In his last decade he developed commercial interests in coconut and cattle production, profiting after 1921 from the support of Governor-General Leonard Wood (q.v.).

Worthen

Worthen's Sparrow *Spizella wortheni* **Ridgway**, 1884
Petrel sp. *Oestrelata wortheni* **Rothschild**, 1902 NCR
[Alt. Phoenix Petrel; JS *Pterodroma alba*]

Charles K. Worthen (1850–1909) was an American illustrator, collector, taxidermist and dealer in natural history specimens. He presented his collection of birds to the Museum of Science, Buffalo, New York. Rothschild (q.v.) named the petrel after Worthen as the latter had '... arranged and managed Mr. Beck's recent trip to the Galapagos Islands.' Beck (q.v.) had collected the holotype.

Wotan

Common Raven ssp. *Corvus corax wotan* Flöricke, 1922 NCR
[JS *Corvus corax corax*]

In Germanic mythology, Wotan is the equivalent of the Norse god Odin. The Anglo-Saxons viewed him as a psychopomp (guider of souls), with whom the raven was frequently associated.

Woznesensky

Sabine's Gull ssp. *Xema sabini woznesenskii* **Portenko**, 1939
[*X. sabini* sometimes regarded as monotypic]

Ilya Gavrilovich Vosnesensky (1816–1871) was a Russian ornithologist who, as a young man, travelled widely. He appears to have lived in St Petersburg working at the Zoological Museum of the Academy of Sciences until 1839. He left Russia (1841) for California where he resided (1842–1848), during which time he visited both Alaska (1843) – at that time owned by Russia – and Kamchatka. Presumably he missed out on the Gold Rush of '49 as he was back in St Petersburg at the Academy of Sciences (1849–1871). (See **Voznesensky**)

Wrangel

California Towhee ssp. *Melozone crissalis wrangeli* **Bonaparte**, 1850 NCR
[Has been treated as a junior synonym of *Melozone crissalis crissalis*, but others have argued it should become the valid name for the subspecies currently known as *M. crissalis petulans*]

Admiral Ferdinand Petrovich Freiherr von Wrangel (1796–1870) was a Baltic German in the Russian navy. He explored in northern Siberia (1820–1824), was Governor of Russian Alaska (1830–1835), and Minister for the Navy (1855–1857). Wrangel Island in the Arctic Ocean off Siberia is named after him, as is a mammal.

Wray

Minivet sp. *Pericrocotus wrayi* **Sharpe**, 1888 NCR
[Alt. Grey-chinned Minivet; JS *Pericrocotus solaris montanus*]

Black-throated Sunbird ssp. *Aethopyga saturata wrayi* Sharpe, 1887
Greater Yellownape ssp. *Chrysophlegma flavinucha wrayi* Sharpe, 1888

Lesser Shortwing ssp. *Brachypteryx leucophris wrayi* **Ogilvie-Grant**, 1906
Long-tailed Sibia ssp. *Heterophasia picaoides wrayi* Ogilvie-Grant, 1910

Leonard Wray (1852–1942) was a British civil servant in Malaya (Malaysia) who explored and collected fauna and flora in the Malay Peninsula. He joined the Perak Civil Service (1881) and became Curator, Perak State Museum, Taiping (1883); State Geologist, Perak (1890); and Director of Museums, Federated Malay States (1904–1908). He was a keen photographer and (WW1) worked as an X-ray technician in auxiliary military hospitals.

Wright

Wright's Flycatcher *Empidonax wrightii* **Baird,** 1858
[Alt. American Grey Flycatcher]

Charles Wright (1811–1885) was an American botanist, teacher and collector who explored the western USA for the Pacific Railroad Company. He supported himself by surveying and teaching, pursuing botany in his spare time. He collected in Texas whilst working for the Boundary Survey Commission. Wright was botanist on the United States North Pacific Exploring Expedition, on board the *Vincennes* (1853–1855), which visited Madeira, Cape Verde, the Cape of Good Hope, Sydney, Hong Kong, several Japanese islands and the western Bering Strait. They returned to San Francisco where Wright was asked to leave the expedition. He proceeded to Nicaragua, where he collected for some months before making his own way to New York. He explored in Cuba (1856–1867) and went with a US Commission to Santo Domingo (1871). Between collecting trips, he spent time at his home in Connecticut and at the Gray Herbarium in Cambridge, Massachusetts. He also served as librarian of the Bussey Institution, Harvard University, for a few months (1875–1876). An amphibian and a reptile are named after him.

Wucherer

Pearly-vented Tody Tyrant ssp. *Hemitriccus margaritaceiventer wuchereri* **P. L. Sclater** & **Salvin**, 1873

Dr Otto Edward Henry Wucherer (1820–1874) was a Portuguese-born German physician and herpetologist. He qualified at Eberhard Karl University, Tübingen, and practised at St Bartholemew's Hospital, London, and also in Lisbon. He discovered the cause of the tropical disease elephantiasis. He left Europe and settled in Bahia, Brazil (1843). He wrote 'Sobre a mordedura das cobras venenosas e seu tratamento' (1867). An amphibian and three reptiles are named after him.

Wulsin

Chinese Blackbird *Turdus wulsini* **Riley**, 1925 NCR
[Alt. Common Blackbird ssp.; JS *Turdus merula mandarinus*]

African Darter ssp. *Anhinga rufa vulsini* **Bangs**, 1918
Greater Vasa Parrot ssp. *Coracopsis vasa wulsini* Bangs, 1929 NCR
[JS *Coracopsis vasa drouhardi*]

Professor Dr Frederick Roelker Wulsin (1891–1961) was an archaeologist, anthropologist, ornithologist and general naturalist. Harvard awarded his bachelor's degree (1913) and PhD (1929). He tutored in anthropology there (1926–1927) and undertook research at Boston University (1932–1941). He was at Tufts College in Massachusetts (1945), becoming Professor of Anthropology (1947–1957). He was in East Africa and Madagascar (1914–1915). In Asia (1921–1924) he led the National Geographic Society's Central China Expedition and collected reptiles and birds in Inner Mongolia, Kansu and Chihli. He was later in India (1924), Belgian Congo and French Equatorial Africa (1927–1928). Later, in Persia (1931), he carried out excavations at the archaeological site of Tureng Tepe. His papers include 'Responses of man to a hot environment' (1943) and 'Adaptations to climate among non-European peoples' (1949). A mammal is named after him.

Wurdemann

Wurdemann's Heron *Ardea wurdemannii* **Baird**, 1858 NCR
[Either a hybrid of *Ardea herodias wardi* x *Ardea herodias occidentalis*, or merely a colour morph of one of these races]

Gustavus Wurdemann (1818–1859) was employed principally as a tidal and meteorological observer in Florida and the Gulf of Mexico (1849–1859) by the US Coast Survey. In his spare time he collected natural history specimens, some of which he sent to the USNM.

Wyatt

Streak-backed Canestero *Asthenes wyatti* **P. L. Sclater** & **Salvin**, 1871

Claude Wilmott Wyatt (1842–1900) was an English ornithologist who collected in Colombia (1870) and co-wrote with Sharpe (q.v.) *A Monograph of the Hirundinidae or the Family of Swallows* (1885).

Wylde

White-browed Scrubwren ssp. *Sericornis longirostris wyldei* **S. A. White**, 1916 NCR
[JS *Sericornis frontalis frontalis*]

Wylde (DNF) was the sub-editor of an Australian newspaper, the *South Australian Register*. The bird's describer, Samuel White, says that Wylde 'assisted me much in ornithological research' but failed to provide any first name(s) for him. We think it was probably C. E. Wylde, who was promoted to Associate Editor in 1928.

Wyman

Wyman's Gull *Larus occidentalis wymani* **Dickey** & **Van Rossem**, 1925
[Alt. Western Gull ssp.]

Luther Everet Wyman (1870–1928) was Curator of Ornithology at the Los Angeles Museum of History, Science and Art. He was interested in natural history from childhood, and made a collection of mounted birds whilst still at school. He was a businessman and a member of the Board of Trade in Chicago for 15 years, but the stress of business led to a nervous breakdown and he left the board and bought an apple ranch. His friend Frank Daggett (q.v.) became Director of the Museum in Los Angeles, and Wyman moved there to supervise palaeontological excavations at the La Brea Tar Pits. He later became Curator of Ornithology. He co-wrote *Field Book of Birds of the Southwestern United States* (1925).

Wynaad

Wynaad's Laughingthrush *Garrulax delesserti* **Jerdon**, 1839
[Alt. Wynaad Laughingthrush]

Wynaad is a region in Kerala state, southern India. Some authors have mistaken it for a person's name and added an incorrect possessive apostrophe.

Wyndham

Horsfield's Bronze Cuckoo ssp. *Chrysococcyx basalis wyndhami* **Mathews**, 1912 NCR; NRM
Yellow-tinted Honeyeater ssp. *Ptilotis flavescens wyndhami* Mathews, 1912 NCR
[JS *Ptilotula flavescens flavescens*]

Named after Wyndham, a town in the Kimberley region of northern Western Australia.

Wyville

Hawaiian Duck *Anas wyvilliana* **P. L. Sclater**, 1878

Professor Sir Charles Wyville Thompson (1830–1882) was a Scottish naturalist. He was Professor of Botany at the University of Aberdeen (1851). Due to his interest in deep-sea biology, he was appointed chief scientist on the *Challenger* Expedition (1872–1876). The *Challenger* called at Hawaii (1875) where its scientists obtained the holotype of this duck.

X

Xanthippe

Foliage-gleaner sp. *Automolus xanthippe* Davidson, 1932 NCR
[Alt. Ruddy Foliage-gleaner; JS *Automolus rubiginosus fumosus*]

Xanthippe was the wife of Socrates (c.470–399 BC). She was reputed to be a difficult woman, and so her name came to signify a shrew or scolding wife. She was apparently considerably younger than her husband; when Socrates was compelled to commit suicide by drinking hemlock, he was at least 70 years old and she was left with their three sons, one of whom was still an infant. The bird's describer gave no reason for his choice of this name. A mammal is also named after her.

Xantus

Xantus's Hummingbird *Basilinna xantusii* **G. N. Lawrence**, 1860
[Syn. *Hylocharis xantusii*]
Xantus's Murrelet *Synthliboramphus hypoleucus* Xantus, 1860
[Alt. Guadalupe Murrelet]

Xantus's Becard *Pachyramphus aglaiae albiventris* G. N. Lawrence, 1867
[Alt. Rose-throated Becard ssp.]
Xantus's (Scrub) Jay *Aphelocoma californica hypoleuca* **Ridgway**, 1887
Xantus's Screech Owl *Megascops kennicottii xantusi* Brewster, 1902
[Alt. Western Screech Owl ssp.]
Xantus's Gnatcatcher *Polioptila melanura abbreviata* **J. Grinnell**, 1926 NCR
[Alt. California Gnatcatcher ssp.; JS *Polioptila californica margaritae*]
Roadside Hawk ssp. *Buteo magnirostris xantusi* **Van Rossem**, 1939 NCR
[JS *Rupornis magnirostris griseocauda*]

Louis Janos (John) Xantus de Vesey (1825–1894) was a Hungarian who was on the staff of William Hammond (q.v.), the collector. Whilst living in the USA (1855–1861), he sent the USNM 10,000 specimens! He is also renowned as a pathological liar. According to Schoenman & Benedek (1976),

'Xantus fled his native Hungary after taking part in the unsuccessful revolt against the Austrian Empire in 1848. A poor but educated and ambitious man, he wrote grandiose accounts of his American exploits. They were published in Hungary where he became famous. His letters make Private Xantus sound like he was in charge. Despite the fact that he plagiarized other travel accounts of the American West, lied about himself, and always claimed to be superior to those around him, Xantus did great work for Baird and the Smithsonian. Xantus once had a photo taken of himself as a US Navy captain, which was published in Hungary. Yet Xantus had never even served in the Navy'. The Austro-Hungarian Empire dispatched an expedition to Siam (Thailand), China and Japan (1868), with Xantus as one of the 18 members. They were charged with collecting botanical and zoological specimens and investigating local ethnography and arts. After leaving Japan he visited Hong Kong, Manila, Singapore, Borneo, Java and Sumatra. He returned to Hungary (1870) with 155,644 specimens in 200 crates. Two reptiles are named after him.

Xavier

Xavier's Greenbul *Phyllastrephus xavieri* **Oustalet**, 1892

Xavier Dybowski (b.1860) was a French explorer who collected fauna and flora in the Congo (1891–1892), presumably as part of a military-aided French survey mission headed by his brother, Jan Dybowski (see **Dybowski, J.-T.**). He wrote an article about an earthquake: 'Tremblement de terre de Turqui observé à Adapazari' (1894).

Xicotencal

White-eared Hummingbird *Trochilus xicotencal* De La Llave, 1833 NCR
[JS *Basilinna leucotis leucotis*]

Xicotencatl (c.1425–c.1522) was a long-lived Tlaxcalan ruler and warrior who opposed the Spanish *conquistadores* in Mexico. At the time of the Spanish conquest of Mexico he was very old and in poor health. The Tlaxcalan historian Diego Muñoz Camargo claimed that Xicotencatl was more than 120 years old, and could only see Cortés if someone else lifted his eyelids for him!. He was also said to have had more than 500 wives and concubines.

Y

Yaeger

Grace's Warbler ssp. *Dendroica graciae yaegeri*
   A. R. Phillips & Webster, 1961
Black-headed Nightingale Thrush ssp. *Catharus mexicanus*
   yaegeri A. R. Phillips, 1991
Streak-backed Oriole ssp. *Icterus pustulatus yaegeri*
   A. R. Phillips, 1995

Lewis D. Yaeger (DNF) was an American collector who assisted the describer, Allan Phillips (q.v.), in Arizona and Sonora.

Yaldwyn

Yaldwyn's Wren *Pachyplichas yaldwyni* Millener, 1988
   EXTINCT
[Alt. Stout-legged Wren]

Doctor John Cameron Yaldwyn (1929–2005) was a New Zealand zoologist who specialised in crustaceans. He edited the detailed report on the distribution of flora and fauna in: *Preliminary Results of the Auckland Island Expedition 1972–1973* (1975). This formed the basis for a programme to rid the islands of non-native mammals, completed in the early 1990s. He was Director of the National Museum of New Zealand, where he founded an archaeozoology department. Yaldwyn was the PhD examiner for Phil Millener, who described the extinct 'wren'. He co-wrote several books and papers, such as *Australian Crustaceans in Colour* (1971) and 'The bird fauna of Niue Island, southwest Pacific, with special notes on the White-tailed Tropicbird and Golden Plover' (1981). The wren became extinct at some time between the arrival of the Polynesians and of Europeans. Amongst other taxa also named after him is the marine fish *Notoclinops yaldwyni*.

Yamada

Tawny Owl ssp. *Strix nivicolum yamadae* **Yamashina**, 1936
[Syn. *Strix aluco yamadae*]

Nobuo Yamada (b.1905) was a Japanese oologist and collector on Formosa (Taiwan) (1936), working for Yamashina (q.v.), with whom he co-wrote 'Nidification of Formosan birds' (1937).

Yamamura

Black-naped Oriole ssp. *Oriolus chinensis yamamurae*
   Kuroda, 1927
[SII *Oriolus chinensis chinensis*]

Miss Yaeko Yamamura (fl.1927) was a Japanese field naturalist and collector on the island of Basilan, Philippines, where her father, Dr Shinjiro Yamamura, was Managing Director of a coconut plantation. The whole family were amateur naturalists and made extensive collections of corals, shells, birds, insects and fishes.

Yamashina

Pagan Reed Warbler *Acrocephalus yamashinae*
   Taka-Tsukasa, 1931 EXTINCT
[Syn. *Acrocephalus luscinius yamashinae*]

Hazel Grouse ssp. *Tetrastes bonasia yamashinai*
   Momiyama, 1928
Lesser Spotted Woodpecker ssp. *Xylocopus minor*
   yamashinai Momiyama, 1928 NCR
[JS *Dendrocopos minor amurensis*]
Eurasian Eagle Owl ssp. *Bubo bubo yamashinai* Momiyama,
   1930 NCR
[JS *Bubo bubo ussuriensis*]
White-backed Woodpecker ssp. *Dryobates leucotos*
   yamashinae Bergman, 1931 NCR
[JS *Dendrocopos leucotos subcirris*]
Brown Booby ssp. *Sula leucogaster yamashinae* **Neumann**,
   1932 NCR
[JS *Sula leucogaster plotus*]
Common Emerald Dove ssp. *Chalcophaps indica yamashinai*
   Hachisuka, 1939 NCR
[JS *Chalcophaps indica indica*]

Marquis Dr Yoshimaro Yamashina (1900–1989) was the second son of Prince Kikumaro Yamashina and developed a keen interest in birds as a child. He was an ornithology graduate at Tokyo University and was awarded his PhD, following research on avian cytology, by the University of Hokkaido. After army service (1932), he built a museum in his backyard to house his collection of bird specimens and books. Wanting to share it with others, he opened it to the public. It later became the Yamashina Institute for Ornithology, moved to larger premises, and was administered under the auspices of the Department for Education. He–received many awards including the Jean Delacour Prize, considered the 'Nobel Prize in Ornithology'. He co-wrote *Handlist of the Japanese Birds*, and wrote *Birds in Japan* (1961). A mammal is named after him.

Yarigueis

Yariguies Brush Finch *Atlapetes latinuchus yariguierum*
Donegan & Huertas, 2006
[Alt. Rufous-naped Brush Finch ssp.]

Lacrimose Mountain Tanager ssp. *Anisognathus lacrymosus yariguierum* Donegan & Avendaño, 2010

These birds are named after the Yariguies, a former indigenous people of Colombia who gave their name to the mountainous Serranía de los Yariguíes area where the birds are found.

Yarrell

Yarrell's Curassow *Crax yarrellii* **E. Bennett**, 1835 NCR
[Alt. Wattled Curassow; JS *Crax globulosa*]
Yarrell's Goldfinch *Carduelis yarrellii* **Audubon**, 1839
[Alt. Yellow-faced Siskin; Syn. *Spinus yarrellii*]
Yarrell's Woodstar *Eulidia yarrellii* **Bourcier**, 1847
[Alt. Chilean Woodstar]

Pied Wagtail ssp. *Motacilla alba yarrellii* **Gould**, 1837

William Yarrell (1784–1856) was an English bookseller, naturalist and amateur ornithologist. He once sent a letter to his friend Leonard Jenyns as he was 'most excited about a new species of swan', which he 'knew was different from the Hooper'. He searched the London markets to purchase any novel animal specimens and he found enough new birds to enable him to publish internal and external descriptions of the swan in a paper for the Linnean Society. Yarrell wrote to Jenyns 'I am almost afraid you have missed another new swan ... the North Americans have a species which they call the Great Grey headed Swan, both the Hooper and the new one are common at Hudson's Bay, and why may not their Great Grey headed Swan occasionally visit us, as well as the others'. This was Bewick's Swan *Cygnus columbianus bewickii* (Yarrell, 1830). He mentioned to his friend that he intended to add to the Philosophical Society's Collection 'one or more of every British Bird's egg that I possess beyond a pair, which I keep for my own drawers'. He wrote *History of British Fishes* (1836) and *History of British Birds* (1843). A fish, Yarrell's Blenny, is also named after him.

Yate

Yate's Sparrow *Passer moabiticus yatii* **Sharpe**, 1888
[Alt. Dead Sea Sparrow ssp.; sometimes considered to be a full species: Afghan Scrub Sparrow]

Colonel Sir Charles Edward Yate, 1st Baronet (1849–1940), was an English soldier and colonial administrator. His career included being the British Political Agent in Oman (1889–1890), Political Officer on the Afghan Boundary Commission, Consul-General in Seistan (1894–1898), and Chief Commissioner in Baluchistan (1900–1904). He wrote *Khurasan and Sistan* (1900).

Yen

Yen's Fulvetta *Alcippe variegaticeps* Yen, 1932
[Alt. Golden-fronted Fulvetta; Syn. *Pseudominla variegaticeps*]

Kwok Yung Yen (also known as Yen Kwokyung) (fl.1915–at least 1951) was a Chinese ornithologist at Sun Yatsen University, Canton (Guangzhou), China (1930–1951). He wrote scientific descriptions of several birds (1934) that had been collected in Malaya (Malaysia) (1905) and lodged in the Selangor State Museum by H. C. Robinson (q.v.), who was the curator there.

Yersin

Yersin's Laughingthrush *Trochalopteron yersini*
H. C. Robinson & **Kloss**, 1919
[Alt. Collared Laughingthrush]

Alexandre Émile Jean Yersin (1863–1943) was a Swiss bacteriologist. Shortly after graduating in Paris (c.1882) he joined Dr Louis Pasteur's team and took French citizenship (1889). Later (1919) he was the Director of the Pasteur Institute in French Indochina (Vietnam), which he established. He was responsible for introducing the rubber tree to Vietnam and for the first quinine plantations there. Yersin discovered the bubonic plague bacillus *Yersina pestis*, which is named after him, and developed an antiserum. He also had a hand in the development of a vaccine for diphtheria. He took an interest in everything around him, which led him to explore in the Vietnamese highlands. He died at Nha Trang, in Annam.

Yokana

Scrub Warbler sp. *Bradypterus yokanae* **Van Someren**, 1919 NCR
[Alt. White-winged Swamp Warbler; JS *Bradypterus carpalis*]

Yokana's (Blue-throated) Sunbird *Anthreptes reichenowi yokanae* **Hartert**, 1921
[Alt. Plain-backed Sunbird ssp.]

Yokana (DNF) was Van Someren's chief collector of bird specimens in Kenya and Uganda.

Young

Golden Whistler ssp. *Pachycephala pectoralis youngi*
Mathews, 1912

The describer, Gregory Mathews (q.v.), was notorious for failing to provide etymological details for taxa that he named. Such is the case here, and we have been unable to identify this 'Young'.

Young, A.

Chestnut-crowned Warbler ssp. *Seicercus castaniceps youngi* **H. C. Robinson**, 1915

Captain Sir Arthur Henderson Young (1854–1938) was a British army officer and colonial administrator in Cyprus (1878–1905). He was in St Vincent (West Indies) (1902–1905) and in the Straits Settlements (Malaysia) (1906–1910) as Colonial Secretary. He became Governor of Straits Settlements and High Commissioner of the Malay States (1911–1920).

Young, C.

Evergreen Forest Warbler ssp. *Bradypterus mariae youngi*
Serle, 1949 NCR
[JS *Bradypterus lopezi camerunensis*]

Charles Gore Young (b.1901) was a colonial administrator who served in Malaya (Malaysia) (1922–1942) and Cameroon (1943–1950). He collected the warbler holotype on Mount Cameroon.

Young, J.

Society Kingfisher ssp *Todiramphus veneratus youngi*
Sharpe, 1892
[Alt. Tahiti Kingfisher ssp.]

John Young (1838–1901) was an English civil servant, ornithologist and collector in the Pacific (1886–1887).

Young, W.

Square-tailed Nightjar ssp. *Crotema fossii youngi*
J. A. Roberts, 1932 NCR
[JS *Caprimulgus fossii welwitschii*]
Bar-throated Apalis ssp. *Apalis thoracica youngi* **Kinnear**, 1936

Rev. William Paulin Young (1886–at least 1945), the son of a Scottish minister, was a missionary to Nyasaland (Malawi) (1911–1931). He was appointed as an unofficial member of the legislative council (1932–1938). He held both the Military Cross and the Distinguished Conduct Medal, which were awarded for outstanding service (WW1). He wrote *A Soldier to the Church* (1919) based on his war experiences. His older brother Rev. Thomas Cullen Young (b.1880) was also a missionary in Malawi.

Yourdin

Yellow-vented Bulbul ssp. *Pycnonotus goiavier yourdini*
Gray, 1847
[Trinomial sometimes amended to *gourdini*]

(See **Gourdin**)

Yvette

Long-tailed Minivet ssp. *Pericrocotus ethologus yvettae*
Bangs, 1921

Yvette Andrews *née* Borup was the wife (1914) of Roy Chapman Andrews (q.v.). She co-led the Asiatic Zoological Expedition for the AMNH to China (1916–1917) and was principal photographer. Yvette and Roy co-wrote the expedition report, *Camps and Trails in China* (1918). They later divorced (1930).

Yzurieta

Cream-winged Cinclodes ssp. *Cinclodes albiventris yzurietae* Nores, 1986

Darío Yzurieta (1931–1996) was an Argentinean naturalist and self-taught artist. He co-authored *Birds of Argentina and Uruguay* (1989).

Z

Zaphiro

Weaver sp. *Sycobrotus zaphiroi* Ogilvie-Grant, 1902 NCR
[Alt. Baglafecht Weaver; JS *Ploceus baglafecht emini*]

(See under **Saphiro**)

Zappey

Zappey's Flycatcher *Cyanoptila cumatilis* Thayer & Bangs,
1909
[Syn. *Cyanoptila cyanomelana cumatilis*]
Zappey's Parrotbill *Sinosuthora zappeyi* Thayer & Bangs,
1912
[Alt. Grey-hooded Parrotbill]

Scaly Francolin ssp. *Pternistis squamatus zappeyi* **Mearns**,
1911 NCR
[*P. squamatus* often regarded as monotypic]

Walter Reeves Zappey (1878–1914) was a collector for
Harvard University's Museum of Comparative Zoology. He
accompanied Ernest Henry Wilson (q.v.) in China (1907–1908),
making collections of birds and mammals. He wrote *Birds of
the Isle of Pines* (1905) with Outram Bangs (q.v.).

Zarudny

Zarudny's Sparrow *Passer zarudnyi* **Pleske**, 1896
[Alt. Asian Desert Sparrow; Syn. *Passer simplex zarudnyi*]
Shrike sp. *Lanius zarudnyi* **Buturlin**, 1908 NCR
[Hybrid: *Lanius collurio* x *L. isabellinus/phoenicuroides*]

Bar-tailed Lark ssp. *Ammomanes cinctura zarudnyi* **Hartert**,
1902
Eurasian Scops Owl ssp. *Otus scops zarudnyi* **Tschusi**, 1903
NCR
[JS *Otus scops pulchellus*]
Zarudny's (Common) Pheasant *Phasianus colchicus
zarudnyi* Buturlin, 1904
Western Rock Nuthatch ssp. *Sitta neumayer zarudnyi*
Buturlin, 1907
Great Reed Warbler ssp. *Acrocephalus arundinaceus
zarudnyi* Hartert, 1907
Fieldfare ssp. *Turdus pilaris zarudnyi* **Loudon**, 1912 NCR;
NRM
Eurasian Collared Dove ssp. *Streptopelia decaocto zarudnyi*
Serebrovski, 1927 NCR
[JS *Streptopelia decaocto decaocto*]

Nikolai Alekseyivich Zarudny (1859–1919) was a Russian
(Ukrainian) zoologist, traveller and ornithologist. He taught at
the Military High School in Orenburg (1879–1892), during
which time he undertook five expeditions through the
Trans-Caspian region (Turkmenistan). He taught natural
history at the Pskov Military School (1892–1906), and in this
period he undertook four journeys through various parts of
Persia (Iran), and was awarded the Russian Geographical
Society's Przhewalski medal. He then worked in Tashkent
(1906), continuing his Middle Asia exploration. He collected
extensively, and his specimens are held by the Zoological
Museum of the Academy of Science. He wrote 'Les reptiles,
amphibiens, et poissons de la Perse orientale' (1903), and
*Third Excursion over Eastern Persia (Horassan, Seistan and
Persian Baluchistan) in 1900–1901* (1916). He was working on
a book about Turkestan birds when he died of accidental
poisoning. Two mammals, two reptiles and several insects
are named after him. (See also **Sarudny**)

Zavattari

Stresemann's Bush-crow genus *Zavattariornis* **Moltoni**,
1938

Orange Bishop sp. *Euplectes zavattarii* Moltoni, 1943 NCR
[Alt. Northern Red Bishop; JS *Euplectes franciscanus
pusillus*]

Professor Edoardo Zavattari (1883–1972) was an Italian zoologist who was the first Director of Zoology at the University
of Rome (1935–1958). He was also a leading Fascist and a
racist academic – being one of the ten leading scientists who
issued a manifesto to justify racism (1938), which was
endorsed by over 300 leading intellectuals.

Zech

Cardinal Woodpecker ssp. *Dendrocopus guineensis zechi*
Neumann, 1904 NCR
[JS *Dendropicos fuscescens lafresnayi*]
White-backed Vulture ssp. *Pseudogyps africanus zechi*
Reichenow, 1904 NCR; NRM

Major Johann Nepomuk Felix Julius Graf von Zech auf
Neuhofen (1868–1914) was Governor of Togo (1904–1910),
then a German colony. After his military career he was sent
to Togo (1895) as part of the administration. He rose from
District Commissioner to Deputy Governor and finally
Governor. He requested retirement from his position on
health grounds and returned to Europe (1910). A mammal is
also named after him.

Zedlitz

Crombec sp. *Sylvietta zedlitzi* **Reichenow**, 1918 NCR
[Alt. Red-faced Crombec; JS *Sylvietta whytii jacksoni*]

Trumpeter Finch ssp. *Bucanetes githagineus zedlitzi*
Neumann, 1907
Yellow-fronted Tinkerbird ssp. *Pogoniulus chrysoconus
zedlitzi* Neumann, 1908 NCR
[JS *Pogoniulus chrysoconus chrysoconus*]
Short-winged Cisticola ssp. *Cisticola brachypterus zedlitzi*
Reichenow, 1909
d'Arnaud's Barbet ssp. *Trachyphonus darnaudii zedlitzi*
Berger, 1911 NCR
[JS *Trachyphonus darnaudii darnaudii*]
Red-billed Firefinch ssp. *Lagonosticta senegala zedlitzi*
Grote, 1922 NCR
[JS *Lagonosticta senegala somaliensis*]
Southern Grey-headed Sparrow ssp. *Passer griseus zedlitzi*
Gyldenstolpe, 1922 NCR
[JS *Passer diffusus diffusus*]

Otto-Eduard Graf von Zedlitz und Trützschler (1873–1927) was
a German ornithologist and explorer in Spitsbergen and
Norway (1900), Tunisia and Algeria (1904–1906 and 1913), and
Eritrea and Ethiopia (1908–1909). He made a trip by Zeppelin
airship to Norway, Bear Island, Spitsbergen and the Arctic
ice fields (1910). He was in the Pripet Marshes (southern
Belarus) during war service (WW1) and was Vice-President
of the German Ornithological Society (1921–1923).

Zelebor

Plumbeous Rail ssp. *Pardirallus sanguinolentus zelebori*
Pelzeln, 1865

Johann Zelebor (1819–1869) was an Austrian zoologist and
taxidermist at the Vienna Natural History Museum. He was
the naturalist on the *Novara* Circumnavigation Expedition
(1857–1859).

Zeledón

Wren-thrush genus *Zeledonia* **Ridgway**, 1889

Zeledón's Tyrannulet *Tyranniscus zeledoni* **G. N. Lawrence**,
1869
[Alt. White-fronted Tyrannulet; Syn. *Phyllomyias zeledoni*]
Canebrake Wren *Cantorchilus zeledoni* Ridgway, 1878
[Syn. *Cantorchilus modestus zeledoni*]
Zeledón's Bush Tanager *Chlorospingus zeledoni* Ridgway,
1905 NCR
[Believed to be a colour morph of *Chlorospingus pileatus*]
Zeledón's Antbird *Myrmeciza zeledoni* Ridgway, 1909
[Alt. Western Immaculate Antbird; Syn. *Myrmeciza
immaculata zeledoni*]

Black-crowned Antpitta ssp. *Pittasoma michleri zeledoni*
Ridgway, 1884
Boat-billed Heron ssp. *Cochlearius cochlearius zeledoni*
Ridgway, 1885
Black-chested Jay ssp. *Cyanocorax affinis zeledoni*
Ridgway, 1899
Zeledón's Manakin *Dixiphia pipra anthracina* Ridgway, 1906
[Alt. White-crowned Manakin ssp.]

José Cástulo Zeledón (1846–1923) was a Costa Rican ornithologist. He had humble beginnings as an apprentice, but
later became an internationally known ornithologist as well
as co-administrator of the drugstore 'Botica Francesa',
which is currently one of the largest private companies in
that country. Advertisements in the press of the time said that
the drugstore was 'managed by naturalists Frantzius and
Zeledón'. He studied under Frantzius (q.v.), and under Baird
(q.v.) and Ridgway (q.v.) at the USNM. He returned to Costa
Rica (1871) as part of a scientific expedition to explore the
forest of Talamanca, and made a very important collection of
birds. The USNM published his 'Catalogue of the birds of
Costa Rica' (1885). His collection of 1,500 birds is in the
National Museum of Costa Rica, which he co-founded. A
reptile is named after him.

Zelich

Zelich's Seedeater *Sporophila zelichi* **Narosky**, 1977
NCR
[Now regarded as a colour morph of Marsh Seedeater,
Sporophila palustris]

Dr Mateo Ricardo Zelich (b.1924) is an Argentine entomologist and ornithologist. He clearly spent a great deal of time
studying the birds around the Uruguay River, as he could
vouchsafe the existence of a particular colony of terns over a
60-year period! He was also a collector of fossils, if not a
palaeontologist, and established a small museum in the town
of Liebig, not far from Columbus, Argentina, to exhibit his
personal collections. He collected the holotype (1969).

Zenaide

Dove genus *Zenaida* **Bonaparte**, 1838

Zenaida Dove *Zenaida aurita* **Temminck**, 1809

Princess Zénaide Charlotte Bonaparte (1804–1854) was the
daughter of Joseph Bonaparte, the elder brother of Napoleon I. She married her cousin (1822), Prince Charles Lucien
Bonaparte (q.v.), the famed ornithologist who named the
dove genus after his wife. Although they had 10 children, four
of whom died in childhood, the arranged marriage was not
happy and the Pope granted Charles Bonaparte permission
to divorce (1854).

Zenker

Zenker's Honeyguide *Melignomon zenkeri* **Reichenow**, 1898
Greenbul sp. *Phyllastrephus zenkeri* Reichenow, 1916 NCR
[Alt. White-throated Greenbul; JS *Phyllastrephus albigularis
albigularis*]

Red-thighed Sparrowhawk ssp. *Accipiter erythropus zenkeri*
Reichenow, 1894
Zenker's Lovebird *Agapornis swindernianus zenkeri*
Reichenow, 1895
[Alt. Black-collared Lovebird ssp.]
Green Turaco ssp. *Tauraco persa zenkeri* Reichenow, 1896
Grey-headed Broadbill ssp. *Smithornis sharpei zenkeri*
Reichenow, 1903
Plain-backed Pipit ssp. *Anthus leucophrys zenkeri*
Neumann, 1906

White-spotted Flufftail ssp. *Sarothrura pulchra zenkeri*
Neumann, 1908
Mackinnon's Shrike ssp. *Lanius mackinnoni zenkerianus*
Grote, 1924 NCR; NRM

Georg August Zenker (1855–1922) was a German botanist and gardener who collected in Cameroon, formerly a German protectorate. He established the settlement which later became the capital, Yaoundé. He had significant land holdings around Bipindi and devoted much time to collecting plants, insects, fish – and apparently human bones, which he is recorded disinterring in the area. He made a particular study of 'pygmies' and other native peoples. Three mammals and a reptile are named after him.

Zenobia

Moluccan Sunbird sp. *Cinnyris zenobia* **Lesson**, 1830 NCR
[Alt. Olive-backed Sunbird; JS *Cinnyris jugularis clementiae*]

Zenobia, Queen of Palmyra (now in Syria) (240–c.274 AD), was a talented and courageous enemy of the Roman Empire. She was captured and taken to Rome as a hostage, but her ultimate fate is uncertain. She may have been executed or, as seems equally possible, she was pardoned, became a Roman matron, married a senator and had a number of daughters.

Zervas

Eurasian Jay ssp. *Garrulus glandarius zervasi* Kleiner, 1940 NCR
[JS *Garrulus glandarius anatoliae*]

A. Zervas (DNF) was a Greek naturalist and conservationist, who worked at the Game Department, Ministry of Agriculture, Athens (1939). He wrote *Game in Greece: Biology and Hunting Methods* (1947).

Ziegler

Ziegler's Crake *Porzana ziegleri* **Olson** & James, 1991 EXTINCT
[Alt. Small Oahu Crake]

Dr Alan Conrad Ziegler (1930–2003) lived on Hawaii for over 30 years and wrote many books about Hawaiian natural history. He spent 15 years as head of the Vertebrate Zoology Division of the Bishop Museum in Hawaii, followed by another 15 years as an independent zoological consultant, and taught at the zoology department of the University of Hawaii. He wrote *Hawaiian Natural History, Ecology, and Evolution* (2002). A mammal is also named after him.

Zietz

Brown Thornbill ssp. *Acanthiza pusilla zietzi* **North**, 1904
Rock Parrot ssp. *Neophema petrophila zietzi* **Mathews**, 1912

Amandus Heinrich Christian Zietz (1839–1921) was a German collector, aviculturist and preparator who worked at the Kiel Museum and emigrated from Germany to Australia (c.1873). He was Assistant Director, South Australian Museum (1888–1909) and a founder member of the RAOU.

Zimmer

Tyrannulet genus *Zimmerius* **Traylor**, 1977

Zimmer's Woodcreeper *Dendroplex kienerii* **Des Murs**, 1855
[Syn. *Xiphorhynchus kienerii*]
Zimmer's Flatbill *Tolmomyias assimilis* **Pelzeln**, 1868
Zimmer's Antpitta *Hylopezus dilutus* **Hellmayr**, 1910
[Syn. *Hylopezus macularius dilutus*]
Zimmer's Tody Tyrant *Hemitriccus minimus* **Todd**, 1925
Zimmer's Tapaculo *Scytalopus zimmeri* **Bond** & **Meyer de Schauensee**, 1940
Zimmer's Swift *Cypseloides cryptus* Zimmer, 1945
[Alt. White-chinned Swift]

Long-winged Antwren ssp. *Myrmotherula longipennis zimmeri* **Chapman**, 1925
Black-chested Mountain Tanager ssp. *Buthraupis eximia zimmeri* **R. T. Moore**, 1934
Barred Hawk ssp. *Leucopternis princeps zimmeri* Friedmann, 1935 NCR; NRM
Jacky Winter ssp. *Microeca fascinans zimmeri* **Mayr** & **Rand**, 1935
Bare-faced Ground Dove ssp. *Metriopelia ceciliae zimmeri* **J. L. Peters**, 1937
Sapphire-spangled Emerald ssp. *Amazilia lactea zimmeri* **Gilliard**, 1941
[Syn. *Polyerata lactea zimmeri*]
Hepatic Tanager ssp. *Piranga flava zimmeri* **Van Rossem**, 1942 NCR
[JS *Piranga hepatica hepatica*]
Natterer's Slaty Antshrike ssp. *Thamnophilus punctatus zimmeri* **O. Pinto**, 1947 NCR
[JS *Thamnophilus stictocephalus stictocephalus*]
Ochre-breasted Brush Finch ssp. *Atlapetes semirufus zimmeri* Meyer de Schauensee, 1947
Olive-faced Flatbill ssp. *Tolmomyias viridiceps zimmeri* Bond, 1947
Tropical Kingbird ssp. *Tyrannus melancholicus zimmeri* Pinto, 1954 NCR
[JS *Tyrannus melancholicus melancholicus*]
Bicoloured Wren ssp. *Campylorhynchus griseus zimmeri* **Borrero** & Hernández-Camacho, 1958
Citrine Canary-flycatcher ssp. *Culicicapa helianthea zimmeri* **Parkes**, 1960
Long-tailed Woodcreeper ssp. *Deconychura longicauda zimmeri* O. Pinto, 1974
Brown-breasted Bamboo Tyrant ssp. *Hemitriccus obsoletus zimmeri* Traylor, 1979

Dr John Todd Zimmer (1889–1957) was an American ornithologist, Curator of Birds at The Field Museum of Natural History, Chicago (1921–1930) and at the AMNH (1930–1957). He wrote *Birds of the Marshall Field Peruvian Expedition* (1930) and *Studies of Peruvian Birds* (1931). He also wrote a catalogue of the Edward E. Ayer Ornithological Library, a part of The Field Museum (1926). He was co-author of volume 8 of *Checklist of Birds of the World* (1979). A mammal is named after him.

Zimmermann

Chinese Crested Tern *Sterna zimmermanni* **Reichenow**, 1903 NCR
[JS *Thalasseus bernsteini*]

Grey-headed Woodpecker ssp. *Picus canus zimmermanni*
Reichenow, 1903 NCR
[JS *Picus canus jessoensis*]

Robert C. Zimmermann (d.1903) was a collector and botanist based at Tsingtao (now Qingdao) in China. At that time, this area was leased by Germany.

Zino

Zino's Petrel *Pterodroma madeira* **Mathews**, 1934
[Alt. Freira, Madeira Petrel]

Paul Alexander Zino, known as 'Alec' (1916–2004), was an amateur ornithologist on Madeira. His son, Dr Francis Zino, known as 'Frank', is a general practitioner and naturalist. The Zino family moved from Genoa to Gibraltar and became British. Later (c.1830) the family business, which involved shipping fruit and vegetables to London, moved to Funchal, Madeira. Father Schmitz (q.v.) discovered the petrel (1903), but by the 1950s it was thought extinct. Alec made recordings of Fea's Petrel *Pterodroma feae* (1967) on Bugio, assuming that its voice might be similar to the Madeira bird's. He played the sound to shepherds in the area where Schmitz had found the petrel – one of them identified it as the souls of shepherds who had died in the mountains! Alec rediscovered nesting birds with eggs (1969) and in 1985 the Freira Conservation Project was set up with Francis as co-ordinator, leading to the purchase of the petrel's breeding area. Through this initiative, the Parque Natural da Madeira successfully applied for funding from the European Union to run a massive recovery programme over a four-year period. Alec Zino was largely responsible for getting the Selvagems turned into a Strict Nature Reserve. Alec and Francis Zino wrote 'Contribution to the study of petrels of the genus *Pterodroma* in the archipelago of Madeira' (1986), and several other publications on Madeira's birds.

Zita

Black-sided Flowerpecker ssp. *Dicaeum sulaense zita*
Harrisson & **Hartley**, 1934 NCR
[JS *Dicaeum monticolum*, NRM]

Inezita 'Zita' Hilda Davis (1904–1952) was the first wife of John Randal Baker (q.v.). She acted as clerical assistant and photographer on his New Hebrides (Vanuatu) Expedition (1933–1934), during which she conducted a love affair with another expedition member, Thomas Harnett Harrisson (q.v.). The Bakers divorced and Zita later married (1937) Richard Crossman, the British Labour politician, cabinet minister and editor of *New Statesman*. Both her husbands were Fellows of New College, Oxford, and colleagues there. She contributed to Baker's series of papers entitled 'The seasons in a tropical Rain forest (New Hebrides)'.

Zoe

Zoë's Imperial Pigeon *Ducula zoeae* **Lesson**, 1826
[Alt. Banded Imperial Pigeon]

Zoë Lesson (DNF) was the first wife of the describer René Lesson (q.v.).

Zotta

Western Barn Owl ssp. *Tyto alba zottae* **Kelso**, 1938 NCR
[JS *Tyto alba tuidara*; Syn. *Tyto furcata tuidara*]

Dr Angel R. Zotta (d.1951) was an Argentinian zoologist. He wrote *Lista Sistemática de las Aves Argentinas* (1944).

Zugmayer

House Crow ssp. *Corvus splendens zugmayeri* **Laubmann**, 1913
White-throated Kingfisher ssp. *Halcyon smyrnensis zugmayeri* Laubmann & Götz, 1926 NCR
[JS *Halcyon smyrnensis smyrnensis*]

Professor Dr Erich Johann Georg Zugmayer (1879–1938) was an Austrian explorer, zoologist, ichthyologist and herpetologist who worked at the Bavarian State Zoological Collection, Munich. He took part in several research trips, exploring the area around Lake Urmia, Persia (Iran), and collecting in Turkestan, western Tibet and Baluchistan. He published accounts of his travels, such as 'Bericht über eine Reise in Westtibet' (1909). Two reptiles and an amphibian are named after him.

Zuloaga

Buff-breasted Sabrewing ssp. *Campylopterus duidae zuloagae* **W. H. Phelps** & **W. H. Phelps Jr**, 1948 NCR
[JS *Campylopterus duidae duidae*]
Bronzy Inca ssp. *Coeligena coeligena zuloagae* Phelps & Phelps, 1959
[May only be an individual variation of *C. coeligena coeligena*]

Dr Guillermo Zuloaga Ramirez (1904–1984) was a Venezuelan geologist and naturalist who was a friend of the Phelps (q.v.) family. His doctorate was awarded by the Massachusetts Institute of Technology. He taught geology in Caracas' Central University and established the Ministry of Mines & Petroleum. He joined Creole Petroleum (1939), then its board of directors (1948).

Zusi

Bogota Sunangel *Heliangelus zusii* **Graves**, 1993

Richard Laurence Zusi (b.1930) is Emeritus Curator, Division of Birds at the USNM, Washington DC. He obtained his master's degree (1953) and doctorate (1959) at the University of Michigan. He taught at the University of Maine (1958–1963). His research interests are the functional anatomy of birds with emphasis on feeding mechanisms, and avian systematics and evolution. In his etymology Graves says: 'I take great pleasure in naming this hummingbird for my friend and colleague, Richard L. Zusi, in recognition of his contributions to the systematics of hummingbirds'. Among his many published papers and longer works is 'Myology of the Purple-throated Carib (*Eulampis jugularis*) and other hummingbirds (Aves: Trochilidae)'. The sunangel is known from a single specimen purchased (1909) in Bogotá, Colombia; the species may now be extinct.

Addenda

Cohn-Haft

Acre Tody Tyrant *Hemitriccus cohnhafti* Zimmer *et al.*, 2013

Dr Mario Cohn-Taft is an American biogeography researcher and bird guide, and an acknowledged expert on Amazonian birds. Dartmouth College, New Hampshire awarded his first degree. He then spent seven years as a research assistant, field assistant or whatever assignments would let him learn more about the birds of various regions, especially in the rainforests of Costa Rica and Panama. He started a master's program at Tulane University and the six-month assignment ended up lasting three years. He entered the Ph.D. program (1990) at Louisiana State University, Baton Rouge, to study rainforest birds, and returned to live in Manaus.

Schultze [revised entry]

Barbet sp. *Tricholaema schultzei* **Reichenow**, 1911 NCR
[Alt. Hairy-breasted Barbet; JS *Tricholaema hirsuta flavipunctata*]
Woodpecker sp. *Mesopicos schultzei* Reichenow, 1912 NCR
[Alt. Elliot's Woodpecker; JS *Dendropicos elliotii johnstoni*]

Dr Arnold Schultze (1875–1948) was a German soldier, geographer and entomologist - a specialist in *Lepidoptera*. He was a cadet in the Brandenburg Field Artillery Regiment (1885), rising to Lieutenant (1896) who transferred to the Prussian Foreign Office (1902–1904) and helped delineate the border between Cameroon and Nigeria. He was assigned to the Railway regiment (1904) but shortly afterwards retired (1905) and joined the Imperial Protection Force in Cameroon. However, ill health forced him to retire from active service that same year. After his retirement from the army he took up the study of geography, natural history and political science in Bonn and was awarded his PhD (1910). As a geographer he participated in the central African expedition of Duke Adolf Friedrich of Mecklenburg and published the geographical part of the scientific results of the expedition. Whilst overseas he collected in the Congo (1907–1910 & 1929–1932), Colombia (1926–1928) and Ecuador (1935–1939). On the return trip from Ecuador (September 1939), just south-west of the Canaries, his ship was sunk by the British cruiser *Neptune*. He and his wife were sent to Dakar and interned for some weeks but were soon released to neutral Portugal because of his poor health. They remained here for the rest of his life. His collections are now in the museums of Stuttgart and Berlin. He published on a number of subjects including the paper: *Die afrikanischen Seidenspinner und ihre wirtschaftliche Bedeutung* (1914). A dragonfly is named after him.

References

Journals
The dates given for journals frequently cover large periods of time. We have consulted many issues in these publications and it would be impractical to list every article consulted.

American Museum Novitates, New York. 1921–2013.
Annales des sciences naturelles: zoologie et paléontologie, Paris. 1864–1915.
Annals and Magazine of Natural History, London. 1841–1966.
Annals of the Transvaal Museum, Pretoria. 1911–2013.
Annuaire du Musée Zoologique de l'Academie Imperiale des Sciences, St. Petersburg. 1896–1932.
The Auk. American Ornithologists' Union. 1884–2013.
Bericht über die Senckenbergischer Naturforschende Gesellschaft, Frankfurt. 1869–1896.
Bonner zoologische Beiträge, Bonn. 1950–2013.
Bulletin du Museum National d'Histoire Naturelle, Paris. 1895–1970.
Bulletin of the American Museum of Natural History, New York. 1881–2013.
Bulletin of the British Ornithologists' Club, London, 1893–2013.
Bulletin of the Museum of Comparative Zoology, Harvard University, Cambridge, Mass., 1863–2013.
The Condor. Cooper Ornithological Society. 1899–2013.
The Ibis, British Ornithologists' Union. 1859–2013.
Jornal de Sciencias Mathematicas, Physicas e Naturaes, Lisbon. 1866–1910.
Journal of the Bombay Natural History Society, Bombay. 1886–2013.
Lori Journal International, Haalderen, The Netherlands. 1999.
Mémoires de l'Academie Impériale des Sciences, St. Petersburg. 1809–1830.
Memoirs of the Queensland Museum, Brisbane. 1912–2013.
Mitteilungen aus dem Naturhistorischen Museum, Hamburg. 1882–1908.
Monatsberichte der Königlichen Preussischen Akademie der Wissenschaften zu Berlin, Berlin. 1856–1881.
The New Zealand Garden Journal (Journal of the Royal New Zealand Institute of Horticulture). Vol. 2, No. 1, March 1997.
Nouvelles archives du Museum de l'Histoire Naturelle, Paris. 1865–1914.
Ornitologiia, Moscow, Vol. 32, 2005.
Philippine Journal of Science, Manila. 1906–1994.
Proceedings of the Zoological Society of London, London. 1830–1965.

Publicaciones ocasionales del Museo de Ciencias Naturales, Caracas, Venezuela. 1962–2013.
Records of the Albany Museum, Grahamstown. 1903–1935.
Records of the Australian Museum, Sydney. 1890–2013.
Records of the South Australian Museum, Adelaide. 1918–2002.
Records of the Western Australian Museum, Perth. 1974–2013.
Revista do Museu Paulista, Sao Paulo. Vol. 23.
Sitzungsberichte der Kaiserlichen Akademie der Wissensschaften, Vienna. 1850–1947.
Stray Feathers: Journal of Ornithology for India and its dependencies, Calcutta. 1872–1899.
Tropical Zoology, Florence. 1988–2013.
Vestnik zoologii, Kiev. 1967–2013.
Videnskabelige Meddelelser fra den Naturhistoriske Forening i Kjöbenhavn, Copenhagen.1849–1912.
The Wilson Bulletin. Wilson Ornithological Society. 1889–2006.
Zoologischer Anzeiger, Berlin/Leipzig. 1878–2013.
Zootaxa, Auckland. 2001–2013.

Dictionaries and general reference works
American Ornithologists' Union. 1983. *Check-list of North American Birds*. 6th Edition. Also 1985, 1987, 1989 and 1998 Supplements. Washington DC.
Banks, R. C. 1988. *Obsolete English Names of North American Birds and their Modern Equivalents*. US Fish & Wildlife Service, Washington DC.
Beolens, B., & Watkins, M. 2003. *Whose Bird?* Christopher Helm, London.
Beolens, B., Watkins, M. & Grayson, M. 2009. *The Eponym Dictionary of Mammals*. Johns Hopkins University Press, Baltimore.
Beolens, B., Watkins, M. & Grayson, M. 2011. *The Eponym Dictionary of Reptiles*. Johns Hopkins University Press, Baltimore.
Beolens, B., Watkins, M. & Grayson, M. 2013. *The Eponym Dictionary of Amphibians*. Pelagic Publishing, Exeter.
Choate, E. A. 1985. *The Dictionary of American Bird Names*. Revised by R. A. Paynter, Jr. Harvard Common Press, Harvard and Boston, Mass.
Clinning, C. 1989. *Southern African Bird Names Explained*. Southern African Ornithological Society, Johannesburg.
Columbia Encyclopedia, Sixth Edition, 2001. Columbia University Press, New York.
del Hoyo, J. *et al.* 1992–2013. *Handbook of the Birds of the World*. Volumes 1–17. Lynx Edicions, Barcelona.